Better Ceramics Through Chemistry V

MATERIALS RESEARCH SOCIETY SYMPOSIUM PROCEEDINGS VOLUME 271

Better Ceramics Through Chemistry V

Symposium held April 27-May 1, 1992, San Francisco, California, U.S.A.

EDITORS:

Mark J. Hampden-Smith
University of New Mexico
Albuquerque, New Mexico, U.S.A.

Walter G. Klemperer
University of Illinois
Urbana, Illinois, U.S.A.

C. Jeffrey Brinker
Sandia National Laboratories
Albuquerque, New Mexico, U.S.A.

MATERIALS RESEARCH SOCIETY
Pittsburgh, Pennsylvania

This work relates to Department of Navy Grant N00014-92-J-1453 issued by the Office of Naval Research. The United States Government has a royalty-free license throughout the world in all copyrightable material contained herein.

Single article reprints from this publication are available through University Microfilms Inc., 300 North Zeeb Road, Ann Arbor, Michigan 48106

CODEN: MRSPDH

Copyright 1992 by Materials Research Society.
All rights reserved.

This book has been registered with Copyright Clearance Center, Inc. For further information, please contact the Copyright Clearance Center, Salem, Massachusetts.

Published by:

Materials Research Society
9800 McKnight Road
Pittsburgh, Pennsylvania 15237
Telephone (412) 367-3003
Fax (412) 367-4373

Library of Congress Cataloging in Publication Data

Better ceramics through chemistry V: symposium held April 27-May 1, 1992, San Francisco, CA, U.S.A./editors Mark J. Hampden-Smith, Walter G. Klemperer, C. Jeffrey Brinker

p. cm. -- (Materials Research Society symposium proceedings, ISSN 1064-8666; v. 271)
Includes bibliographical references and indexes.
ISBN: 1-55899-166-2
1. Ceramics-Chemistry. I. Hampden-Smith, Mark J. II. Klemperer, Walter G. III. Brinker, C. Jeffrey IV. Materials Research Society symposium proceedings on Better Ceramics Through Chemistry five. V. Title: Better ceramics through chemistry 5. VI. Series: Materials Research Society symposium proceedings: v. 271.
TP786.B4852 1992 90-13520
666-dc20 CIP

Manufactured in the United States of America

Contents

*Invited Paper

PART II: PARTICULATE AND POLYMERIC SOLS

*Invited Paper

PART III: CERAMIC THIN FILMS, COATINGS, MEMBRANES, AND FIBERS

*Invited Paper

*Invited Paper

PART IV: AGING, DRYING, AND CONSOLIDATION OF GELS

PART V: CERAMIC COMPOSITES

*Invited Paper

PART VI: HYBRID ORGANIC/INORGANIC MATERIALS

*Invited Paper

PART VII: NOVEL ROUTES TO NON-OXIDE CERAMICS

*Invited Paper

*Invited Paper

PART VIII: VAPOR PHASE ROUTES TO OXIDE CERAMICS

*Invited Paper

Preface

The application of chemical principles and practices to ceramic processing has mushroomed during the past decade, and the Materials Research Society's *Better Ceramics Through Chemistry* symposia have chronicled this growth biennially. The papers in the current volume were presented at the symposium *Better Ceramics Through Chemistry V* of the Materials Research Society Spring Meeting on April 27-May 1, 1992. Financial support was provided by Air Products and Chemicals Inc., Gelest Inc., the Office of Naval Research, and Radiant Technology Inc.

As in the previous *Better Ceramics Through Chemistry* symposia, published as MRS Symposium Volumes 32, 73, 121, and 180, the present volume has as its focus sol-gel processing of oxide ceramics. In addition, however, considerable space is devoted to recent advances in related systems such as metal nitrides, carbides, and borides, polymer precursors, and cement composite materials, as well as other synthesis techniques such as chemical vapor deposition, aerosol routes and electrochemical processing. The volume is designed to provide both a retrospective and a preliminary view of exciting new results of mutual interest to ceramists and chemists.

M.J. Hampden-Smith
W.G. Klemperer
C.J. Brinker

June, 1992

Volume 239—Thin Films: Stresses and Mechanical Properties III, W.D. Nix, J.C. Bravman, E. Arzt, L.B. Freund, 1992, ISBN: 1-55899-133-6

Volume 240—Advanced III-V Compound Semiconductor Growth, Processing and Devices, S.J. Pearton, D.K. Sadana, J.M. Zavada, 1992, ISBN: 1-55899-134-4

Volume 241—Low Temperature (LT) GaAs and Related Materials, G.L. Witt, R. Calawa, U. Mishra, E. Weber, 1992, ISBN: 1-55899-135-2

Volume 242—Wide Band Gap Semiconductors, T.D. Moustakas, J.I. Pankove, Y. Hamakawa, 1992, ISBN: 1-55899-136-0

Volume 243—Ferroelectric Thin Films II, A.I. Kingon, E.R. Myers, B. Tuttle, 1992, ISBN: 1-55899-137-9

Volume 244—Optical Waveguide Materials, M.M. Broer, G.H. Sigel, Jr., R.Th. Kersten, H. Kawazoe, 1992, ISBN: 1-55899-138-7

Volume 245—Advanced Cementitious Systems: Mechanisms and Properties, F.P. Glasser, G.J. McCarthy, J.F. Young, T.O. Mason, P.L. Pratt, 1992, ISBN: 1-55899-139-5

Volume 246—Shape-Memory Materials and Phenomena—Fundamental Aspects and Applications, C.T. Liu, H. Kunsmann, K. Otsuka, M. Wuttig, 1992, ISBN: 1-55899-140-9

Volume 247—Electrical, Optical, and Magnetic Properties of Organic Solid State Materials, L.Y. Chiang, A.F. Garito, D.J. Sandman, 1992, ISBN: 1-55899-141-7

Volume 248—Complex Fluids, E.B. Sirota, D. Weitz, T. Witten, J. Israelachvili, 1992, ISBN: 1-55899-142-5

Volume 249—Synthesis and Processing of Ceramics: Scientific Issues, W.E. Rhine, T.M. Shaw, R.J. Gottschall, Y. Chen, 1992, ISBN: 1-55899-143-3

Volume 250—Chemical Vapor Deposition of Refractory Metals and Ceramics II, T.M. Besmann, B.M. Gallois, J.W. Warren, 1992, ISBN: 1-55899-144-1

Volume 251—Pressure Effects on Materials Processing and Design, K. Ishizaki, E. Hodge, M. Concannon, 1992, ISBN: 1-55899-145-X

Volume 252—Tissue-Inducing Biomaterials, L.G. Cima, E.S. Ron, 1992, ISBN: 1-55899-146-8

Volume 253—Applications of Multiple Scattering Theory to Materials Science, W.H. Butler, P.H. Dederichs, A. Gonis, R.L. Weaver, 1992, ISBN: 1-55899-147-6

Volume 254—Specimen Preparation for Transmission Electron Microscopy of Materials-III, R. Anderson, B. Tracy, J. Bravman, 1992, ISBN: 1-55899-148-4

Volume 255—Hierarchically Structured Materials, I.A. Aksay, E. Baer, M. Sarikaya, D.A. Tirrell, 1992, ISBN: 1-55899-149-2

Volume 256—Light Emission from Silicon, S.S. Iyer, R.T. Collins, L.T. Canham, 1992, ISBN: 1-55899-150-6

Volume 257—Scientific Basis for Nuclear Waste Management XV, C.G. Sombret, 1992, ISBN: 1-55899-151-4

Volume 258—Amorphous Silicon Technology—1992, M.J. Thompson, Y. Hamakawa, P.G. LeComber, A. Madan, E. Schiff, 1992, ISBN: 1-55899-153-0

Volume 259—Chemical Surface Preparation, Passivation and Cleaning for Semiconductor Growth and Processing, R.J. Nemanich, C.R. Helms, M. Hirose, G.W. Rubloff, 1992, ISBN: 1-55899-154-9

Volume 260—Advanced Metallization and Processing for Semiconductor Devices and Circuits II, A. Katz, Y.I. Nissim, S.P. Murarka, J.M.E. Harper, 1992, ISBN: 1-55899-155-7

Volume 261—Photo-Induced Space Charge Effects in Semiconductors: Electro-optics, Photoconductivity, and the Photorefractive Effect, D.D. Nolte, N.M. Haegel, K.W. Goossen, 1992, ISBN: 1-55899-156-5

Volume 262—Defect Engineering in Semiconductor Growth, Processing and Device Technology, S. Ashok, J. Chevallier, K. Sumino, E. Weber, 1992, ISBN: 1-55899-157-3

Volume 263—Mechanisms of Heteroepitaxial Growth, M.F. Chisholm, B.J. Garrison, R. Hull, L.J. Schowalter, 1992, ISBN: 1-55899-158-1

Volume 264—Electronic Packaging Materials Science VI, P.S. Ho, K.A. Jackson, C-Y. Li, G.F. Lipscomb, 1992, ISBN: 1-55899-159-X

Volume 265—Materials Reliability in Microelectronics II, C.V. Thompson, J.R. Lloyd, 1992, ISBN: 1-55899-160-3

Volume 266—Materials Interactions Relevant to Recycling of Wood-Based Materials, R.M. Rowell, T.L. Laufenberg, J.K. Rowell, 1992, ISBN: 1-55899-161-1

Volume 267—Materials Issues in Art and Archaeology III, J.R. Druzik, P.B. Vandiver, G.S. Wheeler, I. Freestone, 1992, ISBN: 1-55899-162-X

Volume 268—Materials Modification by Energetic Atoms and Ions, K.S. Grabowski, S.A. Barnett, S.M. Rossnagel, K. Wasa, 1992, ISBN: 1-55899-163-8

Volume 269—Microwave Processing of Materials III, R.L. Beatty, W.H. Sutton, M.F. Iskander, 1992, ISBN: 1-55899-164-6

Volume 270—Novel Forms of Carbon, C.L. Renschler, J. Pouch, D. Cox, 1992, ISBN: 1-55899-165-4

Volume 271—Better Ceramics Through Chemistry V, M.J. Hampden-Smith, W.G. Klemperer, C.J. Brinker, 1992, ISBN: 1-55899-166-2

Volume 272—Chemical Processes in Inorganic Materials: Metal and Semiconductor Clusters and Colloids, P.D. Persans, J.S. Bradley, R.R. Chianelli, G. Schmid, 1992, ISBN: 1-55899-167-0

Volume 273—Intermetallic Matrix Composites II, D. Miracle, J. Graves, D. Anton, 1992, ISBN: 1-55899-168-9

Volume 274—Submicron Multiphase Materials, R. Baney, L. Gilliom, S.-I. Hirano, H. Schmidt, 1992, ISBN: 1-55899-169-7

Volume 275—Layered Superconductors: Fabrication, Properties and Applications, D.T. Shaw, C.C. Tsuei, T.R. Schneider, Y. Shiohara, 1992, ISBN: 1-55899-170-0

Volume 276—Materials for Smart Devices and Micro-Electro-Mechanical Systems, A.P. Jardine, G.C. Johnson, A. Crowson, M. Allen, 1992, ISBN: 1-55899-171-9

Volume 277—Macromolecular Host-Guest Complexes: Optical, Optoelectronic, and Photorefractive Properties and Applications, S.A. Jenekhe, 1992, ISBN: 1-55899-172-7

Volume 278—Computational Methods in Materials Science, J.E. Mark, M.E. Glicksman, S.P. Marsh, 1992, ISBN: 1-55899-173-5

Prior Materials Research Society Symposium Proceedings available by contacting Materials Research Society

PART I

Molecular Routes to Ceramic Materials

NEW CHEMISTRY FOR THE SOL-GEL PROCESS: ACETONE AS A NEW CONDENSATION REAGENT

SUBHASH C. GOEL*, MICHAEL Y. CHIANG*, PATRICK C. GIBBONS** AND WILLIAM E. BUHRO*
*Department of Chemistry, Washington University, St. Louis, MO 63130
**Physics Department, Washington University, St. Louis, MO 63130

ABSTRACT

A nonhydrolytic sol-gel process that uses acetone as the condensation reagent is described. In the new process, the hydrocarbon soluble $Zn(OCEt_3)_2$ is converted to a rigid, transparent gel and then to ZnO by an aldol-condensation-elimination sequence. The chemical pathway for the transformation has been established, and differs completely from the pathway for the conventional *hydrolytic* sol-gel processing of alkoxide precursors. The new, nonhydrolytic strategy may prove applicable to a variety of metal-alkoxide systems.

INTRODUCTION

In this paper we describe a new, *nonhydrolytic* chemical strategy for the gelation of a zinc dialkoxide and for its subsequent conversion to zinc oxide (zincite). As described below, we expect that this strategy may be applicable to other *basic* metal alkoxides, and with a simple modification to metal alkoxides in general.

Although the *hydrolytic* gelation of alkoxide precursors serves well for silicon and a few other elements [1], the *hydrolytic* gelation of alkoxide precursors of p-, d-, and f-block *metallic* elements is generally difficult to achieve. The hydrolysis behavior of silicon alkoxides and metal alkoxides is known to be very different; the underlying reasons for this difference have been explained in recent reviews by Livage, Sanchez, and Henry [2,3]. Metallic elements are significantly more electropositive than silicon, and can often expand their coordination numbers to allow associative substitution reactions with H_2O. As a result metal alkoxides have higher reactivities towards H_2O than do silicon alkoxides, and upon hydrolysis they readily form precipitates rather than processable gels. This effect should be most extreme for alkoxides of very electropositive, low-coordinate metals.

Several techniques for the successful sol-gel processing of metal alkoxides are now under development [2-4]. One approach is the use of modified alkoxide precursors in which some of the alkoxide ligands are replaced by more hydrolytically stable ligands, typically

 ©1992 Materials Research Society

carboxylates or β-diketonates. Hydrolysis of the resulting mixed-ligand precursors proceeds anisotropically because of the different hydrolysis rates for alkoxide and carboxylate or β-diketonate ligands, and gelation rather than precipitation is frequently observed. Another approach is the use of chemical or physical processes that release H_2O very slowly and therefore promote the slow, controlled hydrolysis of alkoxide precursors. For example, the esterification of acetic acid by alcohols gives slow, in-situ H_2O generation, which has allowed gelation in some metal-alkoxide systems [3]. Therefore, the *hydrolytic* gelation of metal alkoxides can be achieved with appropriate precursor and processing strategies.

We propose that a separate means of countering the extreme hydrolytic reactivity of metal alkoxides is to avoid the use of H_2O entirely. *Nonhydrolytic* processes for converting alkoxides and related precursors to intermediate gels and then to oxide materials should be possible. Condensation reagents other than H_2O could be identified that would provide new condensation pathways and different condensation kinetics. Gels of an entirely new nature could result. Such techniques, once developed, could provide another solution to the metal-alkoxide problem, and would expand the capabilities of sol-gel processing in general.

We have found that acetone - but not H_2O - induces the gelation and subsequent conversion of $Zn(OCEt_3)_2$ to ZnO. The acetone-induced process is the result of a formal aldol-condensation-elimination sequence; the reaction pathway is very different than for conventional, *hydrolytic*, sol-gel chemistry. Our studies have established the identity of the gel-forming intermediate and have defined some of the basic requirements for the extension of this new *nonhydrolytic* sol-gel reaction to the alkoxides of other elements. A preliminary account of our discovery is provided herein.

RESULTS

We previously reported the syntheses of the first soluble zinc dialkoxides (see eq 1) [5]. The 3-ethyl-3-pentoxide complex $Zn(OCEt_3)_2$ and the 2-methoxyethoxide complex $Zn(OCH_2CH_2OMe)_2$ were selected for use in the present study. $Zn(OCEt_3)_2$ is soluble in benzene, toluene, and THF and is thus convenient for solution-phase studies. The molecular structure of $Zn(OCEt_3)_2$ was determined by single-crystal X-ray diffraction and is shown in Figure 1; the details of the structure solution will be published elsewhere. As shown in Figure 1, $Zn(OCEt_3)_2$ is a solid-state trimer having 3- and 4-coordinate zinc atoms. In solution, $Zn(OCEt_3)_2$ shows a slightly higher degree of aggregation ($n = 4.1$) [5]. $Zn(OCH_2CH_2OMe)_2$ is soluble in 2-methoxyethanol, but not in common hydrocarbon or ethereal solvents; its structure is unknown.

$$Zn[N(SiMe_3)_2]_2 + 2ROH \longrightarrow Zn(OR)_2 + 2HN(SiMe_3)_2 \qquad (1)$$

R = CEt_3
CH_2CH_2OMe
$CH_2CH_2OCH_2CH_2OMe$
$CH_2CH_2NMe_2$
$CHMeCH_2NMe_2$
$CH_2CH_2NMeCH_2CH_2NMe_2$

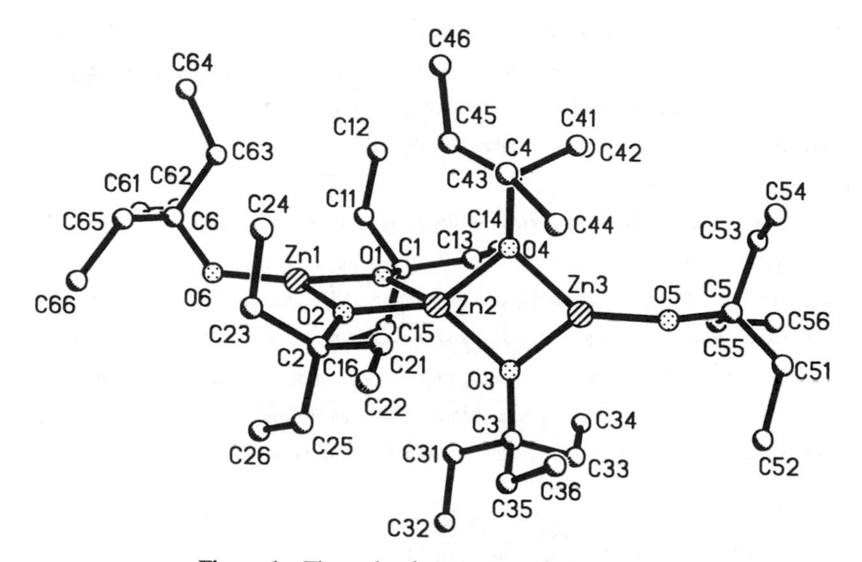

Figure 1. The molecular structure of $[Zn(OCEt_3)_2]_3$.

We first investigated the hydrolyses of $Zn(OCEt_3)_2$ and $Zn(OCH_2CH_2OMe)_2$ under various conditions to determine whether *hydrolytic* gelation would result. The hydrolysis of $Zn(OCEt_3)_2$ in THF by the dropwise addition of a dilute H_2O solution in THF gave a fine, white precipitate immediately. X-ray powder diffraction (XRD) and transmission electron microscopy (TEM) showed that the precipitate consisted of loose aggregates of 20 nm crystalline ZnO particles. Similarly, the hydrolysis of $Zn(OCH_2CH_2OMe)_2$ in 2-methoxyethanol by the dropwise addition of a dilute H_2O solution in 2-methoxyethanol gave a fine, white precipitate. XRD and TEM showed that the precipitate consisted of aggregates of 8 nm primary ZnO particles. In both cases, precipitation of ZnO commenced upon addition of substoichiometric amounts of H_2O. Our results parallel those previously reported by Heistand for the hydrolysis of $[EtZn(O\text{-}t\text{-}Bu)]_4$ [6], in which (monodispersed) spherical aggregates of 15 nm ZnO primary particles were obtained.

The hydrolysis of $Zn(OCH_2CH_2OMe)_2$ was also performed under *basic* conditions by addition of a concentrated aqueous ammonia solution to a 2-methoxyethanol solution of $Zn(OCH_2CH_2OMe)_2$. A colorless, homogeneous solution resulted, which deposited a white precipitate rapidly when the solution was warmed. XRD and TEM revealed that the precipitate consisted of clusters of rod-shaped ZnO crystallites having dimensions of 100 x 1000 nm. The hydrolyses of simple zinc salts under *basic* conditions in the presence of amines or hydroxide ion have been previously shown to give ZnO particles having rod-shaped morphologies [7]. From this and the previous experiments, we concluded that the direct hydrolyses of zinc dialkoxides were unlikely to lead to gelation, but rather only to precipitation. Therefore, the zinc-alkoxide precursors are excellent examples of electropositive, low-coordinate metal alkoxides that are too hydrolytically reactive for conventional sol-gel processing.

However, the addition of acetone to benzene solutions of $Zn(OCEt_3)_2$ resulted in firm, transparent gels within 10 min at room temperature, or within several seconds upon gentle warming. The gels were dissipated by heating above ca. 60 °C for 10 min to give mesityl oxide, Et_3COH, and hydrated zinc oxide according to eq 2. We discovered eq 2 by following up on our observation that $Zn(OCEt_3)_2$ decomposed in acetone-d_6 to give Et_3COD and ZnO while we were attempting to obtain an NMR spectrum. The conversion of acetone to mesityl oxide is formally an aldol-condensation-elimination sequence.

$$4\ \underset{\text{acetone}}{(CH_3)_2C{=}O} + Zn(OCEt_3)_2 \xrightarrow{C_6H_6} [\text{GEL}] \longrightarrow 2\ \underset{\text{mesityl oxide}}{(CH_3)_2C{=}CHC(O)CH_3} + ZnO(H_2O) + 2Et_3COH \quad (2)$$

The aldol condensation is a standard reaction in organic chemistry in which the (often base-catalyzed) dimerization of carbonyl compounds leads to β-hydroxyketones according to eq 3 [8]. (The fate of the original carbonyl compounds is shown by the boxes.) In some cases, aldol condensation is followed by the spontaneous dehydration (H_2O elimination) of the β-hydroxyketones to give α,β-unsaturated carbonyl compounds (see eq 3). This occurs when the resulting α,β-unsaturated carbonyl compounds are particularly stable. The formation of mesityl oxide from acetone in eq 2 *formally* corresponds to the overall process in eq 3, but we will show below that the chemical pathway actually followed by eq 2 is slightly different than eq 3. The aldol condensation is used in organic synthesis for the construction of carbon-carbon bonds and for controlling stereochemistries but to our knowledge it has never before been used in metal-oxide preparations.

basic catalyst

dehydration $-H_2O$ (3)

Under typical conditions $Zn(OCEt_3)_2$ and excess acetone were allowed to react in benzene, toluene, or THF; a typical combination was $Zn(OCEt_3)_2$ (2.00 g, 6.77 mmol), acetone (5 mL), and benzene (50 mL). The amount of excess acetone used was not critical to gel *formation*, but it did affect the *stabilities* of the gels (see below). The resulting transparent, colorless, rigid gels exhibited slow shrinkage and separation from the solvent over ca. 12 h at room temperature. During the shrinkage period the gels became progressively cloudier and ultimately ($>$ 12 h) dissipated to give white precipitates. Chemical and spectroscopic analyses indicated that the as-precipitated powders were hydrated ZnO containing small amounts of residual $OCEt_3$ groups (found: %C, 6.42). XRD and electron diffraction in the TEM gave patterns corresponding to ca. 4 nm zincite crystallites. A TEM micrograph and the electron-diffraction pattern of the as-precipitated material is shown in Figure 2; the primary particle size exhibited in the micrograph was consistent with the XRD data. Additionally, the larger crystallites in Figure 2 clearly showed a rod-shaped morphology, which appeared to be developing even in the smaller crystallites. We concluded that the dissipation of the gels produced basic conditions which favored the rod-shaped zincite morphology; this observation bears on our proposed chemical pathway (see below).

Several other observations were also important for determining the chemical pathway that the acetone-induced sol-gel reaction follows. Spectroscopic monitoring by ^1H NMR indicated that 1.6 mol of Et_3COH had formed at the gel point in eq 2, but that no mesityl

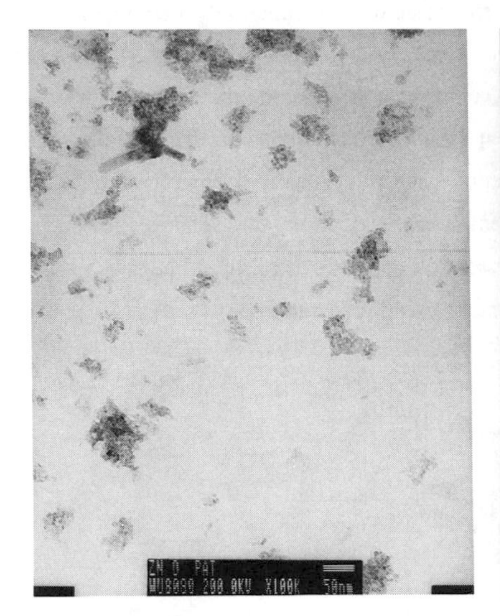

Figure 2. A TEM micrograph (left) and electron-diffraction pattern (right) for the as-precipitated, gel-derived, hydrated zinc oxide powder. The dashed scale bar on the micrograph is 50 nm.

oxide had yet been liberated. Mesityl oxide formed later, during the shrinkage and dissipation of the gels. When $Zn(OCEt_3)_2$ and acetone were allowed to react in a 1:1 stoichiometric ratio a gelatinous material was obtained that became an insoluble solid after solvent removal. Chemical analyses (Zn, C, and H) and IR spectroscopy ($\nu_{C=C}$ = 1638 cm^{-1}) identified the insoluble solid as the polymeric, mono-enolate complex $\{(Et_3CO)Zn[OCMe(=CH_2)]\}_n$ resulting from deprotonation of acetone by a $OCEt_3$ ligand as shown in eq 4. Finally, when $Zn(OCEt_3)_2$ and acetone were allowed to react in a 1:2 stoichiometric ratio a very stable gel was obtained, which did not dissipate in 2 d at room temperature. These observations will be interpreted below.

$$Zn(OCEt_3)_2 + \underset{1:1}{Me_2C{=}O} \longrightarrow 1/n\left[(Et_3CO)Zn{-}O{-}C(Me){=}CH_2\right]_n + Et_3COH \quad (4)$$

DISCUSSION

We believe it is important to establish the chemical pathway for the acetone-induced sol-gel reaction of eq 2 to estimate its potential generality and to indicate how the properties of sols and gels derived from it may be controlled. Knowledge of the detailed pathway should suggest other metal alkoxides for which the new process can also succeed. The identification of the gel-forming intermediate in eq 2 should provide clues as to how the viscosities, stabilities, and other properties of such sols and gels may be altered.

Our proposed chemical pathway for eq 2 is given in Figure 3. The first two steps are the formation of the mono-enolate complex described above, and then the corresponding bis-enolate complex by the stepwise deprotonation of two acetone molecules. The following step is the addition of acetone to the enolate ligands to give *aldolate* ligands. This is the carbon-carbon bond-forming step in the aldol condensation. Although we have not isolated nor directly observed the proposed aldolate intermediate in Figure 3, we have isolated an analogous aldolate complex from a related reaction. As shown in eq 5, $Zn[N(SiMe_3)_2]_2$, which is a functional equivalent of $Zn(OCEt_3)_2$, is able to bring about the aldol condensation of acetone [9]. The structure of the resulting aldolate complex has been verified by X-ray crystallography [9]. Consequently, the steps leading up to the proposed aldolate intermediate in Figure 3 have been substantiated by direct demonstration or by analogy to very similar chemical processes.

$$2Zn[N(SiMe_3)_2]_2 + 4Me_2C{=}O \xrightarrow{-\ 2HN(SiMe_3)_2} [\text{aldolate complex: } (Me_3Si)_2N\text{–}Zn(\mu\text{-O})_2Zn\text{–}N(SiMe_3)_2] \qquad (5)$$

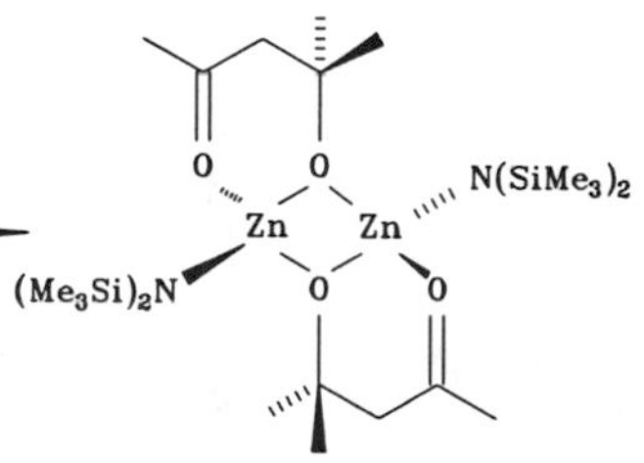

We propose that the aldolate intermediate in Figure 3 is the thermal precursor to ZnO. Keto-enol tautomerization of the aldolate ligands generates enolic protons, which are subsequently transferred to the alkoxide oxygens with the concurrent elimination of mesityl oxide. In these steps aldolate ligands are converted to hydroxide ligands. The resulting $Zn(OH)_2$ is basic and decomposes to hydrated zinc oxide crystallites having rod-shaped morphologies.

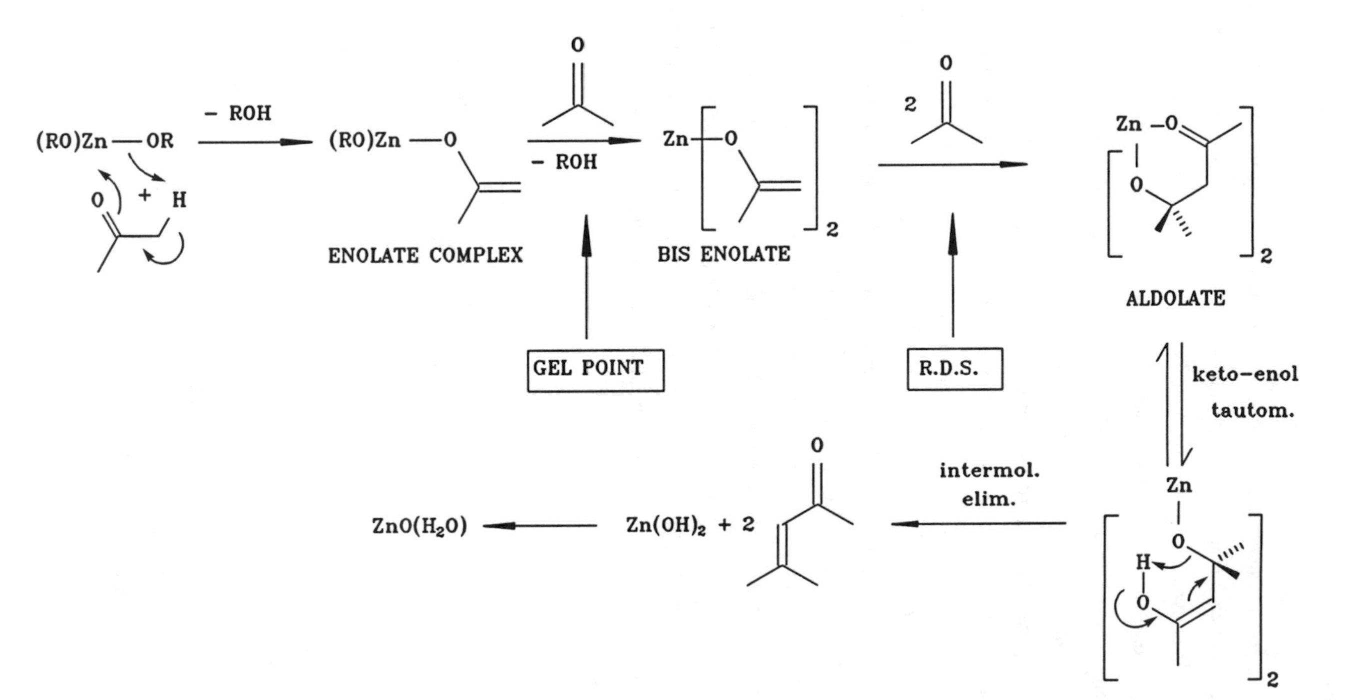

Figure 3. The proposed chemical pathway for the acetone-induced sol-gel reaction of eq 2 (R = $OCEt_3$).

Another reasonable proposal for the eq 2 pathway can be ruled out. Recall that one approach to moderating the hydrolytic reactivity of metal alkoxides is to couple sol-gel processing with chemical reactions that generate H_2O very slowly (see the discussion above). Therefore a reasonable possibility for the eq 2 pathway is the slow, in-situ generation of H_2O by eq 3, followed by the controlled hydrolysis of $Zn(OCEt_3)_2$. If this pathway were actually operating then mesityl oxide would be present at the gel point. Although we have observed large quantities (1.6 mol) of Et_3COH by ^1H NMR at the gel point, mesityl oxide is not initially present in detectable amounts (see above). Mesityl-oxide formation occurs *after* the gel point, during the shrinkage and dissipation of the gel. Because the generation of oxo ligands requires the production of mesityl oxide, we conclude that oxo ligands do not exist in substantial numbers within the gel, but are formed later.

We can now identify the gel-forming intermediates and the rate-determining step in the dissipation of the gel. The presence of 1.6 mol of Et_3COH and the absence of mesityl oxide at the gel point implicate the mono-enolate and the bis-enolate intermediates (see Figure 3) as primary constituents of the gel. We propose that the solid-phase gel network is composed primarily of polymeric enolate intermediates, which are made polymeric by extensive Zn-O-Zn bridging through enolate oxygens (analogously to the structures of polymeric metal alkoxides). We do not exclude the possibility that small numbers of oxo ligands may serve as branch points in the gel network.

The rate-determining step for ZnO formation is affected by the acetone concentration. Thus, when the reaction is carried out with a $Zn(OCEt_3)_2$:acetone ratio of 1:2 the gels are stable for at least 2 d. When a typical excess of acetone is employed the gels dissipate in ca. 12 h. When acetone is used as the reaction solvent, ZnO precipitates in < 15 min. Therefore, the rate-determining step must be the formation of the aldolate intermediate shown in Figure 3 because that is the only step following gel formation that involves acetone. The principal conclusions are that gelation results from the formation of polymeric *enolate* intermediates, and the subsequent conversion to ZnO results from the thermal decomposition of an *aldolate* intermediate.

Is the acetone-induced sol-gel process likely to succeed with other metal alkoxides and with other ketones? The general requirements that emerge from Figure 3 are as follows. (1) The metal alkoxides must be sufficiently *basic* to deprotonate ketones. (2) The resulting enolate intermediates must be stable, polymeric, and have structures appropriate to serve as the network components of gels. (3) The enolate intermediates must react further with ketones to generate unstable aldolate intermediates, which decompose to the corresponding oxides. We have found that other ketones such as 2-

butanone can replace acetone in eq 2, although the gelation times are longer. It seems likely that the stabilities and viscosities of the gels may be affected by the ketone used; more studies are required to state this with certainty.

We have just begun to test the nonhydrolytic eq 2 process with other metal alkoxides. Gel formation is observed in reactions between $Ba(O\text{-}i\text{-}Pr)_2$, which is a basic metal alkoxide, and acetone although the resulting gels are very cloudy and of apparently poor quality. In contrast, no reaction is observed between $Ti(O\text{-}i\text{-}Pr)_4$, which is a non-basic metal alkoxide, and acetone. However, when the less basic O-*i*-Pr ligand in $Ti(O\text{-}i\text{-}Pr)_4$ is replaced by the more basic NMe_2 ligand (i.e., $Ti(NMe_2)_4$), a gel-forming reaction with acetone is observed [10]. Therefore, the generality of the acetone-induced sol-gel process remains to be established although the basicity requirement is apparently confirmed.

A related approach for non-basic metal alkoxides can be suggested. Non-basic metal alkoxides will be incapable of bringing about the aldol condensation of acetone to generate aldolate intermediates analogous to the aldolate in Figure 3. However, it should be possible to introduce aldolate ligands directly by alcohol-interchange reactions between metal alkoxides and diacetone alcohol, $MeCOCH_2CMe_2OH$. Indeed, our preliminary results indicate that $Ti(O\text{-}i\text{-}Pr)_4$ and diacetone alcohol undergo a gel-forming reaction that ultimately affords TiO_2 [10]. We will provide the details of this study in the near future.

CONCLUSIONS

A *nonhydrolytic* sol-gel process has been developed that makes use of acetone as a new condensation reagent. The acetone-induced condensation of $Zn(OCEt_3)_2$ has been shown to proceed by an aldol-condensation-elimination sequence. Polymeric enolate intermediates are the primary constituents of the gel network. Subsequent formation and decomposition of an aldolate intermediate results in ZnO. It is likely that the technique can be extended to other basic metal alkoxides and can be modified for non-basic metal alkoxides. Consequently, this new strategy may ultimately provide an important alternative to conventional hydrolytic sol-gel processing.

ACKNOWLEDGMENTS

Support to W.E.B. was provided by NSF Grant DMR-8914196 and by an NSF Presidential Young Investigator award (1991-1996). The Washington University High-Resolution NMR Service Facility was funded in part by NIH Biomedical Research-Support Shared-Instrument Grant 1 S10 RR02004 and a gift from the Monsanto Co. The Washington University X-ray Crystallography Facility was funded by the NSF Chemical Instrumentation Program (Grant CHE-8811456).

REFERENCES

1. L.L. Hench and J.K. West, Chem. Rev. 90, 33 (1990).
2. J. Livage, M. Henry and C. Sanchez, Prog. Solid State Chem. 18, 259 (1988).
3. C. Sanchez and J. Livage, New J. Chem. 14, 513 (1990).
4. L.G. Hubert-Pfalzgraf, New J. Chem. 11, 663 (1987).
5. S.C. Goel, M.Y. Chiang and W.E. Buhro, Inorg. Chem. 29, 4646 (1990).
6. R.H. Heistand and Y.-H. Chia in Better Ceramics Through Chemistry II, edited by C.J. Brinker, D.E. Clark and D.R. Ulrich (Mater. Res. Soc. Proc. 73, Pittsburgh, PA 1986) pp. 93-98.
7. M.A. Vergés, A. Mifsud and C. Serna, J. Chem. Soc., Faraday Trans. 86, 959 (1990).
8. A.T. Nielson and W.J. Houlihan in Organic Reactions, Volume 16, edited by R. Adams, A.H. Blatt, V. Boekelheide, T.L. Cains, A.C. Cope, D.J. Cram and H.O. House (Wiley, New York, 1968) pp. 1-438.
9. S.C. Goel, M.Y. Chiang and W.E. Buhro, J. Am. Chem. Soc. 113, 7069 (1991).
10. S.E. Bates and W.E. Buhro, unpublished results.

HETEROMETALLIC AGGREGATES AS INTERMEDIATES ON THE MOLECULAR ROUTES TO MULTICOMPONENT OXIDES

LILIANE G. HUBERT-PFALZGRAF
Laboratoire de Chimie Moléculaire, Associé au CNRS, Université de Nice-Sophia Antipolis, 06108 Nice Cedex 2, France

ABSTRACT

Metal alkoxides are often associated with more accessible precursors such as carboxylates, nitrates, ß-diketonates, in chemical routes to multicomponent oxides. The molecular constitution of such solutions has been examined for systems based on niobium or zirconium. The heterometallic aggregates $MNb_2(\mu\text{-OAc})_2(\mu\text{-OR})_4(OR)_6$ (M = Cd, Mg), $PbZr_3(\mu_4\text{-O})(\mu\text{-OAc})_2(\mu\text{-OR})_5(OR)_5$ and $Gd_2Zr_6(\mu_4\text{-O})_2(\mu\text{-OAc})_6(\mu\text{-OR})_{10}(OR)_{10}$ R = iPr have been characterized by X-ray diffraction. Preliminary results on the hydrolysis experiments of these aggregates are given.

INTRODUCTION

Multicomponent oxides represent an important part of advanced materials due to their large palette of technical applications [1,2]. Such quaternary or complex oxides are often based on large transition metals (titanium, zirconium, niobium), lanthanides or heavy main group elements, namely barium or lead. Chemical routes such as the sol-gel process, metal-organic decomposition (MOD) [3] or chemical vapor phase decomposition reactions (MOCVD) [4] are versatile methods for obtaining ceramics in various shapes and thin layers. The first two techniques require the preparation of a solution containing the necessary metal cations in the appropriate stoichiometry. Metal alkoxides $M(OR)_n$ or oxoalkoxides $MO(OR)_n$ are versatile molecular precursors of metal oxides, one of their most attractive features being their solubility in a large variety of solvents and their ability to form heterometallic species, especially by mixing alkoxides of different metals [5]. However, to overcome the difficulty in handling alkoxides and/or in their availability, commonly accessible salts such as carboxylates (often acetates), nitrates [6], halides, hydroxides or ß-diketonates [7) have often been used in conjunction with metal alkoxides [8]. The molecular constitution of such precursor solutions is almost unknown, although a better understanding could allow a better control of the hydrolysis-polymerization process and thus of the properties of the resulting material.

Metal acetates are the most common precursors associated to metal alkoxides. In the presence of niobium isopropoxide, anhydrous cadmium [12], magnesium or lead acetate dissolve readily in toluene or hexane at room temperature over one to two hours, according to eq. I

$$M(OAc)_2 + 2Nb(OiPr)_5 \xrightarrow{\text{hexane}} MNb_2(OAc)_2(OiPr)_{10} \qquad (1)$$

The resulting compounds $M(OAc)_2Nb_2(OiPr)_{10}$ M = Mg (**1**), Cd (**2**), Pb (**3**) obtained in quantitative yields, result from a simple addition reaction. The ν_{as} CO_2 stretching frequency of the carboxylate is shifted to higher values with respect to $M(OAc)_2$ 1590 cm^{-1} for **1** to compare with $Mg(OAc)_2$ (1602 cm^{-1}), and 1574 cm^{-1} for **2** to compare with $Cd(OAc)_2$ (1550 cm^{-1}) and 1560 for **3** (1515 cm^{-1} for $Pb(OAc)_2$), thus excluding the formation of mixed crystals $M(OAc)_2$,$2Nb(OiPr)_5$. The heterometallic character was established for both cadmium and magnesium compounds which were shown to be isostructural by an X-ray structural determination. These trinuclear species display a bent, open-shell structure ($<$NbMNb: 139.66°) with alternating Nb and M atoms, all metals being 6-coordinate (fig. 1). The acetate groups act as assembling ligands and the solid state structure is retained upon dissolution in polar or non-polar solvents, as shown by NMR techniques. The acetate ligands appear equivalent in ^1H as well as ^{13}C NMR while fluxionality is observed for the M(Nb)-OR bonds but these exchange phenomena can be frozen out (at different temperatures depending on the metal M) and three types of OR groups are then detected. Cadmium and lead are metal nuclei giving high resolution spectra (I = 1/2); their sensitivities are 12.3% and 20.6% respectively [13]. Only one signal at high field is detected for both **2** and **3** in the NMR spectrum (δ_{113Cd} = 33.8 ppm for **2**, δ_{207Pb} = 2390 ppm for **3**).

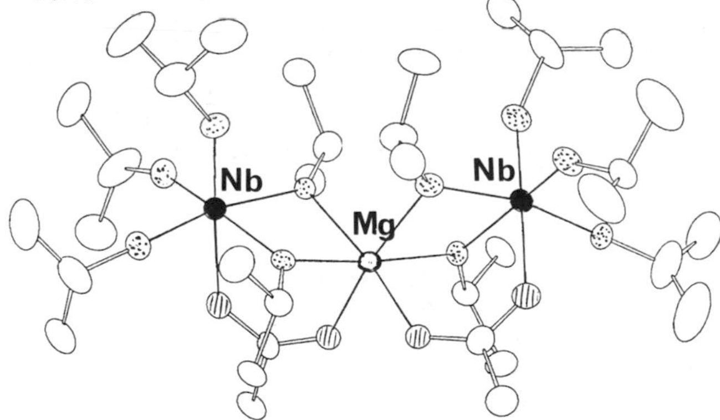

Fig 1: Molecular structure of $MgNb_2(\mu\text{-}OAc)_2(\mu\text{-}O^iPr)_4(O^iPr)_6$. Selected bond lengths and angles (average): Nb-OR_t 1.88, Nb-μOR 2.02, Nb-μOAc 2.17 ; Mg-μOR 2.09, Mg-μOAc 2.04 Å; Nb-O-C (terminal) (163-147°).

We report here some of the results of the investigations of the systems M(II)-$Nb(OR)_5$ M = Pb, Cd, Zn, Ba; M(II)-$M(OR)_4$ M = Ti, Zr; and Ln(III)-$Zr(OR)_4$ Ln = Y, Gd, La. Insight into the molecular composition of homogeneous solutions has been gained by using a variety of spectroscopic techniques, infrared and nuclear magnetic resonance (^1H and ^{13}C as well as metal NMR ^{207}Pb or ^{113}Cd), and in the most favorable cases by single crystal X-ray diffraction. The heterometallic aggregates $MNb_2(\mu-(OAc)_2(\mu-OR)_4(OR)_6$ (M = Cd, Mg), $PbZr_3(\mu_4-O)(\mu-OAc)_2(\mu-OR)_5(OR)_5$ and $Gd_2Zr_6(\mu_4-O)_2(\mu-OAc)_6(\mu-OR)_{10}(OR)_{10}$ R = iPr have thus been unequivocally characterized. Their transformation by hydrolysis-polycondensation reactions has been achieved.

EXPERIMENTAL

All reactions were conducted under inert atmosphere using Schlenk tube techniques. $Nb(OR)_5$ [9] and $[Zr(OiPr)_4(iPrOH)]_2$ [10] were prepared according to the literature. $Ti(OR)_4$ (R = Et, iPr) were commercial products (Aldrich) and were distilled before use. Anhydrous acetates of the different metals were obtained by refluxing the hydrates with acetic anhydride. Nitrates were dried by heating under vacuum.

^{207}Pb and ^{113}Cd NMR spectra were recorded on solutions (concentrations 0.3M for ^{207}Pb and 1M for ^{113}Cd) (Bruker AM-200 spectrometer). The chemical shifts are reported with respect to $M(NO_3)_2$ (M = Pb, Cd) in aqueous solutions as external references, and they are positive to low fields. Infra-red spectra were registered with an IR-455 Bruker spectrometer as Nujol mulls between plates for the air-sensitive derivatives and as KBr pellets for the powder resulting from hydrolysis.

Hydrolysis was achieved at room temperature in THF or isopropanol (0.1 M to 0.05 M) without additives.

The dried powders were characterized by TGA-DTA and X-ray diffraction. X-ray diffraction diagrams were measured using $CuK\alpha_1$ radiation before and after calcination at various temperatures.

RESULTS AND DISCUSSION

The M(II)-$Nb(OR)_5$ System (M = Mg, Cd, Pb)

Metal niobates or tantalates are attractive as electrooptical materials [$LiNbO_3$, [$(Sr,Ba)Nb_2O_6$ (SBN), $Pb(Sc,Nb)O_3$, (PSN)], dielectric ceramics ($CdNb_2O_6$), ceramics for microwave resonators [$PbMg_{1/3}Nb_{2/3}O_3$ (PNM), $Ba(M_{1/3}Ta_{2/3}O_3$ M = Zn (BZT), M = Mg (BMT)] [11]. While alkoxides are generally the source of niobium or tantalum, a variety of salts can in practice be used for the other metals, and have been considered.

1 and **2** are quite stable with respect to elimination of isopropylacetate and further condensation: no evolution is detected after about ten hours in refluxing toluene. A different behavior is observed for the Pb-Nb species **3** : infrared spectroscopy allows monitoring of the formation of the ester ($v_{as}CO_2$ = 1728 cm^{-1}) along with that of a single new metallic derivative (v_{as} CO_2 1586 cm^{-1}), more soluble, probably as a result of a closo structure. Significant differences are found if the reaction between $Nb(OiPr)_5$ and $Pb(OAc)_2$ (2:1 stoichiometry) is achieved directly in refluxing toluene: the formation of an oxo species, poorly soluble in non polar media, is evidenced by the presence of broad absorptions between 800-650 cm^{-1} in the IR due to extended Nb-O-Nb bonds.

The importance of the solvent in the course of the formation of mixed metal acetatoalkoxides is noteworthy. While the reaction between $Cd(OAc)_2$ and $Nb(OiPr)_5$ proceeds at room temperature, no reaction is observed with $[Nb(OEt)_5]_2$, even in refluxing toluene. Addition of small amounts of ethanol allows the reaction to proceed, probably as a result of the formation of the $Nb(OEt)_5(EtOH)$ monomer. The use of a polar solvent which might act as a ligand towards $M(OAc)_2$ appears to be an unfavorable feature on account of its complexation by a metal alkoxide. By contrast with toluene or hexane, no reaction occurs between $Pb(OAc)_2$ and $Nb(OEt)_5$ in THF, as shown by ^{207}Pb NMR, which is a convenient tool for the analysis of the solutions of lead derivatives.[14]

While the reaction between niobium isopropoxide and cadmium, magnesium or lead acetates occurs easily in mild conditions and in hydrocarbons, no reaction was observed with barium acetate, even by heating in the presence of the parent alcohol. The poor reactivity of $[Ba(OAc)_2]_4$ is generally overcome by adding acetic acid.[15] No reaction could however be induced by adding variable amounts of AcOH (up to 15 equivalents per barium), while a large excess led to gelification. Anhydrous zinc acetate also reacts with $Nb(O^iPr)_5$; however, the reaction is less selective than for the other divalent metal acetates; this is most probably related to the nature of anhydrous zinc acetate which is actually an oxide acetate $Zn_4O(OAc)_6$.

Hydrolysis of **1**, for instance in THF, leads to a reactive powder, **1b**, still containing acetate ligands ($v_{as}CO_2$ = 1572 cm^{-1}), although part of them are removed as isopropylacetate ($v_{as}CO_2$ = 1728 cm^{-1}), as shown by the infrared spectrum of the solution. The remaining carboxylates are eliminated by thermal treatment at 400°C as established by TGA. Figure 2 compares the X-ray diffraction patterns of **1b** as well as those of the powder resulting from the hydrolysis of $MgNb_2O(O^iPr)_{10}$ **4**. A pure columbite $MgNb_2O_6$ phase is obtained in both cases, but crystallization is achieved at quite different temperatures: 600°C for **1b**, while more than extra 200°C is necessary for the powder resulting from **4**. The structure of **1**, with alternating Nb and Mg atoms preforms the $[NbO_6]_\infty[MgO_6]_\infty[NbO_6]_\infty$ arrangement of the columbite phase.[16] **4** appears as a good precursor for the formation of stable sols of nanosize particules over a large range of hydrolysis ratios (up to h = 24) in isopropanol.

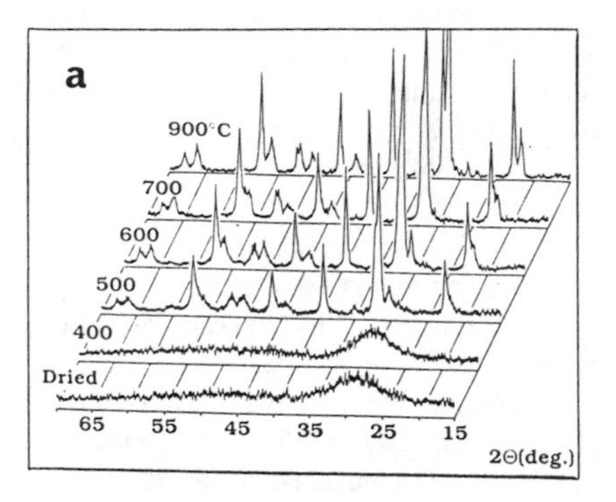

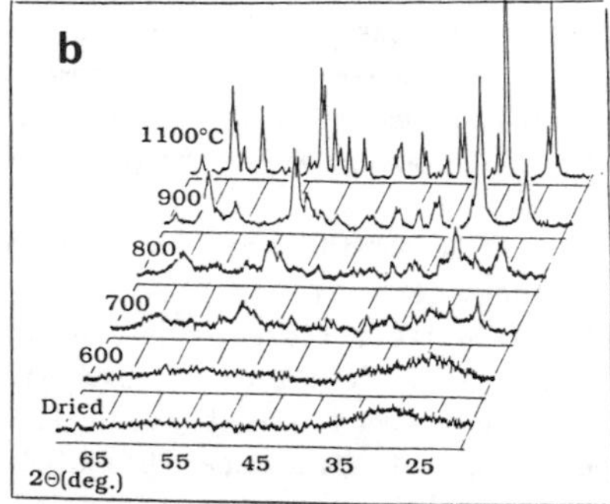

Fig. 2: X-ray diffraction patterns of the powders resulting from the hydrolysis of Mg-Nb mixed metal species; a) hydrolysis of the acetatoalkoxide **1**; b) hydrolysis of the heterometallic alkoxide.

$Pb(II)-M(OR)_4$ Systems (M = Ti, Zr)

Investigation of the Pb(II)-M(IV) system M = Ti, Zr which is also of interest for obtaining many electrooptical ceramics [17,18] has been achieved using titanium or zirconium alkoxides - $Ti(OR)_4$ R = Et, iPr and $[Zr(OiPr)_4(iPrOH)]_2$, respectively - in conjunction with lead alkoxides $Pb_4O(OEt)_6$ or $[Pb(OiPr)_2]_\infty$ or anhydrous acetate.

Lead(II) alkoxide chemistry offers a rich variety of oxo or non-oxo aggregates, depending mainly on the steric bulk of the alkoxide group [19]. $Pb_4O(OEt)_6$ has been shown to be a suitable precursor for the heterometallic Pb-Nb alkoxide $Pb_6O_4(OEt)_4[Nb(OEt)_5]_4$, for which the stoichiometry between the two metals is appropriate to the PNM ceramic formulation [16]. In the case of titanium, the heterometallic Pb-Ti ethoxides display either a 1:2 or a 1:4 stoichiometry (depending on the experimental conditions), and are thus not very attractive for obtaining the $PbTiO_3$ ceramic. Conversion of the $PbTi_2$ species by hydrolysis-polycondensation reactions leads, for instance, to the monoclinic $PbTi_3O_7$ phase, together with red PbO after thermal treatment up to 650°C.

A better match between the ratio of the two metals - as required for the $PbTiO_3$ material - was observed in the case of isopropoxide derivatives [20] The depolymerization of $[Pb(O^iPr)_2]_\infty$ by $Ti(O^iPr)_4$ occurs in refluxing toluene and in the presence of trace amounts of isopropanol in a 1:1

stoichiometry. The resulting mixed metal polynuclear oxoalkoxide **5** shows a single signal at 3888 ppm in the ^{207}Pb NMR spectrum. Its hydrolysis-polycondensation reactions in isopropanol gives successively clear and turbid sols and finally fine precipitates .Thermogravimetric analysis under nitrogen of the amorphous powder obtained for h = 32 shows that the burn off of all organics requires temperatures as high as 475°C. The resulting cristalline material corresponds to the $Pb_2Ti_2O_6$ pyrochlore and the $PbTiO_3$ perovskite phases while the pure perovskite is obtained after 700°C.

Although the reactions between $Pb(OAc)_2$ and titanium or zirconium alkoxides (R = Et, iPr) also occur at room temperature in toluene, they display quite different features by comparison with the system $Nb(OiPr)_5$-$M(OAc)_2$. The formation of oxo derivatives is now spontaneous. Thus the reaction between $Ti(OEt)_4$ and $Pb(OAc)_2$ offers the hexane soluble mixed metal oxoacetatoalkoxide $PbTi_2O(OAc)_2(OEt)_6$ ($v_{as}CO_2$ 1568 cm^{-1}) whose NMR data (^1H, ^{13}C) suggest a triangular framework with hexacoordinated titanium centers. Formation of oxo species was definitely confirmed by the structural characterization of a tetranuclear aggregate $PbZr_3(\mu_4-O)(\mu-OAc)_2(\mu-O^iPr)_5(O^iPr)_5$ **6**, (fig. 3). **6** is obtained by adding lead acetate into a solution of zirconium isopropoxide in toluene. The overall structure of **6**, in which zirconium is 6-coordinate and lead is 5-coordinate, can be considered either as two cubane units (in which one metal -apex- is missing) sharing a face (Zr1, Pb and the oxo ligand O15) or as a distorted tetrahedron. The tetranuclear $PbZr_3$ unit can indeed be viewed as a flattened tetrahedron with the lead in the "apical" position while the zirconium atoms form an isocele triangular unit (distance between O15 and that plane 0.55 Å). The distances Zr1..Zr2 and Zr1...Zr3 are comparable (3.43 Å aver.) but Zr2...Zr3 opens up to 4.10 Å and the oxo ligand is the only one connecting these two metals. The salient feature is thus the coordination of the μ_4-oxo ligand; this ligand is "concave" (Zr2-O15-Zr3 of 134.9°) and drawn towards Zr1. Such a stereochemistry contrasts with most $M_4(\mu_4-X)$ (X = C, N, O) units but is in analogy with that observed for $Ce_4O(O^iPr)_{13}(^iPrOH)$.[21] The coordination sphere of lead is a distorted square pyramid with the oxo ligand in the apical position. Five coordination is scarce for lead in molecular compounds, but a surrounding related to that in **6** is observed for lead-containing high-Tc superconductors. [15b]

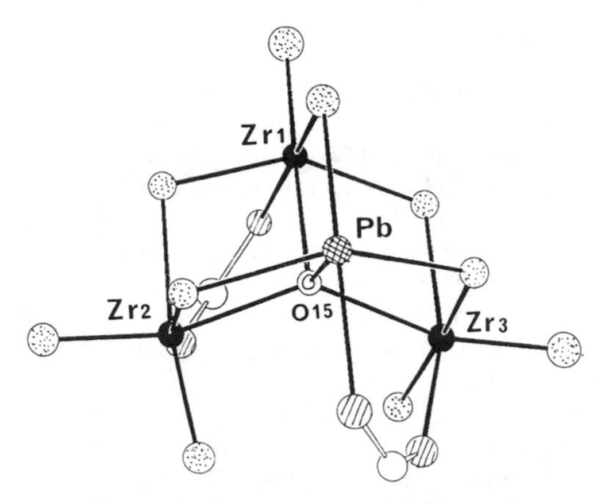

Fig. 3. Molecular structure of $PbZr_3(\mu_4\text{-}O)(\mu\text{-}OAc)_2(\mu\text{-}O^iPr)_5(O^iPr)_5$ (The carbon atoms of the isopropoxide groups have been omitted for clarity) Selected bond lengths (average): Zr-ORt 1.91, Zr-μOR 2.12, Zr-μOAc 2.17, Zr-μ_4O 2.20; Pb-μ_4-O 2.32, Pb-μOAc 2.42, Pb-μ-OR 2.46 Å; Zr...Pb 3.54 Å.

The structure of **6** is non-fluxional and is retained in solution as shown by the non-equivalent isopropoxide as well as acetate ligands (two peaks at 1.95 and 2.00 ppm in a 1:1 ratio).

At first glance, the reactions between group 4 metal alkoxides and $Pb(OAc)_2$ appear dependent on the addition order of the reactants: higher stoichiometric amounts of alkoxide are generally necessary in order to achieve dissolution of $Pb(OAc)_2$ when it is added to a suspension of the lead derivative. However, this observation mainly illustrate the difference in the kinetic of the reactions, indeed ^{207}Pb NMR established that such solutions contain in fact the same Pb species (2560 ppm). Nevertheless the Pb-M products isolated can be different as a result of the variable amount of the zirconium alkoxide present in solution with the original mixed metal species. Thus, while **6** (v_{as} CO_2 1610, 1573 cm^{-1}) derives from a solution where $Pb(OAc)_2$ was added to $[Zr(OiPr)_4(iPrOH)]_2$, a species richer in zirconium, analyzing as $PbZr_4O(OAc)_2(O^iPr)_{14}$ **7** ($v_{as}CO_2$ 1573 cm^{-1}), was obtained when the alkoxide was added to a suspension of lead acetate. Compounds **6** and **7** are formally related by adding a $Zr(OR)_4$ moiety to **6**.

<u>Ln(III)-Zr(OR)$_4$ System</u> <u>Ln = Y, La, Gd</u>

The most commonly used lanthanide alkoxides, isopropoxides, are generally pentanuclear clusters, and although they are reported to form mixed metal alkoxides, for instance with $[Zr(OiPr)_4(iPrOH)]_2$, their structural characterization turned out to be difficult.[22] On the other hand, lanthanide acetates are easily accessible.

Anhydrous lanthanide acetates are less reactive than $M(OAc)_2$ M = Mg, Cd, Pb. Indeed, no reaction was observed between $Ln(OAc)_3$ (Ln = Y, La, Gd) and zirconium isopropoxide in toluene at room temperature. Dissolution of the acetate, however, is in all cases observed by heating. In the case of gadolinium, for instance, a paramagnetic product analyzing as

$Gd(OAc)_3Zr_3O(O^iPr)_{10}$, **8**, was obtained in high yield (85%). IR data show three absorption bands vasCO_2 for the acetate ligands (1670, 1574 and 1544 cm^{-1}), and thus shifted to higher frequencies with respect to $Gd(OAc)_3$ (1539 cm^{-1}).

Compound **8** appears to be an octanuclear heterometallic oxoacetatoalkoxide of formula $Gd_2Zr_6(\mu_4\text{-}O)_2(\mu\text{-}OAc)_6(\mu\text{-}OiPr)_{10}(OiPr)_{10}$. Its molecular structure (Fig. 4) results from two $GdZr_3(\mu_4\text{-}O)(OAc)_2(OR)_{10}$ aggregates having a framework similar to that of $PbZr_3O(OiPr)_{10}(OAc)_2$ assembled via bridging acetato ligands and the two gadolinium atoms. Gd is 7-coordinate while all zirconium centers are 6-coordinate. The tetranuclear unit is more regular than **6** with nearly regular Zr...Zr distances whose values are close to the Zr...Gd one.

In view of the high yield and the absence of ester formation, the origin of the oxo ligand in **8** appears to be the isopropoxo groups. Zirconium tetramethoxyethoxide is more reactive than the isopropoxide derivative and allows reactions with lanthanides acetates to proceed at room temperature giving rise to a compound analyzing as $YZr(OAc)_3(OC_2H_4OMe)_4$ for instance in the case of yttrium and whose characterization is in progress. Reactions achieved in refluxing toluene lead to elimination of 2-methoxyethoxideacetate (va$_s$$CO_2$: 1728 cm^{-1}) and to a loss of solubility of the resulting mixed metal species.

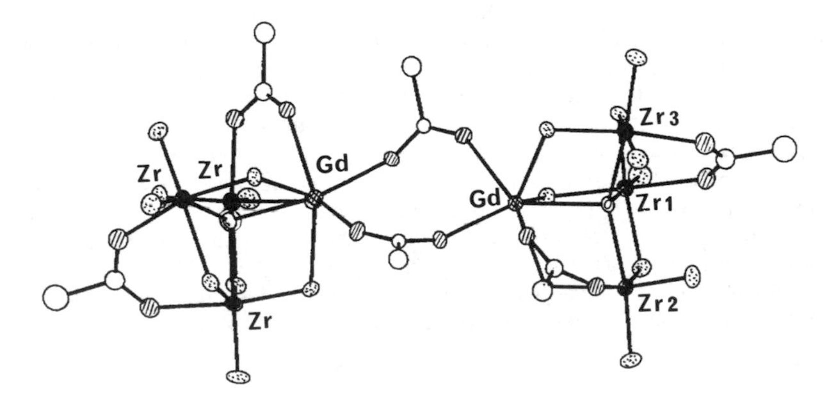

Fig. 4: Molecular structure of $Gd_2Zr_6(\mu_4\text{-}O)_2(\mu\text{-}OAc)_6(\mu\text{-}O^iPr)_{10}(O^iPr)_{10}$. Selected bond lengths: Zr-ORt 1.89, Zr-μOR Zr-μOAc (2.10-2.27), Zr-μ_4-O 2.17 ; Gd-μOR 2.27, Gdμ-OAc, Gd-μ_4O 2.42; Zr..Zr 3.57, Zr...Gd 3.59 and 5.58 Å for Gd..Gd)

Hydrolysis of **7** gives a powder containing acetate ligands (vasCO_2: 1541, vsCO2 1404 cm-1) which are eliminated by thermal treatment at 300°C while much higher temperatures are required the conversion of anhydrous

$Gd(OAc)_3$ into the oxide.[23] After crystallization (starting at 600°C) cubic ZrO_2 containing 25% of Gd_2O_3 (Fig. 5) as a result of the existence of ZrO_2-Gd_2O_3 solid solutions over a large range of compositions is obtained.

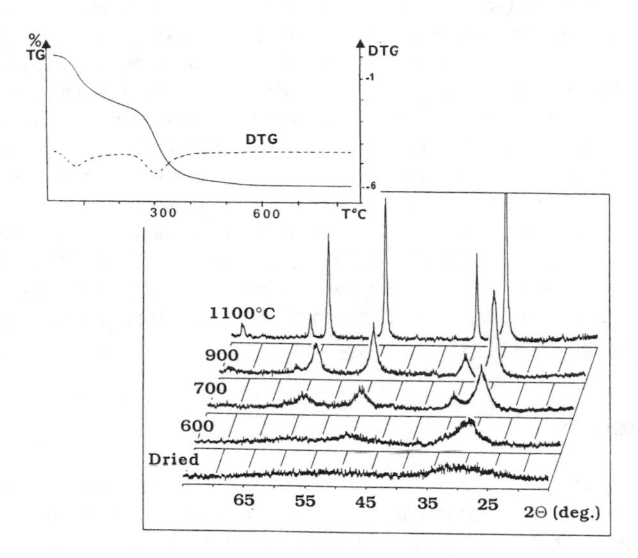

Fig. 5: X-ray diffraction patterns of the powder resulting from the hydrolysis of **8.** The insert shows the thermogravimetric analysis.

Common features for the mixed-metal acetatoalkoxides obtained so far are that the metal introduced as an alkoxide (R = Et or iPr) is predominant and that the as $v_{as}CO_2$ stretching frequencies are generally shifted to higher values with respect to the anhydrous acetate used as the starting material. This suggests that this vibration could be used as a probe for the formation of a mixed-metal species. A common feature of all structures is the hexacoordination of the transition metal, niobium or tantalum. The M-O bond distances follow the pattern Zr-OR < Zr-μ-OR Zrμ_4-O Zr-OAc and M-μOR < M-μOAc < M-μ_4O (M = Pb or Gd). The gadolinium(lead)-oxygen bond lengths are longer than the Zr-O ones as expected on the basis of the larger covalent radius for Gd or Pb. Angles Zr-O-C or Nb-O-C to terminal alkoxide ligands are large and account for the short metal-oxygen bond lengths (< Σrcov). The various metal-oxygen distances are in agreement with the values reported in the literature.

CONCLUSION

Mixed metal acetatoalkoxides can be obtained by reacting niobium isopropoxides and anhydrous acetates $M(OAc)_2$ (M = Mg, Pb, Cd) at room temperature in non polar solvents. Similar reactions are observed for group 4 metal alkoxides and $Pb(OAc)_2$. Solvents such as alcohol or THF generally preclude the formation of such mixed metal acetatoalkoxides. The reaction proceed with the formation of the smallest aggregates that allow the metals to achieve their common coordination numbers regarding the steric demand of the alkoxide ligand and are selective. Niobium, titanium and zirconium tend to be 6-coordinate and this leads to the formation of simple addition products in the case of niobium while hexacoordination can only be achieved via generation of an oxo ligand for the tetravalent metals, titanium and zirconium. The oxo ligand results from the elimination of dialkylether and thus the mixed metal oxo aggregates can be viewed as complexes between acetates and the oxo alkoxides moieties $Zr_3O(OR)_{10}$, $Ti_2O(OEt)_6$. The acetate groups act always as assembling ligands as evidenced by the X-ray structural data or by the difference in the stretching frequencies $\Delta = v_{as}CO_2 - vsCO_2 < 200cm^{-1}$.[24]

ACKNOWLEDGMENTS

The author is grateful to the CNRS (GRECO) for financial support, to Dr F. Chaput (Ecole Polytechnique) for the X-ray diffraction experiments, to B. Septe for the lead and cadmium NMR spectroscopy, and to the contributions and enthusiasm of coworkers whose names are listed in the references.

REFERENCES

1. L.G. Hubert-Pfalzgraf, New J. Chem. 11, 663 (1987).
2. H. Dislich, Angew. Chem. Int. Ed. Engl. 10, 363 (1971).
3. R.W. Vest, T.J. Fitzsimmons, J. Xuu, A. Shaikh, G.L. Liedl, A.I. Schindler and J.M. Honig, J. Solid State Chem. 73, 283 (1988); J.A. Agostinelli, C.R. Paz-Pujolt and A.K. Mehrotra, Physica C, C156, 208-210 (1988); A.S. Shaikaud and G.M. Vest, J. Am. Ceram. Soc. 69, 682-686 (1986).
4. D.C. Bradley, Chem. Rev. 89, 1317 (1989).
5. K.G. Caulton and L.G. Hubert-Pfalzgraf, Chem. Rev. 90, 969 (1990).
6. L.R. Pederson, G.D. Maupin, W.J. Weber, D.J. McReady and R.W. Stephens, Mater. Lett. 10, 437 (1991).
7. C. Sirio, O Poncelet, L.G. Hubert-Pfalzgraf, J.C. Daran and J. Vaissermann, Polyhedron 11, 177 (1992).
8. P.D. Godbole, S.B. Deshponde, H.S. Potdar and S.K. Date, Mater. Lett. 12, 97-101 (1991).
9. D.C. Bradley, B.N. Chakravarti and W. Wardlaw, J. Chem. Soc. 1956, 2381.

10. B.A. Vaartstra, J.C. Huffman, P.S. Gradeff, L.G. Hubert-Pfalzgraf, J.C. Daran, S. Parraud, K. Yunlu and K.G. Caulton, Inorg. Chem. 29, 3126-3131 (1991).
11. F. Chaput, J.P. Boilot, M. Lejeune, R. Papiernik and L.G. Hubert-Pfalzgraf, J. Am. Ceram. Soc. 72, 1355 (1989); O. Renoult, J.P. Boilot, F. Chaput, R. Papiernik, L.G. Hubert-Pfalzgraf and M. Lejeune, Ceramics Today - Tomorrow's Ceramics (Elsevier, London, 1991), p. 1991.
12. S. Boulmaaz, R. Papiernik, L.G. Hubert-Pfalzgraf, J.C. Daran and J. Vaissermann, Chem. Mat. 3, 779 (1991).
13. J.J. Dechter, NMR of metal nucleides, Part I: Main group elements. Progress in Inorganic Chemistry, edited by J.S. Lippard; vol 29, 285 (1982); R.K. Harris, J.J. Kennedy, W. McFarlane, NMR and the Periodic Table, edited by R.K. Harris and B.E. Mann (Academic Press, London 1978), p 366.
14. A. Mosset, I. Gautier-Luneau, J. Galy, P. Strehlov and H. Schmidt, J. Non Cryst. Solids 100, 339 (1988).
15. R. Papiernik, L.G. Hubert-Pfalzgraf and J.C. Daran, J. Chem. Soc., Chem. Commun. 1991, 695; L.G. Hubert-Pfalzgraf, R. Papiernik, M.C. Massiani and B. Septe, MRS Better Ceramics through Chemistry IV, , 180, 393 (1990)
15b. R.J. Cava, B. Batlogg, J.J. Krajewski, L.W. Rupp, L.F. Schneemeyer, T.Siegrist, R.B. VanDover, P. Marsh, W.F. Peck, Jr. P.K. Gallagher, S.H. Glarum, J.H. Marshall, R.C. Farrow, J.V. Waszczak, R. Hull and P. Trevor, Nature, 136, 211 (1988).
16. A.F. Wells , Structural Inorganic Chemistry, 5 ed., Oxford University Press.
17. K. Okazaki, Ceramic Bull. 67, 1946 (1988).
18. M.I. Yanovskaya, E.P. Turevskaya, N.Ya. Turova, M.Ya. Dambekalne, N.V. Kolganova, S.A. Ivanov, A.G. Segalla, V.V. Belov, A.V. Novoselova and Yu.N. Veneutsev, Inorg. Mater., 23, 584 (1987); H. Hirashina, E. Onishi and M. Nakagawa, J. Non Cryst. Sol. 121, 404 (1990).
19. R. Papiernik, M.C. Massiani and L.G. Hubert-Pfalzgraf, Polyhedron 10, 1657 (1991); S.C. Goel, M.Y. Chiang and W.E. Buhro, Inorg. Chem. 29 4640 (1990).
20. L.G. Hubert-Pfalzgraf, R. Papiernik and F. Chaput, J. Non-Cryst. Solids, in the press
21. K. Yunlu, P.S. Gradeff, N. Edelstein, W. Kot, G. Shalimoff, W. streib, B.A. Vaarststra and K.G. Caulton, Inorg. Chem., 30, 2317 (1991).
22. N. Ya Turova, N.I. Kozlova and A.V. Novoselova, Russ. J. Inorg. Chem. 25, 1788 (1980).
23. K.C. Patil, G.V. Chanrashekhar, M.V. George and C.N.R. Rao, Can. J. Chem. 46, 257 (1968)
24. G.B. Deacon and R.J. Philipps, Coord. Chem. Rev., 90, 227 (1980).

FURTHER STUDIES ON ALUMINUM OXOALKOXIDE CLUSTERS

R. A. SINCLAIR, M. L. BROSTROM, W. B. GLEASON AND R. A. NEWMARK.
201-2E-01, 3M CENTER, ST. PAUL, MN 55144

ABSTRACT

Evidence for a tetranuclear aluminum oxoalkoxide cluster $Al_4(\mu_4\text{-}O)(\mu_2\text{-}OBu^i)_5(OBu^i)_6H$ is presented, drawing on data from X-ray, NMR, MS and other spectroscopic and experimental investigations. In principle, these compounds are derived from the well-authenticated aluminum alkoxide tetramers by hydrolytic loss of only one alkoxy group. They contain a single central oxo ligand and only 5-coordinate aluminum atoms. They are the simplest discrete oxoalkoxide structures so far reported for aluminum and may be helpful in understanding the hydrolysis and condensation steps leading to hydrated aluminum oxide sols and gels.

INTRODUCTION

Many of the sol-gel procedures for the production of metal oxides from simple metal alkoxides use high molar ratios of water to alkoxide, leading to extensive hydrolysis and the formation of polymeric oxygen-bridged networks which are difficult to elucidate. A more limited hydrolysis, whereby specific metal alkoxide oligomers would first be formed and then carefully linked together, would allow greater control of crosslinking reactions and would therefore influence final materials properties.

Klemperer has already shown the value of a "building block" approach [1] in sol-gel polymerization. For example, hydrolysis of the cubic octamethyl octasilicate monomer $[Si_8O_{12}](OCH_3)_8$ proceeds through more rigid framework intermediates than those forming in the hydrolysis of simple orthosilicates, such as $Si(OEt)_4$, producing a very high surface area silica xerogel [2]. When $Si(OR)_4$ or $Ti(OR)_4$ is hydrolyzed under low molar ratios of water to alkoxide then more linear polymers are indeed obtained, instead of gel networks, and these can be spun and then converted by heating to the corresponding oxide fiber [3].

Even though the hydrolysis of alkoxides of elements other than silicon is often complicated by the variable coordination number and more electropositive nature of the metal, the importance of a low hydrolysis ratio in isolating definite oligomers is again well recognized [4]. In this way a small number of well-defined crystalline oxoalkoxides have been made and their value as reference structures in understanding mechanism of hydrolysis has been noted [4].

This paper describes a tetranuclear aluminum oxoalkoxide $Al_4O(OBu^i)_{11}H$, which contains only 5-coordinate aluminum atoms and which is structurally related to the more extensively hydrolyzed oxoalkoxide $Al_{10}O_4(OEt)_{22}$[5]. These are the only oxoalkoxides of aluminum for which crystal structures have been reported.

EXPERIMENTAL

All reactions and manipulations were performed under nitrogen using a glove box equipped with a drying train. Well-formed crystals of the aluminum oxoisobutoxide deposited at room temperature from clear solutions of aluminum tri-sec-butoxide in isobutyl alcohol, as previously reported [6]. The alcohol contained about 150 ppm water. Several days were usually required before product formed from a 10% solution. The crystalline product was also obtained by suspending a droplet of water, in the stoichiometric 1:4 ratio, on the underside of the cap used to seal a small glass reaction vial, and then allowing the water vapor to slowly diffuse into the alkoxide solution. Because of the moisture sensitivity of the aluminum oxoalkoxide, appropriate glove box and Schlenk tube procedures were followed in handling samples for analysis [6].

RESULTS AND DISCUSSION

Aluminum oxoisobutoxide is stable for several years at room temperature and is not decomposed even at 130^0C, under nitrogen. It melts sharply at 140^0C. The compound is readily soluble in organic solvents such as toluene and chloroform. The main structural features were established by single crystal X-ray diffraction [6] and are shown in Figure 1 for the metal oxygen framework. The most interesting feature is the absence of the tetrahedral and octahedral metal coordination encountered in aluminum alkoxide tetramers [7]. Instead, all four aluminum atoms are 5-coordinate. In this arrangement, the six terminal isobutoxy and five bridging isobutoxy groups can not be equally distributed amongst the four aluminum atoms. Thus, two of the aluminum atoms are linked to only one terminal alkoxy group, but share three bridging groups, while the remaining two aluminum atoms are each bonded to two terminal and two bridging alkoxy groups. The oxo ligand occupies a central μ_4 - position.

There is strong evidence that a proton is located between two of the terminal alkoxy groups as shown in Figure 1. The internuclear distance between the probable hydrogen-bonded alkoxy oxygen atoms is remarkably short, at only 2.43Å. The two Al-O bonds are elongated (1.819,1.810Å) compared with the other terminal Al-O bonds in the molecule (1.732,1.711,1.709,1.719Å). Also the Al-O-C angles are compressed to 127^0, considerably narrower than the Al-O-C angles found for the other terminal isobutoxy groups (140^0). The exact position of this proton is being determined by neutron diffraction. NMR and IR support for this hydrogen bond is as yet inconclusive. The absence of a ν(O-H) absorption may be due to the O-H-O bridge being highly symmetric.

NMR and FAB MS studies have indicated that some features of the oxoalkoxide species are preserved both in solution and in the vapor phase [6]. For example, a single unresolved resonance in the ^{27}Al NMR spectrum of a toluene solution at 35ppm is further evidence that only 5-coordinate aluminum is present [8,9]. The FAB mass spectrum contains the molecular ion at m/z 927 and relative abundance 5%, while the base peak at m/z 797 is in agreement with loss of $(Bu^i)_2O$ from the molecular ion.

The significance of this particular oxoalkoxide as a possible first stage hydrolysis product for aluminum alkoxides is heightened by comparison with the more extensively hydrolyzed aluminum oxoethoxide reported by N. Ya Turova at Moscow State University [5]. This compound formed in small yield through trace hydrolysis of benzene solutions of $Al(OEt)_3$ oligomers and contains two pentameric aluminum fragments linked by two planar μ_3-O ligands to give a centrosymmetric decameric species, $Al_{10}(\mu_4\text{-}O)_2(\mu_3\text{-}O)_2(\mu_2\text{-}OEt)_{14}(OEt)_8$ (Figure 1).

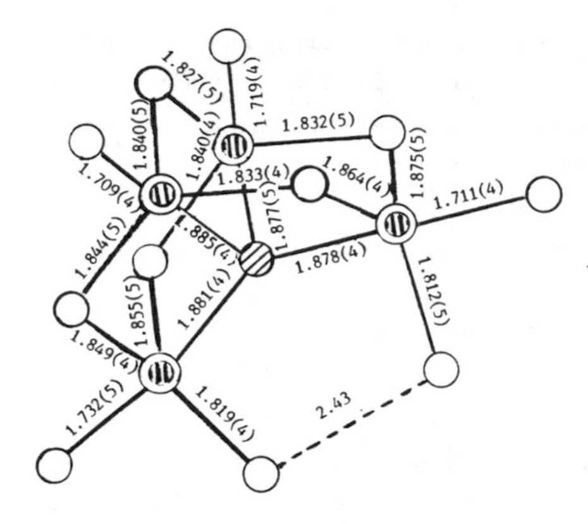

$Al_4(\mu_4\text{-}O)(\mu_2\text{-}OBu^i)_5(OBu^i)_6H$

Figure 1
Structural Relationship
Between
$Al_4(\mu_4\text{-}O)(\mu_2\text{-}OBu^i)_5(OBu^i)_6H$ and
$Al_{10}(\mu_4\text{-}O)_2(\mu_3\text{-}O)_2(\mu_2\text{-}OEt)_{14}(OEt)_8$

○ = Oxygen, R group omitted
◍ = μ_4-Oxygen
● = μ_3-Oxygen
◎ = Aluminum

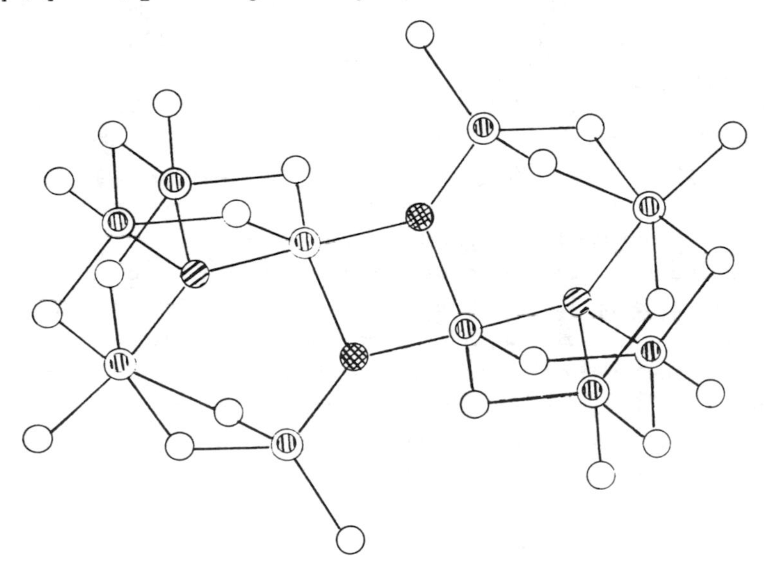

$Al_{10}(\mu_4\text{-}O)_2(\mu_3\text{-}O)_2(\mu_2\text{-}OEt)_{14}(OEt)_8$

The structural unit for the tetranuclear aluminum oxoisobutoxide can be incorporated into these ethoxide pentamers without much modification: three of the aluminum atoms remain 5-coordinate and connected to a μ_4 -oxo ligand, while the fourth aluminum increases its coordination to six. This suggests that a primary oxoalkoxide tetrameric unit may sometimes remain intact during further limited hydrolysis as represented in the following equation:

$$2Al_4(\mu_4\text{-}O)(\mu_2\text{-}OR)_5(OR)_6H + 2Al(OR)_3 + 2H_2O \longrightarrow Al_{10}(\mu_4\text{-}O)_2(\mu_3\text{-}O)_2(\mu_2\text{-}OR)_{14}(OR)_8 + 6ROH$$

Evidence for other molecular aluminum oxoalkoxides and their controlled hydrolytic condensation to higher molecular weight is being sought.

CONCLUSION

The structures determined independently for $Al_4(\mu_4\text{-}O)(\mu_2\text{-}OBu^i)_5(OBu^i)_6H$ and $Al_{10}(\mu_4\text{-}O)(\mu_3\text{-}O)_2(\mu_2\text{-}OEt)_{14}(OEt)_8$, in which 5-coordinate aluminum predominates, contain important similarities. Central μ_4-O ligands are found in both compounds, and the tetranuclear oxoalkoxide clusters appear as two subunits within the decameric framework. Because of the similarity within these structures, and since both compounds are obtained by hydrolysis of aluminum alkoxides at very low hydrolysis ratios, it is tempting to conclude that these species can be primary hydrolysis products in the sol-gel sequence. Further structural studies of hydrolyzed derivatives of aluminum alkoxides are needed to support this idea and to determine whether oxoalkoxide fragments exist in more extended oxide networks.

ACKNOWLEDGMENT

Part of this work was supported by a 3M Leading Edge Academic Program Grant. We wish to thank Professor D. C. Bradley for helpful comment on this research and thank Professors K. Caulton and L. Hubert-Pfalzgraf for sharing their insight into hydrogen-bonded metal alkoxides and oxoalkoxides.

REFERENCES

1. V. W. Day, W. G. Klemperer, V. V. Maing and D. M. Millar, J. Am. Chem. Soc., 107, 8262 (1985).
2. P. C. Cagle, W. G. Klemperer and C. A. Simmons, in Better Ceramics Through Chemistry IV, edited by B. J. J. Zelinski, C. J. Brinker, D. E. Clark and D. R. Ulrich (Mater. Res. Soc. Symp. Proc. 180, Pittsburg, PA 1990), pp. 29-37.
3. K. Kamiya, K. Tanimoto and T. Yoko, J. Mater. Sci. Lett., 5,402 (1986).
4. C. Sanchez, P. Toledano and F. Ribot, in Better Ceramics Through Chemistry IV, edited by B. J. J. Zelinski, C. J, Brinker, D. E. Clark and D. R. Ulrich (Mater. Res. Soc. Symp. Proc. 180, Pittsburg, PA 1990), pp. 47-59.
5. A. I. Yanovskii, N. Ya. Turova, N. I. Kozlova and Yu T. Struchkov, Koord. Khim., 13 (2), 242 (1987).
6. R. A. Sinclair, W. B. Gleason, R. A. Newmark, J. R. Hill, S. Hunt, P. Lyon and J. Stevens, in Chemical Processing of Advanced Materials, edited by L. L. Hench, Wiley Interscience (in press).
7. D. C. Bradley, R. C. Mehrotra and D. P. Gaur, Metal Alkoxides, Academic, London (1978).
8. M. C. Cruickshank, L. S. Dent Glasser, S. A. I. Barri and I. J. F. Poplett, J. Chem. Soc. Chem. Commun., 23 (1986).
9. L. B. Alemany and G. W. Kirker, J. Am. Chem. Soc., 108, 6158 (1986).

STUDIES ON THE FORMATION OF TIN(IV) OXYGEN COMPOUNDS BY SOL-GEL-PROCESS

H. REUTER, M. KREMSER, D. SCHRÖDER AND M. JANSEN

Department of Inorganic Chemistry, University of Bonn, Gerhard-Domagk-Str. 1, W-5300 Bonn 1, FRG

ABSTRACT

Solutions of $Sn(OR)_4$ and $R'Sn(OR)_3$ with different amounts of alkali and alkaline earth metal alkoxides, $M^{I}(OR)$ and $M^{II}(OR)_2$ respectively, have been prepared. Slow evaporation of the solvent resulted in the crystallization of mixed metal alkoxide compounds. Solid state structures have been determined for two examples: $K_3H[Sn(O^iPr)_6]_2$ and $Na_2[^iPrSn(O^iPr)_5]\cdot 5\,^iPrOH$. Solutions as well as solids have been hydrolyzed by moisture. The thermal behaviour of the resulting amorphous powders was investigated by TG, DTA and XRD up to 1000°C in argon atmosphere. The structural evolution during the solid state step mainly depends on the nature and amount of the mono- and bivalent cations. The different influence of both parameters is described.

1. INTRODUCTION

In the past, synthesis of ternary metal tin oxides was usually achieved by the traditional solid state high temperature processes using powdered salt mixtures of the two metals [1]. In contrast, the nowadays well introduced sol-gel-process [2] should allow the preparation of such bimetallic oxides from solutions containing alkoxides of both metals. The solutions are subsequently hydrolyzed to amorphous gels, which are then dehydrated at elevated temperatures to the corresponding mixed metal oxides.

$$\underset{\text{solution}}{\langle Sn(OR)_4\rangle} \xrightarrow[\text{step}]{\text{solution}} \underset{\text{gel}}{[SnO(OH)_2\cdot n\,H_2O]} \xrightarrow[\text{step}]{\text{solid state}} \underset{\text{ceramic}}{SnO_2} \qquad (1)$$

From this point of view, the sol-gel-process has to be divided into two different steps. The first, the **solution step**, starts with the molecular solution of the metal alkoxide and results via the reaction with water in the formation of a gel. The second, the **solid state step**, leads from this gel to the final oxide.

The potential advantages of the sol-gel-process are likely to arise from a better homgeneity, which is attained because all components are mixed in solution on a molecular level, and from a high surface area of the amorphous powders, which lowers processing temperatures compared to the traditional synthetic strategies. However, although processing is very simple, the chemistry involved in the two steps of the sol-gel-process is very complicated, even in those cases where only one metal alkoxide is present in solution, as indicated in equation 1.

In addition, chemistry is also required for the characterization and design of modified and new precursors, as well as for the determination of the physical and chemical properties of the new ceramic materials thus obtained.

During our systematical studies on the structural evolution taking place during the solution step in the binary tin(IV) oxide system, we were able to develop a complete model [3] of this step by modifying the normal sol-gel-process [4]. Lately, our interest has been focussed on the investigation of the solid state step, where we hope to get more detailed information by using bimetallic alkoxide solutions.

In particular, we are interested in isolating solid, crystalline bimetallic precursors from these solutions, uncovering the nature of the bimetallic gels obtained by their hydrolysis, determining the intermediate species formed during thermal treatment and synthesizing new materials based on ternary tin(IV) oxides. Again, we use side by side pure tin(IV) and monoorganotin(IV) alkoxides.

This work reports the first results obtained by the investigation of those tin(IV) precursor systems in mixtures with alkali and alkaline earth metal alkoxides.

2. EXPERIMENTAL

Isopropyltin trichloride was prepared with 89.6 % yield via the reaction of isopropyltriphenylstannane [5] with HCl in toluene [6]. Isopropyltin triisopropoxide was synthezied by the reaction of the corresponding trichloride with potassium in isopropanol [7]. Tin(IV) chloride (Merck) and *n*-butyltin trichloride (Aldrich) were commercially attained and used without further purification. Metalls and solvents were purchased in high purity from standard sources (Riedel-de Haen, Aldrich, Merck), alcohols were dried according to standard procedures [8]. All manipulations during alkoxide syntheses and isolation were carried out under dry argon atmosphere [9], using standard Schlenk techniques [10].

Thermogravimetric and differential thermal analyses were performed in alumina crucibles under argon atmospheres with a NETZSCH STA 429 instrument using heating rates of 5.0°C/min. X-Ray diffraction studies of the amorphous and crystalline powders were undertaken with Cu-K_α radiation and Ge monochromator in transmission mode using a STOE automatic powder diffractometer STADI-P. Single crystals were investigated with an Enraf-Nonius CAD4-diffractometer using Mo-K_α radiation and graphit monochromator.

For the investigation of the solid state reactions all alkoxide solutions were allowed to hydrolyze in moisture during slow evaporation of the solvent and dried at room temperature for one week. All materials thus obtained are fine white powders, which are amorphous as determined by X-ray diffraction (XRD). Their thermal behaviour was determined by thermogravimetry (TG) and differential thermal analysis (DTA), continously from room temperature to 1000°C and discontinously by XRD of powder samples heated for 6h at intermediate temperatures using a conventional furnace.

Detailed synthetic procedures, as well as spectroscopic and single crystal X-ray structural data are being published elsewhere.

3. INVESTIGATIONS IN THE SYSTEM $Sn(OR)_4/KOR$

In our first approach to the synthesis of ternary tin(IV) oxides via sol-gel-process, we studied the $KOR/Sn(OR)_4$-system using isopropanol as solvent. Starting from these bimetallic alkoxide solutions, the sol-gel-process should proceed in the same way as with a tin(IV) alkoxide alone. In addition, the second metal will cause a similiar reaction sequence, which will be more or less independent from the other, as shown in equation 2. Although the bimetallic alkoxide solutions (1) can be prepared in a wide range of different stoichiometries, the final product of the sol-gel-process may be a mixture of hydroxides and oxides (3) or a new ternary phase (4).

$$\begin{matrix} \langle KOR \rangle_m \\ \langle Sn(OR)_4 \rangle \\ \underline{1} \end{matrix} \xrightarrow{H_2O} \begin{matrix} [KOH]_m \\ [SnO(OH)_2 \cdot n\,H_2O] \\ \underline{2} \end{matrix} \xrightarrow{\Delta T} \begin{matrix} m\,KOH\,|\,SnO_2 & \underline{3} \\ K_mSnO_{(2+m/2)} & \underline{4} \end{matrix} \quad (2)$$

Solutions of the foregoing system were synthesized according to equations 3 and 4 by dissolving the appropriate amount of potassium in isopropanol followed by the addition of $SnCl_4$ and filtration of KCl. In detail, solutions were studied with $Sn(OR)_4$ and KOR in a molar ratio of 1:1 (**1a**, m = 1), 1:2 (**1b**, m = 2) and 1:4 (**1c**, m = 4), which correspond, as shown in figure 1, to the hitherto unknown ternary oxide $K_2Sn_2O_5$ (**4a**), as well as to the well known ones, K_2SnO_3 (**4b**) [11] and K_4SnO_4 (**4c**) [12], respectively.

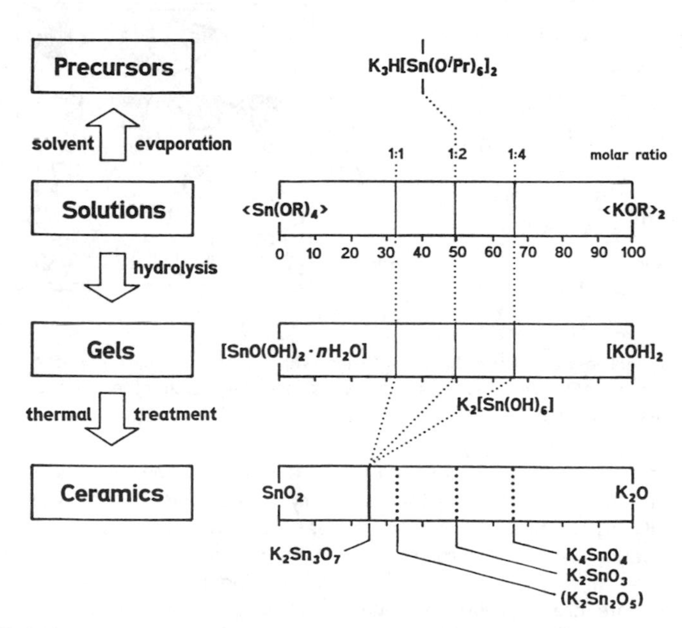

Figure 1: Typical working course of our experiments in the field of bimetallic alkoxide solutions and their integration in the phase system SnO_2/K_2O.

$$(4+m)\,K \quad + \quad (4+m)\ {}^{i}PrOH \longrightarrow \langle KO^{i}Pr\rangle_{4+m} \quad + \quad (4+m)/2\ H_2\uparrow \qquad (3)$$

$$\langle KO^{i}Pr\rangle_{4+m} \quad + \quad SnCl_4 \longrightarrow \langle Sn(O^{i}Pr)_4\rangle\langle KO^{i}Pr\rangle_m \quad + \quad 4\ KCl\downarrow \qquad (4)$$

1a–1c

After slow evaporation of the solvent, all solutions deposited solid phases (**5**) of well defined crystals which were of rhombohedral (**5a**) octahedral (**5b**) and hexagonal prismatic (**5c**) habit, respectively. Until now, only the crystal structure of **5b** could be determined by single crystal X-Ray diffraction [13]. In contrast to the solution with a Sn to K ratio of 1:1, these crystals have the composition $K_3H[Sn(O^{i}Pr)_6]_2$. Obviously, the isolation of a solid bimetallic alkoxide compound is directed more by its insolubility than by the stoichiometry present in the parent solution.

The crystal structure of **5b** consists of octahedral $[Sn(O^{i}Pr)_6]^{2-}$-ions, which are slightly distorted due to their strong interaction with the K^+-ions which link them to two interpenetrating three-dimensional networks, as shown in figure 2. **5b** (3:2, Sn_{oct}) represents the third structure type of a bimetallic M(I)-Sn(IV)-alkoxides, which now is verified unambigously. The two others are $Tl_2[Sn(OEt_6)]$ (2:1, Sn_{oct}) [14] and $K[Sn(O^{t}Bu)_5]$ (1:1, Sn_{tby}) [15]. It is noteworthy that their formation is strongly influenced by the nature of M^+ as well as by the sterical demands of the alkoxide ligands.

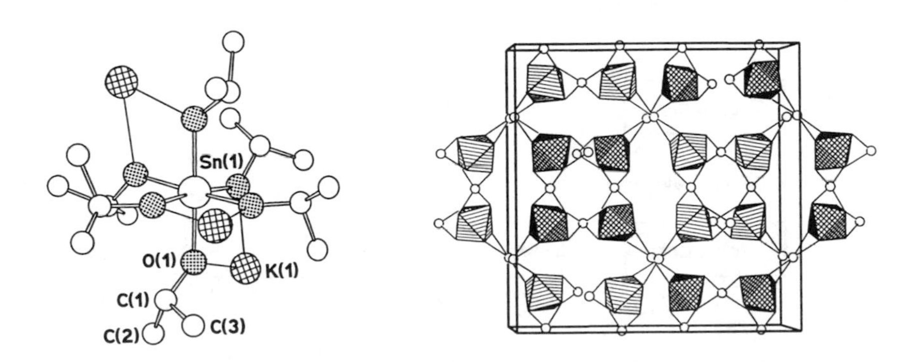

Figure 2: Perspective view of the $[Sn(O^{i}Pr)_6]^{2-}$-ion with its K^+-surrounding (**a**) and of the two three-dimensional cation-anion-networks (**b**) interpenetrating each other in the solid state structure of $K_3H[Sn(O^{i}Pr)_6]$.

At 950°C the thermal treatment of the amorphous powders **2a**–**2c** resulted in the formation of only one and the same crystalline phase, which was identified by XRD as $K_2Sn_3O_7$ [16]. Obviously, the excess of potassium hydroxide present in the final product remains amorphous and does not induce the formation of the ternary phases mentioned above. The results of the more detailed TG, DTA and XRD experiments are summarized in figure 3 using **2c**. In contrast, samples **2a** and **2b** show a contionus but no step-like decomposition and no crystalline intermediates below 400°C.

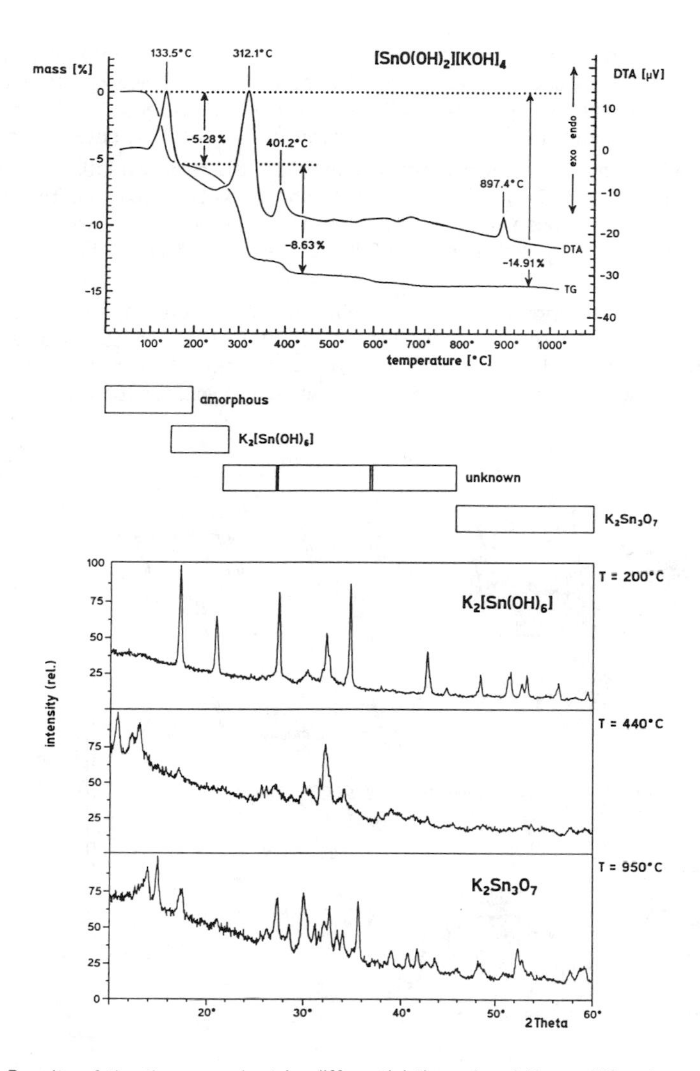

Figure 3: Results of the thermogravimetric, differential thermal and X-ray diffraction studies on the thermal degradation and structural evolution taking place during solid state step of <u>2c</u>.

The total weight loss of <u>2c</u> was found to be 14.91% which is 8.80% greater than expected, using the formulations $[SnO(OH)_2][KOH]_4$ and $\frac{1}{3}(K_2Sn_3O_7|10\,KOH)$ as starting and final points, respectively. In general, this discrepancy is due to the fact, that a small amount of water and solvent is retained during the powder preparation [17]. This was confirmed by elemental analysis, which showed the presence of up to 3% carbon. These *impurities* are usually eliminated during the initial step of the solid state reaction at temperatures between 80-140°C, accompanied by an endothermal effect [18]. However, the organic material is not only vaporized, but also decomposed during this step. In the following, the remaining carbon causes very complicated combustion reactions at elevated temperatures making the complete quantitative interpretation of TG and DTA data impossible.

The XRD experiments show clearly that the structural evolution starts at a temperature of 200° C with the crystallisation of $K_2[Sn(OH)_6]$. At temperatures around 300° C this phase decomposes. Thereafter, several yet unknown crystalline phases are passed through before at 840° C the final phase, $K_2Sn_3O_7$, begins to crystallize.

The unexpected formation of $K_2[Sn(OH)_6]$ stimulated the separate investigation of its structural and thermal properties. Its crystal structure consists, as shown in figure 4, of isolated octahedral $[Sn(OH)_6]^{2-}$-ions linked together by the K^+-ions to layers, which are held together through strong hydrogen bonds [19].

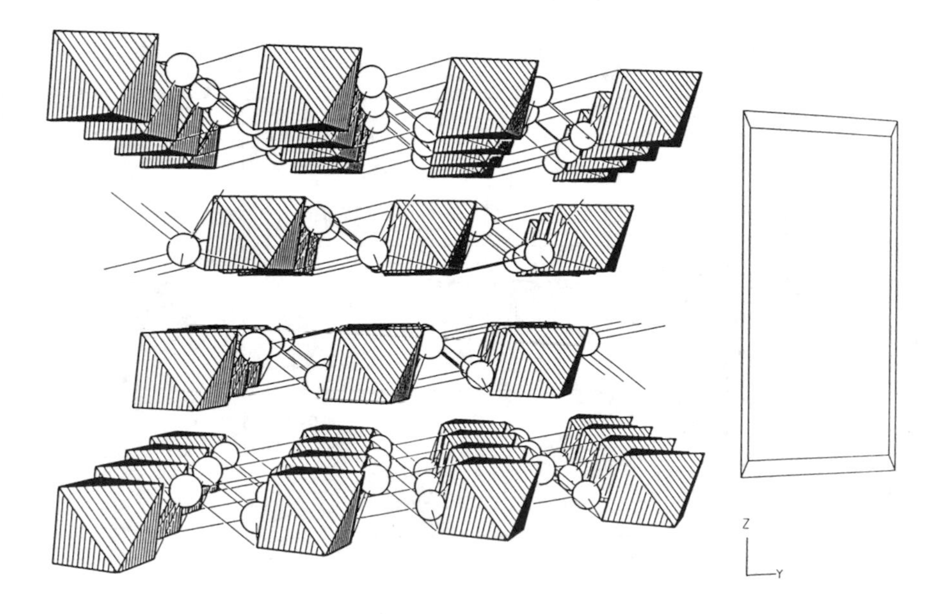

Figure 4: Perspective view on the layer structure of $K_2[Sn(OH)_6]$ as determined by single crystal X-ray diffraction; $[Sn(OH)_6]^{2-}$-octahedra are represented by polyhedra, K^+-ions by small balls [19].

The thermal degradation of $K_2[Sn(OH)_6]$ starts at 321° C with the endothermal loss of nearly 2 of the 3 water molecules, which can be released during the condensation of the six hydroxyl groups. Thereafter, further endothermal degradation steps accompanied by water loss occur at 382° C (~ 2/3) and 600° C (~ 1/3) [19].

This behaviour seems very similiar to those of <u>**2b**</u>. However, $K_2[Sn(OH)_6]$ dehydrates to K_2SnO_3, whereas <u>**2b**</u> forms $K_2Sn_3O_7$ as described above.

4. INVESTIGATIONS IN THE SYSTEM $^{i}PrSn(OR)_3/NaOR$

In a second series of experiments, we have studied the formation of ternary tin(IV) oxides via sol-gel-process using isopropyltin trialkoxides instead of pure inorganic tin(IV) alkoxides. Alkoxide solutions of the Sn and Na components in a molar ratio of 1:2 were obtained according to equations 5 and 6 by the reaction of sodium metal with an excess of isopropanol, followed by the dissolution of an appropriate amount of isopropyltin triisopropoxide.

$$2\,Na + {}^{i}PrOH \xrightarrow{i\text{-}PrOH} \langle NaO^{i}Pr\rangle_2 + H_2\uparrow \qquad (5)$$

$$\langle NaO^{i}Pr\rangle_2 + {}^{i}PrSn(O^{i}Pr)_3 \xrightarrow{i\text{-}PrOH} \underset{\underline{6}}{\langle {}^{i}PrSn(O^{i}Pr)_3\rangle\langle NaO^{i}Pr\rangle_2} \qquad (6)$$

After slowly evaporating most of the solvent, 6 separates out transparent colourless crystals of $Na_2[{}^{i}PrSn(O^{i}Pr)_5]\cdot 5\,{}^{i}PrOH$ (7). In the solid state, 7 exists, as a discrete molecule with the tin atom trigonal-bipyramidally coordinated by the isopropyl group and four isopropoxide ligands in a strongly distorted manner as one can seen from figure 5. Each sodium atom is fourfold coordinated and connected with the tin atom through two bridging alkoxide ligands [20].

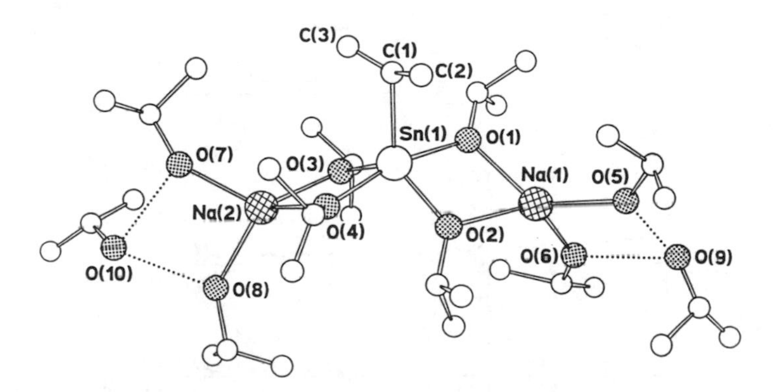

Figure 5. Solid state molecular structure of $Na_2[{}^{i}PrSn(O^{i}Pr)_5\cdot 5\,{}^{i}PrOH$

In order to study the solid state reactions, 6 was hydrolyzed as mentioned above. The fine white powder (8) thus obtained was investigated by TG, DTA and, after thermal treatment at different temperatures, by XRD. The results are shown in figure 6. At 1000°C only one single crystalline phase, Na_2SnO_3 [21], could be determined.

$$\underset{\underline{6}}{\langle {}^{i}PrSn(O^{i}Pr)_3\rangle\langle NaO^{i}Pr\rangle_2} \xrightarrow{H_2O} \underset{\underline{8}}{[{}^{i}PrSnO(OH)][NaOH]_2} \xrightarrow{\Delta T} Na_2SnO_3 \qquad (7)$$

On the basis of equation 7 thermal treatment of <u>8</u> should result in a total mass loss of 22.60 %, which is less (4.81 %) than observed in reality, due to the same reasons as before. Moreover, the interpretation of the TG curve is more complicated because of the thermal decomposition of the isopropyl group directly connected to the tin atom. As seen in figure 6, this endothermal degradation takes place around 300° C with a weight loss of 14.29 %. Theorectically, the loss should be 16.05 %. Again, the difference must be due to carbon remaining in the powder.

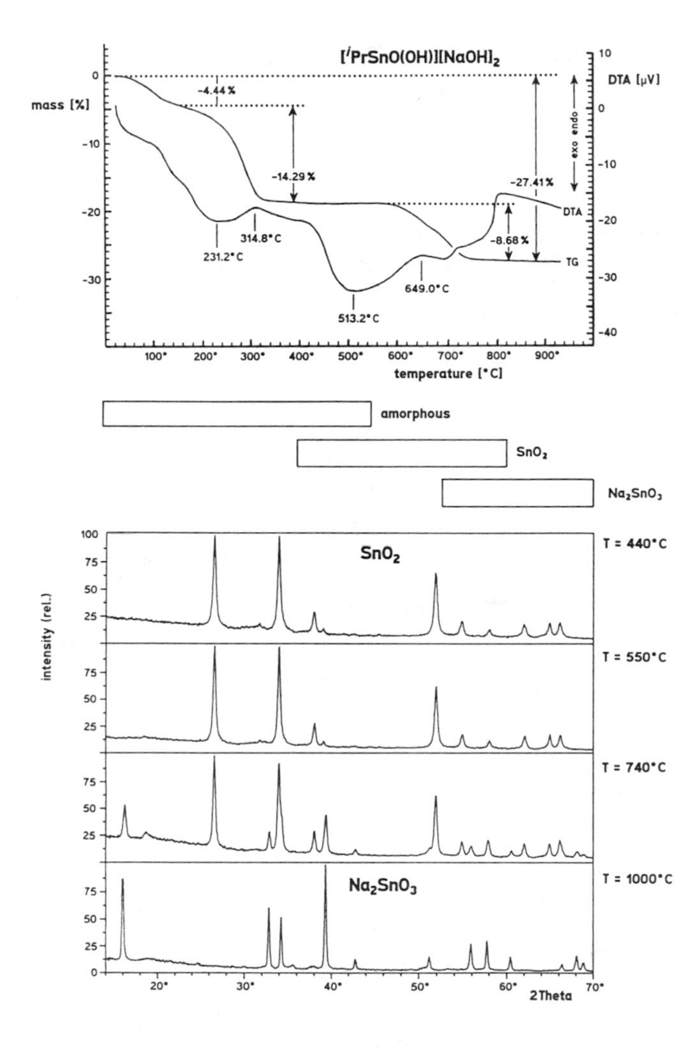

Figure 6: Results of the thermogravimetric, differential thermal and X-ray diffraction studies on the thermal degradation and structural evolution taking place during solid state step of <u>8</u>.

Thereafter, the sample retains a nearly constant weight up to 600°C, whereas at 450°C a strong exothermal reaction occurs. The XRD experiments show, that at this point the crystallization of SnO_2 starts. At 550°C this is still the single crystalline phase which can be observed. However, after a last thermal degradation step at 600°C, a second crystalline phase, Na_2SnO_3, is formed. At 1000° C this is the only phase left.

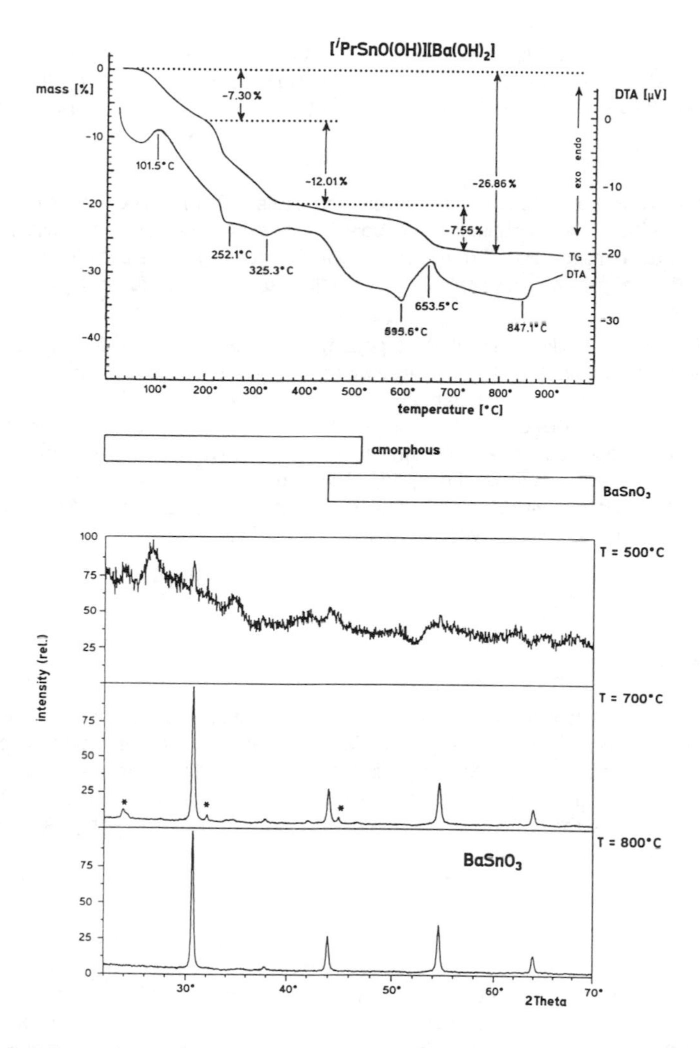

Figure 7: Results of the thermogravimetric, differential thermal and X-ray diffraction studies on the thermal degradation and structural evolution taking place during solid state step of 11.

5. INVESTIGATIONS IN THE SYSTEM $^{n}BuSn(OR)_3/Ba(OR)_2$

Furthermore we have studied the combination of a monoorganotin trialkoxide with a alkaline earth metal alkoxide to check the influence of the bivalent cation on the course of the solid state step of the sol-gel-process. 1:1-Solutions of both alkoxides were prepared according to equations 8-10 by the reaction of potassium metal with an excess of isopropanol, followed by addition of butyltin trichloride and filtration of KCl. Thereafter, the barium metal is dissolved in the resulting solution.

$$3\,K + 3\,^{i}BuOH \longrightarrow \langle KO^{i}Bu\rangle_3 + 3/2\,H_2\uparrow \qquad (8)$$

$$\langle KO^{i}Bu\rangle_3 + {}^{n}BuSnCl_3 \longrightarrow \langle {}^{n}BuSn(O^{i}Bu)_3\rangle + 3\,KCl\downarrow \qquad (9)$$

$$\langle {}^{n}BuSn(O^{i}Bu)_3\rangle + Ba + 2\,^{i}BuOH \longrightarrow \underbrace{\langle {}^{n}BuSn(O^{i}Bu)_3\rangle\langle Ba(O^{i}Bu)_2\rangle}_{\underline{9}} + H_2\uparrow \qquad (10)$$

After evaporating most of the solvent, crystals (10) of a yet unknown structure and composition were obtained from solution 9. This solution was hyrolyzed and the solid-state reactions of the corresponding gel (11) investigated in the same way as in the previous examples. The results are presented in figure 7 at the page before.

Again, the step-like thermal decomposition is very well verified although it is impossible to explain the TG/DTA data quantitatively because of the same reason as before. However, in contrast to the previous example, the powder remains amorphous up to 500°C. At this temperature, the crystallization of $BaSnO_3$ [22] starts. Some dubious reflections (marked by an asterix) of less intensity may arise from an intermediate phase, but this interpretation will remain unsure as long as it is impossible to determine this phase unambigously. At 800°C only those reflections are visible which are due to $BaSnO_3$.

6. CONCLUSIONS

In order to examine the possibilities for synthesizing ternary potassium tin oxides using the sol-gel-process, alkoxide solutions of both metals were prepared in three different stoichiometries. During thermal treatment of the corresponding gels, no signs for the formation of ternary phases with the required stoichiometries could be found, although some crystalline phases, which are yet unknown, were formed in the medium temperature range (350°-800°C). At elevated temperatures (1000°C), only one and the same crystalline phase ($K_3Sn_2O_5$) was observed in all three samples.

The solid state step of sol-gel-process was investigated using TG and DTA. Although the results are difficult to explain in detail, they give a good overview of the different degradation steps and, in combination with XRD, of the structural evolution taking place during thermal treatment. Monoorganotin(IV) alkoxides behave similar to pure tin(IV) alkoxides with the exception that one or two additional degradation steps occur.

Three different ways for the formation of intermediate phases were observed: First, the crystallization of a complex hydroxide ($K_2[Sn(OH)_6]$) at very low temperatures (≦ 200°C) in the presence of an excess of the alkali metal hydroxide. Second, the crystallization of only one component (SnO_2) in a medium temperature range (~ 400°C). Third, the crystallization of a ternary phase ($BaSnO_3$) at elevated temperatures (> 550°C).

Further investigations on a broader basis will be necessary to support these observations.

7. ACKNOWLEDGEMENT

This work is supported by the Deutsche Forschungsgemeinschaft.

8. REFERENCES

1. *"Gmelin Handbuch der Anorganischen Chemie - Zinn"*, Part C3, edited by H. Bitterer, Part C4, edited by E. Koch, (Springer Verlag, Berlin, 1975/1976).
2. C.J. Brinker and G.W. Scherer, *"Sol-Gel-Science - The Physics and Chemistry of Sol-Gel Processing"* (Academic Press, San Diego, 1990).
3. H. Reuter, M. Kremser and D. Schröder in *"Eurogel `91 - Proceedings on the Second European Conference on Sol-Gel-Technology"*, edited by S. Vilminot, (Elsevier Science Publishers, in press).
4. Modification of the sol-gel-process was achieved by use of monoorganotin(IV) compounds.
5. U. Folli, D. Jarossi and F. Taddei, *J. Chem. Soc., Perkin Trans. II* **1973**, 638.
6. H. Puff and H. Reuter, *J. Organomet. Chem.*, **368**, 173 (1989).
7. H. Reuter and D. Schröder, *Chem. Ber.*, submitted for publication.
8. *"Houben-Weyl – Methoden der organischen Chemie"*, I/2, (Georg Thieme Verlag, Stuttgart, 1959), p. 772.
9. D.F. Shriver, *"The Manipulation of Air Sensitive Compounds"*, (Mc Graw-Hill Publishers, New York, 1969), p. 190.
10. *"Handbuch der Präparativen Anorganischen Chemie"*, edited by G. Brauer, 3rd ed., (Ferdinand Enke Verlag, Stuttgart, 1975), p.83.
11. B.M. Gatehouse and D.J. Lloyd, *J. Solid State Chem.*, **2**, 410 (1970).
12. R. Marchand, Y. Piffard and M. Tournoux, *Acta Crystallogr.*, **B31**, 511 (1975).
13. H. Reuter and M. Kremser, *Inorg. Chem.*, submitted for publication.
14. M.J. Hampden-Smith, D.E. Smith and E.N. Duesler, *Inorg. Chem.*, **28**, 339 (1989).
15. M. Veith and M. Reimers, *Chem. Ber.*, **123**, 1941 (1990).
16. M. Tournoux, *Ann. Chim. [Paris]*, **9**, 579 (1964).
17. The discrepancy may also be due to a partial votalization of K_2O as a result of the thermal degradation of an other ternary phase. Its total votalization should give a weight loss of more than 50%.
18. Systematical investigations have shown, that stronger drying conditions (24h at 300°C in vacuum) will reduce the difference between calculated and observed weight loss. In this case, the initial weight loss is much less pronounced than by the drying procedure used in this investigation.

19. H. Reuter, unpublished results.
20. H. Reuter and D. Schröder, *J. Organomet. Chem.*, in preparation.
21. E. Hubbert-Paletta, R. Hoppe and G. Kreuzburg, *Z. Anorg. Allg. Chem.*, **379**, 255 (1970).
22. G. Wagner and H. Binder, *Z. Anorg. Allg. Chem.*, **298**, 12 (1959).

HYDROLYSIS-CONDENSATION OF ALKYLTIN-TRIALKOXIDES

François RIBOT, F. BANSE and C. SANCHEZ
Laboratoire de Chimie de la Matière Condendée (URA 1466). Université Pierre et Marie Curie. 4, place Jussieu. 75252 PARIS. FRANCE.

ABSTRACT

$BuSn(OPr^i)_3$ and $BuSn(OAm^t)_3$ have been synthesized according to a literature procedure. Their hydrolysis have been studied by ^{119}Sn and ^1H NMR spectroscopy. The former precursor yields only a closo-type oxo-hydroxo cation of formula $[(BuSn)_{12}O_{14}(OH)_6]^{2+}$. Investigation of the hydrolysis-condensation products of $BuSn(OAm^t)_3$ reveals expanded coordinenceabout the tin atoms from 4 to 5 and 6 when $H_2O/Sn>1$. For butyltin tri-tert-amyloxide, condensation also promotes the formation of a cluster whose structure is as yet unknown.

INTRODUCTION

The first steps of sol-gel processing are usually performed at low temperature,[1] and therefore allow one to incorporate "fragile" organic compounds inside an oxide-type matrix. The chemical characteristics of silicon make the Si-C bond fairly stable and thus hybrid organic-inorganic networks can be built (ORMOSIL, ORMOCER).[2] For transition metals, networks need to be anchored to one another through complexing functions such as carboxylates, ß-diketonates, and so on.[3]

Tin is a very interesting element because its characteristics make it intermediate between silicon and transition metals. Like transition metals, tin readily expands its coordination sphere from 4 to 5 or 6, and even up to 8.[4] But contrary to transition metals, the Sn-C bond is less ionic and provides a good stability, especially towards nucleophilic agents such as water.[5]

In order to document the possible formation and structure of oxide-type polymers from tin alkoxides precursors containing Sn-C bonds, this work has been started. This paper deals with the characterization by ^{119}Sn and ^1H NMR spectroscopy of the hydrolysis products of two different n-butyl trialkoxides.

EXPERIMENTAL

n-Butyltin trialkoxides were synthesized by reacting stoichiometric amounts of sodium alkoxide with n-butyltin trichloride.[6] 1,2-dimethoxyethane (DME) was used as co-solvent instead of benzene to ease the filtration separative step. All the experiments were carried out with schlenk-line techniques to avoid partial hydrolysis of the compounds. All alcohol solvents were freshly distilled from sodium and DME was distilled from $LiAlH_4$. Sodium isopropoxide was prepared using sodium metal (9.85 g) and sodium tert-amyloxide using sodium hydride (95% : 10.8 g). DME (≈100ml) and isopropyl or tert-amyl alcohol (≈100ml) were used respectively as solvents. A mixture of 25 ml of $BuSnCl_3$ (95%) and 100 ml of DME was mixed by dropwise addition to hot a solution of the respective sodium alkoxide. These mixtures were refluxed for about 4 hours and then allowed to cool. Sodium chloride, formed during the reaction, was removed by filtration. Excess solvents were evaporated under reduced pressure. Finally, n-butyltin trialkoxides were distilled under vacuum.

$BuSn(OPr^i)_3$: B.p.=75°C/0.05 mmHg, white solid, T_f=38°C, $\delta(^{119}Sn)$=-321 ppm (2.5 M in C_6D_6), $J^2(^{119}Sn\text{-}H)$=93 Hz, $J^3(^{119}Sn\text{-}H)$=93 Hz.

$BuSn(OAm^t)3$: B.p.=95°C/0.08mmHg, transparent liquid, $\delta(^{119}Sn)$=-194 ppm, $J^2(^{119}Sn\text{-}H)$=90 Hz, $J^3(^{119}Sn\text{-}H)$=116 Hz.

Hydrolysis experiments were generally performed using a water/alcohol 10% solution.

NMR experiments were performed on a BRUKER AM250 spectrometer (93.276 MHz for ^{119}Sn). ^{119}Sn chemical shifts are relative to tetramethyltin. C_6D_6 or $CDCl_3$ were used as internal lock and their protio impurities as references for ^1H spectra.

RESULTS AND DISCUSSION

Hydrolysis of $BuSn(OPr^i)_3$

Two types of hydrolysis experiments were performed on $BuSn(OPr^i)_3$:

i) One molecule of isopropanol per tin atom was added to $BuSn(OPr^i)_3$. Then hydrolysis (H_2O/Sn=3) was performed with neat water. This experiment gave a turbid suspension from which rectangular x-ray quality crystals grew in a few days. Single crystal X ray diffraction studies[7] indicate the following formula $[(BuSn)_{12}(\mu_3\text{-}O)_{14}(\mu_2\text{-}OH)_6](OPr^i)_2(HOPr^i)_2(H_2O)_2$ (I). An ORTEP drawing of this compound is given in Figure 1. The main feature corresponds to an oxo-hydroxo closo cluster surrounded by n-butyl chains which prevent further condensation. This type of alkyltin-oxo-hydroxo cation, $[(BuSn)_{12}O_{14}(OH)_6]^{2+}$, has already been identified by single crystal X ray diffraction in various hydrolysis compounds of $BuSnCl_3$.[8] The $^{119}Sn\text{-}\{^1H\}$ NMR spectrum of crystals I dissolved in $CDCl_3$ is shown in Figure 2.

ii) Neat $BuSn(OPr^i)_3$ was also hydrolyzed (H_2O/Sn=10) with water in isopropanol (10%) and produced needle shaped crystals (II)in few days.

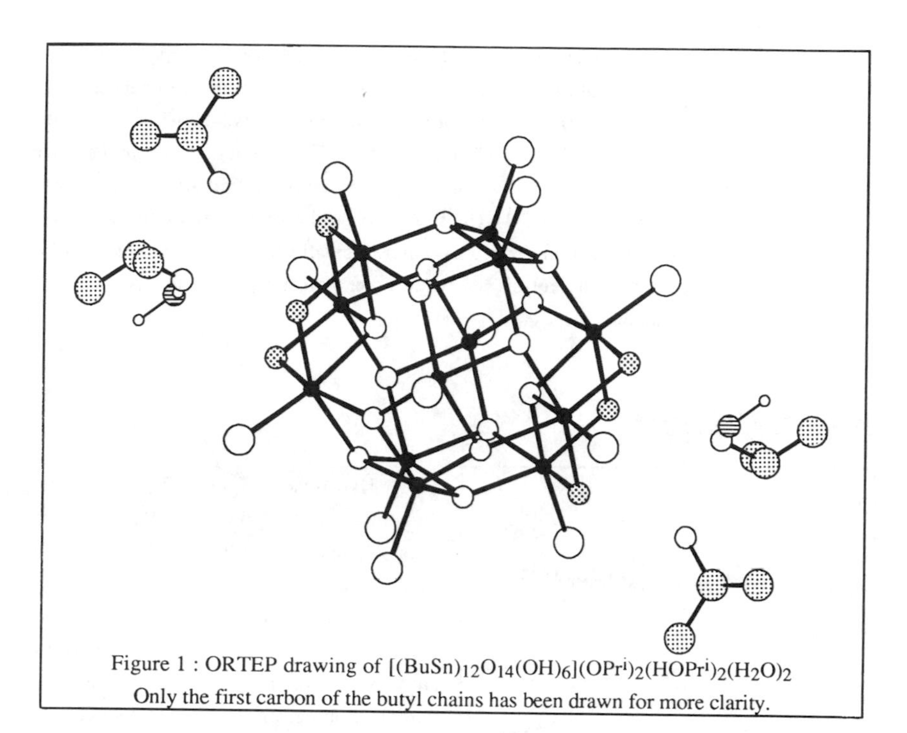

Figure 1 : ORTEP drawing of $[(BuSn)_{12}O_{14}(OH)_6](OPr^i)_2(HOPr^i)_2(H_2O)_2$
Only the first carbon of the butyl chains has been drawn for more clarity.

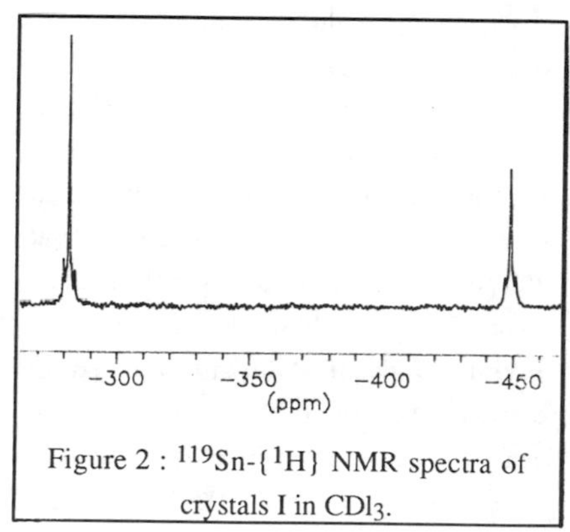

Figure 2 : ^{119}Sn-{^1H} NMR spectra of crystals I in CDI_3.

The ^{119}Sn-{^1H} NMR spectra of I and II are identical. They show two resonances at -281 and -449 ppm with two sets of satellites each. These peaks lie respectively in the range of 5 and 6 fold-coordinate mono-alkyltin.[9] Integration of both resonances on a ^{119}Sn NMR experiment indicates a 1:1 ratio between tin atoms in 5 fold and tin atoms in 6 fold coordination. Actually this spectrum is characteristic of the cluster $[(BuSn)_{12}O_{14}(OH)_6]^{2+}$. Satellites come from three types of scalar coupling between tin atoms bounded by oxo or hydroxo bridges : $J^2(^{119}Sn_P\text{-}^{119/117}Sn_O)$=383 Hz, $J^2(^{119}Sn_P\text{-}^{117}Sn_P)$=177 Hz and $J^2(^{119}Sn_O\text{-}^{117}Sn_O)$=205 Hz (subscript P and O refer to coordination numbers of 5 and 6 respectively).[10] Integration of the ^1H spectrum of crystals II shows only two OPr^i groups and one H_2O for every twelve n-butyl chains. These different amounts of isopropanol and water present in the structure probably explain the differences between crystals I and II.

The $^{119}Sn-\{^{1}H\}$ NMR spectrum (C_6D_6 as internal lock) of a freshly hydrolyzed isopropanol solution of $BuSn(OPr^i)_3$ ($H_2O/Sn=10$, case ii) also exhibits two resonances (-283 and -452 ppm). The poor signal to noise ratio of this spectrum permits identification of only one set of satellites separated by ~375 Hz. Chemical shifts and scalar coupling are very similar to those observed for solutions of crystals I or II. Small variations might arise from solvent effects. Therefore, $[(BuSn)_{12}O_{14}(OH)_6]^{2+}$ seems to be the only species formed upon hydrolysis of $BuSn(OPr^i)_3$ with an excess of water. Actually this cluster might even be obtained for sub-stoichiometric hydrolysis since 1.67 water molecule per tin atom is enough to balance the chemical equation.

Hydrolysis of $BuSn(OAm^t)_3$

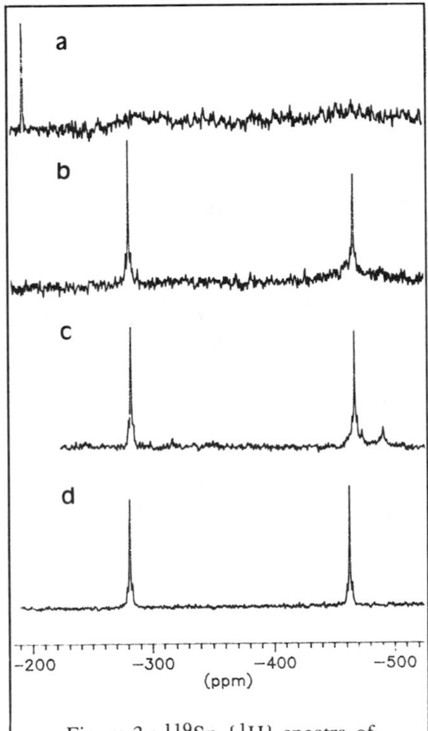

Figure 3 : $^{119}Sn-\{^{1}H\}$ spectra of $BuSn(OAm^t)_3$ hydrolyzed. $H_2O/Sn=1$ (a); $H_2O/Sn=2$ (b); $H_2O/Sn=10$ as prepared (c), dried and dissolved in C_6D_6 (d)

Hydrolysis of neat $BuSn(OAm^t)_3$, with diluted water and $0.5 \leq H_2O/Sn \leq 10$, gave clear solutions, except for $H_2O/Sn=2$ where a small amount of precipitate was observed.

$^{119}Sn-\{^{1}H\}$ NMR spectra of $BuSn(OAm^t)_3$ hydrolyzed with various amounts of water are shown in Figure 3. After hydrolysis, tert-amyl alcohol was added to obtain roughly the same tin concentration (0.3 M) in all the samples. For $H_2O/Sn \leq 1$, only one resonance is observed at -195 ppm. This chemical shift lies in the range of 4 fold-coordinated mono-alkyltin. For $H_2O/Sn \geq 2$, two main resonances are observed at -283 and -468 ppm. These values lie respectively in the range of 5 and 6 fold-coordinated mono-alkyltin. Both of them are flanked by satellites separated by ~395 Hz. The separation and the intensity of these satellites correspond to $J^2(^{119}Sn-^{119/117}Sn)$ scalar coupling.[9-10]

The solution of $BuSn(OAm^t)_3$ hydrolyzed with $H_2O/Sn=10$ was dried under vacuum. The $^{119}Sn-\{^{1}H\}$ spectrum of the resulting white solid, dissolved in C_6D_6, is shown in Figure 3.

The tin spectrum exhibits only two resonances at -282 and -463 ppm. These chemical shifts are quite similar to those observed prior to drying. Small changes in the peak positions can be related to the effects of solvent. Both peaks are flanked by two sets of satellites. Outer satellites are separated by 400 Hz for both resonances. Separations between inner satellites differ from one resonance to the other : 160 Hz at -282 ppm and 180 Hz at -463 ppm. All these values are of the magnitude of $J^2(^{119}Sn\text{-}^{119/117}Sn)$ scalar coupling and are close to those observed for $[(BuSn)_{12}O_{14}(OH)_6]^{2+}$. Integration of the complex multiplets arising from $J^2(^{119}Sn\text{-}H)$ and $J^3(^{119}Sn\text{-}H)$ coupling (^{119}Sn NMR experiment) indicates an equal amount of tin atoms in five fold-coordination and tin atoms in 6 fold-coordination. The ^1H spectrum exhibits characteristic features of n-butyl and tert-amyl groups, especially in the high field region. Careful integration of the signal indicates a ratio n-Bu/OAm^t≈2. This might indicate that all the tert-amyloxy groups are not removed upon hydrolysis or, as in the case of $BuSn(OPr^i)_3$, a charged compound is formed and needs alkoxy groups for its electroneutrality. Up field, three triplets ($C\underline{H}_3$-CH_2-) are observed. One corresponds to a tert-amyl group and the others to two different types of n-butyl chains. These two types of n-butyl chains seem to be in equal quantity and might be related to the presence of tin atoms in 5 and 6 fold-coordination as for compounds I and II.

From all these spectroscopic characterizations two main conclusions can be drawn :

i) For hydrolysis of $BuSn(OAm^t)_3$ with $H_2O/Sn \leq 1$, either the reaction does not occur or a monomeric species of formula $BuSn(OAm^t)_2(OH)$ is the only one formed. This latter compound would hence exhibit a ^{119}Sn chemical shift very close (within 2 ppm) to the one for $BuSn(OAm^t)_3$. Auto-association, which occurs for the first hydrolysis product of $RSnCl_3$ and yields $(RSn)_2(\mu_2\text{-}OH)_2Cl_4(H_2O)_2$ would be prevented by the bulkiness of tert-amyloxy groups.[11-12] This alternative is currently under investigation.

ii) For hydrolysis of $BuSn(OAm^t)_3$ with $H_2O/Sn \geq 2$, tin atoms increase their coordination to 5 and 6. No precipitate or gel is obtained, which seems to indicate that the condensation does not extend to a tri-dimensional polymeric network. Note that hydrolysis, under the same conditions, of $Sn(OAm^t)_4$ quickly yields amorphous precipitates. One may suggest that the major compound formed is a cluster, involving μ_3-O and μ_2-OH bridges, of a yet unknown structure. Attempts are currently being made to crystallize such a compound.

The differences observed between the hydrolysis behavior, may be related to the difference between the degree of association of the precursors. $BuSn(OAm^t)_3$ is a monomeric species while $BuSn(OPr^i)_3$ is prone to dimerization.[13]

CONCLUSION

Hydrolysis-condensation of $BuSn(OPr^i)_3$ in isopropanol yields only compounds containing the oxo-hydroxo cation, $[(BuSn)_{12}O_{14}(OH)_6]^{2+}$, in which tin atoms exhibit cordination numbers of 5 and 6. The ^{119}Sn NMR spectroscopic characteristics of this cluster have been reported.

Hydrolysis of $BuSn(OAm^t)_3$ with more than one water molecule per tin atom increases the cordinence of tin to 5 and 6. The major compound formed upon hydrolysis-condensation, in tert-amyl alcohol, is probably also a cluster whose structure is yet unkown but different from the one with butyltin triisopropoxide.

Alkyltin-trialkoxides exhibit a lower functionality than tin alkoxides because of the C-Sn bond which is not cleaved upon hydrolysis. Therefore, even if the coordination of tin atoms increases to 5 and 6 through μ_3-O and μ_2-OH bridges upon hydrolysis and condensation, the lower functionality of alkyltin-trialkoxides, and likely hydrophobic effects, force the formation of closo-type clusters and limit the spatial extension of the polymeric oxide type network.[14]

To promote the extension of the polymeric oxide type network, the functionality of the precursors can be increased by using a mixture of alkyltin-trialkoxides and tin alkoxides. Such a procedure is under current study.

REFERENCES

1. C.J. Brinker and G.W. Scherrer, Sol-Gel Science: The Physics and Chemistry of Sol-Gel Processing, (Academic Press, New York, 1990).
2. H. Schmidt, H. Scholze and A. Kaiser, J. Non-Cryst. Solids 63, 1 (1984).
3. C. Sanchez, F. Ribot and S. Doeuff, in Inorganic and Organometallic Polymers with Special Properties, edited by R.M. Laine (Kluwer Academic Publisher, New York 1992) pp.267-295.
4. N.W. Alcock and V.L. Tracy, Acta Cryst. B35, 80 (1979).
5. A.G. Davies and P.J. Smith, in Comprehensive Organometallic Chemistry, edited by G. Wilkinson, F.G.A. Stone and E.W. Abel (Pergamon Pres, Oxford, 1982), p.519.
6. D.P. Gaur, G. Srivastava and R.C. Mehrotra, J. Organometallic Chem. 63, 221 (1973).
7. P. Tolédano, F. Banse, F. Ribot and C. Sanchez, Acta Cryst. (submitted).
8. H. Puff and H. Reuter, J. Organometallic Chem. 373, 173 (1989).
9. R.K. Harris, J.D. Kennedy and W. McFarlane, NMR and Periodic Table, (Academic Press, London, 1978), p.348.
10. T.P. Lockhart, H. Puff, W. Schuh, H. Reuter and T.N. Mitchell, J. Organometallic Chem. 366, 61 (1989).
11. C. Lecompte, J. Protas and M. Devaud, Acta Cryst. B32, 923 (1976).
12. H. Puff and H. Reuter, J. Organometallic Chem. 364, 57 (1989).
13. J.C. Kennedy, J.C.S. Perkin II 1977, 242.
14. F. Ribot, P. Tolédano and C. Sanchez, Chem. Mater. 3, 759 (1991).

HYDROLYSIS AND CONDENSATION OF MODIFIED TIN (IV) ALKOXIDE COMPOUNDS TO FORM CONTROLLED POROSITY MATERIALS

CHRISTOPHE ROGER*, M.J. HAMPDEN SMITH* and C.J. BRINKER**
*University of New Mexico, Department of Chemistry and Center for Micro-Engineered Ceramics, Albuquerque, NM 87131,
**Sandia National Laboratories, P.O. BOX 5800, Albuquerque, NM 87185

ABSTRACT

Sol-gel type hydrolysis and condensation has been studied extensively as a method for the control of evolution of microstructure in the formation of inorganic silicates. However, in non silicate systems, the same level of control of microstructural evolution has yet to be demonstrated. In this work, we report the synthesis of mixed ligand tin (IV) alkoxide complexes specifically designed to undergo sequential hydrolysis reactions, where, as a result, control over porosity of the final metal oxide network is expected. A series of tin (IV) alkoxide compounds modified with difunctional carboxylate ligands as templates has been prepared with internuclear tin separations that are determined by the length of the bridging carboxylate group. The first hydrolysis step consists of the removal of the alkoxide ligands to create a three-dimensional network of oxo-bridged tin carboxylate species. In the second step, the bridging groups are removed by acid hydrolysis to leave pores without creating new Sn-O-Sn bonds. The synthesis and characterization of these species and the connection between pore structure (i.e. micro- vs. meso- porosity) as a function of the dimensions of the bridging ligands will be reported.

BACKGROUND

The evolution of microstructure during the sol-gel type hydrolysis and condensation of silicon alkoxides can be controlled as a function of temperature, time, hydrolysis ratio and catalyst.[1] However, microstructural evolution is much harder to control in non-silicon metal alkoxide systems[1,2] because these metal alkoxides often aggregate prior to or during hydrolysis,[3] exhibit a variety of different coordination numbers[2] and hydrolyze much more rapidly than their silicon analogues.[4] Hydrolysis and condensation of metal alkoxides modified with acetic acid have been studied by Sanchez et al.[5] and it has been shown that the carboxylate ligand is retained in the metal coordination sphere after hydrolysis at neutral pH. However, the structural relationship between modifying organic ligands and the final material has not been studied.

One approach to control the properties of sol-gel derived films that has been investigated extensively is the combination of inorganic (especially silicate) and organic polymers to form organically modified silicate (ORMOSIL) materials.[1,6] However, the use of organically modified metal alkoxides to create *porous* metal oxide materials in a general fashion (independent of the metal) has not been carried out prior to this work.

INTRODUCTION

Tailoring the pore structure of a material can be achieved by preparing modified metal alkoxides in which the added ligand is kinetically and/or thermodynamically more stable with respect to hydrolysis and subsequent condensation reactions than the other alkoxides ligands. This can be achieved with a carboxylic acid. The pK_a of the acid is lower than that of the alcohol corresponding to the alkoxide ligand; consequently, the bond between the metal and the carboxylate fragment will be stronger than that of the metal-alkoxide. Moreover, the carbonyl group of the carboxylic ligand bears two lone pairs on the oxygen atom that could interact with the empty d-orbitals of the metal, aiding coordinative saturation at the metal center and so lowering the tendency for aggregation. In Scheme I, hydrolysis of the carboxylic acid modified metal alkoxide (**1**) at neutral pH followed by condensation gives a three-dimensional bridging-oxo structure with the carboxylate retained in the coordination sphere of the metal (**2**). A second hydrolysis step, induced by the presence of acid, removes the carboxylate template (**3**).

Another variation on this process is illustrated in Scheme II, where two metal alkoxide

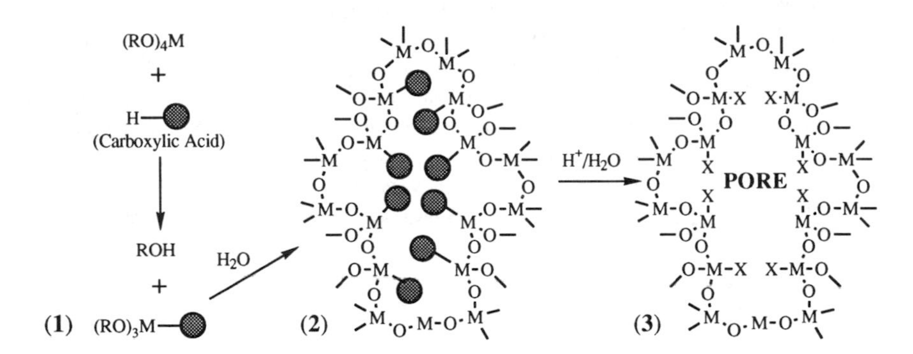

Scheme I. Schematic representation of a general route to porous metal oxides.

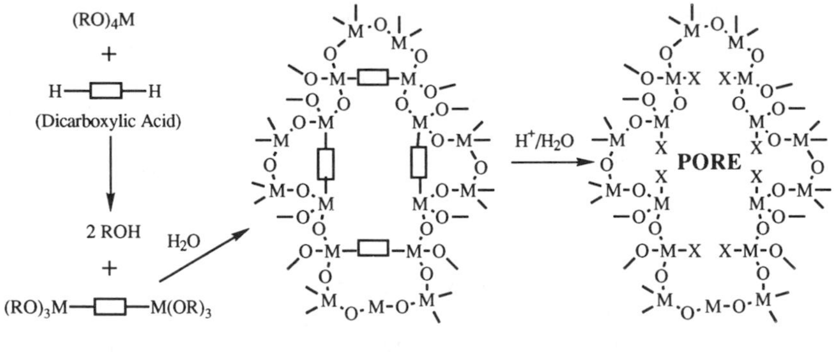

Scheme II. Another schematic representation of a general route to porous metal oxides.

molecules can react with a di-carboxylic acid to form a carboxylate bridged species. An analogous two-step hydrolysis process will first create an analogous three-dimensional network of bridged-oxo metal carboxylates which will then generate a porous structure upon acid hydrolysis. This strategy, without any *thermally-induced* condensation, is likely to preserve the micro- structure created by the templates. These two related processes can be viewed as a general method to porous metal oxides.

In this paper we have investigated the strategies outlined above to determine the relationship between the steric demands of the organic template ligand, the surface area, the pore structure and pore size in the final material. In previous work, we have demonstrated that the tin (IV) alkoxide system is a prototypical example of non-silicon metal oxide and have shown its viability as a model system through structural characterization in solution by NMR spectroscopy and through formation of dense spherical particles and thin films of crystalline cassiterite (SnO_2) via sol-gel methods. Tin (IV) tetra(*tert*-butoxide) has been reacted with different dicarboxylic acids which act as templates to enable the introduction of pore structure in the tin oxide formed.

RESULTS AND DISCUSSION

Tin (IV) tetra(*tert*-butoxide) chosen because it is monomeric in the solid state and in solution[7] reacted in refluxing THF with half an equivalent of succinic acid ($HOOCCH_2CH_2COOH$) to afford a 50/50 mixture of the starting material and a new compound identified as the chelating succinate $Sn(O^tBu)_2[(OOCCH_2)_2]$ (see characterization data below). No evidence for bridging succinate ligand was observed. Reaction of $Sn(O^tBu)_4$ with one equivalent of succinic acid gave the chelating adduct $Sn(O^tBu)_2[(OOCCH_2)_2]$ in almost quantitative yield. The product is slightly soluble in THF or benzene and was characterized by NMR spectroscopy. The proton NMR spectrum shows only one resonance for the two diastereotopic hydrogen atoms (H_a and H_b) of the methylene groups. The low solubility of this material precluded studies to decoalesce these signals at low temperature. The $^{13}C\{^1H\}$ shows a signal at 186.0 ppm characteristic of a carboxylic carbon, the methyl groups of the O^tBu fragment appear as a single resonance (33.2 ppm) with a tin coupling constant of 36 Hz characteristic of terminal alkoxide ligands.[7] Finally, one signal at -613.3 in the ^{119}Sn NMR spectrum is consistent with a six coordinate tin atom.[8] The succinate ligand chelates the tin moiety and the oxygen lone pair of the carboxy group interacts with the d-orbitals of Sn. The structure proposed for this molecule in solution is illustrated in Scheme III.

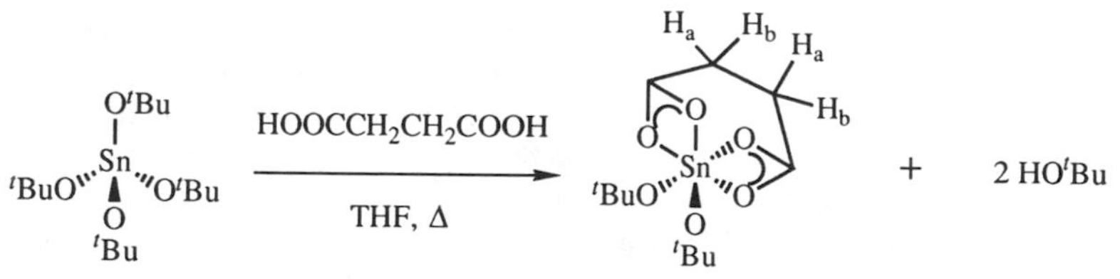

$$Sn(O^tBu)_4 \xrightarrow[\text{THF, }\Delta]{HOOCCH_2CH_2COOH} Sn(O^tBu)_2[(OOCCH_2)_2] + 2\ HO^tBu$$

Scheme III. Synthesis of $Sn(O^tBu)_2[(OOCCH_2)_2]$

As observed in the titanium system[5], the carboxylate ligand is not displaced upon hydrolysis at neutral pH. Addition of an excess of water (10 fold excess per alkoxide ligand) to a THF suspension of $Sn(O^tBu)_2[(OOCCH_2)_2]$ resulted in a white precipitate insoluble in any solvent. The Thermogravimetric Analysis (TGA) confirmed the presence of the succinate ligand in the product (Figure 1). After the loss of adsorbed and structural water (up to 150 °C), the weight lost is equal to that expected from $SnO[(OOCCH_2)_2]$ to SnO_2. The temperature of decomposition is ~ 430 °C. The white powder ($SnO[(OOCCH_2)_2]$) is very hygroscopic and contained some structural water that could not be removed upon drying at 110 °C for 16-24h. Consequently, the elemental analysis corresponds to $SnO[(OOCCH_2)_2]\cdot 0.6\ H_2O$ which is consistent with the TGA data if we consider that adsorbed water is lost on heating at 150 °C.

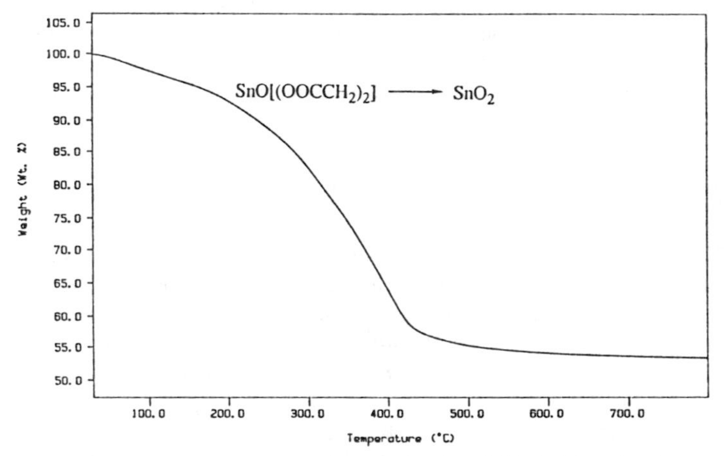

Figure 1. Thermogravimetric analysis of $SnO[(OOCCH_2)_2]\cdot 0.6\ H_2O$

Attempts to remove the succinate ligand thermally at 430 °C for 30 min. resulted in formation of crystalline tin oxide in the cassiterite phase as shown by X-Ray powder diffraction. The crystallite size can be estimated at 10-30 nm from the broadening of the X-ray diffraction peaks. The material possesses a surface area of 4.8 m^2g^{-1}, which means that the particles size is larger than the crystallite size. This experiment demonstrates the need for a less vigorous method to remove the succinate template, permitting retention of the porosity created by the template.

In an alternative strategy, a THF solution of $Sn(O^tBu)_2[(OOCCH_2)_2]$ was hydrolysed overnight with neutral pH water and then nitric acid (69-71%) was added to the suspension. After 1 hour at room temperature, the suspension was filtered, washed with water and acetone, and dried at 110 °C for 16-24h. Elemental analysis of the resulting material does not show any C or N present in the structure. After loss of adsorbed and structural water (up to 150 °C), TGA data reveal a weight loss of 11% corresponding to the transition from $SnO(OH)_2$ to SnO_2. The material possesses a surface area of 110 m^2g^{-1} with 60% of micropore and 40% of mesopore.

Reaction of $Sn(O^tBu)_4$ with half an equivalent of dicarboxylic acids with a rigid backbone (such as oxalic and 4,4'-biphenyl dicarboxylic acids) in refluxing THF gave white insoluble powders. Hydrolysis of these powders at neutral pH gave materials where the TGA shows the presence of one acid molecule for two tins. The product has the general formula (HO)OSn-OOC-R-COOSnO(OH) (R= nothing or biphenyl).

Hydrolysis of the modified tin alkoxide complexes by the oxalic or 4,4'-biphenyl dicarboxylic acids at neutral pH followed by acid hydrolysis gave a material of formula $SnO_{3/2}(OH)$ with surface area of 20 and 77 m^2g^{-1} respectively and 38% micropores for the former (average pore radius 26 Å) and 100% mesopores (average pore radius 40 Å) for the latter.

To demonstrate the generality of this method, $Si(OEt)_4$ and $Ti(O^iPr)_4$ were also modified by a carboxylate ligand[9]. Hydrolysis with water at neutral pH gave white powders of general formula MO(OH)(OOCR) (M= Si, Ti) with surface area of 2.2 and 77 m^2g^{-1} respectively and no microporosity was observed. After acid hydrolysis, $MO(OH)_2$ (M= Si, Ti) was formed with surface area of 994 and 203 m^2g^{-1} respectively. The materials are microporous with significantly lower average pore radii than the material formed after the first hydrolysis.[9]

CONCLUSION

We have demonstrated that upon hydrolysis at neutral pH, the carboxylate modified metal alkoxide compounds retain the carboxylate ligand. However, in these conditions, hydrolysis of the alkoxy groups occurred. The acid template can be removed completely by acid hydrolysis (nitric acid) with no nitrogen incorporation in the final structure. Furthermore, due to low crystallization temperatures of some non-silicon metal oxides, the thermolytic removal of the template results in collapse of the microstructure and formation of crystalline materials.

Although it is too early to correlate the size of the pore to the size of the carboxylate ligand at this stage, the template *creates* the porosity in the final material. Moreover, this work demonstrates that the strategy described above should be a general route to any porous metal oxides.

EXPERIMENTAL

All manipulations were carried out under an atmosphere of dinitrogen. THF was dried and distilled from sodium benzophenone ketyl. C_6D_6 (Isotec Inc.) was degassed and store over molecular sieves (4 Å) under a dinitrogen atmosphere. Water (Aldrich, HPLC grade) and nitric acid (Baker, 69-79%) were used as received. $Sn(O^tBu)_4$ was prepared following a published procedure.[7,10] NMR data were recorded on a Bruker AC-250P NMR referenced to the protio impurities of the deuterated solvent (^1H and ^{13}C NMR) or to external $SnMe_4$ (^{119}Sn NMR), and TGA were obtained on a Perkin-Elmer 7700 instrument.

Representative experiments

In a Schlenk flask, $Sn(O^tBu)_4$ (1.035 g, 2.517 mmol) and succinic acid (0.297 g, 2.517 mmol) were mixed as solids. THF (50 mL) was then added and the suspension was refluxed for 6h. The solvent was then removed, the residue was washed with pentane and dried under an oil pump vacuum to give 905 mg (95% yield) of a white air-sensitive powder of $Sn(O^tBu)_2[(OOCCH_2)_2]$.

^{1}H NMR (C_6D_6, 24 °C, 250 MHz) 1.85 (s, 4H, CH_2), 1.52 (s, 18H, $^4J_{Sn-H}$ = 3 Hz, CH_3).

$^{13}C\{^1H\}$ NMR (C_6D_6, 24 °C, 62.9 MHz) 186.0 (s, COO), 74.6 (s, $\underline{C}CH_3$), 33.2 (s, CH_3, $^3J_{Sn-C}$ = 36 Hz), 26.9 (s, CH_2).

$^{119}Sn\ \{^1H\}$ NMR (C_6D_6, 24 °C, 93.3 MHz) -613.3.

To a Schlenk flask containing a THF (40mL) suspension of $Sn(O^tBu)_2[(OOCCH_2)_2]$ (180 mg, 0.472 mmol), water (170 μL, 9.440 mmol) was added. The suspension was stirred overnight. An aliquot was then taken, filtered, washed with water and acetone, and dried at 110 °C for 20h. The TGA shows a total weight loss (up to 800 °C) of 48% with about 4% weight lost between 25 °C and 150 °C corresponding to a loss of adsorbed and structural water. Analysis Calcld for $SnO[(OOCCH_2)_2]\cdot 0.6\ H_2O$: C, 18.36; H, 2.01. Found: C, 18.74; H, 2.25. The remaining suspension was then treated with 1 mL of nitric acid and stirred for an hour. The powder was filtered, washed with water and acetone, and dried at 110 °C for 20h. The TGA shows a total weight loss of 13% with about 2% weight lost between 25 °C and 150 °C corresponding to a loss of adsorbed and structural water.
Elemental analysis: C, 0.0; N, 0.0.

ACKNOWLEDGMENTS

This work is supported by Sandia National Laboratories at Livermore. We thank the ONR for analytical facilities, the NSF under the chemical instrumentation program for the purchase of an NMR spectrometer and Bill Ackerman for BET surface area measurements.

REFERENCES

1. C.J. Brinker and G.W. Scherer, Sol-Gel Science, The Physics and Chemistry of Sol-Gel Processing, (Academic Press, New York, 1990).
2. M.J. Hampden-Smith, T.A. Wark and C.J. Brinker, Coord. Chem. Rev. 112, 81 (1992).
3. D.C. Bradley and H. Holloway, Can. J. Chem. 40, 1176 (1962).
4. D.C. Bradley, R.C. Mehrotra and D.P. Gaur, Metal Alkoxides, (Academic Press, New York, 1978).
5. S. Doeuf, M. Henry and C. Sanchez, Mat. Res. Bull. 25, 1519 (1990)
6. H. Schmidt in Better Ceramics through Chemistry IV 995 (1991).
7. M.J. Hampden-Smith, T.A. Wark, A. Rheingold and J.C. Huffman, Can. J. Chem. 69, 121 (1991).
8. R. Hani and R.A. Geanangel, Coord. Chem. Rev. 44, 229 (1982).
9. C. Roger, M.J. Hampden-Smith and C.J. Brinker (unpublished results).
10. C.D. Chandler, G.D. Fallon, A.J. Koplich and B.O. West, Aust. J. Chem. 40, 1427 (1987).

POLYNUCLEAR TITANIUM OXOALKOXIDES: MOLECULAR BUILDING BLOCKS FOR NEW MATERIALS?

Y. W. CHEN, W. G. KLEMPERER,* AND C. W. PARK
Beckman Institute of Advanced Science and Technology, Department of Chemistry, and Materials Research Laboratory, University of Illinois, Urbana, IL 61801

ABSTRACT

The $[Ti_7O_4](OEt)_{20}$ molecule, Et = C_2H_5, is very reactive toward ethanol, and its $[Ti_7O_4]$ metal oxide core structure is largely decomposed in <10 minutes. The $[Ti_{16}O_{16}](OEt)_{32}$ molecule, however, has a $[Ti_{16}O_{16}]$ core structure which is relatively stable toward alcoholysis, and solid state ^{17}O MAS NMR experiments using selective ^{17}O labeling techniques show that this core structure is preserved in good yield during sol-gel polymerization.

INTRODUCTION

Sol-gel polymerization of the $[Si_8O_{12}](OCH_3)_8$ cube [1] yields silica xerogels having microstructures, surface areas, and porosities different from those obtained by sol-gel polymerization of $Si(OCH_3)_4$ under the same conditions [2]. These differences have their origin in the different molecular structures of the two precursor molecules (see Figure 1), which cause them to follow different molecular growth pathways during sol-gel polymerization. The larger size and rigidity of the cubic $[Si_8O_{12}]$ core structure generates polymers whose rigidity inhibits extensive crosslinking of the type observed in $Si(OCH_3)_4$-derived gels [2].

This paper is concerned with extending the molecular building block approach to new ceramic materials beyond silica to titania. The molecular building block approach could be successfully applied to silica sol-gel polymerization because the $[Si_8O_{12}]$ core of the $[Si_8O_{12}](OCH_3)_8$ precursor is sufficiently resistant to alcoholysis and hydrolysis to ensure control of the course of the polymerization process [1-4]. The question to be addressed,

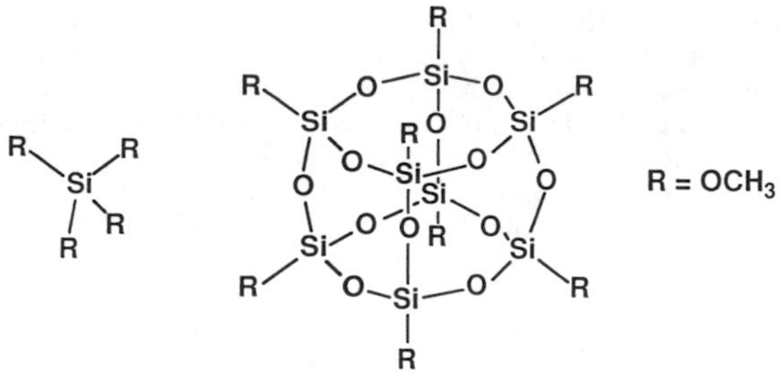

R = OCH_3

$Si(OCH_3)_4$ $[Si_8O_{12}](OCH_3)_8$

Figure 1. Molecular structures of $Si(OCH_3)_4$ and $[Si_8O_{12}](OCH_3)_8$ [1].

therefore, is the stability of the $[Ti_xO_y]$ cores in $[Ti_xO_y](OR)_z$ precursors toward alcoholysis and hydrolysis. Here the $[Ti_7O_4](OEt)_{20}$ [5-7] and $[Ti_{16}O_{16}](OEt)_{32}$ [6,8] molecules, Et = CH_2CH_3, will be considered.

$[Ti_7O_4](OEt)_{20}$

The $[Ti_7O_4](OEt)_{20}$ molecule, Et = C_2H_5, has the C_{2v} metal-oxygen core structure shown at the top of Figure 2 in the solid state according to single-crystal X-ray diffraction studies [5-7]. This molecule can be prepared from $Ti(OEt)_4$ and water isotopically enriched in ^{17}O, H_2O^*, to yield $[Ti_7{}^*O_4](OEt)_{20}$, *i.e.*, material that has been selectively enriched in ^{17}O at the oxide oxygen sites [7]. When the ^{17}O NMR spectrum of this material is measured in toluene solution (Figure 2a), two resonances are observed, one assigned to the triply bridging oxide oxygen atoms (A) and the other to the fourfold bridging oxide oxygen atoms (B), indicating that the $[Ti_7O_4](OEt)_{20}$ metal oxide core structure is stable in toluene solution [7].

When ethanol is added to a solution of $[Ti_7{}^*O_4](OEt)_{20}$ in toluene, the resulting solution displays six major resonances in its ^{17}O NMR spectrum in addition to the two arising from $[Ti_7{}^*O_4](OEt)_{20}$ (Figure 2b), indicating extensive degradation of the $[Ti_7O_4]$ metal oxide core structure. The spectrum bears a close resemblance to the spectrum of $Ti(OEt)_4$ plus 1/2 eq H_2O^* reported in reference [7], suggesting that ethanol acts as a catalyst for converting $[Ti_7O_4](OEt)_{20}$ into an equilibrium mixture of polytitanates.

$[Ti_{16}O_{16}](OEt)_{32}$

The $[Ti_{16}O_{16}](OEt)_{32}$ molecule, Et = C_2H_5, has the S_4 metal-oxygen core structure shown at the top of Figure 3 [6,8]. As was the case for the $[Ti_7O_4](OEt)_{20}$ molecule, it can be prepared from ^{17}O-enriched water to yield $[Ti_{16}{}^*O_{16}](OEt)_{20}$, selectively enriched at its oxide oxygen sites. The ^{17}O NMR spectrum of this material in toluene solution (Figure 3a) displays four resonances corresponding to the four types of symmetry nonequivalent oxide oxygen atoms in the structure.

Unlike the $[Ti_7O_4](OEt)_{20}$ molecule, the $[Ti_{16}O_{16}](OEt)_{32}$ molecule is stable toward ethanol as demonstrated by the ^{17}O NMR spectrum shown in Figure 3b: even after 48 h at ambient temperature, the spectrum of the $[Ti_{16}{}^*O_{16}]$ core structure is clearly retained. This spectrum does not indicate whether the $[Ti_{16}O_{16}](OEt)_{32}$ molecule is reactive toward ethanol, since alkoxide exchange with ethanol implies no net structural change. Such is not the case when other alcohols are employed. Consider, for example, the case of alkoxide exchange with *n*-propanol. This can be probed using ^{13}C NMR spectroscopy as shown in Figure 4. First, the ^{13}C NMR spectrum of $[Ti_{16}O_{16}](OEt)_{32}$ in toluene is measured and seen to display eight distinct resonances in the OCH_2 region corresponding to the eight sets of symmetry-equivalent ethoxide groups present in its structure. Next, alcohol exchange is effected by reacting the alkoxide with *n*-propanol in toluene/propanol and removing all volatiles. Then ^{17}O NMR spectroscopy is used to confirm that the $[Ti_{16}O_{16}]$ core structure has been preserved, and finally, ^{13}C NMR spectroscopy is used to probe the alkoxide ligands present. As shown in Figure 4b, six ethoxide and two propoxide resonances are observed, identifying the product as $[Ti_{16}O_{16}](OEt)_{24}(OPr^n)_8$, Et = C_2H_5 and Pr^n = $CH_2CH_2CH_3$. This result is significant: alcoholysis, hydrolysis, and condensation are closely-related reactions, and given that alcoholysis process with retention of the $[Ti_{16}O_{16}]$ core structure, there is a reasonable prospect that hydrolysis and condensation might proceed in the same fashion.

The experiments just described all indicate good stability for the $[Ti_{16}O_{16}]$ core of $[Ti_{16}O_{16}](OEt)_{32}$ toward alcoholysis, a necessary but not sufficient condition for this molecule to serve as an effective molecular building block for the preparation of new forms of titania. The stability of the titanium oxide core structure toward hydrolysis and condensation under actual sol-gel polymerization conditions can be probed directly using ^{17}O CPMAS solid state NMR spectroscopy to examine $[Ti_{16}O_{16}](OEt)_{32}$-derived xerogels [9]. As shown in Figure 5a, spectra of xerogels prepared from ^{17}O-enriched $[Ti_{16}{}^{*}O_{16}](OEt)_{32}$ precursor and ^{17}O-enriched water display five ^{17}O resonances: one OTi_4 resonance (A), two OTi_3 resonances (B and C), and two OTi_2 resonances (D and E). Significantly different spectra are observed for xerogels prepared from unenriched precursor $[Ti_{16}O_{16}](OEt)_{32}$ and enriched water (Figure 5b) and xerogels prepared from enriched precursor $[Ti_{16}{}^{*}O_{16}](OEt)_{32}$ and unenriched water (Figure 5c). Resonances Band E are observed only when ^{17}O-enriched water is used. Resonances A, C, and D are either not observed or observed only as weak resonances unless ^{17}O-enriched

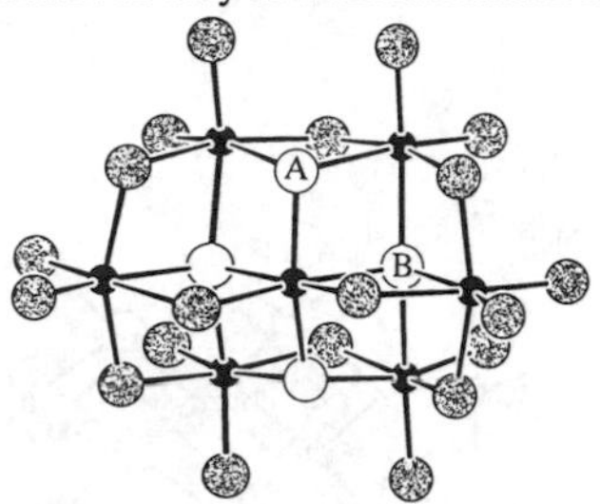

$[Ti_7O_4](OEt)_{20}$

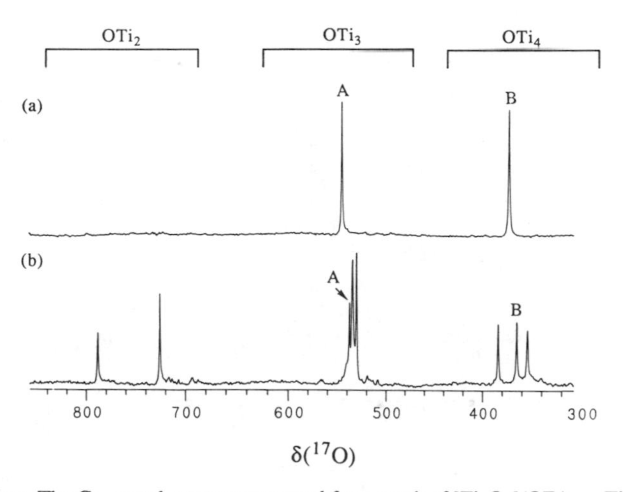

Figure 2. Top: The C_{2v} metal-oxygen structural framework of $[Ti_7O_4](OEt)_{20}$. Titanium atoms are represented by small filled circles, oxide oxygen atoms by large open circles, and alkoxide oxygen atoms by large shaded circles. One member of each set of two symmetry-equivalent oxide oxygen atoms has been labeled. Bottom: 40.7 MHz ^{17}O FTNMR spectra measured at ambient temperature of (a) *ca.* 5 at % ^{17}O-enriched $[Ti_7{}^{*}O_4](OEt)_{20}$ as a 0.17 M solution in toluene and (b) the same solution after diluting with anhydrous ethanol (1:3 v/v ethanol:toluene). Spectrum (b) was measured during the ten minutes immediately following dilution. Chemical shift values relative to fresh tap water are: (a) δ 537 and 365; (b) δ 788, 725, 537, 534, 530, 384, 365, and 354.

$[Ti_{16}{}^{*}O_{16}](OEt)_{32}$ precursor is used, and the chemical shifts for resonances A(376 ppm), C(548 ppm), and D(745 ppm) are close to those observed for the corresponding $[Ti_{16}{}^{*}O_{16}](OEt)_{32}$ resonances in solution, *i.e.*, 382, 554/562, and 749 ppm (see Figure 3). Clearly, there are two types of oxide oxygens present in $[Ti_{16}O_{16}](OEt)_{32}$-derived xerogels, those that are part of the $[Ti_{16}O_{16}]$ core structure and responsible for resonances A, C and D, and those that serve to link these cores together and are responsible for resonances B and E. Although there is some scrambling between these two types of oxide oxygens as evidenced by weak A and C resonances in Figure 5b, the remarkable degree of selectivity observed in Figure 5 indicates the high degree to which the $[Ti_{16}O_{16}]$ core of $[Ti_{16}O_{16}](OEt)_{32}$ is preserved during the course of sol-gel polymerization.

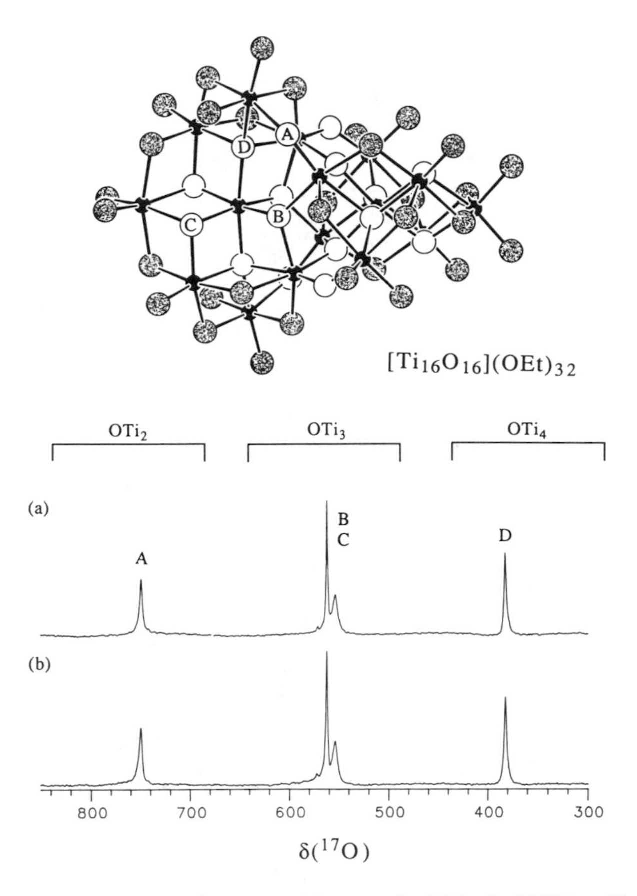

Figure 3. Top: The S_4 metal-oxygen structural framework of $[Ti_{16}O_{16}](OEt)_{32}$. Titanium atoms are represented by filled small circles, and alkoxide oxygen atoms by large shaded circles. One member of each set of four symmetry-equivalent oxide oxygen atoms is labeled: one doubly-bridging oxygen O_A, two triply-bridging oxygens O_B and O_C, and one four-fold bridging oxygen O_D. Bottom: 40.7 MHz ^{17}O FTNMR spectra, measured at ambient temperature, of (a) *ca.* 5 at % ^{17}O-enriched $[Ti_{16}{}^{*}O_{16}](OEt)_{32}$ as a 0.05M solution in toluene and (b) the same solution 48 h after diluting with anhydrous ethanol (1:3 v/v ethanol:toluene). Chemical shift values in (a) and (b), relative to fresh tap water, are: δ 749, 562, 554, and 382.

CONCLUSIONS

1. The metal oxide cores of low nuclearity titanium alkoxides such as $[Ti_7O_4](OEt)_4$ are too kinetically labile to serve as molecular building blocks for the synthesis of new titania materials by sol-gel polymerization.

2. The metal oxide cores of high nuclearity titanium alkoxides like $[Ti_{16}O_{16}](OEt)_{32}$ are sufficiently inert kinetically to serve as molecular building blocks for the sol-gel processing of new forms of titania.

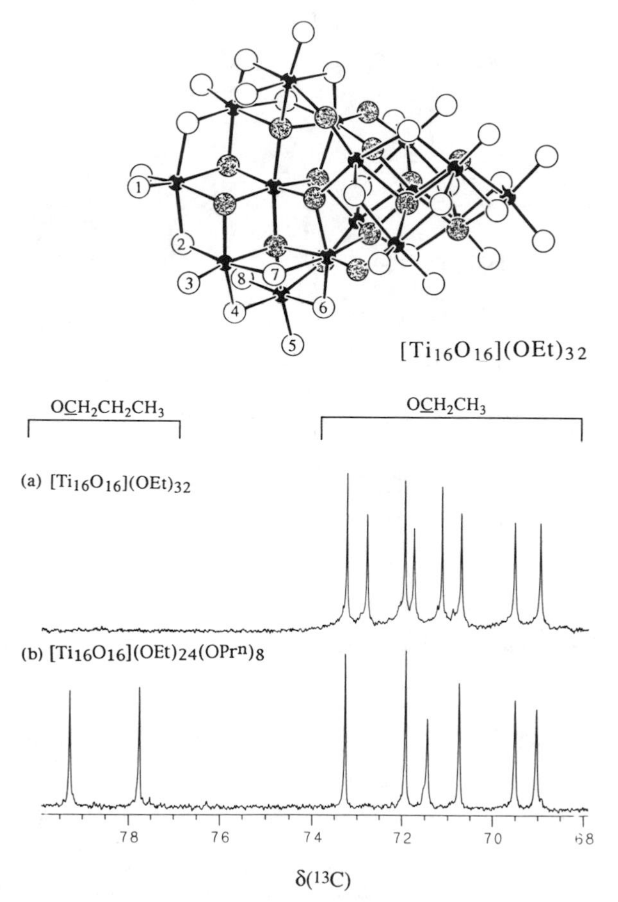

Figure 4. Top: The S_4 metal-oxygen structural framework of $[Ti_{16}O_{16}](OEt)_{32}$ [6,8], with titanium atoms represented by small, filled circles, oxide oxygen atoms by large shaded circles, and alkoxide oxygen atoms by large open circles. One atom from each set of four symmetry-equivalent alkoxide oxygen atoms is labeled. Bottom: 125.8 MHz ^{13}C FTNMR spectra, measured at ambient temperature, of (a) 0.05 M $[Ti_{16}O_{16}](OEt)_{32}$ in toluene and (b) 0.05 M $[Ti_{16}O_{16}](OEt)_{24}(OPr^n)_8$ in toluene. Chemical shift values relative to tetramethylsilane are: (a) δ 73.2, 72.8, 71.9, 71.7, 71.1, 70.7, 69.5, and 68.9; (b) δ 79.3, 77.7, 73.3, 71.9, 71.4, 70.7, 69.5, and 69.0.

3. Solid state ^{17}O CPMAS NMR spectroscopy using selective ^{17}O labeling techniques is a powerful tool for studying molecular growth pathways in titania sol-gel polymerizations.

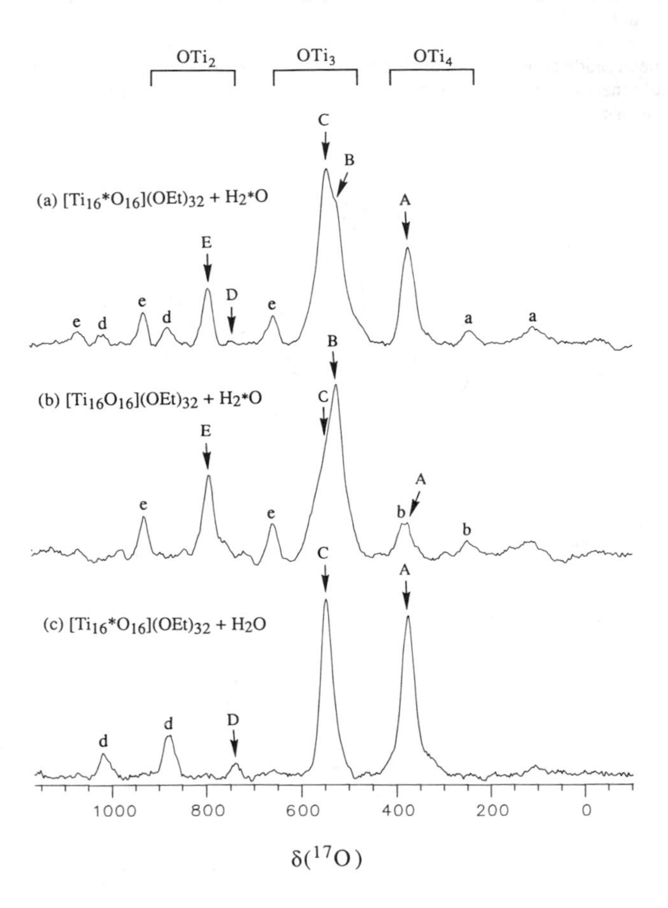

Figure 5. 40.7 MHz MAS solid state ^{17}O NMR spectra of xerogels prepared by first adding a 5mL solution of 150μL H_2O plus 10μL of 12 M HCl in ethanol to a solution of 780 mg $[Ti_{16}O_{16}](OEt)_{32}$ in 5mL tetrahydrofuran to obtain a solution 0.012 M in HCl, 0.844 M in H_2O and 0.032 M in $[Ti_{16}O_{16}](OEt)_{32}$, then allowing the solution to form a clear gel, and finally drying the gel in vacuo. In (a), the sample was prepared from $[Ti_{16}{}^{*}O_{16}](OEt)_{32}$ enriched to 10 at % ^{17}O and water enriched to 45 at % ^{17}O; in (b), $[Ti_{16}O_{16}](OEt)_{32}$ natural abundance in ^{17}O and water 45 at % ^{17}O; in (c), $[Ti_{16}O_{16}](OEt)_{16}$ 10 at % ^{17}O, water natural abundance in ^{17}O. Resonance frequencies and spinning sidebands were identified by varying sample spinning rates: the central resonances are labeled with upper case letters and arrows, and their spinning sidebands are labeled with the corresponding lower case letters. A spinning rate of 5.5 KHz was used in (a) - (c). Chemical shift values relative to fresh tap water are δ 377 ± 2 (A), 526 ± 2 (B), 548 ± 3 (C), 745 ± 4 (D), and 793 ± 3 (E).

ACKNOWLEDGMENTS

This research was supported by the United States Department of Energy, Division of Materials Science, under Contract DE-AC02-76ER01198. We are also grateful to Dr. Zhehong Gan of the University of Illinois School of Chemical Sciences Molecular Spectroscopy Laboratory for assistance in measuring ^{17}O solid state NMR spectra.

REFERENCES

1. V. W. Day, W. G. Klemperer, V. V. Mainz, and D. M. Millar, J. Am. Chem. Soc. **107**, 8262 (1985).
2. P. C. Cagle, W. G. Klemperer, and C. A. Simmons in *Better Ceramics Through Cemistry IV*, edited by B. J. Zelinski, C. J. Brinker, D. E. Clark, and D. R. Ulrich (Mat. Res. Soc. Proc. **180**), Materials Research Society, Pittsburgh, PA 1990, p. 29.
3. W. G. Klemperer, V. V. Mainz, and D. M. Millar in *Better Ceramics Through Chemistry II*, edited by C. J. Brinker, D. E. Clark, and D. R. Ulrich (Mat. Res. Soc. Proc. **73**), Materials Research Society, Pittsburgh, PA 1986, p. 3.
4. C. S. Brevett, P. C. Cagle, W. G. Klemperer, D. M. Millar, and G. C. Ruben, J. Inorg. Orgomet. Polymers **1**, 335 (1991).
5. K. Watenpaugh and C. N. Caughlan, J. Chem. Soc., Chem. Commun. 76 (1967).
6. R. Schmidt, A. Mosset, and J. Galy, J. Chem. Soc., Dalton Trans. 1999 (1991).
7. V. W. Day, T. A. Eberspacher, W. G. Klemperer, C. W. Park, and F. S. Rosenberg, J. Am. Chem. Soc. **113**, 8190 (1991).
8. A. Mosset and J. Galy, C. R. Acad. Sci. Paris Ser. II **307**, 1747 (1988).
9. V. W. Day, T. A. Eberspacher, W. G. Klemperer, C. W. Park, and F. S. Rosenberg in *Chemical Processing of Advanced Materials*, edited by L. Hench and J. West, Wiley, New York, NY 1992, p. 257.

RAMAN AND SURFACE ENHANCED RAMAN SPECTROSCOPY OF TITANIUM CARBOXYLATES

BEATRICE A. VAN VLIERBERGE-TORGERSON, DILUM DUNUWILA AND KRIS A. BERGLUND
Michigan State University, Departments of Chemical Engineering and Agricultural Engineering, East Lansing, MI 48824, USA.

ABSTRACT

Porous stable metal-organic films can be processed under mild conditions by the hydrolysis of carboxylic acid and titanium isopropoxide mixtures. It is of importance to establish the structure of such materials. Raman spectroscopy is a useful technique in determining structural changes in titanium isopropoxide carboxylate solutions, but weak scattering does not allow for study of thin films unless surface enhancement is used. In an attempt to elucidate the relationship between coating solutions and the final structure of a cast film, we discuss Raman spectra of precursor solutions in a variety of conditions.

INTRODUCTION

Titanium alkoxides are widely used for their versatility in producing different adducts by replacement of one or several alkoxy groups [1-3]. Their reaction with carboxylic acids has been extensively studied and the replacement of one or two alkoxy groups was proven to be straightforward whereas the replacement of a third group required a prolonged reflux [4]. The total substitution of the four alkoxy groups by four carboxylates could never be isolated [5]. Furthermore, titanium alkoxides are extremely sensitive to hydrolysis since a white precipitate can be observed with even traces of water [3]. On the other hand, titanium alkoxide - carboxylic acid mixtures with an $R_a \geq 2$ do not readily precipitate upon addition of small quantities of water (R_w=1.5), where R_a and R_w are defined as the ratio of acid to alkoxide and water to alkoxide, respectively. We have used similar mixtures (variety of R_a's and R_w's) to prepare clear solutions that can be spin-cast on glass supports and result in a thin porous film of an unknown titanium network upon drying [6]. The as-obtained film is optically transparent and allows for embedding or binding of a variety of chemicals useful in the design of sensors [7, 8]. Only speculations have been made as to the overall mechanism of these reactions and it is our interest to elucidate the path that leads from a titanium alkoxide precursor to a thin metal-organic film via the hydrolysis of titanium carboxylates. In this report we will selectively study the Raman spectra of a variety of mixtures and relate to the spectrum of the resulting film when applicable. Non-resonance conditions are utilized to avoid the photoreduction of Ti(IV) to Ti(III) centers as has been observed with UV excitation. In the case of film spectra, it is necessary to rely on surface enhancement conditions since the porous network demonstrates very poor scattering owing to its optical transparency. The study covers different alkoxides as precursors (titanium tetraisopropoxide and titanium tetraethoxide), reacted with valeric acid, or butyric acid in one instance, and submitted to mild hydrolysis with varying amounts of water or deuterium oxide. A possible structure of the network is proposed.

EXPERIMENTAL

Valeric (VA) and butyric (BA) acid, titanium isopropoxide (TiPT) and titanium ethoxide (TiOEt) were purchased from Aldrich Chemical Company. Ammonium hydroxide (30%) and nitric acid were from Baker Chemicals. Silver nitrate 99.9+% was from Alfa products, sodium hydroxide GR was from EM Science and the glucose was from the Sigma Chemical Company. All chemicals were used without further purification. The water was reverse osmosis deionized with 18 MΩ x cm resistivity. Silver substrates and film coatings were cast on Corning pre-cleaned single frosted microscope slides. Solutions were studied in Kimble Products capillary melting point tubes sealed with parafilm.

All Raman spectra were recorded on a Spex 1877 triplemate (1800 grooves/mm) with an EG&G OMA-III diode array detector and equipped with a microprobe attachment (Spex-modified Zeiss microscope) with a 180° backscattering configuration using a 40x objective. The excitation source was the 514.5 nm line of a Spectra-Physics Argon laser with a power of 500 mW at the source. The spectrometer slit was set at 100 μm and the filter stage and entrance slits at 5 mm. All spectra were taken at room temperature.

For surface enhancement, a silver coating was prepared according to the method reported by T. M. Cotton and coworkers using Tollen's reaction [9]. Prior to coating, the slides were cleaned in concentrated nitric acid and rinsed in distilled water. The Tollen's reagent was prepared by the precipitation of 50 ml of 2-3% silver nitrate solution with about 50 drops (≈ 2 ml) of 5% sodium hydroxide followed by dropwise addition of 30% ammonium hydroxide (until redissolution of precipitate) in an ice bath. The slides were placed in the Tollen's reagent prior to the addition of 15 ml of a 10% glucose solution. After reaching room temperature, the beaker containing slides and reaction solution was placed in a water bath at 55°C for 1 minute and sonicated (Bransonic 220 sonic bath) for another minute and subsequently rinsed several times with distilled water. The slides were kept in distilled water in the dark until coating (never more than a week delay). At the time of coating, a slide would be removed from the water and thoroughly dried with a hot air dryer, placed in the coating solution for variable amounts of time (15 min to 2 hrs) and spun in a Fisher Scientific benchtop centrifuge for about one minute. The films were kept in the dark until studied.

Solutions were prepared in disposable glass scintillation vials by addition of the required amount of valeric acid to a given volume of alkoxide to reach the desired R_a=[acid]/[alkoxide]. After mixing for a few seconds on a vortex mixer (Fisher Scientific touch mixer), a precise quantity of water was added according to R_w=[water]/[alkoxide] followed by thorough mixing. All reactions were performed at ambient conditions of temperature and light and the spectroscopic response was monitored at different time intervals.

RESULTS AND DISCUSSION

Several spectral windows were investigated in order to characterize the solutions as well as films. The low frequency region provides an insight into Ti-O-Ti or Ti(CO) bonds [10, 11], whereas mid-range frequencies were followed to elucidate the carboxylate's mode of coordination to the metal (1300 to 1800 cm^{-1}) [12, 13]. No attempt was made to investigate below 200 cm^{-1} since the level of Rayleigh scattering is fairly high in this configuration. The high frequency domain only permits an assessment of the disappearance of parent alkoxide and is greatly overwhelmed by the presence of carboxylic acid in excess. For this reason, experiments with deuterated water did not provide the level of information desired. The logical explanation resides in the fact that the amount of added D_2O is always minimal compared to water released during the condensation, thus the deuterium effect in the OH stretch region remains buried under the valeric acid response, especially because of the strong feature due to hydrogen bonding in the concentrations of acid employed (3 to 6 M typically). It is important to note that no color change of the mixture solution was observed upon visible excitation as it is the case with 363.8 nm. [14]. Similarly, no degradation of the film cast on silver substrate could be detected upon exposure to the focused laser beam. We are confident that no photoreduction of the titanium centers is occuring for a visible excitation of 514.5 nm and therefore the spectra recorded are truly representative of the Ti(IV) film mixtures.

Low frequency region: 200 to 1200 cm^{-1}

Literature reports show that a Ti-O-Ti bond vibration is expected around 820 cm^{-1} in IR [10]. Unfortunately in a TiPT/valeric acid/water mixture, this region is heavily obscured by signals due to the released isopropanol at 819 cm^{-1} and a strong vibration of the valeric acid around 825 cm^{-1}. If titanium tetraethoxide is used as a precursor, ethanol is expected to appear with a band at 879 cm^{-1}.

In this case, no definite feature arising from a Ti-O-Ti structure can be observed even though a slight dissymmetry of the valeric acid band at 825 cm^{-1} can be noted. Only when using a TiOEt/butyric/water film solution could we demonstrate evidence of the Ti-O-Ti structure at *ca* . 815 cm^{-1} since the Raman spectrum of butyric acid offers a clean window between 800 and 850 cm^{-1} (figure1). On the same account, a weak feature around 750 cm^{-1} (763 cm^{-1} in IR [10]) can be seen as a shoulder upon mixing and becomes more resolved after 24 hrs. in the TiPT / VA / H_2O mixture for R_a=9 and R_w=1.5.

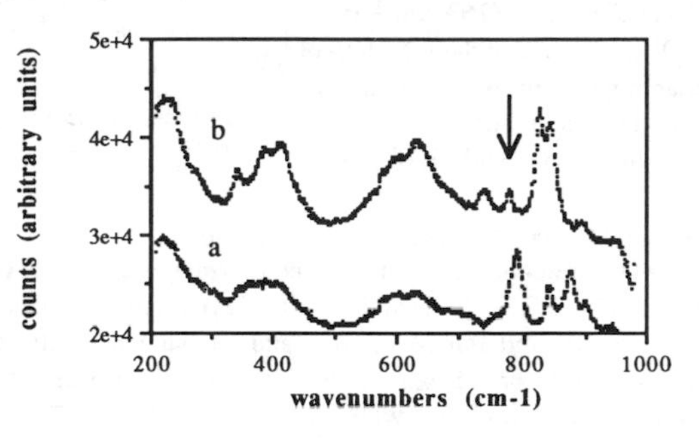

Figure 1. Raman spectra of (a) TiOEt / VA / H_2O and (b) TiOEt / BA / H_2O mixtures with R_a=9 and R_w=1.5

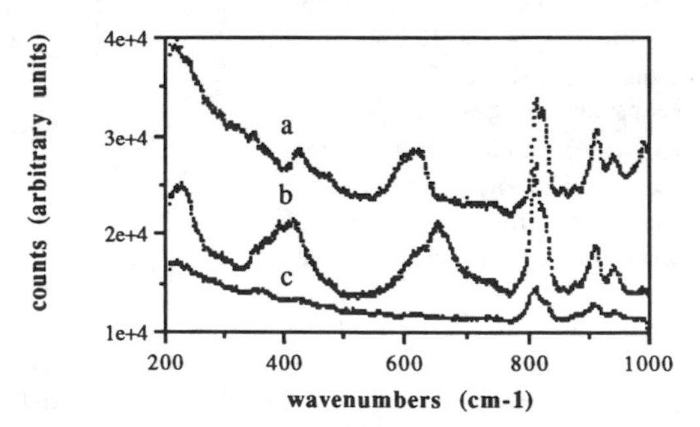

Figure 2. Raman spectra of TiPT / VA / H_2O R_a=9 for (a) R_w=0, (b) R_w=1.5 and (c) R_w=5

In the same region for a fixed R_a=9, but varying R_w's (0 to 10), the titanium isopropoxide doublet at 565 and 612 cm^{-1} (symmetric and antisymmetric Ti-O stretch [15]) undergoes radical changes. Without water, the doublet is changed into a strong assymmetric band with a maximum at 670 cm^{-1} and a shoulder at 630 cm^{-1} whereas the strong feature at 425 cm^{-1} remains. Upon addition of water (R_w=1.5) a shoulder at 380 cm^{-1} appears on the 425 cm^{-1} band while the shoulder at 630 cm^{-1} becomes more pronounced (figure2). For increasing R_w's (5, 7.5 and 10) all spectra are similar: the 380, 425 and 630 cm^{-1} bands remain very weak and broad features acquiring better resolution with time and a weak band at 750 cm^{-1} is present. It is interesting to note that mixtures with very high R_w's (5, 10, 15 and even 30) did not precipitate and remained stable solutions over several months. When D_2O is used in place of water, no shift, attributable to deuterium effect, could be observed in that region. The characteristic (C-O)Ti stretch at 1025 cm^{-1} (1005 cm^{-1} in IR [11]) is very strong in absence of water, but weakens for an R_w=1.5 and practically disappears for $R_w \geq 5$.

When R_w=1.5 is kept constant and R_a varies, some weak non symmetrical vibrations are observed around 425 and 630 cm^{-1} for R_a=2, 3 and 4 (figure 3).

An R_a of 6 brings a shoulder at 670 cm^{-1} and for the highest R_a studied (R_a=9), 670 cm^{-1} dominates with a shoulder at 630 cm^{-1} as observed above. For the high R_a's (6 and 9) a weak structure appears within 24 hrs. at 750 cm^{-1} that can be assigned to Ti-O-Ti bonds (763 cm^{-1} in IR [10]). Once again the vibration of titanium isopropoxide at 1025cm^{-1} remains intense until a sufficient amount of acid is added (R_a=6).

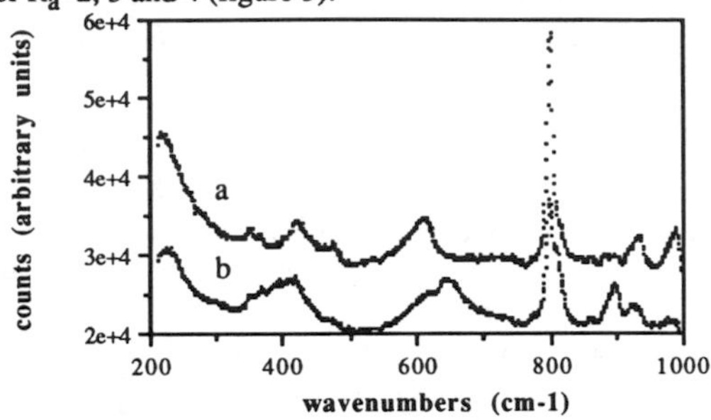

Figure 3. Raman spectra of TiPT / VA / H_2O R_w=1.5 for (a) R_a=2 and (b) R_a=9

Medium frequency range: 1100 to 1900 cm^{-1}

In this region, the emphasis is in determining the mode of binding of the carboxylate anion to the metal [12, 13]. The carboxylic acid symmetric (C-O) stretch is observed at 1310 cm^{-1} (1280 cm^{-1} in IR) and remains strong and unchanged in all conditions. In neat acid the (C=O) antisymmetric stretch exists at 1660 cm^{-1} (1710 cm^{-1} in IR).

When R_a=9 and no water is added, a medium to weak non-symmetrical feature is observed at 1660 cm^{-1}, probably due to the large excess of acid. Upon addition of water (R_w=1.5), a shoulder appears at 1720 cm^{-1} that keeps increasing with higher R_w's (figure 4). This indicates some condensation/polymerization caused by the introduction of water, but it is not clear whether we can distinguish between bidentate chelating or bridging coordination in view of Raman information only. Different authors [16] working with titanium dichloride dicarboxylates have established the coexistence of both modes of coordination through interpretation of IR data. Two models were proposed to accomodate this hypothesis. In the first one, both types of coordination coexist in the same oligomer. In the second case, a mixture of molecules with carboxylates chelated onto the titanium and another form of oligomer with bridging carboxylates between metal centers is speculated. It seems rather unlikely that two different paths could be that clearly distinguished and for this reason the first proposition is preferred.

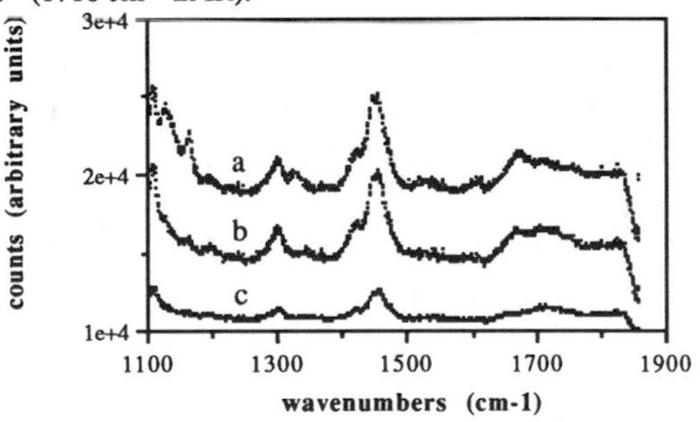

Figure 4. Raman spectra of TiPT / VA / H_2O R_a=9 for (a) R_w=0, (b) R_w=1.5 and (c) R_w=5

This is further substantiated by IR information on a cast film [17] that suggests the existence of both bidentate chelating and bridging in our case as well. On the other hand, for a fixed R_w=1.5, only the 1720 cm^{-1} band is noted for R_a=2, 3 and 4. A shoulder at 1660 cm^{-1} appears for R_a=6 and 9, once again attributable to excess acid (figure 5). The weak shoulder at 1320 cm^{-1} is due to unreacted TiPT for low R_a or R_w.

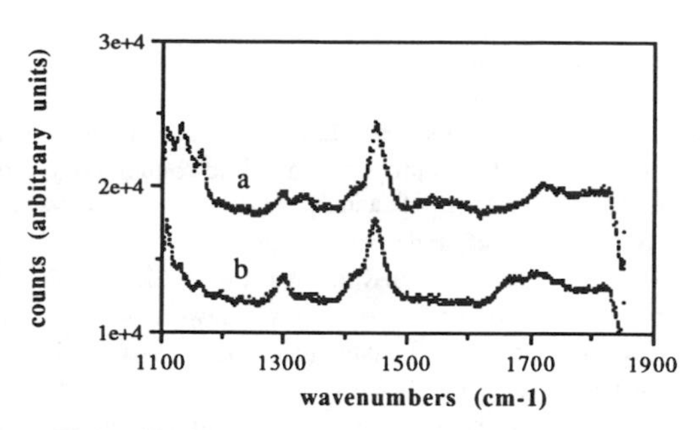

Figure 5. Raman spectra of TiPT / VA / H_2O R_w=1.5 for (a) R_a=2 and (b) R_a=9

High frequencies: 2700 to 3300 cm^{-1}

The different amounts of water in the R_a=9 mixture do not induce any noticeable change in the Raman spectrum. On the other hand, for R_w=1.5 constant, the increasing amount of acid introduced modifies the band at 2970 cm^{-1} from quite resolved (similar to the TiPT signal) at low R_a=2 to becoming a faint shoulder for the highest R_a=9 (as observed in neat valeric acid) (figure 6).

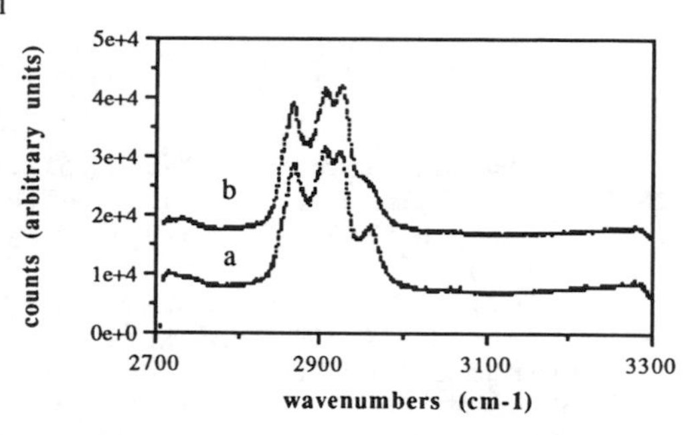

Figure 6. Raman spectra of TiPT / VA / H_2O R_w=1.5 for (a) R_a=2 and (b) R_a=9

Time dependency studies

Two sets of solutions were followed spectroscopically over a 24 hr time period. One experiment involved titanium tetraisopropoxide as the precursor and the second one used titanium tetraethoxide. Both sets had similar R_a's of 9 and R_w's of 1.5. No noticeable change could be observed in these particular conditions between 15 min to 24 hours after mixing of the solutions. Interestingly, spectra were almost identical regardless of the precursor chosen.

Film spectra

As evidenced in a previous communication [18], it is sometimes possible to record Raman spectra of a cast film using a surface enhancement technique. It was shown that the enhancement factor is on the order of 40 for neat titanium isopropoxide. The film study showed that freshly adsorbed films present the same characteristics as the film solution itself. Unfortunately, the signal decreases and eventually disappears over time or for longer casting time. As solvents are driven off and polymerization completes, it can be speculated that the contact of the resulting film with silver islands is impaired. The loss of surface enhancement prevents us from obtaining Raman spectra of completely polymerized films since they possess very poor scattering properties of their own as noted above. It also indicates that reactions may take place after casting.

CONCLUSION

In view of the reported Raman data, consolidated by IR considerations and XPS information [17], we are able to propose a possible connectivity about titanium centers assumed by a film mixture with an R_a=9 and R_w=1.5. The model is in good agreement with the octahedral coordination preferred by titanium (IV) ions.

The oligomer is assumed to consist of randomly placed bridging and bidentate carboxylates, bridging oxo as well as unsubstituted alkoxides. Upon spin-casting, it is likely that further cross-linking occurs either by formation of new Ti-O-Ti bonds or coordination of additional carboxylates through replacement of remaining alkoxides, thus developing a solid interconnected porous network. The resulting film is stable in aqueous neutral solutions compatible with use as a basis for sensing devices.

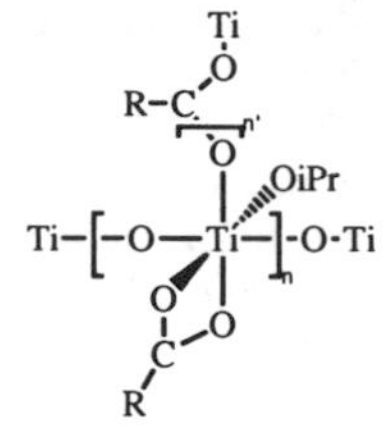

REFERENCES

1- C. Sanchez and J. Livage, *New J. Chem.* 14, 513 (1990).
2- L. G. Hubert-Pfalzgraf, *New J. Chem.*, 11, 663 (1987).
3- D. C. Bradley, R. C. Mehrotra and D. P. Gaur, Metal Alkoxides, (Academic Press, New York, 1978).
4- K. C. Pande and R. C. Mehrotra, *Z. Anorg. Alllg. Chemie*, 291, 97 (1958).
5- K. C. Pande and R. C. Mehrotra, *Z. Anorg. Alllg. Chemie*, 290, 87 (1957).
6- D. Dunuwila, C. D. Gagliardi and K. A. Berglund, unpublished results.
7- R. B. Lessard, M.M. Wallace, W. A. Oertling, C.K. Chang, K. A. Berglund and D. G. Nocera, in Processing Science of Advanced Ceramics, ed. by I.A. Aksay, G.L. McVay and D.R. Ulrich (Mat. Res. Soc. Proc. 155, Pittsburgh, PA, 1989) pp. 109-117.
8- J. I. Dulebohn, B. A. Van Vlierberge, K. A. Berglund, R. B. Lessard, J. Yu and D. G. Nocera, in Better Ceramics through Chemistry IV, ed. by B.J.J. Zelinsky, C.J. Brinker, D.E. Clark and D.R. Ulrich (Mat. Res. Soc. Proc. 180, Pittsburgh, PA, 1990) pp.733-740.
9- F. Ni and T. M. Cotton, *Anal. Chem.* 58, 3159 (1986).
10- V. A. Zietler and C.A. Brown, *J. Phys. Chem.* 61, 1174 (1957).
11- C.G. Barraclough, D.C. Bradley, J. Lewis and I.M. Thomas, *J. Chem. Soc* 1961 2601.
12- R. C. Mehrotra and R. Bohra, Metal Carboxylates, (Academic Press, New York 1983), pp. 48-60 and 233-240.
13- K. Nakamoto, IR and Raman Spectra of Inorganic and Coordination Compounds (Wiley and Sons publishers, New York, 1986), pp. 231-233.
14- C. D. Gagliardi, D. Dunuwila and K. A. Berglund, in Better Ceramics through Chemistry, IV, ed. by B.J.J. Zelinsky, C.J. Brinker, D.E. Clark and D.R. Ulrich (Mat. Res. Soc. Proc. 180, Pittsburgh, PA, 1990) pp. 801-805.
15- M. J. Payne and K.A. Berglund, in Better Ceramics through Chemistry, (Mat. Res. Soc. Proc. 73, Pittsburgh, PA, 1986) pp.627-631.
K. A. Berglund, D. R. Tallant, and R. G. Dosch, in Ultrastructure Processing of Ceramics, Glasses and Composites, ed. by L. L. Hench and D. R. Ulrich (Wiley, New York, 1985), pp. 94-99.
16- J. Amaudrut, *Bull. Soc. Chim.* 7-8, 624 (1977).
17- K. Severin (private communication).
18- B. A. Van Vlierberge-Torgerson, J. I. Dulebohn and K. A. Berglund, Chemical Processing of Advanced Materials, ed. by L. L.Hench (Wiley, & Sons New York, in press).

HYDROLYSIS OF ZIRCONIUM PROPOXIDE BY AN ESTERIFICATION REACTION

I. Laaziz, A. Larbot, C. Guizard, A. Julbe and L. Cot
Laboratoire de Physicochimie des Matériaux, E.N.S.C.M. (C.N.R.S. URA 1312)
8, rue Ecole Normale, 34053 Montpellier Cédex 1, France

ABSTRACT

Zirconium propoxide hydrolysis was performed with the help of an esterification reaction using acetic acid. Products, obtained according to the value of hydrolysis ratio, are successively crystals, sols and gels or precipitates. The similarity of FTIR and NMR spectra for crystals and gels allows to think that the arrangements of ligands around the zirconium atoms are close. The crystalline structure determination permits to precise the environment : acetate groups are always bridging and propoxy groups can be bridging or terminal. Clusters containing 9 zirconium atoms, bridged by oxygen atoms, exist in the structure. The colloid size growth was performed by light scattering. A discussion of the role of acetic acid and the competition between the possible reactions is given.

INTRODUCTION

Because of its good chemical resistance and refractive properties [1], zirconia is an interesting compound for the production of powders, fibers or thin layers. The sol-gel process is a very convenient technique to produce these types of materials. But zirconium alkoxides are easily hydrolyzed [2]. Therefore, we must control the hydrolysis reaction to avoid precipitation of hydrous zirconium oxide.

Hydrolysis, performed by adding acetic acid [3] to the alkoxide solution, leads to the modification of the precursor which affects hydrolysis kinetics and the sol-gel process [4-6].

The objectives of this research are : (a) to define the preparation conditions of zirconia gel from hydrolysis of zirconium alkoxide using esterification reaction, (b) to characterize the molecular structure and the crystalline phase obtained with respect to hydrolysis conditions using X-ray diffraction, NMR and IR spectroscopies.

EXPERIMENTAL

In order to prepare the precursor solution, $Zr(O^{n}Pr)_4$ was mixed with nPrOH. The technique used for the production of water in this solution was an esterification reaction by the addition of acetic acid.

Acetic acid can react with alkoxy groups and with propanol in an esterification reaction. These reactions are in competition and the final product depends on the relative kinetics.

Table I presents the type of products obtained for various times as a function of the temperature and the theoretical hydrolysis ratio h (h=number of AcOH molecules / number of alkoxide molecules). The alkoxide concentration is 2.5 moles per liter of propanol. The results indicate that for the lowest hydrolysis ratios a stable sol was obtained after at least 200 days ; the crystallization and gelation ranges are variable but a precipitate was always obtained under these conditions if h is higher than 2.2. The crystallization and gelation times decrease with increasing temperature. The crystals obtained in each preparation, have identical crystallographic data.

Table I- Results of hydrolysis reaction :
SS = stable sol, C = crystals, G = gel and P = precipitate.

T(°C) \ h	0.2	0.4	0.6	0.8	1	1.5	2	2.5
		SS			C		G	P
7 delay (days)	>200	>200	>100	>100	23	>180	6	1
			SS		C		G	P
20 delay (days)	>200	>200	>200	>200	15	45	2	1
			SS		C		G	P
40 delay (days)	>200	>200	>200	>200	13	55	1	1

The gelation time decreases when the alkoxide concentration is increasing at constant values of h and temperature.

At this stage in the study, it was necessary to determine the true value of hydrolysis ratio to understand the mechanism leading to crystals or gels. Gas chromatography permits the measurement of water molecules produced by the esterification reaction, by controlling the variation of ester in the medium :

$$CH_3\text{-}CH_2\text{-}CH_2OH + CH_3CO_2H \text{ -----> } CH_3CO_2\text{-}CH_2\text{-}CH_2CH_3 + H_2O$$

We have noted that the ester production is important in the first ten minutes and the substitution of propyl groups by hydroxy groups is also more important ; in this case a gel will be obtained. The values reached at the top of the ester concentration curves are the true values of hydrolysis ratio h'. The h' value is a linear function of h according to the following equation : $h'=0.35\ h + 0.09$.

In order to provide a good description of the structure surrounding the zirconium atoms in the sol and in the gel, a comparison between NMR and IR spectra of gels and crystals was performed and the structure of crystals was determined.

RESULTS AND DISCUSSION

Crystal characterization

The crystalline structure was determined by single crystal X-ray diffraction. Analysis of Weissenberg photographs shows a monoclinic unit cell. The crystallographic data are given in table II. The structure was solved by direct methods and the refinements were performed using Shelx 76 programs [7]. The carbon atoms of terminal propyl chains show a significant disorder. This is due to the fact that terminal propyl chains are in "cis" configuration while the bridging ones in "trans" configuration.

Table II- Crystallographic data for $Zr_9(O^nPr)_{18}(OAc)_6O_6$.

	Formula	$Zr_9C_{66}H_{144}O_{36}$	
	Molecular weight	2332.98 g	
Space group	P $2_1/n$	Z	4
a	15.721 (2) Å	d meas.	1.44 (3)
b	28.965 (4) Å	d calc.	1.47
c	23.848 (3) Å	λ Mokα	0.71069 Å
β	103.65 (1)°	μ	0.904 mm^{-1}
v	10552 (2) $Å^3$	R	0.089

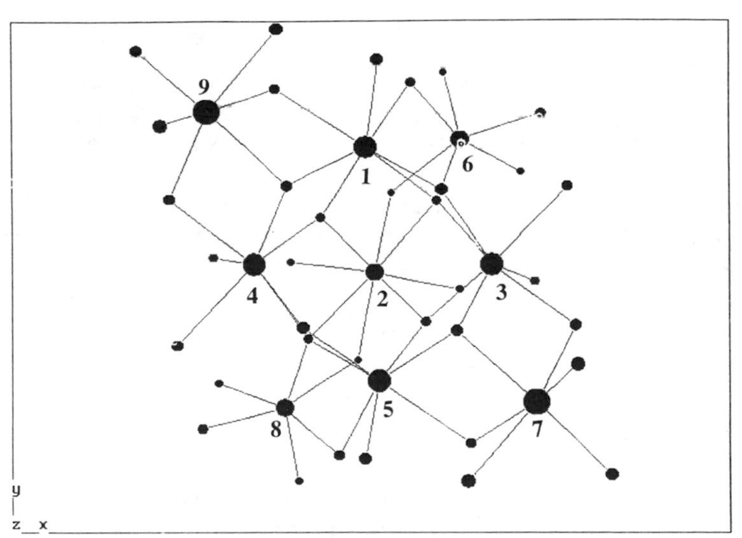

Figure 1 : structural model of $Zr_9(O^nPr)_{18}(OAc)_6O_6$
(● 1 ---> 9 : Zr ; • : O).

So, the last refinement was performed with anisotropic temperature factors for all the atoms except the C atoms of terminal propyl chains for which isotropic temperature factors were used. R factor is equal to 0.089 at the end of the refinement. The structure, represented in figure 1, can be described as a cluster of 9 zirconium atoms . One zirconium atom has a coordination number equal to 8 (oxygen atoms form a square antiprism), 4 others have a coordination number of 7 (monocapped trigonal prism) and the last 4 have a 6-coordinence (octaedral environment). 2 oxygen atoms, noted μ_4-O, bridge 3 zirconium atoms. Zr-O bonds vary from 1.92 to 2.35 Å. Table III shows the environment of zirconium atoms. It should be noted that acetate groups are always bridging and propyl groups can be terminal or bridging.
The structural determination leads to the developped formula : $Zr_9(O^nPr)_{18}(OAc)_6O_6$.

Table III- Environment of zirconium atoms.

Ligands	CH_3-COO Bridging	μ_3-O	μ_4-O	O^nPr Bridging	O^nPr terminal	Coordination Number
Zr_2	xx	xx	xx	xx		8
Zr_1-Zr_5	x	xx	x	xxx		7
Zr_3-Zr_4	xx	xx	x	xx		7
Zr_6-Zr_8	x		x	xx	xx	6
Zr_7-Zr_9	x	x		xx	xx	6

The FTIR spectrum was recorded as dispersed powder in the KBr pellet. Location of bands are reported in table IV. Acetate group has various modes of coordination such as monodentate or bidentate (chelating, bridging or polymeric) [8]. Asymetric and symetric stretching vibration of the COO^- group are located at 1560 and 1412 cm^{-1} respectively, with a $\Delta\nu$ separation equal to 148 cm^{-1} ; this value corresponds to a bridging coordination mode and is in good agreement with those reported in the literature [9].

Table IV- IR band (cm^{-1}) in crystal and gel.

Acetate Group	Crystal	Gel	Propyl group	Crystal	Gel
			ν C-H	2931, 2846	2966,2896
ν_{as} COO	1560	1567	δ C-H	1335	1342
ν_s COO	1412	1419	ν C-O	1046 - 1025	1053 - 1032
δCH_3	1447	1457	ν Zr-O	646, 617	646, 610
			ν Zr-O-Zr	456	470

The ^{13}C NMR spectrum of the crystal was recorded in the solid state. The location of chemical shifts are reported in table V. The difference in chemical shifts for $\underline{C}H_2O$ of the propyl chain shows the existence of two distinct types of propyl chain in the crystal, where the first bridges two zirconium atoms and the second is terminal and bonds with a single zirconium atom.

Table V- ^{13}C NMR chemical shifts (ppm) of solid state $Zr_9(O^nPr)_{18}(OAc)_6O_6$ crystals and gel.

location	crystals	gel
$\underline{C}H_3$ group propoxy	10.1	9.2
$\underline{C}H_3$ acetate group	23.3	23.3
CH_3-$\underline{C}H_2$ propoxy group	25.7	shoulder
CH_3-CH_2-$\underline{C}H_2$ bridging propoxy group	63.7	71.1
CH_3-CH_2-$\underline{C}H_2$ terminal propoxy group	70.7	78
$\underline{C}$-O acetate group	178.9	178.8

Gel characterization

The gel used in this study was obtained in the following conditions : concentration = 2.5 mol.l^{-1} ; theoretical h=2, experimental h'=0.8 , drying temperature = 100°C for 1 hour. For the FTIR spectrum, the location of bands is compared in table IV. A great similarity is observed. The $\Delta\nu$ separation value between ν_s and ν_{as} of COO^- group is equal to 168 cm^{-1} and indicates, as in crystal, a bridging coordination mode.

The comparison of ^{13}C NMR spectra of the gel and the crystals is given in table V. The C of CH_2O in the propyl chain presents two values of chemical shifts ; this fact confirms the two different types of coordination mode : bridging and terminal.

Sol characterization

The study of evolution of particle size in the sol up to the gelation time has been conducted by light scattering. The experiments were conducted with a sol prepared with the following conditions : c=2.5 mol.l^{-1}, h=2 and t=20°C. The gelation time is 48 hours. The particle growth was stopped, for the measurements, by dilution in chloroform. The size varies slowly from 3 to 7 nm in 30 hours. Above this time, the increase of size is faster and the particle radius reaches 18 nm close to the gel time. It can be noted that moving particles with 20 nm in diameter are still present in the gel.

CONCLUSION

Hydrolysis of zirconium propoxide solution in propanol was performed using esterification reaction between acetic acid and propanol. Several products as crystals, gel or precipitate are obtained according to the quantity of acetic acid added in the solution.

The important role of acetic acid as ligand is confirmed. Acetic acid addition reduces the strong reactivity of alkoxides with water.

Zirconium propoxide is modified by acetic acid, by exchange between acetate and propyl groups, and there is a competition between exchange and esterification reactions. The specific kinetics of these reactions depend on the ratio acetic acid over alkoxide.

According to the results, acetic acid reacts firstly with alkoxide and the exchange reaction between propoxy group and acetate happens. The hydrolysis can occur after, depending on free acetic acid in the solution. For the lower values of h, there is not enough produced water in the medium to lead to a gel by hydrolysis and condensation of alkoxide molecules. If h value increases, esterification reaction is more important and the condensation leads to the gel. For the higher values of h, esterification becomes too important, a gel cannot be obtained because of hydrolysis kinetics is too fast and the precitation occurs.

A structural study of $Zr_9(O^nPr)_{18}(OAc)_6O_6$ was made on the crystals and has permitted to describe a cluster of 9 zirconium atoms connected between themselves by oxygen bridge ; acetate ligands are always bridging and propoxy groups can be terminal or bridging.

Zirconium atoms present 8, 7 or 6 coordination number.

A comparative study of crystal and gel conducted with IR and NMR spectroscopies shows the great similarity between the structural arrangement around zirconium atoms. So the cluster of 9 zirconium atoms can be representative of intermediate species between the sol and the gel.

[1] P. Pascal, "Composés Organiques du Zirconium", Nouveau Traité de Chimie Minérale- Tome IX Titane, Zirconium, Hafnium, Thorium, Ed. Masson (1963) 742.

[2] D.C. Bradley, D. G. Carter, Can. J. Chem. 40 (1962) 15.

[3] A.Larbot, J.A. Alary, C. Guizard, L. Cot, J. Gillot, J. of Non Cryst. Solids 104, (1988) 161.

[4] R.C. Fay, Zirconium and Hafnium, "Comprehension coordination chemistry" published by Pergamon Press, edited by G. Wilkinson, R. D. Gillard, G. A. Mc Cleverty, New York, 1987,vol 3, p 32.

[5] R.C. Paul, HS Markhm, Indian J. of Chem., 9 (1971) 247.

[6] J.C. Desikdar, J. Non Cryst. Solids, 86 (1986) 231.

[7] G. M. Sheldrick, Program for Structure Determination, Univ. of Cambridge, England.

[8] N. W. Alcock, V. M. Tracy, T.C. Waddington, J. Chem. Soc. Dalton (1976) 2243.

[9] Von K. H. Thiele, M. Pause, Z. Anorg. Allg. Chem. 441 (1978) 23.

MOLECULAR DESIGN OF CARBOXYLIC PRECURSORS FOR ZIRCONIA

ALLEN W. APBLETT*, JIN LEI, AND GALINA D. GEORGEVA
Tulane University, Department of Chemistry, New Orleans, LA 70118

ABSTRACT

Several carboxylic acid derivatives of zirconium were investigated for their potential utility as precursors for zirconia fibers and thin films. The reactions of $ZrOCl_2 \cdot 8H_2O$ and $ZrCl_4$ with succinic acid, oxalic acid, maleic acid, benzene tetracarboxylic acid, 2-hydroxyiminopropionic acid, 3-hydroxybutyric acid, and β-alanine were performed and the thermal behavior of the products was investigated by thermal gravimetric analysis. The objective was to find preceramic compounds with low pyrolysis temperatures and physical properties suitable for materials processing.

INTRODUCTION

Carboxylic acids play an important role in several materials-oriented processes, including control of hydrolyis mechanisms in the sol-gel process. Often, a carboxylic such as acetic acid proves beneficial in slowing the rate of gellation of metal alkoxide/water systems as well as providing control of the microstructure and improved gel homogeneity [1,2].

The fact that significant improvement in the sol-gel process can be achieved by the addition of carboxylic acids suggests that a practical alternative for the preparation of these materials is metalorganic deposition (MOD)[3]. This technique is a non-vacuum, liquid-based method of depositing thin films. Suitable metalorganic precursors are coated on a substrate and subsequently pyrolyzed to their constituent elements, oxides, or other compounds. Metal carboxylates are the best candidates for the preparation of ceramic oxides and MOD processes for the generation of many oxide-based materials have already been developed: e.g. $BaTiO_3$ [4], ITO [5], SnO_x [6], $YBa_2Cu_3O_7$ [7] and ZrO_2 [8]. Furthermore, methodology has also been developed to achieve epitaxial growth of dense, polycrystalline, zirconia thin films [9].

In many applications, however, carboxylic substituents, while aiding the initial processing of the material, may prove detrimental due to the difficulty of their eventual removal to yield the final product. For example, hydrolyis under standard conditions of most metal carboxylates yields basic metal carboxylates rather than oxides [10]. As well, the removal of carboxylates by pyrolysis is not necessarily achieved cleanly without resort to oxidizing conditions. Metal carboxylates generally decompose via the formation of the metal carbonate and a ketone or to the metal oxide and an acid anhydride [10]:

$$(RCO_2)_2M \longrightarrow MCO_3 + RC(O)R \tag{1}$$

$$(RCO_2)_2M \longrightarrow MO + RC(O)OC(O)R \tag{2}$$

Since removal of the carboxylate requires pair-wise interaction of carboxylates in a random process, eventually a point will be reached where the remaining ones are isolated from each other. These residual carboxylates will then char with further heating leading to carbon incorporation into the product. While an oxidizing atmosphere will, to a large extent, avert this problem, this imposes limitations on eventual applications. Furthermore, the removal of the ketones or anhydrides is not particularly easy where R is a large functional group and high temperatures are often required to completely decompose the metal carboxylates. Clearly, it is important to identify metal carboxylate systems which have desirable processing characteristics. The study reported herein focuses on zirconium carboxylates but the results are applicable to a wide range of metal oxides.

EXPERIMENTAL

All carboxylic acids, benzene tetracarboxylic acid dianhydride, $ZrOCl_2 \cdot 8H_2O$ and $ZrCl_4$ were commercial products and were used as supplied. ZrO(oxalate) [11] and $Zr(oxalate)_2$[12] were prepared by literature routes while 2-hydroxyiminopropionic acid was prepared by a modification of a literature method [13]. CH_2Cl_2 and THF were distilled from P_2O_5 and sodium, respectively. Water was deionized and distilled before use.Thermogravimetric analyses were obtained on a Seiko SSC 5200 TGA/DTA instrument under nitrogen or air atmospheres. Off-gases were trapped and identified by GC/MS spectra on a Hewlett Packard 5995 gas chromatograph/ mass spectrometer. All the zirconium carboxylate complexes with the exception of $ZrO(CO_2CH_2CH_2NH_2)_2 \cdot H_2O$ decompose without melting. The reported decomposition point is the temperature at which the carboxylate ligand begins to decompose. IR spectra were obtained on a Mattson Cygnus 100 FTIR spectrometer. Zirconium analysis was performed gravimetrically upon ignition in oxygen while CH microanalyses were performed by Oneida Research Services Inc., Whitesboro, N.Y. Water content was determined gravimetrically by drying to a constant weight at 105°C.

$ZrO(maleate) \cdot 2H_2O$

$ZrOCl_2 \cdot 8H_2O$ (3.22g, 10 mmol) in water (10 ml) was added to a stirred solution of maleic acid (1.16g, 10 mmol) in water (30 ml). Immediate gellation occurred to yield a translucent mass. Excess water was removed by suction through a medium sintered-glass frit and the gel was dried at room temperature under a stream of flowing dry air. Extensive cracking of the gel occurred in the drying process and the product (2.59 g, Dec. 407°C) was obtained as transparent glassy shards. Analysis: Calculated for $ZrO(maleate) \cdot 2H_2O$; H 2.35 %, C 18.67% Zr 35.45%, H_2O 14.0%. Found; H 2.38 %, C 18.57%, Zr 35.3±0.2%, H_2O 14.1±0.2%

$ZrO(succinate) \cdot 2H_2O$

$ZrOCl_2 \cdot 8H_2O$ (3.22g, 10 mmol) was added to a stirred solution of succinic acid (1.18g, 10 mmol) in water (50 ml). Gentle heating to 50°C caused the solution to solidify to an opaque white gel. The bulk of the water was removed at room temperature under a stream of flowing air followed by heating at 70°C for two hours. Extensive cracking of the gel occurred in the drying process and the product (2.64 g, Dec. 380°C) was obtained as glassy opaque white shards [14].

$ZrO(CO_2CH_2CH_2NH_2)_2 \cdot H_2O$

$ZrOCl_2 \cdot 8H_2O$ (5.30g, 16.5 mmol) and β-alanine (2.92 g, 32.9 mmol) were stirred together in refluxing toluene (100 ml) for 24 hours. Water and HCl was removed from the system by use of a Dean & Stark receiver. The reaction produced a thick, viscous liquid in the bottom of the flask which solidified to a transparent glass upon cooling. The toluene was decanted and the solid was washed with toluene (3 X 10ml). Drying in vacuo cause extensive cracking and the product (4.96g, Dec. 293°C) was obtained as colorless, highly hydroscopic glassy pieces [14].

$[ZrO(O_2C)_2C_6H_2(CO_2)_2ZrO]_x{\cdot}4XH_2O$

$ZrOCl_2{\cdot}8H_2O$ (0.64g, 2.0 mmol) in water (5 ml) was added to a stirred solution of benzene tetracarboxylic acid dianhydride (.21g, 1.0 mmol) in 50:50 water: ethanol (10 ml). The solution slowly gelled over a period of thirty minutes but after several days the gel redissolved to yield a viscous liquid. Removal of the solvent in vacuo produced a colorless glassy solid (0.62 g. Dec. 511°C) [14].

$ZrO(OH)[CO_2C(NOH)CH_3]{\cdot}H_2O$

$ZrOCl_2{\cdot}8H_2O$ (3.22g, 10 mmol) in water (10 ml) was added to a stirred solution of 2-hydroxyiminopropionic acid (1.03g, 10 mmol) in water (30 ml). The solution slowly deposited a microcrystaline solid over a period of several days. Filtration and drying in vacuo produced a white powdery solid (2.18g, Dec. 251°C) [14].

$Zr[CO_2C(CH_3)NO\text{-}]_2$

2-hydroxyiminopropionic acid (2.06g, 20 mmol) was added to a stirred solution of $ZrCl_4$ (2.33g, 10 mmol) in 10:1 CH_2Cl_2 : THF (60 ml). After stirring for 96 hours a white solid (2.67g, Dec. 186°C) was removed by filtration and dried in vacuo [14].

$Zr[CO_2CH2CH(CH3)O\text{-}]$

3-hydroxybutyric acid (2.08g, 20 mmol) was added to a stirred solution of $ZrCl_4$ (2.33g, 10 mmol) in 10:1 CH_2Cl_2 : THF (60 ml). After 5 days a white solid was precipitated by addition of CCl_4 (20 ml). Filtration and drying in vacuo produced a white powder (2.88g, Dec. 275°C) [14].

RESULTS AND DISCUSSION

Reaction of $ZrOCl_2{\cdot}8H_2O$ with organic diacids (succinic, and maleic) leads in all cases to gels presumably consisting of a network of highly-crosslinked polymers having the formula $ZrO(diacid){\cdot}XH_2O$ Thus, rather than chelating, the diacids bridge metal centers and, ultimately, this is detrimental to their eventual removal. The expectation that the use of diacids would lead upon pyrolysis, to formation and volatilization of the cyclic anhydrides (Eq. 3) was only partially realized. While small amounts of succinic and maleic anhydride were detected in the off-gases, the bulk of the decomposition (below the combustion temperature) produced non-volatile linear anhydrides (Eq. 4) as evidenced by bands attributable to the anhydride carbonyls (1750-1850 cm^{-1}) in the infrared spectrum of the solid residue. Therefore, in the absence of oxygen, considerable charring occurred during pyrolysis leading to jet-black products.

$$\text{-Zr(O)O}_2\text{CCH}_x\text{CH}_x\text{CO}_2\text{-} \longrightarrow \text{ZrO}_2 + \text{H}_x\text{C(O)OC(O)CH}_x \quad (3)$$

$$[\text{-Zr(O)O}_2\text{CCH}_x\text{CH}_x\text{CO}_2\text{-}]_n \longrightarrow n\,\text{ZrO}_2 + [\text{-C(O)CH}_x\text{CH}_x\text{C(O)O-}]_n \quad (4)$$

The reaction of $ZrOCl_2{\cdot}8H_2O$ with oxalic acid produces a colorless gel of composition $ZrO(oxalate){\cdot}H_2O$ [11]. Unlike the previous diacids, however, the material decomposed cleanly above 250°C by loss of CO and CO_2.

$$[\text{-OZrO}_2\text{CCO}_2\text{-}]_n \longrightarrow n\,\text{ZrO}_2 + n\,\text{CO} + n\,\text{CO}_2 \quad (5)$$

Zirconium dioxalate decomposes in the same manner, but owing to its high solubility in water, methanol, and THF, it is an ideal MOD precursor for zirconia. Unfortunately, upon long standing these solutions tend to gel due to partial hydrolyis.

Two other zirconium carboxylate systems also have high potential due to their processibilty despite undesirable thermolytic behavior; $ZrO(CO_2CH_2CH_2NH_2)_2 \cdot H_2O$ and $[ZrO(O_2C)_2C_6H_2(CO_2)_2ZrO]_x \cdot 4XH_2O$. The former compound is not only highly soluble in water (> 2g/ml) but it dissolves in its own water of hydration upon heating to 90°C. The viscid melt thus obtained is easily drawn into fibers, or at lower temperatures molded into objects. Upon cooling, it hardens to a translucent solid but must be fired immediately since it deliquesces in moist air. As well, clean removal of the carboxylate is only possible under oxidizing conditions. $[ZrO(O_2C)_2C_6H_2(CO_2)_2ZrO]_x$ was produced under controlled conditions via the slow hydrolysis of benzene tetracarboxylic acid dianhydride since the direct reaction of $ZrOCl_2 \cdot 8H_2O$ with the acid produced an extensively cross-linked polymer. The initial product from the dianhydride was a much more mobile gel which eventually redissolved presumably due to rearrangement to a less cross-linked polymer. The mobility of the gel as well as the high viscosity of the solution attribute excellent processing characteristics to this material. Again, removal of the carboxylate is only possible under oxidizing conditions.

It was found that ideal pyrolysis behavior could best be achieved by careful design of the carboxylate ligands such that they decompose by a low activation-energy route into small, volatile products. Such ideal behavior was observed for 2-hydroxy-iminopropionic acid and 3-hydroxybutyric acid derivatives of zirconium. The former acid yields a crystalline solid with the formula $ZrO(OH)[CO_2C(NOH)CH_3] \cdot H_2O$ upon reaction with $ZrOCl_2 \cdot 8H_2O$. The carboxylic ligand decomposes sharply at 251°C with simultaneous release of acetonitrile and carbon dioxide, as detected by GC/MS. A possible mechanism which accounts for this observation is given in Equation 6.

Me

C O

N C

O O

H HO(O)Zr

$$\longrightarrow ZrO(OH)_2 + MeCN + CO_2 \qquad (6)$$

Unfortunately, the zirconium-containing product obtained had a weight consistent with formation of a complex hydroxide, $Zr_5O_8(OH)_4$. The infrared spectrum of this material exhibited only bands due to hydroxyl and Zr-O vibrations. Further heating to 520°C was required for the final production of zirconia. This decomposition was accompanied only by production of water (GC/MS), again confirming the production of a complex oxide/hydroxide phase. It should be noted that $Zr_5O_8(OH)_4$ was previously identified by Rodd [15]. Its formation was easily avoided by the preparation of zirconium bis(2-hydroxy-iminopropionate) with ligands that chelate through the imino oxygen, reducing the decomposition temperature to 186°C and yielding ZrO_2 directly. Furthermore, the compound is very soluble in a number of organic solvents including THF, CH_3CN, DMSO, and alcohols. Again, production of acetonitrile and CO_2 was observed to occur concurrently, consistent with the proposed mechanism in Equation 7.

$$[Zr(O_2CC(Me)NO)]_2 \longrightarrow ZrO_2 + 2\,MeCN + 2\,CO_2 \qquad (7)$$

In an identical manner, $Zr[CO_2CH2CH(CH3)O\text{-}]_2$ may be prepared by the reaction of $ZrCl_4$ with 3-hydroxybutyric acid. It has similar solubility characteristics and decomposes at somewhat higher temperatures (275°C) to propene, CO_2 (detected by GC/MS) and zirconia. A probable mechanism of this decomposition is;

$$[Zr(O_2CCH_2C(CH_3)O)]_2 \longrightarrow ZrO_2 + 2\,C_3H_6 + 2\,CO_2 \qquad (8)$$

The success of the latter two zirconia precursors is probably best demonstrated by the fact that the ceramic yield is identical (essentially quantitative yield of zirconia) whether oxidizing or inert atmospheres are used in the thermogravimetric analysis. The TGA trace for $Zr[CO_2C(CH_3)NO\text{-}]_2$ is given in Figure 1.

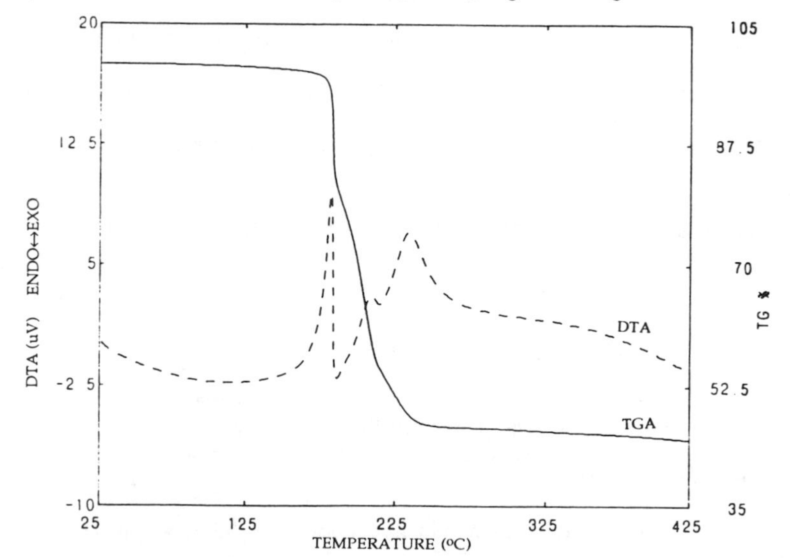

FIGURE 1: TGA trace for $Zr[CO_2C(CH_3)NO\text{-}]_2$ (Heating Rate 4°C/min under N_2)

CONCLUSION

Several zirconium carboxylate systems have been identified that have potentially useful properties for eventual materials production. Thus, stiff or mobile gels, soluble polymers and thermoplastic materials can be prepared by varying the nature of the carboxylate. Furthermore, carboxylate ligands that can decompose in a concerted low-energy pathway to volatile side products have been designed and exhibit great potential for ceramic metal-oxide production.

REFERENCES

1. C. Sanchez, J. Livage, M. Henry, and F. Babonneau, J. Non-Cryst Solids, 100, 65 (1988)

2. S. Doeuff, M. Henry, C. Sanchez, and J. Livage, J. Non-Cryst Solids, 89, 206 (1987)

3. J. V. Mantese, A. L. Micheli, A. H. Hamdi, and R. W. Vest, M.R.S. Bull, 1989 (XIV), 1173.

4. J.J. Xu, A. S. Shaikh, and R. W Vest, IEEE Trans UFFC 36, 307 (1989).

5. J.J. Xu, A. S. Shaikh, and R. W Vest, Thin Solid Films 161, 273 (1988).

6. T. Maruyama and K. Kitamura, Jpn. J. Appl. Phys. 28, L312 (1989).

7. A. H. Hamdi, J. V. Mantese, A. L. Micheli, R.C.O. Laugal, D.F. Dungan, Z.H. Zhang, and K.R. Padmanabhan, Appl. Phys. Lett. 51, 2152 (1987).

8. V. Hebert, C. His, J. Guille, and S. Vilminot, J. Mat. Sci. 26, 5184 (1991).

9. K. T. Miller, F.F. Lange, and D.B.Marshall, J. Mat. Res. 5, 151 (1990).

10. R.C. Mehrotra and R. Bohra, Metal Carboxylates , (Academic Press: New York, 1983) and references therein.

11 E. Lowenstein, Z. Anorg. Allgm. Chem.. 63, 92 (1909).

12. H.S. Gable, J. Am. Chem. Soc., 53, 1276 (1931).

13. V. Meyer and A. Janny, Ber. 15, 1525 (1882).

14. The product was identified as outlined in the general experimental section. The results of the analyis were consistent with the given formulation; agreement between calculated and experimental compositions were similar to those reported for the maleate.

15. R.T. Glazebrook, W. Rosenhein, and E.H. Rodd, British Patent No. 112,973 (29 January, 1917)

PROPERTIES OF ZrO_2 GELS PREPARED BY CONTROLLED CHEMICAL MODIFICATION METHOD OF ALKOXIDE

HISAO SUZUKI*, HAJIME SAITO* AND HIROAKI HAYASHI**
*Toyota Technological Institute, 2-12 Hisakata, Tempaku, Nagoya 468, Japan
**Tsuchiya Co.Ltd., Chiryu Research Laboratory, 224 Higashinamikikita, Yama-machi, Chiryu 472 Japan

ABSTRACT

Polymeric ZrO_2 gels were prepared by the controlled chemical modification method (CCM method) of zirconium-n-propoxide. In this method, the steric hindrance by alkoxide-acetic acid chelation could be used to control the hydrolysis and condensation reaction of the zirconium alkoxide if the amount of hydrolysis water was limited. As a result, polycondensation occurred uniformly in the solution, forming a linear zirconoxane polymer. When the solvent evaporated, the zirconoxane polymer crosslinked with each other and formed polymeric ZrO_2 gels, which were monolithic and transparent. These polymeric gels could be re-dissolved into n-butyl alcohol with acetic acid and mechanical stirring. Heating would enhance the dissolution of the gels. Using the re-dissolved gel solution, dense thin films of ZrO_2 could be obtained by dip-coating procedures without many coating operations.

INTRODUCTION

Recently, wet chemical methods such as the sol-gel processes have attracted much attention in the preparation of the advanced ceramics. Among these, the alkoxide method is especially promising because molecular designs of the ceramic precursors can be achieved [1],[2],[3],[4]. The molecular design of the ceramic precursors could improve the properties of the resulting ceramics significantly [5].

To design ceramic precursors on a molecular scale, precise controls of the hydrolysis and/or polycondensation are essential. To date, many attempts have been done to control the alkoxide reaction with little success because almost all alkoxides except silicon alkoxide hydrolyze very rapidly [1],[2],[3],[6]. In one of the authors' papers, a new method (Controlled Chemical Modification Method or CCM method), in which chemical modifications of an alkoxide was accomplished, was applied to control the reaction of a zirconium alkoxide [7]. This method yielded a monolithic and transparent ZrO_2 gels. These gels were polymeric and finely textured and could be sintered into translucent monolithic polycrystals with very fine pores [8]. If the monolithic gels prepared by the CCM method consisted of linear zirconoxane polymers, these gels should be dissolve in a suitable solvent. After dissolving the gels, high-performance ZrO_2 films could be obtained from the gel solution by dip-coating followed by careful firing. This paper mainly described the precise conditions to prepare these polymeric ZrO_2 gels and the properties of these gels themselves as well as the dense films made from these gels.

EXPERIMENTAL PROCEDURES

Zirconium-n-propoxide and n-butyl alcohol were used as raw material and solvent, respectively. Concentrations of the alkoxide were 0.5 M in all cases. Figure 1 schematically showed the experimental procedures. Previous work indicated that the acetic acid and the moderate amount of water could control the alkoxide reaction in a similar system [7],[8]. Therefore, the acetic acid to alkoxide molar ratio, R_A, was changed from 0 to 2 and water to alkoxide molar ratio, R_W, was changed from 0.5 to 2.

All the solution gelled in two days regardless of different conditions. The actual gelation time of the precursor solution depended upon R_A and/or R_W. After gelation and evaporation of the solvent, n-butyl alcohol and/or acetic acid were added to the gels and then stirred. The gels were found to dissolve in such conditions. In some cases, mixtures of the gels and alcohol were heated with stirring to accelerate the dissolution. Infra-red (IR) spectroscopy was used to identify the chemical bonds in the gels and gel solution. Powder x-ray diffraction technique was performed to identify the crystalline phases in gels after heat-treating up to 600°C for 30 min. in air. Thin films of ZrO_2 were prepared by dip-coating using the re-dissolved gel solution (gel solution) followed by heat-treatment up to 600°C for 30 min. in air. Film thickness was determined by Scanning electron microscope.

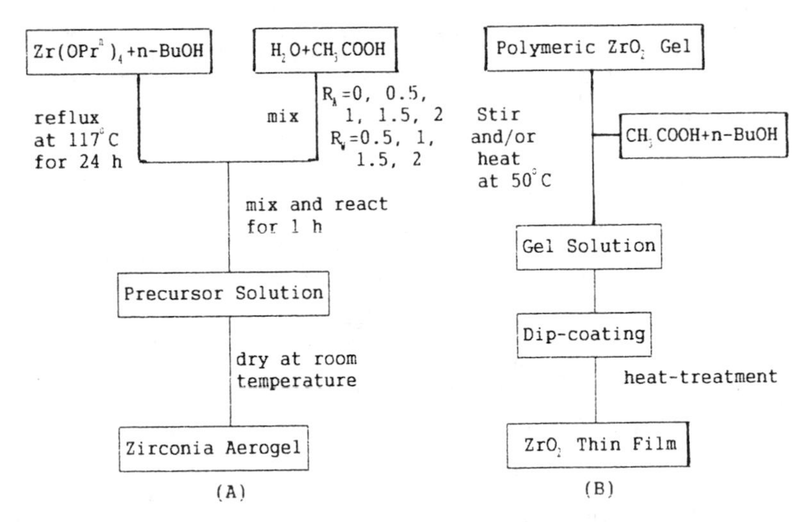

Fig.1 Experimental procedure for preparation of (A)polymeric gels ,(B)gel solutions and thin films

RESULTS AND DISCUSSIONS

Aspects of gels

If the R_A and/or R_W were less than one, precursor solution did not gel. In some cases, precipitation occurred. This showed that the control of R_A and R_W are important to prepare monolithic ZrO_2 gels.

Figure 2 showed some examples of the dried and transparent gels after 3 months of aging. With increasing R_A, gels were more likely to break into pieces (Fig.2). Monolithic and transparent gels formed only under specific conditions (R_A <1.5 and R_W <2 in current study). The addition of excess acetic acid ($R_A \geq 2$) hindered the polycondensation,

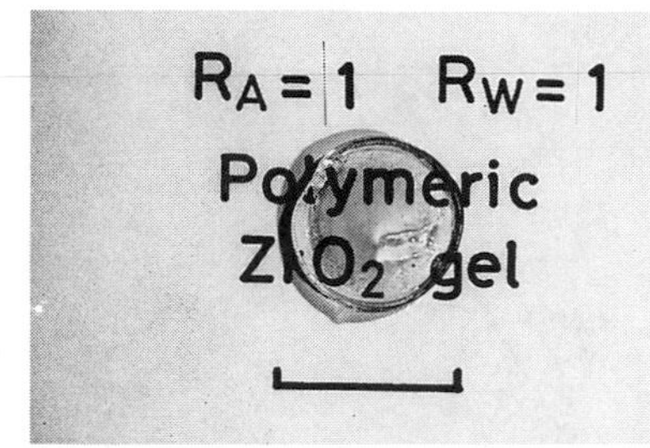

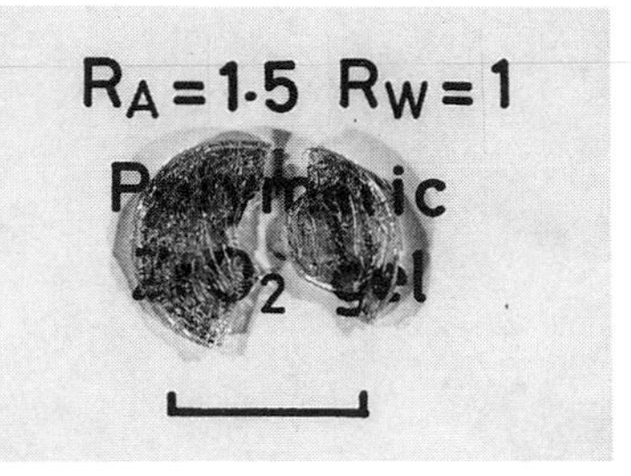

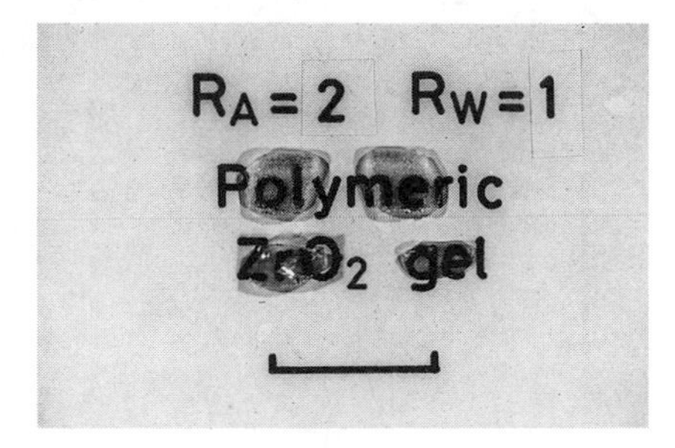

Figure 2. Some examples of the gels prepared by the CCM method.
(A) R_A = 1.0, R_W = 1.0,
(B) R_A = 1.5, R_W = 1.0,
(C) R_A = 2.0, R_W = 1.0.
Bar = 2.5 cm in length

resulting in a longer gelation time. When R_A was too low (<1) and excess water was available (R_W=2), the hydrolysis rate became so high that polycondensation occurred locally. This would result in the precipitation of powders. Therefore, the R_W must be kept below 2. The polymeric gels only formed at specific ranges of R_A and R_W.

Properties of gels

Under specific conditions, the gels prepared by the CCM method would be linear polymeric gels. These gels could be dissolved in a suitable solvent with mechanical forces (stirring). In this study, the re-dissolution in n-butyl alcohol of the monolithic and transparent gels prepared with R_A=1 and R_W=1 or 1.5 were confirmed.

Figure 3 showed the apparent soluble regions in the system of gel-acetic acid-n-butyl alcohol. In the case of the gel prepared with R_A=1 and R_W=1, it could be dissolved in n-butyl alcohol without heating. In contrast, if the gel was prepared with R_A=1 and R_W=1.5, relatively large amounts of alcohol and acetic acid were necessary to dissolve the gel. Otherwise, heating was necessary to dissolve such gels. In a mixture contained more than 70 wt% of the gels, heating was indispensable to completely dissolve the gels. These results showed that the gel with higher R_W consisted of more crosslinked polymer. The higher R_W accelerated the hydrolysis and following polycondensation, resulting in the crosslinked polymer. As already described, excess water (R_W=2) caused rapid hydrolysis, leading to local polycondensation and form colloidal gels. Colloidal gels were insoluble in solvents.

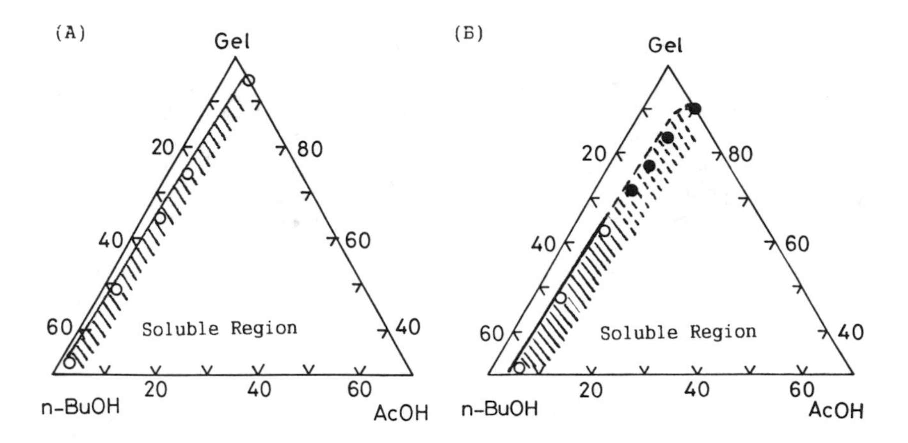

Figure 3. Apparent soluble region in the three component systems.
(A) R_A = 1.0, R_W = 1.0, (B) R_A = 1.0, R_W = 1.5.
○; soluble region, ●; soluble region after heating

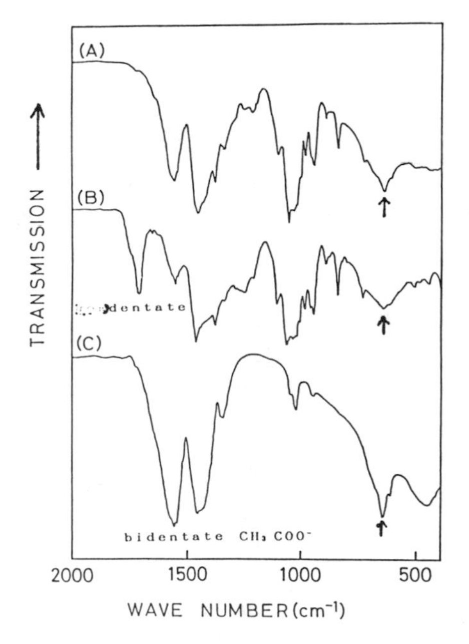

Figure 4. IR spectra of gels and the precursor solution prepared from a gel with R_A=1 and R_W=1.
(A) a gel, (B) the precursor solution prepared from the gel (A),
(C) a gel prepared from the precursor solution (B).

Figure 4 showed the IR spectra of gels and a gel solution prepared by dissolving polymeric gels in n-butyl alcohol. The set of bands around 1560 and 1460 cm^{-1} appeared in all IR spectra. The frequency separation ($\Delta\nu$=100 cm^{-1}) between the ν_{syn}(COO) at 1460 cm^{-1} and the ν_{asym}(COO) at 1560 cm^{-1} was consistent with bidentate acetate ligands. However, its smaller frequency separation suggested chelating acetates rather than bridging acetates occurred [4]. These two bands were observed even after the 500°C firing. Therefore, heat-treatment above 500°C was necessary to obtain polycrystals without residual organic compounds in bulk gels. However, residual carbon could be eliminated at lower temperatures in thin films.

An additional band appeared around 1720 cm^{-1} in the IR spectrum of the gel solution. The band was attributed to the monodentate acetate ligand or free acetic acid [4]. This indicated that the acetic acid acted as a dissolving agent when it was added after gelation of the precursor solution. Furthermore, the broad band around 650 cm^{-1}, corresponding to the Zr-O bond, exhibited higher intensity in the gel prepared from the gel solution than that in the original gel (Fig.4 (C)). This suggested that the polycondensation proceeded to yield more highly polymerized ZrO_2 gel when the polymeric gel was dissolved in a solvent and then gelled again.

Crystallization behavior of gels was investigated by x-ray diffraction (XRD). The polymeric gel was amorphous up to 400°C. Crystallization occurred at 500°C. The crystalline phase was tetragonal because of the small crystallite size (<50 nm). In addition, a very broad hump around (111) peak of the tetragonal zirconia was identified in the XRD pattern. Generally, alkoxide-derived ZrO_2 gels would crystallize around 400°C without amorphous phases [8],[9]. Therefore, the XRD pattern suggested that the amorphous state was stabilized by the highly polymerized Zr-O-Zr network structure and the residual side chain carboxyl group. At 600°C, no hump existed in the XRD pattern. The tetragonal phase was stable up to 800°C.

Preparation of thin film

As already mentioned, the gel solution could be obtained by dissolving polymeric gels in a solvent. In addition, it was possible to prepare thin films from the gel solution by dip- or spin-coating. Figure 5 showed the relation between film thickness after heat-treatment and withdrawal speed for the gel solution prepared from a gel with R_A=1 and R_W=1. Film thickness increased with increasing withdrawal speed. Maximum film thickness attained by one dipping operation was about 0.1 micron. In this case, dense thin films of tetragonal ZrO_2 polycrystals were obtained by heat-treating at 500°C for 30 minutes.

CONCLUSIONS

Transparent polymeric ZrO_2 gels were obtained using CCM method under the specific conditions of R_A=1 and R_W=1 or 1.5. The polymeric gels were soluble in n-butyl alcohol with acetic acid and mechanical stirring. Sometimes, heating was necessary to dissolve polymeric gels prepared with R_W=1.5. The re-dissolubility suggested that the polymeric ZrO_2 gels consisted of linear zirconoxane polymers. As R_W increased, the zirconoxane polymer crosslinked with each other. When R_W was greater than 1.5, colloidal gels formed.

Dense ZrO_2 thin films could be prepared by dip-coating from a solution of ZrO_2 polymeric gels in n-butyl alcohol. Film thickness after

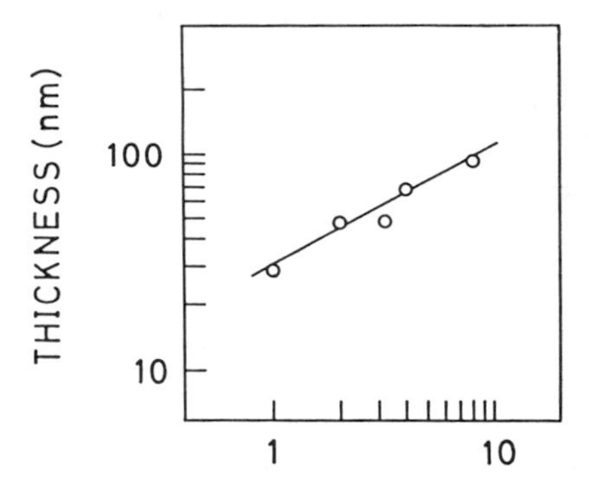

Figure 5. Relation between film thickness after heat-treatment at 500 °C for 30 minutes and withdrawal speed.
We can control the film thickness by changing the withdrawal speed and/or the concentration of the gel.

heat-treatment increased with increasing withdrawal speed up to 0.1 micron. These polymeric gels and films crystallized at 500°C to form the tetragonal zirconia. However, in bulk gels, an amorphous phase remained at this temperature and the crystallization did not complete until 600°C. It concluded that the random network structure of crosslinked zirconoxane polymer as well as the residual carboxyl group stabilized the amorphous state.

These results indicated that the polymeric gels prepared by the CCM method and solutions of these gels are candidates for the precursors of tetragonal zirconia polycrystals with nano-size pores and dense thin films, respectively.

REFERENCES

1. H.Suzuki, H.Saito, in Ceramic Powder Science IV, Ceramic Transactions Vol.22, edited by Shin-ichi Hirano, Gary L.Messing and Hans Hausner (The American Ceramic Society, Inc., Westerville, OH, 1991), p.681

2. H.Dislich, Angew. Chem. 83, 428 (1971)

3. H.Suzuki, K.Ota, H.Saito, Yogyo-Kyokai-Shi 95, 163 (1987)

4. S.Doeuff, M.Henry, C.Asnchez, J.Livage, J.Non-Cryst.Solids 89, 206 (1987)

5. H.Suzuki, H.Saito, T.Hayashi, J.European Ceram.Soc., in press

6. H.Suzuki, Y.Tomokiyo, Y.Suyama, H.Saito, J.Ceram.Soc.Japan, 96, 67 (1988)

7. H.Hayashi, H.Suzuki, H.Sait, J.Ceram.Soc.Japan, 100, 122 (1992)

8. idem, ibid, 100, 240 (1992)

9. D.Kundu, D.Ganguli, J.Mater.Sci.Letter, 5, 293 (1986)

SINGLE-COMPONENT ROUTES TO PEROVSKITE PHASE MIXED METAL OXIDES

CLIVE D. CHANDLER†, M. J. HAMPDEN-SMITH†* AND C.J. BRINKER‡
†Department of Chemistry, University of New Mexico, Albuquerque, NM 87131
*Center for Micro-Engineered Ceramics, University of New Mexico, Albuquerque, NM 87131
‡Inorganic Materials Chemistry, Division 1846, Sandia National Laboratories, Albuquerque, NM 87185

ABSTRACT

Crystalline perovskite phase mixed metal oxides have been prepared at low temperatures from single-component mixed metal-organic precursors specifically designed for this purpose. Pyridine solutions of divalent metal α-hydroxycarboxylates of general empirical formula $A(O_2CMe_2OH)_2$ where A = Pb, Ca, Sr, Ba; Me = methyl, react with metal alkoxide compounds such as $B(OR')_4$, where B = Ti, Zr, Sn with the elimination of two equivalents of alcohol to form species with a fixed A:B stoichiometry of 1:1 according to the equation below.

$$A(O_2C(CMe_2)OH)_2 + B(OR')_4 \longrightarrow A(O_2CCMe_2O)_2B(OR')_2 + 2HOR' \qquad (1)$$

Hydrolysis of these solutions with an excess of water results in homogeneous clear solutions from which white or pale yellow solids can be isolated, by evaporation of the volatile components *in vacuo*. Thermolysis at 350°C in O_2 resulted in formation of crystalline perovskite phase products for the representative examples $PbTiO_3$, $PbZrO_3$, $BaTiO_3$ $CaTiO_3$ and $BaSnO_3$. It is also demonstrated that non-integral stoichiometry phases can be prepared by reactions between controlled stoichiometric precursors. The crystalline phases $PbZr_{0.52}Ti_{0.48}O_3$ and $Ba_{0.6}Sr_{0.4}TiO_3$ were prepared at 350°C by the reactions of the appropriate amounts of the individual, 1:1 stoichiometric mixed metal intermediates $A(O_2CCR_2O)_2B(OR')_2$.

INTRODUCTION

Perovskite phase metal oxide ceramics are attracting increased research interest due to their physical properties which include ferroelectric, pyroelectric, piezoelectric and dielectric behavior and have lead to applications in electro-mechanic transducers, light modulation, charge storage and infra-red detection.[1,2] For these applications, the formation of compositionally and phase pure materials is crucial.

The sol-gel method of hydrolysis and condensation of metal alkoxide precursors is also attractive for formation of such materials due to low processing temperatures and the potential for control over stoichiometry at the molecular level.[3-7] This method has been employed for the formation of $BaTiO_3$, $PbZrO_3$, $PbTiO_3$, and PZT (lead zirconium titanate) species from various metal alkoxide and carboxylate precursors.[2] It has been demonstrated that, on thermal initiation, metal carboxylates can react with metal alkoxide compounds to eliminate an ester group.[8-13] However, it has been proposed[14], that the ester eliminated may be derived from the reaction of adventitious acetic acid with the alcohol solvent. The reaction of metal carboxylates with alcohol solvents has recently been demonstrated in the formation of $Pb_2Zn(OAc)_4(O(CH_2)_2OMe)_4$.[15]

Even in the absence of carboxylate ligands, metal-organic precursors can react prior to, or during hydrolysis to form mixtures of compounds with *undesired* stoichiometries. An example of this problem was recently reported during the formation of $BaTiO_3$ from the double alkoxide, $BaTi(OCH_2CH_2OCH_3)_6$.[16] While hydrolysis of $BaTi(OCH_2CH_2OCH_3)_6$ with an excess of water, resulted in formation of crystalline perovskite phase $BaTiO_3$ at 400°C, partial hydrolysis resulted in formation of the new species $Ba_4Ti_{13}O_{18}(OCH_2CH_2OCH_3)_{24}$. This species could be thermally converted to the crystalline phase $Ba_4Ti_{13}O_{30}$ at 670°C. While this result nicely demonstrates that stoichiometry can be retained on thermolysis of $Ba_4Ti_{13}O_{18}(OCH_2CH_2OCH_3)_{24}$, it also demonstrates that partial hydrolysis can lead to loss of stoichiometric control.

Here, we report the preliminary results on a strategy to control the stoichiometry of different precursors to be that of the final desired phase via formation of "single-component" intermediates.[17-21] It is anticipated that this strategy avoids problems associated with unintended reactions between reactants to form species of undesired stoichiometry which destroys the molecular level homogeneity associated with liquid phase or sol-gel techniques.[3] Furthermore, we can demonstrate that this strategy also accommodates retention of molecular level homogeneity in the synthesis of *non-integral stoichiometry* crystalline metal oxide materials at low temperatures via reactions between controlled stoichiometry precursors.

RESULTS

The preparation and characterization of divalent metal α-hydroxycarboxylates of general empirical formula $A(O_2CR_2OH)_2$ where A = Pb, Ca, Sr, Ba and R = H, Me was recently reported.[22,23] These species are designed to react with metal alkoxide compounds such as $B(OR')_4$, where R' = Ti, Zr, Sn with the elimination of two equivalents of alcohol to form species with a fixed A:B stoichiometry of 1:1 according to equation 1.

$$A(O_2CCR_2OH)_2 + B(OR')_4 \dashrightarrow A(O_2CCR_2O)_2B(OR')_2 + 2HOR' \quad (1)$$

The structure and reactivity of the hydroxyl group in $Pb(O_2CR_2OH)_2$ R = H, Me, has been studied in some detail in the solid state and in solution, and it is likely that the pK_a is lowered compared to aliphatic alcohols by inductive effects and through chelation of the α-hydroxycarboxylate ligands. This provides a thermodynamic driving force for the reaction of equation 1 to proceed to the right.

When the reactions of equation 1 where R = Me are carried out, the reaction products are soluble in pyridine to give homogeneous clear solutions. Furthermore, addition of an excess of water to these rapidly stirred solutions results in initial precipitation followed by rapid re-dissolution to form transparent homogeneous solutions. Evaporation of the volatile components *in vacuo* from these solutions resulted in the formation of either white or pale yellow solids. Thermogravimetric analysis of these solids carried out in oxygen revealed that the weight losses were complete between 300 and 400°C and were consistent with formation of the corresponding metal oxide, ABO_3 (based on $A(O_2CCMe_2O)_2B(OH)_2$). A typical example of the formation of crystalline $BaTiO_3$ formed by this method at 350°C is shown in Figures 1 and 2.

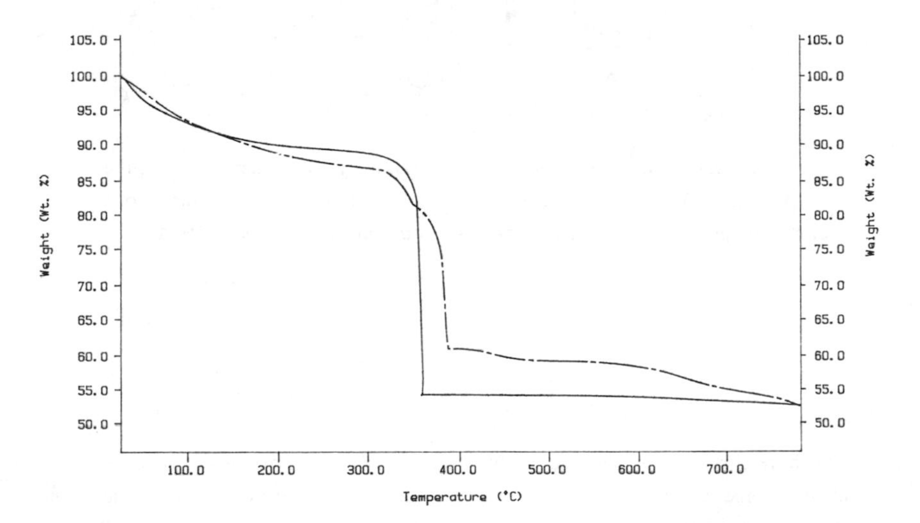

Figure 1: TGA data for the solid hydrolysis product isolated from the reaction between $Ba(O_2CCMe_2OH)_2$ and $Ti(O\text{-}i\text{-}Pr)_4$ in pyridine heated in air (dashed line) and in O_2 (continuous line) atmospheres.

Thermal decomposition of the solid hydrolysis product isolated from the reaction between $Ba(O_2CCMe_2OH)_2$ and $Ti(O\text{-}i\text{-}Pr)_4$ in oxygen resulted in loss of the organic ligands at approximately 350°C corresponding to formation of $BaTiO_3$. On the other hand, thermal decomposition in air resulted in incomplete decomposition by 400°C with a further 5% weight loss corresponding to liberation of CO_2 up to 800°C. X-ray powder diffraction data for this sample heated in oxygen are presented in Figure 2.

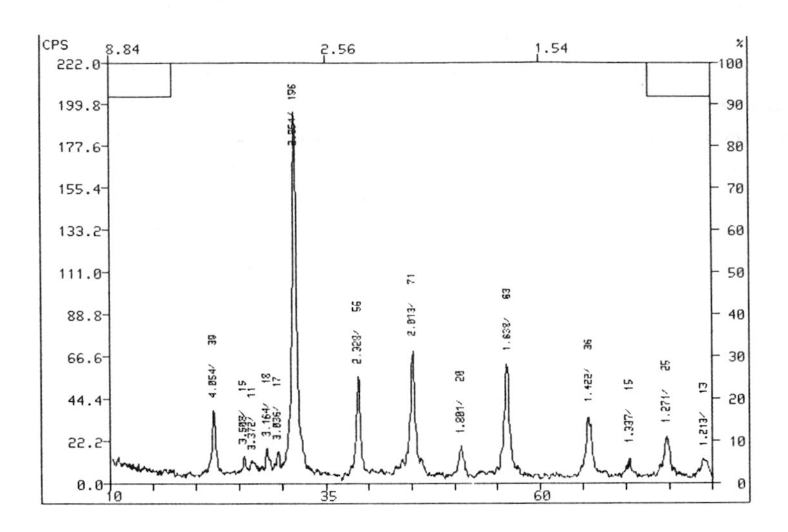

Figure 2: X-ray powder diffraction data for the $BaTiO_3$ formed from the reaction of $Ba(O_2CCMe_2OH)_2$ and Ti(O-*i*-Pr)$_4$ in pyridine followed by hydrolysis, removal of the volatile components *in vacuo* and thermolysis of the powder obtained at 350°C in O_2.

We have also examined the potential of this method for the formation of single phase non-integral stoichiometry materials via reactions between integral stoichiometry precursors. One representative example of each of the two systems, $AB_{1-x}B'_xO_3$ and $A_{1-x}A'_xBO_3$, was investigated. In the first experiment ($A_{1-x}A'_xBO_3$), the solids $Ba(O_2CCMe_2OH)_2$ and $Sr(O_2CCMe_2OH)_2$ were added to a Schlenk flask in the ratio 0.6 : 0.4 respectively. This ratio was chosen because it should be possible to distinguish the formation of a mixture of phases (i.e. $BaTiO_3$ *and* $SrTiO_3$) from the formation of the known crystalline single phase product $Ba_{0.6}Sr_{0.4}TiO_3$ by X-ray powder diffraction. The glycolates were dissolved in pyridine, one equivalent of Ti(O-*i*-Pr)$_4$ was added, the solution hydrolyzed and the volatile components removed *in vacuo*. Thermolysis of the white powder at 350°C for 0.5hr in O_2 gave a white solid that showed $Ba_{0.6}Sr_{0.4}TiO_3$ as the only crystalline phase, Figure 3. A similar reaction between the appropriate amounts of the corresponding PbZr and PbTi solutions yielded crystalline perovskite $PbZr_{0.52}Ti_{0.48}O_3$ as the only crystalline phase.

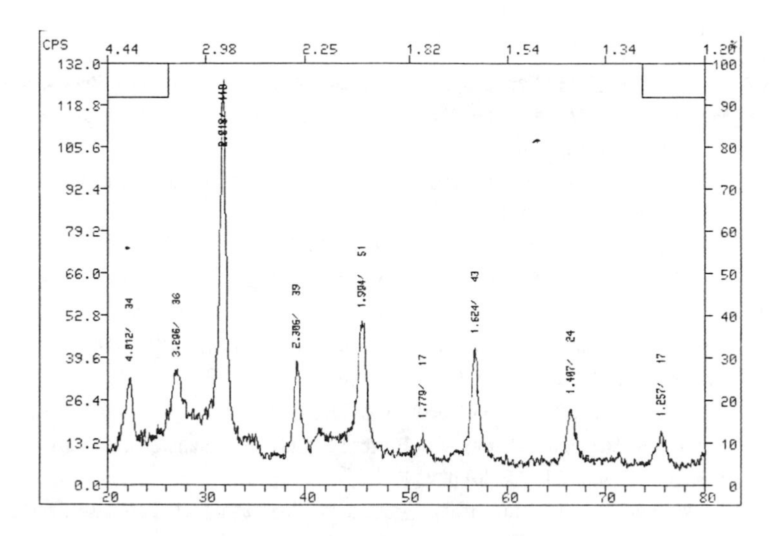

Figure 3: X-ray powder diffraction data for $Ba_{0.6}Sr_{0.4}TiO_3$. See text for details.

CONCLUSIONS

A low-temperature route for the formation of crystalline perovskite phase metal oxides has been developed through the reactions of metal-organic compounds that are specifically designed to react with each other to form intermediates with controlled stoichiometry. As a result of the low temperatures necessary to form crystalline phases, we speculate that the diffusion limited barriers are avoided. This method should be applicable to a wide variety of mixed metal oxide phases with integral and non-integral stoichiometries. While a large number of questions remain to be addressed, to our knowledge, these crystallization temperatures are amongst the lowest reported for the formation of crystalline perovskite phase materials from metal-organic precursors.[24] We are currently investigating the formation of thin films materials via this approach and the generality of these reactions to prepare other complex metal oxides.[25]

ACKNOWLEDGMENTS

We thank the UNM/NSF Center for Micro-Engineered Ceramics for funding this work, the National Science Foundation Chemical Instrumentation program for the purchase of a low-field NMR spectrometer and the Office of Naval Research, Chemistry and Department of Materials Research for the purchase of a 10mm broad-band NMR probe (ONR N00014-91-J-1258). We thank Dongshui Zeng for obtaining the X-ray Powder diffraction data.

REFERENCES

1. A.R. West, in Solid State Chemistry and Its Applications, (John Wiley and Sons, 1989).
2. E.R. Meyers and A.I. Kingon, eds, Ferroelectric Thin Films, Vol 200 (Materials Research Society, Pittsburgh, PA, 1990).
3. C. J. Brinker and G. W. Scherer, Sol-Gel Science. The Physics and Chemistry of Sol-Gel Processing, (Academic Press, 1990).
4. C. Sanchez and J. Livage, Nouv. J. Chem. 14, 513 (1990).
5. J. Livage, J.P. Jolivet and E. Tronc, J. Non-Cryst. Solids 121, 35 (1990).
6. M. Nabavi, S. Doeuff and C. Sanchez, J. Non-Cryst. Solids 121, 31 (1990).
7. J. Livage, M. Henry and C. Sanchez, Prog. Solid State Chem. 18, 259 (1988).
8. D.C. Bradley and I.M. Thomas, J. Chem. Soc., A., 3404 (1959).
9. H.K. Sharma and P.N. Kapoor, Polyhedron 7, 1389 (1988).
10. K.D. Budd, S.K. Dey and D.A. Payne in Better Ceramics Through Chemistry III (Mat. Res. Soc. Proc. 73, Pittsburg, PA 1986) p. 711.
11. S.R. Gurkovich and J.B. Blum in Ultrastructure Processing of Ceramics Glasses and Composites, edited by L.L. Hench and D. R. Ulrich (Wiley-Interscience, New York, 1984) Ch. 12.
12. S.R. Gurkovich and J.B. Blum, Ferroelectrics, 62, 189 (1985).
13. S.R. Gurkovich and J.B. Blum, J. Mater. Sci., 20, 4479 (1985).
14. T.W. Dekleva, J.M. Hayes, L.E. Cross and G.L. Geoffroy, J. Am. Cer. Soc. 71, C-280 (1988).
15. L.F. Francis and D.A. Payne, Chem. Mater. 2, 645 (1990); and references therein.
16. J.F. Campion, D.A. Payne, Chae, H.K., J.K. Maurin, and S.R. Wilson, Inorg. Chem., 30, 3245 (1991).
17. M.J. Hampden-Smith, D. E. Smith and E.N. Duesler, Inorg. Chem. 28, 3399 (1989).
18. M.J. Hampden-Smith, D.S. William and A. Rheingold, Inorganic Chemistry 29, 4076 (1990).
19. M.J. Hampden-Smith, T.A. Wark, L.C. Jones and C. J. Brinker, Proc. Am. Cer. Soc., Symposium V, (Cinncinatti, April 1991).
20. E.A. Gulliver, J.W. Garvey, M.J. Hampden-Smith and A. K. Datye, J. Amer. Cer. Soc. 74, 1091 (1991).
21. T.A. Wark, E.A. Gulliver, M.J. Hampden-Smith and A. L. Rheingold, Inorg. Chem. 29, 4360 (1990).
22. C.D. Chandler, M.J. Hampden-Smith and E.N. Duesler, Inorg. Chem. (submitted for publication 1992).
23. C.D. Chandler, M.J. Hampden-Smith and E.N. Duesler (manuscript in preparation).
24. K. Kiss, J. Magder, M.S. Vukasovich and R.J. Lockhart, J. Amer. Cer. Soc. 49, 291 (1966): T. Fukui, C. Sakurai and M. Okuyama, J. Mater. Res., 7, 791 (1992).
25. C.D. Chandler, M.J. Hampden-Smith, C. J. Brinker and R. Schwartz (work in progress).

SYNTHESIS OF BARIUM TITANATE BY A BASIC pH PECHINI PROCESS

SURESH KUMAR AND GARY L. MESSING
Department of Materials Science and Engineering,
The Pennsylvania State University, University Park, PA.

ABSTRACT

Barium titanate powders and thin films have been processed by the conventional acid pH Pechini process and a modified basic pH Pechini process. Higher pH of the starting solution resulted in significantly finer grains in the sintered thin films. It is proposed that the microstructure refinement is due to the precipitates, formed only in the basic pH solution and resin.

INTRODUCTION

Many devices such as capacitors, transducers, thermistors, etc. are fabricated from $BaTiO_3$ powder. Commercial $BaTiO_3$ powders are manufactured by (1) a conventional mixed oxide method which involves repeated milling and calcination (>1000°C) of $BaCO_3$ and TiO_2 powder mixtures, (2) an oxalate co-precipitation method[1], and (3) a hydrothermal method[2,3]. Solution techniques result in more homogeneous, phase pure, and stoichiometric $BaTiO_3$ powders and fine particle size due to the lower amorphous to crystalline transformation temperature[4] (650-800°C). Solution techniques, however, suffer from high weight losses on calcination.

The Pechini process[5] is based upon an aqueous polyalcohol-citric acid system in which a wide range of metal salts are soluble. For $BaTiO_3$ synthesis, a clear Ti-solution is prepared by mixing Ti-isopropoxide in ethylene glycol and citric acid. $BaCO_3$ is then dissolved into the Ti-solution. Evaporation and thermal polymerization of the final solution results in a solid, amorphous resin, which is calcined at 650-800°C to yield $BaTiO_3$. This method circumvents the problems of segregation or preferential precipitation from solution due to spatial fixation of cations in the resin. Eror and Anderson[6] modified the process by preparing individual cation solutions in ethylene glycol and citric acid and mixing them to obtain the final sol. Other citric acid-based synthesis techniques have been reported[7,8]. Delmon et al.[8] and Lessing[9] reported that the oxide powder morphology from these techniques is dependent on the morphology of the solid, amorphous resin; a porous resin yielding finer powders. Salze et al.[10] varied the amount of various constituents to relate the powder morphology with the solution composition.

Due to the high acid content, the final solution in the Pechini process has a pH in the range of 2-3. During the course of our research, interesting effects were observed on raising the solution pH. In this paper, we report how higher solution pH affects the resin and final microstructure and the resin to $BaTiO_3$ transformation.

EXPERIMENTAL PROCEDURE

An aqueous Ti-solution was prepared by adding 300 ml of titanium isopropoxide into 600 ml of ethylene glycol followed by the addition of 400 g of citric acid. The opaque, white solution was heated for one hour at 80-100°C with continuous stirring. Heating was stopped when no propanol smell could be detected, and the solution became a clear, light yellow color. The Ti-solution was diluted to 1000 ml with distilled water and filtered.

To synthesize $BaTiO_3$, a stoichiometric amount of $BaCO_3$ powder was added slowly to the Ti-solution with continuous stirring at 60-80°C. For a 50 g $BaTiO_3$ batch, three to ten ml of conc. HNO_3 was added to completely dissolve the $BaCO_3$. To examine the effect of solution pH, the pH was increased from 2-3 to 10 by adding a 30% ammonia solution. An acid pH solution and a basic pH solution were then heated at 70-90°C under constant stirring. With continued heating and consequent evaporation, the pH of the basic pH solution continuously decreased, but was raised by further ammonia additions. The solution viscosity increased continuously due to gradual polymerization. The final pH of the basic solution was 6-7 before it turned so viscous that the pH could not be accurately measured. Further heating of the viscous mass for 4 to 6 h at 100-150°C resulted in a clear, transparent, solid, amorphous resin. The acid pH solution was heated similarly to obtain a clear and transparent amorphous resin.

Thermal reactions of the solid resins were analyzed in flowing air with DuPont 951 TGA and 1600 DTA cells interfaced with a TA 2100 computer. The 49% weight loss between 25 and 1000°C and the massive exothermic reaction due to organic oxidation were significantly reduced by pre-calcining the resins for 10 h at 375°C in air. All further analyses were performed on the amorphous, pre-calcined resin, referred to as the "precursor". Thermal analysis samples of 30-35 mg were heated at 10°C/min to the target temperature with an intermediate 10 min hold at 150°C to remove adsorbed moisture. Precursor samples heated to various intermediate temperatures were characterized by x-ray diffraction to identify the crystalline phases. The presence of carbonates was detected by IR spectroscopy. Specific surface area was measured by nitrogen adsorption in a surface area analyzer. Particle density was measured in a helium pycnometer.

For microstructural analysis by SEM, thin films were formed by spin coating the viscous sol onto camphor disks rotating at 6000 to 8000 rpm. After air-drying the films at 85°C for two days, the camphor was sublimed at 150°C to obtain a free-standing, amorphous film of 2-5 μm thickness. The oven-dried films were pre-calcined for 1 h at 300°C in air, before a 5 h calcination at temperatures up to 1100°C in air.

RESULTS

The weight loss on heating the precursor between 25 and 1000°C was 16% (Fig. 1). Significant weight losses are expected only at temperatures >375°C. The weight loss between 400-645°C is due to further oxidation of organics, which is evident from the corresponding large DTA exotherm. The weight loss between 645-705°C is due to a decomposition reaction as indicated by the weak DTA endotherm. No exothermic peak corresponding to $BaTiO_3$ crystallization was found. The acid pH and basic pH precursors had similar thermal behavior indicating that $BaTiO_3$ evolution was not significantly influenced by the solution pH.

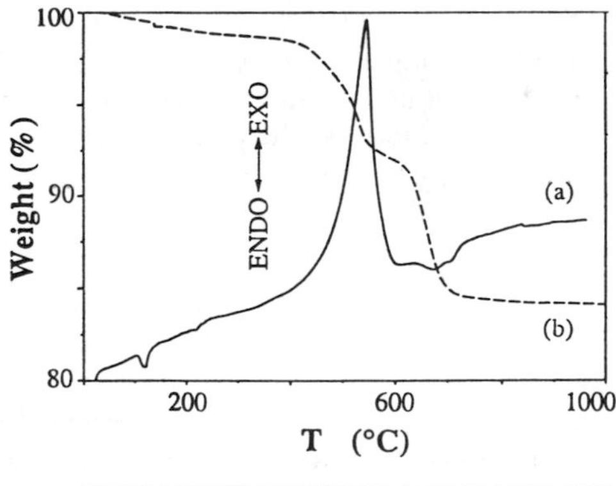

Fig. 1 Thermal analysis of basic pH precursor in air.
(a) DTA
(b) TGA

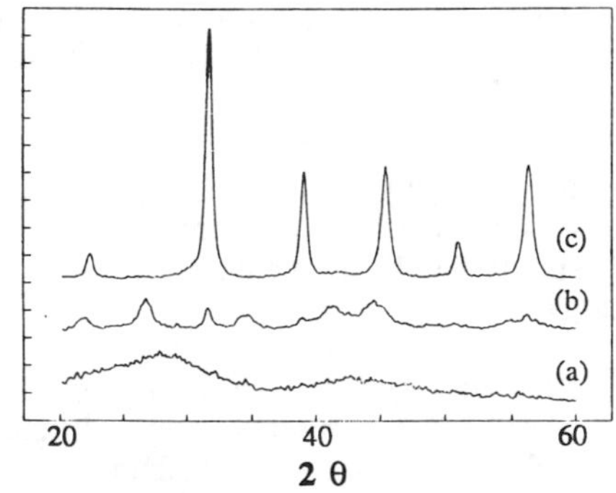

Fig. 2 X-ray diffraction patterns of acid pH precursor.
(a) Precursor - Amorphous
(b) 635°C - Intermediate, $BaTiO_3$
(c) 710°C - $BaTiO_3$

Both basic and acid pH precursors resulted in pure $BaTiO_3$ after calcination at temperatures >700°C and showed identical phase evolution at all temperatures. The x-ray diffraction patterns of the acid pH precursor at temperatures just before and after the endotherm are presented in figure 2. No $BaCO_3$ or TiO_2 was detected by x-ray diffraction at any stage, but carbonate peaks were detected by IR spectroscopy indicating the presence of small amounts of $BaCO_3$ (< 2 v/o). Samples calcined below 635°C showed broad, unidentified x-ray diffraction peaks at 21-22°, 26-28°, and 33-35° 2θ. These peaks disappeared on calcination at higher temperatures indicating a possible intermediate phase, which is now under investigation. Similar unidentified peaks have been reported[11] after low-temperature calcination of Ba,Ti citrates and oxalates. Formation of an intermediate carbonate phase, $Ba_2Ti_2O_5CO_3.CO_2$, has been proposed[12] during calcination of Ba,Ti oxalate.

To investigate the endothermic reaction, the specific surface area and the density of the precursors were measured after heating samples at 10°C/min to temperatures just before (635°C) and after (710°C) the endotherm. The specific surface area increased from 5.7 m^2/g to 19.4 m^2/g during the endotherm. The particle density increased from 4.32 g/cm^3 to 6.03 g/cm^3, a value close to the theoretical density of $BaTiO_3$, during the endotherm. The equivalent particle size as calculated from the specific surface area decreased from 0.25 μm to 0.05 μm. X-ray diffraction showed a significant increase in the $BaTiO_3$ concentration and complete disappearance of the broad, unidentified peaks. From the above results, it is evident that an endothermic decomposition reaction takes place between 635 and 710°C, and most of the $BaTiO_3$ forms as a result of this decomposition.

Microstructural Development

The microstructural development in the basic and acid pH films is shown in figures 3 and 4, respectively. There is no apparent structure in the as-cast acid pH film, but the basic pH film consisted of a structure characteristic of precipitates. After 5 h at 1100°C, the average grain size in the basic pH film was 0.19 μm in contrast with 0.34 μm in the acid pH film. The grain size in these films is considerably finer than previously reported values for sintered powder compacts. The cracks in these films are probably due to drying and/or decomposition stresses. No attempts were made to avoid this problem. The calcined films were dense and well sintered indicating that films of excellent quality could be processed by avoiding the cracks.

It is proposed that the grain size difference between the basic and acid pH films originates from a difference in the x-ray amorphous resins. In contrast with the clear, structureless acid pH film, the basic pH film had a structure consisting of precipitates in an amorphous matrix. During the synthesis of basic pH solution, moderate ammonia additions did not cause any precipitation or turbidity, but precipitation was observed with a large excess of ammonia. Most probably, these precipitates are Ba,Ti citrates[7] or Ti hydroxides[13]. The precipitation of $Ba(OH)_2$ is unlikely due to its high solubility at a pH of 6-7[14]. The microstructure refinement in the basic pH films suggests that the precipitates play an important role in the $BaTiO_3$ grain sizes obtained in the thin films. Experiments are in progress to characterize the precipitates and to determine the chemistry-microstructure relationship in these films.

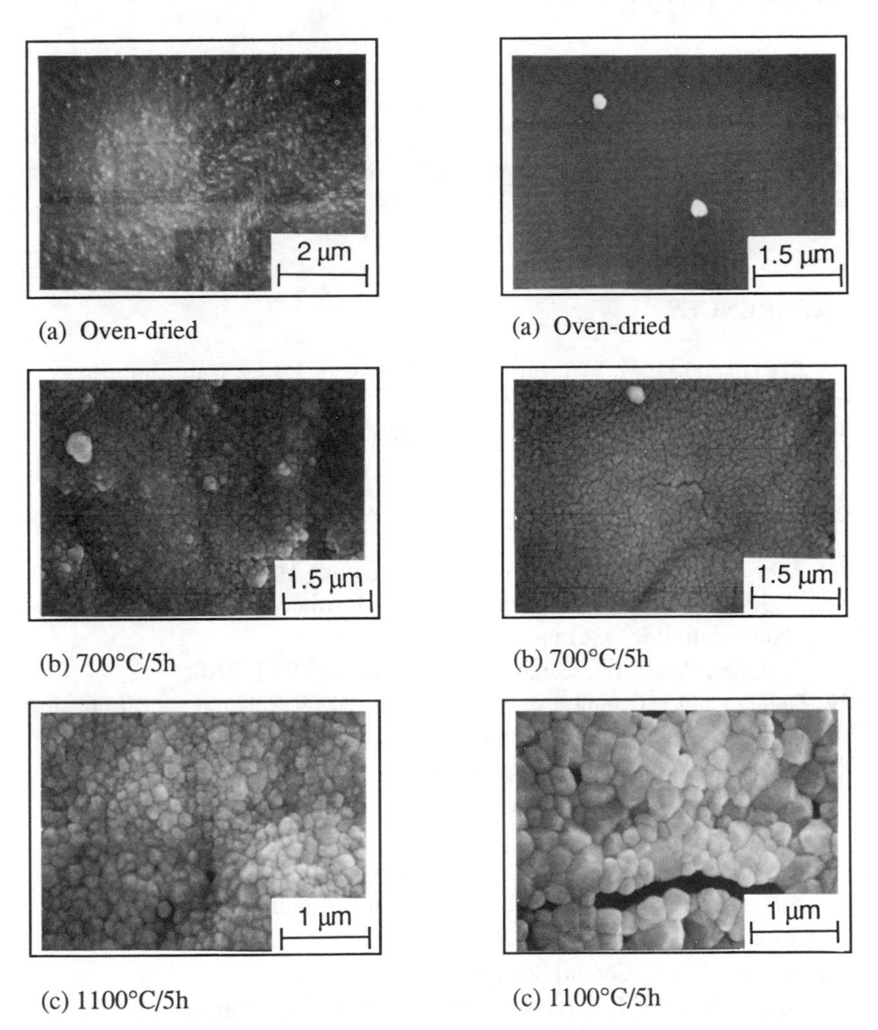

(a) Oven-dried

(b) 700°C/5h

(c) 1100°C/5h

Fig. 3 Microstructure development in basic pH films

(a) Oven-dried

(b) 700°C/5h

(c) 1100°C/5h

Fig. 4 Microstructure development in acidic pH films

CONCLUSIONS

$BaTiO_3$ powders and thin films processed by the conventional acid pH Pechini process and a modified basic pH Pechini process showed identical thermal behavior and crystalline phase evolution. However, the basic pH precursor resulted in a significant grain size refinement. It is proposed that the microstructure refinement is due to the precipitates, formed only in the basic pH solution and resin.

ACKNOWLEDGEMENT

This research was funded by the Ceramics and Electronics Materials Program, DMR-NSF.

REFERENCES

[1]W.S. Clabaugh, E.M. Swiggard, and R. Gilchrist, J. Res. Natl. Bur. Stn., R.P. 2677 56 (5), 289-91 (1956).
[2]S.S. Flaschen, J. Am. Chem. Soc. 77, 6194 (1955).
[3]K. Fukai, K. Hikada, M. Aoki, and K. Abe, Ceram. Int. 16, 285-290 (1990).
[4]P.P. Phule, and S.H. Risbud, J. Mat. Sci. 25, 1169-83 (1990).
[5]M.P. Pechini, U.S. Patent No. 3,330,697 (July 11, 1967).
[6]N.G. Eror and H.U. Anderson in Better Ceramics Through Chemistry II, ed. C.J. Brinker, D.E. Clark, and D.R. Ulrich (Mater. Res. Soc. Symp. Proc. 73, Pittsburgh, PA 1986) pp. 571-78.
[7]B.J. Mulder, Bull. Am. Ceram. Soc. 49 (11), 990-93 (1970).
[8]B. Delmon and J. Droguest in Fine Particles, ed. W.E. Kuhn and J.Ehretsmann (Second Int. Conf., The Electrochem. Soc., Inc., NJ 1974) pp. 242-55.
[9]P.A. Lessing, Bull. Am. Ceram. Soc. 68 (5), 1002-07 (1989).
[10]H. Salze, B. Cales, and P. Odier, Mat. Sci. Mono. (High Tech. Ceram.) 38A 491-99 (1987).
[11]D. Hennings, and W. Mayr, J. Solid State Chem., **26** 329-38 (1978).
[12]H.S.G. Murthy, M.S. Rao, and T.R.N. Kutty, J. Inorg. Nucl. Chem., **37** 891-98 (1975).
[13]F. Schrey, J. Am. Ceram. Soc., **48** 401-05 (1965).
[14]K. Osseo-Asare, F.J. Arriagada, and J.H. Adair in Ceramic Powder Science II, A ed. G.L. Messing, E.R. Fuller, Jr., and H. Hausner (Cer. Trans. **1**, Am. Ceram. Soc., Inc., OH 1988) pp. 47-53.

PREPARATION OF STRONTIUM TITANATE POWDERS BY DECOMPOSITION OF POLYMERIC PRECURSORS

S.L. PESCHKE*, M. CIFTCIOGLU**, D.H. DOUGHTY*** AND J.A. VOIGT***
*New Mexico Institute of Mining and Technology, Socorro, NM
**University of New Mexico, Dept. of Chemical and Nuclear Eng., Albuquerque, NM 87131
***Sandia National Laboratories, Albuquerque, NM 87185

ABSTRACT

The effects of substituting poly(acrylic acid), PAA, and mucic acid for citric acid in the standard Pechini powder synthesis process was studied. Strontium titanate was chosen as the model system for this study. Powders were prepared from precursor solutions in which PAA (average molecular weight = 2,000 and 5000) was substituted for citric acid (up to 70% by weight), and with reduced weight of total organic reagents (50% reduction). The powders prepared with a 30% substitution of PAA for citric acid produced weakly agglomerated submicron particles after calcining. These powders contained a larger fraction of weak agglomerates than the powders prepared by the original Pechini process. As such, the PAA derived powder could be easily fractionated and dispersed with greater than 70% of the calcined powder having particle sizes less than 0.5 μm.

INTRODUCTION

Polymeric precursors are often used in the synthesis of ceramic oxide powders [1,2]. One of the most widely used polymeric routes is the Pechini process [3] in which the desired metal cations are solubilized in water using a hydroxycarboxylic acid (citric acid) as the chelating agent. This solution is mixed with a polyhydroxyl alcohol (ethylene glycol) and heated to dryness to form an amorphous resin that is then pyrolyzed and calcined to form the desired oxide. During the thermal treatment of the solution, esterification reactions occur between the carboxylate groups (citric acid) and the hydroxyl groups (ethylene glycol). The chelation of metals by free carboxyl groups provides a means of intimately mixing a variety of metal cations. It is believed that the high level of mixing is maintained throughout the thermal processing steps and, as a result, homogeneous, high surface area oxide powders are produced at relatively low calcination temperatures.

The ability of hydroxycarboxylic acids to solubilize a wide range of metal cations is one of the strengths of the Pechini process. For example, it is difficult to prepare $SrTiO_3$ precursors by aqueous precipitation techniques because of the relative insolubility of most titanium salts. Titanium is readily solubilized by citric acid as a titanium citrate complex which can then be mixed with a similar complex of strontium to form the desired solution. As a result of the solubilization power of the chelating agent, the Pechini process has found wide laboratory application for the synthesis of many complex oxides [4]. A shortcoming of the process is its tendency to form hard agglomerates during the charring step. The exothermic decomposition of the amorphous resin can cause localized hot spots to develop that lead to agglomerate formation. One of the ways to influence the charring process and the resulting powder properties is to alter the structure of the resin. Recently Tai and Lessing [5] have shown that by altering the ratio of citric acid to ethylene glycol, weakly agglomerated of Sr-doped lanthanum chromite powders can be prepared at a much reduced material cost compared to the original Pechini process.

Little work has been done on how chelating agents other than the commonly used citric acid, CA, affect the processability of powders produced by Pechini type processing [1]. To address this question we chose polyacrylic acid, PAA (average molecular weight = 2000 and 5000), and mucic acid, MA, (also called galactaric acid) in combination with citric acid as chelating agents. PAA was chosen because it contains a high carboxylate content and has a moderate molecular weight. Mucic acid is a dicarboxylic acid with four alcohol groups ($COOH(CHOH)_4COOH$). The high hydroxyl content of mucic acid compared to citric acid (four versus one) provides more branching sites where crosslinking reactions can occur.

We desired to produce powders that could be suspended and used in subsequent colloidal processing studies. Therefore, we sought to maximize the size fraction of a powder that was <1 μm after sonication. The effectiveness of substituting different chelating agents was assessed by comparing the particle size distributions obtained from these experiments with those prepared by the standard Pechini process. The synthesis of $SrTiO_3$ was investigated since it is representative of a wide range of titanate-based electronic ceramic materials.

This work performed at University of New Mexico and Sandia National Laboratories, supported by the U. S. Department of Energy under contract number DE-AC04-76DP00789.

EXPERIMENTAL PROCEDURE

All chemicals were reagent grade and used as received. The chemicals stated purities were: ethylene glycol (Aldrich Chemical Company Inc. and J. T. Baker Inc.) - 99+% and 100%, respectively, tetra-n-butoxy titanate (Aldrich) - 99%, citric acid (Aldrich) - 99.5+%, PAA with molecular weights of 2000 and 5000 (Aldrich) - purities not listed, mucic acid (Aldrich) - 98%, and strontium nitrate (Aldrich and J. T. Baker) - 99+% and 100%, respectively. The tetra-n-butoxy titanate solutions were standardized for weight percent Ti by gravimetric analysis. All metal cation sources (tetra-n-butoxy titanate and strontium nitrate) were analytically weighed to the nearest tenth of a milligram, while all organic reactants were weighed to the nearest hundredth of a gram.

The procedure used to prepare $SrTiO_3$ precursor solutions by the Pechini process [6] is shown in Figure 1. Below we describe the process for preparing 25 grams of $SrTiO_3$ [7]. Ethylene glycol (93.49 gm, 1.51 moles) was measured into a 500 ml Erlenmeyer flask, to which tetra-n-butoxy titanate (46.3655 gm, 0.1363 mole) was added using a disposable plastic syringe. The titanium alkoxide-ethylene glycol solution, while stirred with a magnetic stirrer, was heated for 15 minutes by placing the Erlenmeyer flask in a wax bath held at 75°C. Citric acid (74.08 gm, 0.39 mole) was added gradually to the stirred solution while increasing the temperature to 85°C. This continued for at least 15 minutes or until the citric acid fully dissolved to form a clear yellow solution. Finally, 200-250 ml of distilled water was added. The water dissolved the required quantity of $Sr(NO_3)_2$ (28.8293 gm, 0.1363 mole). Water was removed from the solution by further heating (at temperatures greater than 100°C) until 200 mls of clear yellow precursor solution remained.

A typical modified Pechini procedure for preparing $SrTiO_3$ precursor solutions by substituting equivalent mass of PAA or mucic acid for citric acid is also given in Figure 1. For these preparations, the amounts of citric acid plus substituted organics totaled only 50 weight percent of the citric acid used in the conventional Pechini-based formulation. Furthermore, the procedure utilizing PAA had a 15 weight percent reduction in ethylene glycol. Previous experiments had shown that ethylene glycol could be reduced by this much before solution turbidity occurred. PAA was used either as a powder of average molecular weight of 2,000, or as a 50 weight percent solution in water with an average molecular weight of 5,000. During the preparations, citric acid was always added first and allowed to mix for at least 15 minutes to solubilize the titanate species before adding the PAA or MA. These precursor solutions were heated until only 150 ml remained, instead of the 200 ml left for the original Pechini formulation. Use of the PAA of average molecular weight of 2,000 resulted in solutions appearing clear yellow, whereas the solutions formulated with PAA (ave. mol. wt. 5000) appeared cloudy-white. The mucic acid solutions were similarly prepared except that the standard amount of ethylene glycol was used. These solutions were highly viscous and somewhat cloudy. Table I lists the formulations of precursor solutions used in this work. For identification, each is denoted by a sample letter/number code. Also given are the mole ratios of carboxyl groups to titanium and total metal cations.

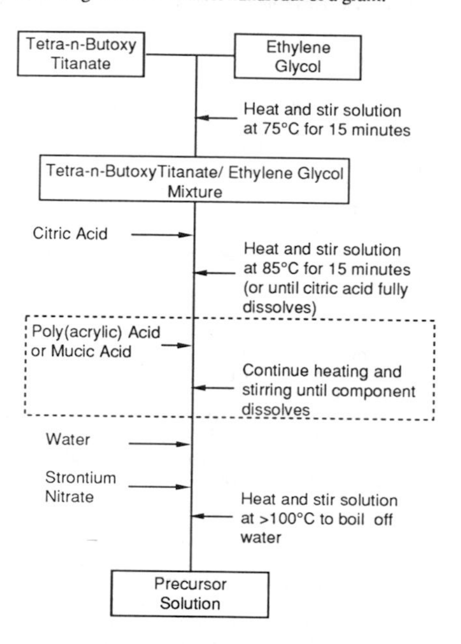

Figure 1. Flowchart for the preparation of precursor solutions (steps enclosed in the dashed box were used in preparing modified Pechini solutions).

Resins were formed by pouring 40 ml of precursor solution into 150 ml Pyrex beakers, and heating on a hot plate for 45 minutes at 150°C. This heating accelerated esterification reactions and eliminated remaining water, producing a viscous, bubbly mass that formed glassy resins upon cooling. The resins were charred at 250°C for 30 minutes in a box furnace following the procedure used by Budd and Payne [6]. The charring process left behind a black ash that yielded the desired oxide powders after calcination. A two step calcination process was used with a lower temperature (400 to 550°C) one hour hold for carbon burnout followed by a higher temperature two hour anneal (550 to 700°C) to crystallize the $SrTiO_3$ phase. See Table I for the calcination schedules used in this study.

Samples of the precursor solutions, the charred resins, and the original components were analyzed for thermal weight loss and decomposition behavior by thermogravimetric analysis, TGA. All samples were heated from room temperature to 900°C at a heating rate of 5°C/minute in 25 sccm air. A sedimentation type particle

Table I. Precursor solution compositions and calcination conditions.

Powder[a]	Type	Moles COOH / Moles Ti	Moles COOH / Moles (Ti+Sr)	Calcination Schedules Used[b]
A-CA	Std. Pechini	8.49	4.25	1a, 2a, 3b
B-CA	Reduced Organics	4.25	2.12	3b
C2000	70%CA/30%PAA	4.10	2.05	3b
D2000	50%CA/50%PAA	4.01	2.01	1, 2, 3,1a, 2a, 3a, 3b
D5000	50%CA/50%PAA	4.01	2.01	1a, 2a, 3b
E2000	30%CA/70%PAA	3.92	1.96	3b
F-MA	70%CA/30%MA	3.75	1.88	3b
G-MA	50%CA/50%MA	3.42	1.71	3b
H-MA	30%CA/70%MA	3.09	1.54	3b

a) 2000 in designator refers to PAA mol. wt. = 2000; 5000 in designator refers to PAA mol. wt. = 5000.
b) 1: 400°C for 1 hr followed by 550°C for 2 hr 1a: 450°C for 1 hr followed by 550°C for 2 hr
2: 400°C for 1 hr followed by 600°C for 2 hr 2a: 450°C for 1 hr followed by 600°C for 2 hr
3: 400°C for 1 hr followed by 700°C for 2 hr 3a: 450°C for 1 hr followed by 700°C for 2 hr
3b: 550°C for 1 hr followed by 700°C for 2 hr

size analyzer (Micromeritics Sedigraph 5100), with a particle size range of 100-0.1 μm, was used to record the particle size distributions of the powders. All the calcined powders were first ground and sieved through a 63 μm sifting pan. The preparation of powder suspensions for particle size analysis involved mixing 1.1-1.2 gm of the sieved powder in 50 ml of distilled water. The particle size distributions were obtained for as-prepared suspensions after 1 hour and 2 hours of sonication in an ultrasonic bath (Fisher Scientific), and after an additional 15 minute treatment using a disrupter horn (Tekmar TSD500, low power setting). Powder suspensions were unstable in water, especially after the ultrasonic treatments. Use of the dispersants Darvan C and Daxad 30 stabilized the powder suspensions. Initially, 10 drops of a dispersant solution were added to a suspension, with more provided if needed. In the event that flocculation persisted, the other surfactant was added. BET surface area (Quantachrome Autosorb-1), scanning electron microscopy (Hitachi S-800) and x-ray powder diffraction analyses were also carried out.

RESULTS AND DISCUSSION

Initial experiments demonstrated that the amount of citric acid and ethylene glycol could be reduced by 50% and 15%, respectively, with little or no effect on the powder properties of the resulting $SrTiO_3$. Using the same calcination schedule (3b - see Table I), the standard Pechini powder (A-CA) and the reduced organic material (B-CA) had surface areas of 15.8 and 20.3 m^2/gm, respectively, with 53% of the particles less than 1 μm after two hours of sonication for both powders. At the reduced organic content, the mole ratio of carboxyl groups to metal ions is 2.1 (see Table I), compared to 4.2 for the standard Pechini method. This reduced carboxyl content appears sufficient to solubilize the cations by chelation. This implies that either the metals are tetrahedrally coordinated (each requiring two carboxyls) or, more likely, octahedrally coordinated, and having other ligands occupy the two open coordination sites. These other ligands could be hydroxyl groups on the citric acid or ethylene glycol, coordinated water molecules, or coordinated nitrate anions. Apparently this modification did not significantly alter the resin structure to cause major differences in calcination behavior of the two formulations. Since one of the goals of this work is to make the process more economical, the reduced level of total organics was used throughout. Table I also lists the ratio of carboxyl groups to metal ions for all samples studied.

The differences in the solution decomposition behavior for the various chelating agent combinations is illustrated by the TGA results given in Figure 2a. The original formulation, the 30%CA/70%PAA-2000 (mol. wt. = 2000), and the 30%CA/70%MA all had completed their decomposition to the oxide by about 600°C. The differences in total weight loss between the samples is explained by the differing amounts of organics and water used as described in the Experimental section. During the charring process the amorphous resin is decomposed into an intimate mixture of strontium carbonate and titanium oxides plus residual carbon. The TGA results of resins charred at 250°C (Figure 2b) when compared to the decomposition of original solution (Figure 2a) show that charring has no effect on the final decomposition temperature for the standard Pechini process. For the PAA substituted material the temperature for complete weight loss was reduced from 600°C to 550°C after the charring step. Since the final amount of weight loss is due to the decomposition reaction of $SrCO_3$ with TiO_2, the lower temperature for complete decomposition indicates an improved level of mixing of the reacting species. Interestingly, the charred MA derived material does not completely decompose until heated above 700°C. This large shift indicates that the powders are less homogeneous after charring.

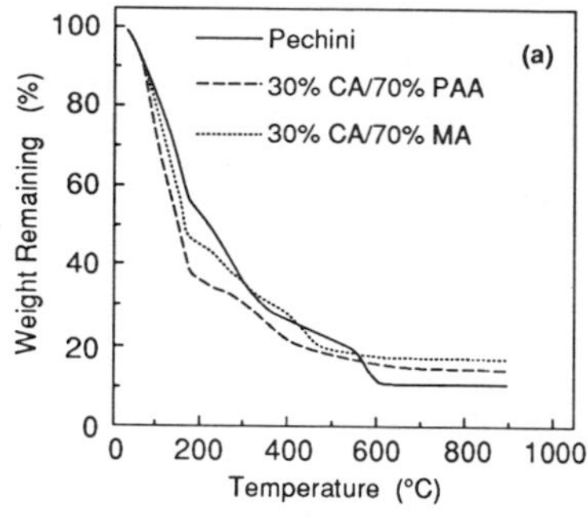

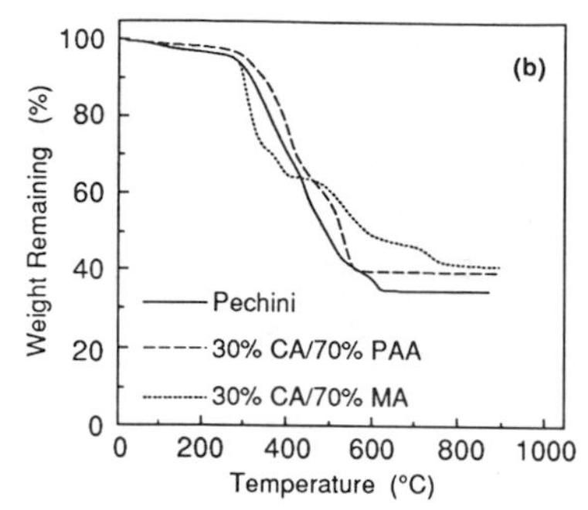

Figure 2. TGA data (heating rate = 5°C/min in 25 sccm air) comparing different chelating agents. (a) as prepared solutions (b) samples after 250°C charring operation

X-ray diffraction analysis of the powders after the calcination step at a given temperature show that the mucic acid-derived powders have a lower $SrTiO_3$ content than the standard Pechini or the PAA modified materials. For example, the x-ray powder diffraction patterns of the original Pechini and the PAA (30%CA/70%PAA-2000) powders show only the diffraction lines for well crystallized $SrTiO_3$ when calcined using the 550/700°C schedule (3b). The mucic acid material (30%CA/70%MA) show weak, very broad diffraction peaks indexed to TiO_2 (anatase) and $SrCO_3$. This result is consistent with the argument that segregation occurs during the charring step of the mucic acid derived powders.

As expected, the surface area the powders decreased with increasing calcination temperature as shown in Figure 3 for the case of the D2000 (50%CA/50%PAA) powder. However, the figure shows that the powders with the higher temperature carbon burnout step (450°C) have larger surface areas than those with the lower temperature burnout (400°C) for a given annealing temperature. SEM photomicrographs in Figure 4 show the increase in crystallite size with increasing calcination temperature for the PAA-derived powder which had the 450°C burnout. The 550°C calcined sample still possesses the shard-like structure characteristic of the charred resin. After calcination at 700°C, the powder is composed of well defined crystallites.

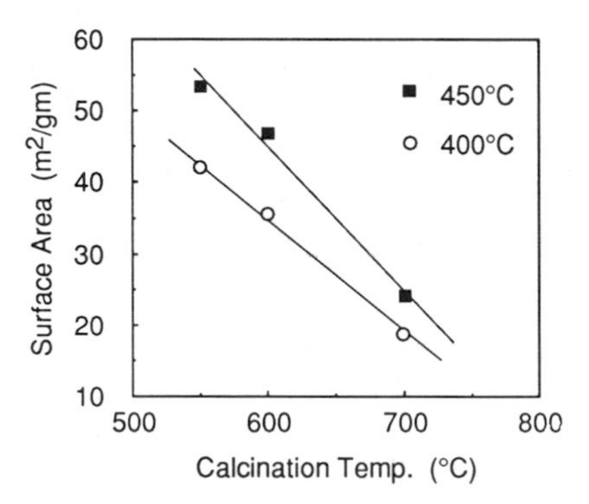

Figure 3. BET powder surface area for powder D2000 (50%CA/50%PAA, PAA mol. wt.=2000) as a function of final annealing temperature and the carbon burnout temperatures of 400 & 450°C.

A quantitative comparison of differences in powder structure from SEM photomicrographs is difficult. A much more quantitative approach is to measure changes in particle size distributions of suspensions after fixed amounts of ultrasonic irradiation [8]. The particle size distributions after two hours of sonication of the powders shown in Figure 4 are reported in Figure 5. The cumulative size distributions are indicative of powders that possess bimodal size distributions. The size of the large particles (>10 μm) is determined by the grinding and sieving operations. If the powders were ground longer or sieved using a finer mesh screen the large particle end of the distribution would be shifted to smaller sizes. Of greater interest is the small particle end of the distribution. For fixed sonication conditions, the mass percent finer than a fixed size gives an indication of the strength of agglomerates larger than that size. In this work, the desired powder would be one in which all particles would be less than 1 μm after sonication. Figure 5 shows that 68% of the particles of the 700°C sample are less than 1 μm after sonication whereas the powders calcined at 550 and 600°C only have 40% of their particles less than one micron. The powder treated at the higher temperature can therefore be assumed to have a larger fraction of weak agglomerates above 1 μm. This approach was used to compare the agglomerate strength of the powders prepared in this study.

Figure 6 shows the effect the use of PAA and MA had on agglomerate strength compared to original Pechini formulated powder. The particle size distribution data given in the figure is for powders that were processed using calcination schedule 3b (550°C/700°C) and sonicated for two hours. The PAA powder, E2000,

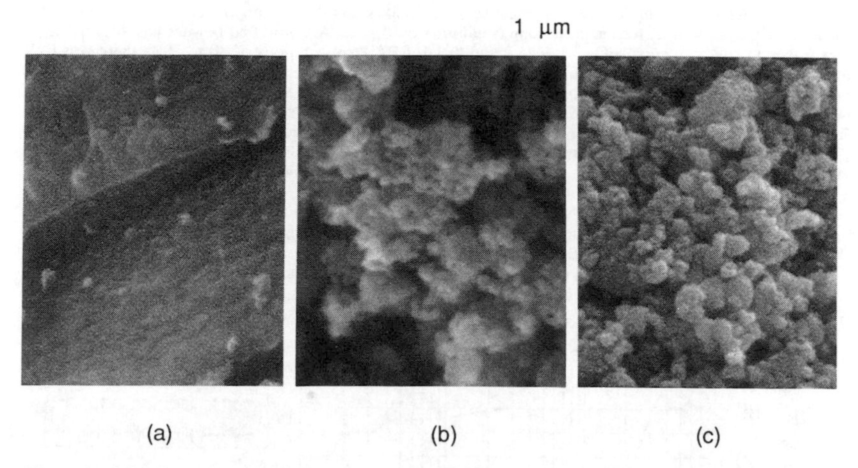

Figure 4. SEM photomicrographs of powder D2000 (50%CA/50%PAA, PAA mol. wt.=2000) calcined using (a) schedule 1a (450°C/550°C), (b) schedule 2a (450°C/600°C) and (c) schedule 3a (450°C/700°C).

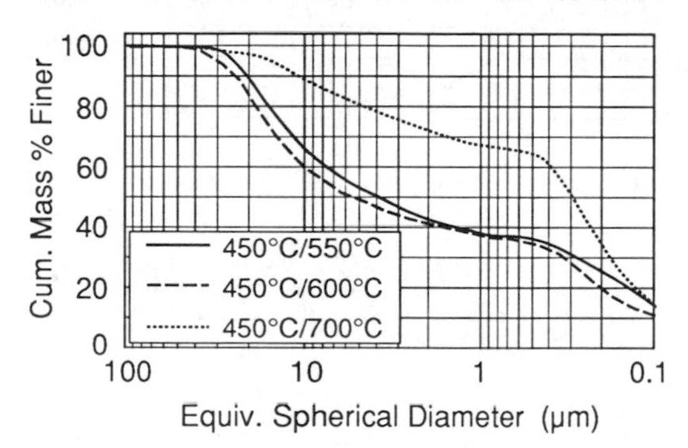

Figure 5. Particle size distribution data for powder D2000 (50%CA/50%PAA, PAA mol. wt.=2000) calcined using schedule 1a (450°C/550°C), schedule 2a (450°C/600°C) and schedule 3a (450°C/700°C). (Powders were sonicated for 2 hours)

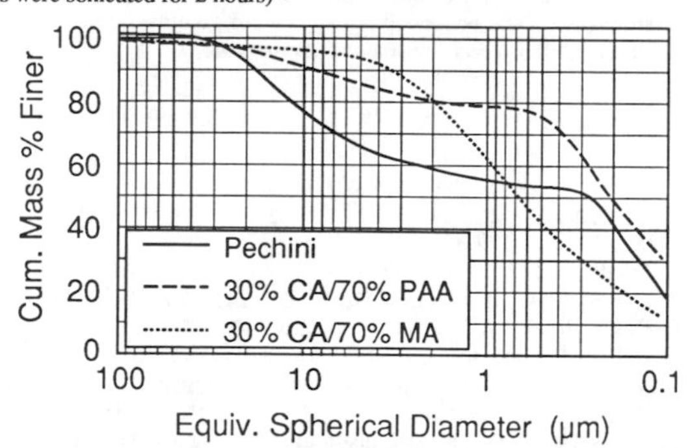

Figure 6. Particle size distribution data for powders A-CA (original Pechini formulation), E2000 (30%CA/70%PAA, PAA mol. wt.=2000) and H-MA (30%CA/70%MA). (Powders were calcined using schedule 3b - 550°C/700°C and sonicated for 2 hours)

(30%CA/70%PAA) has about 80% of its particle mass with sizes less than 1 µm whereas the Pechini material only about 50%. Therefore, for the processing conditions used, the PAA modified powder has a significantly larger fraction of weak agglomerates. It was found that for PAA substitutions of 30 to 70%, there was little effect on powder agglomerate strength characteristics. The higher molecular weight PAA (mol. wt. = 5000) produced powders with stronger agglomerates than those prepared using the Pechini process. For example, powder D5000 (50%CA/50%PAA, PAA mol. wt.=5000) processed using schedule 3b had < 50% of its mass below 1 µm in size after two hours of sonication. Figure 6 also shows that the mucic acid derived material (30%CA/70%MA) has a mass fraction of particles < 1µm similar to the Pechini powder. As with the PAA derived powders, regardless of the ratio of mucic acid to citric acid, the mucic acid derived materials had similar agglomerate strength characteristics.

The weakly agglomerated PAA formulated powders could be easily fractionated to produce suspensions of particles with sizes less than 1 µm. Figure 7 shows particle size distribution data for powder C2000 (70%CA/30%PAA, PAA mol. wt.=2000) in its original suspended state after calcination, after 2 hours of sonication, and after allowing the large particles to settle for 1 hour. The fractionated suspension contains about 70 mass percent of the original agglomerated powder.

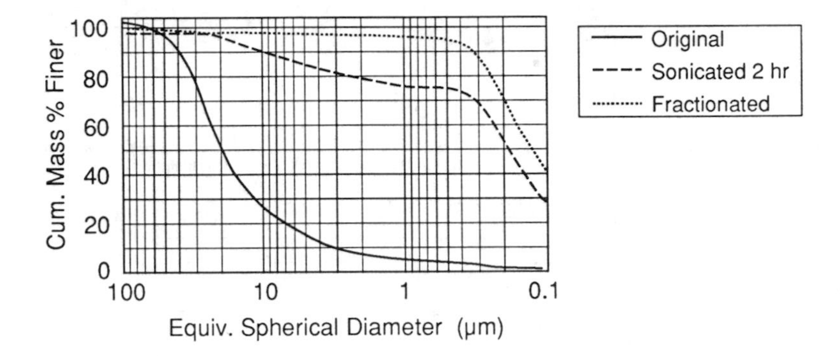

Figure 7. Particle size distribution data for powder C2000 (70%CA/30%PAA, PAA mol. wt.=2000) in as-dispersed state after calcination (Original), original powder suspension after 2 hours of sonication (Sonicated 2 hr) and suspension after 1 hour of sedimentation (Fractionated). The powder was calcined using schedule 3b - 550°C/700°C.

CONCLUSIONS

We have shown that the use of PAA in combination with citric acid as chelating agents in Pechini-type powder synthesis can produce $SrTiO_3$ powders that are weakly agglomerated. Moreover, PAA apparently maintains improved cation mixing during the charring operation over citric acid alone (original Pechini) or the mucic acid/citric acid combination. The better mixing allows for complete conversion to $SrTiO_3$ at lower temperatures producing well crystallized, weakly agglomerated powders.

REFERENCES

1. P. A. Lessing, J. Amer. Ceram. Soc. Bul., 68 1002-7 (1989).
2, M. Zeldin et al., Inorganic and Organometallic Polymers, ACS Symp. Series 360, 1988; L. A. Chick et al., Mat. Sci. Let., 10 6-12 (1990).
3. M. Pechini, U. S. Patent 3,330,697, July 11, 1967.
4. N. G. Eror and H. U. Anderson, in Better Ceramics Through Chemistry II, edited by C. J. Brinker, D. E. Clark and D. R. Ulrich (Mater. Res. Soc. Proc.73, Pittsburgh, PA, 1986) pp. 571-7.
5. Lone-Wen Tai and P. A. Lessing, J. Mater. Res. 7 511-519 (1992).
6. K. D. Budd and D. A. Payne, in Better Ceramics Through Chemistry II, edited by C. J. Brinker, D. E. Clark and D. R. Ulrich (Mater. Res. Soc. Proc.32, Pittsburgh, PA, 1984) pp. 239-244.
7. S. L. Peschke, Masters Thesis, New Mexico Institute of Mining and Technology, 1991.
8. M. Ciftcioglu, J. Amer. Ceram. Soc. Bul., 67 1591-6 (1986).

SYNTHESIS, STRUCTURE, AND CHARACTERIZATION OF $La_{1-x}Ba_xTiO_3$ ($0 \leq x \leq 1$)

Joseph E. Sunstrom IV and Susan M. Kauzlarich
Department of Chemistry
University of California
Davis, California 95616

ABSTRACT

The compounds $La_{1-x}Ba_xTiO_3$ ($0 \leq x \leq 1$) have been prepared by arc melting stoichiometric amounts of $LaTiO_3$ and $BaTiO_3$. Single phase samples can be made for the entire stoichiometry range. The polycrystalline samples have been characterized by thermal gravimetric analysis, X-ray powder diffraction, and temperature dependent magnetic susceptibility. This series of compounds has been studied as a possible candidate for an early transition metal superconductor.

INTRODUCTION

Mixed valent transition metal oxides exhibit a wide range of physical properties that vary with the average valence of the transition metal [1]. New superconducting compounds with transition metals other than copper may be discovered through a systematic study of mixed valent transition metal oxides. Currently, the highest critical temperature observed for a titanium oxide superconductor is 11 K in the mixed valent spinel $LiTi_2O_4$ [2]. The oxygen deficient perovskite $SrTiO_{3-x}$ achieves mixed valency through oxygen deficiency and is superconducting at very low temperatures (T_c = 0.7 K) [3]. Our interest has been in the study of solid solutions of rare earth titanates $RETiO_3$ (RE = La, Ce) with alkaline earth titanates $AETiO_3$ (AE = Sr, Ba) [4]. Synthesis of these types of solid solutions allows one to control the average Ti valence and monitor the properties as a function of the valence and structure [4,5]. $La_{1-x}Sr_xTiO_3$, compounds which are "valence equivalents" of $SrTiO_{3-x}$, were found to not superconduct down to 2 K [4]. This implies that factors other than mixed valency contribute to the superconducting state in titanium perovskite compounds. Since the conduction electron in Ti^{3+}/Ti^{4+} oxides is in a t_{2g} orbital, increased overlap between the Ti t_{2g} and O 2p orbitals is an important variable to consider. The degree of orbital overlap can be altered by the addition of highly electropositive cations.

$BaTiO_3$ crystallizes in a tetragonal structure, and owing to the larger size of the Ba^{2+} ion, one would expect to have different intermediate structures (and therefore different properties) between the end members of the $La_{1-x}Ba_xTiO_3$ series as opposed to $La_{1-x}Sr_xTiO_3$. The larger Ba^{2+} ion will also have a more polarizable electron cloud which would be amenable to Cooper pair formation through better screening of the electrons. Finally, the Ba^{2+} ion is more electropositive than Sr^{2+}, which will make the Ti-O bonding more covalent through decreased competition for electrons.

The structure and properties of both $BaTiO_3$ and $LaTiO_3$ [6] have been studied extensively. Although the synthesis, structure, and thermoelectric properties of $La_{1-x}Ba_xTiO_3$ has been reported, the structure was incorrectly assigned as cubic for $0.15 \leq x \leq 0.6$ [7]. This paper details the synthesis, structure, and properties for the solid solution $La_{1-x}Ba_xTiO_3$.

EXPERIMENTAL

$BaCO_3$ (99.999%) was purchased from Johnson Matthey and dried for 24 hours at 120 °C prior to use. All other materials, and the synthesis of $LaTiO_3$, $AETiO_3$, and $La_{1-x}AE_xTiO_3$ ($0.00 \leq x \leq 1.00$) have been described previously [4].

Oxygen stoichiometry was determined using a DuPont 2100 Thermal Analyzer under conditions previously described [4]. The oxidation products of the samples, determined by X-ray powder diffraction, were $La_2Ti_2O_7$ and $BaTiO_3$. X-ray powder diffraction data were obtained using an Enraf-Nonius Guinier camera or a Siemens D500 diffractometer. A least squares fitting program was used to calculate lattice parameters for the series from measured films (Table II). A Quantum Design SQUID magnetometer was used to make magnetic measurements. The samples with composition $0.00 \leq x \leq 0.05$ were cooled in a field of 1 tesla to 5 K. The magnetization vs. temperature data were taken in an applied field of 10 Gauss (5-170 K) and 1 tesla (170-300 K). For $x = 0.10, 0.20$, the applied field was 1000 Gauss from 5-300 K. At compositions of $x > 0.20$, 1 tesla was applied from 5-300 K. All samples were screened for the Meissner effect at 5 K.

RESULTS AND DISCUSSION

Synthesis and Oxygen Content. By arc melting stoichiometric amounts of $LaTiO_3$ and $BaTiO_3$ under argon, it is possible to make single phase samples for the entire stoichiometry range. Many groups have doped small amounts of La^{3+} into $BaTiO_3$ by reacting La_2O_3 with $BaTiO_3$ in order to study dielectric properties of the material [8]. In a direct solid state reaction of the oxides, the oxygen stoichiometry is too high and little of the $BaTiO_3$ is reduced.

Table I. Oxygen Stoichiometries as Determined by Thermal Gravimetric Analysis.

$LaTiO_{2.99}$	$La_{0.5}Ba_{0.5}TiO_{3.01}$
$La_{0.9}Ba_{0.1}TiO_{2.99}$	$La_{0.4}Ba_{0.6}TiO_{2.98}$
$La_{0.8}Ba_{0.2}TiO_{3.00}$	$La_{0.3}Ba_{0.7}TiO_{2.98}$
$La_{0.7}Ba_{0.3}TiO_{2.99}$	$La_{0.2}Ba_{0.8}TiO_{2.96}$
$La_{0.6}Ba_{0.4}TiO_{2.99}$	$La_{0.1}Ba_{0.9}TiO_{2.97}$

The experimentally determined oxygen stoichiometries are given in Table I. The molar oxygen content of each sample was determined to be approximately 3.00 ± 0.04 using thermal gravimetric analysis. The oxygen content tends to be lower in the Ba rich side of the solid solution. If the molar oxygen content becomes too small (below approximately 2.9), a second phase appears which adopts the hexagonal $BaTiO_{3-x}$ structure type.

Structure. The lattice parameters for the series are shown in Table II. $LaTiO_3$ crystallizes in the $GdFeO_3$ (Pbnm) structure type [6(c)]. The $La_{1-x}Ba_xTiO_3$ solid solution maintains the $GdFeO_3$ structure type until x = 0.20. The x-ray traces for $La_{1-x}Ba_xTiO_3$ (x = 0.05, 0.10, 0.20, 0.30) can be seen in Figure 1 and show the transition to the higher symmetry orthorhombic space group Ibmm. This is in good agreement with the Goldschmidt tolerance factor proposed for this structure type [8].

Table II. Lattice Parameters for $La_{1-x}Ba_xTiO_3$.

x	a(Å)	b(Å)	c(Å)	volume(Å³)
0.00	5.629(2)	5.612(1)	7.914(1)	250.02(12)
0.05	5.593(3)	5.570(3)	7.903(5)	247.24(25)
0.10	5.599(2)	5.586(1)	7.895(3)	246.90(21)
0.20	5.592(2)	5.587(3)	7.891(2)	246.58(15)
0.30	5.588(1)	5.589(1)	7.913(2)	247.30(19)
0.40	5.595(2)	5.590(2)	7.921(2)	247.75(29)
0.50	5.578(4)	5.629(3)	7.920(1)	248.67(21)
0.60	5.590(1)	5.661(5)	7.932(3)	250.98(37)
0.70	3.968(2)			62.47(7)
0.80	3.981(1)			63.09(4)
0.90	3.985(1)			63.29(5)

In addition, the Pbnm space group consists of two substructures: O'-orthorhombic and O-orthorhombic. The O'-orthorhombic is equal to the O-orthorhombic structure with a superimposed Jahn-Teller distortion. The O'-orthorhombic substructure is characterized by a c_o/a_o ratio that is less than $\sqrt{2}$ [9]. The Jahn-Teller distortion for $LaTiO_3$ consists of a compression along the c-axis. In the $La_{1-x}Ba_xTiO_3$ series, the Jahn-Teller distortion is gone by x = 0.05.

Figure 2 shows the cell volume and lattice parameters of the $La_{1-x}Ba_xTiO_3$ ($0.0 \leq x \leq 0.6$) plotted as a function of composition x. The cell volume decreases rapidly until x = 0.20, then begins to increase until x = 0.60. In the $La_{1-x}Sr_xTiO_3$ series [4], the cell volume decreases through the entire series. The cell expansion for x > 0.20 in $La_{1-x}Ba_xTiO_3$ must therefore be due to introduction of the larger Ba^{2+} cation.

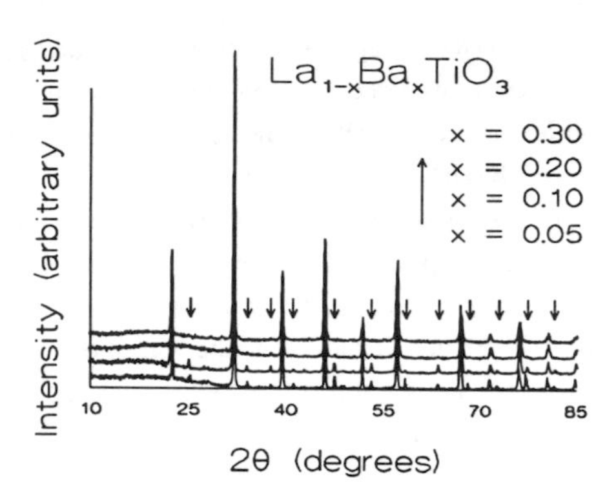

Figure 1. X-ray powder diffraction data for $La_{1-x}Ba_xTiO_3$ for x = 0.30, 0.20, 0.10, 0.05. The arrows indicate reflections due to the Pbnm structure type.

The change in slope of the cell volume at x = 0.20 coincides with the structure change from the Pbnm space group to Ibmm. The samples 0.30 ≤ x ≤ 0.60 are indexed in this space group. As with the $La_{1-x}Sr_xTiO_3$ series, the a and b parameter converge until they are the same within error then diverge. This coincides with a rearrangement of the TiO_6 octahedra. At the x = 0.60 composition, the divergence is at a maximum and the structure changes to higher symmetry at that composition.

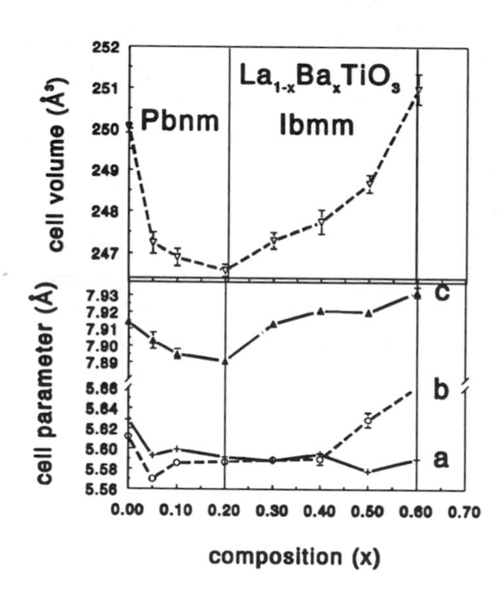

Figure 2. (a) Cell volume vs. composition (b) lattice parameters vs. composition.

The $La_{1-x}Ba_xTiO_3$ ($0.70 \leq x \leq 0.90$) are indexed in the Pm3m space group. Indexing these samples in the tetragonal (P4mm) or orthorhombic (Ibmm) gives lattice parameters which are cubic within calculated error.

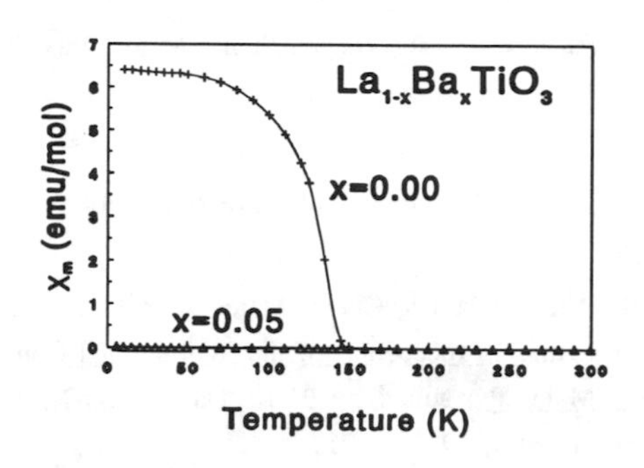

Figure 3. Magnetic Susceptibility vs. Temperature for $La_{1-x}Ba_xTiO_3$ ($x = 0.00, 0.05$).

Magnetic Properties. $LaTiO_3$ is a canted antiferromagnet which shows net ferromagnetic ordering at 150 K. Figure 3 shows that the ordering is destroyed for the $La_{0.95}Ba_{0.05}TiO_3$ sample. We have shown that for the $La_{1-x}Sr_xTiO_3$ sample that the ordering is not destroyed until $x = 0.10$ [4]. In both the Sr and Ba doped $LaTiO_3$, the loss of ferromagnetic ordering coincides with the O'-orthorhombic - O-orthorhombic structure change (loss of Jahn-Teller distortion).

The Pauli susceptibility component of the temperature independent susceptibility is proportional to the density of states (DOS) at the Fermi level. Figure 4 shows the room temperature susceptibilities plotted as a function of composition. The discontinuities in this plot suggest changes in the band structure as a function of x. The increase in susceptibility at the $x = 0.70$ composition is consistent with a change to cubic symmetry.

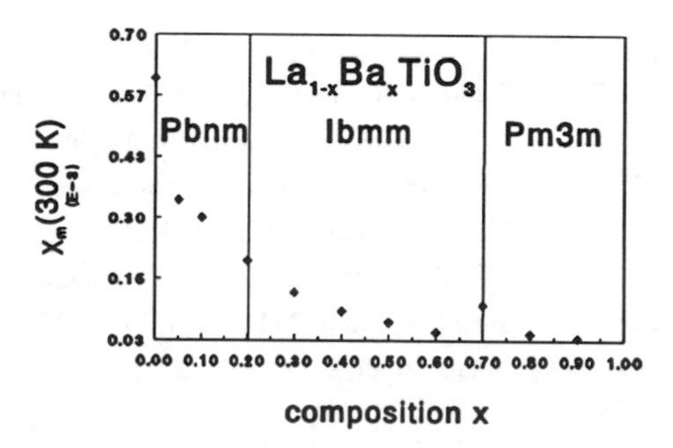

Figure 4. Room Temperature Susceptibilities vs. Composition.

ACKNOWLEDGEMENT

We thank Professor R. N. Shelton for use of the magnetometer, X-ray diffractometer, and TGA. The work was supported by National Science Foundation Solid State Chemistry Grant DMR-8913831.

[1] see for example: (a) B. W. Arbuckle, K. V. Ramanucjachary, Z. Zhang, M. J. Greenblatt, Sol. State Chem. 88, 278 (1990). (b) J. P. Kemp, D. J. Beal, P. A. Cox, J. Sol. State Chem. 86, 50 (1990).

[2] D. C. Johnston, H. Prakash, W. H. Zachariasen, R. Viswanathan, Mater. Res. Bull. 8, 777, (1973).

[3] J. J. Schooley, W. R. Hosler, M. L. Cohen, Phys. Rev. Lett. 12, 474 (1964).

[4] (a) J. E. Sunstrom, IV, S. M. Kauzlarich, P. Klavins, Chem. Mater. 4, 346 (1992). (b) S. M. Kauzlarich, J. E. Sunstrom, IV, P. Klavins, Proceedings of the International Conference of the Chemistry of Electronic Ceramic Materials, edited by P. K. Davies and R. S. Roth (NIST Special Publication 804, Gaithersburg, MD, 1991) p. 217.

[5] (a) Y. Maeno, S. Awaji, H. Matsumoto, T. Fujita, Physica B 165&166, 1185 (1990). (b) M. Higuchi, K. Aizawa, K. Yamaya, K. Kodaira, J. Sol. State Chem. 92, 573 (1991). (c) M. Abbate et. al. Phys. Rev. B 44(11), 5419 (1991). (d) Y. Fujishima, Y. Tokura, T. Arima, S. Uchida, Physica C 185, 1001 (1991).

[6] (a) D. A. Crandles, T. Timusk, J. E. Greedan, Phys. Rev. B 44 13250 (1991). (b) F. Lichtenberg, D. Widmer, J. G. Bednorz, T. Williams, A. Z. Reller, Phys. B82, 211 (1991). (c) M. Eitel, J. E. Greedan, J. Less Common Met., 116, 95 (1986). (d) J. E. Greedan, J. Less-Common Met., 111, 335 (1985). (e) J. P. Goral, J. E. Greedan, J. Magn. Magn. Mater. 37, 315 (1983). (f) D. A. Maclean, J. E. Greedan, Inorg. Chem. 20, 1025 (1980). (g) G. V. Bazuev, G. P. Shveikin, Inorg. Mater. 14, 201 (1978). (h) P. Ganguly, Om. Parkash, C. N. R. Rao, Phys. Status Solidi A 36, 669 (1976).

[9] W. D. Johnston, D. Sestrich, J. Inorg. Nucl. Chem. 20, 32 (1961).

[8] (a) C. A. Kleint, U. Stopel, A. Rost, Phys. Status Solidi A 115, 165 (1989). (b) J. P. Guha, J. Am. Ceram. Soc. 74, 878 (1991). (c) For a more complete review, see F. S. Galasso, Perovskites and High T_c Superconductors, (Gordon and Breach Science Publishers, New York, 1990).

[9] Goodenough, J. B. Prog. Solid State Chem. 5, 145 (1975).

COMBUSTION SYNTHESIS OF Sr-SUBSTITUTED $LaCo_{0.4}Fe_{0.6}O_3$ POWDERS

J. J. KINGSLEY, L. A. CHICK, G. W. COFFEY, D. E. McCREADY AND
L. R. PEDERSON
Pacific Northwest Laboratory, Richland, WA 99352

ABSTRACT

Sr-substituted perovskite $LaCo_{0.4}Fe_{0.6}O_3$ is known to have excellent mixed ionic and electronic conductivity and increased O_2 sorption characteristics. These perovskites are usually prepared by lengthy solid-state reactions of the component oxides at temperatures near 1150°C, and often produce inhomogeneous, multi-phase powders. Presently, it has been prepared by the calcination of combustion-derived fine mixed oxides at 850°C in 6 hrs. Combustion reactions are carried out using precursor solutions containing the corresponding metal nitrates (oxidizers) and glycine (fuel) at 250°C. The metal oxides produced by this process and subsequent calcination were characterized by XRD, TEM and BET surface area analysis.

INTRODUCTION

Mixed ionic and electronic conductors (MIEC's) are considered as potential candidates for many technological applications [1] such as in gas separation [2,3], electrodes in high temperature fuel cells [4,5], catalysts for oxygen evolution [6] and oxygen enrichment in combustion [7]. Among others, Sr-substituted $LaCo_{1-y}Fe_yO_3$ perovskites have been proven to have excellent mixed ionic and electronic conductivity [8]. These oxides are usually prepared by conventional solid state reactions between corresponding oxides or decomposition of solutions containing metal nitrates and acetates. Both of these techniques require high temperature (>1000°C) for the formation of single phase perovskites and often produce inhomogeneous powders. Although many other wet chemical methods such as co-precipitation have been successful in producing single or two component oxides, they have inherent limitations in producing homogeneous ternary or higher multicomponent oxides. Apart from the difficulties in preparing $La_{1-x}Sr_xCo_{1-y}Fe_yO_3$ MIEC's it is noted that their lattice parameters were also not reported. Therefore, the aim of this present study was to prepare single phase and homogeneous Sr-substituted $LaCo_{0.4}Fe_{0.6}O_3$ by a novel technique and to characterize their structure and morphology.

Recently, a novel wet chemical method has been developed [9-14] utilizing exothermic redox decomposition of metal nitrates (oxidizers) and certain organic compounds (fuels). This process can be initiated at temperatures as low as 200°C and produces fine particle oxides in a very short time. The process is simple, auto-catalytic and self-propagating in nature. The process has been used to synthesize a wide variety of single, binary and ternary oxides such as α-Al_2O_3, ruby, CeO_2, ZnO, $Y_3Al_5O_{12}$ (garnet), spinels MAl_2O_4, MCr_2O_4, MFe_2O_4, mullite, t-ZrO_2/Al_2O_3 composite, alkaline earth doped rare earth chromites, manganites, ferrates etc. Also, many have used this process to prepare high Tc cuprate superconductors, however the orthorhombic superconducting phase is formed only after annealing. Presently, we have attempted to use this process to prepare the quaternary oxide, $La_{1-x}Sr_xCo_{0.4}Fe_{0.6}O_3$.

 ©1992 Materials Research Society

EXPERIMENTAL

The precursor solution is prepared by dissolving metal nitrates (oxidizer) and glycine (fuel) in water. The amount of glycine required is determined using the total oxidizing valencies of metal nitrates and the total reducing valencies of the glycine as numerical coefficients to calculate stoichiometric balance. The valencies of the elements used in this calculation are: O (-2), N (0), H (+1), C (+4) and the metal ion, M^{n+} (+n). Thus, one mole of di or tri valent metal nitrate (with total oxidizing valancies -10 and -15) requires 1.11 and 1.66 moles of glycine (with total reducing valency +9).

In a typical experiment, stoichiometric amounts of metal nitrate solutions (standardized by EDTA methods) corresponding to 0.01 mole of $La_{1-x}Sr_xCo_{0.4}Fe_{0.6}O_3$ are mixed in a 150 ml pyrex beaker with the required amount of glycine. The mixture is then decomposed on a hot plate maintained at 250°C. The solution dehydrates and decomposes with ignition producing fine metal oxides. The as-prepared powders are further calcined in a Pt crucible at 850°C. The as-prepared and calcined oxides are characterized by X-ray diffraction, TEM and BET surface area analysis. X-ray diffraction analysis was carried out using Philips APD 3600 Diffractometer with Cu K_α radiation and no internal standard was used for d-spacing values. TEM analysis was made on Philips EM 400 T operated at 120 KeV. BET surface area was measured using nitrogen as adsorbate in Micromeritics 2000 ASAP.

RESULTS AND DISCUSSION

Metal nitrates-glycine compositions with less or more than the stoichiometric amount of glycine are referred to as fuel lean or fuel rich combustion mixtures respectively. Combustion of metal nitrate-glycine precursor mixture for $La_{1-x}Sr_xCo_{0.4}Fe_{0.6}O_3$ produced different oxide phases depending on both the value of x and oxidizer-fuel composition (fuel lean or fuel rich). Fuel lean compositions produced a mixture of oxides with unreacted $Sr(NO_3)_2$ for x=0-0.8. A stoichiometric fuel composition for x=0.2 produced multiphase oxides free of Sr-nitrate. The phases present are identified as similar to $LaCo_{0.4}Fe_{0.6}O_3$ and $Sr_2Fe_2O_5$. For x>0.2, such stoichiometric compositions still produce similar mixed oxides and also increasing amounts of strontium nitrate, $Sr(NO_3)_2$ with x. This could be due to the high decomposition temperature (about 600°C) of $Sr(NO_3)_2$. In order to assist the decomposition of $Sr(NO_3)_2$ and enable Sr^{2+} to form compound oxides, it was attempted to increase the exothermicity and duration of combustion using fuel-rich mixtures. Such compositions on combustion produced mixed oxide phases free of Sr-nitrate. Compositions with 1.5 times the stoichiometric amount of glycine, produced a mixture of phases similar to $SrFeLaO_4$ and $LaCo_{0.4}Fe_{0.6}O_3$ for x=0.2. The amount of fuel was further increased to 2 times the stoichiometric amount to obtain Sr-nitrate free oxide phases for x>0.2. XRD results on these powders indicate that the as-prepared oxides contain a mixture of two phases. The major phase is similar to $LaCo_{0.4}Fe_{0.6}O_3$. The minor phase could possibly be a Sr-rich mixed Fe and Co oxide. The amount of this second phase increased with Sr-content. The XRD patterns of these as-prepared mixed oxides are given in Fig.1.

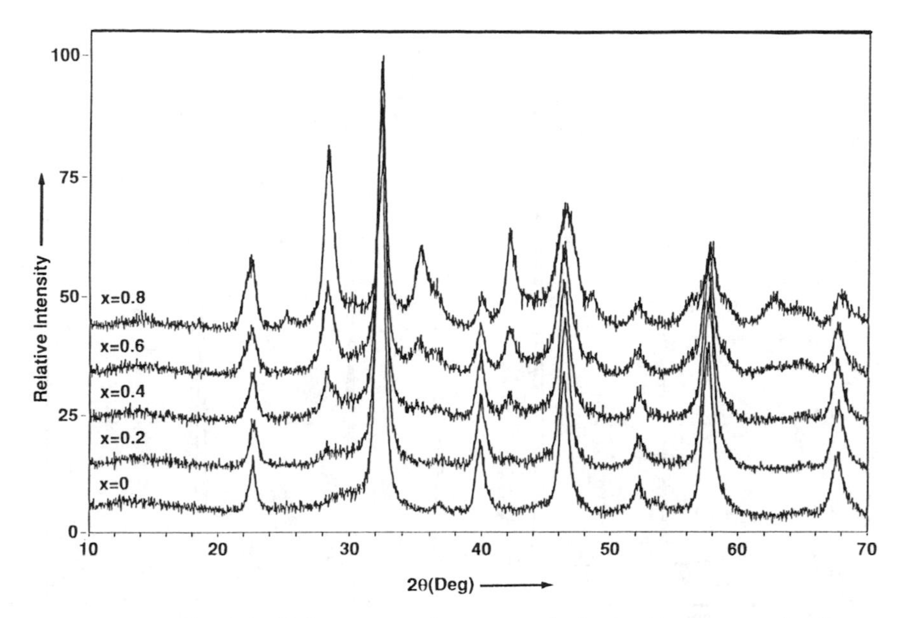

Figure 1. XRD pattern of as-prepared mixed oxides from combustion of metal nitrate-glycine precursors to $La_{1-x}Sr_xCo_{0.4}Fe_{0.6}O_3$

The mixed oxides form single-phase perovskites when calcined at 850°C for 6 hours. The XRD patterns of the calcined oxides are given in Fig.2. Analysis of these patterns indicate the oxides with x=0-0.4 crystallize in an orthorhombic structure, the structure for x=0.0 has been reported as cubic [15]. The peak values for x=0.8 and 0.2 with the corresponding <hkl> are listed in Tables 1 and 2 respectively. The oxides with x=0.6 and 0.8 crystallize in a cubic unit cell, as observed by Matsumoto et al [6]. The lattice parameters calculated from these patterns by least square cell refinement method are given in Table 3. The pseudo-cubic unit cell volume for $La_{1-x}Sr_xCo_{0.4}Fe_{0.6}O_3$ decreases with x and is shown in Fig.3a. The surface area of the as-prepared oxides using fuel rich (2 times stoichiometric glycine) compositions varies from 18-32 m^2/g.

The catalytic activity of perovskite $La_{1-x}Sr_xCo_{0.4}Fe_{0.6}O_3$ (x=0-0.8) in H_2O_2 decomposition was investigated at 70°C, using 10 mg of oxides and 30% H_2O_2(10μl). Preliminary results indicate a linear dependance of relative rate of H_2O_2 decomposition with x in $La_{1-x}Sr_xCo_{0.4}Fe_{0.6}O_3$ and is shown in Fig.3b.

TEM micrographs of the as-prepared mixed oxide and calcined $La_{0.8}Sr_{0.2}Co_{0.4}Fe_{0.6}O_3$ are given in Fig. 4. The as-prepared oxide consists of very fine particles (<100 nm) and transform to single phase and well crystallized perovskite $La_{0.8}Sr_{0.2}Co_{0.4}Fe_{0.6}O_3$ at 850°C with particles of size <1 μm.

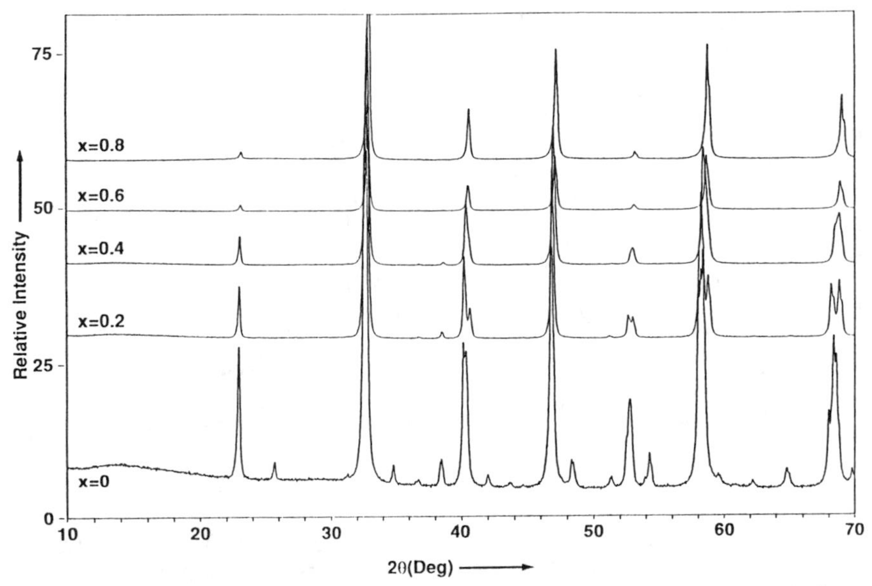

Figure 2. XRD pattern of $La_{1-x}Sr_xCo_{0.4}Fe_{0.6}O_3$ (calcined at 850°C for 6 hrs)

Table 1. The 2θ and <hkl> values for $La_{0.2}Sr_{0.8}Co_{0.4}Fe_{0.6}O_3$

$2\theta_{obs}$	d_{obs}	I/I_o	h k l	d_{calc}
23.14	3.841	2	1 0 0	3.845
32.93	2.718	100	1 1 0	2.719
40.61	2.220	18	1 1 1	2.220
47.25	1.922	42	2 0 0	1.922
53.24	1.719	3	2 1 0	1.719
58.79	1.569	44	2 1 1	1.570
69.03	1.359	23	2 2 0	1.359

Table 2. The 2θ and <hkl> values for $La_{0.8}Sr_{0.2}Co_{0.4}Fe_{0.6}O_3$

$2\theta_{obs}$	d_{obs}	I/I_o	h k l	d_{calc}
22.98	3.867	17	1 1 0	3.868
32.59	2.745	100	0 2 0	2.749
32.82	2.727	98.8	1 1 2	2.727
36.75	2.443	<1	2 1 0	2.439
38.53	2.335	2	1 2 1	2.338
40.22	2.240	29.1	0 2 2	2.236
40.65	2.218	9.4	2 0 2	2.221
46.91	1.935	74	2 2 0	1.934
51.19	1.783	<1	0 3 1	1.782
52.66	1.737	6.9	1 3 0	1.737
53.02	1.726	6.3	2 2 2	1.728
58.25	1.583	59.7	1 3 2	1.583
58.77	1.570	21.1	2 0 4	1.570
68.17	1.374	17.3	0 4 0	1.374
68.80	1.363	19	2 2 4	1.363

Table 3. Lattice parameters for $La_{1-x}Sr_xCo_{0.4}Fe_{0.6}O_3$ (x=0-0.8)

Value o f x	System	Lattice Parameters (Å)
0	Orthorhombic	a=5.468±0.001; b=5.508±0.001; c=7.744±0.002
0.2	Orthorhombic	a=5.442±0.005; b=5.497±0.002; c=7.689±0.007
0.4	Orthorhombic	a=5.442±0.004;b=5.474±0.002; c=7.709±0.006
0.6	Cubic	a=3.8512±0.0002
0.8	Cubic	a=3.8446±0.0003

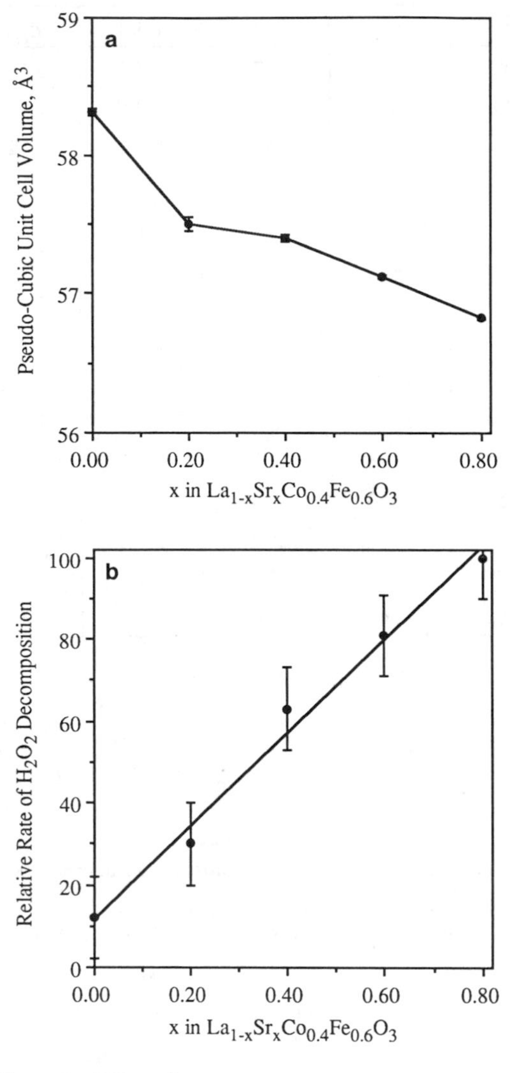

Figure 3. Effect of Sr-Content in $La_{1-x}Sr_xCo_{0.4}Fe_{0.6}O_3$ on
a) Pseudo-cubic unit cell volume
b) Relative rate of H_2O_2 decomposition

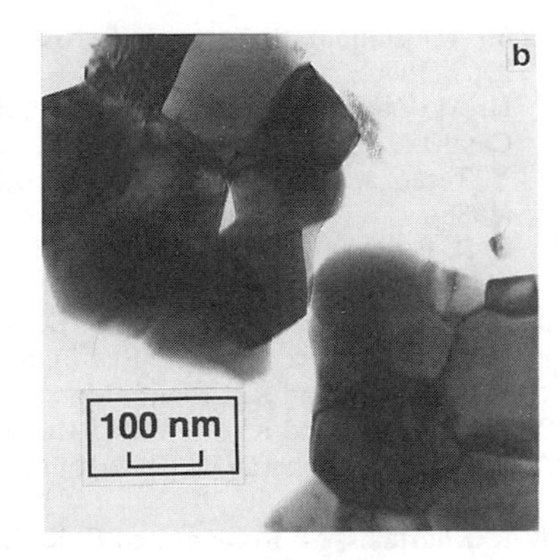

Figure 4. TEM micrographs of $La_{0.8}Sr_{0.2}Co_{0.4}Fe_{0.6}O_3$
a) as-prepared b) calcined (850°C/6 hrs)

CONCLUSIONS

Combustion of metal nitrate-glycine precursor mixtures for $La_{1-x}Sr_xCo_{0.4}Fe_{0.6}O_3$ produced mixed oxide phases. The higher decomposition temperature of strontium nitrate appears to hinder the single phase perovskite formation in the as-prepared powders. Calcination of the as-prepared mixed oxides at 850°C produced single phase perovskite, $La_{1-x}Sr_xCo_{0.4}Fe_{0.6}O_3$ for x=0-0.8. The perovskite $La_{1-x}Sr_xCo_{0.4}Fe_{0.6}O_3$ with x=0-0.4 crystallize in an orthorhombic structure and in a cubic structure for x=0.6 and 0.8. The pseudo-cubic unit cell volume is found to decrease with x in $La_{1-x}Sr_xCo_{0.4}Fe_{0.6}O_3$.

ACKNOWLEDGMENTS

The authors gratefully acknowledge J. E. Coleman and L. A. Charlot for electron microscope analysis. Pacific Northwest Laboratory is operated by Battelle Memorial Institute for the U. S. Department of Energy under Contract DE-AC06-76RLO 1830.

REFERENCES

1. H. L. Tuller, Nonstoichiometric Oxides, Chapt.6 (Academic Press, New York 1981), p.271.
2. Y. Teraoka, H. M. Zhang, S. Furukawa and N. Yamazoe, Chem. Lett. 1743 (1985).
3. Y. Teraoka, T. Nobunaga and N. Yamazoe, Chem. Lett. 503 (1988).
4. O. Yamamoto, Y. Takeda, R. Kanno and M. Noda, Solid State Ionics, 22, 241 (1987).
5. Y. Takeda, R. Kanno, M. Noda, Y. Tomida and O. Yamamoto, J. Electrochem. Soc. 134, 2656 (1987).

6. Y . Matsumoto, S. Yamada, T. Nishida and E. Sato, J. Electrochem. Soc. 127(11) 2360 (1980).
7. H. M. Zhang, Y. Shimizu, Y. Teraoka, N. Miura and N. Yamazoe, Journal of Catalysis, 121, 432 (1990).
8. Y. Teraoka, H. M. Zhang, K. Okamoto and N. Yamazoe, Mat. Res. Bull. 23, 51 (1988).
9. L. R. Pederson, G. D. Maupin, W. J. Weber, D. J. McReady and R. W. Stephens, Materials Letters, 10(9,10), 437-443 (1991).
10. L. A. Chick, L. R. Pederson, G. D. Maupin, J. L. Bates, L. E. Thomas and G. J. Exarhos, Materials Letters, 10(1,2), 6-12 (1990).
11. J. J. Kingsley and K. C. Patil, Materials Letters, 6(11,12), 427-432 (1988).
12. J. J. Kingsley and K. C. Patil, J. Mater. Sci. 25, 1305-1312 (1990).
13. S. Sundar Manoharan, N. R. S. Kumar and K. C. Patil, Mat. Res. Bull., 25, 731-738, (1990).
14. K. Kourtakis, M. Robbins and P. K. Gallagher, J. Solid State Chem., 84, 88-92 (1990).
15. Powder Diffraction File, no: 40-224, Joint Committee on Diffraction Standards.

SYNTHESIS OF FERROELECTRIC PEROVSKITES THROUGH AQUEOUS SOLUTIONS TECHNIQUES

G.Guzmàn-Martel *, M.A.Aegerter* and P. Barboux**
*Instituto de Física e Química de São Carlos, Universidade de São Paulo, Cx. Postal 369, 13560 São Carlos, SP, (Brazil).
**Chimie de la Matière Condensée, Université Pierre et Marie Curie, 75252 Paris (France)

ABSTRACT

The hydrolysis of niobates in aqueous solutions has been studied as a function of pH and concentration. The process has been applied to the coprecipitation of $PbNb_{2/3}Mg_{1/3}O_3$ leading to a low temperature synthesis of this ferroelectric relaxor ceramic. The effect of hydrolysis conditions such as the concentration of bases and acids used, their rate of addition, temperature and the nature of the precursor salts is described. The homogeneity of the precipitates has been characterized through infrared spectroscopy and X-ray diffraction techniques. The perovskite phase appears after heating at 350°C and is obtained as a pure phase above 750°C after one hour heat-treatment. Relaxor ceramics with high dielectric constant can be obtained by sintering above 1000°C. The process also has been successfully applied to the synthesis of other relaxor ceramic compositions such as PZN ($PbNb_{2/3}Zn_{1/3}O_3$), PFN ($PbNb_{1/2}Fe_{1/2}O_3$) as well as $CdNb_2O_6$ and $MgNb_2O_6$ compounds.

INTRODUCTION

A large number of oxide materials based on niobium crystallize in the perovskite or columbite structure. Lead magnesium niobate (PMN) is a typical example of this family. It is a relaxor ferroelectric perovskite whose high dielectric constant and large electrostrictive coefficient provide interesting applications in the field of capacitors and actuators. However, the dielectric properties of this material are very sensitive to the processing of the ceramic because of the formation of a parasitic phase having the pyrochlore structure [1].

Therefore, more reactive precursors are required in order to favor the formation of the perovskite phase relative to the pyrochlore phase. This can be achieved by a better mixing of Mg and Nb as already described by Swartz and Shrout [2] in their two step process:

$$\begin{aligned} MgO + Nb_2O_5 &\rightarrow MgNb_2O_6 \ (1000°C) \\ 3\ PbO + MgNb_2O_6 &\rightarrow 3\ PbNb_{2/3}Mg_{1/3}O_3 \ (800°C) \end{aligned} \quad (1)$$

A better dispersion of MgO also allows a higher reactivity. Such optimized dispersion can be more easily achieved through the use of solution techniques such as sol-gel processes. However as these processes are mainly derived from alkoxide precursors [3], the high cost and difficulty of handling niobium alkoxides make them useless for the synthesis of bulk ceramics whose applications to capacitors yield a low added value.

In this paper we focus on the use of cheap water solution techniques starting from soluble salts such as potassium orthoniobate (K_3NbO_4), and nitrates or acetates of lead and magnesium.

Niobium species are poorly soluble in water at a pH above 7 whereas magnesium and lead hydroxide precipitate above pH 9.5 and pH 5.5 respectively. We therefore first investigate the appropriate pH for coprecipitation. The principle of our process consists in mixing a basic solution, in which niobate species are soluble, with an acidic solution in which Pb and Mg salts are soluble. The solutions are adapted to yield, upon mixing, a solution of intermediate pH (around 11) allowing coprecipitation of the different combinations of the cations.

The process also has been adapted and successfully tested for the preparation of powders of other compositions such as PZN ($PbNb_{2/3}Zn_{1/3}O_3$), PFN ($PbNb_{1/2}Fe_{1/2}O_3$) and compounds such $CdNb_2O_6$ and $MgNb_2O_6$.

Ceramics have been prepared and their dielectric properties and X-ray diffraction patterns are presented.

EXPERIMENTAL

Preparation and characterization of the solutions

The lead and magnesium solutions have been prepared by dissolving commercial nitrate powders (Prolabo) of the respective elements in water. K_3NbO_4 was chosen as the water soluble niobate compound. It was synthesized by first mixing K_2CO_3 and Nb_2O_5 in the appropriate ratio in a ball mill with ethanol and then reacted by solid state reaction at 740°C during 48h. The reaction must be thoroughly completed in order to avoid later on residual carbonate ions. This compound is easily dissolved in water up to a concentration of 0.3M niobium [4] undergoing the following hydrolysis reaction:

$$6K_3NbO_4 + 10\ H_2O \rightarrow 18K^+ + Nb_6O_{19}{}^{8-} + 10\ OH^- \qquad (2)$$

The resulting solution thus provides for each niobium 3 basic equivalents where 10x3/18 are OH^- and 8x3/18 are hexaniobate species. This is demonstrated in figure 1a which shows the titration of a 0.3M solution of K_3NbO_4 by HNO_3 1M. Starting at a high pH, OH^- are first titrated and then, the hexaniobate species precipitate into a niobic acid form in a two steps process. A pH as low as 5 is necessary to neutralize all niobate species and no precipitation occurs until the pH decreases down to 11.

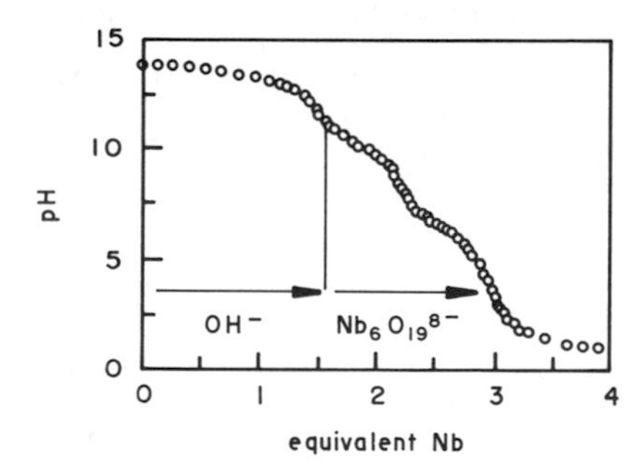

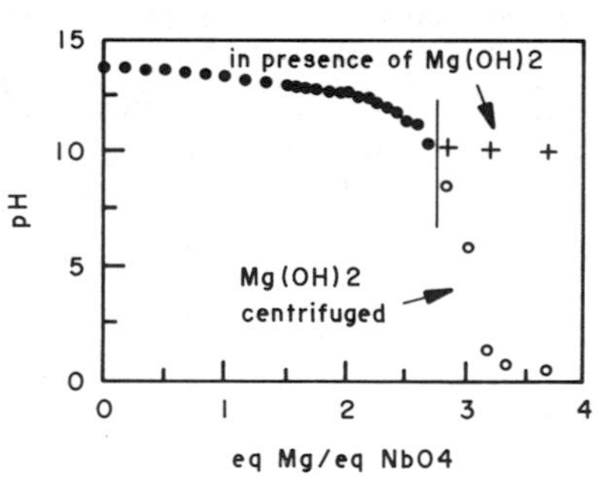

Figure 1. Titration of a 0,3M solution of K_3NbO_4 in water by a) HNO_3 1M b) $Mg(NO_3)_2$ 0,5M.

A slow addition of magnesium nitrate to the orthoniobate solution leads to an immediate precipitation. Because of the presence of $Mg(OH)_2$, which acts as a buffer at pH=10, the solution has to be separated from the precipitate by centrifugation in order to allow the measurement of the final titration (figure 1b). The neutralization point corresponds to a total ratio Mg/Nb=3/2 as can be expected from the precipitation of $5Mg(OH)_2$ + $Mg_4(Nb_6O_{19})$ species. This means that all niobium can be precipitated by magnesium at a pH above 10.

Titration with lead nitrate yields a similar result although $Pb(OH)_2$ is irreversibly precipitated and does not act as a buffer. However, the amount of lead nitrate solution necessary to titrate the niobate is in excess of the expected 3/2 ratio. This is explained by the fact that the addition of lead nitrate salt to an hydroxide solution leads to the formation of a $Pb(NO_3)_2\cdot 3Pb(OH)_2$ composition [6]. In other words, not all the lead acts as an acid towards OH and hexaniobate species since some lead is precipitated with its nitrate counterion according to:

$$4Pb(NO_3)_2 + 6\ OH^- \rightarrow Pb(NO_3)_2\cdot 3Pb(OH)_2 + 6NO_3^- \quad (3)$$

Upon mixing the niobium, magnesium and lead solutions one obtains a mixture of hydroxides and niobates. Since the amount of Pb and Mg corresponding to the PMN composition can only be precipitated with 4 basic equivalents per each niobium and since 1 K_3NbO_4 only provides 3 basic equivalents, at least one KOH group for each Nb should be theoretically added to the solution of hexaniobates. However, at a too high pH, the niobate species do not precipitate. This is demonstrated by the results of the titration curves obtained with solutions in which different amounts of KOH have been added (table I).

Table I: Titration of niobium loss as a function of the final pH:

KOH added/Nb	2.8	0.85	0.6	0.50
final pH	14	12.6	11.5	10.5
Nb loss	35%	25%	1.5%	0% but loss of 10% Mg

The resulting experiment protocol for the preparation of the PMN composition is therefore the following:

A solution containing the appropriate amounts of both Pb and Mg nitrates (Pb/Mg=3, (Pb)=0.5M/l for synthesis of PMN) is rapidly added to a niobium solution (Nb=0.3M, KOH=0.15M); the precipitation readily occurs. The precipitate contains some adsorbed potassium hydroxide or nitrate which can be washed out with distilled water. However, the residual potassium species are better washed out after calcination at 750°C as the first process leads also to a relatively high loss of magnesium hydroxide.

Thermal treatment

The progressive crystallization of the material occurs through endothermic weight losses as shown by DSC and TG measurements (figure 2). Typical X-ray diffraction patterns obtained during the calcination process at different temperatures are shown in figure 3. The dried precipitate is essentially amorphous. Upon heating the P_3N pyrochlore phase appears together with small amounts of the perovskite phase (e.g at 350°C). The unwanted P_3N phase starts to diminish around 700-750°C and at 800°C the X-ray pattern is typical of a well crystallized PMN perovskite with traces of a PbO phase.

A similar process has been applied to the preparation of PZN ($PbZn_{1/3}Nb_{2/3}O_3$), PFN ($PbNb_{1/2}Fe_{1/2}O_3$), $CdNb_2O_6$ and $MgNb_2O_6$. In all cases the magnesium nitrate salt has been substituted by nitrates of the respective elements in adequate proportion. As an example the result of the X-ray diffraction of PZN compound is shown in figure 4. The small amount of the PbO phase which is present can be overcome by precise control of the various concentrations of

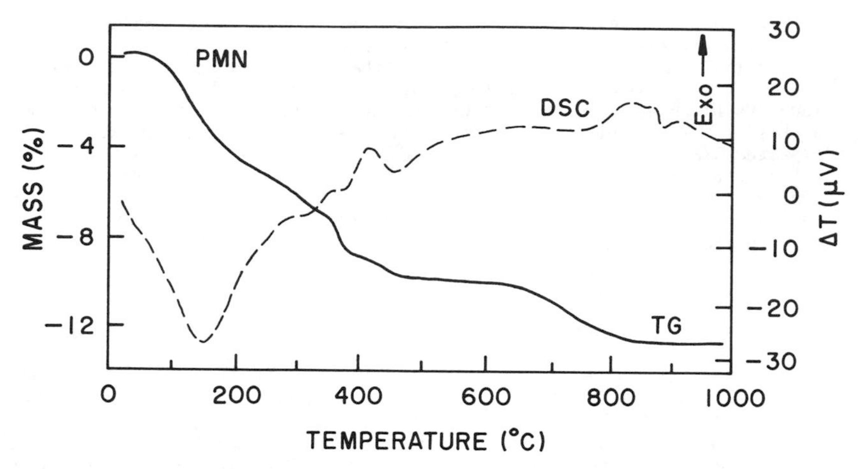

Figure 2. DSC and TG of PMN powder, prepared according to the aqueous process.

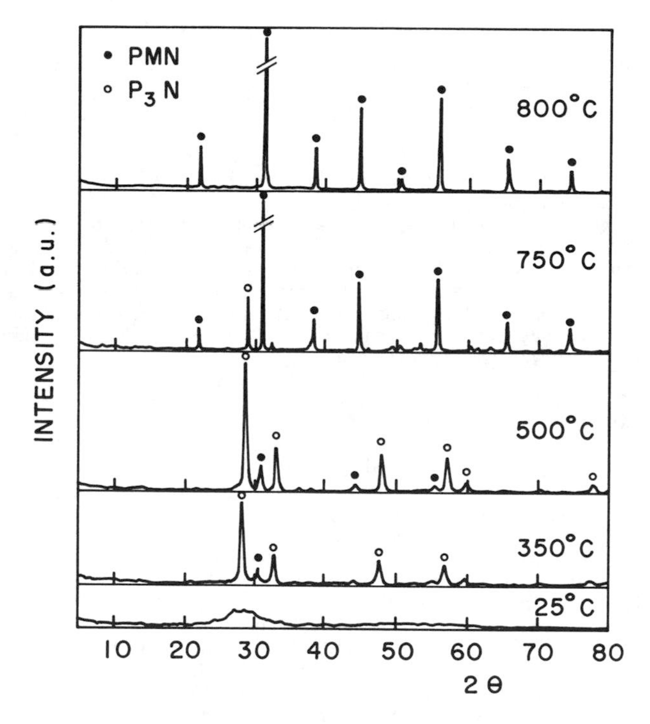

Figure 3. X-ray diffraction patterns of PMN powder measured at 25°C and after heat treatment at 350°C, 500°C, 750°C during one hour respectively.

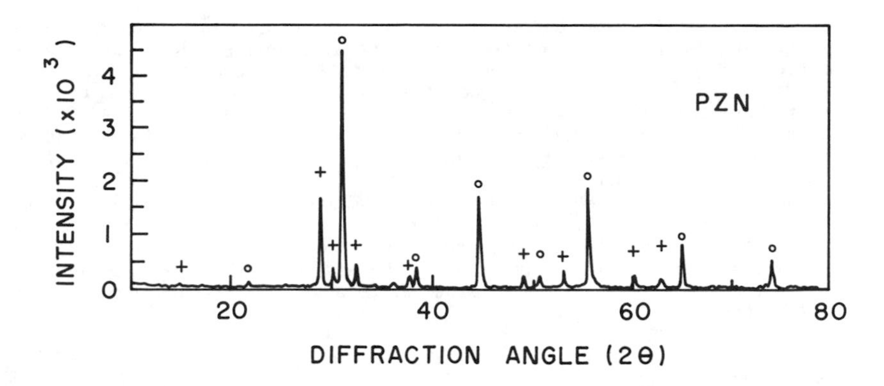

Figure 4. X-ray diffraction pattern of PZN powder calcined at 800°C during one hour (○) PZN, (+) PbO residues.

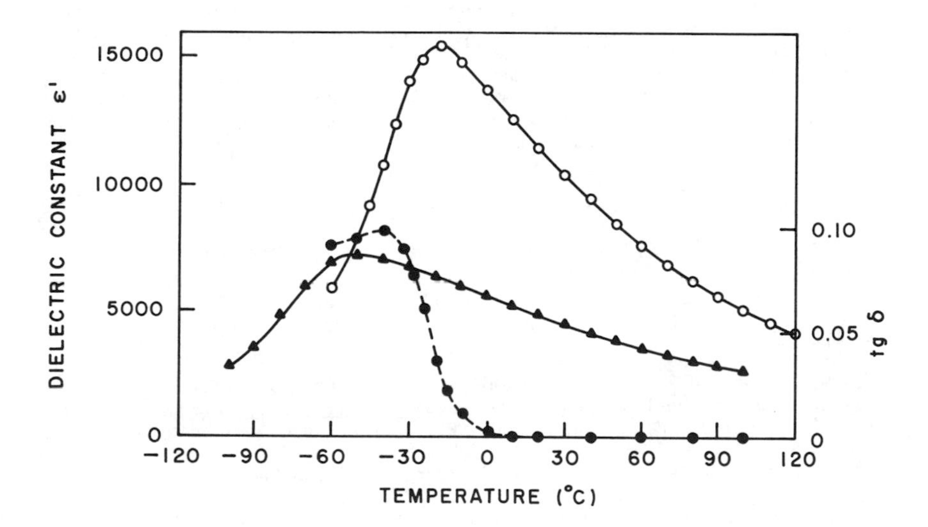

Figure 5. Temperature dependence of the dielectric constant, ϵ' (○,▲) and dielectric loss, tgδ(•), measured at 1kHz of PMN ceramics sintered at 1050°C during 1h. The materials have been prepared with: (○,•) washed powder, (▲) unwashed powder.

the precursor solutions. On the other hand it is known that PbO acts as a good sintering additive and its presence in small concentration should facilitate the process.

Sintering

A heat-treatment at 800°C for half an hour results in the formation of crystalline PMN powders composed of 2 micron size grains. The powders were washed with water to remove the residual potassium present at the grain boundaries of the structure, dried at 150°C and then pressed into pellets under a pressure of $10t/cm^2$. Finally the material was sintered at 1000°C or higher. Figure 5 shows the dielectric constant and dielectric loss measured as a function of temperature at 1kHz for a pure PMN ceramic obtained after sintering at 1050°C during 1 hour. The curves present the typical behavior of the ferroelectric relaxors with a maximum of the dielectric constant slightly higher than 15000 at 250K. The effect of potassium impurities is drastic as observed with a PMN ceramic sintered in the same conditions but prepared without washing the residual potassium. The lower dielectric constant is attributed to the presence of potassium at the grain boundaries of the structure.

CONCLUSION

A cheap process based on aqueous solutions has been derived for the synthesis of PMN ceramics. The pH is a very important parameter for the control of the composition. A too high pH causes a loss of niobium and a too low pH causes a loss of magnesium. As a result PbO is almost always found in excess in the resulting powders. Potassium ions are also coprecipitated and can be remove by washing the powder before or after calcination. The last process avoids the loss of magnesium by dissolution of magnesium hydroxide in water. For well sintered PMN powder, a dielectric constant as high as 15000 has been obtained at 250K. The presence of potassium drastically decreases the dielectric constant of the ceramics. This process was also applied to the synthesis of other niobate containing materials such as PZN, PFN and niobates of Cd and Mg.

ACKNOWLEDGEMENTS

This research has been financed by CNPq, Finep and the Program RHAE - New Materials (Brazil).

REFERENCES

[1] M. Lejeune, J.P. Boillot, Ceramics International, 8 (1982) 99.
[2] S.L. Swartz and T.R. Shrout, Mat. Res. Bull 17 (1982) 1245.
[3] H.U. Anderson, M.J. Pennell and J.P. Guha, Advances in Ceramics, 21 (1987) 91.
[4] G. Guzmàn-Martel, Ph.D. Thesis, University of São Paulo (1991)
[5] M.T. Pope and B.W. Dale, Quart. Rev. 22 (1968) 527
[6] J.L. Pauley and M.K. Testerman, J.Am. Chem. Soc. 76 (1954) 4220.

NEW GROUP 2 COMPOUNDS USEFUL FOR PREPARATION OF THIN FILMS OF ELECTRONIC CERAMICS

William S. Rees, Jr.,* Kerstin A. Dippel, Michael W. Carris, Celia R. Caballero, Debra A. Moreno, and Werner Hesse
Department of Chemistry and Materials Research and Technology Center, Florida State University, Tallahassee, FL. 32306-3006

ABSTRACT

We have prepared examples of several new classes of group 2 compounds, including ether- and amine-substituted metallocenes, inter- and intra-molecular Lewis base stabilized *bis*(β-diketonates), and clam-shell oligioether *bis*(alkoxides), and investigated their use as potential sources in the preparation of ceramic materials from molecular precursors. Examinations have included vapor pressure measurements, hydrolytic, oxidative, thermal and photolytic stability, vapor phase, solution and solid state structures, and evaluation for potential CVD growth of thin films of electronic ceramics. Results to date indicate that intramolecular stabilization is more advantageous than intermolecular stabilization for achievement of optimal CVD source criteria, and that completion of the coordination sphere around the metal atom requires tuning of both ligand spatial and electronic requirements.

BACKGROUND

The recent investigations of the chemistry of the heavier group 2 elements Ca, Sr, and Ba have been driven by the need to develop a basis from which to launch preparations of electronic ceramics.[1] Two specific targets are the ferroelectric $BaTiO_3$ and the superconductor $YBa_2Cu_3O_{7-\delta}$. Until these relatively current chemical studies were motivated by the interest in utilizing controlled methods of materials synthesis, such as sol-gel or chemical vapor deposition (CVD), there was minimal information available about this class of molecules. Specifically, the CVD process demands of high vapor pressure (>10 torr at 100°C) and stable compatibility with oxidizing species in the vapor phase remained unmet by the existing compounds.[2] We have examined the roles of barium coordination number, intermolecular stabilization, intramolecular stabilization, and ligand primary coordination number on each of these two key variables in CVD. The classes of compounds examined include the cyclopentadienides, the β-diketonates, and the alkoxides. Each issue is addressed separately for the bulk of the compounds reported to date.

RESULTS AND DISCUSSION

A comparison of a number of compounds reported to date is presented in **Table I**, with a categorical breakdown of some of the properties of merit for use in CVD processes. One conclusion which emerges quickly from an examination of these entries is the lack of a single precursor which fulfills all the requirements with equal degrees of success. The key issue of vapor pressure is addressed first. In order to maintain a constant flux of chemical delivery by the carrier gas to the growing substrate surface, the condition of saturation equilibrium must be uniformly maintained through growth run time. This is met best in the distillation of a *liquid* source by bubbling the carrier gas through the heated compound. The alternate source states, gases or solids at ambient conditions, have different parameters of concern. To date, no stable, ambient condition gaseous compounds have been reported for barium. Thus, this point is reduced to the problems of solids. They must be sublimed by *percolating* the carrier gas through a heated sample of the compound prior to its introduction into the growth manifold. Such processes of vaporization are no longer equilibrium controlled, and, therefore depend heavily on such variables as particle size (in escence, surface area), long term solid state stability, and partial decomposition/selective volatilization. As a consequence of these difficulties, the employment of liquid sources is *highly* preferable to solids.

As stated, the CVD uses of greatest interest for these precursors are in the formation of metal oxide compositions. Thus, the requirement for oxidizer compatibility is obvious. One of the most volatile compounds reported to date is $Ba(fod)_2$ (fod = 1,1,1,2,2,3,3-heptafluoro-7,7-dimethyloctane-4,6-dionato). This precursor does not exhibit vapor phase oxidative instability; however, it does decompose to BaF_2 upon thermolysis.[1] There is not yet a consensus about the benefits/disadvantages of this primary reaction product. A second, post-deposition H_2O wash, step is required to convert the initial deposit into an oxide structure. Likewise, $Ba(hfac)_2$, $Ba(tfac)_2$, and their tetraglyme adducts also deposit BaF_2 upon thermal decomposition (hfac = 1,1,1,5,5,5-hexafluoropentane-2,4-dionato; tfac = 1,1,1-trifluoropentane-2,4-dionato; tetraglyme = $MeO(CH_2CH_2O)_4Me$).[3]

TABLE I

Compound	Ambient State[a]	Oxygen Sensitivity[b]	Vapor pressure[c]	Stability[d] Condensed	Stability[d] Vapor	Reference
$Ba(C_5H_5)_2$	s	h	~0	– –	?	8
$Ba(C_5Me_5)_2$	s	h	$10^{-4}/200$	–	?	8
$Ba[C_5H(iPr)_4]_2$	s	h	$10^{-3}/100$	√	√	9
$[Ba(C_5Me_5)_2]_2$•Pyrazine	s	h	$10^{-3}/100$	√	√	10
$Ba(C_5H_4CH_2CH_2OMe)_2$	s	h	$10^{-3}/100$	√	√	5
$Ba(fod)_2$	s	l	10/150	√	– –	1
$Ba(acac)_2$	s	l	~0	– –	?	1
$[Ba(tmhd)_2]_{3,4}$	s	l	3/250	– –	√	6
$Ba_5(tmhd)_9(OH)$•$3H_2O$	s	l	5/230	– – –	√	11
$[Ba(tmhd)_2$•$2NH_3]_2$	s	l	5/170	–	√	6
$Ba(tmhd)_2(MeOH)_4$	s	l	~0	– – –	?	15
$Ba(dmmod)_2$	l	l	100/85	√	?	7
$Ba(OMe)_2$	s	l	~0	– –	?	12
$Ba[OCH_2CH_2N(CH_2CH_2OH)_2]_2$•2EtOH	s	l	~0	– – –	?	13
$Ba[O(2,6\text{-}(t\text{-}Bu)_2\text{-}4\text{-}Me\text{-}C_6H_2)]_2$•4THF	s	l	~0	– –	?	14
$HBa_5(O)(OPh)_9(THF)_8$	s	l	~0	– –	?	6
$H_3Ba_6(O)(OtBu)_{11}(OCEt_2CH_2O)(THF)_3$	s	l	~0	– –	?	16
$Ba[O(2,6\text{-}(tBu)_2\text{-}C_6H_3)]_2(HOCH_2CH_2NMe_2)_4$	s	l	~0	– – –	?	17
$Ba_3(OSiPh_3)_6(THF)_{1.5}$	s	l	~0	– –	?	18
$Ba[O(CH_2CH_2O)_2Me]_2$	l	l	$10^{-3}/200$	√	– –	4

Footnotes for Table I

[a] s = solid, l = liquid

[b] h = high, l = low

[c] in mmHg/°C

[d] √ = satisfactory, – = uncertain, – – = unsatisfactory, – – – = very unsatisfactory, ? = unknown

The issue of stability must be subdivided into the vapor phase stability, defined as the ability to recover the starting compound unchanged following vapor transport, and the condensed phase stability of the precursor within the evaporator, either as a liquid or a solid. Numerous precursors contain reactive protic sites and therefore receive low marks in the latter category. Several other compounds which exhibit insufficient vapor pressures (<10^{-3} mm Hg/250°C) to be seriously considered as CVD sources, also have been included for comparison. The component issue of vapor phase stability is particularly troubling in the case of $Ba[O(CH_2CH_2O)_nR]_2$ (**Figure 1**).[4] These ambient temperature liquid compositions have not yet proven themselves to be sufficiently robust for use in CVD.

The first class of compounds examined were the cyclopentadienide complexes of barium.[5] Although incorporation of pendant ether or amine tails onto the ring stabilized the metal center, these compounds proved incompatible for use with CVD of oxide based ceramics (**Figure 2**). In fact, unsurprisingly they are pyrophoric.

Next, the topic of intermolecular versus intramolecular stabilization of Ba(β-diketonato)$_2$ complexes was examined. An ORTEP representation of the structure of $[Ba(tmhd)_2{\bullet}2NH_3]_2$, as determined by single crystal x-ray diffraction, is depicted in **Figure 3**.[6] The parent ligand utilized for the formation of Ba(1,1-dimethyl-8-methoxyoctane-3,5-dionato)$_2$ [Ba(dmmod)$_2$] is represented in **Figure 4**.[7] The coordination number of barium in the dimeric structure is eight, in that for Ba(dmmod)$_2$ nominally it is six. At this time, structural data for Ba(dmmod)$_2$ in the *solid state* are unavailable, due to its ambient condition liquid state. However, in benzene solution it is a monomeric species by cryoscopy.

SUMMARY

The progress in barium chemistry has picked up significantly in the past two years. Much has been uncovered related to the demand to fulfill the metal atom's coordination sphere with 8-10 atoms in order to promote vapor pressure enhancement, decrease solid state cross-catenation, increase solubility in apolar solvents, and increase the stability of the molecules, both in the vapor phase and in the condensed phase. The goals of "ultimate" CVD sources have not yet been realized by *any one* of these new compounds. Therefore, the area still requires significant work. Although the potential exists for Ba(dmmod)$_2$ to be an attractive alternate source to Ba(fod)$_2$, we have not yet grown thin films from it. The utility of $[Ba(tmhd)_2{\bullet}2NH_3]_2$ in thin film growth is comparable to that of $[Ba(tmhd)_2]_{3,4}$.[6]

Figure 1: Proposed structure of $Ba[O(CH_2CH_2O)_3Me]_2$.

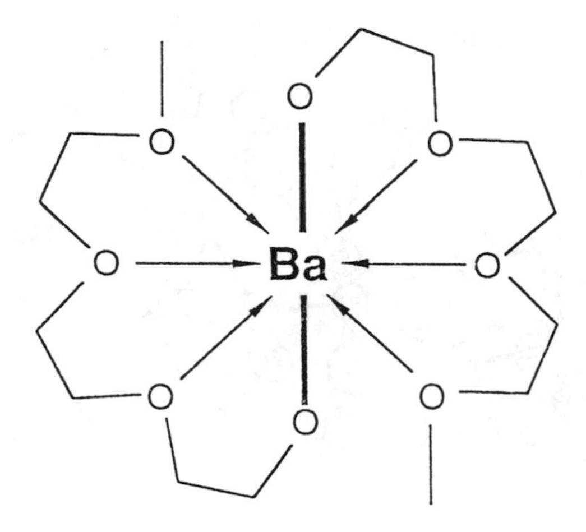

Figure 2: Proposed structure of $Ba(C_5H_4CH_2CH_2OMe)_2$.

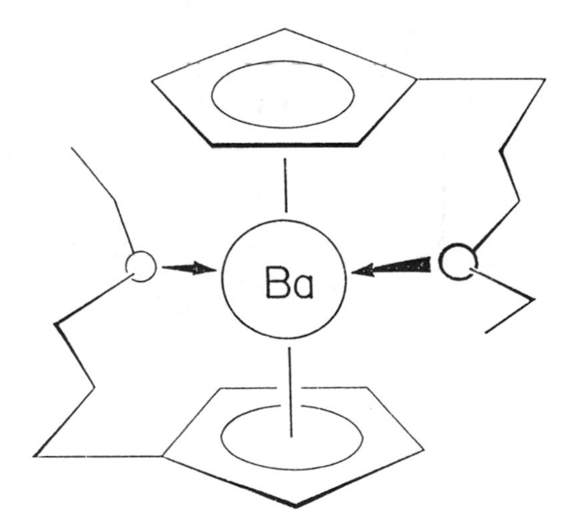

Figure 3: ORTEP representation of $[Ba(tmhd)_2 \cdot 2NH_3]_2$ with thermal elipsoids drawn at the 50% probability level and hydrogen atoms and methyl groups omitted for clarity of viewing.

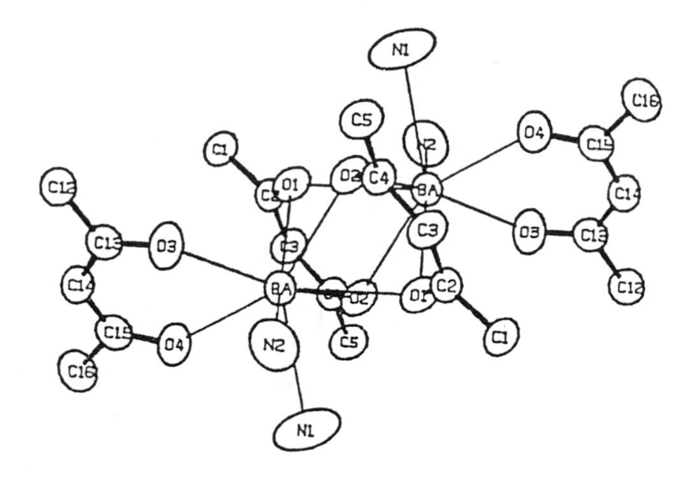

Figure 4: Structural representation of Hdmmod.

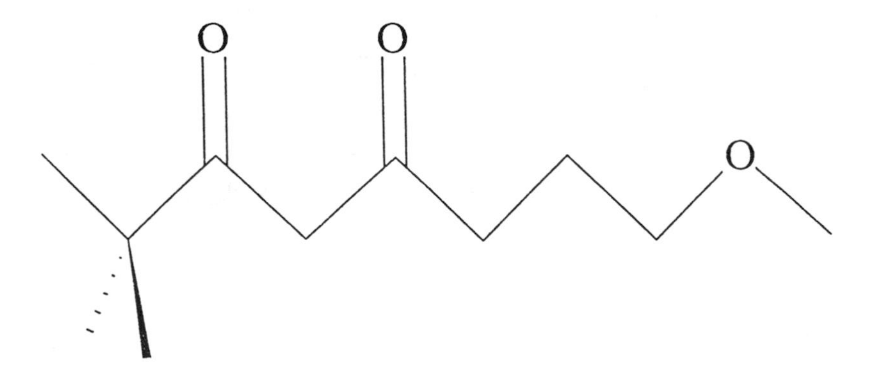

ACKNOWLEDGEMENTS

We gratefully acknowledge the generous financial support provided to this project by DARPA contract number MDA 972-88-J-1006. Both K. A. D. and W. H. acknowledge the Deutsche Forschgungsgemeinschaft for postdoctoral fellowships. We gratefully appreciate Mr. Dick Roche's contribution to the illustrations. Professor T. P. Hanusa (Vanderbilt University) graciously shared a preprint of reference 10 with us.

REFERENCES

1. a. L. M. Tonge, D. S. Richeson, T. J. Marks, J. Zhao, J. Zhang, B. W. Wessels, H. O. Marcy, and C. R. Kannewurf, in *Electron Transfer in Biology and the Solid State: Inorganic Compounds With Unusual Properties, Part III*, M. K. Johnson, R. B. King, D. M. Kurtz, Jr., C. Kutal, M. L. Norton, and R. A. Scott, Eds., ACS Advances in Chemistry Series 226; American Chemical Society: Washington, D.C., **1990**; pp 351-368.
   b. W. S. Rees, Jr. and A. R. Barron, manuscript in preparation.
2. Stringfellow book
3. a. R. Gardiner, D. W. Brown, and P. S. Kerlin, *Chem. Mater.*, **1991**, *3*, 1053-1059.
   b. A. P. Purdy, A. D. Berry, R. J. Holm, M. Fatemi, and K. K. Gaskill, *Inorg. Chem.*, **1989**, *28*, 2779.
   c. C. I. M. A. Spee and A. Macker, in *Science and Technology of Thin Film Superconductors*, R. D. McConnell and S. A. Wolfe, Eds., Plenum: New York, **1989**, pp. 281-294.
4. W. S. Rees, Jr. and D. A. Moreno, *J. Chem. Soc., Chem. Commun.*, **1991**, 1759.
5. a. W. S. Rees, Jr. and K. A. Dippel, in *Ultrastructure Processing of Ceramics, Glasses, Composites, Ordered Polymers, and Advanced Optical Materials V*, L. L. Hench, J. K. West, and D. R. Ulrich, Eds.; Wiley: New York, **1992**, in press.
   b. W. S. Rees, Jr. and K. A. Dipple, *OPPI*, accepted for publication.
6. W. S. Rees, Jr., M. W. Carris, and W. Hesse, *Inorg. Chem.*, **1991**, *30*, 4479.
7. W. S. Rees, Jr., C. R. Caballero, and W. Hesse, *Angew. Chem.*, **1992**, in press.
8. C. J. Burns and R. A. Andersen, *J. Organomet. Chem.*, **1987**, *325*, 31-37.
9. R. A. Williams, T. P. Hanusa, and J. C. Huffman, *J. Am. Chem. Soc.*, **1990**, *112*, 2454-2455.
10. T. P. Hanusa, unpublished result, personal communication.

11. S. B. Turnipseed, R. M. Barkley, R. E. Sievers, *Inorg. Chem.*, **1991**, *30*, 1164.
12. D. C. Bradley, R. C. Mehrotra, and D. P. Gaur, *Metal Alkoxides*, Academic Press: New York, **1978**, p. 50.
13. O. Poncelet, L. G. Hubert-Pfalzgraf, L. Toupet, and J. C. Parson, *Polyhedron*, **1991**, *10*, 2045-2050.
14. P. B. Hitchcock, M. F. Lappert, G. A. Lawless, and B. Royo, *J. Chem. Soc., Chem. Commun.*, **1990**, 1141.
15. A. Gleizes, S. Sans-Lenain, D. Médus, and R. Morancho, *C. R. Acad. Sci., Paris*, **1991**, *312, II*, 983-988.
16. K. G. Caulton, M. H. Chisholm, S. R. Drake, and K. Folting, *J. Chem. Soc., Chem. Commun.*, **1990**, 1349-1351.
17. K. G. Caulton, M. H. Chisholm, S. R. Drake, and K. Folting, *Inorg. Chem.*, **1991**, *30*, 1500-1503.
18. K. G. Caulton, M. H. Chisholm, S. R. Drake, and W. E. Streib, *Angew. Chem.*, **1990**, *102*, 1492-1493.

STABILIZATION OF METAL ALKOXIDES (M = BA, CU) BY ALKANOLAMINES

L.G. HUBERT-PFALZGRAF*, M.C. MASSIANI*, J.C. DARAN** AND J. VAISSERMANN**
*Laboratoire de Chimie Moléculaire, URA CNRS, Université de Nice-Sophia Antipolis, BP 71, 06108 Nice Cédex 2, France
**Laboratoire de Métaux de Transition, URA CNRS, 75230 Paris, France.

ABSTRACT

Alcohol interchange reactions between barium and copper(II) methoxides and triethanolamine give $Ba(teaH\text{-}1)_2$,2EtOH and $[Cu(teaH\text{-}2)]_4$,3tea, respectively. Triethanolamine appears as a tetradentate ligand - chelating in the case of barium, bridging-chelating for copper - and thus ensures high coordination numbers for the metals - eight for Ba and five for Cu - thus decreasing their susceptibility to hydrolysis. The remaining hydroxyl functionality on the coordinated triethanolamine moities favors the formation of solvates with alcohols via H-bonding.

INTRODUCTION

Metal alkoxides $M(OR)_n$ are versatile and thus attractive precursors of oxides, especially for sol-gel processing [1]. However, their practical use is often hampered by their high reactivity towards moisture, leading to difficulty in controlling their hydrolysis. Various additives, also called modifiers or stabilizers, have been used in order to improve the shortcomings of conventional sol-gel methods. Nitrogen donors such as ethylenediamine [2], isopropylamine [3] and alkanolamines (di- or triethanolamine, $OH(C_2H_4)_{x+1}NR'_{2-x}$ (x = 1, 2) [4] are among the most used stabilizers. The latter are also a means of suppressing the precipitation of oxides from the alcoholic solution of metal alkoxides during hydrolysis, and thus providing a rheology appropriate for coating applications [5]. A titanium alkoxide-triethanolamine compound has also been used for the preparation of titanium nitride by pyrolysis. However, the structures of the modified alkoxides have yet to be determined.

The use of classical alkoxides of metals barium and copper is often precluded by the poor stability of barium derivatives (with respect to oxidation and thus to formation of carbon-deficient compounds [6]) and by the unfavorable solubility properties of the copper(II) alkoxides [7]. By contrast with 2-methoxyethanol [8], aminoalcohols could achieve solubilization of Cu(II) [9].

The paper reports the synthesis and characterization of barium and copper(II) alkoxides modified by triethanolamine. Alcohol interchange reactions offered $Ba(teaH\text{-}1)_2$,2EtOH (1) and $[Cu(teaH\text{-}2)]_4$,3tea (2) respectively and both compounds have been structurally characterized by X-ray diffraction.

EXPERIMENTAL

All syntheses and characterizations were carried out as previously reported. X-ray data collection was achieved on a CAD-4 automatic diffractometer using the MoKα_1 radiation. $[Cu(OMe)_2]_\infty$ was obtained according to the literature from copper chloride and lithium methoxide [7], but was shown to be contamined by chlorine and lithium residues.

The barium derivative has been obtained as previously reported [10] and monocrystals of **1** were grown in ethanol.

The copper-triethanolamine derivative was obtained by adding excess triethanolamine to a suspension of $[Cu(OMe)_2]_\infty$ in tetrahydrofurane. After 24 h an insoluble compound (containing chlorine) was eliminated by filtration and **2** was obtained as the second crop by concentration (50%). Crystals suitable for X-ray diffraction were obtained directly from the reaction medium for **2**.

RESULTS AND DISCUSSION

Alcoholysis reactions of the poorly soluble barium or Cu(II) methoxides with excess triethanolamine $N(C_2H_4OH)_3$ (tea) proceed at room temperature -although that the alcohol interchange reactions are much slower than with 2-methoxyethanol-, giving soluble products. Owing to the electropositive character of barium, the same product could also be obtained by directly reacting the metal and triethanolamine in refluxing toluene:THF over 24 h. In addition to the solubility change, the modification of the metal coordination sphere is also shown by the evolution of the infrared spectra especially in the region 1200-400 cm^{-1}.

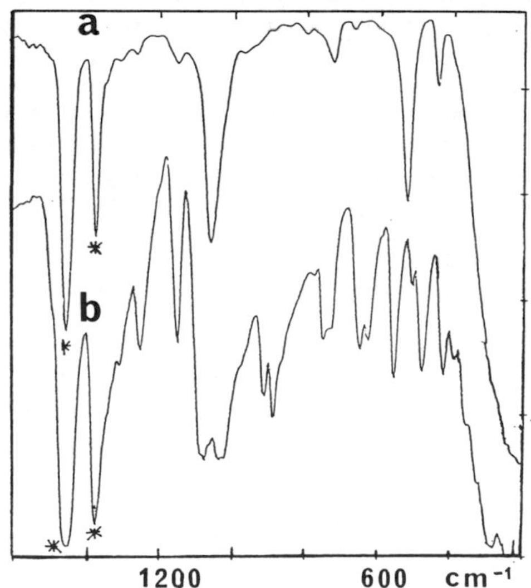

Fig. 1: Infrared spectra in nujol mulls (*): a of copper methoxide, b of **2**

Fig. 1 illustrates that in the case of copper: **1** is characterized by Cu-OR absorption bands at 552, 501, 474, 417 and 386 cm^{-1} while the methoxide has only Cu-OR absorption bands at 525 and 440 cm^{-1}. The barium and the copper(II) derivatives are crystalline products which are both isolated as solvates, as established by X-ray diffraction. Their formulae are $Ba(teaH_{-1})_2,2EtOH$ (**1**) and $[Cu(teaH_{-2})_2]_4,3tea$ (**2**), respectively [11].

Although the barium and copper derivatives are both homoleptic triethanolamine derivatives of divalent metals, their structures are quite different, mainly as a result of the notable difference in size between the two metals (covalent radii 1.17 Å for copper and 1.98 Å for barium) and thus in their coordination number and thus stereochemistry.

Barium generally requires a large coordination number, and in 1 triethanolamine acts as a tetradentate chelating ligand encapsulating the metal which is thus octacoordinated with Ba-O(H) bond lengths ranging from 2.73(1) to 2.78(1) Å and a transannular Ba-N coordination bond of 3.009(3) Å (Fig. 2). While oxoalkoxide have generally been isolated for barium [6], this functional alcohol is able to stabilize barium as a non-oxoalkoxide and the product becomes easy to handle and less sensitive to the atmosphere.

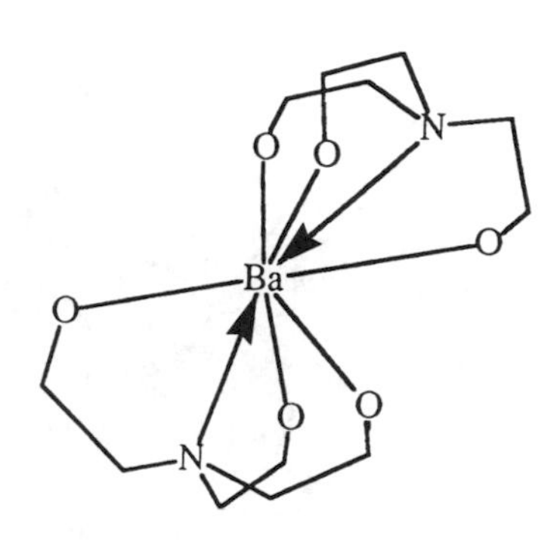

Fig. 2: Molecular structure of the Ba(teaH-1)$_2$ unit. The hydrogen of the hydroxyl groups could not be located.

2 is a paramagnetic compound (ESR in the solid g = 2.43, A = 56.8 G), slightly soluble in toluene, more soluble in THF and in alcohols. It appears in the solid as a centrosymmetric tetramer, solvated by triethanolamine. Two independent molecules of such tetramers are found in the unit cell (monoclinic, space group $P2_1/a$, a = 28.915(4), b = 11.473(7), c = 30.341(9) Å, ß = 116.65(2)°, Z = 4)

Figure 3 represents the molecular structure of one of the tetrameric unit of [Cu(teaH-2)]$_4$,3tea which is based on a eight membered ring. Selected bond lengths and angles are collected in table 1. The four copper atoms, form a regular flattened tetrahedron (Cu---Cu 3.14-3.25 Å, Cu distances + 0.7Å from the average plane), they are connected via one oxygen of the triethanolamine moiety which displays a tetradentate chelating-bridging coordination mode. The geometry around each copper is a distorted square pyramid resulting from one nitrogen and four oxygen atoms. The Cu-O(OH) bond lengths range from 1.85(3) Å to 2.55(3) Å. The apical bonds are much longer than the basal ones, this might be due to Jahn-Teller distorsion as usually observed for 5-coordinate copper(II) and/or to the existence of a Cu-OH coordination bond (the hydrogen atoms could not be localized). These values are in agreement with those observed for [Cu(OC_2H_4OiPr)(acac)]$_4$ [12]. Although the X-ray data are rather poor for **2**, the Cu-O-Cu bridges appear slightly dissymetrical as also observed for Al_2Me_3(teaH-1)$_3$ [13]. The transannular Cu-N bonds lenghts in **2** vary from 1.89(6) to 2.09(6) Å and compare well with the values observed for the Cu-N bonds (basal) in monomeric Cu(II) aminoalkoxides Cu(OR)$_2$ R = $CHMeCH_2NMe_2$

or $C_2H_4NMeC_2H_4NMe_2$ [9], which are among the few structurally characterized and soluble copper(II) alkoxides [14]. Tetrameric derivatives whose structure is based on an eight-membered ring are a common feature of copper(I) chemistry, especially alkoxides or silylamides [15], while cubane-like structures are found for Cu_4O_4 cores based on Cu(II) [12]. The constrained triethanolamine ligand -due to the pyramidal nitrogen atom- appears to favor an eight-membered Cu_4O_4 ring.

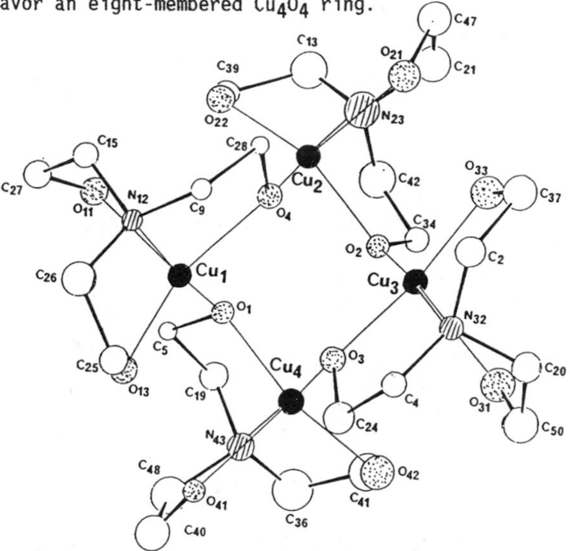

Fig. 3. Molecular structure of one of the $[Cu(teaH\text{-}2)]_4$ units.

Table 1: Selected bond lengths and angles of $[Cu(teaH\text{-}2)]_4$

Bond Lengths (Å)	
Cu1 - O1	1.92(3)
Cu1-N12	2.04(4)
Cu1-O13	2.00(5)
Cu1-O4	2.00(3)
Cu1-O11	2.42(4)
Cu2-O2	2.02(4)
Cu2-O4	1.85(3)
Cu2-N23	1.89(6)
Cu2-O22	1.88(4)
Cu3-O33	1.91(5)
Cu3-O31	2.40(4)
Cu3-O2	1.85(4)
Cu3-N32	2.09(5)
Cu4-O41	2.47(4)
Cu4-O1	1.95(4)
Cu4-O42	2.00(5)
Cu4-N43	2.09(6)

Angles at Cu1 (°)	
N12-Cu1-O1	172.1(17)
O11-Cu1-O1	91.5(15)
O11-Cu1N12	80.8(16)
O13-Cu1-O4	150.1(16)
O13-Cu1-O11	110.8(16)
O4-Cu1-O1	92.2(14)
N12-Cu1-O4	87.4(16)
O11-Cu1-O4	96.5(14)
O13-Cu1-O1	98.5(17)
O13-Cu1-N12	85.5(18)
angles in the ring	
Cu4-O1-Cu1	114.0(16)
Cu1-O4-Cu2	113.8(16)
O4-Cu2-O2	91.2(15)
Cu2-O2--Cu3	112.8(16)
O2-Cu3-O3	94.6(14)
Cu3-O3-Cu4	121.2(16)
O3-Cu4-O1	94.0(14)

A comparable coordination mode behaviour has been observed for $[Ti(teaH\text{-}3)(O^iPr)]_2$ **(3)**, which has been obtained by adding one equivalent of triethanolamine to $Ti(O^iPr)_4$. The higher metal's oxidation state in **3** allows triethanolamine to act as trialcohol with loss of the H of all the hydroxyl functionalities [16].

Both compounds **1** and **2** crystallize as solvates with tea or other alcohol molecules. The existence of solvates is evidenced by the presence of a νOH absorption band in the infrared spectra: 3130 cm^{-1} for **1** and 3100 cm^{-1} for **2**. The broadness of the band is indicative of H-bonding. This was confirmed by X-ray diffraction for **1**, while disorder phenomena associated with a low data to parameters ratio precluded the localisation of the additional tea molecules for **2**. In fact, intermolecular hydrogen bonding (short contacts O--O of 2.54 Å) between the alcohol functionalities on the coordinated triethanolamine and the lattice ethanol molecule is responsible for the formation of an extensive two-dimensional network for **1**, as shown in fig. 4.

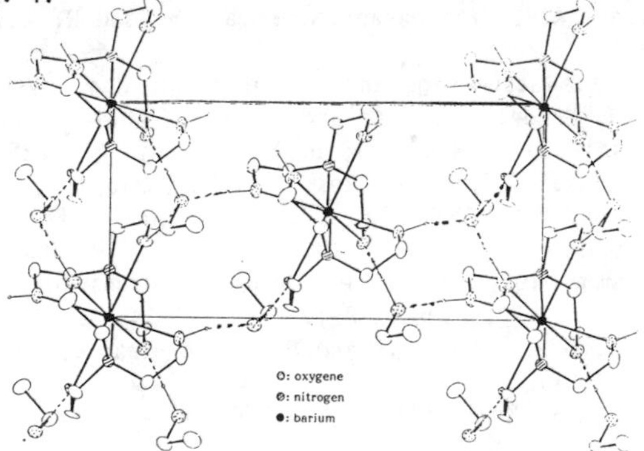

Fig. 4: Two-dimensional hydrogen-bonding network of $Ba(teaH\text{-}1)_2$,2EtOH

Formation of alcohol alkoxide adducts stabilized by intra- or intermolecular H-bonding is a common feature in alkoxide chemistry and this might modify their reactivity [17]. By contrast with most barium alkoxides, which precipitate by hydrolysis, **1** forms stable gels in EtOH by the addition of excess water. The formation of the gel network is probably assisted by H-bonding, as observed in the solid state. The hydrolysis of **2** in EtOH or THF leads to homogeneous solutions.

Triethanolamine is, however, less versatile than 2-methoxyethanol in its coordination modes; only tetradente coordination ones have been characterized. The M-tea bonds are also less labile, and the formation of heterometallic species is less favored than for classical alkoxides or even 2-methoxyethoxides.

CONCLUSION

The polyfunctional alcohol, triethanolamine, is able to achieve depolymerization, and thus solubilization, of insoluble alkoxides such as copper(II) or barium methoxides at room temperature, giving solvated atranes [17] $[Cu(teaH\text{-}2)]_4$,3tea and $Ba(teaH\text{-}1)_2$,2EtOH, respectively. As a tetradentate chelating and/or chelating-bridging ligand, it can lead to high coordination numbers to the metal and thus ensure stabilization toward uncontrolled hydrolysis. The formation of solvates with alcohols, based on extensive intermolecular H-bonding, is favorable for the obtaining of a rheology appropriate to coating applications as well as to the formation of a gel network.

ACKNOWLEDGEMENTS

We thank the CNRS for financial support.

REFERENCES

1. L.G. Hubert-Pfalzgraf, New J. Chem. 11, 663 (1987).
2. S. Katayama and M. Sekine, Better Ceramics through Chemistry IV, 1, 897 (1990); J. Mater. Chem. 1, 1031 (1991).
3. U. Selvarag, A.V. Prasadarao, S. Komarneni and R. Roy, Mater. Let. 12, 311 (1991).
4. N. Tohge, M. Tatsumisago and T. Minam, J. Non Cryst. Solids 121, 443 (1990). T. Monde, H. Kosuka and S. Sakka, Chem. Lett. 287 (1988).
5. Y. Takahashi and Y. Matsuoka, J. Mater. Sci. 23, 2259 (1988). M. Ocana and E. Matijevic, J. Mater. Res. 5, 1083 (1990).
6. K.G. Caulton, M.H. Chisholm, S.R. Drake and K. Folting, J. Chem. Soc. Chem. Commun. 1990, 1349; K.F. Tesh and T.P. Hanusa, J. Chem. Soc. Chem. Commun. 1991, 879. T.P. Thule, S. Raghavan and S.H. Risbud, J. Am. Ceram. Soc. 70, C108 (1987).
7. J.V. Singh, B.P. Baranwal and R.C. Mehrotra, Z. Anorg. Allgem. Chem. 477, 235 (1981).
8. S.C. Goel, K.S. Kramer, P.C. Gibbons and W.E. Buhro, Inorg. Chem. 28, 3620 (1989).
9. S. Goel, K.S. Kramer, M.Y. Chiang and W.E. Buhro, Polyhedron 9, 61 (1990).
10. O. Poncelet, L.G. Hubert-Pfalzgraf, L. Toupet and J.C. Daran, Polyhedron 10, 2045 (1991).
11. teaH-n n = 1, 2, 3 indicates the number of hydroxyl functionalities involved in covalent bonding with the metal.
12. C. Sirio, O. Poncelet, L.G. Hubert-Pfalzgraf, J.C. Daran and J. Vaissermann, Polyhedron 11, 177 (1992).
13. M.D. Healy and A.R. Barron, J. Am. Chem. Soc. 111, 398 (1989).
14. S. Wang, Inorg. Chem. 30, 2252 (1991).
15. H. Chen, M.M. Olmstead, S.C. Shoner and P.P. Power, J. Chem. Soc. Dalton Trans 1992, 451.
16 W.M.P.B. Menge and J.G. Verkade, Inorg. Chem. 30, 4628 (1991); A.A. Naiini, W.M.P.B. Menge and J.G. Verkade, Inorg. Chem. 30, 5009 (1991).
17. B. Vaartstra, J.C. Huffman, P.S. Gradeff, L.G. Hubert-Pfalzgraf, J.C. Daran, S. Parraud, K. Yunlu and K.G. Caulton, Inorg. Chem. 29, 3126 (1990).
18. F. Kober, Zeit. Chem. 20, 49 (1980); T.K. Gar and V.F. Mirovov, Organomet. USSR, 1, 142 (1988).

SOLID STATE STRUCTURES, DECOMPOSITION PATHWAYS, AND VAPOR PHASE BY-PRODUCTS OF Y(acac)$_3$ TYPE OMVPE PRECURSORS FOR THIN FILMS OF YTTRIUM - CONTAINING CERAMIC MATERIALS

William S. Rees, Jr.,* Henry A. Luten, Michael W. Carris, Eric J. Doskocil, and Virgil L. Goedken
Department of Chemistry and Materials Research and Technology Center, Florida State University, Tallahassee, FL.

ABSTRACT

Several useful ceramic materials target compositions contain yttrium, Y_2O_3 and $YBa_2Cu_3O_{7-\delta}$ being the two most widely employed. One known CVD precursor for yttrium-containing thin films is $Y(tmhd)_3$ $\cdot H_2O$ (tmhd = 2,2,6,6-tetramethylheptane-3,5-dionato). We have determined the structure and examined the vapor phase decomposition of this species. A related compound, $[Y(tmod)_3]_2$ (tmod = 2,2,7-trimethyloctane-3,5-dionato), has been prepared, structurally characterized, and studied as an organometallic vapor phase epitaxy (OMVPE) precursor for Y and Y_2O_3 films. Mechanisms of vapor phase decomposition are discussed in terms of solid film deposits and vapor phase by-products. The two precursors are compared to each other with respect to their stability windows, defined as being bound by source volatility on the low side and source stability on the high side.

RESULTS AND DISCUSSION

As one component of an overall project directed at the discovery of a self-commensurate set of precursor chemicals for use in OMVPE growth of $YBa_2Cu_3O_{7-\delta}$, we developed a need to examine the solid state structures, decomposition pathways, vapor phase by-products, and composition of thin films resulting from thermal decomposition of Y(acac)$_3$ type materials. Our recent work with Ba[1] and Cu[2] prompted us to compare the previously utilized "Y(tmhd)$_3$" with "Y(tmod)$_3$". The former has a symmetrical β-diketone ligand, whereas in the latter we have altered the symmetry of the bidentate ligand, the critical observation being to realize that *both compounds have identical empirical formulas* ($YO_6C_{33}H_{57}$). Work in this area from Sievers[3, 4] indicated that, although almost all M(acac)$_n$-type compounds examined were volatile to a certain degree, the greatest vapor pressure was exhibited by those compounds which either contained fluorinated and/or bulkly substituted ligands. The precise utility of fluorine presence in the OMVPE growth of barium-containing electronic materials is not yet determined. However, the demand for a second step, namely a post deposition wash, would appear to be at odds with the rigors of epitaxial growth.[5] Nevertheless, to avoid such a conflict, we chose to explore the alternate option. The well-understood utility of steric hindrance in inhibiting cross-catenation in the solid state structures of oxophillic metals possessing oxygen atom-containing ligands is well-established.[6] The precise role(s) of mono-anionic ligand symmetry, and the important consequences on the vapor pressure of the solids emanating from their complexation with metal cations is less well developed.

The observed increase in vapor pressure upon going from "Y(tmhd)$_3$" to "Y(tmod)$_3$" is depicted in **Fig. 1**. Not having a ready explanation for this behavior, and being unable to refer to the critical *solid state* structures of these two compounds, we turned to the only previously reported structural

 ©1992 Materials Research Society

information relevant to this problem, the GPED structure of $Y(tmhd)_3$.[7] The solid state structures were significantly more applicable to this situation — both precursors are employed at temperatures below their melting points, as solid sources, equilibrating with carrier gasses by percolating (sublimation), not bubbling (distillation). The initial hints of difference in these two species arose from determination of their solution molecular weights. That for "$Y(tmod)_3$" indicated the presence of a dimer, whereas that for "$Y(tmhd)_3$" was close to a monomeric value. Subsequent x-ray diffraction studies showed the correct compositions to be $Y(tmhd)_3 \cdot H_2O$ (**Fig. 3**)[8] and $[Y(tmod)_3]_2$ (**Fig. 4**).[9] The case of $Y(tmhd)_3 \cdot H_2O$ is rather similar to those for "$Ba(tmhd)_2$"[10] and "$Y(O\textit{i}Pr)_3$".[11] Although in the present case the H_2O ligand is innocent and only serving to fill the seventh coordination site on the rather large yttrium ion (radius = 0.893 Å), such is not the case for these other two examples. In the former the correct formulation was discovered to be $Ba_5(tmhd)_9OH \cdot 3H_2O$,[10] and in the latter it was $Y_5(O)(O\textit{i}Pr)_{13}$.[11]

Having the requisite increased vapor pressure source in hand, preliminary CVD experiments have been carried out with $[Y(tmod)_3]_2$. The reactor utilized for deposition is shown in **Fig. 2**, and the relevent data are summarized in **Table I**. At this point, it appears that Y is the initial deposit from all ambients, and it is oxidized subsequently to Y_2O_3 in those experiments utilizing O_2. In order to unambiguously rule out the role of coordinated H_2O in $Y(acac)_3$-type precursors, either advantageous or deleterious upon the CVD process, all the deposition experiments with $[Y(tmod)_3]_2$ also were carried out under hydrous conditions. No distinction could be ascertained between the two sets of experiments, thus, we conclude that, based on all the data available at the present time, H_2O should behave as an innocent ligand in $Y(tmhd)_3 \cdot H_2O$. The full details of each of the x-ray crystal structure determinations, the CVD experiments, and the vapor phase decomposition pathways for these two precursors will be examined in an upcoming publication.[12]

EXPERIMENTAL

$Y(tmhd)_3 \cdot H_2O$. The preparation and characterization of this compound has been described previously.[13]

$[Y(tmod)_3]_2$. To a 250 ml flask containing a magnetic stir bar, 10 g (0.054 mol) of Htmod,[3] and 40 ml of 95% ethanol, 3.1 g (0.055 mol) of KOH in 9 ml of distilled water were added slowly with rapid stirring. A solution of 6.9 g (0.018 mol) of $Y(NO_3)_3 \cdot 3.35\ H_2O$[13(c)] dissolved in 33 ml of distilled H_2O was added drop wise. Upon addition of this solution a white precipitate formed immediately. The mixture was allowed to stir for 12 hr, and then the majority of the ethanol was removed by rotary evaporation. To the remaining solution, 50 ml of distilled H_2O were added and the mixture was again allowed to stir for 12 hr. The crude $[Y(tmod)_3]_2$ was collected by gravity filtration and allowed to air dry. The resultant white solid was dissolved in hexane, dried over anhydrous Na_2SO_4 for 24 hr, gravity filtered, and the filtrate was evaporated to dryness yielding a white powder which was recrystallized from hexane to yield 7.7 g (0.006 mol, 66%) of crude $[Y(tmod)_3]_2$. An x-ray quality crystal was grown from a slowly cooled saturated hexane solution, an ORTEP representation of the determined structure is given in **Fig. 4**.[9] A sublimator equipped with a dry ice cold finger was charged with 7.7 g (0.006 moles) of crude $[Y(tmod)_3]_2$. A thin layer of glass wool was placed on top of the white powder. After evacuation to 1 x 10^{-5} mm Hg, the sublimator was heated to 245°C, the cold finger was cooled to -78°C, and the $[Y(tmod)_3]_2$ was allowed to sublime for 2 hr. The white solid was scraped off the cold finger to yield 6.8 g (0.005 mol, 89%), of purified $[Y(tmod)_3]_2$, based on the initial charge of $[Y(tmod)_3]_2$. Characterization Data: mp

Figure 1. TGA plots (10°C/min, N_2 atm, 45 - 600°C) for: [—] $Y(tmhd)_3 \cdot H_2O$, [----] $[Y(tmod)_3]_2$.

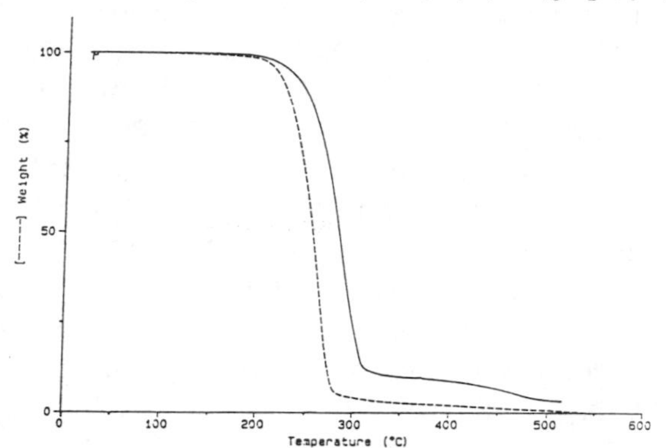

Figure 2. Schematic diagram of the hot-walled reactor used to deposit films from $[Y(tmod)_3]_2$. Key: Z1-Z3, independent temperature zones; S1-S3, substrates to be coated; BH, sublimator heater; PH, pre-heater for deposition gasses; MFC, mass flow controller. Flow through G1 was utilized to prevent premature decomposition of the source for those ambients with which it reacted quickly. G2 makeup was N_2.

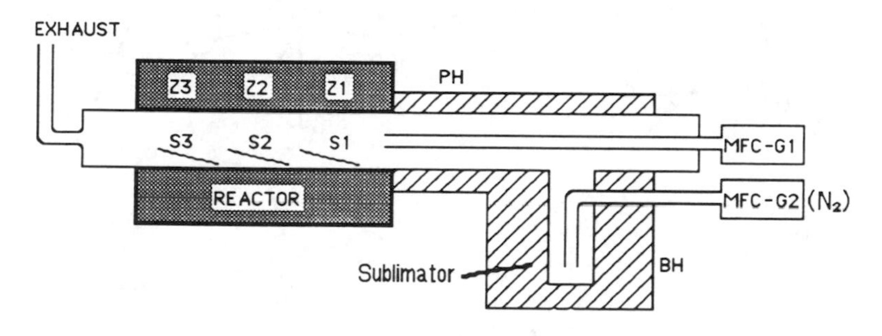

Table I. Effect of Deposition Atmosphere on Thin Film Deposit Composition for $[Y(tmod)_3]_2$.

Atmosphere (G1)	Substrate Temp. (°C)	Flow Rate (sccm) G1	Flow Rate (sccm) G2	Deposit Composition
N_2	480	--	200	Y
O_2	320	100	100	Y_2O_3
H_2	315	20	180	Y
$N_2 + H_2O$	580	--	200	Y
$O_2 + H_2O$	400	100	100	Y_2O_3
$H_2 + H_2O$	460	20	180	Y

Figure 3. (a) ORTEP plot of Y(tmhd)$_3$ • H_2O (50% probability). Hydrogen atoms are omitted for clarity. Y atom is striped, carbon atoms are represented by open circles, oxygen atoms are shaded and numbered. Pertinent interatomic distances [Å (esd)] and angles [deg. (esd)]: Y-O(1) = 2.35 (1), Y-O(2) = 2.24 (1), Y-O(3) = 2.25 (1), Y-O(4) = 2.26 (1), Y-O(5) = 2.22 (1), Y-O(6) = 2.32 (1), Y-O(7) = 2.35 (1); O(1)-Y-O(2) = 73.8 (4), O(1)-Y-O(3) = 128.3 (4), O(1)-Y-O(4) = 156.8 (5), O(1)-Y-O(5) = 89.7 (4), O(1)-Y-O(6) = 78.5 (5), O(1)-Y-O(7) = 80.1 (3), O(2)-Y-O(3) = 76.1 (4), O(2)-Y-O(4) = 126.1 (5), O(2)-Y-O(5) = 80.0 (3), O(2)-Y-O(6) = 142.3 (4), O(2)-Y-O(7) = 117.6 (4), O(3)-Y-O(4) = 72.7 (4), O(3)-Y-O(5) = 125.7 (4), O(3)-Y-O(6) = 141.6 (4), O(3)-Y-O(7) = 78.1 (4), O(4)-Y-O(5) = 83.2 (4), O(4)-Y-O(6) = 78.3 (5), O(4)-Y-O(7) = 97.5 (3), O(5)-Y-O(6) = 74.7 (4), O(5)-Y-O(7) = 155.5 (4), O(6)-Y-O(7) = 81.4 (3). (b) ORTEP plot (50% probability) showing H-bonding interactions between the coordinated H_2O of one Y(tmhd)$_3$ • H_2O and two oxygens of an adjacent molecule. Carbon atoms are omitted for clarity. Pertinent interatomic distances [Å (esd)]: O(7)-O(1′) = 2.71 (2), O(7)-O(6′) = 2.79 (2).

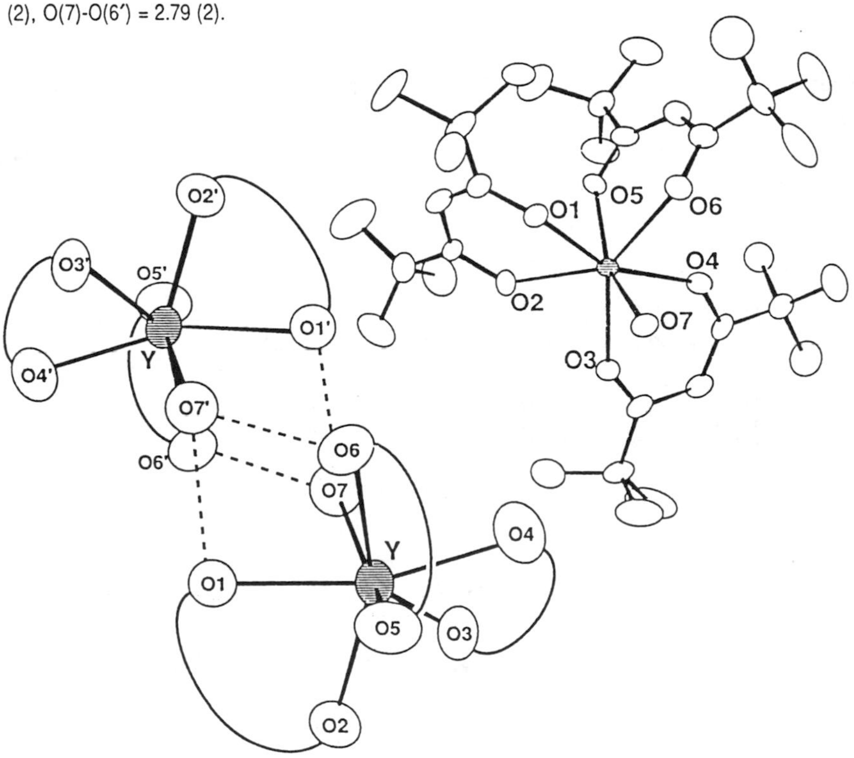

Figure 4. ORTEP plot of [Y(tmod)$_3$]$_2$ (50% probability). Hydrogen atoms are omitted for clarity. Y atoms are striped, carbon atoms are represented by open circles, oxygen atoms are shaded and numbered. Pertinent interatomic distances [Å (esd)] and angles [deg. (esd)]: Y-O(1) = 2.29 (1), Y-O(1′) = 2.42 (1), Y-O(2) = 2.28 (1), Y-O(3) = 2.26 (1), Y-O(4) = 2.27 (1), Y-O(5) = 2.23 (1), Y-O(6) = 2.25 (1); O(1)-Y-O(1′) = 73.8 (3), O(1)-Y-O(2) = 74.7 (3), O(1)-Y-O(3) = 89.9 (3), O(1)-Y-O(4) = 151.9 (3), O(1)-Y-O(5) = 126.6 (3), O(1)-Y-O(6) = 81.8 (3), O(1′)–Y-O(2) = 139.6 (3), O(1′)-Y-O(3) = 77.2 (3), O(1′)-Y-O(4) = 123.1 (3), O(1′)-Y-O(5) = 78.3 (3), O(1′)-Y-O(6) = 120.2 (3), O(2)-Y-O(3) = 78.2 (3), O(2)-Y-O(4) = 79.4 (3), O(2)-Y-O(5) = 141.8 (3), O(2)-Y-O(6) = 78.9 (3), O(3)-Y-O(4) = 74.3 (4), O(3)-Y-O(5) = 126.7 (4), O(3)-Y-O(6) = 157.0 (4), O(4)-Y-O(5) = 80.9 (4), O(4)-Y-O(6) = 103.7 (4), O(5)-Y-O(6) = 74.5 (4), Y-O(1)-Y′ = 106.3 (3). (b) ORTEP (50% probability) showing the Y atoms and oxygen core. Pertinent interatomic distances [Å (esd)]: Y-O(1) = 2.29 (1), Y-O(1′) = 2.42 (1).

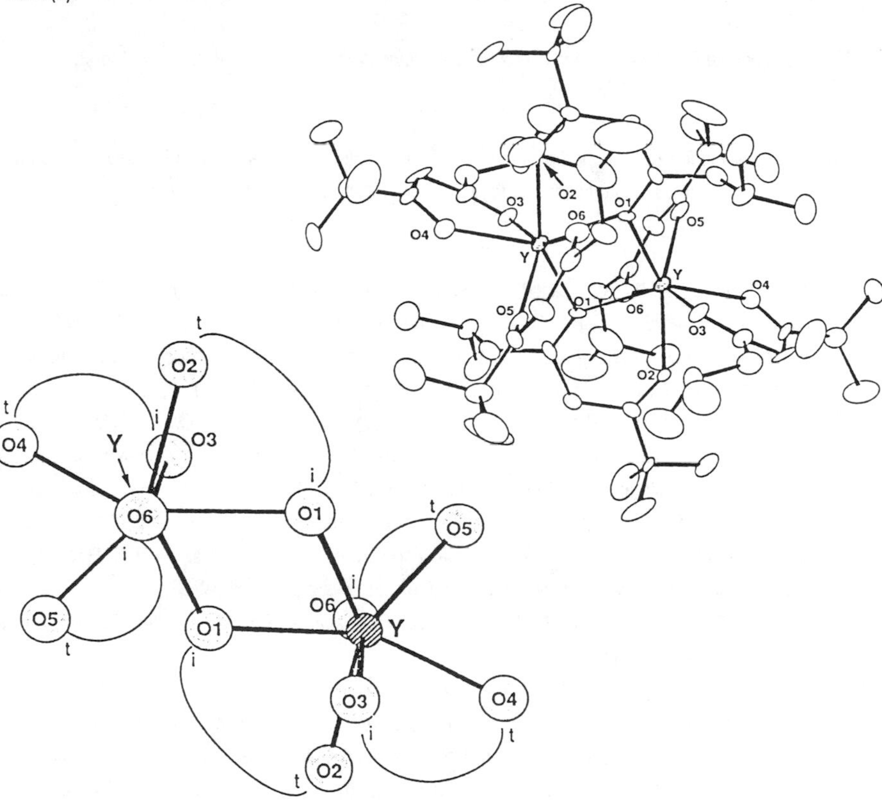

185.8°C. **MW:** (benzene cryoscopy) obs. 1,126 g/mol (calc. dimer: 1,277 g/mol). **MS:** (EI, 70 ev, m/e$^+$, (fragment); L = tmod, M = monomer parent ion) 638(M), 581(M-Bu), 455(M-L), 281($YO_4C_{10}H_8$), 207($YO_4C_4H_6$), 191($YO_3C_4H_6$), 127 (L-Bu), 96($C_6H_8O_2$), 73(OBu), 57(Bu), 41(C_3H_5). **^{1}H NMR:** (300 MHz, positive δ down field referenced to $Si(CH_3)_4$ = 0 ppm utilizing residual $CHCl_3$ = 7.24 ppm in solvent $CDCl_3$) 5.73 [3H, CH], 2.15 [73H, CH_2], 1.30 [3H, CH_3], 1.14 [24H, CH_3], 0.99 [7H, CH_3], 0.93 [19H, CH_3]. **^{13}C NMR:** (75 MHz, positive δ down field referenced to $Si(CH_3)_4$ = 0 utilizing $CDCl_3$ = 77.0 ppm; t, b designate the terminal and bridging end of the bridging ligand respectively) 200.4 [*t*Bu-CO], 198.6 [*t*Bu-CO_t], 194.0 [*t*Bu-CO_b], 192.2 [*t*Bu-CO], 97.0 [CH], 96.3 [CH_b], 50.2 [CH_{2b}], 50.0 [CH_2], 40.2 [ipso C], 39.8 [ipso C_t], 28.1 [($\mathbf{CH_3}$)$_3C_t$], 27.8 [($\mathbf{CH_3}$)$_3$C], 26.3 [CH], 22.7 [($\mathbf{CH_3}$)$_2$CH], 22.5 [($\mathbf{CH_3}$)$_2CH_b$]. **solution IR:** (cm^{-1}, in CCl_4 *vs.* CCl_4 reference) 2948(vs), 2916(s), 2892(m), 2856(m), 1582(vs), 1566(vs), 1544(s), 1528(s), 1524(s), 1504(vs), 1504(vs), 1468(m), 1456(s), 1444(s), 1416(vs), 1396(vs), 1384(s), 1354(s), 1224(m), 1160(m). **UV/VIS:** (nm, hexane *vs.* hexane reference) λ_{max} = 287, c = 1.57 x 10^{-5} **M**, ε = 1.29 x 10^5 **M**$^{-1}$ cm^{-1}.

CVD. The experimental procedure has been reported previously.[2] The reactor is shown in **Fig. 2**.

ACKNOWLEGMENTS

We acknowledge the financial support provided for this project from DARPA contract No. MDA 972-88-J-1006, Mr. D. Roche for the illustrations, and Prof. R. Clark for the use of a TGA instrument.

REFERENCES

1. (a)W. S. Rees, Jr., M. W. Carris, and W. Hesse, *Inorg. Chem.*, **1991**, *30*, 4479. (b) W. S. Rees, Jr. and D. A. Moreno, *J. Chem. Soc., Chem. Commun.*, **1991**, 1759. (c) W. S. Rees, Jr., C. R. Caballero, and W. Hesse, *Angew. Chem.*, **1992**, in press. (d) W. S. Rees, Jr. and M. W. Carris, *Inorg. Syn.*, submitted for publication. (e) W. S. Rees, Jr., M. W. Carris, and W. Hesse, unpublished results. (f) W. S. Rees, Jr. and K. A. Dippel, in *Ultrastructure Processing of Ceramics, Glasses, Composites, Ordered Polymers, and Advanced Optical Materials V*, L. L. Hench, J. K. West, and D. R. Ulrich, Eds.; Wiley: New York, **1992**, in press. (g) W. S. Rees, Jr. and K. A. Dippel, *OPPI*, accepted for publication. (h) W. S. Rees, Jr. and K. A. Dippel, manuscript in preparation, to be submitted. (i) W. S. Rees, Jr., K. A. Dippel, M. W. Carris, C. R. Caballero, D. A. Moreno, and W. Hesse, accompanying manuscript, this volume. (j) W. S. Rees, Jr., Y. S. Hascicek, and L. R. Testardi, accompanying manuscript, this volume. (k) W. S. Rees, Jr. and D. A. Moreno, submitted for publication. (l) W. S. Rees, Jr., *Proceedings of the 4th Florida Microelectronics and Materials Conference*, in press.
2. This is a slight modification of our previous work with CVD of $Cu(acac)_2$ type precursors. See: (a) W. S. Rees, Jr. and C. R. Caballero, in *Chemical Vapor Deposition of Refractory Metals and Ceramics*, T. M. Besmann, B. M. Gallois, and J. Warren, Eds., *Materials Research Society Proceedings*; Materials Research Society: Pittsburgh, Pennsylvania, in press. (b) W. S. Rees, Jr. and C. R. Caballero, *Advanced Materials for Optics and Electronics*, **1992**, *1(1)*, in press. (c) W. S. Rees, Jr. and C. R. Caballero, manuscript in preparation, to be submitted.
3. T. J. Wenzel, E. J. Williams, R. E. Sievers, *Inorg. Synth., Vol. 23*, S. Kirschner, Ed., Wiley-Interscience: New York, **1985**, pp.144- 149.

4. For the preparation of RE(tmhd)$_n$, see: K. J. Eisentraut, R. E. Sievers,. *Inorg. Synth., Vol. 11*, W. L. Jolly, Ed., McGraw-Hill: New York, **1968**, pp. 94-98.
5. (a) L. M. Tonge, *et al.*, in *Electron Transfer in Biology and the Solid State: Inorganic Compounds With Unusual Properties, Part III*, M. K. Johnson, *et al.*, Eds., ACS Advances in Chemistry Series 226; American Chemical Society: Washington, D.C., **1990**; pp. 351-368. (b) W. S. Rees, Jr. and A. R. Barron, manuscript in preparation.
6. For a thorough compilation of structural data on metal acac complexes, see: R. C. Mehrotra, R. Bohra, D. P. Gaur, *Metal β-Diketonates and Allied Derivatives*, Academic: New York, **1978**, pp. 58-194.
7. S. Shibata, K. Iijima, and T. Inuzuka, *J. Molec. Structure*, **1986**, *144*, 181.
8. Complete details of the collection and refinement of diffraction data, and the structure solution will be published elsewhere. Crystallographic data: space group P1, a = 11.553 (3) Å, b = 14.579 (3) Å, c = 14.875 (5) Å, α = 60.45 (2)°, β = 63.10 (2)°, γ = 73.28 (2)°, V = 1937.0 (8) Å^3, Z = 2, ρ_c = 1.095 g-cm^{-3}, R = 7.0%, R_w = 8.8%, data collection at ambient temperature.
9. Complete details of the collection and refinement of diffraction data, and the structure solution will be published elsewhere. Crystallographic data: space group $P2_1/n$, a = 12.993 (3) Å, b = 15.400 (7) Å, c = 18.349 (6) Å, β = 91.50 (2)°, V = 3670 (3) Å^3, Z = 4, ρ_c = 1.156 g-cm^{-3}, R = 10.6%, R_w = 10.3%, data collection at ambient temperature.
10. S. B. Turnipseed, R. M. Barkley, R. E. Sievers, *Inorg. Chem.*, **1991**, *30*, 1164.
11. O. Poncelet, W. J. Sartain, L. G. Hubert-Pfalzgraf, K. Folting, and K. G. Caulton, *Inorg. Chem.*, **1989**, *28*, 263.
12. W. S. Rees, Jr., H. A. Luten, M. W. Carris, E. J. Doskocil, and V. L. Goedken, manuscript in preparation, to be submitted.
13. (a) W. S. Rees, Jr. and M. W. Carris, *Inorg. Syn.*, submitted for publication. (b) G. S. Hammond, D. C. Nonhebel, and C. S. Wu, *Inorg. Chem.*, **1963**, *2*, 73. (c) H. Bergman, in *Gmelin Handbuch der Anorganischen Chemie: Sc, Y, La, und Lanthanide, Teil C 2*, Springer-Verlang: New York, **1974**, pp. 233-234.

AGGREGATION AND HYDROLYSIS REACTIONS OF BISMUTH ALKOXIDES

KENTON H. WHITMIRE, CAROLYN M. JONES, MICHAEL D. BURKART, J. CHRIS HUTCHISON AND ANDREW L. MCKNIGHT
Rice University, Department of Chemistry, P.O. Box 1892, Houston, TX 77251

ABSTRACT

New bismuth alkoxide and oxo-alkoxide complexes have been prepared from the salt metathesis reaction of NaOR with $BiCl_3$. When R = $CH(CF_3)_2$, the product is $[Bi(\mu\text{-}OR)(OR)_2(THF)]_2$ Similar work with R = C_6F_5 has not yielded a simple alkoxide, but complexes of formulation $NaBi_3(\mu_3\text{-}O)(OR)_8(THF)$, $NaBi_4((\mu_3\text{-}O)_2(OR)_9(THF)_2$, $Na_2Bi_4(\mu_3\text{-}O)_2(OR)_{10}$ and $Bi_6(\mu_3\text{-}OR)(\mu_3\text{-}O)_4[\mu_3\text{-}OBi(OR)_4]_3$ have been observed. The reaction of $BiPh_3$ with HOC_6F_5, however, did produce the desired alkoxide which has been characterized as $[Bi(OR)_2(\mu\text{-}OR)(toluene)]_2$ and $[Bi(OR)_2(\mu\text{-}OR)(toluene)]_2 \cdot 2$ toluene. The reaction of this alkoxide with $NaOC_6F_5$ led to the production of $Bi_6(\mu_3\text{-}O)_2(\mu_4\text{-}O)(OR)_{12}$ and $NaBi_3(\mu_3\text{-}O)(OR)_8(THF)_3$. Reaction of $BiPh_3$ with HOC_6F_5 in THF led to the formation of $Bi_6(\mu_3\text{-}OR)(\mu_3\text{-}O)_4[\mu_3\text{-}OBi(OR)_4]_3(THF)_2$. Surprisingly the reaction of $BiEt_3$ and HOR (R = C_6F_5 or Ph) displaced only one Et group to give $[Et_2Bi(\mu\text{-}OR)]_\infty$ which exist as infinite chain polymers with alternating Bi-O-Bi backbones. These spiral chains form chiral helices in the crystal lattice.

SYNTHESES OF BISMUTH ALKOXIDES

The recent discovery of high T_C superconducting copper oxide phases containing bismuth, lead and thallium has led to the investigation of heavy main element alkoxides for use in Sol-gel and CVD syntheses [1-10]. The first homoleptic bismuth alkoxides were reported in the 1960's by Mehrotra and Rai [11] who found that for alkoxides with small R (e.g., methyl, ethyl, etc.) the compounds of formula $Bi(OR)_3$ were insoluble. A mixed alkyl-alkoxide complex had been reported earlier from the work of Calingaert, Soroos and Hnizda [12] where $BiEt_2(OEt)$ was proposed as an intermediate in the oxidation of $BiEt_3$. This complex was independently synthesized by the reaction of $BiEt_2Br$ with NaOEt. More recently soluble and volatile alkoxides have been reported. Evans, Ziller and Hain [13] reported the syntheses of $Bi(O^tBu)_3$ and $Bi(OC_6H_3Me_2\text{-}2,6)_3$. The structure of the latter showed it to be monomeric with a pyramidal bismuth atom. Matchett, Chiang and Buhro [14] at about the same time reported the structure of $Bi(OCH_2CH_2OMe)_3$ which is polymeric. This paper describes studies aimed at producing bismuth alkoxides with intermediate-sized R groups which were intended to allow but also control oligomerization processes.

Following the known salt metathesis methodology for the production of bismuth alkoxides, we found that reaction in tetrahydrofuran (THF) of $BiCl_3$ with 3 equivalents of $NaOCH(CF_3)_2$ (produced in situ from the reaction of the alcohol with NaH) led to the clean production of a trialkoxide of bismuth. The complex crystallizes very easily in large block-shaped chunks which were studied by single crystal X-ray diffraction and shown to have the dimeric structure $[Bi\{OCH(CF_3)_2\}_2\{\mu\text{-}OCH(CF_3)_2\}(THF)]_2$, **1**, (Fig. 1). The coordination geometry about bismuth may be described as square pyramidal. The bridging oxygen groups are asymmetric with Bi-O distances of 2.188(7)Å and 2.688(7)Å. The terminal Bi-O(alkoxide) distances are 2.147(8)Å and 2.088(9)Å. A molecule of tetrahydrofuran is also coordinated to each Bi atom with a Bi-O distance of 2.575(7)Å.

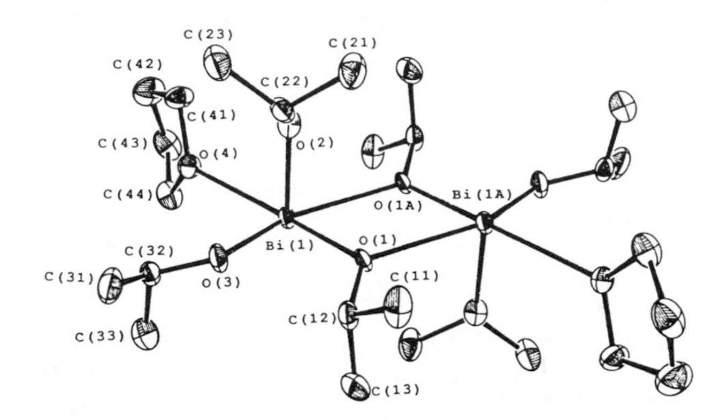

Fig. 1 Structure of **1** (F atoms omitted for clarity).

The reaction of $BiCl_3$ with $NaOC_6F_5$ generated in situ from NaH and HOC_6F_5 in THF proved to be much more complicated than that with $NaOCH(CF_3)_2$. Instead of isolating a simple alkoxide, oxo-bridged species have been observed. The nature of the product is sensitive to the origin and pretreatment of the phenol. When phenol from Aldrich Chemical Co. was used as received, the toluene extract of the product formed in THF yielded two types of crystals: extremely sensitive colorless crystals and nicely-formed yellow crystals. Preliminary X-ray structural analysis indicates the latter to have a formulation of $NaBi_3(\mu_4\text{-}O)(OR)_8(THF)$, **2**. Unfortunately there appears to be an occupational disorder between the Na position and one or more of the Bi sites. The $Bi_3(\mu_3\text{-}O)$ unit appears to be a fundamental building block in these systems (*vide infra*). **2** is one of the first mixed oxo-alkoxide cluster complexes of bismuth. The formation of the oxide ligand probably arises from NaOH or Na_2O present in solutions of the sodium pentafluorophenolate rather than direct hydrolysis of an intermediate alkoxide complex. Other routes which do not use the NaH deprotonation step to be discussed shortly lead to simple alkoxides which can be handled without the formation of the oxide clusters.

X-ray analysis of the colorless crystals has not yet been successful due to problems in transferring the extremely sensitive crystals to the diffractometer. Exposure of toluene solutions of this substance to air, however, led to the formation of pale yellow crystals which have been characterized as $Bi_6(\mu_3\text{-}O)_4(\mu_3\text{-}OR)\{\mu_3\text{-}OBi(OR)_4\}_3$, **3** (Figs. 2 and 3). This complicated aggregate has a similar core structure to that found in hydrolysis products of $Bi(NO_3)_3$ reported by Lazarini [15].

Use of HOC_6F_5 from Strem Chemical Co. without prior drying led to the formation of two clusters with four Bi atoms each. These have been identified as $NaBi_4(\mu_3\text{-}O)_2(OR)_9$ (Fig. 4), **4**, and $Na_2Bi_4(\mu_3\text{-}O)_2(OR)_{10}$, **5** (Fig. 5). Both clusters may be viewed as planar edge-sharing triangles of Bi_4 atoms in which each triangle is bridged by a μ_3-oxo function. The striking difference between the two is that in the first compound the μ_3-O atoms are on the same side of the Bi_4 plane while in the second they are on opposite sides.

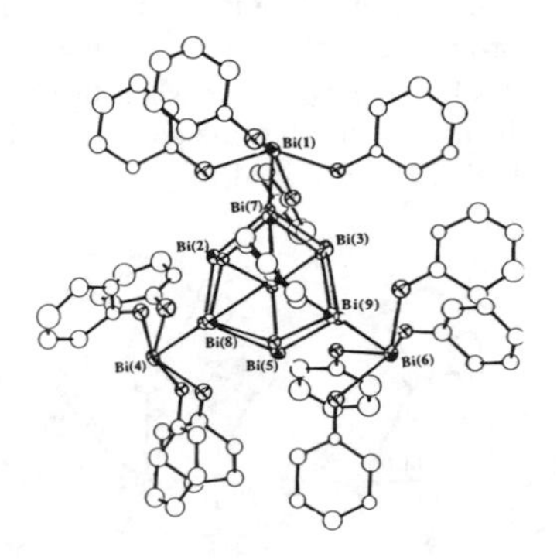

Fig. 2 Top View of the Structure of **3**

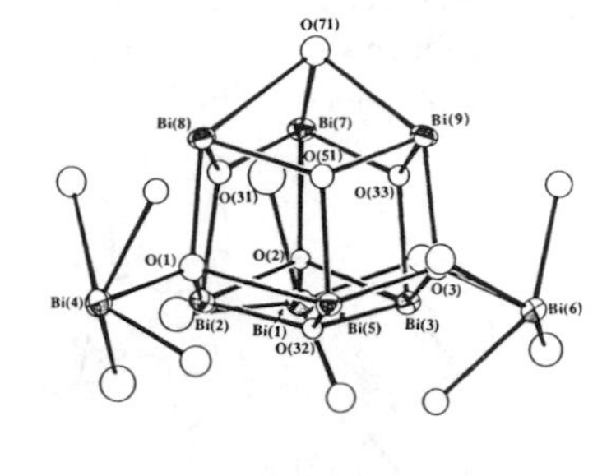

Fig. 3 Side View of the Structure of **3** (C_6F_5 rings omitted for clarity)

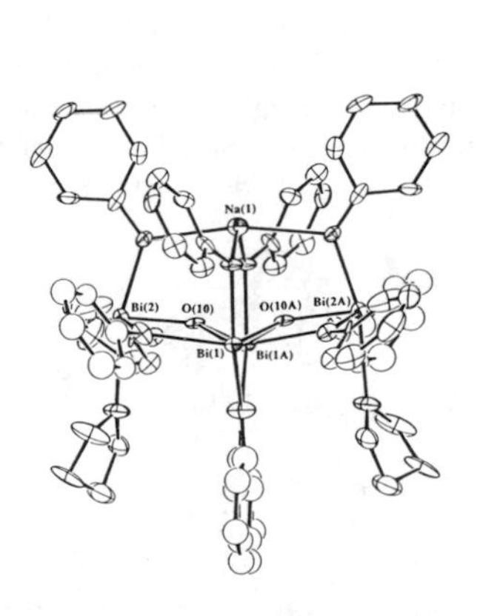

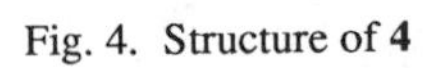

Fig. 4. Structure of **4**

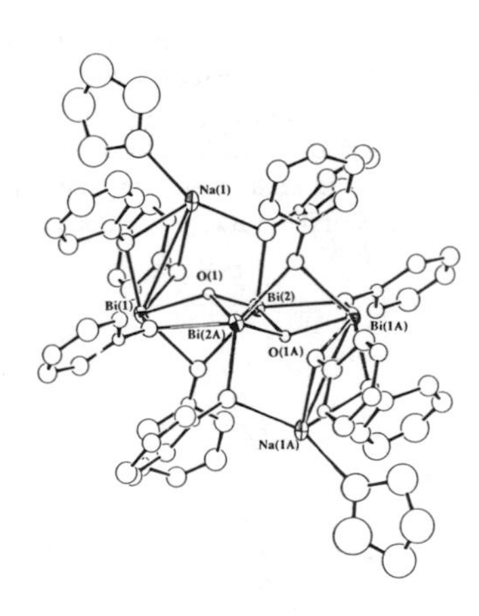

Fig. 5. Structure of **5**

In efforts to circumvent the problems encountered in the reaction of $BiCl_3$ with sodium salts of HOC_6F_5, the solvolysis reactions of organobismuth complexes with aromatic alcohols were probed. This method works well for $BiPh_3$ and HOC_6F_5 which after refluxing 24 hrs in toluene yields the alkoxide $[Bi(OR)_2(\mu\text{-}OR)(toluene)]_2$, **6** (Fig. 6). This compound has been found to crystallize in two different crystal morphologies, a second form possessing additional lattice toluene. The dimer structure is very similar to that observed for R = $CH(CF_3)_2$ but with toluene coordinated in place of THF. Structures of π-arene complexes of bismuth with halide ligands have been reported by Schmidbaur [16] but this is the first example where that has been observed for an alkoxide complex. The toluene is displaced when the complex is dissolved in THF and the complex $[Bi(OR)_2(\mu\text{-}OR)(THF)_2]_2$, **7** (Fig. 7) is obtained.

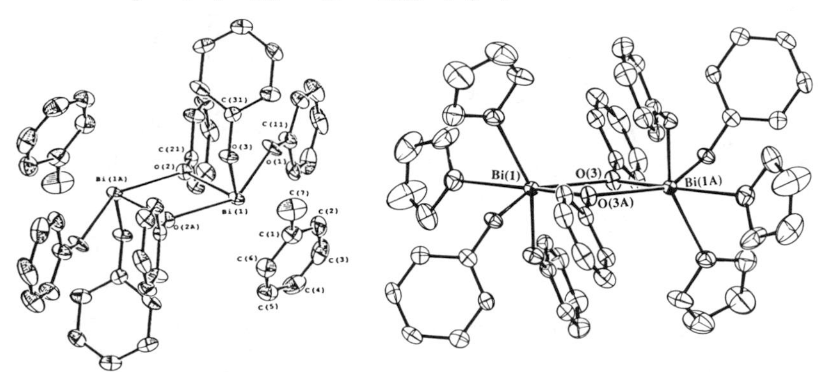

Fig. 6. Structure of **6** Fig. 7. Structure of **7**

6 reacts readily with NaOH but the product of this reaction has not yet been fully characterized. Its reaction with $NaOC_6F_5$ has led to the formation of $NaBi_3(\mu_3\text{-}O)(OR)_8(THF)_3$, **8**, (Fig. 8) and $Bi_6(\mu_3\text{-}O)_2(\mu_4\text{-}O)(OR)_{12}(THF)_2$, **9** (Fig. 9). **8** is very similar in structure to **2**. **9** can be described as two Bi_4 complexes like **4** above fused perpendicular to each other. This supports the hypothesis that some NaOH or Na_2O formed in the synthesis of the $NaOC_6F_5$ solution is carried along leading to oxide formation.

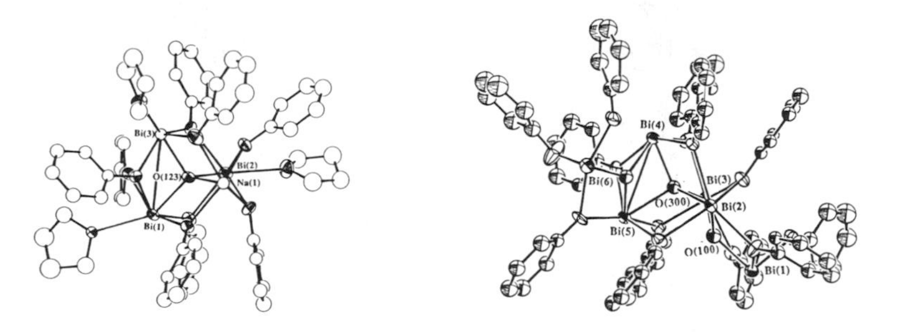

Fig. 8. Structure of **8** Fig. 9. Structure of **9**

When the reaction of $BiPh_3$ with HOC_6F_5 was attempted in THF, crystals of a complex $Bi_6(\mu_3\text{-}O)_4(\mu_3\text{-}OR)\{\mu_3\text{-}OBi(OR)_4\}_3(THF)_2$, **10**, were obtained, probably as a result of adventitious hydrolysis. This complex has essentially the same geometry as **3** but there are an additional two molecules of THF coordinated to the cluster (Fig. 10)

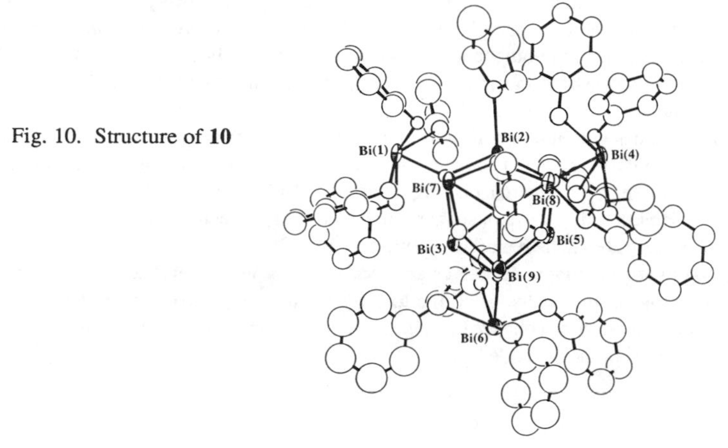

Fig. 10. Structure of **10**

The generality of the organobismuth route was probed by examining the reaction of $BiEt_3$ with HOC_6F_5 and HOPh. This reaction proceeds with replacement of only one Et group to give a complex of formulation $[BiEt_2(OR)]_\infty$, **11**, in both cases. The structures are polymeric with Bi-O-Bi-O- chaings arranged in chiral helices (Figs. 11 and 12). These substances are soluble in organic solvents and give mass spectral data consistent with the monomers. The coordination geometry about the Bi atom is similar to that found for $\beta\text{-}Bi_2O_3$ [17].

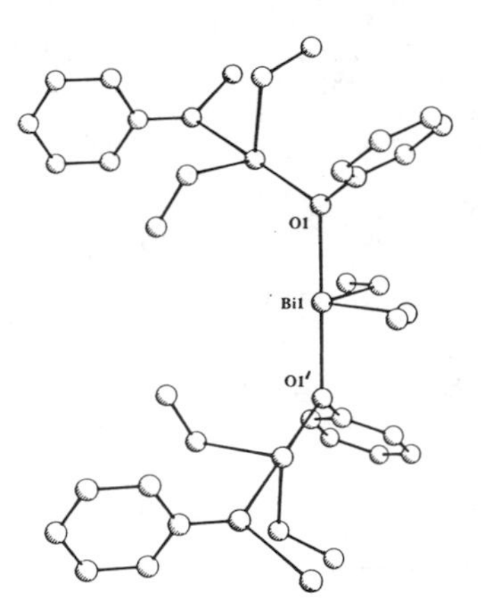

Fig. 11. Polymer Chain Fragment of **11**

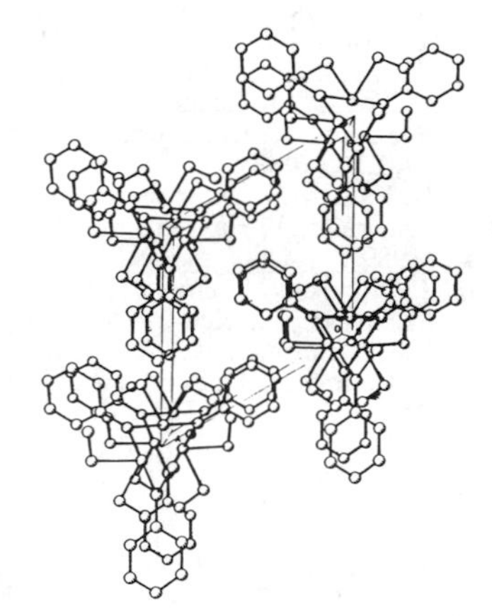

Fig. 12. Packing diagram of **11**

SUMMARY

The salt metathesis reaction between $BiCl_3$ and sodium alcoholates is a simple method of generating bismuth alkoxide complexes but one which has certain synthetic problems, especially in the case where the alcohol is C_6F_5OH. An alternate procedure involving the alcoholysis of organobismuth compounds works cleanly and easily for alcohols with acidic protons. This method avoids the presence of impurities which lead to the formation of oxobismuth clusters.

Bismuth alkoxides with intermediate-sized alcoholate ligands are found as dimers which may function as suitable reagents for the construction of single compound precursors to bismuth-containing high T_C superconducting materials. The presence of Na^+ counterions in several of the bismuth oxide clusters suggests the possibility of cation exchange to achieve heterometallic systems (e.g., the replacement of Na^+ with Cu^+).

The bismuth atoms in these structures are good Lewis acid sites and accept a variety of ligands such as solvent molecules or bridge to oxygen atoms of alkoxides bonded to other metals. This tendency may be enhanced by electron-withdrawing substituents such as fluorine on the organic portion of the alkoxide ligands.

References

1. Chemistry of High-Temperature Superconductors II, edited by D.L. Nelson, and T.F. George (ACS Symposium Series, #377, American Chemical Society, Washington, D.C., 1988).
2. Chemistry of High Temperature Superconductors, edited by D.L. Nelson, M. S. Whittingham, and T.F. George (ACS Symposium Series, #351, American Chemical Society, Washington, D.C., 1987).
3. High Temperature Superconductivity, edited by J.W. Lynn, Springer Verlag, NY, 1990.
4. K. Matsumura, H. Nobumasa, K. Shimizu, T. Arima, Y. Kitano, M. Tanaka and K. Sushida, Jap. J. Appl. Phys., 28, L1797 (1989).
5. T. Nonaka, K. Kaneko, T. Hasegawa, K. Kishio, Y Takahashi, K. Kobayashi, K. Kitazawa and K. Fueki, Jap. J. Appl. Phys., 27, L867 (1988).
6. N. Tohge, M. Tatsumisago and T. Minami, J. Non-cryst. Sol., 121, 443 (1990).
7. S. Shibata, T. Kitagawa, H. Okazaki, T. Kimura and T. Murakami, Jap. J. Appl. Phys., 27, L53 (1988).
8. H. Yamane, H. Kurosawa and T. Hirai, Chem. Lett.(Jap), 1988, 1516.
9. S. Hirano, T. Hayashi, M. Miura and H. Tomonaga, Bull. Chem. Soc. Jpn., 62, 888 (1989).
10. T. Nonaka, M. Green, K. Kishio, T. Hasegawa, K. Kitazawa, K. Kaneko and K. Kobayashi, Physica C, 160, 517 (1989).
11. R.C. Mehrotra and A.K. Rai Indian J. Chem., 4, 537 (1966).
12. G. Calingaert, H. Soroos and V. Hnizda J. Am. Chem. Soc., 64, 392 (1942).
13. W.J. Evans, J.H. Hain, Jr., and J.W. Ziller, J. Chem. Soc., Chem. Comm., 1989, 1628.
14. M.A. Matchett, M.Y. Chiang and W.E. Buhro, Inorg. Chem., 29, 360 (1990).
15. a. F. Lazarini, Acta Cryst., B34, 3169 (1978), b. F. Lazarini, Acta Cryst., B35, 445 (1979).
16. a) A. Schier, J. M. Wallis, G. Muller, and H. Schmidbaur, Angew. Chem. Int. Ed. Engl., 25, 757 (1986); Angew. Chem., 98, 757 (1986).
    b) W. Frank, J. Weber and E. Fuchs, Angew. Chem., 99, 68 (1987).
17. S.K. BLower and C. Greaves, Acta Cryst., C44, 587 (1988).

SOL-GEL SYNTHESIS OF $YBa_2Cu_3O_{7-\delta}$ USING YTTRIUM, BARIUM AND COPPER POLYETHER ALKOXIDE PRECURSORS

CATHERINE J. PAGE, CAROL S. HOUK AND GARY A. BURGOINE
University of Oregon, Department of Chemistry, Eugene, OR 97403

ABSTRACT

Homogeneous solutions of yttrium, barium and copper 2-(2-methoxy)-ethoxyethoxides in 2-(2-methoxy)ethoxyethanol of appropriate 1Y:2Ba:3Cu stoichiometry have been prepared and used in sol-gel synthesis of $YBa_2Cu_3O_{7-\delta}$. To our knowledge, this is the first report of the preparation of a homogeneous solution of yttrium, barium and copper homoleptic alkoxides dissolved only in the parent alcohol. These solutions provide an excellent starting point for sol-gel synthesis of $YBa_2Cu_3O_{7-\delta}$ since the metals are mixed on the molecular scale, and because the solutions are stable with no possibility of ligand exchange to give insoluble products. Hydrolysis and careful heat treatment of gels produced from these solutions yield nearly pure tetragonal $YBa_2Cu_3O_{7-\delta}$ at temperatures as low as 725°C. The preparation of these solutions and their characterization by small-angle x-ray scattering measurements and TEM are reported, as well as preliminary data pertaining to their use in the sol-gel synthesis of $YBa_2Cu_3O_{7-\delta}$.

INTRODUCTION

The synthesis of $YBa_2Cu_3O_{7-\delta}$ by the sol-gel method using alkoxides has been the focus of numerous recent investigations.[1-14] Alkoxide sol-gel synthesis is a complex process which begins with the preparation of a homogeneous nonaqueous solution of metal alkoxides. Subsequent hydrolysis and polycondensation yield an amorphous metal-oxide gel, which can be fired to produce crystalline oxides at temperatures which are low relative to conventional solid-state synthesis. Besides lower processing temperatures, advantages of sol-gel synthesis include higher purity and more homogeneous products, and the ability to form the viscous sol (partially hydrolyzed solution) or gel into various shapes, including fibers and thin films.

The major obstacle to alkoxide sol-gel synthesis of the copper-oxide superconductors has been the insolubility of copper (II) alkoxides of simple alcohols in nonaqueous solvents (i.e. common organic solvents). Some research groups have increased the solubility of copper (II) alkoxides by modification with carboxylic acids,[12] amines,[5,13] or acetylacetonate[6,7] ligands. Such modification usually involves ligand substitution of alkoxide ligands by the modifying agents to give partially or fully-substituted metal complexes. While such complexes are soluble in common organic solvents, they will only undergo partial hydrolysis, leaving organic ligands attached to the metal centers. In the extreme case, fully substituted complexes will not undergo hydrolysis to any extent, and will not be directly incorporated into the metal oxide gel.[5]

Our approach has been to synthesize copper (II) alkoxides made with alcohols that are potentially bi- or tridentate, such as diols or etheralcohols. Such complexes should be more soluble in organic solvents than alkoxides of simple primary or secondary alcohols, and yet they should also undergo hydrolysis. While our work was in progress, it has been confirmed that copper etheralkoxides are indeed more soluble in common nonpolar organic solvents than are the copper alkoxides of simple primary or secondary alcohols.[1-4,8,9] Most of the previous reports of sol-gel syntheses of $YBa_2Cu_3O_{7-\delta}$ using copper (II) etheralkoxides involve mixtures of metal complexes which have different alkoxide ligands (e.g. yttrium and barium isopropoxides with copper 2-(2-methoxy)ethoxy-ethoxide).[1,8,9] In our experience, ligand exchange is relatively rapid in these systems, resulting in precipitation of an insoluble copper alkoxide (e.g. copper isopropoxide precipitates when yttrium or barium isopropoxides are mixed with copper 2-(2-methoxy)ethoxyethoxide). We have therefore restricted our studies to systems where the alkoxide ligand for all three metals is the same, and the solvent is the parent alcohol.

An important feature of etheralkoxide sol-gel systems for synthesis of $YBa_2Cu_3O_{7-\delta}$ is that there is a complexation reaction between barium and copper salts to form a soluble bimetallic complex.[2-4,7] Thus, although the copper (II) etheralkoxides are essentially insoluble in the respective parent alcohol,[9] reaction with the corresponding soluble barium etheralkoxide yields a soluble bimetallic complex in the parent alcohol. We have observed this phenomenon in the case of ethylene glycol alkoxides as well as etheralkoxides.[15] In the case of ethylene glycol alkoxides[15] and 2-methoxyethanol alkoxides,[4] copper is solubilized by the barium salt in a 1:1 ratio in the parent alcohol. However, in the case of 2-(2-methoxy)ethoxyethanol (MEE) and 2-(2-ethoxy)ethoxyethanol (EEE), we have found that the copper alkoxide can be solubilized by the corresponding barium alkoxide in at least a 3:2 copper:barium ratio. This allows the preparation of homogeneous, stoichiometric 1Y:2Ba:3Cu polyetheralkoxide solutions in the parent polyetheralcohol. We report here the preparation of such homogeneous solutions using yttrium, barium and copper MEE alkoxides in MEE. To our knowledge, this is the first report of the preparation of a homogeneous solution of yttrium, barium and copper homoleptic alkoxides of appropriate 1Y:2Ba:3Cu stoichiometry dissolved only in the parent alcohol. These solutions provide an excellent starting point for the sol-gel synthesis of $YBa_2Cu_3O_{7-\delta}$ since the metals are mixed on the molecular scale, and because the solutions are stable with no possibility of ligand exchange to give insoluble products. Preliminary data pertaining to the sol-gel synthesis of $YBa_2Cu_3O_{7-\delta}$ using these solutions are presented.

EXPERIMENTAL

All manipulations of reactants, solvents and alkoxide solutions prior to hydrolysis were carried out under purified nitrogen atmosphere, either using Schlenk glassware and vacuum-line techniques or an inert-atmosphere dry box. Barium metal granules were used as purchased from Alfa. Copper methoxide used in the preparation of the etheralkoxide was prepared from a reaction of copper chloride and excess lithium methoxide,[15] and yttrium oxoisopropoxide was prepared from a reaction of yttrium chloride and potassium isopropoxide.[16] 2-(2-methoxy)ethoxyethanol was dried over sodium and distilled under vacuum.

Copper MEE alkoxide was prepared via alcoholysis of copper(II) methoxide. Cu(II) methoxide was suspended in dry MEE and stirred at ambient temperature for approximately 72 hours. A small amount of solvent was then distilled (under reduced pressure) from the suspension to eliminate methanol liberated via alcoholysis. Concurrent with the distillation, a stoichiometric amount of barium metal[17] (2Ba:3Cu) was dissolved in dry MEE to make a 0.04 molar solution. Addition of the barium solution to the copper suspension resulted in a clear, royal blue solution. The yttrium MEE solution was prepared by dissolving a stoichiometric amount of yttrium oxoisopropoxide ($Y_5O(O^iPr)_{13}$) in dry MEE to make a 0.02 molar solution, from which a small amount of solvent was distilled under reduced pressure to eliminate isopropanol liberated via ligand substitution. Addition of the 2Ba:3Cu solution to the yttrium solution resulted in a clear, dark blue solution.

Small angle x-ray scattering was performed on solutions prior to hydrolysis using a modified Philips x-ray diffractometer with three pin-hole collimators and a position-sensitive detector. Data was analyzed using a Guinier-type analysis[18]. TEM micrographs of solution-phase particles sprayed onto a carbon film were made using a Philips CM-12 TEM/STEM operated at 100 kV with a 70μ aperture.

Hydrolysis of stoichiometric (1Y:2Ba:3Cu) homogeneous MEE solutions yields clear, dark green solutions which gel very slowly. Various water/alkoxide ratios have been employed for hydrolysis. Given below are results from a typical solution which was hydrolyzed in air with a ratio of 5:1 H_2O/metals using a solution composed of 1 part 30% H_2O_2 in water to 9 parts MEE. When gelled in bulk (e.g. in a 50 ml Erlenmeyer flask), a clear green gel separates from the mother liquor within two months.[19] Further separation of the gel from the mother liquor was facilitated by centrifugation, and the concentrated gel was dried in air for ~1 month.

To make the $YBa_2Cu_3O_{7-\delta}$ phase, the dried gel was fired at 725° C in a N_2/O_2 (~95%/5%) atmosphere with an isothermal hold at 350° C for 1 hour on the ramp up. After 18 hours the atmosphere was changed to O_2, and firing at 725° C continued for 6 more hours. The ramp down included an isothermal hold at 550° C for 1 hour. The slow ramp down and isothermal hold at 550°C under flowing O_2 were intended to promote formation of the orthorhombic superconducting phase. Such oxygen annealing is routinely done in solid state preparations of the material to convert the tetragonal $YBa_2Cu_3O_{6+x}$ which is stable above 720°C to the orthorhombic superconducting $YBa_2Cu_3O_{7-\delta}$.[20]

Dried gels and fired samples were characterized by thermogravimetric analysis using a TA Instruments 951 TGA module, and by powder x-ray diffraction using a Scintag XDS-2000 θ–θ powder diffractometer. Magnetic susceptibility measurements were made using a Quantum Design SQUID magnetometer.

RESULTS AND DISCUSSION

Solutions prepared as described above are stable for several months, with no visible precipitate. Preliminary small angle x-ray scattering experiments performed on a solution aged one month suggest the average particle size is less than 40Å. TEM micrographs also indicate a small particle size, as shown below in Figure 1.

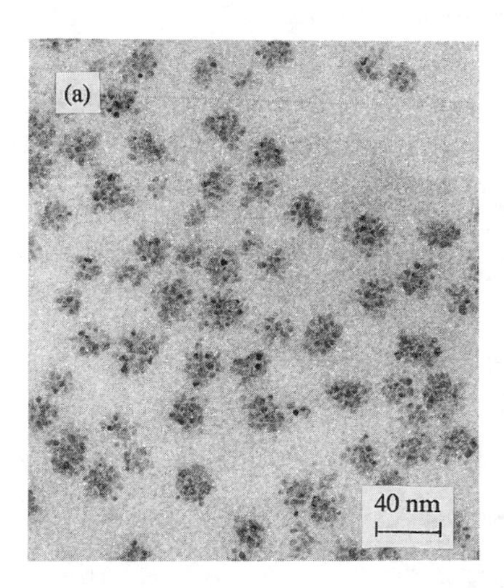

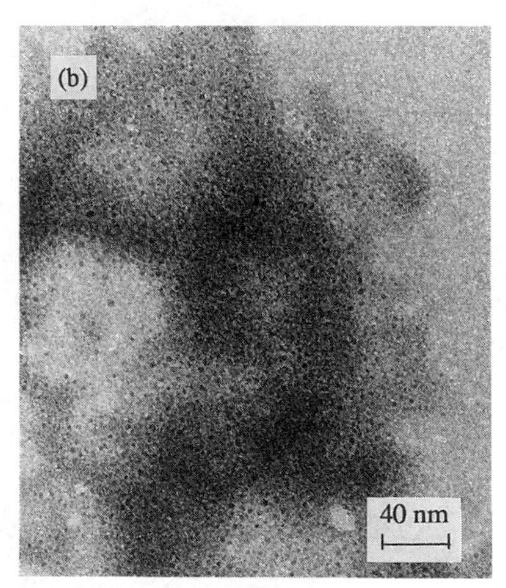

Figure 1: TEM micrographs of an Y-Ba_2-Cu_3 MEE solution sprayed onto a carbon film: a) sprayed in air, and b) sprayed in nitrogen.

The TEM micrographs above confirm that the particle sizes of the species present in the Y-Ba_2-Cu_3 MEE solution are 40 Å or less. The micrograph shown in Figure 1 (a) was made by spraying solution onto the carbon film in air, whereas the micrograph in (b) was sprayed in nitrogen. The clustering seen in (a) is probably due to some hydrolysis which occurred during the spraying process, although the background also shows small particles which are not clustered. Clustering is not seen in any of the micrographs taken of solution sprayed under nitrogen. Thus, the micrograph shown in Figure 1 (b) is probably more representative of the particle size distribution in solution, and indicates a relatively uniform dispersion of very small particles (< 40Å).

Thermogravimetric analysis of the dried gel is shown in Figure 2. Weight loss is essentially complete at 350°C (~38%, possibly corresponding to loss of four solvent molecules per $YBa_2Cu_3O_{7-\delta}$ formula unit). X-ray diffraction of gels fired at 350°C for up to 10 hours show no formation of crystalline phases. After firing to 725°C as described above, x-ray diffraction of the product indicates formation of nearly pure *tetragonal* $YBa_2Cu_3O_{7-\delta}$, as shown in Figure 3.

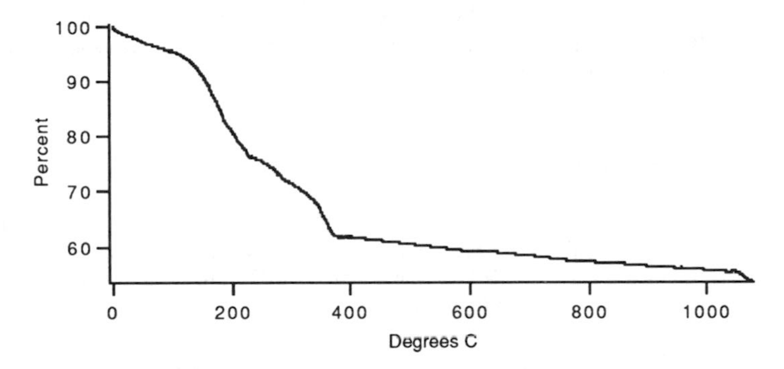

Figure 2: Thermogravimetric analysis of air-dried 1-2-3 MEE gel.

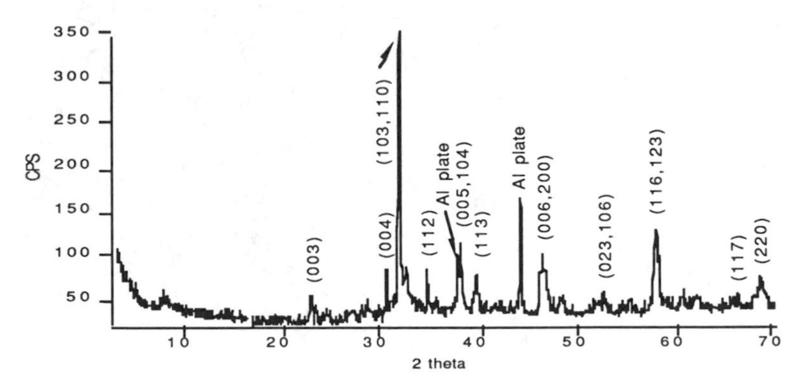

Figure 3: Powder x-ray diffraction for $YBa_2Cu_3O_{7-\delta}$ produced by firing a dried gel as described in the text. Peak indexing is given for the tetragonal phase. Two peaks labeled "Al plate" are from the aluminum sample holder.

Calculation of the Meissner effect from magnetic susceptibility data confirms that less than 5% of the sample is the superconducting orthorhombic phase. The small portion of this phase present had an onset of superconductivity at 75 K. A simple reanneal of the fired product at 550°C under flowing O_2 for 10 hours does not significantly increase the amount of orthorhombic phase relative to the tetragonal phase, as judged by x-ray diffraction. Incomplete conversion of the tetragonal to the orthorhombic phase has also been observed by other researchers preparing this material via sol-gel routes,[2] and via oxalate precipitation.[21] Kordas et al.[2] suggest that such incomplete conversion may be associated with the presence of residual carbon impurity phases. However, this is unlikely in our case because we see no evidence for carbon-containing impurity phases such as barium carbonate in the x-ray diffraction patterns of the product or of the gel fired to intermediate temperatures. More likely may be the explanation of Manthiram and Goodenough,[21] who suggest that oxygen-rich ($\delta < 0.5$ in $YBa_2Cu_3O_{7-\delta}$) tetragonal material will not undergo the transformation to orthorhombic material. These authors suggest that it is necessary to reduce the oxygen content (to $\delta > 0.5$) by firing at 750°C under nitrogen. When we attempted this 'reduction' of our tetragonal material it decomposed. Further studies are underway to determine the oxygen and carbon content of the tetragonal material prepared by this sol-gel method, and to determine routes to effect the transition of this material to the orthorhombic superconducting material.

In summary, we have prepared stable homogeneous solutions of yttrium, barium and copper MEE alkoxides in MEE of appropriate 1Y:2Ba:3Cu stoichiometry. These solutions provide true solution-phase mixing of the metal components, and thus provide an excellent starting point for the sol-gel synthesis of $YBa_2Cu_3O_{7-\delta}$. Hydrolysis and careful heat treatment of gels produced from these solutions yield nearly pure tetragonal $YBa_2Cu_3O_{7-\delta}$ at temperatures as low as 725°C. As with all sol-gel synthetic routes, careful control of many variables, including hydrolysis, gel aging, firing temperatures and firing atmosphere is necessary to develop an understanding of the underlying chemistry. Work is underway in our laboratory to establish the importance of each of these parameters for a reliable sol-gel route to $YBa_2Cu_3O_{7-\delta}$ and results will be presented in a forthcoming publication.

ACKNOWLEDGMENT

This research was sponsored by the National Science Foundation (DMR-8716255). Acknowledgment is also made to the Donors of The Petroleum Research Fund, administered by the American Chemical Society, for partial support of this research (PRF#19769-G3). We thank Eric Shabtach at the University of Oregon for the TEM microscopy, Professor Shankar Rananavare at the Oregon Graduate Institute for performing the small angle x-ray scattering experiments, and Dr. William Ham at the University of Oregon for the magnetic susceptibility measurements, which were performed on a SQUID magnetometer at UC Berkeley. We also thank Dr. Steven Tenhaeff and Craig Love for their help in this study.

REFERENCES

1. H. S. Horowitz, S. J. McLain, A. W. Sleight, J. D. Druliner, P. L. Gai, M. J. VanKavelaar, J. L. Wagner, B. D. Biggs and S. J. Poon, Science 243, 66 (1989).

2. G. Kordas, G.A. Moore, J. D. Jorgensen, F. Rotella, R. L. Hitterman, K. J. Volin and J. Faber, J. Mater. Chem. 1, 175 (1991).

3. G. Kordas, J. Non-Cryst. Solids 121, 436 (1990).

4. G. Moore, S. Kramer and G. Kordas, Mater. Lett. 7, 415 (1989).

5. S. Katayama and M. Sekine, J. Mater. Res. 5, 683 (1990).

6. P. Catania, N. Hovnanian, L. Cot, M. Pham Thi, R. Kormann and J. P. Ganne, Mat. Res. Bull. 25, 631 (1990).

7. N. N. Sauer, E. G. Garcia, K. V. Salazar, R. R. Ryan and J. A. Martin, J. Am. Chem. Soc. 112, 1524 (1990).

8. S. Hirano, T. Hayashi, M. Miura and H. Tomonaga, Bull. Chem. Soc. Jpn. 62, 888 (1989).

9. S. C.Goel, K. S. Kramer, P. C. Gibbons and W. E. Buhro, Inorg. Chem. 28, 3619 (1989).

10. M. W. Rupich, B. Lagos and J. P. Hachey, Appl. Phys. Lett. 55, 2447 (1989).

11. S. Shibata, T. Kitagawa, H. Okazaki, T. Kimura and T. Murakami, Jpn. J. Appl. Phys. 27, L53 (1988).

12. H. Zheng and J. D. Mackenzie, Mater. Lett. 7, 182 (1988).

13. T. Monde, H. Kozuka and S. Sakka, Chem. Lett. 1988, 287.

14. P. Ravindranathan, S. Komarneni, A. Bhalla, R. Roy and L. E.Cross, J. Mater. Res. 3, 810 (1988).

15. C. P. Love, C. C. Torardi and C. J. Page, Inorg. Chem. (1992), in press.

16. Page, C.J., Sur, S.K., Lonergan, M.C., Parashar,G.K., Magn. Reson. in Chem. 29, 1191 (1991).

17. Barium was usually weighed first because it was easier to adjust the mass of copper methoxide powder to obtain a stoichiometric ratio.

18. O. Glatters and O. Kratky, Small Angle X-ray Scattering (Academic Press: NY, 1982), Chapters 5 and 15.

19. Faster gellation can be effected by spreading the solution into a thin layer, as in a Petrie dish. Under these conditions, gellation occurs within a few days.

20. A. Kulpa, A. C. D. Chaklader, G. Roemer, D. L. Williams and W. N. Hardy, Supercond. Sci. Technol. 3, 483 (1990).

21. A. Manthiram and J. B. Goodenough, Nature 329, 701 (1987).

PREPARATION OF BSCCO PRECURSORS BY WATER EXTRACTION VARIANT OF SOL-GEL PROCESS

A. DEPTUŁA*, W. ŁADA* T. OLCZAK* AND A. DI BARTOLOMEO**
* Institute of Nuclear Chemistry and Technology, Warsaw, Poland
** ENEA, C.R.E. Casaccia, Italy

ABSTRACT

The starting acetate sols of molar compositions of Bi:Pb:Sr:Ca:Cu=I–2:O:2:1:2, II–2:0:2:2:3 and III–1.6:0.4:2:2:3 were prepared by NH_4OH addition at various rates, followed by evaporation. Sols were gelled by further evaporation to shard or to microspheres (diameter below 100 μm) by water extraction from the sol emulsion drops formed in 2–ethyl–1–hexanol.The gels were converted to BSCCO phases by thermal treatment. The processes were characterized by XRD and IR examinations. The influence of the sol preparation step connected with formation of polynuclear cations on the following stages of the process was examined.

INTRODUCTION

High temperature superconducting phases in the system Bi–Sr–Ca–Cu–O ($Bi_2Sr_2Ca_1Cu_2O_{8+d}$=BSCCO 2212 with T_c near 80K and $Bi_2Sr_2Ca_2Cu_3O_{10+\delta}$ = BSCCO 2223 with T_c near 110K) were first obtained by Maeda *at al.* [1] by the solid state reaction method. However it is well known [2] that solution techniques, especially sol-gel processes, in many cases are superior to the conventional solid-state reaction reaction. E.g. according to[3] the synthesis of BSCCO 2223 phase has not been possible in this way. Takano *at al.*[4] have found that partial substitution of lead for bismuth effectively facilitates formation of that phase. Recently [8] difficulty in the reproducibility of results has been encountred in Pb dopped 2223 superconductor, due to lead vaporization in the samples prepared by the standard ceramic methods in contrast to those obtained by the solution technique. Mulay *at al.*[9]concluded that also for preparation of the 2212 compounds the sol-gel method is better than any other method.

For the YBCO supercondutor the sol-gel process starting from metal acetate solution was elaborated by Saka and coworkers [5] and after modification used in an earlier work of the present authors [6]. Recently Laine *at al.* [7] noted that metal carboxylates offer a wide variety of bonding arrangements (polymer-like structures for Cu, oxygen bridges between metals for Ca,Sr,Ba carboxylates) and consequently give an opportunity for developing precursors for superconducting ceramics. The present authors also observed [15] that the ability of carboxylic group to bidentate coordination facilitates the formation of gel network in the Ca^{2+}–H_2O–PO_4^{3-} system. Sheppard [10] in the recent paper „Superconductors: Slowly Moving to Commercialization" has considered the sol-gel process of BSCCO preparation, starting from acetate solutions alkalized with ammonia, as one of the most promising. This method has been used in works [11,12,13,] however the rate of ammonia addition to the starting metal acetate (sometimes also mixed with other anion salts) solution has been never studied. This parameter as found by one of the present authors [14] plays an important role in the systems in which polynucler cations are formed during polymerization of simple cations. In the system discussed above Bi^{3+}, Cu^{2+} and Pb^{2+} form polynuclear species [16]. It seemed that also formation of the well known [18] copper ammino complexes can be affected by kinetic phenomena. This work has been undertaken to find the influence of the parameters of the sol preparation on the formation of BSCCO precursors by gelation using water evaporation and water extraction. The latter variant of the sol-gel process [2,17] was succesfully applied [6] to the preparation of medium- sized spherical powders of YBCO superconductors (diameter <50 μm) Such powders of other ceramic materials have important applications [2,17].

EXPERIMENTAL

Concentrated acetate solutions of Sr(0.61M), Ca(1.43M), Cu(0.31M) and Pb(0.38M) were prepared by dissolving commercial reagents. Bi acetate (0.15M) was prepared by dissolution of bismuth hydroxide (freshly precipitated with ammonium hydroxide from Bi nitrate solution) at an excess of acetic acid. The individual solutions were mixed to prepare three staring solutions „SS" of the following molar compositions: I=2Bi:2Sr:1Ca:2Cu; II=2Bi: 2Sr:2Ca:3Cu and III=1.6Bi:0.4Pb:2Sr:2Ca:3Cu. To 50 ml of the solutions, vigorously stirred, aqueous ammonia (13M) was added at rates (A)1.8, (B)0.4 and (C)0.05 ml/min. at controlled pH.The potentiometric titration was continued till the beginning of precipitation.Then to 200–500 ml of SS solution aqueous ammonia was added until pH by 0.1 unit lower than the determined precipitation point(pp) under similar conditions. Visible spectra of those solutions were recorded using a Unicam 1800 Spectrophotometer.

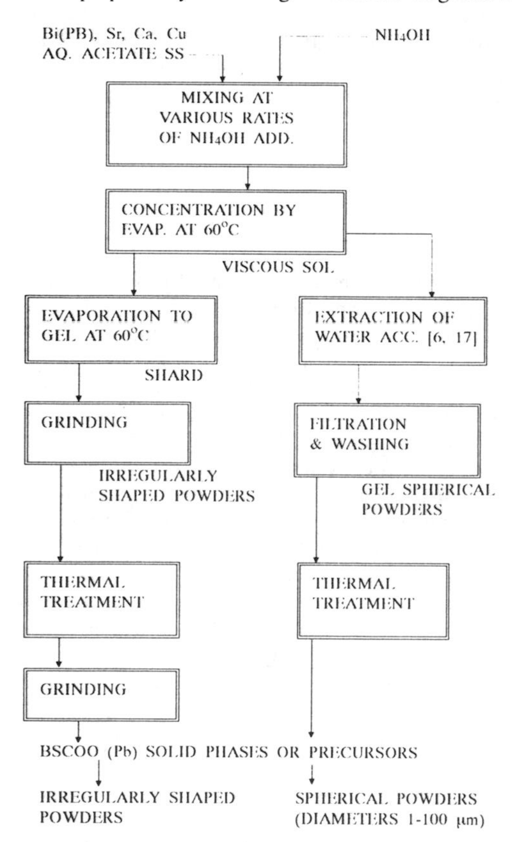

Fig.1. Flow sheet of the sol gel process for preparation of BSCCO(Pb) precursors.

The solutions were concentrated by evaporation of water at 60°C, the viscosity being periodically controlled using Ubbelohde capillaries. One part of the concentrated solution was gelled to medium spherical powders using the process described earlier in [6,17]. The second part was evaporated at 60°C to shard which was then ground in a mortar to irregulary shaped powder. The gels were subjected to thermal treatment. The course of the thermal transformations (TG, DTG, DTA) was studied using a Hungarian MOM Derivatograph. A flow sheet of the process is shown in Figure 1. Products of the thermal treatment of the gels at various temperatures were characterized by X-ray diffraction measurements (Positional Sensitive Detector-Ital Structure) and IR spectra (Perkin Elmer 983) using KBr pellet technique (1/300mg).

RESULTS AND DISCUSSION

It is obvious that all processes of sol preparation should be carried out under conditions under which no precipitates are formed. The potentiometric titration curves of individual solutions shown in Figure 2a point out that first Bi^{3+}(pH 5.3) forms a precipitate, than Cu^{2+}(pH 5.8) and finally Pb^{2+}(pH 10.5).

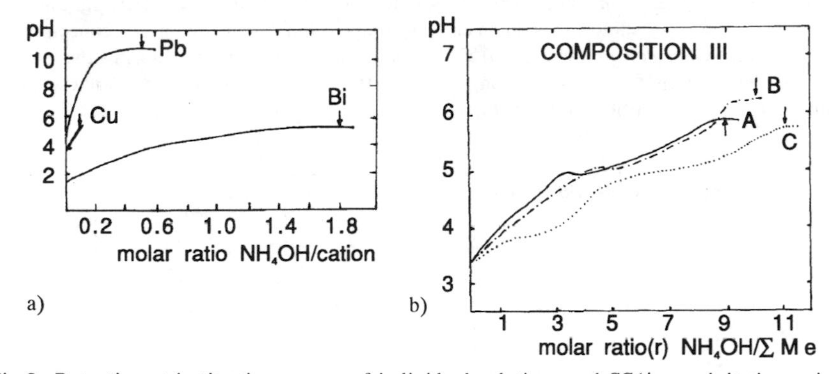

Fig.2. Potentiometric titration curves of individual solutions and SS(↓ -precipitation point).

Table I.
Start of precipitation in SS–NH4OH systems ($r = [NH_4OH] / \Sigma [Me]$)

SS composition	NH4OH feeding rate					
	(A)		(B)		(C)	
	pH	r	pH	r	pH	r
I	5.9	15.4	6.2	16.7	5.8	17.2
II	5.9	10.8	6.3	12.6	5.8	14.5
III	5.9	8.9	6.3	10.2	5.7	10.9

As can be seen in Figure 2b and Table I the formation of the first precipitates from mixed solutions depends on their composition and the rate of ammonia addition. For a given composition the precipitation starts at higher molar ratios(r) of $NH_4OH/\Sigma Me$ and at lower pH value for the lowest ammonia feeding rate. It was also observed that during titration at a low rate the pH after addition of each amount of the reagent drops approx. by 0.2 unit. All the observations seem to indicate that the cations polymerization connected with the formation of –O–,← OH– bridges and H^+ ions is a time consuming process.It was found that the first precipitate contains approx. 74% of Bi which corresponds well to the composition of the Bi0Ac (73.6%Bi). On the basis of the above results, solutions with pH values lower by 0.1 unit than the pH of precipitation were prepared. Visible spectra of those solutions are given in Figure 3.

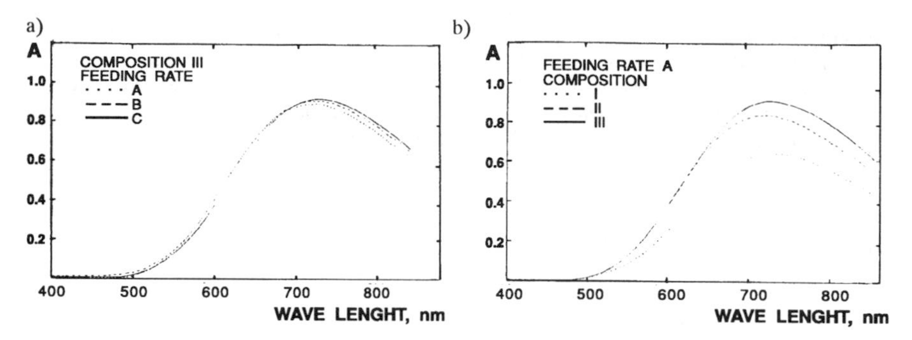

Fig.3. Visible spectra of Bi(Pb)-Sr-Ca-Cu-NH_4OH solutions.

The dark blue solutions of the composition III have the same spectra (Figure 3a).It means that complexation of Cu^{2+} does not depend on the ammonia feeding rate. The spectra in Figure 3b show lower intensity for composition I with lower Cu concentration. All solutions were concentrated at 60^oC. The viscosity was measured during the process. It was observed that the evaporation is more effective for solutions prepared at lower rate of ammonia addition (Figure 4a).

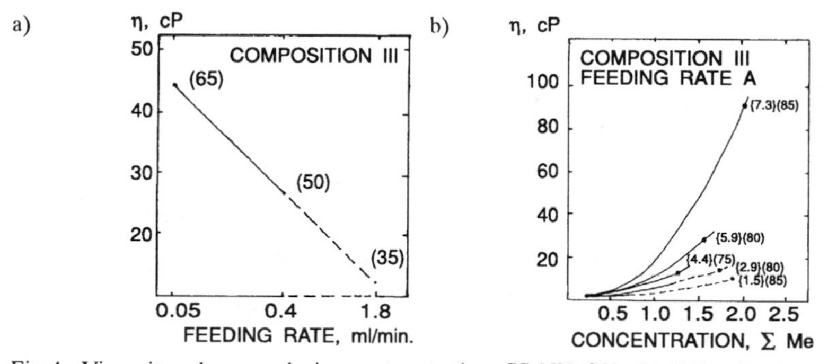

Fig.4. Viscosity changes during concentration SS-NH_4OH. In ()% of evaporated volume,in{}r.

It can be seen that the viscosity increases with the increase in concentration and molar ratio (r) of $NH_4OH/\Sigma Me$ (Figure 4b). The concentration effect for a given composition is the lowest for the solution with r=4.4. For lower r violet crystals appeared during concentration and gelation did not occur. This confirms that polymerization of cations, connected with liberation of molecular water from OH groups (formation of –O– bridges) or from coordinated H_2O (formation of ← OH– bridges) is more effective when reaction is carried out slowly at the first stage. This conclusion is confirmed also by the vicosity changes during concentration of the starting solutions with various molar ratio of $NH_4OH/\Sigma Me$. The viscosity increases with increasing molar ratios i.e. with polymerization. This agrees with Mackenzie's *at al.* [19] data.

After 36–72h all the solutions prepared at maximum pH value (portions 60ml) gelled to dark blue homogeneous gel. The gelation time was shorter for solutions prepared slowly, especially when dopped with Pb. The sols prepared as above (with viscosity 20–80 cP) were gelled succesfully to blue microspheres.

The results of thermal analysis of the gels are shown in Figure 5. XRD patterns and IR spectra of the gel and its calcination products at various temperatures are shown in Figure 6.

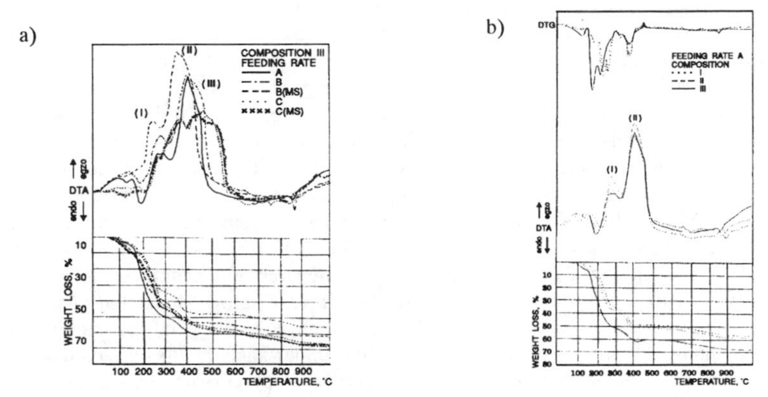

Fig.5. Thermal analysis of various gels (MS-microspheres).

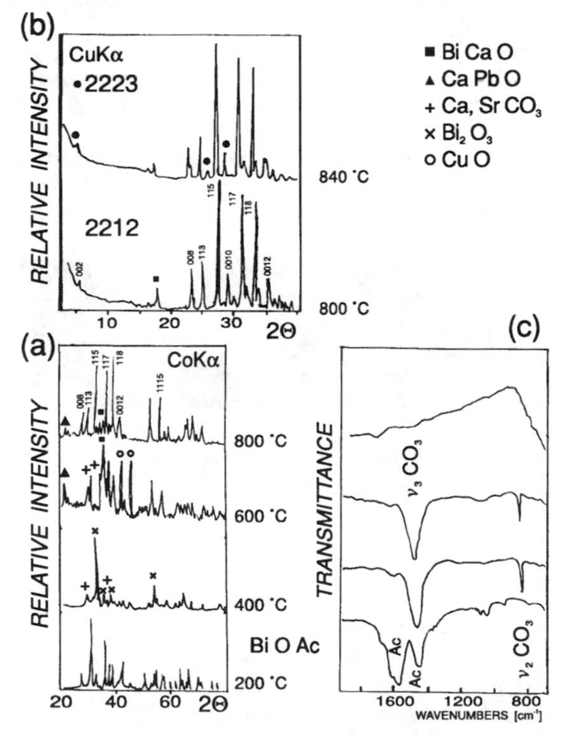

Fig.6. XRD patterns and IR spectra of gel derived products (composition III, feeding rate A, calcination time 24h).

When heating all the gels to 200°C(except for microspheres) weight losses (molecular water and acetates), accompanied by endothermic effects, are observed. At 200°C the product is amorphous with a crystalline BiOAc structure. The same compound was observed above as the first precipitate. The next two exothermic effects at approx. 250°C(I) and 300–400°C (II)can be ascribed to combustion of organics with formation of carbonates of Sr and Ca, and to reactions of metal oxides. These conclusions are supported by the presence of carbonate IR bands at 875 and 1430 $cm^{-1}$for samples calcined at 400–600°C. The last endothermic peak at approx. 850°C, accompanied by a weight loss, can be connected with carbonate decomposition and melting. The weight loss between the last exothermic peak and the endothermic one roughly corresponds the stoichiometry of Sr and Ba carbonates decomposition. The IR spectrum of the sample calcined at 840°C does not show carbonate bands. It can be noted that only for gels prepared at slow (C) feeding rates the third distinct broad exothermic peak appeares at approx. 450°C (III) without appreciable weight loss. This indicates that the final reactions between oxides occur in this case later then for gels prepared by techniques A and B. For a given method of sol preparation DTA and TG curves (Figure 5a) of the spherical powders of the gels (MS) are very similar to those of gels prepared by traditional evaporation methods. All the observations point to the importance of the polymerization grade of cations in the parent sols for thermal properties of precursors. This conclusion confirms the recent data of Wood *at al.* [20] for alumina gels. It can be noted that for the same method of sol preparation (Figure 5b) the composition of the gels does not essentially affect the course of thermal decomposition.

The final calcination of ground shard samples obtained from composition III by technique A at 800°C has shown (Figure 6b) the presence of 2212 BSCCO which starts to convert to phase 2223 at 840°C. Melting was not observed at that temperature. Microspheres were calcined by heating at a rate 1–10°C/min for 24h at 60, 100, 840°C. It was observed that the blue–green transparent microspheres change to transparent brown ones at approx. 150°C. The transparence vanishes at approx. 200°C and the microspheres remain black above 300°C (Figure 7).

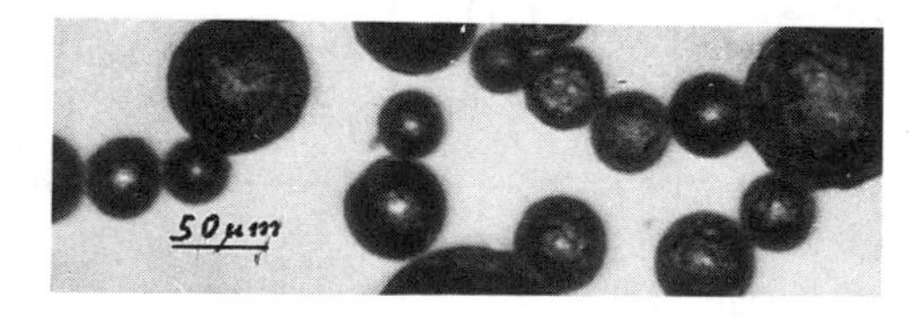

Fig.7 Typical spherical powders of BSCCO(Pb).

CONCLUSIONS

The basic role of formation of sols in the reaction of Bi(Pb)SrCaCu acetate solutions with NH_4OH has been shown. Sols prepared at low ammonia feeding rate gelled more readily in both the applied techniques evaporation to shard or water extraction to microspheres. The thermal characteristics during convertion to BSCCO phases also depends on that parameter. This phenomenon seems to be due to the degree of cations polymeryzation. These preliminary studies do not permit to conclude what grade of cations polymerization is the most advantagous for the formation of BSCCO superconductors. However it seems that more homogeneous, low polymerized gels, facilitate the solid state reaction.

REFERENCES

1. H. Maeda, Y. Tanaka, M. Fukutomi and T.Asano, Jpn.J.Appl.Phys. 27, L209(1988).
2. A. Deptuła, C. Majani, Sol-gel Processes and Their Applications, ENEA Report No. RT/TIB/86/25, ROMA, 1986.
3. R. Ramesh, K. Remsching and J. M. Tarascon, J.Mat.Res. 6, 278, 1991.
4. M. Takano, J. Takada, K. Oda, H. Kitaguchi, T. Miura, Y. Ikeda, Y. Tomii and H. Mazaki, Jap. J.Appl.Phys. 27, L1041(1988).
5. H. Kozuka, T. Umeda, J. Jin, T. Monde and S. Sakka, Bull.Inst.Chem.Res. 66, 80(1988).
6. A. Deptuła, W. Łada, T. Olczak, T. Żółtowski and A. Di Bartolomeo, in Better Ceramics Through Chemistry IV, edited by B. J. J. Zelinski, C. J. Brinker, D. E. Clark and D. R. Urlich (Mat.Res.Soc.Symp.Proc. 180, Pittsburgh, PA, 1990) pp.907–912.
7. R. M. Laine, K. A. Youngdahl, R. A. Kennish, M. L. Hoppe, Z. F. Zhang and J. Ray, J.Mater.Res. 6, 895(1991).
8. R. B. Tripathi and D. W. Johnson, Jr.J.Am.Ceram.Soc. 74, 247(1991).
9. V. N. Mulay, P. V. L. N. Siva Prasad, K. Bhupal Reddy, M. A. Jaleel, J.Mat.Sci.Lett. 9, 1284(1990).
10. L. M. Sheppard, Am.Ceram.Soc.Bull. 70, 1479(1991).
11. K. Tanaka, A. Nozue and K. Kamiya, J.Mat.Sci. 25, 3551(1990).
12. S. Sakka, J. Non-Crystalline Solids 121, 417(1990).
13. D. I. Dos Santos, U. Balachandran, R. A. Guttashow and R. B. Poeppel, ibid. 121, 448(1990).
14. A. Deptuła, Nukleonika 7, 265(1962).
15. A. Deptuła, W. Łada, T. Olczak and A. Di Bartolomeo J. Non-Crystalline Solids, in press (1992).
16. R. E. Mesmer and C. F. Baes, Jr. see ref. [6], pp.85–96.
17. A. Deptuła, J. Rebandel, W. Drozda, W. Łada and T. Olczak, MRS Spring Meeting, San Francisco, April 27–May 1(1992).
18. A. C. Pierre, T. Nickerson and W. Kresic, J. Non-Crystalline Solids 121, 45(1990).
19. R. Xu, E. J. A. Pope, J. D. Mackenzie, ibid. 106, 245(1988).
20. T. E. Wood, A. R. Siedle, J. R. Hill, R. P. Skarjune and C. Goodbrake, see ref. [6], pp.97–116.

THIS WORK HAS BEEN SUPPORTED
BY STATE COMITTEE FOR SCIENTIFIC RESEARCH, POLAND.

Project No. 7.00.63.91.01

FROM CERAMICS TO SUPERCONDUCTORS: RAPID MATERIALS SYNTHESIS BY SOLID-STATE METATHESIS REACTIONS

RANDOLPH E. TREECE, EDWARD G. GILLAN, RICHARD M. JACUBINAS, JOHN B. WILEY, AND RICHARD B. KANER*, University of California, Los Angeles, Dept. of Chemistry and Biochemistry and Solid State Science Center, 405 Hilgard Ave., Los Angeles, CA 90024-1569.

Abstract

The preparation of materials from gas and liquid phase precursor reactions is well documented. However, solid-state precursor routes have remained largely unexplored. This synthetic void led us to develop rapid (< 2 s), solid-state metathesis (exchange) reactions with a very broad range of synthetic applications. An example is the self-propagating reaction:

$$GaI_3 + Na_3As \xrightarrow{\text{Ignite}} GaAs + 3\ NaI$$

where the NaI is simply washed away with water. Analogous reactions allow the preparation of carbides, silicides, chalcogenides (O, S, Se, and Te), and pnictides (N, P, As, and Sb), of main-group, transition-, and rare-earth metals. These methods can also be exploited to produce mixed-metal and mixed-nonmetal solid solutions.

To gain insight into some of the mechanistic factors involved in these reactions, such as initiation, nucleation, and propagation, we have employed a variety of physical characterization methods. This paper will review the range of materials accessible by our synthetic approaches, as well as the control of product crystallinity, phase, and homogeneity through the optimization of reaction conditions.

Introduction

One concern of solid-state chemistry is *controlling* the reactivity of solids. Two classes of solid-state reactions are a) solid-*gas* and b) solid-*solid* processes. In solid-gas reactions, the process can be likened to tarnishing, where the rate of the reaction is severely retarded by the barrier action of the product layer. An example is the oxidation of aluminum in air ($2Al + 3/2\ O_2 \rightarrow Al_2O_3$). The alumina product layer prevents diffusion of both the reactive gas (O_2) to the solid, and diffusion of aluminum atoms from the solid out to the gas.[1] Solid-solid reactions are likewise diffusion limited. In the case of a metathesis (exchange) reaction where $AB + CD \rightarrow AC + BD$, there are two product layers through which the reacting species must diffuse. Early synthetic approaches in solid-gas and solid-solid reactions utilized high temperatures for long durations to overcome diffusion barriers. These high-temperature processes require large amounts of energy, and the products are typically limited to annealed polycrystalline powders of the thermodynamically stable phase. There has been considerable interest in developing synthetic methods which would require less energy input and allow for control of product homogeneity, crystallinity (from amorphous to highly crystalline), and phase (i.e., high-temperature, or metastable forms).

In an effort to overcome diffusion barriers and control the reactivity of solids, investigations have focused on decreasing the particle size and increasing the mixing of the reactants in both precursor and elemental reactions. One elemental method has taken full advantage of the increased reactivity of smaller particles. Soviet researchers, over 25 years ago, discovered that if they made fine enough powders of two metals, a thorough mixture of the two powders could be ignited. They coined the term 'self-propagating high-temperature synthesis' (SHS) for combustion reactions between compacted mixtures of fine elemental powders. SHS reactions are now capable of producing borides, carbides, nitrides, silicides, intermetallics and many other important materials.[2,3]

The SHS reactions are typically ignited with heated tungsten coils, or with laser pulses. It has been noted that in a number transition-metal aluminide, carbide, and boride systems the combustion reactions would not initiate until the least refractory substituent had melted.[3] Melting permits increased surface contact between the reactants at elevated temperatures.

Intimate contact is important because diffusion in solid-solid reactions is a serious impediment to product formation. Another important consideration in SHS reactions, besides initiation, is the adiabatic temperature, T_{ad}. This is the theoretical maximum temperature which the products may attain in any combustion event given the heat of reaction, the heat capacities, enthalpies of fusion, and vaporization.[3]

While SHS methods require little energy input, they still lack effective control over product crystallinity and phase. However, some degree of product control has been established by precursor methods. Molecular-level mixing has allowed stoichiometric and homogeneous ternary metal oxides to be prepared by the precipitation and low-temperature thermolysis of single-source precursors (i.e., $Ba[TiO(C_2O_4)_2] \rightarrow BaTiO_3$).[4] Another low-temperature precursor approach is the sol-gel technique.[5] In this process, a concentrated sol of the reactant oxides and/or hydroxides is converted to a semirigid gel by removal of the solvent. Sol-gel methods have been used to prepare complex oxides and control the formation of noncrystalline solids.

Our group is investigating rapid, solid-state precursor metathesis reactions. In this approach, two solid precursors are mixed together and ignited, to give rapid and self-sustaining reactions. This paper will describe the general approach and wide synthetic application of rapid solid-state precursor reactions. The successful application of this synthetic method with respect to the control of product crystallinity, phase, and homogeneity will also be discussed.

Metathesis Reactions

Wide Synthetic Application. It is now possible to synthesize a very broad range of materials in seconds from self-propagating solid-state precursors. A generalized reaction scheme is shown in Eq. 1,

$$MX_m + m\,AY_n \xrightarrow[\text{2) Wash}]{\text{1) Ignite}} MY_z + m\,AX + [(m \cdot n)-z]\,Y \qquad (1)$$

where M represents a metal, X a halide, A an alkali metal, and Y the nonmetal or metalloid. The precursors are mixed together in a dry box, loaded into a reaction bomb, and ignited by brief, local heating from a hot filament. The byproduct, alkali halide, is removed by simply washing with methanol and/or water, leaving the polycrystalline products. Once the products have been isolated, they are then dried. Materials produced by this method include superconducting (NbN, ZrN, TaN) nitrides,[6] III-V semiconductors (InP, GaAs, InSb, etc.),[7,8] insulating (SiC and BN)[9,10] and magnetic (GdP, SmAs, etc.)[11] materials, transition-metal pnictides (ZrP, NbAs),[12] oxides (Cr_2O_3, CuO, ZrO_2),[13] and dichalcogenides (MoS_2, $MoSe_2$, NiS_2),[14-18] and refractory intermetallic compounds ($MoSi_2$, $NbAl_3$, MoB_2, WSi_2).[19,20] Along with the binary materials, mixed-metal ($Al_xGa_{1-x}As$, $Mo_xW_{1-x}S_2$)[16,21] and mixed-nonmetal ($MoS_{2-x}Se_x$, GaP_xAs_{1-x}, and GdP_xAs_{1-x})[11,18,21] ternary compounds can also be prepared from the appropriate solid-solution precursors.

This method, using solid-state precursors in rapid, self-sustaining reactions, has clear advantages over traditional high-temperature processes. It is possible to synthesize materials in which one of the elements has a high vapor pressure at elevated temperatures, such as nitrides, arsenides, and phosphides. The superconductor ZrN ($T_c = 10.0$ K) is typically prepared under high temperatures and nitrogen pressures, but we have prepared ZrN (T_c onset of 8.5 K, determined by magnetic susceptibility) in seconds from $ZrCl_4$ and Li_3N:[6]

$$ZrCl_4 + 4/3\,Li_3N \longrightarrow ZrN + 4\,LiCl + 1/6\,N_2 \qquad (2)$$

Another example of the utility of this approach in overcoming problems associated with traditional synthesis methods is the rapid preparation of refractory intermetallics. WSi_2 (mp = 2150 °C) is very difficult to prepare by traditional high-temperature methods, due to the refractory nature of W and Si; however, it can be rapidly synthesized by the reaction:[10,19]

$$WCl_6 + 3/2\,Mg_2Si \longrightarrow 3/4\,WSi_2 + 3\,MgCl_2 + 1/4\,W \qquad (3)$$

Initiation Conditions. An important aspect of these reactions is the ignition process. Initiation of the self-sustaining reactions depends on the degree of surface reaction, and propagation of the reaction requires that the heat generated in the initial steps be sufficient to overcome the overall reaction barrier of the bulk mixture. Many of the metathesis reactions self-initiate while others require thermal input, coming from mechanical heating (mortar and pestle), or a hot filament (T < 850 °C). The major factor influencing initiation conditions is phase transition or decomposition points. Vaporization, melting, and/or decomposition of the lowest melting/decomposing precursor increases surface contact, thereby initiating the reaction. The fact that a phase change must occur in one precursor is similar to the need for melting of one element in the SHS reactions; however, the temperatures necessary for initiation in the precursor reactions are much lower.

Since a compounds melting and boiling points are related to its structure, we have investigated the relationship between structure and initiation conditions. The link between structure and initiation temperature is illustrated in the series of reactions:

$$GaX_3 + Na_3P \dashrightarrow GaP + 3\ NaX \qquad (4)$$

where X represents I, Cl, and F. The initial step in the metal iodide and metal chloride precursor reactions appears to be melting and/or vaporization of the metal halide precursors, followed by production of the sodium halide byproduct. Subsequent formation of the III-V occurs within the molten reaction flux. The metal iodides are high vapor-pressure, molecular crystals with boiling/sublimation temperatures less than 360 °C, and reactions with these precursors can be initiated by light grinding with a mortar and pestle. Isostructural gallium chloride melts at 78 °C, and reactions between it and Na_3P self-initiate within seconds upon mixing. Reactions with metal fluorides, however, do not self-ignite. The metal fluorides are three dimensional network solids with melting points greater than 900 °C, and it appears that a phase change in the pnictiding agent (decomp. > 550 °C) initiates the reactions between metal fluoride and sodium pnictide precursors. The absence of any appreciable vapor pressure from the sodium pnictides at room temperature explains the observations of no self-initiating reactions between the fluoride and pnictide precursors. Reactions to form GdN are examples where both of the precursors melt/decompose at temperatures above the temperature of the filament (~ 850 °C). In this case, the reaction between GdI_3 (mp = 930 °C), a 3-dimensional solid, and Li_3N (mp > 950 °C), isostructural with Na_3P, cannot be initiated with the hot wire.

Table I.

Melting/Decomposition Point and Initiation Conditions

MX_3	(mp, °C)	Na_3X	(decomp.,°C)	Initiation Conditions
$GaCl_3$	(78)	Na_3P	(>550)	Stirring Together
GaI_3	(212)	Na_3P	(>550)	Grinding w/mortar & pestle
GaF_3	(>900)	Na_3P	(>550)	Heating w/Hot Wire
GdI_3	(930)	Li_3N	(>950)	*Not Initiated by Hot Wire*

These ignited precursor reactions are very exothermic and quite rapid. The calculated enthalpy, for the WSi_2 reaction shown in Eq. 3, is -306 kcal/mol.[22,23] The driving force is the formation of the salt byproduct. In this reaction, the production of $MgCl_2$ accounts for > 95 % of the heat of product formation. Optical pyrometry indicates that this reaction reaches temperatures in excess of 1000 °C within seconds of initiation, cooling to room temperature in less than 15 seconds. Temperatures measured in the ignition synthesis of GaP, from GaI_3 and Na_3P (Eq. 4), exceed 750 °C, even though the calculated enthalpy is less than half (-138 kcal/mole) of that predicted for the WSi_2 reaction.[22,23] As also observed in the WSi_2 reaction, the formation of the salt byproduct (NaI) provides the thermodynamic driving force (~ 90 % of product ΔH_f) for the production of GaP.

Control of Product Crystallinity. The temperatures generated in these reactions are sufficient to melt or vaporize the reactants and salt byproduct, producing a reaction flux and thereby overcoming solid-state barriers to diffusion. By controlling the conditions of the reaction flux, it is possible to control product crystallinity. The adiabatic temperature, T_{ad} (theoretical upper limit), calculated for WSi_2 (Eq. 3), is 2103 °C. But, the temperature actually achieved may be far less than T_{ad}, due to incomplete reaction and heat loss. It has been shown that larger reaction mixtures lead to greater yields and increased product crystallinity.[6] The larger reaction mixtures provide greater insulation of the molten flux, thereby decreasing heat loss to the surroundings. Improved insulation allows the core of the reaction flux to remain at higher temperatures for longer periods than is permitted with the smaller mixtures.

There is considerable interest in developing techniques which make it possible to produce small particles. Quantum-sized particles of certain semiconductors are desirable for optoelectronic devices, and small particles allow for better densification in high temperature applications. Investigations in our lab have determined that while better insulation of the reaction flux can lead to increased crystallinity, adding a heat sink to the precursor mixture can lead to products with decreased crystallinity.[15] It has been shown that the addition of controlled amounts of inert salt byproduct can lead to a systematic decrease in particle size in certain systems. The effect of the heat sink is to lower T_{ad}, resulting in a reduction of the self-annealing properties the reaction flux.

Control of Product Phase. Closely related to the desire to control crystallinity is the interest in selecting a certain structural phase. For instance, a great deal of research has been applied to the polymorphic ZrO_2 system. For ZrO_2 to be a useful refractory material, it is necessary to stabilize its high-temperature phase at lower temperatures. On heating, normal ZrO_2 undergoes a phase transition at 1100 °C (monoclinic → tetragonal) and again at 2300 °C (tetragonal → cubic).[24] While the tetragonal phase cannot be quenched to room temperature, poorly crystalline cubic/tetragonal (c/t) ZrO_2, has been prepared and isolated by precipitation and sol-gel routes.[25] The challenge has been to prepare a metastable c/t-ZrO_2 phase that will not revert to the stable monoclinic form on heating. It is possible to synthesize and isolate c/t-ZrO_2, from solid-state precursors. After washing the products of the controlled heating of $ZrCl_4$ and Na_2O (Eq. 5) in sealed tubes, an amorphous powder remains. Upon heating to 350 °C for 12 h, X-ray powder diffraction (XRD) patterns reveal the formation of metastable c/t-ZrO_2. The ZrO_2 remains in the metastable phase until heated to above ~ 500 °C, at which point the transformation to monoclinic ZrO_2 becomes apparent.

$$ZrCl_4 + 2\,Na_2O \xrightarrow[\text{2) 5 days}]{\text{1) 220 °C}} ZrO_2\,(\text{amorphous}) + 4\,NaCl \xrightarrow[\text{2) 350 °C / 12 h}]{\text{1) Wash}} \text{c/t-}ZrO_2 \quad (5)$$

The structural transformation caused by heating the amorphous ZrO_2 can be followed in Figure 1. The XRD patterns are of the washed initial products (Fig. 1a), the c/t-ZrO_2 phase formed after heating the amorphous material to > 350 °C (Fig. 1b), and the monoclinic phase coexistent with the c/t-ZrO_2 after heating the metastable material to > 500 °C (Fig. 1c). Work is currently in progress to investigate ternary metal substitution leading to stabilization of c/t-ZrO_2 to higher temperatures.[26]

Phase control is also sought in certain refractory alloys. $MoSi_2$ is used as a furnace element because of its high melting point and resistance to corrosion and oxidation at elevated temperatures. A significant shortcoming to the use of $MoSi_2$ as a high-temperature structural material is its poor ductility. The low-temperature ($T < 1900$ °C) α-form of $MoSi_2$ is the tetragonal $C11_b$ structure, while its high-temperature ß-form is the hexagonal C40 arrangement. It has been proposed that a mixture of α and ß forms might greatly increase ductility. We have been able to make the ß-form of $MoSi_2$ directly in precursor reactions initiated at close to room temperature.[19,20] It may be possible to use this synthetic approach to form other metastable phases and perhaps even prepare new materials which cannot be made in any other way.

Control of Ternary Homogeneity. Solid-state precursors have been used to prepare mixed-metal and mixed-nonmetal solid solutions.[11,16,18,21] A significant challenge in the preparation of ternary solid-solutions is uniform atomic mixing throughout the product. Solid-state precursor reactions make it possible to achieve homogeneous ternary products because they start with well-mixed solid-solution precursors. A comparison was made between two 'ternary' reactions. In the first reaction (Eq. 6), two binary pnictides precursors were physically mixed by thorough grinding before being stirred together with the gallium iodide, and in the second reaction (Eq. 7),

$$GaI_3 + 1/2\,Na_3P + 1/2\,Na_3As \longrightarrow Ga(P, As) + 3\,NaI \quad (6)$$

$$GaI_3 + Na_3P_{0.5}As_{0.5} \longrightarrow Ga(P, As) + 3\,NaI \quad (7)$$

a solid-solution precursor was used. The advantage of using a solid-solution precursor is clearly displayed in Figure 2, where XRD patterns of the products of the two different ternary

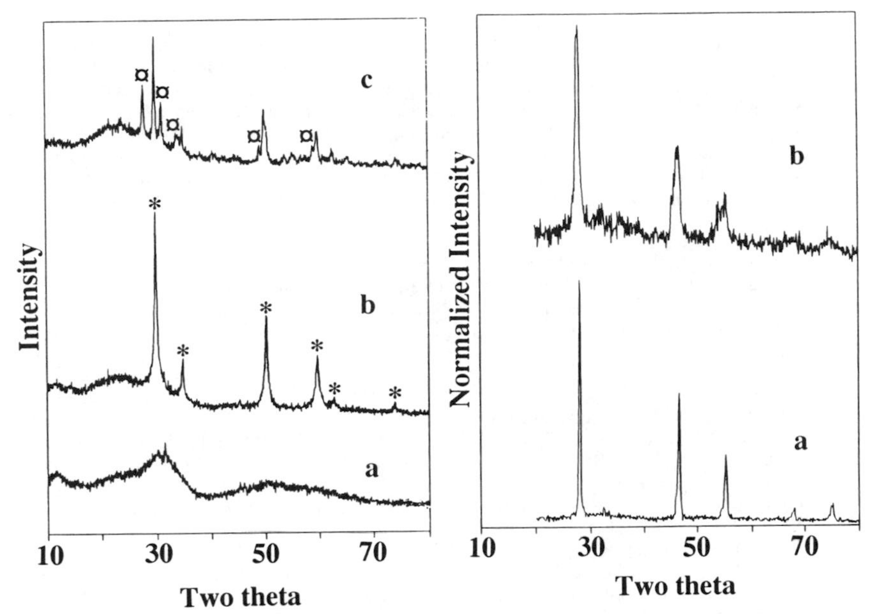

Figure 1. Powder X-ray diffraction pattern of the products described in Eq. 5. The * indicates the c/t-ZrO_2 phase, and the ¤ indicates the monoclinic phase. See text for details.

Figure 2. Powder X-ray diffraction patterns of Ga(P,As) produced in Eq. 6 (Fig. 2b) and Eq. 7 (Fig. 2a). See text for details.

reactions are shown. The diffraction lines of the material represented in the upper pattern (Fig. 2b) are broadened considerably relative to the lower pattern (Fig. 2a). This type of broadening is indicative of significant compositional gradients and variations of regional stoichiometry. Despite good *physical* mixing, the P and As atoms are still very localized within the reaction mixture compared to the homogeneous dispersion possible with atomic-level *chemical* mixing.

Conclusions

The initiation processes are fairly straightforward: the self-sustaining reaction is initiated when at least one of the precursors changes phase (melts, vaporizes, or decomposes). However, the rapid, exothermic nature of these precursor reactions makes it very difficult to observe directly intermediate species or mechanistic steps It is likely though, that shortly after ignition of the reaction, under the extreme conditions present, there may be a variety of species present in the reaction flux, such as the byproduct, precursor molecular fragments, ions in various oxidation states, and free elements. The fact that these highly reactive species are dispersed throughout a high-temperature salt flux accounts for the success of this approach. The diffusion barriers typically associated with solid-solid reactions are significantly decreased. The particle sizes of the reactants has been reduced to essentially atomic dimensions, allowing for rapid product formation. Even the formation of the salt byproduct seems to promote continued reaction by (a) generating a large amount of heat, and (b) becoming a molten host allowing further reactant diffusion and nucleation of the primary product. The reaction ends when the heat has dissipated and the products cool.

Understanding exactly what happens in the reaction flux is necessary for complete control of the process. We are currently investigating the details of the reaction process, but even without that knowledge, many positive aspects of this method are already apparent. (1) The solid-state precursor reactions are easily initiated, rapid, and self-sustaining. This reduces the time and energy put into the reaction process. (2) The species within the flux are extremely reactive and finely dispersed. This enables reactions between refractory materials at temperatures far below their bulk melting points. (3) The maximum temperature achieved by the reaction flux and the amount of time at that temperature affects the yield and crystallinity of products. Larger reaction mixtures are better insulated and lead to greater product crystallinity and yield, whereas, adding an inert heat sink to a reaction mixture leads to decreased product crystallinity by lowering the overall temperature. (4) It is possible to form some high-temperature phases through these reactions, such as cubic/tetragonal ZrO_2 and ß-$MoSi_2$. (5) The ability to use homogeneous, ternary precursors allows for the synthesis of compositionally uniform mixed-metal and mixed-nonmetal solid-solution materials.

Acknowledgments

The authors gratefully acknowledge Dr. P.R. Bonneau, Robert F. Jarvis, Jr., and Lin Rao for their participation in fruitful discussions of this work. This work was supported by the National Science Foundation, Grant No. 8657822, through the Presidential Young Investigator Award Program, and by a David and Lucille Packard Foundation Fellowship in Science and Engineering.

References

1. N.B. Pilling and R.E. Bedworth, J. Inst. Metals *1*, 529 (1923).
2. (a) R.W. Cahn, Adv. Mater. 2, 314 (1990). (b) A.G. Merzhanov and I.P. Borovinskaya, Dokl. Akad. Nauk SSSR (Engl. transl.) *204*, 429 (1972). (c) H.C. Yi and J.J. Moore, J. Mater. Sci. *25*, 1159 (1990).
3. Z.A. Munir and U. Anselmi-Tamburini, Mat. Sci. Reports *3*, 277 (1989).
4. A. Wold, J. Chem. Ed. *57*, 531 (1980).
5. D.W. Johnson, Jr., Amer. Ceram. Soc. Bull. *60*, 221 (1981).
6. E.G. Gillan, R.E. Treece, J.B. Wiley, R.B. Kaner, presented at the American Chemical Society Pacific Conference on Chemistry and Spectroscopy, Anaheim, California, October 1991 (unpublished).
7. R.E. Treece, G.S. Macala, R.B. Kaner, Chem. Mater *4*, 9 (1992).
8. R.E. Treece, G.S. Macala, L. Rao, R.B. Kaner, in preparation.
9. E.G. Gillan, L. Rao, R.B. Kaner, in preparation.
10. Kaner, R.; Bonneau, P.; Gillan, E.; Wiley, J.; Jarvis, Jr., R.; Treece, R.; "Rapid Solid-State Synthesis of Refractory Materials", U.S. Pat., filed 28 January 1991, Notice of Allowance 10 April 1991.
11. R.E. Treece and R.B. Kaner, in preparation.
12. R.F. Jarvis, Jr., and R.B. Kaner, presented at the American Chemical Society Pacific Conference on Chemistry and Spectroscopy, Anaheim, California, October 1991 (unpublished).
13. J.B. Wiley, E.G. Gillan, and R.B. Kaner, unpublished.
14. P.R. Bonneau, R.K. Shibao, and R.B. Kaner, Inorg. Chem. *29*, 2511 (1990).
15. P.R. Bonneau, R.F. Jarvis, Jr., and R.B. Kaner, Nature *349*, 510 (1991).
16. P.R. Bonneau, PhD thesis, University of California, Los Angeles, 1991.
17. P.R. Bonneau, J.B. Wiley, and R.B. Kaner, Inorg. Synth., in press.
18. P.R. Bonneau, R.F. Jarvis, Jr., and R.B. Kaner, Inorg. Chem., in press.
19. E.G. Gillan, L. Rao, R.M. Jacubinas, and R.B. Kaner, presentation at the American Chemical Society Fall Meeting, Washington, DC, August 1992 (submitted).
20. J.B. Wiley and R.B. Kaner, Science *255*, 1093 (1992).
21. R.E. Treece, G.S. Macala, P.R. Bonneau, and R.B. Kaner, in preparation.
22. (a) D.R. Lide, Jr., Ed.,*JANAF Thermochemical Tables* , 3rd ed. (American Chemical Society and American Institute of Physics, Inc.: New York, 1985). (b) R.C.Weast, Ed.,*CRC, Handbook of Chemistry and Physics* ; 64th ed. (CRC Press: Boca Raton, Fl, 1983).
23. O. Kubaschewski and C.B. Alcock, *Metallurgical Thermochemistry*, 5th ed. (Pergamon Press Inc., New York, 1979).
24. A. Wells, *Structural Inorganic Chemistry*, 5th ed.(Claredon Press: Oxford, England, 1986).
25. M.A.C.G van de Graff and A.J. Burggraff in *Advances in Ceramics II*, edited by N. Claussen, M. Ruhle, and A. Heuer (American Ceramics Society, Columbus, OH, 1984) pp 744-765.
26. E.G. Gillan and R.B Kaner, in preparation.

CHARACTERIZATIONS OF SOL-GEL CORDIERITE PRECURSORS

Yun Fa CHEN, Saïd EL HADIGUI and Serge VILMINOT
Groupe des Matériaux Inorganiques,EHICS,1 rue B.Pascal,Strasbourg,France

ABSTRACT

Cordierite powders were prepared using silicon ethoxide, $Mg(NO_3)_2.6H_2O$ and Al $(OBu)_2$etac (etac = ethylacetoacetate) and conditions of hydrolysis have been checked in order pure α cordierite to be obtained. D.T.A. measurements reveal a stepwise crystallization process with intermediate formation of metastable μ cordierite. Infrared studies allow to follow the evolution of the Si-O-Al network with temperature.

INTRODUCTION

Due to its properties, low thermal expansion, low dielectric constant, chemical and thermal stability, cordierite, $Mg_2Al_4Si_5O_{18}$, ceramics are promising materials. The field of their applications deals with catalytic system in automobiles, high temperature furnaces, industrial heat exchangers. It is also considered as an alternative substrate to alumina for electronic applications. The main problem encountered during processing of cordierite powders is that dense ceramics are difficult to obtain without sintering aids, aids that increase the thermal expansion coefficient.

The sol-gel approach has been successfully used by many groups and its success comes from the high reactivity of the resulting powders yielding a strong decrease of the sintering temperature. This low sintering temperature allows the substitution of noble metals currently used as conductors in electronic packaging by copper as it has been demonstrated in our laboratory [1]. The sol-gel processes can be divided in two categories according to the nature of the precursors, either particulate sols or polymeric ones. Particulate sols of boehmite and Ludox have been used, magnesium being introduced as nitrate solution [2]. Polymeric sols involve alkoxide(s) as precursors, either by means of all-alkoxide routes [3-5] or using some soluble metal salt(s) [2,6,7]. Differences in hydrolysis kinetics have to be controlled and most processes involve a silicon alkoxide prehydrolysis.

In our case, we have proceeded to a chemical modification of the aluminum alkoxide by ethylacetoacetate (etac). This paper focuses on the elaboration and characterization of the resulting cordierite powders.

EXPERIMENTAL PROCEDURE

Preparation of gel powders

The silica source in the present work was tetraethyl orthosilicate, TEOS, obtained from Aldrich. Aluminum sec-butoxide (Aldrich) has been analysed for its aluminum content by firing the powder obtained by hydrolysis with a excess water and weighing the resulting α alumina as detected by X-ray diffraction. Magnesium has been introduced as magnesium nitrate, $Mg(NO_3)_2.6H_2O$ (Prolabo). A schematic diagram of the sol-gel process is

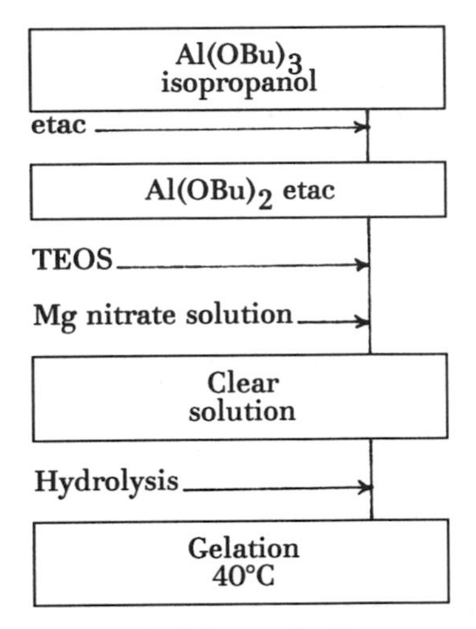

Figure 1. Schematic diagram of sol-gel process

shown in Figure 1. Aluminum butoxide is dissolved in iso-propyl alcohol under stirring. Ethylacetoacetate, (etac), in stoichiometric proportion to replace one OBu group, is dropwise added into the aluminum solution. After a few minutes stirring, TEOS is added, then the magnesium nitrate previously dissolved in the same solvent. A clear solution is obtained even if water has already been introduced by means of hydration molecules. This result is no more observed if we start from Al $(OBu)_3$, the nitrate addition promoting immediate precipitation and clearly illustrates the role of the modifying molecules. Even if hydrolysis has started, the resulting species remain soluble. A first series of samples (COR1) is left at 40°C without any further water addition yielding gelation within two days. A second series (COR2) has been hydrolysed by dropwise addition of an alcohol-water solution with a water amount corresponding to a total H_2O/OR ratio of 1.25 (OR=OBu and OEt). The third series corresponds to a ratio H_2O/OR = 2.5 (COR3). After water addition, the samples are also left at 40°C and monolithic gels are obtained within one hour time.

Drying of the gels has been performed by gradual heating at a rate of 20°C/H to 250°C in air and the powders were left either 2 or 36 hours at this temperature. After drying, the COR1 powders are black whereas the other samples have got a brownish to yellow color. Further calcination at 700°C for two hours results in dark grey powders for COR1 and white ones for COR2 and COR3 respectively.

Measurements of properties

Thermal analysis (TGA-DTA) have been performed on a Setaram apparatus equipped with a heating facility up to 1600°C at heating rates of 4 and 10°C/mn. X-ray diffraction patterns were taken at 1° 2θ per minute using CoK radiation (Siemens D500). Specific areas were measured by nitrogen adsorption at liquid nitrogen temperature (Carlo Erba). The infrared spectra were recorded on a Nicolet FT-IR apparatus on specimens prepared by the pellet method, the powder being mixed with IR-grade KBr and subsequently pressed into a transparent thin disk. For reference a KBr disk was used.

RESULTS AND DISCUSSION

Thermal Analysis

The TGA curves recorded on samples after drying at 40°C reveal, around 150°C, an abrupt weight loss with a strong exothermal effect. Above a sample heated at the same temperature in a furnace, we observe the formation of reddish brown fumes attributed to nitrate decomposition.

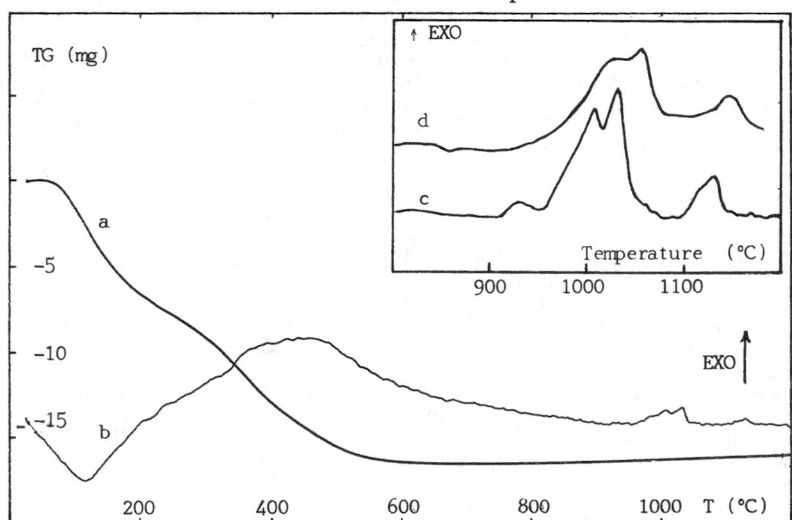

Figure 2. TGA (a) and DTA (b) traces of COR2 samples with 4°C/mn heating rate. D.T.A. traces at 4°C/mn (c) and 10°C/mn (d)

Thermal analysis have therefore been performed on samples after 250°C calcination. The thermograms do not significantly differ according to the hydrolysis conditions. The TGA curve (Figure 2) reveals that the weight loss proceeds within two steps, the first one associated with an endothermal effect (evaporation), the second one with an exothermal effect (combustion). Constant weight is achieved around 600°C. At higher temperature (insert), a succession of exothermal effects is observed on the DTA trace, whose maxima differ with the heating rate, the lower the heating rate, the lower the maxima temperature. Two or three effects appear in the temperature range 920-1080°C and one between 1100 and 1200°C. These effects are related to crystallization as revealed by X-ray diffraction.

X-ray diffraction

In order to get more insight into the effects observed at high temperature on the DTA curve, samples have been heated using the same procedure (4°C/mn), left at the desired temperature for 3 mn and air quenched. X-ray diffraction patterns on the COR2 quenched samples only reveal the presence of μ and/or α cordierite. μ cordierite is the only phase after 900°C treatment. Below this temperature, the samples appear to be amorphous. α cordierite appears after 960°C treatment, μ cordierite being the major phase. Heat treatments at higher temperatures, 1000, 1040 and 1080°C reveal an increase of the crystallization of both phases, μ cordierite being still the major phase. From 1100°C, αcordierite becomes to be the major phase and pure αcordierite is observed on samples after 3 mn treatment at 1200°C. We can therefore conclude that all thermal effects on the DTA curve are related to μ and/or αcordierite crystallization. The presence of multiple effects in the temperature range 920 to 1080°C seems to be related to heterogeneities inducing a stepwise crystallization process.

According to the heating rate influence, the crystallization appears to be kinetically dependent. COR2 samples have been calcined for the same period of time (2 hours) using different heating rates, 1°C/mn to 20°C/mn. X-ray diffraction leads to the following observations after 1100° thermal treatment :

- for heating rates below 4°C/mn, cordierite is the major phase, spinel being also detected, which proportion decreases with increasing heating rate. This result can be related to the fact that spinel is the thermodynamically most stable phase and low heating rate promotes its formation,
- for heating rates between 4°C/mn and 6°C/mn, pure cordierite is observed at 1100°C after two hours treatment,
- for heating rates higher than 6°C/mn, the samples reveal the presence of cordierite as the major phase, spinel and μ cordierite. It is also to be noticed that the μ cordierite diffraction lines are shifted towards smaller d-spacings. This indicates that the μ cordierite phase has no more the required stoichiometry, being displaced towards higher SiO_2 contents.

Samples issued from the different hydrolysis conditions have been calcined at 1100°C for two hours with a heating rate of 4°C/mn. X-ray diffraction patterns reveal :

- the formation of pure cordierite for samples COR1 (no water added) but the powders remain grey even after treatment at higher temperature,
- the same result is observed with COR2 samples but in this case white powders are obtained. The drying time at 250°C does not seem to have any influence,
- when the water amount for hydrolysis is increased, COR3, one observes the presence of small amounts of spinel and mullite. If water is in strong excess, the only phase detected at 1100°C is mullite.

Specific area measurements

TABLE I. Specific area values

Sample	Drying temperature (°C)	Specific area (m^2/g)
COR 2	250	460
	400	480
	500	370
	700	329
	800	35
	850	29
COR 1	250	32
	500	7
	700	12
	850	12

The specific area values of COR1 and COR2 samples are summarized on Table I. For COR2 samples, the maximum value is observed after 400°C calcination. The increase between 250 and 400°C is related to further weight loss in this domain inducing formation of new porosity. The specific area value falls down at 800°C in relation with sintering and crystallization.

For COR1 samples, the values are always very low. Similar results have been observed by KARAGEDOV et al. [4] and can be attributed to the incomplete hydrolysis yielding the formation of pyrolysis products trapped inside the pores.

Infrared studies

COR2 samples have been calcined at different temperatures using the same procedure before measurements. Correct assignment of vibrational frequencies in silicate structures is not at all an easy matter. It has been a subject of controversy for many years now. We mainly focuse on the evolution

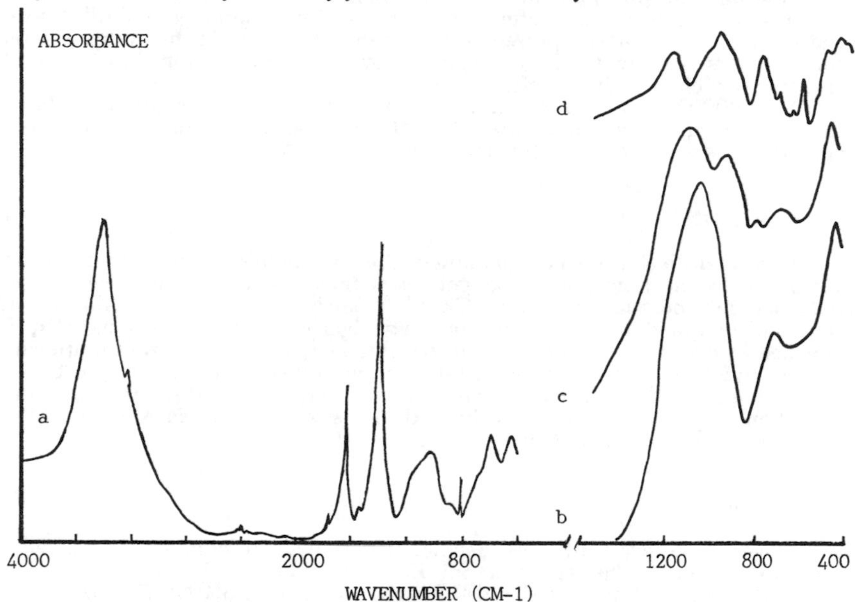

Figure 3. Infrared spectra of COR2 powders after drying at a : 40°C, and calcination at b : 500°C, c : 800°C and d : 1100°C

of spectra in the region between 1200 and 400 cm^{-1}. Outside this region, the major bands correspond to water (3500 and 1600 cm^{-1} domains).

First let us consider the infrared spectrum of the starting powder, that is after 40°C drying (Figure 3). Apart from water bands, one observes the presence of two sharp bands at 1385 and 830 cm^{-1} that have been attributed to NO_3 groups. These bands disappear after thermal treatment above 150°C as expected from the TGA results. Surprisingly, no bands related to etac groups are observed. It has been mentioned that such groups are not hydrolysable. In our process conditions, magnesium nitrate promotes a low pH value that can be responsible for the displacement of etac molecules. The other bands are related to Si-O and Al-O vibrations, the band around 600 cm^{-1} being attributed to AlO_6 groups.

After thermal treatment between 250 and 750°C, three bands are observed as expected in silicate glasses in the 1100, 800 and 500 cm^{-1} regions. The band at 1100 cm^{-1} significantly shifts to a lower frequency and it has been assigned to the M-O bond stretching motion in MO_4 tetrahedron (M=Si or Al). Whereas in glasses obtained by quenching from the liquid phase with compositions close to that of cordierite the 1100 cm^{-1} band is shifted to 1084 cm^{-1} [8], in our case the same band is observed between 1044 and 1012 cm^{-1} according to the thermal treatment. This can be related to the effect of strongly coupled Al-O and Si-O vibration. On the other hand, the band around 600 cm^{-1} related to AlO_6 groups is no more observed on samples calcined above 400°C, indicating that more aluminum ions replace silicon in the tetrahedra unit to form a Si-O-Al network structure. In all samples, the strongest MO_4 bands appear in frequencies greater than 960 cm^{-1} and probably imply a three dimensional network with a highly disordered structure. The shifts observed for the bands in the 800 and 500 cm^{-1} regions have the same origin as for the 1100 cm^{-1} band.

The spectra of samples after 800 and 850°C calcination are similar with the spectra of μ cordierite presented in the literature [5]. At this temperature, the powders appear to be amorphous by X-ray diffraction but infrared spectroscopy clearly reveals their transformation.

After 900°C thermal treatment, new bands start to appear and are related to α cordierite formation. After 1100°C thermal treatment, the infrared spectrum fits very well with the data cited for pure αcordierite [5].

CONCLUSIONS

Pure αcordierite has been obtained after two hours thermal treatment at 1100°C of an amorphous sol-gel precursor issued from hydrolysis-polycondensation reactions using TEOS, magnesium nitrate and aluminum butoxide modified by one mole etac. The hydrolysis conditions have been checked in order pure α cordierite to be obtained. The crystallization appears to be stronghly kinetic dependent and heating rates between 4 and 6°C/mn have to be used. Infrared spectroscopy allowed to follow the thermal evolution of the Si-O-Al network that exhibits strongly coupled Al-O and Si-O vibration in the amorphous phase.

REFERENCES

1.V.Oliver-Broudic, J.Guille, J.C.Bernier, B.S.Han, J.Werckmann, J.Faerber, P.Humbert and B.Carriere, Mat. Sci. Eng. A109, 77 (1989)
2. A.M.Kazakos, S.Komarneni and R.Roy, J. Mater. Res. 5,1095 (1990)
3. M.Nogami, S.Ogawa and K.Nagasaka, J. Mat. Sci. 24, 4339 (1989)
4. G.Karagedov, A.Feltz and B.Neidnicht, J. Mat. Sci. 26, 6396 (1991)
5. H.Suzuki, K.Ota and H.Saito, Yogyo-Kyokai-Shi 95, 25 (1987)
6. H.Vesteghem, A.R.Di Giampaolo and A.Dauger, J.Mat.Sci.Lett.6,1187 (1987)
7. J.C.Broudic, S.Vilminot and J.C.Bernier, Mat.Sci.Eng. A109, 253 (1989)
8.B.N.Roy, J.Amer.Ceram. Soc. 73, 846 (1990).

SOL-GEL PROCESS FROM HETEROBIMETALLIC ALKOXIDES TO INCORPORATE LANTHANIDES IN AN ALUMINA MATRIX

CHAITANYA K. NARULA, Chemistry Department, Research Laboratory, Ford Motor Company, P.O. Box 2053, MD 3083, Dearborn, MI 48121

ABSTRACT

Gels form on addition of seven or more moles of water diluted with *i*-PrOH to one mole of $M[Al(O\text{-}i\text{-}Pr)_4]_3$, M = La, Ce dissolved in *i*-PrOH. Thermal treatment of xerogels derived from $La[Al(O\text{-}i\text{-}Pr)_4]_3$ gives amorphous powders below 900°C. $LaAlO_3$ crystallizes out at 900°C and residual alumina remains amorphous. CeO_2 starts to separate out above 600°C from the xerogels prepared from $Ce[Al(O\text{-}i\text{-}Pr)_4]_3$. Thermally induced structural changes in xerogels prepared from the mixtures of $La[Al(O\text{-}i\text{-}Pr)_4]_3$ and $Ce[Al(O\text{-}i\text{-}Pr)_4]_3$ in 1:1 or 1:3 ratio are different from those prepared from individual precursors. These xerogels remain amorphous below 700°C. Crystalline CeO_2 forms on heating at 900°C but alumina remains amorphous.

Sols are formed on hydrolysis of $M[Al(O\text{-}i\text{-}Pr)_4]_3$ in excess water and acidifying the reaction mixture with acetic acid. Sols can be converted to gels by slow evaporation of volatiles. This method is useful in preparing coatings.

INTRODUCTION

The sol-gel process has received considerable attention as a method for preparing metal oxides [1,2]. This method offers several advantages over conventional synthetic procedures. High surface area materials can be prepared at low temperatures, and the processing does not require extensive heat treatment. Sol-gel processing results in high purity materials, and allows control of microstructure and surface properties. This leads to new applications in thin films and coatings, monoliths, and fiber technology [2].

We are interested in sol-gel processed alumina and silica based materials for various automotive applications. These materials form washcoats for catalyst substrates, and optical coatings for glass. In this presentation, we describe results on sol-gel processed alumina based materials for automotive catalyst substrate applications.

For automotive catalyst applications, washcoat material is required to provide a surface area of 25-30 m^2/g after exposure to 900°C. A suspension of stabilized γ-alumina is used for this purpose. Sol-gel processed alumina can also be employed to prepare washcoat material. Stabilizer addition requires mixing alkoxide precursors to alumina prior to hydrolysis or sequential addition of alkoxides to partially hydrolyzed alumina [1]. A relatively new approach in sol-gel processing from heterobimetallic alkoxides has recently been used to prepare two-component oxides [3,4] in order to achieve uniform distribution of both components. We used this approach to incorporate lanthanides in alumina and have optimized the conditions to prepare gels from heterobimetallic alkoxides, $M[Al(O\text{-}i\text{-}Pr)_4]_3$, M = La, Ce. Gels prepared from these precursors are amorphous and form crystalline materials at high temperatures. Initial results on the evaluation of these materials as washcoat for automotive catalyst are also described.

EXPERIMENTAL

All experimental manipulations for handling the alkoxides were carried out in an inert atmosphere using standard vacuum line techniques. 2-Propanol, *i*-PrOH, was dried over $Na(O\text{-}i\text{-}Pr)$ before use. Commercial $Al(O\text{-}i\text{-}Pr)_3$ and $Al(O\text{-}sec\text{-}Bu)_3$ were distilled to remove oxide impurities. $La[Al(O\text{-}i\text{-}Pr)_4]_3$ and $Ce[Al(O\text{-}i\text{-}Pr)_4]_3$ were prepared by the reaction of $LaCl_3.3\text{-}i\text{-}PrOH$ or $CeCl_3.3\text{-}i\text{-}PrOH$ with $K[Al(O\text{-}i\text{-}Pr)_4]$ as described by Mehrotra et.al.[5].

Sols from $La[Al(O\text{-}i\text{-}Pr)_4]_3$ or $Ce[Al(O\text{-}i\text{-}Pr)_4]_3$ or their mixtures in 1:1 or 1:3 ratios are prepared by adding their solutions in dry 2-propanol to water at 80°C with stirring. The reaction mixture is heated to 90°C to boil off 2-propanol. Sufficient glacial acetic acid is added to obtain a clear sol. The sol is maintained at gentle reflux for 16 h. Gels form on concentrating the sols.

Gels can also be prepared by adding water mixed with *i*-PrOH to a solution of $La[Al(O\text{-}i\text{-}Pr)_4]_3$ or $Ce[Al(O\text{-}i\text{-}Pr)_4]_3$ in *i*-PrOH at -78°C. Gel prepared from $La[Al(O\text{-}i\text{-}Pr)_4]_3$ was dried in vacuum and heated in increments of 100°C and maintained at that temperature for 8hrs. Yellow amorphous powder obtained on drying the gel prepared from $Ce[Al(O\text{-}i\text{-}Pr)_4]_3$ was heated in 100°C increments between 500-900°C with eight hour holding period at each temperature. BET surface area and XRD data of the gels were recorded after each heat treatment.

A clear solution is obtained when $La[Al(O\text{-}i\text{-}Pr)_4]_3$ or $Ce[Al(O\text{-}i\text{-}Pr)_4]_3$ or their mixtures in 1:1 or 1:3 ratios in 2-propanol are reacted with water mixed in 2-propanol at -78°C. Gels form on warming the reaction mixture to room temperature slowly. Volatiles are removed in vacuum to obtain xerogels.

RESULTS AND DISCUSSION

Commercial automotive catalysts are honeycomb structures coated with a washcoat of alumina and stabilizers on which platinum and rhodium metals are deposited. The washcoat decreases the channel volume [Fig.1a] which results in increased back-pressure due to the gas flow. To alleviate this problem and to improve the efficiency of automotive catalyst we have investigated sol-gel processed alumina based materials.

We employed alumina sol prepared by Burggraff's method [6] to coat substrates prior to customizing alumina for automotive catalyst. Compared to commercial catalyst, we achieved higher channel volume [Fig. 1b] without sacrificing surface area [7]. After depositing platinum and rhodium metals, this catalyst was tested on a steady state reactor for oxidation of hydrocarbons and CO and reduction of NO_x. The testing procedure simulates automotive exhaust and is described in detail elsewhere [8]. The efficiency of our catalyst is comparable to a commercial one at 550°C. Testing at 900°C cannot be carried out because the γ-alumina in our formulation was not stabilized. It would undergo phase changes and lose surface area.

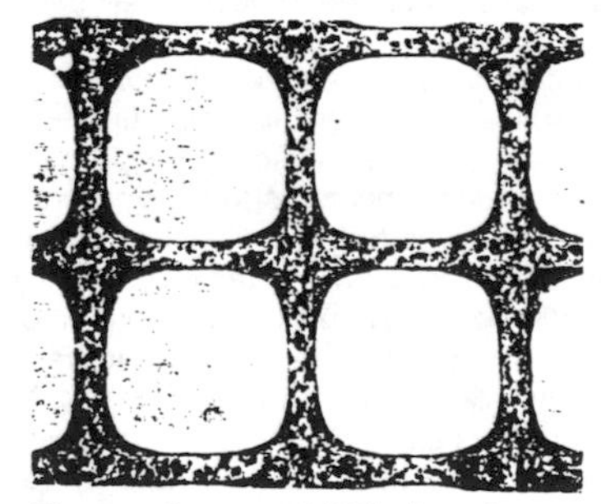

Fig. 1a: Commercial Washcoat

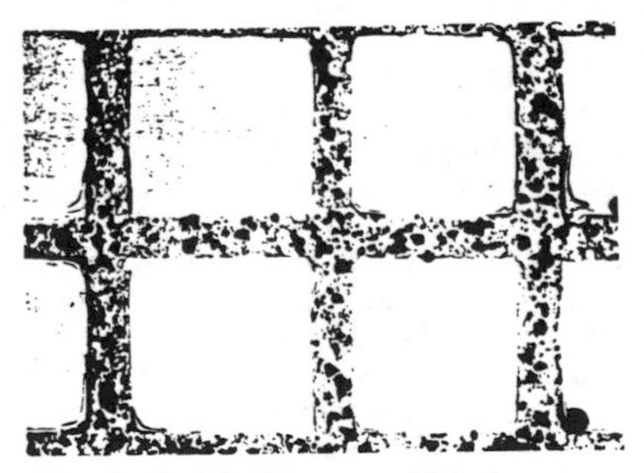

Fig. 1b: Sol-Gel Processed Washcoat

Lanthanum and cerium oxides stabilize γ-alumina and provide improved oxygen storage capacity [9]. This led us to search for suitable methods for the preparation of alumina gels containing lanthanum and cerium oxides. Although a mixture of lanthanum alkoxide and aluminum alkoxide has recently been employed in preparation [10] of $LaAlO_3$, uncertainties in the structures of $Ln(OR)_3$ and substantial loss of products during purification led us to explore other precursors [11]. We selected heterobimetallic alkoxides, $M[Al(O\text{-}i\text{-}Pr)_4]_3$ since they can be prepared in high yield and purity. It has also been suggested that they form M-O-M' species instead of a mixture of metal-oxides on hydrolysis [4,12]. Magnesium aluminate has recently been prepared from $Mg[Al(OR)_4]_2$ [3]. There is also a brief description of the hydrolysis studies on $La[Al(O\text{-}i\text{-}Pr)_4]_3$ and $Zr(O\text{-}i\text{-}Pr)_2[Al(O\text{-}i\text{-}Pr)_4]_2$ [4].

During the course of optimizing gelation conditions, we investigated the reactions of $M[Al(O\text{-}i\text{-}Pr)_4]_3$, [M=La,I; Ce,II] with water. The reactions of I or II with one to six equivalents of water at -78°C in 2-propanol give soluble compounds which are isolated by removal of volatiles in vacuum. Compared to ^1H NMR spectrum of $La[Al(O\text{-}i\text{-}Pr)_4]_3$, [1.23(d) CH_3, 4.24(m) CH], the reaction products exhibit an interesting spectral trend. The reaction product of the equimolar reaction with water shows a complicated spectrum. The spectra simplify as more *i*-PrO groups are replaced. Finally the reaction product with six equivalents of water shows a doublet for CH_3 and a multiplet for CH protons. These data seem to support Mehrotra's suggestion that bridging *i*-PrO groups are preferentially replaced by -OH groups [4].

```
   [ i-Pr   i-Pr ]                           [ H     i-Pr ]
   [  O      O   ]                           [ O      O   ]
M  [     Al      ]    + 6 H2O   --->   M     [    Al      ]    + 6 i-PrOH
   [  O      O   ]                           [ O      O   ]
   [ i-Pr   i-Pr ]3                          [ H     i-Pr ]3
```

Assuming that bridging groups are replaced preferentially, replacement of one bridging *i*-PrO group creates an asymmetrical structure. The incoming second -OH group can occupy any of the residual five positions resulting in symmetrical and asymmetrical structures involving $(OH)_2Al(O\text{-}i\text{-}Pr)_2$ or $(OH)(O\text{-}i\text{-}Pr)Al(O\text{-}i\text{-}Pr)_2$ ligands. These in turn can occupy the cis- or trans-position to $(O\text{-}i\text{-}Pr)_2Al(O\text{-}i\text{-}Pr)_2$ ligands. The introduction of a third -OH group further complicates the structure; the structure becomes simpler after the replacement of six *i*-PrO ligands with -OH groups which creates a symmetrical environment of $(OH)_2Al(O\text{-}i\text{-}Pr)_2$ ligands around lanthanum.

^1H NMR studies of the species isolated from hydrolysis reactions of $Ce[Al(O\text{-}i\text{-}Pr)_3]_4$ are further complicated due to oxidation of Ce^{3+} to Ce^{4+}. Efforts are in progress to grow single crystals of some of the intermediates to obtain structural information. Preliminary data provide some insight into the hydrolysis reaction of $M[Al(O\text{-}i\text{-}Pr)_4]_3$, but further studies are necessary to assign structures to species present in the early stages of hydrolysis. Efforts are underway to follow hydrolysis reactions of $La[Al(O\text{-}i\text{-}Pr)_4]_3$ with EXAFS and XANES to establish the mechanism of gelation.

Gels Derived from $La[Al(O\text{-}i\text{-}Pr)_4]_3$ [I]:

Addition of seven or more equivalents of water in 2-propanol to a solution of I in 2-propanol at -78°C and warming the reaction mixture to room temperature furnishes gels. Thermogravimetric analysis [TGA] of the xerogel prepared by the addition of 12 equivalents of water and removal of volatiles in vacuum shows a weight loss of 25% in the temperature range 50-250°C, and 5% in the 250-400°C region. No further weight loss occurs in the temperature range 400-900°C. After heating the gel to 400°C and maintaining it at that temperature for 8 hours, BET surface area was 221 m^2/g. X-ray powder diffraction [XRD] data shows it to be amorphous. It is necessary to use excess water during preparation of gels, otherwise gels turn black during pyrolysis at ~600°C and lose surface area rapidly. One such sample had a surface area of 5 m^2/g after firing at 950°C.

Sol is prepared by the hydrolysis of I in excess water followed by the addition of acetic acid. This sol forms a gel on evaporation of the volatiles. After drying at 100°C, the transparent gel is collected and powdered. TGA shows a weight loss of 30% in 50-300°C range and no weight loss in the temperature range of 300-900°C.

BET surface area [Fig. 2] decreases slowly as the gel is treated at higher temperatures, but no evidence for the formation of a crystalline phase [Fig. 3] is observed at 700°C from XRD. Crystallization takes place at 900°C and the XRD of the sample matches JCPD pattern # 31-22 for lanthanum-aluminum-oxide, $LaAlO_3$. The BET surface area of this material is 40 m^2/g. It is of interest to note that the classical methods for preparing lanthanum-aluminum-oxide require heating above 1500°C, and the BET surface area of this material is substantially lower than that of the material obtained by our procedure. The residual two equivalents of alumina remain amorphous at 900°C, which could be contributing to the higher surface area of our material.

The transmission electron micrograph [TEM] of the material heated to 900°C shows ~250-300 nm agglomerates. The energy dispersive spectra [EDS] of several individual particles exhibited lanthanum peaks.

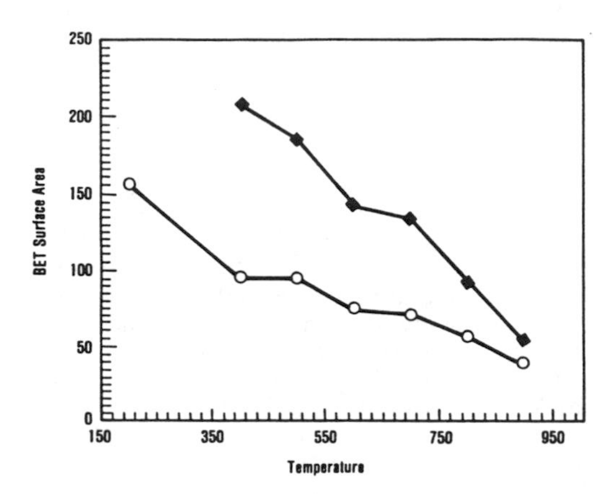

Fig. 2: BET surface area measurements 0-0 gel from I, ■-■ Gel from II

Gels Derived from $Ce[Al(O\text{-}i\text{-}Pr)_4]_3$ [II]:

The reaction of $Ce[Al(O\text{-}i\text{-}Pr)_4]_3$ with seven or more equivalents of water at -78°C forms a gel on warming to room temperature. For this study, a gel was prepared from the reaction of II with excess water and dried in vacuum. TGA of the gel showed a 30% weight loss in the temperature range 50-380°C and no weight loss in 400-900°C range. Pyrolysis of the light-yellow gel at 400°C for eight hours left a yellow powder which was found to be amorphous

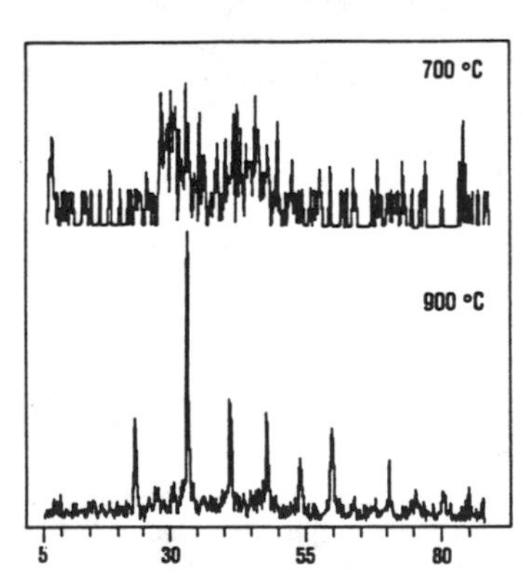

Fig. 3: Changes in XRD of gel derived from I
Crystalline phase is $LaAlO_3$

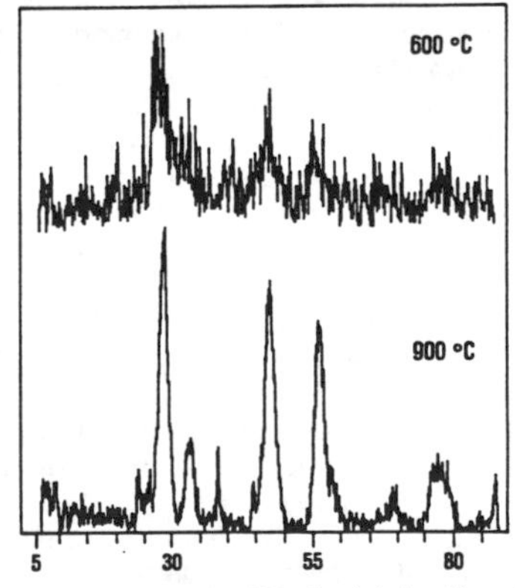

Fig. 4: Changes in XRD of gel derived from II
Crystalline phase is CeO_2

by XRD studies. The BET surface area of this material is 213 m^2/g. BET surface area measurements [Fig. 2] showed that the surface area decreased during heat treatment. XRD studies suggested that the material remained amorphous up to 500°C.

Broad, low-intensity peaks, assignable to poorly crystalline cerium oxide, could be detected in the XRD after heating at 600°C. Further heating at 900°C did not significantly change the intensity or peak width of CeO_2 diffraction peaks. Alumina still remained amorphous. TEM showed particles to be 80-100 nm agglomerates. EDS showed the presence of cerium, aluminum and oxygen in agglomerates.

A sol is also prepared from II by hydrolyzing it in water at 90°C and adding acetic acid. This sol is converted to gel and finally to xerogel by evaporation of solvent. TGA of xerogel shows a weight loss of 30% in the temperature range 50-250°C. The xerogel was pyrolyzed at 400°C for 4 hours. The XRD showed it to be amorphous and BET surface area was found to be 192 m^2/g.

Gels Derived from Mixtures of $La[Al(O\text{-}i\text{-}Pr)_4]_3$ and $Ce[Al(O\text{-}i\text{-}Pr)_4]_3$:

The co-hydrolysis of $La[Al(O\text{-}i\text{-}Pr)_4]_3$ and $Ce[Al(O\text{-}i\text{-}Pr)_4]_3$ in a 1:1, [III], or 1:3, [IV], ratio forms light yellow colored transparent gels which are dried in vacuum to obtain xerogels. TGA of the gels, III and IV, show a weight loss of 30 % in the region 50-400°C and none in the 400-900°C region.

After heating III at 550°C for 4 hours an amorphous yellow solid of BET surface area 159 m^2/g is obtained. Surface area decreases to 110 m^2/g after heating at 700°C for 4 hours and to 46 m^2/g after heating at 900°C. XRD of xerogel III did not show it to be amorphous after heating to 700°C but CeO_2 diffraction peaks are observed after heating at 900°C. No diffraction peaks were observed corresponding to Al_2O_3, La_2O_3, or $LaAlO_3$ as they remain amorphous even on heating at 900°C.

The thermal behavior of gel IV is very similar to III. After heating the material at 550°C for 5 hours, the material is amorphous and the BET surface area is 164 m^2/g. The BET surface area does not change on heating to 700°C but decreases to 119 m^2/g after heating at 900°C. XRD of IV shows weak, broad CeO_2 diffraction peaks after heating at 700°C which are better defined after heating at 900°C. Crystalline Al_2O_3, La_2O_3, $LaAlO_3$ are not formed at this stage.

It should be pointed out that Sato et.al. [9] prepared Ce-La-Al-O materials by heating a mixture of $La(NO_3)_3$, $Ce(NO_3)_3$ and alumina at 500°C. Sato et.al. determined the extent of dissolution of lanthanum oxide in cerium oxide from the shifts of CeO_2 (d=311) diffraction peak in the XRD of La-Ce-Al-O materials. Unlike the observations of Sato et.al., our sol-gel processed La-Ce-Al-O materials show no crystalline phases at 550°C and only microcrystalline CeO_2 is observed after heating to 900°C in XRD.

Sato et. al. also deposited Pt/Rh metal on their La-Ce-Al-O materials and found enhanced oxygen storage as compared to Pt/Rh deposited on a mixture of La_2O_3, CeO_2 and Al_2O_3. The enhanced oxygen storage capacity of Pt/Rh catalyst deposited on these materials was ascribed to interaction between Pt/Rh with CeO_2-La_2O_3 solid solutions. These authors concluded that intimate contact between precious metals and CeO_2 particles is essential for enhanced oxygen storage capacity of cerium oxides. Higher loadings (33% by stoichiometry) and high surface area of La-Ce-Al-O materials prepared by our sol-gel process should provide improved contacts between precious metals and CeO_2. Experiments are in progress to determine the oxygen storage capacity of the materials.

CONCLUSIONS

Gels prepared from heterobimetallic alkoxides, $La[Al(O\text{-}i\text{-}Pr)_4]_3$, $Ce[Al(O\text{-}i\text{-}Pr)_4]_3$ and their mixtures, yield high surface area alumina incorporating lanthanum, cerium or both respectively after drying. Preliminary NMR studies show that lanthanides remain bonded to aluminum in early stages of hydrolyses of alkoxide precursors. During heat treatment of gels, alumina remains amorphous below 900°C in the presence of lanthanum. Although Ce itself separates out as CeO_2, it stabilizes alumina in amorphous state at 900°C. In materials containing La and Ce, La prevents CeO_2 formation upto 700°C and cerium prevents the formation of crystalline $LaAlO_3$. BET surface area of the materials remains high even after heating at 900°C.

ACKNOWLEDGEMENTS

The use of Bruker WM-360 NMR and Philips CM1 2/STEM analytical electron microscope at University of Michigan, Ann Arbor is also gratefully acknowledged. W.L. Watkins provided BET surface area data.

REFERENCES

[1] C.J. Brinker and G.W. Scherer, Sol-Gel Science, Academic Press, London, 1990.

[2] Sol-Gel Technology for Thin Films, Fibers, Preforms, Electronics, and Specialty Shapes, edited by L. Kline, (Noyes Publications, New Jersy, 1988).

[3] K. Jones, T.J. Davies, H.G. Emblem, and P. Parkes, Mater. Res. Soc. Symp. Proc., 73, 111 (1986).

[4] R.C. Mehrotra, Mat. Res. Soc. Symp. Proc., 121, 81 (1988). R. Jain, A.K. Rai, and R.C. Mehrotra, J. Inorg. Chem. (China), 3, 96 (1987).

[5] D.C. Bradley, R.C. Mehrotra, D.P. Gaur, Metal Alkoxides (Academic Press, London, 1978). R.C. Mehrotra and M.M. Agrawal, J. Chem. Soc., Chem. Comm., 469 (1968). R.C. Mehrotra, M.M. Agrawal, and A. Mehrotra, Syn. Inorg. Metal-Org. Chem., 3, 181 (1973). R.C. Mehrotra, M.M. Agrawal, and A. Mehrotra, Syn. Inorg. Metal-Org. Chem., 3, 407 (1973).

[6] A.F.M. Leenars, K. Keizer, and A.J. Burggraaf, Separation with Inorganic Membranes, (Laboratory of Inorganic Science, Twente University of Technology, Enschede, The Netherlands). A.F.M. Leenaars, K. Keizer, and A.J. Burggraaf, J. Membrane Sci., 24, 245 (1985). A.F. Leenaars and A.J. Burggraaf, J. Membrane Sci., 24, 261 (1985). A.F.M. Leenaars, K. Keizer, and A.J. Burggraaf, Chemtech, 564 (1986).

[7] C.K. Narula, W.L. Watkins, and M. Shelef, unpublished results.

[8] H.S. Gandhi, A.G. Piken, M. Shelef, and R.G. Delosh, SAE Transactions, 85, 760201 (1976).

[9] T. Miki, T. Ogawa, M. Haneda, N. Kakuta, A. Ueno, S. Tateishi, S. Matsura, and M. Sato, J. Phys. Chem., 94, 6464 (1990).

[10] N. Watanabe, H. Yamashita, K. Akira, H. Kawagoe, and S. Matsuda, Jpn. Kokai Tokkyo Koho JP 63,175,642. CA 110, 64521 (1989).

[11] R.C. Mehrotra, A. Singh, and U.M. Tripathi, Chem. Rev., 91, 1287 (1991).

[12] R.C. Mehrotra, Coord. Chem. Rev., 31, 67 (1980).

SYNTHESIS OF NEW COMPOUNDS BY ION-EXCHANGE REACTIONS

M.E. Villafuerte-Castrejón, and A. Aragón-Piña.
Instituto de Investigaciones en Materiales, Universidad Nacional Autónoma de México. Apdo. Postal 70-360. México, D. F. 04510.

ABSTRACT

Ion-exchange reactions, particularly with molten salts, are a convenient low temperature route to prepare new materials. Six different complex oxides ($KTaO_3$, $K_2Ti_4O_4$, $NaAlSiO_4$, $Na_2Si_2O_5$, β-$NaAlO_2$ and $K_2Ge_4O_9$) were reacted with different salts (NaCl, KCl, Na_2SO_4, $NaNO_3$, LiCl and $LiNO_3$) by the ion exchange route with molten salts. In this work, new phases and substitution reactions are reported.

INTRODUCTION.

The most widely used synthesis method in solid state chemistry is the direct reaction of a mixture of solid starting materials at high temperatures. Synthesis of solids, their structures and properties are of essential importantance in solid state chemistry. Recent developments in preparative solid state chemistry have shown that it is possible to obtain new solids with novel structures. [1]

Three new routes to solid state synthesis have recently been developed: The precursor method and the two topochemical methods: the redox and the ion exchance reactions.[2][3] These methods have the advantage that the reaction takes place at much lower temperatures than when using normal solid state reaction procedures. This fact is important since at low temperatures the framework is not altered substantially during the change, it means that a topochemical exchange enables synthesis of new metastable phases which cannot be prepared by high temperature routes. [4] [5] [6]

This paper concerns work on six different well known and characterized complex oxides, using the ion exchange route to obtain new phases, polymorphic changes and substitution reactions, at lower temperatures and shorter reaction times than the direct reaction solid state method.

 ©1992 Materials Research Society

EXPERIMENTAL.

The six complex oxides were obtained by conventional solid state reaction. Starting materials were: K_2CO_3 (Baker reagent), Na_2CO_3 (Baker reagent), Ta_2O_5 (Aldrich 99.99), TiO_2 (Baker 99.2%), Al_2O_3 (Analar), SiO_2 (Aldrich 99.995%) and GeO_2 (Baker reagent), mixtures were prepared in ~ 10g amounts by weighing and mixing into a paste with acetone using an agate mortar and pestle, drying and firing in Pt crucibles in an electrical furnace. Temperatures were controlled to ± 10°C. Initial firing was at 600 to 700°C for 3 or 4 hours to expel the CO_2 followed by 950 to 1100°C firing for 3 to 6 days to complete the reactions.

The products of these reactions were identified by X-ray powder diffraction using a Siemens D500 diffractometer, CuKα radiation. Then each one of these complex oxides was reacted with the following reagent grade salts: LiCl, $LiNO_3$, NaCl, $NaNO_3$, Na_2SO_4 and KCl. The salts were selected because little evaporation or decomposition occurs at the melting point.

The experimental method is described as follows: The complex oxides were pelleted. The pellets were transferred to a Pt crucible and covered with one of the selected salts. This was then fired in an electric furnace at the salt's melting point until the reaction reached equilibrium. The salt then was removed by disolving it in water, the pellet was washed and dried in a furnace at 100-200°C, and finally the pellet was analyzed for phase content by x-ray powder diffraction.

RESULTS.

The results of heating experiments of the six oxides with the different fused salts now are discussed in turn:

1.- The products of six exchange reactions are isostructural with the parent oxide.

$$KTaO_3 + NaCl \xrightarrow[\text{3 days}]{900^\circ C} NaTaO_3$$

$$KTaO_3 + Na_2SO_4 \xrightarrow[\text{3 days}]{900^\circ C} NaTaO_3$$

$$KTaO_3 + NaNO_3 \xrightarrow[\text{3 days}]{350^\circ C} NaTaO_3$$

$$\gamma\ NaAlO_2 + LiCl \xrightarrow[\text{3 days}]{700^\circ C} \gamma\ LiAlO_2$$

$K_2Ge_4O_9 + NaCl \xrightarrow[\text{3 days}]{900^oC} Na_4Ge_9O_{20}$

2.- Three ion exchange reactions involving a structural change were observed.

$NaAlSiO_4 + KCl \xrightarrow[\text{3 days}]{900^oC} KAlSiO_4$

$KTaO_3 + LiCl \xrightarrow[\text{15 min}]{700^oC} LiTaO_3$

$\gamma\ NaAlO_2 + LiNO_3 \xrightarrow[\text{3 days}]{300^oC} \beta\ LiAlO_2$

3.- Seven reactions gave different compounds:

$K_2Ti_4O_9 + LiCl \xrightarrow[\text{3 days}]{700^oC} LiTiO_3$

$K_2Ti_4O_9 + LiNO_3 \xrightarrow[\text{2 days}]{300^oC} Li_2TiO_3$

$\alpha\ Na_2Si_2O_5 + LiCl \xrightarrow[\text{3 days}]{300^oC} Li_2SiO_3$

$K_2Ge_4O_9 + Na_2SO_4 \xrightarrow[\text{3 days}]{900^oC} Na_4Ge_9O_{20}$

$K_2Ge_4O_9 + LiCl \xrightarrow[\text{3 days}]{700^oC} Li_2GeO_3$

$K_2Ge_4O_9 + LiNO_3 \xrightarrow[\text{3 days}]{300^oC} Li_6Ge_8O_{19}$

$NaAlSiO_4 + NaCl \xrightarrow[\text{10 days}]{900^oC} Na_8Al_6Si_6O_{24}Cl_2$ Sodalite

4.- Four new phases were found:

$Na_2Si_2O_5 + LiNO_3 \xrightarrow[\text{6 days}]{300^oC} Li_2Si_2O_5$ metastable form

The stable form of $Li_2Si_2O_5$ is well known. The metastable form retains the silicate sheet configuration of the parent $Na_2Si_2O_5$ and is different from that of stable $Li_2Si_2O_5$.

Full structural work was done and it is described in [7].

$NaAlSiO_4 + LiCl \xrightarrow[3\ days]{700^oC}$ new phase

$NaAlSiO_4 + LiNO_3 \xrightarrow[6\ days]{300^oC}$ new phase

$K_2Ti_4O_9 + NaCl \xrightarrow[12\ days]{900^oC}$ new phase

Further work on these three new phases is in progress in order to obtain structural data. X-ray powder diffraction data for these phases are given in Tables I, II and III.

Tables.- I and II. X-ray Powder Diffraction Patterns for the new phases resulting from the reactions:

$NaAlSiO_4$ + LiCl d (A)	I	$NaAlSiO_4$ + $LiNO_3$ d (A)	I
4.77	34	11.78	7
4.41	43	4.25	19
3.52	49	4.12	100
2.69	100	3.79	27
2.67	50	3.70	27
2.4	58	3.18	90
2.36	45	2.93	25
2.34	30	2.514	12
2.33	24	2.292	40
2.012	30	2.010	10
1.79	20	1.76	10
1.56	17	1.52	12
1.54	30		
1.459	10		
1.406	10		
1.372	14		
1.347	7		
1.3006	17		

Table III X-ray powder Diffraction Pattern for the new phase resulting from the reaction of $K_2Ti_4O_9$ + NaCl

d (A)	I	d (A)	I
7.44	100	2.497	30
6.25	95	2.09	60
4.53	7	2.049	28
4.17	15	1.884	8
3.72	12	1.874	10
3.65	12	1.768	4
3.13	12	1.737	14
3.01	7	1.701	4
2.98	20	1.64	14
2.9	100	1.585	5
2.93	15	1.566	7
2.78	6	1.54	10
2.75	6	1.513	15
2.68	40	1.488	10

Conclusions.

Different kinds of compounds have been prepared by ion exchange reactions with molten salts such as the metastable $Li_2Si_2O_5$ which is related structurally to the parent $Na_2Si_2O_5$.

Synthesis of an enterely new class of compounds is one of the most interesting possibilities in the preparation of solids. It can provide new structures, with the promise of new properties and applications. By this method it also is possible to synthesize well-known compounds at lower temperatures with shorter reaction times.

Acknowledgement.

We thank L. Baños for assistance with the X-ray diffraction measurements.

References.

1.- J. Gopalakrishnan. Proc. Indian Acad. Sci. 93, 421-432 (1984).

2.- A. R. West. Solid State Chemistry and its applications. Ed Wiley (1984).

3.- J.M.Honig and C.N.Rao. Preparation and characterization of Materials, Academic Press.

4.- H. Shäfer. Angew. Chem. Internat. Edit. 10, 1, (1971)

5.- C.E. Rice and J.L. Jackel. J. Sol. St Chem. 41, 308-314 (1982).

6.- W.H. Mc Carroll, C.Darling and G.Jakubicki. J. Sol. St. Chem. 48, 189-195 (1983).

7.- R. I. Smith, R.A. Howie, A.R. West, A. Aragón-Piña and M. E. Villafuerte-Castrejón. Acta Cryst. C46, 363-365, (1990).

COMPARATIVE STUDY OF THE SOL-GEL AND METALLO-ORGANIC DECOMPOSITION (MOD) PROCESSES FOR THE PREPARATION OF INORGANIC MATERIALS

GUSTAVO R. PAZ-PUJALT*, W. NIE** AND C. LURIN**
*Corporate Research Laboratories, Eastman Kodak Company, Rochester, NY 14650-2011
**Laboratoire de Science des Materiaux, Kodak Pathé, Chalon-sur-Saône, 71100, France

ABSTRACT

Sol-Gel processing and Metallo-Organic Decomposition are contrasted from the thermodynamics and chemical reactivity points of view. Differences and similarities on precursor requirements, processing characteristics, processing intermediates, and product formation pathways are outlined. Some specific examples are presented and their reaction thermodynamics are compared. Some conclusions regarding thermodynamic versus kinetic control are drawn from these examples.

INTRODUCTION

Oxide ceramics are commonly prepared by solid state reaction and sintering of metal oxide/carbonate mixtures. Because they usually are physical mixtures, they require prolonged grinding/heating cycles. This has led to considerable interest in the preparation of materials by chemical methods in order to achieve stoichiometric control, "atomic level" homogeneity, and to reduce processing times and temperatures. Sol-Gel is by no means a new technology; its progress is very well covered in the literature [1]. Metallo-organic decomposition (MOD) is also an old technology. The decomposition of metal carboxylates was used by ancient peoples to deposit glazes on ceramic vessels. More recently, however, work at various laboratories [2-4] has centered on thin films of high T_c superconductors, ferroelectric ceramics, and other "advanced" materials prepared by this method.

The purpose of the present work is to compare sol-gel and MOD processing as they apply to the preparation of mixed metal oxide materials. Emphasis is placed on the fundamental chemical characteristics of each process rather than on the properties of the resulting products.

In sol-gel processing metal-ligand compounds hydrolyze followed by condensation to produce a network consisting of metal-oxygen-metal bonds.

When different metal precursors are combined and the hydrolysis is carefully controlled the condensation step leads, ideally, to the formation of a mixed metal-oxygen-metal network. This network is usually amorphous. Upon thermal processing the desired mixed oxide material is obtained. In reality hydrolysis is a more complex process than described here. It depends on the type of metal and its properties (charge, electronegativity, coordination number), pH of the hydrolysis medium, and on the nature of the ligand. The hydrolysis of cations is covered in detail by Baes and Mesmer [5]. The chemistry and physics of Sol-Gel processing have been thoroughly covered by Brinker and Scherer [1], Livage et al. describe in depth the sol-gel processing of transition metal oxides [6].

MOD consists of combining precursors of the desired metals in the desired ratios, followed by their thermal decomposition. The decomposition then yields an "atomic level" mixture of the metal oxides/carbonates. Upon thermal treatment the desired mixed metal oxide material is obtained.

$$\text{M–O–L} \rightarrow \text{MO or MCO}_3 + \text{Pyrolysis Products} \quad (1)$$

$$\text{M}'\text{O} + \text{M}''\text{CO}_3 \rightarrow \text{M}'\text{M}''\text{O}_x + \text{CO}_2 \quad (2)$$

where L is an organic ligand and x is determined by the valence state of the metals.

Thermal decomposition products depend on the nature of the cation and not of the ligand, as long as this contains C, H, and O only. Because thermal decomposition is the driving force in this process, MOD precursors must meet certain requirements that will be contrasted to sol-gel precursors in the next section.

PRECURSORS

Sol-gel precursors may be inorganic metal salts (chlorides, nitrates, etc.), or metal linked to an organic ligand as an alkoxide group. Reportedly [1] metal alkoxides are the most widely used precursors for sol-gel research. When referring to sol-gel precursors in this work, unless otherwise noted, they will be metal alkoxides.

The main requirement for sol-gel precursors is that they should be subject to nucleophilic attack by water. The rate of hydrolysis should be controllable by the amount of water added and/or by the action of a catalyst. It is also desirable that the organic hydrolysis product have a relatively low boiling point so that it can be removed from the gel by moderate heating. In addition solubility of the alkoxide in the parent alcohol or other sol-gel compatible solvent is required.

MOD utilizes as precursors metals linked to organic groups, most commonly metal carboxylates, and modified metal alkoxides. In MOD, ligands are separated from the precursor by thermolysis; therefore, compounds that sublime, evaporate, or produce volatile decomposition products are not suitable MOD precursors. Solubility is essential in order to provide the "atomic level" mixing desired. Most commonly, MOD is used for depositing thin films onto a substrate. In order to adjust the rheology of the MOD precursor, it is fundamental that they be soluble in organic solvents because they provide a wide range of physical and chemical properties.

PROCESSING

Hydrolysis

The facile hydrolysis of metal alkoxides invariably leads to metal-oxygen compounds such as hydroxides or hydrated oxides. Most metals are too oxyphilic to let go of oxygen (to be reduced) under such mild conditions as hydrolysis. The acid-base properties of the hydrolytic species depends to a large extent on the electronegativity of the cation. For purposes of comparison of the chemical reactivity and thermodynamic driving forces of sol-gel and MOD, it is convenient to classify hydrolysis products according to their acid base properties.

Thermal Decomposition

The thermal decomposition of metal carboxylates leads to the formation of metal carbonate and the corresponding ketone, both of which decompose further following their own thermodynamic characteristics [7]. Light ketones volatilize readily, whereas larger ones decompose into smaller fragments or undergo complete oxidation. The stability of the metal carbonates is determined by the electropositivity of the cation–the more electropositive the metal the more stable the carbonate. Thus alkali metals, alkaline earths, and lanthanides (including Sc, Y, and La) form stable carbonates that decompose at elevated temperatures (>600°C). Meanwhile, transition metal carbonates decompose readily into the metal oxide and carbon dioxide. Noble and "noble like" metal carboxylates decompose into the metallic form of the element [8].

The formation of a carbonate is independent of the nature of the ligand and is solely determined by the thermodynamic stability of the given oxide in the presence of carbon dioxide [9]. Table I summarizes and contrasts the hydrolysis and decomposition products of sol-gel and MOD precursors, respectively.

Because sol-gel and MOD precursors produce unmistakably different products upon hydrolysis and thermal decomposition, the reactions that lead to the formation of the final mixed metal oxide product obey different energetic pathways.

Table I
Hydrolysis and Decomposition Products for Sol-Gel and MOD Precursors

Element Family	Sol-Gel	MOD
Alkali metal	very strong basic hydroxide	stable carbonate
Alkaline earth	very strong basic hydroxide	stable carbonate
Lanthanides	strong basic hydroxide	stable carbonate
Transition metal	hydroxylated oxides basic, acidic or amphoteric	metal oxides
Noble metals	not common	metallic element

PRODUCT FORMATION

Product formation in sol-gel is the result of the interaction between two or more metal oxides in different stages of hydration. Based on their chemical behavior in the aqueous system, metal oxides may be classified as acidic, basic, or amphoteric [10]. Basic and acidic oxides often combine to produce salts.

$$BaO + Al_2O_3 \rightarrow BaAl_2O_4 \quad (3)$$

The driving force for the formation of mixed metal oxide systems by sol-gel may be viewed in terms of acid-base reactions. MOD, on the other hand, usually consists of solid-state reactions involving mixtures of oxides and carbonates.

Many mixed metal-oxide ceramics of technological interest contain at least one highly electropositive cation. When their preparation is carried out by sol-gel processing, strong basic hydroxides develop as reaction intermediates, whereas MOD processing yields highly stable carbonates. Table II shows some mixed metal oxides, and the intermediates that precede their formation by either sol-gel or MOD processing.

Table II
Hydrolysis and Thermal Decomposition Products for Sol-Gel and MOD Precursors

Material	Sol-Gel	MOD
$LiNbO_3$	$LiOH + HNbO_3$	$Li_2CO_3 + Nb_2O_5$
$BaTiO_3$	$Ba(OH)_2 + TiO_2$	$BaCO_3 + TiO_2$
BaB_2O_4	$Ba(OH)_2 + 2B(OH)_3$	$BaCO_3 + B_2O_3$

Reaction Energetics

Using thermodynamic values available in the literature [11], it is possible to calculate the enthalpies and entropies of reaction for some sol-gel and MOD examples. These quantities describe the bulk. In practice, however, they are affected by surface energy contributions [12]. Nevertheless, they show trends that are expected to be maintained. When thermodynamic quantities are not available, they may be estimated by comparison to analogous compounds or by using empirical relationships based on the electronegativity of the cation [13]. The enthalpy of reaction for the MOD process, for the examples cited, is endothermic. In MOD the reaction is made possible by increasing the temperature to the point where the entropy contribution is sufficient to overcome the positive enthalpy of the reaction, and make the free energy negative, $\Delta G = \Delta H - T \bullet \Delta S$. These examples are thermo-dynamically limited. For the sol-gel process, on the other hand, the reaction is exothermic and spontaneous at ordinary temperatures, therefore kinetically controlled. These findings have been experimentally confirmed by Differential Thermal Analysis [14].

Table III
Comparison of Reaction Enthalpies for MOD and Sol-Gel

$BaCO_3 + TiO_2 \rightarrow BaTiO_3 + CO_2$	$\Delta H = 24310$ cal	MOD
$BaCO_3 + B_2O_3 \rightarrow BaB_2O_4 + CO_2$	$\Delta H = 2.6 \times 10^4$ cal	
$Ba(OH)_2 + TiO_2 \rightarrow BaTiO_3 + H_2O$	$\Delta H = -50045$ cal	SOL-GEL
$Ba(OH)_2 + 2B(OH_4^-)H^+ \rightarrow BaB_2O_4 + 6H_2O$	$\Delta H = -5.8 \times 10^5$ cal	

The examples discussed above represent chemical extremes, i.e., strong bases and stable carbonates versus acidic oxides. When the reactions do not involve such extremes, the difference between MOD and sol-gel becomes less drastic. Consider for example the formation of bismuth cuprate:

$$Bi_2O_3 + CuO \rightarrow CuBi_2O_4 \tag{4}$$

When prepared by the hydrolysis-coprecipitation of the corresponding nitrates under basic conditions, a black powder formed immediately. After washing off excess water with dry ether, a sharp x-ray diffraction (XRD) pattern was obtained [15]. Meanwhile, when the compound was prepared by the codecomposition of carboxylate precursors, the initial decomposition products included alpha-bismuth oxide and copper oxide, which upon mild thermal treatment to remove organics obstructing the crystallization, yielded the mixed oxide as determined by XRD.

Association processes of the type $a + b \rightarrow ab$ almost always lead to very small changes in entropy, therefore, in order to ensure a negative change of free energy, they are inevitably exothermic [16]. The few cases where association processes are endothermic are accompanied by increases in entropy due to vibrational and/or configurational disorder.

An examination of a large number of examples involving highly electropositive cations and acidic oxides for which thermochemical values are reported (Li-Nb, Na-Nb, Ca-Ti, Sr-Ti for example) suggests that the conclusions drawn from the examples discussed in this paper regarding reaction energetics may be broadly generalized for reactions involving this type of reactants for sol-gel or MOD. It is worth noting, for sol-gel, that under actual processing conditions when temperatures are increased in order to drive off hydrolysis products and to dry the gels, pyrolysis of organics may occur leading to the formation of carbon dioxide. This in turn interacts with highly electropositive cations leading to the formation of the corresponding carbonate, thus paralleling MOD and requiring a similar type of thermal treatment for the completion of the process. Strictly speaking, to distinguish sol-gel from MOD, the sol-gel process then must be carried out in a time-temperature regime that precludes the formation of metal carbonates.

CONCLUSIONS

Sol-gel and MOD lead to different process intermediates, based on the position of the cation in the periodic table. Highly electropositive cations lead to stable carbonate intermediates for MOD, and very strong basic hydroxides for sol-gel. When these react with other oxides to produce mixed metal oxides, MOD is generally an endothermic process, whereas sol-gel is exothermic. For transition metals the differences in energetics between sol-gel and MOD become vague. Sol-gel consists of the reaction between metal oxides at different stages of hydration and with different degrees of acidity/basicity, whereas MOD consists of solid-state association reactions. Both processes are expected to be exothermic. Sol-gel processing requires to be carried out at time-temperature coordinates that prevent the formation of metal carbonates. Clearly these generalizations are meant for purposes of comparison. More rigorous study and experimentation are needed in order to arrive at a more fundamental understanding.

ACKNOWLEDGEMENTS

The authors wish to thank the technical contributions of Robert Montesino (Kodak Pathé), Lillie Salter, and Ralph Nicholas (Eastman Kodak). Kay Servais and Karin Barczak are thanked for editing and preparing the manuscript.

REFERENCES

1. C.J. Brinker and G.W. Scherer, *Sol-Gel Science: The Physics and Chemistry of Sol-Gel Processing* (Academic Press, San Diego, CA, 1990).
2. J.V. Mantese, A.L. Micheli, A.H. Hamdi, and R.W. Vest, *MRS Bulletin* **14**(10), 48 (1989).
3. J.J. Xu, A.S. Shaikh, and R.W. Vest, *IEEE Trans. UFFC* **36**, 307 (1989).
4. J.A. Agostinelli, G.R. Paz-Pujalt, and A.K. Mehrotra, *Physica C* **156**, 208 (1988).
5. C.F. Baes and R.E. Messmer, *The Hydrolysis of Cations* (Wiley, New York, 1976).
6. L. Livage, M. Henry, and C. Sanchez, *Progress in Solid State Chemistry* **18**, 183 (1988).
7. R.C. Mehrotra and R. Bohra, *Metal Carboxylates* (Academic Press, London, 1983).
8. G.R. Paz-Pujalt, in *High T_c Superconductor Thin Films*, edited by L. Correra (Elsevier Science Publishers, Amsterdam, 1992).
9. G.R. Paz-Pujalt, A.K. Mehrotra, S. Ferranti, and J.A. Agostinelli, *Solid State Ionics* **32/33**, 1179 (1989).
10. F.A. Cotton and G. Wilkinson, *Advanced Inorganic Chemistry*, 5th ed. (Wiley, New York, 1988), p. 545.
11. *Lange's Handbook of Chemistry,* 5th ed., edited by John A. Dean (McGraw-Hill, New York, 1985).
12. E.I. Tochitsky and N.I. Romanova, *Thin Solid Films* **110**, 55-64 (1983).
13. S. Bratsch, *J. Chem. Educ.* **65**(10), 877 (1988).
14. G.R. Paz-Pujalt, unpublished results.
15. G.R. Paz-Pujalt, *Physica C* **166**, 1772 (1990).
16. A. Novrotsky, in *Solid State Chemistry Techniques*, edited by A.K. Cheetham and P. Day (Clarendon Press, Oxford, 1987).

PART II

Particulate and Polymeric Sols

COMPLEXATION OF Zr(IV) PRECURSORS IN AQUEOUS SOLUTIONS

J. LIVAGE, M. CHATRY, M. HENRY AND F. TAULELLE
Chimie de la Matière Condensée, Université Paris VI, 4 place Jussieu,
75252 Paris, France

ABSTRACT

The sol-gel synthesis of metal oxides can be performed via the hydroxylation and condensation of metal cations in aqueous solutions. The complexation of these ionic species by anions leads to the chemical modification of inorganic precursors at a molecular level. The whole process of hydrolysis and condensation can then be modified allowing a chemical control of the morphology, the structure and even the chemical composition of the resulting powder.

The role of anions during the formation of condensed phases from inorganic precursors in aqueous solutions has to be taken into account. The complexing ability of these anions is described in the frame of the Partial Charge Model as a function of pH and the mean electronegativity of anionic and cationic chemical species. Experimental evidence for the complexation of zirconyl species in aqueous solutions will be given using multinuclear NMR of anions.

INTRODUCTION

The sol-gel synthesis of glasses and ceramics is usually performed via the hydrolysis and condensation of metal alkoxides in organic solvents. The molecular design of these metal-organic precursors provides a control of the morphology and structure of sol-gel materials. The chemical modification of metal alkoxides can be easily performed by adding organic ligands such as carboxylic acids or β-diketones. Complexation occurs via the nucleophilic substitution of alkoxy groups leading to a modification of the structure, functionality and reactivity of these molecular precursors [1].

Metal oxides or hydroxides can also be precipitated from aqueous solutions. This inorganic route has already been used for a long time in industry for the synthesis of alumina, titania or zirconia powders. However little academic work has been published describing the hydrolysis and condensation of metal cations in aqueous solutions [2]. One of the main problems arises from the fact that water behaves both as a ligand and a solvent. A large number of oligomeric species are simultaneously formed. They are in rapid exchange equilibria and it is not possible to predict which one would nucleate the solid phase. The key parameter is usually the pH of the aqueous solutions but foreign anions or cations are often added in order to improve the process. It is well known that spherical hematite particles (α-Fe_2O_3) are obtained when aqueous solutions of $FeCl_3$ are heated under reflux while acicular particles are formed with $Fe(NO_3)_3$ in the presence of phosphate ions. Goethite (α-FeOOH) is precipitated from $Fe(NO_3)_3$ while Akaganeite (β-FeOOH) is formed with $FeCl_3$. The thermo-hydrolysis of $Ti(SO_4)_2$ aqueous solutions leads to anatase TiO_2 while the rutile phase is also formed in the presence of chloride or nitrate [2].

These examples obviously show that anions can play an important role in the formation of condensed phase even if they are not chemically involved in these phases. Anions can lead to the complexation of aqueous cationic precursors (and vice versa). Complexation should actually be considered as the inorganic analogue of the chemical modification of metal alkoxides by organic ligands.

This paper presents a theoretical analysis of the complexation of metal ions in aqueous solutions. It is based on the so-called Partial Charge Model (PCM). Experimental evidence of the complexation of zirconyl precursors by anionic species will be provided by multinuclear NMR.

MODELING

Hydrolysis and condensation

When dissolved in an aqueous medium, metal cations M^{z+} are solvated by water molecules. Because of electron transfers within M-OH_2 bonds, coordinated water molecules become more acidic and spontaneous deprotonation occurs as follows :

$$[M(OH_2)_n]^{z+} + hH_2O \leftrightarrow [M(OH)_h(OH_2)_{n-h}]^{(z-h)+} + hH_3O^+ \quad (1)$$

In dilute solutions, this leads to a whole set of hydrolyzed species ranging from aquo-cations $[M(OH_2)_n]^{z+}$ to oxo-anions $[MO_m]^{(2m-z)-}$. The chemical nature of these hydrolyzed species (or hydrolysis ratio h) mainly depends on the charge of the metal cation and the pH of the aqueous solution [3].

According to the electronegativity equalization principle [4], the deprotonation of coordinated water molecules goes on until the mean electronegativities of the aqueous solution and the hydrolyzed precursor become equal. As shown previously, the mean electronegativity of an aqueous solution decreases linearly with pH as follows [2]:

$$\chi_{aq} = 2.732 - 0.035pH \quad (2)$$

The hydrolysis ratio h can then be calculated in the frame of the Partial Charge Model. The charge of the hydrolyzed species $[M(OH)_h(OH_2)_{n-h}]^{(z-h)+}$ is :

$$z-h = \delta_{aq}(M) + n\delta_{aq}(O) + (2n-h)\delta_{aq}(H) \qquad (3)$$

so that

$$h = [z-\delta_{aq}M(OH_2)_n]/[1-\delta_{aq}(H)] \quad \text{or} \quad h = \Delta_{aq}(z,n,M)/\Delta_{aq}(1,0,H) \qquad (4)$$

where

$$\Delta(z,n,M) = z - \delta[M(OH_2)_n] \qquad (5)$$

Δ_{aq} means that the corresponding species, $M_{aq}{}^{z+}$ and $H_{aq}{}^{+}$, are in equilibrium with the surrounding aqueous solution. Δ_{aq} therefore depends on pH. At low pH Zr(IV) would give rise to the $[Zr(OH)_2(OH_2)_6]^{2+}$ hydrolyzed species (h=2).

However OH groups have a rather negative partial charge (δ_{OH}=-0.06) so that, unless solutions are very dilute (<10^{-4}mol.l^{-1}), condensation occurs via olation leading to the well known cyclic tetramer $[Zr_4(OH)_8(OH_2)_{16}]^{8+}$ the structure of which was determined by X-ray scattering of the solution [5][6] and the solid state [7]. Further polymerization can be obtained by adding a base to Zr(IV) solutions. It leads to the formation of a gelatinous amorphous precipitate. Distribution functions obtained by X-ray and neutron diffraction suggest that this amorphous precipitate is made of plate-like particles the structure of which is close to that of tetragonal zirconia [8]. Crystallization occurs upon heating around 400°C and leads to the metastable tetragonal phase. Crystalline monoclinic hydrous zirconia can be obtained under reflux at low pH [9]. Its formation from tetrameric precursors was described by A. Clearfield [10]. Olation progressively leads to the formation of $[Zr(OH)_4]_n$ sheets while the original square antiprism coordination of zirconium atoms is changed to a cubic coordination. $[Zr(OH)_4]_n$ layers then aggregate into a 3-dimensional fluorite structure via oxolation reactions.

Figure 1. Molecular structure of $[Zr_4(OH)_8(OH_2)_{16}]^{8+}$ precursors
●=Zr, ⊘=OH, O=H_2O

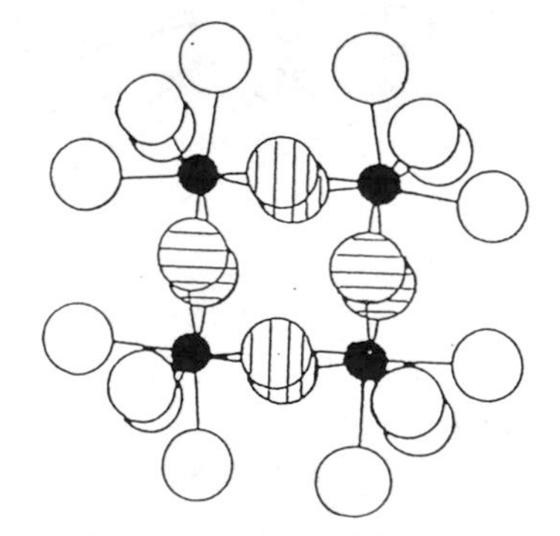

This equation provides a critical pH value for each protonated form $[H_qX]^{(x-q)-}$ of the anion. According to the definition given in relation (5), we can write :

$$\Delta_i(z,n,M) = \Delta_j(z,n,M) + \sigma[M(OH_2)_n](\chi_j-\chi_i) \quad (11)$$

where $\sigma[M(OH_2)_n]$ corresponds to the softness of $M(OH_2)_n$. All Δ_i values can then be expressed with respect to the same Δ_j. If we take H^+ as a reference state, $\chi_p = \chi(H^+) = 4.071$, all χ_q values can be expressed as a function of χ_{aq} and therefore as a function of pH. Calculations lead to :

$$\chi_q = \chi_p - \frac{\Delta_p(z,n-\alpha,M)}{q\sigma(H) + \alpha\sigma(H_2O) + \Delta_p(z,n,M)/(\chi_p-\chi_{aq})} \quad (12)$$

Since χ_{aq} varies linearly with pH (χ_{aq}=2.732 - 0.035 pH), it is possible to express electronegativities as a function of pH, or critical pH* values as a function of electronegativities. This leads to :

$$pH^*_q = \frac{28.6\ \Delta p(z,n,M)}{[\Delta p(z,n-\alpha,M)/\chi_p-\chi_q] - [q\sigma(H) + \alpha\sigma(H_2O)]} - 38.3 \quad (13)$$

An "electronegativity-pH" diagram can then be drawn which could be very useful to describe complexation reactions in aqueous solutions (Fig.2).

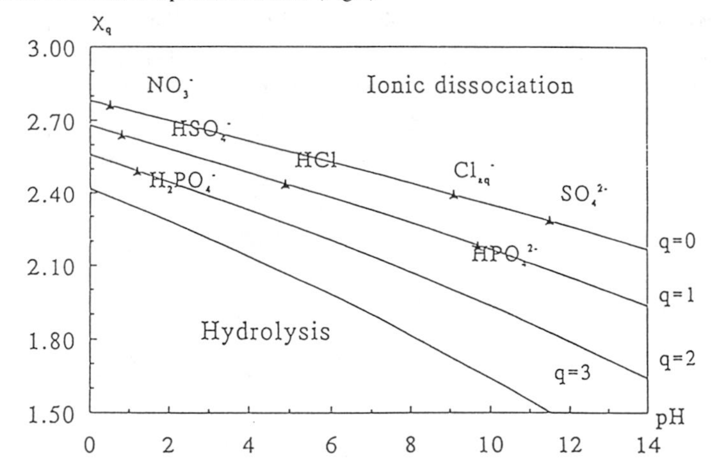

Figure 2. Electronegativity-pH diagram corresponding to the complexation of the tetrameric $[Zr_4(OH)_8(OH_2)_{16}]^{8+}$ precursor

<u>Complexation</u>

Negatively charged anions X^{x-} can react with positively charged hydrolyzed species to give complexes as follows :

$$[M(OH)_h(OH_2)_{n-h}]^{(z-h)+} + X^{x-} \leftrightarrow [M(X)(OH)_h(OH_2)_{n-h-\alpha}]^{(z-h-x)+} + \alpha H_2O \qquad (6)$$

α corresponds to the number of water molecules which have been replaced by one anionic ligand; $\alpha=1$ for monodentate ligands such as Cl^- and $\alpha=2$ for bidentate ligands such as $(SO_4)^{2-}$.

As reported previously, such complexes remain stable only if the following conditions on partial charges are fulfilled [1][2]:

$\delta(X) \leq x$ in order to avoid ionic dissociation in an aqueous medium

$\delta(HX) \geq 0$ in order to avoid the hydrolysis by nucleophilic water molecules

Anionic ligands X^{x-} can be protonated when pH decreases :

$$X^{x-} + q\, H_3O^+ \leftrightarrow [H_qX]^{(x-q)} + q\, H_2O \qquad (7)$$

This leads to (x+1) different species of general formula $[H_qX]^{(x-q)-}$ in which q ranges from 0 to x. The rough formula of the complexed precursor can be written as follows $[M(H_qX)(OH)_h(OH_2)_{n-h-\alpha-q}]^{(z-h-x)+}$ in order to evidence $[H_qX]$ species. The total charge is given by :

$$z-h-x = \delta(H_qX) + \delta[M(OH_2)_{n-\alpha}] - (h+q)\delta(H) \qquad (8)$$

The (x+1) values of q lead to (x+1) critical hydrolysis ratio defined by :

$\delta(H_qX) = q-x$ for ionic dissociation ($0 \leq q \leq x-1$)

$\delta(H_xX) = 0$ for hydrolysis (q=x)

Critical hydrolysis ratio can then be expressed as follows :

$$h_q^* + q = [z-\delta_q(M(OH_2)_{n-\alpha})]/[1-\delta_q(H)] = \Delta_q(z,n-\alpha,M)/\Delta_q(1,0,H) \qquad (9)$$

In an aqueous solution the hydrolysis ratio depends on the pH as shown in relation (4). It is then possible to express h_q^* as critical pH values pH_q^* corresponding to each protonated form of anionic species. If we write h*=h , equations (4) and (9) lead to :

$$\Delta_{aq}(z,n,M)/\Delta_{aq}(1,0,H) + q = \Delta_q(z,n-\alpha,M)/\Delta_q(1,0,H) \qquad (10)$$

According to eq.(13) the critical pH* value, and therefore the complexing range, increases when the mean electronegativity of the anion or the positive charge of the cation M^{z+} decrease and when α increases.

These calculations together with the "electronegativity-pH" diagram provide a theoretical background to predict the complexing ability of anions toward cations. As a general rule, positively charged M_{aq}^{z+} cations tend to attract negatively charged X_{aq}^- anions. Different types of associations can then be described as follows :

a. Both cations M^{z+} and anions X^- are far from each other and behave as free ions surrounded by their own solvation sphere. This occurs in very dilute solutions only.

M — O H H O H H O H H X

M^{z+}_{solv} $(H_2O)_n$ X^-_{solv}

b. Both ions are surrounded by their own solvation sphere (α=0 and q=0) but the corresponding water molecules are bonded via hydrogen bonds. This could be called an ion pair or a third sphere complex in which the anion is the third neighbour only.

M — O H H O H H X

M^{z+}_{solv} X^-_{solv}

c. The anion is bonded to the solvation sphere of the cation (α=0 but q can differ from zero). This could be called a second sphere complex.

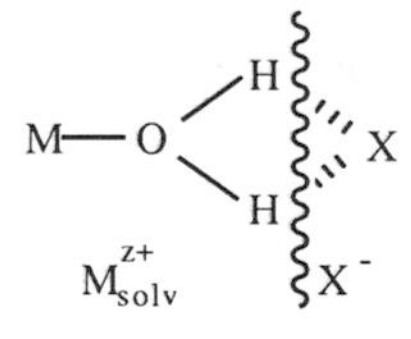

d. The anion is directly bonded to the cation (α and q can be different from zero) giving a first sphere complex M-X.

EXPERIMENTAL EVIDENCE BY MULTINUCLEAR NMR

Many data can be found in literature dealing with the complexation of metal cations by anions. However, measurements have usually been performed on dilute solutions in order to avoid interactions between neigbouring ions. In most cases, such solutions are not concentrated enough to lead to the formation of gels or precipitates. Little information is available on the complexing behavior of anions towards condensed cationic precursors.

Multinuclear NMR provides a convenient method to observe the behavior of anionic species in the solution. Several nuclei are available depending on their nuclear spin. In aqueous solutions, the metal cation can be surrounded by water molecules, OH groups, oxygen atoms or anions. NMR can then be performed at different levels of the aqueous species ; the metal ion when it has a nuclear spin such as ^{27}Al, protons of OH or H_2O and even the anion itself which very often has at least one isotope such as ^{14}N, ^{17}O, ^{19}F, ^{31}P, or ^{35}Cl with a nuclear spin.

The low natural abundance and the large quadrupole moment of the ^{91}Zr isotope does not allow easy NMR observation. Therefore NMR spectra were recorded on anionic species only. All experiments have been performed in acid solutions (pH<7) where the tetrameric square planar polycation $[Zr_4(OH)_8(OH_2)_{16}]^{8+}$ prevails.

According to eq.13, critical pH* values for this precursor are given by :

$$pHq^* = \frac{319.3}{0.5q + 1.4\alpha - [\Delta_p(4,6-\alpha,Zr)]/[4.1 - \chi_q]} - 38.3 \qquad (14)$$

where $\Delta_p(4,6-\alpha,Zr) = -11.17$ for $\alpha=0$, -8.95 for $\alpha=1$ and -6.73 for $\alpha=2$.

NMR spectra were recorded on $(ZrOX_2)_{y/2}$-$(NaX)_{1-y}$ solutions in which X^- is a monovalent anion and $0 \leq y \leq 1$. The total concentration (X^-) remains constant (1mol/l) while the molar fraction of Zr(IV) increases progressively with y. The chemical shift δ, the linewidth $\Delta\nu$ and the peak area S were measured as a function of the molar fraction y. Complex formation leads to a variation of the chemical shift and an increase of the linewidth. The total amount of observed X species is given by S.

<u>Chloride ions Cl^-.</u>

Cl_{aq}^- is not highly electronegative ($\chi=2.3$). Eq.14 shows that monodentate complexes ($\alpha=1$, $q=1$) cannot be formed below pH=5. However second sphere complexes ($\alpha=0$, $q=0$)

could be observed at lower pH in agreement with experiments performed by X-ray scattering of $ZrOCl_2$ aqueous solutions [5].

The ^{35}Cl (I=3/2) NMR spectra of $(ZrOCl_2)_{y/2}$ - $(NaCl)_{1-y}$ solutions are shown in Figure 3. A sharp peak (δ=-0.3ppm, $\Delta\nu$=10Hz) is observed for pure NaCl solutions (y=0). The chemical shift varies slightly while the linewidth increases significantly when zirconium is added, up to δ=4.8ppm and $\Delta\nu$=336Hz for an aqueous solution of $ZrOCl_2$ (y=1). A quasi-linear variation is observed for δ and $\Delta\nu$. The peak area does not vary significantly showing that all Cl species can be seen by NMR.

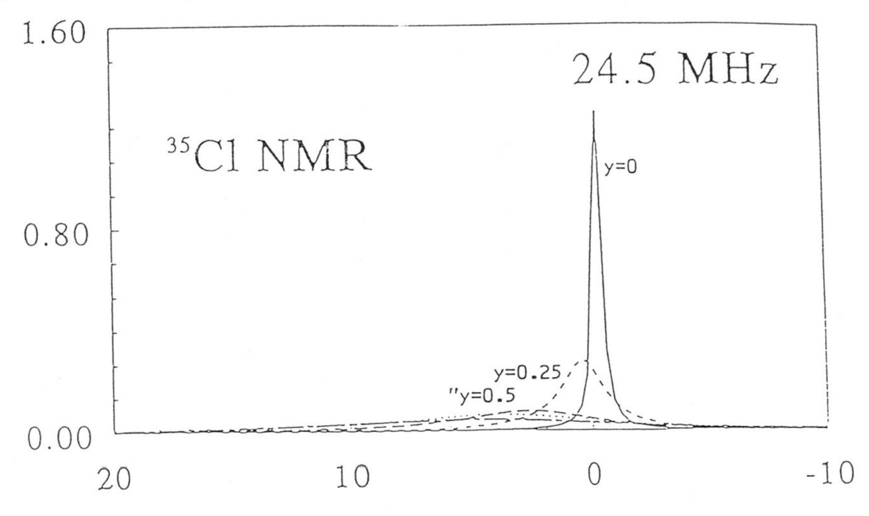

Figure 3. ^{35}Cl NMR spectra of $(ZrOCl_2)_{y/2}$ - $(NaCl)_{1-y}$ solutions

These observations suggest fast exchange reactions of Cl^- ions between two Cl^- species associated either with Na^+ or Zr(IV) species. In both cases, Cl^- ions are almost free and can be observed easily. The strong positive charge of Zr^{4+} leads to an increase of the linewidth due to quadrupolar effects arising from the polarisation of the solvation sphere of Cl_{aq}^-. The complexing ability of Cl^- toward zirconium is very weak and Cl^- behaves as a counter anion for both Na^+ and Zr(IV). It is not bonded to zirconium, in agreement with X-ray diffraction studies showing that $ZrOCl_2$,nH_2O crystals are made of tetrameric cationic species $[Zr_4(OH)_8(OH_2)_{16}]^{8+}$ and Cl^- anions [5][7]. Hydrous zirconia can be obtained at low pH when an aqueous solution of $ZrOCl_2$ is heated under reflux [9][11]. Cl^- anions are released into the solution and are not involved in the polymerization process described by A. Clearfield [10]. Increasing the pH beyond pH=2 leads to the precipitation of amorphous hydrous zirconia ZrO_2,nH_2O. However, according to figure 2, Cl^- anions become complexing between pH=5 and pH=9. This could explain why significant amounts of anions remain in the amorphous oxide when precipitation is performed at high pH. The amount of anions retained by the precipitate is pH dependent, decreasing as the final pH of precipitation increases [10].

Perchlorate ions $(ClO_4)^-$.

Perchlorate anions are highly electronegative (χ=2.857). Complexes (α=1 or α=2) cannot be formed above pH=0. Only second sphere complexes can then be observed.

^{35}Cl NMR spectra of $(ZrO(ClO_4)_2)_{y/2}$ - $(NaClO_4)_{1-y}$ solutions are reported in figure 4. As previously, the linear variation of the chemical shift δ and the linewidth $\Delta\nu$ can be assigned to fast exchange reactions of ClO_4^- between two ion pairs with Na^+ and the tetrameric Zr(IV) precursor. The ClO_4^- tetrahedron is not strongly distorted by Zr^{4+} cations so that quadrupolar effects do not lead to a significant broadening of the ^{35}Cl NMR signal.

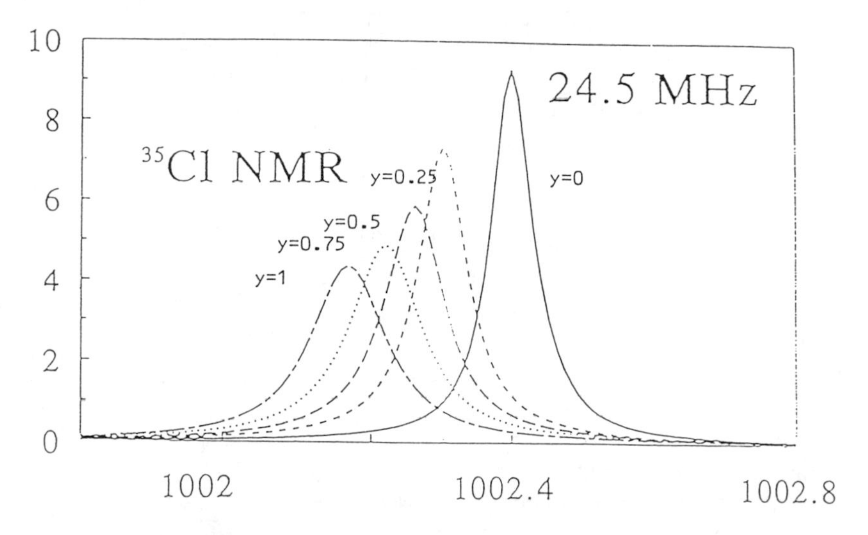

Figure 4. ^{35}Cl NMR spectra of $(ZrO(ClO_4)_2)_{y/2}$ - $(NaClO_4)_{1-y}$ solutions

However, quantitative area measurements show that the number of Cl species detected by NMR decreases when the amount of zirconium increases. About 20% ClO_4^- cannot be detected by NMR in the pure $ZrO(ClO_4)_2$ solution. This could be due to the formation of second sphere complexes "Zr-OH_2-O_4Cl" in which the strong polarization due to Zr^{4+} cations leads to stronger quadrupolar effects. The linewidth becomes too large for NMR signals to be seen. Such an explanation would agree with X-ray scattering experiments showing a peak at 5.25Å which gets larger as the ClO_4/Zr ratio is increased. Such a Zr-Cl distance could actually correspond to second sphere complexes [6].

Nitrate ions $(NO_3)^-$

Nitrates are slightly less electronegative than perchlorates (χ=2.762). Calculations show that bidentate nitrates (α=2, q=0) can form complexes with tetrameric zirconium precursors below pH=2. The ^{14}N (I=1) NMR spectra of $(ZrO(NO_3)_2)_{y/2}$-$(NaNO_3)_{1-y}$ solutions are shown in figure 5. Again both the chemical shift δ and the linewidth $\Delta\nu$ vary linearly with y, suggesting fast exchange reactions between NO_3^- associated with Na^+ and Zr(IV) cations via third sphere complexes. Quantitative measurements show that the amount of ^{14}N which cannot be seen by NMR decreases faster than previously. About 50% nitrates are not visible in the $ZrO(NO_3)_2$ solution. They should be directly bonded to zirconium atoms leading to a strong quadrupolar broadening.

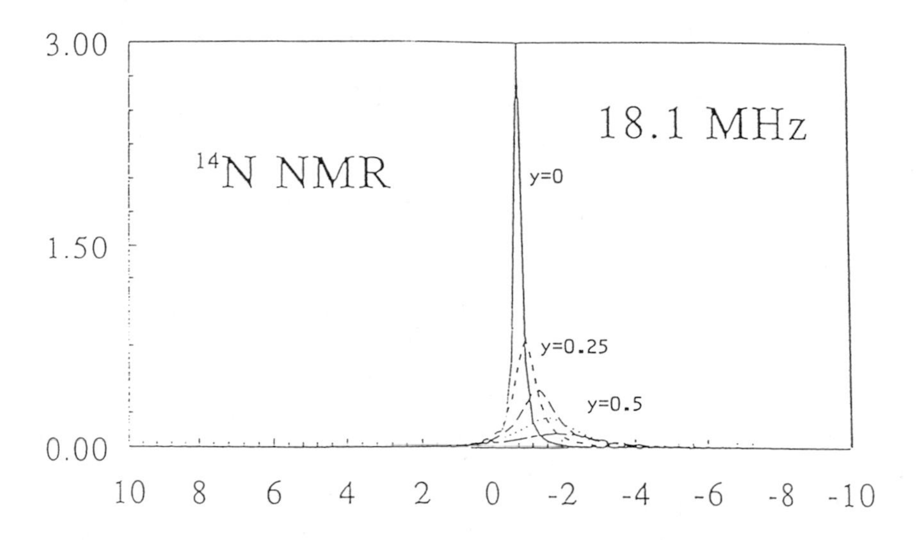

Figure 5. ^{14}N NMR spectra of $(ZrO(NO_3)_2)_{y/2}$-$(NaNO_3)_{1-y}$ solutions

A zirconium hydroxide nitrate $Zr(OH)_2(NO_3)_2{,}4.7H_2O$ can be crystallized from zirconium nitrate solutions at low pH. Its crystal structure was recently determined from X-ray powder diffraction data [12]. It shows that this basic salt is made up of $[Zr(OH)_2(OH_2)_2(NO_3)]^+$ chains in which zirconium atoms are eightfold coordinated by four OH groups, two water molecules and one chelating nitrate. These chains are held together by hydrogen bonds through additional water molecules and nitrate groups located between the chains (Fig.6). In the solution, it may be assumed that half of the nitrates are bonded to zirconium solute species and cannot be seen by NMR because of quadrupolar broadening. The other nitrates just behave as counter anions and are involved in fast exchange reactions as seen by NMR.

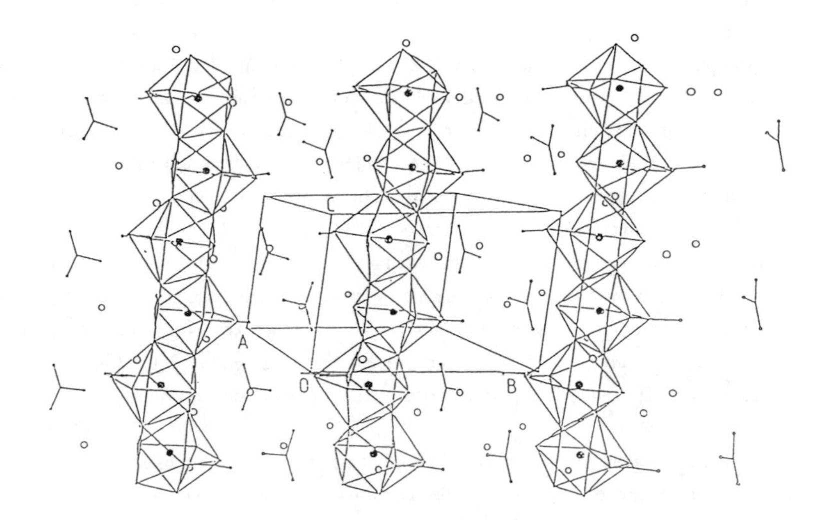

Figure 6. Structure of $Zr(OH)_2(NO_3)_2,4.7H_2O$ according to [12].
O=Zr, O=free H_2O, Y=NO_3^-

CONCLUSION

The complexing behavior of anions toward cations can be predicted in the frame of the Partial Charge Model. Several cases have been described.

i) The anion is not complexing. It just behaves as a counter anion and is not involved in the formation of condensed species. This was the case for Cl^-. $ZrOCl_2,nH_2O$ crystals can be formed in which Cl^- is not bonded to tetrameric cationic species. At high pH, amorphous hydrous zirconia ZrO_2,nH_2O is formed in which Cl^- anions are adsorbed at the surface of the gelatinous precipitate and can be removed by washing.

ii) The anion is strongly complexing. It remains bonded to the cation during the formation of solid phase giving rise to salts rather than hydrous oxides. This is typically the case of phosphates. Phosphoric acid reacts with aqueous solutions of zirconium salts giving an amorphous gelatinous precipitate. The crystallization of the layered α-$Zr(HPO_4)_2,H_2O$ phase occurs progressively when the solution is heated under reflux [13].

iii) The intermediate case corresponds to weakly complexing anions such as nitrates. They can form either first sphere or second sphere complexes and are obviously involved during the first steps of the hydrolysis-condensation process and basic salts are precipitated at low pH. These anions can then be released at higher pH so that amorphous hydrous zirconia is obtained when a base is added.

Sulphates also form complexes with Zr(IV) precursors up to a pH of about 11. They can displace -ol groups from hydrolyzed Zr(IV) species and form sulphate bridges between zirconium atoms. Zirconium sulphate $Zr(SO_4)_2.4H_2O$ is readily crystallized out of moderatly strong sulphuric acid solutions [10]. A large number of basic zirconium sulphates have been described in which $(SO_4)^{2-}$ groups behave as bridging ligands between zirconium atoms [14][15][16]. Moreover, sulphate anions can be used for to control the precipitation of zirconia. Polydispersed ZrO_2 particles are obtained when an acidic aqueous solution of $ZrOCl_2,8H_2O$ is refluxed whereas monodispersed spherical particles are formed upon addition of K_2SO_4 [17]. This could be due to the in-situ generation of sulphato-complexes which act as precursors for nucleation. Protons are formed during the process and non complexing $(HSO_4)^-$ anions are released at a further stage leading to azirconia which does not contain measurable amounts of sulphate.

Such polyvalent anions are very important in the sol-gel synthesis of oxide materials. They prevent or slow down condensation reactions and can be used to control the size and morphology of solid particles. Their precise role is not easy to describe and would require an accurate characterization of all the intermediate species formed during the growth of oxide particles.

REFERENCES

1. J. Livage, M. Henry, C. Sanchez, Progress in Solid State Chemistry, 18, 259 (1988).
2. M. Henry, J.P. Jolivet, J. Livage, Structure and Bonding, 77, 153 (1992).
3. C.F. Baes, R.E. Mesmer, in Hydrolysis of Cations J. Wiley PuB. New York (1976)
4. R.T. Sanderson, Science, 114, 670 (1951).
5. G.M. Muha, P.A. Vaughan, J. Chem. Phys., 33, 194 (1960).
6. M. Aberg, Acta Chem. Scandinavica, B31, 171 (1977).
7. A. Clearfield, P.A. Vaughan, Acta Cryst., 9, 555 (1956).
8. J. Livage, K. Doï, C. Mazières, J. Am. Ceram. Soc., 51, 349 (1968).
9. A. Clearfield, Inorg. Chem., 3, 146 (1964).
10. A. Clearfield, Rev. Pure and Appl. Chem., 14, 91 (1964).
11. J.R. Fryer, J.L. Hutchison, R. Patterson, J. Colloids and Interface Sci., 34, 238 (1970).
12. P. Bénard, M. Louër, D. Louër, J. Solid State Chem., 94, 27 (1991).
13. A. Clearfield, J.A. Stynes, J. Inorg. Nucl. Chem., 26, 117 (1964).
14. D.B. McWhan, G. Lundgren, Inorg. Chem., 5, 284 (1966).
15. P.J. Squattrito, P.R. Rudolf, A. Clearfield, Inorg. Chem., 26, 4240 (1987).
16. M. El Brahami, J. Durand, L. Cot, Eur. J. Solid State Inorg. Chem., 25, 185 (1988).
17. M.A. Blesa, A.J.G. Maroto, S.I. Passaggio, N.E. Figliolia, G. Rigotti, J. Mater. Sci., 20 , 4601 (1985).

FRACTAL GROWTH DURING GELATION

EDWARD J. A. POPE
MATECH, 31304 Via Colinas, Ste 102, Westlake, CA 91362

ABSTRACT

Sol-gel processing has emerged as an important new technology for the fabrication of a wide variety of glass, ceramic, and composite materials. In order to better understand and control the gelation process, a theoretical model has been developed which quantitatively links features of fractal growth with measurable properties of a solution during the sol-gel transition. The result of that effort is the fractal growth model of gelation (FGMG) which expresses the solution viscosity, light scattering, and refractive index changes as a funtion of time and the fractal dimension. In this paper, the application of FGMG to silica gels and multi-component system, such as titanium aluminosilicates, is presented.

INTRODUCTION

Various researchers have observed, using small angle x-ray scattering (SAXS) and small angle neutron scattering (SANS), that the gelation process exhibits characteristics of fractal growth [1]. The application of fundamental concepts of fractal geometry to gelation has resulted in the fractal growth model of gelation (FGMG), which describes viscosity, light scattering, and refractive index changes as a function of time and the fractal dimension [2-6]. This model has been successfully applied to HF-catalyzed silica [2,3,5,6] and base-catalyzed neodymia-silica gels[4]. The significance of this model as a basis for understanding and controlling sol-gel reactions is best illustrated in figure 1, in which fractal growth lies in between the two extremes of linear chain growth and uniform colloidal particle formation. It is this "in between" region where most sol-gel reactions occur. Also presented in figure 1 are the funtionality of polymerization, "f", the degree of cross-linking, "DC", and the theoretical fractal dimension, "D". The observed fractal dimension, even for linear chain structures, tends to be greater than unity due to the tendency of the chains to bend and become entangled.

In table I, the time-dependent property relationships for viscosity, light scattering, and refractive index derived from the FGMG are presented[2,5,6]. By plotting the natural log of the reduced viscosity, eta/eta(0) - 1, versus time, the viscosity growth rate factor, "q_v", can be determined. Similarly, the light scattering growth rate factor, "q_i", can be calculated by plotting the double log of the transmission intensity, "I/I_o", versus time. This assumes, however, that these plots yield a straight line, as would be expected under "ideal" fractal growth conditions.

FGMG has been successfully applied to HF-catalyzed silica, base-catalyzed silica, and Nd-doped silica. Of considerable importance to technologists, however, are complex

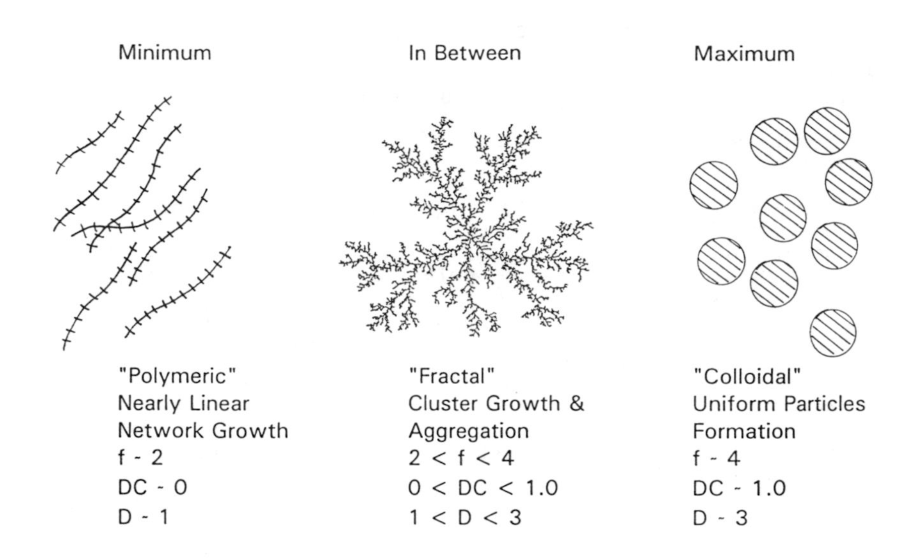

Figure 1: Schematic representation of structural boundary conditions for gel formation.

Table 1: Time dependent property relationships during gelation derived from FGMG.

Property	Equation	Constant	Growth Rate Factor
Viscosity	$\frac{\eta}{\eta_o}-1=A\exp(q_v t)$	$A=\frac{QL}{\rho_o}$	$q_v=(3-D)q_r$
Scattering Cross-section	$S=B\exp(q_I t)$	$B=\frac{3r_o^3 M_o(\frac{n_p}{n_a}-1)}{4\lambda^4 N_a\rho_o}$	$q_I=q_m+(6-D)q_r$
Refractive Index	$\frac{n^2-1}{n^2+2}=\frac{4\pi(N\alpha)_a}{3}+C\exp(q_n t)$	$C=\frac{4\pi((N\alpha)_p-(N\alpha)_a)M_o}{3N_a\rho_o}$	$q_n=q_m+(3-D)q_r$

multicomponant gel systems for such wide-spread potential applications as GRIN optics, non-linear optics, filters, high modulus fibers, magnetic and ferroelectric composites, etc. In this paper, FGMG is applied to multicomponent titanium aluminosilicate gels suitable for high index glasses and GRIN optics.

EXPERIMENTAL

The titanium aluminosilicate gel formula used in this study is slightly modified from that of Caldwell, et al [7]. A solution of 165.8 ml tetraethoxysilane (TEOS) and 382.6 ml ethanol was mixed for 20 minutes, after which 14.4 ml of 0.1 M HCl solution was added to hydrolyze the TEOS. After refluxing for one hour, 13.6 ml titanium-isopropoxide and 28.5 gm aluminum di(sec)butoxide acetoacetic ester chelate (ADAC) was added, followed by continued stirring for one hour. Water was then added to achieve the desired H_2O/TEOS ratio. Solutions were allowed to gel at 23°C. All chemicals were obtained from Morton Thiokol/Alfa products.

Viscosity was measured using a Brookfield model DV-11 digital viscometer. Light scattering was measured with an Aerotech LS10R HeNe laser (25mW) light source and an Aerotech model 71 laser radiometer and an LFI 100-316 detector. Both measurements were plotted simultaneously on a Brookfield model 1202-000 dual pen chart recorder.

RESULTS AND DISCUSSION

The titanium aluminosilicate gel compositions examined contain approximately 5.0 mole percent titania, 5.0 mole percent alumina, and 90.0 mole percent silica. The water mole ratio to TEOS was varied between 4.75 and 8.50, to examine the effect of water content on gelation kinetics and structure. Sample codes, water ratio, gel time, and transparency are given in table two. Both transparency and gelation time are dramatically dependent upon water content, as vividly illustrated in figure 2. The nature of this dependency, however, cannot be inferred from this data alone.

In figure 3A,3B, and 3C, both the reduced viscosity and light scattering data are presented for TSA-2,3, and 4 gels as a function of time,respectively. While the degree of final transparency differs dramatically for these three samples, the most significant change occurs between 10 and 20 minutes after mixing, regardless of total gelation time. The viscosity data follows a nearly ideal fractal growth pattern with no indication of a hierarchical structure, such has been routinely observed in flourine and base-catalyzed silica gels [2-6]. The only significant deviation from a fractal growth pattern occurs during the brief period of dramatic change in light scattering, except for sample TSA-4, in which gelation occurs in under twelve minutes. Similarly, after the initial abrupt change in light scattering, the light scattering behavior also appears fractal for both TSA-2 and TSA-3. Both viscosity and light scattering exhibit fractal characteristics simultaneously, as would be expected.

These preliminary results would suggest the possibility of two largely independent reactions occuring simultaneously in solution. The first reaction, which occurs during the initial ten to twenty minutes, is the formation of large, light-scattering colloidal particles. The second reaction is the formation of a continuous, interconnected gel network. Both reactions appear to be sensitive to water content. Increasing water content affects the size of the colloidal particles, but not the overall kinetics. On the other hand, water content has a dramatic effect upon the kinetics of gel formation.

One possible explanation for the duality of structural evolution in the titanium aluminosilicate system would be phase segregation, in which the colloidal particles are rich in titania and alumina and the network is silica rich. To

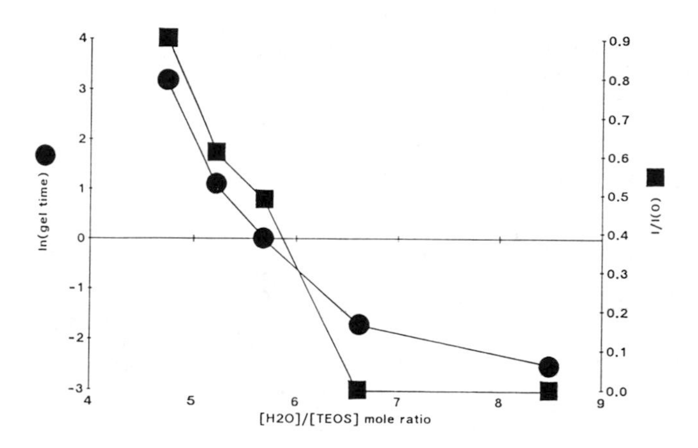

Figure 2: Gel time and transparency of Ti-Al-Si-O gels plotted as a function of water content.

Table II: Gelation time and transparency for seven titanium aluminosilicate gels as a function of water content, aluminum content and titanium content

Sample Code	Alumina (m/o)	Titania (m/o)	Water:TEOS mole ratio	Gel Time (hours)	I/I(0) Final
TSA-1	5.0	5.0	4.75	24.0	0.90
TSA-2	5.0	5.0	5.21	3.0	0.61
TSA-3	5.0	5.0	5.68	1.0	0.49
TSA-4	5.0	5.0	6.61	0.2	0.0
TSA-5	5.0	5.0	8.50	0.08	0.0
TSA-6	5.0	15.0	4.75	0.5	0.0
TS-7	0.0	15.0	4.75	24.0	0.87

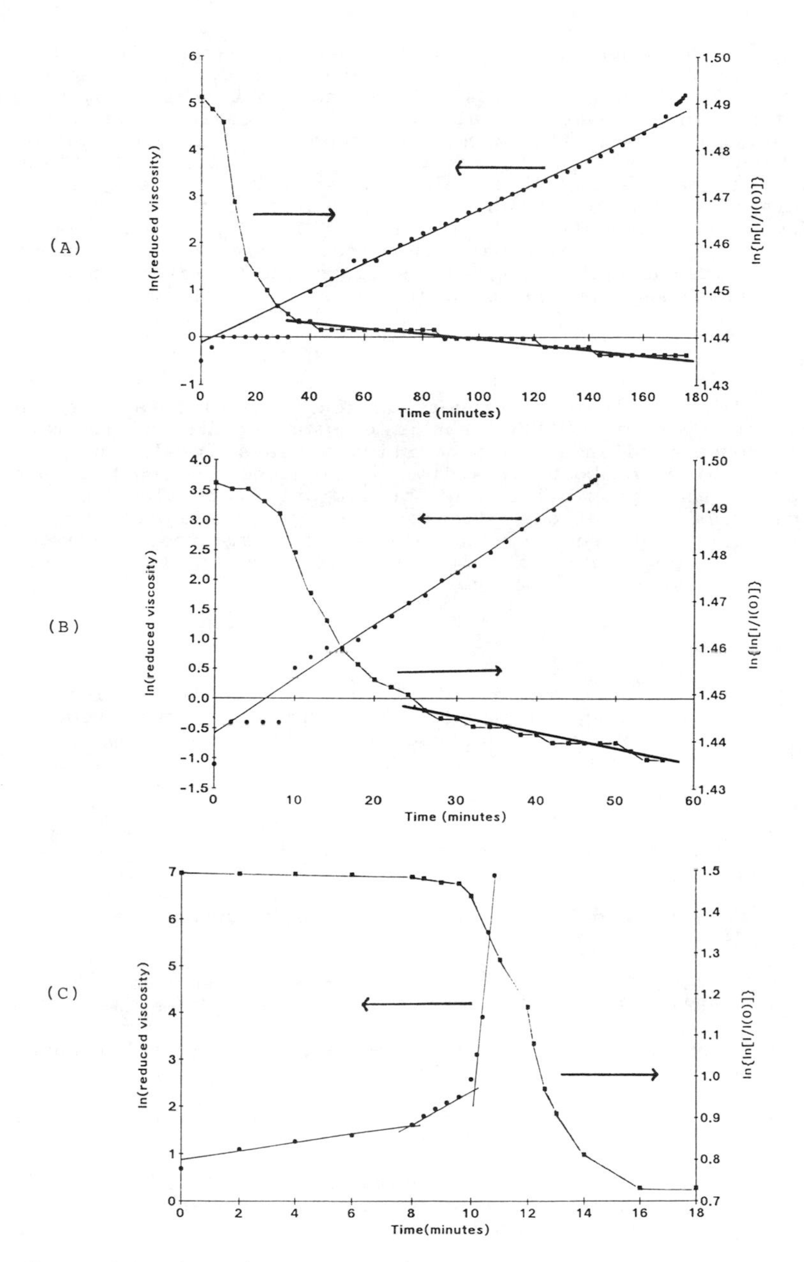

Figure 3: Viscosity and Light Scattering Data plotted by FGMG as a function of time for; A) TSA-2, B) TSA-3, and C) TSA-4.

test this hypothesis, a TSA-1 gel was prepared with three times the titania content, designated TSA-6, as presented in table 2. Both gelation kinetics and light scattering were increased in a manner analogous to increasing the water content. In TS-7, the aluminum component was eliminated, and both gel time and transparency were increased. These results suggest a possible relationship between the titanium precursor, aluminum precursor, and water content in this system. Further study, employing FTIR, NMR, and EXAFS measurements, might elucidate the local chemical environment for the titanium, aluminum, and silicon and, thereby, confirm whether phase segregation is occurring in this gel system.

SUMMARY

In this preliminary investigation, the fractal growth model of gelation (FGMG) has been applied to a complex, multicomponent titanium aluminosilicate glass forming system. The examination of both viscosity and light scattering behavior suggests the possibility of colloidal particle induced compositional variations during the early stages of gel formation. Continued investigation is warranted, however, to examine how intial precursor selection in this system affects final gel homogeneity.

ACKNOWLEDGEMENTS

The author wishes to gratefully acknowledge the assistance of Mr. Alex Almazan in sample preparation and measurements and Mrs. Barbara M. Pope in manuscript preparation.

REFERENCES

1. D.W. Schaefer and K.P. Keefer, Phys. Rev. Lett. 53 (1984) 1383.

2. E.J.A. Pope and J.D. Mackenzie, J. Non-Cryst. Sol., 101 (1988) 198

3. R. Xu, E.J.A. Pope, J.D. Mackenzie, J.Non-Cryst. Sol., 106 (1988) 242.

4. E.J.A. Pope, R.Xu, and J.D. Mackenzie, in "Fractal Aspects of Materials" (Mater. Res. Soc. EA-17: 1988) 159.

5. E.J.A. Pope, in "Scaling in Disordered Materials" (Mater. Res. Soc. EA-25: 1990) 241.

6. E.J.A. Pope and F. Higuchi, Symp. Proc. on Glass and Optical Materials, 93rd Annual Meeting,American Ceramic Society, accepted, in press.

7. J. B. Caldwell, et al., in "Better Ceramics Through Chemistry IV", (Mater. Res. Soc. Proc. 180, Pittsburgh, PA, 1990) 727.

Solvent Quality Effects in Sol-Gel Processing

JOSEPH K. BAILEY
Sandia National Laboratories, Ceramic Synthesis and Inorganic Chemistry Department, Albuquerque, NM 87185.

ABSTRACT

To understand how solvent quality can be used to tailor structure in sol-gel processing of silicon alkoxides, polymerized tetramethoxysilane polymers were synthesized and fractionated to give relatively stable, narrow molecular weight dispersion samples. These polymers had molecular weights ranging from 8000 to 45,000. The solubility parameter range for these polymers is 8.9-14.5 $(cal/cm^3)^{1/2}$. Light scattering confirmed that this range could be used to predict solvent quality. Bulk gels prepared using good versus poor solvents demonstrated that solvent quality can be used to tailor properties of the gels, presumably by modifying the extent of interpenetration of the growing polymers.

INTRODUCTION

The solvent used during the processing of sol-gel derived ceramics can greatly influence the final properties of the gel, such as its surface area and its ability to dry without cracking. Much of the previous research has investigated differences in gel properties when different solvents were used during gel preparation [1-4]. However, little research has investigated the fundamental ability of the solvent to solvate sol-gel polymers. In this work, the polymer science concept of solvent quality is applied to sol-gel processing with the aim of exploiting new methodologies for controlling the structure of silicon alkoxide polymers and gels.

Solvent quality describes the interactions between polymer chain segments and solvent molecules. When the pair-wise free energy of a polymer chain segment and a solvent molecule is lower than that of two polymer chain segments, the solvent is said to be a "good" solvent and the polymer is expanded in solution to maximize polymer-solvent contact. Conversely, when chain segments interact more favorably with other chain segments than with solvent molecules then the solvent is a "poor" solvent [5]. Solvent quality depends on the molecular structure of the polymer: the higher the polymer molecular weight or the greater its extent of branching, the lower its solubility will be. Solvent quality is measured through the solubility parameter, δ, and the second virial coefficient, A_2; the latter can be related to the Flory-Huggins interaction parameter, or more advanced models for solvent-polymer interactions.

The solvent quality affects the radius of gyration of polymer molecules in solution, it determines solubility of the polymers, and it determines phase separation. In a good solvent, the polymers are diffuse and interpenetrating, and may be capable of forming denser films and gels with smaller pores. In a poor solvent, the polymers are more tightly coiled, and should not be as interpenetrating. Figure 1 shows a schematic illustration of the differences which may be possible to induce in sol-gel polymers by solvent quality. Figure 1 implies that linear, or at least weakly branched, polymers are required to allow the flexibility needed for expansion or contraction. Since Kelts and Armstrong [6] demonstrated that acid-catalyzed tetramethoxysilane (TMOS) sols form less cyclized and less branched sols than tetraethoxysilane (TEOS), TMOS was used as a starting alkoxide in this work. A narrow molecular weight distribution was obtained after the polymers were synthesized by using fractionation. With a well-characterized polymer, solvent quality can be quantified. Once the solubility is understood, deciphering the solvent quality effects in the more complex reacting systems can be attempted. Thus, this work is a preliminary attempt to use solvent quality to control structure in sol-gel processing of silicon alkoxides.

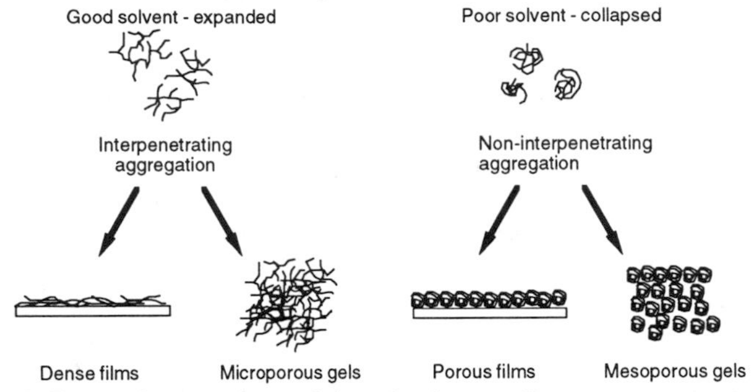

Figure 1: Expected structural control through solvent quality

EXPERIMENTAL

Polymers were prepared by adding substoichiometric quantities of H_2O to TMOS and reacting to near completion. A molar ratio of 1.2 : 1 : 3 of water : TMOS : methanol was used to give a conversion of TMOS to SiO_2 just short of the conversion needed for gelation [6,7]. Hydrochloric acid in a ratio of 0.0005 : 1 of HCl : TMOS was used for catalysis. After the solution was aged at either at room temperature or at 50°C for 3 months or 3 weeks respectively, it was refluxed for 30 minutes and then the solvent was removed by vacuum distillation. The remaining polymerized TMOS, which I have designated p(TMOS), had an oil-like consistency. To ensure that residual methanol and water were removed, an amount of toluene approximately equal in volume to the amount of polymer was added and the solvent was again vacuum distilled. A second amount of toluene was also added and distilled. The resulting polymer solution contained approximately 1.0% by weight of residual toluene. Two batches of p(TMOS) will be discussed in this paper; the first was aged at room temperature for three months and is designated HMW, the second was aged at 50°C for 3 weeks and is designated LMW. These designations refer to high molecular weight and low molecular weight, and will be discussed later.

The fractionation process was performed in two steps. In the first step, 25ml of polymer and 250 ml of pentane were placed in a separatory funnel and shaken well. The polymer-rich phase settled to the bottom and was removed. The polymer phase from the first fractionation step was placed in a clean separatory funnel with 250 ml of pentane and 30 ml of toluene and this mixture was well shaken and the polymer phase was again removed after settling. After the two fractionations, residual toluene and pentane were removed by vacuum distillation. The final yield was approximately 15 ml of polymer. This viscous polymer was then dispersed in the solvent of interest for light scattering and spin coating experiments.

The polymer was characterized by size exclusion chromatography (SEC) using ultra-styragel columns with toluene, calibrated relative to polystyrene. Static light scattering was performed using a light scattering photometer with a 633nm wavelength laser and a scattering angle determined by a 6-7° annulus. The specific refractive index increment (dn/dc) was measured with a laser differential refractometer.

Films were spun cast at 3000 rpm for 30s after the solution had been placed on a stationary substrate. They were analyzed for porosity by a surface acoustic wave (SAW) device technique [9]. Bulk gels were analyzed for porosity by N_2 adsorption/desorption, and morphology was examined by transmission electron microscopy (TEM) using samples of crushed gels supported by lacey carbon TEM grids.

The influence of solvent quality on bulk gel properties was also examined. For bulk gels, 25ml of TMOS, 6.0 ml of H_2O, 10 ml of methanol, and 0.4ml of HCl were mixed with 15 ml of the solvent of interest in a single step sol-gel reaction. After gelation the sols were aged for 24 hours and then the pore fluid was replaced by putting the wet gel in 150ml of toluene for 48 hours, followed by decanting the toluene and adding fresh toluene and allowing another 48 hours for solvent exchange. The pore fluid was exchanged for toluene to eliminate the influence of the surface tension of the solvent on the porosity of the final gel [8]. The gels were then dried at 50°C for 48 hours.

RESULTS AND DISCUSSION

A. Characterization of the polymer and solvent quality

NMR spectroscopy of the unfractionated polymers was performed to test that the TMOS condensation reaction was driven to near completion. The NMR spectra for a p(TMOS) batch with a molar ratio of water:TMOS of 1:1 show that the TMOS had reacted to 49.1% conversion which is near the expected conversion of 50%. Most of the silicon atoms in the polymer were Q2 (linear) sites, however there were also a significant number of Q1 (terminal) and Q3 (branching) sites, indicating lightly branched polymers

The SEC results for the fractionated and non-fractionated samples are shown in Figure 2. The elution time for the peak of the unfractionated sample is 22.12 minutes which corresponds to a molecular weight of 1800 calibrated by polystyrene in toluene. The elution time for the peak of the fractionated sample, which has undergone 2 fractionation steps, was 21.53 minutes which corresponds to a molecular weight of 3400 referenced to polystyrene. The shoulder in the data for the unfractionated sample, which occurs at longer elution times, corresponding to lower molecular weight fractions, has also disappeared from the fractionated samples. The SEC results confirm that the fractionation technique was successful. A polymer sample of higher molecular weight with narrow molecular weight distribution was formed. One could use further fractionation to obtain even higher molecular weights with more narrow distributions, however the yield would be reduced.

A light scattering photometer was used to measure the Rayleigh factor, R_θ, at various concentrations of polymer in the solvent of interest. The relationship between the Rayleigh factor and the molecular weight for small angle scattering when the polymers are much smaller than the wavelength of light is given as:

$$Kc/R_\theta = 1/MW + 2A_2c \quad (1)$$

where MW is the weight average molecular weight, c is the concentration in gm/ml, A_2 is the second virial coefficient, and K is the optical constant equal to $(2\pi^2n^2/\lambda^4N)(dn/dc)^2(1+\cos 2\theta)$ where n is the refractive index of the solvent, and λ is the wavelength of light [5].

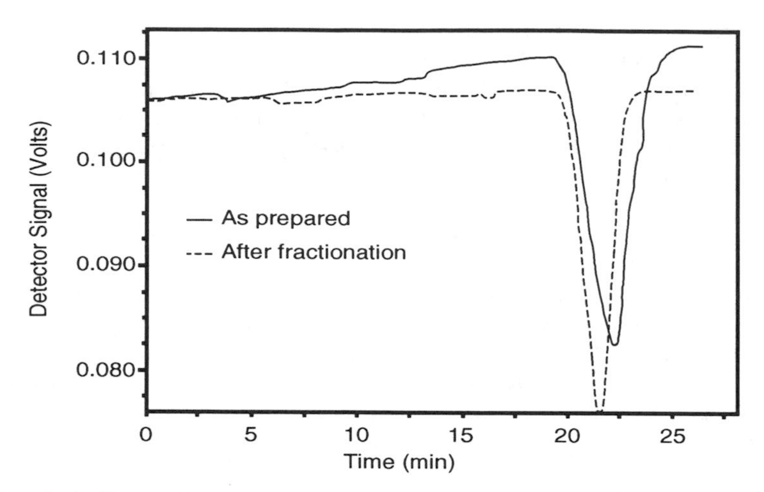

Figure 2: SEC results for fractionated and unfractionated LMW p(TMOS) samples.

Equation 1 suggests that plotting Kc/R_θ vs. c gives a line whose slope is $2A_2$ and the intercept is 1/MW. Figure 3 shows a plot of Kc/R_θ vs. c for the p(TMOS) polymer in three solvents: toluene, methanol, and nitromethane. The data were all taken from a single batch of p(TMOS) which had been twice fractionated and fitted with linear regression. Table 1 summarizes the data obtained for p(TMOS) in these solvents. Table 2 shows the value dn/dc, or specific refractive index increment for p(TMOS) in various solvents measured in this work.

The solubility parameter, δ, for the p(TMOS) polymer was found by experimentally dissolving the polymer in various solvents to be in the range of 8.9-14.5 $(cal/cm^3)^{1/2}$. Thus, nitromethane ($\delta = 12.7$) is expected to be a good solvent, whereas toluene ($\delta = 8.9$) is expected to be a poor solvent [9]. The light scattering data confirm this expectation, since the second virial coefficient is positive for nitromethane, but negative for toluene. The result that nitromethane has a higher molecular weight than toluene is surprising, since the identical polymer sample was used for the light scattering data shown in Fig. 3. It is possible that we are not sampling the entire distribution if toluene is too poor of a solvent to solvate the higher molecular weight polymers . However, the molecular weight of the HMW was 45,000, also determined by light scattering in toluene. Although the HMW result argues that toluene is capable of solvating high molecular weight polymers, solubility is dependent on the degree of branching and cyclization.. Perhaps the LMW samples were more highly branched or more cyclized than the HMW samples. Further NMR studies are underway to determine differences in polymer structure.

B. Effect of solvent quality on film and bulk gel properties

Film casting experiments showed a systematic relationship among the film surface area, pore volume, and molecular weight: films cast from solutions of higher molecular weight polymers had lower surface area and pore volume. This correlation is likely due to the higher degree of condensation of the polymers in the higher molecular weight samples. However, since the light scattering data (Fig. 3) shows a higher molecular weight for better solvents, it is not possible to differentiate the solvent quality effect from the molecular weight dependence.

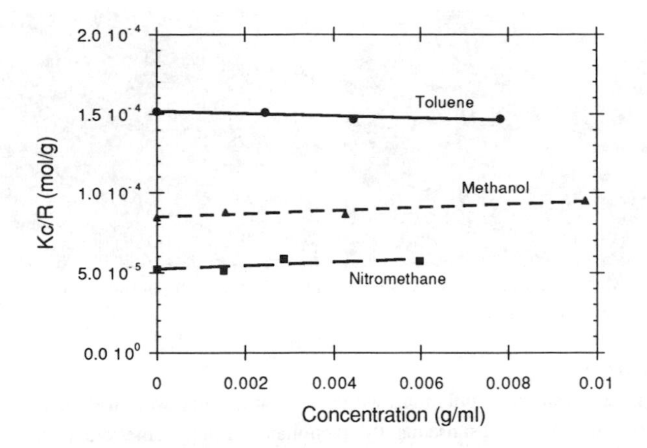

Figure 3: Light scattering results for p(TMOS) plotted as Kc/R_θ versus c.

Table 1
Static Light Scattering (SLS) data for p(TMOS)

	LMW	HMW
MW in Toluene	7,300	45,000
MW in Methanol	12,000	
MW in Nitromethane	19,000	
A_2 in Toluene	-0.66×10^{-3}	-0.64×10^{-3}
A_2 in Methanol	1.01×10^{-3}	
A_2 in Nitromethane	1.07×10^{-3}	
A_2 units - mol cm^3/gm^2		

Table 2
dn/dc for p(TMOS)

Solvent	dn/dc
DMF	-0.0174
Methanol	0.0646
MEK	0.0286
Nitromethane	0.0234
Toluene	-0.0671

Whereas the film casting data does not show dramatic solvent quality effects, the bulk gels made with different solvents do show significant differences. Bulk gels made with toluene had a higher surface area (Langmuir analysis of adsorption data) and a larger pore volume than the gels prepared with nitromethane. Table 3 and Figures 4a,b show the effect of solvent on the bulk gel porosity and morphology. Table 3 shows the pore volume and surface area for gels prepared with three different solvents: toluene, methanol, and nitromethane. In the sol made with toluene, which had the highest surface area, TEM showed porosity on a larger scale than the nitromethane gel which had a finer texture and a correspondingly lower surface area and greater microporosity. These data suggest that the polymers growing in a better solvent were more interpenetrating and could form a denser gel than those grown in a poor solvent, as predicted from Figure 1.

Table 3
Bulk Gel Porosity Data

	Methanol	Nitromethane	Toluene
Langmuir Surface area (m^2/gm)	940	1132	1296
Pore Volume (cm^3/gm)	0.356	0.493	0.669

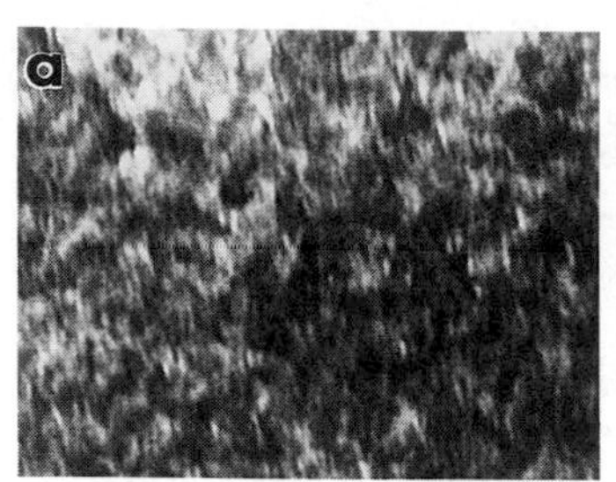

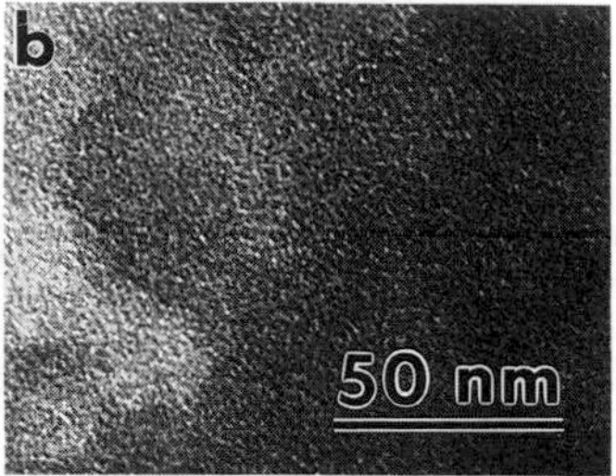

Figure 4: TEM micrographs of the structure of bulk gels a) prepared with toluene b) prepared with nitromethane

CONCLUSIONS

Poly-methoxysiloxane polymers with a low molecular weight dispersion have been synthesized from TMOS and subsequently fractionated with pentane/toluene mixtures. Small angle light scattering was used to measure the weight average molecular weight which varied from 8000 and 45,000. The second virial coefficient was found to be positive in nitromethane and methanol and negative in toluene, confirming that nitromethane is a good solvent for these polymers whereas toluene is a poor solvent. An unusual result is that the polymer molecular weight was found to be higher in good solvents than in poor solvents.

The solvent quality affected the properties of thin films and bulk gels. For films, the influence of solvent quality could not be decoupled from the influence of molecular weight, whereas for bulk gels, the correlation between solvent quality and gel properties corresponds with the ideas shown in Fig. 1. Solvent quality is thus another tool with which the sol-gel chemist can work to tailor material properties.

ACKNOWLEDGEMENTS

C. Jeffrey Brinker and Carol S. Ashley are acknowledged for helpful discussions during this work. R. Weagley, T. Bohuszewicz, and R.A. Assink are thanked for help with SEC, SAW measurements, and NMR respectively. Prof. D.M. Smith and the Center for Micro-Engineered Ceramics at the University of New Mexico are acknowledged for use of the light scattering instruments. The TEM was performed at the Electron Microbeam Analysis Facility at the University of New Mexico Department of Geology and Institute of Meteoritics.

REFERENCES

1. K. C. Chen, T. Tsuchiya, and J. D. MacKenzie, J. Non-Cryst. Solids **81** (1986) 227.
2. T. W. Zerda and G. Hoang, Chem. Mater. **2** (1990) 372.
3. I. Artaki, T. W. Zerda, and J. Jonas, J. Non-Cryst. Solids **81** (1986) 381.
4. T. Katagiri and T. Maekawa, J. Non-Cryst. Solids **134** (1991) 183.
5. P.C. Heimenz, Polymer Chemistry (Marcel Dekker, New York, 1984).
6. L.W. Kelts and N.J. Armstrong, J. Materials Research **4** (1989) 423.
7. J.K. Bailey, C.W. Macosko, and M.L. Mecartney, J. Non-Cryst. Solids **125** (1990) 208
8. R. Deshpande, D.M. Smith, and C.J. Brinker, these proceedings.
9. G.C. Frye and S.J. Martin, Applied Spectroscopy Reviews **26** (1991) 73.
10. E.A. Grulke, in Polymer Handbook, J. Brandrup and E.H. Immergut, editors (Wiley, New York, 1975) VII/519.

A RAMAN SPECTROSCOPIC STUDY OF THE INITIAL STAGES OF HYDROLYSIS OF TETRAETHOXYSILANE

J.D. BARRIE AND K.A. AITCHISON
The Aerospace Corporation, M2/248, P.O. Box 92957, Los Angeles, CA 90009.

ABSTRACT

Multichannel detection Raman spectroscopy was used to follow the hydrolysis and condensation reactions of TEOS/ethanol solutions. Effects of alkoxide concentration, catalyst concentration, water:TEOS ratios, and temperature were studied. It was found that the initial oligomeric composition of hydrolyzed solution was dependent upon the amount of water added to the solution: not only did reaction rates differ, but so to did the reaction products. Higher water concentrations resulted in formation of an oligomer which may contain a trisiloxane ring.

INTRODUCTION

The sol-gel technique has proven to be a valuable route to the synthesis of high purity materials for a number of optical and electronic applications. Many early studies of this process concentrated on the preparation of silicon dioxide from alkoxide precursors such as tetramethoxysilane (TMOS) and tetraethoxysilane (TEOS) [1]. Much is known regarding the behavior of these compounds under a wide variety of experimental conditions as a result of analytical studies using techniques such as Si^{29}-NMR, and Raman and FTIR spectroscopy. Unfortunately, our knowledge of the nature of hydrolysis and condensation reactions of compounds such as TEOS is often confined to specific compositional regimes by limitations in experimental techniques used in the analysis of these reactions. Prolonged data collection times necessitated by the use of single channel detectors in Raman spectroscopy, for example, require the use of idealized experimental conditions giving reduced reaction rates (such as low catalyst concentrations) which are not always representative of the most commonly used conditions. Even more troubling, studies which do not attempt to reduce reaction rates to match their experimental limitations investigate solutions which are changing during the course of the analysis, giving potentially misleading results.

Many studies have successfully applied Raman spectroscopy as a tool for the determination of structure of bulk glasses [2-5], gels and gels evolving toward glasses [6-8], and solution precursors and intermediates in the sol-gel process [9-16]. Critical information regarding the structure of glass surfaces and the nature of the amorphous network have been provided by this technique. Mulder and Damen [10] isolated several oligomeric condensation products of TEOS and utilized Si^{29}-NMR spectroscopy to assign observed Raman bands to these species. Their analysis of the hydrolysis and condensation reactions of TEOS (using single channel detection and low catalyst and water concentrations) formed the basis for the present work.

This study was intended to determine whether Raman spectroscopy using optical multichannel detection is capable of studying sol-gel reactions in time regimes short enough to investigate many of the commonly used experimental conditions. We undertook this investigation not simply to study the reaction pathways of silicon alkoxides, but to develop the capability of addressing the more reactive alkoxide (such as titanium, zirconium, and niobium alkoxides used in the preparation of ferroelectric ceramics via sol-gel processes) and double-alkoxide precursors which are finding increasing application in the preparation of crystalline oxide thin films. Knowledge of the reactivity of these compounds may be critical to the development of crystalline films having proper phase and stoichiometry for device applications.

EXPERIMENTAL

Tetraethoxysilane (TEOS) solutions were prepared with varying ratios of solvent (ethanol), water, and catalyst (HCl) using standard Schlenk-type apparatus. Commercially available TEOS (Alfa, 97%) was purified by distillation. Ethanol (absolute) was dried by refluxing over type-4A molecular sieve and distilled under dry N_2. A standard solution of TEOS in ethanol (1:2 mole ratio of TEOS to ethanol) was prepared and used for all further solutions. Appropriate amounts of this

 ©1992 Materials Research Society

standard solution were placed in septum sealed glass vials for the Raman experiment. Deionized water was mixed with appropriate amounts of dried ethanol and HCl (Mallinckrodt, AR grade) prior to addition to the TEOS:ethanol solution. The water/ethanol/catalyst fraction was added to the stirred TEOS/ethanol fraction by syringe as quickly as possible (<10sec) and data collection was begun immediately thereafter. The relatively large amounts of ethanol used in these experiments resulted in rapid dissolution of the water into the solution, and no evidence of scattering due to immiscibilities was observed.

The concentration of TEOS in each experiment was controlled by variation of the amount of ethanol used in the water fraction, and varied from one to two moles per liter. Water to TEOS ratios were varied from 0.25 to 4 moles of water per mole of TEOS, and the ratio of HCl to TEOS in the solutions was in the range 0.005 to 0.05. Uncatalyzed solutions demonstrated vanishingly slow reaction rates and will not be discussed in this paper. The composition of the reactions discussed in this paper will be denoted by giving the ratio of the reactants in the following order: moles TEOS-moles ethanol-moles water-moles HCl. For example, 1-4-1-0.005 denotes a solution with 4 moles of ethanol, one mole of water, and 0.005 moles of HCl per mole of TEOS. All experimental conditions were examined at least twice to assure repeatability of the characterization.

The Raman experiments were performed using a Coherent model 305 Argon ion laser operating at 514.5 nm, a sample chamber with $\pi/2$ collection optics and a magnetic stirring stage, and an Instruments S.A. U-1000 spectrometer with a Princeton Instruments IRY-700 intensified diode array detector. Typical experimental conditions included: 0.20 W of vertically polarized incident radiation, no polarization analysis, a slit width of 200 μm giving a spectrometer bandpass of 2.1 cm^{-1}, and 1-2 sec collection time for each window of data collected (covering approximately 150 cm^{-1} of the Raman spectrum). For spectra covering more than one window, multiple segments of the spectrum were collected (with overlap) and added together. Data for a single window could be collected at intervals as small as 2 sec, while a complete spectrum covering the range of 100-1600 cm^{-1} typically required about 3 min. Using these collection conditions, no peaks were observed from the glass vials used to contain the solutions.

The data were collected via digital computer and stored on disk for analysis. Deconvolution of overlapping peaks was performed using Jandel Scientific's Peakfit® software, assuming a Voigt lineshape having predominately Lorentzian character. The choice of this lineshape was based upon the observed data for pure TEOS (and TEOS/ethanol solutions) and on the excellent fits determined for the simplest hydrolyzed solutions using this assumption.

RESULTS AND DISCUSSION

When a water/ethanol/catalyst fraction is added to a TEOS/ethanol solution, hydrolysis and condensation reactions begin immediately. The rates of reaction, and the species which are detected via Raman spectroscopy are found to vary with the relative water, TEOS, and catalyst concentrations, and with the temperature of the reactants.

Figure 1 shows the evolution of Raman spectra for a solution of composition 1-4-1-0.05 in the minutes immediately following addition of the water fraction. The TEOS concentration in this solution is 2.1 M. The spectra shown can be interpreted on the basis of assignments made by Ypenburg for the spectra of TEOS [9]. The strong peak at 645 cm^{-1} is attributed to the symmetric SiO_4 stretch of the alkoxide. Upon addition of water and catalyst, this band weakens immediately and a shoulder forms at 670 cm^{-1}. Indications of a second new peak at 701 cm^{-1} are also observed. In less than one minute, however, reactions take place which tend to reduce the intensity of these new bands, and lead to the formation of a band at 604 cm^{-1}. In less than three minutes, additional bands at 575 and 546 cm^{-1} are observed. Each of these bands exhibit appreciable breadth (of the order of 30 cm^{-1} FWHM) and there is extensive overlap.

The peaks at 675 cm^{-1} and 701 cm^{-1} are found at energies consistent with that expected for Raman scattering from TEOS molecules in which one or more of the ethoxy groups have been replaced by hydroxy groups, as the smaller reduced mass of these hydrolyzed species results in higher vibrational energies [16]. In addition, the magnitude of the shifts observed are similar to those found during hydrolysis of TMOS based solutions in which the Raman peaks were assigned to the singly and doubly hydrolyzed monomers and the assignments were confirmed by Si^{29} NMR [16] . By analogy with this work, and taking consideration of the fact that these two peaks evolve sequentially (with the 675 cm^{-1} band preceding the 701 cm^{-1} band), we assign the peaks at 675 cm^{-1} and 701 cm^{-1} to SiO_4 symmetric stretches of triethoxysilanol and diethoxysilanediol,

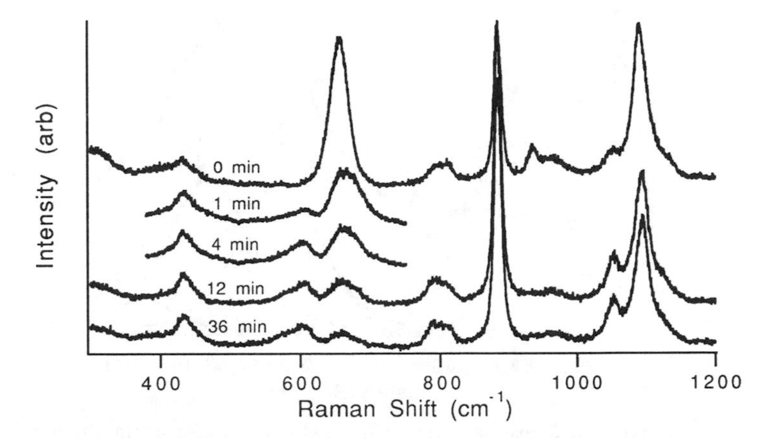

Figure 1: Time evolving Raman spectra of a TEOS solution of composition 1:4:1:0.05. The time elapsed between addition of the water and collection of the spectrum are as noted.

respectively. The assignment of the band at 675 cm^{-1} as arising from the singly hydrolyzed TEOS deviates from the finding of Mulder and Damen's study of TEOS hydrolysis [10], as they did not discern this unresolved peak, and assigned the band observed at 701 cm^{-1} to the singly hydrolyzed TEOS. The excellent time resolution of our spectrocopy was invaluable in isolating this species, as it allowed us to collect spectra resolved both temporally and energetically.

The formation of low molecular weight oligomers via condensation reactions is expected to produce Raman bands at smaller shifts than those produced by TEOS monomers due to the larger reduced mass resulting from the replacement of ethoxy or hydroxy groups by heavier siloxy groups. Our observation of several well defined peaks in the region 520-620 cm^{-1} in the minutes immediately following hydrolysis lead us to conclude that these bands arise from simple siloxane oligomers. On the basis of the work of Mulder and Damen [10], it is clear that at least a substantial fraction of the observed Raman scattering composing these bands is due to hexaethoxydisiloxane (604 cm^{-1}), octaethoxytrisiloxane (575 cm^{-1}), and decaethoxytetrasiloxane (546 cm^{-1}). The breadth of these bands indicates the possibility that some portion of the ethoxy groups on these compounds may be hydrolyzed. At long times (>48 hrs) only the tri- and tetrasiloxane species are still observed in appreciable quantities, with the bands due to higher order chains and rings (expected in the 400-500 cm^{-1} range) obscured by the presence of a band arising from ethanol in the sample. The rapid formation of the condensed species, coupled with the lack of appreciable quantitites of multiply hydrolyzed monomers suggests that the reactivity of these species is significantly greater than that of triethoxysilanol, as had been observed in other studies [17].

When the catalyst concentration of the solution is decreased, the reactions which are observed above are found to slow dramatically. For solutions of composition 1-4-1-0.005, the hydrolysis reactions leading to the formation of diethoxysilanediol are much more evident, and appreciable quantities of this compound (indicated by a Raman peak at 701 cm^{-1}) are observed at times in excess of one hour after the addition of water. Deconvolution of the spectra taken from this composition indicate the presence of an additional peak not observed in the solutions containing higher concentrations of catalyst (Figure 2). This weak peak centered at 628 cm^{-1} is at a higher energy than that expected for the unhydrolyzed dimer, but lower than that expected for any modification of the monomer. On the basis of this, the peak at 628 cm^{-1} is best assigned to the presence of hydrolyzed disiloxanes, although the possibility that this band arises due to ring species such as a hexaethoxycyclotrisiloxane cannot be dismissed.

The effect of increasing the water:TEOS ratio in these solutions (keeping the catalyst concentration constant) is demonstrated in Figure 3. This composition, 1-4-2-0.005, is found to hydrolyze more rapidly than solutions containing a 1:1 molar ratio of water to TEOS, and to produce a different spectrum of monomeric and oligomeric species at longer times as well.

While lesser amounts of water were found to result in a range of silicon containing oligomers

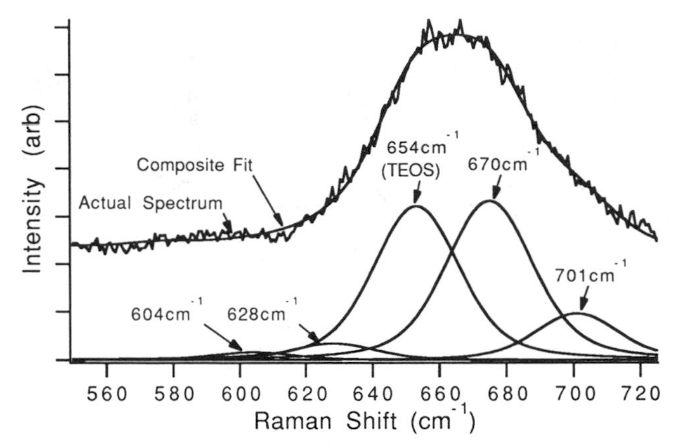

Figure 2: Deconvoluted Raman spectrum of a TEOS solution of composition 1:4:1:0.005 measured 10 min after addition of water.

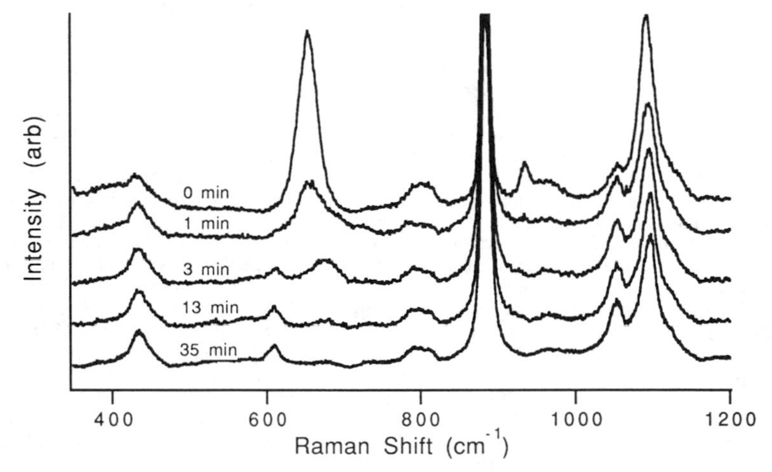

Figure 3: Time evolving Raman spectra of a TEOS solution of composition 1:4:2:0.005. The time elapsed between addition of the water and collection of the spectrum are as noted. Note the peak at 612 cm^{-1} not observed in solutions containing only one mole of water per mole of TEOS.

which gradually coarsened to form a more extended network, increased water contents (2:1 and greater) were found to result in spectra without this complicated range of species. Rather, a pronounced peak at 612 cm^{-1} is observed which dominates the signal in this range of the spectrum. It is apparent that this peak is shifted to slightly higher energies than the peak attributed to unhydrolyzed disiloxanes above, and the linewidth of the peak seems to be reduced as well. This peak is then observed to decrease in intensity in favor of those species which scatter below 550 cm^{-1} and which eventually make up the extended SiO_2 network.

This new peak at 612 cm^{-1} is at a similar energy to the Raman band which has been assigned to trisiloxane rings in silica glass (called D_2 in SiO_2 and typically found around 610 cm^{-1}) [6], and in a variety of crystalline silicates [18]. It is quite possible that these hydrolysis conditions differ enough from those used in previous Raman experiments (which did not observe the formation of cyclics during initial stages of condensation) to induce the formation of these species. Recall that the water-containing fraction was added to the TEOS/ethanol solution quickly, rather than the dropwise fashion commonly used. This deviation from a process which normally allows the water to be consumed almost completely as it is being added may enhance the formation of rings by increasing the extent of hydrolysis of the terminal silanols. Trisiloxane rings have been observed by Si^{29}-NMR measurements on TEOS solutions containing 2:1 molar ratios of water to TEOS [19]. Similar measurements of these compositions would be invaluable for confirming this assignment.

Another interesting feature of spectra taken from solutions which have in excess of two moles of water per mole of silicon is the behavior of the Raman peaks found at 795 and 810 cm^{-1}. The origin of these peaks has been assigned to the SiO_4 asymmetric stretch and a CH_2 rock [9]. Neither of these peaks seems to be affected appreciably in solutions containing water:TEOS ratios of 2:1 or less, and in fact seem to increase somewhat in those containing one mole of water per mole of TEOS. It has been reasoned that the asymmetric stretch of the SiO_4 group is active in the oligomeric species as well [10], and that this is the origin of at least one of these peaks at longer times. Both of these peaks disappear, however, when water:TEOS ratios of 4:1 are used (figure 4). Moreover, they do not decrease in intensity in concert, but rather sequentially. The 810 cm^{-1} peak is the first to disappear, and the 795 cm^{-1} peak follows. The disappearance of these peaks is also observed when additional water is added to a 2:1 solution which has progressed to a stage where little if any monomeric species remain. We assign the vibration at 810 cm^{-1} to the CH_2 rocking motion in the ethoxy groups on TEOS and the several ethoxysilanols formed during hydrolysis on the basis of the following argument. The disappearance of the band at 810 cm^{-1} is the result of the replacement of the remaining ethoxy groups by hydroxy groups or siloxy bridges (at least to the extent that our detection limits for these vibrations is reached). This vibration is not observed in the ethanol which is produced by condensation reactions. The band due to the SiO_4 asymmetric stretch persists for slightly longer times as it can also arise from transitions within disiloxanes, larger oligomers, and even cyclic tri- and tetrasiloxanes. It is likely that the vibrations leading to this band are never fully absent, but that as the solution species coarsen, the band is broadened to the extent that is cannot be detected.

When the original TEOS concentration of the solutions was decreased to 1.0 M by adding additional ethanol (prior to hydrolysis), the sequence of hydrolysis and condensation was observed to follow that observed above for similar water:TEOS ratios, although the rate of the reactions was significantly diminished. The first indications of dimeric species in a 1:10:1:0.005 solution were not observed until 18 min after addition of the water. This is nearly more than three times longer than it took to observed condensed species in a 1:4:1:0.005 solution having twice the initial TEOS concentration. Decreases in reaction rates were also observed when the solution was chilled below room temperature.

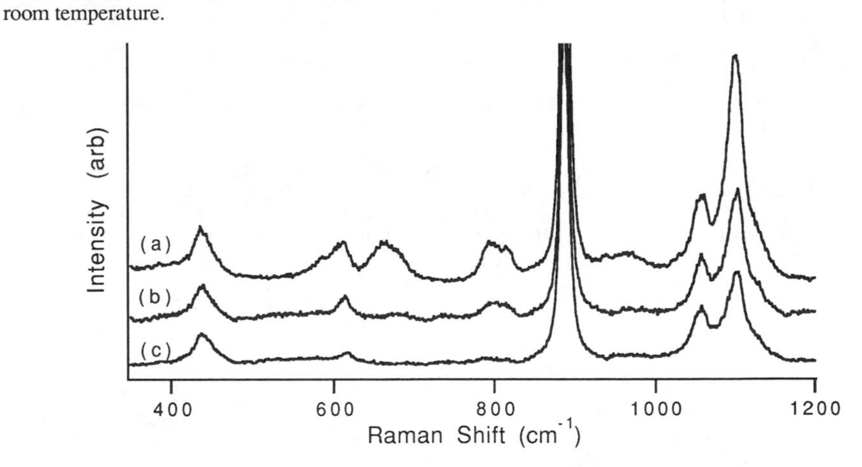

Figure 4: Comparison of Raman spectra of TEOS solutions containing (a) one, (b) two, and (c) four moles of water per mole of TEOS. The time elapsed between addition of the water and collection of the spectrum are 63, 30, and 7 min, respectively. Note the decrease of peaks at 795 cm^{-1} and 810 cm^{-1} in (c).

CONCLUSIONS

The hydrolysis dynamics of TEOS/ethanol solutions using acid catalysis are rapid and complex. The amount of water added to the solution determines not only the extent of the hydrolysis reactions, but also the combination of oligomeric species which form in solution. We observed different hydrolysis/condensation behavior depending upon the amount of water added to the solutions, with higher water:TEOS ratios leading to enhanced concentrations of a compound whose Raman spectrum is similar to that of cyclotrisiloxanes. Changes in catalyst concentration and solution temperature changed the rate of the reactions, but did not seem to alter the reaction products. It is not clear from these studies what the effect of the rate of addition of the water fraction has on the reaction. This type of information can be especially critical in chemical systems in which several alkoxide species are present simultaneously, when double alkoxides are in use, or when alcohols which are not the parent alcohol of the alkoxide are used (e.g., ethanol with titanium isopropoxide).

We conclude that multichannel detection Raman spectroscopy is a valuable technique for the analysis of dynamic reactions which occur during hydrolysis and condensation of sol-gel precursors. The ability to temporally resolve reactions taking place in solutions of nonidealized compositions is invaluable for the understanding of the resultant gel structure.

ACKNOWLEDGEMENTS

The authors appreciate experimental assistance from P.D. Chaffee, J. Ciofalo, and J. Emerson. We thank P.D. Fuqua for helpful discussions. The work was supported by the Aerospace Sponsored Research program.

REFERENCES

1. Brinker, C. Jeffrey, and George W. Scherer, Sol-Gel Science (Academic Press, Inc., San Diego, CA) 1990.
2. Galeener, F.L., Sol. St. Commun. 44, 1037 (1982).
3. Furukawa, Toshiharu, Karen E. Fox, and William B. White, J.Chem. Phys. 75, 3226 (1981).
4. Mulder, C.A.M., J. Non-Cryst. Sol. 95&96, 303 (1987).
5. Tallant, D.R., B.C. Bunker, C.J. Brinker and C.A. Balfe in Better Ceramics Through Chemistry II, edited by C.Jeffrey Brinker, David E. Clark, and Donald R. Ulrich (Mater. Res. Soc. Proc. 73, Pittsburgh, PA 1986) p. 261.
6. Brinker, C.J., R.K. Brow, D.R. Tallant, and R.J. Kirkpatrick, J. Non-Cryst. Sol. 120, 26 (1990).
7. Bertoluzza, Allesandro, Concezio Fagnano, Maria Antonietta Morelli, Vittorio Gottardi and Massimo Guglielmi, J. Non-Cryst. Sol. 48, 117 (1982).
8. Kato, K., J. Mater. Sci. 26, 6777 (1991).
9. Ypenburg, J.W. and H. Gerding, Rec. Trav. Chim. 91, 1245 (1972).
10. Mulder, C.A.M., and A.A.J.M. Damen, J. Non-Cryst. Sol. 93, 169 (1987).
11. Lippert, J.L., S.B. Melpolder, and L.M. Kelts, J. Non-Cryst. Sol. 104, 139 (1988).
12. Balfe, Carol A., Kenneth J. Ward, David R. Tallant, and Sheryl L. Martinez in Better Ceramics Through Chemistry II, edited by C.Jeffrey Brinker, David E. Clark, and Donald R. Ulrich (Mater. Res. Soc. Proc. 73, Pittsburgh, PA 1986) p. 619.
13. Sugahara, Yoshiyuki, S. Okada, K. Kuroda and C. Kato, J. Non-Cryst. Sol. 139, 25 (1992).
14. Zerda, T.W. and G. Hoang, Chem. Mater. 2, 372 (1990), and J. Non-Cryst. Sol. 109, 9 (1989).
15. Zerda, T.W., I. Artaki, and J. Jonas, J. Non-Cryst. Sol. 81, 365 (1986).
16. Artaki, I., M. Bradley, T.W. Zerda, and J. Jonas, J. Phys. Chem. 89, 4399 (1985).
17. Lin, Chia-Cheng and John D. Basil, in Better Ceramics Through Chemistry II, edited by C.Jeffrey Brinker, David E. Clark, and Donald R. Ulrich (Mater. Res. Soc. Proc. 73, Pittsburgh, PA 1986) p. 585.
18. Griffith, W.P., J. Chem. Soc. (A) 1969, 1372.
19. Kelts, Larry W. and Nancy J. Armstrong, in Better Ceramics Through Chemistry III, edited by C.Jeffrey Brinker, David E. Clark, and Donald R. Ulrich (Mater. Res. Soc. Proc. 121, Pittsburgh, PA 1988) p. 519.

SILICON-29 NMR STUDY ON THE INITIAL STAGE OF THE CO-HYDROLYSIS OF TETRAETHOXYSILANE AND METHYLTRIETHOXYSILANE

YOSHIYUKI SUGAHARA, YOICHI TANAKA, SHUJI SATO, KAZUYUKI KURODA, AND CHUZO KATO
Department of Applied Chemistry, Waseda University, Ohkubo-3, Shinjuku-ku, Tokyo, 169 Japan.

ABSTRACT

The hydrolysis and initial condensation processes of the mixtures of TEOS and MTES have been studied by GC-MS and ^{29}Si-NMR in the systems with TEOS:MTES:ethyl alcohol:water:HCl=1:1:14:28:6x10^{-4} and 0.5:1.5:14:28:6x10^{-4}. The dimer $CH_3(HO)_2SiOSi(OH)_3$ possessing both TEOS- and MTES-derived units was identified, indicating that the condensation between a hydrolyzed TEOS-derived monomer and a hydrolyzed MTES-derived monomer was one of dominant dimerization.

INTRODUCTION

Organically-modified ceramics (ORMOCERs) obtained by co-hydrolysis processes have attracted much attention, since they could serve a new type of materials [1,2]. Typically, these ORMOCERs are prepared from organoalkoxysilane possessing an organic modifying group, organoalkoxysilane possessing a polymerizable organic group (like vinyl group), and tetraalkoxysilane (or other metal alkoxides) as an inorganic network former. Most of actual systems, however, are very complicated, so that simple systems have been mainly applied to characterize the structural evolution of heteropolysiloxanes. Lee et al. [3] prepared membranes by the co-hydrolysis of tetra-functional tetramethoxysilane (TMOS) and di-functional dimethyldiethoxysilane (DMDES), and, on the basis of the solid-state ^{29}Si-nuclear magnetic resonance spectroscopy (NMR) results, they proposed the block polymer-like structure where polydimethylsiloxnane and silica-like parts were segregated. Babonneau et al. [4] studied the co-hydrolysis processes of tetraethoxysilane (TEOS)-DMDES mixtures by ^{29}Si-NMR and infra-red spectroscopy (IR). The subsequent paper [5] described further details of ^{29}Si-NMR, as well as IR and Raman spectroscopy results, and the ^{29}Si-NMR signals due to condensation between a TEOS-derived unit and a DMDES-derived one were reported. Hasegawa et al. [6] identified cubic $CH_3Si_8O_{19}^{7-}$ where one $CH_3Si(O_{0.5}\text{-})_3$ unit displaced one of the $Si(O^-)(O_{0.5}\text{-})_3$ units of the $Si_8O_{20}^{8-}$ structure, when TEOS and tri-functional methyltriethoxysilane (MTES) were co-hydrolyzed in the presence of (2-hydroxyethyl)trimethylammonium ion. Recently, Bommel et al. [7] studied the co-hydrolysis processes of the mixtures of TEOS-DMDES, TEOS-MTES, TEOS-phenyltriethoxysilane (PhTES), and TEOS-MTES-DMDES-PhTES by hydrolysis time-condensation time curves, viscosity measurements, and ^{29}Si-NMR. In addition, Glaser et al. [8] characterized the products obtained by the co-hydrolysis of

TEOS-MTES-DMDES mixtures by solid-state ^{29}Si-NMR.

Silicon-29 NMR is one of powerful tools for investigating the hydrolysis and initial polycondensation processes of alkoxysilanes [9] and organoalkoxysilanes [10,11]. Initial polycondensation processes become very important for the co-hydrolysis processes, since the structures in oligomers are preserved in the final products; hence, if co-polymerization is not dominant during the initial polycondensation processes, segregation could not be avoidable.

This paper aims at finding evidence for co-polymerization during the co-hydrolysis process of TEOS and MTES. Although there have been a few ^{29}Si-NMR studies on the co-hydrolysis processes [4,5,7], no ^{29}Si-NMR approaches to identify oligomers possessing units derived from different alkoxysilanes have been reported. Hence, in the present study, we focus dimerization processes, and discuss the identification of the target dimer $CH_3(RO)_2SiOSi(OR)_3$ (where R=H, C_2H_5) which possessing both a TEOS-derived unit and an MTES-derived one.

EXPERIMENTAL

MTES (Tokyo Kasei Chemical Ind. Ltd.) and TEOS (Wako Pure Chemical Ind. Ltd.) were used as received. Other details on materials were described elsewhere [10]. Reactions were conducted in the TEOS-MTES-ethyl alcohol (C_2H_5OH; EtOH)-water (D_2O/H_2O)-HCl system (TEOS:MTES:EtOH:D_2O/H_2O:HCl=1:1:14:24:$6x10^{-4}$ or 0.5:1.5:14:24:$6x10^{-4}$). A trace of chromium(III) acetylacetonate (Cr(acac)$_3$) was also added to shorten the relaxation time of Si-species. After TEOS, MTES, EtOH, and Cr(acac)$_3$ were mixed, D_2O was added dropwise. Then 0.01N-HCl was added similarly, and the mixture was further stirred for 3 min.

Oligomeric species were analyzed by the combination of trimethylsilylation (by the method of Lentz [12]) and gas-liquid chromatography (Shimadzu GC-8A) as described before [13]. The structures of trimethylsilylated derivatives were studied by gas chromatograph-mass spectroscopy (GC-MS; Shimadzu QP-1100EX). Ion impact experiments were carried out with ionizing energy of 20 eV, and the ion source was maintained at 250 °C. The ^{29}Si-{^1H}-NMR spectra were obtained at 79.42 MHz using a JEOL NM-GSX-400 spectrometer. Details of ^{29}Si-NMR measurement conditions were the same as before [10].

RESULTS AND DISCUSSION

The nomenclature of Si environments is followed by the manner similar to the previous papers [10,11,14]. Hence, the TEOS-derived unit of $(HO)_m(EtO)_{4-m-n}Si(O_{0.5}\!\!-\!\!)_n$ is presented as $Q^n{}_{(mOH)}$, and $T^n{}_{(mOH)}$ represents the MTES-derived unit of $CH_3Si(O_{0.5}\!\!-\!\!)_n(OH)_m(OEt)_{3-m-n}$. For the trimethylsilylated derivatives, the number of trimethylsilyl groups attached to silicon is shown in subscript parentheses.

Figure 1 shows a typical gas chromatogram of trimethylsilylated derivatives of the solution in the system with TEOS:MTES:EtOH:H_2O:HCl=1:1:14:24:$6x10^{-4}$. Peaks A-E are assigned on the basis of retention time and GC-MS results (Table I).

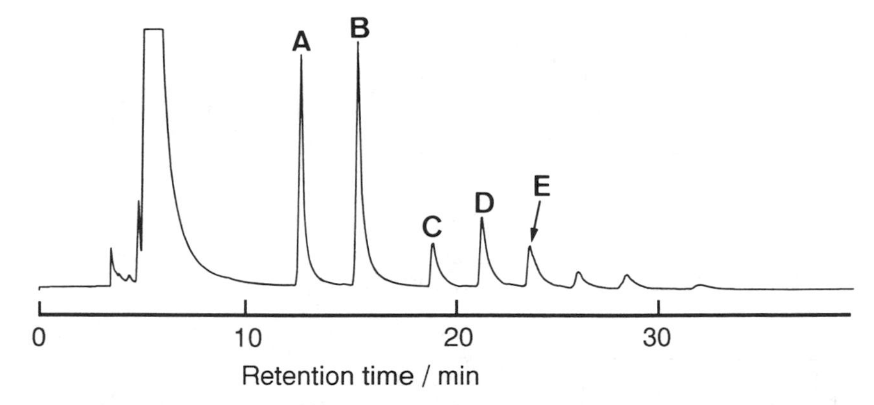

Fig. 1 Gas chromatogram of the trimethylsilylated derivatives of the solution in the system with TEOS:MTES:EtOH:H_2O:HCl=1:1:14:24:6x10^{-4}. The solution was silylated after the reaction for 3 h.

Peaks A and B are due to the trimethylsilylated monomeric species (Since both hydroxyl and ethoxy groups were silylated, all the species of $T^0_{(mOH)}$ and $Q^0_{(mOH)}$ were detected as $T^0_{(3TMS)}$ (Peak A) and $Q^0_{(4TMS)}$ (Peak B)). Peak C is ascribed to the trimethylsilylated dimer derived from MTES only ($T^1_{(2TMS)}T^1_{(2TMS)}$). Peak D can be assigned to the trimethylsilylated dimer possessing both the TEOS- and MTES-derived units ($T^1_{(2TMS)}Q^1_{(3TMS)}$). Peak E is due to the trimethylsilylated dimer derived from TEOS only ($Q^1_{(3TMS)}Q^1_{(3TMS)}$). Other GC peaks having higher retention times should be due to higher oligomers, whose identification is now under investigation. The assignments of GC peaks due to the monomers and the dimers are consistent with the previous report describing the GC results in a similar system [6]. Although condensation during trimethylsilylation cannot be suppressed completely, the presence of peak D strongly suggests that condensation between the TEOS-derived monomer and the MTES-derived one occurred among dimerization.

In order to obtain direct evidence for the presence of the target T^1Q^1 dimer during the co-hydrolysis process, the process was monitored by ^{29}Si-NMR. Figure 2 shows the ^{29}Si-NMR spectra in the system with TEOS:MTES:EtOH:D_2O/H_2O:HCl=1:1:14:24:6x10^{-4}. In the T^0 (-37 - -42 ppm) [11]

Table I The assignments of labeled GC peaks in Fig. 1.

Peak	m/e, $[M-15]^+$ ¶	Structure	Formula (Molecular Weight)
A	295	$T^0_{(3TMS)}$	$CH_3SiO_3[Si(CH_3)_3]_3$ (310)
B	369	$Q^0_{(4TMS)}$	$SiO_4[Si(CH_3)_3]_4$ (384)
C	443	$T^1_{(2TMS)}T^1_{(2TMS)}$	$(CH_3)_2Si_2O_5[Si(CH_3)_3]_4$ (458)
D	517	$T^1_{(2TMS)}Q^1_{(3TMS)}$	$CH_3Si_2O_6[Si(CH_3)_3]_5$ (532)
E	591	$Q^1_{(3TMS)}Q^1_{(3TMS)}$	$Si_2O_7[Si(CH_3)_3]_6$ (606)

¶ No peaks corresponding to molecular ions were observed.

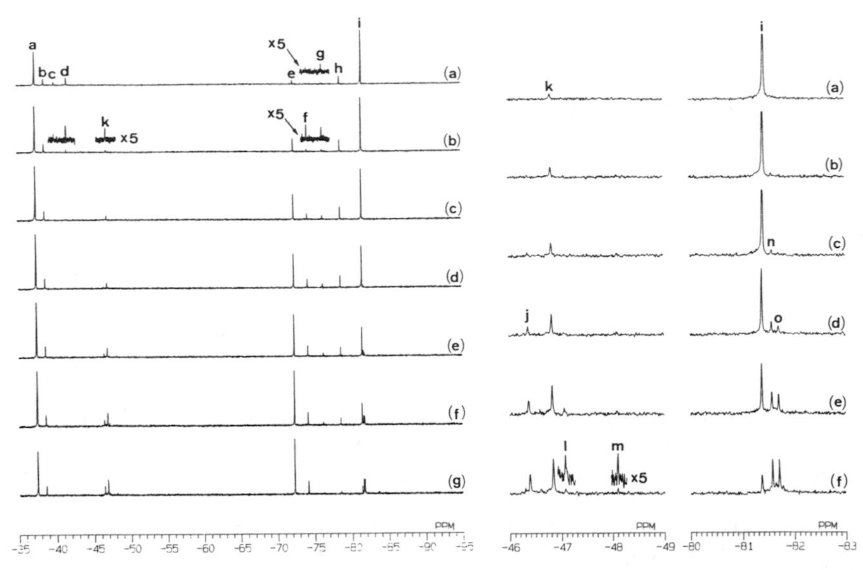

Fig. 2 Silicon-29 NMR spectra of the solution in the system with TEOS:MTES:EtOH:D_2O/H_2O:HCl= 1:1:14:24:6x10^{-4} after (a) 15 min, (b) 30 min, (c) 45 min, (d) 1 h, (e) 1 h 30 min, (f) 2 h, and (g).3 h

Fig.3 Enlarged profile (T^1 and Q^1 region) of Fig.2; after (a) 30 min, (b) 45 min, (c) 1 h, (d) 1 h 30 min, (e) 2 h, and (f) 3 h. Y-axis are expanded twice in all the spectra.

and Q^0 (-72 - -82 ppm) [9] regions, signals a-d and signals e-i were observed. According to the previous studies [7,9], these signals are assigned to monomeric species (signal a: $T^o{}_{(3OH)}$; signal b: $T^o{}_{(2OH)}$; signal c: $T^o{}_{(1OH)}$; signal d: $T^o{}_{(0OH)}$ (MTES); signal e: $Q^o{}_{(4OH)}$; signal f: $Q^o{}_{(3OH)}$; signal g: $Q^o{}_{(2OH)}$; signal h: $Q^o{}_{(1OH)}$; signal i: $Q^o{}_{(0OH)}$ (TEOS)).

The assignments of dimers are based on the distribution of the monomers and the behavior of signal intensities. In the present system, the monomers should be condensed mainly by water-producing condensation (≡Si-OH + HO-Si≡ → ≡SiOSi≡ + H_2O) rather than by alcohol-producing condensation (≡Si-OEt + HO-Si≡ → ≡SiOSi≡ + EtOH), since a large amount of water was present [15]. With respect to the signal intensities, the assignments are based on the assumption that two signals of dimers having different environments of silicon should have similar intensities in all the spectra [10].

Parts of T^1 (-46 - -49 ppm) [11] and Q^1 (-80 - -83 ppm) [9] regions are expanded to discuss the signals due to dimers (Fig. 3), and six signals and signal i were observed. Signal k, which was present even after 15 min (see Fig. 2 (a)), was the strongest among the T^1 signals, and signal j, signal n, and signal o appeared after 1 h. Then signal l and signal m were detected after 3 h. No corresponding Q^2 and T^2 signals were detected even after 3 h, indicating that these signals should

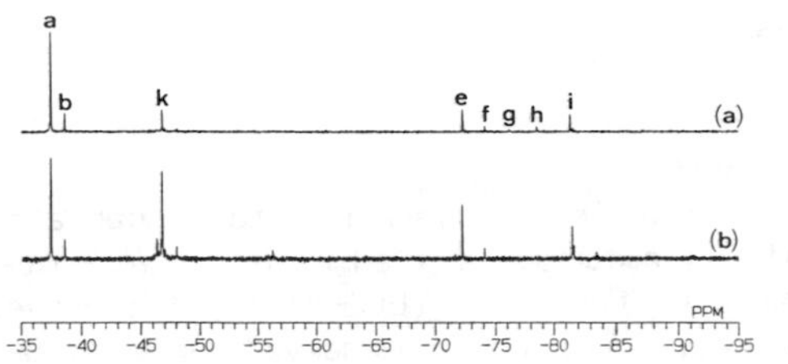

Fig. 4 Silicon-29 NMR spectra of the solution in the system with TEOS:MTES:EtOH:D_2O/H_2O:HCl= 0.5:1.5:14:24:6x10^{-4} after (a) 45 min and (b) 3 h.

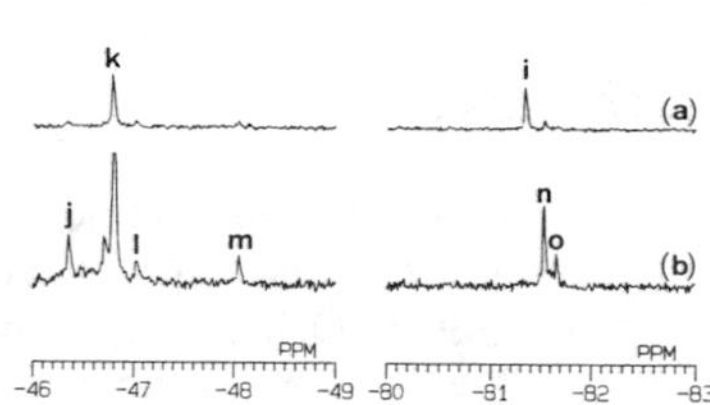

Fig.5 Enlarged profile (T^1 and Q^1 region) of Fig.4; after (a) 45 min and (b) 3 h. Y-axis are expanded twice in all the spectra.

be due to dimers.

When signal k appeared, $T^0{}_{(3OH)}$ was mainly present among the monomers. Hence, it is assigned to $T^1{}_{(2OH)}T^1{}_{(2OH)}$. The signal due to $T^1{}_{(2OH)}T^1{}_{(2OH)}$ also appeared first among the T^1 signals in the hydrolysis and initial polycondensation processes of MTES only [11], and the relative shift from the signal due to $T^0{}_{(3OH)}$ (signal a in Fig. 2) is similar to that in the previous study, which supports the assignment. (The definite chemical shift cannot be evidence because of the strong solvent effect on the chemical shifts of signals.) Signal l and signal m can be assigned to $T^1{}_{(2OH)}T^1{}_{(1OH)}$ (signal l: $T^1{}_{(2OH)}$ unit; signal m: $T^1{}_{(1OH)}$ unit) on the basis of the relative shifts from signal k [11]. This is consistent with the fact that these signals showed similar intensities in Fig. 3 (f).

Signal j, signal n, and signal o exhibit similar intensities in Fig. 3. On the basis of the distribution of the monomers, the formation of two dimers can be probable via water-producing condensation; $Q^1{}_{(3OH)}Q^1{}_{(3OH)}$ and $T^1{}_{(2OH)}Q^1{}_{(3OH)}$. To assign these signals, another system containing a relatively large amount of MTES was studied (TEOS:MTES:EtOH:D_2O/H_2O:HCl=0.5:1.5:14:24:6x10^{-4}) (Figs. 4 and 5). Apparently, signal j and signal n showed similar intensities, indicating the assignment of these signals as $T^1{}_{(2OH)}Q^1{}_{(3OH)}$ (signal j: $T^1{}_{(2OH)}$ unit; signal n: $Q^1{}_{(3OH)}$ unit). Hence, signal o is ascribed to $Q^1{}_{(3OH)}Q^1{}_{(3OH)}$.

The assignments of these six signals are listed in Table II. These NMR results are essentially consistent with the GC results of trimethylsilylated derivatives (Table I). Hence, we conclude that $CH_3(HO)_2SiOSi(OH)_3$ formed in the present

Table II The assignments of the labeled ^{29}Si-NMR signals in Figs. 3 and 5.

Signal	Structure	Formula
j	$\underline{T}^1{}_{(2OH)}Q^1{}_{(3OH)}$	$Me(HO)_2\underline{Si}OSi(OH)_3$
k	$T^1{}_{(2OH)}T^1{}_{(2OH)}$	$Me(HO)_2SiOSi(OH)_2Me$
l	$\underline{T}^1{}_{(2OH)}T^1{}_{(1OH)}$	$Me(HO)_2\underline{Si}OSi(OH)(OEt)Me$
m	$T^1{}_{(2OH)}\underline{T}^1{}_{(1OH)}$	$Me(HO)_2SiO\underline{Si}(OH)(OEt)Me$
i	$Q^0{}_{(0OH)}$	$Si(OEt)_4$ (TEOS)
n	$T^1{}_{(2OH)}\underline{Q}^1{}_{(3OH)}$	$Me(HO)_2SiO\underline{Si}(OH)_3$
o	$Q^1{}_{(3OH)}Q^1{}_{(3OH)}$	$(HO)_3SiOSi(OH)_3$

system as one of the dominant dimers.

CONCLUSION

The GC-MS and ^{29}Si-NMR study of the hydrolysis and initial condensation processes of the TEOS-MTES mixtures showed that the dimer $CH_3(HO)_2SiOSi(OH)_3$, which composed of both the TEOS- and MTES-derived units, formed as well as the dimer having only the TEOS-derived unit and those having only the MTES-derived units. These results indicate that, under the present experimental conditions, the condensation between hydrolyzed alkoxysilanes with different functionalities occurred as one of dominant dimerization reactions.

Acknowledgement

This work was partially supported by Waseda University Grant for Special Research Project. Y.S. thanks Izumi Science and Technology Foundation for financial support. The authors also thank Tsuyoshi Inoue for his experimental assistance.

References

1. H.K. Schmidt, Better Ceramics Through Chemistry IV, edited by B.J.J. Zelinski, C.J. Brinker, D.E. Clark, and D.R. Ulrich (Mater.Res. Soc. Symp. Proc. 180 1990), p. 961.
2. H.K. Schmidt, Inorganic and Organometallic Polymers, edited by M. Zeldin, K.J. Wynne, and H.R. Allcock (ACS Adv. Ser. 360 1988), p.333.
3. Y.T. Lee, K. Iwamoto, H. Sekimoto, and M. Seno, J. Membrane Sci. 42, 169 (1989).
4. F. Babonneau, K. Thornem and J.D. Mackenzie, Chem. Mater. 1, 554 (1989).
5. F. Babonneau, L. Bois, J. Maquet, and J. Livage, Proc. 2nd Eur. Conf. Sol-Gel Tech.(Eurogel'91), edited by S. Vilminot, R. Nass, and H. Schmidt (1992).
6. I. Hasegawa and S. Sakka, Bull. Chem. Soc. Jpn. 63, 3203 (1990).
7. M.J. van Bommel, T.N.M. Bernards, and A.H. Boonstra, J. Non-Cryst. Solids 128, 231 (1991).
8. R.H. Glaser, G.L. Wilkes, and C.E. Bronnimann, J. Non-Cryst. Solids 113, 73 (1989).
9 C.J. Brinker and G.W. Scherer, Sol-Gel Science (Academic Press, New York, 1990), p. 160.
10. Y. Sugahara, S. Okada, K. Kuroda, and C. Kato, J. Non-Cryst. Solids 139, 25 (1992).
11. Y. Sugahara, S. Okada, S. Sato, K. Kuroda, and C. Kato, J. Non-Cryst. Solids, submitted.
12. C.W. Lentz, Inorg. Chem. 3, 574 (1964).
13 I. Hasegawa, K. Kuroda, and C. Kato, Bull. Chem. Soc. Jpn. 59, 2279 (1986).
14. Y. Sugahara, S. Sato, K. Kuroda, and C. Kato, J. Non-Cryst. Solids, in press.
15. R.A. Assink and B.D. Kay, J. Non-Cryst. Solids 99, 359 (1988).

STRUCTURAL CHARACTERIZATION OF GELS PREPARED FROM CO-HYDROLYSIS OF TETRAETHOXYSILANE AND DIMETHYLDIETHOXYSILANE.

FLORENCE BABONNEAU, LAURENCE BOIS and JACQUES LIVAGE.
Chimie de la Matière Condensée, Université P. et M. Curie, 4 place Jussieu, Paris, France.

ABSTRACT

This work is part of a more general study on the structural characterization of organically modified silica gels that are usually prepared from tetrafunctional, trifunctional and difunctional alkoxides. The systems that were investigated in this paper are gels prepared from co-hydrolysis of dimethyldiethoxysilane (DEDMS) and tetraethoxysilane (TEOS). The condensation process was followed by ^{29}Si and ^{17}O liquid NMR and shows clearly evidence for co-condensation reactions. Gels with various DEDMS/TEOS ratios were studied by ^{29}Si and ^{1}H MAS-NMR. The results are clearly correlated to the change in physical properties of the materials, from flexible polymeric gels at high DEDMS content to crosslinked brittle networks at low DEDMS content.

INTRODUCTION

Sol-gel processing of hybrid organic-inorganic materials is currently widely investigated due to the large amount of potential applications of these systems such as transparent matrices for optical devices or fonctionnal coatings[1-3]. Despite this large number of studies, very few are concerned with structural investigation [4,5], certainly due to the complexity of the systems usually derived from tetrafunctionnal, trifunctionnal and difunctionnal silicon alkoxides. Gels have been prepared from tetraethoxysilane (TEOS) and dimethyldiethoxysilane (DEDMS) and lead to a large variety of materials, viscous liquids, flexible or brittle gels. The purpose of this work was to try to understand how the different Si units are connected in the copolymers, and the consequence on the physical properties of the materials. The hydrolysis process has been followed by ^{29}Si and ^{17}O NMR, while gels have been characterized essentially by solid state MAS-NMR spectroscopy (^{29}Si and ^{1}H).

EXPERIMENTAL

The gels were prepared by mixing the various silicon alkoxides in an appropriate ratio. Acidic water (HCl; pH=1) was used for hydrolysis. No other solvent was used. Water is not miscible with the alkoxides but under magnetic stirring, the solution becomes clear within few minutes and is poured into plastic tubes. The cast samples are allowed to gel in open tubes and then air dried at room temperature. For the liquid NMR study, the solution are aged for one week in closed tubes.

NMR experiments were conducted on a MSL 400 Bruker spectrometer. For ^{29}Si liquid experiments, ≈200 scans were used with 5 μs pulsewidth and 10 s delays between scans. For the ^{29}Si MAS experiments, pulsewidth and relaxation delays were respectively 2.5 μs and 60 s. ≈200 scans were acquired per spectrum and tetramethylsilane was used as an external reference. Peaks will be labelled with the conventional D_n and Q_n notation. D and Q respectively refer to $(CH_3)_2SiO$ and SiO_2 units and n is the number of bridging O atoms surrounding Si. Natural abundance ^{17}O NMR spectra were recorded using ≈20 000 scans with 11 μs pulsewidth and 100 ms delays. The ^{1}H MAS-NMR experiments were done with a zirconia rotor using 90° pulses and 8 acquisitions.

RESULTS AND DISCUSSION

Liquid NMR study of the co-hydrolysis process

One objective of this study was to see if ^{29}Si or ^{17}O NMR could be sensitive to co-condensation reactions between D and Q units. DEDMS, TEOS and mixtures of DEDMS/TEOS (1/1 ratio) were hydrolyzed with various amount of water and aged for one week in closed tubes. This experimental procedure allowed to accumulate the NMR spectra for a rather long period with no problem with reaction kinetics.

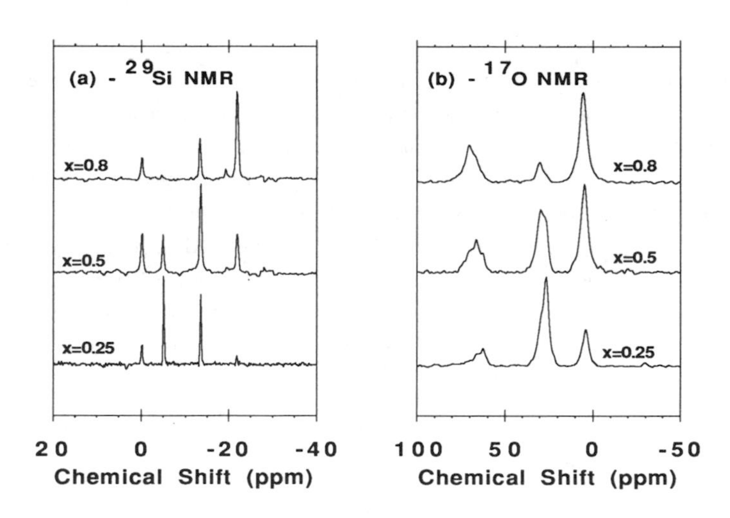

Figure 1 : ^{29}Si NMR (a) and ^{17}O NMR (b) spectra of solutions of hydrolyzed DEDMS with various hydrolysis ratio x=DEDMS/H_2O.

The ^{29}Si NMR results obtained for hydrolyzed solutions of DEDMS are represented in Figure 1a for hydrolysis ratio, x ranging from 0.15 to 0.8. Above 1, the system undergoes a phase separation. The spectra shows three main peaks assigned to D_0 ($\delta\approx$-5 ppm), D_1 ($\delta\approx$-13 ppm) and D_2 ($\delta\approx$-22 ppm) [6]. Their intensity behavior clearly reveals the progress of the condensation reactions when x increases. A smaller peak (δ=-19 ppm) is also present for x greater than 0.5 assigned to cyclic tetramer. Spectra were simulated to extract the percentages of the various species and the amount of oxo bridges formed per silicon estimated ($\%D_1/2 + \%D_2$). This last value agrees well with x showing that all the water was consumed for condensation and that no residual Si-OH groups should be present. This allows to assign the D_0 peak to DEDMS and the D_1 peak to $(CH_3)_2Si(OEt)$-O-.

The same solutions have been studied by ^{17}O NMR (Figure 1b). Three main regions can be defined, corresponding to -$\underline{O}$H groups (0-5 ppm), to -$\underline{O}$Et groups (27-30 ppm) and to Si-$\underline{O}$-Si groups (60-70 ppm) [7,8]. The spectra were simulated. One peak is present in the first region with a chemical shift close to the one of ethanol. Its intensity increases with x. In the second region, the simulation shows the presence of two peaks at 27 and 30 ppm. The peak at 27 ppm corresponds to DEDMS and decreases with x, while the peak at 30 ppm first increases up to x=5 and then decreases. It follows the same evolution as the D_1 peak in ^{29}Si NMR experiments and could be assigned to terminal $\underline{O}$Et groups in a D_1 unit. In the last region, three components can be found at 63, 66 and 70 ppm. The intensity of the 63 ppm peak follows the same evolution as the ^{29}Si NMR peak due to D_1 units and could be assigned to $(CH_3)_2Si(OEt)$-$\underline{O}$- species. The intensity of the two others increases with x. Their chemical shifts are characteristic of oxo bridges between D_2 units in chains (66 ppm) [7] or in cyclic species (70 ppm) [7,8]. The high intensity of the 70 ppm peak for x greater than 0.5 seems to show an important cyclization with formation of D_n species ($n\geq5$).

The same experiments were performed on TEOS with hydrolysis ratio ranging from 0.5 to 2. Extensive data have been published on the ^{29}Si NMR with which the present results are in total agreement [9]. For a low hydrolysis ratio, x=1 (Figure 2b), the main peaks are assigned to Q_1 units, $(OEt)_3\underline{Si}$-O- (δ=-88.8 ppm) and to Q_2 units, $(OEt)_2\underline{Si}$-$(O)_2$- in cyclic species (δ=-94.9 ppm) and in linear species (δ=-96.1 ppm). ^{17}O NMR spectra was also recorded, but not very successfully. For x=0.5, two main peaks are visible at 0.5 ppm ($\underline{O}$H) and 11.8 ppm ($\underline{O}$Et in a Q unit). A small component is present around 30 ppm that could be assigned, according to the ^{29}Si NMR spectrum, to $(OEt)_3Si$-$\underline{O}$-Si≡ species. For x greater than 1, no peaks are visible except those due to $\underline{O}$H and $\underline{O}$Et groups. This feature is certainly related to a large increase in peak linewidths, due to slow motions of the condensed species.

DEDMS and TEOS in a 1/1 molar ratio were co-hydrolyzed with various hydrolysis ratio. The ^{29}Si NMR spectra for x=1.5 (Figure 2c) is compared with the spectra of two solutions : DEDMS/H_2O 1/0.5 (Figure 2a) and TEOS/H_2O 1/1 (Figure 2b). These three preparations correspond to a H_2O/OEt ratio of 0.25. In the D unit region, four new peaks are present in addition to the three main peaks already mentionned ($\underline{D_0}$, $\underline{D_1}$-D and D-$\underline{D_2}$-D). One at -11.5 ppm is assigned to D_1 units bonded to Q units ($\underline{D_1}$-Q) while three other components are present at -16.5, -18.8 and -20.1 ppm. The -18.8 ppm can be due to the cyclic D_4 tetramer, but the other two are certainly related to D_2 units bonded to Q units, (Q-$\underline{D_2}$-Q or D-$\underline{D_2}$-Q). An NMR study is under progress for a better assignment of the peaks. Simulation of the peaks show that at least 50% of the D units could be connected with at least one Q unit. In the Q unit region, additional peaks are also present. One peak is present at -87.9 ppm and could be assigned to a Q_1 unit bonded to a D unit while a new one also appear at -94.2 ppm, assigned to a Q_2 unit ($(OEt)_2\underline{Si}$-$(O)_2$-) bonded to D units.

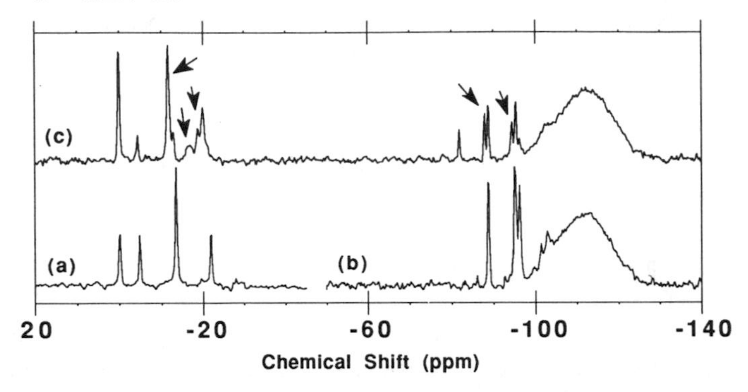

Figure 2 : ^{29}Si NMR spectra of various hydrolyzed systems.
(a) DEDMS/H_2O 1/0.5, (b) TEOS/H_2O 1/1 and (c) DEDMS/TEOS/H_2O 1/1/1.5

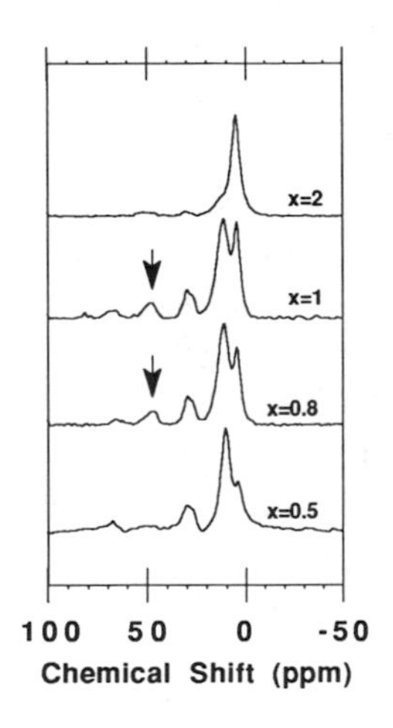

Figure 3 : ^{17}O NMR spectra of hydrolyzed solutions. (DEDMS/TEOS/H_2O 1/1/x)

DEDMS/TEOS systems with x ranging from x=0.5 to 3 were also studied by ^{17}O NMR (Figure 3). Peaks already found for hydrolyzed solutions of DEDMS or TEOS are present assigned to -$\underline{O}$H groups (5 ppm), -$\underline{O}$Et groups in Q units (11.6 ppm) and in D units (26.8 and 29.9 ppm) and oxo bridges between two D units (65-70ppm). The important feature is a new peak around 50 ppm, for x=0.8 and 1. According to the ^{29}Si NMR spectrum (Figure 2c) that show a large number of D units bonded to Q units, a possible assignment for this peak could be oxo bridges between D and Q units. Such a peak was predicted at 59 ppm [8]. An indirect evidence of co-condensation reactions is the disappearance of the NMR peaks for x≥2. For hydrolyzed solutions of DEDMS, the peaks due to oxo bridges between D units are clearly visible. This is certainly due to a high mobility of the linear or cyclic species. Introduction of Q units that can act as crosslinking points lead to the formation of branched species, with a lower mobility, that could not be easily detected.

MAS-NMR characterization of gels prepared from co-hydrolysis of DEDMS and TEOS

Gels have been prepared with various x=DEDMS/TEOS ratios, corresponding to an average functionality f of the precursors ranging from 2.24 to 4. The ratio H_2O/OEt was equal to 1. Above x=9 (f≤2.2), no gels are obtained, only viscous liquids mainly composed with cyclic or linear polydimethylsiloxane species. The ^{29}Si and ^{1}H MAS-NMR of the gels are represented in Figure 4.

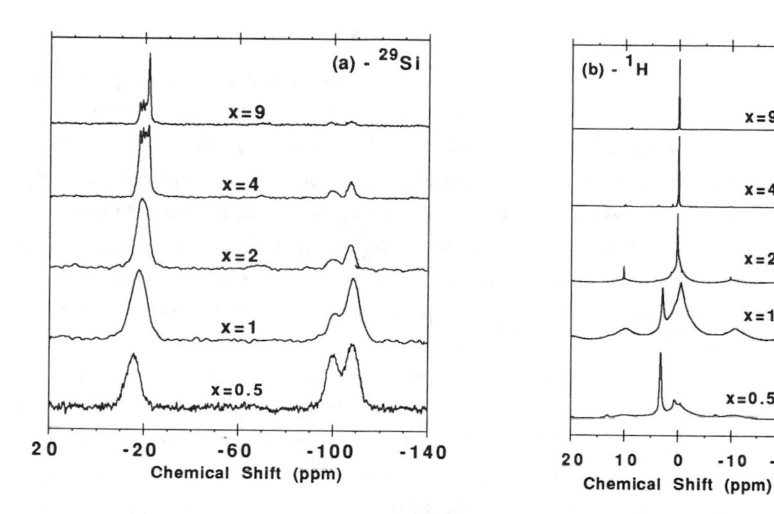

Figure 4 : (a) ^{29}Si and (b) ^{1}H MAS-NMR spectra of gels prepared from various DEDMS/TEOS ratios (x).

The ^{29}Si MAS-NMR spectra for x≥4 (f≤2.4) in the D unit region, presents four sharp peaks at -21.8, -20.4, -19.3 and -18.2 ppm and a shoulder at -17.1 ppm. These peaks were already found in the liquid NMR study of the co-hydrolysis process. The peak at -21.8 ppm is characteristic of D_2 units bonded to D units and is the major peak for high DEDMS content. The peak at -19.3 ppm could be due to cyclic tetrameric species. The other peaks at -20.4, -18.2 and -17.1ppm that increase with the TEOS amount are certainly due to co-condensed species. The small linewidth reflects the high mobility of the species. For x≤4 (f≥2.4), the peak linewidth drastically increases and only one component is now visible. The spectrum is typical of a glassy network. The average chemical shift value varies from -19.2 ppm for x=2 to -17.5 ppm for x=0.5 reflecting a high degree of co-condensation between D and Q units and thus a high degree of crosslinking. The chemical shift values in the Q unit region are also sensitive to this phenomenom. The Q_4 units regularly shift from -110 ppm in a pure silica gel to -107 ppm. The change in mobility of the units when x varies is also reflected by ^{1}H MAS-NMR (Figure 4b). For x≥4, one sharp peak is present at 0 ppm due to methyl groups in D units. The spectrum for x=2 presents two components : in addition to a sharp peak due to very mobile

CH_3-Si groups, a broad component is now present assigned to the same groups but with more restricted motions. Moreover, spinning side bands start to appear showing that dipolar interactions are no longer averaged to zero. For $x \leq 1$, a new peak is present at 3.2 ppm assigned to Si-OH groups. This peak is dominant for x=0.5. This feature is in agreement with a feature already pointed out: the condensation degree of the network increases when the average functionnality of the precursors decreases [4,10].

CONCLUSION

The purpose of this work was to investigate the structure of gels prepared from co-hydrolysis of dimethyldiethoxysilane (DEDMS) and tetraethoxysilane (TEOS). The first objective was to see if ^{29}Si MAS-NMR could be sensitive to co-condensation reactions between the various D and Q units. A prelimary study done on the hydrolysis process using liquid ^{29}Si NMR shows clearly additionnal peaks due to D units bonded to Q units and vice versa. The co-condensation reactions seem to promote a high field shift of the peaks for both units. The most innovative aspect of this work was certainly the use of ^{17}O NMR that in the present case, gave also evidence of an extra peak that could be assigned to oxo bridges between D and Q units. ^{29}Si MAS-NMR spectra of gels prepared with various x=DEDMS/TEOS ratios clearly present shifts in peak position for both D and Q units when the composition varies. According to the results obtained from the liquid NMR study, these shifts can be assigned to the presence of co-condensed species. The increase of such species with the amount of TEOS, leads to a transition from a flexible polymeric material ($x \geq 4$) to a glassy crosslinked network ($x \leq 1$). This variation is perfectly reflected by the evolution of linewidth in both ^{29}Si and ^{1}H MAS NMR spectra. A study is under progress to get precise assignments of the NMR peaks in the liquid state that could be used for a better description of the material in the solid state.

REFERENCES

1. H. Schmidt, J. Non Cryst. Solids, 73, 681 (1985).
2. H. Schmidt, in Better Ceramics through Chemistry IV, edited by B.J.J. Zelinski, C.J. Brinker, D.E. Clark, D.R. Ulrich (Mat. Res. Soc. Symp. Proc. 180, Pittsburgh, PA, 1990) pp 961.
3. S. Diré, F. Babonneau, C. Sanchez, J. Livage, J. Mater. Chem., 2, 239 (1992)
4. R.H. Glaser, G.L. Wilkes, C.E. Bronnimann, J. Non Cryst. Solids, 113, 73 (1989).
5. H-H. Huang, B. Orler, G.L. Wilkes, Macromolecules, 30, 1322 (1987).
6. Y. Sugahara, S. Okada, K. Kuroda, C. Kato, J. Non Cryst. Solids, 139, 25 (1992).
7. U. Scheim, K. Ruhlmann, J.W. Kelly, S.A. Evans Jr, J. Organometallic Chem., 375, 33 (1989)
8. A.R. Bassindale, K.H. Pannell, J. Chem. Soc. Perkin Trans.2, 1801 (1990)
9. J.C. Pouxviel, J.P. Boilot, J.C. Beloeil, J.Y. Lallemand, J. Non Cryst. Solids, 89, 345 (1987).
10. F. Babonneau, L. Bois, J. Maquet, J. Livage, Proceedings of the EUROGEL'91 Conference, Saarbrucken (1991) (in press)

ELECTRONEGATIVITY-BASED COMPUTATION OF ^{29}Si CHEMICAL SHIFTS

HENRY M., GERARDIN C. and TAULELLE F.
Laboratoire de Chimie de la Matière Condensée, University Pierre et Marie Curie, T54-55
4, place Jussieu, 75252 PARIS Cedex 05, FRANCE

ABSTRACT

The Partial Charge Model has been modified to take into account the detailed structure of any molecular sol-gel precursors or inorganic solid networks. Starting from these structure-dependent partial charges, the classical theory of nuclear shielding is applied to compute the electronic cloud compacity $<r^{-3}>_p$, the population unbalance P_u and also the mean excitation energy ΔE. With these three parameters it is possible to explain the chemical shifts variations, spanning from +40 down to -140 ppm, of more than 50 precursors. Depending on the ligands, the well-known upside-down U-curves for series SiX_nY_{4-n} (n=0..4) can be ascribed either to the population unbalance term P_u or to a competition between the two other terms $<r^{-3}>_p$ and ΔE.

1. INTRODUCTION

^{29}Si NMR has proved to be an unvaluable technique for characterizing the various hydrolysis-condensation steps involved during the synthesis of silica-based materials [1][2][3]. Detailed compilations of ^{29}Si chemical shifts data are available [4] which helps the sol-gel chemist for spectra attributions. As ^{29}Si chemical shifts (ranging typically from +50 ppm down to -200 ppm) are very sensitive to local environments, several attempts has been made to correlate them to electronic densities computed from empirical methods [5] as well as from *ab initio* molecular orbital theory [6]. Empirical methods consider only the first coordination sphere of the silicon atom, and are thus not very useful for sol-gel precursors which show mainly second or third coordination sphere shifts. On the other hand, *ab initio* methods are able to take into account the detailed molecular structure to compute accurately the chemical shift tensors. Again, despite this powerness, results are rather meager due to the great number of heavy atoms encountered in sol-gel precursors. The Partial Charge Model (PCM) introduced some years ago [7] does not suffer from such drawbacks as it can be readily modified to take into account the molecular structure of sol-gel precursors. The aim of this contribution is to describe this structure-dependent version of the PCM (SDPCM) and to show how it can be applied to predict the ^{29}Si chemical shifts in various chemical species.

2. THEORY

The classical theory of NMR shielding leads to the following expression for the screening of an atom in a molecule [8][9][10] :

$$\sigma = \sigma_{dia} + \sigma_{para} = \sigma_{dia} - \frac{e^2\hbar^2 <r^{-3}>_p P_u}{3m_e c^2 \Delta E} \qquad (1)$$

In this expression σ_{dia} is the total electronic contribution to the screening whereas σ_{para} comes from the contribution of valence electrons only (p electrons in the case of silicon). Three terms are known to have an influence on the second order paramagnetic contribution namely :

i) The compacity of the p-electronic cloud $<r^{-3}>_p$ is the average over the ground-state wave-function $|\psi_0>$ of the inverse of the cube of the electron-nucleus distance r : $<r^{-3}>_p = <\psi_0|r^{-3}|\psi_0>$. Approximating $|\psi_0>$ as a Slater atomic orbital undergoing a nuclear effective charge $Z^*=Z^*_0(1+fq_{Si})$ leads to : $<r^{-3}>_p = <r^{-3}>_p^\circ(1+fq_{Si})^3$ [5]. In this relation $<r^{-3}>_p^\circ$ is the compacity of the electronic cloud of the neutral silicon atom (q_{Si}=0), f is a screening factor equal to 0.082 within the Slater theory and q_{Si} is the partial charge at the silicon atom.

ii) The population unbalance P_u reflects the anisotropy of the p-electronic cloud within the molecular orbital wave-function $|\psi_0\rangle$. If h_i stands for the polarity of the ith iono-covalent bond ($0 \leq h_i \leq 2$), and if the silicon atom is in tetrahedral coordination, it can be shown that [5] :

$$P_u = \frac{1}{4} \sum_{i=1}^{4} h_i \left[3 - \sum_{j=i+1}^{4} h_j \right] \text{ with } q_{Si} = 4 - \sum_{i=1}^{4} h_i \qquad (2)$$

iii) The mean excitation energy ΔE is the mean difference [11] between the ground-state energy and the various excited states of the molecule. This last term is the polarizability of the bond electronic cloud [11].

The first two terms are obviously dependent on the partial charge at the silicon atom q_{Si}, while the third one does not seem to involve, at first sight, this parameter. This is only partly true as it is known that the optical gap of a binary compound AB strongly depends on the electronegativity difference $|\chi_A - \chi_B|$, the following relationship having been suggested : $\Delta E_{opt} = \sqrt{|10\chi_A - 17.5|} + \sqrt{|10\chi_B - 17.5|}$ [12]. This relationship clearly shows that the higher the electronegativity difference between atoms A and B, the higher the polarity, and also the higher the optical gap ΔE. Consequently, the knowledge of the partial charge distribution seems to be of the utmost importance for computing ^{29}Si NMR chemical shifts. Needless to say such partial charges have to be explicitly dependent on the molecular structure if ones wants to take into account not only first-sphere effects. As our Partial Charge Model was in its original form structure-independent [7], slight modifications are necessary. For the sake of simplicity, let us first consider a diatomic molecule AB. In its structure-independent form the PCM tells us that to a first approximation the electronegativities of atoms A and B vary as : $\chi_A = \chi_A^\circ - kq\sqrt{\chi_A^\circ}$ and $\chi_B = \chi_B^\circ - kq\sqrt{\chi_B^\circ}$, if A is more electronegative than B ($\chi_A^\circ > \chi_B^\circ$ and q>0). At equilibrium $\chi_A = \chi_B$ (electronegativity equalization principle), which allows to express the charge transferred from B to A as a function of the electronegativities of the free atoms χ_A° and χ_B° and of the scale factor k (k=1.36 for Allred-Rochow scale [7], while k=1.56 for Sanderson scale [13]) :

$$kq = \frac{\chi_A^\circ - \chi_B^\circ}{\sqrt{\chi_A^\circ} + \sqrt{\chi_B^\circ}} \Leftrightarrow kq = \sqrt{\chi_A^\circ} - \sqrt{\chi_B^\circ} \qquad (3)$$

Consequently, all what we need to know for computing the partial charge q are the electronegativities of the atoms A et B, and more particularly, we do not need to know the equilibrium distance R_{AB}. Obtaining structure-independent partial charges is here of no help and indeed throw some doubts on the validity of the underlying physics. In fact, the physics of electronegativity equalization upon chemical bond formation is quite right, it is just the approximations made to get relation (3) which are too drastic.

One of the main approximation which is responsible for the structural "amnesy" of the PCM is the neglect of polarization. Following Sanderson, one may interpret the $\pm k(\sqrt{\chi^\circ})q$ term as a size variation undergone by atoms when electrons are added or removed. If this variation is mainly governed by the partial charge q_i owned by an atom i, it is clear that it will also depend to a lesser extent on the surronding partial charges q_j. For the A-B molecule, the electronic cloud of atom A is expanded owing to the presence of an excess electronic charge

-q but also deformed by the presence of a partial charge +q at a distance R_{AB}. The result of such a polarization is a change in the apparent hardness of atom A which is no more equal to $k\sqrt{\chi_A^\circ}$ but must depends also on the values of q and R_{AB}. In order to take into account this environment polarization effect, it can be supposed that it is proportionnal to the polarizing charge q, and inversely proportionnal to the distance R_{AB} [14] :

$$\chi_A = \chi_A^\circ - k\left(\sqrt{\chi_A^\circ}\right)q + k\frac{q}{R_{AB}} = \chi_A^\circ - k\left(\sqrt{\chi_A^\circ} - \frac{1}{R_{AB}}\right)q$$

$$\chi_B = \chi_B^\circ + k\left(\sqrt{\chi_B^\circ}\right)q - k\frac{q}{R_{AB}} = \chi_B^\circ + k\left(\sqrt{\chi_B^\circ} - \frac{1}{R_{AB}}\right)q$$

This allows to rewrite equation (3) as :

$$kq = \frac{\chi_A^\circ - \chi_B^\circ}{\sqrt{\chi_A^\circ} + \sqrt{\chi_B^\circ} - \frac{2}{R_{AB}}} = A$$

The polarity q of the A-B bond, then depends explicitly on the internuclear distance R_{AB}, increasing when R_{AB} decreases. The determination of the scale factor k is possible if the polarity and interatomic distance R_{AB} is known for some reference compounds. Using hydrogen halides HX with X=F, Cl, Br and I, together with Sanderson scale [13] and polarities deduced from experimental dipolar moments $\mu = 4.8 \times q \times d$ (Å), it is possible to show that k=2.416 [15]. For the Allred-Rochow scale the same method lead to k=4.014.

Generalization to polyatomic compounds is straightforward if the last equations are rewritten using a matricial formalism :

$$\begin{bmatrix} k\sqrt{\chi_A^\circ} & \frac{k}{R_{AB}} & -1 \\ \frac{k}{R_{AB}} & k\sqrt{\chi_B^\circ} & -1 \\ 1 & 1 & 0 \end{bmatrix} \begin{bmatrix} q_A \\ q_B \\ \langle\chi\rangle \end{bmatrix} = \begin{bmatrix} -\chi_A^\circ \\ -\chi_B^0 \\ 0 \end{bmatrix} \text{ i.e.}$$

$$\chi_i = \langle\chi\rangle = \chi_i^\circ + k(\sqrt{\chi_i^\circ})q_i + \sum_{j(\neq i)=1}^{n} k\frac{q_j}{R_{ij}} \quad i=1,\ldots,n \qquad (4)$$

$$\sum_{i=1}^{n} q_i = z \qquad (5)$$

In (4) R_{ij} stands for the distance between atom i of charge q_i and atom j bearing a charge q_j, while (5) insures that electric charge will be conserved within the polyatomic compound. A linear system having (n+1) equations for (n+1) unknowns (n partial charges and one mean electronegativity), is thus obtained which can be easily solved on any standard computer, provided that the structure (R_{ij} values) is known.

3. RESULTS AND DISCUSSION

The series studied are of the kind SiX_nY_{4-n} with X,Y=H, CH_3, C_2H_5, F, Cl, OH, OCH_3, OC_2H_5, $N(CH_3)_2$, $Si(CH_3)_3$, $SiCl_3$, SCH_3, SCF_3 and $CH=CH_2$. These series were chosen mainly because some of them show a highly non-monotonous variation of the chemical shift δ with the substitution parameter n ($0 \leq n \leq 4$), and it was the aim of this study to understand the physical origin of such deviations from monotony. Structure-dependent partial charges were computed using the above formalism and Sanderson electronegativity scale (k=2.416) on a 386SX 16 MHz PC. Geometries were taken from literature [16] assuming, in the absence of experimental data in solution, ideal tetrahedral valence-bond (α=109.5°) or torsionnal angles (τ=±60°, 180°). Structural matrix (4) was obtained from molecular coordinates computed using a Z-matrix input [17], and the resulting linear set of equations was solved using the LU decomposition algorithm [18]. The polarities h_i were estimated from the partial charge distributions using a valence-bond formalism:

$$h_i = 1 - \frac{q_{Si} q_i}{\sum_{i=1}^{4} q_i} \qquad (6)$$

where q_{Si} is the partial charge at the silicon atom and q_i the partial charges upon the four surrounding atoms.

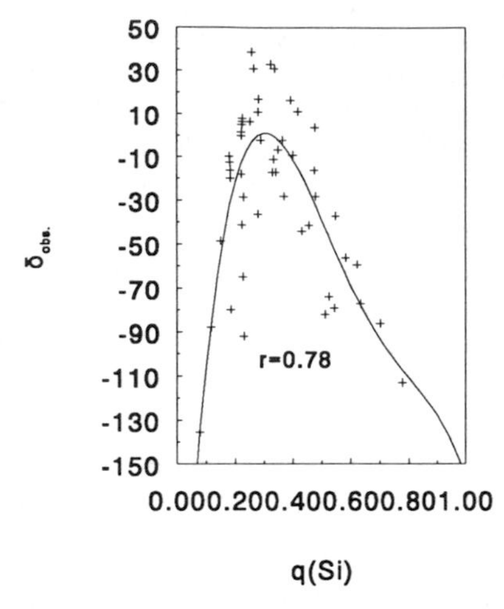

Figure 1 : Experimental chemical shifts versus partial charges.

Figure 1 shows the variation of the chemical shift δ with the partial charge q_{Si}. A typical upside-down U-curve is obtained which can be roughly approximated by a quartic polynomial. The high dispersion of the points reflects the complex variation of the chemical shift with the partial charge predicted by relation (2). However, if each term is taken separately, a much clearer picture emerges. The normalized electronic cloud compacity $R^* = \langle r^{-3}\rangle_p/\langle r^{-3}\rangle_p^\circ = (1+fq_{Si})^3$ varies monotically with q_{Si}, while the normalized population unbalance $P_u^* = P_u/P_u^\circ = (2/3)P_u$ varies approximately quadratically with q_{Si} (fig.2), except for a few mixed species (n≠0,4) which deviate from the theoretical parabola owing to strong anisotropies within the first coordination sphere of the silicon atom. As our SDPCM cannot give a direct estimate of the mean excitation energy ΔE, the following approach have been used to get this crucial parameter. The $SiH_{4-n}F_n$ series was taken as a starting point as it shows a highly non-linear behavior upon the substitution rate n. Knowing that the screening constant must vary as : $\sigma = A - B(1+fq_{Si})^3 P_u^*/\Delta E_n$, we have five experimental points : δ_0, δ_1, δ_2, δ_3 and δ_4 and eight unknowns : A, B, f, ΔE_0, ΔE_1, ΔE_2, ΔE_3 and ΔE_4. Now, if we impose a linear variation of the mean excitation energy with the substitution rate n : $\Delta E_n = [n\Delta E_4 + (4-n)\Delta E_0]/4$, we obtain a non-linear set of five equations with five unknowns A, B, f, ΔE_0 and ΔE_4. This system can be solved by trials and errors leading to the following solution : A=701 ppm, B=1605 ppm, f=0.8732, $\Delta E(SiH_4)$=3.5 eV and $\Delta E(SiF_4)$=9.0 eV. Having this set of parameters, it is now possible to determine from the experimental chemical shift and the computed values of q_{Si} and $P_u^*$of a tetrasubstituted compound SiX_4 , the corresponding value of the mean excitation energy ΔE. Table 1 gives the results of such an adjustment.

Ligand	Me	Et	Cl	OH	OMe	OEt	NMe_2	$SiMe_3$	$SiCl_3$	SMe	SCF_3	C_2H_3
ΔE(eV)	3.87	3.96	6.21	6.3	6.47	6.1	5.05	2.34	3.22	4.43	5.83	3.79

Table 1 : Mean excitation energies values ΔE deduced from the experimental chemical shifts of tetrasubstituted compounds SiX_4.

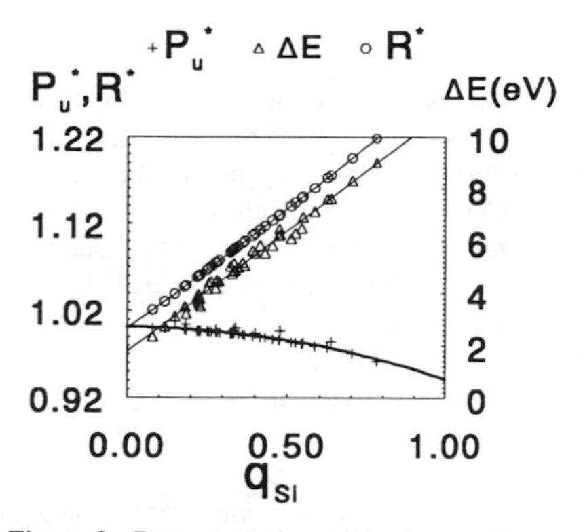

Figure 2 : Deconvolution of fig.1 according to relation (1).

With these values, the mean excitation energies of the mixed compounds can be evaluated using again the approximation $\Delta E_n=[n\Delta E_4+(4-n)\Delta E_0]/4$. Figure 2 shows that a nice linear correlation of mean excitation energies with the partial charge at the silicon atom is obtained : $\Delta E(eV)=9.25q_{Si}+1.77$. Such values appear to be quite reasonnable for silicon compounds, as it is known from optical spectroscopy, that crystalline silicon (q_{Si}=0) have an optical gap of 1.2 eV while for β-quartz it is close to 8.8 eV. Moreover, figure 3 shows that all the investigated chemical shifts can be reproduced with an average error of ±1 ppm from data given in table 1 and from the computed values of q_{Si} (fig.1) and P_u^* (fig.2) using the above values for parameters A, B and f. Solvent effects are not present. Actually, most of the available data are internally referenced, reducing this effect to differential solvation.

Our A value (+701 ppm) which represents the diamagnetic contribution to the ^{29}Si chemical shift, is about 15% lower than the one obtained from *ab initio* calculations for the Si^{4+} ion (+836.9 ppm) [6]. From our B parameter (1605 ppm) we can deduce the following nuclear effective charge for the neutral silicon atom Z^*_0=9.65. This value is about twice higher than the nuclear effective charge computed for an electron within a 3p orbital (Z^*_0=4.285) using Slater rules modified by Clementi and Raimondi [19]. This discrepancy is not very important as it is known that Slater-type orbitals cannot account of the very fine details of the true wave-function of a polyelectronic atom. In particular, it is well-known that using Slater orbitals always leads to much too low values for the compacity $\langle r^{-3}\rangle$ [10]. Experimental values deduced from spin-orbit coupling constants $\langle r^{-3}\rangle_p$=2.31 [20], leads to Z^*_0=11. Our value (Z^*_0=9.65) appears then to be intermediate between these two limiting case. Finally, the f parameter deduced from our analysis (f=0.87) appears to be an order of magnitude higher than the value computed from Slater rules (f=0.082, i.e. 0.35 per valence electron). As it exists no reliable experimental data allowing the measure of the average screening per valence electron in molecules, it is not possible to tell if such a discrepancy is due to the inadequacy of Slater rules in molecules, or to a deficiency of our modelization.

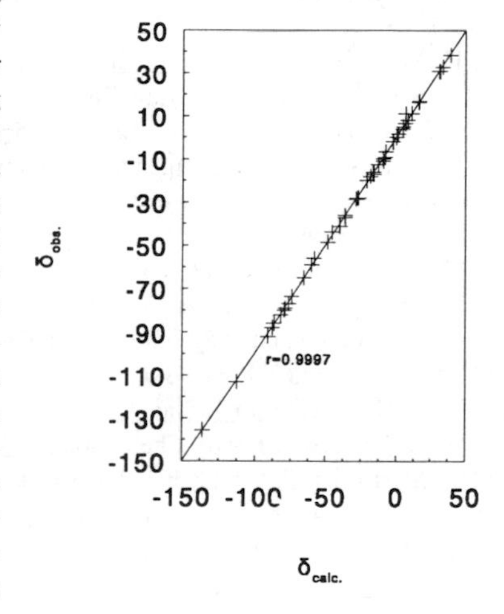

Figure 3 : Correlation between calculated and observed chemical shifts.

4. CONCLUSION

The SDPCM presented here has enabled us to compute within 1 ppm the ^{29}Si chemical shifts of more than 50 molecular precursors. One of the great advantage of our modelization over *ab initio* calculations, is that we are not limited by the number of atoms which could be as great as 53 (such as in $Si[Si(CH_3)_3]_4$) or even infinite provided that a Madelung summation is performed before solving the linear set of equations [21]. One of the main result of our modelization is the theoretical justification of the upside-down U-curves obtained by plotting the ^{29}Si chemical shift versus the partial charge q_{Si} or versus the substitution rate n. Two main phenomena accounts for such a non-linearity :

i) The competition between the electronic cloud compacity $\langle r^{-3} \rangle_p$ and its anisotropy as measured by the population unbalance P_u. This point is not original and has been frequently invoked in the past [5].

ii) The competition between the electronic cloud compacity $\langle r^{-3} \rangle_p$ and its polarizability as measured by the mean excitation energy ΔE. Very few publications stress the importance of this last competition (see e.g. [22]), and it is one of the merit of our model, owing to its intrinsic simplicity, to draw attention upon this crucial point.

5. REFERENCES

[1] I. Artaki, S. Sinha, A.D. Irwin and J. Jonas, J. Non-Cryst. Solids, 72, 391 (1985)

[2]L. Kelts, N. Elffinger and S.M. Melpolder, J. Non-Cryst. Solids, 83, 353 (1986).

[3] J.C. Pouxviel, J.P. Boilot, J.C. Beloeil and J.Y. Lallemand, J. Non-Cryst. Solids, 89, 345 (1987).

[4] H. Marsmann, in NMR 17, Basic Principles and Progress, edited by P. Diehl, E. Fluck and R. Kosfeld, (Springer-Verlag, Berlin, 1986) p. 65.

[5] R. Radeglia, Z. Phys. Chem. Leipzig, 256, 453 (1975).

[6] J.R. Van Wazer, C.S. Ewig and R. Ditchfield, J. Phys. Chem., 93, 2222 (1989).

[7] J. Livage and M. Henry, in Ultrastructure Processing of Advanced Ceramics, Edited by J.D. Mackenzie and D.R. Ulrich, (John Wiley & Sons, New-York, 1988) p. 183.

[8] M. Karplus and T.P. Das, J. Chem. Phys., 34,1683 (1961).

[9] J.A. Pople, J. Chem. Phys., 37, 53 (1962).

[10] C.Y. Jameson and H.S. Gutowsky, J. Chem. Phys., 40, 1714 (1964).

[11]M. Karplus and T.P. Das J. Chem. Phys. 34, 1683 (1961)

[12] N.B. Hannay, in Solid State Chemistry, (Prentice Hall, Englewood Cliffs, 1967) p. 36.

[13] R.T. Sanderson, J. Am. Chem. Soc., 105, 2259 (1983).

[14] W.J. Mortier, S.K. Ghosh and S. Shankar, J. Am. Chem. Soc., 107, 829 (1985).

[15] M. Henry and F. Taulelle, in Proc. Int. Workshop on Application of NMR Spectroscopy to cement science, March 25-26, Guerville, France (1992).

[16] Tables of Interatomic Distances and Configurations in Molecules and Ions, Supplement 1956-1959, Special Publication n°18, (The Chemical Society, London, 1965).

[17] J.E. Nordlander, A.F. Bond IV and M. Bader, Computers and Chemistry, 9, 209 (1985).

[18] W.H. Press, B.P. Flannery, S.A. Teukolsky and W.T. Vetterling, in Numerical Recipes : The Art of Scientific Computing, (Cambridge University Press, Cambridge, 1989), p. 39.

[19] E. Clementi and D.L. Raimondi, J. Chem. Phys., 38, 2686 (1963).

[20] R.G. Barnes and W.V. Smith, Phys. Rev., 93, 95 (1954).

[21] K. Van Genechten, W. Mortier and P. Geerlings, J. Chem. Soc. Chem. Commun., 1278 (1986).

[22]V.G. Malkin, O.V. Gritsenko and G.M. Zhidomirov, Chem. Phys. Letts., 152, 44 (1988).

RAMAN AND INFRARED STUDY OF METAL ALKOXIDES DURING SOL-GEL PROCESS

O. Poncelet, J-C. Robert and J. Guilment
Kodak-Pathé, Centre de Recherches et de Technologie, Chalon-Sur-Saône (71104), France.

ABSTRACT

Organic additives are commonly used in sol-gel chemistry [1]. They can operate as stabilizers of reactive metal alkoxides [2] towards hydrolysis or as drying control chemical additives (such as dimethylformamide). These organic compounds and the by-products of sol-gel reactions can drastically modify the physical properties of final oxide materials, so that it is necessary to optimize the use of these additives. The first steps of hydrolysis of metal alkoxides stabilized by organic additives have been studied by many sophisticated analytical techniques which are difficult to use at technical level and often strongly alter the sol-gel reactions [3]. We chose to use FT-Raman and Infrared spectrocopies, which allow to remain as close as possible of the sol-gel process without altering it. The first steps of hydrolysis of group (IVb) metal alkoxides, more particularly titanium, modified by chlorides, carboxylic acids and alkanolamines have been investigated.

INTRODUCTION

Raman and Infrared spectroscopies are commonly used to study sol-gel materials, mainly during the last steps of the process [4]. The goal of this paper is to promote the use FT-Raman and Infrared spectroscopies to study how chemical additives act to stabilize metal alkoxides. Due to the complex structure and often low molecular symmetry of metal alkoxides, the assignment of various types of ν(M-O) and especially ν(C-O)M bonds has proved to remain rather difficult in IR while classical Raman can have problems with either luminescence or lack of stability of the products. FT-Raman spectroscopy (avoiding fluorescence phenomenom and getting the FT precision) can bring more easily the information.

EXPERIMENTAL

General Procedures.

All reactions were carried out in an atmosphere of prepurified argon at room temperature by using standard inert-atmosphere and schlenk techniques [5]. It can be seen on IR spectra on figure 1 that the alkoxide can be manipulated without hydrolysis under those conditions (no OH band observed at 3400 cm-1). Pure alkoxides and chlorides are from Aldrich or Strem Chemicals; triethanolamine, acetic acid and are respectively from Prolabo and Carlo Erba . All solvents were appropriately dried, distilled prior to use and stored under argon. The spectra were recorded on a Bruker IFS66 spectrometer used either in FT-IR mode with a DTGS detector or in FT-Raman mode (FRA106 accessory) using a Nd-Yag laser and a nitrogen cooled Ge detector. In both cases, we used a resolution of 4 cm-1.

RESULTS AND DISCUSSION

Chemical modification of metal alkoxides

An attractive property [2] of alkoxides is their solubility in a large variety of compositions and solvents including polymers and other alkoxide solutions. This solubility is mainly determined by the polarity of M-O-C bonds, and the degree of oligomerization of the metal alkoxides. The ability of the alkoxide ligands OR to act as terminal or bridging (μ2- or μ3-) groups allows oligomerization to form condensate units in which the metal can reach a higher coordination number. The molecular complexities depend on many parameters which are the nature of the metal alkoxides (increasing with atomic size of the metal), the steric hindrance of R, and also the nature of the solvent used (aprotic or protic).
Many recent works show the importance of the first steps of sol-gel process which is, in fact, a complexation. In this case, it is clear that the degree of oligomerization is an important parameter of this phenomenom which can greatly influence both the hydrolysis rates and the structure of the condensed product.

FT-Raman spectra of group IVb metal alkoxides

IR and Raman spectra of various titanium alkoxides from R =Me to n-Bu, including iPr were recorded and compared to the corresponding alcohols (see figure 1 as an example). From that comparison, it can be easily seen that the ν(Ti-O) band occurs in the 600 cm-1 region and also , for each of the products, a Raman band is observed in the 1150-1200 cm-1 region (see figure 2). This band can be attributed to symmetrical ν(Ti-O-C) mode characteristic of the alkoxy ligand in the coordination sphere.

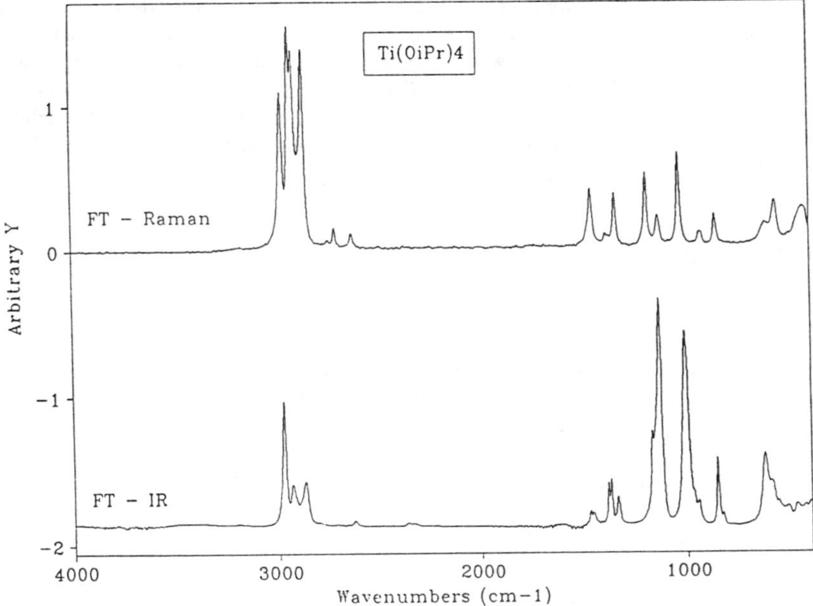

Figure 1 FT-Raman and FT-IR spectra of $Ti(OiPr)_4$

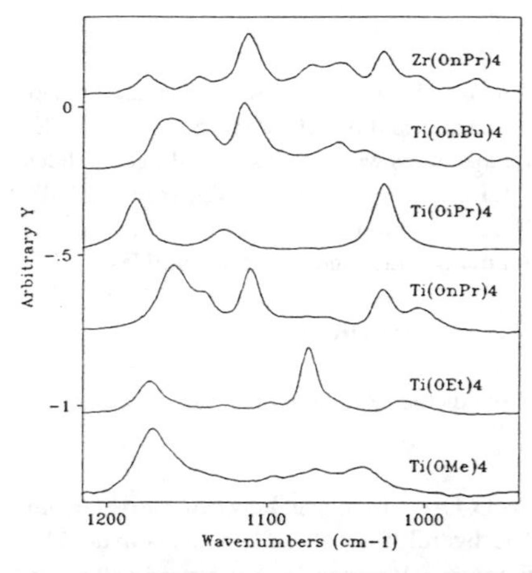

Figure 2 **FT-Raman ν(Ti-O-C) bands of some different titanium (IV) and zirconium (IV) alkoxides**

Alkoxides	ν(C-O)M cm^{-1}	ν(M-O) cm^{-1}
$Ti(OMe)_4$ tetramer	1171	598
$Ti(OEt)_4$ trimer	1173	608
$Ti(O^nPr)_4$	1158	628, 617
$Ti(O^nBu)_4$	1160	619
$Ti(O^iPr)_4$ monomer	1182	593
$Zr(O^nPr)_4$	1175	600, 555

Table 1 **Assignments of FT-Raman ν(M-O-C) and ν(M-O) of some different titanium (IV) and zirconium (IV) alkoxides**

The shift of ν(Ti-O-C) observed can be correlated to the steric hindrance of R and we can show that isopropoxy groups present a stronger steric hindrance effect than n-butoxy (see table 1). The positions of n-butoxy and n-propoxy ν(Ti-O-C) are very similar due, probably , to a same degree of oligomerization.The nature of metal has a strong effect on the shift of ν(M-O-C) for a similar alkoxy group; for $Zr(OnPr)_4$ (covalent radii: 1.45 Å) the ν(Zr-O-C) band is assigned to 1175 cm-1 while for titanium (covalent radii: 1.32Å) the ν(Ti-O-C) band is assigned to 1158 cm-1. This strong shift of Raman bands between two homoleptic metal alkoxides could be useful to study mixtures of alkoxides and /or heterometallic species.

In order to follow an alcoholysis, it appears that this ν(Ti-O-C) band can also be a very useful tool, see for example the figure 3 where then ν(Ti-O-C) band is shifted gradually from 1173 to 1182 cm-1 while substituting one, two of the ethoxy groups with isopropoxy.

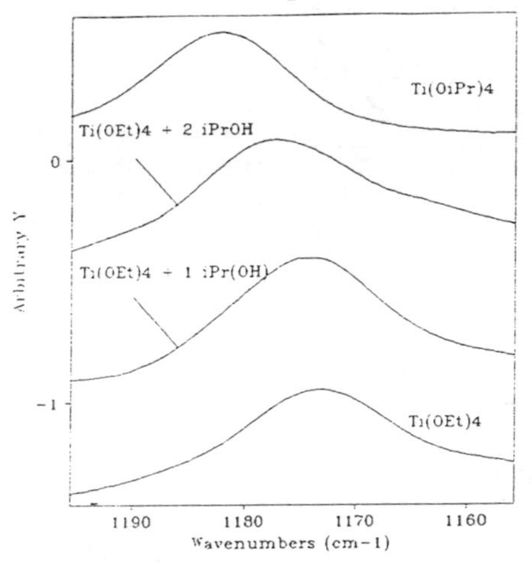

Figure 3 **FT-Raman monitoring of alcoholysis of Ti(OEt)4 by iPROH -ν(Ti-O-C) region-**

Chloroalkoxides.

Metal chloroalkoxides [1] can be used as starting materials in sol-gel process. Generally, they are less sensitive versus hydrolysis than homoleptic alkoxides and therefore more easy to handle. These products can be prepared by reaction of metal alkoxides with halides or hydrogen halides or by reaction of metal chlorides with alcohols. Metal chlorides of group (IVb) react partially with alcohols to form solvated chloroalkoxides. We have chosen to synthetize this kind of compounds by redistribution reactions between titanium alkoxides and titanium chlorides.

$$Ti(OR)_4 + 1/4\, TiCl_4 \longrightarrow 2\, Ti(OR)_3Cl$$

The use of pure products (chloride and alkoxide) avoids the formation of solvated species and simplify the interpretation of the spectra.

Figure 4 shows the influence of chloride on ν-Ti-O-C stretching between two titanium chloroalkoxides ClTi(OiPr)3 and ClTi(OEt)3 and the hydrolysis of these two compounds. The presence of chloride in the coordination sphere of titanium (IV) alkoxides has a strong effect on the ν(Ti-O-C) and ν(Ti-O) bands (see table 2) and a low effect on titanium (IV) ethoxide. These shifts can be due to an increase of molecular complexity (from monomeric species to dimeric or more) in the case of titanium (IV) isopropoxide , while in the case of ClTi(OEt)3 , the molecular complexity remains at least trimeric. The hydrolysis of these two compounds with 4 moles of water/ mole of chloroalkoxides shows a very different behaviour too.
Hydrolysis strongly modifies the FT-Raman spectra of ClTi(OEt)3 in the ν(Ti-O-C) and ν(Ti-O) regions. The bands shift respectively from 1167 to 1133 cm-1 and from 611 to 618 cm-1. This hydrolyzed species remains an alkoxy species which is probably more condensed(the ν(Ti-O) band is larger) or maybe an oxoalkoxide species.
Addition of 4 moles of water to ClTi(OiPr)3 leads to a precipitate which shows a FT-Raman spectra very close to the unhydrolyzed species. In fact, we can observe a 10 cm-1 shift of the ν(C-O) band, so we can think that the coordination sphere of the complexe remains identical to the initial compound but hydrogen bonds can exist between water and OR ligand and lead to precipitation.

Compounds	ν(C-O)Ti cm^{-1}	ν(C-O) cm^{-1}	ν(Ti-O) cm^{-1}
$Ti(OEt)_3Cl$	1167	1075	611
$Ti(OEt)_3Cl$ + H_2O	1133	1076	618
$Ti(OiPr)_3Cl$	1170	1022	617
$Ti(OiPr)_3Cl$ + H_2O	1166	1012	616

Table 2 **Assignments of FT-Raman bands of some chloroalkoxides and their hydrolyzed products**

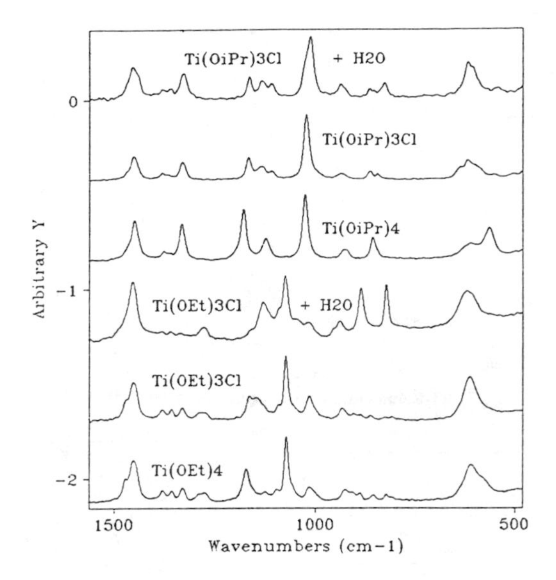

Figure 4 Influence of chloride ion on FT-Raman ν(Ti-O-C) stretching on Chloroalkoxides of titanium and their hydrolyzed products

Chelating agent.

- Carboxylic acids

Acetic acid is often used as additive in sol-gel process. The information given by IR of carboxylatoalkoxide compounds [6] are more often the stretching vibrations of carboxylic groups : νs(COO-) and νa(COO-). The frequency difference Δn is commonly used to determine how the carboxylate ion act - monodentate or bidentate - but do not allow to determine if this acetate ion (bidentate form) is bridging between two titanium or chelating one.
Information provided by FT-Raman concern especially the ν(C-O)M and ν(M-O) region. Fig 5 shows the FT-Raman and Infrared spectra of Ti(OiPr)3OAC. FT-Raman spectrum of Ti(OiPr)4 modified by acetic acid (1:1) shows a strong effect in the ν(Ti-O-C) region, this Raman band shifts from 1182 to 1169 cm-1 while the ν(Ti-O) band shifts from 593 to 610 cm-1 (see table 3). An identical shift of ν(Ti-O-C) band is observed when we introduce chloride ion in the coordination sphere of titanium isopropoxide, however it appears that the shift observed in the ν(Ti-O) region is greater (10 cm-1) for chloride ion than for acetate ion . Acetic acid acts as a chelating agent and increases the degree of coordination , but its stabilizing effect is lower that chloride ion. Figure 6 shows the FT-Raman spectra in the ν(Ti-O) region of Ti(OiPr)3OAc and of the hydrolyzed products. It seems that the hydrolyzed products evolved to TiO2 Rutile.

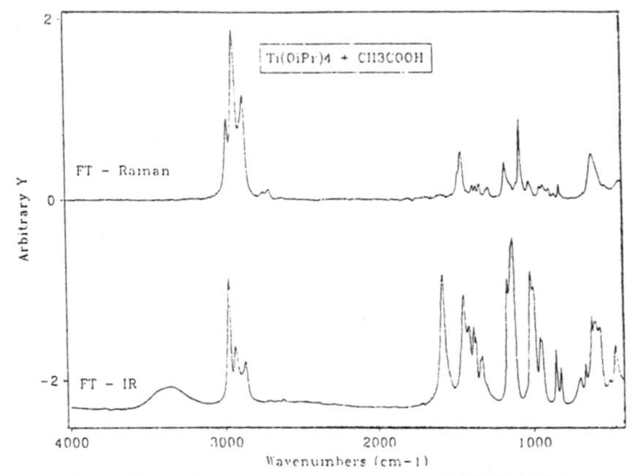

Figure 5 FT-Raman and FT-IR spectra of Ti(OiPr)3OAc

Compounds	ν(Ti-O-C) cm-1	ν(Ti-O) cm-1
Ti(OiPr)3OAc	1169	610
Ti(OiPr)3OAc + 4 H2O	1125	608
TiO_2 Rutile	-	610-448
TiO_2 Anatase	-	639-396

Table 3 Assignments of FT-Raman bands of ν(Ti-O-C) and ν(Ti-O) band of Ti(OiPr)3OAc and the hydrolyzed products

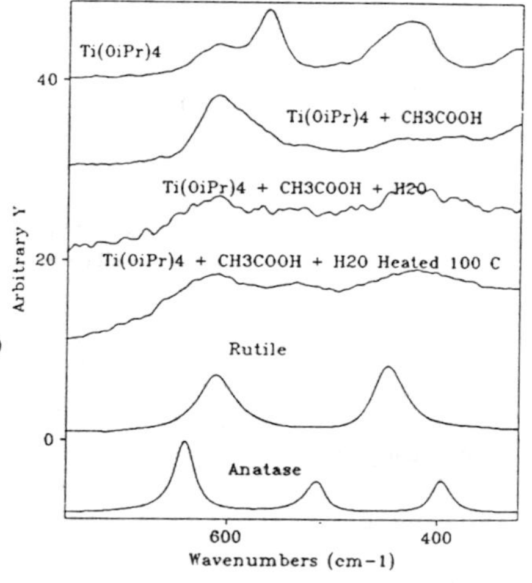

Figure 6 FT-Raman spectra in the ν(Ti-O) region of Ti(OiPr)3OAc and the hydrolyzed products

- Triethanolamine (TEA)

Among the alkanolamines, triethanolamine N(CH2CH2OH)3 is the most efficient ligand due to its possibility to act as a potential tetradentate ligand. Recent structural data have shown that titanium (IV) isopropoxide modified by TEA (1:1) has a dimeric structure where TEA, acting as a tetradentate ligand, is bridging between two titanium [7]. In order to stabilize titanium (IV) alkoxides versus hydrolysis, we have determined the minimal quantity of TEA necessary to obtain stable sols and followed the modification of coordination sphere of titanium (IV) ethoxide and isopropoxide by FT-Raman. It appears that in both cases , titanium (IV) alkoxides are stabilized with 0.6 mole of TEA. The FT-Raman spectra of these stabilized titanium (IV) alkoxides show the apparition of a strong band in the ν(Ti-O) region at 561 cm-1 (see figure 7 and table 4). This fact indicates that we have probably two complexes of titanium with a very similar structure and a same degree of oligomerization, however the difference observed in the ν(Ti-O-C) region indicates that these products keep the alkoxy groups in their coordination sphere. With 0.4 mole of TEA, we can note a strong influence in the ν(Ti-O-C) region but not in ν(Ti-O) region.

compounds	ν(C-O)Ti cm-1	ν(C-O) cm-1	ν(M-O) cm-1
$Ti(OEt)_4$ + 0.4 TEA	1163	1073	607
$Ti(OEt)_4$ + 0.6 TEA	1158	1074	607, 561
$Ti(OiPr)_4$ + 0.6 TEA	1175	1021	615, 559
$Ti(OEt)_4$ + TEA	1129	1077	590, 529

Table 4 Assignments of FT-Raman bands during the stabilization of titanium (IV) alkoxides by triethanolamine

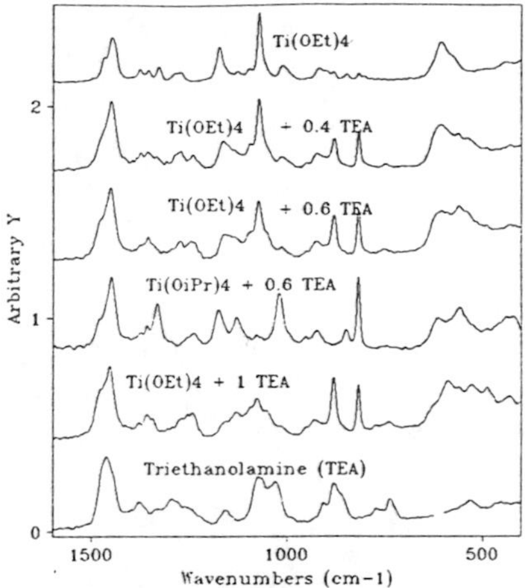

Figure 7 FT-Raman monitoring during the stabilization of titanium (IV) alkoxides by triethanolamine

CONCLUSION

FT-Raman spectroscopy is a promising analytical technique which can provide many information on the first steps of sol-gel process without altering it. This method can bring information on starting materials, mainly alkoxides or modified alkoxides and confirm the significance of the choice of alkoxides in sol-gel technology. The great versatility of FT-Raman spectroscopy allows its use at technical level. The results given by this technique, correlated with other analytical data, would increase our knowledge of the sol-gel process.

REFERENCES

[1] C.J. Brinker, G.W. Scherer, Sol-Gel Science, The Physics and Chemistry of Sol-gel Processing, Academic Press, New-York (1990).

[2] D.C. Bradley, R.C. Mehrotra and D.P. Gaur, Metal Alkoxides, Academic Press, London (1978); L.G. Hubert-Pfalzgraf, New J. Chem., 11, 663 (1987).

[3] J. Livage, M. Henry and C. Sanchez, Progress in Solid State Chemistry, 18, 259 (1988).

[4] C.C. Perry, Xianchun Li and D.N. Waters, Spectrochim. Acta, 47A (9/10), 1487 (1991); J. Chem. Soc. Faraday Trans., 87(15), 761 (1991); A. Takase and K. Miyakama, Jpn. J. Applied Phys., 30(8B), L1508 (1991); M. Aizawa, Y. Nosaka and N. Fujii, J. Non-Cryst. Solids, 128, 77 (1991).

[5] D.D. Perrin and W.L.F. Armarego, Purification of Laboratory Chemicals, Pergamon Press, London (1988).

[6] N. Nakamoto, Infrared and Raman Spectra of Inorganic Compounds, 4e ed., Wiley Interscience, New-York (1986).

[7] W.M.P.B. Menge and J.G. Verkade, Inorg.Chem., 24, 4628 (1991); A.A. Naiini, W.M.P.B. Menge and J.J. Verkade, Ibid., 26, 5009 (1991).

REACTION KINETICS FOR THE HYDROLYSIS OF TITANIUM ISOPROPOXIDE CARBOXYLATE COMPLEXES

CHARLES D. GAGLIARDI*, DILUM DUNUWILA*, BEATRICE A. VAN VLIERBERGE-TORGERSON*, AND KRIS A. BERGLUND*‡
Michigan State University, Departments of *Chemical Engineering and ‡Agricultural Engineering, East Lansing, MI 48824.

ABSTRACT

Titanium alkoxides modified by carboxylic acids have been widely studied as the molecular precursors to ceramic materials. These alkoxide complexes have also been very useful in the formation of stable, porous, optically clear films having many novel applications such as chemical sensors, catalytic supports, and ion-exchange media. To improve the processing of these materials, it is essential to better understand the kinetics of the chemical transformations which occur.

The kinetics of the hydrolysis reaction are studied for selected carboxylic acids using Raman spectroscopy to probe the chemistry of the process. The study has a special emphasis on the titanium isopropoxide-valeric acid system due to the superior quality of these films over other carboxylates. Greater knowledge of the hydrolysis kinetics allows increased control over the quality of the film materials and should be of general interest to those working with modified metal alkoxides.

INTRODUCTION

The chemical modification of titanium alkoxides as sol-gel precursors has led to a better understanding of the sol-gel process, greater control over product characteristics, and the discovery of novel metalorganic materials [1-3]. In particular, much attention has been given to alkoxide modification by acetic acid [1, 3-9] which has proved very helpful to the understanding and development of film materials derived from titanium isopropoxide valerates [10].

The processing of our titanium metalorganic materials can certainly be understood as a special case of a carboxylic acid-modified sol-gel system. The coating solutions incorporate a longer chain carboxylic acid, and a greater quantity of acid, than is normally used. In fact, the carboxylic acid is used in such excess, it functions as both reactant and solvent. The higher acid concentration supports a greater extent of reaction, and the longer acid chain length provides greater steric protection to the titanium from hydrolysis. The acetate group reduces the functionality of the titanium alkoxide because it is harder to remove than the alkoxy group [1], and higher concentrations of longer chain carboxylic acids reduce the functionality even further, producing materials that have excellent resistance to continued hydrolysis after casting. Valeric acid produces the best films, and butyric acid provides the second best, while propionic and hexanoic acid give significantly poorer results [10]. This can be attributed primarily to the steric effect of the acid chain, and to the degree of 'organic character' these chains contribute to the material. Shorter chains provide insufficient protection and longer chains seem to prevent sufficient polymerization as evidenced by the gradual sublimation of soft, octanoic or hexanoic acid-based films previously described [10].

The objective of the current work is to estimate the rates of the hydrolysis and condensation reactions, and to determine the dependence of coating solution stability on the quantity of added water. The experiments focus on the valeric acid system, with some results provided for other carboxylic acids to show the effect of varying the acid chain length.

EXPERIMENTAL

A. Starting Materials

The titanium isopropoxide and carboxylic acids were obtained from the Aldrich Chemical Company and used without further purification. All water was de-ionized, with a resistance of 18 Mohm.

B. Coating Solutions

Except for the sample examined in the rapid mixing cell, all coating solutions were prepared with a total volume of 5 ml, 0.61 ml of titanium isopropoxide, a carboxylic acid to alkoxide ratio of 15, a water to alkoxide ratio of 1, 1.5, 2, or 3, and heptane. In mixing the reactants, water was added after the alkoxide, acid, and heptane had been combined and thoroughly mixed. The hydrolysis was intentionally performed in accordance with the methods we normally use to produce coating materials. To study the earliest effects of hydrolysis, the valeric acid solution with R_W=1.5 was mixed in a quartz rapid mixing cell [11] from two solutions: one providing the alkoxide and 54% of the valeric acid, the other with the remaining acid and the water. The heptane was also distributed between the two solutions to yield the same overall composition used to study the slower reactions.

C. Instrumentation

Raman spectra were collected with a Spex 1877 triple spectrometer equipped with a diode array detector using the 514.5 nm line from a Spectra-Physics argon ion laser producing 200 mW of power at the sample. The solution spectra were taken with a standard cuvette sample holder; the film-porphyrin spectra were taken with the cuvette horizontally positioned and rotated to allow the film to be grazed by the laser beam. The solutions were held in Teflon-stoppered quartz cuvettes or pumped through a rapid mixing cell described elsewhere [11].

RESULTS AND DISCUSSION

Hydrolysis

The first Raman observable manifestation of hydrolysis is the disappearance of the Ti(O-R) band at 1023 cm^{-1}. This structural transformation occurs on a very short time scale. Various stages of this reaction, as shown in Figure 1 for the case of the titanium alkoxide valerate solution, can be resolved with the aid of a rapid mixing cell and a variable speed syringe

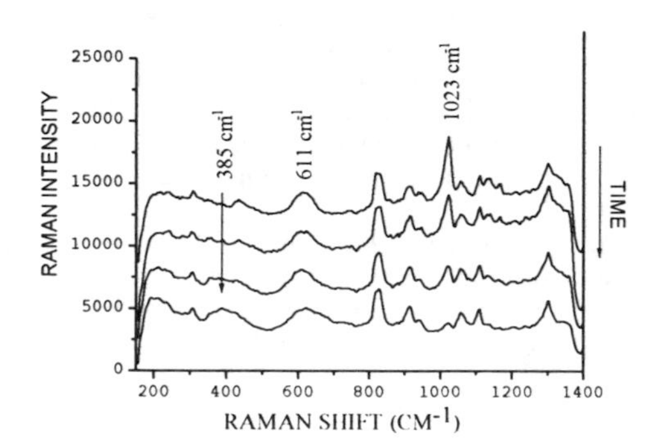

Figure 1. Changes in the Raman spectra during the initial stage of hydrolysis resolved by varying the flow rate in a quartz rapid mixing cell. The top spectrum shows the titanium isopropoxide valerate solution before any appreciable hydrolysis has occurred, and the spectra below show an increasing degree of hydrolysis. The stage of hydrolysis shown in the bottom spectrum is reached within 0.4 seconds after complete mixing.

pump; the faster the reactants are pumped through the cell, the shorter the time between mixing and observation. From these experiments, it is estimated that this reaction takes place within 0.4 seconds of mixing. It is also interesting to note that simultaneous transformations occur in the lower frequency region between 150 and 750 cm^{-1} where the broadening of the band near 600 cm^{-1} and the formation of a band near 385 cm^{-1} can be observed.

Condensation and Solution Stability

Depending on the relative concentrations of reactants, chemical changes can continue to occur in the coating solutions for more than 130 hours or the solutions may stabilize in less than eight hours. The systems studied showed no tendency toward gelation, and the stabilized solutions provided no evidence of detectable colloidal particles. It is no surprise that the initial effect of hydrolysis is the removal of an isopropoxy group, and from the literature, one might also expect that the slower observable transformations would be due to condensation reactions. The modification of titanium alkoxides by acetic acid aids in 'decoupling' the hydrolysis and condensation reactions [1], and it has been noted that sol-gel product characteristics depend upon the relative rates of hydrolysis and condensation, where fast hydrolysis followed by slow condensation produces a polymeric sol [6]. The systems studied here are certainly polymeric in nature, if not merely a solution of oligomers. However, the slow condensation reactions are not necessarily the only slow reactions taking place. Figure 2 shows Raman spectra for aged titanium isopropoxide valerate solutions with various amounts of added water. While the Raman bands above 800 cm^{-1} show no transformation, the growth and reduction of bands at 397 cm^{-1} and 663 cm^{-1} show the creation and consumption of a reaction intermediate. The growth of the bands can be seen for R_w=1.0, where R_w is the water to alkoxide molar ratio. For larger amounts of water (R_w: 1.5 - 3.0), the concentration of the intermediate is already decreasing at the time of the first observation. Thus, the rate of the depletion of the intermediate increases with increasing water concentration, suggesting that the intermediate is being hydrolyzed.

A comparison of various carboxylic acids (propionic, butyric, valeric, and hexanoic) with R_w = 1.5, shown in Fig. 3, indicates that the hydrolysis of the intermediate is somewhat faster for propionic acid than for the other acids. The stability results are summarized in Table I.

Table I. Summary of solution stability

Carboxylic Acid	R_W	Period of Chemical Change
Propionic	1.0	N. A. (precipitates)
	1.5	35 hr
	2.0	9 hr
	3.0	0.45 hr
Butyric, Valeric & Hexanoic	1.0	130 hr
	1.5	80 hr
	2.0	35 hr
	3.0	9 hr

In addition to the chemical changes indicated by the growth and reduction of the bands, there also appears to be reproducible changes in the amount of Rayleigh scattering produced by the sample. These changes may be associated with the formation and depletion of colloidal particles formed by the reaction intermediate(s) which may have bonding patterns similar to those of the hexameric clusters previously isolated for acetic acid systems [1]. However, the only incontrovertible evidence of particle formation occurs for the lowest water condition (R_W=1.0) with propionic acid, which produces a cloudy precipitate. More water or longer acid

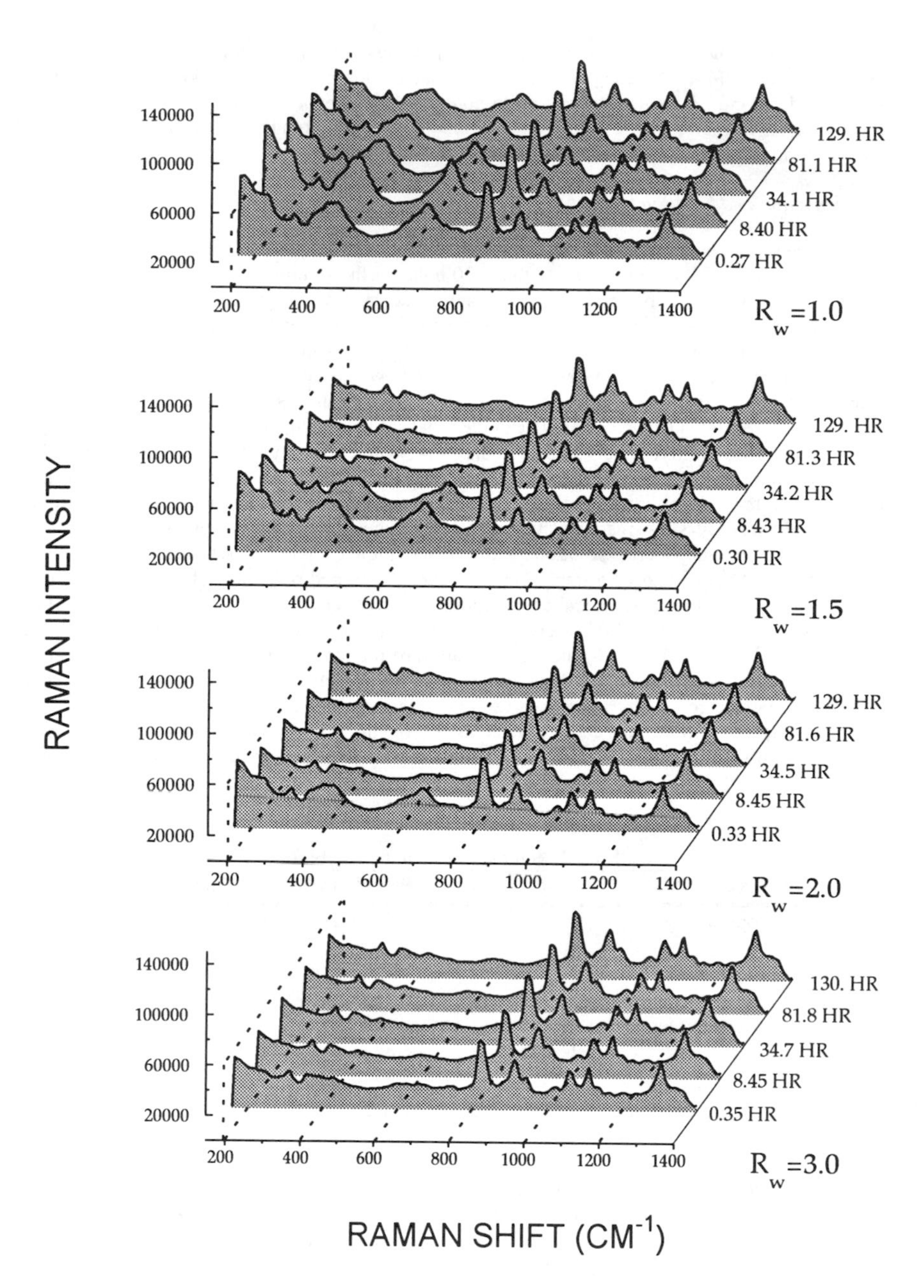

Figure 2. Raman spectra of aged titanium isopropoxide valerate solutions with alkoxide to water ratios (R_W) of 1.0, 1.5, 2.0, and 3.0. Sample age ranged from 0.27 to 130 hr.

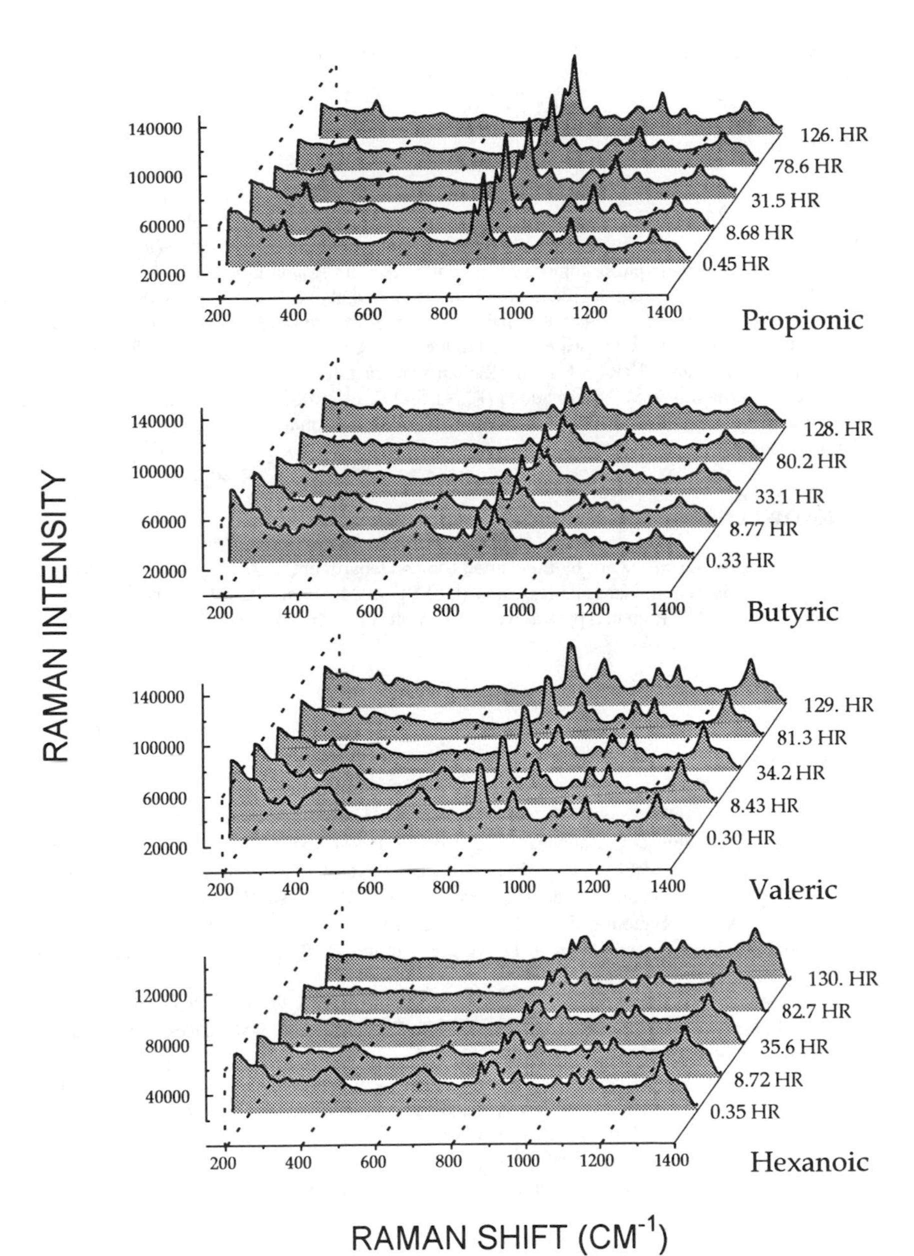

Figure 3. Raman spectra of aged titanium isopropoxide carboxylate solutions with an water to alkoxide ratios (R_w) of 1.5. From top to bottom, the carboxylic acids used were propionic, butyric, valeric, and hexanoic -- all with an acid to alkoxide ratio (R_a) of 15.

chains produce clear solutions with no particles detectable with a Coulter N4 particle size analyzer.

CONCLUSIONS

This Raman spectroscopic investigation of reaction kinetics indicates that the coating solutions do achieve a significant degree of stability for moderate levels of water addition ($R_W \leq 3.0$). Other studies have shown that for an acid to alkoxide molar ratio (R_a) of 9.0, clear, apparently stable, single phase solutions can be formed with R_W values up to 7.5.

Because of the limited degree to which the modified alkoxide can react with water, it is not critical for long-term results that the hydrolysis be 'controlled' by avoiding a local excess of water. Regardless of how the water is introduced, the end result is the same. Furthermore, if sufficient water is added ($R_W \geq 1.5$) the Raman spectra after 130 hours appear to be identical regardless of how much water was added (R_W=1.5, 2.0, or 3.0); within this concentration range, the amount of added water controls the rate of transformation but not the end product characteristics.

ACKNOWLEDGMENTS

The support for this work by the United States Department of Agriculture (USDA) and the Center for Fundamental Materials Research (CFMR) at Michigan State University is gratefully acknowledged. All Raman spectra were taken at the Michigan State University LASER Laboratory.

REFERENCES

1 C. Sanchez and J. Livage, New J. Chem. **14** 513-521 (1990).
2 L. G. Hubert-Pfalzgraf, New J. Chem. **11** 663-675 (1987).
3 C. J. Brinker, G. W. Scherer, *Sol-Gel Science (*Acedemic Press, New York, NY, 1990).
4 S. Doeuff, M. Henry, C. Sanchez, and J. Livage, J. Non-Cryst. Solids **89** 206-216 (1987).
5 C. Sanchez, J. Livage, M. Henry, F. Babonneau, J. Non-Cryst. Solids **100** 65-76 (1988).
6 J. Livage, M. Henry, and C. Sanchez, Prog. Solid St. Chem. **18** 259-341 (1988).
7 C. Sanchez, F. Babonneau, S. Doeuff, A. Leaustic, in *Ultrastructure Processing of Advanced Ceramics,* edited by J. D. Mackenzie and D. R. Ulrich (John Wiley and Sons, New York, NY, 1988).
8 C. Sanchez, P. Tolando, F. Ribot, in *Better Ceramics Through Chemistry IV,* edited by B. J. J. Zelinsky, C. J. Brinker, D. E. Clark, and D. R. Ulrich (Mater. Res. Soc. Proc. **180,** Pittsburgh, PA, 1990) pp.47-59.
9 R. C. Mehrotra and R. Bohra, *Metal Carboxylates (*Acedemic Press, New York, NY, 1983).
10 C. D. Gagliardi, D. Dunuwila, and K. A. Berglund, in *Better Ceramics Through Chemistry IV,* edited by B. J. J. Zelinsky, C. J. Brinker, D. E. Clark, and D. R. Ulrich (Mater. Res. Soc. Proc. **180**, Pittsburgh, PA, 1990) pp.801-805.
11 M. Payne, Master's Thesis, Michigan State University (1986).

USAXS AND X-RAY MICROSCOPY INVESTIGATIONS ON SILICA AND PRECURSORS OF ZEOLITES

Theo P.M. Beelen*, Wim H. Dokter*, Harold F. van Garderen*, Rutger A. van Santen*, Mike T. Browne** and Graeme R. Morrison**
*Eindhoven University of Technology, P.O.Box 513, 5600 MB EINDHOVEN, The Netherlands
** King's College, University of London, Strand, London WC2R 2LS, United Kingdom

ABSTRACT

Combination of SAXS and USAXS measurements provide an extended q-range (0.006-3.0 nm^{-1}) to study fractal growth of both aging silica gel as well as precursors of zeolite-A. Mass (silica) and surface (zeolite) fractals are observed. Scanning transmission X-ray microscopy (STXM) proves to be an extremely useful technique to obtain direct images of wet samples in the 0.1-100 micron range, confirming the SAXS/USAXS results on even larger length scales.

INTRODUCTION

Silica and silica gels are important industrial raw materials. The majority of the processes for silica and silica gel production is based on the acidification of water glass, inducing condensation (polymerization) reactions between silicon tetrahedra:

(XYZ)Si-O- + HO-Si(X'Y'Z') == (XYZ)Si-O-Si(X'Y'Z')

with X,Y,Z,X',Y' and Z' hydroxy or silicic (both monomer and polymer) groups. Due to the tetrahedral-directed bonds ring- and cluster-shaped polymers are favoured resulting in spherical elementary particles [1]. These particles may combine to aggregates [2,3], analogous to the well-studied aggregates formed by hydrolysis of alkoxy silica's; see [4,5] and references cited herein. At sufficient silica concentration the growing aggregates will be interconnected resulting in a network, the gel. This often very fragile network is strengthened during aging. By hydrolysis and condensation reactions silica is redistributed, with dissolution preferably on the periphery and precipitation in the crevices and "necks" of the aggregates [1].

Aging is of crucial importance to the texture of silica. For example, only after aging is the geleous network strong enough to avoid collapse when water is removed by drying, resulting in a mesoporous solid [6]. Because the hydrolysis/condensation reactions are catalyzed by OH^- ions [1,7], the presence of cations and organic templates with polar groups is expected to have a strong influence on the aging processes, analogous to the selectivity in dissolution of silica as shown by NMR [8,9].

During the preparation of zeolites [10] the conversion of the starting silica or silica-alumina gel to the precursors of crystalline zeolites can be considered as an aging process. Investigations on the changing gel structure during aging are necessary to understand the fundamental steps, especially the relations between reaction conditions (pH, concentrations, temperature), addition of cations or templates and the resulting zeolite structure. Recently, scattering by light, X-rays (SAXS) or neutrons (SANS) has proven to be very efficient at providing information concerning the distribution of mass on colloidal scale, irrespective of the phase (solution, sol, gel or amorphous solid) one is dealing with [4,5,11,12]. Exploitation of the fractal properties of aggregates is very helpful in the interpretation of scattering data. Due to the power-law mass density distribution in fractals, from a log-log plot of scattered intensity I versus scattering vector q, the size (radius of gyration) of the fractal aggregates, the size of the primary particles and the fractal dimensionality can be extracted [12].

However, a serious limitation to obtaining the necessary data stems from the often very large q range exhibited by fractal aggregates (in diluted systems up to three decades [11,14]). Combination of SAXS (SANS) and light scattering has been applied [4,11,15] but this last method can not be used for opaque systems (aged silica gels, zeolites). We have chosen an alternative solution: extending X-ray scattering to very small angles. With USAXS (Ultra Small Angle X-ray Scattering) using the synchrotron source at Daresbury the 1-50 nm range of SAXS might be extended to 1000 nm [16,17]. In this paper we show that the SAXS-USAXS combination may be a good alternative for light scattering and SAXS/SANS.

Although the physical basis and the interpretation of scattering spectra from fractal systems is well understood, we are still dealing with an indirect method. A "second opinion" using a direct imaging technique (e.g. microscopy) would give a very valuable confirmation and expansion of the information obtained by scattering. Conventional electron microscopy can not be used because it destroys or changes the tenuous and fragile aggregates by the forced removal of water in the high-vacuum system. A possible solution might be cryo-electron microscopy: extremely fast cooling of the sample preserves the structure of a gelled Ludox sol in the vitreous ice [18].

X-ray microscopy using photons with energies in the range 100 eV to 1 keV fills a niche between optical and electron microscopy [19,20]. The short wavelength of x-rays means that sub-optical resolution can be achieved in specimens that need not to be optically transparent. The dominant beam-specimen interaction is photo-electric absorption which results in elemental contrast even from materials containing adjacent elements in the periodic table; most low atomic number elements have a K or L absorption edge in the soft x-ray region. The possibility to examine species under atmospheric conditions combined with the choice of x-rays in the "water-window" energy range (277-525 eV) is not only attractive for examination of biological material in an aqueous medium, but also very promising for solutions or gels of silicas and zeolites. However, it is only recently that technology has developed sufficiently to allow the manufacture of Fresnel zone plates for use as diffractive focusing elements. In addition, the advent of synchrotron radiation has now provided a sufficiently intense tunable source of radiation in the soft x-ray regime, resulting in the design of several different x-ray microscopes. The King's College scanning transmission x-ray microscope (STXM) was first used on the Daresbury synchrotron radiation source (United Kingdom) in the summer of 1986 and during the past years modifications and improvements have been made [21]. In this paper some results obtained on silicas and zeolites are described, in combination with SAXS-USAXS experiments.

EXPERIMENTAL

Water glass was obtained by dissolution of Aerosil 380 (Degussa) in KOH (Merck,p.a.). Silica gel was prepared by addition of water glass to 1.0 or 0.1 M HCl until pH = 4.0.

Zeolite-A was prepared by addition of a water glass solution (24.7 g water glass (Merck, 27% SiO_2), 2.5 g NaOH, 33.3 g H_2O and 4.85 g SiO_2 (Aerosil 200, Degussa)) to a alumina solution (18.2 g sodium aluminate (Riedel-De Haen; 54% Al_2O_3), 5.2 g NaOH (Merck,p.a.) and 236 g H_2O) and heated at continuous stirring at 80°C during

24 hours. To avoid progress in reaction during the measurements (SAXS, USAXS, X-ray microscopy) samples from the reaction mixture were cooled down to room temperature.

SAXS spectra were measured using synchrotron radiation (λ = 0.154 nm) at station 8.2 at Daresbury Laboratories (United Kingdom). For both solutions and gels cells were used with mylar windows and 1 mm pathway. Spectra were corrected for background and parasitic radiation and checked by extrapolation to $q \rightarrow \infty$ in $I\text{-}q^{-4}$ or $I\text{-}q^{-D}$ plots [22, 23].

USAXS spectra were measured with the Bonse-Hart twin crystal diffractometer system on station 2.2 at Daresbury Laboratory [16, 17]. Assuming infinite slit smearing for the Bonse-Hart system the fractal dimension is determined by adding -1 to the negative value of the slope in log(I)-log(q) plots [24].

X-ray imaging has been performed with the King's College scanning transmission x-ray microscope (STXM) using soft x-ray photons (λ = 3.28 nm) from station 5U2 at the Daresbury Laboratories [19-21]. Images were taken of diluted reaction mixtures from aging silicas or zeolitic precursors. Before scanning, samples were superficially dried (as confirmed by IR-spectra) by exposing small droplets to the air.

RESULTS

To compare the SAXS and USAXS data in figure 1 curves are shown of representative samples of aging silica and precursors of zeolite-A. Because the original spectra showed large differences in scattered intensity, in figure 1 arbitrary intensity units have been chosen and the curves have been shifted along the vertical axis.

The SAXS curve of silica is representative of an aging silica gel [2,6,14]. The growth of the endpoint of the straight line (with constant fractal dimension, D = 1.90), corresponding to the growth of the radius of gyration of the fractal aggregates, did already surpass the low q boundary (0.3 nm^{-1}) at these concentrations after a few minutes, resulting in a straight line over the full q-range during further growth, gelation and aging.

With USAXS the q-range is extended to q = 0.006 nm^{-1}. However, because the Bonse-Hart camera is a scanning system with scanning times in the order of hours, only stable systems (silica in the aging stage, "frozen" zeolitic precursors) can be studied. In figure 1 the USAXS spectrum of an aged silica is shown. Due to the relatively low silica concentration and the extremely high background scattering at the (ultra)low q values, this spectrum showed a rather poor intensity/background ratio.
The only reliable feature to extract from this spectrum proved to be the straight slope, indicating an extended fractality of the aggregates. However, the fractal dimension could not be determined accurately, due to low intensity/background ratio. In figure 1 the USAXS curve has been drawn tentatively assuming the same fractal dimension as found for the SAXS curve (D = 1.90).

Contrary to silica gel samples of the zeolite-A reaction mixture showed the presence of surface fractals with D = 3.6 - 3.7. Only after 7 hours, when by x-ray diffraction in the reaction mixture the presence of small quantities of crystalline zeolite-A could be demonstrated, the fractal dimension approaches D = 4, indicating advancing aging prior to crystallization. Due to the high scattering in the (high silica) zeolitic samples, the intensity/background ratio was much more favourable in the USAXS spectrum, so a reliable log(I)-log(q) plot and fractal dimension could be obtained. Small differences between fractal dimensions in the SAXS and USAXS

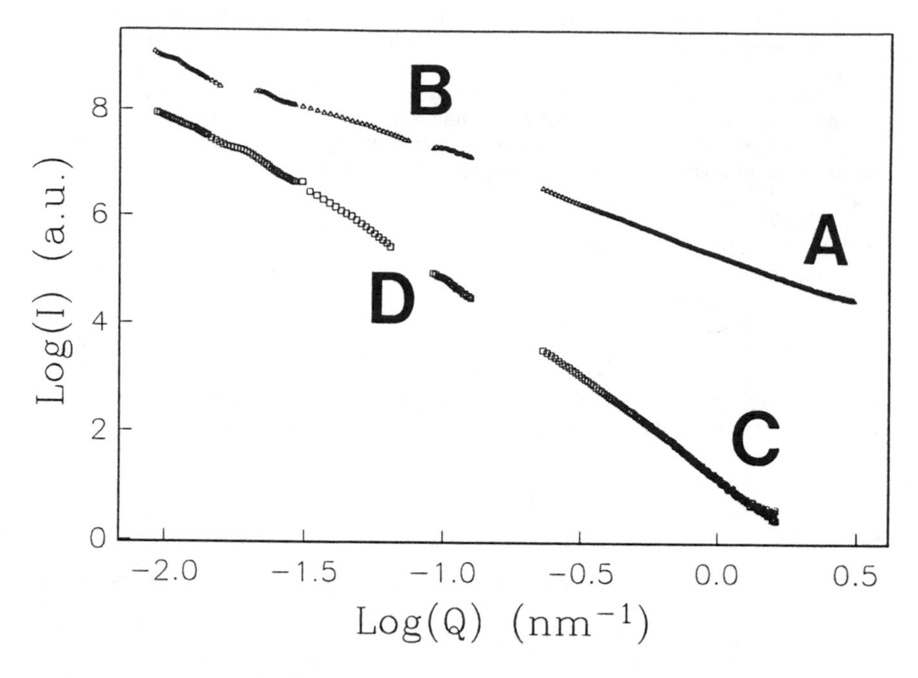

Figure 1. SAXS and USAXS curves of aged silica (A and B) and reaction mixtures of zeolite-A (C and D).

spectrum can be attributed to different samples and different background correction.

The striking difference between the mass and surface fractals of the silica gel and the zeolitic precursor respectively are confirmed by scanning transmission x-ray microscopy (STXM) (figure 2). The (diluted) aged silica shows typical fractal aggregates with high anisotropy as may be expected from reaction-limited cluster-cluster aggregation [25]. Moreover, the fractal range extends even into the micron range, adding another decade to the fractal range of (diluted) silica aggregates. However, one has to be careful because the radius of gyration (indicative of the fractal range in SAXS or USAXS spectra) is much less than the size of the anisotropic aggregates shown in the STXM images. Because the projection of a three-dimensional aggregate to two dimensions will not effect the value of the fractal dimension D for values of D less than 2 [26-28], a rough indication of the fractal dimension of the aggregates may be obtained by applying the relation between mass and radius in fractal systems ($M = a.R^{-D}$) on STXM images. The resulting D values (1.55 - 1.7) are lower than expected, but may be attributed to dilution effects.

The images of the precursor gel of zeolite-A show aggregates of spherical particles. The radius of gyration of the spherical particles has the order of magnitude of 1 micron. According to USAXS the particles behave as surface fractals.

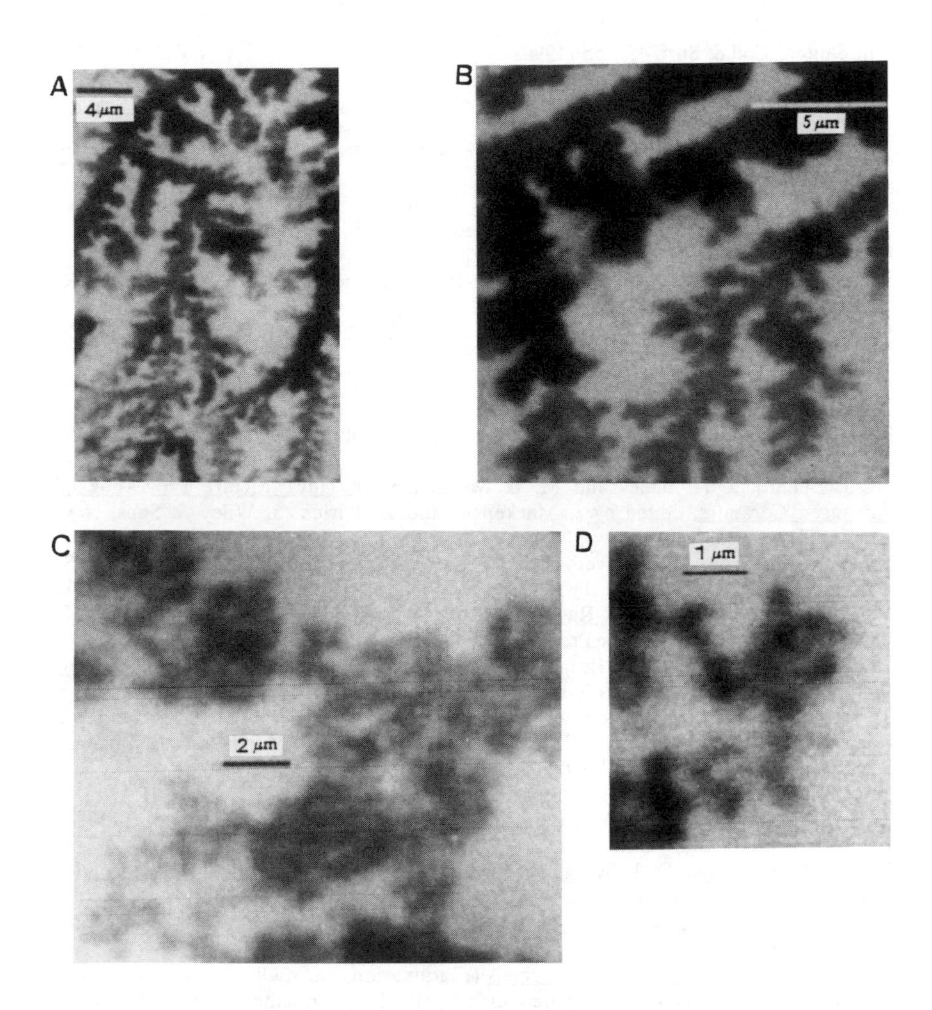

Figure 2. STXM images of diluted solution of aged silica (A and B) and diluted reaction mixture of zeolite-A (C and D).

REFERENCES

1. R. K. Iler, The Chemistry of Silica (J. Wiley & Sons, New York, 1979)
2. T. P. M. Beelen, P. W. J. G. Wijnen, C. G. Vonk and R. A. van Santen, Cat.Letters 3, 209 (1989)
3. P. W. J. G. Wijnen, PhD thesis, Eindhoven University of Technology, 1990
4. J. E. Martin and A. J. Hurd, J.Appl.Cryst. 20, 61 (1987)
5. C. J. Brinker and G. W. Scherer, Sol-Gel Science (Academic Press, Boston, 1990)
6. P. W. J. G. Wijnen, T. P. M. Beelen, C. P. J. Rummens, H. C. P. L. Saeijs, J. W. de Haan, L. J. M. van de Ven and R. A. van Santen, J.Coll.Interf.Sci. 145(1), 17 (1991)
7. E. J. A. Pope and J. D. Mackenzie, J.Non-Cryst.Solids 87, 185 (1986)
8. P. W. J. G. Wijnen, T. P. M. Beelen, J. W. de Haan, L. J. M. van de Ven and R. A.

van Santen, Coll.& Surf. 45, 255 (1990)
9. W. M. Hendricks, A. T. Bell and C. J. Radke, J.Phys.Chem 95, 9519 (1991)
10. H. van Bekkum, E. M. Flanigen and J. C. Jansen, Introduction to Zeolite Science and Practice (Elsevier, Amsterdam, 1991)
11. B. Cabane, M. Dubois, F. Lefaucheux and M. C. Robert, J.Non-Cryst.Solids 119, 121 (1990)
12. J. D. F. Ramsay, Chem.Soc.Rev. 15, 335 (1986)
13. J. Teixeira, in On Growth and Form, edited by H. E. Stanley and N. Ostrowski (M.Nijhoff, Dordrecht, 1986)
14. P. W. J. G. Wijnen, T. P. M. Beelen, C. P. J. Rummens, H. C. P. L. Saeijs and R. A. van Santen, J.Appl.Cryst. 24, 759 (1991)
15. D. W. Schaefer, J. E. Martin, P. Wiltzius and D. S. Cannel, Phys.Rev.Letters 52, 2371 (1984)
16. A. N. North, J. C. Dore, A. R. Mackie, A. M. Howe and J. Harries, Nucl.Instr.Meth.Phys.Res. B47, 283 (1990)
17. G. P. Diakun and J. E. Harries in Synchrotron Radiation in Biophysics, edited by S.S. Hasnain (Ellis Harwood Limited, 1990) p.243
18. J. Bellare, J. K. Bailey and M. L. Mecartney, in Ultrastructure Processing of Advanced Ceramics, edited by J. Mackenzie and D. Ulrich (J. Wiley & Sons, New York, 1988), p.835
19. G. R. Morrison, J. T. Beswetherick, M. T. Browne, R. E. Burge, R. C. Cave, P. S. Charalambous, P. J. Duke, G. F. Foster, A. R. Hare, A. G. Michette, D. Morris, A. W. Potts and T. Taguchi, in X-Ray Microscopy in Biology and Medicine, edited by K. Shinohara et al. (Japan Sci.Soc.Press, Tokyo, 1990) p.99-108
20. G. R. Morrison and R. E. Burge in Synchrotron Radiation in Biophysics, edited by S. S. Hasnain (Ellis Harwood Limited, 1990) p.330
21. G. R. Morrison, S. Bridgewater, M. T. Browne, R. E. Burge, R. C. Cave, P. S. Charalambous, G. F. Foster, A. R. Hare, A. G. Michette, D. Morris and T. Taguchi, Rev.Sci.Instrum. 60(7), 2464 (1989)
22. C. G. Vonk, J.Appl.Cryst. 9, 433 (1976)
23. J. P. Cotton in Neutron, X-ray and Light Scattering, edited by P. Linder and Th. Zemb (North Holland, Amsterdam, 1991) p.24
24. L. A. Feigin and D. I. Svergun, Structure Analysis by Small-Angle X-Ray and Neutron Scattering (Plenum Press, New York, 1987)
25. R. Jullien and R. Botet, Aggregation and Fractal Aggregates (World Scientific, Singapore, 1987)
26. B. Mandelbrot, in Les Objects Fractals (Flammarion, Paris, 1975)
27. A. Tuel, P. Dantry, H. Hommel, A. P. Legrand and J. C. Morawski, Progr.Coll.Polym.Sci. 76, 32 (1988)
28. M. Tence, J. P. Chevalier and R. Jullien, J.Phys. 47, 1989 (1986)

ACKNOWLEDGEMENTS

Financial support was given by the Dutch Department of Economic Affairs (IOP-Katalyse). SAXS-USAXS measurements were performed at the Synchrotron Radiation Source, Daresbury (United Kingdom) under terms of the SERC/NWO Agreement. Special thanks to dr.W.Bras (SRS/NWO) and dr.J.Harries (SRS).

RHEOLOGY OF ZIRCONIA-ALUMINA GELCASTING SLURRIES

A. Bleier, O. O. Omatete, and C. G. Westmoreland
Oak Ridge National Laboratory, P. O. Box 2008, Oak Ridge TN 37831

ABSTRACT

The rheology of zirconia-alumina composite slurries for gelcasting was studied in order to maximize the solids loading. The viscosity and yield stress were controlled by adjusting the pH. This approach allows the solids loading to be maximized for the gelcasting of near-net-shape composites. A strong correspondence exists among the rheological behavior, the surface charge on the particles, colloidal stability, and the maximum solids loading. The best pH conditions for gelcasting composites depends on the specific binary composition.

INTRODUCTION

The effects of suspension properties, such as pH, zeta potential, and the presence of dispersing agents, on the aqueous processing of ceramics are generally understood only in the case of dilute unary suspensions. Yet, most ceramics are composites whose fabrication entails the use of concentrated suspensions. Although the interactions that govern the processing of suspensions with more than one type of solid are complex, they must be properly characterized in order to tailor the microstructures and the final composite properties. In this regard, a major requirement for the fabrication of composites by gelcasting is to predict and to control the rheological behavior of the slurry [1].

The present study addresses this need, in part, by focussing on two criteria for the successful gelcasting of composites that contain zirconia (ZrO_2) and alumina (Al_2O_3): (1) the avoidance of agglomeration and (2) the attainment of a high suspended solids concentration. It stresses the pH-dependent zeta potentials of the oxides and extends recent studies on the particulate interactions that ensure both pourable slurries [2] and uniformly distributed ZrO_2 in an Al_2O_3 matrix [3,4]. Pertinent properties of the powders used are listed in Table I. The best processing conditions for mixtures of these powders [2] and similar Al_2O_3 and ZrO_2 powders [4] exist when the electrophoretic mobilities are positive and adequately high to ensure colloidal stability.

EXPERIMENTAL MATERIALS AND PROCEDURES

The methods used to obtain the properties in Table I are described elsewhere. [2] The colloidal stability of each oxide was assessed by measuring the relative suspension height (RSH) of various, quiescent suspensions of each powder (3.5 vol% ZrO_2, 5.0 vol% Al_2O_3) in cylindrical containers at intervals up to two weeks. [2] Data for suspensions of each oxide are shown in Figure 1A and describe the colloidal stability 1 h after sample preparation. RSH is calculated by normalizing the height of the demarcation between the turbid, lower region and the clear, upper region to the overall height of the solid-liquid mixture. These data were gathered after short times: thus, the RSH values near unity signify that virtually no settling has occurred and that the suspension is colloidally stable, whereas, the low values denote colloidally unstable systems because particulate clusters settle faster than their constituent primary particles. The data in this figure identify three regimes of different colloidal stability, Regions I, II, and III. Comparison of the pH conditions for each region with the zeta potentials (ζ) in Figure 1B, calculated from the electrophoretic mobility [6], reveals that the stability within Regions I and III derives from the high,

Table I: Pertinent Properties of Al_2O_3 and ZrO_2 Powders. [2]

Property	Characterization	
Composition	α-Al_2O_3[a]	t-ZrO_2[b]
Surface Area, m^2 g^{-1}	6.37	6.98
Diameter (D), μm	0.55	0.39
Point of Zero Charge, pH_{pzc}	8.7[c]	7.5[d]
Hamaker Constant (A^*)[e], kT	11.2	34.5[f]

[a]Reynolds, RCHP-DBM; [b]Tosoh, TZ-3Y; [c]Equals isoelectric point, pH_{iep} [5]; [d]The relationship between pH_{pzc} and pH_{iep} has not been studied; [e]For two particles separated by water [4]; [f]Value for m-ZrO_2.

positive and negative surface charge, respectively, and that the instability within Region II results from low surface charge.

The suitability of unary and binary suspensions for gelcasting was assessed by evaluating the maximum solids loading under various pH conditions. For unary systems, this property is the maximum dispersible concentration of a solid that remains sufficiently fluid to ensure that the slurry is pourable and that all the parts of a gelcasting mold can be filled. This concentration was determined for each oxide by incorporating increasing amounts of powder into a suspension until it had prohibitively high viscosity and became pastelike. The maximum loading was also evaluated for binary mixtures of ZrO_2 and Al_2O_3 in which the ZrO_2 concentration was either 20 or 80 vol%. As a reference for suitability in gelcasting and for minimum shrinkage during the drying and sintering stages, 50 vol% is chosen as the minimum acceptable concentration for gelcasting.

Slurry rheology was evaluated by measuring the shear stress (τ) at desired shear rates ($\dot{\gamma}$) with a parallel-plate viscometer (Rheometrics, Model 8400) in a thixotropic loop mode set with four stages: (1) 60 s at 300 s^{-1}, (2) 180 s to reduce $\dot{\gamma}$ to 0 s^{-1}, (3) 180 s to increase it to 300 s^{-1}, (4) 180 s to reduce it again to 0 s^{-1}. This strategy allowed the study of (i) apparent viscosity ($\eta^{app} \equiv \tau/\dot{\gamma}$), Bingham plastic flow with yield stress (τ_B) and constant differential viscosity ($\eta^{dif} \equiv d\tau/d\dot{\gamma}$) when $\tau \geq \tau_B$, (ii) pseudoplastic flow (η^{app} and η^{dif} differ regularly with increasing $\dot{\gamma}$), (iii) hysteresis ($\tau_3 \neq \tau_4$, subscripts specify stages), and (iv) thixotropy (time-dependent τ). [7] Most of the slurries described here were pseudoplastic and exhibited little or no hysteresis.

RESULTS AND DISCUSSION

Maximum Loading. The maximum loading data for ZrO_2 and Al_2O_3 and their mixtures are given in Figure 2. Each unary system (open symbols) has a pH-dependent profile with two regimes (Region I at low pH and a less-distinctive and virtually pH-insensitive Region III at high pH), within which the maximum loading is weakly pH-dependent and is optimized for positively and negatively charged particles (Figure 1), respectively. Between these regimes (Region II), a minimum loading limit occurs for each sign of surface charge at pH 7.0 for ZrO_2 and pH 8.2 for Al_2O_3, values that are in good accord with the pH_{pzc}-values in Table I.

Comparison of the data from Figures 1 and 2 for unary suspensions shows a strong correspondence among the colloidal stability of the particles, their surface charge, and the maximum loading. Regions I and III in Figure 2 compare well with Regions I and III in Figure 1A. Both the loading and the colloidal stability are maximized by achieving the highest zeta potential (Figure 1B).

The maximum loading data for binary systems with either 20 or 80 vol% ZrO_2 are also shown in Figure 2. These systems have strong pH-dependencies, which resemble those of the constituent oxides and typically reside between them. The pH values at the transitions between Region II and Regions I and III are given in Table II, along with the conditions for the lowest loading limit and the most useful pH range for each system. The ζ-potentials for the unary slurries at the pH-transitions between Region II and either Region I or III, noted in Table II,

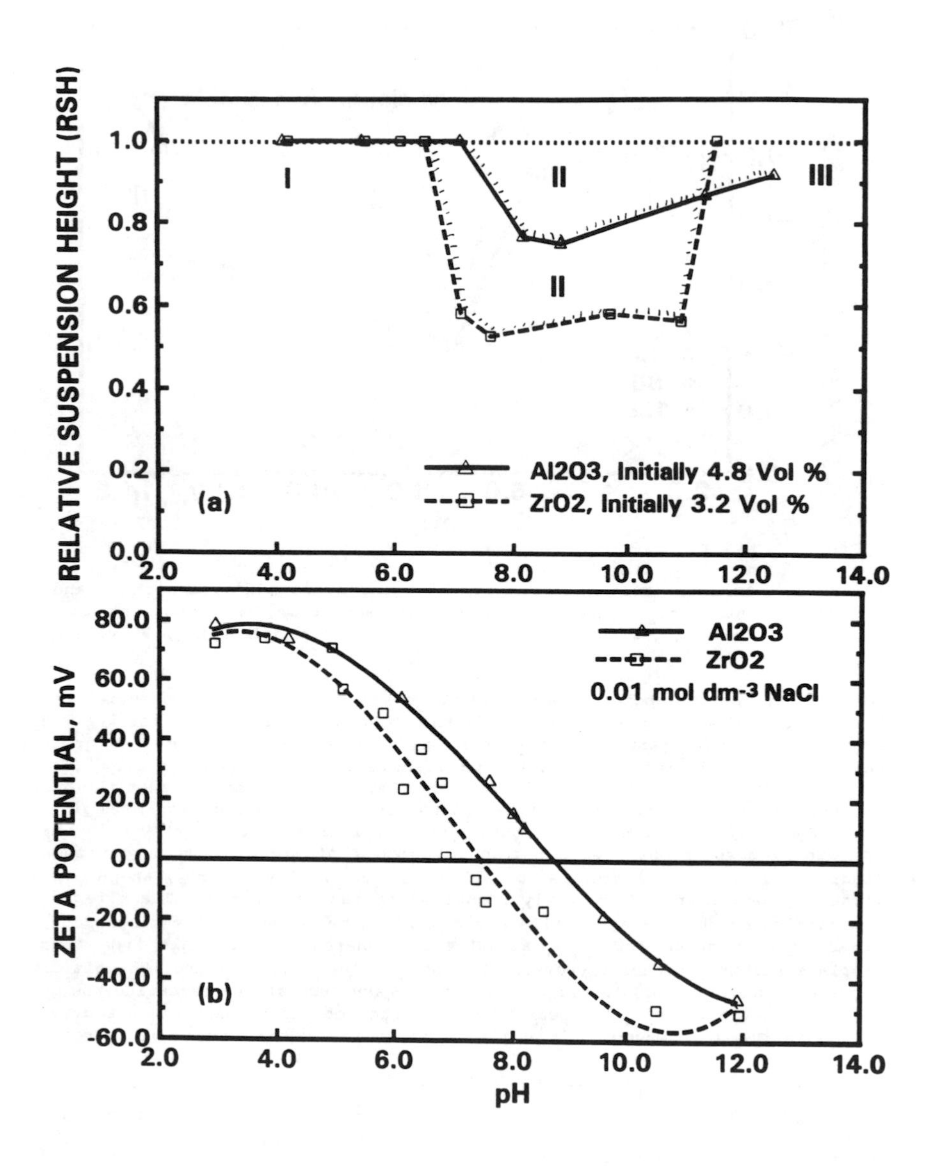

Figure 1: (A) Relative suspension height (RSH) as a function of pH for suspensions of ZrO_2 and Al_2O_3, where hatching denotes particle agglomeration; (B) Zeta potential as a function of pH.

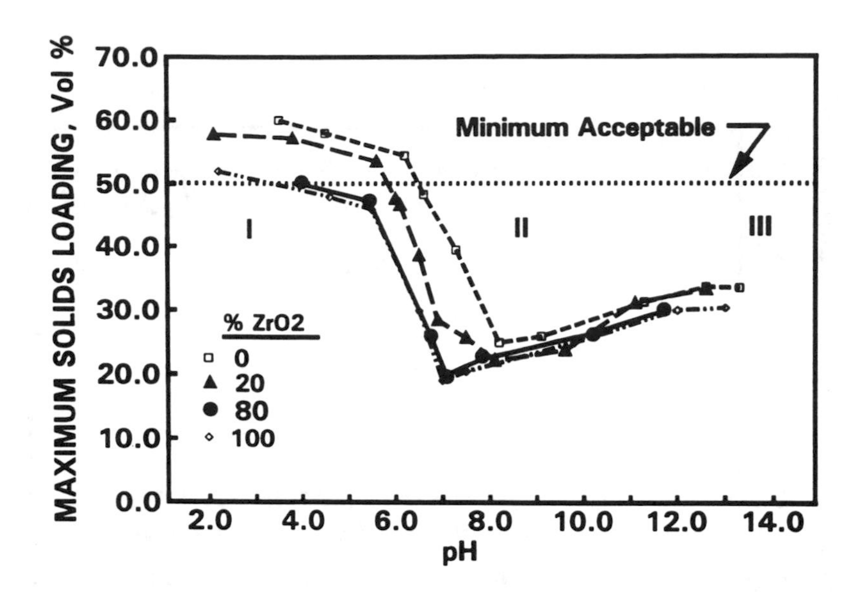

Figure 2: Maximum loadings for ZrO_2, Al_2O_3, and two mixtures with 20 and 80 vol% ZrO_2, based on solids concentration, as functions of pH. Dashed line shows the minimum loading for gelcasting, 50 vol%.

reveal that absolute potentials of ~50 mV comprise a critical ζ-potential range for these oxides. Higher values yield the highest loadings for positively and negatively charged particles of either oxide, and lower ones lead to greatly reduced loading limits. Lastly, the pH-transition between Regions I and II for each mixture agrees well with that for ZrO_2 (Table II), revealing that the lower ζ-potential of ZrO_2 sets the loading limits for the binary slurries in Region I.

Rheology. The composition-dependent rheological nature of 40-vol% unary and binary suspensions prepared within Region I of Figure 2 is summarized in Figure 3. The τ_4-vs-$\dot{\gamma}$ profile for an Al_2O_3 slurry (0 vol% ZrO_2) at pH 4.2 is virtually Newtonian and is nearly identical to that at pH 6.2. The effects of composition on the τ_4-vs-$\dot{\gamma}$profile are clarified by comparing the other data in Figure 3 with those for Al_2O_3 at pH 4.2. Whereas the 20-vol% ZrO_2 slurry displays slightly higher τ_4-values at most $\dot{\gamma}$-values, the 80- and 100-vol% ZrO_2 slurries exhibit pseudoplasticity, with the unary ZrO_2 slurry revealing the most extreme behavior. Thus, the near-Newtonian rheology of a unary Al_2O_3 slurry is increasingly made pseudoplastic as the relative ZrO_2 concentration is increased.

Table II: pH-Dependent Gelcasting Behavior of ZrO_2-Al_2O_3 Slurries. [2]

Composition[a] (Vol% ZrO_2)	Stability Transitions[b] (I ↔ II)	(III ↔ II)	Lowest Maximum Loading	Useful Range[c]
0[d,e]	pH 6.2	pH ≈ 12.6	pH ≈ 8.2	pH ≤ 6.5
20	pH 5.6	pH ≈ 12.6	pH ≈ 8.1	pH ≤ 5.8
80	pH 5.5	pH ≥ 12.0	pH ≈ 7.0	pH ≤ 3.3
100[f]	pH 5.5	pH ≈ 12.0	pH ≈ 7.0	pH ≤ 3.3

[a]Based on total solids; [b]Figure 2; [c]At least 50 vol% solids; [d]100 vol% Al_2O_3; [e]ζ = 52 and -47 mV for Al_2O_3 at low and high pH transitions, respectively; [f]ζ = 49 and -46 mV for ZrO_2 at low and high pH transitions, respectively.

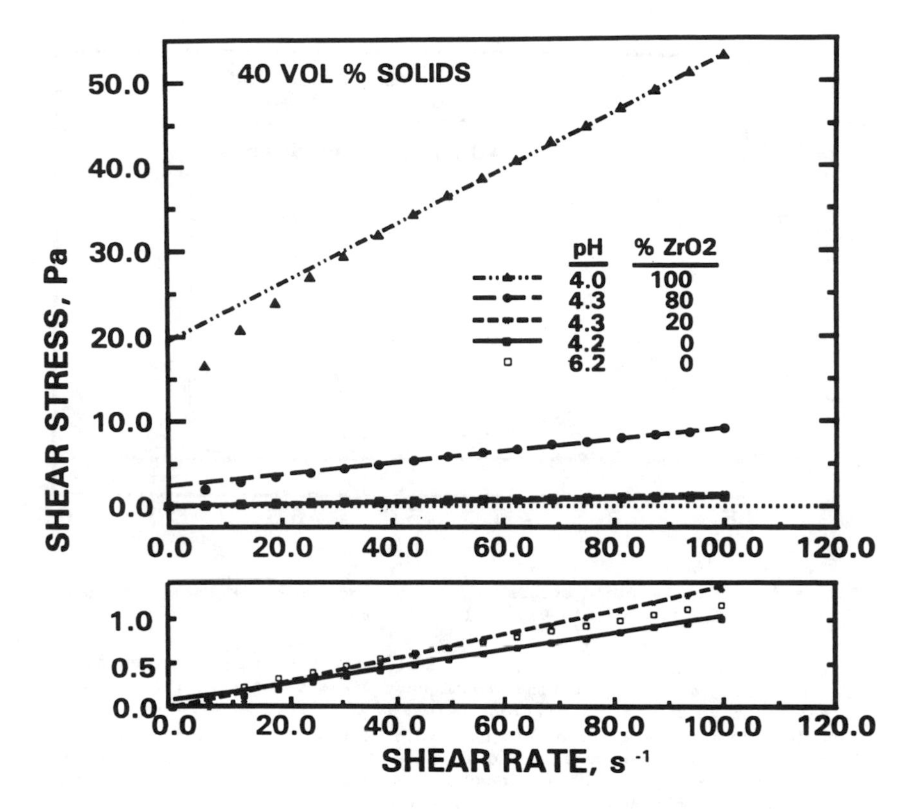

Figure 3: Shear stress in the fourth stage of the thixotropic loop (τ_4) as a function of shear rate ($\dot{\gamma}$) for the compositions depicted within Region I of Figure 2. The line identified with each profile shows the Bingham relationship ($\tau = \tau_B + \eta^{dif}\ \dot{\gamma}$) that best describes it.

Although the RSH data in Figure 1A suggest that ZrO_2 and Al_2O_3 have similar colloidal stability, the loading limits for each oxide component within Region I (Figure 2) imply that ZrO_2 is the less stable one in the concentrated systems. Importantly, the ζ-potential of ZrO_2 is less than that of Al_2O_3 (Figure 1B), and the ratio of the Hamaker constant ($\underline{A}^*$) for ZrO_2 to that for Al_2O_3 is ~3.1, while the analogous ratio of their particle sizes ($\underline{D}$) is 0.71 (Table I). Consequently, the electrical double layer free energy that adds to stability (repulsion) basically scales with the term, $\underline{D}\zeta^2$, [4a,8] and is less for ZrO_2, while the van der Waals free energy that promotes agglomeration (attraction), depends linearly on the term, $\underline{A}^*\underline{D}$, [9] and is greater for ZrO_2. Thus, the colloidal stability of ZrO_2, the less stable component, controls the rheological trends in Figure 3.

The pH-dependent rheological nature of 40-vol% unary and binary suspensions prepared within Region I of Figure 2 is summarized in Figure 4. The τ_B-values of a unary Al_2O_3 slurry (0 vol% ZrO_2) are virtually constant over the range pH 4.2 to 6.2, being nearly zero due to the nearly Newtonian behavior described earlier (Figure 3). The effects of composition on the τ_B-values are revealed by a comparison of the other data sets in Figure 4 with the one for the Al_2O_3 slurries. The data for slurries containing 20 vol% ZrO_2 reveal essentially the same pH-dependence as the ZrO_2-free ones below approximately pH 5.7, but exhibit a higher Bingham yield stress near pH 6. Also, the 80- and 100-vol% ZrO_2

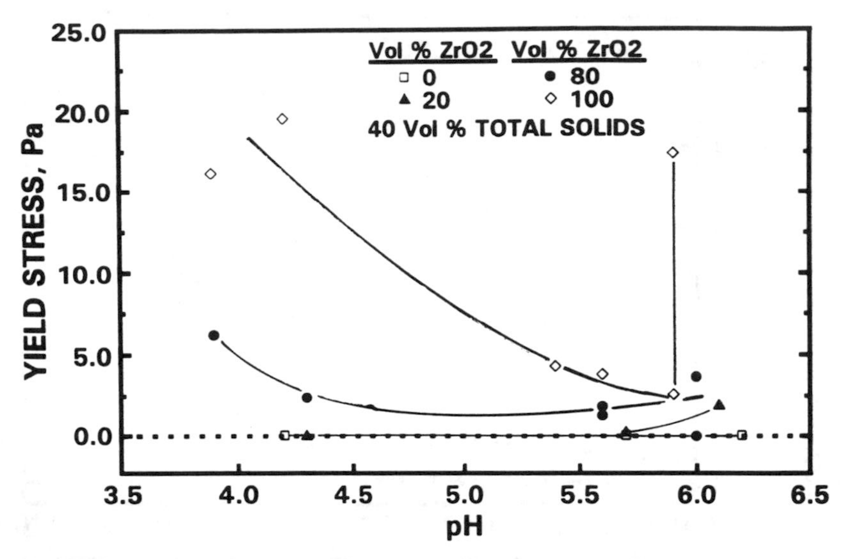

Figure 4: Bingham yield stress (τ_B) as a function of pH for the compositions depicted within Region I of Figure 2.

slurries exhibit notably higher τ_B-values over the entire pH range depicted, with the unary ZrO_2 slurry exhibiting the highest yield stresses. Thus, an increase in the relative ZrO_2 concentration increases the deviation from Newtonian behavior and induces a higher Bingham yield stress throughout the range, pH 4.2 to ~6. According to Figures 2 and 4, this pH range is a critical portion of Region I because the conditions near pH 6 approach those that limit the solids loading for the 80- and 100-vol% ZrO_2 slurries.

The explanations given earlier for the effects of composition on the τ_4-vs-$\dot{\gamma}$ profiles in Figure 3 also appear to apply generally to Figure 4. The dominant role of ZrO_2, via its relative colloidal stability, is inferred from the effect of composition on the Bingham yield stress over the pH range depicted in Figure 4 and the effect of pH as the pH_{pzc}-value, 7.5 (Table I, Figure 1B), is approached. However, the pH-dependence of the Bingham yield stress, τ_B, at low pH (Figure 4) seems incompatible with the ζ-potential profiles in Figure 1B. That is, the τ_B-values of the 80- and 100-vol% ZrO_2 slurries increase as the ζ-potential of each type of oxide increases near pH 4. This behavior, mainly that of τ_B, is apparently the result of electroviscous effects, in which an enhanced resistance to flow occurs when the electrical double layers deform in the process (primary effect) or when these layers increase the effective particle volume and distort the particle trajectories under a shear force (secondary effect). [7c,10] The behavior exhibited by the data in Figure 4 will be addressed by future research on the pH-dependence of τ_B and the effects of particle size on rheology.

CONCLUSIONS

The data presented here clearly show that high colloidal stability is needed for acceptable rheological properties of unary and binary slurries of ZrO_2 and Al_2O_3. The slurry pH is seen to be a critical processing parameter for maximizing the solids loading because it directly affects the absolute and relative zeta potentials that ensure stability. Although the colloidal stability of each oxide component affects the overall rheology, the lower stability of ZrO_2

dominates and increases the pseudoplasticity and Bingham yield stress in ZrO_2-rich systems. Lastly, strong electroviscous effects found in ZrO_2-rich slurries are not found in Al_2O_3-rich slurries at the same total solids loading.

ACKNOWLEDGEMENT

Research sponsored by the U.S. Department of Energy, Division of Materials Sciences and Assistant Secretary for Conservation and Renewable Energy, Office of Industrial Technologies, Industrial Energy Efficiency Division, under contract DE-AC05-84OR21400 with Martin Marietta Energy Systems, Inc.

REFERENCES

1. O. O. Omatete, M. A. Janney, and R. A. Strehlow, Am. Ceram. Soc. Bull. 70 (10), 1641-1649 (1991).
2. O. O. Omatete, A. Bleier, C. G. Westmoreland, and A. C. Young, Ceram. Eng. Sci. Proceed. 12 (9-10), 2084-2094 (1991).
3. (a) S. Baik, A. Bleier, and P. F. Becher, in *Better Ceramics through Chemistry II*, edited by C. J. Brinker, D. E. Clark, and D. R. Ulrich (MRS, Pittsburgh, 1986) pp. 791-800; (b) A. Bleier and C. G. Westmoreland, in *Better Ceramics through Chemistry III*, edited by C. J. Brinker, D. E. Clark, and D. R. Ulrich (MRS, Pittsburgh, 1988) pp. 145-154; (c) A. Bleier and C. G. Westmoreland, in *Better Ceramics Through Chemistry IV*, edited by B. J. J. Zelinski, C. J. Brinker, D. E. Clark, and D. R. Ulrich (MRS, Pittsburgh, 1990) pp. 185-190.
4. (a) A. Bleier and C. G. Westmoreland, J. Am. Ceram. Soc. 74 (12), 3100-3111 (1991); (b) A. Bleier, "Secondary Minimum Interactions and Heterocoagulation Encountered in the Aqueous Processing of Alumina-Zirconia Ceramic Composites", Colloids and Surfaces, In press.
5. (a) A. Bleier and C. G. Westmoreland, in *Interfacial Phenomena in Biotechnology and Materials Processing*, edited by Y. A. Attia, B. M. Moudgil, and S. Chander (Elsevier, Amsterdam, 1988) pp. 217-236; (b) W. C. Hasz, *Surface Reactions and Electrical Double Layer Properties of Ceramic Oxides in Aqueous Solution*, M.S. Thesis (MIT, Cambridge, 1983) 165 p.
6. (a) D. C. Henry, Proc. R. Soc. Lond. A133, 106-129 (1931); (b) A. L. Smith, in *Dispersion of Powders in Liquids*, 3rd ed., edited by G. D. Parfitt (Applied Science Publishers, London, 1981) pp. 99-148; (c) R. J. Hunter, *Zeta Potential* (Academic, London, 1981) p. 71.
7. (a) R. J. Hunter, *Foundations of Colloid Science*, Vol I (Clarendon, Oxford, 1987) pp. 76-89; (b) R. J. Hunter, *Foundations of Colloid Science*, Vol II (Clarendon, Oxford, 1989) pp. 992-1052; (c) Th. G. M. van de Ven, *Colloidal Hydrodynamics* (Academic, London, 1989) 582 p.
8. (a) J. Th. G. Overbeek, in *Colloid Science*, edited by H. R. Kruyt (Elsevier, Amsterdam, 1952) pp. 245-277; (b) Reference [7a], pp. 395-449.
9. H. C. Hamaker, Physica, 4, 1058-1072 (1937).
10. D. H. Everett, *Basic Principles of Colloid Science* (Royal Society of Chemistry, London, 1988) pp. 123-124.

PRODUCTION OF SPHERICAL POWDERS OF INORGANIC COMPOUNDS BY WATER EXTRACTION VARIANT OF SOL-GEL PROCESS

A. DEPTUŁA, J. REBANDEL, W. DROZDA, W. ŁADA, T. OLCZAK,
Institute of Nuclear Chemistry and Technology, Warsaw, Poland

ABSTRACT

A method of preparation of spherical powders (with particle diameters <100μm) of metal oxides,their homegeneous mixtures or compounds,composites and metals has been elaborated.Depending on the nature of cations, the starting solutions were prepared by extraction of anions with Primene JMT from salt solutions, by addition of dopants to the sols or by complexing with acetic acid or/and ammonia. The sols or broths were then emulsified in 2–ethyl–1–hexanol containing SPAN 80. Drops of the emulsion were solidified by extraction of water with the solvent. The process was carried out continuously in a laboratory and/or in a pilot plant.

INTRODUCTION

It is a little known fact that Heard[1] is the real inventor of the sol-gel process for the preparation of round particles of inorganic oxides. More than ten years later wide research programs were started in connection with production of nuclear ceramic fuels [2–5], especially in the form of oxide or carbide microspheres, generally with diameters more than 100μm. The fuels were planned to be used in High Temperature Gas Cooled Reactors (HTGR) as well as for low energy vibrocompaction of nuclear ceramics directly in the fuel rod. Very soon the sol-gel processes of production of those materials have attained technological maturity and some pilot plants have been constructed[6]. Relatively less attention has been paid to the preparation of inorganic materials in the form of medium-sized spherical powders with diameters from several to 100μm. The advantages of such powders over irregularly shaped ones are: (a) fluidity (b) low errosion effects and (c) small hydraulic resistance for liquids.

There are two basic families of the sol-gel processes [6]. The first is the classical process in which gels are formed from inorganic colloidal solutions (nearly all the mentioned above nuclear ceramic materials were prepared in this way). In the second process gels are prepared from metallorganic compounds, in particular from metal alkoxides.However if the latter process is applied to preparation of spherical powders ultrafine submicrone powders are obtained. They do not exhibit the advantages mentioned in points (a) and (c).Quite recently very interesting attempts to join both the processes (in order to prepare medium spherical powders) have been described by Sherif and Shyu [7].

In contrast to vast information concerning the equipment for production of sol-gel derived coarse microspheres [1–6] and also some submicron spherical powders [8,14] there are no available data on the equipment for production of medium spherical powders [9,10,11,12].

In the present paper the variant of the sol-gel process, elaborated in IChTJ in Poland, used for continuous production of medium-sized spherical powders of inorganic compounds is described.

EXPERIMENTAL

A simplified flowsheet of the elaborated process is shown in Fig. 1. For preparation of sols using continuous operation the extraction process of anions from metal salt solution seems to be the optimal. A commercial product Primene JMT (principaly $C_{18}H_{37}NH_2$) diluted 1:1 with petroleum was used as extractant.The equipment for continuous extraction is shown in Fig. 2. The external surfaces of the mixers and settleres,supplied by Glass Factory Wołomin (Warsaw), were covered with transparent heating layers. The layers permit the process to be carried out at elevated temperatures with visual control of its progress. The equipment used under laboratory conditions (feed rate of starting solution „SS" 1–2 l/h) differed from that used in the pilot plant (feed rate of SS 10–20 l/h) essentialy only with respect to the tube diameters of the mixer-settler. The practical volume ratio of organic to aqueous phases at which no precipitation of hydroxides took place was found experimentally. The extractant was regenerated by contacting with solution of Na_2CO_3.

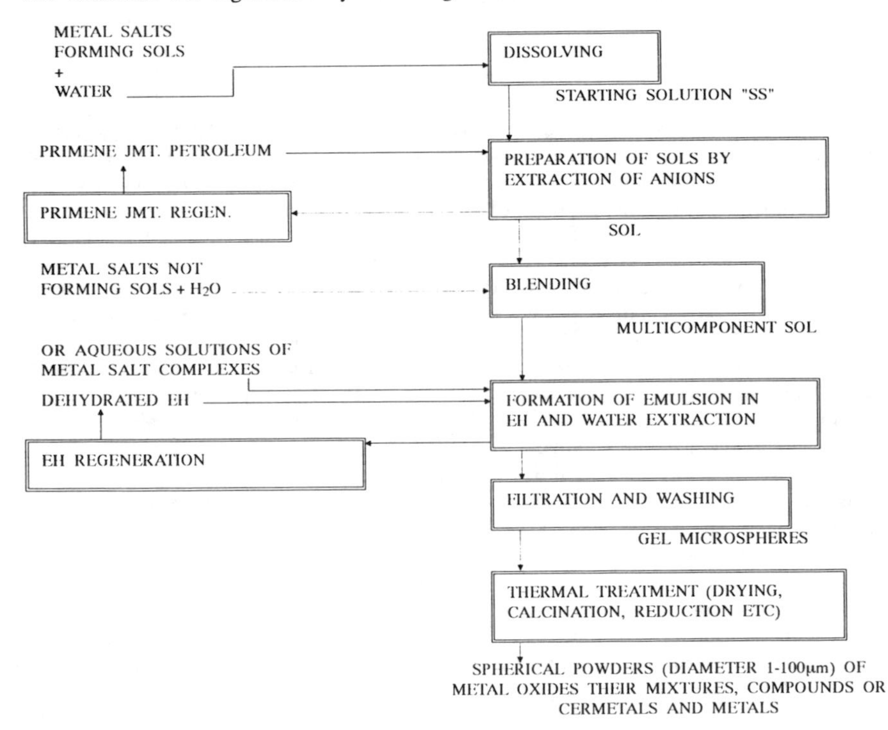

Fig.1. Simplified flowsheet of the IChTJ sol-gel process.

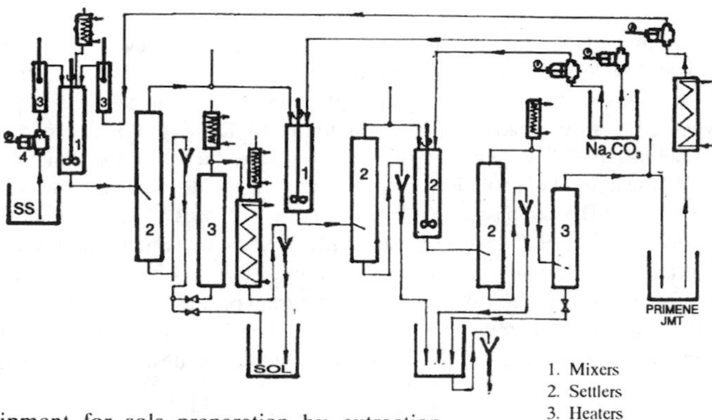

Fig.2. Equipment for sols preparation by extraction.

The gelation step was based on the 0RNL method of water extraction from the sol drops dispersed in an organic solvent [2–6]. Some modifications concerning the formation of sol emulsion in the solvent by vibrating capillaries and the vacuum regeneration process of the solvent were introduced [13]. In cases when cations did not form concentrated hydroxide sols, a complex solution (broth) were used for gelation. The equipment for gelation is shown in Fig. 3.

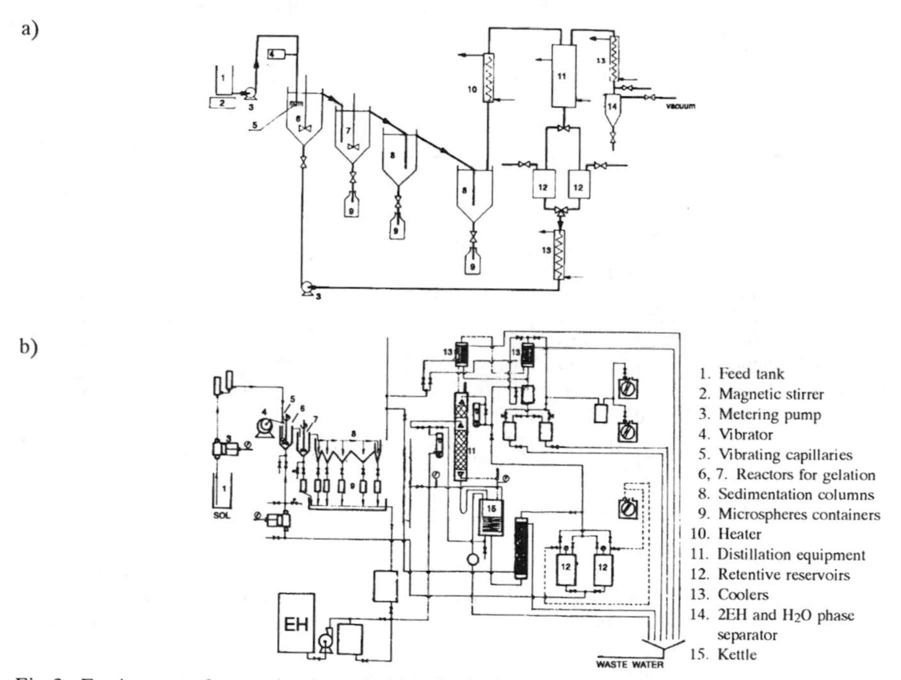

Fig.3. Equipments for production of gel spherical powders (a-laboratory 0.5-1 l of sol/h; b-pilot plant 5-10 l of sol/h)

First, an emulsion of the sol in 2–ethyl–1–hexanol, with an addition (1–2 % vol.) of the surfactant SPAN–80(EH), was prepared. Drops of emulsion were formed by introducing the sol(by means of vibrating capillaries) under the surface of a vigorously stirred solvent–EH. At the same time, the dehydrated EH (4g H_2O/l) in an amount of about 100 volumes per 1 volume of the sol was introduced to the reactor. A part of the water, contained in the emulsion drops was extracted to EH and the suspension of the gel particles in EH was passed to sedimentation columns. The supernant liquid (EH saturated up to 10–14 g H_2O/l) was sucked into a vacuum regeneration system and the water was evaporated at approx. 70°C. The dehydrated EH was recycled. The gelled particles were periodically removed from the bottom of the sedimentation column and washed with a suitable solvent. They were then submitted to thermal treatment under conditions and atmospheres suitable for preparation of the desired final product.

RESULTS AND DISCUSSION

Two principles seem to be important in preparation of sols: (1) production of highly concentrated sols (2) preparation of sols with the smallest possible anion:cation molar ratio. The detailed parameters for this step are summarized in Table I.

TABLE I.
Preparation of metal hydroxides sols by anion extraction process.

No.	Starting solution	Metal conc. M	Extraction parameters (step, temperature °C)			Final sol	
			I	Digestion	II	Metal conc. M	anion / Me molar ratio
1.	$AlCl_3$ or	2.5-					
	$Al(NO_3)_3$	-2.7	80	100,0.5h	80	3.5-4.0	0.1-0.4
2.	Al salts	1.5-,					
	$TiCl_4$(15%)	-2.5	20	100,0.5h	80	1.8-4.2	0.1-0.4
3.	Al salts	1.1-					
	$TiCl_4$(44%)	-1.7	20	100,0.5h	80	2.0-2.4	0.3
4.	$TiCl_4$	2.2	30			3.0	0.3
5.	$ZrO(NO_3)_2$	1.2	20			1.3	0.8
	or $ZrOCl_2$	2.2	20			2.5	1.3
6.	$Fe(NO_3)_3$	1.5	20			1.7	0.3
	or $FeCl_3$	3.0	20			3.4	0.6
7.	$CrCl_3$	2.1	20		80	2.8	0.5

Note. In ( ) % on a total oxide basis.

It can be seen that for the preparation of alumina sols two steps of extraction at elevated temperatures (with boiling of the aqueous phase between the steps) are necessary. This procedure is advantage out also for alumina-titania and chromia sols.It was observed that yttria sol cannot be prepared by extraction of anions because precipitation of hydroxides takes place upon addition of the extractant.

The above listed sols were gelled and calcined to respective oxides. In order to prepare mixed oxides or compounds, metal salts were added to the sols listed in Table I.

Sols prepared by complexation with acetic acid or/and ammonia were gelled as well. Examples of the prepared compounds are shown in Table II. In Fig. 4 microphotographs of some samples are shown.

TABLE II.
Sperical powders of metal oxides, their compounds, metals and cermetals prepared by sol-gel process.

No.	Starting sol	Dopant	Calcination °C	atmosphere	Final composition % *
1.	Alumina sol	Zr sol	900	air	Al_2O_3-ZrO_2(15-45%)
2.	" "	Cr "	900	"	" -Cr_2O_3(15%) green
			1650	"	" violet
3.	" "	Fe "	700	"	" -Fe_2O_3(7.5%)
4.	" "	" "	800	H_2	" -Fe(5%)
5.	" "	$MgCl_2$+YCl_3	900	air	" -MgO(0.3%)-Y_2O_3(0.3)
6.	" "	$(NH_4)_2MoO_4$	700	"	" -MoO_3(7.5%)
7.	" "	"	1100	H_2	" -Mo(5%)
8.	" "	$VOCl_3$	850	air	" -V_2O_5(15%)
9.	" "	$Ni(NO_3)_2$	900	"	" -NiO(10-50%)
10.	" "	"	900	H_2	" -Ni(5-40%)
11.	" "	$LiNO_3$	650	air	$LiAlO_2$
12.	" "	Li,Na,Mg salts	1400	"	β and β" alumina
13.	Zirconia sol	$CaCl_2$	800	"	CaO stabilized ZrO_2
14.	" "	YCl_3	800	"	Y_2O_3 " "
15.	" "	$CeCl_3$	800	"	CeO_2 " "
16.	Fe "	$Ni(NO_3)_2$	800	"	Fe_2O_3-NiO(10-30%)
17.	" "	"	900	H_2	Fe-Ni(25%)
18.	" "	YCl_3	800	air	$Y_3Fe_5O_{12}$
19.	" "	Zn,Mn nitrat.	1200	"	$Zn_{0.43}Mn_{0.5}Fe_{2.1}O_4$
20.	Titania "	$K_4Fe(CN)_6$	140	"	Ti hexacyanoferrate
21.	Ammonium wolframate		900	"	WO_3
22.	Natrium silicate →			gelation →	
	reaction with HCl →		900		SiO_2
23.	Y,Ba,Cu-Ac-NH_4OH	sol	900	(see[15])	YBCO superconductors
24.	Bi,Sr,Ca,Cu-Ac-NH_4OH	"	840	(see[16])	BSCCO "
25.	Ca^{2+}-Ac-PO_4^{3-}		900	(see[17])	Hydroxyapatite

* ( ) % on a total oxide basis.

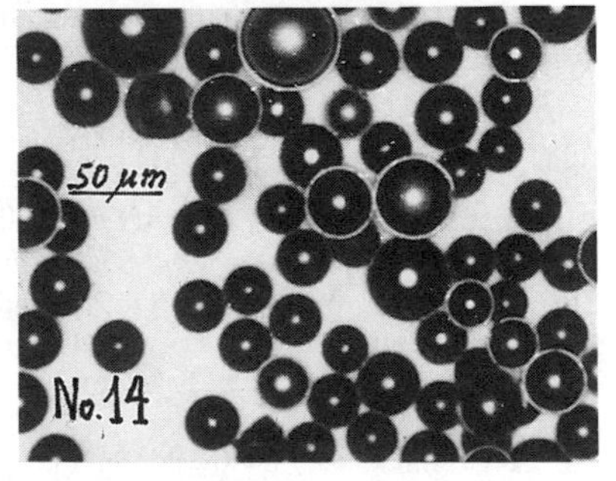

Fig.4. Microphotographs of some spherical powders.

Some of the spherical powders listed in Table II were used for:
(a) coating using plasma spraying technique [18,19,20],
(b) as a support for Pt catalyst [22],
(c) as a sorbent for ^{137}Cs for primary coolants of nuclar reactors [21].
These applications confirm the expected advantages of spherical powders over irregularly shaped ones.

CONCLUSIONS

An efficient method of production of spherical powders (diameter 1–100μm) of various inorganic compounds has been elaborated.It has been found that complex solutions (e.g. with ammonia and/or acetic acid) can be used as staring solution.

REFERENCES

1. Lewellyn Heard, U.S. Patent No.2 505 895 (2 May 1950).
2. I Processi Sol-gel per la Produzione di Combustibili Ceramici (Proc.Int.Nucl.Symp. Turin, Oct. 2–3, 1967) Comitato Nazionale Energia Nucleare, Rome,1968.
3. Sol-gel Processes for Ceramic Nuclear Fuels, (Proceedings of a Panel sponsored by IAEA) Vienna, 6–10 May 1968.
4. Symposium on Sol-gel Processes and Reactor Fuel Cycles, Gatlinburg, 4–7 May 1970, CONF–700502.
5. Sol-gel Processes for Fuel Fabrication, (Proceedings Panel, organised by IAEA, Vienna, 21–24 May, 1973), IAEA–161.
6. A. Deptuła and C. Majani, Sol-gel Processes and Their Applications, ENEA Report No. RT/TIB/86/25, Roma, 1986.
7. F.G. Sherif and L.J. Shyu, J.Am.Ceram.Soc. 74 375(1991).
8. J.H. Jean, D.M. Goy and T.A. Ring, Am.Ceram.Soc.Bull. 66, 1517(1987)
9. J.L. Wooddhead and D.L. Segal, Chemistry in Britain, April, p.310 (1984).
10. G. Wilson and R. Heatcote, Am.Ceram.Soc.Bull. 69, 1137(1990).
11. A.C. Pierre, Am.Ceram.Soc.Bull. 70, 1281(1991).
12. C.W. Turner, Am.Ceram.Soc.Bull. 70, 1487(1991).
13. A. Deptuła, H. Hahn, J. Rebandel, W. Drozda and B. Kalinowski, Polish Patent No.83 484 (20 September,1977).
14. E. Matijević, Annu. Rev. Materials Sci. 15, 483(1985).
15. A. Deptuła, W. Łada, T. Olczak, T. Żółtowski and A. Di Bartolomeo, in Better Ceramics Through Chemistry IV, edited by B.J.J. Zelinski, C.J. Brinker, D.E. Clark and D.R. Urlich (Mat. Res. Soc. Symp. Proc. 180, Pittsburgh, PA, 1990) pp. 907–912.
16. A. Deptuła, W. Łada, T. Olczak and A. Di Bartolomeo, MRS Spring Meeting, San Francisco, CA, April 27–May 1(1992).
17. A. Deptuła, W. Łada, T. Olczak and A. Di Bartolomeo J.Non-Crystalline Solids, in press (1992).
18. L. Golonka, L. Pawłowski, in European Hybrid Microelectronics Conference (Proc., Avignon, 1981) p. 373.

19. L. Pawłowski, Ch. Martin, P. Fauchais, 10–th Thermal Spraying Conference (Proc., Essen, May 1983) p. 167.
20. L.M. Vincent, Report SECMLG 10873/LMV, Fontanay aux Roses (1978).
21. A. Deptuła, B. Bartoś, A. Matyjek, Nukleonika 30, 121(1985).
22. A. Deptuła, presented at V-th Polish-Italian Seminar, Poland, Jeziorowskie, 1976 (unpublished).

NOVEL PRECIPITATION ROUTE TO SILICA GRAIN

Barbara Simms and Tom Gallo, Akzo Chemicals Inc., Precursors/Polymeric Materials, Dobbs Ferry, NY.

ABSTRACT

We describe a novel precipitation route to silica grain that lies in the interface between sol-gel and Stöber-type silica. The use of acetic acid as a catalyst for TEOS hydrolysis provides access to a precipitation window in which partially hydrolyzed TEOS and TEOS monomer, when reacted with aqueous ammonia, combine to form pumice-like silica particles in up to 90% yield as SiO_2. Precipitated particles exhibit narrow particle size distributions that may be controlled for average particle sizes from 50μ to 400 μ. SEM micrographs show that the particles are agglomerates of small particles, which is consistent with the high degree of observed macroporosity.

INTRODUCTION

A recent Chemical Abstracts search revealed 386 references to sol-gel chemistry of tetraethyl orthosilicate, TEOS. Most of this literature discusses acid or acid-base catalyzed routes to gel monoliths, or base catalyzed routes to Stöber-type [1] silica powders. These references reflect the commonly accepted belief that silica particles greater than 5 μ in size cannot be precipitated from TEOS hydrolysis mixtures. Thus, patents on the preparation of polymicron silica grain via sol-gel routes typically claim process steps for comminution [2], screening [3], or by capillary drying action [4] breaking apart a TEOS alcogel or xerogel to achieve large, uniform, silica particles. Emulsion routes can also provide large silica particles, but require the addition of hydrocarbon solvents and additives to the reaction mixture [5].

These references rely upon a mineral acid hydrolysis catalyst under conditions which favor full hydrolysis of the TEOS prior to onset of the gelation phase or base addition. Limited comparable data exists on the effects of weak acids such as acetic acid on the sol-gel chemistry of TEOS [6]. By studying the hydrolysis behavior of TEOS with acetic acid, we have discovered a general route to the precipitation of highly porous, narrow particle range, particulate silica with an available average particle size of from 50μ to 400μ.

EXPERIMENTAL

Reagents were as follows: Akzo Silbond Condensed TEOS (98% TEOS monomer, 1.5% ethanol, 100 ppm water, and less than 1% TEOS dimer by gas chromatography area %), Ethanol U.S.P (100%) , reagent grade acetic acid (99.9%), reagent grade aqueous ammonia (28 wt.%), deionized distilled H_2O.

Experiment 1

Hydrolysis mixtures were prepared as shown in Table I. Duplicates of each hydrolysis mixture were prepared for a total of six hydrolysis mixtures: 1.1a, 1.2a, 1.3a, 1.1b, 1.2b, 1.3b. The mixtures were prepared by combining the TEOS, ethanol, acetic acid, and water, into each of six 4oz glass jars with plastic screw caps. The jars containing mixtures 1.1a, 1.2a, and 1.3a were capped and placed in a shaker bath at 35 °C. The shaker bath was set at speed 8, which provided vigorous shaking action of approximately 150 Hz. The mixtures were allowed to react for 1.5 hours After 1.5 hours of reaction time 14.5 ml of 0.0355 M NH_3 were added to each jar. Similarly, the jars containing mixtures 1.1b, 1.2b, and 1.3b were capped and placed in a shaker bath at 33 °C set at speed 8.. The mixtures were allowed to react for 1.5 hours After 1.5 hours of reaction time, 14.5 ml of 0.355 M NH_3 was added to each jar. Following base addition, all reaction mixtures were replaced in the shaker bath at speed 8 at 33°C for an additional 18 hours. A total of 8.77 mol of H_2O per mol TEOS was added for each reaction mixture in the two-step hydrolysis. The precipitated products were filtered and dried in a convection oven at 130°C for 4 hours prior to sieve analysis. The yield of SiO_2 for the precipitated powders was as follows: 1.1b, 69%; 1.2b, 61.3%; 1.3a, 55.9%.

Table I. Hydrolysis mixtures for Experiment 1. (*) indicates that reaction mixture yielded precipitated product.

reaction mixture	TEOS (g)	ethanol (g)	acetic acid (g)	$H_2O(1)$ (g)	$H_2O(2)$ (g)	NH_3 (g)	T°C
1.1a	30.00	15.00	3.33	6.67	14.50	8.92E-03	35
1.2a	30.00	16.11	2.22	6.67	14.50	8.92E-03	35
1.3a*	30.00	17.22	1.11	6.67	14.50	8.92E-03	35
1.1b*	30.00	15.00	3.33	6.67	14.50	8.92E-02	33
1.2b*	30.00	16.11	2.22	6.67	14.50	8.92E-02	33
1.3b	30.00	17.22	1.11	6.67	14.50	8.92E-02	33

mol ratio TEOS:other	TEOS	Ethanol	Acetic Acid	$H_2O(1)$	H_2O total
1.1 (a,b)	1	1.92	0.43	2.76	8.77
1.2 (a,b)	1	2.06	0.29	2.76	8.77
1.3 (a,b)	1	2.21	0.14	2.76	8.77

Experiment 2

Hydrolysis mixtures were prepared as shown in Table II. Progress of the hydrolysis was monitored by disappearance of TEOS and increasing ethanol concentration by gas chromatography. Reaction mixtures 2.1-2.3 were prepared in glass jars and heated at speed 8 in a shaker bath as in Experiment 1. Reaction mixtures 2.4 were prepared in each of three 500 ml round-bottom three-neck Pyrex reaction flasks each equipped with a Heller mechanical stirrer, a water-jacketed condenser, and an addition funnel. The impeller speed for the stirred reactions was 198 rpm. The Teflon impeller blade measured 19 by 75 mm. The mixtures were hydrolyzed prior to g.c. analysis for the hydrolysis times shown in Table II. Immediately following the removal of a 1μl sample for g.c. analysis, aqueous ammonia was added to the hydrolysis mixtures. Shaker bath mixtures 2.1-2.3a and 2.1-2.3b received 14.5 ml of 0.355 M NH_3. Shaker bath mixtures 2.1c, 2.1b, and 2.3c received 14.5 ml of 0.0355 M NH_3. Reaction mixtures 2.4a and 2.4b received 90 ml of 0.355 M NH_3 while 2.4c received 90 ml of 0.0355 M NH_3. Following base addition, all reaction mixtures were mixed at 32°C for an additional 18 hours. A total of 8.77 mol of H_2O per mol TEOS was added for each reaction mixture in the two-step hydrolysis. The precipitated products were filtered and dried in a convection oven at 130°C for 4 hours prior to sieve analysis. The yield of SiO_2 for the precipitated powders was 90%. Precipitated powders contained 0.8% carbon, 1.7% hydrogen, 92% SiO_2.

Table II. Hydrolysis mixtures for Experiment 2. (*) indicates that reaction mixture yielded precipitated product. Hydrolysis time is time elapsed prior to ammonia addition. G.C. results are for samples taken immediately prior to NH_3 addition.

reaction mixture	mol ratio TEOS: acetic acid	hydrolysis time, min	ethanol g.c. %	TEOS g.c.%	mol ratio TEOS: NH_3	T°C
2.1a	0.43	90	60.9	35.7	5.15E-03	32
2.1b	0.29	90	48.3	47.3	5.15E-03	32
2.1c	0.14	90	44.8	52.4	5.15E-04	32
2.2a	0.43	180	81.7	17.2	5.15E-03	32
2.2b	0.29	180	61.5	35.4	5.15E-03	32
2.2c	0.14	180	52.1	44.4	5.15E-04	32
2.3a	0.43	240	93.3	5.6	5.15E-03	32
2.3b	0.29	240	80.6	17.3	5.15E-03	32
2.3c	0.14	240	64.7	31.7	5.15E-04	32
2.4a	0.43	120	75.6	22.6	5.15E-03	32
2.4b*	0.29	120	58.3	38.5	5.15E-03	32
2.4c*	0.14	120	60.0	39.2	5.15E-04	32

RESULTS AND DISCUSSION

The six hydrolysis mixtures in Experiment 1 yielded three precipitation products. Thus reaction mixtures 1.3a, 1.1b, and 1.2b yielded easily filterable white silica powder while the remaining mixtures produced gels. A sieve particle size analysis of these products is shown in Table III.

Table III. Sieve Analysis for Precipitated powders 1.3a, 1.1b, 1.2b, 2.4c

Particle Size, μ	1.1b	1.2b	1.3a	2.4c
>500	41%	0%	0%	8%
+180/-500	52%	46%	8%	77%
+106/-180	7%	35%	17%	13%
+45/-106	0	19%	48%	2%
-45	0%	0%	27%	0%

Of the reaction mixtures in Experiment 2 only reaction mixtures 2.4b and 2.4c yielded precipitated product. These reaction mixtures differed from the shaker bath mixtures or mixture 2.4a in the % of total TEOS hydrolyzed prior to base addition, as shown in Table II. The better mixing available with mechanical stirring at 198 rpm compared to the shaker bath accounts for the faster hydrolysis observed in the stirred reaction vessels. The particle size distributions for powders harvested from reactions 2.4b and 2.4c were similar. The particle size distribution for 2.4c is shown in Table III.

Porosity data for the product from reaction mixture 2.4c is presented in Table IV with comparable data for comminuted xerogel [2]. All of the precipitated

Table IV. Incremental Pore Volume Distribution (Desorption) for 2.4c.

Average Diameter, Å	% of Maximum Pore Volume	
	precipitated (2.4c)	gel
11.3	0.83%	0.51%
12	0.54	0.58
13	0.86	0.50
14	0.51	0.76
14.8	0.47	0.76
15.4	0.23	0.80
16.9	1.08	1.20
19.3	0.93	3.45
22.8	1.60	6.86
30.5	4.10	26.9
47.3	5.49	41.6
74.4	5.02	0.86
116.4	14.0	0.42
242.1	10.2	0.31
379.9	39.0	0.33
930.5	14.5	0.16
1355.5	0.29	0.45

products have similar morphology, pore size distributions, and residual carbon content. The xerogel used for comparison was prepared by a two-step HNO3 hydrolysis/NH_3 gelation process [2]. Percent carbon for the precipitated gel is 0.8% compared to 2.7% for the comminuted gel. Figure 1a shows the sponge-like appearance of the precipitated silica grain. Figure 1b reveals that each grain is composed of many smaller particles.

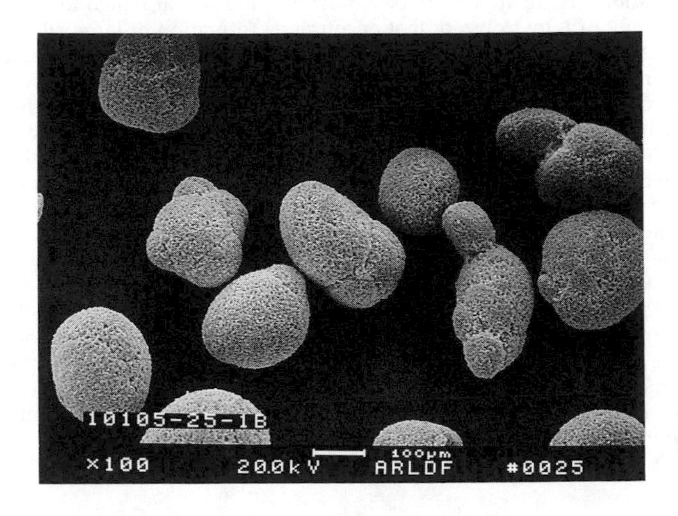

Figure 1a. SEM micrograph of precipitated silica.

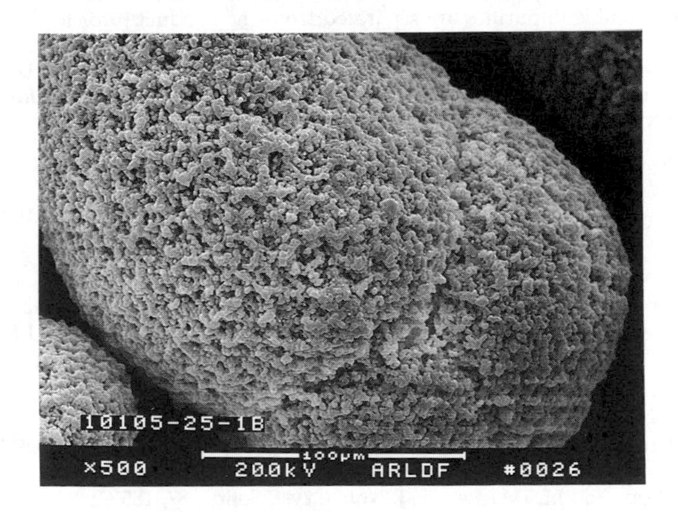

Figure 1b. Magnification shows sub-particle microstructure.

Results from Experiment 2 show a link between extent of hydrolysis of TEOS, and ratio of ethanol to TEOS in the reaction mixture, to the phenomenon of precipitation. There appears to be a narrow window, at which the TEOS is approximately 61% hydrolyzed, that encourages formation of large silica agglomerates upon addition of aqueous ammonia. That a certain percent of TEOS remains unreacted upon harvesting the silica powder is evident from yields ranging from 52% to 90% for the five precipitation products discussed here. The gels contain a high percentage of desirable macropores, but this is probably due more to the agglomeration mechanism than to any other factor. The similarity of these powders to Stöber silica in porosity and carbon content may be due in part to the higher pH of the initial hydrolysis reaction mixtures, as well as the 40% of TEOS which is hydrolyzed under basic conditions.

The two-step hydrolysis is critical to the formation of uniform silica particles. Other authors have experimented with acetic acid catalyzed gels without observing particle formation [6]. A single patent reference [7] describes formation of 50μ x 1μ needles from acetic acid catalyzed TEOS hydrolysis, but we obtained only loose gels when we attempted to repeat the patent examples. We postulate an agglomeration mechanism for the particle formation that depends on the presence of both TEOS monomer and TEOS oligomers at high pH, conditions provided by the two-stage acid-base hydrolysis.

CONCLUSIONS

We have discovered a general route to highly porous, uniform silica powders that lies in the interface between sol-gel and Stöber-type silica. The TEOS-derived powders can be easily harvested by conventional filtration, so that unhydrolyzed TEOS and soluble impurities are separated from the product prior to further processing. Following calcination, the ultra high-purity synthetic quartz grain is suitable for semiconductor glass applications. In addition its commercial attractiveness, the precipitation process described offers fertile ground for investigating cross-over mechanisms between acid and base hydrolysis of TEOS.

REFERENCES

1. Werner Stöber, Arthur Fink, and Ernst Bohn, J. Col.Interf. Sci., 26, 62 (1968).
2. Barbara L. Simms and R. Chihang Chu, U.S. Patent No. 5 017 354 (21 May 1991).
3. Sophia R. Su, and Leo F. Fitzpatric, U.S. Patent No. 4 943 425 (24 July 1990).
4. Paul M. Schermerhorn, Michael P. Teter, and Robert V. Vandewoestine, U.S. Patent No. 4 789 389 (6 Dec. 1988).
5. Eiji Hattori , Toru Tagawa, and Yasuo Oguri, U.S. Patent No. 4 865 829 (12 Sept. 1989).
6. E.J.A. Pope and J.D. Mackenzie, J. Non. Cryst. Solids, 87, 185 (1986).
7. Shin-Etsu Chemical Co., Ltd., Japanese Patent Laid-Open No. 184514 (19 July 1990).

THEORETICAL AND EXPERIMENTAL INVESTIGATIONS OF THE GROWTH OF SILICA AND TITANIA PARTICLES IN LOW MOLECULAR WEIGHT ALCOHOLS

Michael T. Harris, Osman A. Basaran, and Charles H. Byers
Chemical Technology Division, Oak Ridge National Laboratory, Oak Ridge, TN 37831

ABSTRACT

Silica and titania particles are readily synthesized by the hydrolysis of alkoxides in various alcohols. The present research studies experimentally and models the chemistry and growth kinetics of silica and titania particles in low molecular weight alcohols (methanol, ethanol, 1-propanol, 1-butanol). The theoretical model is based on solving the hydrolysis and condensation kinetic expressions and applying the method of moments to expedite the solution of the governing population balance equation.

INTRODUCTION

An understanding of the chemistry and physics of homogeneous precipitation from metal alkoxide solutions is essential to the production of good powders and to the development of theoretical models that describe this process. Numerous studies have been done on both silicon alkoxide and titanium alkoxide systems. Hydrolysis and condensation kinetics have been studied and major reaction species have been identified. The formation and growth of submicron particles have also been studied. As a result of the hydrolysis and condensation kinetic data and particle growth studies, it has been proposed that the formation of monodisperse silica and titania particles is governed by (i) the type polymer (e.g., highly crosslinked) formed during condensation and hydrolysis (TPF) [1], (ii) homogeneous nucleation (HN) [2,3], (iii) controlled aggregation (CA) [4,5,6], or (iv) the precipitation and collapse of polymeric silicate species to form dense particles and the deposition of the polysilicates onto the surface of the dense particles [7,8]. For brevity, the last mechanism will be termed the collapsed polymer-polymer deposition model (CPPD).

Investigators [2,8,9,10] have shown that spherical silica particles are formed in all aliphatic and branched low molecular weight alcohols during the ammonia catalyzed hydrolysis of TEOS. There is disagreement, however, as to the effect of solvent on the ultimate particle size. Tan *et al.* [9] suggested that the size of the alcohol solvent has little effect on particle size, while Harris and Byers [10] and Matsoukas and Gulari [3] observed an increase in particle size as the molecular weight of the solvent increased. The effect of different alcohols on the ultimate size of silica particles is not well understood.

In contrast to the case of formation of silica from alkoxides, the solvent has a profound effect on the shape and morphology of particles formed by the hydrolysis of "true" metal alkoxides such as titanium ethoxide and zirconium butoxide [11]. Harris et al. [11,12] have shown that spherical particles are formed in normal alcohols, ethanol and propanol. However, a ramified, nearly gel-like, precipitate is formed in tert-butanol. This effect is due primarily to the occurrence of alcoholysis. Nabavi et al. [13] state that alcoholysis results in the formation of oligomeric alkoxides in normal alkoxides and of monomeric alkoxides, due to steric hindrance, in the bulky tert-butyl alcohol. Furthermore, these authors [13] suggest that the difference in morphology and size distribution of titania particles results because nucleation and growth rates are decoupled

 ©1992 Materials Research Society

for trimeric species, such as $Ti(OEt)_4$ and $Ti(OBu^n)_4$, whereas they are coupled for monomeric titanium alkoxides $Ti(OBu^t)_4$. In addition to determining the effect of alcoholysis on the particle morphology, a model to predict condensation/precipitation and particle growth kinetics is also required.

The objective of this research is to evaluate the effects of precipitation kinetics and colloidal stability on the synthesis of submicron-monodisperse particles by metal alkoxide hydrolysis in normal alcohols (e.g., methanol, ethanol, 1-propanol and 1-butanol). Silica particles synthesized from TEOS are used as model systems for situations in which the hydrolysis reaction is easily controlled, while condensation is fast. Titania particles synthesized from titanium alkoxides serve as model systems when hydrolysis reaction is facile, while condensation is slow [14].

EXPERIMENTAL

Methanol, ethanol, 1-propanol, and 1-butanol were used as solvents and were of analytical reagent quality. The TEOS (Fisher Scientific Co.) was the purified grade. Titanium ethoxide from Alfa Products was used without further purification. Triple-distilled water was used in all studies. The ammonia (Mallinckrodt, Inc.) was of analytical grade.

A Tracor 550 Gas Chromatograph equipped with a flame–ionization unit was used to follow the production of ethanol. The chromatograph was operated with a column temperature of 130°C and a detector temperature of 175°C. A glass column packed with 1% SP1000 on 80/100 Carbopac C was used to separate the species involved in the reactions.

Absorbance readings for determining monomeric silicic acid by the molybdate method [12] and total soluble titania by the peroxide method were obtained with a Perkin–Elmer Model 200 Spectrometer at 400 nm.

THEORETICAL

A model for the formation of monodisperse silica by the base-catalyzed hydrolysis of TEOS must include consideration of both the hydrolysis and condensation reactions, which are written as follows:

$$Si(OC_2H_5)_4 + 4H_2O \xrightarrow[OH^-]{K_H} Si(OH)_4 + 4C_2H_5OH\ , \tag{1}$$

$$Si(OH)_4 \underset{K_D}{\overset{K_C}{\rightleftharpoons}} SiO_2 \downarrow +\ 2H_2O. \tag{2}$$

If ammonia is the catalyst, the rate expression for the hydrolysis reaction for dilute TEOS solutions is [15]:

$$\frac{\mathrm{d}[TEOS]}{\mathrm{d}t} = -K_H[H_2O]^{1.5}[NH_3]^{0.9}[TEOS], \tag{3}$$

where K_H is the hydrolysis rate constant.

The condensation-dissolution rate expression is given by

$$\frac{\mathrm{d}[Si(OH)_4]}{\mathrm{d}t} = -K_C([Si(OH)_4] - [Si(OH)_4]_e) - \frac{\mathrm{d}[TEOS]}{\mathrm{d}t}, \tag{4}$$

where K_C is the condensation rate constant and the subscript e stands for the equilibrium concentration.

In addition to the hydrolysis and condensation reactions, the following solvent dissociation and acid-base reactions are also possible:

$$2HS \rightleftharpoons H_2S^+ + S^-, \tag{5}$$

$$NH_4^+ + HS \rightleftharpoons NH_3 + H_2S^+, \tag{6}$$

$$NH_3 + H_2SiO_2(OH)_2 \rightleftharpoons NH_4^+ + HSiO_2(OH)_2^-, \tag{7}$$

where HS represents the treatment of ethanol plus water as one solvent (ethanol + water) [16].

Equilibrium expressions for Eqs. 5-7 and electroneutrality allows the computation of $[NH_4^+]$ which is the major cation that controls the Debye length. Solving for $[NH_4^+]$, the following expression is obtained

$$\begin{aligned}[NH_4^+]^3 + [NH_3]K_{Si}[NH_4^+]^2 - (K_{Si}[NH_3][TSi] + 10^{-7.5}[NH_3])[NH_4^+] \\ - 10^{-7.5}[NH_3]^2 K_{Si} = 0.\end{aligned} \tag{8}$$

In the case of titanium ethoxide, the hydrolysis reaction is very fast and condensation is the rate limiting step. During condensation, partial charge analysis suggests that the following alcoxolation and hydrolysis-alcoxolation reactions are possible:

$$Ti - -OR + HO - -Ti \rightarrow Ti - -O - -Ti + ROH \text{ and/or} \tag{9}$$

$$Ti - -OR + H_2O + RO - -Ti \rightarrow Ti - -O - -Ti + 2ROH. \tag{10}$$

Previous studies have shown that the induction period, t_i, varies inversely with the square of the concentration of total soluble titania $[TST]^2$ and inversely with the concentration of water $[H_2O]$. Thus, the rate expression for the [TST] is given by:

$$\frac{\mathrm{d}[TST]}{\mathrm{d}t} = K_{CT}[H_2O]([TST] - [TST]_e)^2 \tag{11}$$

where K_{CT} = condensation rate constant and $[TST]_e$ = equilibrium total soluble titania concentration.

Silica and titania particle formation and growth involve a plethora of possible reaction combinations. However, if several of the reaction combinations are lumped into one expression, such as a nucleation rate expression, the problem becomes more tractable. Thus nucleation is defined as the rate at which dense primary particles are formed by the removal of total soluble silica (i.e. silicic acid) or total soluble titania.

Once these nuclei are formed, it is postulated that growth occurs by controlled aggregation. These postulates are tested by a detailed model which simultaneously solves the hydrolysis and condensation rate expressions (Eqs. 3 and 4 for silica) and the population balance equations that govern the coagulation of Brownian particles with electrostatic repulsive forces [4,17,18]. The population balance equation is solved by the method of moments as outlined by Santacesaria et al. [17]. However, the model used here assumes that the aggregate and primary particles obey lognormal size distributions. Furthermore, an average collision efficiency factor, $\overline{\beta}_E$, is used.

RESULTS AND DISCUSSION

The variation of the silicic acid concentration during the hydrolysis of 0.1M TEOS in methanol, 1-propanol and 1-butanol is shown in Figures 1 and 2. The water and ammonia concentrations were 6.0 and 0.7M, respectively. These solutions were stirred throughout the reaction. Similar to the results of Bogush and Zukoski [4], the silicic acid concentration increases to a maximum value and subsequently decreases to an equilibrium concentration.

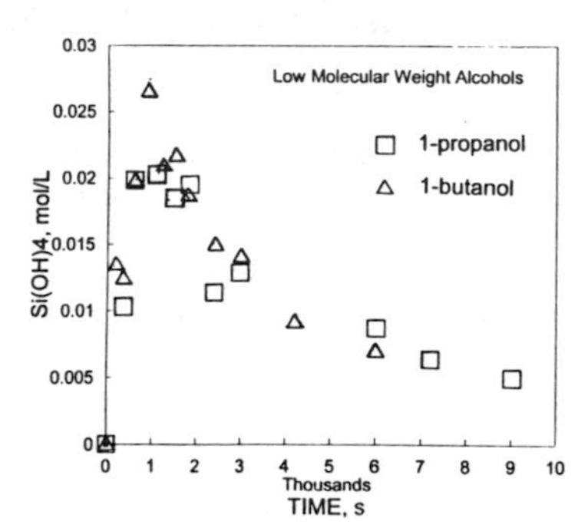

Figure 1. Experimental silicic acid concentration: TEOS hydrolysis in 1-propanol and 1-butanol (0.1M TEOS, 6.0M H_2O, 0.7M NH_3).

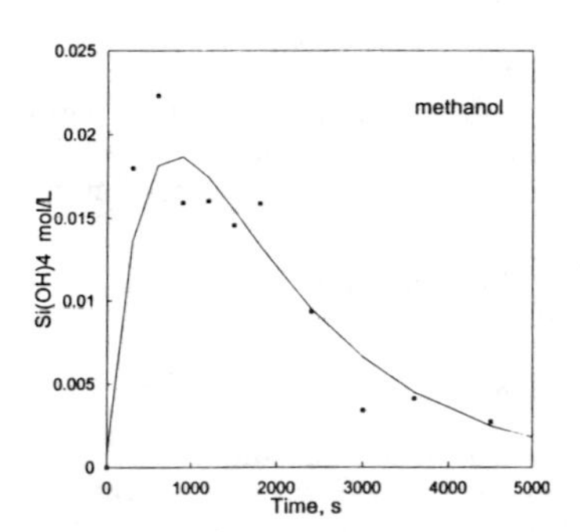

Figure 2. Predicted and experimental silicic acid concentration: TEOS hydrolysis in methanol (0.1M TEOS, 6.0M H_2O, 0.7M NH_3).

Figure 2 shows the experimental and predicted silicic acid concentration for silica synthesis in methanol. The hydrolysis and condensation rate constants were 0.0036 and 0.13 min^{-1}, respectively.

Figure 3 shows the predicted (population balance model) and experimental particle growth kinetics for TEOS hydrolysis in methanol and 1-propanol. The parameter values include a Hamaker constant of 1.0×10^{-21}J, a zeta potential of -46 mV, a primary particle size of 20 nm, and $K_{Si} = 8 \times 10^{-5}$. The dielectric constants of methanol and

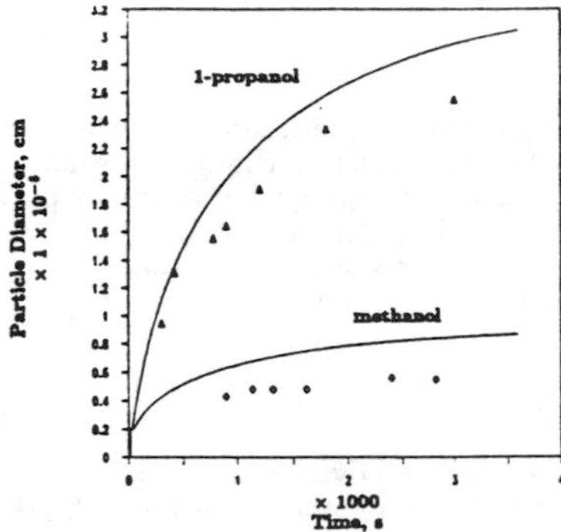

Figure 3. Predicted and experimental particle growth kinetics: TEOS hydrolysis in methanol and 1-propanol (0.1M TEOS, 6.0M H_2O, 0.7M NH_3).

1-propanol solutions were 43 and 24, respectively. The hydrolysis and condensation rate constants for TEOS hydrolysis in 1-propanol were 0.0031 and 0.092 $(mol/L)^{-2.4}min^{-1}$.

There is semi-quantitative agreement between the predicted and experimental particle growth kinetics. Both the model and data indicate that particle size increases with an increase in the size of the alcohol solvent. In 1-butanol, the particles grew to a size of 0.5 to 0.6 μm. The main difference between the solvents is the dielectric constant, which decreases with an increase in the alcohol's molecular weight. Solvent effects can therefore be explained by the controlled aggregation model which suggests that the particle size should increase with a decrease in the dielectric constant of the solvent.

The total soluble titania concentration is given in Figure 4. A fit of the data to Eq. 11 is also shown. The condensation rate constants were 3.8 and 4.2 min^{-1} $(mol/L)^{-1}$ for the hydrolysis of TIE with 0.4 M H_2O and 1.3M H_2O in ethanol. Thus, the condensation of TIE appears to be governed by an hydrolysis-alcoxylation reaction.

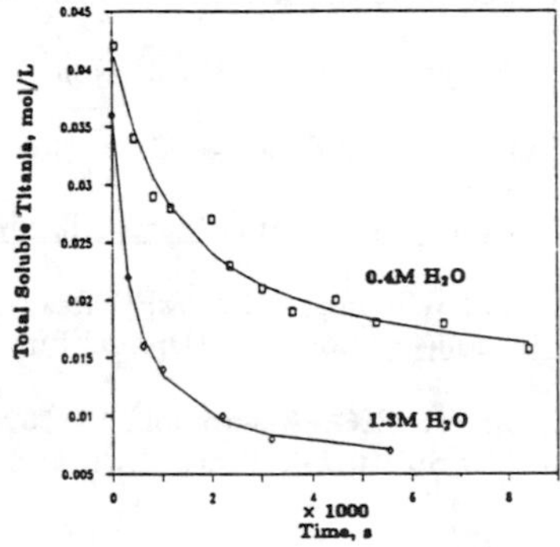

Figure 4. Total soluble titania concentration: TIE hydrolysis in ethanol (0.04M TIE).

These rate constants were used in the chemical kinetic-population balance model to show the rapid growth of titania particles to a particle size of approximately 0.1 to 0.2 μm. Unlike the silica particles, these "larger" particles continued to aggregate, albeit at a slower rate, because of insufficient repulsive forces for colloidal stabilization. Similar results were obtained for the hydrolysis of TIE in 1-butanol.

CONCLUSIONS

The ultimate size of silica particles in methanol, 1-propanol and 1-butanol is governed by the controlled aggregation mechanism. Higher molecular weight alcohols have lower dielectric constants which promote aggregation to larger particle sizes.

The condensation of titanium alkoxides is governed by a hydrolysis-alcoxylation type reaction. Aggregative growth models show that titania particles rapidly grow to 0.1 to 0.2 μm particles in ethanol and 1-butanol.

ACKNOWLEDGMENTS

This research was sponsored by the Division of Material Sciences, Office of Basic Energy Sciences, U.S. Department of Energy under contract DE-AC05-84OR21400 with Martin Marietta Energy Systems, Inc.

REFERENCES

[1] Iler, R.K., The Chemistry of Silica, (John Wiley and Sons, New York, 1988).
[2] Barringer, E.A. and Bowen, H.K., Langmuir, 1, 414 (1985).
[3] Matsoukas, T. and Gulari, E., J. Colloid Interface Sci., 124, 252 (1988).
[4] Bogush, G.H. and Zukoski, C.F., J. Colloid Interface Sci., 142, 19 (1991).
[5] Look, J.L., Bogush, G.H., and Zukoski, C.F., Faraday Discussion Chem. Soc., 90, 345 (1990).
[6] Bailey, J.K. and McCartney, M.L., Mat. Res. Soc. Symp. Proc., 180, 153 (1990).
[7] Bailey, J.K. and McCartney, M.L., in press.
[8] Stober, W., Fink, A. and Bohn, E., J. Colloid Interface Sci., 25, 62 (1968).
[9] Tan, C.G., Bowen, B.D. and Epstein, N., J. Colloid Interface Sci., 118, 290 (1987).
[10] Harris, M.T. and Byers, C.H., J. Non-Cryst. Solids, 103, 49 (1988).
[11] Harris, M.T., Byers, C.H., and Brunson, R.R., Mat. Res. Soc. Symp. Proc., 121, 287 (1988).
[12] Harris, M.T., Brunson, R.R., and Byers, C.H., J. Non-Cryst. Solids, 121, 397 (1990).
[13] Nabavi, M., Doeuff, S., Sanchez, C., and Livage, J., J. Non-Cryst. Solids, 121, 31 (1990).
[14] Bradley, D.C., Mehrotra, R., and D.P. Gaur, Metal Alkoxides, (Academic Press, New York, 1978).
[15] Byers, C.H., Harris, M.T., and Williams, D.F., I&EC Res., 26 (9), 1916 (1987).
[16] Tremillon, Chemistry of Nonaqueous Solvents, (Dreidel Publishing Company, Dordrecht, Holland, 1974).
[17] Feeney, P., Napper, D.H., Gilbert, R.G., Macromol., 17, 2520 (1984).
[18] Santacesaria, E., Tonello, M., Storti, G., Pace, R.C., and Carra, S., J. Colloid Interface Sci., 44, 111 (1986).

THE FAST SOL-GEL SYNTHETIC ROUTE TO SUPPORTED GLASS FILMS: SYNTHETIC FEATURES, SCOPE, APPLICATIONS AND MECHANISTIC STUDIES

Y. Haruvy† and S.E. Webber
Department of Chemistry and Biochemistry and Center for Polymer Research, The University of Texas at Austin, Austin, TX 78712, USA
† On sabbatical-leave from Soreq NRC, Yavne 70600, Israel.

ABSTRACT

The fast sol-gel synthetic route to glass films is facilitated by a solvent-free reaction of trialkoxysilane monomers at 60-80°C and near-stoichiometric water-to-siloxane ratios. The viscous polymer formed within several minutes is spin-cast onto a support (typical thickness 10-25µm), cured within a few hours at 60-70°C or 10-40 min. under an intense electric field at room temperature, and can be characterized on the same day. Multi-layered glass-film assemblies of varying indices of refraction and a total thickness up to 80µm are prepared by the same fast process with slight modifications.

Guest molecules such as laser dyes for 2D lasers and chromophores for second-harmonic-generation are incorporated into these glass films at high concentrations by direct addition to the reaction mixture. Their absorbance and fluorescence spectra are typical of monomeric chromophores up to concentrations of $\sim 10^{-2}$M. Crystallites that may appear at higher concentrations exhibit unusual polymorphism. The smoothness of such films enables surface anchoring of iso-environment photoactive molecules and investigation of their photophysics.

^{29}Si-NMR studies of the hydrolysis pattern in our acid-catalyzed fast sol-gel reaction prove that uniform hydrolysis of the monomer, necessary for a single-phase polymerization, is attained only when the hydrolysis is fast enough to become kinetically controlled, e.g. at elevated temperature or at high acid-concentration. Our reaction pattern and the data of others can be interpreted in terms of a single mechanism of hydrolysis in which two different rate-determining-steps can prevail, depending predominantly on the reactivity of the alkoxysilane and the ratio of the acid-catalyst to the (apparently-inert) proton-binding solvent.

INTRODUCTION

The sol-gel synthetic routes to glasses and ceramics have been extensively investigated and comprehensively reviewed [1-5], alkoxysilanes being the most investigated precursors. Chromophores incorporated into sol-gel produced glassy materials have exhibited promising characteristics for use in non linear optics such as laser systems [6-8]. Of special interest were supported thin films embodying such chromophores, aiming at applications such as surface laser systems and second harmonic generation (SHG) [9-10].

Tetraalkoxysilanes are the precursors for inorganic glasses capable of encaging large molecules (e.g. chromophores, enzymes). Cracking and fragmentation are the common obstacles in the fabrication of such glass articles and films, supported films in particular, due to extensive volume-contraction which accompanies the expulsion of solvent and condensation products [11]. Prolonged and complex processes as well as additives or surfactants [12-13] are necessary to facilitate crack-free supported glasses of desired thickness and optical properties.

In this work we present a fast and convenient synthetic sol-gel route to glass thin-films attached to a rigid support and discuss its key features and the special properties of the glass produced. The processes and scope of embodying guest-molecules in the fast sol-gel derived glasses are further discussed as well as the process-tuning to facilitate multilayered films and waveguides. Incorporation, chemical binding and electric-poling of donor-acceptor chromophores into the fast sol-gel matrix, aiming at SHG devices, or onto its surface, aiming at tethered DA systems are further reviewed. Finally, the ^{29}Si-NMR investigation of the hydrolysis pattern unique to the fast sol-gel reaction is described and the conclusions derived thereafter are discussed.

EXPERIMENTAL

Alkoxysilanes [14], p-nitroaniline and 4,4' diaminodiphenylsulfone were purchased from Aldrich, and laser dyes were purchased from Lambdachrome. All the materials were used without further purification. The experimental set-ups were previously described in detail: sol-gel reaction set-up and preparation of support glasses [15], electric poling set-up and SHG measurements [16] and NMR spectroscopy [17]. In general, the sol-gel reactions, under acid (HCl) or base (Me_2NH) catalysis, were carried out in disposable vials immersed in a thermostated bath and magnetically stirred. The progress of the hydrolysis and condensation reactions was monitored gravimetrically and when the reaction product turned viscous it was spin-cast on a precleaned preheated support. The samples were then left to cure until their surface was no longer sticky, and their thickness was measured as previously described [14].

OBSERVATIONS, RESULTS AND DISCUSSION

Fast Sol-Gel Synthetic Principles

The three primary principles on which the crack-free fast sol-gel synthesis is established are volume minimization, stress relaxation and phase homogenization.

Volume Minimization. The first objective in cracking elimination was minimization of the volume of the reactants. As shown in Table I, we can approach this goal by eliminating solvent, selecting the smallest alkoxide-groups and near-stoichiometric water-to-siloxane ratio. We may also evaporate part of the alcohol produced before casting the film and further decrease the volume contraction of the drying gel. A fringe benefit of this approach is rate-enhancement of all the sol-gel reactions due to increase in concentration of the reactants.

Table I. Calculated Contraction during Sol-Gel Preparation of Glass

Reactants	Volume Contraction (a)	Longitudinal Contraction
$Si(OEt)_4/H_2O/EtOH$ 1 : 4 : 4 (m/m)	7.7	2.0
$Si(OMe)_4/H_2O$ 1 : 2 (m/m) (b)	3.1	1.5
$Si(OMe)_4/H_2O$ 1 : 2 (m/m) -2MeOH	2.1	1.27
$MeSi(OMe)_3/H_2O$ 1 : 1.5 (m/m) -2MeOH	1.5	1.14

(a) for calculation convenience all densities are taken as 1. (b) single-phase is formed within 30 s.

Stress Relaxation. Film cracking and detachment from the support were substantially reduced by minimizing the volume of the reactants yet not eliminated since in supported films the entire volume contraction is restricted to one dimension. Small film-sections free of cracking or detachment have been observed upon slow and cautious drying, implying the occurrence of some stress-relaxation, probably via bond-reformation and segment-relocation processes. Obviously, to eliminate cracking and detachment we need to enhance stress relaxation.

Three principle processes can account for stress-relaxation in such polymer: Si-atom relocation, O-Si-O-Si-O segment relocation, and relocation of gel-particles [15]. Surfactants [9, 12] probably facilitate particles-segregation and relocation thereby, while drying control agents [13] probably allow sufficient time for segment relocation processes, thus allowing stress-relaxation throughout the drying stage. Since our goal was an additive-free glass, we preferred an alternative to these earlier approaches: to impart the sol-gel the capability for Si-atom relocation, thus directly maintaining gel-reformation and stress-relaxation. The precondition for such an inherent mobility is that a substantial part of the Si atoms, at least part of the time, will be bound to the polymer by only two substituents, while the other two remain free to move, as illustrated in Figure 1.

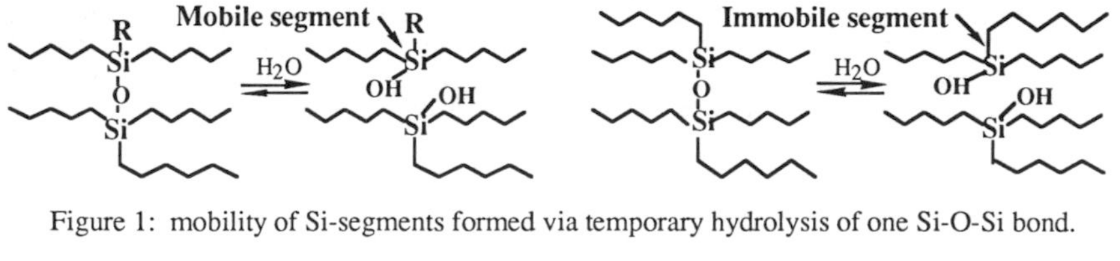

Figure 1: mobility of Si-segments formed via temporary hydrolysis of one Si-O-Si bond.

Tetraalkoxysilanes can form four bonds to the matrix, making the probability to find two bonds simultaneously open extremely low. To allow relaxation we must limit the number of bonds Si–gel to a maximum of **3** and thence **2** upon temporary rehydrolysis (mobile segment). To impart this mobility, we need to block one substituent, preferably with a small-volume group. The smallest stable single blocked alkoxysilane is MTMS [$MeSi(OMe)_3$], the sol-gel process of which was investigated long ago for cladding laser rods but abandoned being "too slow" [18]. The small methyl substituent is sufficient for our objective to block one Si-bond, yet it does not introduce an organic-polymer modification of the silicate [8] which is not necessary for our purposes. Crack-free supported films (up to 25μm thick) were attained right from the first sol-gel preparations with this monomer and were stable and shock-insensitive throughout the drying stage.

Phase Homogenization. Sol-Gel reactions of MTMS under acid catalysis and near-stoichiometric water-to siloxane ratios (MR) were found typically free of phase-separation [15]. It did occur *after* the film-casting when the sol-gel was prepared at *room temperature*, manifested in a hydrophobic liquid expelled from the gel and later solidifying into an opaque layer. This implied that a substantial amount of unhydrolyzed monomer was still present in the gel that separated from the cast film following rapid evaporation of the methanol from its surface.

MTMS hydrolysis at low temperatures apparently did not produce a uniform population of semi-hydrolyzed intermediates, but rather a mixture rich in $MeSi(OH)_3$ and $MeSi(OMe)_3$, the latter being expelled from the gel as it undergoes condensation. Such hydrolysis pattern can be argued in terms of increasing Si reactivity with increasing number of hydroxy groups, yet an ^{29}Si-NMR study is necessary to verify this [17]. The practical question is how to manipulate the molecules to undergo uniform partial hydrolysis regardless their reactivity. Since the latter reflects the reaction energy of activation we achieve the kinetically controlled reaction pattern by raising the temperature or by enhanced catalysis, thus attaining primarily the desired semi-hydrolyzed products. Indeed, the post-hydrolysis phase separation was eliminated by carrying out the sol-gel reactions at 70-80°C. This principle of temperature manipulation of the hydrolysis pattern to maintain phase-homogenization after the film-casting (at room temperature) has not been suggested previously.

Special Properties of the Polymethysiloxane (PMSO) Glass

Hydrolyzed MTMS monomers carry three hydroxy groups and an intact methyl group each that allow amphiphilicity of the surface [22]. The wet PMSO surface can undergo rearrangement to match the environment it is facing, as demonstrated in Table II. It is evident that PMSO surface exposed to air while drying turns to be a methyl-rich hydrophobic surface.

Table II. PMSO–Water Contact Angles

Contact Surface	Contact Angle (deg) (a)
Air	80 ± 2
Glass (b)	68 ± 2
Parafilm (b)	86.5 ± 0.5 (c)
Cellophane (b)	19 ± 3

(a) using a 200 μl droplet
(b) PMSO film mechanically separated from the contact-surface when cured.
(c) compared to 105-114 deg. for paraffin.

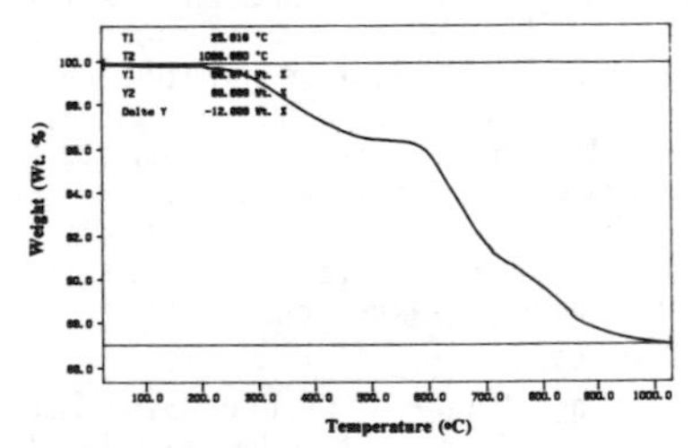

Figure 2. TGA spectrum of the PMSO glass

Sol-gel fast reactions at low MRs are known to proceed preferably via a linear polymerization pattern [19]. The MTMS monomer allows also formation of linear ladder-like polymer segments with pendant methyl-groups [20] which are hindered from cross-linking. These segments probably endow the mechanical stability and flexibility of the resultant glass films and allow facile and rapid curing of relatively thick supported films. This explains also the unique capability of PMSO glasses to embody both hydrophilic and hydrophobic guest molecules at high concentrations. It imposes, however, some difficulty to prepare multilayered film assemblies and waveguides as will be discussed later.

The PMSO supported-films (ca. 25 μm thick) exhibit extraordinary thermal stability: two thermal cycles between liquid nitrogen and room temp. are needed for cracks to develop slowly. Their TGA at the range 28-1028 °C displayed in Figure 2 above shows weight loss of 12.9%:

ca. 4% loss up to 500 °C can be attributed to continued condensation and loss of methanol and the rest to oxidation of the methyl-groups (10.4% at most) or their pyrolysis (to carbon or silicon-carbide) evident by the dark coloration of the samples at the end of the experiment.

Scope of the Fast Sol-Gel Synthesis

The fast sol-gel route is not limited to the MTMS monomer. Testing the principle of "three hydrolyzable substituents" we have applied it to mixtures of tetramethoxysilane (TMOS) and dimethyldimethoxysilane (DMDMS) employing exactly the MTMS reaction and drying procedures and keeping the water-to-siloxane molar ratio at 1.5. All the films severely cracked as long as the ratio TMOS : DMDMS was above 1, whereas at ratios ≤ 1 all films were crack-free, implying that it is the *average* number of hydrolyzable substituents that must be kept ≤ 3. Further, we have extended the fast sol-gel reaction of trimethoxysilane [21]. Crack-free drying of the supported film (on a flexible-support) followed by oxidation (and condensation) of the Si-H groups resulted in glass-films that were free of organic-absorption-peaks and improved optical quality.

A wide spectrum of guest molecules could be incorporated in the fast sol-gel derived matrices and embodied as discrete molecules at very high concentrations by incorporating them in the reaction mixture. This usually has a marginal effect on the fast-sol-gel process itself, e.g. for laser dyes at a typical loading of 0.2 wt% in the monomer (ca. 10^{-2} M in the dry glass). In general, proper dissolution of the guest molecules in the polymerizing siloxane results in glass that exhibits absorbance and fluorescence spectra typical of discretely embodied species. Thus, donor-acceptor type molecules such as p-nitroaniline (PNA) and PNA-allyl-derivatives were embodied in the glass-films at loadings as high as 10-20% retaining the optical quality for the latter. Further, guest molecules which interact with acid-catalysts were successfully incorporated into the glass employing a modified sol-gel synthetic route using the strong base dimethylamine as catalyst [15].

Metal ions (e.g. Cu, Tl, Ce) were also easily incorporated into the fast sol-gel glass films. Since these ions induce sol-gel precipitation they were incorporated at a later stage of the reaction and additional polymerization was allowed for a few seconds only. The resulting glass-films exhibited the typical absorbance of the metal salts while retaining the optical clarity of the glass.

Scope of the Fast Sol-Gel Applications

Multilayered glass-films and waveguides were attained by successive casting of several gel-layers, applying the same sol-gel and drying processes of a single-layer for films up to 50μm thick. A 10% increase in the water-to-siloxane ratio is crucial to ensure hydrophilicity of the films' surface and sufficient interlayer-adherance and eliminate upper-layer gliding and formation of stripes [22]. For an efficient wave-guide we must produce a higher index of refraction in the middle layer so that the other two act as cladding layers. This higher index of refraction was attained by the incorporation of dyes or aromatic monomers (e.g. phenyltriethoxysilane). In such waveguides dye-fluorescence intensities in front-face vs. right-angle measurements differ only by a factor of 10 despite an area-factor of 1000 between the fluorescing surface areas, since most of the light intensity was indeed guided through the dye-layer and only a small portion came out through its surfaces.

Laser features of the three-layered assemblies were measured using a fiber optic which was rotated with the respect to the film edge, facilitating a comparison of the "bulk" fluorescence emitted from the surface of the film with that emitted from its edge under the same conditions, as illustrated in Figure 3.

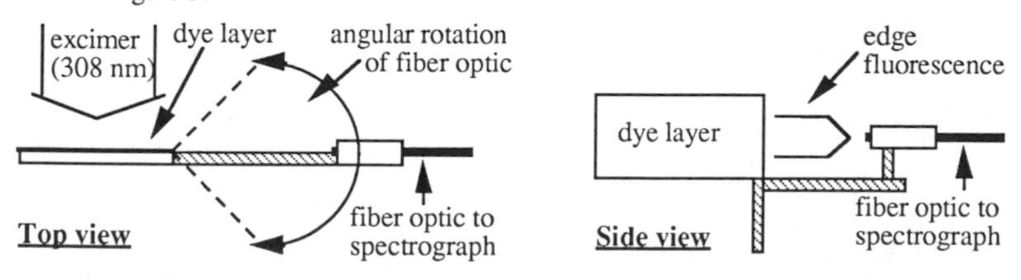

Fig. 3. Optical set-up for the measurement of band-narrowing and lasing-angle width.

The emission that emerged parallel to the film surface had an angular half-width of ca. 6°, somewhat red-shifted and more symmetric than the bulk fluorescence, despite the non-polished edges of our films. These results are very similar to those of Knobbe et al. [23] for dyes dispersed in composite sol-gel/organic polymer glasses and are typical of dye lasing in the waveguide cavity.

Second harmonic generation (SHG) is facilitated by donor-acceptor (DA) molecules organized in a non-centrosymmetric distribution [24]. This can be attained by embodying DA molecules in a host matrix, then orienting them to produce a non-centrosymmetric pattern, most conveniently by poling them with an intense electric-field [24, 25]. Samples of PMSO-embodied PNA cast on a metallic support were poled (while fresh) until their surface appeared cured [16]. Thin films (1-2 μm) with ca. 3% PNA were optically clear and exhibited SHG signal 2-10 times stronger than a comparable powdered urea, whereas negligible SHG was observed without poling. Thicker films (10-20 μm) demonstrated SHG with or without poling and strong light-scattering by PNA-crystallites inside the PMSO.

The molecular dispersity of PNA in the thin films and their efficient poling-orientation were argued in terms of a threefold effect of this thickness: a tenfold increase in the electric field across the thinner film; accelerated curing due to faster escape of the methanol; accelerated field-induced curing due to the stronger field (see discussion later on). The peculiar SHG observed in thick non-poled (oven-cured) films was attributed to non-centrosymmetric PNA polymorphic crystallites formed in the PMSO sol-gel films together with its regular centrosymmetric crystalline form [24]. PNA-allyl derivatives facilitate crystallite-free loading of 15-20% in the PMSO, as well as chemical binding through HSi-monomers, yet their poling is more difficult [26] and is currently being investigated.

Accelerated field-induced curing was observed in PMSO gels subjected to the high electric field used for the SHG studies. A typical sol-gel recipe film became crack-free cured under these conditions 100x faster: within 10-20 min. at room temperature [27]. This dramatic curing strongly depends on the water-to-siloxane ratio in the sol-gel recipe and its mechanism is believed to involve bond-orientation, ion-migration and surface corona-electrolysis [28] and is currently being studied for other condensation polymerization processes (e.g. polyesters, polyimides).

Tethered donor-acceptor molecular assemblies A perfectly smooth surface comprising amino groups was prepared by the fast sol-gel method using mixtures of methyltrimethoxysilane and 3-aminopropyltrimethoxysilane. Donor molecules diisothiocyanate derivatives were anchored to the glass surface via thiourea groups. Tethered donor-acceptor assemblies were then prepared using a diamino or dialcohol long-chain tether-molecule and an isocyanate-derivatized acceptor molecule [29]. The photophysics of siloxane-anchored naphthalene-tether-anthracene assemblies is being studied to determine the correlation between the tether-length and their energy-transfer dynamics.

Mechanistic Studies of MTMS Hydrolysis Using ^{29}Si NMR

Our synthetic observations and the corresponding mechanistic interpretations [15] comply with those of Pouxviel et al. [30], yet are apparently in contradiction to others [31-32]. This conflict suggests three key questions: what is the reaction mechanism; what is the rate determining step; and is the hydrolysis rate increasing [30] or decreasing [32] with successive hydrolysis ? We could interpret the results of various investigators in terms of the S_N2-Si mechanism [4: p. 130] in which different rate controlling steps manifest under different experimental conditions [17]:

$$(RO)_3Si{-}OR \underset{(1)}{\overset{H^+}{\rightleftharpoons}} (RO)_3Si{-}\overset{+}{O}HR \underset{(2)}{\overset{H_2O}{\rightleftharpoons}} \overset{+\delta}{H}\overset{}{O}R\cdots Si(OR)_3\cdots O\overset{+\delta}{H_2} \underset{(3)}{\overset{-ROH}{\rightleftharpoons}} (RO)_3Si{-}\overset{+}{O}H_2 \underset{(4)}{\overset{-H^+}{\rightleftharpoons}} (RO)_3Si{-}OH$$

The number of alkyl-groups predominates the reactivity of alkylethoxysilanes [32] implying that the protonation (1) controls the overall hydrolysis rate due to differences in inductive effect. The number of alkoxy-groups predominates the hydrolysis rate of tetramethoxysilane [Assink et al., 31] and hence the reaction rate decreases with degree of hydrolysis (pattern I). An opposite order (pattern II) was observed for tetraethoxysilane [Pouxviel et al., 30]. This pattern of increasing rate with hydrolysis stage (predicted by Schmidt [33]) should be attributed to the slower hydrolysis of ethoxysilanes (due to steric hindrance) that allows the nucleophilic attack (2) to become thermodynamically-controlled and thus a rate controlling step.

In our ^{29}Si-NMR measurements [17] we have found that the fast-reacting $EtSi(OMe)_3$ exhibits the hydrolysis pattern I whereas the slow-reacting $MeSi(OEt)_3$ exhibits the hydrolysis-pattern II in which large portion of the monomer remains unhydrolyzed. Hence, although the different hydrolyzed species of MTMS itself can not be resolved, it becomes reasonable that its hydrolysis will result in a substantial residue of unhydrolyzed monomer under relatively slower reaction conditions (i.e RT). This hydrolysis will follow the desired hydrolysis pattern I with negligible residue of unhydrolyzed monomer under conditions of accelerated hydrolysis (80°C) essential for the phase-homogeneity of the fast sol gel reaction.

ACKNOWLEDGEMENT

The authors would like to acknowledge the financial support of the State of Texas Advanced Technology Program (Grant numbers 003658-394 and 003658-245A).

REFERENCES

1. Better Ceramics Through Chemistry, edited by C.J. Brinker, D.E. Clark and D.R. Ulrich, Mater. Res. Soc. Symp. Proc., 32 (1984); 73 (1986), 121 (1988).
2. Glasses and Glass Ceramics from Gels, Proc. Int. Workshop, *J. Non-Cryst.Solids*, 48 (1982); 63 (1984); 82 (1986); 100 (1988); 121 (1990).
3. Sol-Gel Technology for Thin Films, Fibers, Preforms, Electronics and Specialty Shapes, edited by L.C. Klein (Noyes Publ., Park Ridge, NJ, 1988).
4. C.J. Brinker and G.W.Scherer, Sol-Gel Science: the Physics and Chemistry of Sol-Gel Processing (Academic Press, San-Diego, CA, 1990).
5. Science of Ceramic Chemical Processing, Edited by L.L. Hench and D.R. Ulrich (Wiley, New-York, NY, 1986).
6. R. Reisfeld, J. Phys. Coll. C7 (suppl. 12) , 48, 423 (1987).
7. F. Salin, G. Le Saux, F. Georges and A. Brun, Optic. Let., 14, 785 (1989).
8. E.T. Knobbe, B. Dunn, P.D. Fuqua, F. Nishida and J.I. Zink, in Proc. 4th Int.Conf. Ultrastructure of Ceramics, Glasses and Composites, (Wiley, New-York, 1989).
9. R. Reisfeld, V. Chernyak, M. Eyal and A. Weitz, Proc. SPIE, 1016, 240 (1988).
10. E. Toussaere, J. Zyss, P. Griesmar and C. Sanches, Non Linear Optics, 1, 000 (1991).
11. S. Sakka, K. Kamiya, K. Makita, and Y. Yamamoto, in Ref. 2b, p. 223: c.f. Table I.
12. D. Avnir, V.R. Kaufmann and R.Reisfeld, R., J. Non-Cryst. Solids, 74, 395 (1985).
13. L.L. Hench, in Ref. 5 p.52 (1986).
14. Y. Haruvy, A. Heller and S.E. Webber, Supramolecular Architectures in two and three dimensions, edited by T. Bein (ACS Symp. Ser., Washington, 1992) in press.
15. Y. Haruvy and S.E. Webber, Chem. Mater., 3, 501 (1991).
16. Y. Haruvy, J. Byers and S.E. Webber, Polym. Prepr., 32, 134 (1992).
17. Y. Haruvy, J.O. Wallin, B.A. Shoulders and S.E. Webber, Chem. Mater., submitted (1992).
18. H. Dislich and A. Jacobsen, Angew. Chem., 12, 439 (1973).
19. C.J. Brinker, K.D. Keefer, D.W. Schaefer, R.A. Assink, B.D. Kay and C.S. Ashley, in Ref. 2b, p. 45.
20. S. Sakka, Y. Tanaka and T. Kokubo, in Ref. 2c, p. 24.
21. Y. Haruvy and S.E. Webber, Chem. Mater., 4, 89 (1992).
22. Y. Haruvy, A. Heller and S.E. Webber, Proc. SPIE, 1590, 59 (1991).
23. E.T. Knobbe, B. Dunn, P.D. Fuqua and F. Nishida, Appl. Opt., 29, 2729 (1990).
24. Non-linear Optical Properties of Organic and Polymer Materials, Edited by D.J. Williams (ACS Symp. Ser., 233, Washington, 1983).
25. P.N. Prasad B.A. Reinhardt, Chem. Mater., 2, 660 (1990).
26. J. Byers, Y. Haruvy and S.E. Webber, in progress (1992).
27. Y. Haruvy and S.E. Webber, US Pat. Appl. 07 / 707,140 (1992).
28. Y. Haruvy, Q. Hibben and S.E. Webber, in progress (1992).
29. J. Chan, Y. Haruvy and S.E. Webber, in progress (1992).
30. J.C. Pouxviel and J.P. Boilot, J. Non-Cryst. Solids, 89, 345 (1987).
31. R.A. Assink and B.D. Kay, J. Non-Cryst. Solids, 99, 359; 104, 112; 107, 35 (1988).
32. H. Schmidt, H. Scholze and A. Kaiser, in Ref. 2b, p. 1.
33. H. Schmidt, A. Kaizer, M. Rudolph and A. Lenz, in Ref. 5, p. 87.

STABILITY OF AQUEOUS SUSPENSIONS OF HIGH SURFACE AREA ZIRCONIA POWDERS IN THE PRESENCE OF POLYACRYLIC ACID POLYELECTROLYTE

MICHEL M.R. BOUTZ, R.J.M. OLDE SCHOLTENHUIS, A.J.A. WINNUBST AND A.J. BURGGRAAF
University of Twente, Faculty of Chemical Technology, Laboratory for Inorganic Chemistry, Materials Science and Catalysis, P.O. Box 217, 7500 AE Enschede, The Netherlands.

ABSTRACT

In this paper results of efforts to prepare stable aqueous suspensions of high surface area (> 100 m^2/gr) yttria stabilized tetragonal zirconia powders are presented using a low molecular weight ammonium polyacrylic acid deflocculant. Zetapotentials, viscosities, agglomerate sizes and sedimentation volumes have been measured to find the optimum pH and deflocculant concentration. It has been found that the optimum pH-value coincides with the pH at which the polyacrylic acid is fully dissociated. Using deflocculant concentrations below the optimum value leads to a highly unstable system, while concentrations above the optimum value influence the stability in a much weaker way.

INTRODUCTION

For many structural applications it is desirable to obtain very fine-grained (<0.5 μm), dense tetragonal zirconia ceramics. Grain growth during sintering can be restricted by densifying at low temperatures. To obtain good sinterability at these low temperatures ultra fine powders with a low degree of agglomeration are required. Such powders can be produced by a gel precipitation technique using metal chlorides as precursor chemicals [1,2]. Nanocrystalline (crystallite size 8-10 nm) powders are obtained with this so-called chloride-method. After isostatic compaction (400 MPa) densities ≥ 95% of the theoretical one can be reached during free sintering at 1100-1150°C.

However, the flowability of these powders is generally poor and relatively low densities (45%) result from dry (isostatic) pressing at 400 MPa. Density gradients introduced into the green compacts during dry pressing may lead to differential sintering, ultimately causing severe cracking of samples. This problem becomes more severe in the case of samples with larger diameters. More homogeneous green microstructures with higher densities can be made by using spray dried powders or by using colloidal techniques like slip casting and pressure casting for green compact formation. Stable aqueous suspensions are needed for these purposes.

In this paper results of efforts to stabilize high surface area (> 100 m^2/gr) yttria stabilized tetragonal zirconia powders are presented using a commercial grade, low molecular weight polyanionic deflocculant. By measuring several suspension characteristics (ζ-potential, viscosity, agglomerate size and sedimentation volume) an optimum pH and deflocculant concentration have been sought.

EXPERIMENTAL PROCEDURE

The tetragonal zirconia powders used in this investigation have the chemical composition $0.95ZrO_2 0.05YO_{1.5}$ (denoted as ZY5). Synthesis took place by the previously mentioned chloride-method [2]. After calcination at 500°C the specific surface area of this powder as measured by the BET-method using N_2 (Micromeritics 2400 ASAP) equals 110-120 m^2/gr. The used deflocculant was obtained from Allied Colloids Ltd. and has the commercial name Dispex A40. Dispex A40 is a water solution of the ammonium salt of a low molecular weight polycarboxylic acid. Infrared absorption measurements indicated that this polycarboxylic acid is polyacrylic acid (PAA) . The solid content of Dispex equals 40%; deflocculant concentrations mentioned below are based on dry Dispex on a dry weight base (dwb) of ZY5.

Zetapotentials were calculated from electrophoretic mobilities (Malvern Zetasizer IIc) using the Helmholtz-Smoluchowski equation. These measurements can only be performed on very dilute suspensions. Aqueous suspension with 1 wt% solids and the desired Dispex concentrations were prepared in polyethylene flasks. HNO_3 and NaOH solutions were used to adjust the pH to the desired value. Double distilled and deionized water has been used throughout this investigation. The suspensions were rolled with Teflon marbles and the pH was adjusted until it remained constant within 0.1 pH units during one night. This procedure typically took several days. An average of five samples was taken from each suspension for electrophoretic mobility measurements and 5-6 measurements were performed per sample to ensure accuracy. Suspensions were prepared at pH 5, 7 and 9 with different Dispex concentrations. One series of suspensions were prepared, in which the Dispex concentration was varied without fixing the pH-value.

Suspensions with 35 wt% solids (8 vol%) and various amounts of Dispex (pH not fixed) were used for viscosity, particle size and sedimentation volume measurements. To attain pH equilibrium these suspensions were rolled with Teflon marbles during one week. Sediment volumes were measured by pouring the suspensions into graduated 10 ml cylinders. After one week and one month the sedimentation cake heights were measured. Viscosities were measured using a rotary viscosity meter (Reoscan 30). The following procedure was adapted: first the shear rate was increased from 0 to 64 sec^{-1} in two minutes, then the suspension was sheared at 64 sec^{-1} during 30 sec and finally the shear rate was decreased to zero again in two minutes. The torque necessary to rotate the cylinder at a certain angular velocity is recorded continuously and the viscosity is calculated from these two parameters. For particle size determination by dynamic light scattering (Horiba laser diffraction) the 35 wt% solids suspensions were strongly diluted.

RESULTS AND DISCUSSION

Both the surface-charge of yttria stabilized tetragonal zirconia polycrystals (Y-TZP) and the dissociation of polyacrylic acids strongly depend on pH. The point of zero charge (pzc) is situated at ± pH 7 for Y-TZP [3]; at more acidic pH-values the surface charge is positive, while at pH-values above the pzc a negative surface charge exists. For polyacrylic acids the fraction (α) of functional groups which are dissociated (i.e. COO^-) and those which are not dissociated (i.e. COOH) will vary with pH, as shown by Ceserano et al. [4] for the case of polymethacrylic acid (PMAA). They found that PMAA is neutral at pH-values ≤ 3.5, while it is fully dissociated at pH-values ≥ 8.5. Thus at pH-values below 7 the surface of Y-TZP and the surface-active component of Dispex (PAA) are oppositely charged, while at more basic pH-values both charges have a negative sign.

To investigate the influence of pH on the interaction of Dispex with zirconia, ζ-potentials have been measured as a function of Dispex concentration at pH-values below, near and above the pzc of zirconia (i.e. at resp. pH 5, 7 and 9). These data are presented in fig. 1. It can be seen that at pH 5.1 the ζ-potential decreases from +35 mV in the absence of Dispex to +18 mV for 3 w/o Dispex. At pH 7 the ζ-potential decreases rapidly with increasing Dispex concentration. In the absence of Dispex the ζ-potential is very low (-3 mV), with small additions of Dispex it decreases rapidly and finally saturates at -40 mV for concentrations above 1 w/o. At pH 9 the dependence of the ζ-potential on Dispex-concentration is very similar to the one at pH 7. The ζ-potential also saturates at -40 mV, but the rate at which it decreases is less than at pH 7 and the saturation value is reached at ± 1.5 w/o.

From the data presented in fig. 1 it is clear that at pH 5 the repulsive forces acting between solid particles in the suspension will decrease with increasing Dispex concentration. At this pH the fraction of dissociated carboxylgroups of PAA will be quite low [4] and adsorption onto the positively charged zirconia surface only leads to a charge neutralization, thereby lowering the suspension stability. At pH 9 the polymer is fully dissociated and adsorption onto the negatively charged zirconia surface now imposes a large negative charge on the zirconia particles.

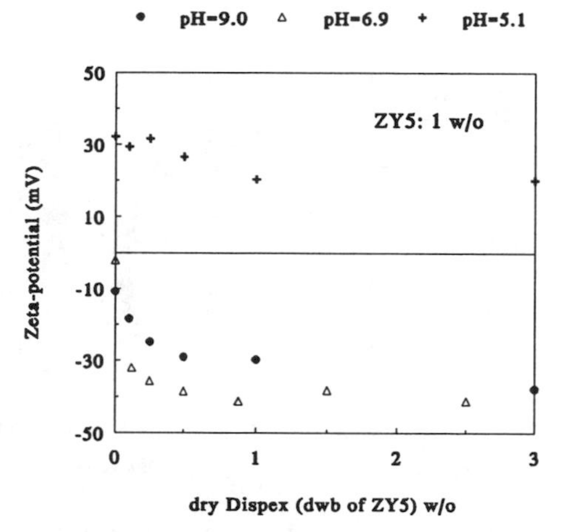

Figure 1 Zetapotential of ZY5 as a function of Dipsex concentration at pH 5, 7 and 9.

Adsorption of the negatively charged polymer takes place at the (minority) of positive surface sites. The faster decrease of the ζ-potential with increasing Dispex concentration at pH 7 compared to pH 9 can be explained by the larger surface concentration of positive sites and the absence of an energetic barrier for adsorption, since the surface of zirconia is effectively uncharged at pH 7.Since Dispex A40 is an ammoniumsalt solution having a pH equal to 8.5, suspensions will become increasingly basic upon Dispex addition. To stabilize the pH below 8.5 nitric acid needs to be added, thereby enhancing the counterion-concentration. This in turn leads to a decrease in the double layer thickness of the suspended particles accompanied by a decrease in stability. This is especially the case for suspensions with higher solid contents and smaller particle sizes, since these two parameters determine the average interparticle distance.

Therefore, the stability of suspension containing 35 wt% solids has been studied as a function of Dispex-concentration without fixing the pH-value. The dependence of both the ζ-potential and the pH of the suspensions on the Dispex concentration can be found in fig. 2. The ζ-potential decreases from +15 mV in the absence of Dispex to -40 mV for a 3 w/o concentration. Simultaneously, the pH changes from 5.5 to 8.7 for 3 w/o Dispex. Sedimentation heights (expressed as relative heights, i.e. the percentage of the total height of the suspension in the graduated cylinders occupied by the sediment cake) after one week and one month settling are shown in fig. 3 for 0, 0.05, 0.1, 0.5 and 3.0 w/o Dispex. From this figure it is clear that very small (0.05 - 0.1 w/o) additions strongly destabilize the suspensions and reasonable stability is only found for larger concentrations ($\geq$ 0.5 w/o).

The rheological character of the suspensions also changes drastically upon Dispex addition. In the absence of Dispex the 35 wt% suspensions behave like a Newtonian liquid, i.e. the viscosity is independent of the applied shear rate. Suspensions with Dispex additions all show a time-dependent dilatant (=rheopectic) behaviour; i.e. the viscosity increases with time and with increasing shear rate. Viscosities measured while increasing the shear rate (η_{up}) thus always have smaller values than those measured while decreasing the shear rate (η_{down}). In fig. 4 the average viscosity ($= \frac{1}{2}\{\eta_{up} + \eta_{down}\}$) is shown as a function of Dispex concentration for a shear rate of 5.5 sec^{-1}. Upon addition of 0.05 w/o Dispex the viscosity increases from < 5 mPasec to > 700 mPasec, by further increasing the concentration to $\geq$ 0.5 w/o the viscosity decreases again to $\pm$ 6 mPasec. The area between the increasing and decreasing shear rate curves in the shear stress - shear strain diagram is a measure for the rheopexy of the suspensions; this area is going through a minimum for a concentration of 2 w/o Dispex.

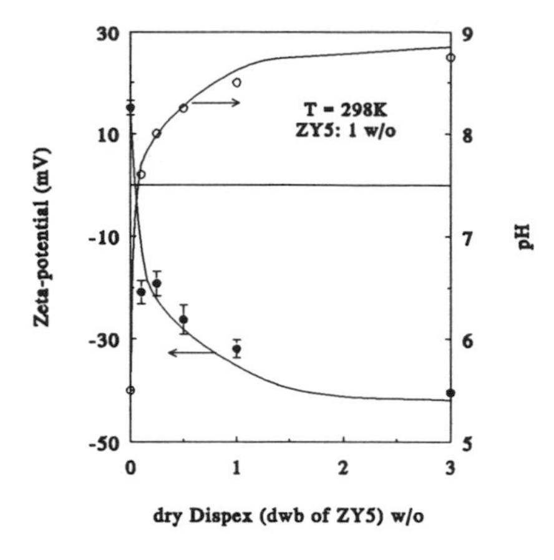

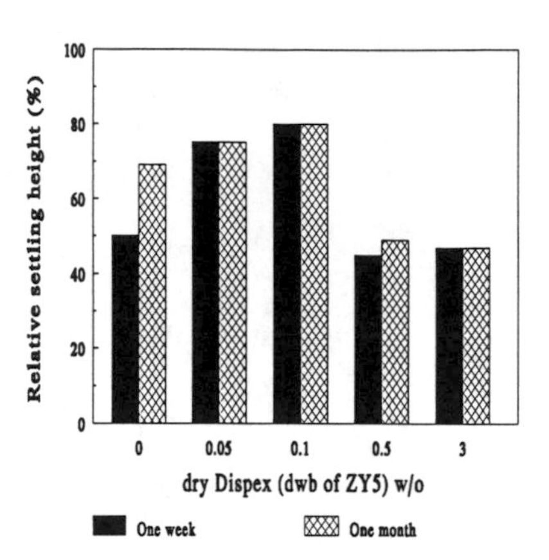

Figure 2 Zetapotential and pH of 1 w/o ZY5 suspensions as a function of Dispex concentration.

Figure 3 Relative sedimentation heights after 1 week and 1 month settling of 35 w/o ZY5 suspensions with different amounts of Dispex.

The trend in agglomerate size as a function of Dispex concentration is very similar to the trend in viscosity. By adding 0.1 w/o Dispex the agglomerate size increases from 1 μm to 7 μm, the agglomerate size then goes through a minimum (0.5 μm) for 0.5 w/o and then slowly increases to 1 μm for 3 w/o Dispex.

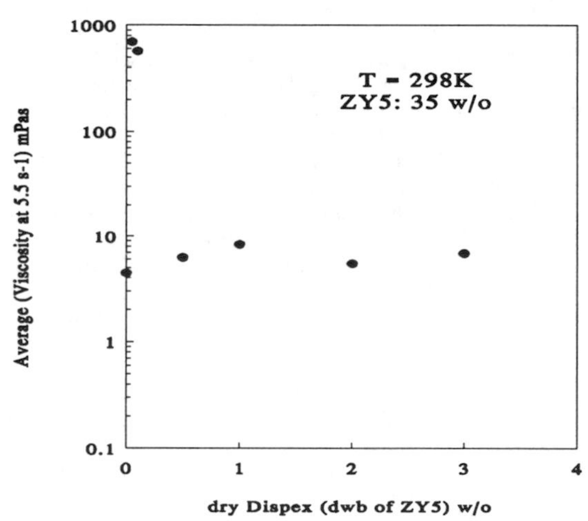

Figure 4 Average viscosity (shear rate 5.5 sec^{-1}) of 35 w/o ZY5 suspensions as function of Dispex concentration.

The effect of varying the Dispex concentration on the characteristics of the suspensions as described above can be interpreted as follows. For very small concentrations (0.05 - 0.1 w/o) of deflocculant only incomplete adsorption occurs. By adsorption of the negatively charged deflocculant onto the zirconia particle a patch is formed with a negative charge. Flocculation then occurs because of the interaction of this negatively charged patch on one particle with a bare (positively charged) patch on a second particle. This so-called patch charge flocculation is favoured by low molecular weight polymers with high charge densities and leads to the observed increase in agglomerate size and viscosity in the investigated suspensions. By increasing the concentration of Dispex the pH rises above 8.5 and the now fully dissociated PAA can form a monolayer on the zirconia particles imposing a strong negative charge on the particles. The ζ-potential reaches its saturation value (-40 mV) and viscosity, agglomerate size and sedimentation volume go through a (broad) minimum. For the investigated high surface area zirconia powders this optimum concentration is situated at 1.5 - 2.0 w/o. A further increase in deflocculant concentration only increases the concentration of nonadsorbed polymer. The observed (gentle) increase in agglomerate size might be explained by depletion flocculation [5]. Viscosity also increases slightly due to this flocculation and the increase in the concentration of polymer in solution.

CONCLUSIONS

It has been found that sufficiently stable aqueous suspensions of high surface area ($> 100\ m^2/gr$) zirconia powders can be prepared by using a polyanionic deflocculant (Dispex A40). The interaction of the surface-active component of Dispex (PAA) with the suspended zirconia particles strongly depends on both pH and deflocculant concentration. Basic pH-values (≥ 8.5) are favourable for suspension stability. Under these conditions PAA is fully dissociated and adsorption of the negatively charged polymer imposes a strong negative charge on the solid particles. In combination with this electrostatic stabilizing effect a steric stabilization interaction also might be contributing to the suspension stability [4].

An optimum deflocculant concentration exists, where the ζ-potential reaches its maximum value and viscosity and particle size go through a minimum. Very small concentrations lead to strong flocculation, while concentrations above the optimum value have a much weaker destabilizing effect.

ACKNOWLEDGEMENTS

Akzo Chemicals B.V. Amsterdam is gratefully acknowledged for financial support of this investigation.

REFERENCES

[1] K.Haberko, Ceramurgia Int. 5 (4), 148 (1979).
[2] W.F.M. Groot Zevert, A.J.A. Winnubst, G.S.A.M. Theunissen and A.J. Burggraaf, J. Mater. Sci. 25, 3449 (1990).
[3] C.Simon, Ceramics Today, Tomorrow's Ceramics, ed. by P.Vincenzini (Elsevier Science Publishers, 1991), 1043.
[4] J.Ceserano III, I.A.Aksay and A.Bleier, J. Am. Cer. Soc. 71 (4), 250 (1988).
[5] D.H.Napper, Polymeric stabilization of colloidal dispersions (Academic Press Inc., London, 1983), 332.

PRODUCTION OF SOLS FROM AGGREGATED TITANIA PRECIPITATES

John R. Bartlett and James L. Woolfrey

Advanced Materials Program, Australian Nuclear Science and Technology Organisation, Private Mail Bag 1 Menai N.S.W. 2234. Australia

ABSTRACT

Titania and titania/zirconia sols, with solids loadings exceeding 1000 g dm^{-3}, have been prepared on a 10 kg scale by chemical methods involving the hydrolysis of an appropriate mixture of alkoxides, followed by peptisation with dilute nitric acid. The rate of peptisation of the hydrolysates was determined by static light-scattering and photon-correlation spectroscopy, enabling the hydrodynamic radius, the radius of gyration and the fractal dimension of colloidal species to be monitored as a function of peptisation time.

The peptisation kinetics were influenced by a range of factors including the initial solids loading, the acid concentration, the reaction temperature and the age of the alkoxide hydrolysate. Hydrolysate peptisation is first order with respect to concentration of acid and exhibited a non-integer reaction rate order (complex mechanism) with respect to solids concentration. Sols produced from freshly-precipitated hydrolysate peptised at a faster rate than aged precipitates but slowly re-aggregated after peptising, yielding "equilibrium" aggregate sizes often exceeding 100 nm. This effect was not observed in sols produced from aged hydrolysate. These differences are interpreted using DLVO theory.

INTRODUCTION

A typical sol-gel process can be represented schematically by :

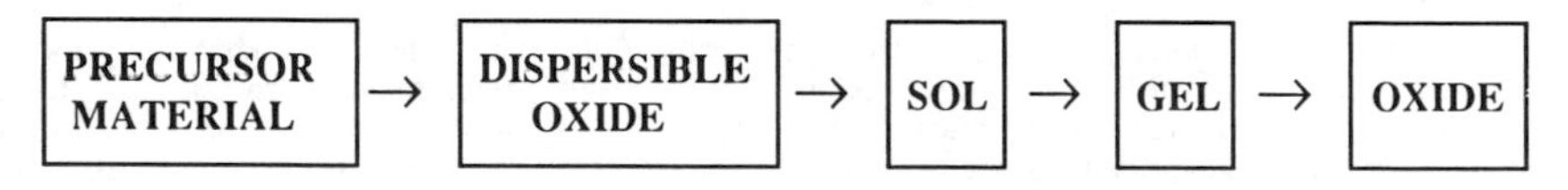

The precursor material (either an inorganic salt or alkoxide solution) is chemically processed to form hydrous metal oxides or hydroxides. A critical step in sol-gel processing is the de-aggregation of the agglomerated oxide or hydroxide particles by peptisation with an appropriate acid or alkali. The repulsive electrostatic-forces generated between adjacent particles, due to the formation of charged surface sites and an associated electric double-layer, promote de-aggregation and formation of a stable sol. The initial conversion of the precursor material into a colloidal dispersion is an important component of the cost of sol-gel processing.

In a previous study [1], we examined the effects of selected peptising agents and the acid:titania ratio on the kinetics of peptisation of TiO_2 precipitates produced via alkoxide hydrolysis. In this study, we investigate the influence of solids loading and hydrolysate ageing on the rate of peptisation. This work is part of a continuing program to obtain a detailed understanding of the processes involved in producing sols on a scale suitable for industrial processing [2].

EXPERIMENTAL

Sample preparation

The method used to prepare titania sols has been described previously [1]. Briefly, pure tetraisopropyltitanate (TPT) was rapidly added to water at ambient temperature, producing large flocs which settled readily. The settled product was washed, diluted to the required concentration in water (typically 0.1 to 2 mol dm^{-3}), and finally peptised in tightly capped teflon containers at 333 K with nitric acid (0.1 moles HNO_3 per mole TPT). During peptisation, the suspensions were slowly agitated with a magnetic stirrer. Samples were withdrawn at selected intervals and rapidly quenched to ambient temperature, to enable the peptisation kinetics to be monitored by photon-correlation spectroscopy. Selected hydrolysate slurries were also aged at ambient temperature for periods ranging from 1 day to several weeks, prior to peptisation at 333 K.

Sample characterisation

The z-averaged hydrodynamic diameter of the colloidal aggregates was determined via photon-correlation spectroscopy, using either a Malvern Autosizer IIC or a Nicomp 170 Computing autocorrelator interfaced (via optic fibres) to a Wyatt Technologies Dawn-F static light scattering spectrometer. All data reported below were obtained using HeNe laser irradiation ($\lambda = 632.8$ nm) and a detector located at either 79.5 or 90 ^{o} to the incident beam. Z-averaged diameters were obtained from the measured autocorrelation functions using the method of cumulants [3]. Radii of gyration and fractal dimensions of selected colloidal species were obtained by monitoring the angular dependence of scattered light simultaneously at 15 discrete angles in the Dawn-F spectrometer.

RESULTS

The influence of TiO_2 concentration on the rate of hydrolysate peptisation is illustrated in Figures 1 and 2. Peptisation kinetics were found to obey the rate law :-

$$\frac{d}{dt}(\Delta P) = -k[\mathrm{TiO_2}]^{0.52}[\mathrm{HNO_3}]\Delta P$$

where $\Delta P = (P_t - P_\infty)$ and P_t is the z-averaged aggregate size (determined by photon-correlation spectroscopy) at time t.

The effects of ageing the hydrolysate in aqueous solution at ambient temperature, prior to peptisation, are shown in Figure 3. Sols produced from freshly-precipitated hydrolysates peptised at a faster rate than aged precipitates (Figure 3) but slowly re-aggregated after achieving a minimum hydrodynamic diameter, yielding "equilibrium" aggregate sizes often exceeding 100 nm. This effect was not observed in sols produced from hydrolysates aged for more than 21 days. Model DLVO calculations of the barrier to aggregation in these systems are summarised in Figure 4.

Variations in the static intensity of light scattered by selected TiO_2 sols, as a function of scattering wave-vector (Q), are illustrated in Figure 5. The fractal dimension calculated from this data is 1.50 ± 0.05.

DISCUSSION

Effect of solids loading on peptisation kinetics

Peptisation of flocculated oxide particles involves two distinct chemical processes :

(a) disruption of weak chemical or van der Waals bonds between the individual crystallites, breaking down aggregates or agglomerates (de-aggregation), and

(b) evolution of an electrostatic charge, or steric layer, on the surface of the de-aggregated particles, which inhibits re-aggregation and maintains the dispersed state.

In an earlier study [1], it was shown that for a fixed concentration of solids, the rate of peptisation in the TiO_2/HNO_3 system could be written as :

$$\frac{d}{dt}(\Delta P) = -k[HNO_3]^n \Delta P \tag{1}$$

where $n = 1.1 \pm 0.2$ (*i.e.* the reaction is first order with respect to acid concentration). However, for variable $[TiO_2]$, a more complete expression for the rate of peptisation is :

$$\frac{d}{dt}(\Delta P) = -k[TiO_2]^m[HNO_3]\Delta P \tag{2}$$

If the HNO_3:TiO_2 mole ratio has a fixed value, r (which was equal to 0.1 in this study), then

$$[HNO_3] = r[TiO_2] \tag{3}$$

and hence

$$\frac{d}{dt}(\Delta P) = -k[TiO_2]^m r[TiO_2]\Delta P \tag{4}$$

i.e.

$$\frac{d}{dt}(\Delta P) = -k'[TiO_2]^{m+1}\Delta P \tag{5}$$

If the values of r and $[TiO_2]$ are fixed during peptisation, Equation (5) may be integrated to yield

$$\Delta P = \Delta P_0 e^{-(k''t)} \tag{6}$$

where

$$k'' = k'[TiO_2]^{m+1} \tag{7}$$

Hence, a plot of $\ln(k'')$ against $[TiO_2]$ for slurries with varying $[TiO_2]$ will yield a line of slope $(m+1)$, assuming a fixed value for r.

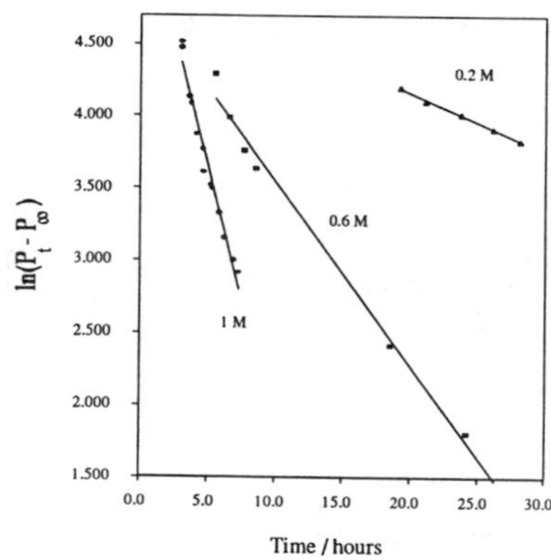

Figure 1. Variations in $\ln(\Delta P)$ with time for selected values of $[TiO_2]$.

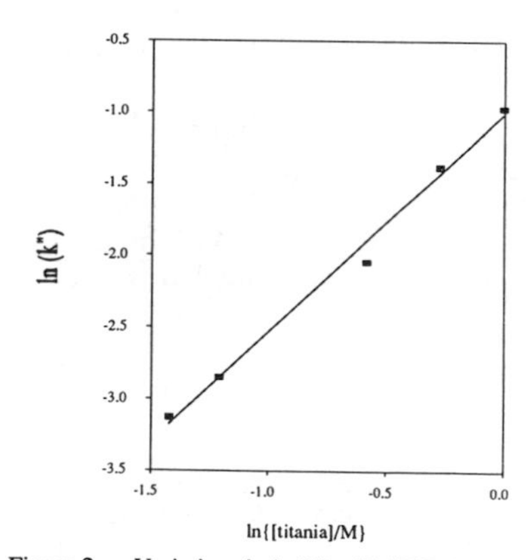

Figure 2. Variations in $\ln(k'')$ with $[TiO_2]$.

Variations in $\ln(\Delta P)$ with time for selected TiO_2/HNO_3 dispersions, with fixed values of r and varying $[TiO_2]$, are illustrated in Figure 1. Decreasing $[TiO_2]$ is accompanied by a decrease in the rate of peptisation, and peptisation obeys a pseudo first order rate law, as expected from Equations 5 and 6 [1]. The value of m deduced from a plot of k'' against $[TiO_2]$, Figure 2, is 0.52 ± 0.03.

The apparent reaction-rate order of reactants in chemical processes provides an insight into the nature of the reaction mechanism. In particular, integer reaction-rate orders are consistent with simple reaction mechanisms, whereas non-integer values reflect complex mechanisms involving multiple steps, steady-state and/or quasi-equilibrium (competing) processes. The experimental observations, Figure 3, suggest a complex reaction mechanism based on sequential de-aggregation and re-aggregation steps, such as

$$[TiO_2]_i:[TiO_2]_j \xrightarrow{k_P(i,j)} [TiO_2]_i + [TiO_2]_j \qquad (8)$$

$$[TiO_2]_i + [TiO_2]_j \xrightarrow{k_A(i,j)} [TiO_2]_i:[TiO_2]_j \qquad (9)$$

where $k_P(i,j)$ and $k_A(i,j)$ denote the rates of de-aggregation and re-aggregation, respectively, of TiO_2 aggregates composed of i and j monomers (crystallites). Such steps may occur simultaneously.

Re-aggregation of peptised sols

An interesting feature of the reactions occurring during peptisation of fresh hydrolysates (for $[TiO_2] > 0.5$ mol dm^{-3}) is that the dispersions re-aggregate slowly during peptisation, reaching final stable aggregate sizes that are significantly higher than the minimum values initially achieved, Figure 3. This phenomenon is not observed in sols produced from aged hydrolysates, which do not re-aggregate after peptising.

The DLVO theory of colloid stability, originally developed by Derjaguin, Landau, Verwey and Overbeek [4], has proven successful in modelling the interaction potential in colloidal systems such as titania. In this theory, the total interaction potential energy between two colloids is given by :

$$V_T(s) = V_R(s) + V_A(s) \qquad (10)$$

where the repulsive ($V_R(s)$) and attractive potentials ($V_A(s)$) are calculated from [5,6] :

$$V_R(s) = \frac{\pi \varepsilon_0 \varepsilon_R r_1 r_2}{r_1 + r_2} \zeta^2 \ln\left\{1 + e^{(-\kappa s)}\right\} \qquad (11)$$

$$V_A(s) = -\frac{A}{6}\left\{\frac{2r_1r_2}{y} + \frac{2r_1r_2}{y + 4r_1r_2} + \ln\left(\frac{y}{y + 4r_1r_2}\right)\right\} \qquad (12)$$

$$\kappa = \sqrt{\frac{2Ie^2z^2}{\varepsilon_0 \varepsilon_r k_B T}} \qquad (13)$$

Here, s is the distance between the surfaces of the interacting colloids of radii r_1 and r_2, A the effective Hamaker constant for the particles in a given medium, ε_0 the permittivity of free space (8.854×10^{-12} F/m), ε_r the relative dielectric constant of the medium, ζ the zeta potential of the colloids (assumed to be the same for both species), κ the Debye-Hückel parameter, I the ionic strength, z the valence of the ionic species in solution, e the electronic charge (1.602×10^{-19} C), k_B Boltzmann's constant (1.381×10^{-23} J/K), T the absolute temperature and $y = s^2 + 2r_1s + 2r_2s$.

Model DLVO calculations of the total interaction potential between two colloids with varying radii in aqueous solution at 333 K are illustrated in Figure 4. For a population of colloids with radii of 45 nm, *i.e.* $r_1 = r_2 = 45$ nm (the minimum aggregate radius observed in the

TiO_2/HNO_3 system for $[TiO_2] = 1.0$ mol dm^{-3}), the calculated barrier to aggregation exceeds 30 kT, and hence the sol is kinetically stable. However, decreasing the value of r_2 (with r_1 held constant at 45 nm) is accompanied by a corresponding decrease in the height of the barrier, which falls below 10kT when $r_2 < 5$ to 10 nm. Under such conditions, flocculation can occur.

Variations in the aggregate size and polydispersity of sols prepared from fresh hydrolysates are compared in Figure 3. When the z-averaged aggregate size achieves its observed minimum value (diameter = 90 ± 2 nm), the size polydispersity remains relatively high, implying that the sol contains a mixture of small and large particles. As discussed above, the low potential energy barrier between relatively small (*i.e.* $r_1 < 10$ nm) and large aggregates ($r_2 \geq 45$ nm) favours flocculation, with an associated increase in the measured z-average diameter. It is anticipated that aggregation would continue until all small colloids had been consumed, yielding a final stable equilibrium aggregate size that is somewhat higher than the minimum value, Figure 3. As predicted, complete consumption of the small aggregates is also reflected in a substantial decrease observed in the size polydispersity.

A different behaviour is evident for sols produced from hydrolysates aged for 21 days prior to peptisation, Figure 3. The polydispersity decreases rapidly during peptisation, and reaches a low equilibrium value before peptisation is completed. This behaviour is consistent with a relatively narrow size distribution, implying that there are no small colloids present in the ensemble when peptisation is complete. Consequently, no re-aggregation (as reflected by an increase in z-averaged aggregate size) would be expected in this system, as demonstrated in Figure 3.

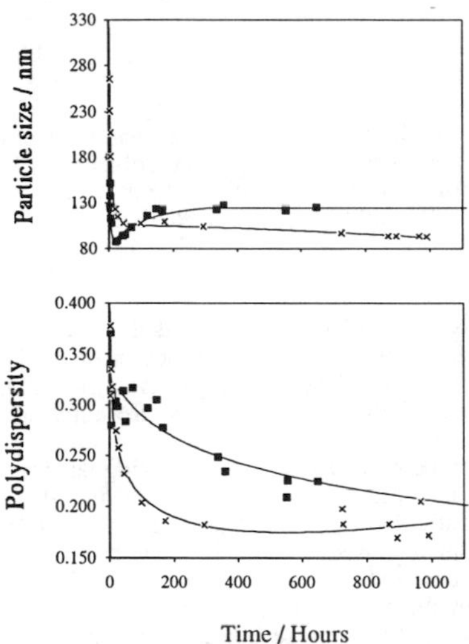

Figure 3. Variations in aggregate size (top) and polydispersity (bottom) for fresh (■) and heavily aged (21 days) hydrolysate (×).

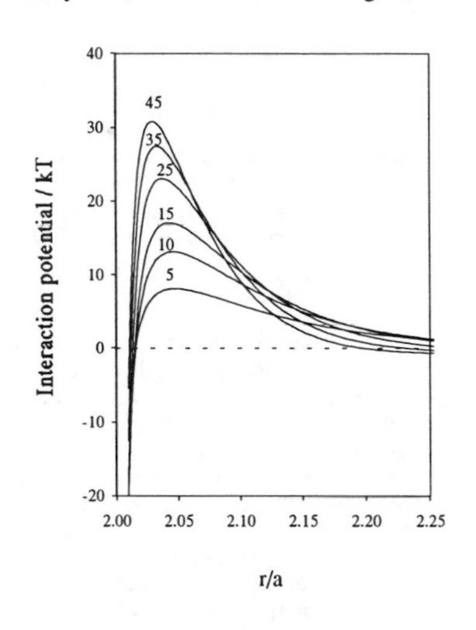

Figure 4. Interaction potential between TiO_2 colloids, with varying radii. Conditions : r_1 = 45 nm; r_2 = 5 to 45 nm; A = 5 x 10^{-20}J; I = 20 mol m^{-3}; ζ = 60 mV; T = 333 K; r = centre-to-centre separation distance between colloids; a = $0.5(r_1+r_2)$.

Fractal structure of aggregates

The concept of fractal geometry has proven to be a powerful tool for elucidating the structure of aggregates in a variety of colloidal systems, including silica [7-9], polystyrene [10] and branched polymers and gels [11]. In a typical experiment, the static intensity of laser light scattered by the sample is monitored as a function of scattering wave-vector, yielding the associated static structure-factor from which the radius of gyration, r_g, and the mass-fractal dimension, d_f, are obtained. The intensity of light scattered by a fractal aggregate is given by [12]:

$$I(Q) = KP(Q)S(Q) \tag{14}$$

where the form factor and inter-particle structure factor (*P(Q)* and *S(Q)*, respectively) are

$$P(Q) = \left(n_p - n_s\right)^2 \left\{3\frac{\sin(Qr_0) - Qr_0\cos(Qr_0)}{(Qr_0)^3}\right\}^2 \tag{15}$$

$$S(Q) = 1 + \frac{d_f \Gamma(d_f - 1)\sin((d_f - 1)\tan^{-1}(Q\varepsilon))}{(Qr_0)^{d_f}\left(1 + 1/\left[Q^2\varepsilon^2\right]\right)^{(d_f-1)/2}} \tag{16}$$

and

$$Q = \frac{4\pi n_s}{\lambda}\sin\left(\frac{\theta}{2}\right) \tag{17}$$

Here, Q is the scattering wave-vector, θ the scattering angle with respect to the incident beam, λ the wavelength of the incident light (*in vacuo*), n_s the refractive index of the solvent, n_p the refractive index of the scattering species, r_0 the radius of the crystallites from which the aggregates are composed (determined via TEM in this study) and ε is related to r_g via

$r_g = \varepsilon\left[\sqrt{\left(d_f(d_f+1)\right)}\right]$.

The static intensity of light scattered by a fully peptised (*i.e.* "equilibrated") TiO_2 sol, produced from freshly precipitated hydrolysate, is illustrated in Figure 5 as a function of Q. The values of r_g and d_f calculated from this data are 140±1 nm and 1.50±0.05, respectively, while the corresponding hydrodynamic radius (r_h), measured via photon-correlation spectroscopy in a separate experiment, is 59±1 nm.

The magnitude of d_f provides insight into the geometrical structure and packing density in colloidal aggregates. For example, infinitely narrow rods exhibit a fractal dimension of 1.0 (*i.e.* mass $\propto r_g^{1.0}$), while a value of 3.0 is characteristic of fully dense spheres (*i.e.* mass $\propto r_g^{3.0}$).The value of 1.50 obtained for peptised TiO_2 aggregates in this study is consistent with an open, extended, chain-like aggregate structure, with short branches [11].

The ratio $R_r = r_h/r_g$ is also strongly influenced by the aggregate structure of the scattering species. In particular, it can be demonstrated [7] that $R_r = 1.29$ for dense spheres, while a value of 0.79±0.04 is typically found for linear, random-walk polymers. Values between 0.54 and 0.72 have been reported for aggregated silica spheres [7,8], reflecting significant deviations from spherical geometry in these systems. However, the value obtained in this study for TiO_2 aggregates is even lower ($R_r = 0.42$), indicating even greater deviations from spherical geometry. This latter result is consistent with the extended, chain-like aggregate structure deduced above for the TiO_2 colloids.

CONCLUSIONS

Titania sols have been produced on a 10 kg scale (oxide basis) by a method involving the hydrolysis of pure alkoxides with pure water, and peptisation of the resulting hydrolysate at 333 K with dilute nitric acid. These sols could be routinely concentrated to 1000 g dm^{-3}.

1. The kinetics of peptisation were monitored by static light-scattering and photon-correlation spectroscopy. The rate of peptisation was dependent on the HNO_3:TiO_2 mole ratio and on the

concentration of solids, and obeyed the following rate law :

$$\frac{d}{dt}(\Delta P) = -k[\mathrm{TiO_2}]^{0.52}[\mathrm{HNO_3}]\Delta P$$

The non-integer exponent implies a complex reaction-mechanism.

2. The z-averaged aggregate size in sols produced from fresh hydrolysates decreased from an initial value in excess of 10 µm to less than 100 nm during peptisation, and then slowly increased, reaching final equilibrium values in the range 50 to150 nm (depending on the concentration of solids). In contrast, sols prepared from aged hydrolysates did not re-aggregate following peptisation. These differences were interpreted using DLVO theory.

3. The TiO_2 aggregates produced by peptising fresh hydrolysate with HNO_3 exhibited a fractal structure, with a fractal dimension of 1.50±0.05. The value obtained for the ratio r_h/r_g was 0.42, where r_g is the radius of gyration (determined via static light-scattering) and r_h the hydrodynamic radius (measured using photon-correlation spectroscopy). These results are consistent with an open, extended, chain-like aggregate structure.

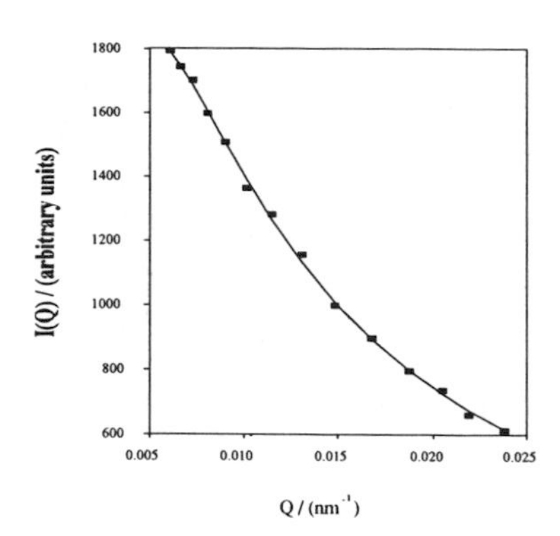

Figure 5. Static light scattering intensity as a function of scattering wave-vector for fully peptised TiO_2/HNO_3 sols prepared from fresh hydrolysate. $[TiO_2] = 0.8$ mol dm^{-3}. The least-squares fit was obtained using Equations 14 to 17.

ACKNOWLEDGMENTS

The authors gratefully acknowledge helpful discussions with Jim Woodhead, Dr Rose Amal and Dr David Waite. This work was funded by a grant from the Australian Energy Research Development Council.

REFERENCES

1. J.R. Bartlett and J.L. Woolfrey in "Chemical Processing of Advanced Materials (Proceedings of the 5th Ultrastructure Meeting, February, 1991, Orlando, Florida)", (L.L. Hench, Ed.), Chapter 23 (in press).
2. J.R. Bartlett and J.L. Woolfrey, Mat. Res. Soc. Symp. Proc. 180, 191-196 (1990).
3. B. Burn and R. Pecora, "Dynamic Light Scattering", (Wiley, New York, 1976).
4. E.J.W. Verwey and J. Th. G. Overbeek, "Theory of the Stability of Lyophobic Colloids", (Elsevier, Amsterdam, 1948).
5. P.C. Hiemenz, "Principles of Colloid and Surface Chemistry", (2nd Edition, Marcel Dekker, New York, 1986).
6. R. Hogg, T.W. Healy and D.W. Fuerstenau, Trans. Faraday Soc., 62, 1638-1651 (1966).
7. P. Wiltzius, Phys. Rev. Lett., 58, 710-713 (1987).
8. E. Ziemath, M.A. Aegerter, J. Moreira, M. Figueiredo, and J. Zarzycki, Mat. Res. Soc. Symp. Proc., 121, 311-316 (1988).
9. D.W. Schaefer, J.E. Martin, P. Wiltzius and D.S. Cannell, Phys. Rev. Lett., 52, 2371-74 (1984).
10. C. Cametti, P. Codastefano and P. Tartaglia, J. Colloid Interf. Sci., 131, 409-422 (1989).
11. J.E. Martin and A.J. Hurd, J. Appl. Cryst., 20, 61-78 (1987).
12. R. Amal, J.A. Raper and T.D. Waite, J. Colloid Interf. Sci., 140, 158-168 (1990).

PART III

Ceramic Thin Films, Coatings, Membranes, and Fibers

STRESS AND PHASE TRANSFORMATION PHENOMENA IN OXIDE FILMS: REAL-TIME SPECTROSCOPIC MEASUREMENTS

GREGORY J. EXARHOS, AND NANCY J. HESS
Pacific Northwest Laboratory, Richland, Washington 99352

ABSTRACT

In situ optical methods are reviewed for characterization of phase transformation processes and evaluation of residual stress in solution-deposited metastable oxide films. Such low density films most often are deposited as disordered phases making them prone to crystallization and attendant densification when subjected to increased temperature and/or applied pressure. Inherent stress imparted during film deposition and its evolution during the transformation are evaluated from phonon frequency shifts seen in Raman spectra (TiO_2) or from changes in the laser-induced fluorescence emission spectra for films containing rare earth (Sm^{+3}:$Y_3Al_5O_{12}$) or transition metal (Cr^{+3}:Al_2O_3) dopants. The data in combination with measured increases in line intensities intrinsic to the evolving phase are used to follow crystallization processes in thin films. In general, film deposition parameters are found to influence the crystallite ingrowth kinetics and the magnitude of stress and stress relaxation in the film during the transformation. The utility of these methods to probe crystallization phenomena in oxide films will be addressed.

INTRODUCTION

Stress is manifest in ceramic films and arises from thermal properties mismatch between film and substrate materials and the inherent microstructure which develops as a result of the deposition process and attendant and post processing procedures. The film stress may be beneficial or detrimental depending upon its magnitude and the particular application intended for the film. For example, a slight compressive stress improves adhesion to the substrate. However, excessive stress can result in buckling of the film, and marked degradation in interfacial bonding leading to delamination. In addition, deposition onto thin substrates can cause deformation of the substrate. The magnitude of interfacial stress can be mediated through various pre and post processing treatments such as laser annealing where both an increase in stress at the substrate surface and stress relaxation in films deposited thereon have been reported. **[1,2]** An empirical correlation between stress and thermal expansion coefficient also has been reported **[3]** and other physical properties such as refractive index are influenced by film stress. Therefore, knowledge of residual film stress and how it is modified during film processing is paramount to understanding and modeling film properties.

Residual stress in films is comprised of interfacial stress, arising from film substrate interactions, and inherent stress related to the film microstructure and associated chemistry. Deposition temperature and ion energy constitute physical processes which can alter film stress during PVD or IAD processing of films. **[4,5]** Stress also can be introduced chemically in solution-deposited ceramic films. Previous work has shown that the stress state of silica or titania films deposited using sol-gel routes is dependent upon solution pH, and the presence of complexing agents added to the

 ©1992 Materials Research Society

solution from which the film is cast. **[6,7]** Subsequent thermal processing converts organic precursors to the oxide, induces transformation to a new phase, and alters the stress state of the film.

Thermally-induced transformation phenomena in oxide films deposited by one of two solution routes are treated in this review. Sol-gel processing is a well established method for the preparation of anti-reflective coatings, multi-layer structures and fibers in addition to large optical parts. **[8,9]** Using this method, controlled hydrolysis and condensation of a metal organic precursor, such as an alkoxide, generate a polymeric oxide which can be deposited as a film using dip coating or spin casting procedures. Modification of the solution equilibria through addition of complexants or control of pH, influences the localized chemical bonding which affects the density and refractive index of the deposited amorphous film. **[6,7,10]** The tendency of such films to transform to a new phase upon heating or compression is similarly controlled. This technique produces single component metal oxide films of high optical quality and also is appropriate for deposition of multi-component metal oxide phases provided that the organic precursors are mutually soluble in an appropriate solvent. A second solution-based deposition method requires glycine complexation of metal nitrate precursors in water. **[11]** The glycine acts to complex the metal cations, enhance solubility, and frustrate precipitation of low solubility cation species to maintain the target stoichiometry. Following spin casting, thermal treatment of the amorphous precursor induces an oxidation-reduction reaction which removes both nitrate and glycine from the evolving oxide matrix. This procedure is particularly amenable to the preparation of multi-component metal oxide films such as doped garnet (Sm^{+3}:$Y_3Al_5O_{12}$) and ruby (Cr^{+3}:Al_2O_3). The stress state of these films changes throughout processing. The evaluation and regulation of the residual film stress allows control over the physical properties of the film and the phase stability.

A number of techniques currently are used to evaluate stress in thin films. Perhaps the best known method requires deposition of the film on a thin substrate. Stress can then be evaluated from the extent of deformation exhibited by the substrate.**[12-14]** A second principal method involves X-ray diffraction measurements using the "$\sin^2 \psi$" method where changes in the lattice constant are determined directly. Based upon an appropriate model, the lattice strain can be converted to a stress. **[15-18]** These methods are not easily applied to *in situ* analysis since they require in one case a thin substrate and in the second case, considerable time for data collection and analysis.

Optical spectroscopic probes present an attractive alternative for determination of stress in thin films. Raman methods have been used for a number of years to evaluate stress and stress homogeneity in crystalline dielectric films. **[19-21]** Deviations in measured phonon frequencies from values characteristic of an unstressed single crystal correlate with residual film stress. **[22]** Through careful choice of laser excitation wavelength and scattering geometry, non-destructive depth profiling of the film stress can be achieved. **[23,24]** The principal advantages of this method include rapid measurement time and the ability to acquire the data *in situ* while the film is being deposited or subsequently modified by exposure to conditions (temperature and/or pressure) sufficient to induce transformation to a different phase. Transformation rate constants can be evaluated from time dependent increases in line intensity of the evolving phase in analogy to XRD methods. **[25]**

A second optical method relies on laser-induced fluorescence of cation dopants in oxide matrices and is the basis for the remote determination of static pressure in diamond anvil cell measurements. **[26]** The method is based upon a correlation between red shifts in the fluorescence emission wavelength (ruby, samarium-doped garnet) and pressure for selected emission lines. **[27]** The fluorescence spectrum also serves to uniquely identify a specific crystalline phase. Therefore, time-resolved fluorescence measurements can be used in much the same way as Raman data to quantify stress in films and characterize phase transformation processes.

In this paper, Raman spectroscopy and laser-induced fluorescence measurements are used to evaluate stress in oxide films and follow in real time the dynamics of thermally induced phase transformations. The methodology, advantages and limitations of each method are reviewed and related experiments are described for evaluating the interfacial stress which may differ from the residual stress usually measured in dielectric films. Transformation phenomena in films produced via sol-gel routes (TiO_2) and the glycine-nitrate method (Sm^{+3}:$Y_3Al_5O_{12}$, and Cr^{+3}:Al_2O_3) will be discussed. Empirical correlations between phase stability and processing parameters are described based upon the spectroscopic data.

EXPERIMENTAL

Solution Preparation of Thin Ceramic Films

Films Deposited Using Sol-Gel Methods

Solutions containing varying amounts of $TiCl_4$, $Ti(OC_2H_5)_4$, $Ti(OC_4H_9)_4$, ethanol, butanol, HCl (ethanolic solution), LiCl, and H_2O were prepared at room temperature with the compositions listed in **Table I**. Previous results, based upon Raman and fluorescence measurements, indicated that marked changes in solution species occurred as a function of solution reaction time. **[7,28]** Therefore, following mixing, the solutions were covered and allowed to equilibrate at a temperature of 293 K for different periods of time in order to evaluate the influence of reaction time and solution composition on the phase stability and stress state of the processed films. Amorphous films 200 nm in thickness were formed by spin casting (1 min, 1000 to 3000 rpm) equilibrated solutions onto cleaned 2.54 cm diameter fused silica flats. Two specific subjects were investigated: (1) effect of solution chemistry on thin film phase composition, and (2) effect of solution equilibration time on phase stability (crystallization kinetics). The stress state of the film was evaluated in each of these experiments from shifts in Raman active phonon frequencies relative to a single crystal standard. **[22]**

Films Deposited from Metal Nitrate/Glycine Aqueous Solutions

A new route to the production of thin ceramic films has been developed in which metal nitrates and a low molecular weight amino acid are intimately combined in aqueous solution, deposited as an amorphous layer, and thermally crystallized. Glycine behaves as a cation complexant and generates an amorphous matrix throughout which cations are homogeneously distributed. This method has been used previously to prepare yttria-stabilized zirconia films as well as films of the high T_c superconductor $YBa_2Cu_3O_{7-x}$. **[11]** Samarium doped garnet (Sm^{+3}:$Y_3Al_5O_{12}$) and ruby (Cr^{+3}:Al_2O_3) films have been derived from aqueous precursor solutions containing the requisite metal nitrate salts in the proper proportion and glycine (NH_2CH_2COOH). The ratio of glycine to total metal cation content had a typical value of 3, and cation dopant concentrations ranged from 0.05 to 0.25 weight percent. Aqueous solutions were concentrated by evaporation to facilitate spin casting operations. Cleaned silica substrates were coated with precursor solutions at spin rates of 8000 rpm. Thermal treatment of the as-deposited amorphous films at temperatures near 1125 K (garnet) and 1475 K (ruby) were required to generate the target crystalline phases. Film thickness averaged 100 nm. The crystalline phase and isothermal crystallization kinetics have been evaluated through measurements of the laser-induced fluorescence.

Table I. Sol-Gel Solution Composition, Phase Composition and Film Stress

Titania Films Deposited Immediately after Mixing
Heat Treated to 525 K for 6 Hours
(A = Anatase; R = Rutile)

	System	Composition (M/l) Ti	H_2O	HCl	LiCl	$E_g(cm^{-1})^a$	Phase
	(C_2H_5OH + $TiCl_4$)						
1.	No Additions	0.50	0.00	0.00	0.00	146	80% A, 20% R
2.	HCl Added	0.50	2.80	0.60	0.00	-	100% R
	(C_2H_5OH + $Ti(OC_2H_5)_4$)						
3.	No Additions	0.54	0.00	0.00	0.00	143	100% A
4.	HCl Added	0.54	11.0	1.20	0.00	144	100% A
5.	2x HCl Added	0.54	11.0	2.40	0.00	146	100% A
6.	HCl + LiCl	0.54	11.0	1.20	1.08	147	100% A, Trace R
7.	2x HCl + LiCl	0.54	11.0	2.40	1.08	148	60% A, 40% R
	(C_4H_9OH + $Ti(OC_4H_9)_4$)						
8.	No Additions	0.54	0.00	0.00	0.00	-	Amorphous
9.	LiCl Added	0.54	0.00	0.00	1.08	-	Amorphous
10.	HCl Added	0.54	11.0	2.40	0.00	144	100% Anatase
11.	HCl + LiCl	0.54	11.0	2.40	1.08	149	50% A, 50% R

[a] Raman mode frequency used to evaluate film stress. A value of 143 cm^{-1} is found for an unstressed single crystal of anatase.

Optical Measurements

Film thickness, homogeneity, and refractive index of amorphous and crystallized films were determined by means of optical microscopy, transmission spectroscopy, and ellipsometry.

Time-resolved Raman spectroscopy was used to follow the isothermal crystallization kinetics of amorphous titania films deposited on silica substrates. A copper resistance furnace, designed to hold samples at 65° relative to the optical axis of the spectrometer, was mounted in the sample chamber of a SPEX model 1877 triple spectrometer (1200 groove/mm grating in the dispersion stage). Raman scattered light excited by 100 mW of 488 nm radiation from a CW argon ion laser was dispersed onto a Princeton Instruments liquid nitrogen cooled CCD detector (1152 x 278 diodes). Slitwidths were maintained at 250 μm for all measurements. Sample temperatures were determined by means of a thermocouple in contact with the coated substrate and were maintained to $\pm$2 K using a proportional temperature controller. The spectrometer was calibrated on the basis of the known room temperature Raman frequencies of a single crystal anatase standard. Spectra were recorded as a function of time immediately following insertion of the amorphous titania sol-gel coated substrates into the preheated furnace. Typical exposure times averaged 1 sec, allowing collection of sufficient data for complete characterization of the transformation process.

Changes in phonon frequencies in response to an applied hydrostatic pressure and at various temperatures were measured for titania (anatase) crystals constrained in a heated diamond anvil cell (Merrill-Bassett design) containing an alcoholic pressure medium. The sample was located within a 250 μm cavity in an inconel gasket which was sandwiched between the diamond anvil faces. Pressures to several hundred kilobar could be achieved and were determined from the wavelength shift of a samarium doped garnet pressure calibrant. **[27]** In this experiment, a Raman microprobe (Spex Model # 1482) was used to image the sample, deliver the probe excitation, and collect the backscattered Raman emission.

The same instrumentation was used to measure laser induced fluorescence emission from the samarium garnet and ruby films excited by 100 mW of 488 nm or 514.5 nm radiation respectively from the ion laser. Time resolved intensity measurements during thermally induced crystallization of these films also were obtained by means of the CCD detector. The fluorescence lifetimes of the principal emission lines were measured using a phase modulation technique. **[29]** Here, the incident excitation was modulated at 250 Hz and the phase shift in the fluorescence emission at a spectrometer selected wavelength was synchronously detected by means of a photomultiplier tube and lock-in amplifier. The time resolution for this measurement was 1 sec.

ANALYTICAL METHODS

Raman Spectroscopy of Dielectric Films

Phase Characterization
Temperature and Pressure Effects

The utility of Raman spectroscopy to identify particular phases of a material is illustrated in **Figure 1** which depicts the vibrational spectra of anatase (C), rutile (B) and amorphous (A) sol-gel films all deposited on silica substrates. The rutile and anatase phases are easily distinguishable from one another. In the amorphous TiO_2 film, features similar to the rutile phase are apparent, however, linewidths have increased by a factor of three and the overall Raman intensity is diminished by a factor of 50.

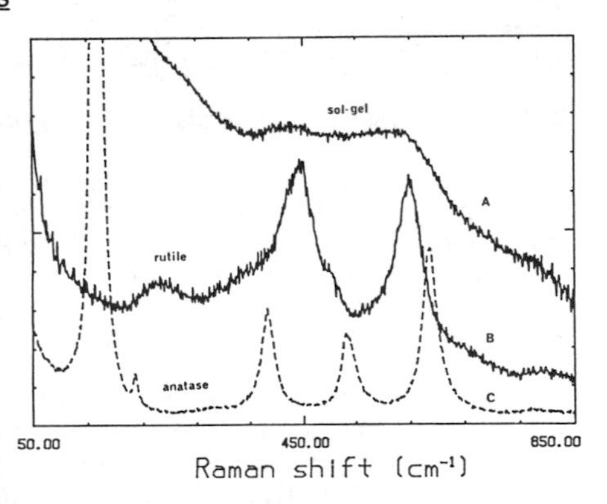

Figure 1. Raman spectra of TiO_2 films on silica.

Vibrational mode frequencies shift in response to an applied temperature or pressure as a result of anharmonicity in the vibrational potential energy. For a simple anharmonic oscillator, the square of the resonance frequency, $\omega^2(T)$, is predicted to vary linearly with temperature, T, according to equation 1. **[30,31]**

$$\omega^2(T) = \omega_o^2 \, [1 - (g_o^2/2f_o^2)k_BT] \quad . \tag{1}$$

Here, ω_o is the low temperature limit of the resonance frequency, k_B is Boltzmann's constant, and f_o and g_o are respectively the harmonic and first anharmonic contributions to the potential energy. Raman line intensities also exhibit a temperature dependence owing to thermal partitioning between the ground and excited vibrational states for each individual mode. The ratio of Stokes, I_s, to anti-Stokes, I_a, Raman intensities for each mode is proportional to the Boltzmann factor allowing temperatures to be computed from intensity measurements according to equation 2. **[32]**

$$I_s \, [\omega + \omega(T)]^4 / I_a \, [\omega - \omega(T)]^4 = \mathbf{S}(\omega) \, \exp[\hbar c \omega(T)/k_BT] \quad . \tag{2}$$

$\mathbf{S}(\omega)$ is the response function of the spectrometer used for the measurement, $\hbar$ is Planck's constant, c is the speed of light, and ω is the frequency of the probe laser.

The observed shifts of vibrational mode frequencies, ω_i, with applied pressure also arise as a result of anharmonicity in the vibrational potential energy. Assuming that the mode frequency is a function of volume, a change of volume (ΔV) in the material as a result of contraction or expansion produces a relative change in frequency ($\Delta\omega_i$) for each allowed mode as shown in equation (3),

$$\Delta\omega_i/\omega_i = -\gamma \Delta V/V \quad , \tag{3}$$

where the proportionality constant γ is called the mode Grüneisen parameter. **[33,34]** The variation of the mode frequency with respect to pressure then can be written as,

$$(\partial\omega_i/\partial P)_T = -\gamma\omega_i/V \, (\partial V/\partial P)_T = \gamma\kappa\omega_i \quad , \tag{4}$$

where κ is known as the isothermal compressibility of the material. Therefore, both pressure and temperature can affect vibrational mode frequencies. Conversely, a measurement of the vibrational frequencies can be used to evaluate temperature and pressure of a sample optically. The correspondence of pressure to residual stress in thin films is the basis for optical measurements of this quantity. **[28]**

Isothermal Crystallization Kinetics in Amorphous TiO_2 Films

Based on **Figure 1**, normalized Raman line intensities can be correlated with the extent of crystallization of amorphous films since the scattering cross section for the crystalline phase is significantly larger than that from the amorphous phase. **[34]** The evolving line intensity of the 143 cm^{-1} mode can be used to determine the time-dependent fractional crystallinity, $F(t)_T$, of an amorphous TiO_2 film during isothermal annealing. **Figure 2** illustrates a collection of successive spectra measured on a sample (# 5 in Table 1) annealed at 623 K which have been ratioed against a spectrum of the pristine amorphous film prior to crystallization. Measured spectra confirm ingrowth of the anatase phase under these conditions. **Figure 3** shows the normalized intensity increase for the 143 cm^{-1} mode as a function of time, $I_{143}(t)/I_{143}(t_\infty)$, where $I_{143}(t)$ is the time dependent intensity and $I_{143}(t_\infty)$ is the mode intensity of the completely crystallized film. The observed sigmoidal ingrowth behavior is represented quite well by equation (5) which has been used by Avrami to describe the crystallization kinetics of amorphous phases. **[35]**

$$F(t)_T = 1 - \exp(-k_{T,N}t^N) \ , \quad (5)$$

where $k_{T,N}$ is the temperature-dependent crystallization rate constant, t is time, and N is the critical growth exponent. Crystallization kinetics have been measured for titania sol-gel films as a function of solution processing time prior to coating deposition, and annealing temperature. Results are used to assess the stability of deposited films and follow phase transformation phenomena.

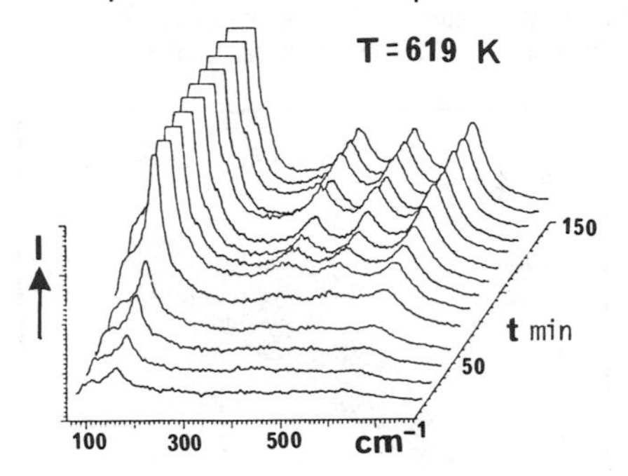

Figure 2. Time-Resolved Raman Spectra of an amorphous sol-gel deposited TiO_2 film showing growth of an anatase phase.

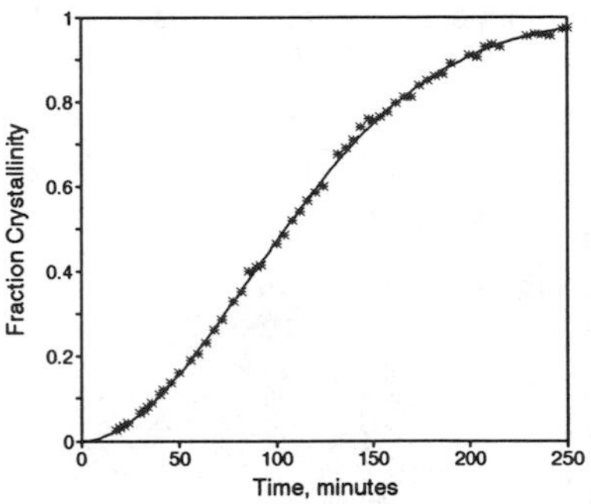

Figure 3. Normalized intensity data (143 cm^{-1} mode) and best fit (solid line) to equation 5.

Frequency Contour Analysis to Determine Film Stress and Temperature

As mentioned previously, expansion and contraction of a material induce changes in vibrational mode frequencies and this observation can be used to determine sample temperature and residual stress provided that the mode frequency shifts with temperature and pressure are known. In order to independently determine these variables the response to T and P of two uncoupled modes is required. The accuracy by which stress and temperature in a film may be determined improves if the two chosen modes exhibit contrasting behavior in their temperature and pressure responses. The anatase phase of TiO_2 is particularly amenable to these restrictions. **Figure 4a** depicts the vibrational frequency surface in pressure-temperature space for the high frequency E_g mode of an anatase single crystal constrained within a heated diamond anvil cell. The surface was constructed on the basis of equation 6, which is an empirical analytical representation of the measured pressure and temperature dependence of the mode frequency.

$$\omega_i = \omega_i^o + a_iP + b_iT + c_iPT \ . \quad (6)$$

Here, ω_i is the frequency of the mode of interest, and a_i, b_i, and c_i are fitting parameters. The intersection of a horizontal plane, constructed at a fixed frequency, with this surface defines a curve of constant frequency. There exist a number of T,P combinations which lie on this curve. Therefore, measurement of a single frequency is not sufficient to uniquely determine T and P. However, a simultaneous frequency measurement for an independent mode with contrasting behavior would allow a unique determination of T and P. **Figure 4b** shows the surface for the low frequency E_g

mode of anatase. Notice that in contrast to the high frequency mode, the frequency surface is increasingly positive at higher temperatures and pressures. A contour construction showing curves of constant frequency for both modes in P,T space appears in **Figure 4c**. The crossing points of these curves identify particular pressures and temperatures corresponding to the measured frequencies of the two independent modes. A numerical solution of equation 6 also can be used to extract the desired P,T information. When this formalism is applied to measurements of a thin film, the calculated pressure correlates with the cumulative residual stress in the film. The uncertainty in P and T determined by this method is related to the accuracy of the single crystal standard data and the contrast in curvature for the two modes in the contour plot. At temperatures exceeding 673 K, the estimated uncertainty in calculated T and P values exceeds ± 10% due to a softening of the low frequency mode.

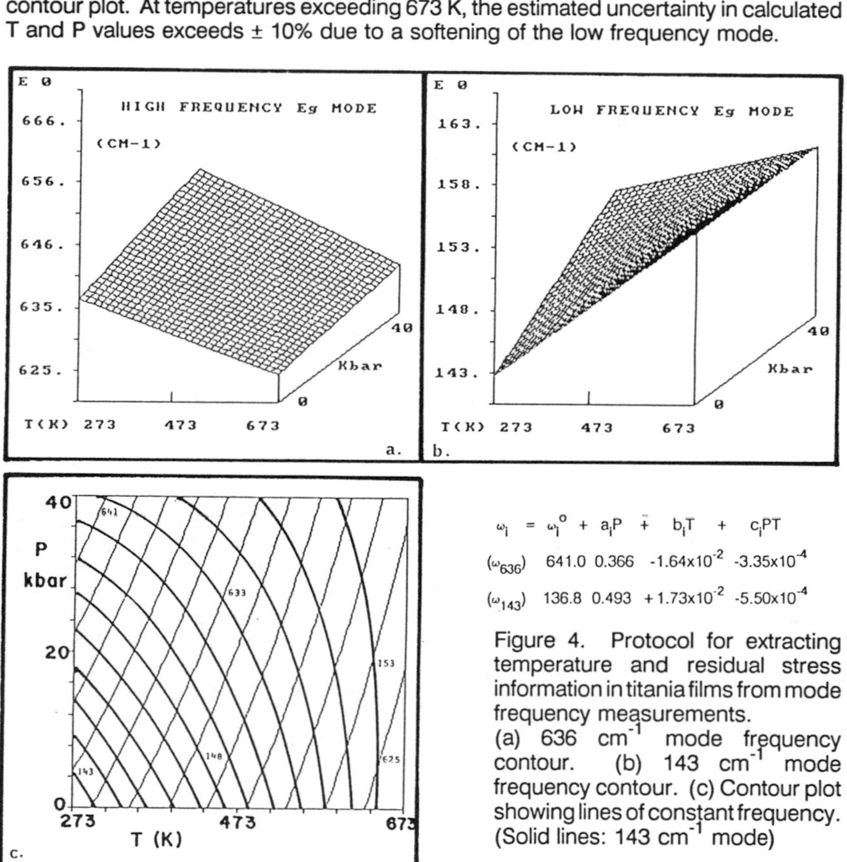

$$\omega_i = \omega_i^o + a_iP + b_iT + c_iPT$$

	ω_i^o	a_i	b_i	c_i
(ω_{636})	641.0	0.366	-1.64×10^{-2}	-3.35×10^{-4}
(ω_{143})	136.8	0.493	$+1.73\times10^{-2}$	-5.50×10^{-4}

Figure 4. Protocol for extracting temperature and residual stress information in titania films from mode frequency measurements. (a) 636 cm^{-1} mode frequency contour. (b) 143 cm^{-1} mode frequency contour. (c) Contour plot showing lines of constant frequency. (Solid lines: 143 cm^{-1} mode)

Fluorescence Spectroscopy of Dielectric Films

Phase Characterization, Temperature and Pressure Effects

Introduction of fluorescent cation dopants in ceramic oxides allows localized chemical structures to be probed as a result of the distinctive fluorescence exhibited

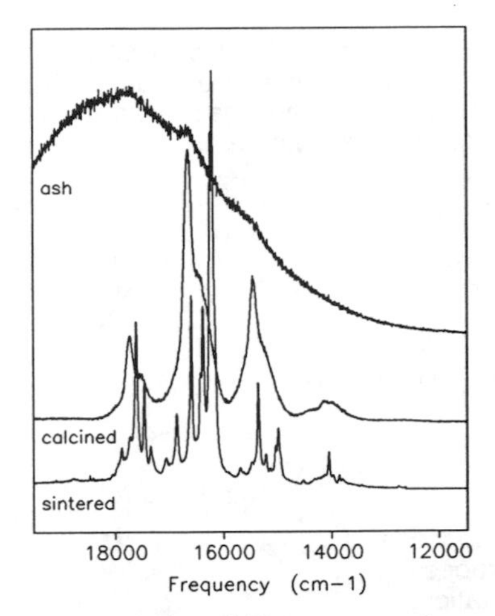

Figure 5. Induced Fluorescence from Sm^{+3}:$Y_3Al_5O_{12}$ irradiated at 488 nm as a function of processing temperature.

by the cation upon laser excitation. An example of the sensitivity of this technique is illustrated by measurements of samarium garnet (Sm^{+3}:$Y_3Al_5O_{12}$) prepared using the glycine-nitrate method. **Figure 5** depicts several laser induced fluorescence spectra (488 nm) of the initial material soon after preparation. The broad features in the spectrum are indicative of Sm^{+3} in a disordered environment. Following calcining at 973 K, the residual organic material is removed as signified by the disappearance of the broad fluorescence background. Further processing at temperatures above 1125 K generates the garnet structure with its own distinctive fluorescence. These results demonstrate the complexity which can accompany transformation processes in complex oxide materials. Recent x-ray diffraction data suggests that a transient perovskite phase also may be present below 1125 K.

The frequency and lifetime of the laser induced fluorescence in samarium-doped garnet are found to vary independently with pressure and temperature respectively as summarized in **Figure 6. [36]** These correlations suggest that this material can be used to optically probe ambient temperature and pressure. When deposited as a tin film, measured changes in fluorescence frequency can be related to film stress. In addition, introduction of a thin interfacial layer of this material at the interface between any film and substrate will be an effective probe of interfacial stress.

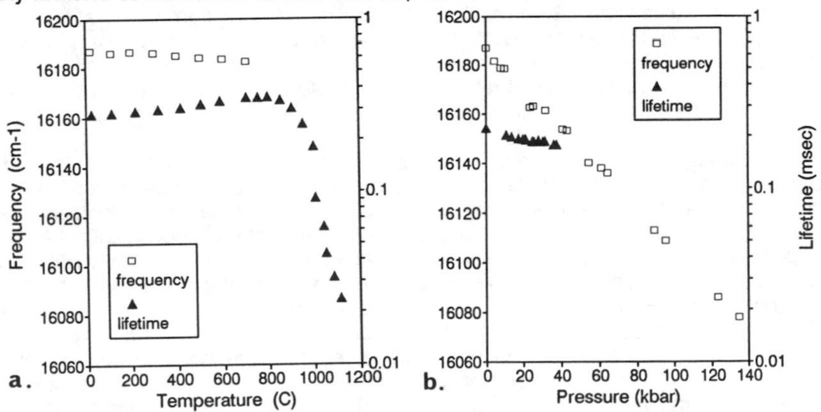

Figure 6. Variation in laser induced fluorescence frequency and lifetime with temperature (a) and pressure (b) for samarium doped garnet. (488 nm excitation)

Sample Temperatures from Fluorescence Lifetime Measurements

An example showing how temperature can be determined at a dielectric film-substrate interface involves preparation of a bilayer structure consisting of a thin layer of ruby on silica onto which titania is deposited. A schematic of this structure along with the measured response of the ruby R_1 fluorescence lifetime is illustrated in **Figure 7**. Simultaneous collection of both the red ruby fluorescence and the Raman scattered light is possible allowing an independent verification of temperature.

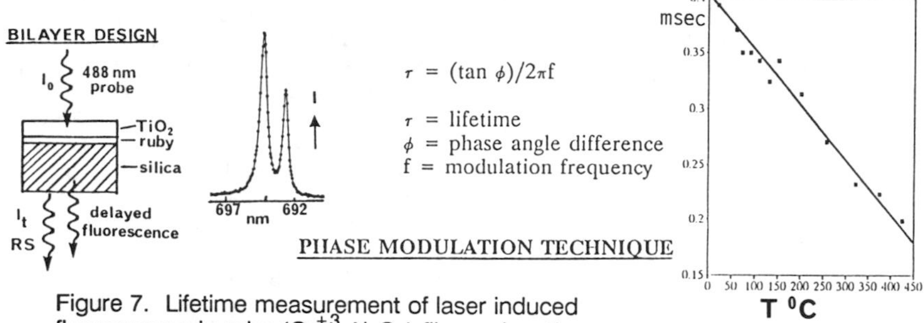

Figure 7. Lifetime measurement of laser induced fluorescence in ruby (Cr^{+3}:Al_2O_3) film probes the interfacial film temperature. (488 nm excitation)

IN SITU MEASUREMENTS OF TRANSFORMATION PHENOMENA

Isothermal Crystallization of Sol-Gel Deposited Titania

The properties and transformation kinetics of sol-gel deposited titania films are influenced by the solution chemistry and processing time. Table 1 shows that residual film stress and the stable crystalline phase which evolves during annealing depends upon the composition of the precursor solution. Increased solution equilibration time also affects the transformation kinetics and tends to retard crystallization. **[7]** Increased gelation time in low pH solutions also tends to lower the refractive index of spin cast amorphous titania films. Both results suggest that long solution equilibration times lead to the generation of increasingly larger polymeric titanium-oxygen n-mers. During spin casting, a larger fraction of free volume is incorporated into the film thereby lowering the measured index of refraction. Crystallization is likewise impeded since rearrangement of the higher titanium-oxygen polymers is restricted sterically. **[34]**

For all sol-gel deposited titania films subjected to isothermal annealing, a sigmoidal ingrowth of the crystalline phase has been observed which can be described in terms of the Avrami growth model. Measurements at different annealing temperatures indicate an activation energy for this process of 34 kcal/mole. However, critical Avrami exponents range from about 2 (hour equilibration) to 1.5 (week equilibration). The relatively small Avrami exponents are not consistent with the observed spheroidal microstructure. However, a modification to the Avrami formalism which incorporates diffusion-limited mass transport does predict critical exponents in accord with these measurements. During crystallization, the film densifies in the vicinity of the nucleation centers thereby restricting mass transport to the nucleation center. This concept is based upon the observed microstructure and presence of void shells around the growing spheroidal crystallites.

In addition to measuring Raman intensities to probe the crystallite ingrowth kinetics, associated Raman frequencies for each allowed mode also are available. For a titania film crystallizing at 625 K, both E_g mode frequencies are found to decrease during the course of film crystallization. Raman frequencies measured in real time and the corresponding temperature and film stress data determined using the formalism discussed previously are shown in **Figure 8**. Raman measurements demonstrate that compressive stress in the coating is relieved during the crystallization process. The degree of stress relaxation is related to the film deposition parameters and particularly to the temperature at which the substrate is held during film deposition.

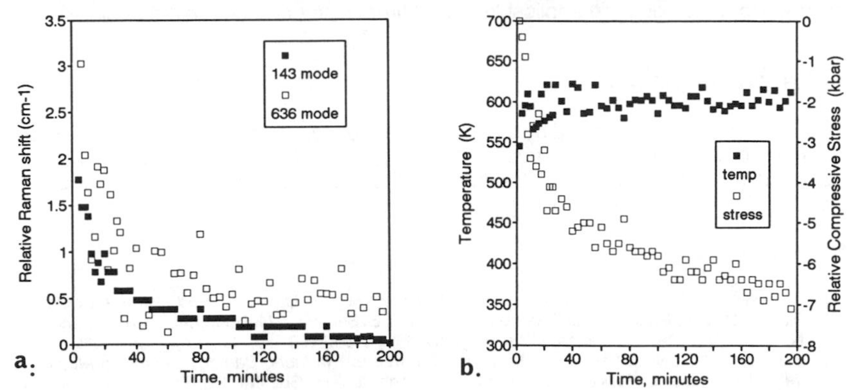

Figure 8. (a) Relative changes in the E_g mode frequencies for a titania sol-gel film at 625 K during isothermal crystallization. (b) Relative change in compressive film stress and temperature with time.

Fluorescence Measurements During Formation of Garnet

Fluorescence intensity changes may be used to follow ingrowth of a particular crystalline phase. For Sm^{+3}-doped yttria and alumina containing materials prepared using the glycine-nitrate process, formation of the garnet phase is identified from the observed fluorescence spectrum as seen in **Figure 5**. By following the intensity of the fluorescence specific to the garnet phase relative to the fluorescence from the initial amorphous phase the ingrowth kinetics may be studied as shown in **Figure 9**. The garnet phase evolves during heating at 1275 K after an induction time of about 10 min. Typical Avrami ingrowth kinetics are seen where the critical exponent in this case is 1.7 and the crystallization rate constant is 0.037 min^{-1}. The decrease in fluorescence of the amorphous precursor phase also is illustrated along with the measured processing temperature.

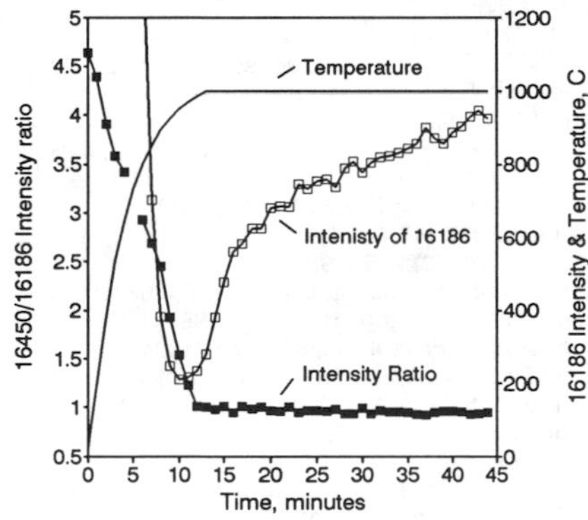

Figure 9. Evolution of fluorescence characteristic of Sm^{+3}:$Y_3Al_5O_{12}$ upon annealing of the amorphous precursor.

SUMMARY

Two optical methods for real-time evaluation of crystallite ingrowth kinetics and associated changes in interfacial stress in titania, ruby, and samarium-doped garnet films have been developed and are based upon time-resolved Raman spectroscopy and laser-induced fluorescence measurements. The contour analysis method (Raman measurements) appears to work well for titania films probably because the two E_g modes chosen for the analysis exhibit such contrasting behavior as a function of applied temperature and pressure. The fluorescence method appears to be general for all materials and is a good candidate method for evaluating film-substrate interfacial stress. Both methods are applicable to real-time *in situ* analysis of materials.

ACKNOWLEDGEMENTS

This work has been supported by the Materials Sciences Division of the Office of Basic Energy Sciences, U.S. Department of Energy. Pacific Northwest Laboratory is operated by Battelle Memorial Institute for the U.S. Department of energy under Contract DE AC06 76 RLO 1830.

REFERENCES

[1] A.J. Weber, A.F. Stewart, G.J. Exarhos, and W.K. Stowell in *Laser Induced Damage in Optical Materials: 1986*, **NIST** Special Publication 752, ed. H.E. Bennett, A.H. Guenther, D. Milam, and B.E. Newnam (U.S. Dept. Commerce, Washington, DC) pp. 542-556 (1988).
[2] B.J. Pond, J.I. DeBar, C.K. Carniglia, and T. Raj in *Laser Induced Damage in Optical Materials: 1988*, **NIST** Special Publication 775, ed. H.E. Bennett, A.H. Guenther, B.E. Newnam, and M.J. Soileau (U.S. Dept. Commerce, Washington, DC) pp. 311-319 (1989).
[3] G.J. Exarhos and K. M. Crosby, Proc. Soc. Photo-opt. Instrumen. Eng. **1438**:324 (1990).
[4] D.S. Rickerby, B.A. Bellamy, and A.M. Jones, Surf. Eng. **3**:138 (1987).
[5] F.L. Williams, R.D. Jacobson, J.R. McNeil, J.J. McNally and G.J. Exarhos, J. Vac. Sci. Tech. **A6(3)**:2020 (1988).
[6] M. Guglielmi, G. Scarinci, N. Maliavski, A. Bertoluzza, C. Fagnano, and M.A. Morelli, J. Non. Crys. Sol. **100**:292 (1988).
[7] W.S. Frydrych, K.F. Ferris, and G.J. Exarhos in *Atomic and Molecular Processing of Electronic and Ceramic Materials*, ed. I.A. Aksay, G.L. McVay, and T.G. Stobe, (Mat. Res. Soc. Pittsburgh) pp. 147 (1988).
[8] I.M. Thomas, Optics News **1986(7)**:18 (1986).
[9] S. Sakka, Ceram. Bull. **64(11)**:1463 (1985).
[10] T. Assih, A. Ayral, M. Abenoza, and J. Phalippou, J. Mat. Sci. **23**:3326 (1988).
[11] L.R. Pederson, L.A. Chick, and G.J. Exarhos, U.S. Patent #4,880,772 (1989).
[12] C.M. Drum and M.J. Rand, J. Appl. Phys. **39**:4458 (1968).
[13] J.D. Finegan and R.W. Hoffman, J. Appl. Phys. **30**:597 (1959).
[14] B. Haworth, C.S. Hindle, G.J. Sandilands, and J. R. White, Plas. Rubber Process. Appl. **2**:59 (1982).
[15] P.M. Ramsey, H.W. Chandler, and T.F. Page, Thin Solid Films **193/194**:223 (1990).
[16] H. Behnken and V. Hauk, Thin Solid Films **193/194**:333 (1990).
[17] A. J. Perry, M. Jagner, P.F. Woerner, W.D. Sproul, and P.J. Rudnik, Thin Solid Films **43/44**:234 (1990).
[18] D.I. Ma, S.B. Qadri, M.C. Peckerar, and D. McCarthy, Thin Solid Films **206**:18 (1991).
[19] G.J. Exarhos and D.M. Friedrich, Microbeam Analysis **1987**:147 (1987).
[20] T. Nishioka, Y. Shinoda, and Y. Ohmachi, J. Appl. Phys. **57(2)**:276 (1985).
[21] K. Kubota, M. Nakayama, H. Katoh, and N. Sano, Solid State Commun. **49(2)**:157 (1984).
[22] N.J. Hess and G.J. Exarhos, Proc. Soc. Photo-opt. Instrumen. Eng. **1055**:194 (1989).
[23] K. Yamazaki, M. Yamada, and K. Yamamoto, Jpn. J. Appl. Phys. Pt.1 **23(6)**:681 (1984).
[24] M.B. Stern, T.R. Harrison, V.D. Archer, P.F. Liao, and J.C. Bean, Sol. St. Comm. **51(4)**:221 (1984).
[25] G.J. Exarhos, and W. M. Risen, Jr., J. Am. Ceram. Soc. **57(9)**:401 (1974).
[26] J.D. Barnett, S. Block, and G.J. Piermarini, Rev. Sci. Instr. **44**:1 (1973).
[27] N.J. Hess and G.J. Exarhos, High Press. Res. **2**:57 (1989).
[28] W.S. Frydrych, G.J. Exarhos, K.F. Ferris, and N.J. Hess, Proc. Mat. Res. Soc. **121**:343 (1988).
[29] K.T.V. Grattan, R.K. Selli, and A.W. Palmer, Rev. Sci. Instrumen. **59(8)**:1328 (1988).
[30] P. Bruesch in *Phonons: Theory and Experiments I. Lattice Dynamics and Models of Interatomic Forces, Springer Ser. in Sol. St. Sci.* ***34***, ed. P. Fulde (Springer, Berlin) pp. 152 ff (1982).
[31] G.J. Exarhos and J. W. Schaaf, J. Appl. Phys. **69(4)**:2543 (1991).
[32] G.J. Exarhos and J. W. Schaaf, Proc. Soc. Photo-opt. Instrumen. Eng. **1055**:185 (1989).
[33] W.F. Sherman, Bull. Soc. Chimique **9/10**:I-347 (1982).
[34] G.J. Exarhos and M. Aloi, Thin Solid Films **193/194**:42 (1990).
[35] M. Avrami, J. Chem. Phys. **7(12)**:1103 (1939); **8(2)**:212 (1940); **9(2)**:177 (1941).
[36] N.J. Hess and D. Schiferl, J. Appl. Phys. **71(5)**:2082 (1992).

CHEMICAL PROCESSING OF FERROELECTRIC NIOBATES EPITAXIAL FILMS

SHIN-ICHI HIRANO, TOSHINOBU YOGO, KOICHI KIKUTA,
HISANOBU URAHATA, YASUHIDE ISOBE, TOSHIYUKI MORISHITA,
KOJI OGISO AND YASUHIRO ITO
Department of Applied Chemistry, Nagoya University,
Furo-cho, Chikusa-ku, Nagoya 464, Japan

ABSTRACT

Recent achievements in the chemical processing of ferroelectric niobates epitaxial films are reported. The molecular level designing of the precursor solution was stressed as well as the control of key processing factors. Examples are presented for $LiNbO_3$/Ti, K(Ta,Nb)O_3 and (Sr,Ba)Nb_2O_6 films from metal alkoxide precursors. Crystalline epitaxial films with stoichiometry could be synthesized through control of the intermediate state of starting molecules in solution. Water vapor stream was found to play an important role in crystallizing gel films to desired phases at relatively lower temperatures.

INTRODUCTION

Ferroelectric thin films have been receiving attention because of their potential for emerging applications, which include multilayer hybrid capacitors, nonvolatile memories, pyroelectric detectors, surface acoustic wave (SAW) devices, electrooptic, nonlinear optic, pyroelectric and photorefractive devices. A number of niobates are candidate ferroelectric thin films, as listed in Table 1, which have been extensively studied [1-12].

Among several processing methods, sol-gel processing has the potential for good homogeneity, ease of chemical composition control, high purity, low temperature processing, and large area and versatile shaping over physical methods. In addition to these advantages, the sol-gel method provides the capabilities of reproducible coating thickness and hybridization of functionalities. However, more research is required to establish the key processing factors which affect the microstructure-properties relationship, leading to better understanding and control of ceramic processing through chemistry. This paper describes the sol-gel processing of niobate thin films of $LiNbO_3$/Ti, K(Ta,Nb)O_3 and (Sr,Ba)Nb_2O_6 while focusing on the key control factors which are critical for the fabrication of high quality sol-gel films.

PROCESSING AND CHARACTERIZATION OF Ti DOPED $LiNbO_3$ FILMS

$LiNbO_3$ based optical waveguides have been processed by the proton substitution or the thermal diffusion of Ti metal in $LiNbO_3$ single crystals. In the conventional Ti diffused waveguide, Sugii et al. reported about the lattice contraction along the **a** axis, which causes misfit dislocations and cracks in the thermally diffused layer [13]. The wave guide channels thus fabricated have been known to smear due to the inhomogeneous thermal diffusion of Ti depending upon defects in crystals. The sol-gel processing does provide the

 ©1992 Materials Research Society

Table I Ferroelectric Niobate Thin Films

Materials	Phenomena	Applications
Related Perovskite Structure		
$LiNbO_3$ (LN) ($LiNbO_3$/Ti)	Piezoelectric Electrooptic	Pyrodector, SAW Waveguide device, SHG Optical modulator
K(Ta,Nb)O_3 (KTN)	Pyroelectric Electrooptic	Pyrodetector Waveguide device Frequency doubler Holographic storage
Pb($Mg_{1/3}Nb_{2/3}$)O_3 (PMN)	Dielectric	Memory, Capacitor
Tungsten-bronze Structure		
(Sr,Ba)Nb_2O_6 (SBN) (Pb,Ba)Nb_2O_6 (PBN) $(Sr,Ba)_{0.8}R_xNa_{0.4}Nb_2O_6$ (R: Cr,Zn,Y) (K,Sr)Nb_2O_6 (KSN) (Pb,K)Nb_2O_6 (PKN) $Ba_2NaNb_5O_{15}$ (BNN)	Dielectric Pyroelectric Electrooptic	Memory Pyrodetector Waveguide device

homogeneous solid solution of $LiNbO_3$/Ti at relatively low temperatures [7].

Figure 1 illustrates the experimental procedure of Ti doped $LiNbO_3$ films on sapphire substrates. In this work, refractive index of the Ti doped $LiNbO_3$ films prepared at 600°C were examined with an automated ellipsometer (MIZOJIRI OPT. Co. Ltd., DVA-36LD, λ=632.8 nm). Dielectric properties of $LiNbO_3$ films on Pt/sapphire C were also measured with LCR meter at 100 kHz.

Partial hydrolysis of double alkoxide precursors was found to be effective to crystallize the gel films at lower temperatures. Crystallization was enhanced by the heat-treatment of gel films in flow of O_2/water vapor mixture, which was found to accelerate the reaction among alkoxy-groups in gel films, leading to the avoidance of up-taking free carbons during crystallization.

The XRD profile of $LiNbO_3$/Ti films on sapphire C substrates crystallized at 550°C is shown in Fig. 2. The films with highly preferred orientation to (006) plane were successfully synthesized on the buffered layers of preapplied

LN films. The films prepared on A and C-plane of sapphire substrates show the epitaxial growth, which was confirmed by RHEED and pole figure of the $LiNbO_3$/Ti film on sapphire C-plane (0001). The FWHM of the Rutherford back scattering (RBS) spectra indicates that Ti exists homogeneously in the $LiNbO_3$ film without reaction with sapphire substrate.

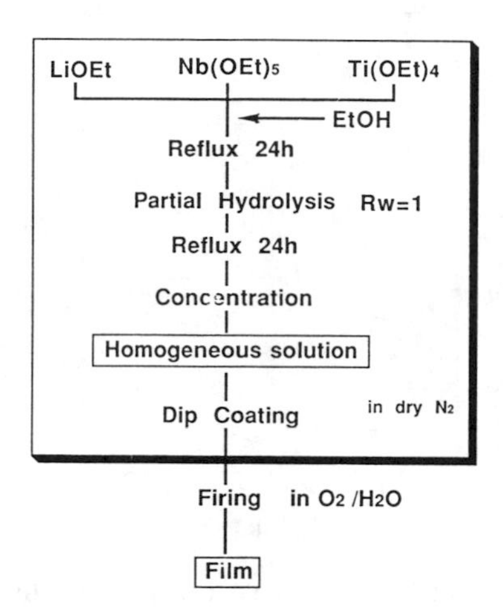

Fig. 1 Processing scheme for Ti-doped $LiNbO_3$ films.

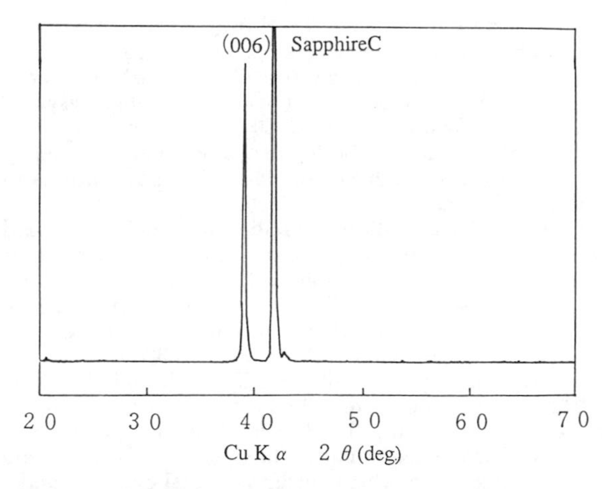

Fig. 2 XRD profile of Ti-doped $LiNbO_3$ thin film on sapphire (C) substrate (Buffer layer is used).

Dielectric constants of 3.0 mol% Li_2TiO_3 doped $LiNbO_3$ films were 45 and 7 for the epitaxial film and non-preferred film, respectively. Figure 3 shows the change of refractive index of the Ti doped $LiNbO_3$ films. A 2.5 mol% Li_2TiO_3 doped $LiNbO_3$ film showed the slightly higher refractive index. Further Ti doping induced the abrupt increase in these values, which makes it possible to fabricate the optical waveguide as proposed in the previous paper [6].

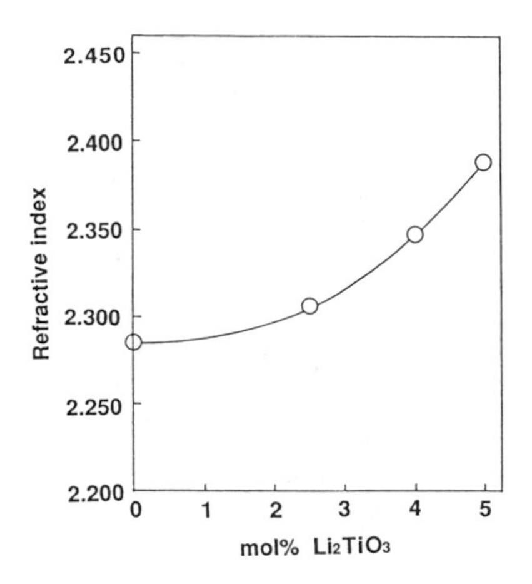

Fig. 3 Refractive index of Li_2TiO_3 doped $LiNbO_3$ thin films on $LiNbO_3$ (z) substrates.

PROCESSING OF $K(Ta,Nb)O_3$ FILMS

Potassium tantalate niobate ($K(Ta,Nb)O_3$ (KTN)) of perovskite structure has been known to have a ferroelectric character. Since the Curie temperature of KTN can be varied by controlling the Ta:Nb ratio, the key ferroelectric properties are compositional-dependent [14]. KTN single crystals have been grown by a modified Kyropoulos method. However, the growth of large KTN single crystals is difficult due to being a compositional inhomogeneity. The demand for KTN thin films has been increasing for applications in electrooptic devices and the pyroelectric device.

Potassium ethoxide, tantalum ethoxide and niobium ethoxide were used as starting materials. A certain amount ratio of tantalum ethoxide and niobium ethoxide was reacted in absolute ethanol to coordinate metal bondings in B-site of perovskite structure. Then a stoichiometric ratio of potassium ethoxide solution was added to react at refluxing temperature for 24 h.

The changes of coordination states in a KTN original precursor and partially hydrolyzed precursor were followed by NMR and FT-IR measurements. In the ^{13}C NMR spectra, bridging metal-oxygen bonds in the original precursor diminished to break-up, indicating that the condensation-polymerization reaction of metal alkoxides takes place. The behavior was also confirmed in FT-IR spectra as shown in Fig. 4, in which bridging metal-oxygen bonds and carbon-oxygen bonds were broken-up to form oligomers.

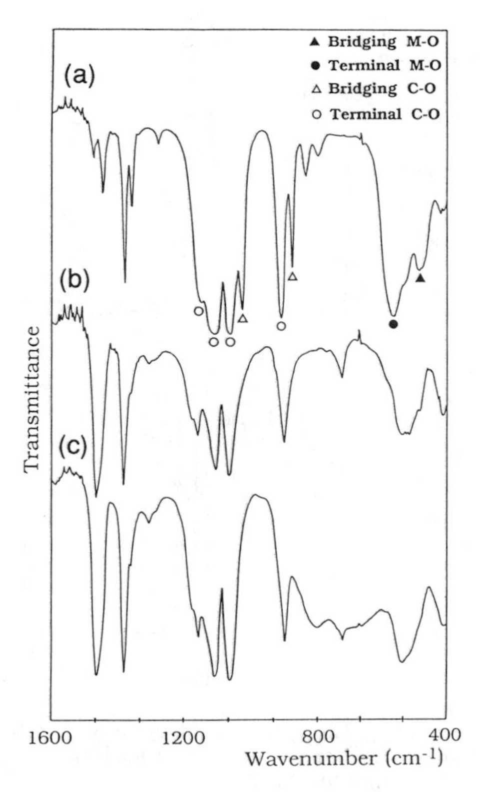

Fig. 4 FT-IR spectra of niobium ethoxide and KTN precursors
(a) $Nb(OEt)_5$, (b) KTN precursor, (c) partially hydrolyzed KTN precursor.

An MgO (100) plate was chosen as a substrate because of the similar oxygen packing with KTN (100). XRD profiles of the KTN thin films crystallized at 675°C are shown in Fig. 5. The formation of pyrochlore phase was observed after heat treatment without flow of water vapor (Fig. 5a). A mixture of water vapor and oxygen gas during calcination has a pronounced effect on the elimination of remaining organic components, leading to the formation of the preferred orientation of the perovskite phase (100) as shown in Fig. 5b. A similar effect was observed in $LiNbO_3$ [2-7]. Continuing the water vapor treatment hinders the further formation of the perovskite phase during the crystallization process, which might be attributable to potassium reacting with the hydroxy group and disturbing the network of the oxygen-metal bond at 675°C.

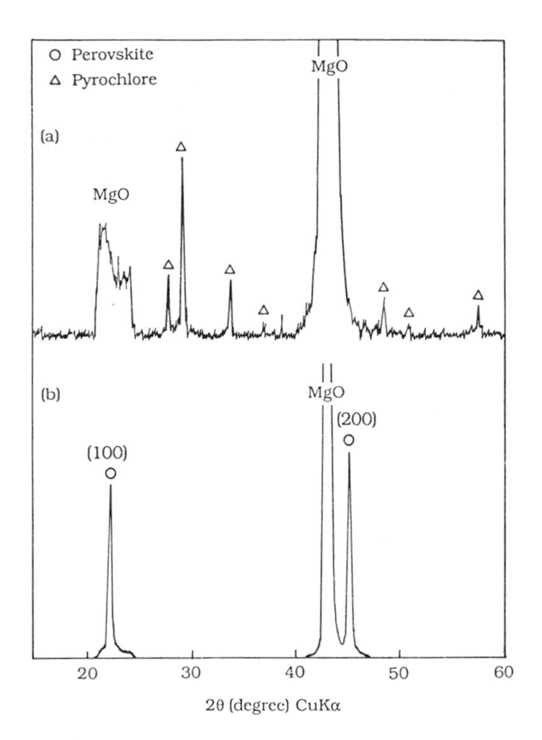

Fig. 5 XRD profile of $KTa_{0.65}Nb_{0.35}O_3$ thin films on MgO (100) substrates crystallized at 675°C after calcination through (a) oxygen at 300°C, (b) oxygen/water vapor at 300°C.

PROCESSING OF STRONTIUM BARIUM NIOBATE (SBN)

SBN thin films have been received a great attention as integrated devices. Neurgaonkar et al. [15] synthesized the epitaxial SBN films on SBN substrates by the liquid phase epitaxy in vanadium oxide flux. However, the precise control of composition is usually difficult in the flux growth.

In this work, SBN films were synthesized on MgO (100) substrates through the molecular designed metallorganic compounds. Figure 6 shows the experimental procedure to proceed the coordination reaction of metal alkoxides. Ethoxyethanol (EGMEE) was found to be an effective stabilizing agent of alkoxides. NMR spectra of the alkoxide solution treated with EGMEE indicated that the alkoxide ligands on metals were partially substituted for EGMEE, affording the stable homogeneous solution.

The XRD profiles of films crystallized at 670°C on an MgO (100) substrate is shown in Fig. 7, which indicates the preferred orientation of the (100) plane on MgO (100) substrates. An edge-on SEM photograph of the film is shown in Fig. 8. The film is very smooth and crack-free. The film quantities were affected by the concentration of the coating solution, withdrawal speed of substrate, and the heating rate.

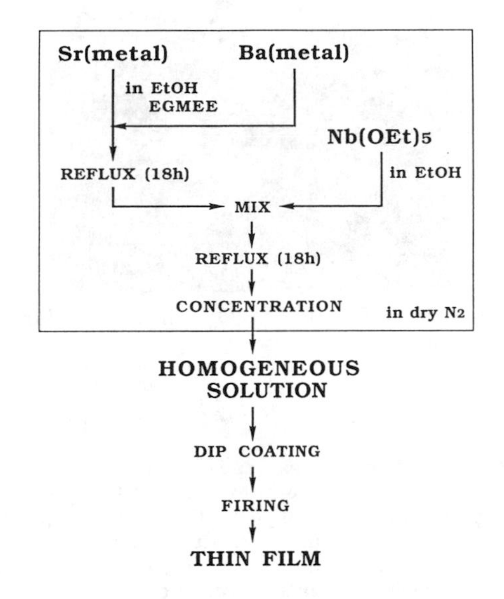

Fig. 6 Processing scheme for SBN thin films

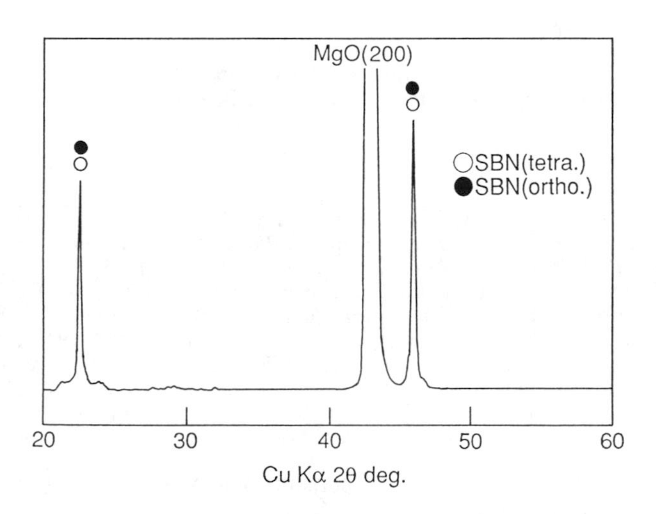

Fig. 7 XRD profile of $Sr_{0.5}Ba_{0.5}Nb_2O_6$ thin films on MgO (100) substrates crystallized at 700°C.

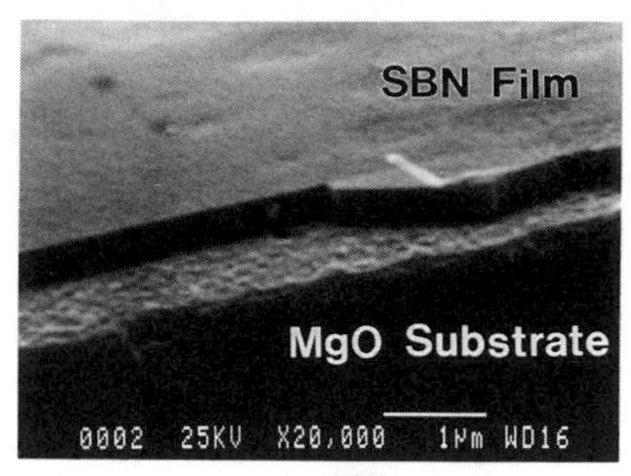

Fig. 8 SEM photograph of $Sr_{0.5}Ba_{0.5}Nb_2O_6$ thin film on MgO(100) crystallized at 700°C for 1h.

CONCLUSION

The chemical processing does realize a promising route to synthesize ferroelectric films, which lead to the integration of functionalities. The concept described in this paper can be applied to the chemical processing of ceramic films with controlled conditions and orientations at relatively low temperatures. The crystallization of films could be promoted by the controlled partial hydrolysis of metal alkoxides in intermediate coordination state and then the pre-heat treatment of alkoxy-derived films in flow of water vapor and oxygen gas mixture.

ACKNOWLEDGEMENT

This work was partly supported by Grant-in-Aid for scientific Research on Priority Areas, The Ministry of Education, Science of Culture.

REFERENCES

[1] S.Hirano and K.Kato, *Adv. Ceram. Mater.* **5** 142 (1987).
[2] S.Hirano and K.Kato, *J. Non-Cryst. Solid* **100** 538 (1988).
[3] S.Hirano and K.Kato, *Adv. Charm. Mater.* **3** 505 (1988).
[4] S.Hirano and K.Kato, *Bull. Chem. Soc. Japan* **62** 429 (1989).
[5] S.Hirano and K.Kato, *Solid State Ionics* **32/33** 765 (1989).
[6] S.Hirano and K.Kato, *Mater. Res. Soc. Symp. Proc.* **155** 181 (1989).
[7] S.Hirano, K.Kikuta and K.Kato, *Mater. Res. Soc. Symp. Proc.* **200** 3 (1990).
[8] K.Kikuta, W.Sakamoto and S.Hirano, *Ceram. Trans.* **12** 717 (1990).
[9] S.Hirano, T.Yogo and K.Kikuta, *J. Ceram. Soc. Japan* **99** 1026 (1991).
[10] S.Hirano, T.Yogo, K.Kikuta, K.Kato, W.Sakamoto and S.Ogasahara, *Ceram. Trans.* **25** 19 (1992).
[11] S.Hirano, T.Yogo, K.Kikuta, T.Morishita and Y.Ito, *J. Amer. Ceram. Soc.* **75** in press (1992).
[12] S.Hirano, T.Yogo, K.Kikuta and K.Ogiso, *J. Amer. Ceram. Soc.* **75** in press (1992).
[13] K.Sugii, M.Fukuma and H.Iwasaki, *J. Mater. Sci.* **15** 523 (1978).
[14] S.Triebwasser, *Phys. Rev.* **114** 63 (1959).
[15] R.R.Neurgaonker and E.T.Wu, *Mater. Res. Bull.* **22** 1095 (1987).

CHEMICAL PROCESSING AND PROPERTIES OF NANOCRYSTALLINE $BaTiO_3$

Z. XU, H.K. CHAE*, M.H. FREY AND D.A. PAYNE
Department of Materials Science and Engineering, Materials Research Laboratory and Beckman Institute, University of Illinois at Urbana-Champaign, Urbana, IL 61801

ABSTRACT

Barium titanate ($BaTiO_3$) thin layers were prepared by sol-gel processing. Details are reported for the synthesis route from methoxyethoxide precursors. Partially hydrolyzed solutions were spin-cast onto metallized silicon substrates, and a multilayering technique was used to develop sub-micron structures. Information is reported on the thermal processing conditions used and the development of structure. Crystallization started at 600°C and was fully developed by 700°C. The room temperature structure was cubic, and the grain size was 25-50 nm. Details are reported for the dielectric properties. Ferroelectricity was not observed. Dielectric constant (200-300) increased with increasing grain size, and was stable with respect to temperature, field and frequency. The properties are attractive for potential integrated capacitor applications.

INTRODUCTION

In this paper we report on the sol-gel processing of $BaTiO_3$ dielectrics with dense nanocrystalline microstructures. Our interest lies in the integration of thin layers with silicon and the development of new devices. We expect unusual properties for $BaTiO_3$ at the size and scale of the nanostructural level where cooperative long-range ordering forces should diminish. In fact, we shall demonstrate an absence of ferroelectric properties for $BaTiO_3$ at the 25-50 nm size level, with no polarization-reversal or Curie-Weiss characteristics. The resulting stability with respect to voltage and temperature make nanocrystalline $BaTiO_3$ dielectrics attractive for capacitive applications. Emphasis is placed on a materials chemistry approach to the fabrication of sub-micron structures by sol-gel processing.

EXPERIMENTAL PROCEDURE

Barium and titanium alkoxide precursors (methoxyethoxides) and complexes were prepared by a method originally developed by Campion et al. [1,2]. Briefly, a solution of barium methoxyethoxide $[Ba(OCH_2CH_2OCH_3)_2]$ was prepared by reaction of excess 2-methoxyethanol ($CH_3OC_2H_4OH$, Aldrich, 99.99%) with Ba metal (Alfa, 99.99%) under dry nitrogen. Similarly, titanium methoxyethoxide $[Ti(OCH_2CH_2OCH_3)_4]$ was prepared by the exchange reaction of titanium isopropoxide (Aldrich, 99.99%) with excess methoxyethanol. Repeated distillation and vacuum drying gave a clear viscous liquid. Finally, a solution of

*Present address: Korea Institute of Science and Technology, Seoul 130-650, Korea

titanium methoxyethoxide in methoxyethanol was added to barium methoxyethoxide and refluxed. For concentrations less than 0.5M in $BaTiO_3$ stable clear brown solutions were obtained. The solutions were partially hydrolyzed prior to the spin-casting of thin layers. For partial hydrolysis, a barium titanate methoxyethoxide solution was combined with a 2-methoxyethanol solution which contained 2 moles of water per mole of $BaTiO_3$. The concentration of the final solution was adjusted to 0.2M in $BaTiO_3$ prior to spin-casting. Under these conditions, dense crack-free thin layers were formed.

Thin layers were deposited onto silicon substrates by a multilayering spin-casting technique. The substrates consisted of (100) Si wafers with a thermally grown SiO_2 layer (~700 nm) which was sputter-coated with Ti (~20 nm) followed by Pt (~150 nm). Spin-casting was carried out in a dust-free environment in a clean-hood. The precursor solution was syringed through a filter (0.2 µm) onto the substrate, and spin-cast at 3000 rpm for 50 seconds. After each deposition, the specimens were either heat-treated on a hot plate (300°C, 1 m) or rapidly treated in a box furnace (750°C, 1 m) following an established method [3]. The deposition and heat treatment procedure was repeated until the desired thickness was achieved. A final heat treatment was necessary to complete crystallization.

Phase development was determined by X-ray diffraction (XRD), and scanning electron microscopy (SEM) was used to determine layer thickness and grain size. The interface between $BaTiO_3$ and the metallized silicon substrate was probed by scanning Auger electron spectroscopy (SAES). Details within the microstructure were resolved by transmission electron microscopy (TEM).

Gold electrodes ($4.5x10^{-4}$ cm^2) were sputter-deposited through a mask onto the surface of $BaTiO_3$ dielectrics. Values of dielectric constant (K) were calculated from measurements of capacitance (C) obtained from an impedance analyzer (HP 4192A) at 25mV/µm a.c. oscillation level. A d.c. bias voltage (V) was applied with field strengths (E) up to 300kV/cm to obtain C-V and dielectric saturation characteristics. Also, the current (I)-V characteristics were monitored to dielectric breakdown. A hot-stage was used to determine the temperature dependence of dielectric properties. All contact measurements were made through use of micromanipulators and an optical microscope.

RESULTS AND DISCUSSION

Figure 1 gives XRD data for $BaTiO_3$ thin layers which were formed from partially hydrolyzed precursors and heat-treated for 1h. Crystalline $BaCO_3$ could be detected after 500°C but was absent above 650°C. This is in concurrence with previous FTIR data [1]. $BaTiO_3$ started to crystallize by 600°C and was fully developed by 700°C. No tetragonal peak splitting could be detected for $BaTiO_3$, even after 750°C for 1 h (Fig. 1 (b)). A pseudocubic lattice parameter of 0.397(6) nm was calculated at 25°C.

Figure 2 illustrates the evolution of microstructure, as observed from fired surfaces, in the SEM. After heat treatment at 400°C for 1h (Fig. 2 (a)) the $BaTiO_3$ layer was dense, smooth and featureless. After heat treatment at 750°C for 1 h the layer was also dense, smooth, but crystallized into a nanostructure (~25 nm). Prolonged heat treatment at 750°C for 10h did not promote any significant grain growth. Larger grain sizes at the nanostructure level were obtained by heat

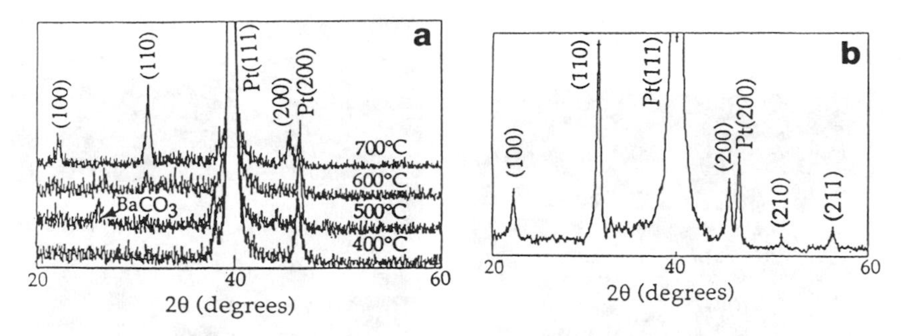

Figure 1. (a) XRD data for $BaTiO_3$ thin layers heat treated at different temperatures; (b) typical data for a thin layer heat treated at 750°C for 1h.

treatment at higher temperatures, e.g., ~40 nm at 850°C, 1h (Fig. 2 (c)) and ~50 nm at 950°C, 15m (Fig. 2 (d)). The latter condition developed some incipient discontinuous grain growth, but for all cases, SAES analysis did not detect any significant reaction between $BaTiO_3$ and the Pt interface.

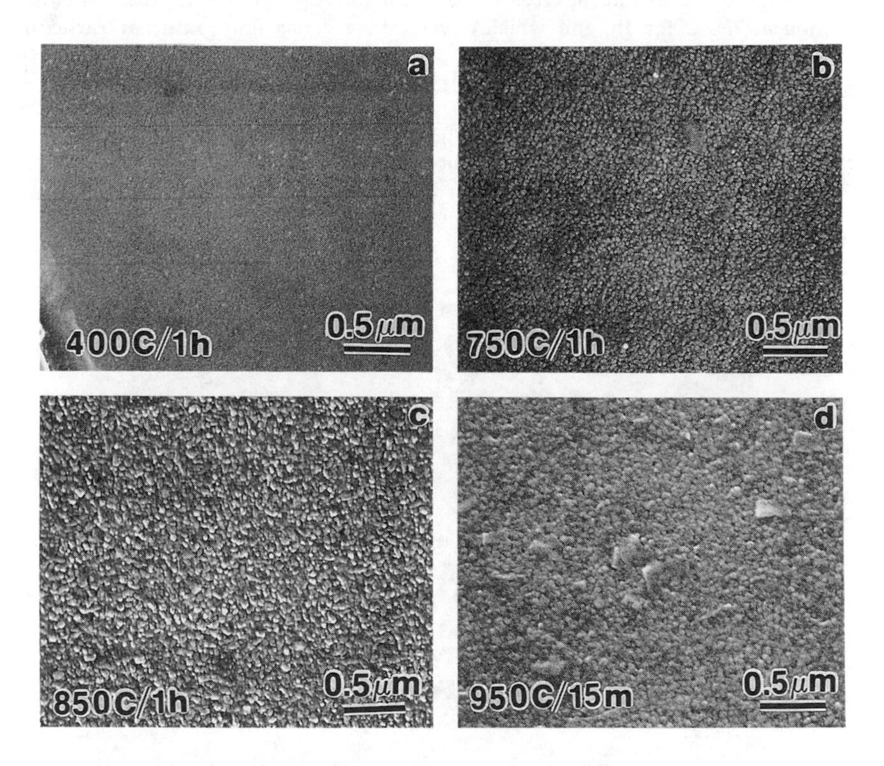

Figure 2. Microstructures of $BaTiO_3$ thin layers heat treated at: (a) 400°C/1h; (b) 750°C/1h; (c) 850°C/1h; (d) 950°C/15m.

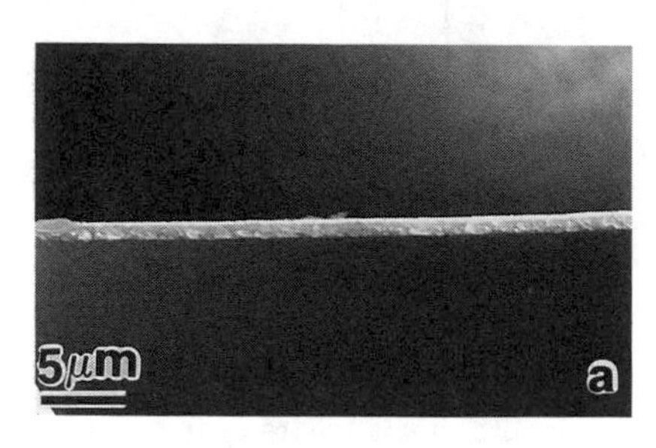

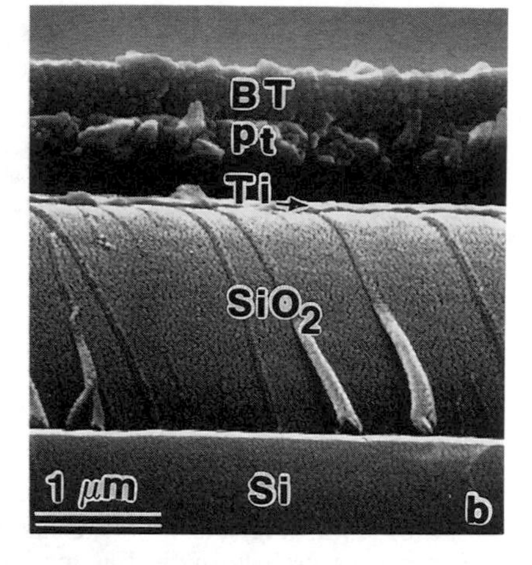

Figure 3. SEM photomicrographs of a cross-section of a thin layer heat treated at 750°C for 1h.

Figure 3 illustrates a cross-section of a $BaTiO_3$ thin layer, which was heat treated at 750°C for 1h, and which developed a dense fine grain nanostructure. Figure 3 (a) illustrates the uniformity of the layer thickness, and Fig. 3 (b) the integrity of the Pt/Ti metallized barrier between the $BaTiO_3$ dielectric and the Si/SiO_2 substrate. TEM analysis (Fig. 4) confirmed the grain size was ~25 nm after heat treatment at 750°C for 1h (cf., Fig. 3 (b)). Ferroelectric domains were not observed in the fine grain nanostructure. Occasional twin structures were observed, as indicated by arrows in Fig. 4, but they did not disappear on heating in the hot-stage of the TEM to 500°C. They were thought to be growth twins, not transformation twins.

Figure 4. TEM photomicrograph of the nanostructure in a $BaTiO_3$ thin layer heat treated at 750°C for 1h.

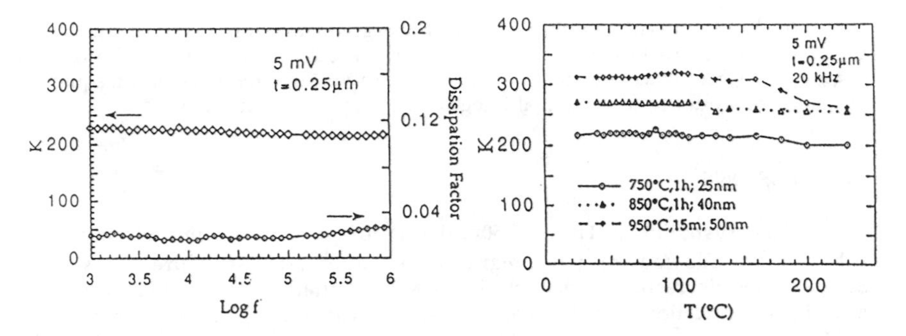

Figure 5. Frequency dependence of dielectric constant and dissipation factor for a thin layer (750°C/1h).

Figure 6. Temperature dependence of dielectric constant for thin layers heat treated at different temperatures.

The dielectric properties for $BaTiO_3$ capacitors integrated on Si are illustrated in Figs. 5-8. Figure 5 indicates good dielectric quality (dissipation factor < 0.02) for thin layers crystallized at 750°C for 1h, with no dispersion in the measured frequency range. The calculated value of dielectric constant was K=230 at room temperature, i.e., for the 25 nm grain size. Prior to crystallization, the dielectric constant of the dense amorphous layer was K=25. The temperature dependence of K for crystalline material is illustrated in Figure 6, which indicates the absence of a dielectric anomaly at 130°C where the ferroelectric Curie-point (T_c) should normally occur. On cycling ± 100°C around this temperature, Curie-Weiss behavior was not observed. Figure 6 also indicates that values of dielectric constant increased with increasing heat-treatment conditions, i.e., as the grain size grew from 25 to 50 nm. Typical values of dielectric constant were K=270 at 40 nm and K=310 at 50 nm. Ferroelectric hysteresis loops were not observed by polarization-reversal measurements, only linear dielectric behavior. Figure 7 indicates weak saturation of dielectric constant with increasing applied field strengths up to

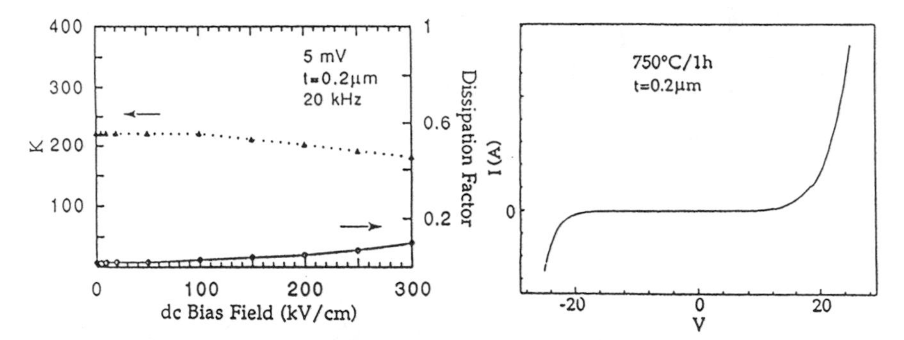

Figure 7. dc bias dependence of dielectric properties for a thin layer heat treated at 750°C for 1h.

Figure 8. I-V curve for a thin layer heat treated at 750°C for 1h.

300kV/cm. Current-voltage characteristics are given in Figure 8, and indicate a breakdown strength in excess of 1 MV/cm for a 0.2 μm thin layer. The overall response characteristics are stable with respect to voltage, temperature, frequency, etc., and are attractive for potential integrated capacitor applications.

CONCLUSIONS

Nanocrystalline $BaTiO_3$ (25-50nm) was found to be nonferroelectric. The sol-gel route allowed for the integration of sub-micron structures on silicon. Stable dielectric response characteristics were suitable for a wide variety of capacitive applications, including, decoupling capacitors for VLSI circuits and DRAM's.

ACKNOWLEDGEMENTS

The research was supported by the U.S. Department of Energy under Contract No. DE-DMR-FG02-91ER45439. The use of facilities in the Center for Microanalysis of Materials at the University of Illinois is greatly acknowledged. We are grateful for the continued interest and advice of Professors W.G. Klemperer and D.L. Wilcox in the research.

REFERENCES

1. J-F Campion, D.A. Payne, H.K. Chae and Z. Xu, Ceram. Trans. **25**, 477 (1991).
2. J-F Campion, J.K. Maurin, D.A. Payne, H.K. Chae and S.R. Wilson, Inorg. Chem. **30**, 3244 (1991).
3. L.F. Francis and D.A. Payne, J. Am. Ceram. Soc., **74**, 3000 (1991).

OPTIMIZATION OF SOL-GEL DERIVED PZT THIN FILMS BY THE INCORPORATION OF EXCESS PBO

G. TEOWEE, J.M. BOULTON AND D.R. UHLMANN
Arizona Materials Laboratories, Department of Materials Science and Engineering, University of Arizona, Tucson, AZ 85721.

ABSTRACT

A series of PZT precursor solutions was prepared which incorporated excess PbO to give the composition $Pb_{1+x}Zr_{0.53}Ti_{0.47}O_{3+x}$, where $0 \leq x \leq 0.3$. These solutions were spin coated on platinized Si wafers and fired at elevated temperatures up to 750C for 30 mins. After crystallization into single-phase perovskite, the films were studied using XRD, optical microscopy and electrical characterization techniques (hysteresis loops and dielectric properties). It was found that the presence of excess PbO significantly improved the PZT films in terms of phase assemblage, microstructure and electrical properties. Under optimized conditions, films with dielectric constants of around 3000 can be obtained.

INTRODUCTION

PZT films are widely used (or proposed to be used) in a gamut of electrical, optical and electrooptic applications including infrared sensors [1], neural networks [2], waveguides [3], optical switches [4], second harmonic generation [5], optical data storage [6] and ferroelectric memory [7].

There are many techniques to deposit PZT films which can be classified into either physico-chemical or wet chemical approaches. These techniques include magnetron sputtering [8], d.c. diode sputtering [9], ion beam assisted sputtering [10], evaporation [11], laser ablation [12], plasma spraying [13], chemical vapor deposition [14] and sol-gel [15].

The main issues pertaining to these methods in obtaining PZT films include film-substrate interaction, precision of stoichiometry control, degree of homogeneity, control of microstructure and compositional doping, among others. PZT thin films behave differently compared to their bulk ceramic counterparts; usually their properties are inferior to those of bulk samples [16, e.g.].

Incorporation of excess PbO has been carried out in bulk ceramics and its effects studied; it not only serves as a liquid phase sintering aid [17], but also compensates for lead volatility loss at elevated firing temperatures. The excess PbO aids in eliminating porosity, resulting in transparent PLZT [18] and in Bi-doped PZT [19]. In sintered PZT ceramics, PbO loss results in lower values of dielectric constant and the d_{33} piezoelectric coefficient [20], and also induces the formation of secondary phases such as ZrO_2, $La_2Zr_2O_7$ and $La_2Ti_2O_7$ [18] in PLZT ceramics. The control of the PbO content is often the most important processing issue in PbO-containing systems.

EXPERIMENTAL

The preparation of the precursor solutions has been described in a previous paper directed to stoichiometric PZT 53/47 [5]. In the current study, excess PbO (in the form of additional Pb acetate trihydrate in the precursor solutions) was incorporated in a series of thin films with nominal compositions $Pb_{1+x}Zr_{0.53}Ti_{0.47}O_{3+x}$ where $0 \leq x \leq 0.3$. Spin coating of these 1.0 M solutions was performed on platinized Si wafers at 2000 rpm for 20s. Multiple coatings with intermediate firing at 500C for 30 mins. were used to achieve thick coatings. The films were fired at 650-750C for 30 mins. to obtain the desired single-phase perovskite. XRD and optical microscopy were performed on the films to study their phase assemblages and morphologies. Electrical characterization, using a Hewlett Packard 4192A impedance analyzer, was carried out on Pt-PZT-Pt capacitors formed by sputtering top Pt electrodes through a physical shadow mask on the films. Hysteresis loops were obtained using a Sawyer-Tower circuit.

RESULTS AND DISCUSSION

Thicknesses of the films (3 coatings) after firing at 700C are displayed in Fig. 1 as a function of PbO content. The thickness is seen to increase with increasing PbO content, reflecting the higher solid contents of the solutions with increasing amounts of PbO. XRD scans indicated that PZT films with excess PbO up to 20 mole % fired above 650C exhibited single-phase perovskite. PZT films with 15 mole % excess PbO were free of pyrochlore when heated as low as 600C. In contrast [21], PZT films with no excess PbO have to be fired above 725C to eliminate the pyrochlore phase and achieve single-phase perovskite (see Table I). No secondary phases, i.e., Pb, pyrochlore or PbO, were detected in the PZT films with varying amounts (2.5-20 mole %) of excess PbO fired between 650C and 750C. In contrast, for powders derived from gels containing excess PbO, an extra PbO phase was observed in powders fired between 650C and 750C; while in PZT powders with no excess PbO, only single-phase perovskite powders were observed when fired above 500C.

Table I. Phases Detected in PZT 53/47 Films with Excess PbO After Firing

Temperature (C)	Amount of Excess PbO 0 mole %	15 mole %
500	Pyrochlore	Pyrochlore + Perovskite
600	Pyrochlore + Perovskite	Perovskite
700	Pyrochlore + Perovskite	Perovskite
725	Perovskite	Perovskite

The dielectric properties of the PZT thin films are shown in Fig. 2 as a function of firing temperature. Adding excess PbO in general enhanced the dielectric constant and dielectric loss, whereas tan δ remained fairly constant. Films with high dielectric constants also tended to display large dielectric losses as expected, since dielectric loss = tan $\delta \cdot \epsilon_r$. Fig. 2a shows that the dielectric constant of all films with excess PbO

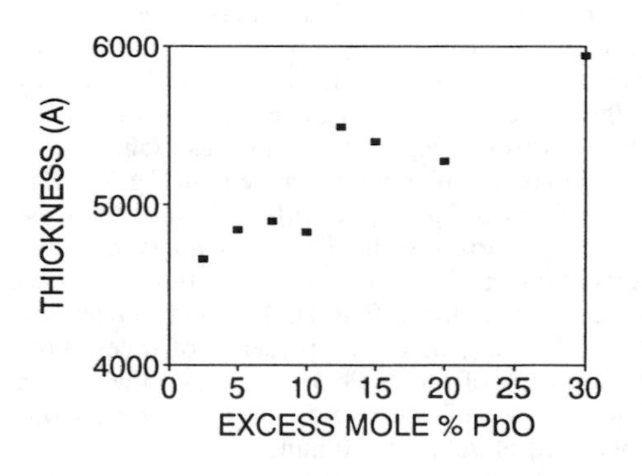

Figure 1. Thickness of PZT 53/47 films as a function of excess PbO (films fired at 700C).

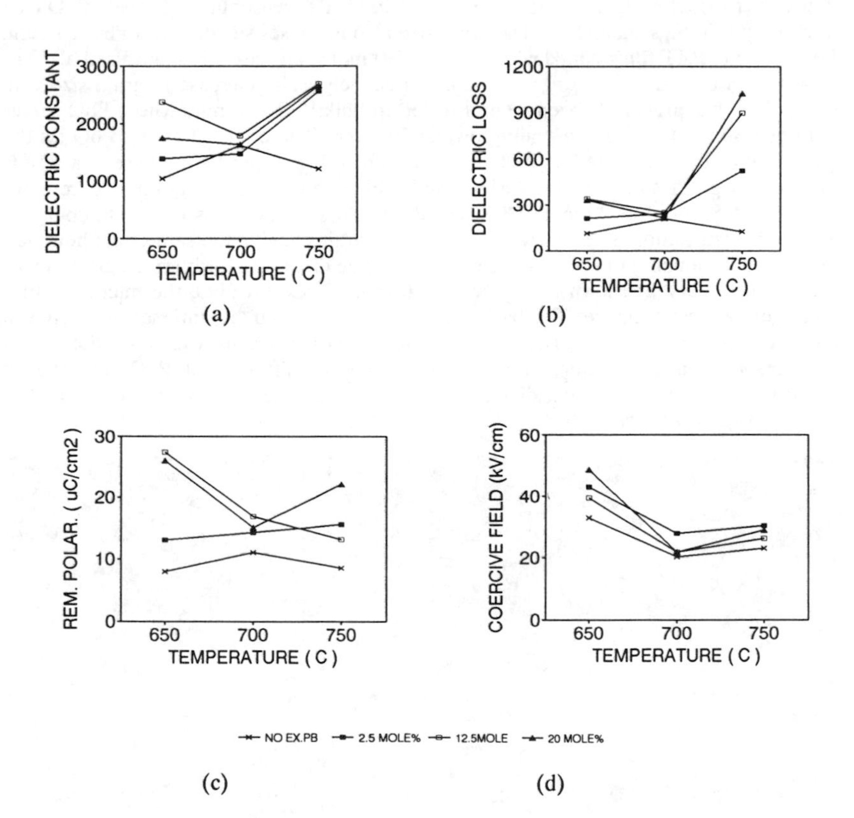

Figure 2. a) Dielectric constants b) Dielectric losses c) Remanent polarizations, and d) Coercive fields of PZT films as a function of excess PbO and firing temperature.

heated at 750C were about 2500. Regardless of the amount of excess PbO, the dielectric constant is much higher than in a film with no excess PbO ($\epsilon_r \approx 1200$). Thus, adding excess PbO aids films which are treated at higher temperatures where PbO volatility poses a serious problem. However, the more PbO that is added, the higher is the accompanying dielectric loss (Fig. 2b) after firing at higher temperatures (750C).

Adding excess PbO increased the remanent polarization, as seen in Fig. 2c. The highest P_r (27.4 $\mu C/cm^2$) was achieved by firing a film doped with 12.5 mole % excess PbO at 650C for 30 mins. The remanent polarization in films with no excess PbO remained fairly constant with firing temperature; while no consistent pattern was seen for films containing excess PbO. The values of coercive field (E_c) tended to decrease when fired at higher temperatures (Fig. 2d). In general, the presence of excess PbO raised the value of E_c regardless of the amount of excess PbO incorporated when the films are fired at 650-750C. The lowest value of coercive field (~20 kV/cm) was achieved in the film with no excess PbO fired at 700C for 30 mins.

Film microstructures were studied by optical microscopy. Clearly seen is the dense microstructure and the lack of "rosettes" in PZT films containing excess PbO (see micrographs in Figs. 3a and b). The grain size also increases steadily with PbO content. For example, PZT films doped with 2.5 and 20.0 mole % excess PbO and fired at 700C exhibited grain sizes of ~1 μm and 5 μm respectively. The increase in grain size with excess PbO has previously been demonstrated in bulk PZT ceramics where PbO serves both as a sintering aid and a grain growth promoter. The effect of excess PbO on the grain size of the films fired at 700C is illustrated in Fig. 4. Initially, there is a steady increase in grain size with increasing excess PbO; however, the grain size remains constant at 5 μm after 20 mole % excess PbO. At smaller levels of excess PbO, the grain size distribution is narrow with most of the grains having the same size; however, increasing the amount of PbO broadens the grain size distribution with some grains twice the size of others, as shown in Fig. 3. At 30 mole % excess PbO, the microstructure undergoes a drastic change. The basic grains can be observed, but interspersed between these grains are smaller clusters of percolated darker grains which resulted in a heterogeneous microstructure resulting from massive diffusion of PbO into the Pt substrate. Dielectric constant, dielectric losses and remanent polarizations seemed to increase with grain size.

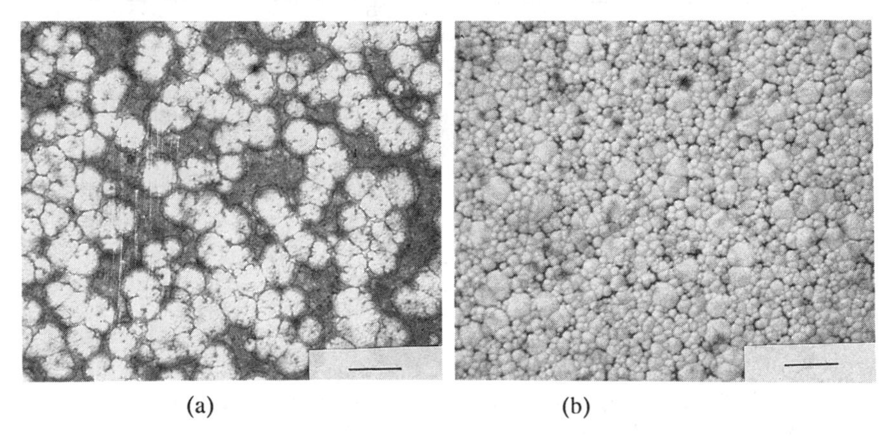

(a) (b)

Figure 3. Optical micrographs of PZT 53/47 films fired at 700C. a) no excess PbO b) 15 mole % excess PbO (Bar = 10 μm).

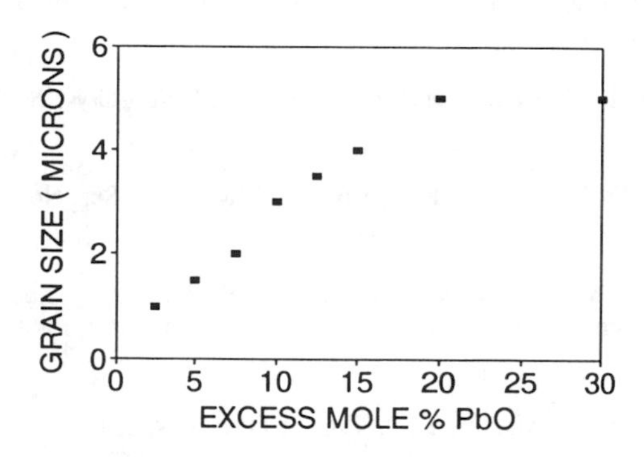

Figure 4. Grain size of PZT 53/47 films as a function of excess PbO (films fired at 700C).

In summary, it can be seen that the preparation of PZT films by sol-gel methods is not a simple process. Oxidation of organic species occurs upon firing which can lead to localized reducing conditions within the films resulting in the reduction of PbO to metallic Pb. The Pb can then react with the Pt substrates, which effectively act as Pb-sinks, resulting in PbO-deficient films. Aside from this interfacial reaction, PbO loss due to volatility can occur from the film/air interface. PbO volatility is exacerbated in films due to their inherent high surface area. The incorporation of excess PbO, therefore, aids PZT film processing by compensating for these PbO losses. In addition, it also acts as a grain growth promoter and as a sintering aid. The large grain sizes, and the subsequent enhanced mobility of the domain walls, are responsible for the superior dielectric properties of the films observed here.

CONCLUSIONS

The incorporation of excess PbO in the range of 2.5 to 20.0 mole % has proven to be beneficial in affecting the phase assemblages, microstructures and properties of PZT thin films. The presence of excess PbO promotes the formation of single-phase perovskite films at lower temperatures (~600C) compared to those (~725C) in films with no excess PbO. Porous "rosettes" were eliminated altogether with excess PbO: the PZT films were denser and exhibited larger grain sizes with increasing amounts of excess PbO. The electrical properties were also improved with excess PbO in the films. Dielectric constants as high as 3000 were achieved in a PZT 53/47 film with 7.5 mole % excess PbO fired at 750C. This is the highest value of ϵ_r obtained to date of any PZT thin film.

ACKNOWLEDGMENT

The financial support of the Air Force Office of Scientific Research is gratefully acknowledged.

REFERENCES

1. M. Okayama and Y. Hamakawa, Ferroelectrics, 118, 261 (1991).

2. L.T. Clark, R.U. Grondin and S.K. Dey, Proc. of 1st IEE Intl. Conf. Artificial Neural Networks, London, 1989.

3. E.R. LaSerra, Y. Charbouillot, P. Baudry and M.A. Aegerter, J. Non-Cryst. Solids, 121, 323 (1990).

4. H. Higashino, H. Adachi, K. Setsune and K. Wasa, Inst. Phys. Conf. Ser. No. 130, 23 (1989).

5. G. Teowee, J.M. Boulton, W.M. Bommersbach and D.R. Uhlmann, "Second Harmonic Generation From $PbTiO_3$-Based Ferroelectric Thin Films," J. Non-Cryst. Solids, in press.

6. C.E. Land, M.A. Butler and S.J. Martin, Tech. Dig. IEDM 1989, 251.

7. C.A.P. de Araujo, L.D. McMillan, B.M. Melnick, J.D. Cuchiaro and J.F. Scott, Ferroelectrics, 104, 241 (1990).

8. K. Sreenivas and M. Sayer, J. Appl. Physics, 64, 1484 (1988).

9. A. Okada, J. Appl. Phys., 48, 2905 (1977).

10. R.N. Castellano and L.G. Feinstein, J. Appl. Phys., 50, 4406 (1979).

11. M. Oikawa and K. Toda, Appl. Phys. Lett., 29, 491 (1976).

12. S. Otsubo, T. Maeda, T. Minamikawa, Y. Yonezawa, A. Morimoto and T. Shimizu, Jpn. J. Appl. Phys., 29, L133 (1990).

13. B. Malric, S. Dallaire and K. El-Assal, Mater. Lett., 5, 246 (1987).

14. M. Okada, T. Tominaga, T. Araki, S. Katayama and Y. Sakashita, Jpn. J. Appl. Phys., 29, 718 (1990).

15. K.D. Budd, S.K. Dey and D.A. Payne, Br. Ceram. Proc., 36, 107 (1986).

16. D.R. Uhlmann, G. Teowee, J.M. Boulton and B.J.J. Zelinski, Mat. Res. Soc. Symp. Proc., 180, 645 (1990).

17. G.S. Snow, J. Amer. Ceram. Soc., 56, 479 (1973).

18. B. Song, D. Kim, S. Shirazaki and H. Yamamura, J. Amer. Ceram. Soc., 72, 833 (1989).

19. G.S. Snow, J. Amer. Ceram. Soc., 57, 272 (1974).

20. I.K. Lloyd, M. Kahn and S. Lang, Cer. Trans., 11, 390 (1989).

21. G. Teowee, J.M. Boulton, S.C. Lee and D.R. Uhlmann, "Electrical Characterization of Sol-gel Derived PZT Films," Mater. Res. Soc. Proc. 243, in press.

ELECTRICAL AND OPTICAL PROPERTIES OF SOL-GEL PROCESSED $Pb(Zr,Ti)O_3$ FILMS

S.D. Ramamurthi, S.L. Swartz, K.R. Marken, J.R. Busch, and V.E. Wood
Battelle, 505 King Avenue, Columbus, OH 43201

ABSTRACT

Lead-zirconate-titanate, $Pb(Zr_{0.53}Ti_{0.47})O_3$, films were produced by the sol-gel method from alkoxide and acetate precursors in a 2-methoxyethanol solvent system. The PZT films were deposited on platinized silicon and single-crystal $SrTiO_3$ substrates for electrical and optical characterization, respectively. The processing parameters, especially excess PbO content and annealing conditions, were shown to have a significant effects on the properties of PZT films. Epitaxial PZT films deposited on $SrTiO_3$ waveguided over 10 mm distances with propagation losses as low as 5.9 dB/cm at 783 nm and a linear electro-optic effect was also demonstrated.

INTRODUCTION

Lead-zirconate-titanate (PZT) is a useful ferroelectric material for a variety of electronic and electro-optic applications [1]. In recent years, thin-film device geometries for PZT and other such ferroelectric compositions have gained interest due to their versatility in microdevice applications [2]. Electronic applications of PZT thin films may include non-volatile memories, dynamic random access memories, capacitors, micro-actuators, and IR sensors. Electro-optic applications, on the other hand, may include optical waveguide modulators, optical interconnects, memories and displays.

PZT thin films can be produced by a variety of deposition methods including sol-gel, sputtering, laser ablation, CVD (chemical vapor deposition) and modification of these methods [3]. In this work, PZT films were produced by the sol-gel method with the focus on the requirements for non-volatile semiconductor memories and optical waveguide devices. In this paper, the effects of a variety of process variables on the electrical and optical properties are described. The process variables included excess PbO content, type of substrates, buffer and electrode layer thicknesses, and annealing conditions.

EXPERIMENTAL PROCEDURE

The PZT solution was prepared using $Pb(OOCH_3)_2{\cdot}3H_2O$ (J.T. Baker), $Zr(O\text{-}nC_3H_7)$ (Strem Chemicals), and $Ti(O\text{-}iC_3H_7)_4$ (Aldrich) precursors and HPLC grade 2-methoxyethanol, $CH_3OC_2H_4OH$ (Aldrich) solvent [4]. The details of the solution preparation are provided elsewhere [5]. Some of the relevant modifications in PZT solution preparation are discussed here. Control over PZT stoichiometry improved when stock solutions of each of the precursors were used to prepare PZT solutions. In preparing the lead precursor solution, multiple distillations were carried out to ensure complete removal of water. Stable PZT solutions could also be prepared by refluxing and vacuum-distilling zirconium and titanium stock solutions together and subsequently repeating the procedure with the lead-zirconium-titanium mixture. In vacuum distillations, if the concentration of the solution increased dramatically, gelation occurred. Stable PZT solutions were formed by either using the vacuum distillation approach or by using the procedure described in references 4 and 5 .

The hydrolyzed PZT solution was spin coated on a variety of substrates, and the procedure is described schematically in Figure 1. Each layer was spin coated at 2500 rpm, providing 600-700Å of film thickness (after annealing) per deposition. The as-deposited films

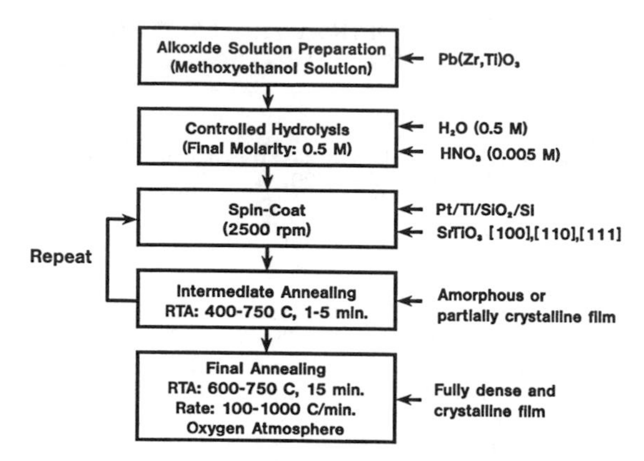

Figure 1. PZT film fabrication procedure by sol-gel method.

were annealed in a Rapid Thermal Annealer (Process Products Corporation, RTM-2016) at 100-1000°C/minute. Four layers (≈2500Å) or eight layers (≈6000Å) of PZT were deposited on $SrTiO_3$ substrates for optical characterization. Four layers (2500-2800Å) of PZT were deposited on platinized silicon wafer for electrical property measurements.

The platinized silicon substrates for electrical characterization were prepared by e-beam sputtering of SiO_2, titanium, and platinum with desired thicknesses. The deposition of SiO_2 and Ti intermediate layers were required to prevent reaction of Pt with Si during annealing and to promote adhesion. In two samples, the SiO_2 buffer layer was thermally grown instead of using the e-beam deposition method. For optical characterization, $SrTiO_3$ crystals with [100], [110], and [111] orientations (Commercial Crystals, Naples, FL) were used as substrates for PZT film deposition.

Electrical properties, including coercive field (Ec), dielectric constant (K), remanent polarization (Pr), and fatigue, were measured using a RT-66A system (Radiant Technologies). For these measurements, the top gold electrodes were sputter deposited (1 mm diameter). The refractive indices and waveguiding distances were measured with a prism-coupling method described elsewhere [5]. The electro-optic effect in a PZT film on $SrTiO_3$ was demonstrated using a waveguide Bragg diffraction method described subsequently in this paper.

X-ray diffraction (XRD, Rigaku-DMAX-II) was used to characterize the crystallinity and orientation of the PZT films. The microstructure and film thickness were analyzed by scanning electron microscopy (SEM, JEOL-840).

RESULTS AND DISCUSSION

The electrical and optical properties of the sol-gel processed PZT films are described below.

Electrical Properties:

For electrical characterization, PZT films were deposited on $Si/SiO_2/Ti/Pt$ substrates. The ferroelectric properties including, Pr, Ec, and K as a function of process variables are summarized in Table 1. The electrical property values were obtained by averaging 9-18 measurements over three different locations on the film. Columns 2 and 3 in Table 1 provide the targeted and measured (SEM) layer thicknesses of various substrate

Table 1. The electrical properties of PZT films deposited on platinized silicon electrodes.

Sample	Substrate SiO_2/Ti/Pt (Å)	SEM SiO_2/Ti/Pt/PZT (Å)	PbO (Excess in %)	IA[a] (°C/min)	FA[b] (°C/min)	Pr ($\mu C/cm^2$)	Ec (kV/cm)	K
A1	1000/1000/1000 e-beam SiO_2	900/800/700/2600	6	400/5	650/30	30.3	144.0	1233.0
B1	1000/500/1500 e-beam SiO_2	600/400/900/2700	6	400/5	650/30	28.0	196.8	1186.5
C1	1000/1000/1000 thermal SiO_2	1600/1200/800/2550	6	400/5	650/30	17.0	437.5	464.0
A2	1000/1000/1000 e-beam SiO_2	1200/900/800/2600	10	400/5	650/30	26.3	98.3	1056.5
B2	1000/500/1500 e-beam SiO_2	700/400/1000/2600	10	400/5	650/30	22.8	76.0	1048.5
C2	1000/1000/1000 thermal SiO_2	1600/1200/800/2800	10	400/5	650/30	27.3	89.3	1124.5
B3	1000/500/1500 e-beam SiO_2	600/400/1000/2650	10	400/1	650/30	25.5	83.3	1115.5
B4	1000/500/1500 e-beam SiO_2	-	10	400/5	625/30	24.5	89.5	1050.0
B5	1000/500/1500 e-beam SiO_2	-	10	400/5	600/30	1.1	63.3	80.0

a: intermediate annealing
b: final annealing

layers. The process variables in Table 1 include PbO content, Pt electrode layer thickness, intermediate annealing time and final annealing temperature. As an example, the hysteresis loop of sample B2 is shown in Figure 2. In Table 1, with the exception of samples C1 and B5, the Pr values of the samples ranged from 22-31 $\mu C/cm^2$, K ranged from 1048-1233, and Ec was between 63-197 kV/cm. The maximum fluctuations were observed in the coercive field values. Hence, most of the process variable effects were analyzed as a function of Ec.

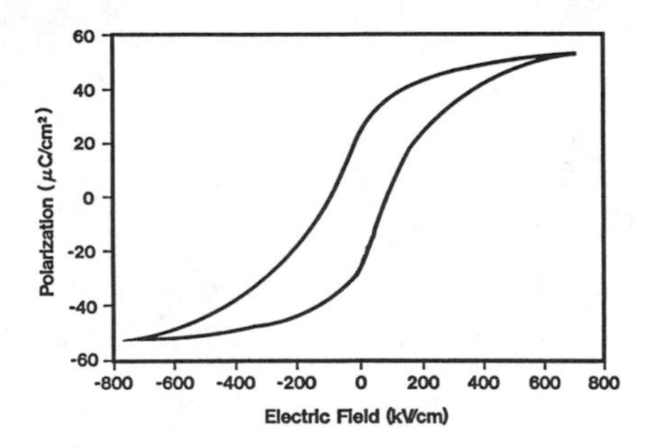

Figure 2. A ferroelectric hysteresis loop for PZT film deposited on platinized silicon (sample B2).

The effect of excess PbO content on coercive field is shown in Figure 3a. In all three types of substrates, the Ec values decreased with an increase in excess PbO content. The control of lead stoichiometry in sol-gel PZT films requires careful solution processing, and has a profound effect on crystallization [6]. The tendency for lead deficiency is due either to incomplete removal of water from the lead precursor or to PbO evaporation during

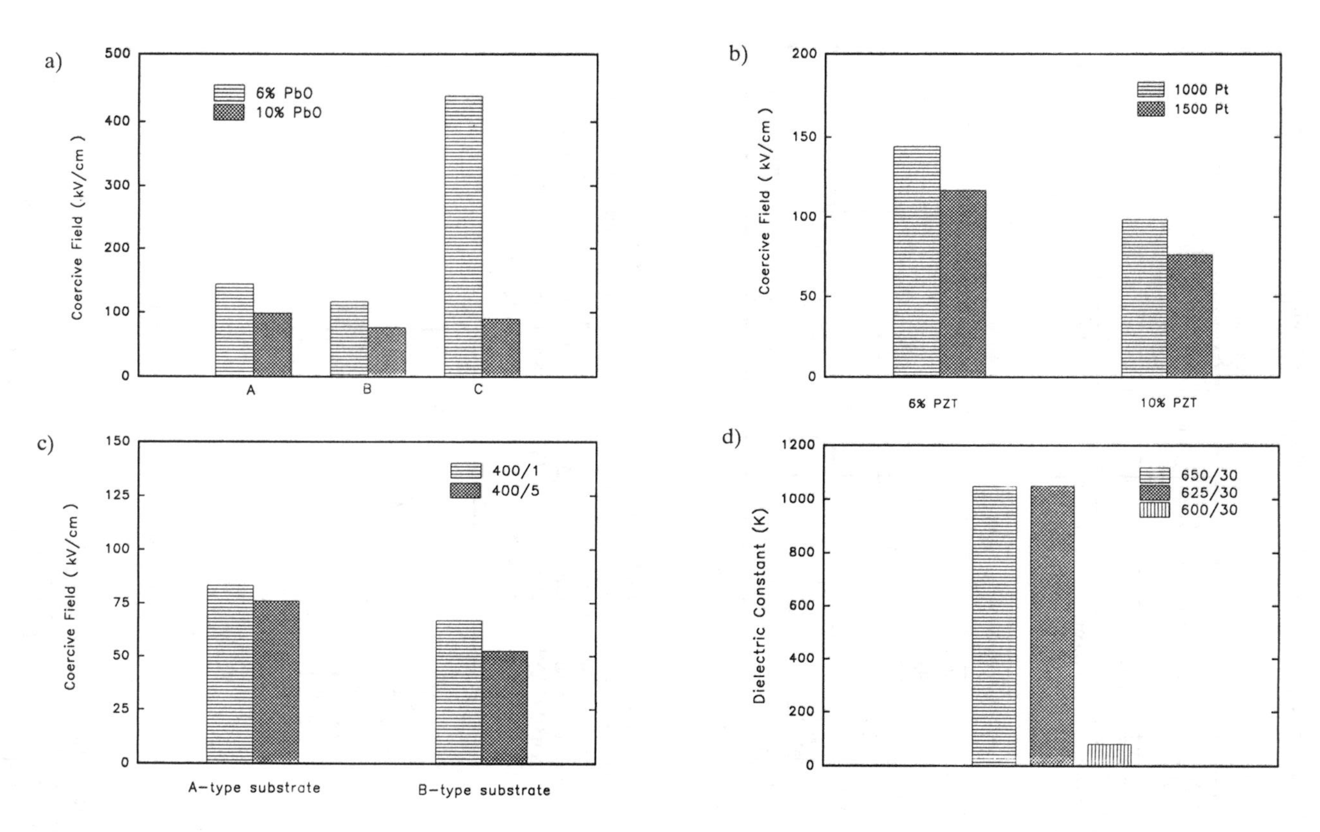

Figure 3. The variation in coercive field (Ec) or dielectric constant (K) as a function of a) PbO content, b) Pt electrode layer thickness, c) intermediate annealing time, and d) final annealing temperature.

annealing. The addition of excess lead compensates for these losses and results in highly crystalline perovskite films. The variation in Ec with the platinum electrode layer thickness is shown in Figure 3b. Interdiffusion between the electrode and barrier layers (Pt, Ti, SiO_2) during film processing is common [7]. The interdiffusion increases and PZT crystallinity decreases with increasing porosity of the top platinum layer. Hence, denser and thicker Pt layers are required to improve crystallinity and minimize interdiffusion, resulting in enhanced electrical properties. The effect of intermediate annealing (400°C for 5 min. verses 400°C for 1 min.) on Ec is shown in Figure 3c. Longer intermediate annealing resulted in lower Ec values, possibly due to increased density and homogeneity of the PZT films. The effect of final annealing on the dielectric constant is shown in Figure 3d. Here, the coercive field was an inappropriate variable because sample B5 did not crystallize completely and did not exhibit ferroelectric hysteresis. Figure 4 shows the x-ray diffraction patterns of samples, B2, B4, and B5 annealed at 600°C, 625°C, and 630°C for 30 min., respectively. As shown in Figure 4 sample B5 contained a poorly crystalline pyrochlore phase which reduced the dielectric constant significantly. The crystallinity and electrical properties indicate that the final annealing at 625°C for 30 minutes is sufficient to densify and crystallize the PZT film.

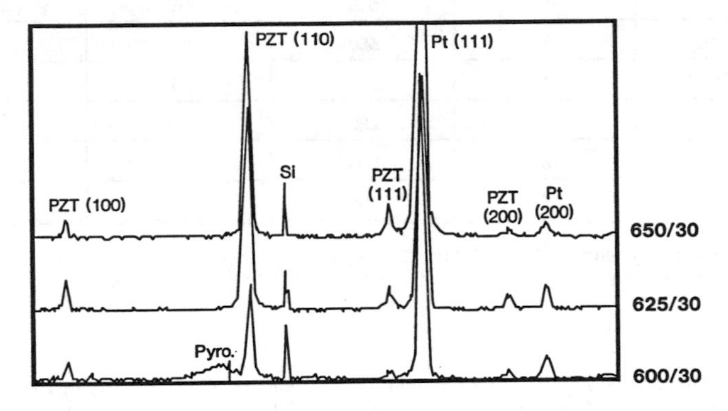

Figure 4. X-ray diffraction patterns of PZT films processed at 600°C, 625°C, and 630°C.

In summary, the electrical properties of PZT films were improved by increasing the excess PbO content from 6% to 10%, by depositing denser and thicker (>800Å) Pt electrode layers, using longer intermediate annealing times, and final annealing temperatures of 625°C or higher.

Optical Properties:

Ferroelectric films for optical waveguide applications must satisfy the following:

- Absence of defects and optical inclusions,
- Grain sizes smaller than the operating wavelength,
- Higher refractive index than the substrates,
- Thickness range between 0.2-1.0μm, and
- Crystallographic orientation and/or epitaxy.

The refractive indices of PZT films (> 2.48) are significantly higher than $SrTiO_3$ (2.39). In an earlier study, epitaxial PZT films were produced on $SrTiO_3$ ([100], [110], and [111]) crystals [5, 8]. Hence, optical waveguiding was achieved in PZT films when a minimum thickness of about 3000Å (4 layers) was obtained. To study the effects of annealing conditions on the optical properties, the annealing temperature and time were varied from

400°C for 1 min. (A-type) to 400°C for 1 minute plus 750°C for 5 min. (B-type). The optical properties, including refractive indices and waveguiding distances using the prism-coupling method are summarized in Table 2. The refractive index of PZT films increased as the film thickness increased and the intermediate annealing program changed from A-type

Table 2. The optical properties of PZT films deposited on $SrTiO_3$ substrates.

Substrate Orientation	Annealing Conditions	Film Thickness (SEM; nm)	Refractive Index (633 nm)	Waveguiding Distance (mm)
[100]	A[a]	280	2.48	< 1
[100]	B[b]	610	2.50	8
[100]	B	510	2.55	10
[100]	B	550	2.55	> 10
[110]	A	280	2.50	2
[111]	A	280	2.49	< 1
[111]	B	500	2.56	--
[111]	B	550	2.55	10

a: 400°C for 1 min. (intermediate); 700°C for 30 min. (final)
b: 400°C for 1 min. plus 750°C for 5 min. (intermediate); 750°C for 15 min. (final)

to B-type. These effects result from the high homogeneity and high density of PZT films associated with thicker films and with higher intermediate annealing temperature. High quality PZT films were produced with refractive indices as high as 2.56 and waveguiding distances as long as 10 mm (limited by substrate size) at 633 nm wavelength. The propagation losses for PZT films were 15 dB/cm (TE_0) and 7.0 dB/cm (TE_1) at 633 nm and 5.9 dB/cm (TE_0) at 783 nm.

The electro-optic effect was also demonstrated for PZT film waveguides using a Bragg grating modulator [9]. The output of the diffracted beam in a two prism setup was measured as a function of the driving voltage waveform (see Figure 5a). The output (Figure 5b) shows the behavior expected for Pockels electro-optic modulation [9].

In conclusion, high quality PZT films were deposited on platinized silicon ($Si/SiO_2/Ti/Pt$) substrates, and important properties for non-volatile memory applications were measured:

- Coercive fields (Ec) as low as 63 kV/cm,
- Remanent polarizations (Pr) as high as 31 $\mu C/cm^2$, and
- Dielectric constants (K) as high as 1233.

For optical waveguide applications, the PZT films deposited on $SrTiO_3$ substrates showed:

- Refractive indices (n) as high as 2.56,
- Waveguiding distances as long as 10 mm,
- Low propagation losses (5.9 dB/cm at 783 nm), and
- Linear electro-optic effects.

a)

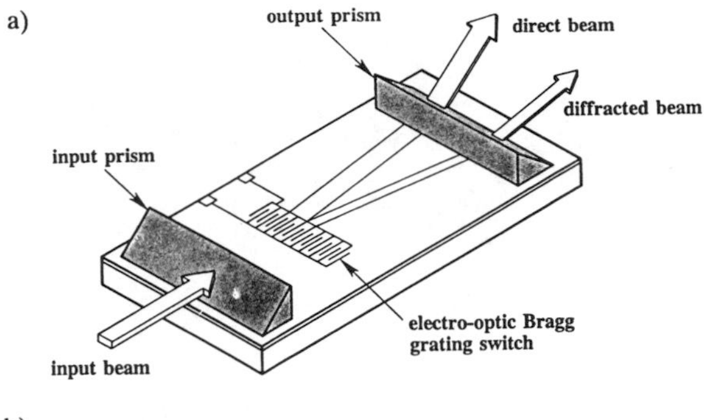

b)

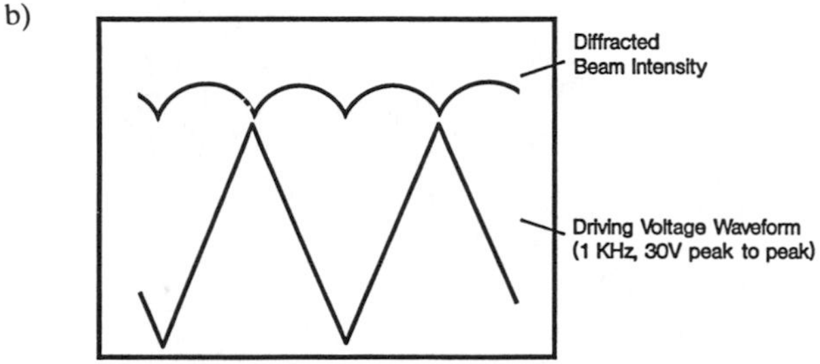

Figure 5. a) A two-prism setup for measuring the output diffracted beam intensity as a function of applied voltage. b) The results of the guided wave electro-optic modulation.

ACKNOWLEDGEMENT

This project was supported by Battelle Columbus Internal Research and Development funds. The authors wish to acknowledge Steve Bright for sol-gel experimental work.

REFERENCES

1. S.L. Swartz, IEEE Transactions on Electrical Insulation, 25(5), 1990, 935.
2. P.K. Tien, in Integrated Optics, ed. D. Marcuse, IEEE Press, New York, 1972, 12.
3. See review by: R.A. Roy, K.F. Etzold, and J.J. Cuomo, Mat. Res. Soc. Symp. 200, 1990, 141.
4. K.D. Budd, S.K. Dey and D.A. Payne, Proc. Br. Ceram. Soc. 36, 1985, 107.
5. S.D. Ramamurthi, S.L. Swartz, V.E. Wood, and J.R. Busch, Cer. Trans., 25, 1992, 353.
6. B.A. Tuttle, R.W. Schwartz, D.H. Doughty, and J.A. Voigt, Mat. Res. Soc. Symp. 200, 1990, 159.
7. S. Matsubara, T. Sakuma, S. Yamamichi, and Y. Miyasaka, Mat. Res. Soc. Symp. 200, 1990, 243.
8. S.H. Rou, A.I. Kingon, S.D. Ramamurthi, and S.L. Swartz, in "Defect Structures in Crystalline Electronic Oxides", Mat. Res. Soc. Symp. Proc. 1992.
9. R.G. Hunsperger, Integrated Optics: Theory & Technology (Springer-Verlag, Berlin, 1982), section 8.6.1.

ELECTRICAL PROPERTIES OF CRYSTALLINE AND AMORPHOUS $Pb(Zr_xTi_{1-x})O_3$ Thin Films PREPARED BY THE SOL-GEL TECHNIQUE

Yuhuan Xu, Chih-Hsing Cheng, Ren Xu and John D. Mackenzie
Department of Materials Science and Engineering, University of California,
Los Angeles, CA 90025

ABSTRACT

$Pb(Zr_xTi_{1-x})O_3$ (PZT) solutions were prepared by reacting lead 2-ethylhexanoate with titanium n-propoxide and zirconium n-propoxide. Films were deposited on several kinds of metal substrate by dip-coating. Crystalline PZT films and amorphous PZT films were heat-treated for 1 hour at 650°C and at 400°C, respectively. Electrical properties including dielectric, pyroelectric and ferroelectric properties of both crystalline and amorphous PZT films were measured and compared. The amorphous PZT thin films exhibited ferroelectric-like behaviors.

INTRODUCTION

In the past twenty years ferroelectric thin films have been extensively studied[1] and a variety of novel devices have been demonstrated[2]. Many deposition techniques have been employed for the preparation of ferroelectric thin films.[1] The sol-gel method is one of the most promising techniques with the following advantages: good homogeneity, ease of composition control, low sintering temperature, large area thin films, possibility of epitaxy and lower cost than other techniques. In the last year, we studied ferroelectric $LiNbO_3$ thin films prepared by sol-gel technique and discovered that the amorphous $LiNbO_3$ films have ferroelectric-like behavior.[3]

Ferroelectric lead zirconate titanate (PZT) is a main industrial product among piezoelectric ceramic materials that have excellent piezoelectric properties. The dielectric, piezoelectric and ferroelectric properties of PZT can be modified by doping to suit a variety of applications.[4] The preparation of PZT thin films by sol-gel method began in 1984. Presently, many researchers have been involved in this field.[5-8] The purpose of this work is to find out whether the amorphous PZT films have similar behavior too.

PREPARATION OF PZT THIN FILMS

$Pb(Zr_xTi_{1-x})O_3$ thin films with Zr/Ti = 52/48 were prepared by sol-gel technique. The PZT solution was prepared by reacting lead 2-ethylhexanoate with titanium n-propoxide and zirconium n-propoxide.[9,10] The flow chart of the processing is shown

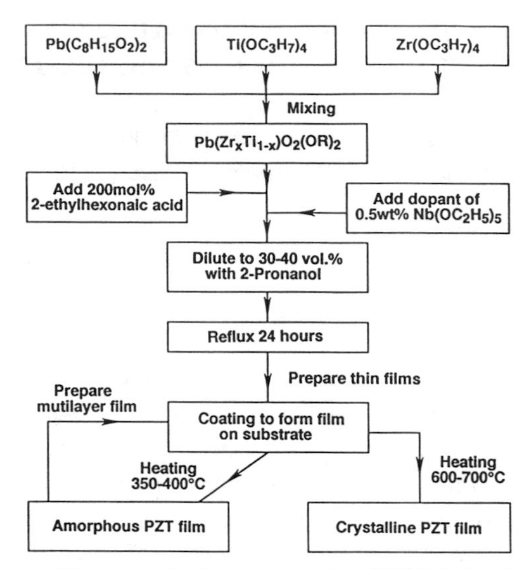

Fig. 1 The flow chart of the processing for the preparation of PZT thin films by the sol-gel method.

in Fig. 1. 2-ethylhexonaic acid (200 mol%) and $Nb(OC_2H_5)_5$ (0.5 wt%) were added into the solution to control the hydrolysis conditions and to improve the piezoelectric property of PZT films[4], respectively. Films were deposited on several kinds of metal substrate (including Pt, Ti and Au passivated silicon wafer) by dip-coating. These samples were separated to two batches and heat-treated for 1 hour between 350—400°C and between 650—700°C, respectively. The average film thickness for one single dipping was around 1000Å. Thicker films could be obtained by multiple dipping and heating. Fig. 2 shows the X-ray diffraction patterns. The films heated at temperature above 650°C possess crystalline perovskite structure and the films heated at temperature below 400°C are still amorphous. These results were also confirmed by the experiments of high-resolution TEM and electron-diffration. Fig. 3 shows the SPM micrographs of both crystalline PZT film (a) and amorphous PZT film (b). The average grain size of crystalline PZT is about 500Å.

ELECTRICAL PROPERTIES

For the measurement of electrical properties Au-electrodes were sputtered on crystalline PZT films and amorphous PZT films with thickness of around 0.3 μm on metal substrates in oder to form metal/PZT film/metal sandwich structure samples. By using a HP 4192A Impedance Analyzer, dielectric permittivity and dissipation loss factor as the functions of frequency (from 10 Hz to 10^7 Hz) and temperature (from room temperature to 150°C) were measured, respectively. The permittivity of amorphous PZT is much lower than that of crystalline PZT. In both crystalline PZT and amorphous PZT samples

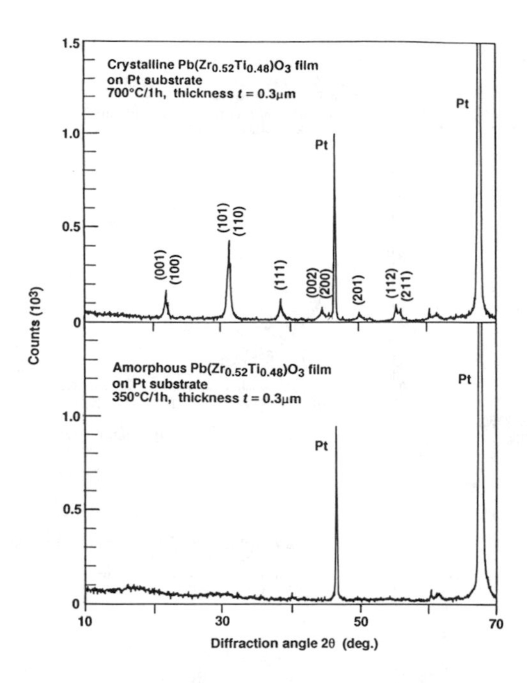

Fig. 2 X-ray diffraction patterns of crystalline PZT (upper) and amorphous PZT (lower) thin films.

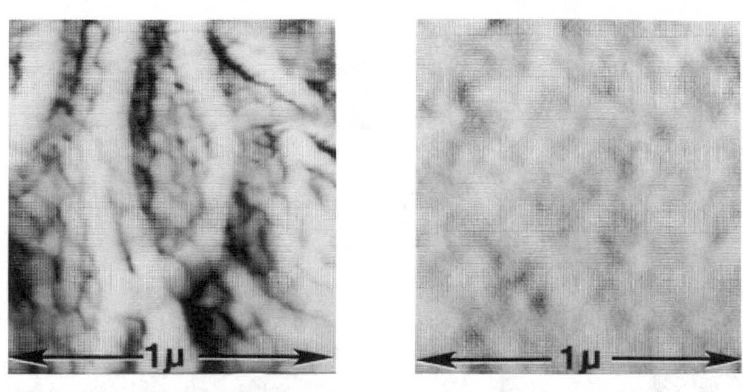

Fig. 3 Micrographs of PZT thin film on Au substrate. Picture area is 1 X 1 μm^2. (a) Crystalline PZT (650°C / 1h, thickness $t = 0.3\mu m$); Average grain size is 500Å. (b) Amorphous PZT (400°C / 1h, thickness $t = 0.3\mu m$); No grain boundaries were observed. (Pictures were taken by Topomatrix TMX2000 SPM system.)

the dielectric permittivity and loss factor have a piezoelectric resonance peak at about 1 MHz and a peak of loss factor caused by dipole relaxation at about 12 MHz[11,12] (see Figs. 4, 5). From the loss factor of crystalline PZT obvious interface-polarization effect caused by grain-boundaries can be observed.

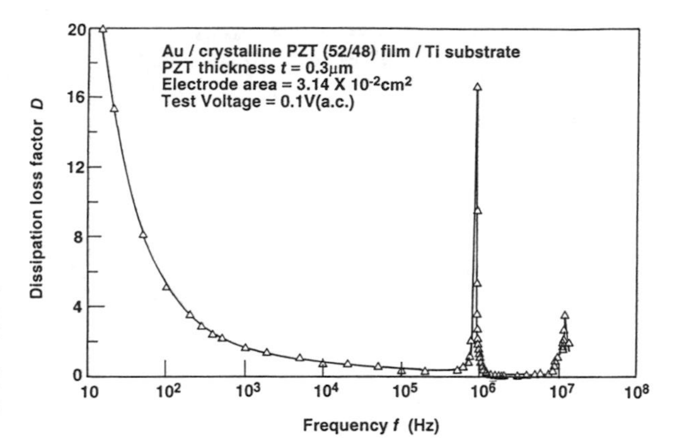

Fig. 4 Dielectric loss changes with frequency in crystalline PZT thin film. A strong peak at 1Mc is caused by piezoelectric resonance absorption and another peak at 12Mc is caused by dipole relaxation.

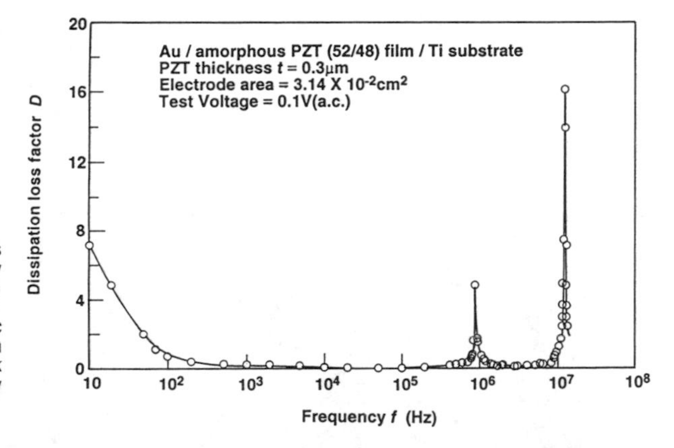

Fig. 5 Dielectric loss changes with frequency in amorphous PZT thin film. A peak at 1Mc is caused by piezoelectric resonance absorption and another strong peak at 12Mc is caused by dipole relaxation.

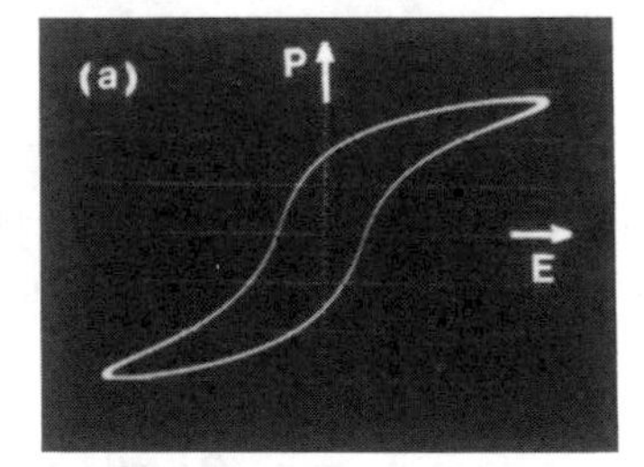

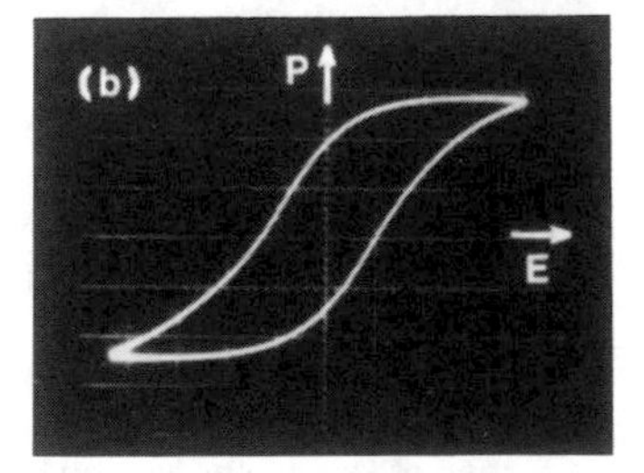

Fig. 6 *P-E* hysteresis loops of Au / PZT (52/48) thin film(0.3μm) / Ti substrate. (a) Crystalline PZT; Scale of *P*: 21μC/cm^2/div. and scale of *E*: 14kV/mm/div.. (b) Amorphous PZT; Scale of *P*: 2μC/cm^2/div. and scale of *E*: 6.5kV/mm/div..

Pyroelectric current was measured by a picoammeter (417 Keithly Instruments) connected in series with the sample, while the sample was heated at a moderate rate from room temperature to 150°C. Before the measurement both crystalline samples and amorphous samples were poled at 150°C with a dc electric field of 7 kV/mm. The pyroelectric coefficients *p* of the two kinds of sample are listed in Table 1. For the amorphous PZT samples the pyroelectric current change with time were observed. After poling 2, 44, 162 hours, the pyroelectric current (which was measured by heating the sample at the same heating rate in the three heating cycles) become stable. After heating up the poled amorphous PZT sample to 250°C the pyroelectric current disappeared owing to the sample was depolarized. Therefore, it was identified that the pyroelectric effect in the amorphous PZT was caused by the contribution of poled dipoles.

Ferroelectric hysteresis loops of both crystalline and amorphous PZT thin films were observed by a modified Sawyer-Tower bridge with an adjustable compensation condenser. From the saturated loops (Fig. 6) the remanent polarization p_r of 31.5 $\mu C/cm^2$ and 3.2 $\mu C/cm^2$ as well as the coercive field E_c of 14 kV/mm and 7.8 kV/mm in crystalline PZT and amorphous PZT samples, respectively, were obtained. For comparison, electrical properties of crystalline PZT and amorphous PZT thin films are summarized in Table 1.

From above experiments it should be identified that there are some permanent dipoles, which may be the distorted Ti-oxygen octahedra and Zr-oxygen octahedra, in the network of amorphous structure. These dipoles can be oriented by applying a high electric field. After poling a permanent polarization, which can be reversed by extranal field, were formed in the amorphous PZT films.

Table I Electrical Properties of Crystalline and Amorphous $Pb(Zr_{0.52}Ti_{0.48})O_3$+0.5wt%$Nb_2O_5$ Thin Films at Room Temperature

Properties	Crystalline PZT (700°C/1h)	Amorphous PZT (400°C/1h)
Film's thickness *t*	0.3μm	0.3μm
Average grain size (Å)	500	no grains were observed
Dielectric permittivity ε (25°C)	1850 (1kc) 927 (100kc)	160 (1kc) 49 (100kc)
Resistivity (d.c.) ρ (Ω-cm)	3.1×10^8	1.4×10^7
Pyroelectric coefficient *p* (nC/cm^2K)	18.3	13.8
Remanent polarization P_r ($\mu C/cm^2$)	31.5	3.2
Coercive field E_c (kV/mm)	14.0	7.8
Breackdown strength E_b (kV/mm)	127	53
Optical refractive Index *n*	2.5 (t > 0.35μm)	between 2.0 and 2.2

CONCLUSIONS

1) PZT thin films on metal substrates were prepared by the sol-gel technique. It was confirmed that the PZT films annealed at temperatures above 650°C and below 400°C were crystalline and amorphous, respectively.

2) The dielectric permittivity of amorphous PZT is much lower than that of crystalline PZT. In both crystalline PZT and amorphous PZT thin films evident piezoelectric resonance peaks and dipole relaxation were observed.

3) The amorphous PZT thin films have pyroelectric effect and *P-E* hysteresis loop. This ferroelectric-like behavior indicated that there is a permanent polarization, which can be reversed by external field, in the amorphous PZT films.

4) The advantages of ferroelectric-like amorphous materials are low processing temperature, low dielectric permittivity and good transparency without grain-boundaries.

ACKNOWLEDGEMENTS

This work was supported by the Air Force Office of Scientific Research, Directorate of Chemical and Materials Science under Grant No. AFOSR-91-0096. The authors wish to thank Mr. Huddee Ho, TopoMetrix Co., Santa Clara, CA, for the Scanning Probe Microscope pictures in this work.

REFERENCES

1. E.R. Myers and A. I. Kingon (editors), Feroelectric thin films, (Mater. Res. Soc. Proc. **200**, Pittsburgh, PA, 1990).
2. Yuhuan Xu and John D. Mackenzie, Integrated Ferroelectrics, 1, 17 (1992).
3. Ren Xu, Yuhuan Xu and J. D. Mackenzie, presented at the 3rd International Symposium on Integrated Ferroelectrics, Colorado Springs, Colorado, 1991(to be published in Ferroelectrics).
4. Yuhuan Xu, Ferroelectric Materials and Their Applications, 1st ed. (North-Holland Elsevier Science Publishers, Amsterdam, 1991) p. 101.
5. Sandwip K. Dey and Rainer Zuleeg, Ferroelectrics, **108**, 37 (1990).
6. S.S. Dana, K.F. Etzold and J. Clabes, J. Appl. Phys., **69**, 4398 (1991).
7. Noboru Tohge, Satoshi Takahashi and Tsutomu Minami, J. Am. Ceram. Soc., **74**, 67 (1991).
8. K. R. Udayakumar, S. F. Bart, A. M. Flynn, J. Chen, L. S. Tavrow, L. E. Cross, R. A. Brooks and D. J. Ehrlich, in Proc. IEEE Micro Electro Mechanical Systems, (Nara, Japan, 30 Jan.--2 Feb. 1991), p. 109.
9. K. C. Chen, A. Janah and J. D. Mackenzie, in Better Ceramics Through Chemistry II, edited by C. J. Brinker, D. E. Clark and D. R. Ulrich, (Mater. Res. Soc. Proc., **73**, Pittsburgh, PA, 1986), p. 731.
10. R. A. Lipeles, D. J. Coleman and M. S. Leung, in Better Ceramics Through Chemistry II, edited by C. J. Brinker, D. E. Clark and D. R. Ulrich, (Mater. Res. Soc. Proc., **73**, Pittsburgh, PA, 1986), p. 665.
11. I. Bunget and M. Popescu, Physics of Solid Dielectrics, 1st ed. (Elsevier Science Publishers, Amsterdam, 1984), p. 316.
12. G. A. Smolenskii (Editor-in-Chief), Ferroelectrics and Related Materials, 1st ed. (Gordon and Breach Science Publishers, New York, 1984), p. 443.

PREPARATION OF $BaTiO_3$ AND $PbTiO_3$ THIN FILMS ON $BaPbO_3$ SUBSTRATES BY THE SOL-GEL METHOD AND THEIR PROPERTIES

M. KUWABARA, T. KURODA, S. TAKAHASHI, AND T. AZUMA
Kyushu Institute of Technology, Department of Applied Chemistry,
Tobata, Kitakyushu 804, Japan

ABSTRACT

Barium titanate, lead titanate and barium-lead titanate thin films were prepared by the sol-gel method using metal alkoxides on metallic conducting $BaPbO_3$ ceramic substrates. Thin films were deposited on the polished surfaces of the substrates by spin coating solutions containing $Ba(OC_2H_4OCH_3)_2$, $Pb(OCOCH_3)_2$ and $Ti(O\text{-}iC_3H_7)_4$ and methoxyethanol. The gel films were then fired at appropriate temperatures to yield ceramic thin films. A 0.50 μm-thick $BaTiO_3$ film showed a dielectric constant of 280 and tan δ of 3% at 1kHz with no distinguishable peak of dielectric constant at any temperature between 20°C and 200°C. Both $PbTiO_3$ (0.75 μm thick) and $Ba_{0.7}Pb_{0.3}TiO_3$ (0.2 μm thick) films, on the other hand, showed a distinguishable peak of dielectric constant at their Curie point with a rather deformed P-E hysteresis loop at room temperature.

INTRODUCTION

There have been many attempts to produce ferroelectric thin films of barium titanate[1-4], lead titanate[5,6], lead zirconium titanate (PZT)[7,8], and so forth, by using various techniques of dry and wet methods. Some methods have yielded such ferroelectric thin films of good quality, but few successful preparations of barium titanate thin films with good ferroelectric properties at room temperature have been reported so far. Ferroelectric thin films have been deposited on various kinds of substrates, and meanwhile the deposition of thin films on platinized silicon wafers has recently been the main stream from the viewpoint of practical use. However, there still be a contamination problem by platinum in the thin films deposited on platinized silicon wafers, and the contamination may result in degraded ferroelectric properties particularly for thin films with a thickness < 1μm. A better selection of substrate materials and their surface states to be used as well as a more significant improvement of the film forming processes should thus be required to prepare thin films with ferroelectric properties of much higher quality. In this connection, the utilization of the sol-gel processing in conjunction with the searching of suitable substrate materials may be one of the most promising ways to achieve the aim particularly from the view points of the easiness and reproducibility of the film forming process.

It is a principal advantage of the sol-gel method to be able to use a variety of metalloorganic compounds as starting materials. By choosing suitable metalloorganic compounds and the solution system to produce starting sol solutions, which are successively submitted to dip coating or spin coating, it may be possible to improve the structure and electronic properties of the thin films obtained. In this paper, we report the microstructures and dielectric properties of barium titanate, lead titanate and barium-lead titanate thin films, which were deposited on metallic conducting barium lead oxide, $BaPbO_3$, ceramics by the sol-gel method using metal alkoxides. Dense $BaPbO_3$ ceramics were probably first used as a substrate for the preparation of ferroelectric thin films in this study.

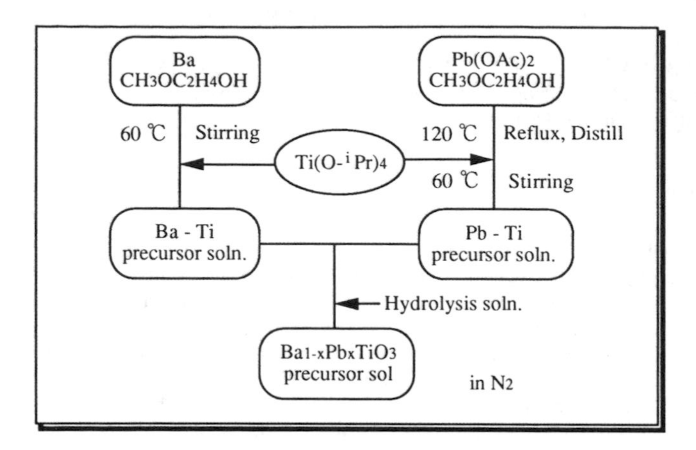

Fig. 1. Schematic of the sol-gel processing used.

EXPERIMENTAL

The procedure used in this study for producing starting sol solutions with the compositions $Ba_{1-x}Pb_xTiO_3$ (x=0, 0.3 and 1; designated BTO, BPT-30 and PTO, respectively) is briefly illustrated in Fig. 1. Methoxyethanol was used as an alcoholic medium in this sol-gel processing. In the present procedure, a barium titanate precursor solution was prepared simply by mixing barium methoxyethoxide and titanium isopropoxide in methoxyethanol at 60°C, while a lead precursor solution was produced using partially alkoxylated lead acetate and titanium isopropoxide, following the way that has appeared in the literature.[9] A hydrolysis solution was added to the mixtures containing exactly required amounts of barium titanate and lead titanate precursor solutions (or to the respective single component precursor solutions) to prepare the barium-lead titanate starting (including BTO and PTO) sol solutions. The obtained sol solutions were coated by spin coating on polished surfaces of metallic conducting $BaPbO_3$ ceramics, which were used as a substrate which provided the function of an electrode. The gel thin films prepared on the substrates were then fired at 800°C for 1h to yield $BaTiO_3$ and at 750°C for 30min to produce both $PbTiO_3$ and $Ba_{0.7}Pb_{0.3}TiO_3$ films all in air.

The temperature dependence of dielectric constant and dissipation factor, tan δ, was obtained for their films using an impedance analyzer (YHP-4192A) at a frequency in the range of 1kHz to 100kHz by a two-probe method with Au sputtered films as a counter electrode. P-E hysteresis loops were also obtained for the films using a Sawyer-Tower circuit at a frequency of 60Hz or 1kHz. SEM technique was used to examine structures, film thickness and surface morphologies of the thin films. The crystal phase of the thin films and their crystallographic orientations were examined by X-ray diffraction (XRD) analysis.

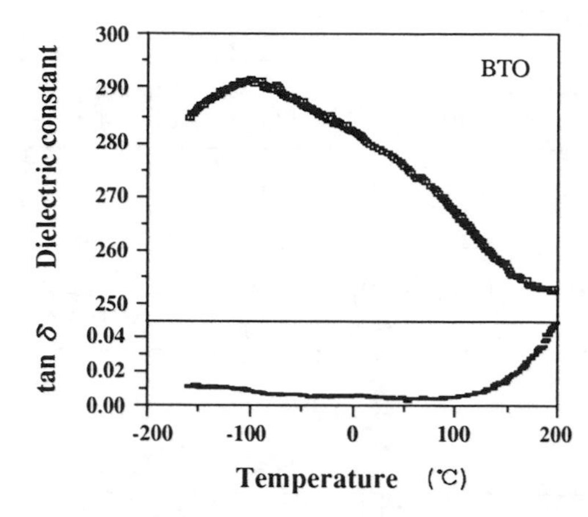

Fig. 2. Temperature dependence of dielectric constant and tan δ for a BTO thin film prepared.

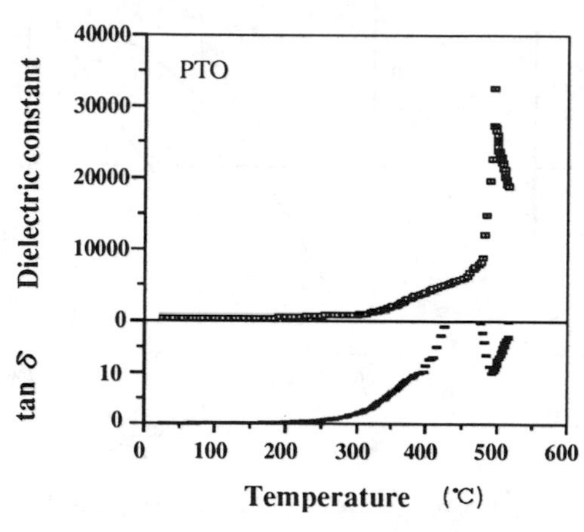

Fig. 3. Temperature dependence of dielectric constant and tan δ for a PTO thin film prepared.

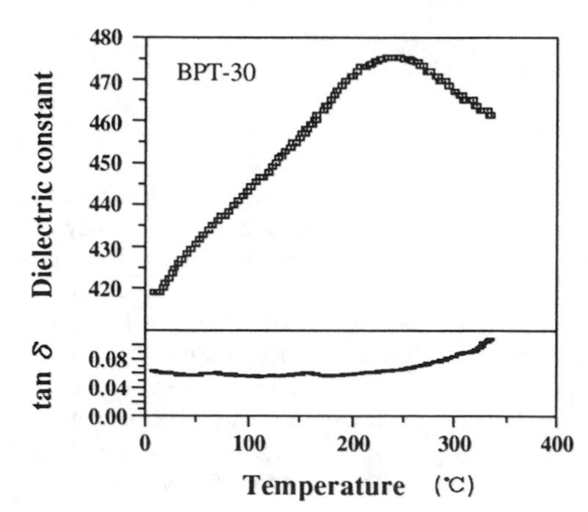

Fig. 4. Temperature dependence of dielectric constant and tan δ for a BPT-30 thin film prepared.

Table I. Specifications of the Ceramic Thin Films Prepared in the Present Study

Sample	Firing Temp. (°C)	Firing Time (h)	Grain size (μm)	Thickness (μm)	ε_r (20°C, 1 kHz)	tan δ (%) (20°C, 1 kHz)	P_r (μC/cm^2) E_c (kV/cm)
BTO	800	1.0	< 0.05	0.50	280	3.0	------------
BPT-30	750	0.5	~ 0.1	0.20	480	3.6	Pr=7.4, Ec=120 (1 kHz)
PTO	750	0.5	0.1~0.2	0.75	310	7.2	Pr=4.9, Ec=53 (60 Hz)

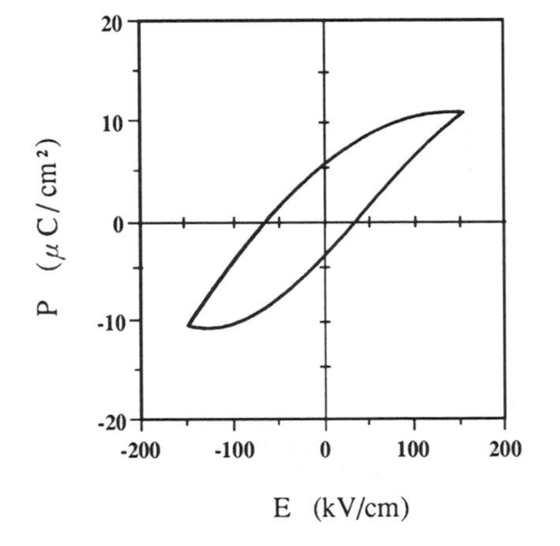

Fig. 5. P-E hysteresis loop for sample PTO (60Hz).

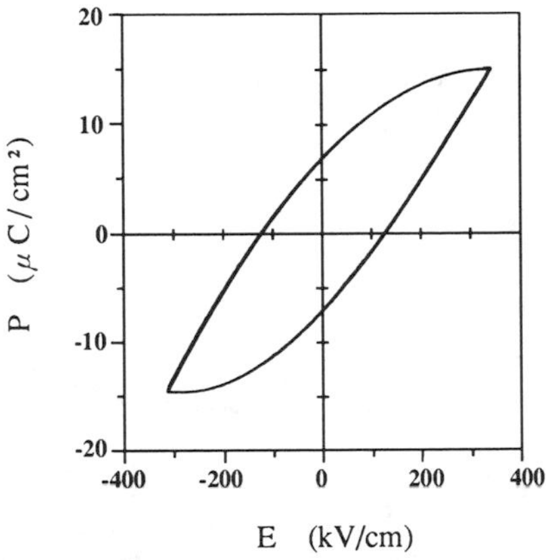

Fig. 6. P-E hysteresis loop for sample BPT-30. (1kHz).

RESULTS AND DISCUSSION

Figures 2 through 4 show the temperature dependence of dielectric constant and tan δ for $BaTiO_3$ (BTO), $PbTiO_3$ (PTO) and $Ba_{0.7}Pb_{0.3}TiO_3$ (BPT-30) thin films, respectively. By XRD analysis no phases other than perovskite can be detected in these thin films and no special crystallographic orientation was observed for BTO and BPT-30 films, while a considerable orientation in the (101) plane was observed for the PTO film. Figures 5 and 6 show P-E hysteresis loops for PTO and BPT-30 films, respectively. The P-E hysteresis loop for the PTO film was obtained at 60Hz and that for the BPT-30 film was obtained at 1kHz. They showed a slight change in its shape of getting slim with increasing frequency from 60Hz to 1kHz. It is obvious that these P-E hysteresis loops show unsaturated characteristics. Saturated P-E hysteresis loops could not be obtained for these films because of an insufficient limiting voltage of the instrument used. Since these films have been confirmed to be highly resistive at room temperature (though their exact resistivities were not obtained), evidently, P-E hysteresis loops for these films resulted from ferroelectricity but not resistive loss. Specifications of these thin films, including the data of dielectric properties and remnant polarizations Pr measured at room temperature, are shown in Table I. In this table, a value of Pr for the BTO film is not listed, because no definite P-E hysteresis loop was observed for the film at room temperature.

The dielectric constant-temperature curve for a BTO thin film measured at 10kHz and a oscillation (OSC) level of 0.1V is shown in Fig. 2. It was confirmed that the characteristic showed no significant frequency dependence in the frequency range of 1kHz to 100kHz. A broad peak of dielectric constant was observed around -100°C, however no characteristic peaks were observed at the Curie point (≈120°C) or at the orthorhombic to tetragonal phase transition (≈0°C). This suggests that the BTO film obtained may be paraelectric at room temperature. This observation is supported by the lack of a P-E hysteresis loop. It has not been decided yet whether the broad dielectric peak appearing around -100°C is connected with a ferroelectric phase transition (the rhombohedral to orthorhombic phase transition of barium titanate normally located≈-80°C), but an experiment for P-E hysteresis loop observation below -100°C is now under way. The lack of ferroelectricity in this film is considered to be connected with its very small grain size of less than 0.05 μm.

The dielectric constant-temperature characteristic for the PTO thin film (measured at 1kHz and OSC=0.1V) was, on the other hand, found to show a characteristic peak at the Curie point (≈490°C), as shown in Fig. 3. It can be recognized that this film showed an unusually high value of dielectric constant above 300°C. However the high dielectric constant above 300°C is associated with a very large value of tan δ, which probably is due to a significant decrease in resistivity at high temperatures. Another distinguishable but broad peak is seen in the dielectric constant-temperature characteristic for the BPT-30 thin film (at 100kHz, OSC=0.1V), as shown

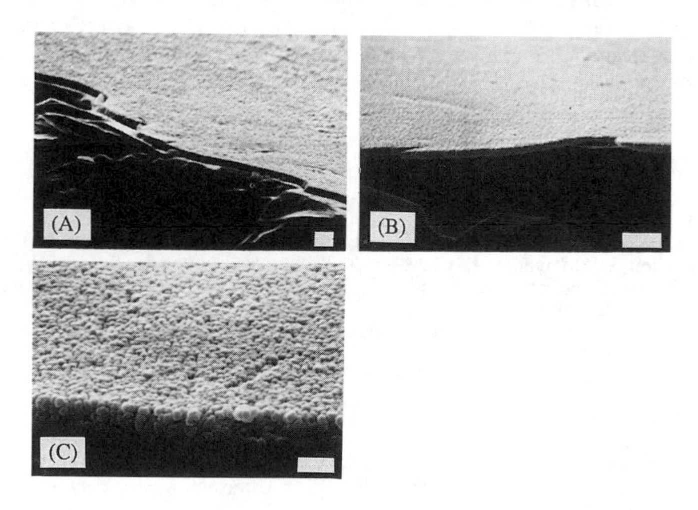

Fig. 7. SEM photographs of (A) BTO, (B) BPT-30 and (C) PTO ceramic thin films, deposited on $BaPbO_3$ ceramic substrates. Bars=1 μm.

in Fig. 4. The peak temperature corresponds to the Curie point of BPT-30 (≈250°C)[10]. The appearance of the broad peak at the Curie point in this film is likely connected with the grain size effect the same as in the case of the BTO film.

SEM morphologies shown in Fig. 7 may be of help to understand the dielectric properties obtained for the films. This figure demonstrates that fine-grained, dense films were deposited on BPO ceramic substrates for both BTO and BPT-30 compositions, while larger-grained film was produced for the PTO composition. This figure also shows that all the films were produced to have smooth surfaces and no microcrackings, with no appreciable interface layers between the films and the substrates. The grain sizes and morphologies of these thin films may interpret well their dielectric properties shown in Figs. 2 through 4, and it has been clearly shown that metallic conducting BPO ceramics can be used as a substrate for producing barium titanate-based thin films without causing any significant degradation of dielectric properties.

CONCLUSION

Ceramic thin films of BTO, PTO and BPT-30 have been prepared on metallic conducting $BaPbO_3$ ceramic substrates by the sol-gel method using metal alkoxides. A BTO film with a grain size <0.05 mm showed a broad peak of dielectric constant around -100°C but with no appreciable peak of dielectric constant at the Curie point (≈120°C), which indicates that this film is paraelectric at room temperature. PTO and BPT-30 films, on the other hand, showed a characteristic peak of dielectric constant at their Curie point. Although the quality of ferroelectricity in these films could not be evaluated by hysteresis loop measurements, it may be possible that these films were produced to be ferroelectric at room temperature from the appearance of dielectric constant peak at their Curie point. A broad peak of dielectric constant at the Curie point in the BPT-30 film may be interpreted on the basis of its grain size effect. This study clearly demonstrates that metallic conducting $BaPbO_3$ can be used as a substrate with the function of an electrode for producing barium titanate-based thin films without causing no significant degradation of dielectric properties.

REFERENCES

1. V.S. Dharmadhikari and W.W. Grannemann, J. Appl. Phys., 53, 8988 (1982).
2. J. Fukushima, K. Kodaira, and T. Matsushita, Am. Ceram. Soc. Bull., 55, 1064 (1976).
3. J.F. Campion, D.A. Payne, H.K. Chae, and Z. Xu in Ceramic Powder Science IV; Ceram. Trans. 22, edited by S. Hirano, G.L. Messing, and H. Hausner (Am. Ceram. Soc., Ohio, 1991) pp. 477-489.
4. T. Tuchiya, T. Kawano, T. Sei, and J. Hatano, J. Ceram. Soc. Jpn., 98, 743 (1990) [in Japanese].
5. J.B. Blum and S.R. Gurkovich, J. Mater. Sci., 20, 4479 (1985).
6. K.D. Budd, S.K. Dey, and D.A. Payne in Better Ceramics Through Chemistry II, edited by C.J. Brinker, D.E. Clark, and D.R. Ulrich (Mater. Res. Soc. Proc. 73, Pittsburgh, PA, 1986) pp. 711-716.
7. J. Fukushima, K. Kodaira, and T. Matsushita, J. Mater. Sci., 19, 595 (1984).
8. S.K. Dey, K.D. Budd, and D.A. Payne, IEEE Trans. Ultrasonics, Ferroelectrics, and Frequency Control, 35, 80 (1988).
9. S.D. Ramamurthi and D.A. Payne, J. Am. Ceram. Soc., 73, 2547 (1990).
10. B. Jaffe, W.R. Cook Jr., and H. Jaffe, Piezoelectric Ceramics, (Academic Press, London, 1971), p. 94.

LOW TEMPERATURE FORMATION OF FERROELECTRIC THIN FILMS

Chi Kong Kwok and Seshu B. Desu

Department of Materials Science and Engineering
Virginia Polytechnic Institute & State University
Blacksburg, VA 24061

ABSTRACT

Two novel processes have been developed to lower the transformation temperature of ferroelectric lead zirconate titanate (PZT) thin films with high Zr/Ti ratio. One process makes use of high pressure and the other process uses seeding to encourage the transformation.

A previous study has shown that nucleation is the rate-limiting step for the perovskite formation. Therefore, any process that enhances the kinetics of the nucleation will likely decrease the transformation temperature. In this process, a very thin seeding layer, which has a low effective activation energy for perovskite formation, is used to provide nucleation sites needed for the low temperature perovskite formation. In this study, the pyrochlore to perovskite phase transformation temperature of $PbZr_xTi_{1-x}O_3$ films of high Zr/Ti ratio (e.g. x = 53/47) can be lowered by as much as 100°C.

INTRODUCTION

Recently, ferroelectric thin films have been actively researched in the area of nonvolatile memory applications. Lead zirconate titanate (PZT), one of the most studied ferroelectric materials, shows promising electrical properties in the memory applications. The commonly used composition is about 53% Zr and 47% Ti (53/47), which is located at the morphotropic phase boundary. Films of this composition possess the very good ferroelectric properties required for fast and reliable memory switching.

In order to integrate PZT films into the existing semiconductor processes, many process problems have to be solved. One of the problems is the high temperature post-deposition annealing needed to form the desirable perovskite phase. This annealing is required because most of the as-deposited films are amorphous, and they form an intermediate nonferroelectric pyrochlore phase before the formation of the perovskite phase. The transformation temperatures for initial perovskite formation (T_i^{per}) and complete perovskite formation (T_c^{per}) of the PZT films are functions of the compositions and of the type of substrate used for deposition. Typical annealing temperatures for the 53/47 films are from 650°C to 750°C. At these annealing temperatures, interdiffusion among the PZT films, the contact electrodes, and the underlying metallization becomes a genuine concern; furthermore, thermal stress developed during the high temperature annealing may affect the long term reliability of the device.

The objective of this study is to develop novel processes of thin film deposition which require a lower perovskite transformation temperature.
In this study, three different approaches for low temperature perovskite transformation are investigated and they show different degrees of success. These methods are : (1)

(1) high pressure synthesis and (2) a two-step seeding process.

Viskov and co-workers reported that many thallium-containing compounds would not transform from pyrochlore structure to perovskite structure even annealed at 800°C but these compounds transformed to perovskite phase at 600°C under a pressure of 40-60 kbar [1]. Pyrochlore structures in general have more open crystal structures when compared to perovskite structure. In the case of PZT materials, the pyrochlore phase has a specific volume about six percent larger than that of the perovskite structure. Consequently, increasing the pressure encourages the pyrochlore phase to transform to the perovskite phase.

Previous kinetic study by the authors was performed by following the nucleation and growth of the perovskite phase using scanning electron microscopy [2]. Analysis of the kinetic data showed that the activation energies of the nucleation and growth of the 53/47 PZT perovskite phase are 441 kJ/mole and 112 kJ/mole, respectively. Therefore, the transformation from the pyrochlore phase to the perovskite phase is apparently nucleation controlled. Since the nucleation is the rate-limiting step of the transformation, the kinetics of transformation can be expedited when a large amounts of nucleation sites are created on the substrate. The purpose of the two-step process is to provide a large number of such favorable nucleation sites by the seeding layer.

Seeding processes have been demonstrated to enhance the kinetics of crystallization of gel particles [3] and to encourage the formation of epitaxial films [4]. In this paper, we describe a high pressure process and a two-step process which utilize the seeding to produce ferroelectric films at low transformation temperatures. These processes can, in principle, be applied to other material systems that undergo a similar polymorphic phase transformation.

EXPERIMENTAL

1. Precursor and film preparation

PZT thin films were fabricated from sol-gel precursors (0.4M) of lead acetate, titanium isopropoxide, and zirconium n-propoxide dissolved in glacial acetic acid and n-propanol. The proper amounts of Zr and Ti alkoxides were premixed in the presence of propanol and acetic acid before the addition of lead acetate. These solutions were hydrolyzed with appropriate amounts of water and were further diluted with propanol and acetic acid to form the final precursors. The details of the precursor preparation are similar to the one suggested by Yi and Sayer [5]. Seven precursors with different Zr and Ti compositions were prepared and the Zr/Ti ratios are 0/100, 30/70, 40/60, 53/47, 65/35, 75/25, and 100/0. The compositions of the films made by these precursors were determined by an electron microprobe. The compositions of each of the films prepared by the sol-gel method were within 1 atomic percent of their target compositions.

Thin films were deposited by spin coating on single crystal sapphire and Pt-coated Si substrates. The thicknesses of the films were controlled by the spin speed and the concentration of the precursor. After spin coating, the films were placed on a hot plate and the organic solvents were baked out at 150°C for 5 minutes. The film thickness can be built up by repeating the spin coating and the baking. Final annealing was performed in a tube furnace in air.

2. High pressure process

In this study, PZT samples having film thicknesses of about 350 nm and area of

0.6 cm by 1.5 cm were inserted into platinum capsules, filled with pure oxygen and then sealed by electrical welding. Each capsule was placed into a superalloy high pressure container. Each entire setup was then slided into a tube furnace and pressurized by a pressure line to the desired pressure.

3. Two–step process

In this study, the first step involved depositing a very thin layer of lead titanate (PT) onto a sapphire substrate. The spin speed used was 9000 rpm for 30 seconds. The film was then baked at 150°C for 5 minutes and annealed at 500°C for 15 minutes in air. The final PT film thickness was about 45 nm. At this stage, the PT film had already been transformed into the perovskite structure. We anticipate that a thinner seeding film, for instance, a film of few atomic layer thick, would still be effective in this study.

The second step involved depositing the PZT film with desired composition and thickness onto the PT–coated substrate. In this study, Zr/Ti composition of 53/47 and thickness of 300nm were chosen based on previous experiments which have shown that PZT films having this combination produced very good ferroelectric properties. PZT films were obtained by spinning and baking twice with a spin speed of 1500 rpm. Final annealings were performed in a tube furnace at temperatures ranging from 500°C to 750°C. These films were then characterized by x–ray diffraction.

RESULTS AND DISCUSSION

In most of the thin film deposition techniques, such as the sol–gel method, metalorganic decomposition (MOD), and sputtering, as–deposited PZT films generally have amorphous structures. Post–deposition annealing is needed to produce the desirable ferroelectric perovskite phase. The amorphous structure first transforms into a metastable pyrochlore phase, and this pyrochlore phase transforms into perovskite phase at a higher annealing temperature. The pyrochlore phase has an oxygen–deficient fluorite structure and it is non–ferroelectric. The details of the pyrochlore to perovskite transformation have been studied by transmission electron microscopy [6]. The transformation of the pyrochlore phase to the perovskite phase can be monitored by x–ray diffraction. For the 53/47 PZT films, the perovskite phase has sharp and well–defined major peaks at 2θ of 31.3°, 38°, and 55.5°. The pyrochlore phase has very broad major peaks at 2θ of 29.5°, 34.2°, and 49.2°. Figure 1 shows the x–ray diffraction spectra of PZT 53/47 films annealed from 500°C to 650°C for a fixed annealing time of 15 minutes. At 500°C, only the pyrochlore phase is found. At 550°C, both the perovskite phase and the pyrochlore phase are present in the film; and at 600°C, the transformation is completed, and only a single perovskite phase is present. Thus, in this case, the T_i^{per} is about 525°C and the T_c^{per} is 600°C for the 53/47 PZT films.

Similar experiments have been performed on films with compositions ranging from pure lead titanate (PT) to pure lead zirconate (PZ); the results are summarized in Fig. 2. Figure 2 illustrates the presence of the perovskite and pyrochlore phases as a function of annealing temperatures and compositions. The T_i^{per} and T_c^{per} of the Zr–rich phases are much higher than that of the Ti–rich phases. The T_c^{per} of the PT film is 500°C, which is 100°C lower than that of the 53/47 PZT film.

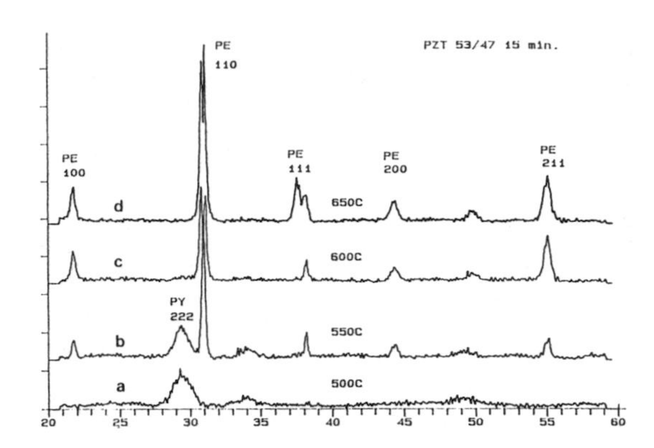

Figure 1. XRD patterns of unseeded 53/47 PZT films annealed at (a) 500°C; (b) 550°C; (c) 600°C; and (d) 650°C for 15 minutes. The symbols PE and PY indicate perovskite phase and pyrochlore phase, respectively.

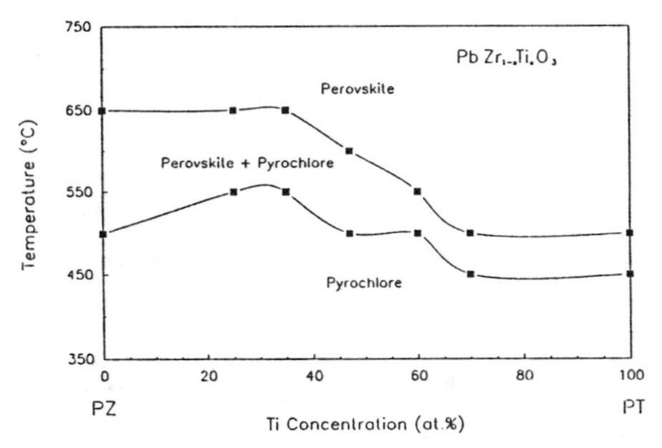

Figure 2. Initial and complete perovskite transformation temperatures as a function of Zr and Ti composition in the $PbZrO_3$–$PbTiO_3$ solid solution.

High pressure process

PZT films of 53/47 composition were annealed at 450°C for 5 hours in the same setup under both normal pressure and elevated pressures of 50 and 100 bar. For the PZT films annealed at normal atmospheric pressure, only pyrochlore phase was observed. However, as the pressure increased to 50 and 100 bar, only the perovskite phase was observed. The theoretically predicted condition for the conversion of pyrochlore phase to perovskite phase as specified by Viskov [1] is such that if the ratio of the (Volume of perovskite/ Volume of pyrochlore) raised to the third power is less than 0.40, then, it is possible to convert pyrochlore to perovskite. In the case of 53/47 PZT, the ratio of cell volume to the third power is about 0.396 which does satisfy the above condition. This process confirms that the pyrochlore to perovskite transformation involved a volume decrease.

Two–step seeding process

For the two–step process in this study, PT was chosen as the first deposited layer. Subsequently, 53/47 PZT films were deposited on top of the PT films and annealed at temperatures ranging from 450°C to 600°C in 25°C intervals at a fixed annealing time of 15 minutes. These two–layer films wil be called PT–PZT films hereafter.

The results of T_c^{per} as a function of annealing temperature and time are summarized in Figs. 3 and 4. Figure 3 denotes that at a fixed annealing time of 15 minutes, the T_c^{per} of the PT–PZT film is 525°C. As the annealing time is increased to 1 hour, the T_c^{per} is further decreased to 500°C, as seen in Fig. 4. Hence, the transformation temperature of the 53/47 PZT films is decreased by 100°C due to the presence of the seeding layer.

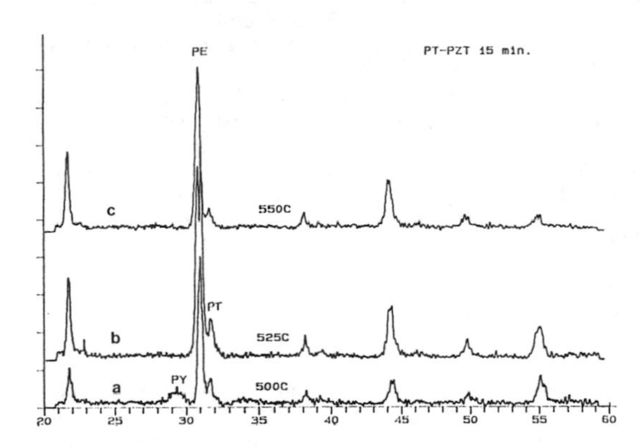

Figure 3. XRD patterns of the seeded 53/47 PZT films annealed at (a) 500°C; (b) 525°C; (c) 550°C for 15 minutes.

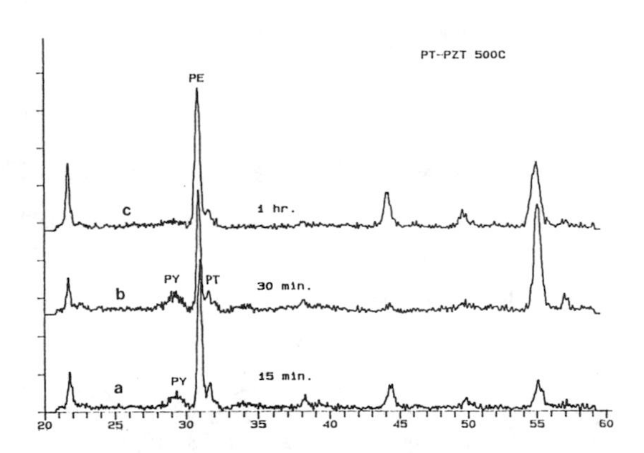

Figure 4. XRD patterns of the seeded 53/47 PZT films annealed at 500°C for (a) 15 minutes; (b) 30 minutes and (c) 1 hour.

In the PZT thin film system, previous work has shown that nucleation is the rate–limiting step so the lowering of T_c^{per} by the two–step process is most likely related to the lowering of nucleation activation energy of the perovskite formation. The two energy barriers that oppose nucleation are the strain energy and the surface energy that incur during transformation. However, the strain energy should not change whether or not the PZT film is formed in the presence of a PT seeding layer. Thus, the major effect of the two–step process is the possible decrease of surface energy, which in turn, increases the kinetics of the nucleation.

The lowering of the surface energy can be explained by the lattice matchings of the PZT/PT and PZT/sapphire interfaces. Both the PT and the 53/47 PZT phases have a tetragonal "perovskite" structure. The lattice parameters, a and c, for the PT and the 53/47 PZT structures are 0.3899, 0.4036 nm and 0.4153, 0.4146 nm, respectively [7]. The sapphire substrate has a rhombohedral structure but is commonly represented by a hexagonal unit cell. The lattice parameters, a and c, are 0.4758 and 1.2991 nm [7]. For the PZT/PT interface, the lattice mismatch in the a direction is about 3.4 % and less than 0.2 % in the c direction. For the PZT/sapphire interface, there are no common matching planes. If the lattice mismatch is small, the interfaces could be coherent or semicoherent. Either coherent or semicoherent interfaces will have a much lower surface energy when compared to an incoherent interface. It is plausible that a seeding layer, which has a crystal structure similar to that of the deposited film and has a small lattice mismatch, can provide the preferred nucleation sites and thus increase the formation kinetics of the deposited film.

CONCLUSIONS

We have demonstrated two different processes which can lower the pyrochlore to perovskite transformation temperature of the 53/47 PZT films by 100°C. For polymorphic transformation from an open structure to a compact strcucture, high pressure invariably encourages low temperature transformation and this is confirmed in this study. The underlying principle of this two–step process is that the nucleation rate of the film of desired composition and thickness can be controlled by the introduction of the seeding layer. The results from x–ray diffraction supports this supposition.

ACKNOWLEDGMENTS

This work is partially supported by a project from DARPA through ONR.

REFERENCES

1. A. Viskov, E. Zubova, K. Burdina and Y. Venevtsev, Soviet Physics – Crystallography, **15**(5) 932 (1971)
2. C. Kwok and S. Desu, Ceramic Transactions: Ferroelectric Films, 25, A. Bhalla, ed., ACS, Westerville, OH, 73 (1992)
3. M. Kumagai and G. Messing, J. Am. Ceram. Soc. **67** (11) C230 (1984)
4. K. Miller and F. Lange, J. Mater. Res. **6** (11) 2387 (1991)
5. G. Yi and M. Sayer, Ceramic Bulletin, **70**(7) 1173 (1991)
6. C. Kwok and S. Desu, Ceramic Transactions: Ferroelectric Films, 25, A. Bhalla, ed., ACS, Westerville, OH, 85 (1992)
7. Powder Diffraction File, Inorganic Phase, JCPDS (International Center for Diffraction Data, Swarthmore, PA 1986) #6–452, #33–784 and #10–173.

MICROSTRUCTURAL VARIATIONS IN SOL-GEL PROCESSED LITHIUM NIOBATE THIN FILMS

VIKRAM JOSHI*, GRACE K. GOO**, AND MARTHA L. MECARTNEY**
*University of Minnesota, Dept. of Chemical Engineering and Materials Science, Minneapolis, MN 55455
**University of California at Irvine, Materials Science and Engineering Program, Irvine, CA 92717

ABSTRACT

$LiNbO_3$ thin films were deposited by dip coating Li-Nb alkoxide solutions onto silicate glass substrates and single crystal sapphire substrates. Microstructural characterization using transmission electron microscopy (TEM) showed significant differences in film microstructures dependent on the initial solution chemistry. Fully crystalline films could be obtained after heat treatments at 400°C in air. The grain size and porosity were dependent on the amount of water of hydrolysis in the alkoxide sol. The higher the water content, the larger the grain size and porosity. Crystallization studies of nucleation and growth of $LiNbO_3$ for films heat treated from 300-600°C indicated that higher temperatures or long soak times generated large facetted grain structures. Single crystalline films were obtained on (0001) sapphire substrates.

INTRODUCTION

Lithium niobate is a promising material for integrated optic devices due to its fast response time and low absorption. Applications include usage in dielectric waveguides, Q-switches, integrated optics and many others. The interest in preparing the lithium niobate by sol-gel routes is due to the homogeneity and high purity that can be achieved in the final product. Thin films can be easily prepared by sol-gel processing, at relatively low temperatures as compared to traditional film preparations [1-6]. Work performed by Nashimoto and Cima [6] on the development of $LiNbO_3$ thin films using X-ray techniques has shown that the amount of water of hydrolysis plays an important role in the resulting film structure. Other studies have shown that crystallization of the amorphous lithium niobate gel initiated at temperatures as low as 200°C [7]. Changes in particle morphology were observed at higher processing temperatures. The goal of this study is to observe, via TEM, the effects of varying processing temperature, soak time, amount of water of hydrolysis, and substrate with respect to the microstructural development of lithium niobate thin films .

EXPERIMENTAL PROCEDURE

The procedure for obtaining the lithium niobate complex alkoxides are given in reference [3]. Thin films were dip coated onto substrates and heating was performed at 10°C/min. in air. To study the effect of water of hydrolysis, three different alkoxide to water ratios were used; namely, 1:0, 1:1, and 1:2. To study the effect of temperature, samples with a 1:1 alkoxide to water ratio were used with soak temperatures of 300, 400, 500, and 600°C for 60 minutes. To study the effect of soak time, samples with a 1:1 alkoxide to water ratio were used and held at 400°C for 10, 30, 60, 120, and 240 minutes. All these samples were prepared using glass substrates and were quenched in air following the heat treatment. To study the effect of different substrates on thin film growth, single crystal sapphire (0001 orientation) substrates were used.

A modified TEM specimen preparation technique was developed in order to eliminate any damage or artifacts caused by ion milling. To study films in plan view, conventional TEM specimen preparation involves the deposition of the film followed by mechanical thinning and ion milling from one side. The ion milling process may result in introducing artifacts or film damage. Peel off of the film from the substrate was a significant problem. For the technique used in this study, the substrates were initially thinned to perforation before film deposition. A similar technique has been employed previously for studying nucleation and growth of thin films grown by vapor phase deposition [8]. Figure 1 shows schematics for the two different techniques for TEM sample preparation of thin films. It can be seen that the specimen prepared by using the modified technique produces a thicker film in the electron transparent region. Figure 2 shows TEM micrographs comparing the microstructure of the films synthesized with similar processing conditions (1:1 alkoxide to water ratio, 400°C, 60 min.) but with the two different TEM specimen preparation techniques described above. The grain size for both films is similar; however, the film deposited on the pre-thinned substrate shows cracking. The perforation in the pre-thinned substrate acts as an edge, and dip coated films in general are not uniform or continuous at the edge of a substrate.

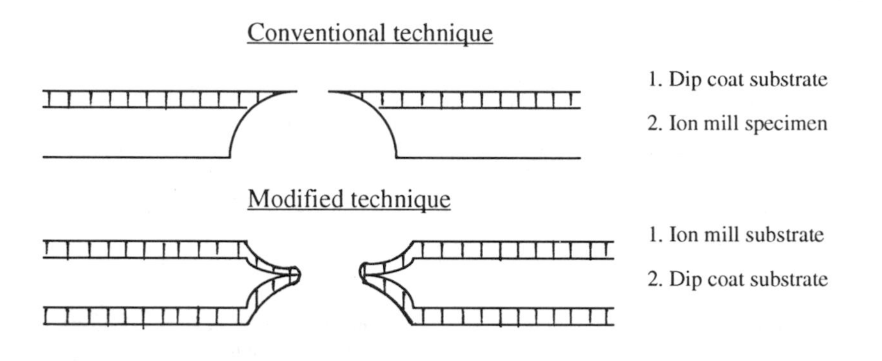

Figure 1: Schematic illustration of the two different techniques for TEM sample preparation of thin films.

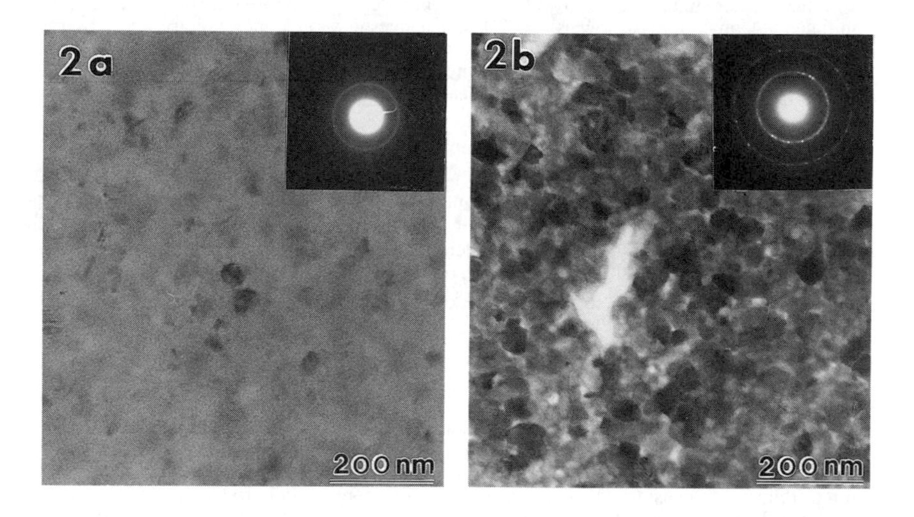

Figure 2: Microstructural comparison of $LiNbO_3$ thin films synthesized with similar processing conditions (1:1 alkoxide to water ratio, 400°C, 60 min.) but with the two different TEM specimen preparation techniques; (a) conventional, (b) modified.

RESULTS AND DISCUSSION

Effect of alkoxide to water ratio

Figure 3 shows the TEM results with corresponding SAD patterns for varying the alkoxide to water ratio for samples treated at 400°C for 60 minutes on glass substrates. TEM specimens for these films were prepared by conventional technique. The sample with 1:0 alkoxide to water ratio is amorphous with no pores visible. The sample with 1:1 alkoxide to water ratio gives extremely refined crystallites on the order of 30 nm, and the film is very dense. As the alkoxide to water ratio increases (1:2), the size of the crystallites increases (150 nm) as well as the formation of porous network in the thin film. The results indicate that not only is hydrolysis critical for crystallization but the amount of water also plays an important role in determining the grain size and porosity of the film.

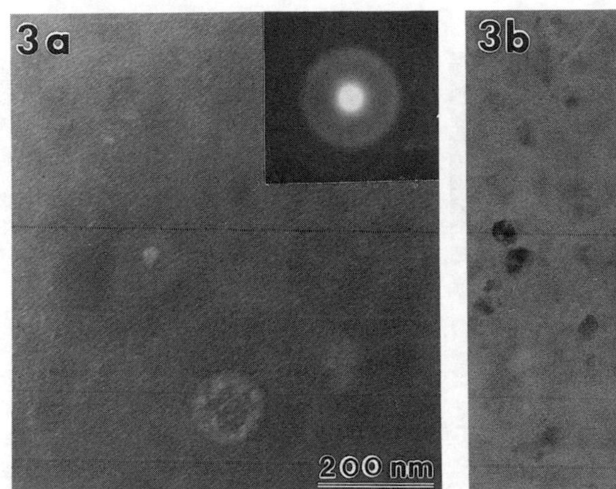

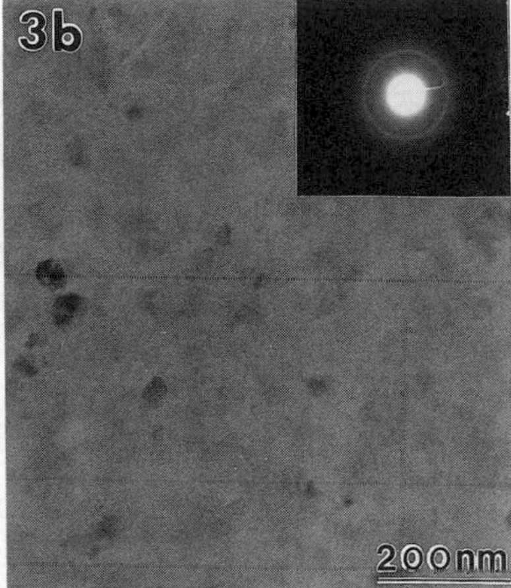

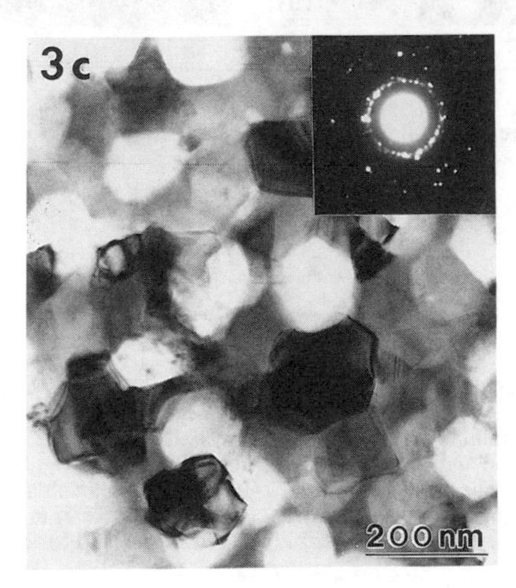

Figure 3: Film heated at 400°C for 60 min; (a) 1:0, (b) 1:1, and (c) 1:2 alkoxide/water ratio.

Effect of soak temperature

Figure 4 shows the TEM micrographs of thin films of 1:1 alkoxide to water ratio processed at 400°C and 600°C for 60 minutes. At 300°C, the film overall was still amorphous, but by 400°C, the films were completely crystallized with crystallite size on the order of 30 nm. And as the temperature increased, the crystallites increased in size to about 125 nm and a porous network formed similar to that obtained at a high water to alkoxide ratios. At 600°C, Figure 4b shows the development of facetted particles. This is due to the domination of the low surface energy planes consistent with previous studies on the processing of $LiNbO_3$ powders at 600°C [7].

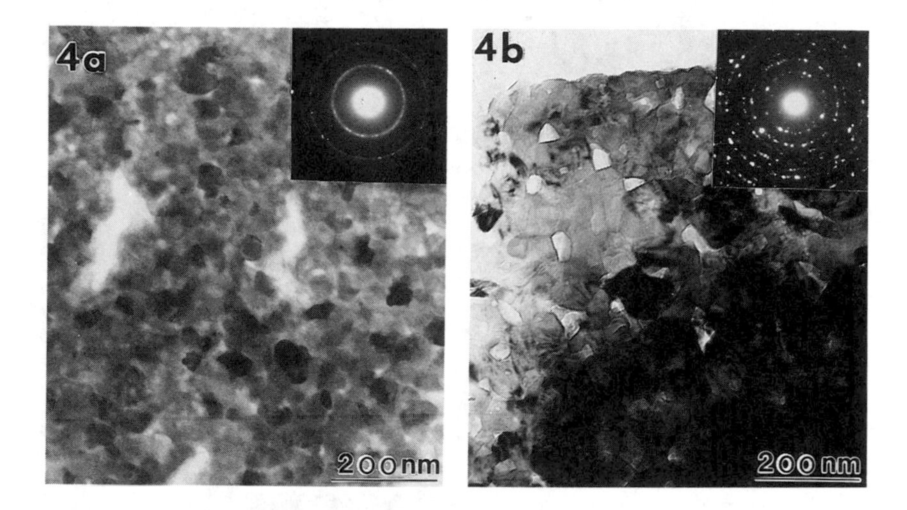

Figure 4: Film heat treated at (a) 400°C and (b) 600°C for 60 min. for 1:1 alkoxide/water ratio.

Effect of soak time

The effect of soak time is shown in Figure 5 for selected samples with 1:1 alkoxide to water ratio held at 400°C. The film held for only 10 minutes appeared featureless and amorphous. After 30 minutes the film began to crystallize. As soak time increased, the size of the crystallites increased and a porous network forms (compare Figure 5a held for 60 min. and 5b held for 240 min). The grain size for the 240 min. sample is on the order of 150 nm which is similar to those obtained from films with a 1:2 alkoxide to water ratio heat treated for 1 hr. at 400°C (Fig. 3c), but the porosity is lower in the 1:1 sample held for 240 min.

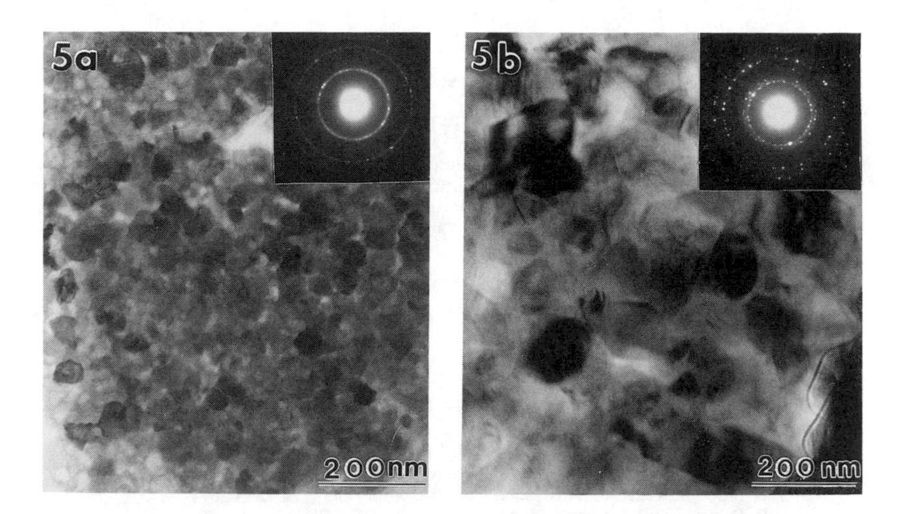

Figure 5: Film heat treated at 400°C for (a) 60 min. and (b) 240 min. for 1:1 alkoxide/water ratio.

Effect of substrate

Epitaxial film growth for lithium niobate was attempted using (0001) single crystal sapphire. Films were coated and treated at 400°C for 60 minutes for 1:1 alkoxide to water ratio. TEM examination of the film (1:1, 60 min, and 400°C) shows that the sapphire substrate yielded epitaxial growth over large areas. The SAD pattern in Figure 6 shows single crystal nature of the film and the zone axis is [0001].The film were highly defective however, with significant strain, dislocations and twinning evident.

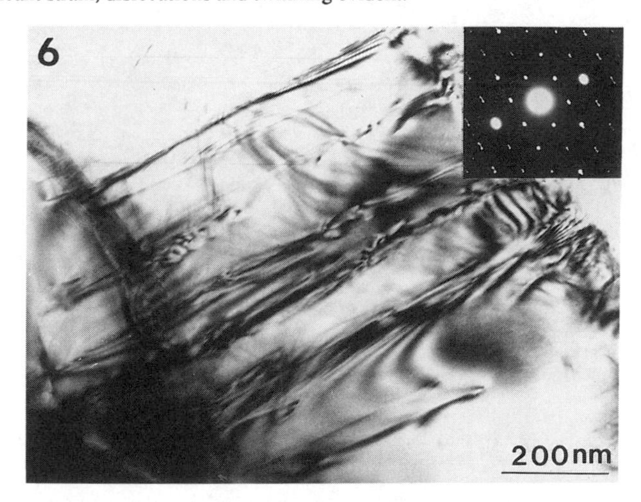

Figure 6: $LiNbO_3$ film on single crystal sapphire (0001) substrate, heat treated at 400°C for 60 min. for 1:1 alkoxide/water ratio.

CONCLUSION

The above results indicate that the film morphology for $LiNbO_3$ can be tailored very explicitly by varying one or more of the following parameters: amount of water of hydrolysis, film treatment temperature, film treatment time. The higher the degree of hydrolysis, the larger the grain size and larger the pore size. Long times or high temperatures induced the formation of facetted large grains. Epitaxial film growth can be obtained on (0001) single crystal sapphire. More investigations are in process to achieve a defect-free epitaxial $LiNbO_3$ film.

ACKNOWLEDGMENTS

This work has been supported through a grant from the Air Force Office of Scientific Research under contract number 49620-89-C-0050.

REFERENCES

1. S. Hirano and K. Kato, J. Non-Cryst. Solids 100 , 538 (1988).
2. N.P. Castings, F. Duboudin, and J. Ravez, J. Mater. Res. 3 , 557 (1988) .
3. S. Hirano and K. Kato, Advanced Ceramic Materials 2 , 142 (1987).
4. D.P. Partlow and J. Greggi, J. Mater. Res. 2 595 (1987).
5. D.J. Eichorst and D.A. Payne, in Better Ceramics Through Chemistry IV, eds. B.J.J. Zelinski, C.J. Brinker, D.E. Clark, and D.R. Ulrich, (Mat. Res. Soc. Proc. 180, Pittsburgh, PA, 1990) p. 669.
6. K. Nashimoto and M.J. Cima, Mat. Lett. 10, 348 (1991).
7. V. Joshi, G.K.Goo, and M.L.Mecartney, in Synthesis and Processing of Ceramics: Scientific Issues, eds. W.E. Rhine, T.M.Shaw, R.J. Gottschall, and Y. Chen (Mat. Res. Soc. Proc. 249, Pittsburgh, PA, 1992), p.459.
8. M.G. Norton, S.R. Summerfelt, and C.B. Carter, App. Phys. Lett. 56 , 2246 (1990).

THE GROWTH OF SINGLE CRYSTAL-LIKE AND POLYCRYSTAL $KNbO_3$ FILMS VIA SOL-GEL PROCESS

Chih-Hsing Cheng, Yuhuan Xu, and John D. Mackenzie
Department of Materials Science and Engineering
University of California, Los Angeles

Jie Zhang, and LeRoy Eyring
Center For Solid State Sciences
Arizona State University

ABSTRACT

Transparent and uniform thin films of crystalline $KNbO_3$ were synthesized on different substrates such as $SrTiO_3$, ZrO_2, MgO and Pt by the dip-coating method using a double alkoxides solution. The film was epitaxially crystallized at the temperature around 750°C for 3 hours in air on $SrTiO_3$ single crystal substrate. The structure of the films has been studied with X-ray and electron diffraction, and transmission electron microscopy.

Different molar ratios of 2-ethylhexanoic acid (2EHA) to alkoxides were applied to control the hydrolysis rate of the sol-gel solution. Dense films were obtained after modification of the alkoxides solution.

INTRODUCTION

Potassium niobate ($KNbO_3$) is one of the great potential materials for a variety of applications such as frequency doublers, optical waveguides, and holographic storage systems[1-2], due to its high electro-optic and nonlinear coefficients and good photorefractive properties. Recently the demand for the high quality and transparent ferroelectric thin films is increasing with the advances in the integrated optics. Fabrication of these films by sol-gel processing[3], including $BaTiO_3$[4], PZT[5], $LiNbO_3$[6], SBN[7], etc. has been the subject of much study. In addition, single crystal-like $LiNbO_3$ films have been successfully prepared using bimetallic alkoxides by S. Hirano et al.[8] and R. Xu et al.[9]. Epitaxial thin films of $KNbO_3$ have also been deposited on MgO single crystal substrate by ion beam sputtering technique[14]. However, only a few investigations of sol-gel processing of $KNbO_3$ have been reported[10-12]. Single-phase $KNbO_3$ powder[13] and polycrystalline films[12] were successfully achieved. There have been no reports of the growth of epitaxial $KNbO_3$ films via sol-gel processing in the past.

This work was carried out to study the crystallization of $KNbO_3$ films on various substrates. The paper also focuses on the effects of microstructure of thin films versus the modified double alkoxides solution.

EXPERIMENTAL

Figure 1 illustrates the flow chart of the preparation of coating solutions and films. A potassium ethoxide solution was prepared by dissolving potassium metal in anhydrous ethanol. Niobium ethoxide, 99.99% (liquid), was obtained from Alfa Chemicals. These potassium and niobium ethoxides were mixed and refluxed for 24 hours to ensure the formation of the double alkoxides solution[6]. The 2-ethylhexanoic acid (2EHA) was added into the double alkoxides solution to control the hydrolysis reactions. Then the solution was refluxed for 6 hours to make the modifying reaction completely. After being concentrated by distilling off the excess ethanol, the stock solution was ready for the dip-coating purpose. All reactions were carried out under an environment of dry N_2 box.

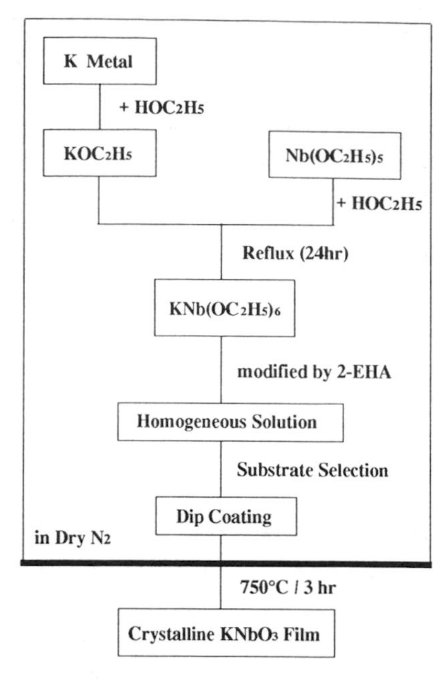

Fig. 1 Flow chart of synthesis of KNbO3 film by sol-gel method

SEM photographs of the microstructure of the films with and without modifying by organic acid, 2EHA, are shown in Figure 2. Both films were deposited on MgO substrates and calcinated at 750°C for 3 hours in air. Dense and transparent $KNbO_3$ films were obtained after the alkoxides solution was modified by 2EHA. Crack size around few μm range could be detected easily in the film without modifying by 2EHA.

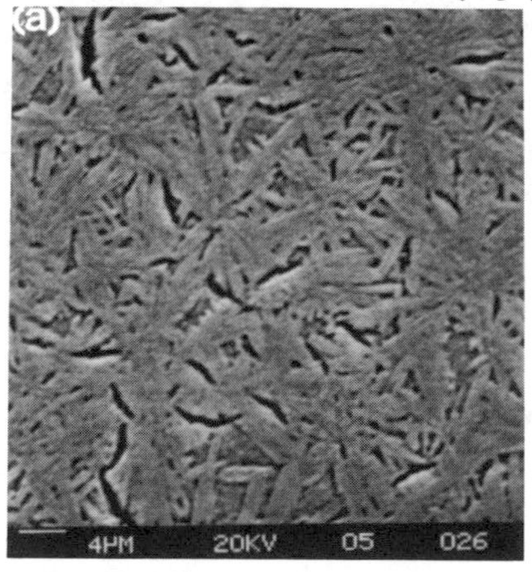

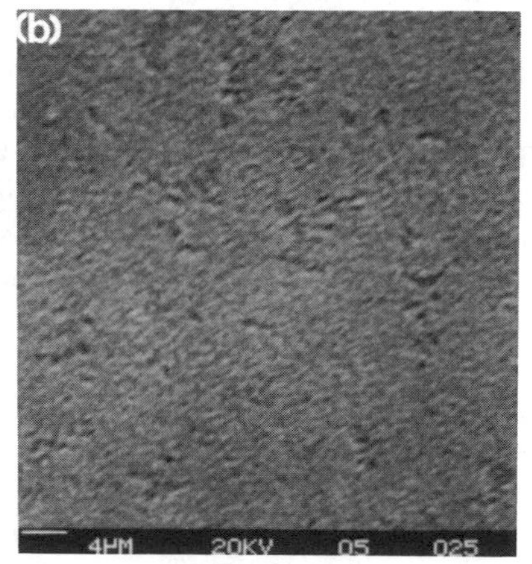

Figure 2. SEM for $KNbO_3$ films on MgO substrates, film fired at 750°C for 3 hours in air. (a) without any dopant, and (b) the alkoxides solution modified by 2EHA.

Optically polished $SrTiO_3$, MgO and ZrO_2 wafers were used as substrates. These substrates were cleaned by absolute ethanol and acetone before the dipping procedure. Thicker film was achieved by multiple coating and heat treatment. Each layer of film was heat treated crystallized before the next deposition. Thickness of films was measured by the Dektek depth profilometer.

XRD patterns and electron diffraction were used to exam the crystallization of films. The structure of film/substrate interface was investigated using high resolution transmission electron microscopy (HRTEM). Morphologies of the films were studied by SEM.

RESULTS AND DISCUSSIONS

A 2EHA modified $KNbO_3$ alkoxides solution was used as the coating solution. The solution is stable for about two months in a dry box. Single-phase $KNbO_3$ film began to crystallize at 700°C and became highly crystallized at 750°C. XRD patterns of the films showed complete crystallization and strongly preferrential orientation of the films along [110] direction on $SrTiO_3$ (100) substrate (Figure 3). These films exhibited epitaxial crystallization on $SrTiO_3$ and no other "potassium-deficient" phase. However, X-ray diffraction data can not distinguish between an epitaxial film from a single crystal film. For further characterization of the epitaxial growth of $KNbO_3$ film on $SrTiO_3$ substrate, the HRTEM of the interfacial lattice fringe was carefully studied.

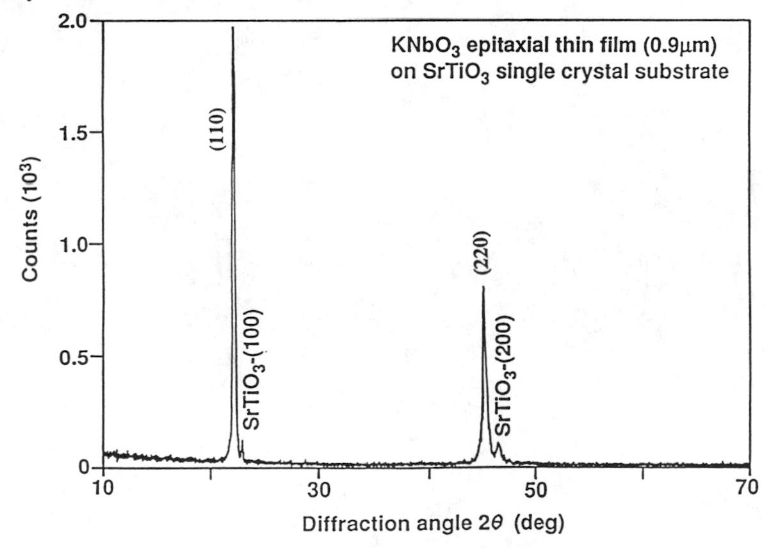

Figure 3. XRD of $KNbO_3$ film on $SrTiO_3$ single crystal substrate. The heat treatment temperature is 750°C for 3 hours.

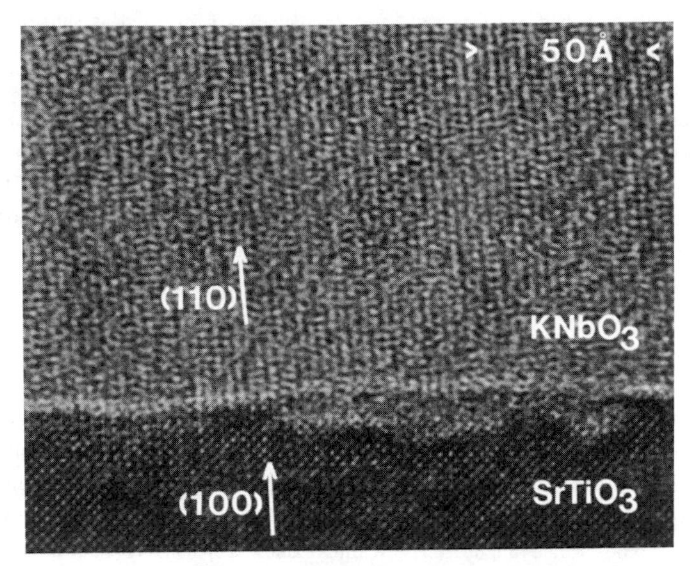

Fig. 4 HRTEM of the $KNbO_3/SrTiO_3$ epitaxial interface.

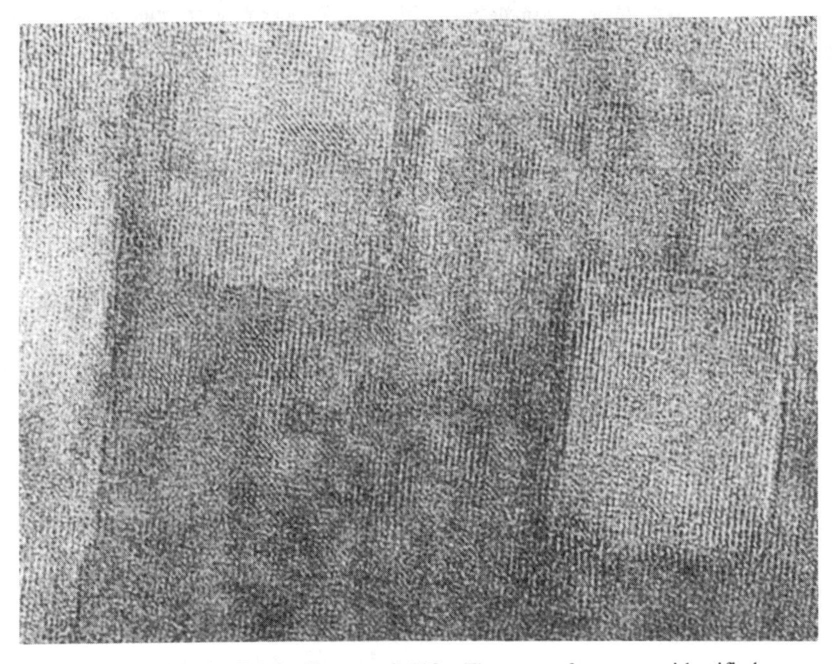

Fig. 5 HRTEM of a $KNbO_3$ film onto $SrTiO_3$. The square features are identified as ferroelectric domains.

Figure 4 shows the HRTEM immage of lattice fringe of the $KNbO_3$ film on $SrTiO_3$. Across all observable interfaces, neither grain boundaries nor misallignment of lattices were detected. Preliminary analysis shows that the immage of lattice fringe and the electron diffraction patterns correspond to that viewed from the direction perpendicular to the (110) plane. No other orientation was observed through the whole interfaces. Thus the film is believed to be single crystal.

Figure 5 gives another HRTEM of a locally $KNbO_3$ film. The three large structures noted in the micrograph are attributed to incoherent ferroelectric domains similar to those that have been reported previously by Lin et al.[15] and M.S. Ameen et al.[14]. Generally, domain walls could be identified only for high quality films.

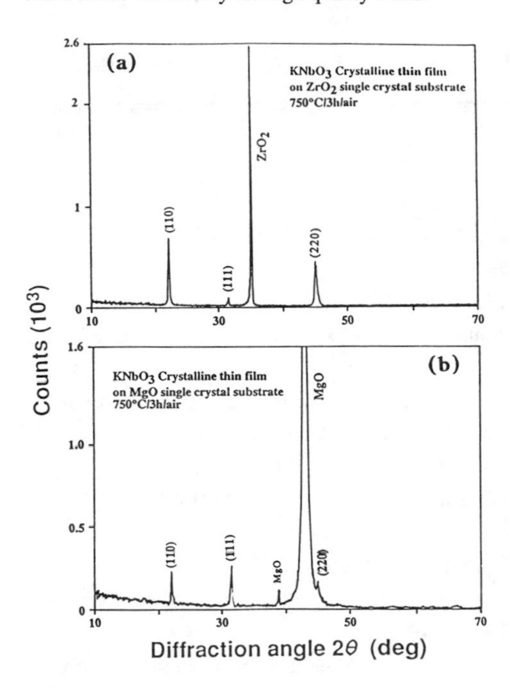

Fig. 6 XRD of $KNbO_3$ film on (a) ZrO_2, and (b) MgO substrates.

Crystallization of the films on MgO and ZrO_2 substrates is shown in Figure 6. Films on each MgO and ZrO_2 substrates were polycrystalline. Table I summerizes the influence of substrate on the crystallization of $KNbO_3$ films. The smallest mis-match between the films and substrates is along the c axis of $SrTiO_3$.

CONCLUSIONS

(1) Transparent $KNbO_3$ thin films on various substrates have been fabricated by sol-gel method.

(2) Single-phase $KNbO_3$ films were prepared on various substrates such as $SrTiO_3$, MgO and ZrO_2 via sol-gel process.

(3) Epitaxial growth of single crystal-like $KNbO_3$ film was achieved on $SrTiO_3$ single crystal substrate.

ACKNOWLEDGEMENT

This work was supported by the Air Force Office of Scientific Research, under Grant No. AFOSR-91-0096, and by the NSF under Grant No. DMR-8820017. The authors also thank Dr. Ren Xu's help in this work.

Table I. Summary of $KNbO_3$ film on various substrates.

Material	Crystal system	Lattice parameters(Å)	Lattice mismatch w.r.t. $KNbO_3$	Growth of $KNbO_3$ (750°C/3 h)
$KNbO_3$	Ortho.	a = 5.695 b = 5.721 c = 3.973		
$SrTiO_3$	cubic	3.905	||a 31% ||c 1.7%	single crystal-like
MgO	cubic	4.208	||a 26% ||c 5.9%	polycrystalline
ZrO_2	cubic	5.070	||a 11% ||c 27%	polycrystalline

REFERENCES

(1) Yuhuan Xu, *Ferroelectric Materials and Their Applications*, North-Holland, N.Y. (1991).

(2) P. Gunter, *Phys. Rep.*, **93 (4)**, 199-299 (1982).

(3) Yuhuan. Xu, and John.D. Mackenzie, *Integrated Ferroelectrics*, **1**, 17-44 (1992).

(4) J. Fukushima, K. Kodaira, A. Tsunashima and T. Matsushita, *Yogyo Kyoaishi,* **83**, 204 (1975).

(5) E. Wu, K.C. Chen and J.D. Mackenzie, *Better Ceramics Through Chemistry,* (eds. C.J. Brinker, D.E. Clark and D.R. Ulrich), North-Holland, N.Y., 169 (1984).

(6) S. Hirano and K. Kato, *Advanced Ceramic Materials*, **2(2)**, 142 (1987).

(7) Y. Xu, C.J. Chen, R. Xu, and J.D. Mackenzie, *Phys. Rev.*, B, **44**, 35-41 (1991).

(8) S. Hirano and K. Kato, *Mat. Res. Symp. Proc.*, **155**, 181 (1989).

(9) R. Xu, Ph. D. Thesis, University of California, Los Angeles, 1991.

(10) E. Wu, A.X. Kuang, and J.D. Mackenzie, Proc. of the Sixth IEEE International Symposium on Applications of Ferroelectrics, 388-390 (1986).

(11) S.L. Swartz, P.J. Melling, and C.S. Grant, in Optical Materials: Processing and Science, (eds. D.B. Poker and C. Ortiz), *Mat. Res. Soc. Proc.*, **152**, Pittsburgh, PA, 1989, pp. 227-232.

(12) M. Amini and M.D. Sacks, *Mat. Res. Soc. Symp.* **180**, 675-683 (1990).

(13) A. Nazeri, A.X. Kuang and J.D. Mackenzie, *J. of Materials Science,* **25**, 3333-3337 (1990).

(14) M.S. Ameen, T.M. Graettinger, S.H. Rou, H.N. Al-Shareef, K.D. Gifford, O. Auciello, and A.I. Kingon, *Mat. Res. Soc. Symp. Proc.*, **200**, 65-76 (1990).

(15) P.J. Lin, L.A. Bursill, *Phil. Mag.* **48**, 251 (1983).

PREPARATION AND CHARACTERIZATION OF SOL-GEL THIN FILMS OF $KTiOPO_4$.

MARK A. HARMER and MARK G. ROELOFS
Du Pont, Central Research and Development, Experimental Station, P.O.Box 80328, Wilmington, DE 19880-0328

ABSTRACT.

In this paper we describe a novel route for the formation of thin films of potassium titanyl phosphate ($KTiOPO_4$ or KTP) via a sol-gel route. The ability to form films of high optical quality depends upon the precursor chemistry developed. Formation of stoichiometric films is especially challenging in three component systems and a number of precursors have been investigated. The basic chemistry, processing and resulting microstructure will be described. Transparent, dense and continuous films have been prepared with grain sizes typically of 0.3µm or less depending upon the processing conditions. Some textured growth patterns are observed although the films are essentially polycrystalline. Using the prism coupling technique the films have been found to exhibit waveguiding.

INTRODUCTION.

Recently, there has been growing interest in the application of sol-gel chemistry for the formation of a range of thin film materials. Ferroelectric thin films, for example lead zirconate titanate, barium titanate and lithium niobate [1,2], have been fabricated by the sol-gel technique. Such materials are of interest for a range of applications from piezoelectric or electro-acoustic transducers, ferroelectric memory cells, to optical waveguides in electro-optical devices. Of particular interest is to gain a greater understanding of the precursor chemistry and to relate this to the final microstructure to investigate correlations between processing, properties and microstructure.

We report herein a novel route for the preparation of potassium titanyl phosphate ($KTiOPO_4$ or KTP) using a sol-gel route. KTP is a non-linear optical material which is being investigated for a number of non-linear optical applications, and in particular for frequency doubling the 1 µm radiation of Nd lasers [3]. The materials properties also make it attractive for various electro-optic applications, such as modulators and Q switches and recently low-loss optical waveguides were developed for KTP suggesting that this material is promising for integrated-optic applications. Most of these applications have been developed using single crystal KTP. It is believed that the formation of thin film KTP may broaden the number of device applications.

To the authors knowledge this is the first example of a sol-gel route to KTP. Sol-gel is particularly attractive for several reasons ranging from the good control of stoichiometry leading to high purity, the ability to forms films without the need for sophisticated apparatus, often lower processing temperatures, the ability to change the ratio of metal ions and if necessary to add dopants in a controlled way. In the case of the three component system investigated in this system, the choice of the precursor chemistry is crucial to form stoichiometric films. In this paper we report on the basic precursor chemistry and describe our initial results on the relationship between final microstructure and processing conditions.

EXPERIMENTAL.

A sol-gel route for the formation of KTP was developed based upon the aqueous hydrolysis and condensation of the alcoxides, potassium ethoxide, titanium isopropoxide and the phosphorous source was a mixed phospho ester, n-butyl phosphate. The phosphorus source, available commercially (Gelest Inc.) was used as supplied and consists of a 1:1 mixture of $OP(OH)(OC_4H_9)_2$ and $OP(OH)_2(OC_4H_9)$, with hydroxyl groups present on the phosphorous. The P content was determined by ICP analysis. Titanium isopropoxide was vacuum distilled prior to use. The solvent used was isopropanol. KTP precursor dried gels were investigated using thermogravimetric analysis (TGA) in conjunction with mass spectral analysis to indentify evolved decomposition products.

Thin films were prepared by spin coating the precursor solution on to a range of substrates. Substrates which have been used include polycrystalline quartz discs and the single crystal substrates silicon, lanthanum aluminate ($LaAlO_3$) and lanthanum gallate ($LaGaO_3$). Films were prepared by applying the precursor solution to the substrate and spinning the substrate at 750 rpm for about 15 seconds and then increasing the spin speed to 2000 rpm for 2 minutes. A single spin resulted in a final film thickness of about 100 nm as described below. Films were crystallized by calcination to about 750 C. Thicker films were obtained using multiple coating/calcination steps. Calcined films were characterized by x-ray diffraction, scanning electron microscopy and in the case of a silicon substrate, the film thickness was measured using an Rudolf AutoEL ellipsometer. A single coating on a single crystal silicon substrate with the KTP precursor solution leads to an amorphous film of about 180 nm in thickness as measured by ellipsometry. Upon calcination to 750 C, crystallization of the film occurs, with a final film thickness of about 100 nm. Waveguiding and the effective index of refraction of the optical modes for films of various thicknesses were measured using the prism coupler technique with a 45-90-45 ^{0} rutile prism and a green He-Ne laser [4].

RESULTS AND DISCUSSION.

The synthesis of the KTP precursor solutions and the formation of both thin film and bulk KTP is summarized in Figure 1. Hydrolysis and condensation was affected by the slow addition of 4 moles of water per mole of titanium added as a dilute solution over a period of about 30 minutes followed by heating the solution to about 60 C overnight. High concentrations of water, for example, 10 moles per titanium leads to precipitation. The resultant clear yellow solutions were stable for several weeks and were used to prepare both bulk and thin film KTP. Extended heating of these solutions leads to the formation of clear gels as shown in Figure 2, which on subsequent drying lead to clear glass pieces, typical of sol-gel derived reactions. The successful formation of KTP thin films (and bulk) is very dependent upon the precursor chemistry developed. Alternative sources of P, such as H_3PO_4 and $OP(OC_2H_5)_3$, were not successful due to the formation of insoluble potassium phosphates in one case and in the latter case the phospho ester is very slow to hydrolyse and simply volatilizes upon heating resulting in complete loss of the phosphorous. The mixed P complex with both OH and butoxo functional groups, leads to stoichiometric KTP, and most likely provides stability by coordination via the hydroxides to Ti forming Ti-O-P linkages with additional solubility due to the presence of the butoxo groups. The resultant clear gels and dried glasses indicate homogeneous mixing between the various reagents.

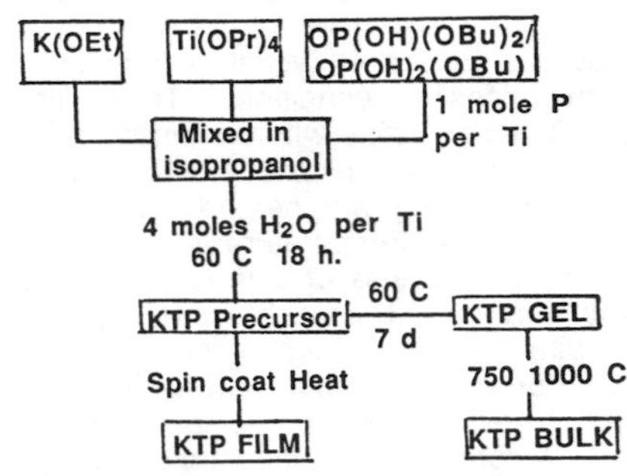

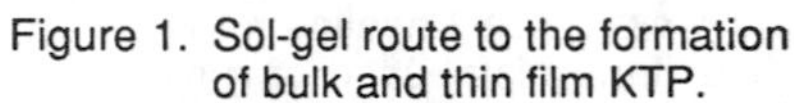
Figure 1. Sol-gel route to the formation of bulk and thin film KTP.

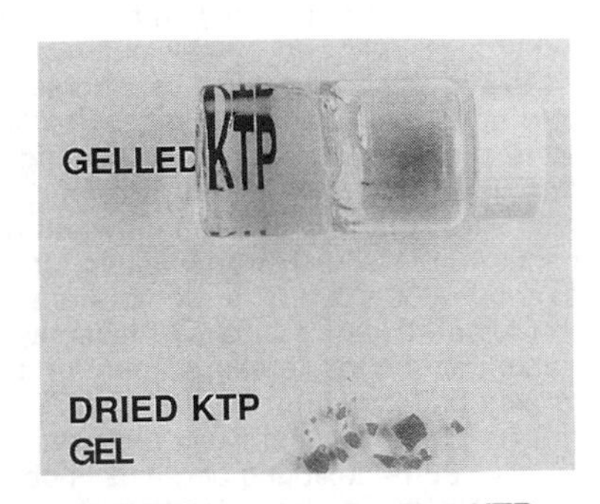

Figure 2. Examples of gelled KTP precursor and dried gel.

Crystalline KTP was prepared in bulk form by the calcination of the dried gels in air. Figure 3 shows the x-ray diffraction of material calcined at different temperatures. The material remains amorphous up to about 500 C with crystallization occurring above about 700 C. The diffraction pattern obtained is consistent with the formation of phase pure KTP. A single phase orthorhombic KTP is obtained [5]. Thermogravimetric analysis of the dried gels indicated that the decomposition of the dried gels was completed by ca. 650 C, with two well defined events occurring at about 350 and 630 C, with a final conversion yield of about 60% by weight (under nitrogen). Mass spectral analysis was used to look at the volatiles produced during heating. A complex mixture of organics and water is evolved including products of bimolecular reactions for example dibutylether at the higher temperature (630 C), and at the low end butene is most prominent and also some butanol and butyraldehyde. Products derived from isopropanol and ethanol are not prominent. The mass spectral analysis is consistent with complete hydrolysis of both the ethoxide and isopropoxide groups on the K and Ti centers, with incomplete hydrolysis of the butoxo groups on the P. This is to be expected since it is well known that phospho esters are very slow to hydrolyse. Bulk samples of crystalline KTP were found to exhibit non-linear optical (NLO) activity, with values of NLO about 1000 times that of quartz, which is typical for polycrystalline KTP.

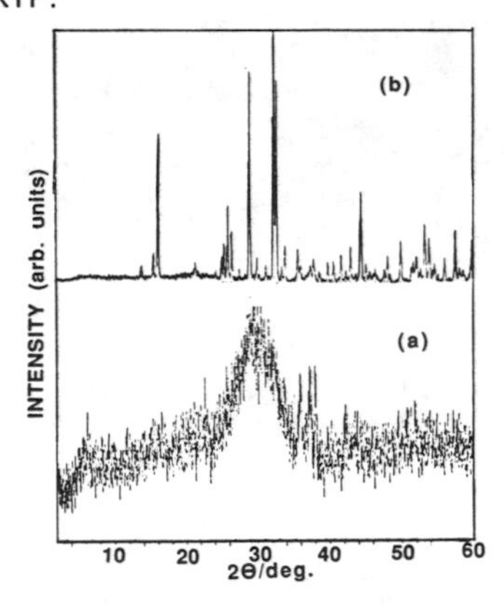

Figure 3. X-ray diffraction pattern of bulk KTP calcined in air for 10 hours at (a) 500 C, and (b) 750 C.

Crystalline KTP in thin film form were prepared by spin coating on to various substrates, followed by calcination. The microstructure of a series of crystalline KTP thin films on quartz substrates is shown in Figure 4. What is particularly striking is the variation of microstructure with processing conditions. These films were prepared using multiple coating and calcination steps where in most cases the final film thickness is about 0.7 μm. Figure 4 (a) and (b) shows the microstructure of a thin film grown at a very slow heating rate, heated at 0.2 C per minute up to 750 C and annealed for 10 hours. Higher temperature annealing (for example 900-1000 C to promote grain growth) causes loss of both K and P from the films shown by energy dispersive x-ray analysis. Using slow heating rates generates a considerable amount of texture within the film. Some almost 'star like' growth patterns are observed where grains appear to point out from a number of growth centers, most likely representing growth from a low number of nucleation points. Some texturing of grains is observed. A considerable amount of porosity is evident in these films. In sharp contrast, films grown at fast heating rates (>10 C per minute) appeared very dense with more uniform distribution of grain sizes, Figure 4(c), a film calcined to 750 C. In the latter case the grain sizes are in the range of 0.1 to 0.3 μm. Figure 4(d) shows a thin film also grown at a fast heating rate but calcined at 800 C showing a similar microstructure but with a slightly larger grain size. The resulting films are dense, continuous, transparent and appear reasonably uniform. A more detailed investigation of the the effect of processing conditions is in progress.

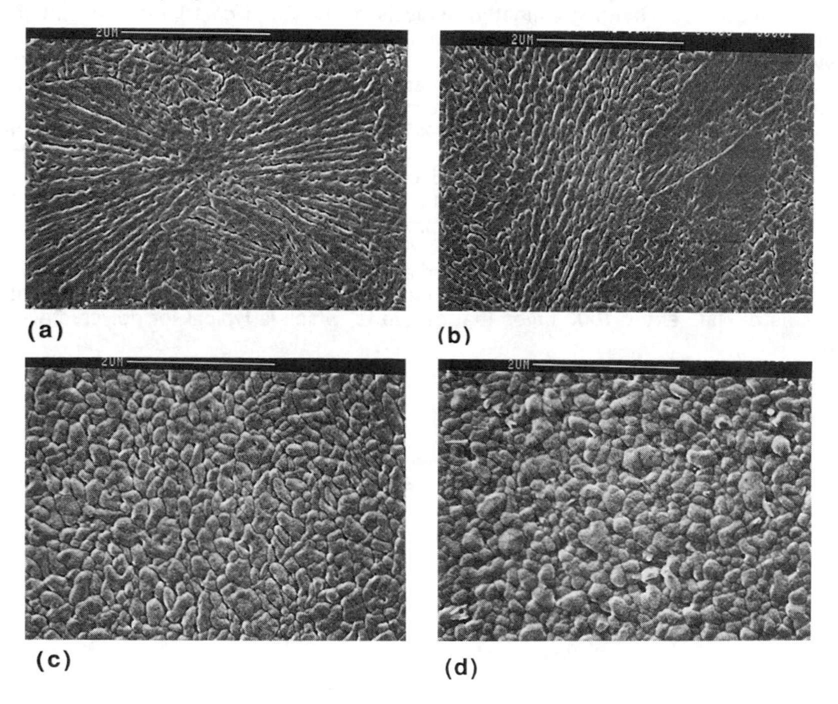

Figure 4. SEM of thin film KTP on quartz substrates developed by (a) and (b) slow heating at 0.2 C per minute to 750 C, (c) rapid heating at 10 C per minute to 750 C and (d) rapid heating at 10 C per minute to 800 C.

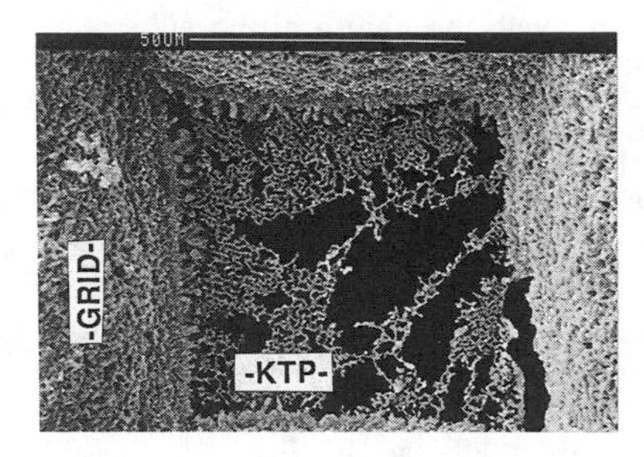

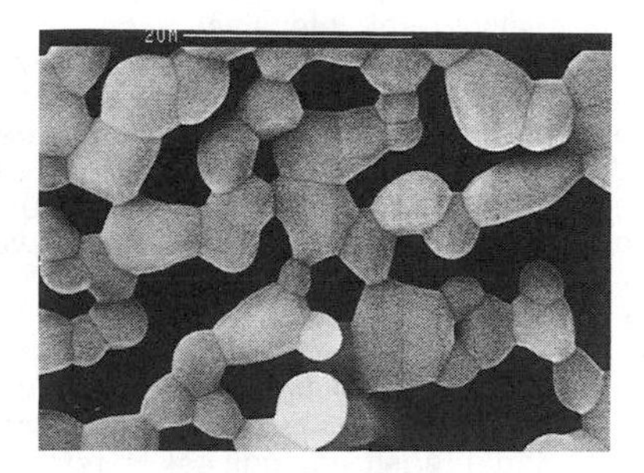

Figure 5. SEM of free standing films of polycystalline KTP grown on a TEM grid.

Interestingly, we have also found we can prepare free standing films of KTP by direct spinning of the precursor on to an TEM copper grid, followed by calcination to 750 C, Figure 5. The calcined film maintains enough integrety to be self supporting and this technique will be expoited for detailed TEM studies. The grain size appear larger than the films grown on the substrate, probably since the growing films is not pinned to the substrate. It is felt this kind of approach may prove useful for studying the grain boundaries and growth patterns of collections of grains.

The crystallinity of the sol-gel derived thin films on various substrates was assessed using X-ray diffraction. Figure 6 shows the X-ray diffraction of thin film KTP grown on a number of substrates (heating rates 10 C per minute). In general the diffraction pattern corresponds to the bulk sample and literature values for KTP [5], with no evidense for additional phases. It is interesting to note however, that

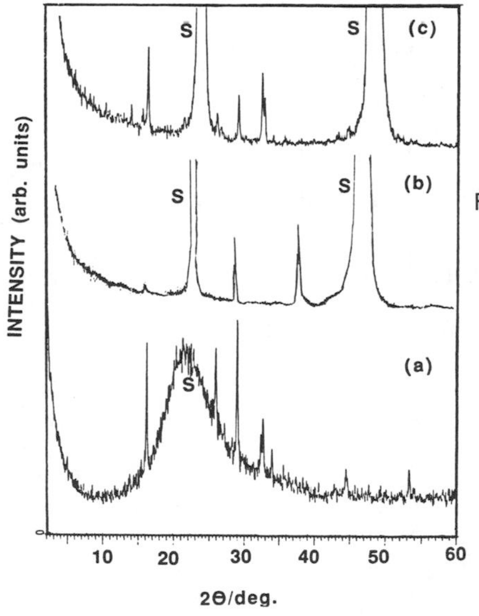

Figure 6. X-ray diffraction of thin film KTP grown on (a) polycrystalline quartz, (b) single crystal $LaAlO_3$ and (c) single crystal $LaGaO_3$.

the relative peak intensities varies considerably with the nature of the substrate. Films grown on polycrystalline quartz, show different intensities than the bulk sample indicating some preferred growth patterns. This was also found in the optical measurements as described below. Films grown on the single crystal substrate $LaAlO_3$ show pronounced differences in the growth patterns however more detailed work is needed to fully understand the growth patterns as a function of substrate. The low intensities however tend to suggest that these films are mainly polycrystalline.

Waveguiding was observed in the sol-gel derived KTP thin films (measured for an ca. 0.7 µm film grown on a quartz substrate) using the prism coupling method. The film supported two optical modes for both TE and TM polarizations, with the measured indices of refraction were estimated as $n_{f,TE}$=1.74 and $n_{f,TM}$=1.77. These values are slightly lower than the calculated index for dense polycyrstalline KTP which is 1.816 [4]. A more detailed report on the waveguiding properties will appear separately [4]. The difference between these two values is presumably due to the presence of voids in the film some of which are observed by SEM . In addition, we note that the index for TM polarization was slightly higher than for TE, suggesting a slight preference for the **c**-axis of the crystallites to be normal to the plane of the film.

CONCLUSIONS.

In summary we have developed a novel route to thin film KTP. The processing of such films depends very strongly upon the chemistry and processing conditions. With slow growth rates for example very striking growth patterns are observed. Polycrystalline films on quartz substrates have been developed to exhibit waveguiding.

REFERENCES.

1. D. P. Partlow and J. Greggi, J. Mater. Res. 2, 595 (1987).

2. A. H. Carim and B. A. Turtle, J. Am. Cer. Soc. 74, 1455 (1991).

3. J. D. Bierlein and H. Vanherzeele, J. Opt. Soc. Am. B **6**, 622 (1989).

4. A. Yariv and P. Yeh, *Optical Waves in Crystals* (John Wiley & Sons, New York, 1984), p. 416.

5. Natl. Bur. Stand. (U.S.) Mongr. 25, (1984) 21 110.

6. M. A. Harmer and M. Roelofs, submitted to J. Mat Sci.

SYNTHESIS OF CUPRATE BASED HIGH-Tc SUPERCONDUCTORS BY POLYMERIZED COMPLEX METHOD

MASATO KAKIHANA*, MASAHIRO YOSHIMURA*, HIROMASA MAZAKI**, HIROSHI YASUOKA**, LARS BÖRJ ESSON***
*Research Laboratory of Engineering Materials, Tokyo Institute of Technology, Nagatsuta 4259, Midori-ku, Yokohama 227, Japan
**Department of Mathematics and Physics, The National Defense Academy, Hashirimizu 1-10-20, Yokosuka 239, Japan
***Department of Physics, Chalmers University of Technology, S412 96 Göteborg, Sweden

ABSTRACT

We report on a novel precursor technique to produce highly pure and homogeneous cuprate based high-Tc superconductors based on the formation of a polymer metal complex in which various metal ions in desired stoichiometric ratio can be uniformly dispersed. Application of the polymerized complex method is demonstrated on the ceramic synthesis of polycrystalline compounds viz. $YBa_2Cu_3O_y$, $YBa_2Cu_4O_y$, $Bi_{1.6}Pb_{0.4}Sr_2Ca_2Cu_3O_y$ and $Bi_2Sr_2Ca_{0.8}Y_{0.2}Cu_2O_y$. The superconducting properties were studied with complex magnetic susceptibility measurements which exhibited sharp superconducting transitions without any indication of two-phase character. The materials fabricated by this method showed improved properties compared to those for ceramic samples of previously reported synthesis procedures.

INTRODUCTION

The occurrence of high-Tc superconductivity in the series of cuprate based oxide ceramics that have been discovered for the last five years has generated intense research activities worldwide aiming at elucidating the origin of the superconductivity at high temperatures. Current research is directed both to the precise characterization of the fundamental physical properties of the new superconducting materials and to the development of new and useful synthetic methods. Clearly the understanding of the unique properties of these compounds, as well as their technological application, depends crucially upon the progress of the synthetic method. To date the vast majority of the powder synthesis of these compounds have been carried out using the conventional ceramic method usually involving a mixed-oxide process which first requires mechanically mixing the oxides and/or carbonates and then calcining the resulting mixture at an appropriate temperature. Several disadvantages are evident, such as inhomogeneous mixing, non-uniformity of particle size and shape, high impurity contents, lack of reproducibility, necessity of repeated grinding and firing because of incomplete reaction. There have been many studies whose goal is developing alternative wet chemical techniques to obtain a better homogeneous precursor material with well interspersed elements before heat treatments at high temperatures. Several wet chemical routes have been developed to reach this goal, such as sol-gel formation based on hydrolysis of mixed metal alkoxides [1] and on coprecipitation of mixed metal-hydroxides [2]. Problems have still arisen with nonhomogeneity owing to cation segregation possibly occurring during hydrolysis or coprecipitation process, which is difficult to be controlled for solutions with more than two metallic ions.

The present "polymerized complex method" is a modified version of the polymeric precursor technique which was first developed by Pechini [3] in 1967 to address the problem mentioned above. The original patent for the process [3] indicates the formation of a polymeric resin produced through polyesterification between metal chelate complexes using alpha-hydroxycarboxylic acids (e.g. citric acid, glycolic acid and lactic acid) and a polyhydroxy alcohol such as ethylene glycol. The general idea is to obtain a polymer precursor comprising randomly coiled macromolecular chains throughout which the cations are atomistically distributed. Heating of the resin at high temperatures causes a breakdown of the polymer. Low cation mobility in the highly viscous polymeric resin may lead to the assumption that only little segregation of the various cations occurs during the pyrolysis. After Pechini, in 1980's the polymerized complex method has been successfully used to prepare many other ceramic compounds such as lead magnesium niobates [4], $LaMnO_3$ [5,6], $BaTiO_3$ [5] and $SrTiO_3$ [7,8]. However, it has never received much attention as a synthetic method for cuprate based high-Tc superconducting ceramics until now.

The aim of this paper is to demonstrate the advantages of the polymerized complex method in the synthesis of highly pure cuprate based high-Tc ceramic superconductors such as $YBa_2Cu_3O_{7-d}$, $YBa_2Cu_4O_{8-d}$, $Bi_2Sr_2Ca_{0.8}Y_{0.2}Cu_2O_{8+d}$, and $Bi_{1.6}Pb_{0.4}Sr_2Ca_2Cu_3O_{10+d}$. Emphasis is placed, on the one hand, to the remarkable simplicity of procedures eliminating steps such as careful aging of gel and fine control of pH as usually required for sol-gel methods, and on the other hand, to outstanding purity of the product as is revealed by superconducting transition widths much sharper than those for ceramic samples of previously reported synthesis procedures.

POLYMERIZED COMPLEX METHOD

The polymerized complex technique used in this study includes three important major steps: (1) formation of stable citrate complexes uniformly dispersed into ethylene glycol and water, (2) fabrication of polymerized complexes which are formed through dehydration among the hydroxyl groups of ethylene glycol and citric acid with chelated metal ions, and (3) pyrolysis of the polymerized complex precursor. A schematic picture of chemical reactions involved in making polymerized complexes is shown in Fig.1. The amount of each raw material (nitrate and carbonates are preferred) to make a required composition is mixed with citric acid solution in a 1:1 molar ratio of metal ions and citric acid, and 200 ml of ethylene glycol on each 150 mmol of cations are added to the mixture. Heating at moderate temperatures (around 140 °C to 190 °C) causes dehydration reactions that occur among hydroxyl groups of ethylene glycol and citric acid [9]. Prolonged heating at 190 °C evaporates the excess water and ethylene glycol from the solution, which will promotes polymerization occurring through dehydration (or esterification) process among low molecular weight oligomers and ethylene glycol. It is, however, important to notice that actual chemical reactions involved in this process are quite complicated as the solutions are highly complex mixtures involving not only complex formation in a great variety of way but also numerous relatively strong intermolecular interactions. For example, citric acid (CA) is a polyfunctional ligand which can coordinate to metal ions in different ways; e.g. one of three carboxylic acid groups in CA acts simply as a bidentate ligand like M_2 in Fig.1, CA can coordinate to a metal ion through the hydroxyl group and one of the carboxylic acid groups forming either a five membered chelate ring (like M_1 in Fig.1) or six membered chelate ring (like M_3 in Fig.1). In addition to this, metal-citrate complexes thus formed can be stabilized in the presence of ethylene glycol, as it possesses two alcoholic hydroxyl functional groups with strong complexation affinities to metal ions. Nevertheless, such interactions between ethylene glycol and metal ions are not taken into consideration in Fig.1. Of ideal is to have a polymeric product in which different cations are randomly distributed with a desired stoichiometry as shown in Fig.1. Note the remarkable simplicity of the method described above as no rigorous experimental condition such as precise control of pH is involved. The polymeric product thus obtained is heated to about 350-400 °C in a beaker to remove as much of the organics as possible. The resultant fine powder is then ground , transferred to an Al_2O_3 crucible, calcined, reground, pelletized, sintered and annealed for various duration as given in TABLE 1. Detailed heat treatment conditions optimized for obtaining single-phase superconductors are described elsewhere [10-13].

TABLE 1: Summary of the thermal process for cuprate based superconductors

Sample	Calcination		sintering							
			step 1		step 2		step 3		step 4	
	T/ °C	*t*/h	*T*/ °C	*t*/h	*T*/ °C	*t*/h	*T*/ °C	*t*/h	*T*/ °C	*t*/h
Y123 *	940	12	950	15	400	12				
Y124 **	800	20	820	30	820	60				
Bi,Pb-2223***	800	36	850	36	850	36	850	36	850	72
Bi-2212****	840	12	865	12						

*$YBa_2Cu_3O_y$, **$YBa_2Cu_4O_y$, ***$Bi_{1.6}Pb_{0.4}Sr_2Ca_2Cu_3O_y$, ****$Bi_2Sr_2Ca_{0.8}Y_{0.2}Cu_2O_y$

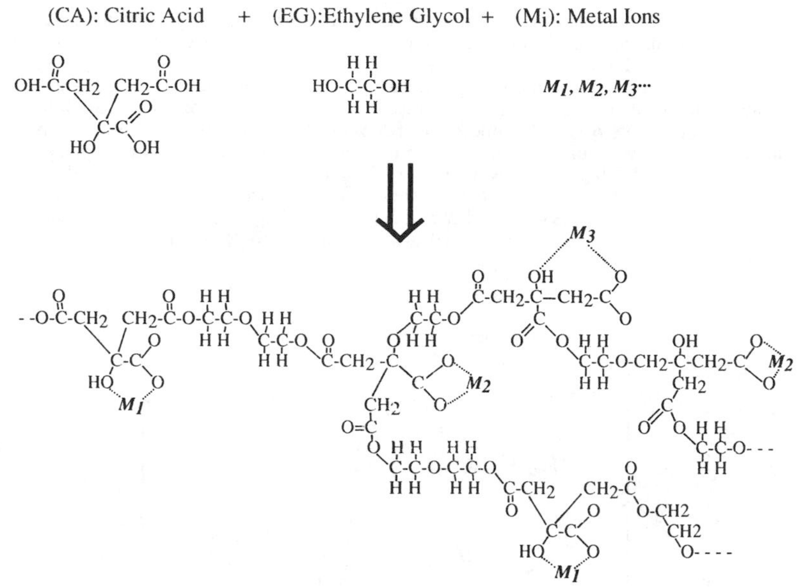

(CA): Citric Acid + (EG):Ethylene Glycol + (M_i): Metal Ions

$M_1, M_2, M_3 \cdots$

Fig.1. Schematic of chemical reactions involved in polymerized complex method.

RESULTS

A variety of measurements have been performed on the cuprate based high-Tc superconductors prepared by the present polymerized complex method with a view to either establishing the phase purity of the products or characterizing their superconducting properties. Both X-ray diffraction and Raman scattering measurements have shown that our samples are virtually of single phase. It should be stressed that Raman spectroscopy is quite capable of detecting amorphous impurity phases [14], which are very hard to be detected by the X-ray diffraction method because of their diffuse diffraction. Moreover Raman spectroscopy is a very sensitive tool for detecting traces of non-superconducting impurity phases in high-Tc superconductors, since the insulating or semi-conducting compounds have much larger Raman scattering cross-sections than the metallic superconductors, as has been demonstrated in a previous paper [10]. A general scheme for impurity detection in ceramic superconductors is given elsewhere [15], where emphasis was placed to the importance of employing an appropriate analytical technique in ensuring the phase purity according to types of impurity phases (i.e. crystal vs. amorphous or metallic vs. semiconducting, insulator).

The superconducting transition was measured in terms of complex magnetic susceptibility, $\chi=\chi' - i\chi''$. Our measuring system consisted of a Hartshorn bridge and a temperature control system, which makes it possible to measure simultaneously χ' and χ'' of a specimen in an ac magnetic field $h(t) = h_0 \sin(2\pi f t)$ as a function of temperature, where h_0 is the amplitude of the applied magnetic field and f is the frequency. In the present work, we used h_0=100 mOe and f=132 Hz. A full description of the measuring system for complex magnetic susceptibility was reported in our previous paper [16].

The superconducting transitions of the Y123 and Y124 as measured in terms of complex susceptibility are shown in Fig.2. Note the different temperature scale in the two figures. The profile of χ is typical of that of a conventional disk sample of oxide superconductors [17]. The diamagnetic transition occurs in a two-step manner representing the bulk superconductivity and inter granular coupling, respectively (Such a two-step transition is clearly visible more remarkably in Y123 than in Y124, because of an expanded temperature scale in Fig.2 (a).) The onset of the diamagnetic transition, *T*c(onset), was observed at 92.6 K for Y123 and at 82.6 K for Y124 (shown by an arrow in the figure) followed by a sharp transition corresponding to the weak-link coupling between the bulk superconducting grains. The overall diamagnetic transition is almost completed by about 88 K for Y123 and 70 K for Y124 which is indicative of good connectivity between the grains. Besides, a single peak appearing on the χ'' curve implies that the sample contains only one superconducting phase. It should be pointed out that the superconducting transition of our Y124 sample prepared by the polymerized complex method *under 1 atmosphere oxygen pressure* is much sharper, ΔTc(10-90 %)=4.5 K, than for ceramic samples of previously reported synthesis procedures (ranging from 28 K < ΔTc(10-90%) < 38 K [18,19]), which we attribute to the high purity and compositional homogeneity of our sample.

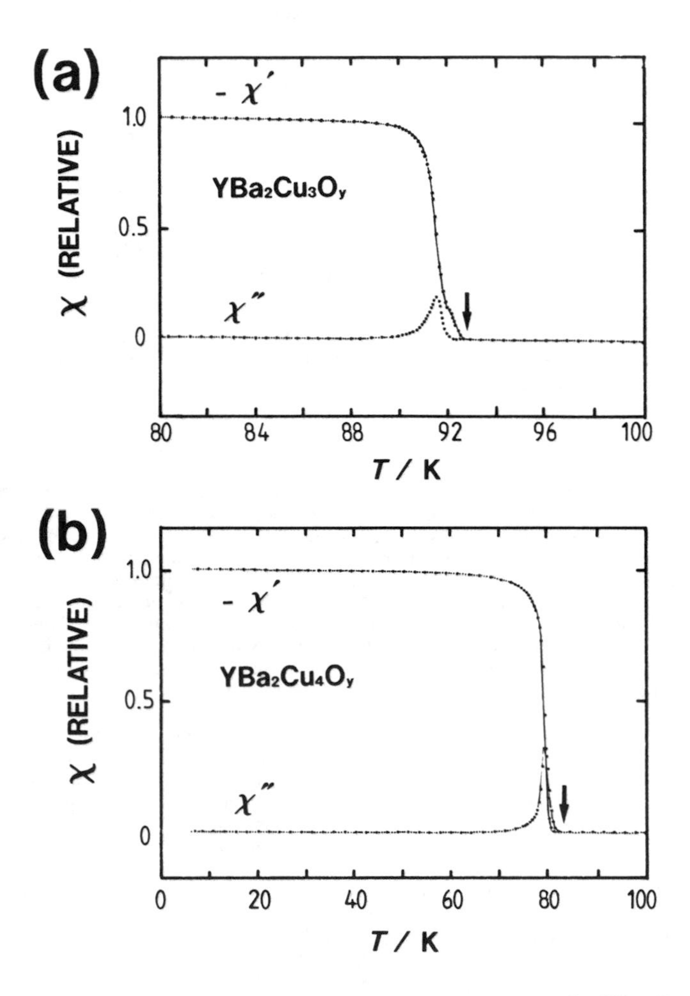

Fig.2 Real (χ') and imaginary (χ'') components of χ vs. temperature for $YBa_2Cu_3O_y$ (a) and $YBa_2Cu_4O_y$ (b) prepared by the polymerized complex technique. The amplitude of the applied field is 100 mOe.

The overall superconducting properties of $Bi_{1.6}Pb_{0.4}Sr_2Ca_2Cu_3O_{10+d}$ (Bi,Pb-2223) and $Bi_2Sr_2Ca_{0.8}Y_{0.2}Cu_2O_{8+d}$ (Bi2212) were examined by complex magnetic susceptibility measurement as shown in Fig.3. The real part, χ', exhibits a typical two-step transition manner representing the bulk superconductivity and weak coupling of superconducting grains at their boundaries. The onset temperature, *T*c(onset), of χ' is 108.0 K for Bi,Pb-2223 and 91.6 K for Bi2212 (shown by an arrow in the figure). The diamagnetic signal is almost completely saturated at a temperature lower than *T*c(onset) by 10 K; the transition width (10-90%) of the full diamagnetism being 5.2 K for Bi,Pb-2223 and 4.0 K for Bi2212. A single loss peak with a maximum at about 103 K for Bi,Pb-2223 and at about 87 K is observed in the χ'' part of the susceptibility, which indicates that the sample is a monophase superconductor. It is worthwhile to notice that no appreciable signal could be detected in the χ'' curve for Bi,Pb-2223 between 80 and 95 K, as it can prove that the sample predominantly contains the Bi,Pb-2223 superconducting phase without the Bi,Pb-2212 superconducting phase of lower Tc's (ranging from 80-90 K) which is generally present as a parasitic minor impurity even after extending heating to 785 h [20]. It is generally acceptable that the broadening of superconducting transition reflects the presence of extraneous phases such as bulk secondary phases or some grain boundary phases and inhomogeneities of the material. We therefore attribute the narrow superconducting transition and a single χ'' peak demonstrated in Fig.3 to the high purity and compositional uniformity of the samples obtained by our polymerized complex technique.

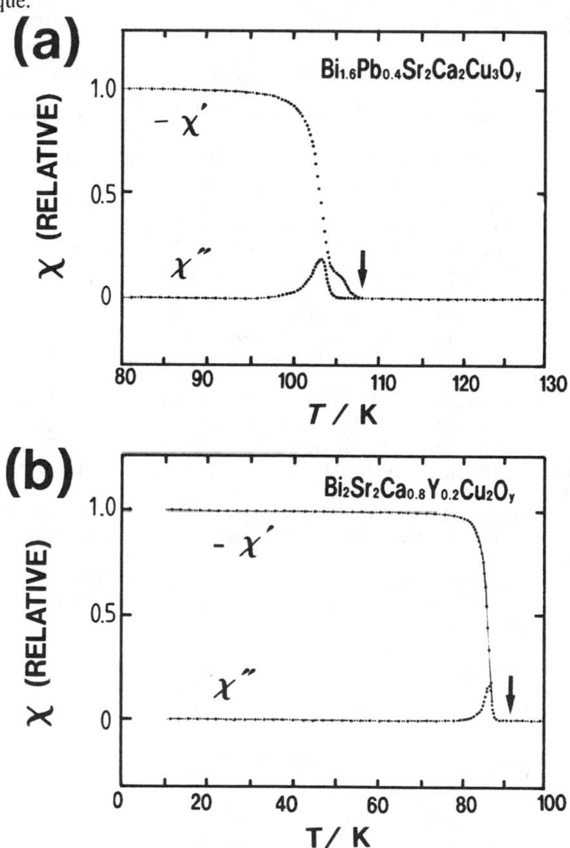

Fig.3 Real (χ') and imaginary (χ'') components of χ vs. temperature for $Bi_{1.6}Pb_{0.4}Sr_2Ca_2Cu_3O_y$ (a) and $Bi_2Sr_2Ca_{0.8}Y_{0.2}Cu_2O_y$ (b) prepared by the polymerized complex technique.

CONCLUSION

We have developed the polymerized complex precursor technique for the fabrication of some typical copper based high-Tc superconductors. The materials exhibited sharper superconducting transitions than those for earlier samples prepared either by the conventional powder-mix technique or by solution techniques based on coprecipitation or sol-gel techniques, which we attributed to the higher compositional homogeneity and purity of our samples. The present technique is advantageous over the recent solution techniques from the viewpoint of its remarkable simplicity and less skillfulness required for procedures. The polymerized complex method can be used to great advantage in the synthesis of complicated compounds such as cuprate based high-Tc superconductors in their pure form, which are definitely needed for better understanding of the high-Tc superconductivity in these fantastic ceramic oxides.

Acknowledgement

This work was partially supported by the Swedish National Science Research Council and the Swedish Board for Technical Development. Part of this work was done at the Center for Ceramic Research in Tokyo Institute of Technology.

REFERENCES

1. P. Ravindranathan, S. Komarneni, A. Bhalla, R. Roy and L. E. Cross, J. Mater. Res. *3*, 810 (1988).
2. P. Barboux, J. M. Tarascon, L. H. Greene, G. W. Hull and B. G. Bagley, J. Appl. Phys. *63*, 2725 (1988).
3. M. Pechini, U. S. Patent No. 3 330 697 (11 July 1967).
4. H. U. Anderson, M. J. Pennell and J. P. Guha, in *Advances in Ceramics, Ceramic Powder Science*, edited by G. L. Messing, K. S. Mazdiyasni, J. W. McCauley and R. A. Haber (Am. Ceram. Soc. *21*, Westerville, OH, 1987), p.91.
5. N. G. Eror and H. U. Anderson, in *Better Ceramics Through Chemistry II*, edited by C. J.Brinker, D. E. Clark and D. R. Ulrich (Mater. Res. Soc. Proc. *73*, Pittsburgh, PA 1986) p.571-577.
6. P. A. Lessing, Am. Ceram. Soc. Bull. *168*, 1002 (1989).
7. K. O. Budd and D. A. Payne, in *Better Ceramics Through Chemistry*, edited by C. J. Brinker, D. E. Clark and D. R. Ulrich (Mater. Res. Soc. Proc. *32*, Elsevier New York, 1984) p.239-244.
8. S. G. Cho, P. F. Johnson and R. A. Condrate SR, J. Mater. Sci. *25*, 4738 (1990).
9. K. J. Saunders, *Organic Polymer Chemistry*, (Chapman-Hall, London, 1973), Chapt. 10.
10. M. Kakihana, L. Börjesson, S. Eriksson and P. Svedlindh, J. Appl. Phys. *69*, 867 (1991).
11. M. Kakihana, M. Käll, L. Börjesson, H. Mazaki, H. Yasuoka, P. Berastegui, S. Eriksson and L. G. Johansson, Physica C, *173*, 377 (1991).
12. H. Mazaki, M. Kakihana and H. Yasuoka, Japan. J. Appl. Phys. *30*, 38 (1991).
13. M. Kakihana, M. Yoshimura, H. Mazaki, H. Yasuoka and L. Börjesson, J. Appl. Phys. *70* (15 April, 1992) in press.
14. M. Kakihana, L. Börjesson and S. Eriksson, Physica B, *165&166*, 1245 (1990).
15. M. Kakihana, M. Yoshimura, H. Mazaki and H. Yasuoka, Report of Res. Lab. of Eng. Materials, Tokyo Inst. of Tech. *17*, 63 (1992).
16. T. Ishida and H. Mazaki, Phys. Rev. *B20*, 131 (1979).
17. H. Mazaki, M. Takano, R. Kanno and Y. Takeda, Japn. J. Appl. Phys. *26*, L780 (1987).
18. K. Kourtakis, M. Robbins, P. K. Gallagher and T. Tiefel, J. Mater. Res. *4*, 1289 (1989).
19. H. Murakami, S. Yaegashi, J. Nishio, Y. Shiohara and S. Tanaka, Japn. J. Appl. Phys. *29*, L445 (1990).
20. G. Jakob, M. Huth, T. Becherer, M. Schmitt, H. Spille and H. Adrian, Physica B, *165&166*, 1667 (1990).

APPLICATION OF ETHANOLAMINE METHOD FOR THE PREPARATION OF INDIUM OXIDE-BASED SOLS AND FILMS

YASUTAKA TAKAHASHI, HIDEO HAYASHI and YUTAKA OHYA
Department of Chemistry, Faculty of Engineering, Gifu University, Yanagido, Gifu 501-11, Japan.

ABSTRACT

It is found that ethanolamine method, recently developed by us, is very useful to form stable, concentrated (0.4-0.5M) indium oxide-based sols from which very uniform, transparent In_2O_3 and ITO films are obtained by dip-coating. Indium isopropoxide and acetate can be used as the starting materials. The optical and electrical properties of these films were examined. A transparent ITO film with the highest conductivity (3030 S/cm, after annealing at 0.05 torr-600°C for 60 min) was obtained from the acetate at 600°C. The films from the alkoxide had lower conductivities probably due to the impurities included in the alkoxide.

INTRODUCTION

The needs to develop economical fabrication process of transparent and electrically conducting films have recently much increased because of recent requirement for the larger display, heat-mirror and opto electric devices. Among them Sn-doped In_2O_3 (ITO) film seems to be one of the most promising materials, and its preparation has mainly been studied with sputtering and chemical vapor deposition. Sol-gel technique (dip- or spin-coating) also is considered to be one of interesting preparation methods of oxide thin films because of its low cost of fabrication. So far some trials to obtain ITO films using the sol-gel technique have been reported, and the starting materials are limited to indium salts [1], acetylacetonates [2] or long chain carboxylates [3], partly because of lack of enough solubility [4] and higher cost of the alkoxides which are common to the sol-gel process. As a result, the morphologies and hence the conductivities of ITO films which were prepared by sol-gel process are slightly less satisfactory, except for those reported by Arfsten et al. [5].

Ethanolamine method developed by us can be applied for the preparation (dip-coating) of films of various oxides such as Al_2O_3, Y_2O_3, TiO_2, ZrO_2, VO_x, Nb_2O_5, Ta_2O_5, Fe_2O_3, as well as various mixed oxides such as YIG, YAG, YVO_4 etc. [6-10]. Ethanolamines possess following two important effects:

(1) The addition of ethanolamines to the alcoholic solutions of alkoxides suppress their hydrolysis giving homogeneous, stable sols without the occurrence of gel formation or precipitation. Thus, the sols show ease of handling, good shelf stability (sometimes over 2 years) and can be applied to the dip-coating of uniform oxide films.

(2) The presence of ethanolamines enhances the solubilities of the alkoxides into alcoholic solvents. Even metallic salts such as acetates can be dissolved into alcoholic solvents in the presence of suitable ethanolamines.

Ethanolamines such as diethanolamine (DEA), triethanolamine (TEA) and, sometimes, monoethanolamine (MEA) can be used as the a sol stabilizer. But, types of ethanolamines that are effective depend on the kind of alkoxides used. At this stage the reasons are unclear. Therefore, the experimental selection of most suitable ethanolamines for the alkoxides is to be done. Indium and related alkoxides may be one of the cases.

In this work, examining whether or not the ethanolamine method developed by us is applicable to the indium alkoxide system to give stable indium alkoxide sols and investigating the properties of the films of indium oxide or ITO prepared from ethanolamine-stabilized sols are initially aimed at. Some trials to obtain more useful material than the alkoxide showed that indium acetate was a possible compound that gave stable In_2O_3-based sols. The results of these investigations are reported here.

EXPERIMENTAL

Starting materials --- Commercial indium triisopropoxide of 99.9% purity (High Purity Chemicals Co. Ltd.) was used as obtained, which is reported to contain the impurities: Ca, 0.001%; Fe, 0.001%; Na, 0.08%; Si, 0.005%; Zn, 0.001%. Commercial Guaranteed Reagent Grade ethanolamines (MEA, DEA and TEA) and 3N tin tetraisopropoxide (High Purity Chemicals Co. Ltd.) were used without further purifications. Extra Pure Reagent Grade isopropanol was used as a solvent after dehydration by aluminum isopropoxide and distillation under nitrogen.

Indium acetate was prepared by refluxing a suspension of 4N $In(OH)_3$ in a mixture of acetic anhydride and n glacial acetic acid under nitrogen for 24 hrs followed by filtration of the white, crystalline precipitates and drying. The indium content in the product indicated that it had the composition of $In(OAc)_2(OH)\cdot H_2O$ or $In(OAc)(OH)_2\cdot AcOH$.

Preparation of the dip-solution of In_2O_3 and ITO --- Into the suspension of the indium alkoxide in isopropanol, was added TEA equimolar to the alkoxide under nitrogen. The orange colored alkoxide slowly dissolved to afford yellow colored, almost transparent solution within 24 hrs. After filtration of a small amount of residues, the filtrate was used for the dip-coating of In_2O_3 films. 0.4M solutions thus formed are so stable that no precipitates or gellation took place even after the addition of excess water (5-6 molar times) to the solution. More concentrated solutions can be obtained, but they had a gel formation tendency during the storage. The dip-solutions for ITO were similarly prepared by adding known amount of tin isopropoxide. The doping quantity of Sn in ITO is described in terms of Sn mol%, namely Sn/(Sn+In), in this paper.

Using indium acetate, similar homogeneous isopropanol solution was obtained, where DEA was found to be more effective. The addition of DEA (the molar ratio of DEA/acetate=2) accelerated the dissolution of the acetate in isopropanol forming a clear, stable solution. Tin alkoxide was used to prepare tin-doped solution.

Instant gel formation or precipitation of white needle crystals from the solution, when DEA was replaced by TEA, was observed.

Dip-Coating of In_2O_3 and ITO films and their characterizations --- The dip-coating of the films was performed by soaking a suitable substrate plate, pulling-up at constant speed (6 cm/min), drying at 110°C in air for 15 min and firing at a given temperature in air for 30 min. Normally this coating cycle is repeated 5 times using Corning Co. #7059 glass and silica glass plates as substrate. TiO_2(50nm)-precoated glass plates which were obtained by the dip-coating using titanium isopropoxide sol were further used to examine the substrate effect. The crystal modifications of the coated films or powders derived from the solution were determined by XRD. The thickness and the refractive indices of the films were evaluated from the interference bands observed in the visible spectra or were directly measured from the TEM images of the cross section. Average film thickness values per coating cycle are found to be 60-70nm from 0.4M alkoxide, 71-79nm from 0.5M alkoxide, and 65-68nm from 0.5M acetate solution. The films thicker than 80 nm per coating cycle tended toward having fine cracks. Therefore, in the case of alkoxide, 0.4M solution was mainly used.

The conductivities of the films in air were measured by 4-probe method using an conventional instrument (Rolesta Model TCP-400, Mitsubishiyuka Co. Ltd.). The carrier density and mobility for several films were measured by Van der Pauw's method. The room temperature conductivity of the In_2O_3 films were interestingly found to be increased with time during which the samples are stored at room temperature in air after the preparation. A typical example of this behavior of an In_2O_3 film on #7059 glass is shown in Fig. 1. Immediately after firing it was 6.3 S/cm, increasing rapidly at the initial stage and levelled off to 40 S/cm within five to ten days. The conductivities of ITO films also had a similar but weak dependence on time. Therefore, the conductivity values described in this paper have been cited from those of leveling-off data.

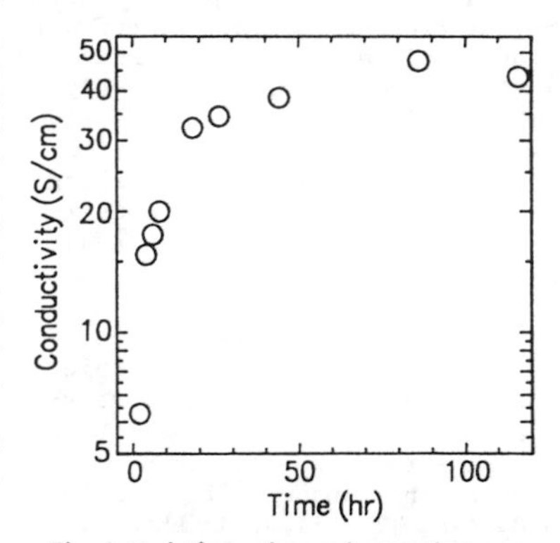

Fig. 1. Variations of room temperature conductivity of an In_2O_3 film on a glass plate with the time of storage at room temperature in air.

RESULTS AND DISCUSSION

(1) The effect of ethanolamines.

Indium isopropoxide is almost insoluble to isopropanol, but showed dissolution in the presence of TEA equimolar to the alkoxide to afford 0.4M isopropanol solution, and the formed solution showed good stability on addition of excess water (5-6 molar times the alkoxide) without precipitation or gel formation. This solution was seen to posess good shelf stability, and used for the dip-coating of In_2O_3 films, indicating that the ethanolamine method is very useful for the case of indium alkoxide, too. DEA had a similar but weak effect, and no concentrated solution could be obtained with DEA. MEA showed no such positive effects.

As described in the Experimental section, the alkoxide contained a large amount of sodium impurity, which could affect the conductivity of n-type semiconductor like In_2O_3 in this case. So, $In(OH)_3$ as a starting material was further examined. The hydroxide was completely insoluble in alcoholic solvents even in the presence of ethanolamines. But, it was found that indium acetate obtained by the reaction between the hydroxide and acetic anhydride in acetic acid was soluble to isopropanol solvent in the presence of TEA or DEA giving a homogeneous solution. However, when TEA was used, the solution showed rapid gel formation, or crystalline product, TEA complex of indium acetate, was separated. On the other hand, 0.5M isopropanol solution of the acetate obtained using DEA (the molar ratio of DEA/acetate=2) showed good stability, and hence this solution was used for the dip-coating of In_2O_3 films.

The addition of tin isopropoxide up to 10 mol% relative to indium alkoxide or acetate showed no harmful effects on the stability of the solutions, and ITO films were obtained from the tin-doped solutions.

It is supposed that these interesting behaviors of the ethanolamines can be attributed to the formation of stable aminoglycolates by the reaction with metallic derivatives (especially alkoxides). This has already been confirmed from the analysis of the reaction products derived from the Ti and Nb alkoxides [11].

(2) The optical and electrical properties of the In_2O_3 and ITO films.

Crystalline In_2O_3 films were obtained above about 400°C from both the alkoxide and acetate solutions. Their XRD patterns were very similar to those of powder diffractions, showing no particular crystal orientation, different from sputtered films [12]. They were very uniform and transparent (optical transmittance is higher than 90%), and regular interference bands were observed in the visible spectra. Fig. 2 shows the wavelength dispersion of the refractive indices of these In_2O_3 films evaluated from the interference bands. The refractive indices at wavelength 600nm for the films obtained from the alkoxide and the acetate solutions are estimated to be 1.75 and 1.93, respectively. The acetate yielded the films with a higher refractive index. Nevertheless, the index seems to be lower than that (n=2.1) observed for sputtered In_2O_3 films [13], indicating that the film is not fully densified. Mean crystallite sizes evaluated from the half width of the most intensive (222) XRD diffraction peaks of both films are 24-26 and 33-35nm, respectively, which also were confirmed by TEM observation.

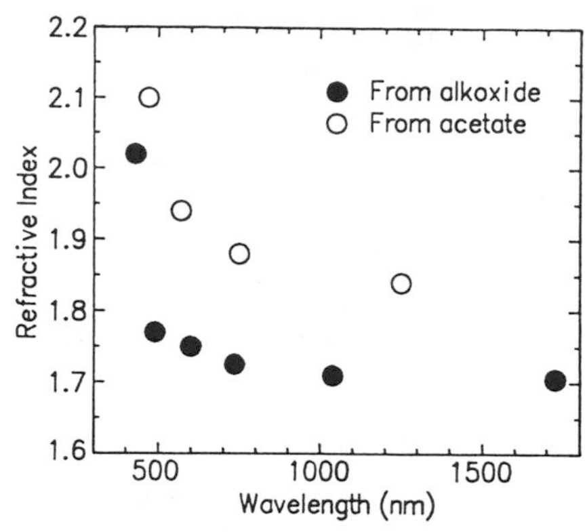

Fig. 2. Refractive index dispersions of In_2O_3 films prepared at 600°C from alkoxide and acetate.

The starting materials also affected the conductivities of the In_2O_3 and ITO films, as shown in Table 1. The films from the acetate had the conductivities twice as high as those from the alkoxide. This difference could be ascribed to the impurities included in the starting materials rather than to the structural effect of the indium compound. The films prepared on TiO_2-precoated glass plate showed twice higher conductivities. The deposition temperatures above 600°C gave less conductive films. The conductivities were much increased by annealing in vacuo (see later).

Table 1. Examples of conductivities (S/cm) of In_2O_3 and ITO (5mol% Sn doped) prepared from alkoxide and acetate.

Deposition Temp. (°C)	Substrate	From alkoxide In_2O_3	From alkoxide ITO	From acetate In_2O_3	From acetate ITO
600	Glass	40–50	180 670*	71	400 3030*
600	TiO_2–coated glass	80	330	---	630
800	SiO_2 glass	30	200	---	63

* After annealing at 600°C–0.05 torr for 60 min.

As described in Experimental Section, the room temperature conductivity of the In_2O_3 films were interestingly found to be increased with time during which the samples are stored at room temperature in air after the preparation. The carrier must be generated by oxygen vacancies. In this case, decrease in conductivity by the reaction with oxygen in air during the storage can be expected. But the expected result was contradictory. Possibility that film distortion induced during the firing and slow relaxation play an important role cannot be overruled.

Fig. 3 shows the effect of the quantity (mol%) of tin dopant on the conductivities of formed ITO films on #7059 glass from the TEA-stabilized alkoxide solution. Highest conductivity (180 S/cm) was found around 5 mol%, almost consistent with the case

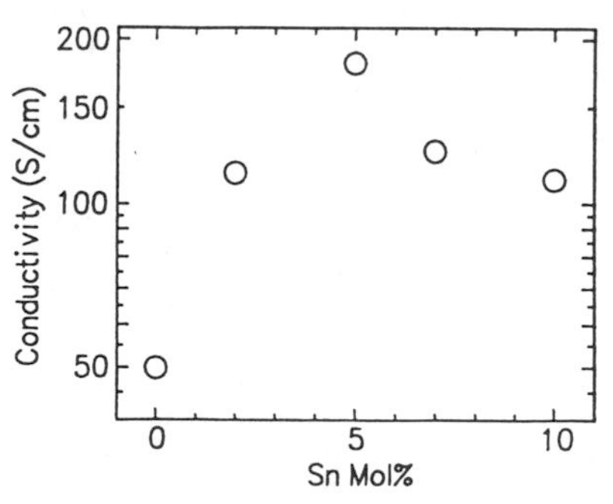

Fig. 3. Effect of Sn mole% on the conductivity of ITO films prepared at 600°C on glass plates from alkoxide.

of PVD and CVD, although the conductivity itself seems to be lower at least by one order of magnitude than that of films prepared by sputtering. Similar trend was found in the case of ITO films derived from the acetate solution, and the maximum conductivity at 5 mol% of Sn-doping were 400 (on bared glass) and 630 S/cm (on TiO_2-precoated glass), suggesting that the starting materials and the substrate had a small but important effect on the conductivity of ITO films.

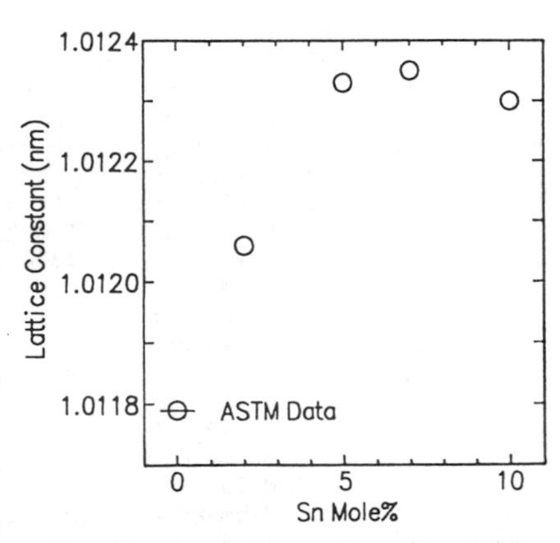

Fig. 4. Relationship between Sn mole% and lattice parameter of ITO powders obtained at 700°C from alkoxide.

The relationship between the lattice parameter of ITO powders and Sn-doped quantity is shown in Fig. 4. It is found that Sn-doping up to 5 mol% induces lattice expansion, and above 5 mol% the lattice parameter remains constant. At 7-10 mol% doping the XRD peak due to segregated SnO_2 was observed at 26.6°, suggesting that the solubility limit of SnO_2 in In_2O_3 at 600-700°C exist around 10 mol%, namely Sn/(In+Sn)=0.05, which is somewhat lower than the value 20 mol% reported for a higher temperature 1825K [14].

Typical TEM images of the cross section of the ITO films prepared by 5 times coating on glass substrates from alkoxide and acetate solutions are shown in Figs. 5a and 5b, respectively. From both Figures it is found that the films are composed of regular five layers with each layer thickness of 70-80 nm. The crystallite size of ITO from the alkoxide was smaller than that from acetate, consistent with the case of In_2O_3. In Fig. 5a four dark lines exist between the layers, and in Fig. 5b rather bright lines. The presence of the dark lines are due either to heavily concentrated impurities at the surface of each layers during the heat treatment or to an unique crystallite arrangement near each surface tops. In any event it is important to note that there are some discontinuities between the layers.

The most conductive ITO films with 5 mol% Sn-doping obtained from

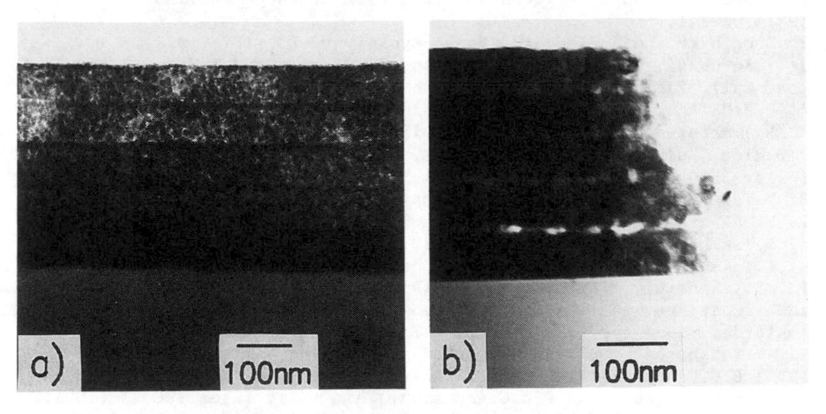

Fig. 5. TEM images of the cross sections of ITO films prepared from alkoxide (a) and acetate (b)

acetate solution were further heat-treated at 600°C-0.02 to 0.05 torr for 1 hr, and the conductivities of about 3000 S/cm were obtained as already shown in Table 1. A typical visible spectrum of the heat-treated film is shown in Fig. 6 along with that of the film before the heat-treatment. Strong transmittance reduction above 1200 nm wavelength is observed due to plasma reflectance. The carrier density and mobility at room temperature were $9.10 \times 10^{20}/cm^3$ and $20.8\ cm^2/V \cdot S$, respectively. The conductivity is roughly comparable to that of PVD ITO films.

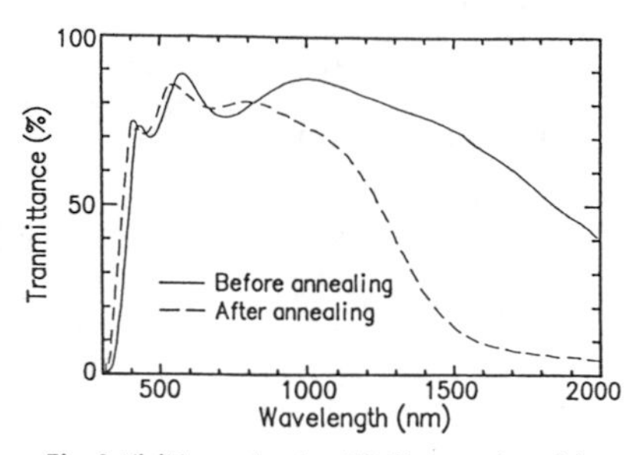

Fig. 6. Visible spectra of an ITO film on a glass plate before and after annealing at 600°C–0.05 torr.

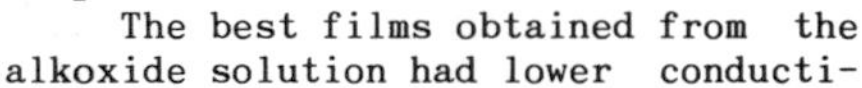

The best films obtained from the alkoxide solution had lower conductivities of 500-1000 S/cm even after annealing at reduced pressures. From these observations it can be concluded as follows:

(1) Ethanolamine method is applicable to prepare stable In_2O_3-based sols from which uniform In_2O_3 and ITO films can be dip-coated.

(2) ITO films with high conductivity can be prepared using ethanolamine stabilized sols derived from the acetate.

(3) Purity of starting materials for the preparation of highly conductive films is important.

(4) Efforts to overcome the discontinuity between each layers of the multi-layered films are to be taken.

ACKNOWLEDGMENT

The authors gratefully acknowledge Mr. Masaaki Kanamori for the measurement of carrier density and mobility of In_2O_3 and ITO films.

REFERENCES

1. T.Furusaki and K.Kodaira, in High Performance Ceramic Films and Coatings, edited by P.Vicenzini (Elsevier Science Publishers, Amsterdam, 1991), p.241.
2. S.Ogihira and K.Kinugawa, Yogyo-Kyokai-Shi **90**, 157 (1982).
3. T.Furusaki, K.Kodaira, M.Yamamoto, S. Shimada and T.Matsushita, Mater. Res. Bull. **21**, 803 (1986)..
4. D.M.Mattox et al., Thin Solid Films **204**, 25 (1991).
5. N.J.Arfsten, R.Kaufman and H.Dislich, in Ultrastructure Processing of Ceramics, Glasses and Composites, edited by L.L.Hench and D.R.Ulrich (Wiley-Interscience Publishers, New York, 1984), p.189.
6. Y.Takahashi and Y.Matsuoka, J. Mater. Sci. **23**, 2259 (1988).
7. Y.Takahashi and Y.Wada, J. Electrochem. Soc. **137**, 267 (1990).
8. Y.Takahashi, Kagaku-Kogyo **34**, 482 (1988).
9. Y.Takahashi and K.Yamaguchi, J. Mater. Sci. **25**, 3950 (1990).
10. Y.Takahashi, Y.Ohchi and T.Sasaki, in High Performance Ceramic Films and Coatings, edited by P.Vicenzini (Elsevier Science Publishers, Amsterdam, 1991), p.127, and the references cited therein.
11. Y.Takahashi et al., unpublished results.
12. J.C.C.Fan and F.J.Bachner, J. Electrochem. Soc. **122**, 1719 (1975).
13. C.H.L.Weijtens and P.A.C.van Loon, Thin solid Films **196**, 1 (1991).
14. J.L.Bates, C.W.Griffin, D.D.Marchant and J.E.Garnier, Am. Ceram. Soc. Bull. **65**, 673 (1986).

GROWTH OF ORIENTED TIN OXIDE THIN FILMS FROM AN ORGANOTIN COMPOUND BY SPRAY PYROLYSIS

ISAO YAGI* AND SHOJI KANEKO**
* Kawai Musical Instruments Mfg. Co., Research & Development Center, Hamamatsu 430, Japan
** Shizuoka University, Dept. of Materials Science & Technology, Hamamatsu 432, Japan

ABSTRACT

Tin oxide films were grown from di-n-butyltin diacetate on a heated glass substrate by a pneumatic spraying system. The effects of various film growth parameters, i.e. solvent, solution feed rate, film thickness, and film growth rate on the microstructures of the films were studied by X-ray diffraction and scanning electron microscopy. The SnO_2 films of the (200) plane were grown by the optimum growth parameters.

INTRODUCTION

Chemical techniques of making thin solid films have recently been increasingly studied. Spray pyrolysis, one of the known techniques, has been used for coating ornamental films and anti-abrasive films on glass materials [1]. In 1966, CdS films for solar cells were formed by this technique [2]; later, films of noble metals, metal oxides and chalcogenide compounds were formed because of the technique's simplicity of apparatus and good productivity on a large scale [3].

The effects on the microstructure of the films have seldom been studied in detail. Moreover, tin tetrachloride has been mostly used as a starting source compound [4-6]. We, therefore, studied the relationship between the microstructures of the films and source compounds by using some organotin compounds, and revealed that the preferred orientations of the films were greatly affected by the starting source compounds [7]. We tried to grow the SnO_2 thin film of highly preferred orientation of (200) plane from di-n-butyltin diacetate, which had indicated the preferred orientation of (200) plane [7], by studing the effects of various film growth parameters, *i.e.* solvent, solution feed rate, film thickness, film growth rate, and substrate temperature on the microstructure of the films.

EXPERIMENTAL

Sample Preparation

Di-n- butyltin diacetate of NITTO KASEI, more than 95% in purity (DBTDA), was used. All solvents were dried. The spray pyrolysis apparatus is represented schematically in Fig. 1. The DBTDA organic solution was atomized by a pneumatic spraying system of about 1.0 kg/cm^2 air pressure. The droplets were transported onto a heated, Corning 7059 glass substrate of 25X25X1 mm in size, by the spraying air. The solution was atomized not consecutively but intermittently: the substrate temperature was lowered by the spraying air, so that it took several tens of second for the next spray until the substrate temperature recovered. The time of one spray was one second. The film growth parameters studied were solvent, solution feed rate (R_f), film thickness (T_f), film growth rate (G_f), and substrate temperature (T_S).

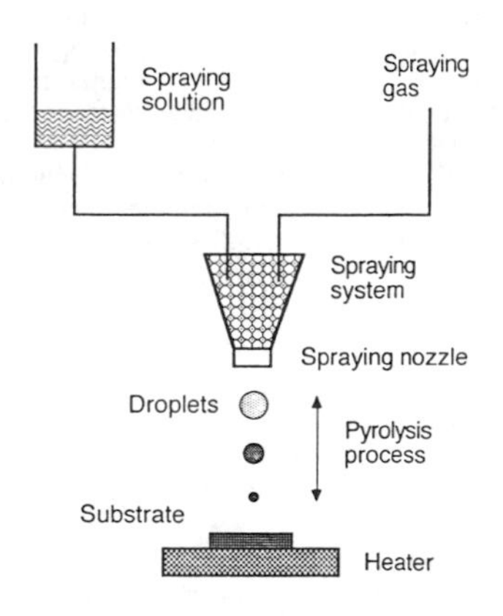

Fig. 1 Schematic diagram of spray pyrolysis apparatus

 ©1992 Materials Research Society

Characterization of The films

The crystal structures of the films were measured by X-ray diffraction. CuKa radiation was used with a Shimazu XD-610 diffractometer. The preferred orientation of the films was evaluated by texture coefficient calculated from the following equation [8],

$$TC(hkl) = \frac{I(hkl) / I_0(hkl)}{1/N\,(\sum_N I(hkl) / I_0(hkl))}$$

where TC is the texture coefficient of the (hkl) plane, I is the measured intensity, I_0 is the JCPDS standard intensity of SnO_2 (cassiterite), and N is the reflection number (N=31). The higher deviation of the texture coefficient from unity indicates the higher preferred orientation of the films. The surface morphologies of the films were observed by using a JOEL JSM-T300 scanning electron microscope.

RESULTS AND DISCUSSION

Effects of Solvents

The XRD profiles of tin oxide films grown from various organic solvents are represented in Fig. 2. Benzene and ethanol were adopted to investigate the different effects between non-polar and polar solvents because these have almost the same boiling point. Every diffraction peak of the film grown from the benzene solution is much broader than that from the ethanol solution (Fig. 2 a and c). This indicates the film grown from benzene solution probably has either smaller crystallite or lower degree of crystallinity than that from ethanol solution. The result can be ascribable to the degree of polymeric structures of DBTDA in organic solvents: DBTDA is less polymeric in benzene than in ethanol. Ethanol is a more preferable solvent to grow crytalline tin oxide films.

1-butanol and methanol also were adopted to investigate the effects of solvents which have different boiling points. The boiling points of 1-butanol, ethanol, and methanol are 117°C, 78°C, and 65°C respectively. The XRD profile of the film grown from 1-butanol solution at a solution feed rate 1.0 ml/s (Fig. 2 b') is different from that from ethanol solution (Fig. 2 c). The XRD profile of the film grown at a solution feed rate 0.6 ml/s (Fig. 2 b), however, is the same that from the ethanol solution. The difference is ascribable to the declination of the substrate temperature during growth at the higher solution feed rate (Rf=1.0 ml/s); because the boiling point of 1-butanol is much higher than that of ethanol, hence, there is little difference between 1-butanol and ethanol as a solvent. The XRD profile of the film grown from methanol solution is very different from those from the other solvents, *i.e.* the diffraction peak of Sn_2O_3 [$SnO{\cdot}SnO_2$] phase at about 31° two theta in diffraction angle. The reason for the formation of Sn_2O_3 phase is probably as follows: The pyrolysis of DBTDA gives both SnO_2 and SnO. Normally SnO reacts quickly with oxygen in air to form SnO_2. In the methanol solution, however, Sn_2O_3 is also formed under the condition of less oxygen because oxygen reacts with methanol to give aldehyde in the presence of SnO_2 at elevated temperature. Consequently, ethanol is the most preferable solvent to grow (200) plane oriented SnO_2 thin films.

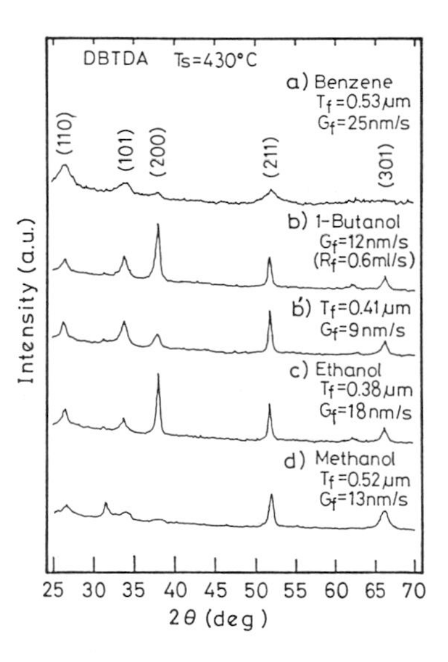

Fig. 2 X-ray diffraction profiles of tin oxide films from various solvents

Effects of Film Thickness

The XRD profiles of tin oxide films of various film thicknesses (T_f) are represented in Fig. 3. The intensities of the diffraction peaks of (101), (200), and (301) planes increase with increasing film thickness from 0.19mm up to 0.85mm. The intensity of the (211) plane increases with increasing film thickness to 0.38mm, but hardly increase with increasing film thickness from 0.38mm to 0.85mm. In contrast, those of (110) and (310) planes do not increase at all. Consequently, (200) plane grows dominantly parallel to the surface of substrate.

Effects of Film Growth Rate

The XRD profiles of tin oxide films grown at various growth rates (G_f) are represented in Fig. 4. Most of the diffraction peaks become sharp with decreasing growth rate. The intensity of the diffraction peak of (200) plane increases obviously with decreasing growth rate. This indicates the lower growth rate results in higher degree of crystallinity in the growth of SnO_2 thin films.

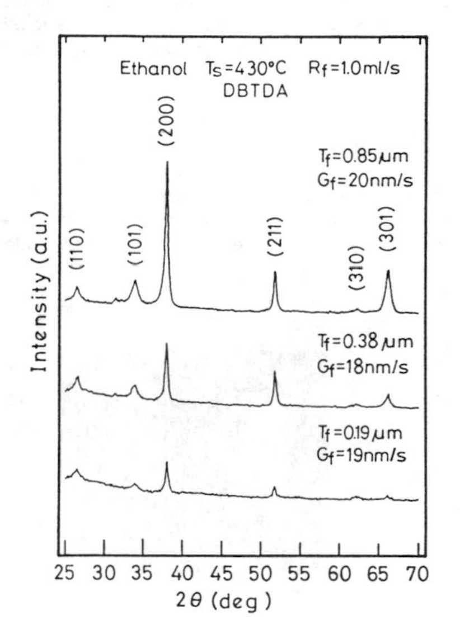

Fig. 3 X-ray diffraction profiles of tin oxide films of various film thicknesses

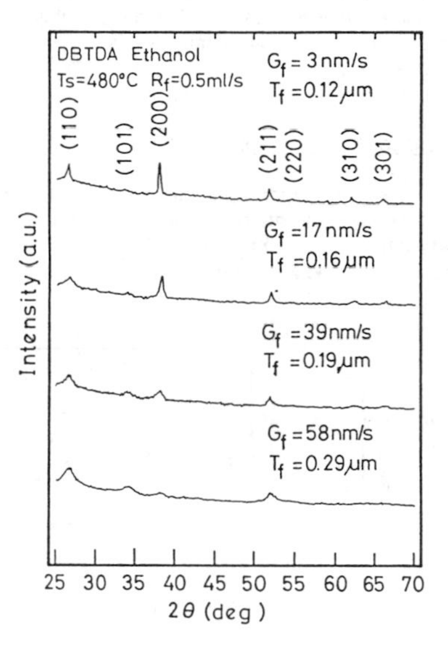

Fig. 4 X-ray diffraction profiles of tin oxide films grown at various growth rates

Effects of Substrate Temperature

The texture coefficients of (110), (200), and (211) planes of SnO_2 thin films grown at various substrate temperatures (T_S) are represented in Fig. 5. The texture coefficient of the (200) plane is the lowest value at a substrate temperature 340°C, and greatly increases with raising substrate temperature. In contrast, that of the (110) plane is the highest value at 340° C, and decreases with raising substrate temperature, then reaching a very low value at 480°C. The texture coefficient of the (211) plane is very low at any substrate temperature, but also slightly decreases with increasing substrate temperature. The results indicates the higher temperature is preferable to the growth of preferred orientation of the (200) plane of SnO_2 thin films.

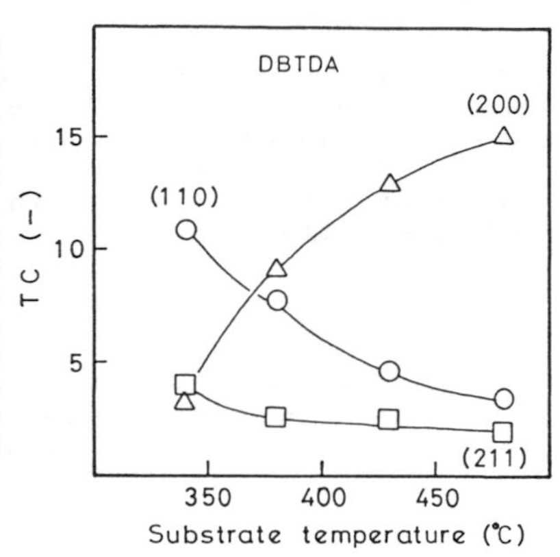

Fig. 5 Texture coefficients of SnO_2 films grown at various substrate temperatures

Highly Preferred Orientation of (200) Plane

The XRD profile of tin oxide thin films grown from ethanol solution at a growth rate 3nm/s and a substrate temperature 480°C showed highly preferred orientation of (200) plane (Fig. 6). The relative intensities of the diffraction peaks of (110), (211), (310), and (301) planes are 10, 4, 8, and 3 respectively.

Surface Morphology of The Film

The SEM micrograph of tin oxide thin film grown from ethanol solution at a growth rate 17nm/s and at a substrate temperature 430°C is represented in Fig. 7. The surface of tin oxide thin film is very smooth and has no pinholes. The measuring bar indicates 0.1 μm. The particle size of tin oxide is estimated to be 2 to 3 hundred nanometers.

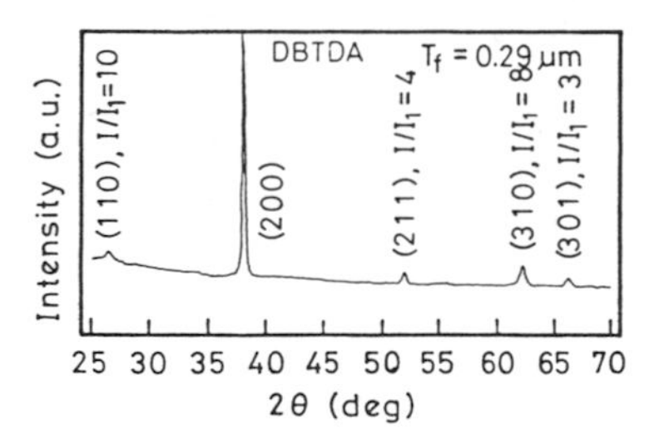

Fig. 6 X-ray diffraction profiles of tin oxide thin films of highly preferred orientation of (200) plane

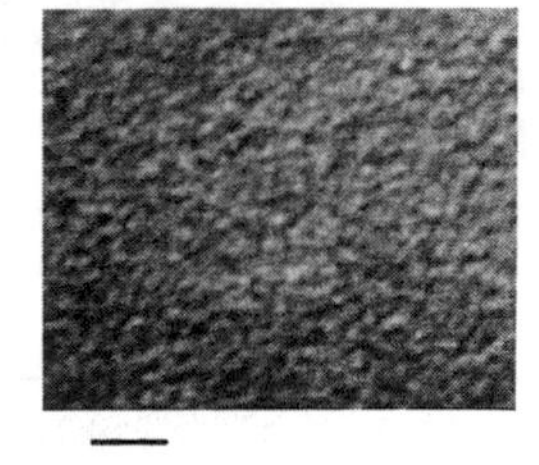

Fig. 7 Micrograph of tin oxide thin film grown from ethanol solution at 430°C

SUMMARY

The growth of tin oxide thin films from di-n-butyltin diacetate by spray pyrolysis was studied in terms of various film growth parameters, *i.e.* solvent, solution feed rate, film thickness, film growth rate, and substrate temperature. The following was revealed:

(1) The tin oxide film grown from benzene solution probably has either small crystallite or low degree of crystallinity; those from alcohol solution have higher degree of crystallinity.

(2) Ethanol is the best solvent to grow the (200) plane oriented SnO_2 thin films. 1-butanol has almost the same effects as ethanol. The Sn_2O_3 phase exists in the tin oxide film grown from methanol solution.

(3) The preferred orientation of (200) plane becomes dominant with increasing film thickness and with decreasing film growth rate.

(4) The preferred orientation of (200) plane becomes dominant with increasing substrate temperature; in contrast, that of (110) plane becomes dominant with decreasing substrate temperature.

(5) The growth of the highly preferred orientation of (200) plane of SnO_2 thin film was successful. The film has a smooth surface and consists of many particles of 2 to 3 hundred nanometers in size.

References

[1] J.M. Mochel, U.S. Patents, No. 2 564 706; 2 564 707; 2 564 708; 2 564 710; 2 564 987(1951).
[2] R.R. Chamberlin and J.S. Skarman, J. Electrochem. Soc. 113, 86(1966).
[3] J.B. Mooney and S.B. Radding, Ann. Rev. Mater. Sci. 12, 81(1982).
[4] C. Agashe, M.G. Takwale, B.R. Marathe, and V.G. Bhide, Sol. Energy Mater. 17, 99(1988).
[5] M. Fujimoto, Y. Nishi, A. Ito, T. Mishuku, H. Iida, and S. Shirasaki, Jpn. J. Appl. Phys. 27, 534(1988).
[6] M. Fujimoto, T. Urano, S. Murai, and Y. Nishi, *ibid.* 28, 2587(1989).
[7] I. Yagi, S. Fukushima, F. Imoto, and S. Kaneko, HYOMEN KAGAKU 12, 316(1991).
[8] C. Barret and T.B. Massalski, Structure of Metals (Pergamon, Oxford, 1980), p.204.

SOL-GEL MULTILAYERS APPLIED BY A MENISCUS COATING PROCESS

Jerald A. Britten and Ian M. Thomas

Lawrence Livermore National Laboratory, Livermore, CA 94550

ABSTRACT

We describe a meniscus coating method to produce high-laser damage threshold, silica/alumina sol-gel multilayer reflectors on 30+ cm substrates for laser-fusion applications. This process involves forcing a small suspension flow through a porous applicator tube, forming a falling film on the tube. A substrate contacts this film to form a meniscus. Motion of the substrate relative to the applicator entrains a thin film on the substrate, which leaves behind a porous, optical quality film upon solvent evaporation. We develop a solution for the entrained film thickness as a function of geometry, flow and fluid properties by an analysis similar to that of the classical dip-coating problem. This solution is compared with experimental measurements. Also, preliminary results of multilayer coating experiments with a prototype coater are presented, which focus on coating uniformity and laser damage threshold (LDT).

INTRODUCTION

Sol-gel processing for the manufacture of multilayer high reflectors (HR's) for high-power laser applications offers the possibility of reducing the cost of these optics, which are presently made by electron-beam evaporation techniques. Laser damage-resistant single-layer antireflective coatings on large optics have been made by sol-gel dip-coating for several years [1]. In recent work, Thomas [2] has demonstrated damage-resistant alumina-silica sol-gel multilayer HR's applied by spin-coating on 5.1 cm diameter substrates. These mirrors reflect 1.06 micron light and exceed LDT (40 J/cm^2 at 10 ns) and reflection (>99%) specifications required by our application in laser-driven inertial confinement fusion (ICF) experiments.

The next generation of ICF driver lasers will be of square segmented aperture and will require turning mirrors of up to 50 cm in length. A process to apply sol-gel multilayer coatings must be scaleable to this size. Dip coating multilayers is not an attractive method due to the potential intercontamination of the incompatible sols when dipping between large tanks, and due to the large amounts of suspension required. Recently we have studied the feasibility of spin-coating multilayers on large substrates 20-30 cm in diameter [3]. We have discovered that the highly non-Newtonian rheology of one of the suspensions (alumina) used in our process results in a radial film thickness decrease from center to edge that cannot be compensated by the reflection bandwidth. This difficulty, as well as our requirements for square or rectangular cross-sections have prompted a search for an alternative coating method.

One alternative, known as meniscus coating or laminar film coating, offers promise for our application. This process operates by creating a thin liquid film over the surface of a porous metal applicator tube by forcing the coating suspension through the tube wall. A meniscus is formed by contacting this film with a substrate, the applicator is dropped down about 1 mm and then the substrate is moved relative to the applicator. This movement entrains a very thin liquid film on the substrate which leaves behind an optical quality coating of solid particles as the carrying fluid evaporates. The coating fluid collects in a trough surrounding the applicator and is continuously filtered and recirculated. A schematic of the coating process and meniscus geometry is shown in Figure 1. This process is very similar in principle to dip coating, but minimal contact between coating sols occurs and very small suspension volumes are required.

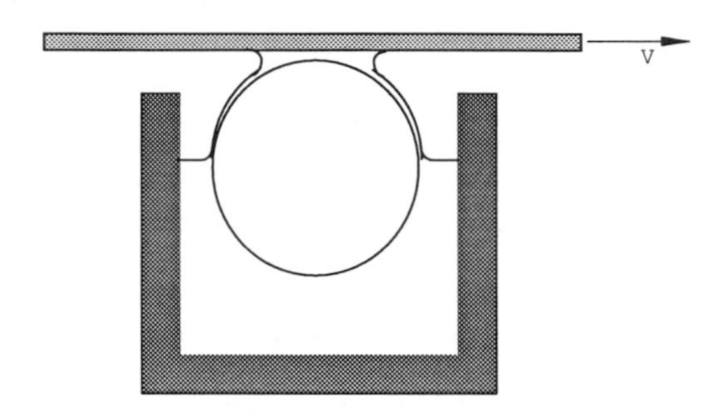

Figure 1. Schematic of meniscus coating process.

The following presents a brief description of the fluid dynamics of meniscus coating, and then describes preliminary multilayer coating experiments which have demonstrated the feasability of this process.

FLUID DYNAMICS OF MENISCUS COATING

The thickness of a liquid film on a surface emerging from a quiescent pool was first analyzed by Landau and Levich (cf. [4]) and later by Wilson [5] among others. Their analyses are based on the perturbation that the entrained film flow imposes on a stationary meniscus, in the limit of a small capillary number $\mu V/\sigma$, where μ is the fluid viscosity, V is the substrate translation rate and σ is the surface tension. The shape of a meniscus in the neighborhood of a falling film flow has been analyzed by Ruschak and Scriven [6] for the case where the downstream film thickness $h_f = (3\mu Q/\rho^2 g \cos\alpha)^{1/3}$ is much smaller than the radius of curvature of the meniscus. In the above, Q is the specified mass flow rate per unit length of the tube, ρ the liquid density, g the gravitational acceleration, and α the angle of inclination of the falling film away from vertical. For such cases, the meniscus is also quasistationary and the falling film appears as a boundary layer.

We have modeled the essential features of the fluid dynamics of meniscus coating for a Newtonian fluid in an idealized geometry, by combining the solutions mentioned above to match the curvature of the meniscus as it stretches from the very thin falling film flow on the applicator to the even thinner entrained film on the substrate. The simplified geometry is shown in Figure 2. The major simplification for modeling purposes is the well-defined angle α.

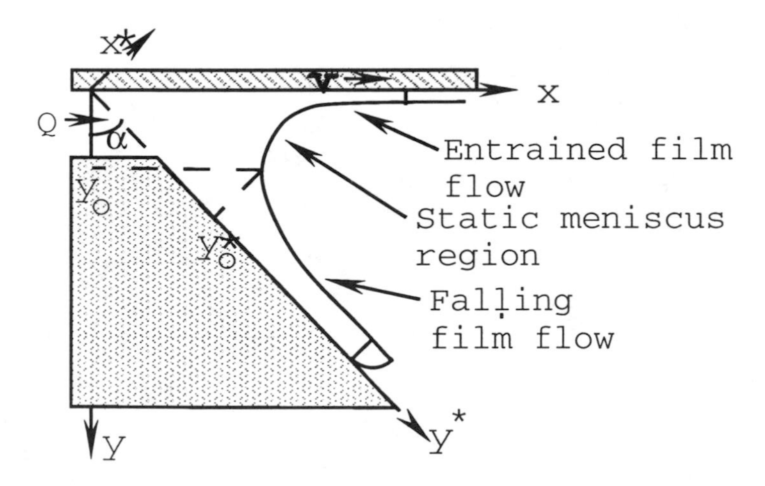

Figure 2. Idealized geometry of meniscus coating process used in fluid flow analysis.

The details of the analysis are given in another paper [7] but the result for the entrained film thickness is:

$$h_\infty = \frac{1.3375\left(\frac{\mu V}{\rho g}\right)^{2/3}\left(\frac{\rho g}{\sigma}\right)^{1/6}}{\left[2(1+\sin\alpha) - 1.1444\cos^{4/3}\alpha\left(\frac{\rho g}{\sigma}\right)^{1/3}\left(\frac{3Q\mu}{\rho g\cos\alpha}\right)^{2/9}\right]^{1/2}} \quad (1)$$

Except for the term multiplied by 1.1444 in the denominator, eq. (1) is identical to the leading-order term in the dip-coating film thickness equation developed by Wilson [5]. This small correction (about 10%) represents the stretching perturbation of the meniscus curvature due to the falling film. Equation (1) shows that the major effect of the falling film is geometrical, and that larger curvatures (larger α) result in thinner films. In the actual coating apparatus in which the applicator is a tube, (see Figure 1) an increase in the flowrate can decrease slightly the angle of attachment of the meniscus to the falling film. Thus, the geometrical effect will be a weak function of the flowrate as well.

We tested the theory by measuring the dependence of film thickness on substrate translation speed. A prototype of a commercial apparatus (CAVEX meniscus coater; Union Carbide Specialty Coating Systems) was used to deposit a series of single coats of silica sol at varying speeds. The optical thickness of the dried film was measured with a Perkin-Elmer Lambda-9 spectrophotometer. These measurements are shown in Figure 3. The best fit to a power-law model of the h_∞ -vs- V data gives an exponent of 0.658, which is very close to the theoretical value of 2/3.

COATING EXPERIMENTS

The preparation and properties of the silica and alumina suspensions used in this study have been described elsewhere [2]. The machine used to apply the above calibration layers was fitted with a dual dispense system and used for applying multilayers onto 30-cm square panes of 0.5 cm thick window glass and also onto 15 cm square fused silica substrates.

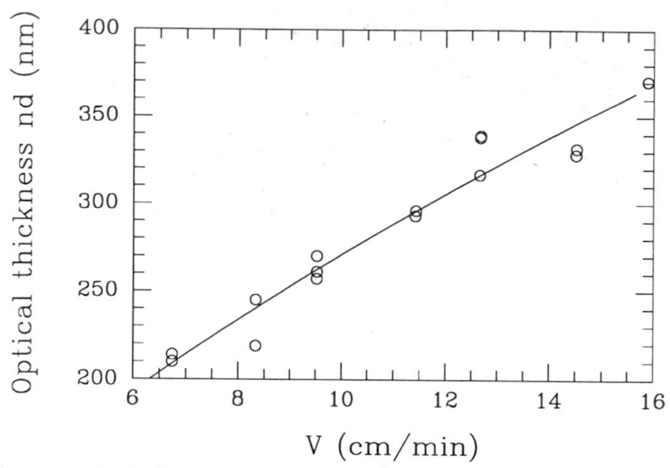

Figure 3. Optically measured film thickness as a function of translation rate for SiO_2 single layers. Solid line is best fit of $nd = AV^p$, with $A = 59.5$, $p = 0.658$.

Preliminary experiments showed that single–layer films of the alumina sol were very uniform. Films from the silica sol, however, exhibited banding or striations parallel to the applicator tube, caused by vibrations or drive-induced irregularities in the applicator tube translation rate. The silica films were also splotchy and spatially nonuniform in all directions if the translation rate was too high. The reason for this is that the alumina sol is more viscous so that films are initially thicker and vibration effects are damped. It also gels and becomes essentially immobile at relatively low solids volume fractions [3]. The silica sol on the other hand remains quite fluid until volume fractions approaching about 50%, and precipitates, rather than gels, as the solvent evaporates. Faster translation rates correspond to thicker films which dry more slowly and are more highly influenced by random air currents, which can accelerate drying locally. When this occurs, it appears that the receding liquid can pull solvent and suspended particles away from the dry areas, resulting in a splotchy film. Thus it was concluded that very slow translation rates give better film uniformity and are less influenced by drying irregularities. The desired optical film thickness for slow translation rates can be obtained by increasing the initial solids fraction of the sol. Banding occurred in all experiments, but we are confident that vibrations can be eliminated in future optimized machines.

Several quarterwave silica/alumina multilayer coatings of from 10-36 layers were made to characterize reflection, uniformity and laser damage characteristics. Many of the coatings, particularly ones consisting of a large number of layers, suffered from particulate contamination due to lack of a clean-room environment and lack of a filter system for the recirculated sol. The sol also tended to accumulate particulates over time due to a mechanism to be discussed later.

Figure 4 shows a transmission spectrograph for a 20-layer coating measured at three locations on a 15-cm fused silica substrate. This coating was designed to reflect 1.06 μ light. The number of layers is not sufficient to provide 99+% reflection due to the relatively small index difference between the silica and alumina layers (1.22 / 1.44) but it is sufficient to address coating uniformity issues. Figure 4 shows that we are able to control substrate translation rates to deposit quarterwave multilayer stacks of the desired optical thickness.

Calculations with a commercially available thin-film design software package show that the minimum transmission for a 1.22/1.44 10-pair multilayer is about 10.6%. Figure 4 shows about 14% as the actual value. The difference is probably due to small calibration errors or to densification of the individual material layers caused by capillary forces, which could act to reduce the index difference of the multilayer slightly.

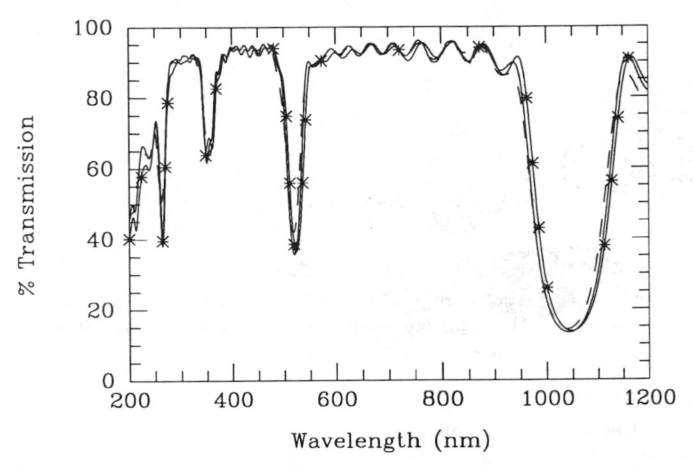

Figure 4. Transmission spectra for a 20-layer silica/alumina HR on a 15-cm fused silica substrate at three random locations.

Figure 5 shows the uniformity of a 20-layer coating applied onto a 30 cm square substrate. The uniformity is plotted as 1 - $\lambda_{max}/ < \lambda_{max} >$, where λ_{max} is the measured wavelength of maximum reflection locally, and $< \lambda_{max} >$ is the average of this over all locations measured. Twenty-two spot measurements were made with the spectrometer 2.5 cm apart along two lines; one parallel to and one perpendicular to the direction of translation of the substrate. The maximum shift in λ_{max} is seen to be about 2%, with the average being less than 1%. This coating exhibited the banding structure in the silica layers described earlier. However, due to the width of the reflection band and the averaging effect of many layers, these bands, which are subtle variations in the wavelength of reflected harmonics of the quarterwave, appear not to influence the primary peak reflection uniformity significantly. In fact, the reflection uniformity of this HR meets the requirements for some of our applications.

The biggest problem we experienced was the accumulation of particulates, due largely to a self-contamination mechanism arising from the multilayer coating process. Where the meniscus detaches from the substrate, a thick film or drip line is formed. This leads to a thick deposit upon solvent evaporation. This crust continues to build up with subsequent applications and then begins to craze. Wetting by the applicator tube can then detach small particles from this rind which remain on the applicator tube surface or fall with the flow back into the suspension, where in general they do not resuspend. These particles, large with respect to the original suspended ones, can then be entrained and transferred onto the substrate during subsequent coating applications. Amelioration of this problem requires active cleaning of the knife edges on both the leading and trailing edges after every coating. An advanced coating machine under development will have such a feature, and will also have the capability for continuous filtration of the coating sol.

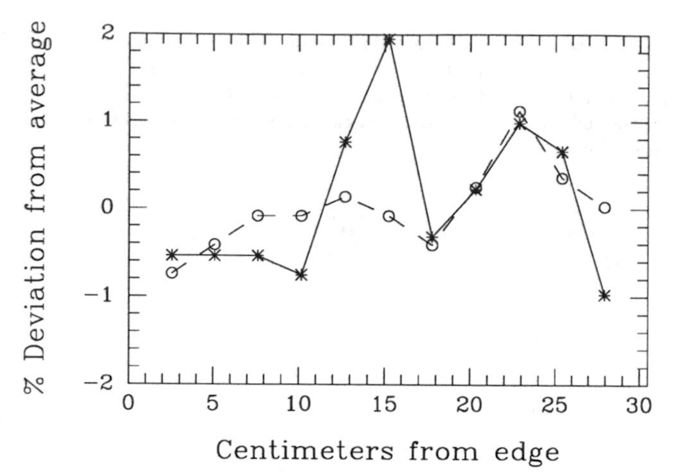

Figure 5. Deviation of λ_{max} from average value at 22 locations on a 30-cm square window glass substrate coated with a 20-layer silica/alumina quarterwave stack: —— Perpendicular to applicator; – – parallel to applicator.

LASER DAMAGE STUDIES

The LDT for meniscus-coated HR's should be in principle the same as for spin-coated ones, which are quite good. We performed laser damage tests on a bare 15 cm fused silica substrate, single-layer coatings of both alumina and silica, and a 20-layer multilayer (that with transmission spectrum shown in Figure 4) at our damage facility. This facility is described in [8]. Test results are shown in Table 1. Both of the single layer films tested higher than the bare substrate, which shows that there is nothing intrinsic to the coating process itself that is of concern with respect to LDT. The multilayer, however, failed at about 22 J/cm^2, a value that is somewhat better than e-beam coated HR's before their LDT is improved by laser conditioning [9]. The coating, although spatially uniform, was dirty due to the reasons described above. These results suggest that elimination of the particulate problem as discussed above will also fix the laser damage problem.

Table 1

Laser damage test results for single-layer SiO_2 and AL_2O_3 coatings and a $(SiO_2/AL_2O_3)^{10}$ multilayer on 15 cm silica substrates applied by meniscus coater. LDT in J/cm^2 for 1064 nm, 10 ns pulse

coating	LDT	comments
bare substrate	>57.4	front surface
silica single layer	>84.6	50 artifacts/mm^2 $< 30\mu$
alumina single layer	67.3	no artifacts
silica/alumina 20 layers	21.8	2000 artifacts/mm^2 $< 30\mu$ 28-38 J/cm^2 massive damage

The meniscus coating process would be more attractive if the number of layers required for 99% reflection can be reduced. The silica coating has about the lowest index possible for an oxide, but several high-index materials are potential candidates. The major limitation is low LDT. The alumina sol has several desirable properties other than high LDT, including good adhesion and binding characteristics to produce films which do not craze or scatter. We are actively investigating other oxide suspensions, binders etc. that may increase the index of the high-index layer without degrading LDT or adhesion performance.

CONCLUSIONS

The potential for meniscus coating of large-area sol-gel multilayer laser reflectors has been demonstrated. The fluid dynamics of this coating process has been analyzed and found to differ from the familiar dip-coating problem only in a small term which corresponds to the stretching of the meniscus by the falling film to which it is attached. Coating uniformity over 30 cm scales has been demonstrated, and mechanisms for self-contamination of the coating by particulates have been identified. These will be alleviated in future coating machine designs. The laser-damage threshold of meniscus-coated HR's appears to be adequate for our purposes once the particulate problem is overcome.

ACKNOWLEDGEMENTS

This work was performed under the auspices of the U.S. Department of Energy by Lawrence Livermore National Laboratory under contract number W-7405-Eng-48. The assistance of Hendrik Bok in setting up the meniscus coating apparatus is greatly appreciated.

REFERENCES

1. Thomas, I.M., Applied Optics 25 1481 (1986)
2. Thomas, I.M., Applied Optics 28 4013 (1989)
3. Britten, J.A. and Thomas, I.M., J. Applied Physics 71 972 (1992)
4. Levich, V.G., Physicochemical Hydrodynamics (Prentice-Hall, Englewood Cliffs, NJ, 1962), p. 675
5. Wilson, S.D.R., J. Engineering Math. 16 209 (1982)
6. Ruschak, K.J. and Scriven, L.E., J. Fluid Mech. 81 305 (1977)
7. Britten, J.A., submitted to Chem. Eng. Communications. (1992)
8. Morgan, A.J., Rainer, F., DeMarco, R.P., Kozlowski, M.R. and Staggs, M.C., in Laser Induced Damage in Optical Materials: 1989 Proc. Boulder Damage Symposium Nov 1-3, 1989, SPIE vol 1438 47 (1990)
9. Wolfe, C.R., Kozlowski, M.R., Campbell, J.H., Rainer, F., Morgan, A.H., and Gonzalez, R.P., in Laser Induced Damage in Optical Materials: 1989 Proc. Boulder Damage Symposium Nov 1-3, 1989, SPIE vol 1438 360 (1990)

PHASE TRANSFORMATIONS IN SOL-GEL YAG FILMS

R. S. Hay
Wright Laboratory, Materials Directorate
WPAFB OH 45433

ABSTRACT

Diphasic yttrium-aluminum garnet (YAG) sols were gelled across TEM grids to make films. The films were heat-treated up to 300 hours between 800°C to 1150°C. Microstructure and phase evolution were observed by TEM. YAG fraction and grain size, matrix phases and grain size, nuclei/area, and film thickness were measured. Yttrium-aluminum monoclinic and transition alumina appeared at 800°C. YAG nucleated between 800°C and 950°C. Above 850°C nucleation was site-saturated. Final nuclei density averaged 0.3/μm^3. Between 850°C and 1000°C YAG growth had ~$t^{1/2}$ dependence and Q ~ 280 kJ/mole. Below 850°C nucleation was continuous and growth had ~$t^{0.85}$ dependence. Above 1000°C YAG growth had ~$t^{1/4}$ dependence and the matrix grains coarsened with ~$t^{1/4}$ dependence. Unreacted areas were more abundant in thinner films because the nuclei/area and reaction rate decreased. YAG growth was often accompanied by formation of 20 – 100 nm subgrains. Above 1100°C there was some reaction to YAG by a different process.

I. INTRODUCTION

Diphasic YAG ($Y_3Al_5O_{12}$) sols made by controlled hydrolysis of aluminum and yttrium isopropoxides were used to coat fibers[1,2]. They were similar to alumina sols made by Yoldas[3,4]. In a preliminary study[5] it was found that nucleation and growth of YAG was in some ways similar to that of α-alumina from boehmite gels[6,7]. Extensive subgrain formation accompanied the growth of YAG. These subgrains dominated microstructure evolution at higher temperatures.

Little work has been done on the kinetics of YAG formation from diphasic gels, but such studies have been done for other ceramics[8]. This work follows the preliminary study by focusing on the kinetics of nucleation and growth of YAG. Most of the data comes from TEM observations. The kinetics are compared with materials that exhibit similar behaviour. A more thorough analysis and descripton of the work is given elsewhere[9].

II. EXPERIMENT TECHNIQUE

YAG sols were made by a procedure described elsewhere[5,9]. Drops of the sols were placed on 150 mesh nickel TEM grids and gelled across the spaces to make films. The films were heat-treated in air with a 50°C/min heating rate in a Netzsch differential thermal analysis/thermogravimetric analysis (DTA/TGA) equipped furnace. Heat-treatments of 10 minutes to 300 hours at 50°C intervals from 800°C to 1150°C were done.

Film thickness (δ) was measured from Fizeau fringe intensity in white light. Thicknesses less than 60 nm could not be measured this way, so TEM methods were used. Thickness varied from 25 nm to 400 nm; most were between 100 and 200 nm. Measurement error was estimated to be +/- 25 nm. Ion-milling and carbon coatlng were not needed for TEM observation. Standard and high resolution TEM techniques were used with a JEOL 2000FX operating at 200 kV. Measurements of YAG area fraction were made from dark-field images. Twenty to thirty different areas with measured thickness were photographed for each temperature-time. Where possible, YAG grain size was measured and grains/area were counted to measure nucleation rate. Matrix phases were identified from SAD ring patterns. Matrix grain size histograms were made from 200 grain diameter measurements.

III. RESULTS AND DISCUSSION

In the thinner films spherulitic YAG crystals grew as branched, curved cylinders with pores in between, but in thicker films they grew as a continuous sheet (Fig. 1). Wetting phases were not observed at grain impingement sites, at the reaction interface, or on matrix grain boundaries by HRTEM. No impurities were found at grain impingement sites. A preferred orientation of YAG was not observed. Subgrains formed at the reaction front. Their density was higher in thicker films treated at higher temperatures. At 800°C subgrains were absent, but high strain and defect concentration were manifested as variable and wavy diffraction contrast.

The reaction had characteristics of spherulitic growth[10,11]. Such growth is promoted by difficult nucleation, small diffusion coefficients, and multicomponent systems, and is characterized by radially symmetric crystal habit with non-crystallographic branching. This described YAG growth, except the radial symmetry was often weak, and the non-crystallographic branching could have been caused by a Rayleigh instability in thin film single crystals[12].

YAM (yttrium-aluminum monoclinic, $Y_4Al_2O_9$) and transition-alumina diffraction rings were visible at 800°C (Fig. 2). Crystallization was gradual. YAM was identified from prominent rings at .291 and .301 nm[13]. Below 900°C individual spots were diffuse and superposed on a diffuse ring. This came from 50 – 100 nm grains with diffraction contrast suggestive of high defect content. With time these grains seem to have evolved to less defective, smaller grains. The formation of YAM from aluminum and yttrium oxide

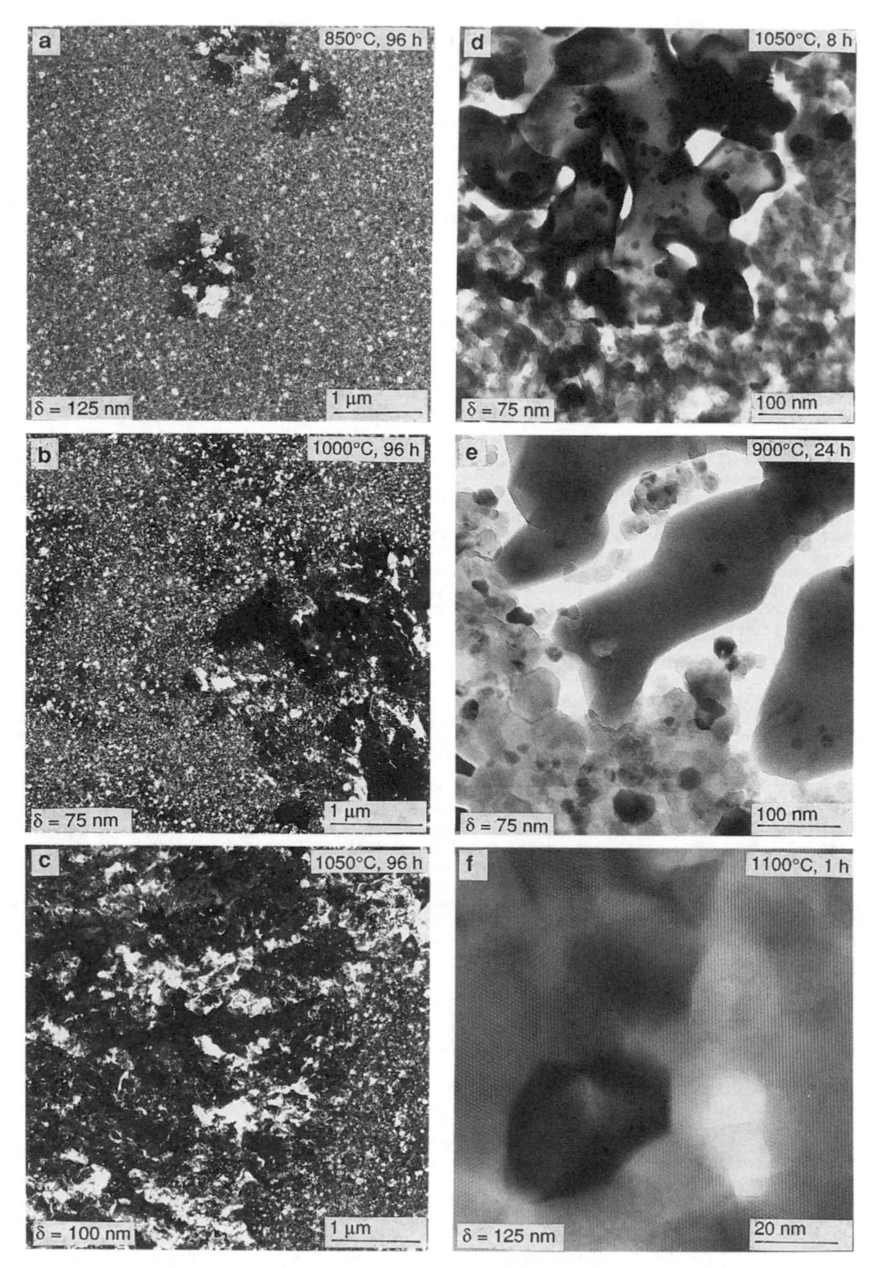

Fig. 1 a-c) Spherulitic YAG growth for t = 96 h (dark field); d-e) the reaction front, showing non-crystallographic branching, defect formation, and inclusion of matrix between branches (bright field); f) subgrains formed near the reaction front in a slightly thicker film (bright field, $\sim[111]_{YAG}$).

prior to formation of YAG was consistent with relative immobility of yttrium compared to aluminum and oxygen.

The transition aluminas have similar structures so their specific identification was difficult[13]. The rings were diffuse compared to YAM rings. Θ-alumina identification was tentatively made from the .154 nm {$\bar{3}$13} and .245 nm {111} rings. However, Θ-alumina is not usually found below 1000°C[13]. Above 1000°C the rings became less diffuse, and at 1100°C the relative intensities gradually changed, perhaps from development of preferred orientation.

A transient ring with weak intensity appeared at .269 nm (Fig. 2). An even weaker ring sometimes appeared at .321 nm. At higher temperatures they appeared and disappeared earlier. They could have been YAG[13] or $YAlO_3$ garnet[14,15] rings. Evidently a small amount of fine grained garnet crystallized with YAM and transition-alumina, and later disappeared. Its transient nature suggests it is $YAlO_3$ garnet. Its presence roughly coincided with the nucleation of spherulitic YAG.

After 1 hour at 1150°C, 100 – 300 nm YAG grains appeared in the few areas of the matrix not consumed by growth of spherulitic YAG. Apparently, areas not consumed by growth of spherulitic YAG reacted to YAG by a different mechanism. Such growth was identified as abnormal grain growth in a preliminary study[5]. It is now apparent that spherulitic YAG formed from reaction between YAM and transition alumina, but may gradually take the character of abnormal grain growth if the matrix reacts to YAG by a different mechanism.

Above 1000°C the matrix grains coarsened. Coarsening at 1050°C – 1150°C was described by:

$$d_{av} - d_o = k_o t^n \exp^{-Qm/RT} \qquad (1)$$

where Q_m = 200 +/- 30 kJ/mole, k_o = .08 (m/s^n), n = 0.23 – 0.25, t is time, RT has the usual meaning, and d_o was estimated to be 10 nm (Fig. 3).

The normalized grain size distributions were computed above 950°C (Fig. 4). YAM (65 vol%) and transition-alumina (35 vol%) grew simultaneuously, and in addition there was porosity that coarsened from 10 nm interconnected pores to 20 nm isolated pores at 1000°C. The matrix grain size distribution was log-normal, which is usually associated with grain growth[17], but grain size dependence on $t^{1/4}$ is associated with interface[18] or surface[19] diffusion control of coarsening. Classical coarsening theory[18] and a new theory[20] both predict normal distributions skewed toward large sizes. Neither match the observed log-normal distribution. None deal with simultaneous coarsening of two phases.

Nucleation occurred in random locations, without preference for visible defects. Nuclei/volume was roughly constant, so nuclei/area scaled with film thickness (δ). From 800°C to 950°C nuclei/volume increased with time, but after 3 hours at 950°C or 24 hours at 850°C it was site-saturated (Fig. 5). Final nuclei/volume varied between .15/μm^3 and .40/μm^3. Reliable counts after high temperature – long time heat-treatments could not be made because of grain impingement.

Nuclei may form by reaction of YAM and transition alumina with a volume loss of 7.7%. Nuclei might also form from decomposition of the transient fine-grained $YAlO_3$ garnet[16], with a volume increase of 11%, or reaction of $YAlO_3$ garnet with transition alumina, with a volume increase of 3.5%. The transient appeareance of $YAlO_3$ garnet during nucleation is circumstantial evidence in favor of the latter mechanism. If so, nucleation in this system is not analogous to alumina or mullite.

Site-saturation of nucleation, independent of heat-treatment, is similar to that observed for α-alumina in boehmite gels[6,7]. It was suggested that nucleation of a polymorphic phase transformation was related to particle packing in the gel[7]. It is unclear how particle packing would influence nucleation in diphasic gels. There seems to be no general and well accepted explanation for sparse nucleation and site-saturation in sol-gel derived ceramics.

Spherulitic YAG grew as circular grains until impingement, some to 10 μm in diameter (Fig. 1). Growth was two-dimensional. Since nuclei/area scaled with δ a modified expression for the Avrami kinetics of nucleation and growth was used[21]:

$$\log\{\ln[1/(1-f)]\} = \log(k\delta t^{n_A'}) \qquad (2)$$

where f is the area fraction, t is time, n_A' is an area growth exponent, and k is a constant. The effect of δ on f is the same as t on f. However, spherulitic grains also grew faster *after* nucleation in thicker films.

Avrami plots were prepared. k and n_A' were calculated before and after nuclei site-saturation. If there was no significant difference they were lumped together. The thickness effect on growth was found by normalizing the data with the linear best fit for k and n_A', and replotting against δ:

$$\log\{\ln[1/(1-f)]\} f \equiv \delta^{1.4} \qquad (3)$$

Correlation was weak (.51), probably from errors in measurement of δ (Fig. 6). Avrami plots were prepared again with the empirical correction for δ to f from (3) normalized to δ of 100 nm (Fig. 7). The correlation coefficients were compared to those from the uncorrected plots. A large improvement was found at higher temperatures.

The linear growth exponents (n' = n_A'/2), activation energies (Q_r), and rate constants for linear growth between 850°C and 1000°C were found from Avrami plots:

$$800°C: \quad n' \sim 0.85 \qquad (4a)$$

$$850°C - 1000°C: \; n' \sim 0.5, \; Q_r \sim 280 \text{ kJ/mole}, \; k_1 \sim 2200 \qquad (4b)$$

$$1050°C - 1100°C: \; n' \sim 0.25, \; Q_r' \sim 120 \text{ kJ/mole} \qquad (4c)$$

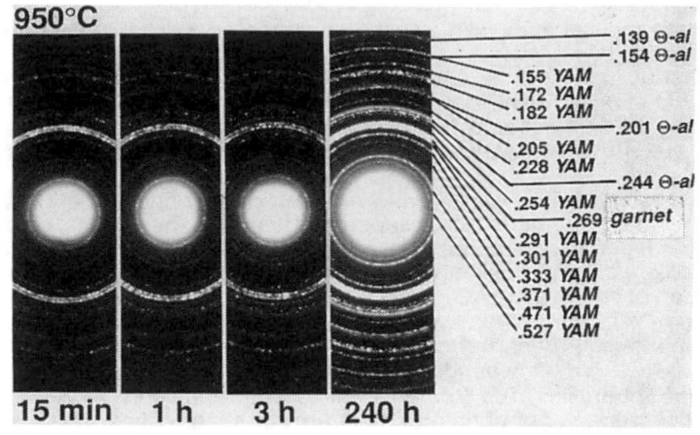

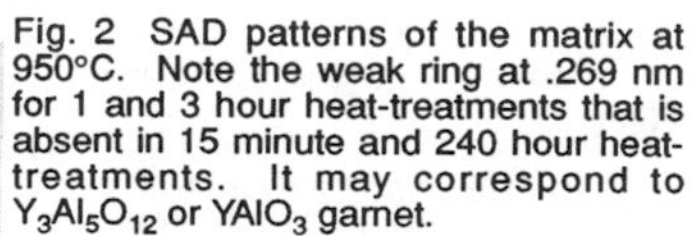

Fig. 2 SAD patterns of the matrix at 950°C. Note the weak ring at .269 nm for 1 and 3 hour heat-treatments that is absent in 15 minute and 240 hour heat-treatments. It may correspond to $Y_3Al_5O_{12}$ or $YAlO_3$ garnet.

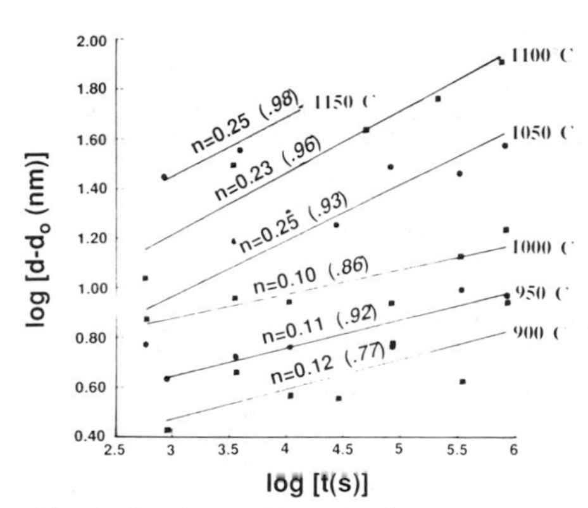

Fig. 3 Average matrix grain size vs. time. Note the change in slope between 1000°C and 1050°C.

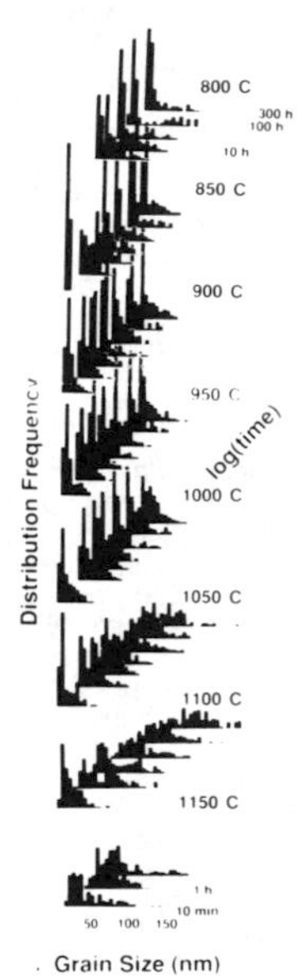

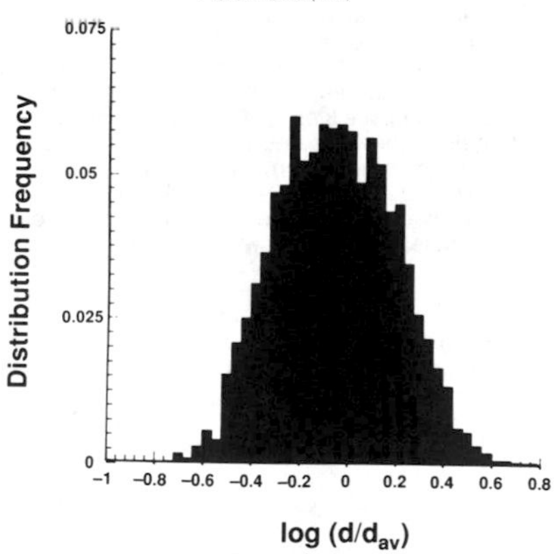

Fig. 4 Matrix grain size distributions vs. T & t (top), and the log-normal matrix grain size distribution at 950°C and above (bottom).

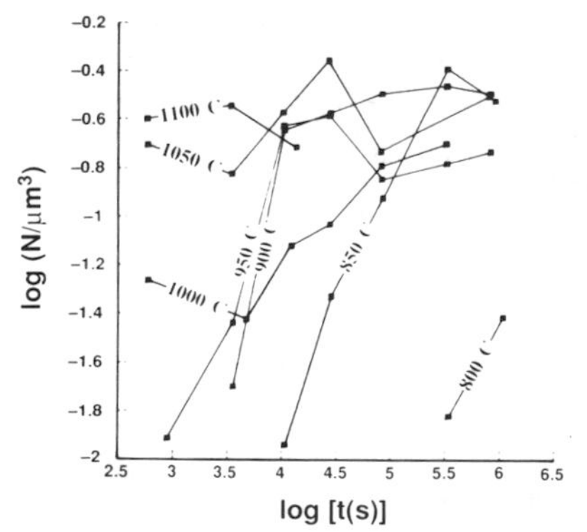

Fig. 5 Nuclei (N) density vs. time. The average for site saturation was about $0.3/\mu m^3$.

The kinetics between 850°C and 1000°C was consistent with diffusion control and nuclei site-saturation[21]. The higher n' at 800°C was consistent with diffusion control with constant nucleation rate[21]. n' decreased by ~1/4 and the matrix grain growth exponent (n) increased by ~1/4 between 1000°C to 1050°C.

IV. GROWTH MECHANISMS

Growth of spherulitic YAG must involve diffusion and reaction between grains of YAM and transition alumina along a reaction front (Fig. 8). Reactants do not have to diffuse through the product layer, so the rate determining flux should be constant and growth should be interface controlled. However, for complete reaction the growth rate (G) must be inversely proportional to the matrix grain size (d). Therefore:

$$G = d\Gamma/dt = (n'k_2t^{n'-1}/d)\exp(-Q_r/RT) \qquad (5)$$

where Γ is the spherulite radius and k_2 is a constant that incorporates the pre-exponential term for diffusion and the driving force[22]. n' should be one but it was observed to be ~0.5 for constant d and after site-saturation.

The growth rate *should* be constant, but instead decreases with time by –1/2. Such growth is found when impurities collect at the reaction front[23]. The impurities must inhibit the reaction and be mobile compared to G; otherwise they are incorporated into the spherulite. The reaction is rate limited by diffusion of this impurity from the front. If they diffuse along grain boundaries G should scale with 1/d. The impurities build up as the reaction progresses and should concentrate where spherulites impinge. Impurity concentrations at impingement sites were not observed, so if this mechanism caused diffusion control trace amounts of impurity were responsible.

From (5) it is apparent that if d changes, G must change. Substituting (1) into (5):

$$\Gamma = \frac{k_2t^{1/2}\exp(-280/RT)}{.08t^{1/4}\exp(-200/RT) + d_o} \qquad (6)$$

where R is in units of kJ/mole°C. Two growth regimes are evident. When grain growth is insignifiacnt equation (5) can be recovered. When grain growth is significant:

$$\Gamma = .000275t^{1/4}\exp(-80/RT) \qquad (7)$$

The predicted time dependence suggests that matrix grain growth may lower n' by increasing the required diffusion distance or reducing the number of boundaries an impurity can use for fast diffusion paths. However, the change in G between 1000°C and 1050°C was abrupt. The change in matrix grain growth in this interval was also abrupt. Alumina phase transformations were suggested to be rate controlling for mullite growth at ~ 1000°C[24], but this is controversial[25,26]. In these YAG films no phase transformations were observed for matrix alumina (Fig. 2), but phase identification was ambiguous. The activation energy for mullite growth was ~1000 kJ/mole[25,26], nearly four times that of YAG, so it seems unlikely that the mechanisms are similar. Activation of another diffusion mechanism above 1000°C might also have caused an abrupt change in mechanism.

Possible explanations for dependence of growth rate on thickness are: 1) surface energy effects; 2) phase distribution effects, and; 3) gel structure effects. These will be discussed in a forthcoming publication[9]; in summary, all are difficult to test and none seem satisfactory. The most important effect of thickness was that growth of spherulitic YAG was inhibited in thin films, so phase evolution in films and bulk gel are different. However, it is clear that the $\delta^{1.4}$ dependence can only be valid over a limited range; otherwise growth rates in thick samples would approach unrealistically large values.

VI. CONCLUSION

Microstructure and phase evolution can be studied by TEM in electron transparent sol-gel derived films without specimen preparation. Spherulitic YAG formed in diphasic aluminum oxide – yttrium oxide films. Its growth can be described by Avrami kinetics between 800°C and 1100°C. Film thickness and simultaneous coarsening of the matrix grains affected growth. Explanations for many of the observations were incomplete or inconsistent. For matrix grain growth, grain size distributions suggest grain boundary migration was operative, but the time dependence of growth suggested coarsening was operative. For spherulitic growth linear time dependence was expected, but parabolic time dependence was observed. Explanations based on impurity diffusion control are plausible, but independent evidence for their existence was lacking. A correlation between YAG nucleation and the transient appearance of a metastable garnet phase was found. Difficult nucleation with site-saturation was similar to that observed in alumina and mullite, but metastable garnet was not present in those systems. The reasons for site saturation are unknown.

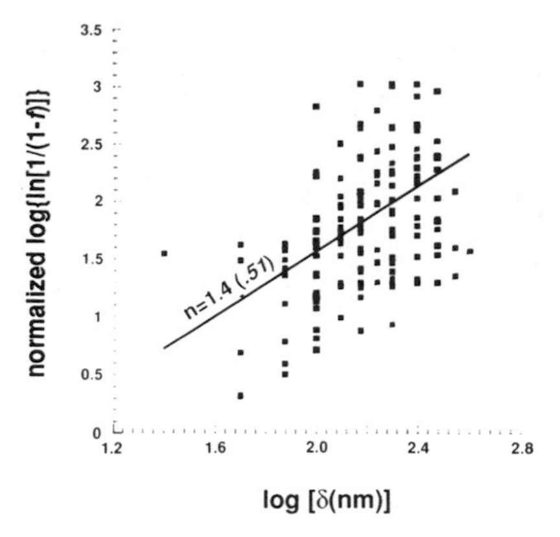

Fig. 6 Normalized Avrami plot of area fraction spherulitic YAG vs. film thickness (δ).

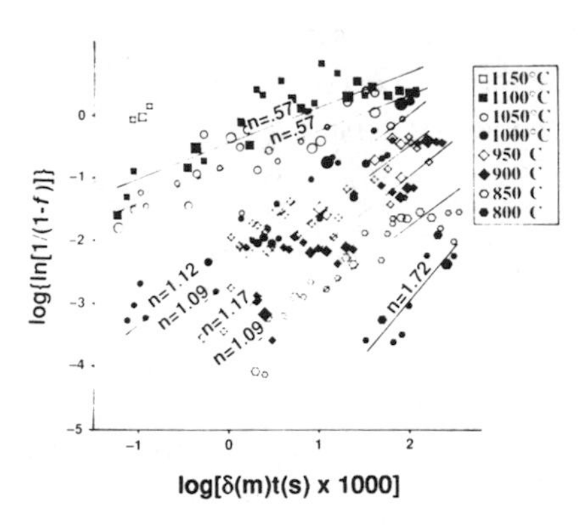

Fig. 7 Avrami plot of thickness corrected area fraction of spherulitic YAG vs. time. The slopes (n) correspond to n_A' used in the text, where $n_A'/2 = n'$.

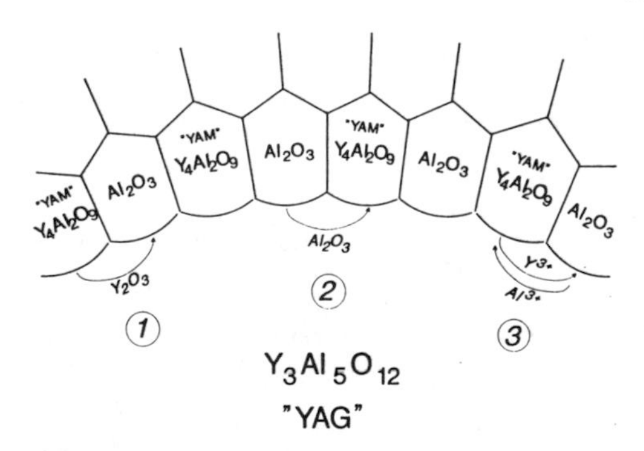

Fig. 8 Simplified schematic of the reaction between YAM and Al_2O_3 that forms spherulitic YAG. The reaction interface *should* not change with time. There are three possible transport mechanisms. Of these (2) seems most likely.

REFERENCES

[1] R. S. Hay and E. E. Hermes, *Ceram. Eng. Sci. Proc.* **11**, 1526 (1990).
[2] R. S. Hay, *Ceram. Eng. Sci. Proc.* **12**, 1064 (1991).
[3] B. E. Yoldas, *Am. Ceram. Soc. Bull.*, **54**, 286 (1975).
[4] B. E. Yoldas, *Am. Ceram. Soc. Bull.*. **54**. 289 (1975).
[5] R. S. Hay, E. E. Hermes, and K. A. Jepsen, *Ceramic Transactions – Ceramic Thin and Thick Films*, ed. by B. V. Hiremath, Am. Ceram. Soc., **11**, 243 (1990).
[6] F. W. Dynys and J. W. Halloran, *J. Am. Ceram. Soc.* **65**, 442 (1982).
[7] L. Pach, R. Roy, and S. Komareni, *J. Mater. Res.* **5**, 278 (1990).
[8] R. Roy, Y. Suwa, S. Komareni, *Science of Ceramic Chemical Processing*, L.L. Hench and D.R Ulrich, eds., John Wiley and Sons, p. 247 (1986).
[9] R. S. Hay, to be published.
[10] H. D. Keith and F. J. Padden, *J. Appl. Phys.*, **34**, [8] 2409 (1963).
[11] A. Spry, *Metamorphic Textures*, Pergamon Press (1969).
[12] D. J. Srlovitz and S. A. Safran, *J. Appl. Phys.*, **60**, 247 (1986).
[13] R-S. Zhou and R. L. Snyder, *Acta Cryst.* **B47**, 617 (1991).
[14] I. Warshaw and R. Roy, *J. Am. Cer. Soc.*, **42**, 434 (1959).
[15] M. L. Keith and R. Roy, *Am. Min.* **39**, [1&2] 1 (1954).
[16] O. Yamaguchi, K. Takeoka, K. Hirota, H. Takano, A Hayashida, *J. Mat. Sci.* **27**, [5] 1261 (1992).
[17] C. S. Pande, *Acta metall.* **35**, [11] 2671 (1987).
[18] A. J. Ardell, *Acta metall.* **20**, 601 (1972).

[19] P. Wynblatt and N. A. Gjostein, *Acta metall.* **24**, 1165 (1976).
[20] R. T. DeHoff, *Acta metall. mater.* **39**, [10] 2349 (1991).
[21] J. W. Christian, *The Theory of Transformations in Metals and Alloys*, 2nd Ed., Pergamon Press, Oxford (1975).
[22] H. Schmalzried, *Solid State Reactions*, 2nd Ed., Verlag Chemie (1981).
[23] H. D. Keith and F. J. Padden, *J. Appl. Phys.*, **35**, [4] 1286 (1964).
[24] J. C. Huling and G. L. Messing, *J. Am. Ceram. Soc.*, 74, [10] 2374 (1991).
[25] W. Wei and J. W. Halloran, *J. Am. Ceram. Soc.*, 71, [7] 581 (1988).
[26] D. X. Li and W. J. Thomson, *J. Mater. Res.*, **5**, [9] 1963 (1990).

STRUCTURAL AND ELECTRICAL PROPERTIES OF Mn-Co-Ni-O THIN FILMS PREPARED BY DIP-COATING TECHNIQUE

SHOJI KANEKO,* NAOTO MAZUKA,* AND TAMOTSU YAMADA**
* Shizuoka University, Hamamatsu 432, Japan
** Nissei Electric Co., Ltd., Iwata-gun, Shizuoka 438-02, Japan

ABSTRACT

Mn-Co-Ni-O thin films of metal atomic ratio 3.0 : 1.9 : 1.0 were prepared on glass substrates from methanol solutions of the corresponding metal ß-diketonates by dip-coating. As-prepared films were heated at 900°C for 1 h mostly after being calcined at 450°C for 5 min. The film thickness increased with increasing concentration of the solution as well as the number of lifting times. However, the effect was not apparent with the sample prepared without calcination. The prepared films were observed to crystallize into a complicated spinel phase by the heating process at 900°C for 1 h. The surface of the dense film composed of particles of about 0.2 μm diameter, was almost even. The thermal and aging responses of electric resistance showed the film to be a good material as a thermistor.

INTRODUCTION

Mn-Co-Ni-O, transitional metal complex compounds have been used as a negative temperature coefficient (NTC) thermistor,[1)-4)] however, thin film thermistor having desirable thermal response and resistivity has not been available yet. We attempted the formation of Mn-Co-Ni-O thin film by means of a dip-coating using the corresponding metal ß-diketonate complexes as starting materials. A metal ß-diketonate complex has a molecular structure as shown below:

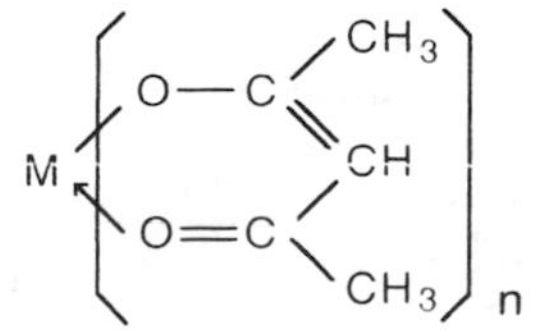

$$M\left[\begin{array}{l} O-C(CH_3)= \\ \quad\quad CH \\ O=C(CH_3)- \end{array}\right]_n$$

(M: metal, n: integral number)

This compound is known not to dissolve so easily into solvent but may appear as the following features when used here: (1) smooth conversion to metal oxide by thermal decomposition due to direct metal-oxygen bond, (2) small volume change at the decomposition due to small functional group, and (3) handle with ease due to non-hydrolysis in organic solvent.

EXPERIMENTAL

Formation of film

For a dipping solution 3 wt% methanol solution of Mn, Co, and Ni

Table I. Dissolution of Metal Acetylacetonates into Various Solutions

	$Mn(acac)_2(H_2O)_2$	$Co(acac)_2(H_2O)_2$	$Ni(acac)_2(H_2O)_2$
methanol	o	o	o
ethanol	o	o	x
acetylacetone	Δ	Δ	Δ
acetone	o	Δ	Δ
acetic acid	x	x	o
benzene	o	Δ	x
toluene	Δ	Δ	x

o : dissolved, Δ : partially dissolved, x : not dissolved

acetylacetonates of metal atomic ratio 3.0 : 1.9 : 1.0 was prepared on the basis of dissolution test shown in Table I, and then thin film was formed on a glass substrate (Fig. 1). The metal content of thin film formed was determined by means of an atomic absorption spectrophotometry after being dissolved into HCl solution.

Characterization of film

The film thickness was measured by a probe contact method and/or a X-ray fluorescence method. The identification of the crystalline phase of the film was carried out by XRD. The film surface was observed by SEM and EPMA. The resistance of the film was measured in a silicone oil kept at 25°C(R_{25}), 100°C(R_{100}), or 200°C(R_{200}) after a Ag-Pd electrode was painted on the film. The specific B

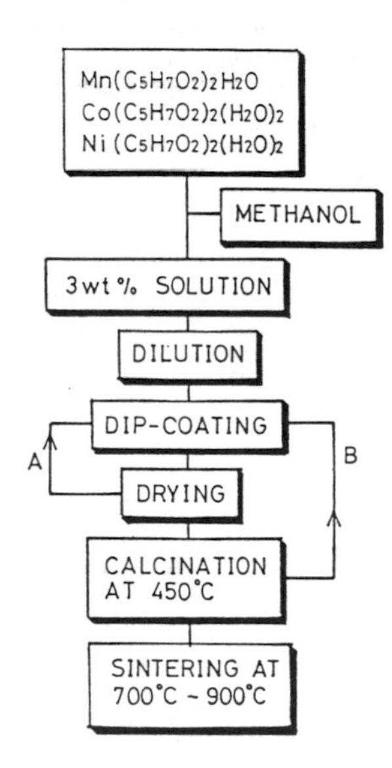

Fig. 1. Preparation method of Mn-Co-Ni-O thin film

constant on a thermistor was calculated by the following equation:

$$B_{T1/T2} = \frac{\ln R_{T1} - \ln R_{T2}}{(T_1 + 273)^{-1} - (T_2 + 273)^{-1}} \qquad (1)$$

where, R_{T1} and R_{T2} are the resistances (ohm) of the film at temperatures T_1(°C) and T_2(°C), respectively.

RESULTS AND DISCUSSION

Film formation

The thickness of the thin film formed via route A in Fig. 1 did not increase so much with the lifting times, as shown in Fig. 2. This could be explained as a result of the dissolution of the film deposited. On the other hand, the film formed via route B increased with an increase in the number of lifting times, making the effectiveness of calcining very clear. It was also proved that the film thickness increased with both the concentration of the dipping solution and the lifting rate. The comparison of the metal contents of the thin film formed and the source dipping solution was carried out : The difference of the composition between the solution and the thin film, tabulated in Table II, was probably due to sublimation of cobalt acetylacetonate judging from its DTA and TG curves; this ca. 30% reduction of the mole ratio of cobalt component of the film compared with that of the dipping solution should be considered in the film formation, despite of the unnecessary consideration on the bulk thermistor of the same composition.[5]

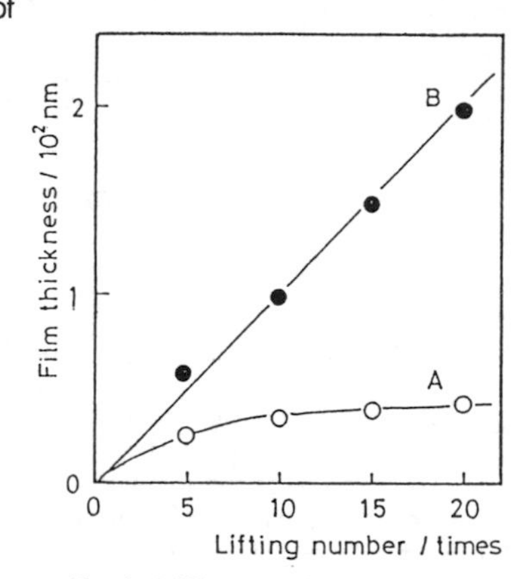

Fig. 2. Difference of film thickness between the two preparation processes

Table II. Comparison of Composition between Dipping Solution and Thin Film

Mn : Co : Ni (mole ratio)	
Solution	Thin film
3.0 : 1.9 : 1.1	3.0 : 1.4 : 1.2
3.0 : 2.5 : 1.1	3.0 : 1.6 : 1.1
3.0 : 3.0 : 1.1	3.0 : 1.9 : 1.2

Crystallization of film

Figure 3 shows the XRD profiles of three thin films of about 800 nm thick as-prepared, and heat-treated at 700°C and 900°C. It seems that the crystallization of the thin film is almost complete by 700°C, and the d spacings given from the XRD profiles of the films all are in the middle of the corresponding d spacings of the cubic $MnCo_2O_4$[6] and $NiMn_2O_4$[7] spinels. Therefore, the crystalline phase of the thin film has been identified to be a solid solution of the two spinels.

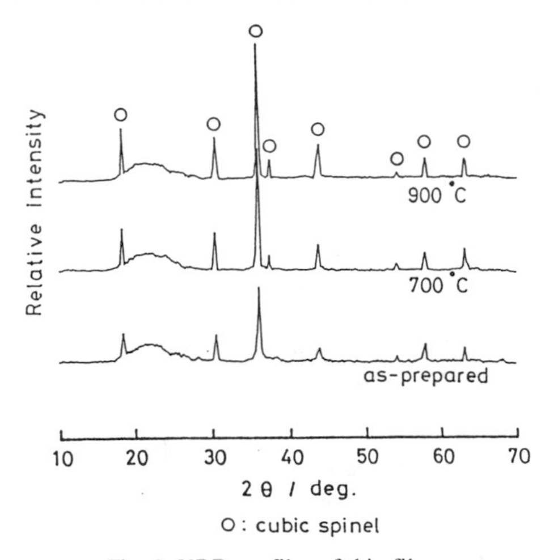

Fig. 3. XRD profiles of thin films

Surface morphology of film

The surface of the thin films of about 200 nm thick was observed by SEM (Fig. 4). As-prepared film was of even surface without crack or abnormal particle growth. On the other hand, the occurrence of crack observed in some places of the film that had been heat-treated at 700°C or 900°C was probably due to the difference of thermal expansion coefficients between the film and the substrate. It is desirable that heat-treatment process should be controlled carefully for the formation of film of quality. These heat-treated films of particles below 0.2 μm diameter were dense and of rather smooth surface.

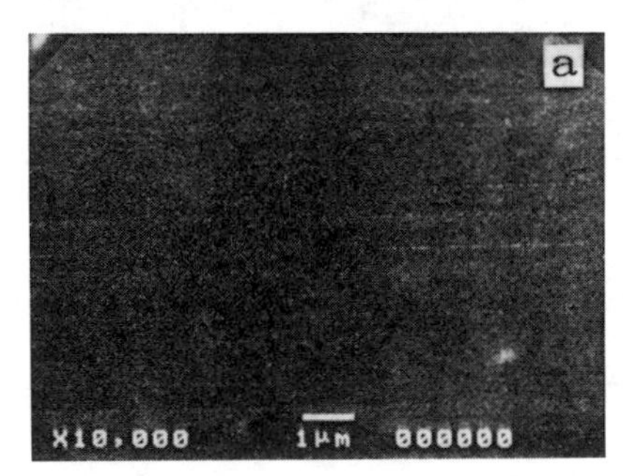

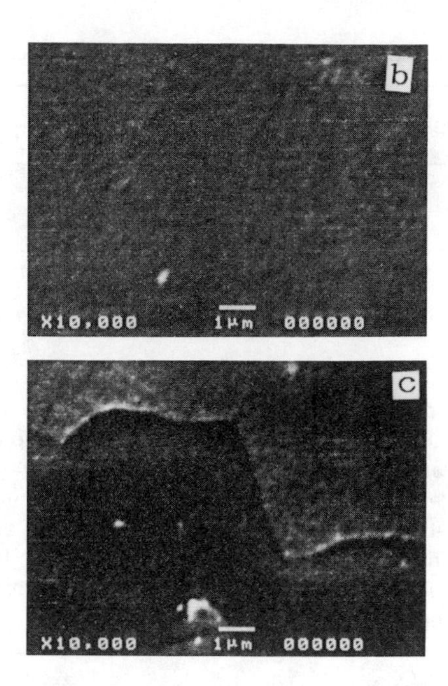

Fig. 4. Microstructure of Mn-Co-Ni-O thin films
a) as-prepared, b) 700°C, and c) 900°C

Electrical properties of film

The temperature dependence of electrical resistance of the film heat-treated at 900°C was non-linear in accordance with that of B constant in Equation (1), as shown in Fig. 5. Although the film thickness affected the electrical resistance of the film, the two $B_{100/200}$ constants and $B_{25/100}$) were all close because of the same composition (Table III). That is, the control of electrical resistance of the film is possible by the change of film thickness. The electrical resistance of the film was largely stable by aging at 150°C for 300 h after the slight increase at the beginning period of heating probably due to the reduction of oxygen vacancy of film by oxidation in air.

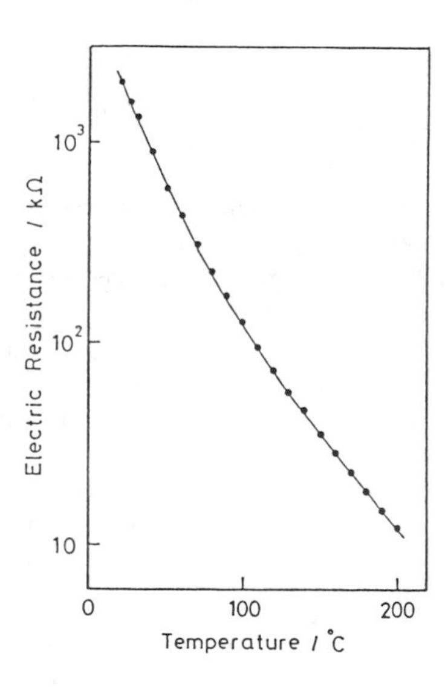

Fig. 5. Temperature vs. electrical resistance of thin film

Table III. Effect of Film Thickness on Electrical Properties

Film thickness / nm	R_{200}	R_{100}	R_{25}	$B_{100/200}$	$B_{25/100}$
	/ kΩ			/ K	
100	102	1060	(13000)	4132	(3718)
200	22.8	240	3200	4152	3842
300	12.0	123	1612	4109	3817

SUMMARY

The present work on the preparation of Mn-Co-Ni-O thin films from the corresponding metal ß-diketonates mixed-solution by dip-coating has been summarized as follows:

(1) The metal ß-diketonates dissolved into methanol to some extent (ca. 3 wt%) and decomposed thermally to convert to the complex-oxide at 450°C.

(2) A Mn-Co-Ni-O thin film of a good composition for thermistor has been prepared from the mixed metal ß-diketonates solution at higher cobalt content compared with the prescribed composition.

(3) The thin film prepared here has a complex spinel phase; a solid solution of $MnCo_2O_4$ and $NiMn_2O_4$.

(4) The B constant of thin film on thermistor is of higher value than that of the bulk of the same composition.

REFERENCES

1) T. Meguro, T. Sasamoto, T. Yokoyama, T. Yamada, Y. Abe, and T. Togai, Yogyo Kyokai Shi, 95, 336 (1987).

2) A. Feltz, F. Schirrmeister, J. Tpfer, and M. Steinbr ck, "Euro-Ceramics", 2. 231 (1989).

3) Nitya Vittal, R. C. Aiyer, C. R. Aiyer, M. S. Setty, S. D. Phadeke, and R. N. Karekar, J. Appl. Phys., 64, 5244 (1988).

4) R. Metz, R. Legros, A. Rousset, J. P. Caffin and, A. Loubiere, Silicate Industriels, 3-4, 71 (1990).

5) S. Kaneko, N. Mazuka, and T. Yamada, unpublished data.

6) JCPDS, Powder Diffraction File No. 1-1110.

7) JCPDS, Powder Diffraction File No. 23-1237.

MOLECULAR RECOGNITION ON ACOUSTIC WAVE DEVICES: MODIFIED ZEOLITE-SILICA THIN FILMS WITH TAILORED ADSORPTION PROPERTIES

YONGAN YAN,[a] THOMAS BEIN,[a,*] KELLY D. BROWN,[b] RAY FORRISTER,[b] and C. JEFFREY BRINKER[c]

[a] Department of Chemistry, Purdue University, West Lafayette, IN 47907, USA

[b] Department of Chemistry, University of New Mexico, Albuquerque, NM 87131, USA

[c] Sandia National Laboratories, P. O. Box 5800, Albuquerque, NM 87185, USA

ABSTRACT

Microporous thin films composed either of zeolite crystals embedded in sol-gel derived glass or of a molecular coupling layer, zeolite crystals and a porous silica overlayer, were formed on the gold electrodes of Quartz Crystal Microbalances (QCM). The microporosity of the thin films was characterized by *in situ* nitrogen and vapor sorption isotherms. Both preparation methods result in thin films with substantial microporosity. Selective adsorption based on molecular size exclusion from the microporous films could be achieved.

INTRODUCTION

Growing efforts are being devoted towards the design of chemical microsensors for environmental monitoring and industrial processing [1]. We have recently developed microsensors with molecular sieving functions in chemically selective layers on piezoelectric devices [2,3,4]. The successful design of microporous thin films retaining the molecular sieving capabilities of glass-embedded zeolites requires that several issues be addressed:

- The glass matrix should not introduce additional, undesired porosity.
- The glass matrix should not clog the zeolite pores after deposition, i.e., the zeolite porosity should be accessible from the gas phase.
- The attachment technique should result in a mechanically stable film that provides effective coupling between the zeolite crystals and the oscillating QCM.

In situ nitrogen adsorption data of different zeolite-containing glass films were used to explore some of these aspects. In an extension of the glass-based composite approach we have also explored the attachment of zeolite crystals on sensor substrates by molecular means, i.e., by using a bifunctional coupling layer that binds to the substrate as well as to the zeolite crystals. As an example of such a system, we discuss a chemical microsensor with combined molecular sieve and selective surface interactions, based on novel silicalite-silica composite thin films on the active area of QCMs. The function of this sensor will be discussed in the context of selective responses towards ethanol.

In addition to the high sensitivity and selective size exclusion offered by molecular sieve films, the nature of the chemical surface interactions is important. In the thin films described below, interference from water (2.65Å) which is smaller than ethanol (4.3Å), could be minimized through hydrophobicity of the molecular sieve and the matrix (silica). The interplay of size exclusion and surface affinity thus provides highly selective adsorption suitable for sensor applications.

EXPERIMENTAL SECTION

The QCM crystals used in the present study are AT-cut 5-9 MHz piezoelectric resonators, with 0.20 cm^2 gold electrodes deposited on chromium underlayers on opposite sides of the crystal. The sensitivity of a 6 MHz QCM according to the Sauerbrey relationship [5] is 12.3 ng Hz^{-1} cm^{-2}.

The sol-gel derived, dip-coated composites were prepared from a suspension of silicalite (MFI) and Na-faujasite (FAU) in acid- or base-catalyzed (A2 and B2, respectively) TEOS sols as described previously [2b, 6]. Silicalite is a crystalline molecular sieve of composition SiO_2 with a pore system of zig-zag channels along *A* (free cross-section ca. 5.1 x 5.5 Å), linked by straight channels along *B* (5.3 x 5.6 Å), while FAU is composed of truncated Si/Al octahedra (sodalite cages) that form supercages with 7.5 Å circular openings and 13 Å internal diameter, with typical composition $Na_{58}Al_{58}Si_{134}O_{384}$ [7].

The preparation of the microporous layer for the ethanol sensor (MFI-EtOH) involved two steps: First, silicalite crystals (about 3 µm diameter) were chemically anchored to the QCM gold electrodes via a thiol-organosilane coupling layer [4]. The silicalite crystals bonded to the QCM electrodes were then further coated with an amorphous, porous silica layer, prepared via sol-gel processing from $Si(OEt)_4$ [8], so that stable microporous thin film composites were obtained. Nitrogen sorption at liquid nitrogen temperature on films where the microporosity is presaturated with ethanol revealed a nitrogen monolayer capacity of 0.0028 g/(g film), corresponding to only 9.8 m^2/g external surface area.

The selectivity and sensitivity of the composite film coated QCM devices was determined by measuring the dynamic vapor sorption isotherms in a computer controlled helium vapor flow system. QCM frequency changes of 0.1 Hz could be detected with a Keithley 775A frequency counter interfaced to a 386-based personal computer. Equilibrium of the vapor sorption was assumed when the frequency changes were less than 1.0 Hz in 30 seconds.

RESULTS AND DISCUSSION

Uncoated and glass-coated QCMs. Nitrogen adsorption isotherms of uncoated QCM crystals are Type II and indicate surfaces with no porosity or with macropores. A polished 9 MHz QCM crystal shows a BET surface area of 2.0 cm^2/cm^2 and a C constant of 20. This deviation from a perfectly flat surface is attributed to microscopic roughness of the devices. If the crystal is dip-coated with A2 sol and heated in air at 350°C to result in a coating mass of 1.55 $\mu g/cm^2$, the BET surface area does not increase. However, coating the crystal with a B2-sol derived glass film results in some porosity of the film when heated at the same conditions. The BET surface area increases to 10 cm^2/cm^2.

The conditions of the heat treatment have a profound effect on the porosity of the glass films. Calcination in oxygen at 450 °C further increases the surface area of the films and introduces some microporosity. For instance, if a B2-derived film (45 $\mu g/cm^2$) is treated at 450 °C, the BET surface area increases to 19 cm^2/cm^2 (corresponding to 42 m^2/g; derived from the desorption branch), and micropore filling with 0.019 g N_2/g film is observed. In the sorption branch, the micropore filling is 0.010 g N_2/g. The difference between the adsorption and desorption branches found in several isotherms is probably associated with tortuous porosity of diameter close to that of nitrogen, such that hysteresis results.

Adsorption on Zeolite Sol-Gel Composites. Silicalite (MFI) and faujasite (NaY) type zeolites were embedded in different sol-gel derived glasses by dip-coating a suspension of the zeolites in the corresponding sol. The microporosity and external surface areas of these films were determined from α-plots of the nitrogen adsorption isotherms taken on the QCMs.

In general, the isotherms of the zeolite-glass composites show distinct microporosity as indicated by the initial steep rise of the adsorption at very low partial pressures. However, the isotherms differ from ideal Type I (Langmuir) adsorption which would be expected for zeolite crystals. After the initial steep rise the films are not saturated but continue to adsorb nitrogen in a fashion typical for many porous solids with Type IV isotherms. The continued adsorption indicates the presence of meso- and macroporosity that must result from combining the zeolites with the glass matrix.

Figure **1** shows a comparison of adsorption isotherms of silicalite (MFI) embedded in A2 glass and heated to 350 and 450 °C. Porosity parameters obtained from these isotherms are listed in Table 1. At about the same total coating weight, the adsorption isotherm resulting from lower temperature heating shows clearly much less adsorption in the micropore region (0.015 g N_2/g film at p/p_o=0.10) than that resulting from high temperature heating (0.047 g N_2/g film at p/p_o=0.10). Even if the desorption branch with the hysteresis is evaluated, higher

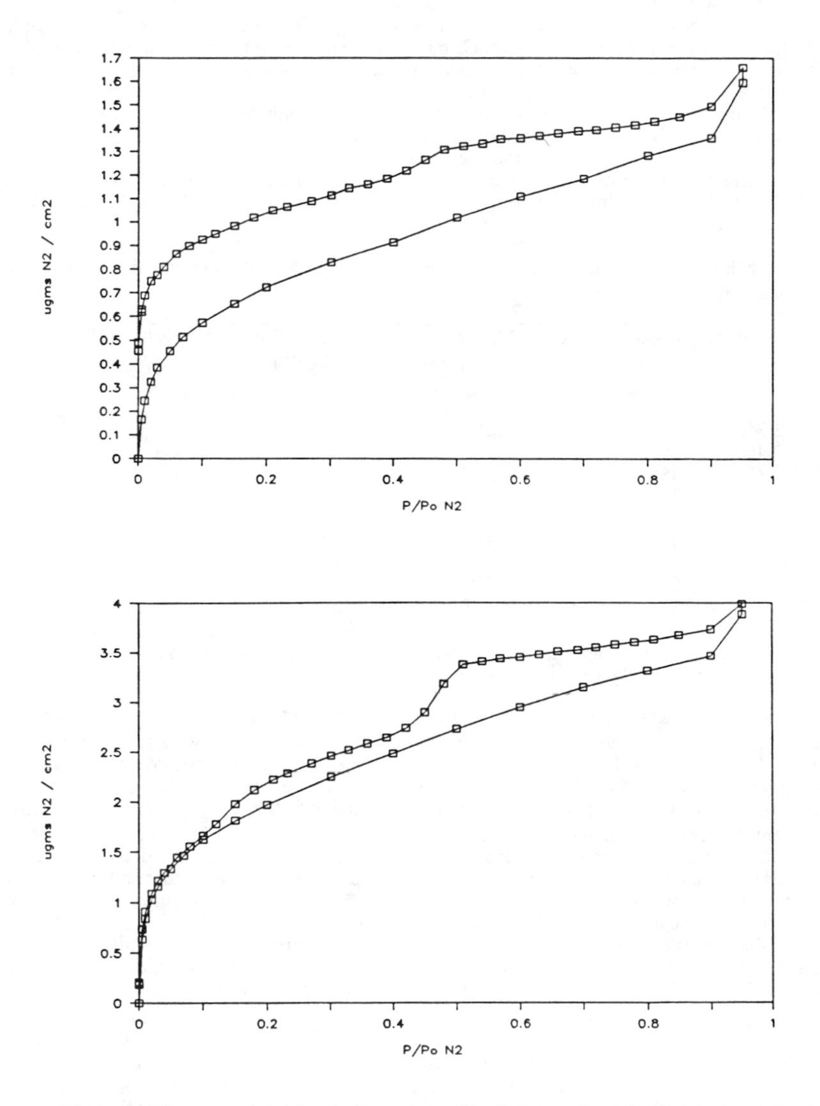

Figure 1. Nitrogen adsorption isotherms on silicalite crystals embedded in A2-derived glass films calcined at 350 °C (top), and calcined at 450 °C (bottom).

microporosity results from heating at higher temperature (0.021 vs. 0.014 g N_2/g). The other important difference between the two isotherms relates to the desorption behavior. After heating at lower temperature, the desorption branch is offset by a strong hysteresis similar to that observed with some of the glass films themselves (see above) which is probably associated with a tortuous pore network of diameters close to that of nitrogen. On heating to 450°C in oxygen, the hysteresis in the low pressure region all but disappears, and some minor hysteresis in the mesopore region remains. Apparently the oxygen treatment removes residual organics from the films such that access to the zeolites is more facile.

The microporosity of these composite films is substantially less than that expected for a film consisting of 100% accessible silicalite (0.17 g N_2/g). Since the weight fraction of the zeolite in the films is not known, it cannot be determined exactly how much of this decrease in microporosity must be attributed to potential pore clogging, ineffective acoustic coupling, or dilution with the glass film. Previous spectroscopic studies have established that zeolite porosity is accessible in similar films [2b].

Table 1. Microporosities and external surface areas of zeolite-glass composite films from nitrogen adsorption isotherms.[a]

Sample	Coating Mass $\mu g/cm^2$	Microporosity g N_2/g film	External Surface m^2/g film	Hysteresis at $p/p_o<0.1$
MFI-A2-350	39	0.014	36	+
MFI-A2-450	35	0.021	110	-
MFI-B2-350	45	0.029	62	+
MFI-B2-450	34	0.029	138	-
FAU-A2-350	55	0.011	64	-
FAU-A2-450	45	0.017	60	-
FAU-B2-350	51	0.020	35	+
FAU-B2-450	41	0.027	55	-

[a] Evaluated from α-plots of the desorption branches; when no hysteresis is present (-), the data are similar to those derived from the adsorption branches.

When the silicalite crystals are embedded in a B2-derived glass film, a comparison of the adsorption isotherms obtained after calcination at 350 and 450 °C show very similar trends as with the A2 glass (Table 1). Again, the adsorption isotherm resulting from lower temperature heating shows clearly much less adsorption in the micropore region (0.027 g N_2/g film at p/p_o=0.10) than that resulting from high temperature heating (0.059 g N_2/g film at p/p_o=0.10). The strong hysteresis disappears only on calcination at 450 °C. The B2-derived glass matrix results in more porous composites than the A2-derived films, as reflected in the higher microporosities and external surface areas, determined at both calcination temperatures. This is consistent with the known behavior of base-catalyzed TEOS sols which retain greater porosity on condensation due to their more compact aggregate structure.

Films derived from faujasite crystals show a different behavior with respect to calcination temperature, compared to silicalite. Figure 2 shows nitrogen sorption isotherms of FAU in A2-derived films. At both temperatures, no hysteresis is observed in the micropore region, but the microporosity still increases on heating to 450 °C, from 0.011 to 0.017 g N_2/g film. Similarly, little hysteresis is observed if a B2-derived FAU composite is heated to 350 °C, in contrast to the silicalite B2-film. As shown in Table 1, the microporosity of the FAU-B2 films is higher than that of the A2-derived films, but the external surface areas remain similar.

These results suggest that the sols do not affect the access to the larger FAU pores as much as they limit access into silicalite pores at low calcination temperatures where the glass may still contain residual organics. Thus, the adsorption behavior of the FAU films is not much affected by the different calcination treatments. Relative to the bulk micro-porosity (0.25 g N_2/g), that of the FAU-containing films is reduced more than that of the MFI films, possibly due to increased penetration of the sol into the larger FAU pores.

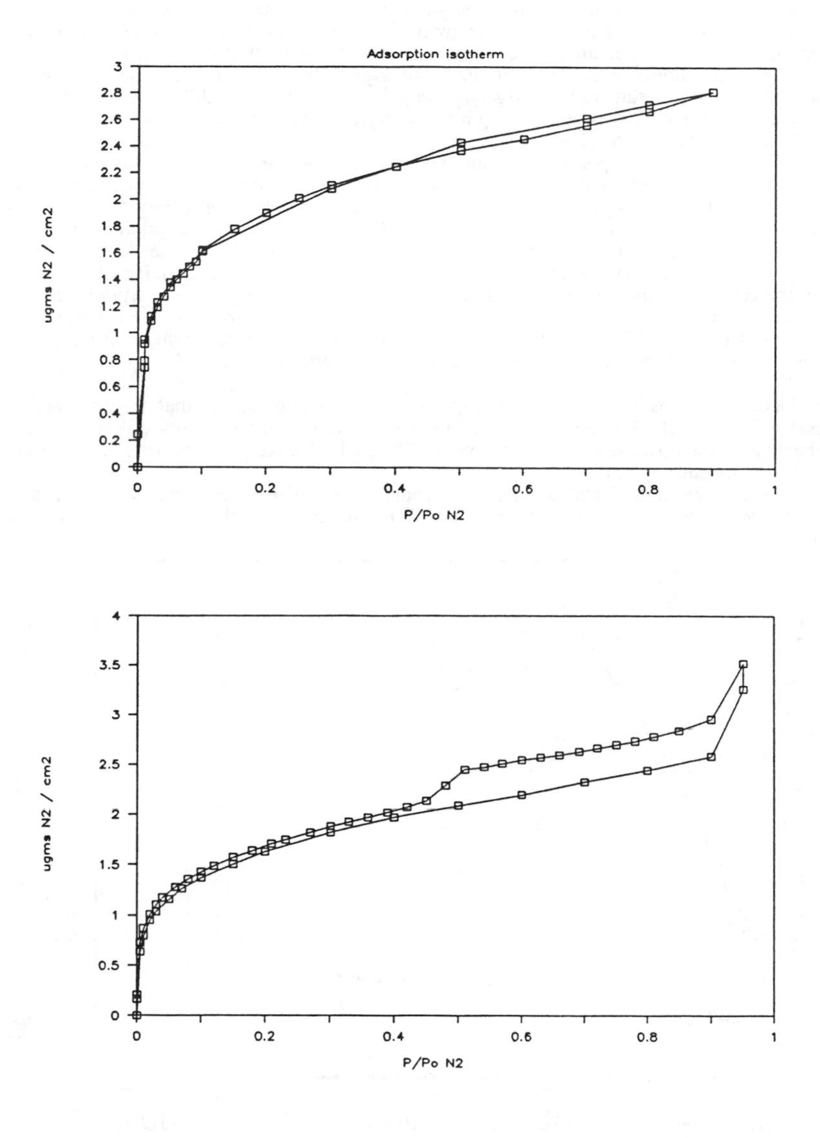

Figure 2. Nitrogen adsorption isotherms on faujasite crystals embedded in A2-derived glass films calcined at 350 °C (top), and calcined at 450 °C (bottom).

Ethanol selectivity. The selective vapor responses of a QCM coated with the microporous, hydrophobic silicalite composite layer (MFI-EtOH) are illustrated by the sorption of pure vapors of ethanol (ca. 4.3 Å diameter), 2,2,3-trimethylpentane (iso-octane, ca. 6.2 Å) and water (2.65 Å) in the range of 0-2000 parts per million (ppm) in moles (Figure **3**). The QCMs were pre-degassed at 170 °C in a helium flow for 20 min. The sorption of ethanol (top curve) shows the largest and fairly linear response as a function of vapor concentration. A 50.4 ppm ethanol vapor concentration results in 0.024 μg sorption per μg coating layer, 309 ppm cause further mass increase to 0.050 μg/μg, and 722 ppm to 0.076 μg/μg, which is about 1000 times higher than the alcohol response of a bare QCM.

In contrast, the response of the microdevice to iso-octane (bottom line) exhibits an almost negligible change with increasing vapor concentration. As the concentration of iso-octane is varied from 200 ppm to 700 ppm, the amount adsorbed changes from 1.66 ng/μg to 1.87 ng/μg, which is a mass increment of only 0.6% of that of alcohol sorption at the same concentrations. The results show that novel sensing layers with highly effective molecular sieving functions and very low external surface areas can be designed. Sorption of the small water molecules (middle line) shows also a small response. Although water sorption slightly increases with increasing concentration (at the two concentrations given above, the capacity is respectively 12.2 and 15.5 ng/μg), it already approaches a rectilinear isotherm at these low concentrations. This behavior differs drastically from the sorption of ethanol.

The microsensor design discussed in this communication shows that highly selective responses can be achieved when microporous adsorption is combined with tailored surface interactions in composite films. Exploration of additional selective interactions in microporous glass composites is in progress.

Funding from the National Science Foundation (Division of Materials Research) and from the Department of Energy (New Mexico WERC program) for this research is gratefully acknowledged.

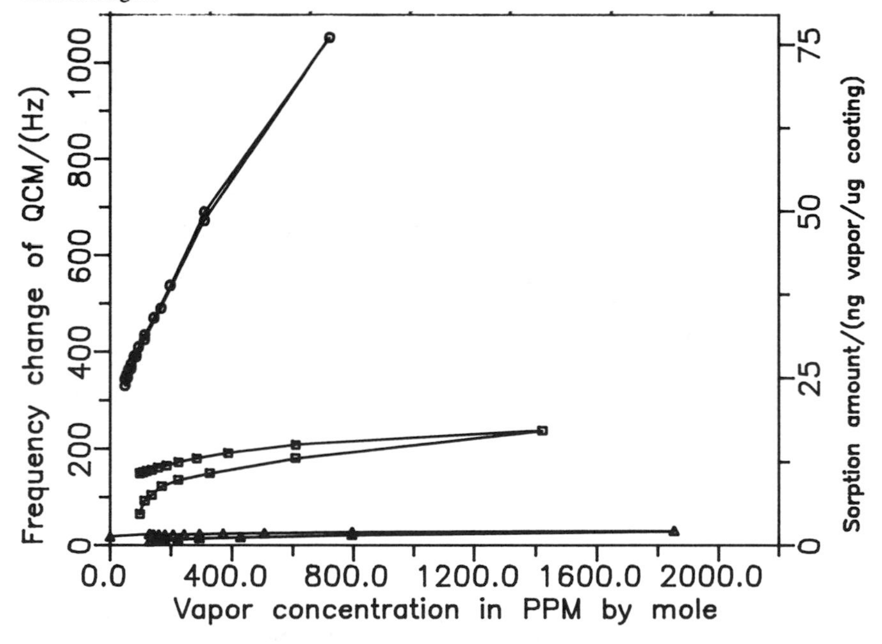

Figure 3. Sorption isotherms of ethanol (O), water (□) and isooctane (Δ) on a QCM coated with 168 μg/cm^2 of MFI-EtOH composite at 20 °C.

REFERENCES

1 For recent reviews see: (a) J. Janata, Anal. Chem. 62, 33r (1990). (b) R. C. Hughes, A. J. Ricco, M. A. Butler and S. J. Martin, Science 254, 75 (1991).

2 (a) T. Bein, K. Brown, G. C. Frye and C. J. Brinker, U. S. Patent Application 07/580,373, allowed on Jan. 23, 1992. (b) T. Bein, K. Brown, G. C. Frye and C. J. Brinker, J. Am. Chem. Soc. 111, 7640 (1989).

3 T. Bein, K. Brown and C. J. Brinker, Stud. Surf. Sci. Catal. 49, P. A. Jacobs, and R. A. Van Santen, Eds.; (Elsevier, Amsterdam, 1989), p. 887.

4 Y. Yan and T. Bein, J. Phys. Chem., in press.

5 $\Delta f = -2f_o^2 \Delta m A^{-1}(\rho_q \mu_q)^{-1/2}$, where Δf is the frequency shift, f_o the parent frequency of the QCM, Δm the mass change in g, A the piezoelectrically active area in cm^2, ρ_q the density (2.648 g cm^{-3}) and μ_q the shear modulus (2.947 $x10^{11}$ dynes cm^{-2}) for AT-cut quartz. See: G. Sauerbrey, Z. Physik 155, 206 (1959).

6 C. J. Brinker and G. W. Scherer, Sol-Gel Science, (Academic Press, San Diego, 1990), p. 110.

7 (a) E. M. Flanigen, J. M. Bennett, R. W. Grose, J. P. Cihen, R. L. Patton, R. M. Kirchner and J. V. Smith, Nature 271, 512 (1978). (b) W. M. Meier and D. H. Olson, Atlas of Zeolite Structure Types, 2nd Edition, (Butterworths, 1987).

8 T. Bein and Y. Yan, U.S. Patent in preparation.

STRUCTURE OF THIN SOL-GEL COATINGS NEAR THE SURFACE OF SUBSTRATES

HIROSHI HIRASHIMA AND TAKAO KUSAKA
Keio University, Faculty of Science and Technology
3-14-1, Hiyoshi, Kohoku-ku, Yokohama 223 Japan

ABSTRUCT

The sol-gel method is one of the most appropriate technologies for the preparation of functional thin oxide coatings. However, few studies on the structure near the interface between sol-gel coatings and substrates have been reported. In this study, the structures of thin TiO_2 gel coatings were investigated by XPS in order to elucidate the interaction between the coatings and the glass substrates. TiO_2 gel coatings were prepared from Ti-alkoxide by hydrolysis and dip-coating on silica glass plates. Thickness and refractive index measurements were made by ellipsometry. Changes in O1s spectra for the coatings were observed at locations near the interface. Dynamic microhardness measurements for the coatings and the coated glasses were made. These results suggest the formation of a Ti-O-Si bond at the interface.

INTRODUCTION

Thin films of transition metal oxides have interesting physical properties, e.g. high electrical conductivities, high dielectric constants, high refractive indices, etc.. They can be applied to microelectronics and integrated optics. The sol-gel process is one of the most appropriate technologies for the preparation of thin oxide films. Many works on the preparation, structure and physical properties of sol-gel coatings have been published. The structure near the interface, between the coating and the substrate, is important for the application and preparation of sol-gel coatings. However, only a few papers on the structure of the interface have been published [1, 2]. The formation of chemical bonds between silica sol-gel coatings and glass substrates has been suggested [1]. The authors previously reported the preparation, densification and crystallization of transition metal oxide sol-gel coatings, such as V_2O_5, TiO_2, Nb_2O_5 and Ta_2O_5 [3]. In this study, the structure of TiO_2 coatings near the interface has been investigated. X-ray photoelectron spectra, XPS, and dynamic microhardness of TiO_2 coatings were measured.

EXPERIMENTAL PROCEDURES

TiO_2 gel coatings were prepared from $Ti(iso\text{-}C_3H_7O)_4$, 99.9 % and supplied by Soekawa Rikagaku Co., Tokyo, by hydrolysis in ethanol solution and dip-coating. The concentration of $Ti(iso\text{-}C_3H_7O)_4$ was 0.2 mol/l. The amount of H_2O used for hydrolysis was 4 times theoretical. HCL, 0.05 mol to 1 mol $Ti(iso\text{-}C_3H_7O)_4$, was added as a catalyst. Clear sols were obtained were used for dip-coating. Silica glass plates were used as substrates. The withdrawal rate was about 6 cm/min. Dipping was repeated after drying at 90°C to increaee the thickness. After dipping and drying, the coatings were heat treated at various temperatures in the range from 200 to 900°C for 30 min. The thickness and refractive index of the coatings were measured by an automatic ellipsometer (Shimadzu, AEP-100). X-ray

 ©1992 Materials Research Society

diffraction measurements of the coatings after heat treatment were made using a rotating sample holder for thin films (Rigaku, 2651A1). X-ray photoelectron spectra of the coatings were measured after Ar ion etching using MgKα radiation (JEOL, JPS-90SX). The accelerating voltage for Ar ion was 400 V, and the etching rate was about 4 nm/min. The binding energy, BE, was calibrated using a vapor-deposited Au film on the sample surface (Au4f lines). The O1s spectra were deconvoluted using Gaussian lineshapes after smoothing, background subtraction and satellite removal.

Dynamic microhardness measurements were made using a trigonal diamond pyramid indenter (Shimadzu, DUH-200), and calculated according to

$$DH = \alpha \frac{P}{D^2} \tag{1}$$

where DH is dynamic microhardness, α is a constant, α = 37.838 for trigonal pyramid indenter, P is the load, and D is the indent depth. P is loaded with an electric balance. D was measured continuously for increasing P, in the range from 0 to 20 g. The loading rate was 0.72 g/s. The distribution in DH values for the coatings and coated glass were made in the depth direction. The distribution in microhardness for sub-micron range can be easily measured in short time by this method.

RESULTS AND DISCUSSION

Thickness and refractive index values for TiO_2 gel-derived coatings varied remarkably with increasing temperature of heat treating in the range 200 to 400°C (Fig. 1). Gel coatings heat treated at temperatures less than 300°C were amorphous to X-ray diffraction (Fig. 2). Diffraction peaks for crystalline TiO_2, Anatase, were found for coatings heat treated at 400 to 900°C. Significant increases in values of refractive index for heat treatment up to 400°C may be attributed to densification of the coatings by evaporation of solvent and to crystallization of the gel. On the other hand, refractive index value for coatings hardly changed after heat treatment from 500 to 700°C. The increase in the refractive index after heat treatment at 800°C may be attributed to densification by sintering. The phase transition from Anatase to Rutile was not observed after heat treating at 900°C.

An example of the distribution in DH near the interface between gel coatings and the glass substrate is shown in Fig. 3. The thickness of the coating was small, about 100 nm. Therefore, very small loads, less than a few g, were needed to measure DH for the coatings, and the reproducibility of the measurement was not good. Evaluation of DH for the coatings was

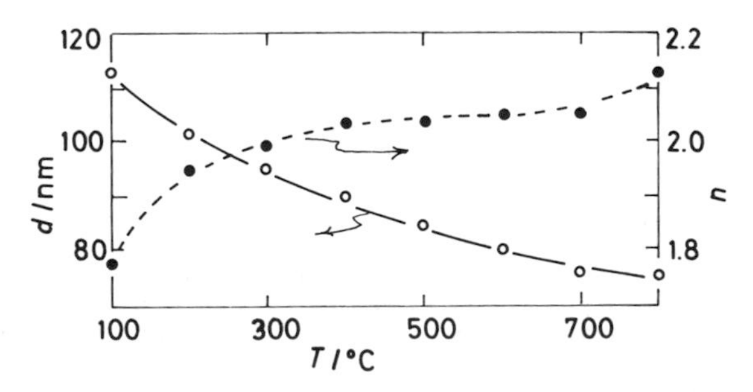

Fig. 1 Effect of heat treatment temperature on thickness, d, and refractive index, n, for TiO_2 gel coatings. (o; d and •; n)

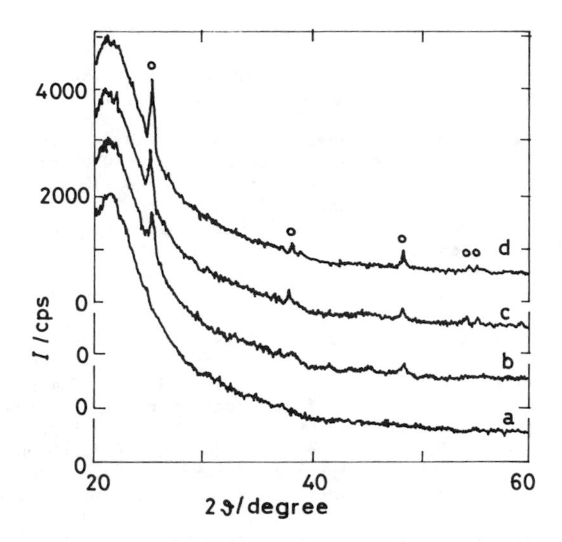

Fig. 2 X-ray diffraction patterns for TiO_2 gel coatings after heat treatment at (a) 300°C, (b) 400°C, (c) 500°C and (d) 900°C, for 30 min. (CuKα, o; Anatase)

difficult. On the other hand, reproducibility of DH measurements for the silica glass plate was good, about 5 % or less, because of relatively high DH values. DH of the silica glass substrates, at 1 μm from the interface, is shown as a function of the thickness for the coatings (Fig. 4). DH values of the substrate increased remarkably when the thickness of the coating was more than 100 nm. DH is affected by elastic and inelastic factors for solids. The increment in DH for the coated glass near the interface suggests

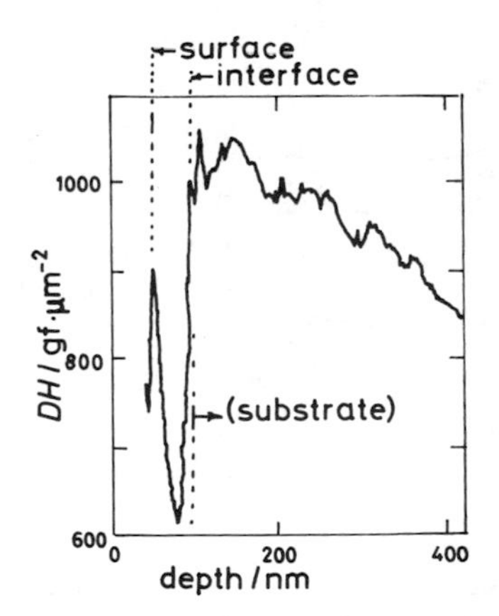

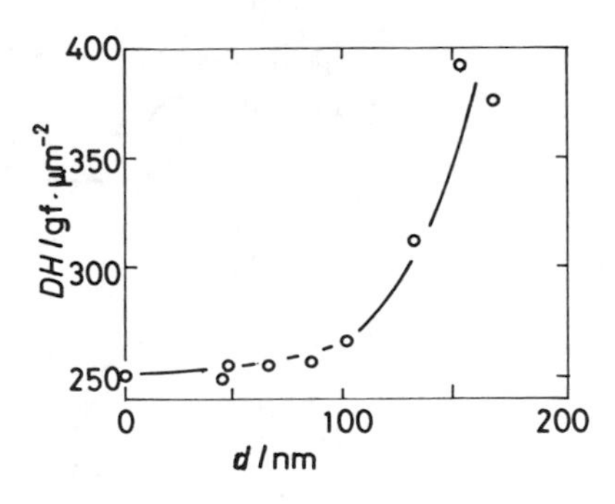

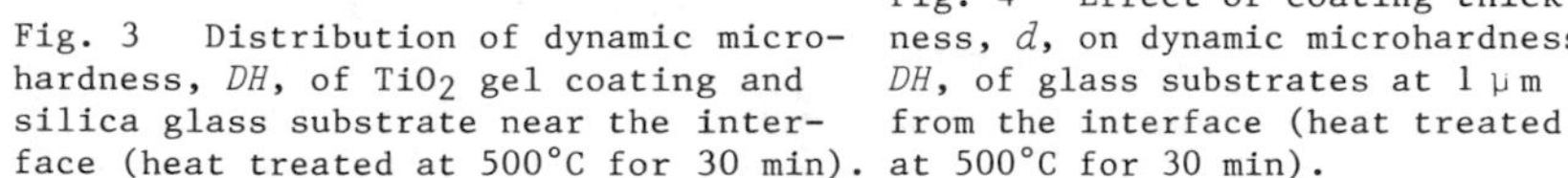

Fig. 3 Distribution of dynamic microhardness, *DH*, of TiO_2 gel coating and silica glass substrate near the interface (heat treated at 500°C for 30 min).

Fig. 4 Effect of coating thickness, *d*, on dynamic microhardness, *DH*, of glass substrates at 1 μm from the interface (heat treated at 500°C for 30 min).

that any kind of stress induced by the gel coating exists in the glass substrate in this region. Stress in gel coatings has been reported [4]. The stress in the substrate may be caused by coating thicker than about 100 nm. This result also suggests good adhesion for the gel coating to the glass substrate.

Fig. 5 shows an example of changes in XPS for TiO_2 gel coatings determined by Ar ion etching. After Ar ion etching, low binding energy peaks for Ti2p spectra, which may be attributed to the low valent Ti ions, were observed. This result shows that Ti^{4+} ion in the coating is reduced by Ar ion etching. Therefore, it is difficult to discuss the coordination state of Ti ion in the coating from Ti2p spectra.

O1s spectra of the TiO_2 coatings consist of 2 or more peaks (Fig. 6). Near the surface of the coating, O1s spectra consist of 2 peaks, Peak 1, BE $\cong$ 530 eV, and Peak 2, BE $\cong$ 532 eV (Fig. 6a). These two peaks were also observed for crystalline TiO_2 (Anatase). This result coincided with the X-ray diffraction data for the gel coating heat treated at 500°C (Fig. 2c). Peak 1 can be attributed oxygen in the Ti-O-Ti bond [5]. The peak area for Peak 2 increased in the inner layer of the coating (Fig. 7). The binding energy of monovalent oxygen, such as OH^- and O^-, has been reported to be about 532 eV, and is similar to that of Peak 2. This result shows that the concentration of residual OH^- is high in the inner layer, and/or that O^- is formed by reduction of TiO_2 by Ar ion etching. It is not yet clear which is dominant.

Near the interface, where the Si2p spectrum was observed, Peak 2 became smaller and then disappeared (Fig. 6c). This result suggests the structure of the TiO_2 gel coatings near the interface is different from Anatase or Anatase-like structure. For the interface layer, both Si and Ti signals were observed. The interface layer was about 20 nm thick. This was larer than the surface roughness of the glass substrate, which was about 6 nm.

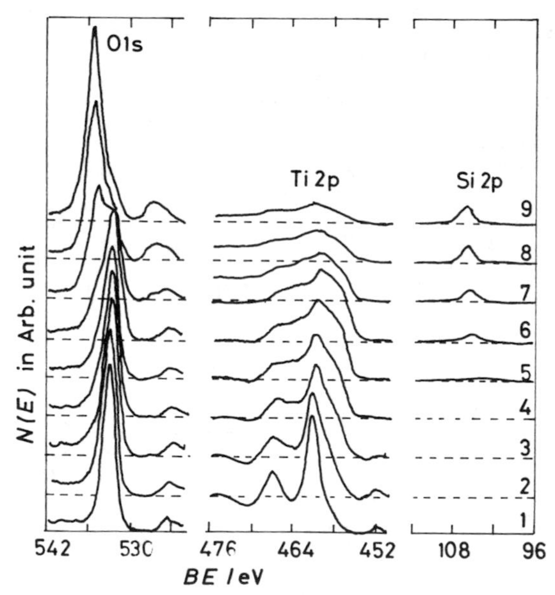

Fig. 5 XPS of TiO_2 gel coatings, heat treated at 500°C for 30 min, after Ar ion etching for (1) 0 s, (2) 90 s, (3) 180 s, (4) 270 s, (5) 360 s, (6) 450 s, (7) 540 s, (8) 630 s, and (9) 720 s.

These results suggest that an interface layer, which has a different structure from crystalline TiO_2, is formed, and that a Ti-O-Si bond is formed in this layer. The good adhesion of the coating, suggested by DH measurements, may be attributed to the formation of a Ti-O-Si bond. The formation of a Si(in coating)-O-Si(in glass) bond at the interface between silica gel coating and glass substrate has been suggested [1]. However, further investigations are necessary to confirm the formation of a Ti-O-Si bond. Peak 3, BE ≅ 533 eV, was also observed in this region. This peak can be attributed to bridging oxygen in silicate glasses [7].

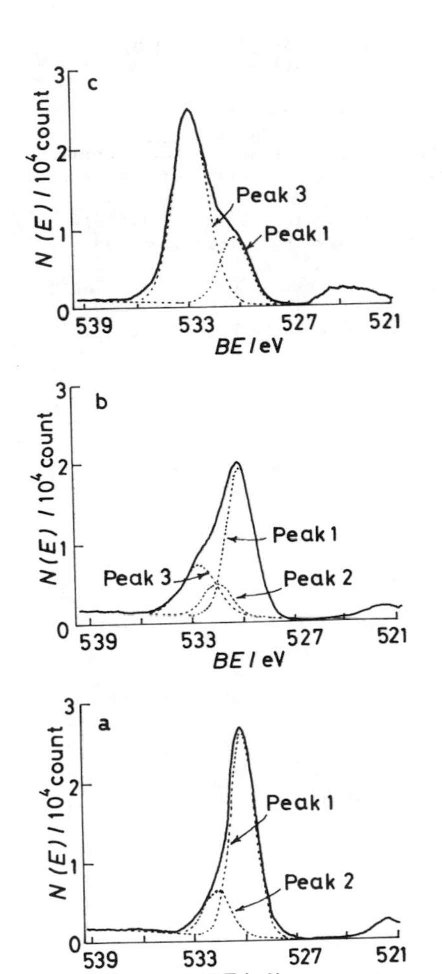

Fig. 6 O1s spectra of TiO_2 gel coatings, heat treated at 500°C for 30 min, after Ar ion etching for (a) 0 s, (b) 450 s, and (c) 630 s.

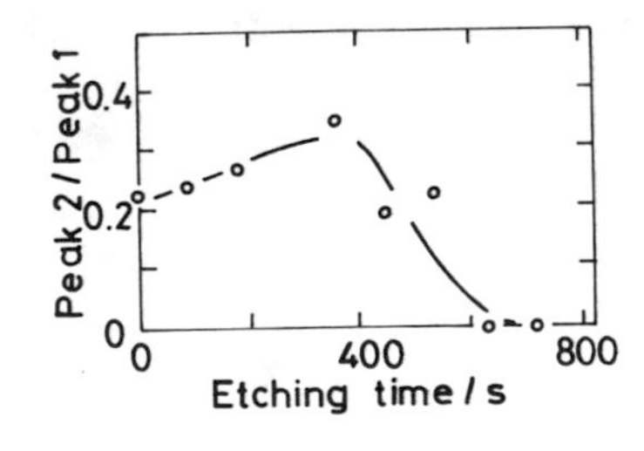

Fig. 7 Effect of etching time on the peak area ratio of [Peak 2]/[Peak 1] of TiO_2 gel coating heat treated at 500°C for 30 min.

CONCLUSIONS

(1) TiO_2 gel coatings crystallize at temperatures higher than 400°C.
(2) Dynamic microhardness of the glass substrates was increased for TiO_2 gel coatings thicker than 100 nm.
(3) Ols peak profiles of XPS show that the structure of the TiO_2 gel coating near the interface is different from the structure in the outer layer. The formation of a Ti-O-Si bond is suggested.

REFERENCES

1. T.A. Michalske and K.D. Keefer, in Better Ceramics Through Chemistry III, edited by C.J. Brinker, D.E. Clark, and D.R. Ulrich (Mater. Res. Soc. Proc. 121, Pittsburgh, PA 1988)pp.187-197.
2. B.D. Fabes and W.C. Oliver, J. Non-Cryst. Solids, 121, 348 (1990).
3. H. Hirashima, R. Muratake, T. Yamashita, and T. Chiba, in Better Ceramics Through Chemistry IV, edited by B.J.J. Zelinski, C.J. Brinker, D.E. Clark, and D.R. Ulrich (Mater. Res. Soc. Proc. 180, Pittsburgh, PA 1990)pp.611-616.
4. K.A. Cerqua and J.E. Hayden, J. Non-Cryst. Solids, 100, 471 (1988).
5. Y. Kaneko, M. Yamane, and Y. Suginohara, Yogyo-Kyokai-Shi, 89, 599 (1981).
6. L.M. Moroney, R.St C. Smart, and M. Wyn Roberts, J. Chem. Soc., Faraday Trans. I, 79, 1769 (1983).
7. Y. Kaneko and Y. Suginohara, Nippon-Kinzoku-Gakkai-Shi, 42, 285 (1978).

LASER DENSIFICATION OF SOL-GEL DERIVED TiO_2 - THIN FILMS.

N. ARFSTEN, B. LINTNER, M. HEMING, O. ANDERSON AND C.R. OTTERMANN

* SCHOTT Glaswerke, P.O.B. 2480, D-6500 Mainz, FRG

Abstract

Laser substrate heating is discussed as an alternative for the densification of sol - gel thin films. Homogeneous films of n(550 nm) = 2.30 could be obtained, keeping the bulk substrate temperature below 80 °C.

Introduction

One of the main applications of sol - gel technology today, is the coating of large area (up to 3 x 4 m) flat - glass plates, e.g. with solar reflective or anti - reflective coatings [1]. The production process involves the dipping of a thoroughly precleaned glass plate into the coating solution, withdrawal of the plate under controlled atmospheric conditions at a constant speed and thermal conversion of the adherent film into an oxide layer. The thermal conversion is performed in an oven process at temperatures up to 400 - 500°C. The speed - limiting factor of this process is the maximum attainable cooling rate of the flat - glass plate.

To overcome this limitation, we have investigated the possibility of laser densification as an alternative process [2]. In the current contribution, we report the results of CO_2 - laser densification of titania thin films.

Experimental

Alcoholic solutions containig $TiCl_2(OC_2H_5)_2$ [3] were spin - coated on clean float and borosilicate glass (TEMPAX) substrates with a thickness of 3 mm. As - spun film thicknesses were on the order of 100 – 200 nm. The laser - densification was performed at 10.6 μm under controlled atmospheric conditions using a BLS60C CO_2 - laser with two crossed galvano mirror scanners for two dimensional beam scanning. The maximum deflection speed of the laser beam is $\approx$ 2 m/s. The focused beam (TEM_{00} - mode, 60 Watt integral output power, $d_{1/e}$ = 0.5 mm) delivers up to 30 kW/cm^2 to the sample. Refractive indices and layer thicknesses were derived from transmission and reflection measurements. TEM - pictures of the TiO_2 - films (lifted off in 0.2% HF) were taken on a ZEISS EM 10 CR electron microscope at 100 keV acceleration voltage. SIMS and IBSCA [4] were used to obtain information about the elemental depth profiles. Hydrogen depthprofiles were calculated from NRA experiments [4]. Grazing incidence X - ray diffraction experiments on a modified SIEMENS D - 500 were performed to obtain information concerning the film microstructure.

Laser substrate heating

Large area substrates cannot be densified with a single laser shot. Instead a scanning laser beam system has to be used. To control the temperature distribution within the substrate, it is necessary to know its functional dependence on the process parameters: applied power, beam profile and velocity. To calculate the time dependent temperature distribution under a scanning laser beam (fig. 1), one has to solve the differential equation:

$$\frac{\delta T}{\delta t} - D \cdot \nabla^2 \cdot T = \frac{Q(x, y, z, t)}{C_p} \qquad (1)$$

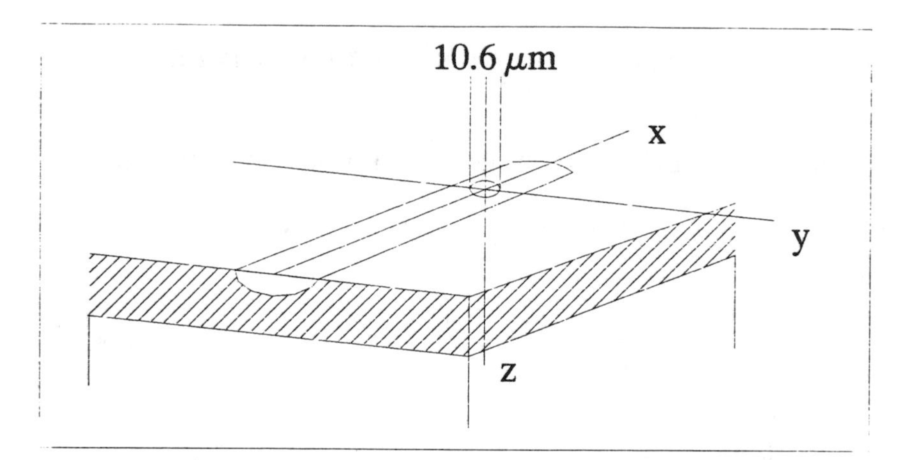

Figure 1: Schematic of the sol - gel laser densification via CO_2 - laser beam scanning in x - direction.

where D is the thermal diffusivity ($5 - 6.5 \cdot 10^{-3}$ cm^2/s) and C_p is the specific heat per unit volume ($1.7 - 1.9$ $J \cdot K^{-1} \cdot cm^{-3}$). With the laser operating in the TEM_{00} - mode and scanning in x - direction, the absorbed power density Q reads:

$$Q(x, y, z, t) = \frac{P(1-R)}{2\pi r^2} \cdot e^{-[((x-vt)^2+y^2)/2r^2]} \cdot \frac{e^{-\beta z}}{\beta} \tag{2}$$

where P is the total applied power, R the reflection coefficient of the substrate/film surface ($\approx 10 - 20\%$ [4]), v the velocity and r the radius of the beam. β is the extinction coefficient of the substrate, which amounts to ≈ 1000 cm^{-1} for silicate glasses. Thus, most of the energy is absorbed within a depth of $10\mu m$, whereas only $< 1\%$ of the energy is absorbed within the sol - gel film. For temperature distribution calculations, it is sufficient to assume, that the energy is deposited in an infinitesimal thin top - layer.

Eq. (1) has been solved using Green's function technique [5] under the assumption, that K and D are temperature independent[1]. The temperature along the scanning beam in depth of 0, 0.05 and 0.1 times the beam radius for two different beam velocities is shown in fig. 2. The absolute temperature is given in units of thermal conductivity times beam radius to absorbed power. The in - depth distribution is mainly governed by the ratio of the velocity of the heat source to the velocity of the generated heat wave (rv/D). To restrict the temperature to a depth of 10% of the beam radius, this ratio should be > 500. This operation mode was used throughout the investigation.

The laser heating affects the substrate itself:

- As shown in fig. 3, heating the surface of a TEMPAX - substrate up to 650°C with local heating times on the order of 0.6 ms is already sufficient to cause a decrease of the leached surface layer thickness, a dehydration of the substrate surface and an increase of the Na^+ - concentration at the substrate surface.
- Increasing the deposited energy beyond a threshold value, we observe surface melting with the accompanying formation of wall - type, relief structure (see fig. 4). The wall height increases with delivered energy. In the case of float glass, the creation of the relief structure occurs only in a very narrow parameter

[1] We found, by using the T -dependent K - data of TEMPAX [7] that this procedure overestimates the surface temperature in the T - interval of 300 – 700 °C by 10 – 15%.

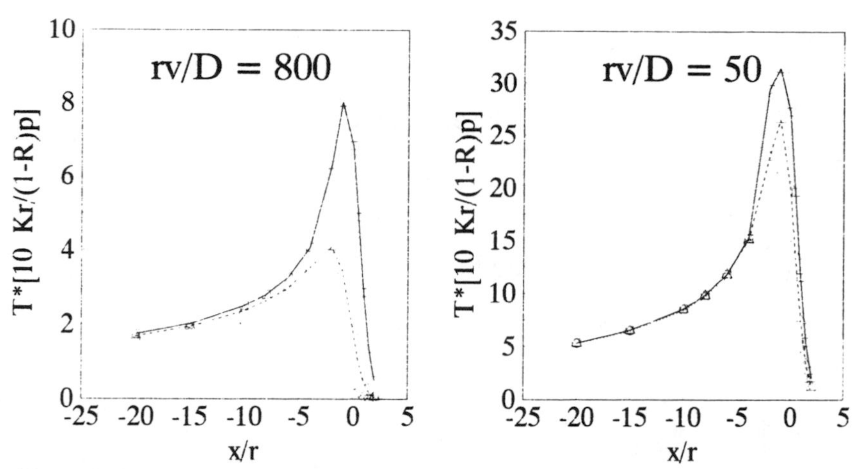

Figure 2: Temperature distribution along the scanning laser direction for the ration rv/D = 800 (left) and 50 (right). The laser currently irradiates the spot x/r = 0. The temperature is shown in units of Kr/(1-R)P. K: thermal conductivity. The temperature along the scan is shown for depths according z/r = 0, 0.05 and 0.10.

window. Due to the larger thermal expansion coefficient, the material tends to crack and we observed spots of evaporated glass components on the surface close to the laser track.

- On the other hand, we could not detect any stress relief in pre - hardened flat glass substrates; even for parameters, which induce surface cracks.

TiO_2 - densification: Optical properties

Taking the refractive index as an indicator for the density of the films, we were able to obtain visibly homogeneous films up to n (550 nm) = 2.30. Compared to the as - spun films, the thickness decreased by ≈ 60%. Interferometric measurements yield a rms surface roughness of 0.7 ± 0.2 nm. Thus, the scanning mode did not induce any artificial surface topology. Although the films withstood higher deposited energy densities, peeling of substrate surface layers prevented the formation of highly densified films.

The dispersion relation of the laser densified films is rather similar to those for titania deposited by other methods (see fig. 5).

The refractive index of the densified material closely correlates with its extinction coefficient k. At n = 2.2, k - values of ≈ 0.01 are observed. Again, this correlation seems to be independent, whether the material is densified in an oven process or using a scanning laser beam (see fig. 5).

Structured films for decoration purposes have been obtained using a simple mask technique and washing off the undensified parts with appropriate solutions.

TiO_2 - densification: Film microstructure

The short local densification times have a drastic effect on the film microstructure. Typical TEM - pictures of comparable oven and laser densified films (d = 85-87 nm, n = 2.24-2.26 at 550 nm) are shown in fig.6. Whereas crystallites up to 50 nm diameter have been observed in oven densified materials, the laser densified material is x - ray amorphous. The electron diffraction diagram shows indications of the anatase structure, with maximum crystallite size of 5 – 10 nm.

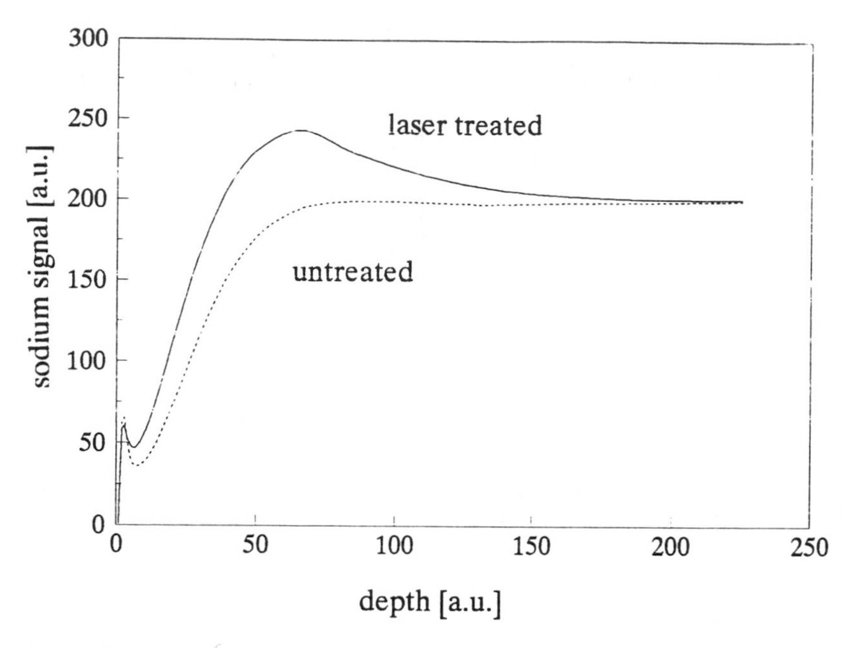

Figure 3: SIMS - Na - depth profiles of a TEMPAX glass substrat. The laser treatment (650°C surface temperature, 0.6 ms local heating time) reduces the leached substrate surface layer and increases the Na - concentration at the substrate surface.

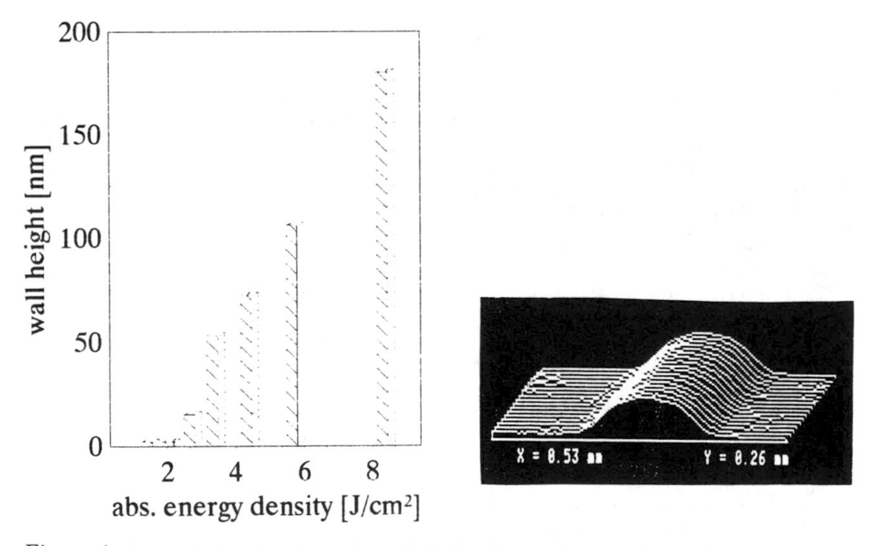

Figure 4: Laser - induced surface melting: Relief height as a function of total deposited energy in a single track (left) and relief structure along the laser track at 3 J/cm^2 (right, relief height 72 nm). Substrate glass was TEMPAX.

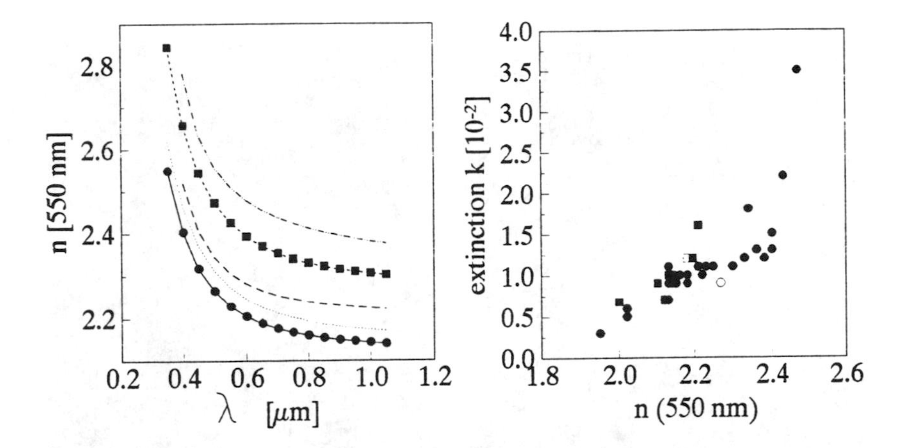

Figure 5: Left: Dispersion relation of densified titania thin films compared with those deposited by other methods.: From top to bottom Ion plating, laser densified (cracked), reactively evaporated, oven densified, laser densified. Data taking partially from [8]. Substrates with densified films of n = 2.4 exhibited a large number of cracks. Right: Extinction coefficient vs. refractive index for a variety of laser and oven densified titania thin films. Open/closed circle: Oven/laser densified on TEMPAX; open/closes squares: Oven/laser densified on float glass.

Sodium depth profiles indicate, that no Na^+ - diffusion occurs into the titania thin films during the laser treatment.

H/Ti ratios depends largely on the laser parameter (see fig. 7). With increasing n, the hydrogen content decreases [8]. A typical value for films densified to n = 2.23 is H/Ti = 0.16. Chlorine contaminations are on the order of 1 - 3% as determined from RBS measurements.

Conclusion

The results of this investigation demonstrate:

- With 10.6 μm radiation it is possible to densify titania thin films up to densities comparable to those obtained in a conventional oven process. The local densification occurs on a several orders of magnitude faster time - scale, than in an oven process.
- Homogeneous, smooth films are obtained by line scanning.
- The restriction of the heated substrate zone to a depth of about 10% of the beam radius keeps the bulk glass temperature below 80 °C. The treatment of pre - hardened glass substrates without stress relief is possible.
- Structured films for decoration purposes are easily obtainable, by using mask techniques and washing off the undensified film.
- Extinction coefficients of the titania films are correlated with the refractive index; they do not differ significantly from those obtained by conventional densification, in spite of the fast densification process.
- Due to the short local heating times, we do observe a Na - enrichment at the substrate surface, but no diffusion of Na into the thin film.

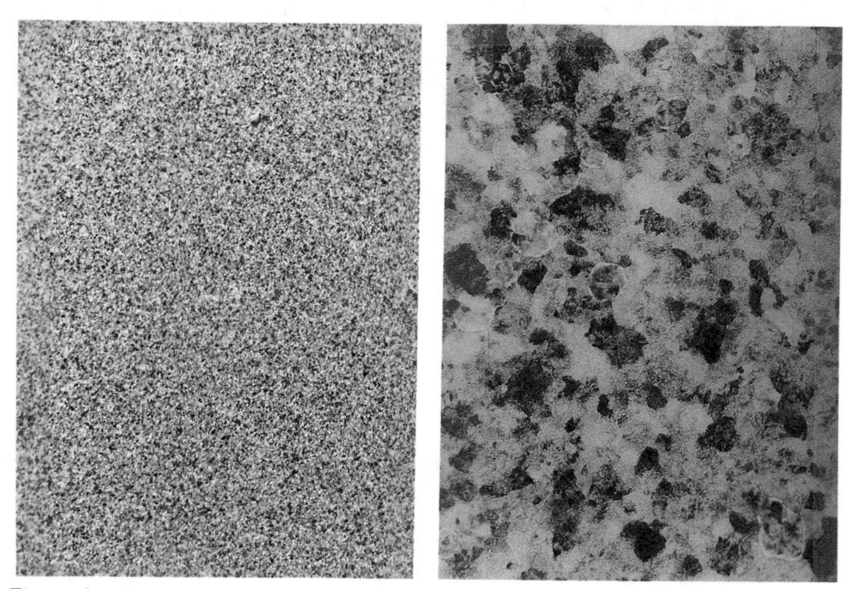

Figure 6: TEM - pictures of laser (left) and oven (right) densified titania thin films with comparable optical properties.

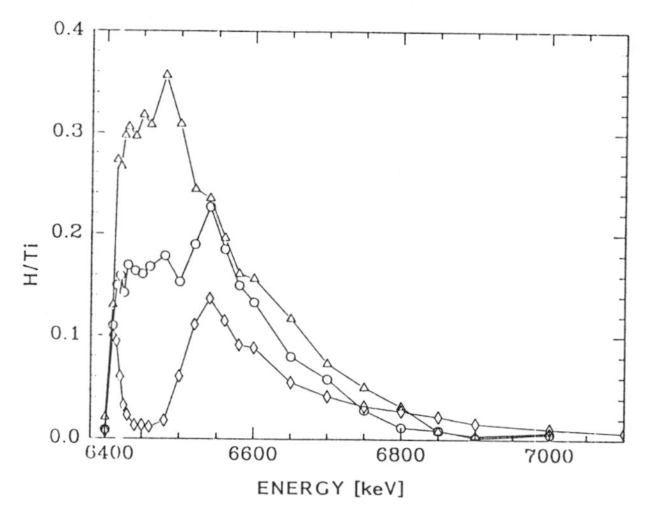

Figure 7: NRA - hydrogen depth profiles for three different laser treated samples. The hydrogen concentration decreases with increasing refractive indices: 2.15, 2.23 and 2.43 (cracked sample).

- Titania crystallizes in the anatase modification; the average grain size is a factor of 5 - 10 smaller, than those obtained in the oven process.

- The hydrogen content closely correlates with the refractive index.

Thus laser densification is a viable alternative to the conventional oven process, especially for the coating of temperature sensitive materials, like laminates, and for the production of structured coatings.

References

[1] H. Dislich in: Sol-gel technology for thin films, fibers, preforms, electronics, and specialty shapes. L.C. Klein, ed., Noyes Publ., 1988, Ch. 4.

[2] N. Arfsten, H. Piehlke, E. Hußmann: DE 37 44 368, 1987.

[3] H. Schröder, in: Physics of Thin Films (G. Hass, ed.). 5, 87 (1969).

[4] H. Bach. Glastechn. Ber. 56,1 (1983); ibid. 56,29 (1983).

[5] C. Buerhop, B. Blumenthal, R. Weissmann, N. Lutz and S. Biermann. Appl. Surf. Science 46, 430(1990).

[6] H.E. Cline, T.R. Anthony. J. Appl. Phys. 48, 3895(1977).

[7] Y.S. Toulonkian, R.W. Powell, C.Y. Ho, P.G. Klemens, in: Thermophysical properties of materials, Plenum Press. Vol. 12, 922(1970).

[8] K. Bange, C.R. Ottermann, O. Anderson, U. Jeschkowski, M. Laube and R. Feile. Thin Solid Films. 197, 279(1991)

CRYSTALLINE THIN AND/OR THICK FILM OF PEROVSKITE-TYPE OXIDES BY HYDROTHERMAL ELECTROCHEMICAL TECHNIQUES

Masahiro YOSHIMURA
Research Laboratory of Engineering Materials, Tokyo Institute of Technology, Nagatsuta, Midori, Yokohama, 227, Japan

ABSTRACT

Well crystallized polycrystalline perovskite-type oxide thin and/or thick films were prepared at sufficiently low temperatures by newly developed "hydrothermal electrochemical techniques", where metals were electrochemically oxidized and reacted with some components in hydrothermal solutions. $BaTiO_3$ films 70 to 300 nm thick were formed in $Ba(OH)_2$ solution at 100-200℃ under saturation vapor pressure on the Ti substrate and Ti deposited glass substrates. The electrical current enhanced to thicken their films. $SrTiO_3$, $BaFeO_3$ and $LiNbO_3$ films were also prepared.

INTRODUCTION

The perovskite-type oxides such as $BaTiO_3$, $SrTiO_3$, $PbTiO_3$, etc. have attracted much attention in electroceramics: capacitors, thermistors sensors, actuators, optoelectronic materials, etc. Recent demands to miniaturize and/or upgrade require the fabrication of their thin films. Fabrication methods of thin films are conventionally divided into two categories 1)dry method such as vaccum deposition, sputtering, ion plating and CVD, etc. and 2)wet method: dip, spray coating of organic/inorganic precursors, and/or sol/gel method [1, 2].

These methods generally require a high temperature above 500 ℃ or higher during and/or after fabrication of the film, because as-deposited films would be amorphous thus need the heating to crystallize, or even in mixture states of precursors which must be heated to react to produce the perovskite-type compounds. Those heating would bring about pinholes, cracking and/or peeling in films and/or reaction with substrates. Furthermore these conventional processing methods are two-step or multi-steps: that is film formation and production of compounds, or production of compounds in powders then film formation, or film formation in amorphous then crystallization.

During developing the hydrothermal processing to fabricate well crystallized fine powders having homogeneous composition, size and shape, we have applied anodic oxidation under hydrothermal processing[3, 4]. The anodic oxidation is generally used for the fabrication of amorphous oxide thin film of bulb metals (Ta, Nb, Ti, Al, etc.). It was believed, however, that only simple oxides could be fabricated. We have experienced that hydrothermal reactions using high temperature solutions can provide well crystallized double oxide materials which might have reacted with the component(s) in the solution. Thus we have combined these two methods expecting that cations in the solution can be incorporated into the anodic oxide film to yield double oxides in the crystallized form on the substrate. we have named this new technique as hydrothermal electrochemical technique [5] where thin film of double oxides as perovskite can be prepared at relatively low temperature, 200℃ or below, by *in situ* reaction on the substrate. This paper summarizes this hydrothermal electrochemical technique.

HYDROTHERMAL ELECTROCHEMICAL METHODS

This technique provides the film of double oxides on a metal, alloy or oxide as an electrode substrate which react with cation component(s) in the high-temperature solution chemically and/or electrochemically by passing electrical current. Therefore this method may involve following reactions including the originally considered one [5]:

 ©1992 Materials Research Society

1. Hydrothermal reaction between anodically oxidized film with solution component(s),
2. Electrochemical reaction with solution component (s),
3. Electrolytic oxidation of dissolved species and hydrothermal reaction,
4. Hydrothermal reaction between anodically dissolved species and solution component(s),
5. Electro-discharge reaction between anodic oxidized film with solution component(s),
6. Electro-deposition of charged oxide particles.

Thus these methods possess following characteristics
1) Complex oxide thin film can be prepared.
2) The films are prepared at relatively low temperatures.
3) Complex oxide films are directly (in situ) fabricated on the substrate.
4) The film is crystalline.
5) The growth rate of the film is relatively high.

PREPARATION OF $BaTiO_3$THIN FILM

$BaTiO_3$ thin film was formed on Ti substrate when it was treated hydrothermally in $Ba(OH)_2$ solution passing electrical current [5] . Figure 1 illustrates the electrochemical cell to fabricate the $BaTiO_3$ thin film. A Ti plate is used as the anode with the DC power supply, but two Ti plates as electrodes can be used with AC power supply. The produced film has lustrous with various beautiful colors, blue, violet, gold, etc., depending on the experimental conditions. Adhesion of the film on the metal substrate was so strong that no exfoliation was observed after bending the specimen plate. According to the X-ray diffraction analysis (Fig. 2), only poorly crystallized anatase was detected in the product at a room temperature. Well crystallized $BaTiO_3$ was observed on the Ti substrate treated above 100℃ even after a short period as 30 min.

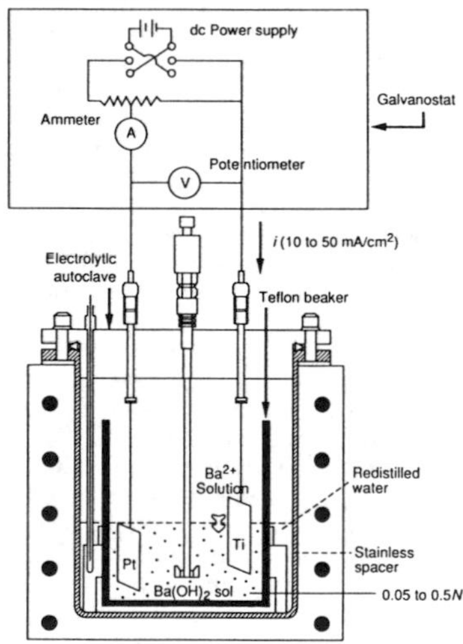

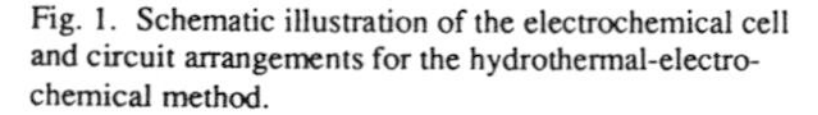
Fig. 1. Schematic illustration of the electrochemical cell and circuit arrangements for the hydrothermal-electrochemical method.

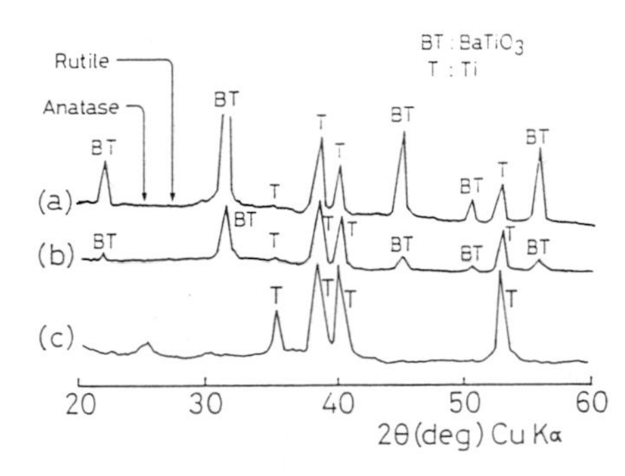

Fig. 2. XRD patterns of the films produced by hydrothermal-electrochemical method in 0.5N-$Ba(OH)_2$: (a)200℃, 30 min. (b)100℃, 30 min, (c)room temperature, 60 min.

The symmetry of the $BaTiO_3$ was considered to be cubic since no obvious peak splitting, which is typical for the tetragonal and other forms, was observed. In addition no preferred orientation of $BaTiO_3$ grains was detected when a polycrystalline Ti plate was used as a substrate.

The thickness of the film increases with the increase of (1) treating temperature, (2) treating time, (3) Ba concentration, and (4) current density. As shown in Fig. 3 [6, 7], it increased from 70 to 300 nm with treating time from 10 to 60 min. under the conditions at

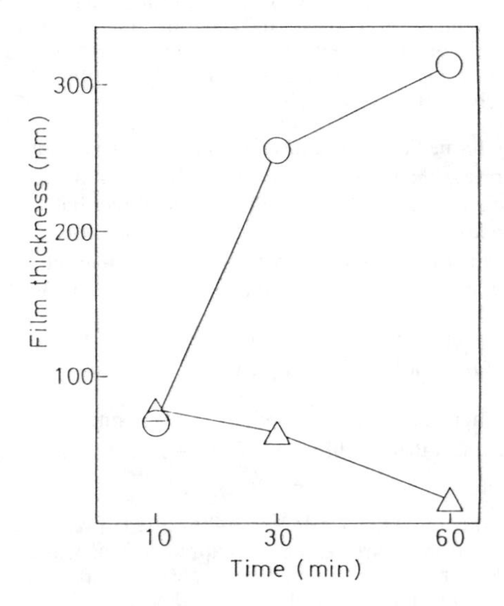

Fig. 3. Changes of the thickness of $BaTiO_3$ film against the treatment time, estimated from the weight change. Films were formed at 200℃ in 0.5N-$Ba(OH)_2$ with current density of 15(○) and 0 (△) mA/cm^2.

200℃ with the current density of about 15 mA/cm^2 in an 0.5N - $Ba(OH)_2$ solution. The thickness was estimated from the weight gain of the substrate, which was almost comparable with the results of SEM observation. Thin film of $BaTiO_3$ was also formed in $Ba(OH)_2$ solutions under a pure hydrothermal condition without any electric current. It is considered to be formed due to the presence of oxide layer on the Ti metal substrate because the surface of Ti metal is generally covered by the oxide layer. It was less than 70 nm, and it decreased with time suggesting probable dissolution in the $Ba(OH)_2$ solution. At higher temperatures under hydrothermal conditions, $BaTiO_3$ could be grown to μm size with tetragonal symmetry [8], and moreover an epitaxially grown $BaTiO_3$ film could be prepared on the $SrTiO_3$ substrate [9, 10].

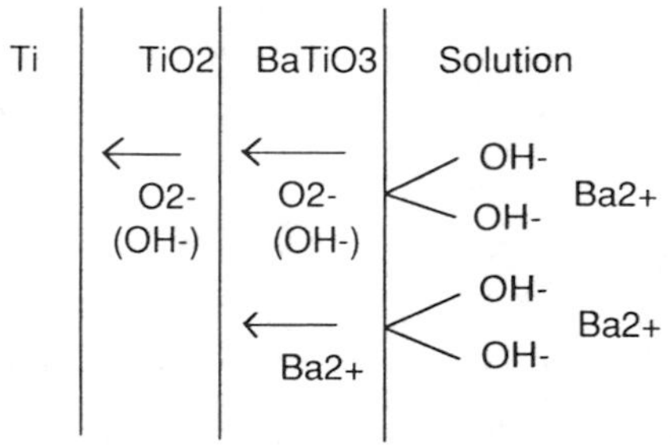

Ti | TiO2 | BaTiO3 | Solution

← O2- (OH-) | ← O2- (OH-) | OH- OH- Ba2+

← Ba2+ | OH- OH- Ba2+

Fig. 4. A proposed growth mechanism of $BaTiO_3$ film by the hydrothermal electrochemical method (see text).

It was observed that OH^- ions are essential to form $BaTiO_3$ film because no $BaTiO_3$ was prepared in any solutions except a strongly alkaline one. The electrical current was not essential to form $BaTiO_3$ but it was very effective to thicken the film. From the above results, we proposed a tentative growth mechanism of $BaTiO_3$ film as shown in Fig. 4 in this hydrothermal electrochemical method. This is:

(1) $BaTiO_3$ is firstly formed at the interface between the Ti substrate and the solution, where chemisorbed OH^- ions at the surface of Ti (probably with a oxide layer) can attract Ba^{2+} ions, even though an electrical field is not preferable to attract Ba^{2+} ions to the anode.
(2) Once $BaTiO_3$ layer was formed, Ba^{2+} ion must diffuse into interior due to its concentration gradient, thus it requires a little high temperature as 100℃. This Ba^{2+} diffusion might be enhanced by the diffusion of O^{2+} (and/or OH^-) assisted by electrical current.
(3) Anodic oxidation of Ti to form TiO_2 proceeds even at low temperature as a room temperature because O^{2-} (and/or OH^-) can be incorporated into TiO_2 layer by the electrical current.
(4) The formation front of $BaTiO_3$, therefore, must be the interface of $BaTiO_3/TiO_2$. This is corresponding with the fact that the diffusion of Ti^{4+} is more difficult than O^{2-} (and Ba^{2+}) in TiO_2 (and $BaTiO_3$) [11].

According to this model, TiO_2 layer might exist between the surface $BaTiO_3$ layer and the Ti substrate when an electric current was applied. This was confirmed by thin film X-ray diffraction analyses changing the incidence angles. An appreciable TiO_2 layer was not detected in the products by simple hydrothermal methods, whereas an anatase layer could be detected in the products by hydrothermal electrochemical method as illustrated in Fig. 5 [12]. The relative increase of anatase peak intensity to $BaTiO_3$ one when the incidence angle was increased from 0.5° to 1.5° indicates that the anatase layer exist beneath the $BaTiO_3$ layer.

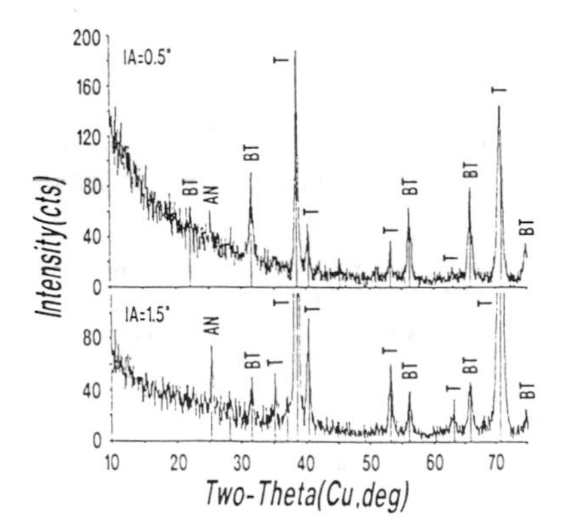

Fif. 5. Diffraction profiles taken at the incidence angles of 0.5° (upper) and 1.5° (lower) of the GIAB geometry for the $BaTiO_3$ thin film prepared by the hydrothermal-electrochemical method. Notation are BT ($BaTiO_3$), T (Ti) and AN (anatase).

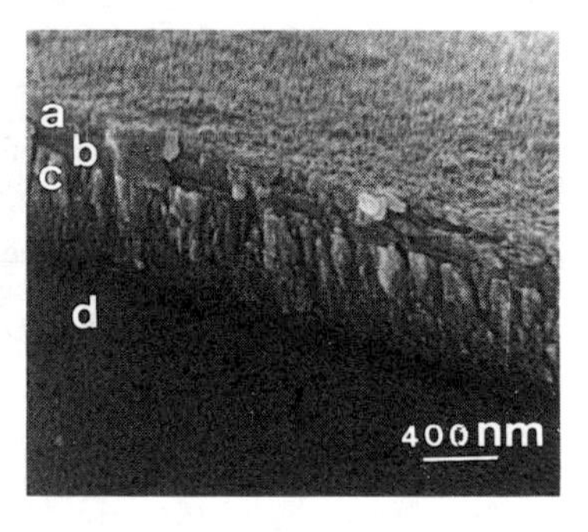

Fig. 6. SEM micrograph of the fractured surface of $BaTiO_3$ thin film prepared on a Ti-sputtered glass substrate at 200℃ with the current density of 10 mA/cm^2 for 30 min in 0.1N-$Ba(OH)_2$. a : $BaTiO_3$, b : TiO_2 , c : sputtered Ti , d : glass substrate,

The intermediate layer of TiO_2 was also detected in the SEM observation of the cross section of the $BaTiO_3$ film prepared from sputtered Ti layer on a glass substrate (Fig.6) [13]. This figure also demonstrates that the formed film consists of fine grains of $BaTiO_3$ with 30 nm in size. We could prepare similar $BaTiO_3$ thin films on flexible polymer film as a substrate using similar technique [14]. The hydrothermal-electrochemical technique is noticable to provide such multi-layered film with crystallized double oxide layer(s) at sufficiently low temperatures.

$SrTiO_3$ THIN FILM

Similar to the formation of $BaTiO_3$, $SrTiO_3$ thin films could be prepared when $Sr(OH)_2$ solutions were used under hydrothermal conditions, (Fig. 7) [15]. The thickness seemed to increase with increase in treating time but less dependent with the concentration of the Sr solutions. The electron microscopic observation revealed that the film consists of grains with approximately 10 nm in size.

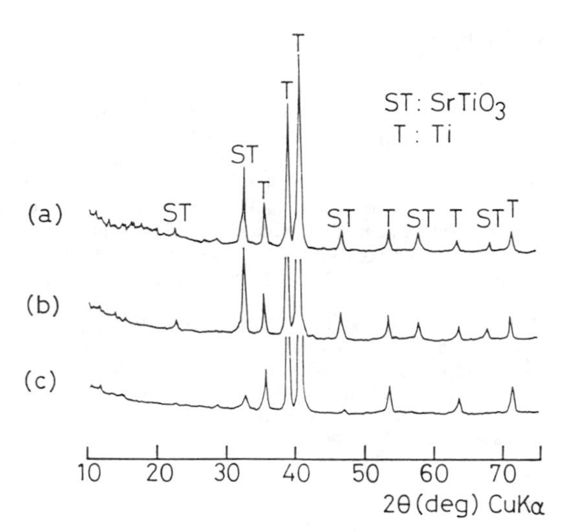

Fig. 7. XRD patterns of the $SrTiO_3$ films produced by hydrothermal-electrochemical method in various $Sr(OH)_2$ solutions at 200℃ with the current density of approximately 15 mA/cm^2: (a) 5N for 90 min., (b) IN for 90 min., and (c) IN for 30 min.

$BaFeO_3$ FILM

When Fe substrate was used for the hydrothermal electrochemical method in strong alkaline solution, $Ba(OH)_2$ + NaOH or KOH, $BaFeO_3$ films [16] were formed at temperatures above 90℃ . $BaFeO_3$ was not formed without electrical currents, without alkaline additives such as NaOH or KOH, or below 90℃.

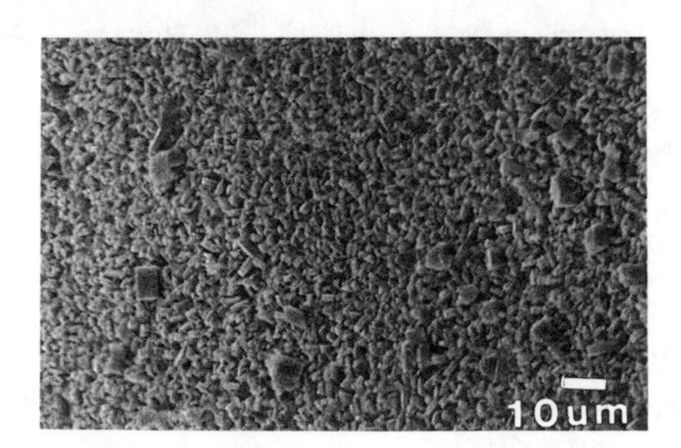

Fig. 8. SEM photograph of the surface of $BaFeO_3$ film prepared by hydrothermal electrochemical method for a FE substrate in 0.25M $Ba(OH)_2$ - 0.5M NaOH at 200℃ with the current density of 18 mA/cm^2 for 60 min.

Since the film consists of grown crystals of $BaFeO_3$ with 2 - 3 μm in size, Fig. 8, the $BaFeO_3$ seems to be precipitated from strong alkaline solutions. In fact, anodic polarization curve during the procesing suggested that the Fe substrate appears to be in overpassivity state, where Fe would be dissolved in the solution as ferrate ions. The lattice parameters of the produced $BaFeO_3$ were a=0.5694(1) nm, c=2.789(4) nm, v=7.829(1) nm^3 and c/a=4.898 for a hexagonal perovskite lattice, which agrees with previous data [17] for the powder sample prepared by a solid state reaction. It is very interesting that $BaFeO_3$ was firstly formed in the film at low temperatures of 90 - 200℃.

OTHER PEROVSKITE FILMS

The film of $LiNbO_3$ can be prepared using the hydrothermal (electrochemical) techniques as follows: When a Nb substrate was treated electrochemically in LiOH solutions at RT-200℃ , anodic films of amorphous Nb_2O_5 were formed at a first stage. An electric discharge (sparking) occurred when the applied voltage exceeded the breakdown voltage of the film, then the reaction between the amorphous Nb_2O_5 and the Li ion in the solution occurred to yield $LiNbO_3$ grains to form a porous film [18]. A dense $LiNbO_3$ film [19] could be prepared by a simple hydrothermal method without electric currents from a Nb substrate in 0.1M LiOH solution at 400℃ under 60 MPa for 1h. Thin films of $PbTiO_3$ and the perovskite of Ba-Nb-Oxide were also prepared using this hydrothermal electrochemical techniques.

As described above, the hydrothermal electrochemical techniques is a unique method which allows us to prepare many double oxides such as perovskite in the form of thin and/or thick films even at low temperatures below 200℃. Some of them seem to be metastable and/or new phases. Moreover it is advantageous to prepare well crystallized film due to the hydrothermal solution. Thus we hope this method will provide a new area of research and development in ceramics science and technology.

ACKNOWLEDGEMENT

The author would like to thank the members in our research group, Dr. N. Ishizawa, Dr. E-S. Yoo, Mr. M. Hayashi, Mr. K. Kajiyoshi,Mr. H. Kimura, Mr. M. Nakatsu, Mr. H . Banno, Mr. H. Yamaguchi, Mr. S. Mizunuma, Mr. U-S. Cho, Dr. M. Kakihana, and Dr. M. Yashima, to participate these research, and also to the New Technology Agency, Japan and the Murata Scientific Foundation for parttial financial support for this research.

REFERENCE

1. Ceramic Thin and Thick Films, Ceramic Trnsactons, vol. 11, edited by B. Hiremath, (American Ceramic Society, OH, 1990).

2. Ferroelectric Films, Ceramics Transactions, vol. 25, edited by A. S. Bhalla and K. M. Nair, (American Ceramic Society, OH, 1992).

3. S. E. Yoo, M. Yoshimura and S. Sōmiya, Sintering '87, Proceedings vol. 1, edited by S. Sōmiya, M. Shimada, M. Yoshimura, and R. Watanabe, (Elsevier, 1988), pp. 108-13.

4. S. E. Yoo, M. Yoshimura and S. Sōmiya, J. Mater. Sci. Lett., 8 [5] (1989), PP. 530-32 .

5. M. Yoshimura, S. E. Yoo, M. Hayashi and N. Ishizawa, Jpn. J. Appl. Phys., 28 [11] (1989), L2007-09 .

6. M. Yoshimura, S. E. Yoo, M. Hayashi and N. Ishizawa, Microelectronic Systems, Ceramic Transactions, vol. 15, Ed. K. M. Nair, R. Pohanka and R. C. Buchanan, (The Am. Ceram. Soc. 1990), PP.427-436.

7. N. Ishizawa, S. E. Yoo, M. Hayashi and M. Yoshimura, Ferroelectric Thin Films, (Mater. Res. Soc. Proc. 200 Pittsburgh, PA 1990), pp. 57-62.

8. K. Kajiyoshi, N. Ishizawa, and M.Yoshimura, J. Am. Ceram.Soc., 74 [2], (1991), pp. 364-74 .

9. K. Kajiyoshi, N, Ishizawa, and M. Yoshimura, Jpn. J. Appl. Phys., 30 [1B] (1991) pp. L120-123

10. K. Kajiyoshi, N. Ishizawa, and M. Yoshimura, pp. 271-278 in Ref. [2].

11. A. Garcia - Verduch and R. Lindner, Arkiv fr Kemi, 5 , (1953) 313-16.

12. N. Ishizawa, M. Hayashi, H. Banno, S. Mizunuma, H. Yamaguchi, S. E. Yoo, and M. Yoshimura, Report of Res. Lab. Eng. Mater., Tokyo Inst. Tech., 16, (1991), pp.9-16.

13. M. Hayashi, N. Ishizawa, S. E. Yoo, and M. Yoshimura, J. Ceram. Soc. Jpn., 98 [8] (1990), pp.931-33 .

14. N. Ishizawa, H. Banno, M.Hayashi, S. E. Yoo and M. Yoshimura, Jpn. Appl. Phys., 29 [11], (1990), pp. 2467-72 .

15. S. E. Yoo, M. Hayashi, N. Ishizawa and M. Yoshimura, J. Am. Ceram. Soc., 76 [8] (1990), pp. 1561-63 .

16. M. Yoshimura, M. Nakatsu, S-E. Yoo and N. Ishizawa, to be published in Japan J. Appl. Phys.

17. M. Zanne and C. Gletzer , Bull. Soc. Chem. France, 3 [5], (1971), pp.1567-70. 1567 - 70 (1971)

18. M. Yoshimura, S-E. Yoo., H. Kimura, M. Hayashi, and N. Ishizawa, Proc. of Fall Meeting of the Ceramics Society of Japan, (1989), pp. 28-29.

19. M. Yamaguchi, M. Kakihana, M. Yashima, and M. Yoshimura, Proc. of 30th Meeting of Basic Ceramics Science, (1992), pp. 103.

ELECTROPHORETIC DEPOSITION OF SOL-GEL-DERIVED CERAMIC COATINGS

YINING ZHANG,* C. JEFFREY BRINKER,† AND RICHARD M. CROOKS*
*Department of Chemistry and UNM/NSF Center for Micro-Engineered Ceramics, University of New Mexico, Albuquerque, NM.
†Center for Micro-Engineered Ceramics, University of New Mexico, Albuquerque, NM, and Sandia National Laboratories, Albuquerque, NM.

ABSTRACT

The physical, optical, and chemical characteristics of electrophoretically- and dip-coated sol-gel ceramic films are compared. The results indicate that electrophoresis may allow a higher level of control over the chemistry and structure of ceramic coatings than dip-coating techniques. For example, controlled-thickness sol-gel coatings can be prepared by adjusting the deposition time or voltage. Additionally, electrophoretic coatings prepared in a four-component alumino-borosilicate sol display interesting optical characteristics. For example, the ellipsometrically-measured refractive indices of electrophoretic coatings are higher than the refractive indices of dip-coated films cast from identical sols, and they are also higher than any of the individual sol components. This result suggests that there are physical and/or chemical differences between films prepared by dip-coating and electrophoresis.

INTRODUCTION

Electrophoresis refers to the migration of charged molecules, particles, or polymers in an electric field[1]. Electrophoretic deposition of ceramic precursors onto conductive substrates is an alternative to existing techniques which include dip-coating, spin-coating, spraying, and chemical vapor deposition[2-4]. Although dipping and spinning are the most extensively investigated sol-gel coating methods, electrophoretic coating techniques permit additional control over the coating process[5,6]. For example, the voltage or current, coating time, and various Faradaic electrode processes can be used to prepare complex coatings that are difficult or impossible to fabricate using conventional sol-gel coating methods.

The primary goal of this work is to provide a preliminary survey of differences between the physical, optical, and chemical characteristics of electrophoretically and dip-coated sol-gel films.

EXPERIMENTAL

Aluminoborosilicate sols (71% SiO_2 - 18% B_2O_3 - 7% Al_2O_3 - 4% BaO, equivalent wt% oxides) prepared from hydrolyzed mixtures of the alkoxides and aged for different times (0, 3, 5, 7, 10, and 14 days) at 50 C were used to prepare sol-gel thin films by dip-coating and electrophoretic deposition. For electrophoretic experiments designed to compare coating weights, thin Al foils, 1-1.5 cm^2, oriented parallel to each other and separated by 1-1.5 cm, were used as the anode and cathode. The electrodes were immersed in an unaged sol, and a constant voltage or current was applied between the electrodes. Following electrolysis, the anode and cathode were withdrawn from the sol, rinsed with ethanol to remove the adherent dip-coated overlayer, dried overnight at 20-25 C, and weighed. For each set of experimental conditions, a reference sample was prepared by dip-coating only. Dip-coated substrates were not rinsed with ethanol.

Experiments were also carried out to compare the optical characteristics of dip- and electrophoretically-deposited films. The substrates for these experiments were 2000 Å-thick

Au films deposited onto 1 x 2 cm diced Si wafers prepared with a 50 Å Cr adhesion layer. For these experiments, electrophoretic coatings were formed by applying a constant cell current of 5 x 10^{-3} mA/cm^2 for 2 to 10 min. Following deposition, the electrodes were withdrawn from the electrophoresis medium, dried overnight at 20-25 C, and weighed. In some cases, the substrates were rinsed with ethanol immediately following electrophoresis to remove the dip-coated overlayer. Ellipsometry was used to measure the thicknesses and refractive indices of both dip- and electrophoretically-coated films[7].

RESULTS AND DISCUSSIONS

Coating Weights

Tables I and II compare the total weight of films prepared by dip-coating and electrophoresis at constant potential and constant current, respectively. Several important conclusions can be drawn from these data. First, the anode-coating weights for the electrophoretically-prepared films are significantly higher than either the cathode or dip-coated substrate weights, indicating that electrophoretic deposition takes place only at the anode under the conditions used here. Second, more of the sol resides on the electrophoretically-coated anodes than on the dip-coated substrates, which demonstrates that substrates can be coated more heavily by electrophoretic deposition than by dip-coating. Third, the weight of the electrophoretic coatings increases with coating time and with the cell voltage or current.

Figure 1 shows how the current resulting from constant voltage electrophoresis decreases as a function of time. The current decreases because the growing film continually increases the total cell resistance, eq 1, where i_{cell} is the electrophoretic current; V_{cell} is the constant

$$i_{cell} = (V_{cell} - V_{film})/R_{soln} \quad (1)$$

applied cell voltage; V_{film} is the voltage drop in the film, which is equal to the product of the current and the film resistance; and R_{soln} is the resistance of the deposition solution, which can be considered constant in these experiments. The rate of current decrease is a function of the physical and chemical properties of the coatings, since the density, porosity, thickness, and the chemical nature of the film control changes in the film resistance. The film will cease to grow when $V_{film} = V_{cell}$.

Table I Comparison of coating times and weights for films prepared by constant voltage electrophoresis and dip-coating from an unaged four-component sol. The anodes and cathodes were rinsed in ethanol immediately following removal from the deposition solution to remove the dip-coated contribution to the weight. The electrode separation was 1 cm.

	Exp. 1	Exp. 2	Exp. 3	Exp. 4	Exp. 5
Area of Al Foil (cm^2)	1.5	1.5	1.0	1.0	1.0
V_{cell} (V)	5	5	6	8	10
Coating Time (min)	5	10	15	15	15
Anode Coating Weight (g)	0.0018	0.0027	0.0011	0.0016	0.0020
Cathode Coating Weight (g)	0	0.0002	0	0	0
Dip-Coating Weight (g)	0.0004	0.0005	0.0002	0.0002	0.0002

Table II Comparison of coating time and weight for films prepared by constant current electrophoresis and dip-coating from an unaged four-component sol. All electrodes were 1.0 cm^2 Al foils separated by 1.0 cm. The anode and cathode were rinsed in ethanol immediately following removal from the deposition solution to remove the dip-coated contribution to the coating weight.

	Exp.1	Exp. 2	Exp. 3	Exp. 4	Exp. 5
i_{cell} (mA)	0.25	0.5	0.75	1.0	1.25
Coating Time (min)	10	10	10	10	10
Anode Coating Weight (g)	0.0017	0.0029	0.0064	0.0101	0.0120
Cathode Coating Weight (g)	0	0	0	0	0
Dip-Coating Weight (g)	0.0003	0.0004	0.0005	0.0003	0.0005

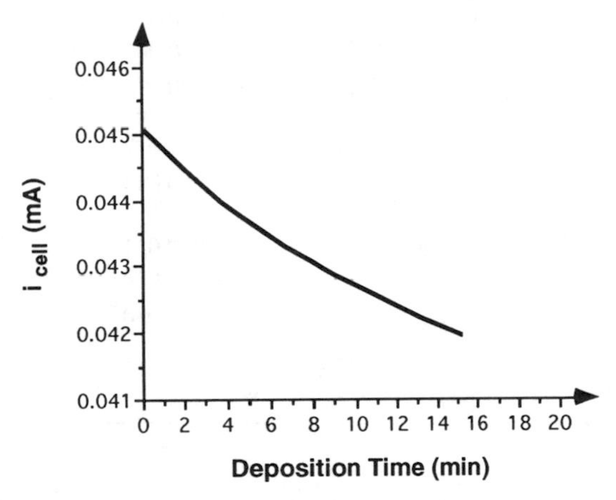

Figure 1 Cell current as a function of deposition time for a constant voltage (4 V) electrophoretic deposition of an unaged, four-component sol onto a 1.0 cm^2 Al foil.

Thickness and Optical Properties

The refractive indices and thicknesses of dip- and electrophoretically-deposited films were measured by ellipsometry. The results obtained from dip-coated films cast from sols aged for different times are shown in Table III. As expected, the refractive indices of the films are close to that of SiO_2 (n = 1.45)[8], since it is the primary component of the sols. Longer aging times increase the sol cluster sizes, resulting in formation of thicker and more porous films[9]. Aged films with higher porosities have lower refractive indices since they are less dense than films prepared from unaged sols. The refractive indices measured for electrophoretically-deposited coatings are much larger than those obtained for dip-coated films, Table IV. This effect may arise from density or other morphological differences that exist within the films, from chemical differences, or from a roughened Au/coating interface.

Electrophoretically-deposited coatings are probably more dense than dip-coated films for two reasons. First, particle deposition is directed towards pinholes in the films, where the

Table III Ellipsometrically measured refractive indices and thick-nesses of dip-coated films prepared from four-component sols aged for various lengths of time at 50 C.

Aging Time (days)	Refractive Index	Thickness (nm)
5	1.394±0.015	132±2
7	1.359±0.003	144±5
14	1.324±0.004	148±4

Table IV Refractive indices and thicknesses of films prepared by constant current (i_{cell} = 5 x 10^{-3} mA/cm^2) electrophoresis from four-component sols aged for various lengths of time at 50 C. Refractive indices and thicknesses were obtained for electrophoretically deposited films covered with a dip-coated layer, except in one case.

Aging Time (days)	Coating Time (min)	Refractive Index	Thickness (nm)
0	2	2.334±0.333	101±21
0	5	1.870±0.005	134±2
0	10	1.417±0.009	163±4
3	2	2.028±0.285	115±21
3	5	1.993±0.018	119±3
5	2	2.148±0.040	242±15
5	10	1.812±0.014	347±4
7	5	1.504±0.022	266±1
7	10	1.647±0.028	288±3
7*	5	1.653±0.030	69±2

*In this case only, the substrate was rinsed with ethanol after being withdrawn from the sol to remove the adherent dip-coated overlayer.

electric field is highest, resulting in low defect-density coatings. Second, film density also might increase as a result of the finite impact velocity of the particles on the electrode surface. The velocity, v, is approximated by eq 2[2], where q is the charge on the particle, E is the

$$v \propto qE/\eta r \qquad (2)$$

magnitude of the electric field, η is the sol viscosity, and r is the hydrodynamic radius of the particle.

Additional morphological differences in electrophoretically-deposited films may arise because particles with different size-to-charge ratios will selectively deposit on the substrate at different rates and with different velocities. Such selective deposition should result in compositionally distinct strata within the coatings, which might induce a distribution of different refractive indices across the film thickness.

Chemical differences between dip- and electrophoretically-deposited coatings may also be present. For example, polymerization reactions could be driven by the finite impact velocity of the mobile particles against the substrates. Finally, Faradaic electrode reactions may also induce chemical reactions within electrophoretically-deposited films that change either their physical or chemical nature.

To determine if compositional variations within the films arise from time-dependent variations in the deposition conditions, we measured the change in V_{cell} as a function of time for a constant current electrophoretic deposition. The voltage required to maintain the current at a fixed value should increase linearly with time, according to eq 3[10], where υ is the

$$V_{cell} = V^0_{cell}\,(1 + \upsilon N Z i_{cell} t / \sigma A D) \qquad (3)$$

particle volume, N is the particle density of the sol, Z is the electrophoretic mobility of the particle given by eq 2 and Z=v/E, D is the separation between the electrodes, t is the coating time, V^0_{cell} is the initial cell voltage, σ is the film conductance, and A is the geometrically projected film area. Figure 2 illustrates how V_{cell} changes as a function of coating time for a sol that was aged for 5 days. The results indicate that there are three distinct linear regions in the voltage versus time plot, rather than one as predicted by eq 3, and confirm our expectation that there are chemically or physically distinct regions within the electrophoretically-deposited coatings.

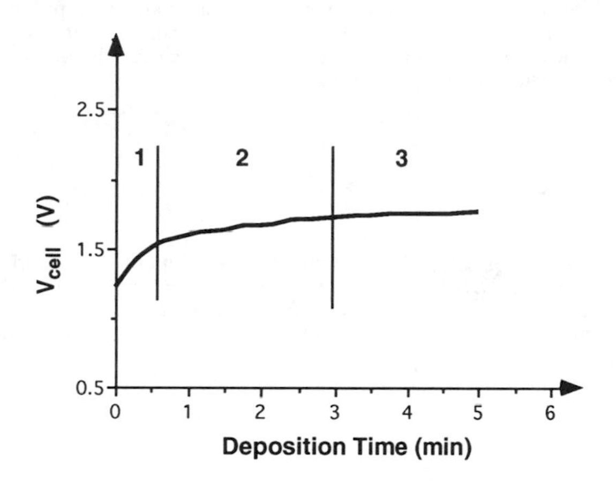

Figure 2 V_{cell} versus deposition time for constant current electrophoretic deposition of a four-component sol aged for 5 days at 50 C; $i_{cell} = 5 \times 10^{-3}$ mA/cm^2.

To insure that the surprisingly high refractive indices do not arise from macroscopic heterogeneities in the electrophoretically-deposited films, we compared SEM micrographs of dip- and electrophoretically-deposited films deposited from a sol aged for 7 days. However, we did not observe any significant differences between the film morphologies, and therefore we believe that the unique optical properties of the electrophoretically-coated films arise from real and controllable differences in chemical and physical structure rather than from cracked or otherwise poorly formed coatings.

CONCLUSIONS

Our results show that electrophoretically-deposited sol-gel films are chemically and physically different than dip-coated films prepared from identical sols. The most surprising result is that the refractive indices of electrophoretically-deposited coatings are much higher than expected. It is interesting that some refractive indices measured for the electrophoretically-deposited films are even larger than the maximum value of the highest refractive index component of the sol (BaO, n = 1.98)[8]. At present, we do not fully understand this unanticipated result, however, the data are reproducible and independent measurements of the refractive indices yield identical results.

We speculate that the unusual optical characteristics of the electrophoretically-deposited films arise from structural and chemical gradations brought about by the following conditions which are unique to electrophoresis. First, chemically distinct species within this four-component sol deposit on the substrate at different rates depending on their size-to-charge ratio. Second, more mobile species impact the electrode with a higher velocity, which increases the film density and may promote chemical reactions, such as polymerization, that are not induced by dip-coating. Third, Faradaic reactions occurring at the electrode-film interface may cause chemical reactions, which impart unique characteristics to the film.

Details of the structural and chemical properties which lead to the interesting optical properties of these films are presently under investigation. Although the mechanism of electrophoretic deposition of ceramic precursors from multicomponent polymeric sols is not well understood at the present time, our preliminary results clearly demonstrate that electrophoresis can lead to a much higher level of control over the chemistry and structure of ceramic coatings than alternative deposition methods.

ACKNOWLEDGMENTS

The authors thank Ms. Carol Ashley for preparing the sol solutions. This work was funded by the Department of Energy, Division of Basic Energy Sciences, and the UNM/NSF Center for Micro-Engineered Ceramics, a collaborative effort supported by NSF (CDR-8800352), Los Alamos and Sandia National Laboratories, the New Mexico Research and Development Institute, and the ceramics industry.

REFERENCES

1. C. Gutierrez, J. Mosley, and T. Wallace, J. Electrochem. Soc. 109, 923 (1962).
2. C.L. Foss, M.S. Thesis, Massachusetts Institute of Technology, Sept. 1982.
3. W.J. Dalzell and D.E. Clark, Ceram. Eng. Sci. Proc. 7, 1014 (1986).
4. D.E. Clark, W.J. Dalzell and D.C. Folz, Ceram. Eng. Sci. Proc. 9, 1111 (1988).
5. L.E. Scriven in Better Ceramics Through Chemistry III edited by C.J. Brinker, D.E. Clark, and D.R. Ulrich (Mat. Res. Soc., Pittsburgh, Pa., 1988), p. 717.
6. W.J. Dalzell, Jr. M.S. Thesis, The University of Florida, 1988.
7. Ellipsometric measurements were made using a Rudolf 43603-200E manual ellipsometer configured with a He-Ne laser at a 70° angle of incidence and a 45°/135° azimuth of the quarter-wave compensator. The complex refractive index of a naked Au substrate measured prior to coating, and an algorithm developed at Sandia National Laboratories were used to calculate film thicknesses and refractive indices (Tardy, H. L. "ELLIPSE User's Manual and Program Reference"; Report No. 89-0008; Sandia National Laboratories:Albuquerque, NM).
8. CRC Handbook of Chemistry and Physics edited by D.R. Lide (CRC Press, Inc.), 71th Edition 1990-1991.
9. C.J. Brinker and G.W. Scherer, Sol-Gel Science (Academic Press, INC. San Diego, CA 1990) p. 357.
10. A. Sussman and T.J. Ward, RCA Review, 42, 178 (1981).

CORROSION RESISTANT ZrO_2 SOL-GEL COATING ON STAINLESS STEEL

M. ATIK*,** AND M.A. AEGERTER**
* Laboratoire de Science des Matériaux Vitreux, Université des Sciences et Techniques du Languedoc, Place E. Bataillon, 34060 Montpellier, France
** Instituto de Física e Química de São Carlos, Universidade de São Paulo, Cx. Postal 369, 13560 São Carlos, SP, Brazil.

ABSTRACT

Corrosion resistant Sol-Gel ZrO_2 coatings deposited by dip coating technique on 316 L stainless steel sheets have been fabricated utilizing sonocatalysed precursors sols prepared from zirconium alkoxide $Zr(OC_3H_7)_4$, isopropanol, glacial acetic acid and water. Their composition varied between 0,025 to 0,9 mol/l ZrO_2. The coatings have been characterized by X-ray diffraction, IR optical reflection, ellipsometry and scanning electron microscopy. When densified under a slow heat treatment thin coatings exhibit outstanding corrosion resistance as no weight change has been observed after 24 h chemical attack in H_2SO_4 aqueous solution at 81°C and 10 h oxidation treatment in air at 800°C. Thick films as well as films densified under rapid heat treatments tend to crack and are less corrosion resistant.

INTRODUCTION

Zirconia (ZrO_2) is a well known material which has superior properties such as high mechanical strength, chemical durability, alkali resistance, heat resistance against oxidation and refractoriness. Its use to prepare coatings on glass substrates has been reported with emphasis directed toward the understanding of the hydrolysis, polymerization and sintering processes as well as the obtention of good optical properties such as transparency and high reflection [1-4].

Very few reports have been published concerning the obtention and the characterization of ZrO_2 Sol-Gel thin films deposited on metallic substrates [5,6]. The stability and the optical appearance of ZrO_2 films coated on stainless steel sheets have been found strongly dependent of the choice of the sol precursors; the use of zirconium alkoxides leads to less stable films than Zr tetraoctylate and Zr tetrakis(acetylacetonate).

The oxidation resistance was tested in air at 800°C showing that the weight increase of a 130 nm thick coating of ZrO_2 on stainless steel sheet was not less than one half of that of the substrate alone after a 10h test [6].

Several works were recently presented at the 6th International Workshop on "Glasses and Ceramics from Gels" studying the properties of mullite coating on carbon steel [7], borosilicate on mild steel [8], methyltrialkoxysilane (MTOS) on steel [9] and ZrO_2 on stainless steel [10].

This paper presents a systematic study of the durability against chemical attack and thermal oxidation of zirconia thin coatings prepared from zirconium propoxide, isopropanol and water sols which have been previously sonocatalyzed and then deposited by dip coating technique on 316L stainless steel sheets. The corrosion resistance of the films was evaluated by measuring the weight change as a function of time under different heat and chemical treatments.

EXPERIMENTAL

Zirconium sols were prepared by dissolving zirconium propoxide $Zr(OC_3H_7)_4$ in isopropanol (C_3H_7OH) and adding under ultrasonic irradiation (sonicator W385 Heat Systems-Ultrasonics, Inc., 20 kHz) glacial acetic acid and excess water to complete the hydrolysis. The ultrasonic procedure was carried on until obtaining clear and transparent sonosols. The resulting sols were then left undisturbed for four weeks at room temperature.

The substrates were 316L stainless steel sheets, degreased ultrasonically in acetone. They were dipped into the solutions and withdrawn at a speed of 10 cm/min and then dried at 40°C for 15 minutes. The samples were then heat treated at 400°C for 1 hour in air to remove the organic residues.

Two tests have been applied for measuring the oxidation and corrosion resistance:

Test A: the samples were first densified at various temperatures and then placed in a 15% H_2SO_4 aqueous solution at 72-89°C for different periods of time (up to 32 hours). The determination of the total chemical corrosion was determined by measuring the weight loss of the samples due to a slow leaching of the coatings.

Test B: the oxidation test was carried out in air typically at 800°C for 2 to 10 hours. The oxidation was determined by measuring the weight increase of the coated sheets resulting from the formation of various oxide compounds.

RESULTS AND DISCUSSION

The thickness of ZrO_2 films deposited on 316 stainless steel and densified after a heat treatment in air at 800°C for 9 h is shown in figure 1 as a function of the ZrO_2 sol concentration. The values have been measured by ellipsometry (Rudolf S 2000); they increase linearly with the concentration of ZrO_2.

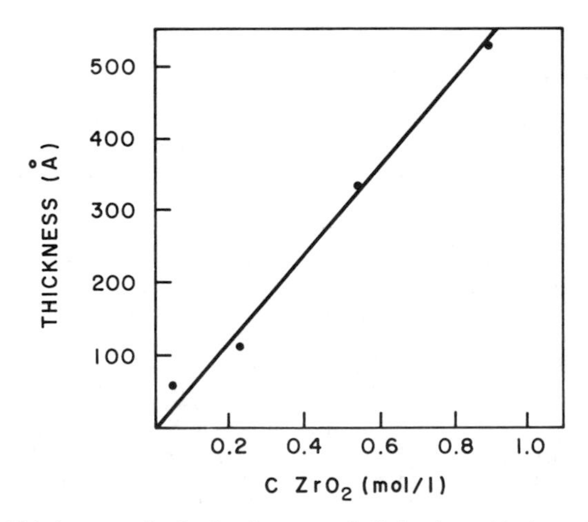

Fig. 1. Thickness of single layers of ZrO_2 deposited on both sides of 316L stainless steel sheet by dip coating technique (v=10 cm/min) and then densified in air at 800°C during 9 h versus ZrO_2 sol concentration.

Measurements of reflection in the IR region of unsintered samples show characteristic OH bands at ~3600 cm^{-1}, Zr-O-C groups at 1476.8 and 1452.4 cm^{-1} and Zr-O-Zr at 665.7 cm^{-1}. During sintering the OH and Zr-O-C bands disappear and the Zr-O-Zr band increases strongly. X-ray diffraction data has confirmed that the films densified at 800°C have the tetragonal ZrO_2 structure (111 peak at d= 2.98Å).

The effect of air oxidation on samples prepared with different ZrO_2 sol concentration (and consequently different film thickness) is shown in figure 2. All the samples have been heat treated at 800°C during 6 h but their sintering has been realized either under a rapid heating process (RHP, direct introduction of the samples in preheated furnace) or a slow heating process (SHP, 5°C/min from 25 to 800°C with 1 h holding time at 400°C to complete the elimination of the organic compounds). The rate of oxidation, observed as a weight gain, increases with the ZrO_2 sol concentration for RHP while no weight change is observed using SHP. Apparently the oxidation resistance does not depend of the solution concentration or film thickness but is probably linked to the films morphology and their structural homogeneity. Thermal stresses, cracks, etc. are likely to develop during a rapid heating, particularly for thicker coatings.

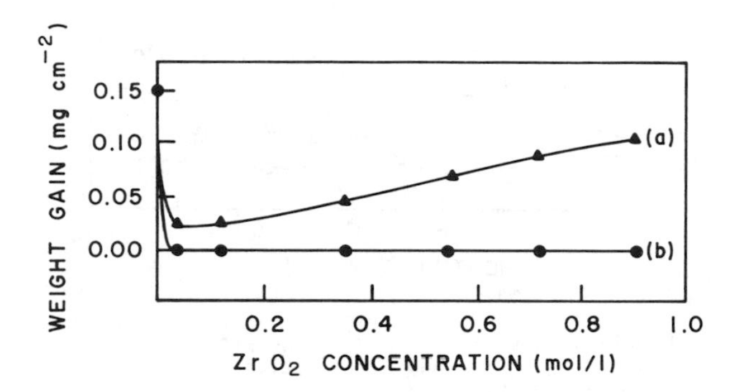

Figure 2. Weight increase of ZrO_2 coated 316L stainless steel measured after air oxidation treatment at 800°C/6 hrs versus ZrO_2 sol concentration a) rapid heating process b) slow heating process (see text).

The effect of time on the oxidation treatment is shown in figure 3. Curve a, b and c show the weight gain of uncoated samples at 950, 870 and 800°C, which is seen to increase with the temperature. ZrO_2 coating of 35 nm thickness tested at 800°C sintered under RHP (curve d) shows a reduction of the oxidation rate and confirms that this fast densifying process introduces defects which impede a perfect protection. The effect is more pronounced for thicker coatings.

The samples sintered under SHP do not exhibit any weight gain up to 8 h heat treatment (thickness of 35 nm) and up to at least 10 h for smaller thicknesses confirming the above conclusions.

Figure 4 shows two SEM micrographs taken for 5,5 nm thick ZrO_2 sintered using both processes. The surface rapidly heated and kept only 10 minutes in the furnace at 800°C shows clearly regions where oxidation can take place easily. On the contrary the surface slowly heated and kept 9 hours at 800°C is smoother and is better protecting the substrate.

Figure 5 shows a series of measurements made at high temperature indicating that the films have however a limited protection and do not protect the stainless sheet indefinitely. These results clearly show that the films only delay the oxidation process and that, after a certain time or for higher temperatures, the films are not efficient at all.

The chemical corrosion in H_2SO_4 aqueous solution in shown in figure 6 where the weight loss of the samples is plotted as a function of time for different experimental conditions.

The weight loss of uncoated samples (curve a, b, c) increases with time and temperature. The weight loss of coated samples sintered under SHP shows a drastic reduction which depends of the thickness of the films. Thick coatings ($\geq$ 55 nm) provide protection up to 10 h and thin coatings (<40 nm)

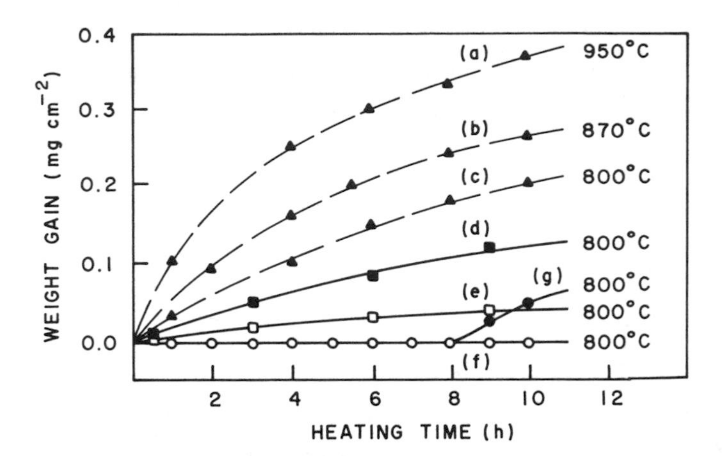

Figure 3. Weight gain versus oxidation time in air. a), b), c) uncoated substrates tested at 950, 870 and 800°C respectively, d) 35 nm ZrO_2 coated sample sintered under RHP and tested at 800°C, e) idem for 5,5 nm coating, f) and g) 5,5 and 35 nm respectively coated samples sintered under SHP and tested at 800°C.

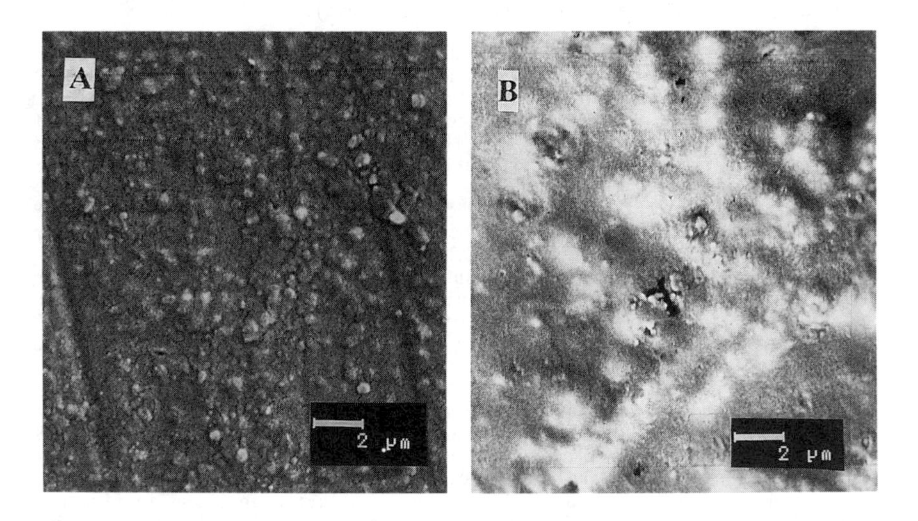

Figure 4. SEM micrographs of 5,5 nm thick ZrO_2 coating a) sintered under SHP and kept at 800°C during 9 h; b) sintered under RHP and kept 10 min at 800°C.

do not present any chemical attack up to 24 h at 81°C. No test has been realized for coatings sintered under RHP. The figure once again shows that the

SHP leads to much more homogeneous coatings. However thicker coatings, even densified under SHP, present a chemical corrosion after a certain time of treatment.

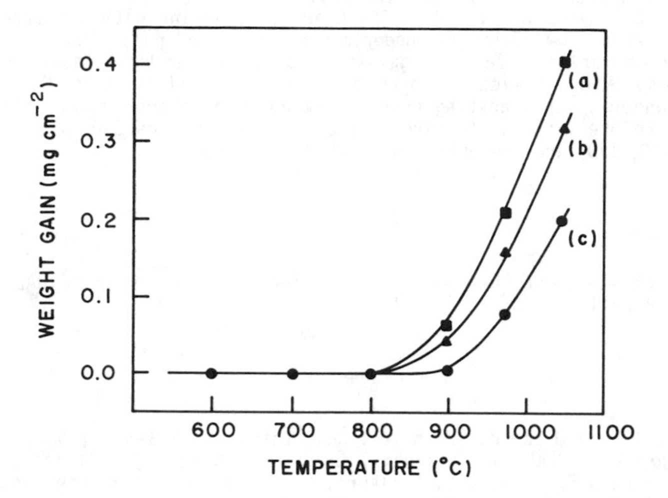

Figure 5. Weight gain of 5,5 nm ZrO_2 coating sintered under SHP, measured after oxidation in air at different temperatures. Time of heat treatment a) 9h, b) 6h, c) 3h.

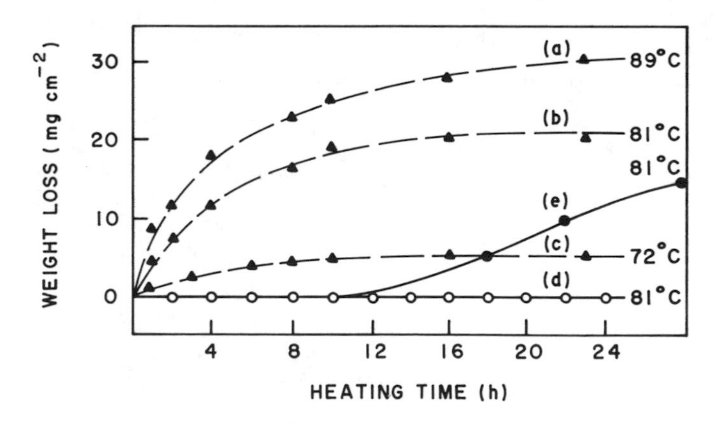

Figure 6. Weight loss of samples as a function of immersion time in 15% H_2SO_4 aqueous solution. a), b) and c) uncoated samples tested at 72, 81 and 89°C respectively d) ZrO_2 coated substrate sintered under SHP and tested at 81°C with films having a thickness <40 nm; e) same as d) but for a film of 55 nm.

CONCLUSION

Sol-gel ZrO_2 coatings obtained from sonocatalyzed hydrolysis of Zr propoxide have been deposited on 316L stainless steel sheet by dip coating technique. The films densified at 800°C are crystalline with a tetragonal structure. When slowly sintered under a 5°C/min heated rate they provide an outstanding corrosion protection against air oxidation up to at least 10 hours at 800°C and H_2SO_4 chemical attack up to at least 24 h at 81°C. Films densified under a rapid heating process present cracks and thermal stresses and are therefore less protective. Other systems such as SiO_2,Al_2O_3, SiO_2-Al_2O_3 and SiO_2-TiO_2 are presently under study.

ACKNOWLEDGEMENTS

This research has been financed by CNPq, Finep and the Program RHAE - New Materials (Brazil).

REFERENCES

1. D. Ganguli and D. Kundu, J. Mater. Sci. Lett., 3, 503-504 (1984)
2. N. Tohge, A. Matsuda and T. Minami, Chemistry Express, 2, 141-144 (1987).
3. O. De Sanctis, L. Gomez, N. Pellegri, C. Parodi, A. Marajofski and A. Duran, J. Non Cryst. Solids, 121, 442-448 (1989).
4. L. Yang and J. Cheng, ibid., 112, 442-448 (1989).
5. K. Sugioka, H. Tashiro, K. Toyoda, H. Murakami and H. Takai, J. Mater. Res., 5 2835-2840 (1990).
6. K. Izumi, M. Murakami, T. Deguchi and A. Morita, J. Am. Ceram. Soc., 72, 1465-68 (1989).
7. A.R. Di Giampaolo, M. Puerta, J. Lira and H. Ruiz, Proc. VIth International Workshop on Glasses and Ceramics from Gels, Sevilla (SP), 6-11 october 1991, Abstract CP 12.
8. M. Guglielmi, D. Festa, P.C. Innocenzi, P. Colombo and M. Gobbin, ibid., Abstract CP 15.
9. K. Izumi, H. Tanaka, Y. Uchida, N. Tohge and T. Minami, ibid., Abstract CP 20.
10. M. Atik and M.A. Aegerter, ibid., Abstract N 1.

PROTECTIVE ZIRCONIA THIN FILMS ON METAL SUBSTRATES.

ANDREA TOMASI*, PAOLO SCARDI** & FABIO MARCHETTI*
*Institute for Scientific and Technologic Research, Trento, Italy;
**Department of Materials Engineering, University of Trento, Trento, Italy.

ABSTRACT

Stabilized Zirconia thin films were deposited by R.F. Magnetron Sputtering onto iron substrates. Deposition parameters such as gas pressure and R.F. voltage were optimized to obtain homogeneous and well adherent protective coatings. The morphology and structure of films before and after thermal treatments were investigated by Scanning Electron Microscopy (SEM), X-Ray Diffraction (XRD) and Thermogravimetrical Analysis (TGA).

Auger Electron Spectroscopy (AES) depth profiling techniques were used to obtain the chemical distribution of the elements in order to study the influence of two important parameters on the adhesion of the ceramic: the presence of a thin layer of native metal oxide and the residual surface roughness after polishing.

The effectiveness of coatings to protect the metal substrate against high temperature oxidation in air was evaluated by thermogravimetrical analysis at 773K. The protective action of the ceramic film was well demonstrated.

INTRODUCTION

Owing to their electrical and optical properties, stabilized zirconia thin films have been proposed for several applications in the field of optical or electronic devices, such as laser mirrors, gyros, broadband interference filters [1,2,3], oxygen sensors and buffer layers for integrated circuits [4,5]. Even so, the most interesting features of zirconia materials are probably those connected with their thermomechanical properties. Zirconia is a good thermal insulator with high toughness and thermal shock resistance, also in severe environmental conditions. The monoclinic (m) phase, which is thermodynamically stable at room temperature, undergoes a transformation to tetragonal (t) (1170°C) and cubic (c) (2370°C) before melting (2680°C). The volume change associated with the t-m transformation (about 3%) can be exploited to achieve very high toughness in structural ceramics, but can have a destructive effect on zirconia coatings subjected to thermal treatments. To avoid this effect, t or c phases can be stabilized in a large temperature range, even below room temperature, by the addition of a number of oxides, including MgO, CaO and several rare earths oxides [6].

Fully stabilized zirconia's have thermal expansion coefficients in the range $12\text{-}14\times10^{-6}$ ($°C^{-1}$), very close to that of many metals and high temperature alloys; the good thermal matching with metals makes these materials suitable to produce ceramic coatings. In fact, thick coatings of stabilized or partially stabilized zirconia prepared by Plasma Spray [7] have been proved effective as a thermal barrier, protecting metal surfaces of engine components. By allowing a higher working gas temperature, zirconia coatings can improve the efficiency of thermal engines, particularly gas turbines.

Thin protective films with such properties, would be quite useful but little work has been reported [8,9,10,11]. Physical Vapour Deposition (PVD) techniques are ideally suited to make thin, homogeneous, smooth and well adherent films. Moreover, the film microstructure can be optimized for a given application by opportunely changing process parameters [12,13]. In this work ceramic thin films were deposited on metal substrates by magnetron sputtering, using a commercially available stabilized Zirconia target. Iron was employed as a substrate to test the

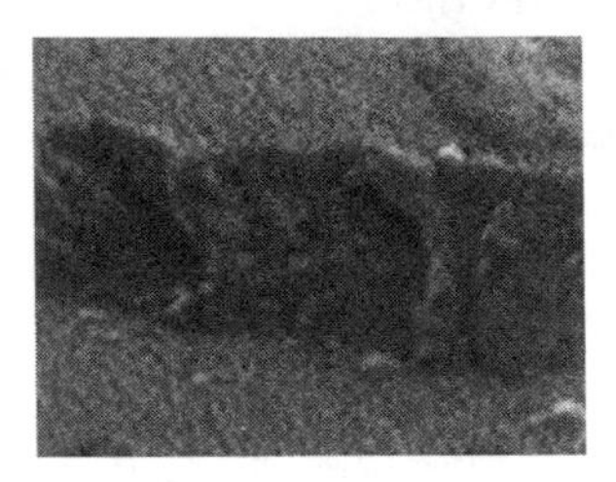

Fig. 1. Cross section of 300nm zirconia film on silicon fractured in LN_2 (51KX).

effectiveness of the films to coat and to protect a very reactive metal surface at high temperature. The oxidation kinetics were followed by TGA, SEM, AES and XRD.

EXPERIMENTAL

Fully stabilized (12.5mol%CaO) zirconia coatings with nominal thickness of 100 and 300 nm were deposited using a r.f.-Magnetron sputtering system, Z400 (Leybold), equipped with 7.5cm targets. Iron substrates about $1cm^2$ by 0.2cm thick, were finished with silicon carbide 4000 papers or polished with 0.1μm diamond before the standard cleaning cycle. Substrates were heated (350°C) in high vacuum and etched by backsputtering and the system was pumped down to 10^{-7} mbar before the deposition started.

The true film thickness, measured, for each deposition batch, by an Alphastep 200 profilometer (Tencor), was 115nm and 325nm. SEM (Cambridge stereoscan 200) and XRD (Rigaku D-max IIIB) were used to study, respectively, the morphology and phase composition of as prepared and thermally treated samples. All thermal treatments were performed at 773K with 10°/min heating and cooling rate. 24 hour treatments were performed using a Mettler TGA, to follow the weight increase due to oxidation. AES (Perkin Elmer PHI 4200) was used to analyze samples as prepared and after thermal treatment up to 2 hours, in order to study the behaviour of the film and the first stages of iron oxidation.

RESULTS AND DISCUSSION

After testing a preliminary set up [14], the r.f. potential and the Argon working pressure were adjusted to 1.5kV and $2x10^{-3}$ mbar respectively, in order to obtain the desired dense microstructure. In fact, it is well known that low working pressure promotes the formation of dense films, corresponding to zone T in the Thornton model [13], with a compressive intrinsic stress playing a favourable role against cracking. High crystallinity achieved using low sputtering pressures [12] also increases the thermal stability. Even though it has been pointed out that a dense microstructure is more sensitive to heat shock [12], this should not be the case for very thin films with good elastic coupling with the metal substrate such as stabilized zirconia on iron or steels. On the contrary, the advantages of a dense structure with respect to oxygen diffusion and high temperature corrosion in general are quite evident.

The microstructure of our zirconia films is visible in Figure 1, showing the LN_2 fracture surface of a 300nm layer deposited onto silicon substrate.

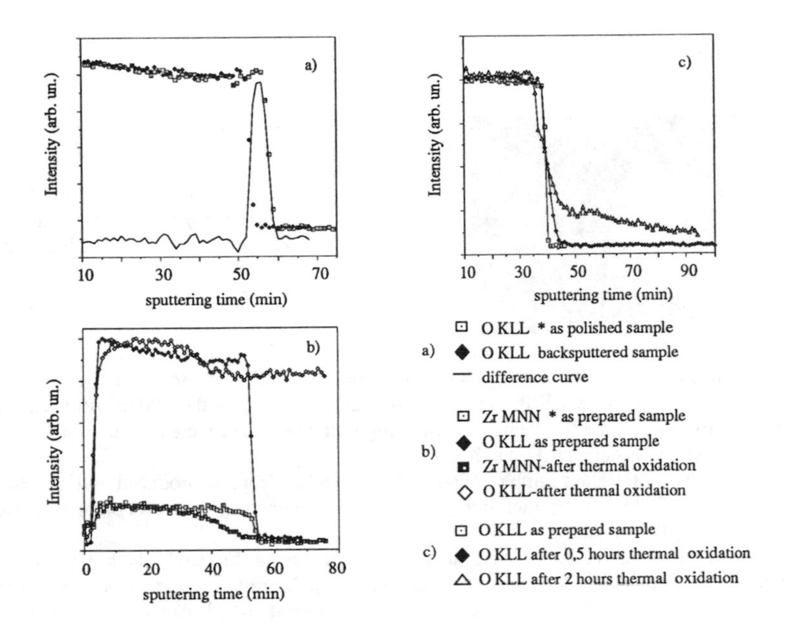

* The notation is relative to the specific Auger transition used in the depth profile

Figure 2. AES depth profiles. 100nm zirconia film as prepared, etched and not (a); 100nm film as prepared (etched) and after 1 hour at 773K (b); 300nm film onto etched iron as prepared, after 1/2 hour and 2 hours at 773K (c).

100nm films were deposited both on backsputtered and on as polished (4000 paper) iron samples, in two separate batches, to study the effect of the more or less thin native oxide layer, almost always present on metal surfaces, on film adhesion and effectiveness against oxidation. For the sample deposited without etching, AES depth profiling (Figure 2a) showed the presence of the thin iron oxide layer as an oxygen peak at the interface. But, as shown by TGA (Figure 3) and XRD, after thermal treatments the film protective action did not seem to be affected by this thin layer.

After 1 hour at 773K, AES (Figure 2b) showed the film to be still intact, with the same sharp interface, but iron oxide started to form below it. The morphology of the sample surface

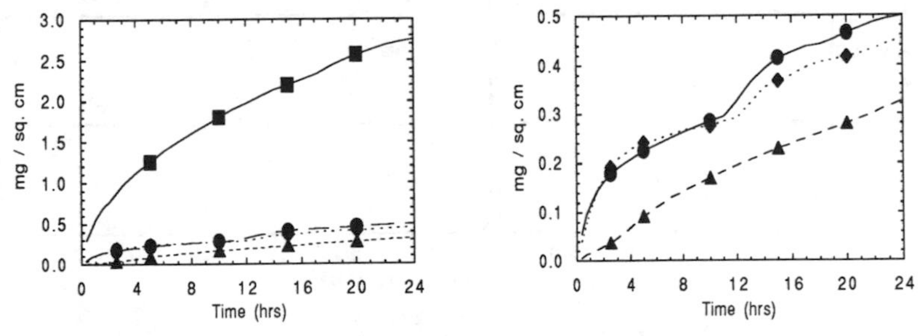

Figure 3. TGA of bare iron substrate ■, 100nm film (ecthing: yes ◆, not ●), 300nm film ▲ (left); expanded scale (right).

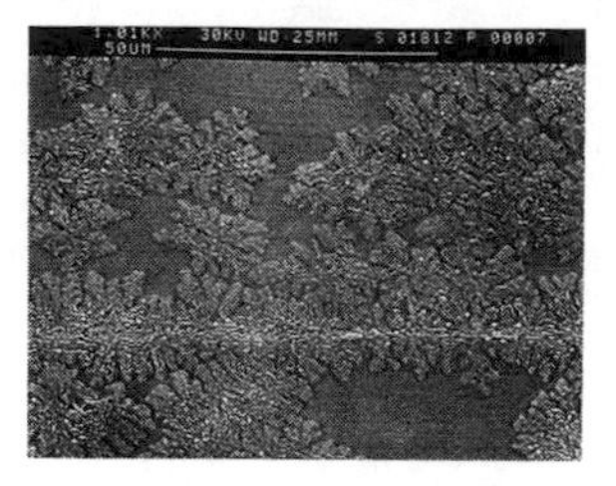

Fig. 4. SEM micrograph of the surface of 100nm coated iron sample after 1 h at 773K.

is shown in Figure 4. After 2 hours, the film started to crack because of iron oxide growth. However, in spite of its small thickness, the overall behaviour of the 100nm film was good. The protective action can be evaluated comparing data for weight increase due to oxidation of coated and bare iron samples, Figure 3.

XRD patterns of 100nm samples deposited onto etched iron, as produced and treated 1, 2 and 24 hours at 773K, are reported in Figure 5. No difference was found for samples prepared without etching the substrates.

A modified version of the Rietveld method [15,16] was used to study the kinetics of oxide formation. The procedure consists of fitting a wide angular range of the experimental pattern using analytical functions for peaks and a polynomial for background, allowing the simultaneous refinement of both structural (structure factor, atomic positions inside cell, temperature factor, occupancy, lattice parameters) and microstructural parameters (crystallite size, microstrain, preferred orientation, phase percentages) for the present phases.

Figure 6 shows the graphic output of our program for the sample treated 24 hours. Four phases were present: cubic zirconia, α-iron, Fe_2O_3 and Fe_3O_4.

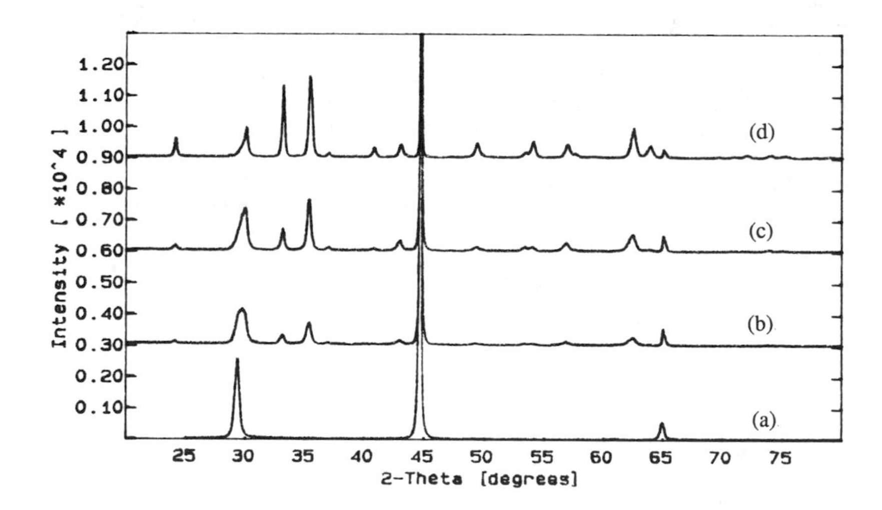

Figure 5. XRD patterns of 100nm films on iron: as prepared (a), after 1 hour (b), 2 hours (c) and 24 hours (d) at 773K.

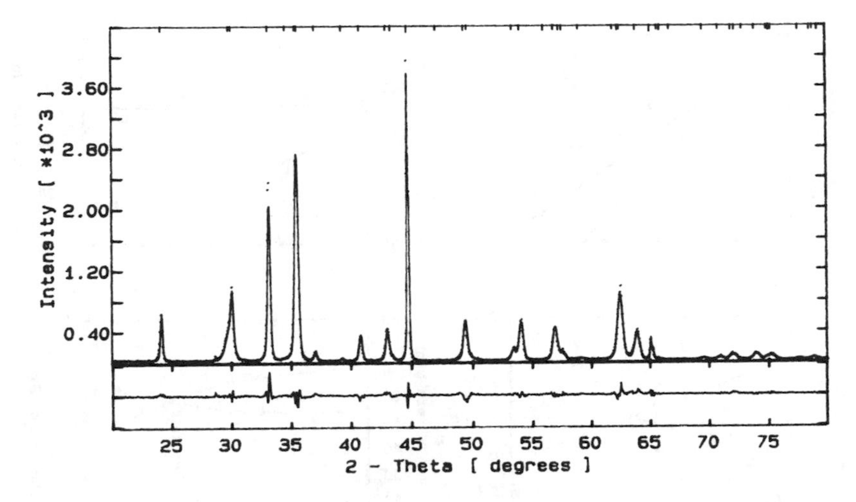

Figure 6. XRD experimental (dots) and modelled (line) spectrum of 100nm sample treated 24 hours at 773K. The residual (exp. minus theor) is shown below.

The results of the modelling routine, for all 100nm films on etched iron, are reported in Figure 7. As shown in Figure 7a, zirconia crystallite size, 34nm, shrunk while microstrain increased after thermal treatments, probably because of iron oxide formation that started to crack the film. The same model can account for the decrease of zirconia lattice parameter, Figure 7b. The very large value for the as prepared film, 0.5270nm, with respect to the value for 12.5mol%CaO-stabilized zirconia powder, 0.513nm [17], was due to a high compressive stress in the film plane. In fact, for the adopted Bragg-Brentano geometry, the in-plane compressive stress corresponds to an increase in the measured interplanar distances of planes parallel to the surface. On the contrary, α-iron lattice parameter was almost constant, very close to the standard value of 0.28664nm. The zirconia film was highly oriented along [111] direction. The reduction of preferred orientation, expressed by the PGF parameter in Figure 7c (0=max, 1=no orientation), started only after 2 hours treatment, confirming that crack formation was absent until then. In Figure 7c the ratio between the relative phase percentages of Zirconia and iron, which should be a constant through all analyses, it is also reported. The fact that a decreasing trend was found indicates that iron oxides grew between the substrate and the film, reducing the intensity diffracted by the substrate.

Iron oxides crystallite size increased and microstrain decreased slowly as a function of treatment time, Figures 7d, 7e. Only Fe_3O_4 showed a small preferred orientation along [022] direction, almost absent after 24 hour treatment. The lattice parameters for iron oxides were much higher than standard values (a=0.50356nm, c=1.37489nm for Fe_2O_3 and a=0.8396nm for Fe_3O_4); the high stress relaxed after 24 hours, Figures 7f, 7g. Figure 7h reports the relative phase content of iron oxides. The values reported were obtained after a normalization to the phase percentage of zirconia, that is supposed to be a constant among all the analyses. The results indicate a higher content of Fe_3O_4 after 1 and 2 hours, when the film is still continuous, while Fe_2O_3 is higher after 24 hours.

300nm films, in spite of the increased stress due to greater thickness, withstood thermal treatments and were more effective against oxidation, as shown by TGA , Figure 3. XRD (not shown) revealed the presence only of Fe_3O_4 in the first 2 hours of treatment at 773K. This

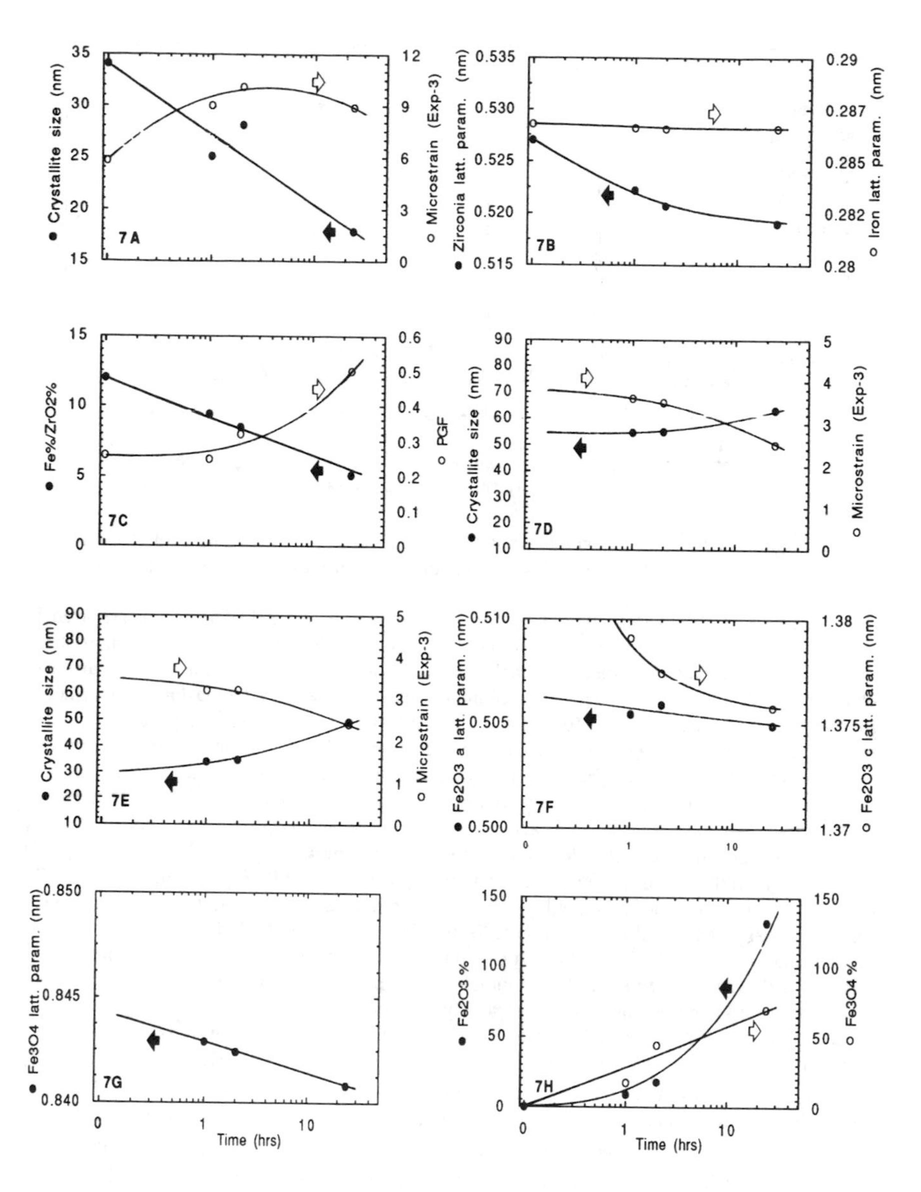

Fig. 7. Results of refinements for 100 nm samples: zirconia crystallite size and microstrain (7A), iron and zirconia lattice parameters (7B), PGF and percentage ratio (7C), crystallite size and microstain for Fe_2O_3(7D) and Fe_3O_4(7E), lattice parameters for Fe_2O_3 (7F) and Fe_3O_4(7G), phase content of Fe_2O_3 and Fe_3O_4(7H).

resulted from a decreased oxygen diffusion through the film in respect to 100nm coatings. AES, Figure 2c, showed clearly that iron oxide formed beneath the film; both after 1/2 hour and 2 hour treatments the interface remained sharp and film thickness was unchanged, confirming film stability.

CONCLUSION

Stabilized Zirconia thin films were deposited by r.f. magnetron sputtering onto iron substrates. Process parameters were adjusted to obtain a dense well crystallized film, to be effective in protecting the very reactive metal surface. In spite of the dense structure and large compressive intrinsic stress all coatings were adhered well and did not show any spalling effect due to thermal treatments. The presence of a thin native oxide layer on the metal surface and the substrate finishing (4000 paper or 0.1 μm diamonds) did not seem to affect adhesion and film properties. All 100nm films were protective against oxidation at 773K, but a 300nm thickness increased the protective action, reducing the oxidation rate, without worsening the thermomechanical properties.

ACKNOWLEDGEMENTS

This work was supported by CNR - Progetto Finalizzato Materiali Speciali per Tecnologie Avanzate. The authors wish to thank Mr. Nicolodi for technical support in film deposition and Dr. Lutterotti for the help in XRD data processing.

REFERENCES

[1] W.H. Lowdermilk, D. Milam and F. Rainer, Thin Solid Films, **73**, 155 (1980).
[2] W.T. Pawlewicz, D.D. Hays and M. Martin, Thin Solid Films, **73**, 169 (1980).
[3] K.V.S.R. Apparao, N.K. Sahoo and T.C. Bagchi, Thin Solid Films, **129**, L71 (1985).
[4] Y. Miyahara, K. Tsukada and H. Miyagi, J. Appl. Phys, **63**, 2431 (1988).
[5] W. Shi, J. Shi, W. Yao, Zh. Qi, Sh. Tang, G. Zhou, Thin Solid Films, **203**, 267 (1991).
[6] V.S. Stubican, G.S. Corman, J.R. Hellmann and G. Senft, in *Advances in Ceramics*, edited by N. Claussen, M. Ruhle and A. Heuer (The American Ceramic Society, Columbus, 1984), Vol. 12, p.96.
[7] R.J. Bratton and S.K. Lau, in *Advances in Ceramics*, edited by A. Heuer and L.W. Hobbs (The American Ceramic Society, Columbus, 1981), Vol. 3, p.226.
[8] R. Di Maggio, P. Scardi and A. Tomasi, Mat. Res. Soc. Symp. Proc., **180**, 481 (1990).
[9] R. Di Maggio, P. Scardi, A. Tomasi and F. Marchetti, Proceeding of the European Ceramic Society Second Conference, Augsburg, Germany (1991). In press.
[10] A.S. Kao and C. Hwang, J. Vac. Sci. Tachnol. A, **8** [4] 3289 (1990).
[11] F.C. Toriz, A.B. Thakker and S.K. Gupta, Surf. Coat. Technol., **39-40**, 161 (1989)
[12] M. Mesbah, A. Boyer, E. Groubert, J. Mat. Sci., **27**, 83 (1992).
[13] J.A. Thornton and D.W. Hoffman, Thin Solid Films, **171**, 5 (1989).
[14] P. Scardi and A. Tomasi. Unpublished work.
[15] L. Lutterotti, P. Scardi and P. Maistrelli, J. Appl. Cryst, **25** (1992).
[16] P. Scardi, L. Lutterotti, R. Di Maggio and P. Maistrelli, Mat. Sci. Forum, **79-82**, 233 (1992).
[17] M. Ruhle and A.H. Heuer, in *Advances in Ceramics*, edited by N. Claussen, M. Ruhle and A. Heuer (The American Ceramic Society, Columbus, 1984), Vol. 12, p.14.

PLASMA SPRAYED ALUMINA-TITANIA COATINGS

T.J. Steeper,[a] A.J. Rotolico,[b] J.E. Nerz,[b] W.L. Riggs II,[c] D.J. Varacalle, Jr.,[d] and G.C. Wilson[d]
[a]Savannah River Laboratory, Aiken, SC; [b]Metco/Perkin-Elmer, Westbury, NY;
[c]TubalCain Company, Loveland, OH; [d]Idaho National Engineering Laboratory,
EG&G Idaho, Inc., Idaho Falls, ID

ABSTRACT

This paper presents an experimental study of the air plasma spraying (APS) of alumina-titania powder using argon-hydrogen working gases. This powder system is being used in the fabrication of heater tubes that emulate nuclear fuel tubes for use in thermal-hydraulic testing. Experiments were conducted using a Taguchi fractional-factorial design parametric study. Operating parameters were varied around the typical spray parameters in a systematic design of experiments in order to display the range of plasma processing conditions and their effect on the resultant coatings. The coatings were characterized by hardness and electrical tests, surface profilometry, image analysis, optical metallography, and x-ray diffraction. Coating qualities are discussed with respect to dielectric strength, hardness, porosity, surface roughness, deposition efficiency, and microstructure. Attempts are made to correlate the features of the coatings with the changes in operating parameters.

INTRODUCTION

The plasma-spray process is one of the most versatile and rapid methods of applying coatings, and its applications continue to increase.[1] The process is commonly used to form ceramic coatings. Plasma spray guns typically use a non-transferred dc plasma torch configuration with gun powers ranging from 10 to 100 kW. The basic design of the plasma gun consists of a cone-shaped cathode inside a cylindrical anode, which forms a nozzle. An inert gas (usually argon, argon-hydrogen, or argon-helium) flows through the space between the electrodes where it is ionized to form a plasma. Powders injected into the plasma are then accelerated and melted by its high temperature. The molten droplets (i.e., splats) are propelled onto the substrate where they solidify and accumulate to form a thick, well-bonded coating. For quality coatings, the particles must absorb enough heat from the flame to melt thoroughly, but not so much that they overheat, oxidize, and/or vaporize. At the same time, the particles must travel fast enough to adhere well once they strike the substrate.

Some of the more common coating functions obtainable with the plasma-spray process include wear resistance, heat and oxidation resistance, corrosion resistance, electrical or thermal conductivity or resistivity, restoration of dimension, and clearance control. A multitude of operational parameters control the quality of the coating, making it difficult to find the optimum processing conditions for any specific application.[2,3]

As coating property requirements become more sophisticated, a better knowledge of the underlying principles is necessary for improved process control and coating quality. Plasma spraying has been developed to a large extent by empirical means, while the scientific research lags behind. This work attempts to further the scientific understanding of the physical mechanisms involved in the formation of ceramic coatings by determining which processing parameters affect the structure and properties of the coatings. Former work in this area centered on the use of super-fine ceramic alumina-titania powders [4] using argon-hydrogen working gases and alumina-titania powders sprayed with nitrogen-hydrogen working gases.[5]

Thermal spray technology is being used to fabricate heater tubes for use in thermal-hydraulic experiments.[6] These tubes are heated with a high-amperage dc power source to simulate nuclear fuel tubes. The heater tubes are fabricated using a multilayered, graded coating system (metal bond coat, ceramic insulator, metal conductor, ceramic insulator, aluminum skin). Plasma spraying is used to fabricate the bond coat, conductor, and two insulator layers, and a twin-wire electric arc system is used to fabricate the heater skin. For this application, the capability of thermal spray processes to apply very thin layers of insulating materials well bonded to an aluminum base tube, metallic conductor, and aluminum skin is crucial to the success of the heater tubes.

Plasma-sprayed alumina-titania coatings have been studied for the ceramic insulator of the heaters. These coatings have been used for varied applications in the automotive, transportation, aerospace, and aircraft industries. The fabricated coatings are extremely wear resistant, heat resistant, resistant to most acids and alkalies, and have high dielectric strength. In addition, bond strength is high, interparticle strength is high, and the finish is very smooth. The coatings exhibit little evidence of through porosity.

The alumina-titania coating used for the heater application must survive thermal cycling from thermal-hydraulic testing, which induces the tendency for spalling and destruction of the insulator coating. The coating must be thin enough so the heater will match the prototypical heat storage and transfer of a nuclear fuel tube. At the same time the coating must have sufficient dielectric strength to insulate the electrical conductor from the aluminum skin and aluminum substrate. If the electrical resistance at any point in the insulator layer is very low or zero (electrical short), either the insulator layer was damaged during spraying (i.e., cracking) or an insulator coating was too thin, allowing tendrils of molten or vaporized metal to follow the interconnected porosity and thus short through the insulator layer.

For this application, porosity, electrical resistance, and cracking of the ceramic insulator are the most important microstructural features that must be controlled for the construction of durable heaters.

EXPERIMENTAL PROCEDURE

A Taguchi-style,[7] fractional-factorial L8 design of experiment was employed to evaluate the effect of seven plasma processing variables on the quantitatively measured attributes. The quantitative Taguchi evaluation of the plasma spray process displays the range of measured coating attributes attainable, and it statistically delineates the impact of each factor on the measured coating characteristics across all combinations of other factors. The Taguchi analysis was done using measured responses with software on a personal computer.[8]

Experiments ATA01 through ATA08 represent the eight runs evaluated with the Taguchi L8 approach. The experiments are detailed in Table 1. Each variable has a high and low level selected to band around the nominal settings (i.e., Experiment ATA09). The parameters varied were current, primary gas flow, secondary gas flow, powder feed rate, spray distance, traverse rate, and substrate cooling.

A Metco MBN plasma spray system with a 9MB gun was utilized for this study. The primary gas was argon; the secondary gas was hydrogen; and the powder carrier gas was argon sprayed at 0.793 scmh (28 scfh) (console flow) for all nine experiments. The powder injection was external to the torch and directed perpendicular to the flow. An x-y manipulator ensured the standoff distance and repeatability in the experiments. A y-step of 0.0032 m (0.125 in.) and four passes were used to fabricate each of the coatings.

Commercially available thermal-spray powder (Metco 130 alumina-titania) was plasma sprayed onto 6061 aluminum plates (51 x 63 x 3 mm) cooled by air jets on the back side. Short tubes, as shown in Figure 1, were also coated. The substrates were grit blasted with No. 30 alumina grit before spraying. The substrates were then coated with a nickel-aluminum bond coat of approximately 75 μm (3 mils), which was plasma sprayed at the manufacturer's recommended process parameters.

Table 1. Alumina-Titania Thermal Spray Experiments ATA01 through ATA09

Experiment	Current (A)	Primary Flow scmh(scfh)		Secondary Flow scmh(scfh)		Feed Rate kg(lb)/h	Distance mm(in.)	Traverse mm(in.)/s	Cooling yes/no
ATA01	450	1.70	60	0.28	10	1.36(3.0)	88.9(3.5)	457.2(18)	N
ATA02	450	1.70	60	0.28	10	2.27(5.0)	114.3(4.5)	660.4(26)	Y
ATA03	450	2.27	80	0.425	15	1.36(3.0)	88.9(3.5)	457.2(26)	Y
ATA04	450	2.27	80	0.425	15	2.27(5.0)	114.3(4.5)	457.2(18)	N
ATA05	500	1.70	60	0.425	15	1.36(3.0)	114.3(4.5)	457.2(18)	Y
ATA06	500	1.70	60	0.425	15	2.27(5.0)	88.9(3.5)	660.4(26)	N
ATA07	500	2.27	80	0.28	10	1.36(3.0)	114.3(4.5)	660.4(26)	N
ATA08	500	2.27	80	0.28	10	2.27(5.0)	88.9(3.5)	457.2(18)	Y
ATA09	500	2.27	80	0.425	15	2.27(5.0)	114.3(4.5)	457.2(18)	Y

MATERIALS CHARACTERIZATION RESULTS

Table 2 lists the coating characterization results for this study. The attributes evaluated were ceramic thickness using optical microscopy, superficial hardness using a Rockwell 15N test, microhardness using a Vickers test, porosity using image analysis, dielectric strength, deposition efficiency, electrical resistance, and coating roughness using image analysis.

The coating thicknesses, as revealed by optical metallographic observations at 200X magnification, are listed in Table 2. Average thicknesses from 12 measurements of the alumina-titania layers ranged from 71 to 175 μm (2.8 to 6.9 mils), reflecting the influence of the various spraying parameters.

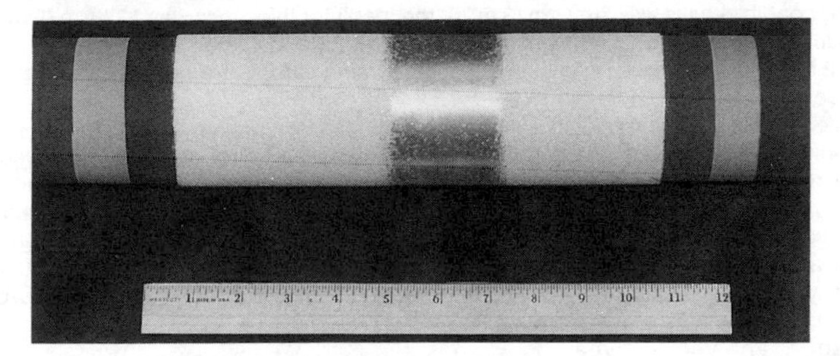

Figure 1. Heater tube.

Table 2. Coating Characterization Results for Experiments ATA01 through ATA09

Experi-ment	Thickness mils	Thickness μm	Hard-ness[a]	Hard-ness[b]	Porosity %	Dielectric Strength V/mil	Deposition Efficiency %	Roughness μm	Resistance (kΩ) Tube	Resistance (kΩ) Plate
ATA01	6.4	163	60	1009	8.71	230	78.7	1.66	65	900
ATA02	4.4	112	55.3	1070	8.82	197	74.4	1.70	132	8400
ATA03	3.8	97	57.3	1100	6.53	160	72.0	2.09	76	25
ATA04	5.75	146	65.6	1160	9.99	254	61.7	1.70	89	3200
ATA05	6.1	155	76.5	1050	4.60	178	86.2	1.90	125	2600
ATA06	6.1	155	65	1090	7.21	111	83.3	1.70	37	24
ATA07	2.8	71	51.5	903	7.13	208	51.5	2.10	23	57
ATA08	6.1	155	62.7	1095	9.96	236	51.3	2.08	62	650
ATA09	6.9	175	68.5	1120	9.89	292	61.8	1.73	100	3000

a. Superficial Rockwell 15N hardness measurement; b. Vickers microhardness values.

Porosity measurements for the coatings, as revealed by image analysis, are listed in Table 2. A Dapple Image Analyzer with a Nikon Epiphot metallograph was used for the metallurgical mounts. Image analysis procedures were first tested for sensitivity to parameter variance. The average porosity of the ceramic coatings ranged from 4.6 to 10%.

Superficial Rockwell hardness and Vickers microhardness measurements were taken on the coatings. The superficial Rockwell hardness measurement was taken normal to the deposit using the 15N method. Vickers microhardness measurements were taken on a cross section through the coating. Ten measurements were taken and averaged. The superficial Rockwell hardnesses ranged from 51.5 to 76.5; the microhardness measurements ranged from 903 to 1160.

Dielectric strength was determined using an Associated Research AC Hypot Model 4030. The test was conducted by applying an increasing voltage across the coating surface to the aluminum substrate using a 1-mm-diameter probe. Average values from three measurements ranged from 111 to 292 V/mil.

Deposition efficiency for the experiments was determined with conventional techniques by measuring the amount of sprayed ceramic deposited for an allotted time. The deposition efficiencies ranged from 51.3 to 86.2%.

Surface roughness was determined with image analysis. The data from each image were mathematically treated according to ANSI Standard B46.1, which indicates that roughness is calculated as the average departure y from the mean height in a given region. The average departure y was determined for 20 frames, and the 20 frames were averaged to yield the final measured roughness. The roughness of the coatings ranged from 1.66 to 2.10 μm (higher values are rougher).

The electrical resistance of the insulator coating was determined by spraying the first three heater coatings on 6061 aluminum tubes (diameters representative of the actual fuel tubes) and coupons, and then measuring the electrical resistance from the conductor to the base metal. The bond coat thickness was 76.2 μm (3 mils), the insulator thickness was 127 μm (5 mils), and the conductor thickness was 356 μm (14 mils). The tube insulator resistances ranged from 23 to 132 kΩ. This procedure was then duplicated using sample coupons. The plate insulator resistances ranged from 24 to 8400 kΩ.

By determining the best values for coating attributes, an improved coating for this particular application can be selected. This coating would have, in order of greatest importance, high insulator resistance to minimize power leakage, low porosity to enhance the thermophysical properties of the coating, high dielectric strength for a voltage barrier, low thickness for prototypical heat transfer of the actual fuel tubes, high hardness for toughness, rough surface finish for adhesive strength, and high deposition efficiency for cohesive strength. Using this philosophy, coatings ATA05 and ATA02 are the best coatings produced in this study. Coating ATA07 was mediocre for this application.

Image analysis revealed variances in the microstructures (i.e., porosity, cracking, unmelted particles) for the experiments. Figures 2, 3, and 4 illustrate the microstructures for coatings ATA02, ATA05, and ATA07, respectively. No macroscopic cracking (segmented cracking that appears either parallel or normal to the substrate) was apparent in the body of any of the coatings within the test matrix.

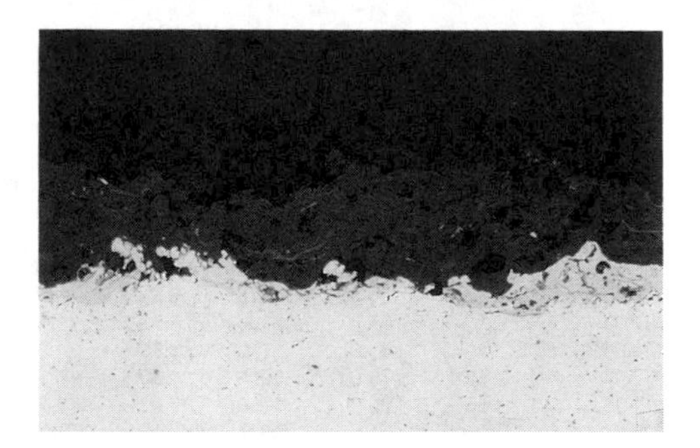

Figure 2. Microstructure of as-sprayed coating ATA02 (200X).

Porosity as small as 1 μm was revealed at a magnification of 500X. Surface examinations revealed splat-like formations, indicating that the material was predominantly molten upon deposition (see Figure 5, coating ATA05).

Coating ATA05 appears to be one of the smoothest of the as-sprayed samples, indicating the presence of molten conditions at deposition, which was verified by surface profilometry. This coating was produced with a primary argon gas flow of 1.7 scmh (60 scfh) and a secondary flow of 0.425 scmh (15 scfh) at 500 A. This was the highest energy case for the experimental scheme (i.e., power-to-flow volume ratio). The experiment also had the highest flow ratio of secondary hydrogen to the total flow. The addition of hydrogen to the working gas enhances plasma-particle heat transfer. The coating had the lowest porosity and high values of hardness and tube resistance. It is considered the best coating produced in this study. Coating ATA02 is also considered a good coating, despite the high porosity, with high values of tube resistance and dielectric strength.

Coating ATA07 had the lowest tube resistance, hardness, and deposition efficiency for the series and was also the thinnest coating. This coating was produced with a primary argon gas flow of 2.27 scmh and a secondary flow of 0.28 scmh at 500 A. This coating was produced with the lowest energy for the experimental scheme and the lowest ratio of secondary flow to the total flow. Therefore, higher power-to-volume ratios and higher ratios of secondary hydrogen flow to the total flow will result in more molten particles, better powder deposition, and lower porosity for this application.

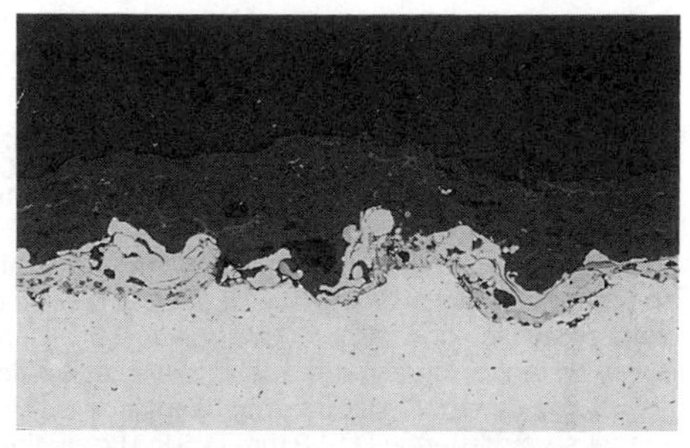

Figure 3. Microstructure of as-sprayed coating ATA05 (200X).

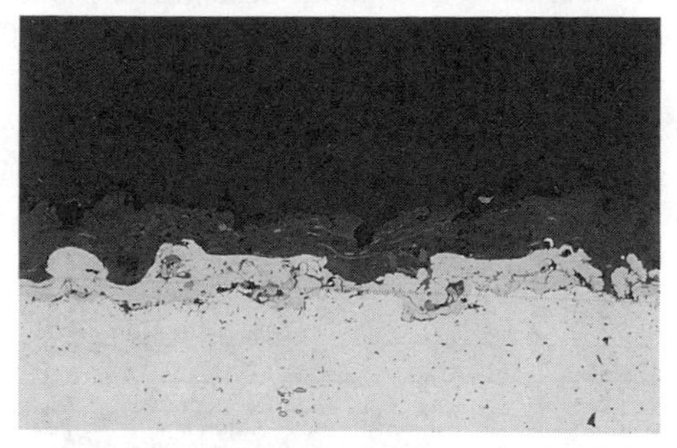

Figure 4. Microstructure of as-aprayed coating ATA07 (200X).

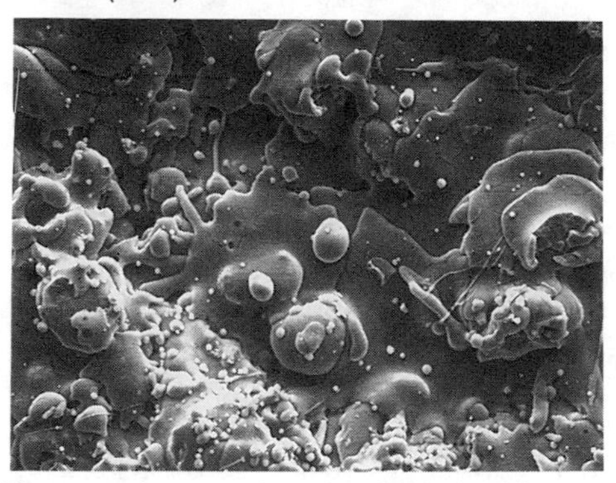

Figure 5. As-sprayed coating ATA05 (500X).

DISCUSSION OF TAGUCHI FRACTIONAL-FACTORIAL EXPERIMENT DESIGN

Taguchi-type fractional-factorial testing is an efficient means of determining broad-based factor effects on measured attributes. This methodology statistically delineates the impact of each variable on the measured coating characteristics across all combinations of other factors.

The spray tests were conducted and evaluated once, and all data points were considered in the analysis of variance (ANOVA) calculations. The rho percent (ρ%) calculation indicates the influence of a factor or parameter on the measured response, with a larger number indicating more influence. The ANOVA calculations guide further experimentation by indicating which parameters are the most influential on coating attributes. This information is extremely useful in developing new coating specifications.

The optimum coating for this application in (as shown in Table 3 in order of priority) would have high insulator resistance, low porosity, high dielectric strength, low thickness, rough surface finish, high hardness, and a high deposition efficiency. Table 3 illustrates the results of the Taguchi analysis.

The Taguchi evaluation indicated that substrate cooling was the most dominant contributor to tube resistance at 39.6 ρ%, with cooling resulting in higher resistance. Secondary influences were spray distance (20.1 ρ%), current (16.0 ρ%), and primary argon flow (14.3 ρ%), with longer spray distance, lower current, and lower argon flow resulting in higher resistance.

Spray distance was the most significant contributor to plate resistance at 35.0 ρ%, with longer spray distance [114.3 mm (4.5 in.)] resulting in higher resistance. Secondary influences were current (18.3 ρ%), powder feed rate (16.5 ρ%), and primary argon flow (14.0 ρ%), with lower current, higher powder feed rate, and lower argon flow resulting in higher resistance.

Powder feed rate was the most significant contributor to lowering porosity at 42.4 ρ%, with lower feed rate resulting in lower porosity. Secondary hydrogen flow was 20.7 ρ% with the higher flow rate resulting in lower porosity.

Dielectric strength was most influenced by traverse rate at 40.8 ρ%, with the slower traverse rate resulting in higher dielectric strength. Other contributors were secondary and primary flow rate at 23.4 and 16.7 ρ%, with the lower flow of hydrogen and higher flow of argon resulting in higher dielectric strength.

Table 3. Results of the Taguchi Analysis

	Processing Factor						
Desired Attribute	Amperes ρ%/Level[b]	Argon flow ρ%/Level[b]	H_2 flow ρ%/Level[b]	Feed rate ρ%/Level[b]	Spray distance ρ%/Level[b]	TR[a] ρ%/Level[b]	Cooling Y/N
1 High Resistance tube	16.0/L	14.3/L	2.44/H	1.16/H	20.1/H	6.4/L	39.6/Y
2 High Resistance plate	18.3/L	14.0/L	3.8/L	16.5/H	35.0/H	0.3/H	12.1/Y
3 Low Porosity	13.9/H	9.5/L	20.7/H	42.4/L	1.8/H	6.7/H	5.1/Y
4 High Dielectric Strength	9.7/L	16.7/H	23.4/L	0.4/H	8.3/H	40.8/L	0.9/N
5 Low Thickness	0.6/L	20.7/H	4.2/L	10.5/L	11.2/H	52.4/H	0.4/Y
6 High Surface Finish	17.9/H	46.0/H	1.0/H	14.7/H	0.8/H	2.8/H	16.8/Y
7 High Hardness (Rockwell)	9.3/H	11.7/L	36.8/H	0.3/H	0.5/H	38.5/L	2.9/Y
8 High Micro-hardness	12.3/L	0.5/H	31.7/H	37.8/H	3.7/L	6.9/L	7.1/Y
9 High Deposition Efficiency	2.0/L	71.3/L	21.5/H	3.0/L	1.3/L	0.1/H	0.7/Y

a. TR = traverse rate; b. Level = parameter level, H = high, L = low.

Coating thickness buildup was dominated by traverse rate (52.4 ρ%). Increasing the traverse rate will result in thinner coatings. Other equal contributors were primary flow rate, powder feed rate, and spray distance.

Surface finish is most influenced by the primary argon flow rate (i.e., 46.0 ρ%). Current (17.9 ρ%), substrate cooling (16.8 ρ%), and powder feed rate (14.7 ρ%) were secondary influences. Using the higher values for the process parameters with substrate cooling will result in a rougher coating, which is desired for better bonding in the multilayered-coating heater design.

Superficial hardness increase (i.e., Rockwell) was equally influenced by a slower traverse rate (38.5 ρ%) and higher hydrogen flow (36.8 ρ%). Microhardness increase (i.e., Vickers) was most influenced by a larger powder feed rate (37.8 ρ%) and a larger hydrogen flow rate (31.7 ρ%).

Deposition efficiency was most influenced by the primary-to-secondary flow split. Lower argon flow (71.3 ρ%) and higher hydrogen flow (21.5 ρ%) resulted in higher deposition efficiency.

Selection of the optimum levels of the design factors can produce an improved coating for this particular application. This coating would have high insulator resistance, low porosity, high dielectric strength, low thickness, high hardness, rough surface finish, and high deposition efficiency. This coating can be obtained by using a current of 450 A, a primary argon gas flow of 1.7 scmh (60 scfh), a secondary hydrogen flow of 0.425 scmh (15 scfh), a powder feed rate of 1.36 kg/h (3.0 lb/h), a spray distance of 114.3 mm (4.5 in.), a traverse rate of 660.4 mm (26 in.)/s, and substrate cooling. For this study, Experiment ATA05 had the closest process parameters to these optimized parameters.

The Taguchi evaluation employed in this study is significant because it directs further experimentation that considers the most important process or coating attributes and the process parameters that affect these attributes. The most important attributes may differ for the same material in different applications, and the baseline data generated in this study can be used to develop specific confirmation runs that approach the desired application attributes.

SUMMARY AND CONCLUSIONS

An experimental study of the plasma spraying of alumina-titania powder using argon-hydrogen working gases has been presented. Experiments employed a Taguchi fractional-factorial approach with typical process parameters. The coatings were characterized by hardness tests, electrical tests, surface roughness, image analysis, and optical metallography. Coating qualities were determined with respect to insulator resistance, dielectric strength, hardness, porosity, deposition efficiency, and microstructure.

The alumina-titania coating thicknesses, reflecting influences of spraying parameters, ranged from 71 to 175 μm. Insulator tube resistance ranged from 23 to 132 kΩ. Insulator plate resistance ranged from 24 to 8400 kΩ. Porosity for the coatings, as revealed by image analysis, ranged from 4.6 to 10%. The superficial Rockwell hardnesses ranged from 51.5 to 76.5, and the microhardness measurements ranged from 903 to 1160. Dielectric strength measured for the coatings ranged from 111 to 292 V/mil. Deposition efficiencies ranged from 51.3 to 86.2%. Surface roughness ranged from 1.66 to 2.10 μm.

The Taguchi evaluation indicated that powder feed rate and hydrogen flow were the most significant contributors to porosity. Spray distance dominated the insulator plate resistance. Substrate cooling dominated the tube resistance. Surface finish was most influenced by the primary argon flow. Coating thickness buildup was dominated by traverse rate. Deposition efficiency was most influenced by the primary-to-secondary flow split. Dielectric strength was most influenced by traverse rate. Rockwell hardness was equally influenced by traverse rate and secondary hydrogen flow. Vickers hardness was influenced by powder feed rate and secondary hydrogen flow. An improved coating for this particular application can be obtained by using a current of 450 A, a primary argon gas flow of 1.7 scmh (60 scfh), a secondary hydrogen flow of 0.425 scmh (15 scfh), a powder feed rate of 1.36 kg (3.0 lb)/h, a spray distance of 114.3 mm (4.5 in.), a traverse rate of 660.4 mm (26 in.)/s, and substrate cooling. Experiment ATA05 had the closest process parameters to these parameters for this study.

The objective of this and future work is to optimize alumina-titania coatings. After baseline data are generated on factors that influence coating characteristics, other important characteristics must be quantitatively evaluated using a similar design of experiment approach. Then the design engineer wishing to utilize an alumina-titania coating can review the ranges of coating characteristics generated through the plasma spray process. Coatings of known structure can be applications-tested to prioritize the characteristics that impact the coating performance in a specific application. From this methodology, processing parameters can be adjusted, optimized, and confirmed. A realistic specification can be made for the coating as sprayed, and ultimately, the specification will be transferred back to the control parameters only.

The procedure described in this paper will assist in selecting and optimizing operational parameters for future alumina-titania plasma spray processing experiments and applications. Future work is needed in analytical modeling to obtain a better understanding of the process.

ACKNOWLEDGMENTS

The technical input provided by V. L. Smith-Wackerle, INEL, is gratefully acknowledged. The work described in this paper was supported by the U.S. Department of Energy, Assistant Secretary for Defense, under DOE Idaho Field Office Contract No. DE-AC07-76ID01570.

REFERENCES

1. E.D. Kubel, Advanced Materials and Processes, 132 (6) 69-80 (December 1987).
2. E. Pfender, "Fundamental Studies Associated With the Plasma Spray Process," Proceedings of the National Thermal Spray Conference, (Orlando, FL, September 14-17, 1987).
3. P. Fauchais, "State of the Art for the Understanding of the Physical Phenomena Involved in Plasma Spraying at Atmospheric Pressure," Proceedings of the National Thermal Spray Conference, (Orlando, FL, September 14-17, 1987).
4. T.J. Steeper, D.J. Varacalle, Jr., G.C. Wilson, A.J. Rotolico, J.E. Nerz, and W.L. Riggs II, "A Taguchi Experimental Design Study of Plasma Sprayed Alumina-Titania Coatings," Thermal Spray Coatings: Properties, Processes, and Applications, (American Society of Metals National Thermal Spray Conference, Pittsburgh, PA, May 1991).
5. T.J. Steeper, D.J. Varacalle, Jr., G.C. Wilson, A.J. Rotolico, J.E. Nerz, and W. L. Riggs II, "A Design of Experiment Study of Plasma Sprayed Alumina-Titania Coatings," Thermal Spray: International Advances in Coating Technology, (American Society of Metals International Thermal Spray Conference, Orlando, FL, June 1992).
6. T.J. Steeper, D.J. Varacalle, Jr., G.C. Wilson, and V.T. Berta, "Use of Thermal Spray Processes to Fabricate Heater Tubes for Use in Thermal-Hydraulic Experiments," Thermal Spray Coatings: Properties, Processes, and Applications, (American Society of Metals National Thermal Spray Conference, Pittsburgh, PA, May 1991.
7. G. Taguchi and S. Konishi, Taguchi Methods: Orthogonal Arrays and Linear Graphs, (ASI Press 1987).
8. Culp, R.F., SADIE, (1989).

A Microporous Silica Membrane

Wilhelm F. Maier* and Herbert O. Schramm
Department of Chemistry, Universität Essen and Max-Planck-Institut für Kohlenforschung, D-4330 Mülheim 1, Federal Republic of Germany

ABSTRACT

The separation properties of a microporous silica membrane, prepared be e-beam evaporation of quarz onto a mesoporous support membrane is reported.

While polymer membranes are already of increasing technological impotance, there are only few applications of inorganic membranes yet. Inorganic membranes can be thermally regenerated and have a high thermal and chemical resistance. The potential application of inorganic membranes are equilibrium shift of high temperature reactions, gas and liquid separation and heterogeneous catalysis - applications where the ceramic membranes supplement rather than substitute the successful polymer membranes. The central problem limiting the application of inorganic membranes is the separation mechanism, which is fundamentally different from the separation mechanism of polymer membranes. In polymer membranes the separation is largely guided by differences in interaction or solubility of the feed molecules with the polymer matrix.

Gas separation in ceramic membranes is dominated by the pore size (see scheme 1). In large (macro) pores the gas diffusion can be modeled by the laws of laminar or turbulent flow. The permeability is proportional to the pressure an the square of the pore radius and independent of molecular properties. There is no separation at large pore diameters. In pores smaller than the medium free path way of the gas molecules Knudsen flow is observed. Here molecules interact more often with the pore wall than with other gas molecules and their diffusion becomes dependent on their flow speed, which is proportional to the pore diameter and inversely proportional to the square root of the molecular weight of the gas molecule. The permeability is inversely proportional to the square root of the molecular weight. The separation limit is given by the Knudsen separation factor α. Due to this dependence there are no impressive separation factors achievable with Knudsen flow conditions, separation follows the molecular weight and is most efficient with molecules of very different molecular weight. H_2, for example, will flow through such a pore 4 times faster than O_2.

At pore diameters of 4-5 times the molecular dimension or smaller the molecules spend more time interacting with the pore walls than in free flow inducing the molecules into

Scheme 1: Differences in gas permeation in rigid pores of decreasing diameter at a given temperature. Arrows indicate medium free pathway

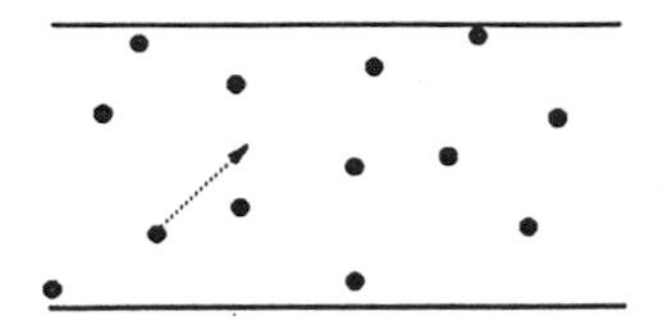

Gas flow at large pore diameters:
no separation - regular gas diffusion

$$F_0 \sim \frac{\varepsilon c_1 r^2 P}{\tau \eta}$$

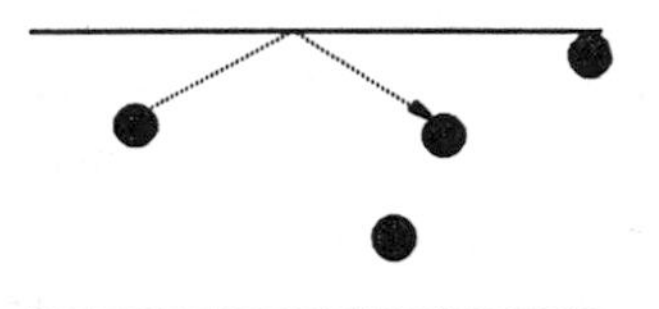

gas flow at smaller diameters (2 - 30 nm):
Knudsen flow - the lighter the faster

$$F_0 \sim \frac{\varepsilon c_2 r}{\tau M^{1/2}} \qquad \alpha = \sqrt{\frac{M_2}{M_1}}$$

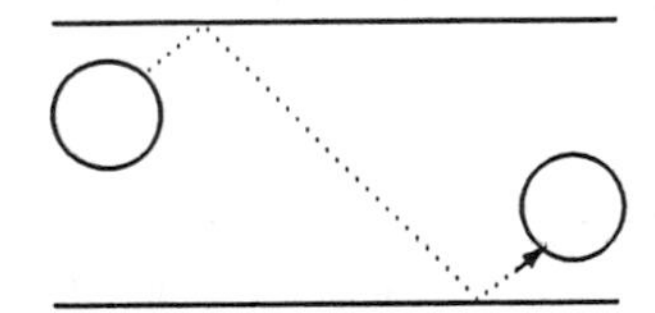

gas flow at very small diameters (< 2 nm):
surface diffusion - transport of adsorbed molecules

$$F_0 \sim \frac{\varepsilon c_3 D}{\tau r}\left(\frac{dX}{dp}\right)$$

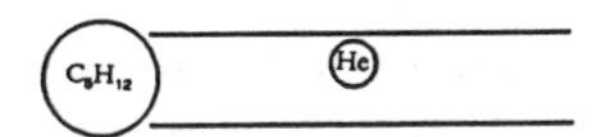

pores of molecular size:
molecular sieving

F_0	= permeability	P	= mean pressure in the system
c_n	= constants	η	= viscosity
ε	= porosity	D	= surface diffusion coefficient
τ	= tortuosity	X	= portion of monolayer, that occupies the surface
r	= mean pore radius	p	= pressure
M	= molar mass	α	= separation factor

an adsorbed state. Diffusion on the pore surface or the pore wall may become more significant. For highly condensible gases or gases that adsorb more easily due a favorable interaction with the surface diffusion may also be seen at larger pore diameters. The permeability for adsorbed molecules is dependent on the surface coverage and the surface diffusion coefficient and inversely proportional to the pore radius.

Since the surface coverage will be the larger the smaller the vapor pressure for a given molecular class the surface diffusion efficiency will be inversely proportional to the molecular weight. This means, that upon reducing the pore size the larger (more adsorbed) molecules start to diffuse better through the pores than the smaller ones. For a given mixture of two molecules at given separation conditions there will be a pore size, where there is no separation at all when Knudsen flow and surface diffusion are operating and cancel each others separation effect. Surface diffusion has already been observed in 1966 on Vycor glas.[1] Condensable gases (isomeric pentanes and other hydrocarbons) were found to diffuse better through Vycor membranes than hydrogen and noble gases.[2] On mesoporous alumina membranes there is Knudsen behavior with light gases and increasing permeability for hydrocarbons with size.[3] On silica membranes Knudsen flow is evident for inert gases, while some condensable gases show increased permeation.[4]

When pores get even smaller, molecules of a certain size become too large to enter the pores. At this point the separation is dominated by molecular sieving and only molecules smaller than the pore diameter can penetrate the membrane. Clearly, the smaller the medium pore diameter the more sensitive is the membrane to pin holes and other membrane defects. The ease of preparation of defect free membranes will therefore determine the potential success of inorganic membranes in the future.

Because of the rather poor separation factors at Knudsen flow and the separation reversal at conditions of surface diffusion inorganic membranes need a very narrow pore size distribution and have to be tailor made for individual applications. The separation by molecular size appears to be the most promising and attractive goal for the development of inorganic membranes. Ilias and Govind have stated in 1989,[5] that the availability of inorganic membranes with pores in the range of 1 nm or less will open new areas in membrane reactor technology and gas separation.

While microporous membranes are being prepared most commonly from zeolites by embedding in a polymer or by Sol-Gel procedures, we have found that such membranes can readily be prepared by e-beam evaporation. When silica is evaporated at 10^{-5} mbar and condenses onto a cold surface, the resulting film is microporous with a narrow pore size distribution and a mean pore diameter of 0,6 ± 0,2 nm.[6] The pore size distribution was determined from the nitrogen adsorption isotherm by t-plot methods and by the method of Horvath and Kawazoe.[6] Such films have been used already in heterogeneous catalysis to shield Pt films as the catalyst metal from direct interaction with organic substrates before the microporous structure had been identified.[7] It was concluded, that hydrogen diffuses through these membranes effectively, gets activated at the silica-Pt interface and the activated hydrogen returns to the outer surface in a spillover fashion to react with the organic substrate.[8]

We have studied the use of these microporous silica films as membranes. To obtain a sufficient flow it is essential to keep the separating membrane very thin, which requires the use of a support membrane. As support membranes alumina disc membranes (Anotec, diameter 46 mm) with a pore diameter of 20 nm and porous silica capillaries (Bioran, Schott) with a medium pore diameter of 10 nm (outer diameter 0.4 mm, length 120 mm) were used.

The BIORAN capillaries were mounted in a deposition chamber, so that they could be rotated continuously (40 rpm) during deposition and pretreatment. To clean the capillaries they were treated by glow discharge at about 1 torr for 15 min before silica deposition. At a pressure of 10^{-5} to 10^{-4} torr the silica film of various thickness was deposited by electron beam evaporation of fused quarz, deposition rate 10 nm/min. The capillary was mounted in a simple stainless steel t-connector with swage lock fittings by silicon rubber or by epoxy resin. After hardening over night the permeation of single gases and the separation of gas mixtures was studied in with these capillaries (see figure 1). The permeate and the retentate flow were monitored by a on-line quadrupole and by flow meters. The total permeate flow was about 1% of the feed flow, which was usually about 20 ml/s on a single capillary. Table 1 lists some separation factors obtained from the permeation of selected gases. There is little difference between the separation factors obtained with the support alone and the support with the silica membrane except for the largest molecule isobutane. All gases except isobutane show Knudsen behavior. At this point it is not clear, whether the highly reduced permeation of isobutane is due to pore condensation causing a hindered permeate flow or wether there is already a hindered free flow due to the molecular size (molecular sieving). The reduced permeation of the larger molecule is evidence against surface diffusion, which should bring about enhanced flow of the larger molecule. With an equimolar gas mixture of isobutane and hydrogen

Figure 1: Schematic single capillary reactor for the permeation studies

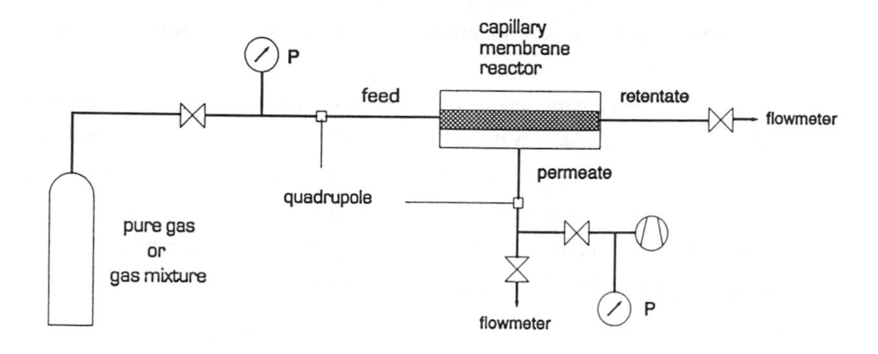

Table 1: separation factors of hydrogen for the gases indicated obtained from the permeation rate of single gases at a pressure of 0.1 bar and a permeation of 21 ml/cm²s. (Capillary length 10 cm, film thickness 500 nm).

	He	CH_4	N_2	CO	Ar	i-Bu
Knudsen-limit:	1,4	2,8	3,7	3,7	4,5	5,4
support only	2,3	2,1	3,5	3,6	4,3	5,0
silica film	2,1	2,8	3,3	3,2	4,0	320 ± 19

the separation factor dropped to 26 ± 3 on the microporous membrane, clearly better than the Knudsen limit but not as large as the flow difference of the individual gases.

The use of the alumina support membrane confirmed the trend already observed with the capillary support. The separation membrane was deposited contiuously at a deposition rate of 15 nm/min on the alumina membrane by e-beam evaporation at 10^{-5} to 10^{-4} torr until a thickness of 520 nm was obtained. Without any further treatment the membrane was mounted in a simple filtration equipment. The permeation of He and of hydrogen were determined independently. The H_2/He separation factors exceeds the Knudsen factor with and without the microporous silica membrane indicating a special interaction of He with the support membrane. All other gases show Knudsen behavior.

Attempts to improve the separation quality of the membranes by clogging the pores with various chlorosilanes have failed. Although we have obtained a further reduction in total flow by a factor of 2-3, there was no improvement in separation factors. Despite the large separation factor for isobutane we are convinced, that the flow of gases through these membranes is still dominated by defects. We hope to be able to demonstrate superior separation of small molecules in the close future on defect free inorganic membranes.

Acknowledgement: We are grateful for the support by the Bundesministerium für Forschung und Technologie (BMFT), project number O3C257 5.

References:

1. S.-T- Hwang, K. Kammermeyer Can. J. Chem. Eng. (1966) 82.

2. A.L. Wright, A.S. Berman, S. Prager J. Phys. Chem. 82 (1978) 2131.

3. F. Suzuki, K. Onozato, Y. Kurokawa J. Noncryst. Solids 94 (1987) 160.

4. R.J.R. Uhlhorn, M.H.B.J. Huis In't Veld, K. Keizer, A.J. Burggraaf J. Mater. Sci. Lett. 8 (1989) 1135.

5. S. Ilias, R. Govind JAIChE 31 (1989) 18.

6. W.F. Maier, H. Schramm, M. Wiedorn Angew. Chem. 103 (1991) 1523; Int. Ed. 30 (1991) 1509.

7. A.B. McEwen, R.H. Fleming, S. Baumann, W.F. Maier Nature 329 (1987) 531; J.M. Cogen, K. Ezaz-Nikpay, R.H. Fleming, S. Baumann, W.F. Maier Angew. Chem. **1987**, 99, 1222.

8. W.F. Maier, Angew. Chem. 101 (1989) 135.

THERMAL STABILITY OF SUPPORTED TITANIA MEMBRANES

KRISHNANKUTTY-NAIR P. KUMAR, V.T. ZASPALIS, F.F.M. DE MUL*, KLAAS KEIZER AND ANTHONIE J. BURGGRAAF
Laboratory of Inorganic Chemistry, Materials Science and Catalysis, Faculty of Chemical Technology and *Faculty of Applied Physics, University of Twente, P.O. Box 217, 7500 AE Enschede, The Netherlands.

ABSTRACT

The beneficial effects of the support constraint on the improved thermal stability of supported titania membranes were studied by following the anatase to rutile phase transformation in supported and unsupported titania membranes. This was studied using Raman spectroscopy and XRD. Supported membranes showed a higher transformation temperature, about 150°C higher, (slower rate of transformation) compared to the unsupported ones. Unsupported membranes showed a slight thickness dependence on the phase transformation temperature, but less significant compared to the difference in phase transformation behaviour between supported and unsupported membranes of similar thicknesses.

INTRODUCTION

Ceramic membranes are gaining more and more importance because of their high thermal stability compared to organic membranes. When a ceramic membrane is subjected to high temperatures, the following thermally activated process can occur: viscous sintering, crystallization and grain growth followed by densification and consequent change in pore size and size distribution. At high temperatures, support-membrane interaction is also important in the case of supported membranes. Certain types of ceramic membranes, especially titania and alumina, can also undergo phase transformations from one crystalline form to another. These transformations can often be accompanied by a volume change causing the membrane to crack or peal off from the support. The transformation can also cause a drastic reduction in the porosity, and an increase in the mean pore diameter of the membranes [1]. However, very few studies have been carried out concerning the thermal stability of supported membranes [1-4]. It has been shown that [1] the major cause of surface area and porosity reduction in unsupported titania membranes is the enhanced sintering of the rutile phase during the transformation. For all practical purposes, it can be assumed that a membrane with higher anatase to rutile transformation temperature will have a better stability. Therefore understanding the phase transformation behaviour will give an indirect indication of the textural stability of the membranes. In this paper, we report the phase transformation behaviour of supported and unsupported titania layers.

EXPERIMENTAL

Supported titania membranes and unsupported gel layers were produced from a titania sol synthesized by the hydrolysis of titanium isopropoxide. Titanium isopropoxide (TIP), diluted with isopropyl alcohol (IPA), is hydrolysed to give a white precipitate of titanium oxyhydroxide. This step is followed by filtration, washing and peptisation with HNO_3 to give a stable sol. The hydrolysis reaction is very complex and the reaction produces polycondensates with different chemical compositions which depend upon the hydrolysis conditions like water to alcohol ratio, type of catalyst and the reaction temperatures. Unsupported gel layers of three different thickness were formed by pouring controlled amounts of 0.3 mole titania sol in a glass petri dish. The excess water in the sol was allowed to evaporate to form a hydrogel layer. The formed layer is dried and peeled off from the glass dish. The thickness of the layers were 7 μ (USM-7), 15 μ (USM-15) and 50 μ (USM-50). Supported membranes (designated as SM-6) with 6 μ thickness are formed on a porous α-alumina support by dipping one side of the support in the sol for few seconds. A layer will be formed on the support by a slipcasting mechanism [2].

The membranes are dried at 40°C and 60% relative humidity and calcined at different temperatures. For both supported and unsupported membranes, all calcinations were carried out at a very slow heating rate of 0.2°C/min up to 450°C, then cooled down to room temperature at a rate of 0.2°C. These membranes are further calcined at different temperatures ranging from 450 to 1200°C with a heating rate of 1.7°C/min. They were kept at the set temperature for 8 h, and then furnace cooled by switching off the power. All of the calcination temperatures discussed in this paper correspond to the second dwell period. A detailed procedure for making sols and supported titania membranes is given elsewhere [2].

Thermal decomposition characteristics were measured by a Polymer Lab. Thermal Science DTA at a heating rate of 10°C/min in air. The DTA measurements were performed using platinum sample cups with α-alumina as the reference. All the measurements were carried out in flowing air with a flow rate of 20 ml/min. The Raman spectra were recorded using a SPEX-Triplemate spectrometer with microscope attachment (objective 50X ; LASER spot size = 2 μm; focal depth=5 μm) equipped with an EGG multichannel detector. The laser source was a Spectra Physics Ar^+-Laser operating at a wavelength of 514.5 nm, with 25 mW LASER power focused at the spot. Typical measuring times were 1-20 sec.

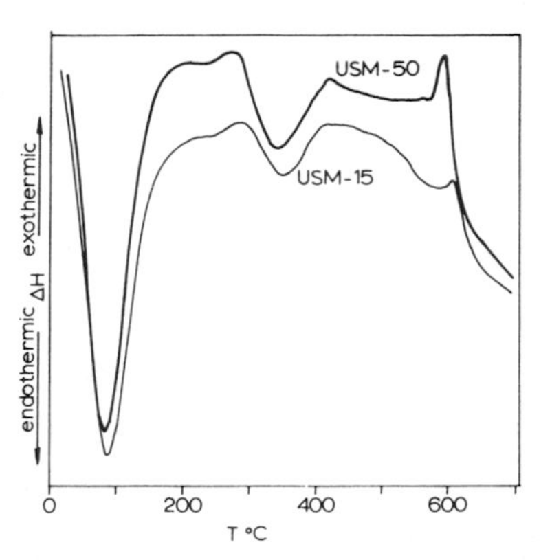

Fig.1 DTA traces of USM-50 and USM-15.

RESULTS AND DISCUSSIONS

DTA traces of USM-15 and USM-50 are illustrated in Fig.1. There are two endothermic peaks, one at c.a. 95°C and the second one at c.a. 350°C. The first peak represents the removal of physically adsorbed water from pores of the gel. The second peak represents decomposition of residual nitrates and the organics. There is a third peak in the range of 580 to 600°C, that corresponds to the enthalpy released during the anatase to rutile phase transformation. The details about the thermal behaviour of sol-gel derived titania membranes are reported else where [1]. We can clearly see a small shift in the peak corresponding to the phase transformation temperature. Furthermore, the area under the peak is higher in the case of thicker membranes, although all of the thermal analysis were done with same sample size. This result shows that processes of phase transformation and recrystallization are sluggish in the case of thin membranes. This fact can be seen from the X-ray step scan spectra of USM-50 and USM-15 given in Fig.2. At 600°C, USM-15 has a clear peak corresponding to anatase at 25.4°(2θ). In the case of USM-50 the peak corresponding to anatase is relatively week compared to USM-15. The strong peaks (for both USM-50 and 15) at 27.5°(2θ) represent the rutile phase.

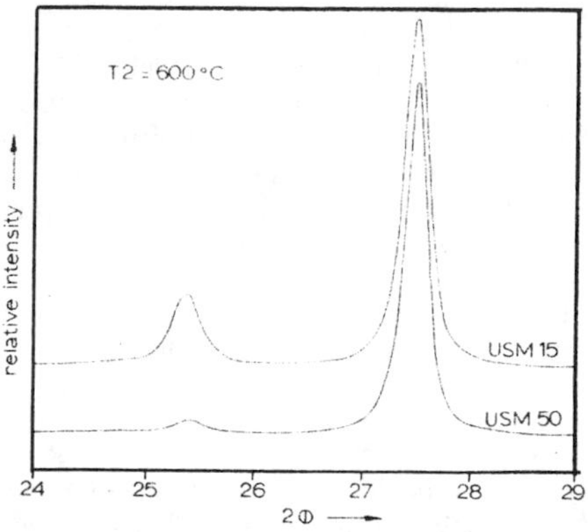

Fig.2 X-ray step scan spectra of USM-50 and USM-15 calcined at 600°C.

Anatase and rutile polymorphs of titania give distinct and characteristic Raman spectra. In addition, Raman spectra will give information about transformations occurring at or very near to the surface such as surface residual stress controlled nucleation. The most important capability is that the measurement of Raman intensities of different phases can be used to quantify the amount of phase present without much standardization [5]. However care should be taken to use integrated intensities of the Raman bands, for quantitative calculations, in order to avoid band broadening effects like differences in crystallite size.

Raman spectra that were recorded for SM-6 and USM-50 (the spectrum of USM-50-W is exactly similar to that of USM-50) are illustrated in Fig.3 a & b. All the samples were heated at 700°C for 8 h at the same time in the same furnace. Typically, the supported membranes show an anatase spectrum. All the peaks marked (▾) belong to anatase while the ones marked (•) belong to rutile. The crystallographic unit cell of anatase is body centred (space group D_{4h}^{19}, $I4_1/amd$) and contains two primitive unit cells, each of which contains two formula units of TiO_2. According to the factor group analysis [6], six modes, (A_{1g} + $2B_{1g}$

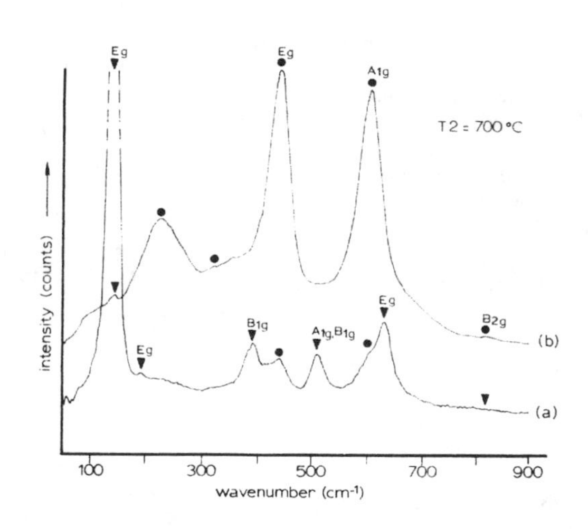

Fig.3 Raman spectra of (a) SM-6 and (b) USM-50 calcined at 700°C for 8 h.

+ $3E_g$) are Raman active. In the present case, all of the six bands are observed. Only the first overtone of the first-order mode with B_{1g} at 398 cm^{-1} (796 cm^{-1}) is not observed. However, the observed peaks have a slight shift (up to 5 cm^{-1}) with respect to the published results [6]. This shift is probably due to the residual stress in the supported membrane [7] or due to the difference in oxygen stoichiometry [8-10]. Moreover, the relative intensities of the peaks are not in agreement with the published data [6]. This difference is believed to be due to the crystallographic ordering of titania primary particles in the membranes. This aspect is currently under investigation.

In Fig.3 (See curve b), a Raman spectrum of an unsupported titania layer (USM-50) is illustrated. This material is predominantly rutile at this temperature. The rutile structure has two titania molecules in the unit cell with the space group D_{4h}^{14} ($P4_2/mnm$) [6]. According to group theory four modes are Raman active, (A_{1g}+ B_{1g} + B_{2g} + E_g). Similar to the anatase spectrum, the rutile spectrum also has slight band shifts. The small band at 142 ± 2 cm^{-1} is probably due to the traces of anatase phase in the sample [8]. The relatively large band at 230 ± 3 cm^{-1} is not included in the set of first order bands predicted by the group theory [6]. All of the other bands correspond to the predicted bands for the rutile system. The spectrum of an unsupported layer without HNO_3 (USM-50-W) is exactly identical to the one illustrated in Fig.3 (See curve b). This observation clearly shows that the difference in the phase transformation behaviour of supported and unsupported titania membranes is due to something other than the slight difference in chemistry.

Fig.4. shows the fraction of rutile phase, calculated from Raman intensities, that is present in the membranes after calcination at different temperatures. The fraction of rutile present (C_R) was calculated on the basis of the semi quantitative equation [5] given below:

$$C_R = K (I_R^{447} \backslash(I_R^{447} + I_A^{144})) \quad (1)$$

K is a constant which is approximately equal to unity while I_R^{447} and I_A^{144} are the integrated Raman intensities of 447 cm^{-1} band of rutile and the 144 cm^{-1} band of anatase, respectively. The area of analysis was a circle of 10 μ diameter, corresponding to $\approx 10^9$ grains (if we assume the grain size of titania to be ca. 10 nm).

Fig.4 a, b and c represent plots for unsupported membranes with 50, 15 and 7μm, thick respectively. The 50μ thick membrane has a slightly lower transformation rate compared to the thinner ones (15 and 7μ). However, the supported membrane shows a considerably higher transformation temperature compared to both the unsupported membranes. Initially, this difference in phase transformation behaviour of USMs with thickness was believed to be due to the residual water in these gels. However, as mentioned above, this effect is not significant.

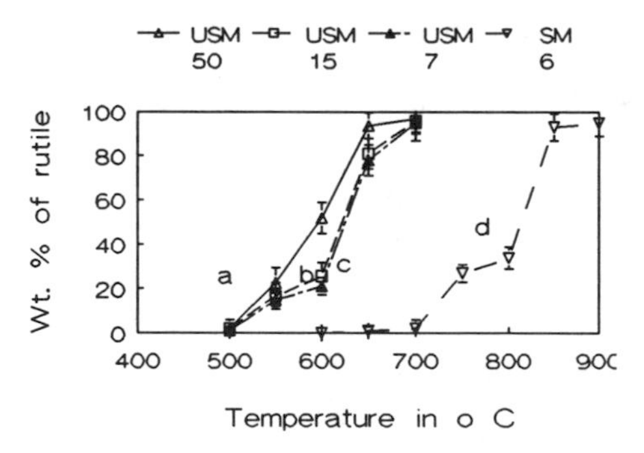

Fig.4 Fraction of rutile present vs calcination temperature (8 hrs soaking time) of titania membranes.

As mentioned earlier there is a volume change of about 8% during the anatase to rutile phase transformation. In the case of supported membranes this -ve volume change is prevented by the rigid support. The in plane stress, σ_x, developed due to this volume change can be expressed as [10]:

$$\sigma_x = (E/(1-\nu))(\Delta V/3V) \quad (2)$$

where E is the Young's modulus, ν is the poisson's ratio and ($\Delta V/V$) is the volumetric strain, which is about 0.08. Assuming a E of 50 GPa for porous titania (~ 30% porosity) we obtain a value of 1777 MPa for the in plane stress. This can be taken as the upper limit, however we should consider stress relaxation due to creep and can assume a value which is half of the calculated value. This stress value is in agrement with the observed Raman shifts of 143 cm^{-1} mode of anatase phase in the partially transformed supported membranes measured at room temperature. The peak shift was always found to be -3 to -5 cm^{-1} indicating a tensile stress of 990 to 1660 MPa (assuming a shift of 3 cm^{-1}/GPa [10]).

From the Clapeyron equation [11], the change in the phase transformation temperature, ΔT, can be written as

$$\Delta T = (T\Delta V / \Delta H) \Delta P \quad (3)$$

where T is the equilibrium transformation temperature, that is assumed to be 700 °C; ΔV is the volume change accompanied with the anatase-rutile transformation, - 1.73 cm^3/mole; ΔH is the enthalpy of transformation, 6.51 x 10^3 J/mole [12]; ΔP is the pressure. If we apply the stress developed during phase transformation calculated using equation (2) (ΔP =889 MPa) in equation (3) we will get a ΔT value of 230°C. This can obviously explain the shift in the transformation temperature.

It has been seen from the FE-SEM micrographs that the rutile crystallites are at least 4 to 5 times larger than the anatase crystallites. Based on this it not unreasonable to think that the critical nuclei size of rutile is bigger than the primary crystallite size of anatase. This means that for forming a single rutile nuclei, more than one anatase crystallite are required. Therefore the number of nearest neighbour in contact (coordination number) determines partly the nucleation rate. The growth rate is also affected in the same way. This is another mechanism which supplements the slower rate of anatase to rutile phase transformation in supported membranes.

SUMMARY

Titania that was produced through the method mentioned in this paper will undergo anatase to rutile phase transformation in the temperature range of 550 to 600 °C. This temperature range is in the lower side of the range of temperatures reported in the literature, because of the very small particle size and high surface area compared to traditionally prepared anatase. Supported membranes showed a higher transformation temperature (slower rate of transformation) compared to the unsupported ones. Unsupported membranes showed a slight thickness dependence on the phase transformation temperature, but less significant compared to the difference in phase transformation behaviour between supported and unsupported membranes of similar thicknesses. This difference in phase transformation behaviour is primarily attributed to the large stress which is developed due to the negative volume change during the anatase-rutile transformation in a constraint environment. The lower level of coordination in the supported membranes has a direct influence on the nucleation and growth kinetics in the anatase to rutile transformation. The common practice of evaluating the textural characteristics of unsupported membranes and using the findings directly to understand the properties of supported membranes should be reconsidered. Laser Raman spectroscopy is a useful tool to study the stability of the membranes.

REFERENCES

1. K.P. Kumar, K. Keizer and A.J. Burggraaf, "Thermal Stability of Ceramic Membranes: I Textural Evolution and Phase Transformation in Supported Titania membranes, Submitted to J. Am. Ceram. Soc.

2. V.T. Zaspalis, "Catalytically Active Ceramic Membranes: synthesis, properties and reactor applications", Ph.D thesis, University of Twente, Enschede, The Netherlands (1990).

3. A. Larbot, J.P. Fabre, C. Guizard and L. Cot, "New Inorganic Membranes: Titania and Zirconia Membranes", J. Am. Ceram. Soc., **72** [2], 257-261 (1989).

4. R.J.R. Uhlhorn, "Ceramic Membranes for Gas Seperation: synthesis and transport properties", Ph.D Thesis, University of Twente, Enschede, The Netherlands (1990).

5. D.R. Clarke and F. Adar, " Measurement of the Crystallographically Transformed Zone Produced by Fracture in Ceramics Containing Tetragonal Zirconia, " J. Am. Cer. Soc., **65** [6] 284-288 (1990).

6. U. Balachandran and N.G. Eror, "Raman Spectra of Titanium Dioxide," J. Solid State Chem., **42,** 276-282 (1982).

7. J.H.L. Vonken, C. Lijzenga, K.P. Kumar, K. Keizer, A.J. Burggraaf and B.C. Bonekamp,"A new method for the measurement of stress in thin drying layers produced during the formation of ceramic membranes" J. Mat. Sci., 27, 472-78 (1992)

8. J.C. Parker and R.W. Siegel, " Raman Microprobe Study of Nanophase TiO_2 and Oxidation-Induced Spectral Changes, " J. Mater. Res., **5** [6], 1246-1252 (1990).

9. C.A. Melendres, A. Narayanaswammy, V.A. Maroni and R.W. Siegel, " Raman Spectroscopy of Nanophase TiO_2," J. Mater. Res., **4** [5], 1246-1250 (1989).

10. G.J. Exarhos and N.J. Hess, "Interfacial Stress and Phase Transformation Phenomena in Oxide Thin Films", in this proceeding

11. R.A. Swalin, The Thermodynamics of Solids, John Wiley and Sons, New York, 1972.

12. JANAF Thermochemical Tables, ACS, AIP, NBS, Third Edition, Part II, (Chase et al. eds.) U.S.A, 1985

PREPARATION AND CHARACTERIZATION OF MICROPOROUS SOL-GEL DERIVED CERAMIC MEMBRANES FOR GAS SEPARATION APPLICATIONS

R.S.A. DE LANGE, J.H.A. HEKKINK, K. KEIZER AND A.J. BURGGRAAF
University of Twente, Faculty of Chemical Technology, Laboratory of Inorganic Chemistry, Materials Science and Catalysis, P.O. Box 217, 7500 AE Enschede, The Netherlands

ABSTRACT

Mesoporous alumina membranes were modified with microporous silica. The polymeric silica sols consist of very weakly branched polymeric molecules with a fractal dimension of 1.3 and Guinier radius of ≈ 2.0 nm. The resulting top layer material is microporous with a porosity of 40% and a layer thickness of 60 nm. Gas transport is activated with an activation energy of 11 kJ/mol for hydrogen and molecular sieve like separation factors have been obtained. From the mechanism of gas transport the conclusion can be drawn that the pores are of molecular dimensions (0.7 - 1 nm diameter).

INTRODUCTION

The preparation of coatings and thin films, for e.g. magnetic, optical, electrical and separation applications, is a rapidly expanding field in the ceramic technology. The sol-gel process is a commonly used technique for the preparation of these films because of a number of advantages, compared to conventional techniques, such as high purity, high homogeneity and the possibility to tailor the microstructure (pore size, porosity and pore size distribution) [1].

Especially because of this last property, the sol-gel process is very important for the formation and modification of ceramic membranes. Commercially, inorganic membranes have nowadays applications in many areas such as food and beverage processing, biotechnology and water treatment. Emerging areas, however, are the use of inorganic membranes in high temperature gas separation and (catalytic) membrane reactors with the aim to enhance productivity and selectivity of processes which are limited by thermodynamics. For these applications, research efforts are directed towards the decrease of pore size to radii < 1 nm [2].

In the present paper, the sol-gel synthesis and modification of ceramic membranes and the relation between sol morphology and the obtained microstructure of the calcined membrane material is discussed. Finally the effect of modification on gas transport and separation properties will be given.

Sol-Gel Synthesis

With the sol-gel process, the morphology of sols can be directed towards particulate (colloidal) sols or polymeric sols by controlled hydrolysis and condensation of alkoxides. Important process parameters are water content, type of catalyst and the used precursor. In turn, the nature of the sol morphology determines very strongly the microstructure of the dried and calcined material. Polymeric systems consisting of weakly branched oligomers or polymers with low fractal dimensions ($D_f << 2$), result in microporous ($r_{pore} < 1$ nm, IUPAC definition [3]) materials [4]. Contrary to this, particulate or highly branched systems with a high fractal dimension, result in mesoporous systems.

In order to obtain weakly branched polymers, the hydrolysis and condensation processes have to be carried out in the presence of an acid catalyst, with a relatively low water content. Under these conditions the hydrolysis rate is fast with respect to the condensation rate and the hydrolysis rate decreases as the number of OR-groups decreases with the progress of hydrolysis [4,5]. In the very early stages of the process, at $\approx t/t_{gel} < 0.01$, condensation involves the most basic species, i.e. monomers or end groups of chains. The condensation process under these conditions can be described as a reaction limited cluster-cluster aggregation process [6]. The resulting sol consists of very open fractal structures.

Gas Transport Through Porous Inorganic Membranes

Gas transport mechanisms through mesoporous (pore diameter 2 nm < d_{pore} < 50 nm [3]) membranes are Knudsen diffusion, surface diffusion and multi-layer diffusion/capillary condensation [7]. At high temperatures and for non-sorbable gases, however, the Knudsen diffusion component, equation (1), determines the transport.

$$F_{Kn,0} = \frac{2.\epsilon.\mu_{Kn}.\overline{v}.\overline{r}}{3RT} \quad with \quad \overline{v} = \sqrt{\frac{8RT}{\pi M}} \tag{1}$$

Where $F_{0,Kn}$ is the Knudsen permeability (mol/m^2.s.Pa), ϵ is the porosity, μ_K is a shape factor equal to $1/\tau$, where τ is the tortuosity, R is the gas constant (=8,31 J/mol.K), T is the absolute temperature (K), r is the modal pore radius (m), v is the average molecular velocity (m/s) and M is the molecular mass of a gas molecule (kg/mol).

From equation (1) it can be seen that i) the transport rate decreases as function of temperature and ii) the maximum separation factor α^* (= $F_{Kn,0A}/F_{Kn,0B}$) is equal to the square root of the ratio of the molecular masses. For many applications, however, these separation factors are not high enough! A method to enhance the separation factor is modification with a microporous system.

Several techniques and systems have been reported in literature to achieve microporous membranes, these include sol-gel modification [8-11] and CVD modification [12-14] of mesoporous glass or ceramic membranes and more recently the development of zeolitic membranes [15]. The separation properties are molecular sieve like and contrary to Knudsen diffusion, the permeability increases with increasing temperature.

The exact mechanism of activated transport, or micropore diffusion, is not completely clear yet, only some qualitatively pictures have been reported [9,13,16,17]. These pictures show that in these pores of molecular dimensions, the size of the molecules and specific molecule-wall interactions are responsible for differences in potential energy barriers and therefore in transport rates. These types of interactions only take place as the pore diameter is not larger than approximately 4-5 times the molecular diameter.

EXPERIMENTAL

Mesoporous γ-alumina membranes (50% porosity, top layer thickness of 7 μm, 3 nm pore diameter and sharp pore size distribution) were prepared by dipcoating of α-alumina supports (40% porosity, mean pore radius 160 nm) with a colloidal AlOOH (boehmite) dip solution, using a procedure which has been extensively described in previous publications [18,19]. Polymeric silica sols were prepared by hydrolysis and condensation of tetra-ethyl-ortho-silicate (TEOS, Merck, p.a.grade) in ethanol with HNO_3 as catalyst. The used molar ratios of the silica sol are TEOS:ethanol:water:HNO_3 = 1 : 3.8 : 6.5 : 0.09.

Non-supported silica membranes were prepared by pouring the polymeric solution in polypropylene dishes. The liquid films were dried overnight in ambient conditions. Supported silica modified membranes were prepared by a dipcoating procedure of γ-alumina membranes in polymeric silica solutions.

In order to investigate the correlation between the sol morphology and the final microstructure, Small Angle X-ray Scattering (SAXS) experiments were performed using synchrotron radiation (SERC Synchrotron Radiation Services, Daresbury UK). The obtained spectra were corrected for background and parasitic scattering.

Microstructure characterization of calcined non-supported silica was performed with nitrogen adsorption using a volumetric adsorption unit (Carlo Erba Sorptomatic 1900).

Layer thickness of supported silica modified membranes was investigated with Scanning Auger Microscopy (SAM, Perkin Elmer PHI 600) and X-ray Photo electron Spectroscopy (XPS, Kratos XSAM 800). Sputter profiles were obtained by surface analysis between succeeding sputtering (by argon bombardment) steps.

Gas permeation measurements were performed in equipment described elsewhere [18]. For these experiments single gases have been used. For the separation measurements, gas mixtures have been used.

RESULTS

In figure 1, a typical log(I)-Log(Q) plot for silica polymeric solutions is given for the high Q-range, where Q is the scattering vector defined as:

$$Q = \frac{2.\pi}{\lambda}\sin(2.\theta) \qquad (2)$$

A clear transition is present from the Guinier region to the linear fractal region. For mass fractals, the slope -X in the fractal region is equal to the fractal dimension D_f if polydispersity can be neglected. This is a very reasonable assumption if the sol is far from the gelation point [20]. As gelation times are in the order of 1 to 2 weeks for the prepared sols, polydispersity is neglected. Some first results for this system are summarized in table I.

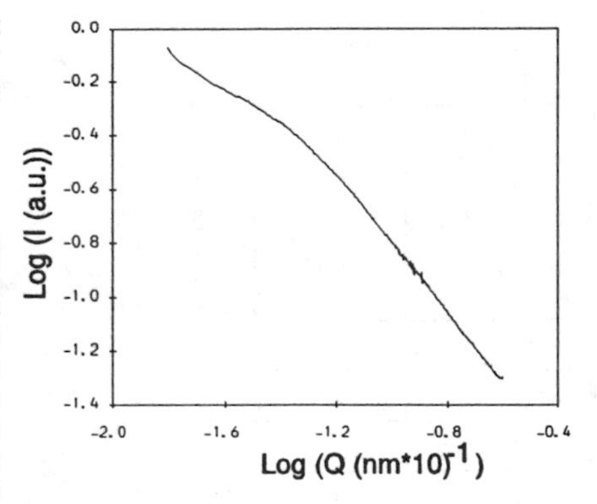

Figure 1 : Log(I)-Log(Q) plot silica reacted for 3 hours

The fractal dimension and the aggregate size increase as a function of reaction time. After one hour of reaction the scattering intensity is too low to analyze the spectrum. The reactions are still in the initial phase, the scattering centra probably consist of very small oligomers. After two hours of reaction a Guinier radius of only 1.5 nm is calculated with a slope of -1. This is the lowest possible value for a connected fractal [20] and means that perfectly linear polymeric molecules are formed. After three hours of reaction, which is the standard reaction time for the sols for membrane modification, the slope is -1.3 and the Guinier radius is ≈ 2.0 nm. This can be assigned to very weakly branched, chain-like structures.

Table I : Fractal dimension D_f and Guinier radius R_g of silica sols determined with SAXS as function of reaction time

Time (h)	D_f	R_g (nm)
1	N.D.#	N.D.
2	1	1.5
3	1.30 ± 0.05	2.0

: Not detectable

For the formation of microporous films from sols, it is important that the fractal dimension is smaller than 1.5. Then it is possible for the polymeric molecules to penetrate into each other during the film formation process [21]. This process can be considered as a forced gelation! The calcined film then contains pores in the molecular sieve range. This will be shown by permeation measurements. Probably a very low fractal dimension (≈ 1) is not preferable because then the porosity will be too low.

In figure 2 some nitrogen adsorption isotherms are given for *non-supported* silica calcined at different temperatures. The isotherms are all of type I and can be ascribed to microporous (r_{pore} < 1 nm) materials [3]. Unfortunately, in this pore size region, there is no generally accepted model to calculate pore sizes from adsorption isotherms as is the case for mesoporous materials. In table II the porosities are given for an extended series of experiments with silica. From these data it is clear that the porosity does not decrease dramatically as function of temperature till 600°C. Densification takes place by continued condensation reactions and structural relaxation processes [1]. Above this temperature strong densification takes place because viscous sintering starts at these temperatures. Important to note is that the material remains microporous! The conclusion from these experiments has to be that the thermal stability of these membrane toplayer materials is limited to ≈600°C.

Table II : Thermal stability non-supported silica

Calcining conditions	Porosity (%)
400°C, 3 h.	41
400°C, 3 h., 500°C, 3 h.	31
400°C, 3 h., 600°C, 3 h.	28
700°C, 3 h.	11
800°C, 3 h.	0
500°C, 12 h.	35
600°C, 12 h.	21

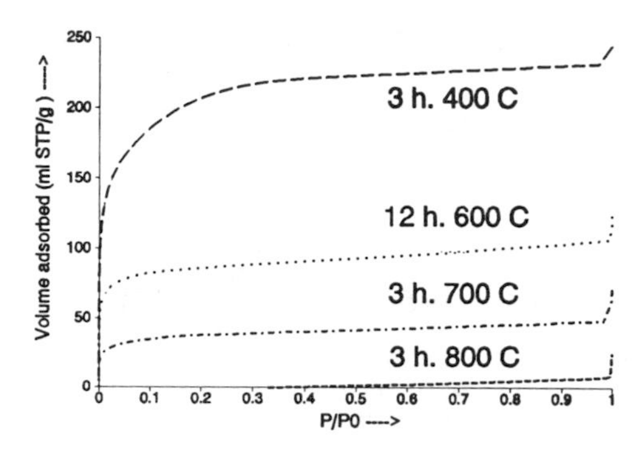

Figure 2 : Nitrogen adsorption isotherms non-supported silica (heating rate 25°C/hr in air)

When translating these results to membrane properties however, one has to bear in mind that the sol-gel transition for *supported* membranes takes place in 10-30 seconds instead of hours for *non-supported* layers. There is more time for condensation reactions to take place, what will result in a larger and more branched polymers. For *supported* silica a denser material is expected with smaller pores.

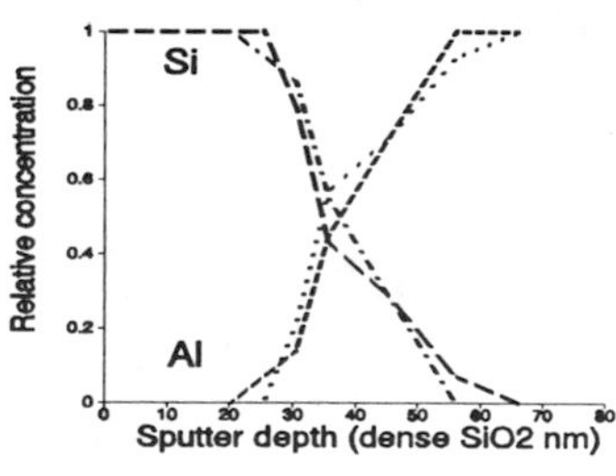

Figure 3 : SAM sputter profile SiO_2 modified membrane

However, because of the extremely thin silica layer compared to the support thickness, what will be shown in the next paragraph, no gas adsorption experiments can be performed with supported membranes.

In figure 3 a SAM sputter profile for silica modified γ-alumina membranes is given. The silica modification is present as a layer of ≈ 30 nm on top of the γ-alumina membrane and a layer of ≈ 30 nm in the pores of the γ-alumina. With XPS a comparable profile has been found. The sputter depth is calculated using a calibrated sputter rate for *dense* silica of 2.55 nm/min. So an error is made in the calculation of the absolute layer thickness for porous silica. The order of magnitude, however, is correct, this is supported by the permeation data.

H_2 and CO_2 permeabilities as function of temperature for a silica modified membrane calcined at 400°C are given in figure 4 and 5 respectively. In all cases P_{low} is 0 bar. The permeability F_0 is calculated by dividing the flux (mol/m^2.s) by the pressure difference ΔP (Pa) over the membrane. For both gases, transport is activated with apparent activation energies of 11 kJ/mol for hydrogen and 2 kJ/mol for carbon-dioxide. This activation energy is composed of an activation energy of diffusion and an activation energy of adsorption.

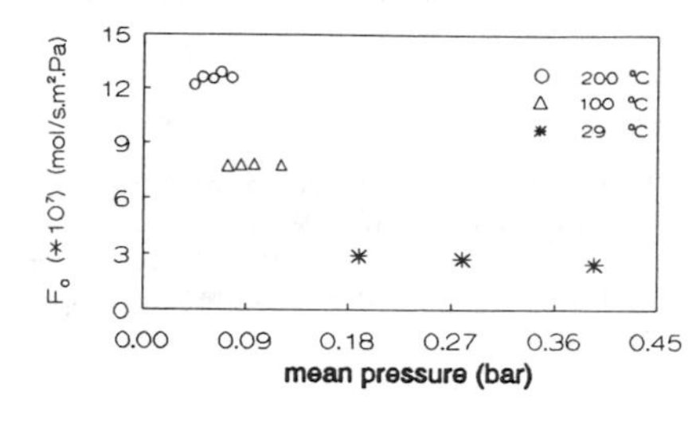

Figure 4 : H_2-permeability silica membrane, E_{act} = 11 kJ/mol (12.3 kJ/mol corrected for support)

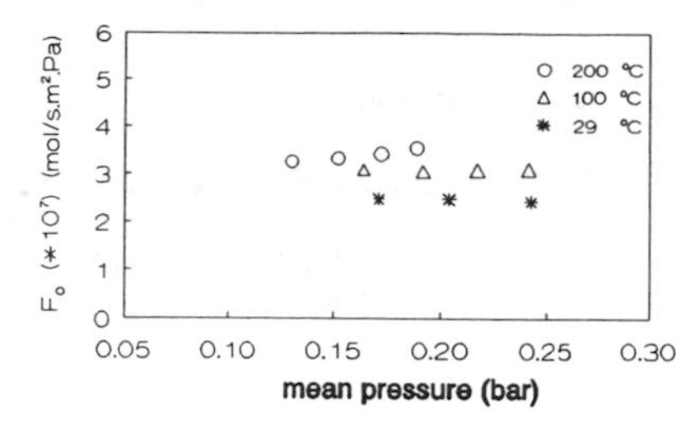

Figure 5 : CO_2-permeability silica membrane, E_{act} = 2 kJ/mol (2.7 kJ/mol corrected for support)

Also it has been found that the gas permeability for these membrane types decreases in the order He > H_2 > CO_2 > CH_4 > C_3H_6 [8,9,22]. According to Knudsen theory, lighter gases should diffuse faster. However, He permeates faster than H_2 and CO_2 permeates faster than CH_4, even though they are heavier molecules. The order of permeability appears to be strongly related to the size of the molecules and also specific molecule-wall interactions will play a role (especially CO_2). The obtained permeabilities are very high compared to published literature data; factors 10-20 higher than for sol-gel modified membranes [10-12] and 100-200 times higher than for CVD-modified membranes [13-14]. This can be attributed to the very thin modified toplayer. In table III some separation results are shown for silica modified membranes and silica/titania modified membranes calcined at 400°C prepared in our laboratory.

Table III : Separation factors and permeabilities of fastest permeating component for microporous membranes.

System	T (°C)	α	F_o*10^7 (mol/m^2.s.Pa)
SiO_2	260	H_2/C_3H_6 : 200	H_2 : 20
	100	CO_2/CH_4 : 20-50	CO_2 : 10
SiO_2/TiO_2	150	H_2/CH_4 : 200	H_2 : 1

As expected from the permeation results, a large improvement of the separation factors has been achieved. The separation factors are molecular sieve like; compared to Knudsen-type separation factors, they are more than one order of magnitude higher. Separation experiments showed to be very sensitive to membrane quality; few pinholes or cracks have severe consequences for separation factors. A strict control of the membrane formation is therefore necessary.

CONCLUSIONS

It is shown that sol-gel modification is a very effective technique for the synthesis and modification of ceramic membranes. Polymeric silica sols were prepared with a fractal dimension of 1.3 and a Guinier radius of ≈2.0 nm, which, after drying and calcination, lead to very thin (60 nm) microporous toplayers. The thermal stability of silica modified membranes is limited to ≈ 600°C; above this limit viscous sintering takes place.

High gas permeabilities and separation factors have been obtained with these membranes. This combination makes these membranes excellent candidates for gas separation applications. The mechanism of gas transport leads to the conclusion that the pore sizes of the modified toplayers are of molecular dimensions (0.7 - 1 nm diameter).

ACKNOWLEDGEMENTS

The authors express their sincere appreciation to T.P.M. Beelen, W. Dokter and H.F. van Garderen of the University of Eindhoven and W. Bras of the SRS, Daresbury for many valuable scientific discussions. A.J.H. van den Berg of the Centre for Materials Research of the University of Twente is gratefully thanked for performing SAM and XPS measurements and Shell Research Amsterdam for their financial assistance.

REFERENCES

[1] C.J. Brinker and G.W. Scherer, Sol-Gel Science: The Physics and Chemistry of Sol-Gel Processing, (Academic Press, San Diego, 1990).
[2] R.R. Bhave, Inorganic Membranes; Synthesis, Characteristics, and Applications, (Van Nostrand Reinhold, New York, 1991).
[3] K.S.W. Sing, D.H. Everett, R.A.W. Haul, L. Moscou, R.A. Pierotti, J. Rouquérol, T Siemieniewska, Pure & Appl. Chem., 57 (4), 603 (1985).
[4] C.J. Brinker, K.D. Keefer, D.W. Schaefer, R.A. Assink, B.D. Kay and C.S. Ashley, J.Non-Crystalline Solids, 63, 45 (1984).
[5] I. Strawbridge, A.F. Craievich and P.F. James, J. Non-Crystalline Solids, 72, 139 (1985).
[6] P.Meakin, in On Growth and Form, edited by H.E. Stanley and N. Ostrowsky (Martinus Nijhof, Dordrecht, 1986), p. 111.
[7] R.J.R. Uhlhorn, K. Keizer and A.J. Burggraaf, J.Membr.Sci., 46, 225 (1989).; ibid., 66 (2&3), 259 (1992).
[8] R.S.A. de Lange, J.H.A. Hekkink, K. Keizer and A.J. Burggraaf, Key Engineering Materials, 61 & 62, 77 (1991).
[9] R.J.R. Uhlhorn, K. Keizer and A.J. Burggraaf, J. Membr.Sci., 66 (2&3), 271 (1992).
[10] S. Kitao, H. Kameda and M. Asaeda, Membrane, 15 (4), 222 (1990).
[11] H. Ohya, T. Hisamatsu, H. Fujimoto, S. Sato and Y. Negishi, Key Engineering Materials, 61 & 62, 353 (1991).
[12] S. Kitao and M. Asaeda, Key Engineering Materials, 61 & 62, 267 (1991).
[13] T. Okubo and H. Inoue, AIChE Journal, 35 (5), 845 (1989).
[14] G.R. Gavalas, C.E. Megiris and S.W. Nam, Chem.Eng.Sci., 44 (9), 1829 (1989); M. Tsapatsis, S. Kim, S.W. Nam and G. Gavalas, Ind.Eng.Chem.Res., 30 (9), 2152 (1991).
[15] E.R. Geus, A. Mulder, D.J. Vischjager, J. Schoonman and H. van Bekkum, Key Engineering Materials, 61 & 62, 57 (1991).
[16] J.E. Koresh and A. Soffer, Sep.Science Techn., 18, 723 (1983); Chem.Soc.Farad. Trans.I, 82, 2057 (1986).
[17] A.J. Burggraaf, H.J.M. Bouwmeester, B.A. Boukamp, R.J.R. Uhlhorn and V.T. Zaspalis, in Science of Ceramic Interfaces, edited by J. Nowotny (Elsevier Science Publishers, Amsterdam, 1991), p. 525.
[18] R.J. van Vuren, B.C. Bonekamp, K. Keizer, R.J.R. Uhlhorn H.J. Veringa and A.J. Burggraaf, in High Tech Ceramics, edited by P. Vincenzini (Elsevier Science Publishers, Amsterdam, 1987), p. 2235.
[19] A.F.M. Leenaars and A.J. Burggraaf, J.Colloid and Interf.Sci., 105 (1), 27 (1985).
[20] J.E. Martin and A.J. Hurd, J.Appl.Cryst., 20, 61 (1987).
[21] B.B. Mandelbrot, The Fractal Geometry of Nature, (Freeman, San Francisco, 1982).
[22] J.D. Way and D.L. Roberts, Sep.Sci.Technol., 27 (1), 29 (1992).

IMOGOLITE AS A MATERIAL FOR FABRICATION OF INORGANIC MEMBRANES

JEFFREY C. HULING *, JOSEPH K. BAILEY **, DOUGLAS M. SMITH *
AND C. JEFFREY BRINKER **
*UNM/NSF Center for Micro-Engineered Ceramics, University of New Mexico, Albuquerque, NM 87131
**Sandia National Laboratories, Ceramic Synthesis and Inorganic Chemistry Department, Albuquerque, NM 87185

ABSTRACT

Imogolite is a structurally microporous tubular clay comprising one-dimensional pore channels that are 0.8 - 1.2 nm in diameter, depending on composition. The microporous structure of natural and synthetic imogolite has been investigated by nitrogen adsorption as a function of outgassing temperature. A significant increase in adsorption at low relative pressure ($P/P_0 \sim 10^{-6}$) after 275°C outgassing reflects a high concentration and narrow distribution of 0.8 - 0.9 nm diameter pores (i.e., the imogolite tubes) and supports the potential use of imogolite in inorganic membrane applications.

INTRODUCTION

The use of membranes is widespread in such diverse applications as food processing, industrial waste treatment, gas separations and biomedical processing. Although the majority of commercially used microporous or ultramicroporous membranes are based on organic polymer films, inorganic membranes potentially have several outstanding advantages including chemical, mechanical and thermal stability. Unfortunately, it has proven difficult to prepare supported inorganic membranes with sufficiently small pore sizes (0.2 - 2.0 nm) and with the absence of cracks necessary for such applications as gas separation, dialysis, and reverse osmosis. One strategy to control the pore size of ceramic membranes is to utilize microporous inorganic materials, such as imogolite, having structures that comprise one-dimensional pore channels.

Imogolite is a naturally occurring, hydrated aluminosilicate mineral that derives considerable microporosity from a nanoscale tubular crystal structure [1]. In contrast to the layer silicates kaolinite or halloysite, in which one oxygen per SiO_4 tetrahedron is shared with the gibbsite layer and the remaining three oxygens are involved in SiO_4 corner-sharing, in imogolite the SiO_4 tetrahedra are isolated and effectively inverted, sharing three oxygens with the gibbsite layer and leaving one oxygen as part of a silanol group. The strain associated with this increased Si-O-Al bonding is accommodated by bending the gibbsite layer, thus creating a tube having distinct external (Al-OH) and internal (Si-OH) surface characteristics (Figure 1). The composition can be written $(HO)_3Al_2O_3SiOH$, which reflects the arrangement of atoms encountered on passing from the exterior to the interior of the tube. The internal and external tube diameters are ~1 nm and ~2.5 nm, respectively.

Most papers on synthetic imogolite are found in the soil science and clay mineralogy literature and deal with chemical processing conditions and characterization (e.g., TEM, solid state NMR) of the tubular imogolite structure. While the tremendous potential of imogolite in materials science has been discussed [2], the actual practical application of imogolite has been limited to intercalation studies in which the outer diameter of the imogolite tube is used to control the basal spacing (i.e., microporous structure) of pillared clays [3-6]. Our work, however, is primarily concerned with the inner diameter and internal (Si-OH) surface of the imogolite tubes. By embedding imogolite in a dense sol-gel silicate matrix, supported inorganic membranes will be produced for gas separation, ultrafiltration and catalysis in which transport occurs only through the cylindrical imogolite channels. Unlike pore channels in amorphous gels and crystalline zeolites, these micropores will not be interconnected in three dimensions and will not vary in diameter along the path of diffusion.

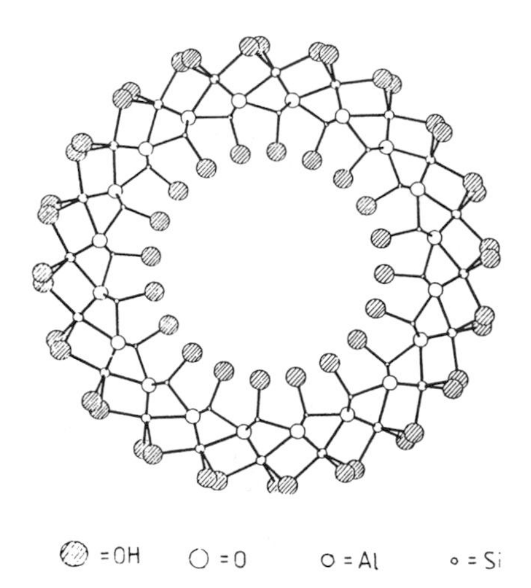

Figure 1.

Schematic depiction of the cross-section of the imogolite tube structure. Structure proposed by Cradwick et. al. [1]; schematic from MacKenzie et. al. [7].

Under ambient atmospheric and temperature conditions, the imogolite tubes are filled with water. Since this clearly renders the tubes ineffective as channels for gas transport, it is important to know both the minimum temperature at which this water is removed and the maximum temperature that can be reached before the material begins to dehydroxylate and the tube structure begins to break down. Of these distinct processes, only the latter has been addressed in the literature. According to thermogravimetric analysis [7], solid state NMR [7], and X-ray diffraction [8], dehydroxylation begins at ~300°C.

There have been several adsorption studies using imogolite [9-11], although none have directly measured its micropore structure. The effect of prior heating was investigated by Egashira and Aomine [9], who found that the surface area of imogolite was maximized after heating to 300°C. Adams [10] outgassed imogolite at 200°C and indirectly characterized its micropore structure by measuring the accessible micropore volume as a function of adsorbate molecular size. To measure the pore size distribution of imogolite more directly and more precisely, we have performed adsorption experiments on both natural and synthetic imogolite at low relative pressures of nitrogen ($10^{-6} < P/P_0 < 0.99$) over a range of outgassing temperatures.

In this paper we present our preliminary results on the synthesis of phase-pure imogolite and the characterization of its micropore structure.

EXPERIMENTAL PROCEDURE

Farmer and Fraser [12] described a method for synthesizing imogolite in very dilute (millimolar) solutions of $Al(ClO_4)_3$ and $Si(OH)_4$. The imogolite yield was found to be maximized --relative to boehmite and other non-tubular products--by maintaining low precursor concentrations, starting with an excess of $Si(OH)_4$ (Si/Al = 0.6, vs. the stoichiometric ratio of 0.5), and by an initial, temporary (i.e., immediately reversed) increase in solution pH to partially neutralize the aluminum salt and promote imogolite nucleation. Significant amounts of imogolite were formed in the sol by heating at 95 - 100°C for 1 day, although heating for 5 days gave the best results. [The yield of synthetic imogolite was estimated by measuring the volume of gel obtained by flocculating a sol at pH ~ 10 with NH_4OH and centrifuging.]

Our approach modifies that of Farmer and Fraser by "seeding" the formation of imogolite. By adding a portion of a previously processed synthetic imogolite sol to a mixture of 2.5 mM $Al(ClO_4)_3$ and 1.25 mM $Si(OH)_4$, we have found that the pH adjustment step can be avoided, excess orthosilicic acid does not increase the imogolite yield, and imogolite formation is nearly complete after heating at 95 - 100°C for 2 days. The seeding sol, added in a 1:4 ratio to the precursor solution, is ultrasonically dispersed prior to its addition to increase its concentration of potential imogolite growth sites.

Natural imogolite was provided by K. Wada, following a chemical purification treatment [13] to remove organic matter, extractable oxides, and, to a lesser extent, non-tubular amorphous silicates.

Nitrogen adsorption experiments were performed at 77K using the static volumetric method (Micromeritics ASAP 2000M adsorption analyzer) at $10^{-6} < P/P_0 < 0.99$. BET surface areas were determined from adsorption at five relative pressures in the range 0.05 to 0.30. Micropore area and micropore volume were determined from t-plots interpreted using the Harkins and Jura analysis. Differential pore volume plots were determined using the Horvath-Kawazoe model [15] with the Saito-Foley modification [16] to reflect the cylindrical rather than slit-shaped geometry of the imogolite micropores. Total pore volume was estimated from nitrogen uptake at $P/P_0 \sim 0.99$.

RESULTS AND DISCUSSION

The SEM micrograph in Figure 2 shows synthetic imogolite "fabric" in which the visible "threads" are tube bundles, each containing many individual (~2.5 nm diameter) imogolite tubes. This fabric structure is similar in appearance to the gel films of natural imogolite that form on volcanic ash due to weathering [14]. The natural imogolite films, however, typically have more open, web-like structures. In both cases the imogolite tubes are highly aligned: In natural imogolite the 10 - 30 nm width of the bundles limits the range of this ordering [13], whereas in the more dense synthetic fabric the imogolite tubes are ordered over hundreds of nanometers. This can be seen in the TEM micrograph in Figure 3, which also shows that the tight packing of the synthetic imogolite appears to exclude all extraneous, non-tubular material from the fabric structure. Electron diffraction confirms that the repeat distance perpendicular to the lines in Figure 3 is ~ 2.6 nm, corresponding to the outer diameter of the imogolite tubes.

The nitrogen adsorption study of imogolite performed by Adams [10] gave a Type I isotherm characteristic of microporous materials. Experimentally, such a result often indicates that considerable adsorption occurs at relative nitrogen pressures lower than that of the first measurement, which is typically $P/P_0 \sim 0.01$. To determine the micropore structure of natural and synthetic imogolite more precisely, we determined the adsorption isotherms beginning at much lower pressures, $P/P_0 \sim 10^{-6}$, and following outgassing at incrementally increasing temperatures.

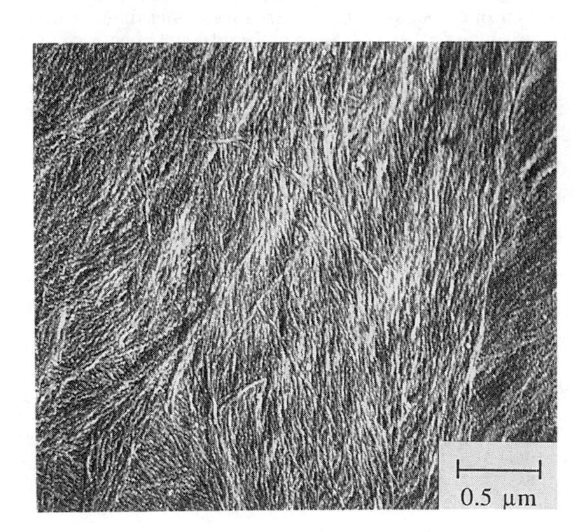

Figure 2.

SEM micrograph of synthetic imogolite "fabric." Visible features are tube bundles, each containing numerous individual imogolite tubes.

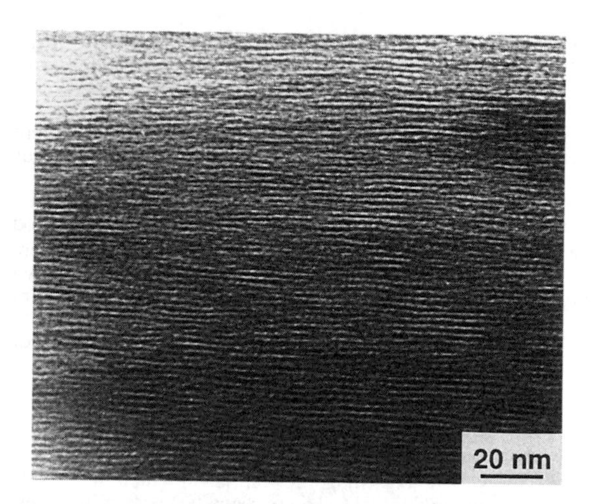

Figure 3.

TEM micrograph of synthetic imogolite "fabric." Note the highly ordered parallel configuration of imogolite tube packing.

Results of the adsorption studies show that the pore structure of the natural material is essentially constant as outgassing temperature is increased from 110°C to 250°C. After outgassing at 250°C, the differential pore volume plots (Figure 4) show that both the natural and synthetic samples have peaks at pore diameters of 0.8 - 0.9 nm. The natural imogolite, however, has a much greater fraction of its total pore volume contributed by pores with diameters > 1 nm (also see Table I). This is probably indicative of non-tubular material not separated from the natural imogolite during its chemical purification treatment. The peaks at 0.8 - 0.9 nm are consistent with adsorption at the ends of imogolite tubes where some water loss has occurred, or at the channels between tightly packed tubes.

Increasing the outgassing temperature leads to significant changes in adsorption for both the natural and synthetic imogolites. The isotherms in Figure 5 reveal that after outgassing at 275°C there is considerable adsorption at very low relative pressures ($P/P_0 \sim 10^{-6}$) that does not occur after 250°C outgassing. This characteristic feature--analogous to a step function in the isotherm--corresponds to newly acquired access to the tubes or inter-tube channels following the removal of water. It is therefore more pronounced for the synthetic material, which contains more of the tubular imogolite phase. Adsorption in or between the ordered imogolite tubes is most dramatically illustrated by the magnification of the peaks at ~0.8 nm in the differential pore volume plots (Figure 6). As shown in Table I, the increases in total (BET) surface area and total pore volume of the natural material that are caused by increasing the outgassing temperature from 250°C to 275°C are entirely accounted for by increases in the micropore area and micropore volume, respectively.

Table I. Nitrogen adsorption results for natural and synthetic imogolite.

	Natural		Synthetic	
T_{outgas}	250°C	275°C	250°C	275°C
BET surface area (m^2/g)	265	318	287	293
Micropore area (m^2/g)	60	112	195	183
Micropore volume (cm^3/g)	0.0258	0.0503	0.0913	0.0849
Total pore volume (cm^3/g)	0.1709	0.1933	0.1670	0.1746
$\frac{\text{Micropore volume}}{\text{Total pore volume}}$ (%)	15.1	26.0	54.7	48.6

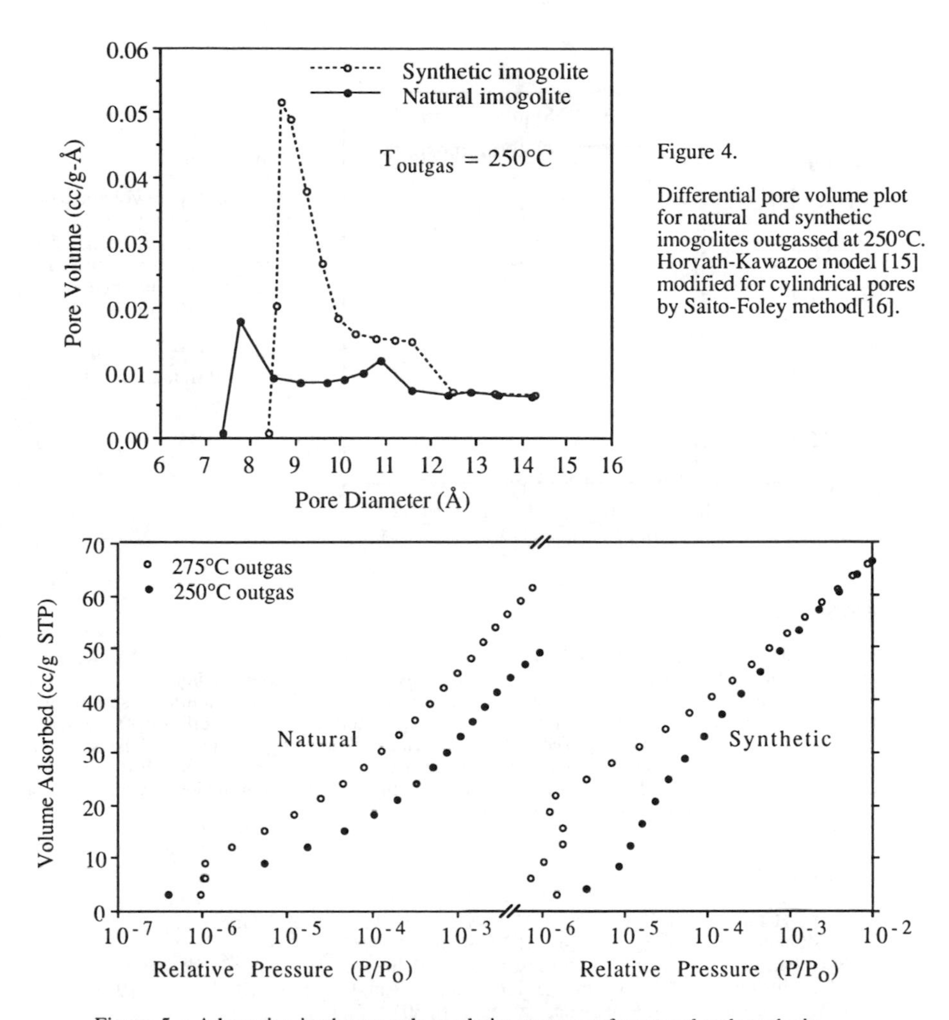

Figure 4.

Differential pore volume plot for natural and synthetic imogolites outgassed at 250°C. Horvath-Kawazoe model [15] modified for cylindrical pores by Saito-Foley method[16].

Figure 5. Adsorption isotherms at low relative pressures for natural and synthetic imogolites outgassed at 250°C and 275°C.

CONCLUSIONS

Imogolite synthesis has been simplified and accelerated by seeding dilute aluminosilicate solutions heated at 95 - 100°C. The micropore structure of phase-pure synthetic imogolite--and the change in this structure with increasing outgassing temperature--has been directly characterized by nitrogen adsorption at low relative pressures ($10^{-6} < P/P_o < 0.99$). Following outgassing at 275°C, there is considerable adsorption of nitrogen at $P/P_o \sim 10^{-6}$. This corresponds to adsorption in cylindrical pores of ~0.8 nm diameter, which is consistent with access to the imogolite tube channels upon removal of water. Such demonstration of exceptionally well defined microporosity confirms that imogolite is indeed a promising material for use in inorganic membrane applications.

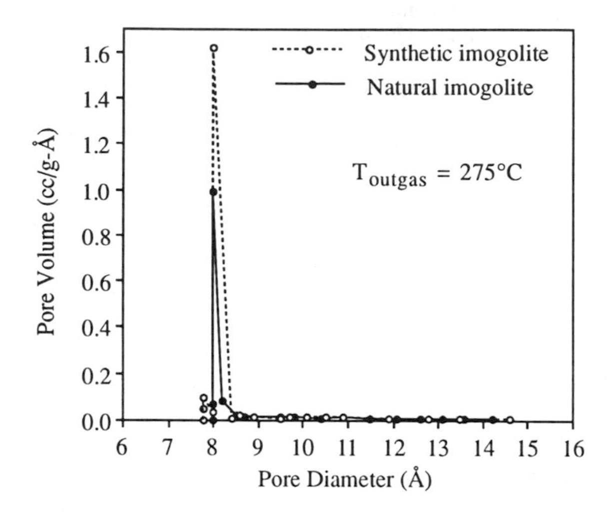

Figure 6.

Differential pore volume plot for natural and synthetic imogolites outgassed at 275°C. Horvath-Kawazoe model [15] modified for cylindrical pores by Saito-Foley method[16].

Note that thescale of pore volume differs from that in Figure 4.

ACKNOWLEDGEMENTS

Supported in part by the UNM/NSF Center for Micro-Engineered Ceramics, which is funded by the NSF (CDR-8803512), Sandia and Los Alamos National Laboratories, the New Mexico Research and Development Institute, and the ceramics industry. Additional support was provided by Sandia National Laboratories, a U.S. Department of Energy facility operated under contract #DE-AC-04-76DP00789. The authors gratefully acknowledge the technical assistance of William Ackerman of UNM in performing the adsorption experiments. The TEM work was performed in the Electron Microbeam Analysis Facility in the Department of Geology and Institute of Meteoritics at the University of New Mexico.

REFERENCES

1. P.D.G. Cradwick, V.C. Farmer, J.D. Russell, C.R. Masson, K. Wada, and N. Yoshinaga, Nature Phys. Sci. 240, 187 (1972).
2. V.C. Farmer, M.J. Adams, A.R. Fraser, and F. Palmieri, Clay Minerals 18, 459 (1983).
3. I.D. Johnson, T.A. Werpy, and T.J. Pinnavaia, J. Am. Chem. Soc. 110, 8545 (1988).
4. L.M. Johnson and T.J. Pinnavaia, Langmuir 6, 307 (1990).
5. L.M. Johnson and T.J. Pinnavaia, Langmuir 7, 2636 (1991).
6. T.A. Werpy, L.J. Michot, and T.J. Pinnavaia, in Novel Materials in Heterogeneous Catalysis, edited by R.T.K. Baker and L.L. Murrell (American Chemical Society, Washington, D.C., 1990), pp. 119-128.
7. K.J.D. MacKenzie, M.E. Bowden, I.W.M. Brown, and R.H. Meinhold, Clays and Clay Minerals 37, 317 (1989).
8. S.J. van der Gaast, K. Wada, S.I. Wada, and Y. Kakuto, Clays and Clay Minerals 33, 237 (1985).
9. K. Egashira and S. Aomine, Clay Science 4, 231 (1974).
10. M.J. Adams, J. Chromatog. 188, 97 (1980).
11. K. Wada and T. Henmi, Clay Science 4, 127 (1972).
12. V.C. Farmer and A.R. Fraser, in International Clay Conference 1978, edited by M.M. Mortland and V.C. Farmer (Elsevier Science Publishers, Amsterdam, 1979), pp. 547-553.
13. K. Wada, Am. Mineral. 52, 690 (1967).
14. K. Wada, Clay Minerals 8, 487 (1970).
15. G. Horvath and K. Kawazoe, J. Chem. Engr. Japan 16, 470 (1983).
16. A. Saito and H.C. Foley, AIChE Journal 37, 429 (1991).

SOL-GEL DERIVED PZT FIBERS

J.M. BOULTON, G. TEOWEE AND D.R. UHLMANN
Arizona Materials Laboratories, Department of Materials Science and Engineering, University of Arizona, Tucson, AZ 85712.

ABSTRACT

Sol-gel routes were developed to prepare densified PZT fibers. The effects of the degree of hydrolysis and the addition of an organic polymer to the precursor sol on fiber forming ability were investigated. Results on the crystalline and microstructural development of gels and fibers are presented. The effects of the incorporation of excess PbO and sintering atmosphere are also discussed, particularly in relation to densification.

INTRODUCTION

Sol-gel methods have been used to prepare a range of ceramic fibers [1]; and these methods are utilized commercially to manufacture refractory alumina and aluminaborosilicate fibers [2,3]. Several ferroelectric fibers have been prepared by wet chemical techniques, namely $LiNbO_3$ [4], $BaTiO_3$ [5], $PbTiO_3$ [6-8] and PLZT [9]. Ferroelectric (FE) fibers are attractive for composites in sensor applications, such as hydrophones and transducers [10-11].

With regard to lead zirconate titanate (PZT), conventional melt-processing routes to fibers cannot be used due to incongruent melting and the volatility of PbO at elevated temperatures. Alternative routes include replication methods by impregnating activated carbon fibers in a PZT acetate-methoxyethanol derived solution [11], the extrusion of a slurry of PZT powder in a binder phase [12], and topotactic reaction of hydrated TiO_2 fibers [13]. In this study, we report a sol-gel route to PZT fibers based on a lead acetate precursor solution. We have used similar solutions to prepare PZT thin films with excellent FE properties [14,15].

EXPERIMENTAL

Precursor solutions were derived from $Pb(OAc)_2 \cdot 3H_2O$, $Zr(O^nPr)_4$ and $Ti(O^iPr)_4$ [14]. Methanol was used as the solvent and distilled water was added to give the ratio 6 moles of H_2O: 1 mole of Pb. Anhydrous solutions were also investigated; these were prepared by using a high boiling point alcohol and distilling off the alcohol-water azeotrope [16]. Both $PbZr_{0.53}Ti_{0.47}O_3$ and $PbZr_{0.53}Ti_{0.47}O_3 \cdot 0.15PbO$ compositions were investigated. Fibers were also prepared by dispersing 53/47 single-phase perovskite powder prepared via precipitation and spray-drying (diameter ~0.5-2 μm) in the precursor sol (wt. of added powder/wt. of oxide equivalent of sol=2.5). Polyvinylpyrrolidone was added to the precursor sols to aid in fiber-forming. The precursor sols were concentrated by rotary evaporation to clear, viscous liquids. Fibers were drawn by inserting and withdrawing a glass rod from the sol.

RESULTS AND DISCUSSION

Fibers exceeding 1 m in length could readily be drawn from concentrated precursor solutions. Fiber formation was notably easier from hydrolyzed solutions compared to those without water, indicating the effects of hydrolysis-condensation on fiber-forming ability. The addition of the organic polymer, PVP, also aided fiber formation; PVP has previously been used to aid in the formation of $LiNbO_3$ fibers from alkoxide solutions [17]. The speed of withdrawal of the glass rod from the precursor sol basically determined the fiber diameter which could be varied from ~5 to 200 μm. With practice, fibers with diameters in the range of 10-20 μm could be routinely prepared.

After drawing and gelation, the fibers were dried. The dried gels were investigated by DTA, TGA and XRD. Thermal analysis results on the $PbZr_{0.53}Ti_{0.47}O_3$ gel in O_2 are shown in Fig. 1. These results indicate that weight loss is almost complete by 350C. In DTA, a large exotherm was observed at 300C, associated with decomposition of organics. A smaller exotherm was also seen at 457C; this reflected the crystallization of pyrochlore (based on XRD of DTA samples before and after the exotherm). No exotherm attributable to the subsequent crystallization/transformation to perovskite was observed.

XRD results on $PbZr_{0.53}Ti_{0.47}O_3$ and $PbZr_{0.53}Ti_{0.47}O_3 \cdot 0.15PbO$ gels fired in O_2 at various temperatures for 30 mins. are shown in Fig. 2. In these examples the gels were heated using the following schedule: 5C/min to 180C, 1C/min to 350C and then 5C/min to the desired temperature. This schedule, which was used for firing the fibers, ensured slow and complete burn-out of the organic components prior to crystallization. For the $PbZr_{0.53}Ti_{0.47}O_3$ gel, both pyrochlore and perovskite phases were observed on firing to 450C and 550C, with single-phase perovskite being obtained on firing to 650C. The addition of excess PbO hindered the formation of perovskite at lower temperatures; perovskite was not observed until 550C. However, single-phase perovskite was still obtained at 650C. As discussed later, the addition of excess PbO has beneficial effects on the microstructure of fibers fired at higher temperatures.

The degree of hydrolysis and heating rate had significant effects on the early crystalline development of the gels. In the gel prepared without added water (i.e., the water for hydrolysis derived from the atmosphere), the formation of perovskite is hindered. The gel fired at 450C only shows pyrochlore, with pyrochlore being the major phase at 550C and single-phase perovskite again being formed at 650C. Faster heating rates promoted perovskite formation at lower temperatures. Perovskite was the major phase on firing to 450C at a constant heating rate of 5C/min; however, firing to 650C was still required to form perovskite as the sole phase.

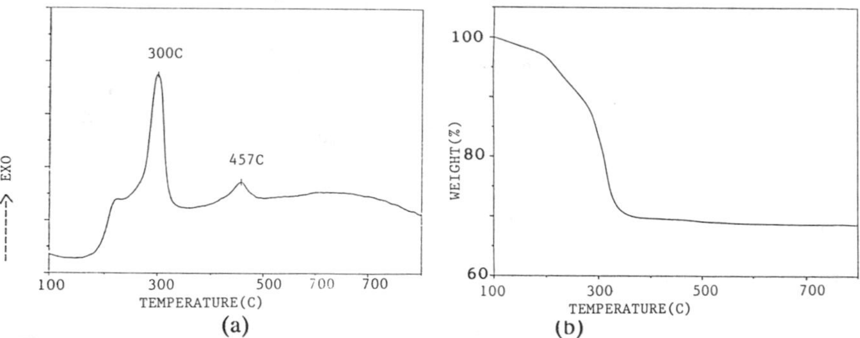

Figure 1. DTA(a) and TGA(b) of $PbZr_{0.53}Ti_{0.47}O_3$ gel in O_2 [6 moles of H_2O:1 mole of Pb].

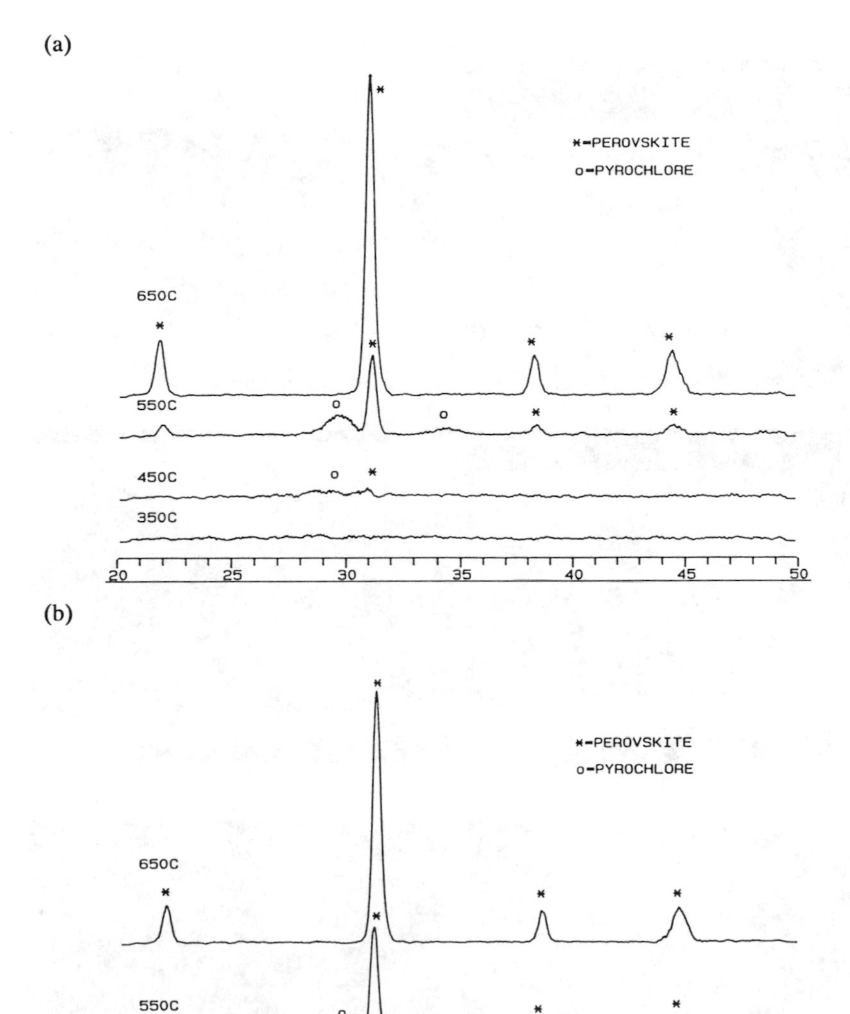

Figure 2. XRD patterns of (a) $PbZr_{0.53}Ti_{0.47}O_3$ and (b)$PbZr_{0.53}Ti_{0.47}O_3 \cdot 0.15PbO$ gels fired at various temperatures in O_2 for 30 mins [6 moles of H_2O:1 mole of Pb]. x-axis = 2θ.

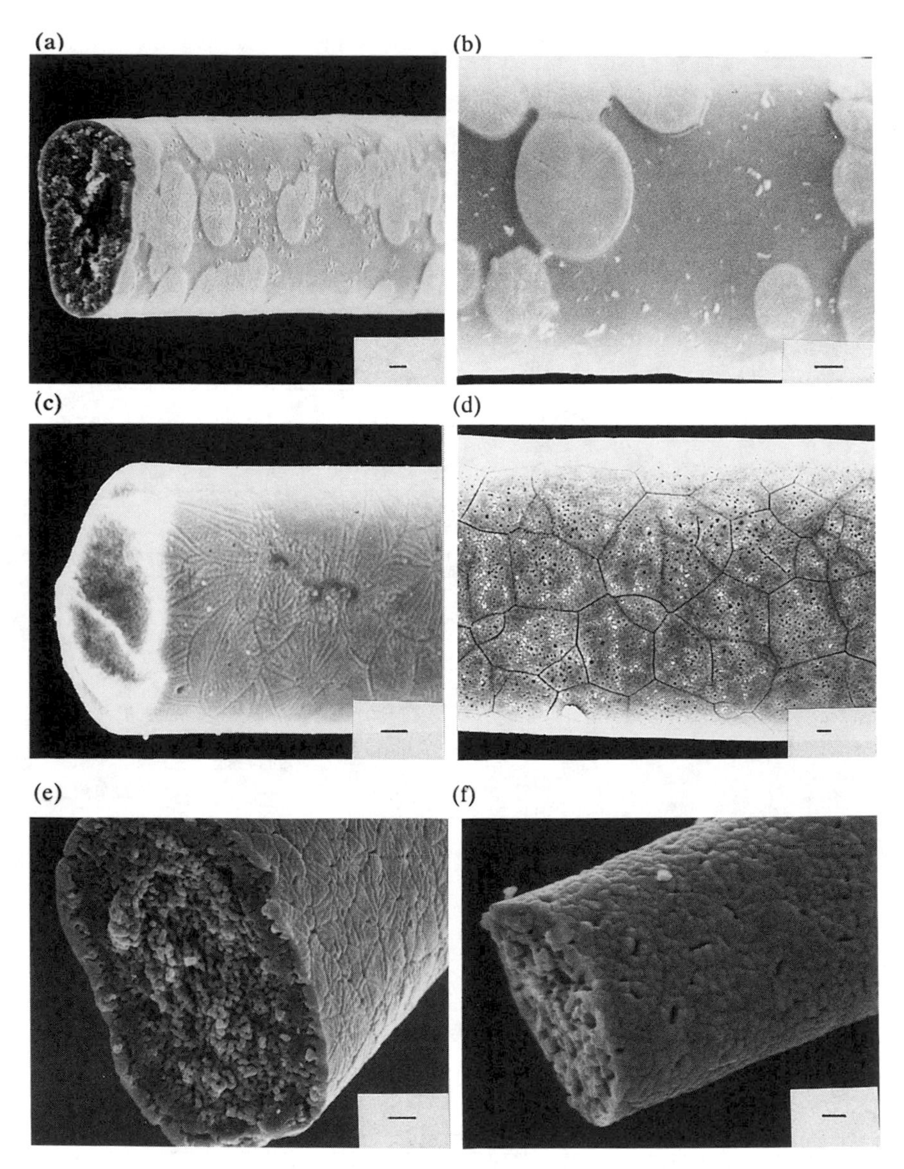

Figure 3. SEM micrographs of PZT fibers fired under various conditions [6 moles of H_2O:1 mole of Pb in all cases]: (a) $PbZr_{0.53}Ti_{0.47}O_3$, 800C/30 mins in O_2 (b) $PbZr_{0.53}Ti_{0.47}O_3 \cdot 0.15PbO$, 750C/30 mins in O_2 (c) $PbZr_{0.53}Ti_{0.47}O_3 \cdot 0.15PbO$, 800C/30 mins in O_2 (d) $PbZr_{0.53}Ti_{0.47}O_3 \cdot 0.15PbO$, 900C/30 mins in O_2 (e) $PbZr_{0.53}Ti_{0.47}O_3 \cdot 0.15PbO$, 500C/30 mins in O_2 and 800C/30 mins in He (f) $PbZr_{0.53}Ti_{0.47}O_3 \cdot 0.15PbO$, 500C/30 mins in O_2 and 850C/120 mins in He. Bar = 1 μm in all cases.

Fired PZT fibers were investigated by SEM. Figure 3(a) shows the microstructure of a $PbZr_{0.53}Ti_{0.47}O_3$ fiber fired at 800C for 30 mins. The interior of the fiber is porous; and the surface shows large crystalline rosettes of the order of ~3-4 μm with a central nucleation site from which radial growth occurs. Rosette structures have frequently been encountered in PZT thin films [18]. The incorporation of excess PbO had beneficial microstructural effects which we have also observed in PZT thin films [19]. The excess PbO acts as a sintering aid and compensates for PbO volatility at elevated temperatures. On firing to 750C (Fig. 3(b)) large isolated rosette structures are observed; but on firing to 800C (Fig. 3(c)) a dense microstructure is obtained. Rapid grain growth occurs on firing at elevated temperatures. Fig. 3(d) shows a fiber fired at 900C with a grain size approaching 10 μm. This grain growth likely results from liquid phase sintering by PbO (m.p. 886°C). Isolated pores can be seen within the grains, likely resulting from PbO volatilization at these high temperatures. Dense, small-grained fibers could be formed by firing in O_2 to 500C for 30 mins. (to complete organic decomposition) and then in He to the desired temperature. Fig. 3(e) and Fig. 3(f) show fibers fired in He at 800C/30 mins and 850C/2 hrs respectively.

Incorporating PZT powder into the precursor sol gave rise to thicker fibers (50-100 μm). These fibers were relatively porous after firing, even in He, as exemplified in Fig. 4(a) for a fiber fired at 800C for 30 mins in He. This excessive porosity results from large-scale shrinkage of the rigid gel matrix around the powder particles on firing, without coalescence of the entire fiber, giving rise to large voids. Sol-gel processing can also be used to prepare hollow ceramic fibers by entrapping air bubbles into the precursor sol. This has previously been demonstrated for TiO_2 fibers [20]. Fig. 4(b) shows a hollow $PbZr_{0.53}Ti_{0.47}O_3$ fiber fired in O_2 at 750C for 30 mins.

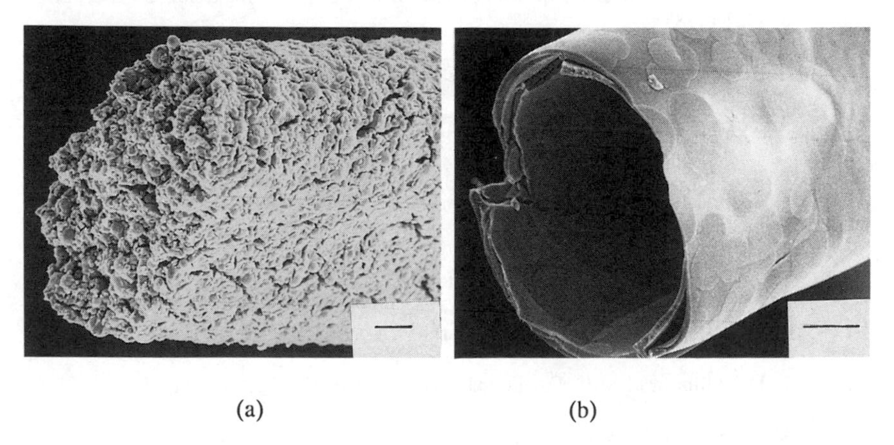

(a) (b)

Figure 4. (a) $PbZr_{0.53}Ti_{0.47}O_3$, 500C/30 mins in O_2 and 800C/30 mins in He, spray-dried PZT powder dispersed in precursor sol. (b) Hollow $PbZr_{0.53}Ti_{0.47}O_3$ fiber, 750C/30 mins in O_2. Bar = 10 μm in both cases.

CONCLUSIONS

PZT fibers can readily be drawn from viscous solutions based on Pb acetate and Zr and Ti alkoxides. The addition of water and polyvinylpyrrolidone facilitate fiber formation. The PbO content, the degree of hydrolysis and the heating rate affect the initial phase development of gels fired at low temperatures (≤550C). These differences in phase assemblage are mitigated when the gels are fired at higher temperatures (650C) where only single-phase perovskite is observed. The firing atmosphere affects the grain size and densification of the fibers; for example, a He furnace atmosphere gave fibers with a fine grain size offering an enhanced driving force for densification. Incorporation of excess PbO is also beneficial in terms of microstructural development by acting as a sintering aid and by compensating for PbO volatility at high temperatures.

ACKNOWLEDGMENT

The financial support of the Air Force Office of Scientific Research is gratefully acknowledged.

REFERENCES

1. D. Segal, Chemical Synthesis of Advanced Ceramic Materials, (Cambridge University Press, Cambridge, 1989).

2. M.J. Morton, J.D. Birchall and J.E. Cassidy, Brit. Pat. No. 1360197 (19 June 1970).

3. H.G. Sowman and D.D. Johnson, Ceramic and Engineering Science Proceedings 6, 1221 (1985).

4. S. Hirano, T. Hayashi, K. Nosaki and K. Kato, J. Am. Ceram. Soc. 72, 707 (1989).

5. T. Yoko, K. Kamiya and K. Tanaka, J. Mater. Sci. 25, 3922 (1990).

6. H. Janusson, C.E. Millar and S.J. Milne in Better Ceramics Through Chemistry IV, edited by B.J.J. Zelinski, C.J. Brinker, D.E. Clark and D.R. Ulrich (Mater. Res. Soc. Proc. 180, Pittsburgh, PA 1990) pp. 421-424.

7. S.I. Aoki, S.C. Choi, D.A. Payne and H. Yanagida in Better Ceramics Through Chemistry IV, edited by B.J.J. Zelinski, C.J. Brinker, D.E. Clark and D.R. Uhlrich (Mater. Res. Soc. Proc. 180, Pittsburgh, PA 1990) pp. 485-490.

8. L. Del Olmo and M.L. Calzada, J. Non-Cryst. Solids 121, 424 (1990).

9. V.K. Seth and W.A. Schulze, Ferroelectrics 112, 283 (1990).

10. T.R. Gururaja, D. Christopher, R.E. Newnham and W.A. Schulze, Ferroelectrics 47, 193 (1983).

11. D.J. Waller, A. Safari, R.J. Card and M.P. O'Toole, J. Am. Ceram. Soc. 73, 3503 (1990).

12. Y. Chida, T. Nishimura and Y. Oguri, European Patent Appln. No. 248432 A2 (9 Dec. 1987).

13. T. Nishi, S. Fujitsu, M. Miyayama, K. Koumoto and H. Yanagida, Sogo Shikensho Nenpo (Tokyo Daigaku Kogakubu) 46, 143 (1987).

14. G. Teowee, J.M. Boulton, W.M. Bommersbach and D.R. Uhlmann, "Second Harmonic Generation From $PbTiO_3$-Based Ferroelectric Thin Films," J. Non. Cryst. Solids, in press.

15. G. Teowee, J.M. Boulton, S.C. Lee and D.R. Uhlmann, "Electrical Characterization of Sol-Gel Derived PZT Films," Mater. Res. Soc. Proc. 243, in press.

16. S.R. Gurkovich and J.B. Blum, Ferroelectrics 62, 189 (1985).

17. S. Hirano (private communication).

18. L.N. Chapin and S.A. Myers, in Ferroelectric Thin Films, edited by E.R. Myers and A.I. Kingon (Mater. Res. Soc. Proc. 200, Pittsburgh, PA 1990) pp. 153-158.

19. G. Teowee, J.M. Boulton and D.R. Uhlmann, "Optimization of Sol-Gel Derived PZT Thin Films By The Incorporation of Excess PbO," paper presented at this meeting.

20. M. Aizawa, Y. Nakagawa, Y. Nosaka, N. Fujii and H. Miyama, J. Non-Cryst. Solids 124, 112 (1990).

PART IV

Aging, Drying, and Consolidation of Gels

MECHANICS OF GELS

GEORGE W. SCHERER
Du Pont Co., Central R & D, Experimental Station 356/384, Wilmington, DE 19880
USA

ABSTRACT

The response of wet gels to applied stresses is discussed, with emphasis on the role of flow of the pore liquid. Even when the network of the gel is purely elastic, the gel exhibits time-dependent behavior that resembles viscoelasticity, which results from fluid flow. The permeability of the gel can be determined from measurement of this time-dependence. Fluid flow also influences the thermal expansion behavior of gels, and can cause severe stresses to develop during supercritical drying, if the autoclave is heated too rapidly.

INTRODUCTION

To understand drying and cracking of gels, it is first necessary to understand their mechanical properties, and this in turn requires consideration of the role of fluid flow within the gel. When a strain is imposed on a wet gel (say, by deflecting a rod in three-point bending), the stress required to maintain a constant deflection is found to vary with time [1]. The reason is that the load is initially supported in part by the liquid and in part by the solid; over time, the liquid flows until the pressure in the liquid drops to zero, and all of the load is transferred to the solid network. Consequently, even if the network is purely elastic, the stress-strain response is time-dependent. In addition, there may be a true viscoelastic response owing to stress-induced breaking and reforming of bonds in the network [2]. If a gel is heated, the thermal expansion of the liquid (which is usually much greater than that of the solid network) causes stretching of the network; gradually the network retracts, and squeezes the liquid out of the network [3]. Thus, the gel shows a time-dependent thermal expansion behavior, which has been shown to play a critical role in stress development in supercritical drying [4]. In this paper, these effects are reviewed, and it is shown how measurement of the kinetics of the response of a gel can be used to determine the permeability of the network.

In the following analyses, the gel is treated as a porous medium containing fluid, so we begin by discussing the appropriate constitutive equations, originally developed by Biot. Then we define the permeability, and qualitatively explain its influence on mechanical behavior. This behavior is illustrated for the case of elastic and viscoelastic networks under applied loads, and in response to a change in temperature.

CONSTITUTIVE EQUATIONS

Although the liquid and solid components of a gel are intimately associated, they are separate phases, and their mechanical behavior has been successfully described using a model originally developed for porous solids [5]. The use of this model for organic gels has been discussed by Johnson [6]. If the network of the gel is elastic, the constitutive equations are of the form [7,8]

$$\varepsilon_x = \frac{1}{E_p}\left[\sigma_x - \nu_p\left(\sigma_y + \sigma_z\right)\right] - \frac{P}{3K_p} \tag{1}$$

where ε_x is the strain in the x-direction, σ_x, σ_y, and σ_z are the components of total stress, P is the stress in the liquid (equal to the pressure, but opposite in sign); E_p, ν_p, and $K_p = E_p/[3(1-2\nu_p)]$ are respectively Young's modulus, Poisson's ratio, and the bulk modulus of the network. The constitutive parameters E_p and ν_p must be measured while $P = 0$ (*i.e.*, by a very slow application of a load), so that the network responds as if no liquid were present. Eq. (1) differs from the constitutive equation for a conventional elastic material only by the term in P, so the gel is simply elastic when $P = 0$. However, if a load is applied suddenly, the liquid cannot instantly escape from the network, so it must support part of the load, and $P \neq 0$. The *total* stress σ_x represents the sum of the forces exerted on the solid and liquid phases on a cross-section of a gel, whereas the *network* stress $\tilde{\sigma}_x$ represents the force on the solid phase per unit area of gel. These quantities are related by [7,8]

$$\sigma_x = \tilde{\sigma}_x + (1-\rho)P \tag{2}$$

where ρ is the volume fraction of solids in the gel.

PERMEABILITY

Liquid is assumed to move relative to the network according to Darcy's law [8]:

$$J = \frac{D}{\eta_L}\nabla P \tag{3}$$

where J is the flux (volume per unit area of gel per time), D is the permeability of the network (units of area), η_L is the viscosity of the liquid, and ∇P is the pressure gradient driving the flow. The permeability increases with the porosity $(1 - \rho)$ and pore radius (r_p) of the network. Simple models [8] suggest a dependence of the form

$$D \sim (1-\rho) r_p^{\,2} \tag{4}$$

but the distribution of pore sizes, and the way in which they are interconnected, must be taken into account in general. Gels have high porosity, but such small pores that the permeability is quite low.

When a load is applied to a gel, the liquid begins to flow in such a way that the stress P goes toward zero. The same behavior is exhibited by a sponge: if a saturated sponge is suddenly squeezed in one direction, the liquid instantly moves in a direction perpendicular to the applied strain; however, the liquid cannot instantly escape from the pores of the sponge, so it stretches the solid network. That is, friction prevents the liquid from squirting out of the sponge, and leaving the network to be compressed uniaxially. Instead, the instantaneous response of the sponge is the same as for an incompressible body: it stretches laterally under a uniaxial compression, maintaining a (nearly) constant volume. The liquid then gradually drains from the network until $P = 0$, while the network retracts laterally. The rate of this response depends on the permeability, which is a measure of the rate at which liquid can move through the network.

MECHANICAL RESPONSE OF AN ELASTIC NETWORK

Consider a rod of gel mounted on a three-point bending fixture immersed in liquid (so that no evaporation can occur from the gel). The rod is suddenly bent to a fixed deflection and held in that position, and the load W required to sustain the deflection is measured as a function of time. The results of such an experiment are presented in Fig. 1, where the sample is a silica gel that has been washed in ethanol so that only the pure solvent is present in the pores. The load relaxes rapidly at first as the liquid flows out of the rod, then arrives at a plateau and remains constant. For a gel with an elastic network, it has been shown [1] that the characteristic time for load decay in three-point bending is

$$\tau_b \equiv \frac{\eta_L a^2 (1-2\nu_p)}{DG_p} \tag{5}$$

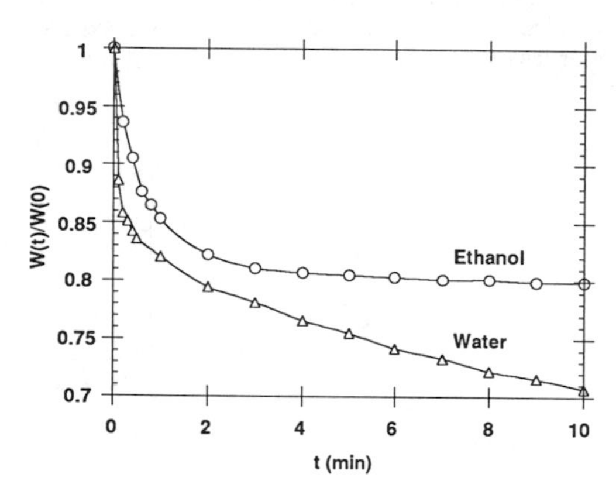

a)

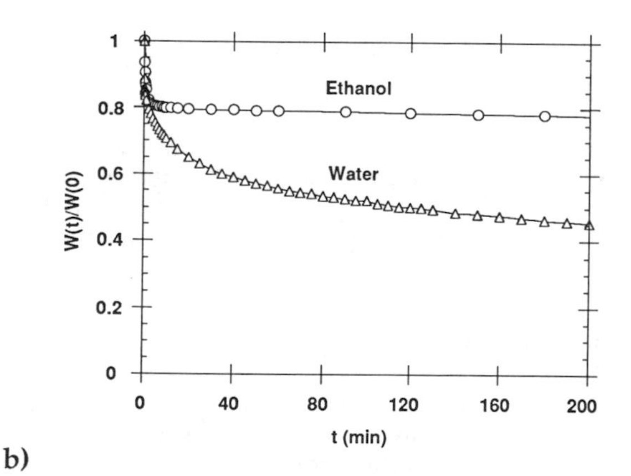

b)

Fig. 1 Load W as function of time t needed to sustain constant deflection in rod of silica gel containing pure ethanol or pure water (data from ref. 1). The initial relaxation (a) results from flow of liquid; sample containing ethanol sustains constant stress in long term (b), while gel containing water relaxes continuously.

where a is the radius of the rod and $G_p = E_p/[2(1+\nu_p)]$ is the shear modulus of the network. The load decay is found to be in excellent agreement with the predicted form:

$$\frac{W(t)}{W(0)} = \frac{2(1+\nu_p)}{3} + \frac{8(1-2\nu_p)}{3} \sum_{n=1}^{\infty} \beta_n^{-2} \exp\left(-\beta_n^2 t / \tau_b\right) \qquad (6)$$

where β_n is the nth root of Bessel's function of the first kind of order one. Beam-bending experiments performed on silica gels of various diameters and containing various solvents show, as indicated in Fig. 2, that the dependence of τ_b on a and η_L is in agreement with eq. (5).

This type of measurement can be used to determine D from τ_b. Moreover, the initial strain occurs at constant volume (since the liquid remains in the gel), so the load and deflection can be used to calculate the *shear* modulus, G_p; the final load yields Young's modulus (as in a beam-bending measurement on an ordinary elastic material), and Poisson's ration can be calculated from G_p and E_p. Thus, one experimental curve yields four important properties of the gel.

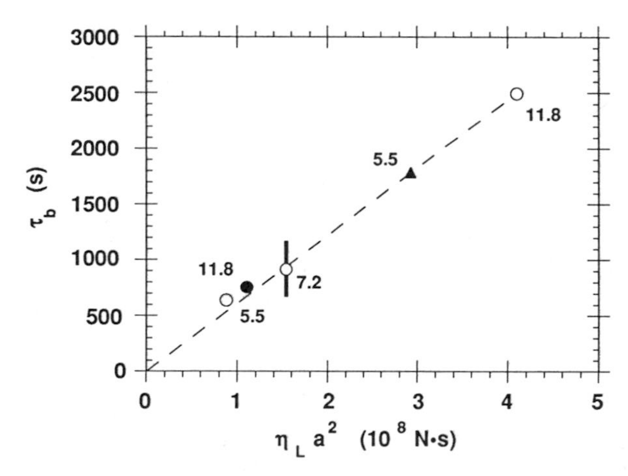

Fig. 2 Relaxation time for load decay versus $\eta_L a^2$, where a is rod radius and η_L is viscosity of liquid in pores of gel. Samples tested in ethanol (o, 1.2 mPa•s), acetone (● , 0.32 mPa•s), and isobutanol (▲, 3.9 mPa•s); numbers are diameters of rods in millimeters. Vertical bar represents data obtained using thermal expansion method. Data from ref. 1.

MECHANICAL RESPONSE OF A VISCOELASTIC NETWORK

In contrast to the results obtained in alcohol, when the silica gel contains water, as shown in Fig. 1, the initial relaxation is followed by a steady decrease in the load. In ethanol, the gel network is elastic, and the only load decay is that produced by fluid flow; in water, the load continues to decrease due to a true viscoelastic effect, as the water attacks strained siloxane bonds and thereby permits stress relaxation. The relaxation in aqueous solvents is found to be nonexponential, with a relaxation spectrum containing a wide range of relaxation times [2]. Analyzing the spectrum is difficult, because the viscoelastic and hydrodynamic effects are concurrent. For the base-catalyzed gels depicted in Fig. 1, the hydrodynamic relaxation is fast compared to the viscoelastic effect; the opposite is true for the acid-catalyzed systems studied in ref. 2, because the latter have much lower permeability, so τ_b is long compared to the average viscoelastic relaxation time.

THERMAL EXPANSION

If the temperature of a gel rises rapidly, the expanding liquid stretches the gel network; at high enough heating rates, the thermal expansion coefficient of the gel is equal to that of the liquid in the pores. If the temperature is raised and then held constant, the gel first expands, then contracts to its initial dimension; the contraction occurs as the liquid drains from the network. The characteristic time for this process is [3]

$$\tau \equiv \frac{\eta_L a^2 \beta}{D K_p} = \frac{\tau_b}{2(1-\nu_p)} \tag{7}$$

where $\beta = (1+\nu_p)/[3(1-\nu_p)]$. As shown in Fig. 3, when a gel is subjected to a complicated thermal cycle, its dimensional changes can be accurately described by the theory presented in ref. 3, where the only free parameter is τ. The figure shows that the gel expands when the temperature is raised, as the expanding liquid stretches the network; then, while the temperature is held constant, the network contracts back to its original dimension ($\Delta h/h_o = 0$). Similarly, the gel contracts when the temperature decreases, but the network expands back to $\Delta h/h_o = 0$ when the temperature is held con-

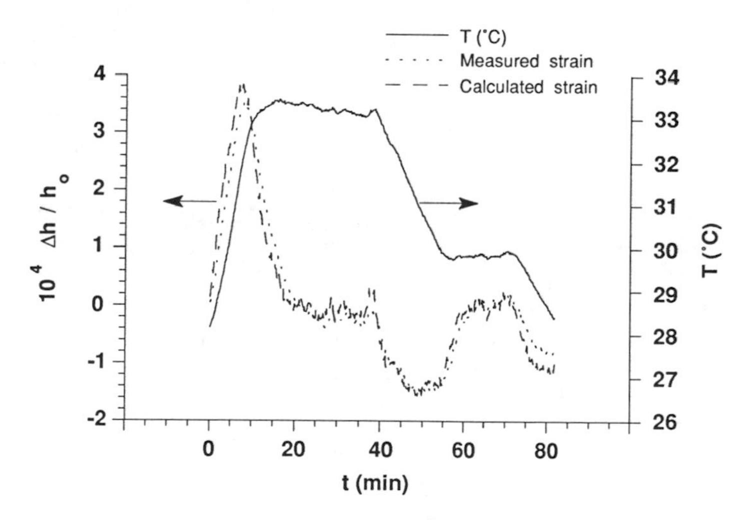

Fig. 3 Measured and calculated axial strain ($\Delta h/h_o$) for gel rod with diameter of 7.2 mm and length of 9.5 cm subjected to thermal cycle shown by solid curve. Calculated curve is one-parameter fit of measured strain to theoretical curve given in ref. 3; parameter is τ. Data from ref. 3.

stant. Given K_p and v_p, this type of measurement can be used to determine the permeability of the gel; as shown in Fig. 2, the results are in very good agreement with those obtained from beam-bending experiments.

Some gels exhibit syneresis, a spontaneous contraction produced by condensation reactions within the network [8]. That shrinkage is superimposed on the thermal strains when the gel is subjected to a temperature cycle, as shown in Fig. 4. The gel is an acid-catalyzed silica gel with a low permeability, so that its characteristic time τ_b is much greater than that of the base-catalyzed gel in Fig. 3. When the data are compared to the prediction for an elastic gel, there is a systematic disagreement, resulting from the continual contraction (syneresis) of the network. That is, the calculation predicts that the sample will return to its original dimension during an isothermal hold, but it actually relaxes to a smaller dimension. When a constant strain rate is superimposed on the calculated strain, the fit (Fig. 4b) is excellent. This type of experiment can thus be used to measure both the permeability and the rate of syneresis; however, a separate measurement (such as beam-bending) is needed to obtain the constitutive parameters, so that D can be calculated from τ_b. Since beam-bending is not sensitive to

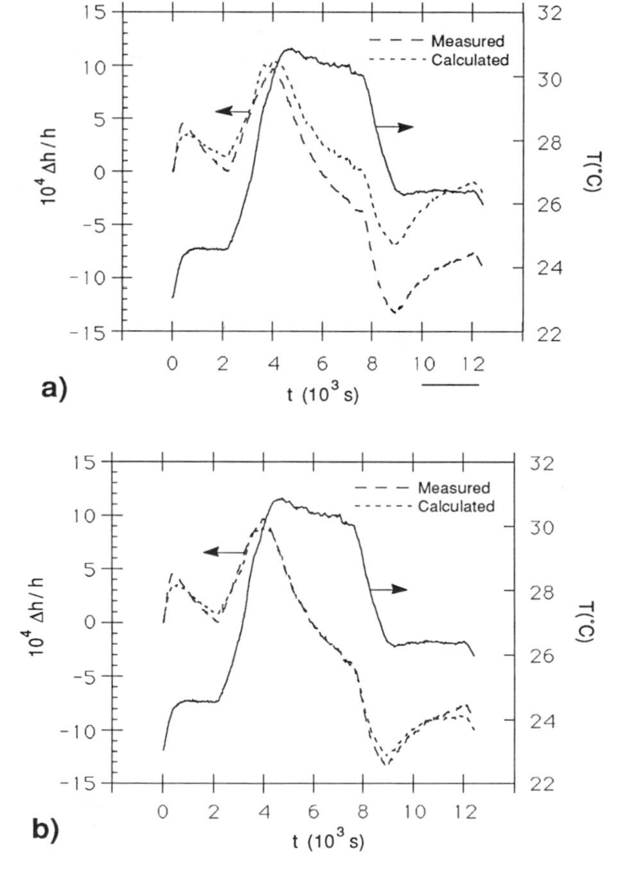

Fig. 4 Measured and calculated axial strain ($\Delta h/h$) for acid-catalyzed silica gel rod subjected to thermal cycle shown by solid curve. Calculated curves are fits of measured strain to theoretical curve given in ref. 3, using (a) one parameter, τ, or (b) two parameters, τ plus a constant syneresis strain rate, $\dot{\varepsilon}_s$.

syneresis, it is advantageous to apply both techniques to characterize the gel.

The expansion of the liquid stretches the network of the gel, and in extreme cases this can lead to dangerous tensile stresses in the gel. For example, it is sometimes seen that gels crack during heating in the autoclave, in the preparation of supercritically dried aerogels [9]. This is most likely to occur when the gels are large and the rate of heating is fast. In such cases, it has been shown [4] that the calculated thermal stresses exceed the strength of the gel. Very high axial stresses develop if a long cylindrical gel is in contact with the walls of the vessel, because the liquid must flow along the length, rather than the radius, to escape from the gel. In that case, the gel fractures into slices normal to its axis of rotation. If the same type of gel is heated without being in contact with the walls, it does not crack at the same heating rate; it will crack at much higher heating rates, but the cracks will run in random directions [9]. These phenomena are in accord with predictions of a stress analysis based on the constitutive equations presented above [4].

CONCLUSIONS

Wet inorganic gels exhibit two types of time-dependent response to applied stresses and strains: transient effects produced by fluid flow within the pores, and true viscoelastic behavior caused by chemical attack of the pore fluid on strained bonds in the network (a phenomenon also known as chemorheology [10]). The crucial distinction between the mechanical behavior of wet gels and ordinary elastic or viscoelastic materials is the role of the liquid within the network. The movement of the liquid relative to the solid results in delayed responses of the gel to applied forces and changes in temperature. This behavior is accurately described by a model originally developed for porous materials, such as soil. The model has been extended to describe drying stresses [7,8], and thermal strains [3] and stresses [4]. On the basis of the model, it is possible to extract the properties of the gel (including permeability, rate of syneresis, and elastic moduli) from measurements of load relaxation [1] and/or thermal strain [3] in wet gels, and these methods are much easier than conventional measurements of those properties.

REFERENCES

[1] G.W. Scherer, J. Non-Cryst. Solids, 142 [1-2] 18 (1992)
[2] G.W. Scherer, S.A. Pardenek, and R.M. Swiatek, J. Non-Cryst. Solids, 107 (1988) 14-22
[3] G.W. Scherer, H. Hdach, and J. Phalippou, J. Non-Cryst. Solids, 130, 157 (1991); correction 136, 269 (1991)
[4] G.W. Scherer, "Stress development during supercritical drying", accepted by J. Non-Cryst. Solids (Proc. Third Int. Symp. on Aerogels, Wurzburg, Sept. 1991)
[5] M.A. Biot, J. Appl. Phys., 12, 155 (1941)
[6] D.L. Johnson, J. Chem. Phys., 77, 1531 (1982)
[7] G.W. Scherer, J. Non-Cryst. Solids, 109, 171 (1989)
[8] C.J. Brinker and G.W. Scherer, Sol - Gel Science (Academic Press, NY, 1990)
[9] T. Woignier and J. Phalippou, Univ. Montpellier, France, private communication
[10] K. Murakami and K. Ono, Chemorheology of Polymers (Elsevier, New York, 1979)

THE EFFECT OF GEL AGING ON THE PHYSICAL AND OPTICAL PROPERTIES OF A GEL-DERIVED GRIN GLASS

Mark A. Banash, Tessie M. Che, J. Brian Caldwell, Robert M. Mininni, Paul R. Soskey, Victor N. Warden and Hui H. Chin, EniChem America, 2000 Cornwall Road, Monmouth Junction, NJ 08852

Variations in gel aging times and temperatures are shown to affect the optical properties of a gel-derived GRIN lens. These properties include the variation of the total index change as well as the shape of the index profile. However, these optical properties do not appear to correlate with the amount of modifier removed during the leaching step. Also, regardless of aging, gels appear identical with respect to calcining response and composition of pore liquid.

Introduction

Sol-gel methods have been used successfully to make optical GRIN glass [1]. Recent work on radial gradient-index (GRIN) glass rods in the TiO_2-Al_2O_3-SiO_2 system [2,3] uses an acid to leach a multicomponent silicate gel to create a composition gradient of TiO_2. This gel is then fixed, dried and sintered to make gradient-index glass. One of the difficulties encountered with this process is that identically formulated gels do not always behave identically during the leaching step. In most cases the gel survives the acid bath but in some instances the gel is severely weakened by the process, sometimes even to the point of dissolving. One possible cause of this inconsistency is that the gels are not aged identically. In this study several aging parameters are varied to determine the subsequent effects on the gel's survivability, its chemical and physical properties, and the optical properties of the resultant gradient-index glass product.

Experimental Procedure

A TiO_2-Al_2O_3-SiO_2 system containing 20% titania by mole weight was synthesized using an established procedure [2]. This TMOS based system has a gelation time of 4-5 hours at room temperature. The aging variations carried out in this study are listed in table I - these include whether gelation occurs at room or elevated temperatures , how long the gel is heated after gelation, and the length of time the gel ages at room temperature before it is used. Gel samples are identified by the classification X-Y-Z, where X is the number of days the gel sits at room temperature after casting, Y is the subsequent number of days the gel is heated at 70°C and Z is the number of days the gel ages at room temperature before further processing. After aging, the gels were leached for 24 hours at ambient temperatures in an aqueous acid medium to create the gradient in titania index modifier. Each gel was then fixed, dried, calcined and finally sintered to glass. Gels were calcined in a box furnace under air. The temperature was raised from 25 °C to 450°C at a rate of 5 degrees/min. The xerogels were then sintered by placing the samples on a platinum gauze "boat" inside a quartz tube in a Lindberg tube furnace. The samples are heated from 450 °C to 550°C under flowing air and then to 760°C under flowing He. The end product is a transparent GRIN rod.

Sample ID	Final Leachant Ti Concentration (mg/l)	Δn	Profile Comments
0-1-1	1646	Sample Broke	
0-1-7	1552	0.027	Highly Aberrated
0-1-14	1727	0.022	Highly Aberrated
0-2-1	1426	0.028	Aberrated
0-2-7	1445	0.028	Highly Aberrated
0-2-14	1589	0.034	Highly Aberrated
1-1-1	1694	0.025	Aberrated
1-1-7	1616	0.023	Aberrated
1-1-14	1698	0.027	Slightly Aberrated
1-2-1	1573	0.025	Aberrated
1-2-7	1547	0.025	Aberrated
1-2-14	1624	0.028	Least Aberrated

Table I - Elemental and Optical Results of Aging Study on TiO_2-Al_2O_3-SiO_2 GRIN system

Pore Liquid Infra-red Analyses

Pore liquid which evolves from the gels during syneresis was analyzed for differences in composition by infra-red spectroscopy using a Mattson Cygnus 100 FTIR spectrometer. It can be postulated that differences in the extent of gel network formation during aging would result in different proportions of the elimination products in the pore liquid. However, as evident in the IR spectra shown in Figure 1, there are no discernible differences between liquids from the shortest and longest aging cycles.

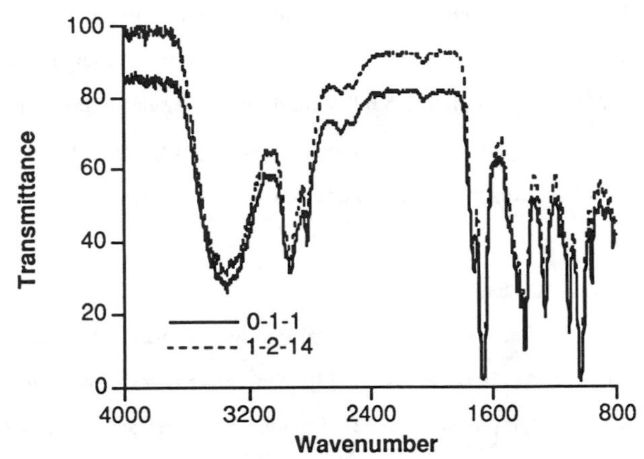

Figure 1 - Infra-Red Spectra of Pore Liquids from Differently Aged Gels

Thermogravimetric Analyses

Thermogravimetric (TGA) analyses were performed using a Perkin Elmer TGA7 analyzer to determine if there were any differences in solvent removal from two differently aged gels during calcining. Both gels were aged 1 day at 25°, then 1 day at 70°C. GEL 1 was then allowed to sit for 3 days and GEL 2 for 14 days at room temperature. Table II shows the weight loss of these two gels. No major differences in the order or extent of solvent loss were observed in the thermogravimetric plots. The plots themselves display a one-step process - such a process resists analysis of percent loss of individual solvent components.

Sample Preparation	GEL 1	GEL 2
As is	80.1%	83.3%
Additional 2 hrs at 100°C	77.9%	76.3%
Additonal 285 hrs at 100°C	22.9%	23.5%

Table II - Weight loss at 450°C from TGA analysis on 20% TiO_2-Al_2O_3-SiO_2 system.

Elemental Analyses of Leaching Solutions

Atomic Absorption analysis was performed to measure the concentration of titanium in each leaching solution after 24 hours. The instrument employed was a Perkin

Elmer Model 3100 AA unit using the 363 nm absorption line of Ti. This line is linear over the 0 to 250 mg/l concentration range and was calibrated using standard techniques. Samples of leaching solution (0.5 ml) were drawn at regular intervals over the 24 hour leach. Prior to analysis, the samples were diluted by a factor of ten using a 0.1% KCl aqueous solution . Results are given in Table I.

Optical Characterization

Refractive index profiles were measured by Mach-Zehnder interferometry which has been described in detail elsewhere [2]. Refractive index profiles for Samples 0-1-7 and 1-2-14 are displayed in Figure 2. The changes in refractive index (Δn) for lenses made from different gels are given in Table I.

Discussion

With the exception of Sample 0-1-1 which cracked during drying at 70°C, all gels survived processing. The thermal analyses provided no detectable differences between the differently aged gel systems. Previous IR work with similar gel formulations have revealed compositional differences in the pore liquids of differently aged gels. These differences are seen as significant variations in peak height as well as in the appearance of new peaks or the disappearance of old ones. No such effects are seen here, indicating that compositional differences are slight. As shown in Table I, variations in aging do effect the amount of Ti removed from the gel by the leaching solution. However, these amounts do not correlate directly with the observed index changes of the respective lenses.

For many GRIN lens applications, the optimal shape of the refractive index profile is a parabola. The derivative of a parabolic profile is linear with respect to radial position, thus analysis of the linearity of the profile derivative can provide a method for identifying optical quality GRIN lenses. For an acceptable lens, this linearity must extend over an appreciable amount of the lens radius. The derivative plot of a good quality GRIN lens (Sample 1-2-14) shows little deviation from linearity over approximately 90% of its radius, in comparison to that for a poor quality GRIN lens (Sample 0-1-7) which is linear over less than 70% (Figure 2) . The optical quality of all lenses is summarized in Table I. It appears that the systems which were not allowed to gel and initially age at room temperatures (0-Y-Z series) resulted in GRIN lenses with optical defects and/or poor index profiles. In the 1-Y-Z series, only Experiment 1-2-14 resulted in an good quality GRIN lens. The poor quality of the 0-Y-Z lenses may be due to coarsening or phase transformation effects which are known to occur more rapidly at higher gelation temperatures [4]. Leaching of more disordered gel structures may result in the aberrations observed in the index profiles of the 0-Y-Z series.

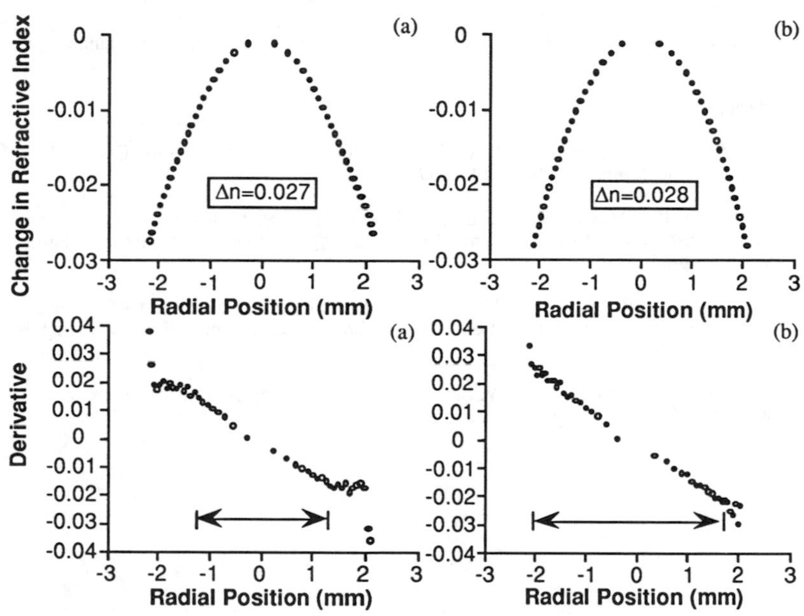

Figure 2 - Refractive Index Profiles and Corresponding Derivative Plots for a) Sample 0-1-7 and b) Sample 1-2-14. Arrows indicate range of useful profile.

Conclusions

GRIN lenses made via a sol-gel process show variations in the total index change and index profile shape as a result of differences in gel aging parameters. Optical evaluation of GRIN lenses provides a sensitive method for measuring the effects of aging on the microstructure of TiO_2-Al_2O_3-SiO_2 gels. The results suggest that enhancing aging, either by heating the gel or by allowing it to sit for extended periods, improves the optical qualities of the resultant glass. Work is presently being carried out to correlate AA results with the shape and magnitude of the index change using both interferometry and EDS to prepare Ti maps of lenses. Additional IR analyses to determine exact pore liquid composition are also being performed. Future planned studies include surface adsorption and rheological measurements to provide more insight into the effects of aging on the properties of these gel systems.

Acknowledgements

The authors thank Ian McCallum for his assistance in the preparation of the manuscript.

References

1. T.M. Che, J. B. Caldwell and R.M. Mininni, Proc. SPIE - Int. Soc. Opt. Eng. Vol. 1328 - Sol-Gel Optics (1990) pp.145-159.

2. J.B. Caldwell, T.M. Che, R.W. Cruse, R.M. Mininni, R.N. Nikles, V.N. Warden and M.A. Banash, Mater. Res. Soc. Symp. Proc. Vol. 180 - Better Ceram. through Chem. IV (1990) pp. 727-732.

3. M.A. Banash, J.B. Caldwell, T.M. Che, R.M. Mininni, P.R. Soskey and V.N. Warden, Proc. SPIE - Int. Soc. Opt. Eng. Vol. 1590 - Submol. Glass. Chem. Phys. (1991) pp. 8-13.

4. C. Jeffrey Brinker and George W. Scherer in Sol-Gel Science (Academic Press, Inc., San Diego, CA, 1990) pp. 356-405.

SOL-GEL FILMS WITH TAILORED MICROSTRUCTURES*

DEBORAH L. LOGAN, CAROL S. ASHLEY AND C. JEFFREY BRINKER**
Ceramic Synthesis and Inorganic Chemistry Department 1846
Sandia National Laboratories, P.O. Box 5800, Albuquerque, NM 87185-5800
**also UNM/NSF Center for Micro-Engineered Ceramics,
University of New Mexico, Albuquerque, NM 87131

ABSTRACT

The refractive index and porosity of silicate films formed by dip-coating were tailored by varying the aging time of a two-step acid- and base-catalyzed sol to control the size and structure of the polymeric species prior to coating. ^{29}Si NMR showed that there was a 2% increase in the total number of bridging oxygens with sol age, consistent with a cluster-cluster growth mechanism. However, several percent monomer remained regardless of aging. Dip-coated films were characterized by ellipsometry before and after heating to moderate temperatures at three rates. Sol aging prior to film deposition leads to greater film porosity, consistent with the concept of aggregation of fractal clusters, but only after a heat treatment to remove the organic species associated with incompletely hydrolyzed monomers from the pores.

INTRODUCTION

Organic dye-doped thin films could be useful for applications such as chemical sensors [1,2], photochromic switches, tunable lasers [3-5], luminescent solar collectors [5] and optical data storage [6]. These applications require a rigid transparent matrix so the dye can exhibit a photochromic response, a low temperature synthesis technique to prevent decomposition of the dye, and, often, control of the thin film microstructure. Compared to conventional organic polymer matrices, sol-gel-derived inorganic glass matrices exhibit superior chemical, thermal, mechanical and optical properties and can be processed at low temperatures compatible with the dye molecules.

Another advantage of sol-gel processing is the ability to tailor the microstructure of the film for applications such as optical switches that require a dense matrix or to create the porosity needed for a chemical sensor. While there are many ways to affect the competition between phenomena that promote a more compact or a more porous structure in a dip-coated film, e.g., by changing the composition or dip-coating conditions [7], it is also possible to tailor the microstructure of films obtained from a single sol composition by changing the size and/or structure of the entrained inorganic species [8].

---

*This work was performed at Sandia National Laboratories, supported by the U.S. Department of Energy under Contract # DE-AC04-76-DP00789.

We can exploit the fractal nature of the inorganic polymers formed in the sol to tailor the microstructure of the resulting dip-coated film. When the sol is aged the fractal precursors grow larger, which affects the porosity of the film in two ways. Because the density of a mass fractal object scales with its size r as $1/r^{(3-D)}$, where the mass fractal dimension D is less than 3, larger fractals are more porous. In addition, as cluster size increases the probability of clusters intersecting each other increases. This means that larger clusters will interpenetrate less, or are more opaque to each other, so aggregation of larger clusters leads to more porous films as illustrated schematically in Figure 1.

Fig. 1

Cluster-cluster aggregation of larger, mutually-opaque clusters from aged sols leads to more porous films.

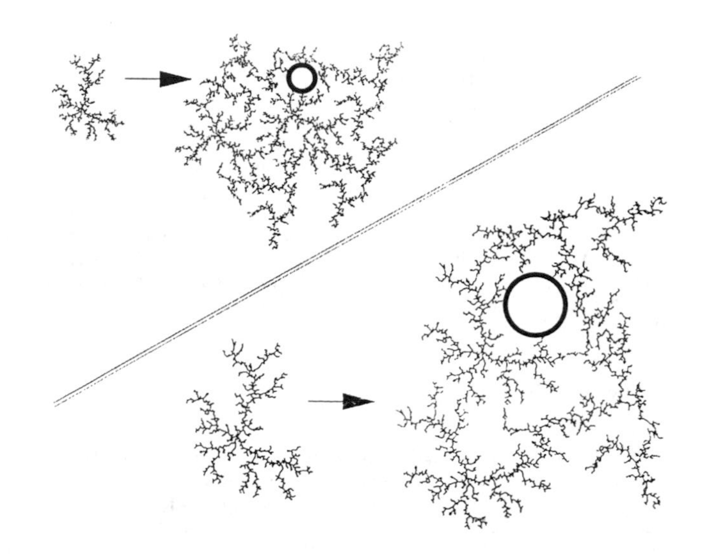

EXPERIMENTAL

Silica sols were prepared by a two-step process involving acid-catalyzed hydrolysis of tetraethoxysilane (TEOS) in ethanol followed by addition of base as described by Brinker et al. [9,10]. SAXS studies indicate that the silicate polymers formed by this process are mass fractals with mass fractal dimension $D \approx 2.4$ [11]. The oxide concentrations of the sols were varied from ~2.4 to 4.8 wt% by dilution with ethanol. After dilution, the sols were aged in an explosion-proof oven at 50°C, and samples were removed at intervals up to the gel point. Other samples were permitted to gel and were sheared to a coatable consistency by either 30 minutes of ultrasonic agitation or by the combination of dilution with an equal volume of ethanol and exposure to ultrasound. Films were applied to Si substrates by dip-coating at rates varying from 12.7 to 20.8 cm/min in a nitrogen atmosphere. After 20 minutes drying under a heat lamp at 65°C, the films were fired in air at moderate temperatures (usually up to 400°C). The samples were fired on edge in quartz trays, and the oven ramp varied from 2°C/min to an estimated 50°C/min (the latter achieved by placing the quartz trays directly on ZrO_2 pallets in the oven at temperature). Ellipsometry was used to determine the refractive index and thickness before and after firing.

RESULTS AND DISCUSSION

Based on the concept of aggregation of mutually opaque fractals, we expect that sol aging, which increases the size of the fractal clusters, should increase film porosity and decrease refractive index. The expected effect did not appear until after heat treatment, however, as Figure 2 shows. The NMR results (Figure 3) show that the unaged sol is 2% unhydrolyzed and partially hydrolyzed TEOS monomer (Q^0), and with aging the monomer increases to 7%, presumably due to limited siloxane bond hydrolysis. The total number of bridging oxygens increases 2% with aging, however. Presumably monomer produced by dissolution is more than compensated for by additional polymerization of higher Q species (cluster-cluster growth). The films exhibit little variation in porosity as deposited because monomer is able to fill in the pores of fractal aggregates. The expected increase in film porosity with sol aging appears only after heat treatment to burn off the organic species associated primarily with unhydrolyzed or partially hydrolyzed monomer and to a lesser extent with more highly condensed polymers.

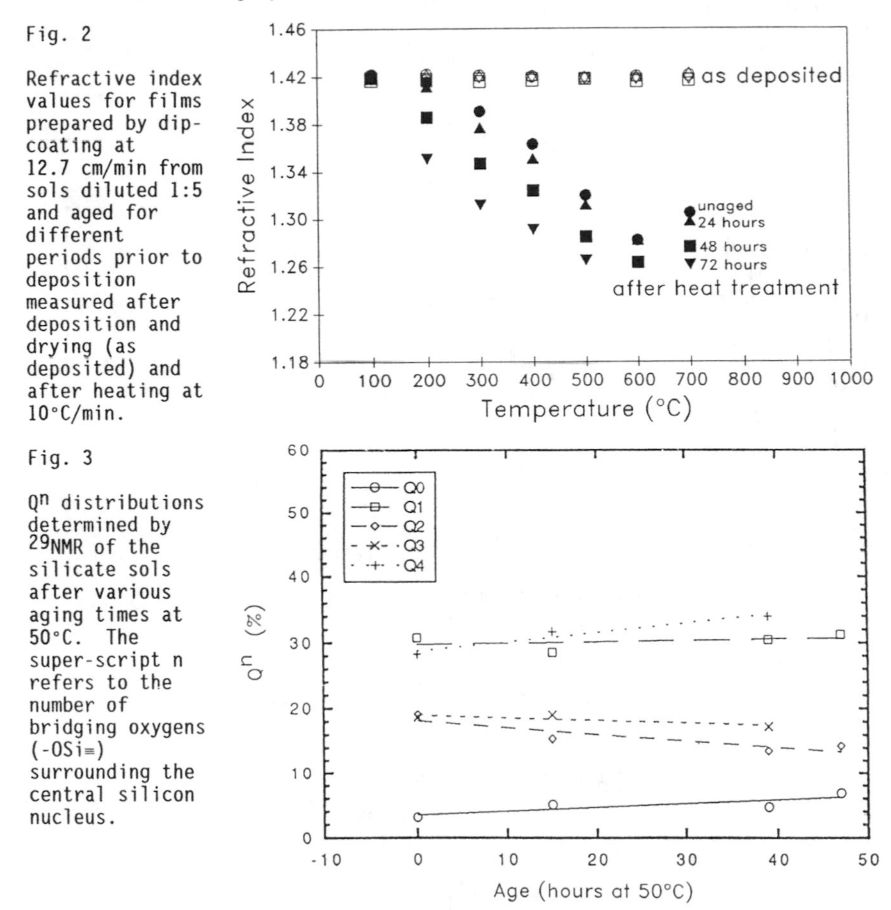

Fig. 2

Refractive index values for films prepared by dip-coating at 12.7 cm/min from sols diluted 1:5 and aged for different periods prior to deposition measured after deposition and drying (as deposited) and after heating at 10°C/min.

Fig. 3

Q^n distributions determined by ^{29}NMR of the silicate sols after various aging times at 50°C. The super-script n refers to the number of bridging oxygens (-OSi≡) surrounding the central silicon nucleus.

Comparing the percent shrinkage data (calculated from ellipsometric measurements) in Figure 4 to the TGA results in Figure 5, the samples pass through three stages when heated, as described in Table 1. In the 200°C to ~550°C range, there is a competition between the removal of organic species associated mainly with incompletely hydrolyzed monomer, which tends to increase the film porosity, and continued condensation reactions, which tend to decrease the film porosity. The former effect dominates and is more pronounced at higher heating rates because there is less time for condensation reactions to occur. The corresponding reduction in refractive index at higher heating rates is shown in Figure 6.

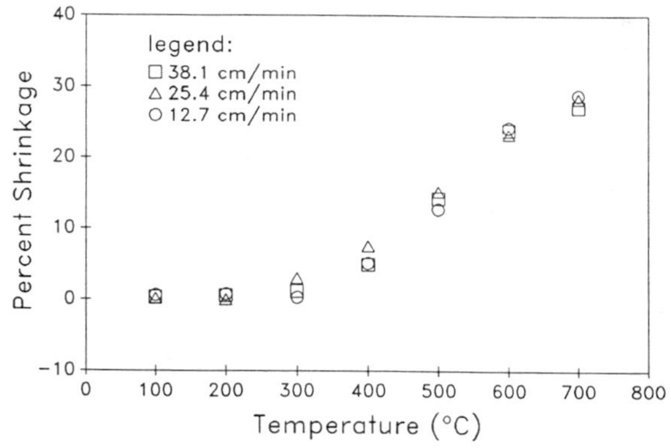

Fig. 4

Percent shrinkage for samples diluted 1:4, aged 15 hours, and heated 10°C/min, calculated from ellipsometric measurements.

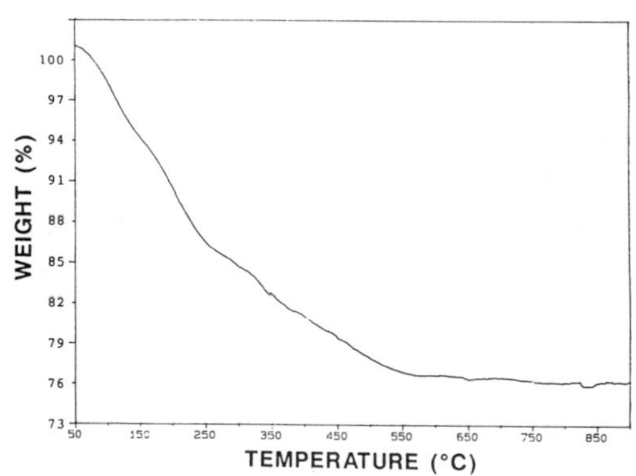

Fig. 5

TGA results for a thin film sample prepared from a sol diluted 1:5 and aged 72 hours prior to deposition, heated at 10°C/min in air.

Table 1. Three Stages of Heat Treatment

Temperature Range	Observation	Process(es)	Effect on Porosity
room temperature to 200°C	weight loss without shrinkage	desorb H_2O and organics	little change
200°C to ~550°C	weight loss and shrinkage	burn off organics	↑
		continued condensation reactions	↓
~550°C to 700°C	shrinkage without weight loss	sintering	↓

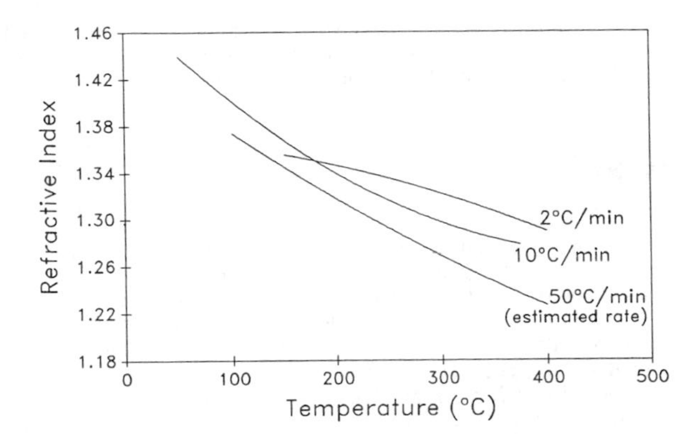

Fig. 6

Refractive index values versus temperature at three heating rates for films prepared from sols aged 94 hours (past the gel point), broken up with ultrasound and dip-coated at 25.4 cm/min.

Figure 7 shows that dilution has little effect on refractive index either before or after heat treatment because the data for all dilutions scale in same manner with t/t_{gel}. More dilute sols are easier to age and dip-coat, however. A sol diluted 1:2 (sol:ethanol) by volume gels within 13 hours, but a sol diluted 1:4 or 1:5 takes several days, which makes it easier to age samples to near the gel point for dip-coating. The 1:5 dilution makes the most uniform films for more consistent ellipsometric measurements, and the 1:5 gel is the easiest to break up with ultrasound to a coatable consistency.

Fig. 7

Refractive index values versus normalized gel time for samples dip-coated at 25.4 cm/min, before and after heating to 400°C for 10 min.

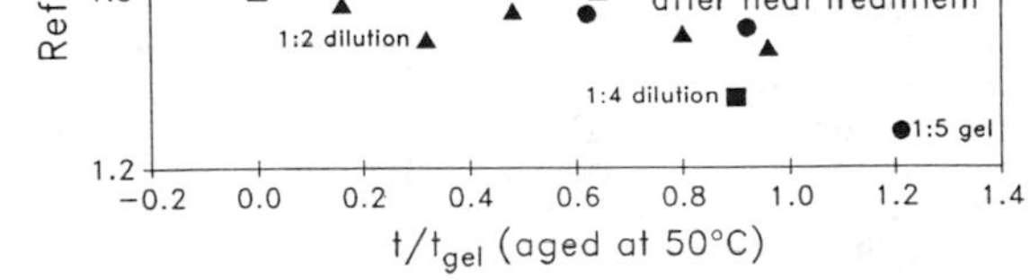

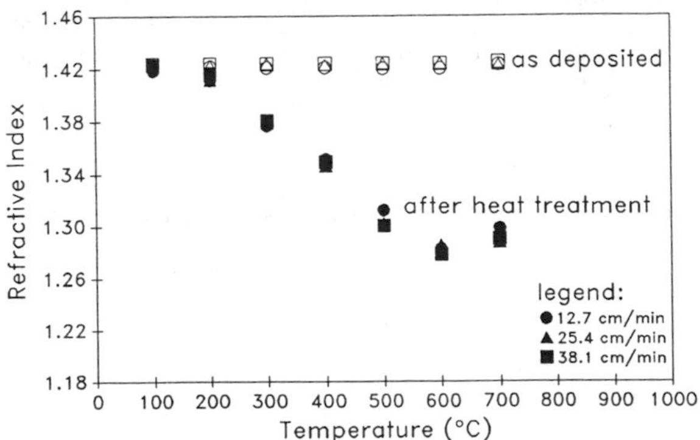

Fig. 8

Refractive index values versus temperature for samples diluted 1:5, aged 24 hours, and heated at 10°C/min.

Figure 8 shows that dip-coating withdrawal speed has no impact on refractive index. Dip-coating withdrawal speed does affect thickness but has no impact on shrinkage (Figure 4).

SUMMARY

Aging a two-step acid- and base-catalyzed sol to increase the size and opacity of the fractal precursors produced more porous silicate films. A heat treatment was required, however, because organic species associated primarily with incompletely hydrolyzed monomers had to be burned out of the pores. We are currently investigating a three-step process, using a second acid catalysis step to fully hydrolyze the monomer prior to adding base and aging, in order to produce films with controlled porosity that do not require heat treatment, and therefore are compatible with organic dyes.

ACKNOWLEDGMENTS

The authors thank R.A. Assink for the NMR spectra and D.C. Goodnow for the TGA results.

REFERENCES

1. R. Zusman, C. Rottman, M. Ottolenghi and D. Avnir, J. Non-Cryst. Solids 122, 107-109 (1990).
2. V. Chernyak, R. Reisfeld, R. Gvishi and D. Venezky, Sensors and Materials 2 (2), 117-126 (1990).
3. B. Dunn, E. Knobbe, J.M. McKiernan, J.C. Pouxviel and J.I. Zink in Better Ceramics Through Chemistry III, edited by C.J. Brinker, D.E. Clark and D.R. Ulrich (Mater. Res. Soc. Proc. 121, Pittsburg, PA, 1988) pp. 331-342.
4. R. Reisfeld in Sol-Gel Optics, SPIE Proceedings 1328, 29-39 (1990).
5. R. Reisfeld in Sol-Gel Science and Technology, edited by M.A. Aegerter, M. Jafelicci Jr., D.F. Souza and E.D. Zanotto (World Scientific, Singapore, 1989) pp. 323-345.
6. D. Levy, S. Einhorn and D. Avnir, J. Non-Cryst. Solids 113, 137-145 (1989).
7. C.J. Brinker, A.J. Hurd, G.C. Frye, P.R. Schunk and C.S. Ashley, J. Ceram. Soc. of Japan 99 (10), 862-877 (1991).
8. C.J. Brinker and G.W. Scherer, Sol-Gel Science (Academic Press, San Diego, CA, 1990) pp. 799-811.
9. C.J. Brinker, K.D. Keefer, D.W. Schaefer and C.S. Ashley, J. Non-Cryst. Solids 48 47-64 (1982).
10. C.J. Brinker, K.D. Keefer, D.W. Schaefer, R.A. Assink, B.D. Kay and C.S. Ashley, J. Non-Cryst. Solids, 63, 45-59 (1984).
11. D.W. Hua and D.L. Logan, unpublished results.

PORE SIZE AND SURFACE TEXTURE MODIFICATION OF SILICA VIA TRIALKYLSILYLATION

Duen-Wu Hua, Douglas M. Smith
UNM/NSF Center for Micro-Engineered Ceramics, University of New Mexico, Albuquerque, NM 87131

ABSTRACT

Surface modification of silica via silylation is often undertaken to change surface chemistry and hence, adsorption characteristics but little attention is paid to its use for pore morphology modification, the subject of this work. Modification of two silica samples was conducted by trimethylsilylation. The degree of silylation/ surface coverage was monitored by TGA and elemental analysis. Pore structure was studied by nitrogen adsorption and condensation, and small angle x-ray scattering (SAXS). Previously, we demonstrated that surface area, surface texture, pore size distribution, and total pore volume are indeed changed in a controlled fashion due to silylation of mesoporous silicas. Generally, the mean pore size is smaller, the pore size distribution is narrower, and the pore surface is smoother after silylation. However, questions remain concerning the effect of small-scale surface roughness and the fractional surface coverage on the effective pore size. This study concentrates on the silylation of smaller pore material ($r < 40$ Å). Significant pore size distribution narrowing and smoothing of the pore surface with increasing surface coverage was noted.

INTRODUCTION

Surface modification of porous silica has been widely studied and used in various fields [1]. It is well known that the active -OH groups on the surface of silica can react with alcohol or alkylchlorosilane and change the pore chemistry. However, the use of silylation for modifying pore size and size distribution as well as the physical texture of the pore structure is seldom considered. Many studies have been devoted to the silylation of silica gels [2-5] concerning the change of adsorption properties, the effect of large R groups, and the fractal nature of the resulting gels but no particular attention was paid to the effect of silylation on pore morphology. Previously, we had studied [6] the effect of silylation on various mesoporous silicas, such as phase separated glasses (Vycor and control pore glass: CPG). The results showed that the pore surface gets smoother with silylation, and the pore sizes decrease as a function of R group size and coverage. In principle, silylation should be particularly effective for microporous material. This study compares the silylation of a smaller pore silica xerogel with our previous results. The effect of the silylating agent on the surface area, pore size, and pore surface roughness are also investigated.

EXPERIMENTAL

Samples used in this study are sol-gel derived water washed silica xerogels, one from the 2-step acid/base catalyzed reaction of TEOS: tetraethyl orthosilicate (commonly denoted as B2 gel) and one from a 2-step acid/acid catalyzed reaction (denoted as A2 gel) [7]. Davis and co-workers [8] have shown that the washing of these gels in ethanol removes unreacted and partially reacted silicates which can cause phase separation if the wet gel is placed in water. Washing in water promotes further condensation/polymerization reactions leading to a decrease in surface area and increased pore volume. For this work, both wet gels are washed 3 times with ethanol to remove unreacted monomers/oligomers and 3 times with water, each wash cycle was about 24 hours [8]. Drying at 393 K for two days, the final xerogel with primarily a Si-OH surface is obtained.

Silylation was conducted by placing of dried silica into a benzene solution with various trimethylchlorosilane (TMCS) concentrations. After shaking for 24 hours at room temperature, TMCS/silica was filtered, washed with benzene and then acetone, dried at 413 K for 2 days, and stored in a desiccator [2]. The degree of silylation was calculated from elemental analysis and represented by carbon contents of the samples. The blank carbon value was subtracted from the sample treated only with benzene and acetone.

Pore structure analysis was conducted using nitrogen adsorption/condensation at 77 K. Samples were outgassed at 403K overnight before analysis. Surface area was calculated from the adsorption branch of the isotherm in the range of $0.05 \leq P/P_0 \leq 0.3$, and the pore size distribution was calculated from the desorption branch. Small angle x-ray scattering was performed at Oak Ridge National laboratory with a pinhole camera using a 64x64 2 dimensional position sensitive detector and on the Kratky U-slit setup with a linear position sensitive detector at the University of New Mexico (CMEC), the wavelength used are 1.542 Å (Cu-K_α), the q range covered is about 0.005 to 0.6 $Å^{-1}$. Surface area were obtained by using the relative intensity from the higher q region of the scattering curves.

RESULTS AND DISCUSSIONS

Nitrogen adsorption data and TMS (trimethylsilyl group) surface coverage (calculated from the carbon contents by elemental analysis) for both gels are given in Table I. Surface coverage (in terms of the fraction of the surface covered by the TMS group) is calculated from the carbon contents by elemental analysis and the estimated size of the TMS group (0.38 nm^2 by Bondi's model [9]).

Table I. Nitrogen adsorption results for various trialkylsilylated B2 gels.

	SA (m^2/g)	PV (cm^3/g)	C %	TMS group/g	Coverage
B2M1	539	0.897	0.5	0	0%
B2M2	525	0.809	2.24	2.91×10^{20}	20.5%
B2M3	461	0.693	4.91	7.38×10^{20}	52.0%
B2M4	436	0.673	5.71	8.71×10^{20}	61.4%
A2M1	615	0.761	0.15	0	0%
A2M2	572	0.679	2.75	4.35×10^{20}	26.9%
A2M3	493	0.574	5.25	8.53×10^{20}	52.7%
A2M4	471	0.546	5.58	9.08×10^{20}	56.1%

The nitrogen surface area and total pore volume decrease with increasing TMCS concentration (higher surface coverage). This is a similar effect as observed for phase separated glasses with larger pores like CPG-75 [6]. The water washed B2 xerogels should have the same surface smoothing effects towards TMCS. With higher TMCS concentration, the surface gets smoother and the pores get smaller. As the starting pore size decreases, this effect should become more obvious.

The surface area of A2 and B2 silica gels are plotted in Figure 1. As can be clearly seen, the surface area of both gels decrease as the TMCS concentration increases and reach a minimum. This result suggests the presence of small scale surface roughness in the materials. Notice that the A2 gel requires a much lower TMCS concentration to reach a minimum. This is reasonable since the A2 gel has a smaller pore structure.

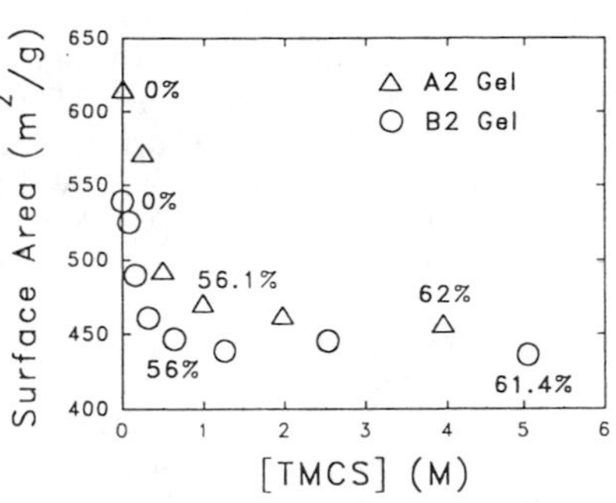

Figure 1. Surface area for TMCS silylated A2 & B2 gels.

Figure 2 gives pore size distributions of TMCS silylated B2 gels and the effect is clear. Silylation appears to selectively react in the larger pores which results in a significant narrowing of the pore size distribution. This may be the result of steric hindering of the silylation reaction in smaller pores. Even at only 20% surface coverage, most large pores have been eliminated.

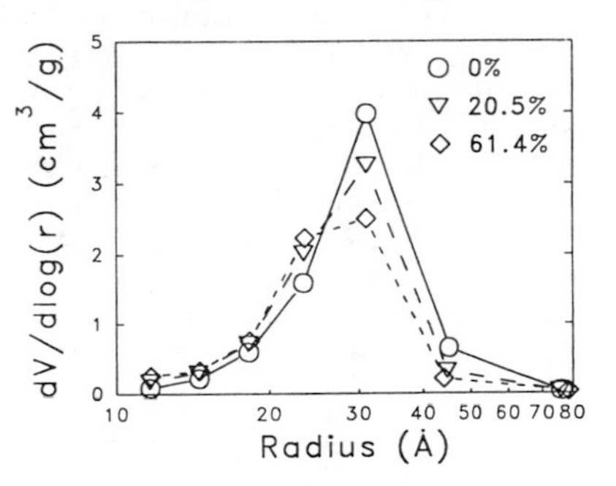

Figure 2. Pore size distribution for TMCS silylated B2 gels.

Comparatively, the pore size distributions of A2 xerogel have a different shape (Fig. 3) but the same effect of silylation. It has smaller pore sizes to begin with (≈ 24Å), and the sily -lation also narrows down the pore size distribution and reduces the pore size.

It is interesting to note that with increasing extent of silylation, the large peak at ≈ 24 Å decreases in both size and magnitude but the peak at ≈ 17 Å increases in magnitude and is invariant in size. We attribute this peak to the tensile strength limit of liquid nitrogen at 77 K since it corresponds to P/P_0 = 0.42 in the Kelvin type analysis [10]. This implies that the peak is representative of the volume of the pores with size ≤ 17 Å rather than a narrow peak corresponding to an actual distribution as narrow as narrow as indicated. The use of xerogels with micropore sizes (such as A2 gel dried from mother liquor or aprotic solvents are under way.

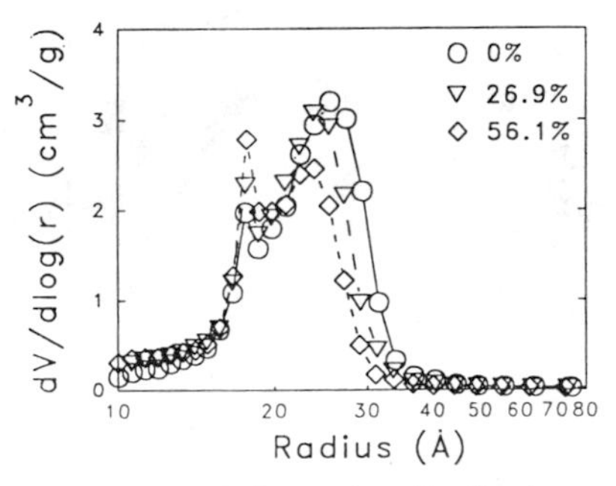

Figure 3. Pore size distribution for TMCS silylated A2 gels.

Small-scale pore surface roughness is another important parameter for many practical applications. In a previous paper, Pfeifer et al. [3] showed that by using nitrogen adsorption data measured for a sample with various thicknesses of a preadsorbed film such as water [11], one could derive the surface fractal dimension. For a fractal surface, the measured film surface area, A, decreases with increasing adsorbed film volume, V, as:

$$A \propto V^{(2-D)/(3-D)} \qquad (1)$$

in which D=2 will be a flat surface. Thus a log-log plot of A versus V should be linear and have a slope -M (M>0), which yields the dimension of the surface from $D_s = 2 + [M/(M+1)]$.

From the difference between the measured pore volume of the silylated sample and the pore volume of the blank, the film (TMS) volume is found. Figure 4 is a $\log(A/A_0)$ vs. $\log(PV_0\text{-}PV_p)$ plot for A2 and B2 gels silylated with TMCS. We calculate the surface fractal dimension for TMCS silylated gel as 2.26 for B2 gel and as 2.16 for A2 gel. This indicates that the pore surface for A2 is smoother than B2 gel.

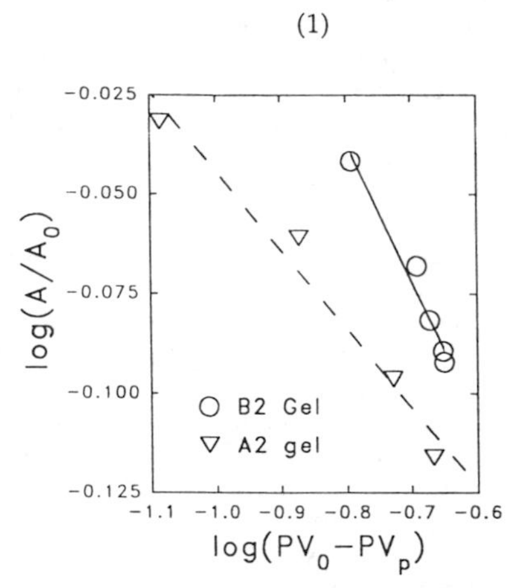

Figure 4. Surface area vs. film volume of TMS on A2 & B2 gels.

Basically, our SAXS results support the nitrogen adsorption data. Figure 5 shows representative SAXS curves for TMCS silylated B2 gels. We found that all B2 samples exhibited power law scattering for scattering vector between $0.2 < q < 0.4$ Å^{-1}. However, only the highly silylated sample (61.4%)

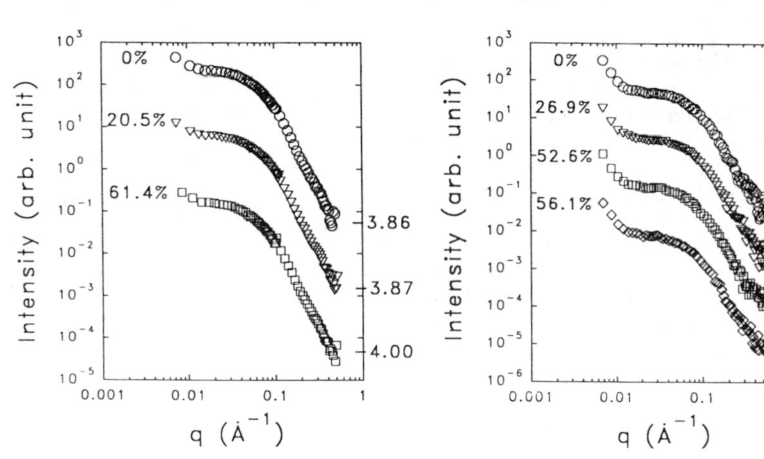

Figure 5. SAXS curves of TMCS silylated B2 gel

Figure 6. SAXS curves of TMCS silylated A2 gel

showed Porod behavior (slope -4 for a smooth surface); the others deviate to various degrees. As for the A2 gels (Fig. 6), the range is about 0.1 to 0.3 Å^{-1}, and the trend is the same as B2 gels. However, the Porod slope of the highly silylated A2 gel (56.1%) only approaches -4 (≈ 3.92), which suggests that the surface is not completely smoothened by silylation. This is probably due to the range that we use for the Porod analysis. From Hurd et al. [12,13], the specific surface area of a smooth surface can be measured by normalizing to the total scattered intensity Q :

$$Q = 2 \pi I_0 (\Delta\rho)^2 V \varphi (1-\varphi) \qquad (2)$$

$\Delta\rho$ is the difference in electron density, V is the illuminated volume and φ is the volume fraction of the sample, then the specific surface area σ_0 is:

$$\sigma_0 = S/V\rho_0 = \pi \varphi (1-\varphi) \lim(q^4 I)/\rho_0 Q \qquad (3)$$

where ρ_0 is the mass density of the loosely packed sample and $\lim(q^4I)$ is the Porod constant. σ_0 can be converted to a surface area that is measured with a resolution r as:

$$S(r) = \sigma_0 r^{2-D} \qquad (4)$$

where D is the surface fractal dimension, and r is about 4 Å for a nitrogen adsorption experiment.

The surface area calculated from the SAXS curves of Figure 5 using the large q range (0.2 - 0.4 $Å^{-1}$) is about 590 m^2/g for plain B2 gel and drops to 553 m^2/g with medium TMCS silylation. The final value of 538 m^2/g is obtained for the highly silylated gel. The calculated surface area and related parameters for different xerogels are listed in Table II. The decreasing Porod slope clearly shows that with a higher concentration of silylating agent, the surface gets smoother.

Table II. Surface area and related parameters calculated from SAXS.

	SA (m^2/g)	slope	Coverage	N_2 SA (m^2/g)
B2M1	590	-3.86	0%	539
B2M2	553	-3.87	20.5%	525
B2M4	538	-4	61.4%	436
A2M1	641	-3.73	0%	615
A2M2	592	-3.78	26.9%	572
A2M3	549	-3.86	52.7%	493
A2M4	506	-3.92	56.1%	471

When highly silylated with any silylating agent, we can observe Porod behavior (smooth surface with slope -4) for the sample. Compared to the surface area obtained from nitrogen adsorption (BET) data, all of the SAXS surface area have the same decreasing trend, but the magnitudes are larger. This difference is probably caused by assuming the surface to be smooth, the q range chosen for calculations; and/or the fact that SAXS observes all pores which includes closed porosity that is inaccessible to BET analysis.

CONCLUSION

The change in pore morphology, size, and surface texture for two different porous silica xerogels as a function of silylation has been measured. The pore size distribution may be considerably narrowed and the mean pore size is typically reduced by a dimension comparable to the R group diameter (4 - 6Å). Silylation appears to selectively modify the large pores in the size distribution which is possibly a result of less steric hindrance to the silylation reaction in larger pores. Silylation also smooths the surface of the pores and may be used as a probe of the surface fractal dimension.

ACKNOWLEDGEMENTS

This work is supported by the UNM/NSF Center for Micro-Engineered Ceramics, a collaborative effort supported by NSF (CDR-8800532), Los Alamos and Sandia National Laboratories, the New Mexico Research and Development Institute, and the ceramics industry. The authors gratefully thank Dr. J. S. Lin of Oak Ridge National Laboratory in his helping with the SAXS instrument. The help of CMEC staff: W. Ackerman (nitrogen adsorption), J. Anderson (sample preparation) and R. Ju (elemental analysis) is also appreciated.

REFERENCES

1. R.K. Iler, The Chemistry of Silica, (Wiley & Sons, New York, 1979).
2. Y. Nakamura, M. Shinoda, K. Danjo, K. Iida and A. Otsuka, Advanced Powder Technology 1 (1), 39 (1990).
3. P. Pfeifer, G.P. Johnston, R. Deshpande, D.M. Smith and A.J. Hurd, Langmuir 7, 2833 (1991).
4. T. Okubo, H. Inoue, AIChEJ, 34, 1031 (1988).
5. P.W. Schmidt, D. Avnir, D. Levy, A. Hober, M. Steiner and A. Roll, J. Chem. Phys. 94, 1474 (1991).
6. D.W. Hua and D.M. Smith, Langmuir, submitted.
7. C.J. Brinker, K.D. Keefer, D.W. Schaefer and C.S. Ashley, J. Non-Cryst. Solids 48, 47 (1982).
8. P.J. Davis, C.J. Brinker and D.M. Smith, J. Non-Cryst. Solids, in press.
9. A. Bondi, J. Phys. Chem. 68, 441 (1964).
10. S.J. Gregg and K.S.W. Sing , Adsorption, surface area, and porosity (Academic Press, New York, 1982), p. 154.
11. S.B. Ross, D.M. Smith, A.J. Hurd and D.W. Schaefer, Langmuir 4, 977 (1988).
12. A.J. Hurd, D.W. Schaefer and A.M. Glines, J. Appl. Cryst. 21, 864 (1988).
13. A.J. Hurd, D.W. Schaefer, D.M. Smith, S.B. Ross, A.L. Mehaute and S. Spooner, Phys. Rev. B, 39 (13), 9742 (1989).

PORE STRUCTURE EVOLUTION OF SILICA GEL DURING AGING/ DRYING: EFFECT OF SURFACE TENSION

Ravindra Deshpande, Douglas M. Smith and C. Jeffrey Brinker; UNM/NSF Center for Micro-Engineered Ceramics, University of New Mexico, Albuquerque NM 87131 USA.

ABSTRACT

Two-step acid/base (B2) and acid/acid (A2) catalyzed silica gels have been aged in ethanol or water baths followed by various aprotic solvents with a range of surface tensions. The physical and chemical structures of xerogels dried from these aprotic solvents were studied by a series of techniques (nitrogen adsorption, elemental analysis, TGA, SAXS)and compared to the corresponding structures of low temperature (CO_2) and high temperature (ethanol) aerogels. The aprotic solvents help to isolate the effects of pore fluid surface tension during drying since they do not react with the gel surface. B2 xerogels showed a linear decrease in surface area with increasing surface tension. Pore volume and pore size followed a similar trend. Micropore analysis on A2 xerogels showed an increase in micropore volume and surface area with increase in surface tension, whereas total surface area and pore volume showed an opposite trend. Thus, by varying surface tension and aging, one is able to independently control surface area, pore volume, and pore size.

INTRODUCTION

Before drying, the structure and chemistry of a wet gel may be considerably altered by varying the aging conditions such as time, temperature, pH and pore fluid [1,2,3,4]. In previous studies of the effects of alcohol and water, final pore fluid was found to be the dominant factor in xerogel surface areas and pore volume. A decrease in surface area upon drying was observed resulting from condensation during drying and/or surface tension induced collapse. The effect of surface tension on xerogel structure could not be isolated, however because of accompanying reactions of the solvent with the gel such as reesterification reactions or alcoholysis of siloxane bonds. Deshpande and co-workers [5] have studied the effect of pore fluid surface tension on base catalyzed silica gels and demonstrated that significant changes in pore structure could be obtained by changing surface tension.

The effect of surface tension (γ) during solvent exchange and drying is manifested in the form of capillary pressure (P_c) [6]:

$$P_c = -(2\,\gamma_{LV}\cos\theta)/r_{pore} \qquad (1)$$

where θ is contact angle and r_{pore} is the pore radius. The tension in the liquid is supported by the solid network, which in turn experiences compression. Shrinkage depends on the response of the gel to the maximum capillary stress generated [6]. Aprotic solvents do not participate in condensation or reesterification reactions, so that the effects of surface tension and capillary stresses on xerogel structure are more clearly elucidated.

The supercritical drying process for making aerogels avoids liquid-vapor menisci during drying, thus eliminating capillary pressure and essentially preserving the wet gel structure. The most common methods of making aerogels involve either directly removing pore fluid above its critical point (for ethanol T_c= 516 K, P_c = 63 bar) or low temperature replacement of pore fluid with liquid CO_2 following by the removal of CO_2 above its critical point (T_c = 304 K, P_c = 73 bar).

Here we use physical and chemical characterization of xerogels and aerogels to study the effects of surface tension on their pore structure. Different aprotic solvents exhibiting a wide range of surface tension were used as the drying fluids to partially isolate the effects of surface tension without affecting the chemistry of the silica gel.

EXPERIMENTAL

Silica gels were prepared via a two-step acid/base (B2) or two-step acid/acid (A2) catalyzed hydrolysis/condensation of tetraethylorthosilicate (TEOS) [7]. After initial aging in the mother liquor for 22 hours at room temperature, the wet gels were placed in a large excess of ethanol. In order to remove any unreacted monomer from the gel network [2] the gels were washed with pure ethanol in five 24 hour steps, using fresh ethanol for each successive step. These gels were then washed in various aprotic solvents with a range of surface tensions. Exchange of the ethanol pore fluid was achieved by washing in an aprotic solvent for 48 hours, with intermediate removal of excess solvent from the top of the container, followed by addition of fresh aprotic solvent. The various aprotic solvents employed are listed in Table 1 along with water and ethanol used for comparison as a final pore fluid.

Table 1: Surface tension of different pore fluids used at room temperature.

Final pore fluid	Surface tension, dyne/cm	Final pore fluid	Surface tension, dyne/cm
Supercritical	0	Acetone	23.7
Liquid CO_2(283 K)	2	Cyclohexane	25.3
Liquid CO_2 (273 K)	4	Acetonitrile	29.3
Liquid CO_2 (268 K)	7	Nitromethane	32.7
Ethanol	22.7	1:4 dioxane	33.6
Tetrahydrofuran	23.1	Water	72

A similar series of samples was prepared by washing five times with ethanol and then once with water prior to exchange with the final aprotic solvent (or water). Exchange of water pore fluid with aprotic solvents was achieved by washing as described above. Only aprotic solvents miscible with water were used in this set. A similar drying scheme was followed for all the samples. The gels were dried at 323 K for 48 hours and then at 383 K for 48 hours.

Two A2 and B2 aerogel samples were also prepared. The first was made by heating the gel washed five times with ethanol above the critical temperature (T~ 525 K) in an autoclave and removing the supercritical ethanol while maintaining the temperature above the critical temperature. The second sample was prepared at low temperature by exchanging ethanol with liquid CO_2 and then removing CO_2 above its critical temperature (T~ 309 K). Additional samples were prepared by exchanging ethanol with liquid CO_2 but the CO_2 was removed below its critical point so that the surface tension experienced by the drying gel is that of liquid CO_2.

Nitrogen sorption at 77 K was used to obtain surface areas and pore volumes of xerogels. A five point BET analysis [$0.05 < P/P_0 < 0.3$, N_2 molecular cross-sectional area=0.162 nm^2] was conducted to obtain surface areas and a single condensation point [$P/P_0 = 0.995$] was used to determine the total pore volumes. Because of large upper pore size, aerogel pore volumes were determined via measuring the bulk (by mercury displacement at ambient pressure) and skeletal density (by helium pycnometry).

Small angle x-ray scattering (SAXS) data were collected using a Rigaku 12 kW rotating anode x-ray generator (Cu-Kα, λ=0.1532 nm) with Kratky U-slit optics and a position sensitive detector. The data were then corrected for slit collimation in order to evaluate the radius of gyration and the fractal dimension by analysis of the Guinier and Porod regions, respectively.

RESULTS AND DISCUSSION

The effect of pore fluid surface tension on A2 and B2 xerogel areas are presented in Figure 1 along with surface areas for B2 aerogels. We observe that for aprotic solvents by increasing the surface tension of the pore fluid, the surface area of B2 xerogels decrease in a roughly linear manner. There is little effect of the intermediate water wash.

The surface area of ethanol washed A2 samples are significantly less than B2 samples and there is little dependence on surface tension. With an intermediate water wash, we observe a strong surface tension dependence and increased surface area for each value of surface tension employed. For both A2 and B2 samples, surface areas of samples dried in water are considerably higher than expected from extrapolation of the trend for intermediate water wash samples to 72 dyne/cm.

Figure 2 illustrates the effect of surface tension on the pore volume of A2 and B2 xerogels prepared with and without an intermediate water wash prior to final solvent exchange. The effect of surface tension on pore volume is similar to that observed for the surface area results. For the B2 samples without a water wash, the pore volume changes by over a factor of four for the surface tension range studied and the pore volume is inversely proportional to surface tension. For the samples dried from liquid carbon dioxide, the porosities are over 80% and approach those obtained for aerogels [6]. The effect of a water wash is to significantly increase the pore volume (by a factor of ~1.5).

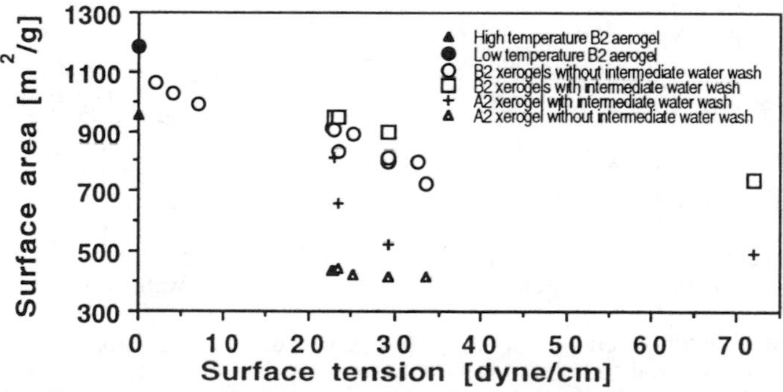

Figure 1 Effect of pore fluid surface tension on xerogel surface area.

Micropore analysis (for pore radius ≤ 15 Å) was carried out on the A2 xerogels since in comparison to B2, A2 is relatively weak structure and is expected to undergo more capillary stress induced collapse [8]. Micropore volume and surface areas for A2 samples with and without water wash prior to final aprotic solvent are shown in Figure 3, 4 and 5,6 along with total pore volume and surface area, respectively. Micropore volume and surface area for samples without intermediate water wash contribute over 75 percent to total pore volume and surface area, illustrating the compliant nature of the ethanol washed A2 gel and the greater extent of capillary stress induced collapse. Samples with an intermediate water wash exhibit an increase in micropore volume and surface area with increasing surface tension, whereas total pore volume and surface area show an opposite trend. An intermediate water wash resulted in hydrolysis of surface ethoxide groups and subsequent condensation leading to a stiffer network. As a result of relatively stronger network, there was gradual increase in the micropore contribution to the total pore volume with increasing surface tension.

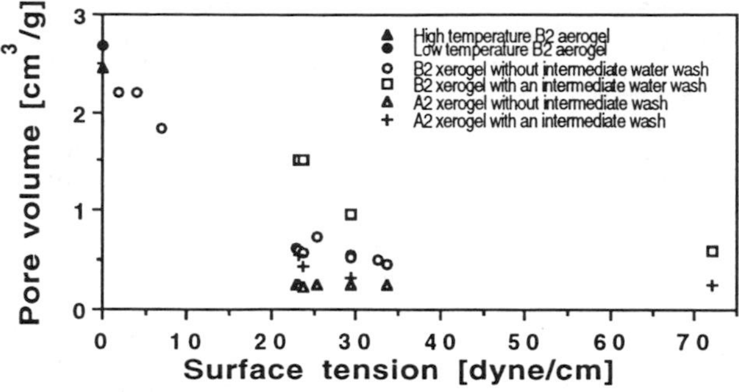

Figure 2 Effect of pore fluid surface tension on xerogel pore volume.

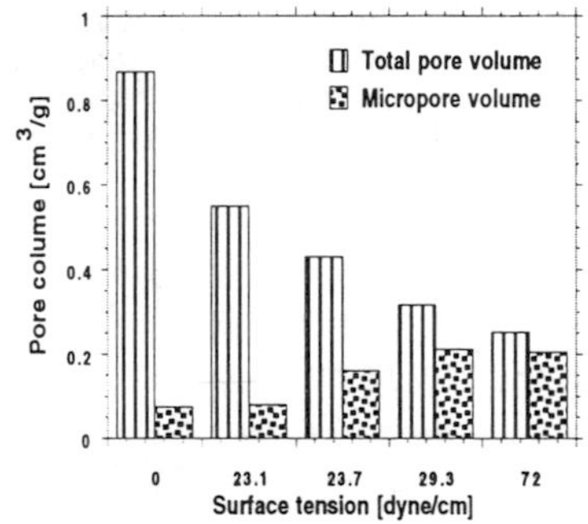

Figure 3 Effect of surface tension on micropore and total pore volume of A2 xerogels with intermediate water wash.

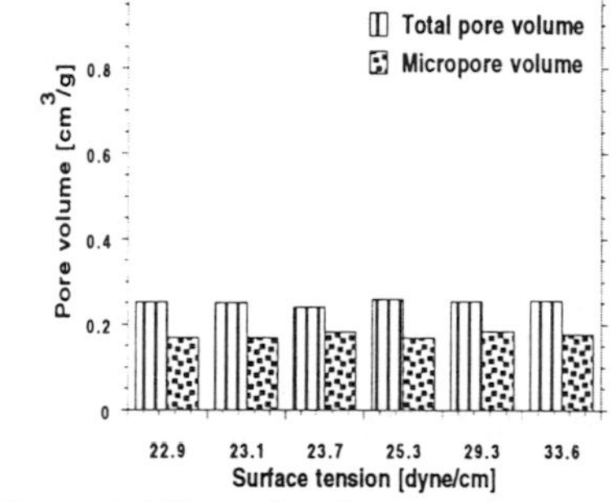

Figure 4 Effect of surface tension on micropore and total pore volume of A2 xerogels without intermediate water wash

Additional information concerning structure can be extracted from the average pore radius (defined as the ratio of twice the pore volume to surface area). Figure 7 shows the effect of surface tension on average pore radius of A2 and B2 xerogels with and without an intermediate water wash. For B2 samples without an intermediate water wash, the average pore radius varied by a factor of four over the surface tension range studied. The use of an intermediate wash changed the average pore radius by a factor of approximately two over the same surface tension range. A2 xerogels with and without an intermediate water wash exhibited smaller average pore radius compared to corresponding B2 sample dried from pore fluid of same surface tension.

SAXS curves for A2 xerogels with and without an intermediate water wash are shown in Figures 8 and 9 respectively. We observe a consistent decrease in pore size and scattering feature size with increase in surface tension employed.

The microstructure of a xerogel reflects the original wet gel structure and its extent of collapse during drying by the capillary pressure. As seen from

Equation 1, increasing the surface tension causes a linear increase in P_c, so increasing the surface tension should cause an increase in the extent of

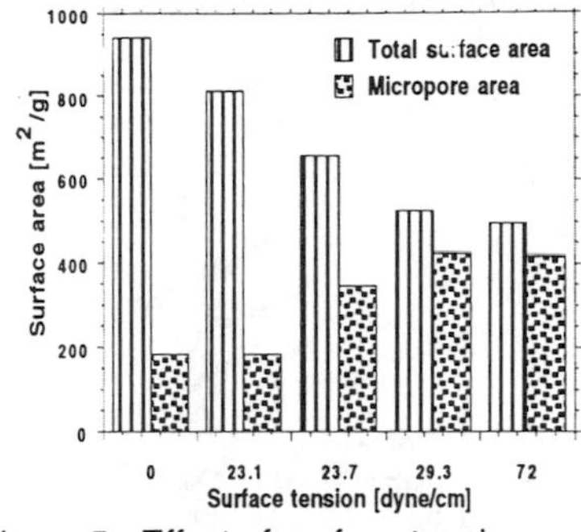

Figure 5 Effect of surface tension on micropore and total surface area of A2 xerogels with intermediate water wash.

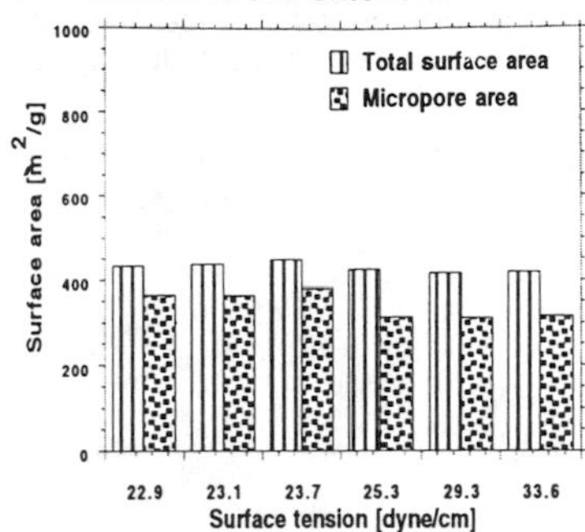

Figure 6 Effect of surface tension on micropore and total surface area of A2 xerogels without intermediate water wash.

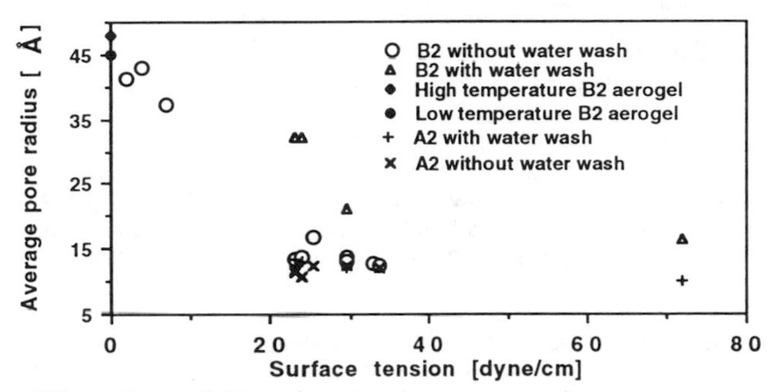

Figure 7 Effect of pore fluid surface tension on xerogel average pore radius.

collapse of the gel. This collapse is resisted by the stiffening of the gel that results from condensation reactions that accompany drying as reactive -OH and -OR groups are forced in close proximity. Even in the absence of further condensation reactions, shrinkage would be resisted by an increase in viscosity of the gel network as it is concentrated by the evaporation of solvent.

The surface area, pore volume, and pore size of the B2 xerogels suggest that the gel network is progressively collapsed as the capillary pressure is increased through an increase in the pore fluid surface tension. The effect of an intermediate water wash is to hydrolyze surface ethoxide groups (that results from ethanol washing). This promotes the condensation reactions that resist collapse leading to greater pore volume and slightly increased surface areas.

Compared to the B2 gels, the ethanol washed A2 gels are less highly crosslinked and presumably more fully ethoxylated due to the acid catalyst. Consequently A2 gels are more compliant than B2 gels and condensation reactions are inhibited during drying. For the range of pore fluid surface tension studied, the capillary pressure is sufficient to collapse the gels to such an extent that the increasing viscosity of the network ultimately limits further shrinkage. There is little dependence of surface area and pore volume on surface tension.

The intermediate water wash hydrolyzes the surface ethoxide groups, promoting further condensation reactions that resist complete collapse of the gel. We observe increased surface areas and pore volumes and a strong dependence of these properties on pore fluid surface tension.

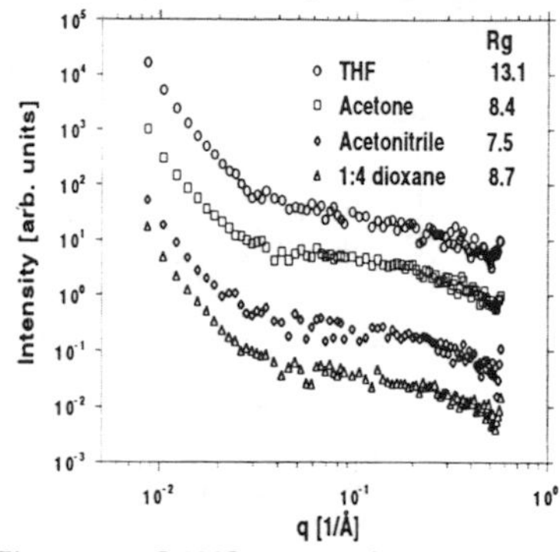

Figure 8 SAXS curves for A2 gels without intermediate water wash.

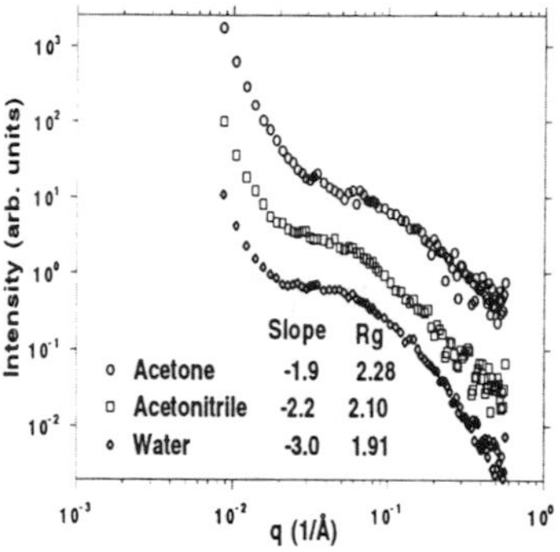

Figure 9 SAXS curves for A2 gels with intermediate water wash.

CONCLUSIONS

The linear dependence of surface area, pore volume, and pore size on surface tension illustrates that for base catalyzed gels the capillary stress dominates the extent of structural collapse during drying. For more compliant ethanol washed acid-catalyzed gels, the extent of structural collapse was greater and the surface area and pore volume depended on the extent of hydrolysis of gel surfaces. Our results show that the xerogel microstructure is a result of the competition of capillary pressure and the rigidity of gel network.

ACKNOWLEDGEMENTS

This work has been supported by UNM/NSF Center for Micro-Engineered Ceramics. The authors thank W. Ackerman and Dr. D.W. Hua of UNM/NSF CMEC for nitrogen condensation and SAXS results and Dr. G.W. Scherer of Du Pont for his many helpful discussions.

REFERENCES

[1] D.M. Smith, C.L. Glaves, P.J. Davis, and C.J. Brinker, In Situ NMR Study of Gel Pore Structure During Drying and Aging in Better Ceramics Through Chemistry III edited by C.J. Brinker, D.E. Clark, and D.R. Ulrich (Mater. Res. Soc. Proc. 121, Pittsburgh, PA 1988) pp.657-662.

[2] P.J. Davis, C.J. Brinker, D.M. Smith, J. Non-Crystalline Solids, in press.

[3] P.J. Davis, C.J. Brinker, D.M. Smith, R.A. Assink, J. Non-Crystalline Solids, in press.

[4] P.J. Davis, R. Deshpande, D.M. Smith, C.J. Brinker, R.A. Assink, J. Non-Crystalline Solids, in press.

[5] R. Deshpande, D.W. Hua, D.M. Smith, and C.J. Brinker, J. Non-Crystalline Solids, in press.

[6] C.J. Brinker and G.W. Scherer, Sol-Gel Science: The Physics and Chemistry of Sol-Gel Processing, (Academic Press, San Diego, 1990).

[7] C.J. Brinker, K.D. Keefer, D.W. Schaefer, and C.S. Ashley, J. Non-Crystalline Solids, 48 , 47 (1982).

[8] C.J. Brinker and G.W. Scherer. J. Non-Crystalline Solids 70 301. (1985).

INTERPRETATION OF AEROGEL SHRINKAGE DURING DRYING.

BHARATH RANGARAJAN and CARL T. LIRA
Department of Chemical Engineering, Michigan State University, East Lansing, MI 48824, USA.

ABSTRACT

A variety of silica aerogels have been prepared by the hydrolysis of TEOS, and dried using supercritical CO_2. The shrinkage which occurs during the drying process is dependent on the gel formulation and the extent of aging of the gels in their pore liquor. Such aging normally results in an increased density, modulus and pore size of wet gels. Upon drying the corresponding aerogels show the opposite behavior for modulus and density, which decrease with the extent of aging. Both drying and aging shrinkage were not observed for base-catalyzed gels, and were very small for HF-catalyzed gels. The use of formamide resulted in reduced drying shrinkage and a slightly larger amount of syneresis. Drying shrinkage is associated with the presence of micropores. Shrinkage during drying has been observed using a high pressure view cell and it was found that most of the shrinkage occurred during depressurization. An explanation consistent with the above is proposed.

INTRODUCTION

Aerogels are a fascinating class of extremely low density porous solids and have been the subject of several international conferences [1]. They are of great interest due to their unusual material properties. Several properties of aerogels such as elastic modulus, modulus of rupture and thermal conductivity scale with density [2,3]. Characterizing and understanding the shrinkage during drying will enable better control over the density of the final material, and will allow a tailoring of the aerogel properties.

In theory the supercritical drying process should eliminate the vapor-liquid interface and result in an undistorted structure. However shrinkage on the order of 10-15% is not uncommon in biological, organic and inorganic samples [4-8]. Scherer, et al. [3,8-10], have recently studied the high-temperature (or supercritical methanol/ethanol) drying process to understand the reasons for shrinkage. In this study we look at the shrinkage which occurs in the low-temperature (or supercritical CO_2) drying process for silica gels. Several different systems have been examined in this study. Some of the factors examined are the effects of formamide, the type of catalyst, and the silica content and aging in pore liquor.

For the purposes of this discussion the term *alcogel* will be used to refer to wet gels, and *aerogel* to the gels dried by the supercritical process. Gels dried in air are called *xerogels*. The *age* (or aging) of a gel is defined as the time between gelation (the macroscopic gel point) and the initiation of the drying process, i.e. removal of the gel from pore liquor. The age of an aerogel is the age of the alcogel from which it is prepared. Shrinkage during aging in pore liquor is called *syneresis or aging shrinkage*. Shrinkage during drying is called *drying shrinkage*.

EXPERIMENTAL

Table I gives the formulations of the different gels used in this study. The molar ratio of water to TEOS (R) is maintained at 10, except for type 8. Gels without formamide were made by replacing the formamide, in type 1-7 gels, with an equal volume of ethanol. The various components were mixed and stirred together for 10 minutes in a Pyrex vessel with a Teflon coated spinbar. Temperature was not controlled during mixing, and a slight rise in temperature was noticed. The sol was poured into acrylic cuvettes of 1.05 cm width and syringes of 1.15 or 1.39 cm diameter, where it was allowed to gel and age at room temperature. After aging the alcogel was removed from its residual liquor, and its dimensions and weight were measured. Details are presented elsewhere [7].

Before drying the gels were soaked in absolute ethanol for at least a day. The alcogels were transferred into a high pressure vessel, and flushed with liquid CO_2 for 4 hours (10 +/- 0.4 MPa, 16-20°C). At this time most of the ethanol is removed. The vessel is heated to 40°C (+/- 1°C) and flushed with supercritical CO_2 under the same conditions for 2 hours. The system was then depressurized slowly over a period of 2 hours. The dimensions of the specimen were observed during the supercritical drying process using a high pressure view cell. The effluent gases were sampled and analyzed by gas chromatography. More descriptions about the system, drying process and measurements are given in an earlier paper [6].

Moduli of several alcogels and aerogels were measured using a procedure described by Scherer [11]. A three point bend test was performed on a United SFM Test System with a 500 gram load cell, on samples with a length to diameter ratio of 13.5 and at downdrive rates ranging from 0.06 mm/min to 30 mm/min. A line was fit to the data with a regression coefficient of 0.999 and moduli were determined.

Table I. Summary of gel formulations used in this study.

Gel Type / Compound	1	2	3	4	5	6	7	8	9
TEOS (ml)	15	25.9	22.9	19.8	16.8	13.8	8.4	21.1	15
Ethanol (ml)	12.5	12.5	12.5	12.5	12.5	12.5	12.5	22.0	12.5
Water (ml)	12.1	20.9	18.5	15.8	13.6	11.1	6.8	6.8	12.1
Formamide(ml)	12.5	12.5	12.5	12.5	12.5	12.5	12.5	0.0	0.0
Catalyst (ml)	2	3.5	3.1	2.6	2.2	1.8	1.1	0.3	0.044
Type	A	A	A	A	A	A	A	HF	B
% TEOS (v/v)	27.7	34.4	33.0	31.2	29.2	26.7	20.3	42.0	39.6

Catalyst A was 70 (w/w) nitric acid. Catalyst B was a 29% (w/w) solution of NH_4OH.

RESULTS

Most of the results presented here pertain to gels of type 1, and unless otherwise stated, this is the system we will be discussing. Figure 1 shows the effect of aging on the dimensions of alcogels and of aging and drying on the dimensions of aerogels. The dimensions reported are the width or diameter of the gel and are expressed as a percentage of the corresponding dimension of the container that they were gelled in. The mass of the aerogel as a fraction of the mass of the sol used to make the alcogel remains constant with aging time at about 10.7% with a standard deviation of 0.9%, indicating that the changes which occur are structural and textural, and that the density changes are due to shrinkage. With aging in pore liquor the skeletal density of the alcogels increased from 0.11g/cc to 0.17 g/cc. The skeletal density of the alcogel is the mass of the solid material, which is the aerogel mass divided by the alcogel volume. Aerogels however showed a dramatic decrease in density with age from 0.32 to 0.18 g/cc [Figure 5, ref 7]. Figure 2 shows the dependence of density on age for five other gel types. In addition to the dependence of densities on silica contents, aerogel densities decrease with age. Figure 3 shows shrinkage for differing silica contents for three different gels, of similar age, both in the presence

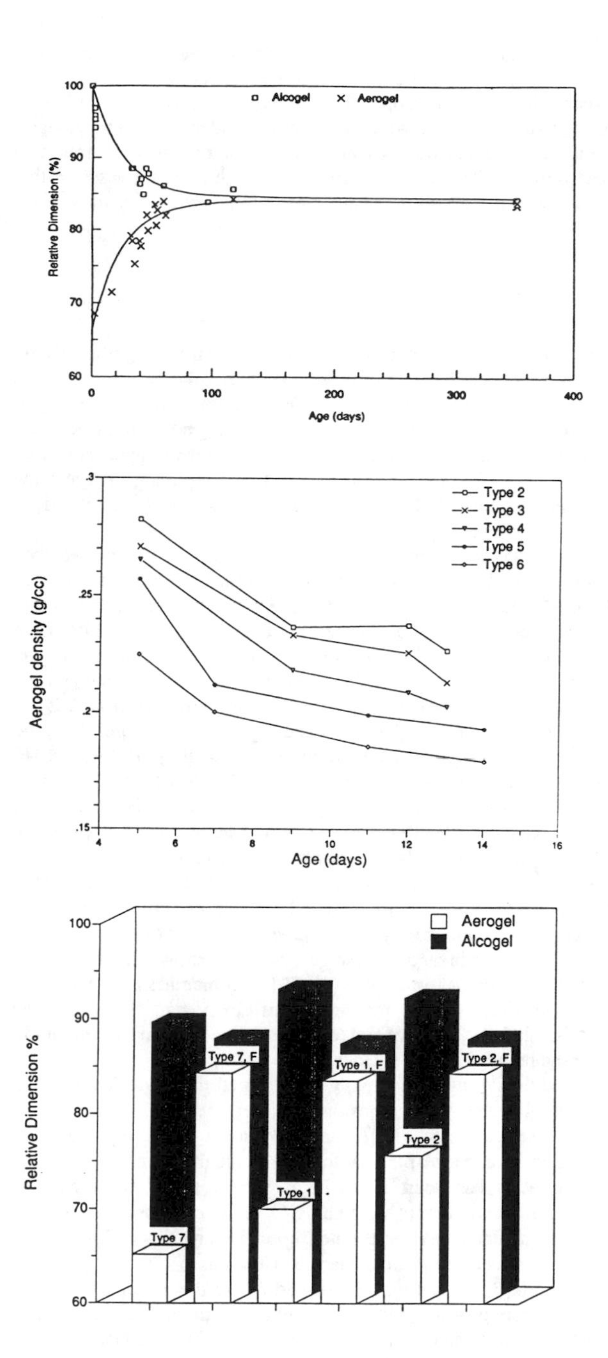

Figure 1. The effects of aging and drying on the dimensions of type 1 alcogels and aerogels. The dimensions are normalized on the basis of the container in which the gel was cast. The difference between 100% and the top line is the syneresis or aging shrinkage of alcogels. The difference between the alcogel curve and aerogel curve is the drying shrinkage. The total shrinkage is the sum of these two.

Figure 2. The effects of aging and drying on the aerogel density for gels of type 2-6. See ref. [7] for more details on type 1 systems.

Figure 3. The effect of formamide and silica content on the aging and drying shrinkage of gels of type 7, 1, and 2.

and absence of formamide. SEM micrographs of the fracture surface of three different aerogels, of type 1, have been presented earlier [7]. It was noted that the microstructure of the gel kept coarsening over a period of a year although macroscopic shrinkage ceased in two months. Very young gels have a large amount of fine structure, which is absent in older gels. The alcogels become noticeably stronger as they age. Some measured values of elastic modulus are presented in Figure 4. Sample dimensions were monitored during drying in a high pressure vessel with a view port. During the drying process it was noticed that most of the shrinkage took place during the depressurization step [6].

DISCUSSION

It can be very clearly seen from Figure 1 that young type 1 gels shrink significantly on drying, and that old alcogels show very little drying shrinkage. With age the alcogels undergo syneresis and become more dense. Since modulus scales with density, the moduli of these gels increase as they shrink [14-16]. Reasonable estimates of their moduli are made using these data and compared with experimental measurements in Figure 4. Details about the applicability of this correlation are discussed elsewhere [7]. It is worth noting that for young aerogels the modulus is about 300 times greater than that of the corresponding alcogel, and a modulus 20 times larger than that of a significantly aged aerogel [7].

If all the alcogels, of a given composition, were subjected to the same stresses they would achieve the same final normalized dimension, density and modulus. In fact if allowed to age in pore liquor the alcogels do just that, and show the same levels of syneresis. The fact that young aerogels are more dense implies that they have experienced larger stresses which have compressed them to this extent. SEM micrographs [7] show that the fine structure of the gel disappears with age, due to coarsening and ripening [17]. Dumas et al [18] have studied the effect of aging on pore size and volume, and show an increase in mean pore size from 2.2-2.5 nm to 6-7.8 nm. They have also studied base-catalyzed gels and show that these gels have no microporosity. As a part of this study we have dried base-catalyzed gels aged 5-30 days, and HF catalyzed gels aged 3-20 days. The base-catalyzed gels are very weak (G~0.1 MPa), with an elastic modulus 10-20 times smaller than acid-catalyzed alcogels [14], but show no measurable aging or drying shrinkage. The HF catalyst acts similar to a base and helps form a gel without many small pores. These gels show less than 4% linear shrinkage on drying. This leads us to the conclusion that the drying stress is related to the pore size, and that the smaller the pores the larger the drying stresses. This conclusion is further strengthened by the data shown in Figure 3, where the effect of formamide on syneresis and drying shrinkage is shown. It may be seen that gels with formamide show a lower drying shrinkage. These gels however show a larger amount of syneresis. This is consistent with the role of formamide [19-20]. Formamide accelerates the condensation reactions and forms gels with larger and more uniform pores. The enhanced condensation rate accounts for the slightly larger extent of syneresis. The larger and more uniform pores are responsible for the reduced drying shrinkage.

In order to better understand the drying shrinkage, we monitored the gel as it was being dried, in a high pressure view cell. Effluent gases from the cell were sampled and analyzed. Details of these experiments are presented elsewhere [6] and only some of the highlights are presented here. Most of the drying shrinkage took place at the end of the drying process, during depressurization. A similar observation has been made by Boyde et al [5] for biological specimens. Following initial displacement of ethanol with liquid CO_2, the ethanol concentration in the effluent fell to less than 10 ppm. Increases were noted upon heating to 40°C and upon depressurization, but at all times the concentration of ethanol in CO_2 was below 150 ppm, far below the solubility limit of 10000 ppm [6]. A run was performed where the vessel was flushed with liquid CO_2 for about 48 hours before heating to 40°C. Again on heating a small amount of ethanol was evolved, but on depressurization no ethanol was detected. In this case also,

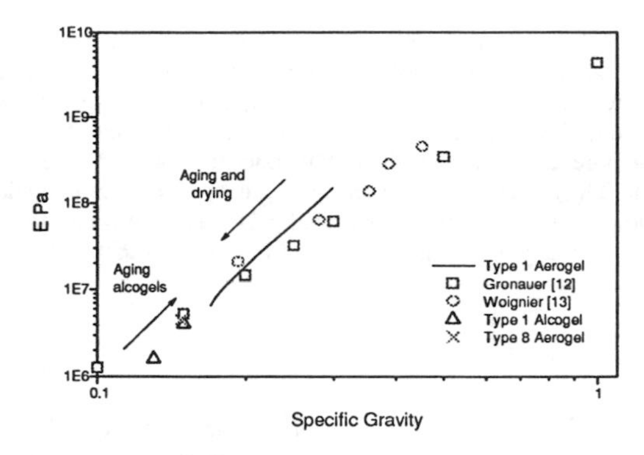

Figure 4. A comparison of the estimated [7] elastic modulus of type 1 aerogels with measured values for type 1 alcogels, type 8 aerogels and several values reported in the literature [12, 13] as a function of density. The effects of aging on alcogels, and of aging and supercritical drying on aerogels are indicated by the arrows.

significant shrinkage was measured, and the aerogel showed a total shrinkage of 26% instead of 30% for another gel of similar age dried with the usual procedure.

In the following section we will examine different probable causes of drying shrinkage. It should be kept in mind that the drying shrinkage is related to the presence of small pores, and occurs on depressurization, and that any explanation for drying shrinkage should explain these facts. The stresses experienced by young gels are at least an order of magnitude greater than those experienced by older gels. Some of the questions addressed are whether the drying shrinkage is due to accelerated syneresis, relaxation, incomplete diffusion, or a manifestation of adsorption.

Accelerated syneresis : The hypothesis here is that during the drying process we are accelerating syneresis due to an increased temperature. This is of extreme significance in the high-temperature process where the temperature is raised above 250°C. Heating the alcogels in ethanol at 40°C for 24-48 hours did not result in any shrinkage. If this were the mechanism, all of the drying shrinkage would have occurred soon after raising the temperature to 40°C, instead of during depressurization. Also all aerogels of a given composition would have a similar final density, due to similar shrinkage.

Relaxation : The assumption underlying this argument is that there exists some sort of stress which is causing collapse of the matrix, and that this collapse is retarded by the slow rate at which liquid can escape from the gel. Acid-catalyzed gels have small pores and extremely low permeabilities of about 1 nm^2 [21]. The argument then is that if the ethanol in the gel is replaced by liquid or supercritical CO_2, a fluid with a lower viscosity, the network can relax at a faster rate, hence the shrinkage. If this argument were true the drying shrinkage should begin soon after liquid CO_2 replaced the ethanol instead of during depressurization. If CO_2 was adsorbed in the micropores its mobility could be reduced and the situation would be different. In that case the rate at which CO_2 can escape is reduced and other forces due to adsorption become important. Relaxation alone does not explain the timing of the shrinkage. When coupled with adsorption, relaxation forces could become very significant. This point will be brought up in the section dealing with adsorption.

Diffusion : Incomplete diffusion will result in the pores of the gel being filled with a liquid (EtOH) instead of supercritical CO_2 when the temperature is raised to 40°C. This would result in capillary stresses and gel collapse. Unsteady state diffusion in a single pore has been modeled by treating the pore as a cylinder of length 2L. The pore is modeled as initially filled with pure fluid A (EtOH), and at the pore mouth there is a pure fluid B (CO_2). The diameter of the pore does not explicitly enter the classical diffusion equation, but the size of the pore does affect the diffusion coefficient D. For CO_2 diffusing into EtOH D is ~3×10^{-5} cm^2/s, and the self diffusion coefficient for CO_2 is 2×10^{-4} cm^2/s [22], both values at 18-20°C. Therefore as liquid

CO_2 diffuses in and replaces the ethanol the diffusion coefficient will increase and the process is auto-accelerating. The diffusion equation was solved assuming equimolar counter diffusion and a constant diffusion coefficient. Figure 5 shows the solution to this equation over a range of parameters. For a given diffusion coefficient and pore length the time required, for the concentration of A at the center of the pore to be 0.1 % of the initial concentration of A, is given on the vertical axis. For a gel of width 0.8 cm (L=0.4 cm) a processing time of 4 hours in liquid CO_2 seems reasonable for this diffusion process. We have also dried gels which are 3 mm thick with similar processing conditions. The drying shrinkage for this gel is similar to that of larger specimens.

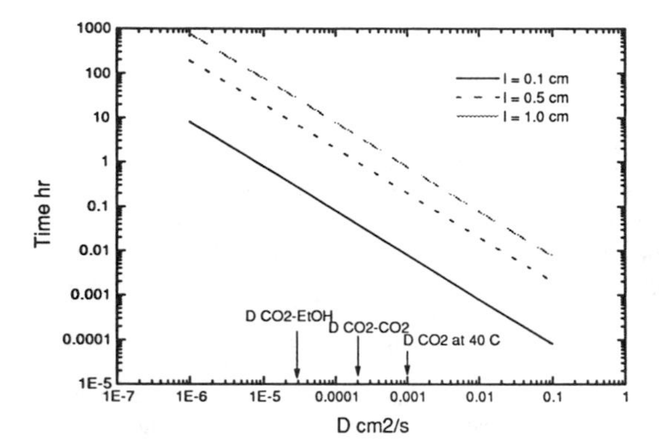

Figure 5. The time required for the centerline concentration of a fluid A, in a pore of length 2L, to reach 0.1% of its initial concentration, when a pure fluid B is at the pore mouth. (C(0,t) = 0.001)

$$\frac{\partial C}{\partial t} = D\frac{\partial^2 C}{\partial x^2}$$

$$x = 0, \frac{\partial C}{\partial x} = 0; x = L, C = 0$$

$$t = 0, C = 1$$

In micropores the diffusion coefficients are lowered significantly [23], and could easily be one hundredth of the bulk value, when the pore diameter is on the order of two molecular diameters. However electron micrographs show large pores and voids separating particles and the micropores cannot be very long. Assuming that a micropore is one micron in length the time required for hindered diffusion ($D \sim 10^{-10}$ cm^2/s) is about 20 seconds. Adsorption of molecules on the surface of the gel could significantly increase the time required for diffusion in micropores. Fluid molecules migrating through a pore are continually stopped at the pore walls, and after some time they resume their migration and are replaced by other fluid molecules. For a fluid A the time required for diffusion (without adsorption) of A through a capillary of 1 nm diameter and 10 microns length is insignificant when compared to the 5000 seconds required for diffusion with adsorption if fluid A physically adsorbs on a solid with a heat of adsorption of 10 - 12 kcal/mol [24]. This analysis leads us to believe that the time allowed for diffusion alone is reasonable. With adsorption a time of 4 hours is reasonable but may not be long enough to eliminate all the ethanol. This is probably why there is a small amount of ethanol evolved on heating. Extended flushing with liquid CO_2 for 48 hours does not eliminate shrinkage. If incomplete diffusion alone was the dominant mechanism, greater shrinkage during heating would be expected when an interface is likely to form.

Adsorption : The final hypothesis presented here is that there could be a stress related to adsorption. Adsorbed liquids can result in relatively large forces especially in solids with micropores. Israelachvilli [25] has measured large forces on the order of 100 bar between mica plates, which oscillate with plate separation and are primarily repulsive. Grand Canonical Ensemble Monte Carlo and Canonical Ensemble Monte Carlo simulations on the effects of adsorption on the forces between two walls, separated by molecular distances, also show oscillatory forces [26,27]. The magnitude of these forces increases as the separation between the walls is reduced [26-28]. These forces would cause the gel to remain in a swollen state as long as there existed a liquid-like layer in the micropores. Since liquid CO_2, supercritical CO_2 and ethanol are adsorbed [6,29] on the gel, the displacement of one fluid by another would not result in collapse. However desorption of CO_2 is likely to reduce the adsorption stresses and result in compression. Also the adsorption force may become attractive as the pressure is reduced. Such an explanation would explain why the shrinkage occurs on depressurization. The phenomenon of swelling on adsorption and shrinkage on desorption has been seen by many researchers [31-32 and references therein]. Since younger gels have smaller pores, the adsorption stresses in these gels would be higher. On aging the processes of dissolution and re-precipitation result in a removal of the microporosity of the gel. Since the adsorption forces are strongly dependent on pore size, they are not as large in aged gels. A greater understanding of the fundamentals of high pressure adsorption is needed to be able to make quantitative predictions of adsorption and related stresses.

Our analysis indicates that adsorption in micropores seems to be the primary cause of drying shrinkage in silica gels, whether through modified diffusion or through adsorption forces. Adsorption can cause attractive forces as well as prevent relaxation by disjoining forces, depending on the pressure and the pore size. The appeal of this mechanism is enhanced by the fact that it could explain the similar shrinkage on depressurization of biological specimens during critical point drying. Most of the biological specimens dried are on the order of 1~2 mm or less, and diffusion becomes less of a problem. CO_2 does adsorb on biological materials [32] and the mechanism for drying shrinkage could be the one described.

SUMMARY AND CONCLUSIONS

Aging alcogels in pore liquor results in reduced drying shrinkage. Shrinkage during drying has been quantified and correlated with the structural (modulus) and textural (pore size) changes taking place during aging. Drying stresses are shown to be non-constant and depend on pore size. It is shown that by controlling the pore size, by aging or the use of HF or a base catalyst, shrinkage can be reduced or even eliminated. Several different explanations for shrinkage are proposed and examined. Stresses due to adsorption seem to be the most likely cause of shrinkage.

REFERENCES

1. *First, Second and Third International Symposium on Aerogels,* ed J. Fricke, Wurzburg, Germany (1986), eds Vacher R., Phalippou J., Pelous J., Woignier T., Montpellier, France, (1988), ed J. Fricke, Wurzburg, Germany (1990).
2. Fricke J., in *Aerogels,* ed. Fricke J., Springer Verlag, Berlin, 1, (1986).
3. Hdach H., Woignier T., Phalippou J., Scherer G.W., *J. Non Cryst. Solids,* 121, 202, (1990).
4. Mulder C.A.M., van Lierop J.G., in *Aerogels,* ed. Fricke J., Springer Verlag, Berlin, 68, (1986).
5. Boyde A., Bailey E., Jones S.J., Tamarin A., *IITRI/SEM,* Vol. 1., 507, (1977).
6. Rangarajan B., Lira C.T., *J. Supercritical Fluids,* 4-1, 1,(1991).
7. Rangarajan B., Lira C.T., *J. Non-Cryst. Solids,* 136, 111, (1991).
8. Woignier T., Phalippou J., Hdach H., Scherer GW, in *Better Ceramics Through Chemistry IV,* eds. Zelinsky B.J.J., Brinker C.J., Clark D.E., Ulrich D.R., 1087, (1990).
9. Scherer G.W., to be published in *J. Non-Cryst. Solids,* (1992).
10. Pauthe M., Quinson J.F., Hdach H., Woignier T., Phalippou J., Scherer G., *J. Non Cryst. Solids,* 130, 1, (1991).
11. Scherer G.W., to be published in *J. Non-Cryst. Solids,* (1992).

12. Gronauer M., Kadur A., Fricke J., in *Aerogels,* ed. Fricke J., Springer Verlag, Berlin, 167, (1986).
13. Woignier T., Phalippou J., Vacher R., in *Better Ceramics Through Chemistry III,* eds. Brinker C.J., Clark D.E., Ulrich D.R., 697, (1988).
14. Scherer G.W., *J. Non-Cryst. Solids,* 109, 183, (1989).
15. Scherer G.W., Pardenek S.A., Swiatek R.M., *J. Non-Cryst. Solids,* 107 [1], 14, (1988).
16. Scherer G.W., in *Better Ceramics Through Chemistry III,* eds. Brinker C.J., Clark D.E., Ulrich D.R., 179, (1988).
17. Iler R K, *The Chemistry of Silica,* Wiley, New York, (1979).
18. Dumas J., Quinson J.F., Serughetti J., *J. Non Cryst. Solids,* 125, 244, (1990).
19. Hench L.L. in *Science of Ceramic Chemical Processing,* eds. Hench L.L., Ulrich D.R., Wiley, New York, 52, (1986).
20. Brinker C. J., Scherer G. W., in *Sol-Gel Science,* Academic Press, San Francisco, (1990).
21. Scherer G.W., Swiatek R.M., *J. Non-Cryst. Solids,* 113, 119, (1989).
22. *'The Properties of Gases and Liquids',* 4th edition, Reid R.C., Prausnitz J.M., and Poling B.E., McGraw Hill, 607, (1987).
23. Mavrovouniotis G.M., Brenner H., *J. Colloid Interface Sci,* 124, 269, (1988).
24. *'The Dynamical Character of Adsorption',* de Boer J.H., Clarendon Press, Oxford, 37, (1953).
25. Israelchvili J.N., *Chem. Scr.,* 25, 7, (1985).
26. Snook I.K., van Megan W., *J. Chem. Soc., Faraday Trans. 2,* 77, 181, (1981).
27. van Megan W., Snook I.K., *J. Chem. Soc., Faraday Trans. 2,* 75, 1095, (1979).
28. Ash S.G., Everett D.H., Radke C., *J. Chem. Soc., Faraday Trans. 2,* 69, 1256, (1973).
29. Strubinger J.R., Parcher J.F., *Anal. Chem,* 61, 951, (1989).
30. Serda P.J., Feldman R.F., in *The Solid Gas Interface vol 2,* ed Flood E.A., Marcel Dekker Inc, New York, 729, (1967).
31. Flood E.A., Heyding R.P., *Can. J. Chem.,* 32, 660, (1954).
32. Nakamura K., Hoshino T., Ariyama H., in *Proceedings of the 2nd International Symposium on Supercritical Fluids,* ed Mc Hugh M.A., John Hopkins University, Baltimore, Maryland, 185, (1991).

PREPARATION OF LOW-DENSITY AEROGELS AT AMBIENT PRESSURE

Douglas M. Smith, Ravindra Deshpande and C. Jeffrey Brinker;

UNM/NSF Center for Micro-Engineered Ceramics, University of New Mexico, Albuquerque NM 87131.

ABSTRACT

Low density aerogels have numerous unique properties which suggest a number of applications such as ultra high efficiency thermal insulation. However, the commercial viability of these materials has been limited by the high costs associated with drying at high pressures (supercritical), low stability to water vapor, and low mechanical strength. Normally, critical point drying is employed to eliminate the surface tension and hence, the capillary pressure, of the pore fluid to essentially zero. However, we show that by employing a series of aging and surface derivatization steps, the capillary pressure and gel matrix strength may be controlled such that gel shrinkage is minimal during rapid drying at ambient pressure. The properties (density, surface area, pore size, SAXS) of aerogel monoliths prepared from base catalyzed silica gels using this technique, supercritical CO_2 drying, and supercritical ethanol drying are compared. An additional advantage of this approach is that the final gels are hydrophobic.

INTRODUCTION

Aerogels are extremely porous materials representing the lower end of the density spectrum of man-made materials. Densities as low as 0.003 g/cm^3 have been reported for silica aerogels [1]. Aerogels with porosities in the range of 85% to 98% may be transparent or translucent. They exhibit strong Rayleigh scattering in the blue region and very weak in the red region. In the infrared (IR) region of electromagnetic spectrum, the radiation is strongly attenuated by absorption leading to application in radioactive diode systems, which effectively transmit solar radiation but prevent thermal IR leakage [2]. Unusual acoustic properties (sound velocity as low as 100 m/s) suggests use in impedence matching and other acoustic applications. Aerogels with a refractive index of 1.015 to 1.06 cover the region not occupied by any gas or solid. Thereby resulting in applications in high energy physics [2]. Fricke [3] has described several aerogel applications based on their insulating properties such as the reduction of heat losses through windows, energy-effective greenhouses, and solar ponds for long-term energy storage.

Since aerogels are made by sol-gel processing, their microstructure may be tailored to optimize any property for a specific application. Various precursors, including metal alkoxides, colloidal suspensions, and a combination of both under several mechanisms of gelation may be used to synthesize gels [4]. By varying aging conditions such as time, temperature, pH, and pore fluid, the parent wet gel micro-structure may be altered [4,5,6,7,8].

Considerable shrinkage and reduction in surface area and pore volume is usually observed during drying by evaporation. This may be a result of continuing condensation reactions between terminal -OH groups and/or surface tension induced collapse. The capillary pressure generated during drying is related to the pore fluid surface tension, γ_{LV}, and contact angle, θ, between the fluid meniscus and pore wall as follows,

$$P_c = -(2\,\gamma_{LV}\cos\theta)/a \qquad (1)$$

where a is the pore radius. The conventional supercritical pore fluid extraction method of aerogel synthesis involves elimination of capillary pressure to avoid the deformation of the wet gel during drying.

Aerogels were first reported by Kistler almost 60 years ago [8]. However recent advances in sol-gel processing technology along with increasing environmental concern caused by alternate insulation material manufacturing has regenerated interest in energy conservation and thermal insulation applications. The most common method of making aerogels involves directly removing the pore fluid above its critical point (for ethanol T_c= 516 K, P_c=63 bar). An alternate low temperature method involves replacing the pore fluid with liquid CO_2 and then removing CO_2 above its critical point (T_c = 304 K, P_c = 73 bar).

Aerogels suffer from several drawbacks for commercial applications. An important disadvantage is the high processing cost associated with supercritical drying. For the high temperature ethanol extraction process, significant chemical and physical changes in gel structure can occur as a result of the greatly accelerated rates of aging and changes in the equilibrium behavior of various reactions whereas, the low temperature carbon dioxide exchange process is limited to only pore fluids miscible in liquid CO_2. Additionally, condensation of moisture within the hydrophilic porous network leads to loss of insulating properties and capillary stress induced collapse of the aerogel [9].

In this study we explore an alternative process to reduce aerogel processing cost by designing a route at ambient pressure. We use a series of aging and surface derivatization steps to alter the gel matrix strength and capillary stresses developed respectively, thereby minimal shrinkage is observed during rapid drying at ambient pressure.

EXPERIMENTAL

Silica gels were prepared via a two-step acid/base catalyzed hydrolysis procedure (refered to as B2 in [11]) and used in our previous studies [5,6,7,8]. Gels were cast in 7.8 mm diameter and 6 cm long plastic tubes. The details of aging and washing the samples to improve the gel matrix strength and to remove any unreacted monomer from the network and of surface chemical modification are described elsewhere [10]. Samples were washed with excess of aprotic solvents prior to drying. Samples were dried at room temperature for 24 hours, followed by 323 K for 24 hours. These monoliths were denoted as ambient pressure aerogels.

Two B2 conventional aerogel samples were also prepared. The first was made by heating a gel previously washed five times with ethanol above the critical temperature (T~ 525 K) in an autoclave and removing the supercritical ethanol. A second sample was prepared at low temperature by exchanging ethanol with liquid CO_2 and then removing CO_2 at 308 K.

Nitrogen sorption at 77 K was used to study pore structure. Samples were outgassed under vacuum at 373 K for at least two hours and five point BET analysis [$0.05 < P/P_0 < 0.3$, N_2 molecular cross-sectional area=0.162 nm^2] was conducted to obtain surface areas. Pore size distributions were calculated from the desorption branches of the isotherms. Because of large upper pore size, aerogel pore volumes were determined via measuring the bulk (by mercury displacement at ambient pressure) and skeletal density (by helium pycnometry).

Small angle x-ray scattering (SAXS) data were collected using a Rigaku 12 kW rotating anode x-ray generator (Cu-Kα, λ=0.1542 nm) with Kratky U-slit optics and a position sensitive detector. The data were then corrected for slit collimation in order to evaluate the radius of gyration and the fractal dimension by analysis of the Guinier and Porod regions, respectively.

To probe the quantity and nature of the surface groups (SiOH, SiOR) TGA/DTA was conducted in air to 923 K at the rate of 10 K per minute using an Omnitherm TGA/DTA.

Water intrusion experiments were carried out in a Autoscan-33 mercury porosimeter to determine the hydrophobic nature of the surface. The ambient pressure aerogel was sealed in rubber cots with water and pressurized from 0 to 30 kpsia.

Changes in the dimensions of the gel rod and corresponding fluid loss during drying at room temperature (~298 K) was measured using a Fotonic Sensor [12] and Fisher Scientific XD-400D balance. A polished brass disk was used as a reflector with the optical probe. The probe was calibrated at the start of experiment using a micrometer stage. The calibration in air indicated that 1 V change in output of optical probe corresponded to 334 μm displacement.

RESULTS AND DISCUSSION

The unique properties of the aerogels are a direct result of their ultrafine microstructure. Table 1 shows the surface areas measured via nitrogen/BET analysis for various gel types. The decrease in surface area for the xerogel as compared to the low temperature aerogel is a result of condensation reactions of surface SiOH and SiOR groups during drying [7]. The lower surface area for high temperature aerogel is a result of greatly accelerated aging at this higher temperature [5]. We believe the reduction in surface area for the ambient pressure aerogel is a result of blocking small-scale surface roughness with the surface modification agent.

Table 1 Surface area for various B2 gel types.

Gel type	Surface area, m^2/g	Gel type	Surface area m^2/g,
Xerogel	950	Low temperature aerogel	1185
Ambient pressure aerogel	869	High temperature aerogel	954

Figure 1 compares the pore volume of an ambient pressure aerogel with a conventional B2 aerogel and a B2 xerogel. Bulk densities corresponding to each sample are indicated on top of respective calumns. Final pore volumes are a result of competing effects including capillary stress induced collapse, which tends to reduce the pore volume and condensation/polymerization reactions which tend to stiffen the matrix resisting the collapse. The low pore volume of the xerogel is a result of capillary stress induced collapse. For conventional aerogels, supercritical drying leads to large pore volumes, since the samples experience very low capillary stresses. The large pore volume of the ambient pressure aerogel results from reduced capillary stresses due to contact angle modification and resistance to pore collapse offered by the surface organic groups. (Since the pore volume are obtained via density measurements, it should be noted that the introduction of organic groups on the surface leads to slight reduction in the skeletal density)

Additional information on the pore structure can be extracted from the pore size distributions. Figure 2 compares the PSDs of a xerogel, conventional aerogel, and ambient pressure aerogel. Both ambient and conventional aerogels exhibit similar pore size distributions and large areas under the curves (corresponding to high pore volumes) compared to the xerogel pore size distribution.

Small angle x-ray scattering results support the similarity in pore structure for conventional and ambient pressure aerogels. Figure 3 presents the SAXS curves for low temperature and ambient pressure aerogels. For both curves, the linear Porod region extends to similar q value and yield similar radius of gyration (R_g) values indicating similar scattering feature size. The more negative Porod slope observed for the ambient pressure aerogel illustrates smoothing of the surface by the surface modification agent.

Thermal analysis of the aerogels established the surface chemical nature. The high temperature aerogel showed a ~14 percent weight loss with a sharp exotherm at around 513 K, which corresponds to removal of the surface ethoxide groups, however for low temperature aerogels this loss was only 8 percent indicating increased esterification with high temperature processing. A second weight loss occured at around 673 K due to terminal -OH groups. For the high temperature aerogel it was 2 percent, while for the low temperature aerogel it was 5 percent. The ambient pressure aerogel exhibited a ~10 percent weight loss at around 543 K with a sharp exotherm, which corresponds to a loss of surface organic groups. These gels showed no weight loss at near 673 K indicating an absence of terminal -OH groups.

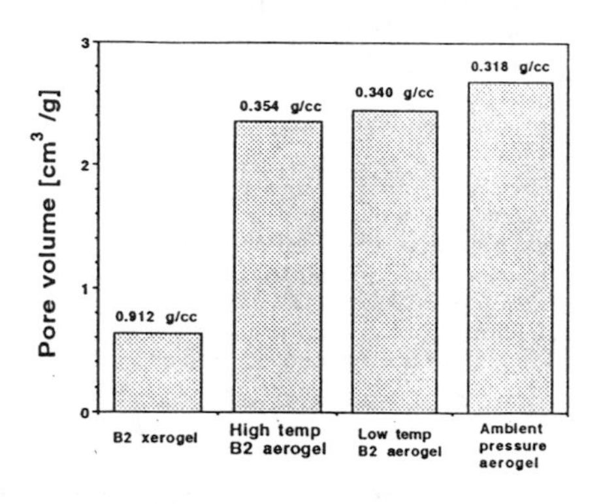

Figure 1 Pore volumes of various gel types

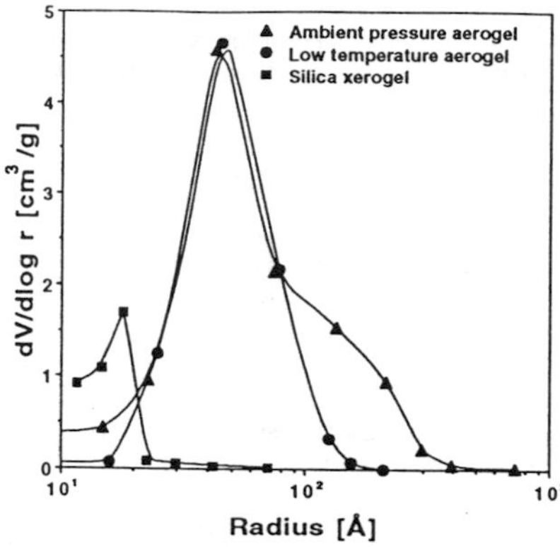

Figure 2 Comparison of Pore size distributions

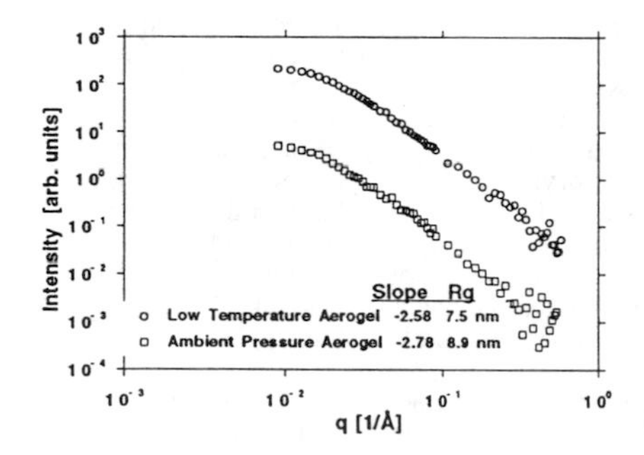

Figure 3 SAXS curves for conventional and ambient pressure aerogels.

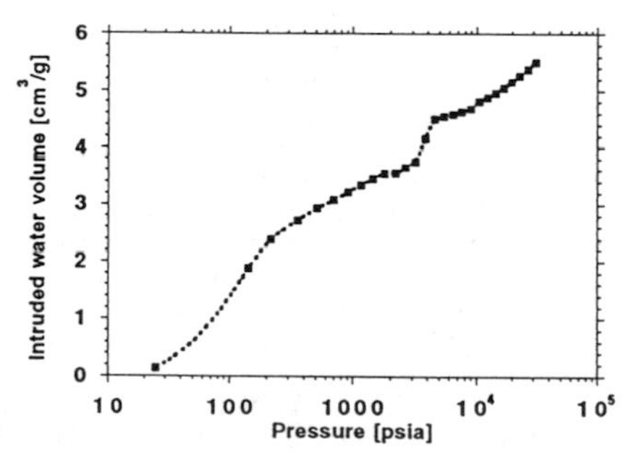

Figure 4 Water intrusion curve for ambient pressure aerogel

To demonstrate the hydrophobic nature of the ambient pressure aerogels, water intrusion experiments were carried out (Figure 4). Apparent intrusion at pressures less than 100 psia is a result of filling around the sample/water container. Pressures over 100 psia were required to force water inside pores. Whereas both low and high temperature aerogels shattered immediately upon exposure to water.

To understand the effect of capillary stresses on shrinkage during drying, length and weight of a ambient pressure aerogel with acetone as final pore fluid was continuously monitored. These results are presented in Figure 5a. Figure 5b shows the length of the sample and percent acetone loss as function of fluid coverage on the gel surface. The sample shows continuous shrinkage illustrating the effect of capillary stresses, however as shown in Figure 5b, when the fluid coverage is about a monolayer, considerable expansion is observed. We believe that when the fluid coverage reaches a monolayer most menesci are eliminated and the structure expands due to stress relaxation of surface organic groups. It is important to note that the maximum shrinkage observed for these aerogels is significantly smaller than that for a B2 xerogel drying under similar conditions of temperature and final pore fluid.

Similar experiments were carried out for samples washed with different aprotic solvents to vary the surface tension during drying. Figure 6 shows percent axial shrinkage during drying as function of percent fluid loss. With increasing surface tension more shrinkage was observed. Cracking of the sample was observed for the highest surface tension fluid used. It is important to note that under isothermal conditions, the drying rate is different for these samples. The drying rate is expected to affect the shrinkage and subsequent recovery. Experiments are in progress to address this issue.

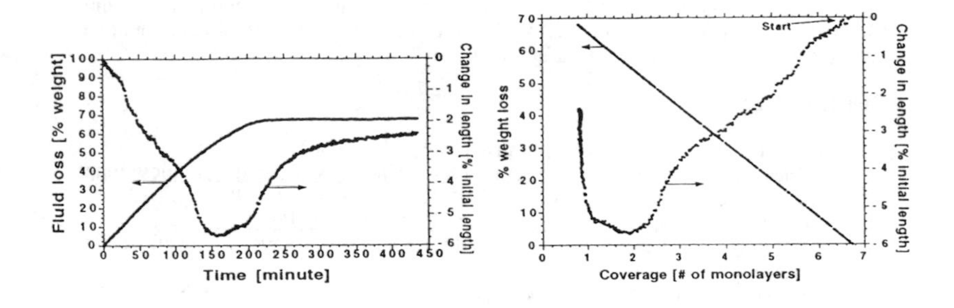

(a) (b)

Figure 5 Shrinkage and fluid loss from ambient pressure aerogel during drying with acetone as final pore fluid.

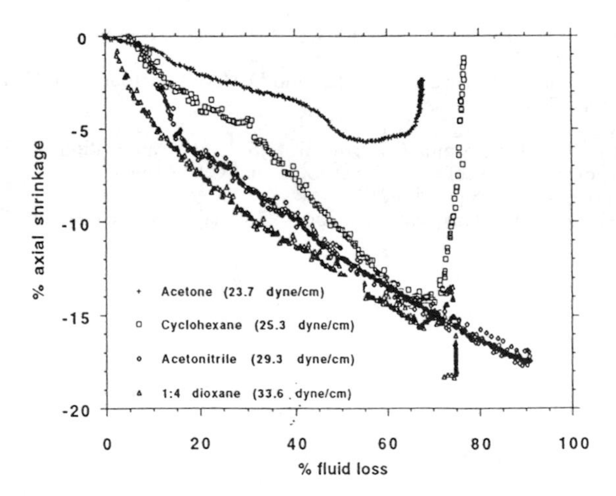

Figure 6 Shrinkage and fluid loss for samples dried from aprotic solvents with different surface tension.

CONCLUSIONS

Low density silica aerogels were prepared by surface chemical modification of B2 silica gels at ambient pressure. Micro-structural analysis of ambient pressure aerogels shows striking similarities with the aerogels prepared by conventional supercritical processing routes. The hydrophobic nature of ambient pressure aerogels gave them structural stability against humidity.

ACKNOWLEDGEMENTS

This work was supported by the UNM/NSF Center for Micro-Engineered Ceramics which is a collaborative effort of the National Science Foundation (CDR-8803512), Los Alamos and Sandia National Laboratories, the New Mexico Research and Development Institute, and the ceramics industry. The authors would like to thank J. Anderson, W. Ackerman, and Dr. D.W. Hua of CMEC for sample preparation, nitrogen sorption, and SAXS analysis respectively and Dr. G.W. Scherer of Du Pont for loan of equipment and software for dimensional change measurements.

REFERENCES

[1] L.W. Hrubesh, Chemistry and Industry, 24, 824 (1990).
[2] J. Fricke and R. Caps in Ultrastructure Processing of Advanced Ceramics edited by J.D. Mackenzie and D.R. Ulrich (Wiley, New York, 1988), p. 613.
[3] J. Fricke in Sol-Gel Technology for Thin Films, Fibers, Preforms, Electronics and Speciality Shapes edited by L.C. Klein (Noyse Publications, Park Ridge NJ, 1988), p. 226.
[4] C.J. Brinker and G.W. Scherer, Sol-Gel Science (Academic Press, San Diego, 1990).
[5] D.M. Smith, C.L. Glaves, P.J. Davis, and C.J. Brinker, In Situ NMR Study of Gel Pore Structure During Drying and Aging in Better Ceramics Through Chemistry III edited by C.J. Brinker, D.E. Clark, and D.R. Ulrich (Mater. Res. Soc. Proc. 121, Pittsburgh, PA 1988) pp.657-662.
[6] P.J. Davis, C.J. Brinker and D.M. Smith, Journal of Non-Crystalline Solids, in press.
[7] P.J. Davis, C.J. Brinker, D.M. Smith and R.A. Assink, Journal of Non-Crystalline Solids, in press.
[8] R. Deshpande, D.W. Hua, C.J. Brinker and D.M. Smith, Journal of Non-Crystalline Solids, in press.
[9] S.S. Kistler, Nature, 127, 741 (1931).
[10] R. Deshpande, D.M. Smith, C.J. Brinker, U.S. Patent Application, 1992.
[11] C.J. Brinker, K.D. Keefer, D.W. Schaefer, and C.S. Ashley, J. Non-Crystalline Solids, 48 ,47 (1982).
[12] G.W. Scherer, H. Hdach, and J. Phalippou, Journal of Non-Crystalline Solids, 130 (1991) 157.

PREPARATION AND PHYSICAL PROPERTIES OF V_2O_5 AEROGELS

Hiroshi Hirashima and Kazumi Sudoh
Keio University, Faculty of Science and Technology
3-14-1, Hiyoshi, Kohoku-ku, Yokohama 223 Japan

ABSTRACT

Monolithic aerogels and aerogel coatings of hydrated V_2O_5 containing 0 to 10 mol% GeO_2 were prepared from metal alkoxides by hydrolysis and autoclave drying. The bulk aerogels of high porosity, more than 90 %, were obtained. On the other hand, the porosity of the aerogel coatings, about 50 nm in thickness, was about 50 %. The aerogels consisted of micro fibrils, less than 20 nm in diameter. Diffraction peaks of crystalline vanadium oxides were not observed for the as-dried aerogels. Dynamic microhardness of the bulk aerogels was very low, but increased with addition of GeO_2. Dc electrical conductivity of the bulk aerogels was lower than that of the bulk xerogels by one order of magnitude, and was isotropic. The aerogels were surface fractal with $D_s \cong 2.7$.

INTRODUCTION

Hydrated V_2O_5 gels are known to be semiconducting and to have high and anisotropic conductivity [1]. Dense thin films of V_2O_5 gels can be applied to antistatic coatings, switching devices, electrochromic films, electrodes for Li-batteries, etc.. For some applications, such as catalysts and sensors, microporous gels with high surface area are appropriate. Many works on silica and silicate aerogels have been reported. However, few papers on aerogels of transition metal oxides have been published. V_2O_5 xerogels are known to have a fibrous structure [1, 2]. On the other hand, silica aerogels usually consist of nano-particles [3]. Preparation and physical properties of fibrous aerogels are also interesting. In this study, V_2O_5 aerogels were prepared from vanadyl ethoxide by hydrolysis and autoclave drying. The authors previously reported the preparation of V_2O_5 xerogels from vanadyl alkoxides [4] and their microstructure [2]. Effects of additive oxides on physical properties of V_2O_5 xerogels have been also reported [2, 5]. Electrical conductivity decreased with addition of TiO_2 or GeO_2, but scratch hardness was improved by these additives. In this study, effect of GeO_2 addition on physical properties of V_2O_5 aerogels was also discussed.

EXPERIMENTAL PROCEDURES

V_2O_5 wet gels containing 0 to 10 mol% GeO_2 were prepared from vanadyl triethoxide and germanium tetraethoxide, 99.9 %, supplied by Soekawa Rikagaku Co., Tokyo, by hydrolysis in ethanol solutions without any catalysts [4, 5]. Vanadium content of the solutions was about 2.5 wt% as V_2O_5. After aging at room temperature for 1 to 3 weeks, the wet gels were dried supercritically using an autoclave at 255°C and 210 atm for 1 h. The wet gels were not washed before drying. Bulk density, shrinkage after drying, and X-ray diffraction of the dried gels were measured. Specific surface area was measured by BET method with N_2. Dc conductivity of the bulk aerogels were measured by a 2-terminal method using vacuum evaporated

Au films as electrodes. Dynamic microhardness of the bulk aerogels was measured using a trigonal diamond pyramid indenter (Shimadzu, DUH-200) [6].

The wet gels were ultrasonically peptized in ethanol and used for preparation of gel coatings. The gel coatings were prepared by dip-coating [2, 5]. Soda-lime-silicate glass plates were used as substrates. Immediately after dipping, the coated glass plates were put into an autoclave and supercritically dried. Thickness and refractive index of the coatings were measured by ellipsometry.

RESULTS AND DISCUSSION

Syneresis was observed during aging of the V_2O_5 wet gels. The volume shrinkage during aging for 1 to 3 weeks was about 25 to 30 %. The volume shrinkage during supercritical drying was less than 10 %. The shrinkage of the gel containing 10 mol% GeO_2 by autoclave drying was scarcely observed. On the other hand, the volume change of the gel without addition of GeO_2 was about 10 %. The shrinkage of the gels during drying under an atmospheric pressure was more than 90 % in volume. The bulk density of the supercritically dried gels was 0.07 to 0.1 g/cm^3, and was about 5 % of the bulk density of the xerogels or smaller. The porosity of the V_2O_5 aerogels, which was estimated from bulk densities of the aerogels and the dense xerogels, was 90 to 96 %. When the wet gel was supercritically dried without aging, the volume shrinkage was larger, and the bulk density of the dried gel was 0.5 to 1 g/cm^3. The porosity was estimated to be 50 to 75 %. This result suggests that the gel structure changes during aging and the stiffness of the gel increases.

The aerogels were black in color and brittle. The fraction of reduced V ion, $[V^{4+}]/[V_{total}]$, determined by wet chemical analysis was about 0.20 to 0.25. These values are similar to those of the xerogels. The molar ratio of $[H_2O]/[V_2O_5]$ in the aerogels, which were kept in a dessicator with silica gel, was about 0.5. The gels have a fibrous structure and consist of micro fibrils, less than 20 nm in diameter [2].

Broad and weak diffraction peaks were observed for the aerogels (Fig. 1c). The diffraction patterns of the aerogels were different from those of any crystalline vanadium oxides, such as V_2O_5, VO_2 and V_2O_3. Diffraction

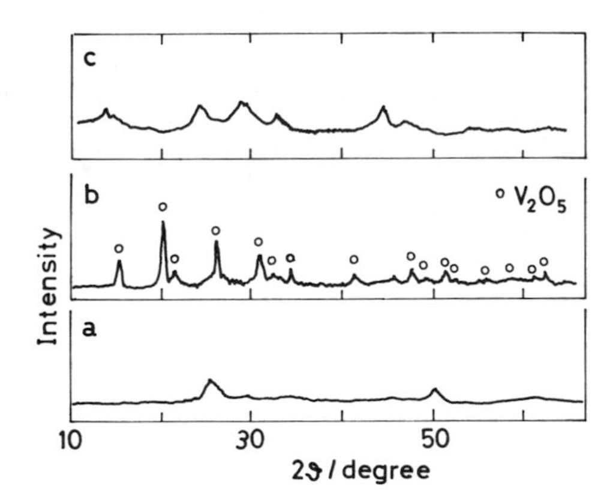

Fig. 1 X-ray diffraction patterns of V_2O_5 gels, (a) V_2O_5 xerogel, dried at 200°C for 1 h, (b) V_2O_5 xerogel, heat treated at 250°C for 1 h, and (c) V_2O_5 aerogel, as dried. (CuKα)

peaks of crystalline V_2O_5 (orthorhombic) were observed for the xerogels after heat treatment at 250°C for 1 h (Fig. 1b). Although heated at 255°C during supercritical drying, the aerogels contain a certain amount of H_2O. These results suggest that the hydrated V_2O_5 gels are stabilized under high pressure during supercritical drying.

The specific surface area of the V_2O_5 aerogels was 140 to 200 m^2/g. These values are much lower than that of silica aerogels which is often reported to be more than 500 m^2/g [7]. This result is attributed to the difference in microstructure of the aerogels; V_2O_5 aerogels consist of micro fibrils and silica aerogels consist of granular particles. However, the further discussion should be needed to clarify the reason for the low specific area of the V_2O_5 aerogels.

Dynamic microhardness of the aerogels was very low in comparison with that of usual glasses (Table 1). However, dynamic microhardness of the aerogels increased with addition of GeO_2. Mechanical strength of the V_2O_5 aerogels without additives was too low to prepare the samples for conductivity measurement. Only the sample containing 10 mol% GeO_2 was able to be prepared. Dc conductivity of the bulk aerogels was lower than that of the bulk xerogels by about one order of magnitude (Fig. 2). Dc conductivity of the bulk aerogel was isotropic as well as that of the bulk xerogels. The

Table I Dynamic microhardness of aerogels

Sample	Dynamic microhardness
V_2O_5 aerogel	3.73 gf/m^2
$90V_2O_5 \cdot 10GeO_2$ aerogel	10.99
Microscope slide glass	160
Silica glass	250

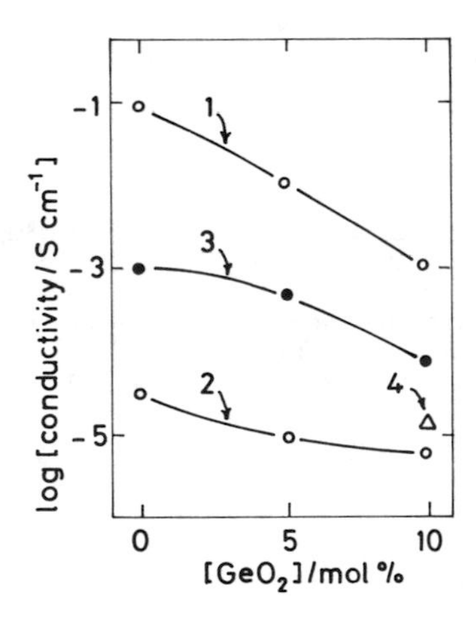

Fig. 2 Dc conductivity of V_2O_5 gels at 25°C, (1) xerogel coating, parallel to the surface, (2) xerogel coating, perpendicular to the surface, (3) bulk xerogel, and (4) bulk aerogel.

low conductivity of the aerogels may be attributed to their high porosity.

Small angle X-ray scattering of the bulk aerogels was measured using Kratky U slit (Rigaku, type 2223). The power low was valid in Porod region, where the scattering vector was about 0.07 to 0.1 A^{-1}. The slope was about -3.3 . The aerogels may be surface fractal with $D_s \cong 2.7$. This value is near to that of acid-catalized silica aerogel [8].

Transparent thin coatings, about 50 nm in thickness, were obtained by supercritical drying of the wet V_2O_5 gel coatings. Refractive index values of the coatings were about 1.4, and they were much lower than those of the V_2O_5 xerogel coatings, 1.8 to 2.0 (Table II). This result shows the low density of the supercritically dried coatings. The porosity of the aerogel coatings was estimated from the refractive indeces according to the following equation [9],

$$q = \frac{N_e^2 - 1}{N_e^2 + 2} \cdot \frac{N_f^2 + 2}{N_f^2 - 1} \qquad (1)$$

where q = volume fraction of substance with refractive index N_f and N_e = apparent refractive index of the film. The estimated porosity of the aerogel coating was about 50 %, and was much lower than that of the bulk aerogels. The low porosity of the aerogel coatings may be attributed to the following reasons; (1) the wet gel coatings were dried without aging, and (2) densification of the gel coating took place during dip-coating because of rapid evaporation of solvent from thin coatings.

Table II Refractive indices, n, and porosity of V_2O_5 coatings

Sample	n	Porosity*
Xerogel coating	1.84	11 %
Aerogel coating	1.41	50

* Porosity of the coatings were estimated using Eq. (1) with assumption, n of the dense xerogel = 2.0

CONCLUSIONS

(1) V_2O_5 aerogels were obtained by supercritical drying of wet gels prepared from vanadyl ethoxide. The porosity of the bulk aerogels was more than 90 %.

(2) Dc conductivity of the bulk aerogels was lower than that of the bulk xerogels by one order of magnitude.

(3) Microhardness of the aerogels was very low and increased with addition of GeO_2.

(4) V_2O_5 aerogel coatings, about 50 nm in thickness, were prepared by supercritical drying of the wet gel coatings. The porosity was about 50 %.

Acknowledgement

The authors gratefully acknowledge Prof. M. Yamane of Tokyo Institute of Technology and his colleagues, Dr. A. Yasumori and Mr. M. Iwasaki, for their help in preparation of aerogels, and also acknowledge Mr. S. Sasaki of Keio University for his help in XRD and SAXS measurements.

REFERENCES

1. J. Livage, in Better Ceramics Through Chemistry, edited by C.J. Brinker, D.E. Clark, and D.R. Ulrich (Mater. Res. Soc. Proc. 32, Pittsburgh, PA 1984)pp.125-134.
2. H. Hirashima, S. Kamimura, R. Muratake, and T. Yoshida, J. Non-Cryst. Solids, 100, 394 (1988).
3. C.M. Lampert and J.H. Mazur, in Aerogels, edited by J. Fricke (Springer-Verlag, Berlin, 1986)pp.154-159.
4. H. Hirashima, K. Tsukimi, and R. Muratake, Seramikkusu-Ronbun-Shi, 97, 235 (1989).
5. H. Hirashima and K. Sudoh, J. Non-Cryst. Solids, 121, 68 (1990).
6. H. Hirashima and T. Kusaka, (in this volume)
7. J. Zarzicki and T. Woignier, in Aerogels, edited by J. Fricke (Springer-Verlag, Berlin, 1986)pp.42-48.
8. D.W. Schaefer, C.J. Brinker, J.P. Wilcoxon, D.-Q. Wu, J.C. Phillips, and B. Chu, in Better Ceramics Through Chemistry III, edited by C.J. Brinker, D.E. Clark, and D.R. Ulrich (Mater. Res. Soc. Proc. 121, Pittsburgh, PA, 1988)pp.691-696.
9. R.M.A. Azzam and N.M. Bashara, Ellipsometry and Polarized Light, paper back edition (North-Holland, Amsterdam, 1987)p.359.

PART V

Ceramic Composites

LEUCITE-POLLUCITE GLASS CERAMICS: A NEW FAMILY OF REFRACTORY MATERIALS WITH ADJUSTABLE THERMAL-EXPANSION

R.L. Bedard, R.W. Broach, and E.M. Flanigen, UOP Research and Development, Tarrytown, NY

ABSTRACT

Novel ceramic compositions can be produced from carefully chosen, modified, and processed zeolites. The use of zeolites as precursors for leucite-pollucite glass ceramics having adjustable thermal-expansion behavior is described. The facile sintering properties of the zeolite-derived precursors allow production of these high-temperature materials at several hundred degrees below the solidus temperatures. A representative leucite-pollucite ceramic was studied by high temperature x-ray powder diffraction and synchrotron powder diffraction to determine the nature of the crystal structure and its relationship to the leucite and pollucite end members of the family.

INTRODUCTION

Leucite and pollucite, which are refractory phases with identical aluminosilicate frameworks, have potassium and cesium cations contained in intracrystalline tunnels, respectively. The use of leucite ($K_2O:Al_2O_3:4\ SiO_2$) is primarily in applications where a high thermal-expansion coefficient (about 28 ppm°C^{-1}) is needed, such as in dental ceramic applications [1], porcelain coatings for metals, and metal-ceramic seals. The extremely low thermal-expansion coefficient (as low as 1.5 ppm°C^{-1}; essentially 0 above 200°C) of pollucite ($Cs_2O:Al_2O_3:4\ SiO_2$) may be useful in low thermal-shock structural applications and optical materials. The high melting points of leucite (1693°C), Rb-leucite (1675°C), and pollucite (>1900°C) have probably hindered the ceramic application of the stoichiometric compositions.

The crystallography of leucite and pollucite phases has been previously reported [2,3]. The K-leucite end member has a tetragonal unit cell with a=13.08 Å and c=13.78 Å (at 25°). This cell transforms at about 660°C to a cubic cell with a=13.51 Å. The Rb end member is also tetragonal at room temperature and has lattice parameters of a=13.29 Å and c=13.75 Å. The structural transformation of the Rb form to a cubic cell, with a=13.55 Å, occurs at 310°C [2]. The Cs form, known as pollucite, is cubic at all temperatures with a cell constant a=13.68 Å at room temperature [2].

Ceramic bodies consisting of leucite-pollucite solid solutions have not been reported, although the mutual solubility of Rb and Cs in pollucite and K and Rb in analcite zeolites, prepared as powders by hydrothermal methods, has been described by Barrer and coworkers [4,5]. The use of zeolite precursor powders allows the formation of ceramics containing any solid solution in a vast range of compositions containing K, Rb, and Cs cations. All of these compositions are accessible by ion exchanging the appropriate zeolite to give the desired cation ratio.

EXPERIMENTAL

The NaY zeolites, with SiO_2/Al_2O_3 ratios of 3.5 and 5, were commercially available powders from UOP and were ion exchanged with K, Rb, and Cs by conventional ion exchange procedures [6]. Compositions of the zeolites are presented in Table I.

TABLE I. Compositions of Zeolite Powders

#	Zeolite	Anhydrous Oxide Ratios
I	KY(3.5)	0.02 Na_2O: 0.91 K_2O: 1.0 Al_2O_3: 3.4 SiO_2
II	KY(5)	0.96 K_2O: 1.0 Al_2O_3: 5.0 SiO_2
III	K,CsY(5)	0.78 K_2O: 0.22 Cs_2O: 1.0 Al_2O_3: 5.0 SiO_2
IV	Rb,KY(5)	0.02 Na_2O: 0.04 K_2O: 0.87 Rb_2O: 1.0 Al_2O_3: 4.9 SiO_2
V	Cs,KY(5)	0.70 Cs_2O: 0.23 K_2O: 1.0 Al_2O_3: 5.0 SiO_2

Glass ceramic bars were made by precalcining the zeolite powders for 1 hr at 1050°C, then by screening out agglomerates of greater than 250 μ, pressing the resulting amorphous powder in a steel die (5000 psi), and sintering the bars at 1250°C for 4 hr.

The reported coefficients of thermal-expansion (CTE's) have been corrected with a standard alpha-Al_2O_3 reference sample. Lattice parameters of the leucite-pollucite ceramic examined by high temperature x-ray powder diffraction were calculated by least squares analysis of data collected with an added internal Si standard, assuming a 3 ppm°C^{-1} CTE for Si. Synchrotron powder diffraction data was collected on beam line X-7A at the Brookhaven NSLS. The ceramic sample was ground into a powder and sealed in a 1.0 mm glass capillary. A Ge(111) double crystal monochromater, a Ge(220) analyzer crystal, and a Kevex solid state detector were used. Slits were set at 8 mm wide and 1 mm high before the sample and 8 mm wide after the sample. The x-ray synchrotron wavelength of 1.306441(17)Å was calibrated using a Si standard.

RESULTS AND DISCUSSION

The formation of leucite-pollucite ceramics from zeolite precursors is a facile process because of the homogeneous cation distribution in the precursor zeolite powders. Achieving the same degree of homogeneity by other methods is not possible, except perhaps with prolonged glass melting. The refractory nature of high-Cs glasses, coupled with volatilization losses of Cs_2O from the melts [7], make compositional control and formation of these materials by conventional glass ceramic processes difficult.

Figures 1A through D show x-ray powder diffraction patterns of the ceramics produced by sintering zeolites powders II through V. The x-ray pattern of the leucite ceramic made from zeolite sample I, KY(3.5) was nearly identical in profile to the x-ray pattern from sample II, KY(5.0), except with only about half as much crystallinity. The x-ray data indicates that the zeolite-derived glass ceramics each contain one crystalline phase. Unambiguous indexing of the ceramic made from powder III was not possible with x-ray data collected in-house, and a high-resolution synchrotron study was necessary.

The CTE's, from 25-700°C, for the K, Rb, and Cs materials have been reported by Taylor and Henderson [2] as 27.6 ppm°C^{-1}, 17.0 ppm°C^{-1}, and 3.8 ppm°C^{-1}, respectively. The thermal-expansion properties of leucite-pollucite glass ceramics are strongly dependent on the cationic composition of the starting zeolite powder as well as the SiO_2/Al_2O_3 ratio. Figure 2 presents some of the wide range of thermal-expansion behaviors that can be demonstrated by the zeolite-derived leucite-pollucite materials. The leucite produced from KY(5.0) displays a rapid volume increase at about 400°C. This increase is caused by the well-known tetragonal-to-cubic phase transition of leucite.

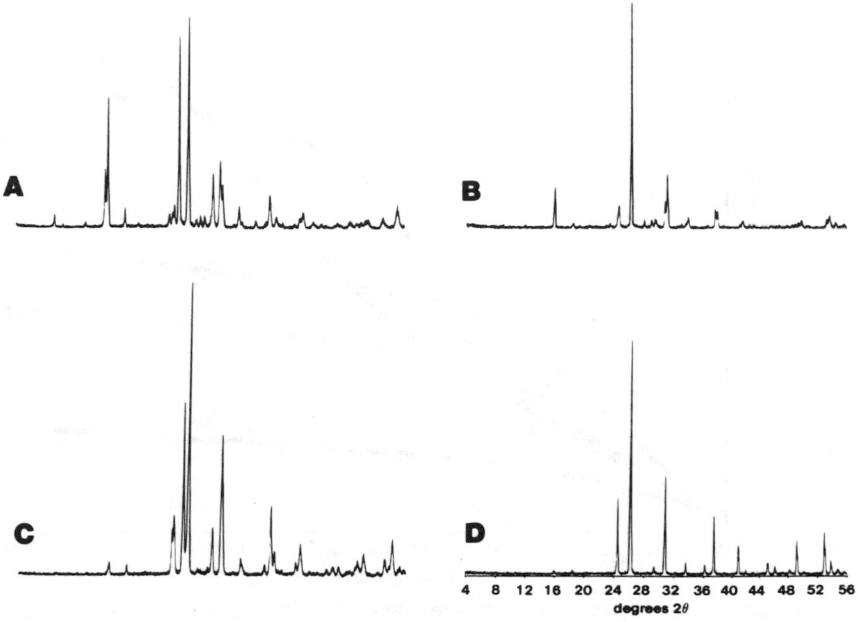

Figure 1. X-ray Powder Diffraction Patterns of Glass Ceramics Derived from Powders II through V

The CTE's in the 50-700°C temperature interval for the five samples of this study were as follows: I. 28 ppm°C^{-1}; II. 27 ppm°C^{-1}; III. 14 ppm°C^{-1}; IV. 16 ppm°C^{-1}; and V. 4.5 ppm°C^{-1}. Discontinuities are observed in the thermal-expansion curves of ceramics made from powders III, IV, and V (Figure 2). These discontinuities in the thermal-expansion curves are usually attributed to the completion of the "unfolding" of the aluminosilicate framework and the resulting onset of low thermal-expansion behavior [2]. The ceramic sample made from powder IV, the $Rb_{0.87}K_{0.04}$ zeolite, displays an unusually large CTE of greater than 30 ppm°C^{-1} in the 25 to 300°C range. The thermal-expansion of the ceramic made from powder I, the KY(3.5), is fairly smooth and continuous until a 670°C discontinuity.

Synchrotron Diffraction of a Leucite-Pollucite Glass Ceramic

The nature of the crystal structures of the ceramics containing intermediate levels of Cs was investigated by a high resolution synchrotron diffraction study of a representative sample made from powder III (cation composition $K_{0.78}Cs_{0.22}$). The indexed diffraction data is given in Table II and representative diffraction peaks are shown in Figure 3. The x-ray pattern was indexed on a hexagonal cell with dimensions a=18.92025(79)Å and c=11.73997(77), with a cell volume of 3651.1(5)Å^3. The material represents a new crystalline modification in the leucite-pollucite series. The previously known phases in this ceramic family have all been either tetragonal or cubic.

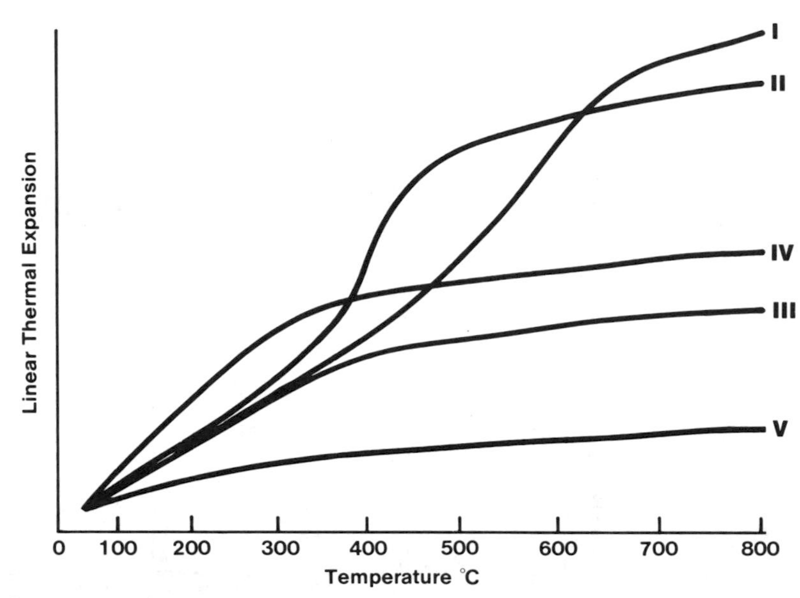

Figure 2. Dilatometer Traces Showing Thermal Expansion Behaviors of Glass Ceramic Bars Derived from Zeolite Powders I through V

TABLE II. Synchrotron Diffraction Data: $K_{0.78}Cs_{0.22}$ Solid Solution

h k l	d (Å)	Int.	h k l	d (Å)	Int.
1 0 2	5.5273	554	4 2 1	2.9988	508
2 1 1	5.4842	2036	1 0 4	2.8890	1680
2 0 2	4.7750	544	5 0 2	2.8648	1440
2 2 0	4.7392	298	5 1 1	2.8589	2614
1 1 3	3.6173	775	4 1 3	2 6419	515
3 1 2	3.5963	901	5 2 0	2.6283	1251
3 2 1	3.5837	1896	4 0 4	2.3872	1713
4 0 2	3.3628	14311	4 4 0	2.3686	1582
2 2 3	3.0171	810			

The relationship of the new leucite-pollucite solid solution structure to the cubic pollucite structure is straightforward. The hexagonal structure represents a trigonal distortion, or slight elongation (about 1%), of the cubic cell along the body diagonal or [111] direction. The hexagonal cell axes a and c correspond to the face diagonal and one-half the body diagonal of the cubic cell.
One possible explanation of this trigonal distortion is an ordering

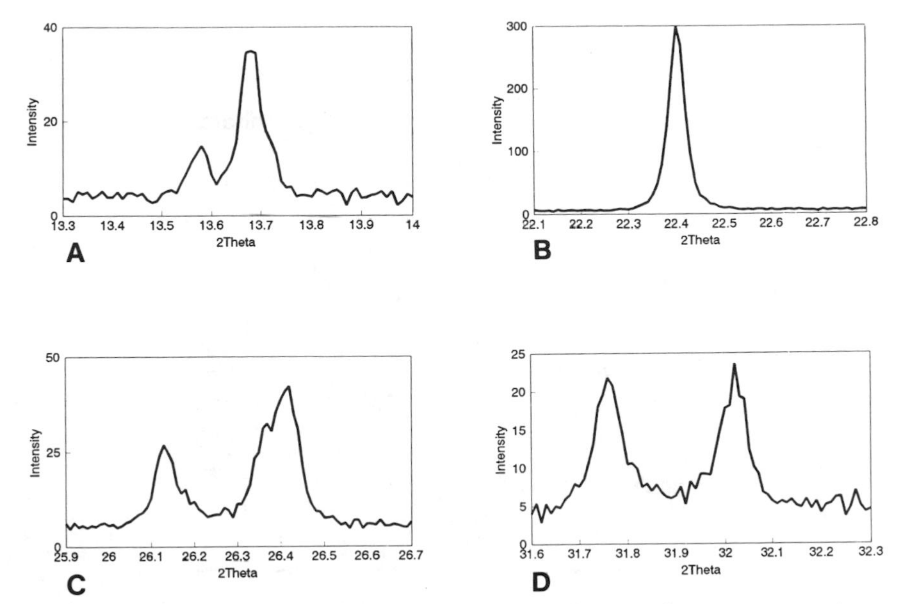

Figure 3. Selected Expanded X-ray Diffraction Peaks from Synchrotron Powder Diffraction Pattern of Hexagonal Solid Solution

of the K and Cs ions to produce a distinct phase rather than a solid solution. Another ceramic having a cation composition of $K_{0.70}Cs_{0.29}$ shows a nearly identical, apparently hexagonal x-ray pattern. These data are consistent with a homogeneous K-Cs solid solution in these materials rather than a distinct hexagonal phase. The range of K-Cs compositions over which the hexagonal structure is present is not yet known. The ceramic made from powder V is clearly cubic, however (Figure 1), indicating that the hexagonal structure is not present at a 0.70 Cs fraction.

<u>High Temperature X-ray Diffraction of Leucite-Pollucite Solid Solution</u>

The same sample that was studied by synchrotron powder diffraction was mounted as a thin layer on the platinum ribbon stage of a high temperature powder diffraction device on a Seimens D-500 diffractometer with a θ-θ source-detector configuration. The temperature was adjusted to the desired level and the diffraction pattern was recorded. Temperatures studied included 25, 350, 450, 550, and 650°C.

The data indicated a gradual shifting of diffraction peaks from the hexagonal positions at room temperature to the characteristic cubic positions of high leucite (Figure 4). The symmetry transformation was complete between the temperatures of 450 and 550°C. No apparent hexagonal splitting occurred in the diffraction peaks at 550 and 650°C. The resolution of this instrument was not good enough to determine the lattice parameters of the hexagonal phase at various temperatures. The diffraction pattern at 650°C was sharp enough to allow indexing of the peaks to give a unit cell of

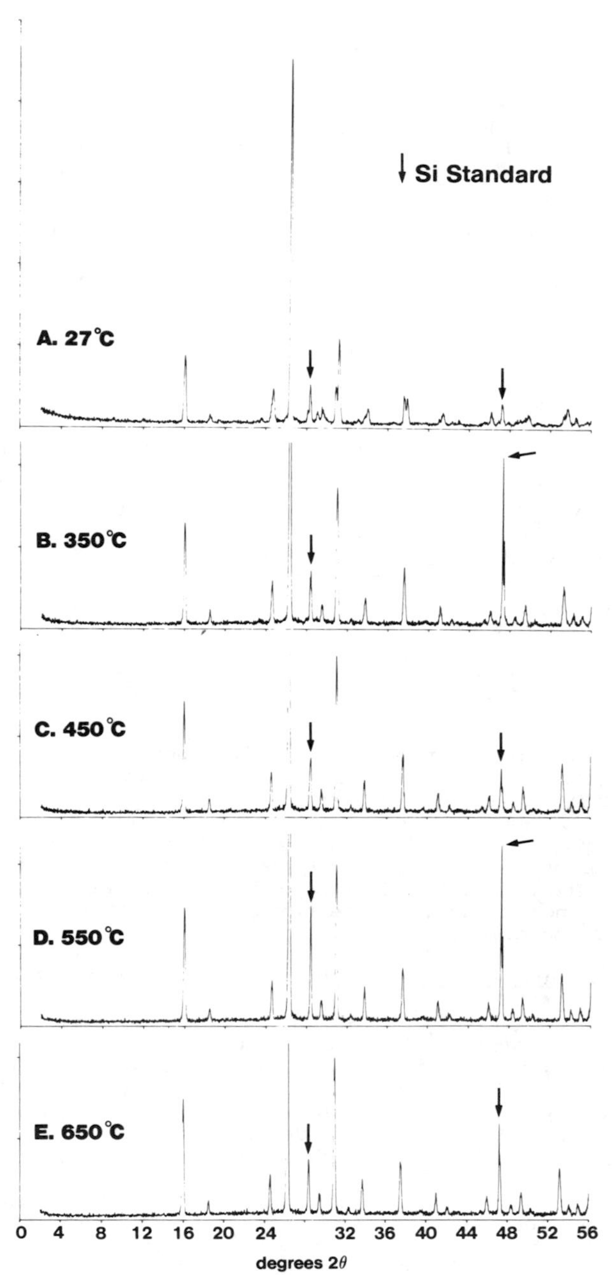

Figure 4. High Temperature X-ray Powder Diffraction Data for Hexagonal Solid Solution

13.5627(15)Å and a cell volume of 2494.8(8)$Å^3$. Comparison with the volume at room temperature from the synchrotron data (based on volumes of the asymmetric units) indicates a volume expansion of 2.5% in the 25 to 650°C temperature range. The CTE of the crystalline portion of the leucite-pollucite glass ceramic is calculated to be 13 ppm°C^{-1}. The CTE of the ceramic part made from powder III, measured by dilatometry, was 14 ppm°C^{-1}, which was in close agreement with the crystallographic value.

SUMMARY

Leucite-pollucite solid solutions can be made over a large compositional range from ion-exchanged zeolite powders. A new hexagonal structure modification has been discovered in intermediate K-Cs solid solutions. The hexagonal structure transforms to the familiar cubic high-leucite structure at elevated temperatures. A wide variety of thermal-expansion behaviors can be engineered by adjusting the SiO_2/Al_2O_3 ratios and cation compositions of the precursor zeolites. The final properties of these glass ceramics can be tailored to numerous potential applications

ACKNOWLEDGEMENTS

This research was carried out (in part) at the National Synchrotron Light Source, Brookhaven National Laboratory, which is supported by the U.S. Department of Energy, Division of Materials Sciences and Division of Chemical Sciences.

REFERENCES

1. C. Hahn and K. Teuchert, Ber. Dt. Keram. Ges., **57**, 208-214 (1980).
2. D. Taylor and C.M.B. Henderson, The American Mineralogist, **53**, 1476-1489 (1968).
3. D.W. Richerson and F.A. Hummel, J. Am. Cer. Soc., **55**, 269-273 (1972).
4. R.M. Barrer and N. McCallum, J. Chem. Soc., 4029-4035 (1953).
5. R.M. Barrer, J.W. Baynham, and N. McCallum, J. Chem. Soc., 4035-4041 (1953).
6. R.L. Bedard and E.M. Flanigen, U.S. Patent 5,071,801 Dec. 12, 1991.
7. G.H. Beall and H.L. Rittler, Adv. Cer., 301-312 (1982).

SINTERING PROCESS AND MECHANICAL PROPERTIES FOR OXIDE-BASED NANOCOMPOSITES

Koichi Niihara, Atsushi Nakahira and Masahiro Inoue
The Insititute of Scientific and Industrial Research, Osaka University, 8-1, Mihogaoka, Ibaraki, Osaka 567, Japan.

ABSTRACT

Oxide-based Al_2O_3/SiC nanocomposites, in which the nanometer-sized nonoxide SiC particles are dispersed mainly within the matrix Al_2O_3 grains, were fabricated by the normal powder metallurgical methods. High resolution transmission electron microscopic observation showed that the nanometer sized SiC particles were directly bonded to the matrix Al_2O_3 grain without any reacting layers. The nano-sized SiC dispersions gave the significant improvement of mechanical properties such as fracture strength over 1000 MPa and no degradation of its strength up to 1000°C for Al_2O_3/SiC nanocomposites, was observed. The sintering process and the correlation between mechanical properties and micro/nanostructure will be discussed for this nanocomposite.

INTRODUCTION

Oxide-ceramics have been used in various engineering fields because of many desirable properties such as high Young's modulus, wear resistance, good chemical inertness and easy processing. However, their applications are still limited by some disadvantages such as relatively low fracture toughness and strength. Therefore, the reinforcement of oxide ceramics have been tried by incorporating the particle, whisker and fiber with the micrometer size into the matrix grain boundary. Recently, ceramic-based nanocomposites have attracted much attention as the material of structural components since the studies by Niihara and his colleagues[1-4]. They reported that Al_2O_3/SiC, MgO/SiC, Al_2O_3/Si_3N_4, Al_2O_3/TiC, mullite/SiC, $Al_2O_3/SiC/ZrO_2$ nanocomposites could be successfully fabricated by the conventional powder metallurgical techniques such as hot-pressing, pressureless sintering and sintered/HIPing [5,6]. The mechanical properties of these nanocomposites were found to be significantly improved by the dispersion of nanometer-sized particles within the matrix grain. Therefore, the understanding of sintering mechanism in the fabrication processes and mechanism of strengthening and toughening is very important in the advance of tailoring microstructures of ceramic composites.

The purposes of this paper are to clarify the sintering mechanism and intragranular structure for the Al_2O_3/SiC nanocomposite and to discuss the nanostructure/ mechanical properties relationship for understanding the roles of intragranular dispersions.

EXPERIMENTAL PROCEDURE

The starting materials were highly pure γ-Al_2O_3 from Asahi Chemical Co. and β-SiC from Ibiden Co.. The γ-Al_2O_3 powder had the irregular shape less than 50nm. The β-SiC powder had the spherical particle mostly below 100nm and in part some hundred nm from transmission electron microscopic observation. Al_2O_3/SiC containing 5, 11, 17 and 33 vol% were mixed by the ball milling method in ethanol for 12 hrs using Al_2O_3 balls. After drying, these mixtures were further ball-milled in dry conditions for 24 hrs. The mixed Al_2O_3/SiC powders were hot-pressed in a graphite die at 1500 - 1800°C under N_2 atmosphere with applied pressure of 30MPa.

The sintered specimens were subjected to the measurements of Young's modulus, fracture toughness and fracture strength. Test-pieces with dimension of 3x4x36 mm were used for measuring the mechanical properties. Young's modulus was measured by the flexural vibration method. Fracture toughness was estimated by indentation microfracture method[7]. Fracture strength was done using the 3-point bending method with a span of 30 mm at a crosshead speed of 0.5mm/min up to 1500°C in air.

Crystalline phases were identified by X-ray diffraction analysis. The density was measured by Archimedes immersion method. The microstructure was observed by scanning electron microscopy(SEM) and transmission electron microscopy(TEM).

RESULTS AND DISCUSSION

Sintering process and microstructure

Figure 1 shows the transmission electron micrographs for Al_2O_3/5vol%SiC nanocomposites hot-pressed at 1000, 1300 and 1600°C. At 1000°C, as seen in Fig. 1(A), the SiC particles with the diameter of approximately 200nm are dispersed in approximately 50nm Al_2O_3 particles. Fig. 1(B) shows that the SiC particles are surrounded by 200nm Al_2O_3 as Al_2O_3 matrix grains coarsen at 1300°C. The neck of Al_2O_3 grains are stably grown and in part the closed pores are observed around SiC particles until this hot-pressing temperature. At 1600°C, the SiC particles are trapped within fine-grained Al_2O_3 matrix, as seen in Fig.1(C). These results show the Al_2O_3/SiC nanocomposites were successfully prepared by sintering technique. These closed pores around SiC were eliminated with the increase of the hot-pressing temperature. Furthermore, Al_2O_3/SiC nanocomposites were observed to be completely dense without trapped pores within Al_2O_3 matrix grains, whereas for monolithic Al_2O_3 the intragranular pores were often observed within Al_2O_3 grains. This difference suggests that the dispersion of intragranular SiC particles within Al_2O_3 matrix grain inhibited the mobility of grain boundary.

However, it is noted that smaller SiC than some tens nm were already dispersed within the small Al_2O_3 grains even at 1100-1200°C (not shown in this figure). These results mean that some small SiC particles less than 100 nm are trapped within Al_2O_3 matrix during the grain growth Al_2O_3 before the densification is completely finished. The critical radius of intragranular SiC particle is thought to be dependent strongly on the hot-pressing temperature. Therefore, it may be concluded that the intragranular SiC is caused by the grain boundary break-away from the pinning force of SiC during matrix grain growth.

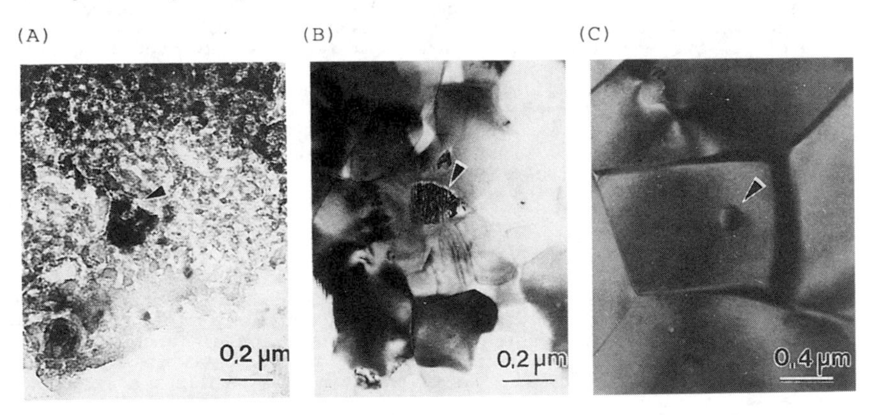

Figure 1 Transmission electron micrographs of Al_2O_3/SiC nanocomposites hot-pressed at (A)1000°C, (B)1300°C and (C)1600°C

Mechanical Properties

Figure 2 gives the variation of Young's modulus with the volume fraction of SiC for Al_2O_3/SiC nanocomposites. Young's modulus for the fully dense Al_2O_3/SiC nanocomposites increase with the SiC content. This change of Young's modulus is in fairly good agreement with the linear mixed rule. However, Al_2O_3/SiC nanocomposite with lower relative density showed the lower Young's modulus than expected values.

Figure 3 shows the effect of SiC content on the fracture toughness for Al_2O_3/SiC nanocomposites. As seen in Fig. 3, the dispersion of SiC gives the increase in fracture toughness from 3.0 to 4.7MPam$^{1/2}$ with only 5 vol% SiC and keeps the constant value up to 33 vol% SiC. This increase of fracture toughness for Al_2O_3/SiC nanocomposites is attributable to the direct interaction of SiC particles with the crack front, the crack deflection, and/or the shielding effect due to residual stress caused by the difference of thermal expansion coefficient between Al_2O_3 and SiC during cooling from the fabrication temperature[8].

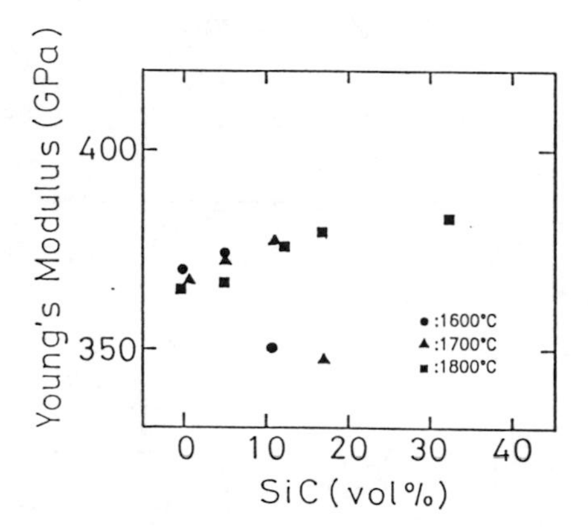

Figure 2 Variation of Young's modulus with the SiC content for Al_2O_3/SiC nanocomposites.

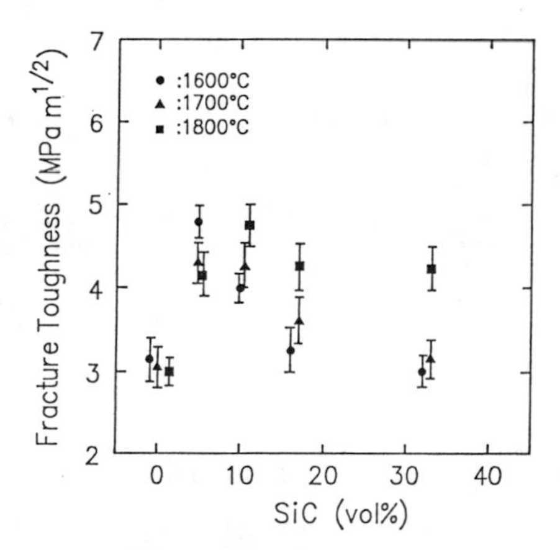

Figure 3 Variation of fracture toughness with the SiC content for Al_2O_3/SiC nanocomposites.

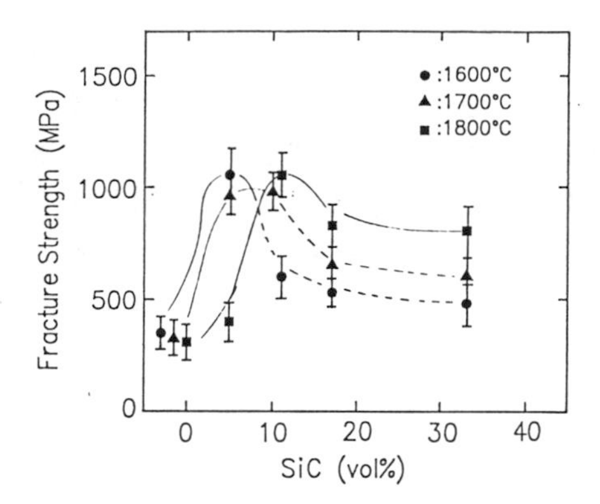

Figure 4 Variation of fracture strength with the SiC content for Al_2O_3/SiC nanocomposites.

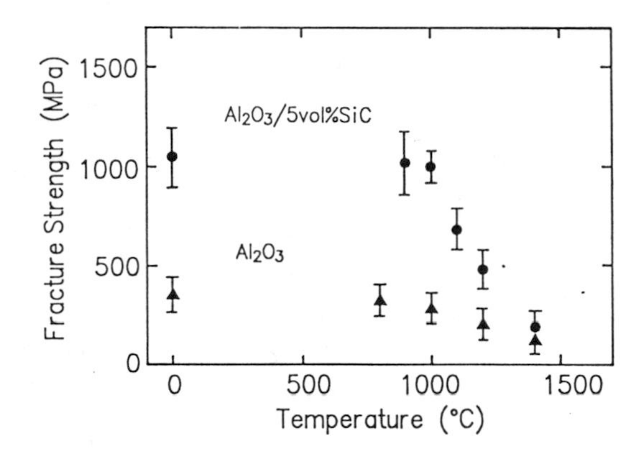

Figure 5 Temperature dependence of fracture strength for Al_2O_3/SiC nanocomposites.

Fracture strength of Al_2O_3/SiC nanocomposites at room temperature is illustrated in Figure 4. Fracture strength is dependent on the relative density and decrease with lowering relative density, as shown by dashed lines in figure. However, the fully dense Al_2O_3/SiC nanocomposites show the maximum strength over 1000MPa at only 5 vol% SiC dispersions and give the slightly-decreased strength up to 33 vol% SiC. It is striking finding that the further improvement in strength was achieved to 1500MPa by thermal treatment in air or inert atmosphere at 1200 - 1300°C for Al_2O_3/5vol%SiC nanocomposite.

Another surprizing finding is that the high strength of this nanocomposites was independent of the grain size of the Al_2O_3 matrix. The detailed results will be given elsewhere[9].

This tremendous increase in fracture strength cannot be explained only by the increase in fracture toughness. The observation of fracture surface showed the completely transgranular fracture mode for Al_2O_3/SiC nanocomposites, whereas for monolithic Al_2O_3 the intergranular fracture was caused by the thermal expansion mismatch between a-axis and c-axis of Al_2O_3[10]. Therefore this notable increase in strength for Al_2O_3/SiC nanocomposite is thought to be contributed to the decrease of fracture origins caused by the refinement of matrix Al_2O_3 grain and/or the compressive stresses around SiC particle which inhibit the grain boundary fracture of matrix.

Temperature dependence of fracture strength for Al_2O_3/SiC nanocomposites is shown in Figure 5, compared with that for monolithic Al_2O_3. Monolithic Al_2O_3 exhibited the gradual degradation of strength over 600-800°C, above which temperature the fracture surface mode was completely intergranular. Therefore, the degradation in this temperature range is thought to be caused by the slow crack growth due to the grain boundary sliding and cavity formation/growth[11]. On the other hand, Al_2O_3/SiC nanocomposites showed a notable improvement in strength up to 1000°C. The fracture mode at 1000°C as well as room temperature was found to be transgranular. Thus, it may be concluded that the significant improvement in high-temperature strength of Al_2O_3/SiC nanocomposites is achieved through inhibition of grain boundary fracture.

CONCLUSION

The Al_2O_3/SiC nanocomposites were successfully fabricated by the usual powder metallurgical methods. The transmission electron microscopic observation of sintering process and the mechanical tests gave the following results;

1. The smaller intragranular SiC less than 100 nm within Al_2O_3 was confirmed to be consolidated at 1100 - 1200°C, whereas relatively large SiC particles with approximately 200 nm were found to be trapped within Al_2O_3 grains above 1400°C.
2. SiC particles inhibited the grain boundary movement and resulted in microstructural stability in sintering process.
3. Young's modulus increased linearly with the SiC content. Fracture toughness and strength were remarkably improved by incorporating the finer SiC particles into matrix grains; 70% increase in fracture toughness and 200% increase in fracture strength.
4. High-temperature strength of Al_2O_3 was also improved by the dispersion of nanometer-sized SiC particles within the matrix grains.

REFERENCES

1. K. Izaki, K. Hakkei, K. Ando and K. Niihara, in Ultrastructure Processing and Advanced Ceramics, eds by J.D. Mackenzie and D.R. Ulrich, 891-900 (John Wiley & Sons, NY, 1988).
2. F. Wakai, Y. Kodama, S. Sakaguchi, N. Murayama, K. Izaki, and K. Niihara, Nature, 344, 421-422(1990).
3. K. Niihara, K. Suganuma, A. Nakahira and K. Izaki, J. Mater. Sci. Lett. 10, 598-599(1990)
4. K. Niihara, K. Izaki and T. Kawakami, J. Mater. Sci. Lett., 10, 112-114(1990).
5. K. Niihara and A. Nakahira, Proc. of MRS Meeting on Advanced Composites, 129-134(Plenum, NY, 1988).
6. K. Niihara, Electronic Ceramics, 9, 44-48(1989).
7. K. Niihara, A. Nakahira and T. Hirai, J. Amer. Ceram. Soc., 67, C1-C2(1984).
8. J. Selsing, J. Amer. Ceram. Soc., 44, 419(1961).
9. K. Niihara, J. of The Ceram. Soc. Jpn., 99, 974-982(1991).
10. R.W. Rice, Proc. Brit. Ceram. Soc., 20, 205-215(1972).
11. A.G. Evans and A. Rana, Acta Metall., 28, 129-139(1980).

AMORPHIZATION AT THE METAL-CERAMIC INTERFACE IN $Nb - Al_2O_3$ COMPOSITE THROUGH EXTENSIVE INTERPHASE DIFFUSION,

Abhijit Ray and Shyam K. Samanta
Mechanical Engineering Department, Plasticity Laboratory, University of Nevada-Reno, NV-89557.

ABSTRACT

A novel, low temperature and pressureless bonding technique has been developed to fabricate Al_2O_3 coated Nb composite. The nature of the interface bond has been characterized by extensive electron microscopic investigation, and a phenomenological model has been developed to explain the mechanism of bond formation. In the present technique a suspension (in distilled water) of fine Al_2O_3 powder (mean particle size of $0.2\mu m$) was applied on polished and cleaned Nb blocks. The coatings were air dried and the coated metals were heat treated at $1100°C$ to achieve bonding between the ceramic and the metal. The process was carried out in an argon atmosphere. To reduce the oxygen impurity content in argon, the gas was passed over Mg turnings at $300°C$ before its entry into the bonding apparatus. Control of oxygen impurity in the bonding atmosphere in conjunction with the use of fine ceramic particles led to enhancement of interfacial diffusion. Scanning electron microscopy of the bonded specimens showed a consolidated layer in the ceramic phase close to the metal-ceramic interface. Transmission electron microscopy of the metal-ceramic interfacial region r0evealed a distribution of Al_2O_3 particles in an amorphous phase containing varying amount of Al, Nb, and O. Prior treatment with Mg was to reduce the oxygen impurity content below the equlibrium oxygen partial pressure (at $1100°C$) of Al_2O_3. This would result in loss of oxygen from the surface of the ceramic particles. Thus a region, with high concentration of $O^{=}$ vacancy would form along the surfaces of the particles. As a result of excessive oxygen depletion, further continuation of this process would lead to degeneration of the crystal structure of the regions along the surfaces of the Al_2O_3 particles. These regions along the Al_2O_3 particle surfaces may then act as easy diffusion paths for Nb. Extensive diffusion of the metal into the ceramic phase along these paths would lead to the formation of an amorphous phase along the ceramic particle surfaces. Eventually, this phase will occupy the voids in between the ceramic particles, that are close to the metal-ceramic interface, and thus bond them together with the metal.

INTRODUCTION

The requirement that the modern engineering materials be used in increasingly more severe working conditions has given a major emphasis on the development of materials with tailored properties. While, even a few decades ago, the primary means of tailoring the material properties was alloying of different metals, presently material properties are also modified through joining of dissimilar materials to form a composite material. Recent surge in the use of composite materials in various engineering applications led to extensive research in this area. One of the most potent of all these materials is the metal-ceramic composite. There exist a number of methods for fabrication of metal-ceramic composites.

The hot isostatic pressing (HIP)[1] technique is frequently used to form laminated metal-ceramic composites.

Ceramic overlay coatings[2] can be developed through physical vapor deposition, thermal spraying, laser chemical vapor deposition and sol-gel[3] processing techniques. In addition, oxide coatings on alloys can be developed through selective oxidation. An example may be the development of Al_2O_3 or Cr_2O_3 enriched coatings on the alloy surfaces containing more than 10% aluminum or 16% chromium respectively through controlled oxidation[1].

A novel processing technique has been developed[4], taking a completely different approach, to produce a thin coating of Al_2O_3 on Nb. While in all of the existing techniques interphase diffusion is increased by the use of elevated processing temperature in conjunction with improved contact between the metal and the ceramic, in the present technique, in addition to improved contact between the two phases, increased defect concentration in the ceramic phase (due to composition control of the bonding atmosphere) resulted in enhanced interphase diffusion.

EXPERIMENTAL

Niobium pieces were polished with 600 grit abrasive paper and coated with a suspension (in distilled water) of Al_2O_3 powder (0.2 μm mean particle size). These specimens were bonded at 1100°C in an argon atmosphere using a special bonding apparatus. The unique feature of this apparatus is its built-in ability to cool the specimen from the bonding temperature with an argon gas flow while the specimen is inside the apparatus, thereby preventing any possibility of oxidation of the metal. The argon gas was purified by passing it over magnesium turnings at 300°C prior to its entry in the bonding apparatus. Such a process would reduce the oxygen content of the gas to a very low level. Bonding and cooling of the specimens were followed by appropriate specimen preparation for scanning and transmission electron microscopy.

RESULTS

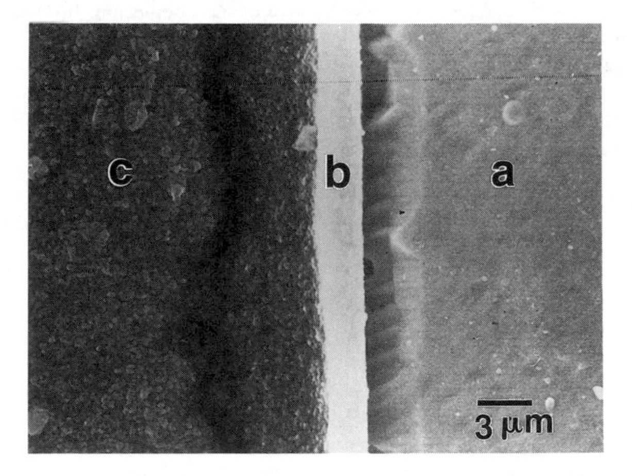

Figure 1. Scanning electron micrograph of a cross sectioned specimen bonded for 4 hours at 1100°C showing: (a) the metal (niobium); (b) a consolidated layer at the metal-ceramic interface; (c) the loosely bonded alumina particles away from the metal-ceramic interface.

A consolidated layer (Fig.1) in the ceramic phase close to the metal-ceramic interface has been observed in the scanning electron micrographs of the bonded specimens. In the transmission electron micrograph (Fig.2) a phase containing varying amount of aluminum, oxygen and niobium has been observed close to the metal-ceramic interface. Figure 3 shows a diffused ring pattern that has been obtained from the interfacial phase shown in Fig.2. Such patterns are typically produced by amorphous phases in the diffraction mode. Figure 4 shows the energy dispersive X-ray spectrum obtained from the interfacial amorphous phase. As can be seen, oxygen, aluminum and niobium are the major elements present in the interfacial amorphous phase. Other elements seen in these spectra are deposited on the interfacial region during ion beam milling of the specimens. Thus, iron, nickel and chromium are deposited from the stainless steel cathode of the ion mill; copper and molybdenum from the specimen holder of the ion mill; source of argon is the argon ions used to thin the specimen in the mill.

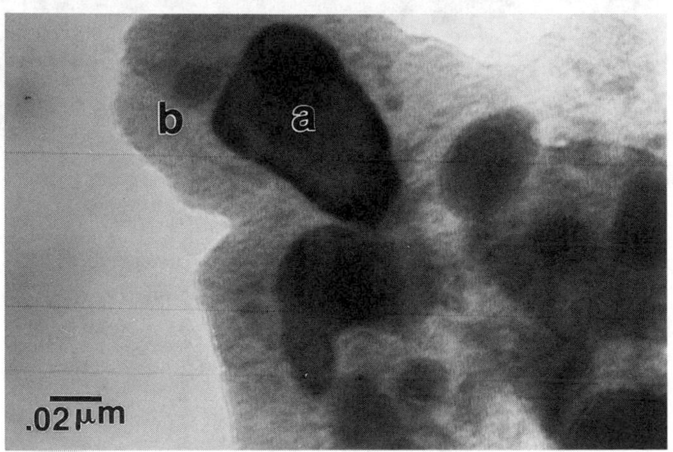

Figure 2. Transmission electron micrograph of a specimen bonded for 4 hours at 1100°C. Specimen preparation for transmission electron microscopy was done through sectioning parallel to the metal-ceramic interface. (a) an alumina particle at the metal-ceramic interface; (b) amorphous phase around the alumina particle.

DISCUSSION

Aluminum is known to form a number of gaseous aluminum oxides in addition to solid Al_2O_3. Table I shows the different reactions leading to the formation of gaseous aluminum oxides from Al_2O_3, the corresponding change in the standard free energies (for one gram mole of O_2) at the bonding temperature ($1100°C$) and the relations between the equilibrium partial pressures of these oxides and oxygen. Further, for these reactions consideration of stoichiometry in the gaseous phase at the gas-ceramic interface (e.g., oxygen partial pressure, p_{O_2}, is $\frac{1}{4}{}^{t}h$ of p_{AlO}, the partial pressure of AlO for the reaction $2Al_2O_3 \rightleftharpoons 4AlO + O_2$) enable us to determine the partial pressures of oxygen present in equilibrium with each of these oxides, assuming that none of the other oxides are present in the gaseous phase (see table II). In the presence of all of the gaseous aluminum oxides, partial pressure of oxygen, p_{O_2}, in the gas phase will be the same as the maximum oxygen partial pressure that can exist at the gas-ceramic interface.

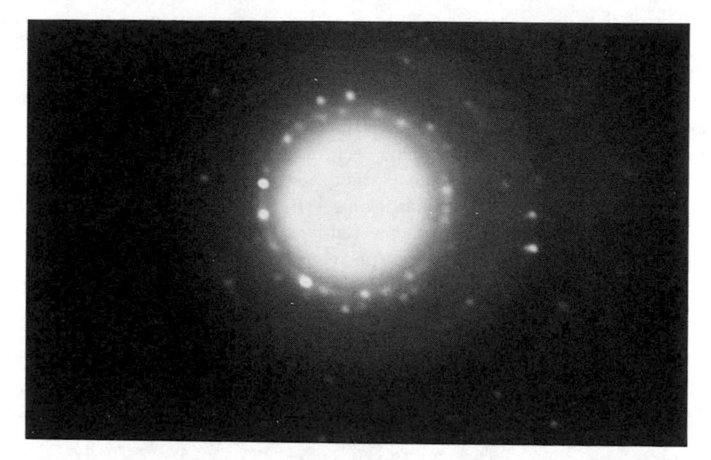

Figure 3. Diffraction pattern obtained from the amorphous phase shown in figure 2.

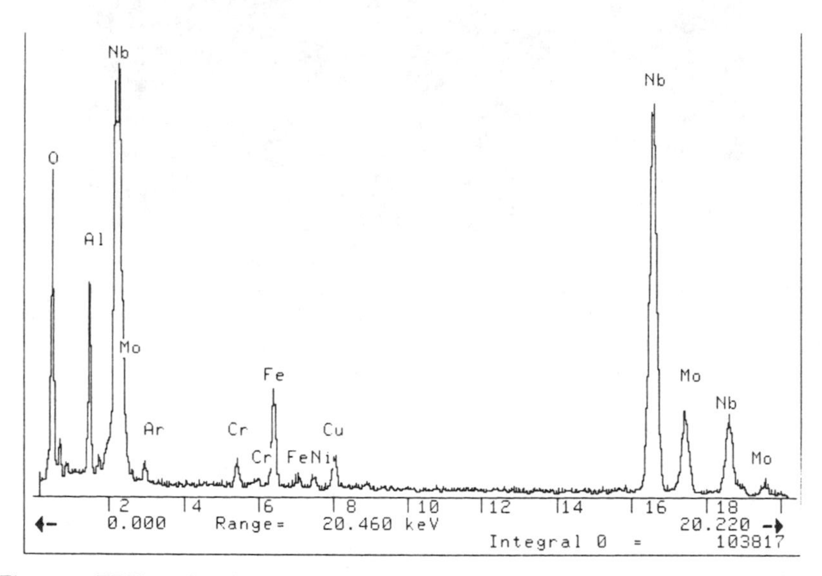

Figure 4. EDX results obtained from the amorphous phase shown in figure 2.

Therefore, from table II we can say, the partial pressure of oxygen at the interface will be 1.39×10^{-19} atmosphere. Using Langmuir's equation[5] we can relate the rate of loss of

oxygen, Z_{O_2} $(moles/cm^2.hr)$, from the surfaces of the aluminum oxide particles and the oxygen partial pressure, p_{O_2}, at the gas-ceramic interface.
$Z_{O_2} = 1.59\times10^5 \frac{p_{O_2}}{\sqrt{M_O T}}$
where M_O is the molecular weight of oxygen and T is the the temperature (K). Substitution for p_{O_2}, M_O and T yields
$Z_{O_2} = 2.93\times10^{-20}$
Further, Z_{O_2} can be related to the rate of increase of vacancy concentration on the oxide particle surfaces using the relation $\dot{V}_O = \frac{2NZ_{O_2}}{M}$, here N is the Avogadro's number and M is the average number of lattice sites in a square cm on an Al_2O_3 particle surface. The lattice parameter of Al_2O_3 on the basal plane (the most densely packed plane) is ~ 5 Å[6]. This is the minimum lattice parameter in this crystal. The lattice parameter has a maximum value of ~ 12 Å in a direction perpendicular to the besal plane. For the purpose of our calculation we choose an average value of ~ 8 Å. Therefore, in an area of 1 square cm the number of lattice sites is, $M \sim 1.56 \times 10^{14}$. Substituting $M = 1.56\times10^{14}$, $N = 6.023\times10^{23}$ and $p_{O_2} = 1.39\times10^{-19}$ in the above expression,

$$\dot{V}_O = 2.26\times10^{-10}/sec$$

Al_2O_3 forms Schottky type disorder. The equilibrium defect concentration in this material is $\sim 10^{-12}$ at the bonding temperature[7]. Therefore, as result of the control of the oxygen impurity in the bonding atmosphere, shortly after the initiation of the bonding cycle, at the calculated rate of vacancy generation, the surface vacancy concentration will be a few orders of magnitude higher than the equilibrium defect concentration of the Al_2O_3 particles. Considering diffusion in the ceramic phase as a process of discrete and random jumps of ions to neighbouring vacant sites, it is shown[8] that the diffusivity of Nb on the surface of an Al_2O_3 particle can be expressed as

$$D_{Nb,s} = \alpha C_{Nb,s}(\frac{2}{3}\dot{V}_O t - C_{Nb})^2$$

where α is a constant involving L, the number of cationic sites nearest to a diffusing Nb ion; d, the jump distance for a cation; ν, the jump frequency and a constant κ_s, the ratio of the concentrations of diffusing cations and that of available cationic vacancy. $C_{Nb,s} = \frac{C_{Nb}}{C_{Nb}+C_{Al}}$, where C_{Nb} is the concentration of Nb ions and C_{Al} is the concentration of Al ions on the surface of a ceramic particle. The above expression suggests that the diffusivity of Nb on the surface of an aluminum oxide particle will increase with the square of the available vacancy concentration, $(\dot{V}_O t - C_{Nb})$, and as a result of the enhanced interphase diffusion a consolidated layer is formed at the metal-ceramic interface (observed in the scanning electron micrographs of the bonded specimen). Similar expressions for diffusivities of oxygen, aluminum and niobium ions are also derived[6]. The diffusion coefficients are found as,
$D_{O,i} = \alpha_a C_{v,a}^2$
$D_{Al,i} = \alpha_c(1 - C_{Nb,c})C_{v,c}^2$
$D_{Nb,i} = \alpha_c C_{Nb,c} C_{v,c}^2$
where $D_{O,i}$, $D_{Al,i}$ and $D_{Nb,i}$ are the diffusivities of oxygen, aluminum and niobium ions respectively. α_a and α_c are similar constants as α in the previous expression while subscripts a and c refers to the anionic and cationic sublattices. $C_{Nb,c}$ is equivalent to $C_{Nb,s}$ and $C_{v,a}$ and $C_{v,c}$ are the available anionic and cationic vacancy concentrations. In an Al_2O_3 particle, close to the metal-ceramic interface, simultaneous diffusion of aluminum and niobium ions in the cationic sublattice will reduce the available cationic vacancy

concentration. As a result, the mass transport in the cationic sublattice will be slow compared to that in the anionic sublattice. The resulting deviation in stoichiometry in these ceramic particles will lead to a break down of their crystal structure and thus produce an interfacial amorphous phase enriched in oxygen, aluminum and niobium.

Table I. The reactions and associated free energies of gaseous aluminium oxides.

Reactions	Molar free energies	Partial pressure relations
$Al_2O_3 \rightleftharpoons O_2 + Al_2O$	238.972 (kcal/mole)	$9.08\times10^{-39} p_{O_2}{}^{-1} = p_{Al_2O}$
$2Al_2O_3 \rightleftharpoons O_2 + 4AlO$	577.148 (kcal/mole)	$1.33\times10^{-92} p_{O_2}{}^{-1} = (p_{AlO})^4$
$2Al_2O_3 \rightleftharpoons O_2 + 2Al_2O_2$	409.236 (kcal/mole)	$7.14\times10^{-66} p_{O_2}{}^{-1} = (p_{Al_2O_2})^2$
$2Al_2O_3 + O_2 \rightleftharpoons 4AlO_2$	408.048 (kcal/mole)	$1.104\times10^{-65} p_{O_2} = (p_{AlO_2})^4$

Table II. The equilibrium oxygen partial pressures of the gaseous aluminium oxides.

Oxides	Oxygen partial pressure
Al_2O	9.528×10^{-20} (atmosphere)
AlO	1.39×10^{-19} (atmosphere)
Al_2O_2	1.213×10^{-22} (atmosphere)
AlO_2	3.51×10^{-23} (atmosphere)

ACKNOWLEDGEMENT

This work has been supported by the United States National Science Foundation Grant MSM 8714377.

REFERENCES

1. M. Ruhle, M. Backhaus-Ricoult, K. Burger and W. Mader, Ceramic Microstructure, vol. 21, edited by J.A. Pask and A.G. Evans (London, New York: Plenum Press), p.295, 1986.
2. S.R.J. Saunders and J.R. Nicholls, Material Science and Technology, vol.5, p.780.
3. M.G. Hocking, V. Vasantasree, and P.S. Sidky, Metallic and Ceramic Coatings: Production, High Temperature Properties and Applications, (Longman Scientific and Technical, 1989), pp.49-200, ibid, p.293.
4. S.K. Samanta, and A. Ray, Proc. Adv. Mats. 1, 61 (1991).
5. Langmuir, Jones, and Mackay, Phys. Rev. 30, 201 (1927).
6. G.V. Samsonov, The Oxide Handbook, (New York: IFI/Plenum, 1973), p.23.
7. W.D. Kingery, H.K. Bowen, D.R. Uhlmann, Introduction to Ceramics, 2nd ed. (New York: John Wiley & Sons, 1976), pp.142, ibid, p.238.
8. A. Ray, and S.K. Samanta, Submitted to Phil. Mag., 1991.

MODELING AND EXPERIMENTAL INVESTIGATIONS OF PRESSURE INFILTRATION OF SOL-GEL PROCESSED 3-D CERAMIC MATRIX COMPOSITES

HSIEN-KUANG LIU AND AZAR PARVIZI-MAJIDI
Center for Composite Materials and Department of Mechanical Engineering, University of Delaware, Newark, DE, 19716

ABSTRACT

A parabolic rate kinetics model has been developed and verified by experiments for the pressure infiltration of textile ceramic matrix composites using sol-gel processing with added solid particles. Darcy's law and a global permeability of the whole system were adopted in the model. The permeability of the whole system consisting of fiber preform, compaction, and filter was derived from the individual permeabilities of the fiber preform and particle compaction inside the preform which were based on the Kozeny-Carman equation. Experiments were conducted using 3-D angle interlock woven carbon fiber preforms and silica sol containing silica particles of 0.25μm and 0.5μm sizes. The total infiltration time was inversely proportional to the constant processing pressure and increased significantly with reducing the solid particle size by a factor of two. This is because the permeability of particle compaction layer is the dominant term and it is proportional to the square of particle size.

INTRODUCTION

Sol-gel processing has attracted wide interest for fabrication of ceramics and ceramic composites due to its low temperature processing and versatility in terms of the materials and microstructures obtained. However a major problem in sol-gel processing is excessive shrinkage during drying which leads to extensive matrix cracking in composites where the stiff fiber preform constrains matrix deformation. Liu and Parvizi-Majidi [1] studied the control of shrinkage by the addition of solid particles to the sol. The mixture of sol and particles suspension was infiltrated into a 3-D angle interlock woven preform by pressure infiltration. Drying cracks were reduced and density of the composite increased. Processing can be better understood if kinetics of pressure infiltration is studied. Lange [2] proposed the parabolic rate kinetics of pressure filtration of alumina slurry. Danforth [3,4] also proposed parabolic rate kinetics of filtration for bimodal mixtures of colloidal silica and monosized colloidal silica.

This work investigates kinetics of pressure infiltration of mixture of silica sol and silica solid particles into a 3-D angle interlock woven preform. The relationship among permeabilities of material system, processing time, processing pressure, and size of particles is being studied.

EXPERIMENTAL PROCEDURES

Infiltrate and preform

The infiltrate combines both a sol and a suspension. The silica sol was prepared using a modification of the recipe given by Klein [5] in which tetraethyl-orthosilicate (TEOS), ethanol, deionized water and 7wt% HNO_3 were mixed at a volume ratio of 1 :1 :1.6 :0.06. The solid content in the sol was 7 wt%. Colloidal suspensions of silica particles with particle sizes of 0.25μm and 0.5μm were prepared by a method analogous to that used by E.I. DuPont de Nemours & Co. for making Stober silica suspension [1,6,7]. Particle sizes were examined by transmission electron microscopy (TEM) and are shown in Figure 1. The infiltrates were prepared by mixing appropriate volumes of the sol and the particle suspensions. Table 1 gives the characteristics of the infiltrates used.

The fiber preform was a 3-D angle interlock woven fabric[1] of PAN-based type I carbon fiber with a fiber volume fraction of 48% and thickness of 0.245". It was cut into a circular disk of 2.43± 0.06" diameter. Fibers had a diameter of 7.2μm.

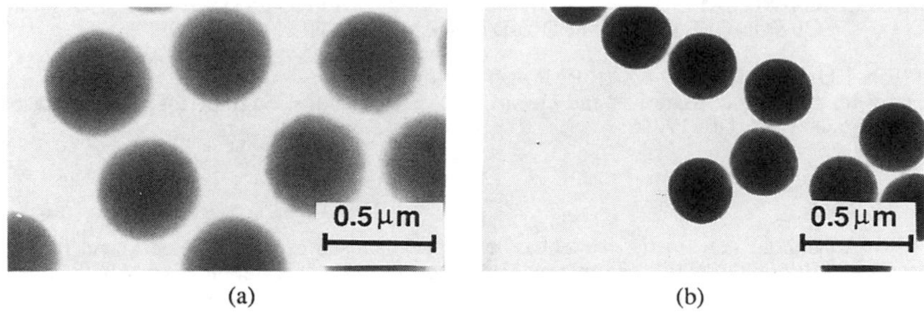

(a) (b)

Figure 1 TEM micrographs of silica particles in suspension (a) 0.5μm (b) 0.25μm.

Table I Volume fractions and viscosities of the two infiltrates used in experiments.

Infiltrate	v_1	v_2	v_c	v_0	Viscosity (cP)
Sol + 0.25μm particles	0.036	0.0498	0.6	0.0429	2.2cP
Sol + 0.25μm particles	0.036	0.0577	0.59	0.0477	2.34cP

v_1 : volume fraction of solids in sol; v_2 : volume fraction of particles in suspension
v_0 : volume fraction of solids in infiltrate; v_c : volume fraction of solids in particle compaction

Pressure Infiltration

The preform was placed in a pressure infiltration apparatus as shown in Figure 2. The infiltrate was pushed through the thickness direction of preform, i.e., perpendicular to warp and weft yarns, using an Instron testing machine. The pressure infiltration apparatus was specifically designed to obtain data on the matrix deposition[1].

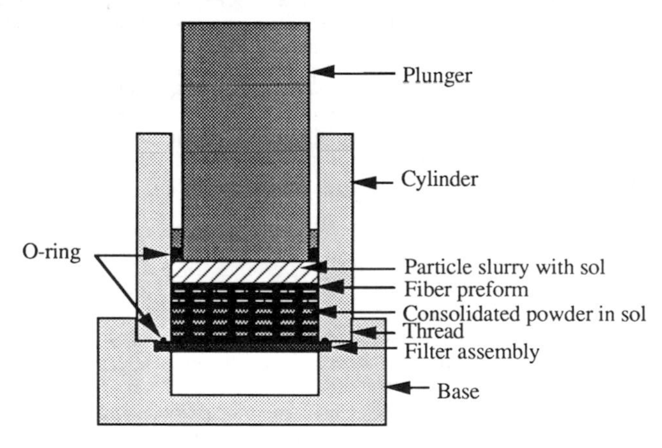

Figure 2. Schematic of pressure infiltration apparatus.

The preform was placed above the filter assembly and 300 ml of the infiltrate was poured into the cylinder and then pushed through the preform by pressing down the plunger. A hydraulic Instron (type 1321) set at constant load mode was used to provide pressure. To prevent filter paper from rupture, infiltration was initially carried out at a low level of pressure. Once a significant layer of powder compaction had formed inside the preform, the pressure was raised to the desired level. Displacement of the moving plunger was recorded during infiltration. The infiltration experiments were terminated after different processing times.

Pressure exerted on the infiltrate was calculated by dividing the applied load by the cross-sectional area of the inner circle of cylinder. After infiltration, the specimen was ejected and dried in an oven at 50°C for 18 hours. The weight of specimen was measured before and after drying and the composite green density was calculated from its dimensions and weight after drying. In order to facilitate cutting and specimen preparation for microscopy, the composite green body was infiltrated with epoxy by soaking it in epoxy and then curing at 150°C for 2 hours. It was then cut and polished down to 1μm finish for examination of the microstructure and the thickness of the particle compaction layer in preform by optical microscopy.

MODELING

Flow through porous media can be described by Darcy's law[8]

$$J = \frac{\kappa_x P_x}{\mu \delta} \tag{1}$$

where J is flow rate per unit area; κ_x is effective permeability of the porous medium; P_x is the pressure difference across the medium; μ is viscosity of fluid; and δ is the thickness of the medium. For the present problem, balancing the moving distance of plunger "d" and the fluid flowing through the porous medium by integrating equation (1) over the processing time t gives

$$d = \sqrt{\frac{2\kappa_e P\ x\ t}{\mu(L_f+L_1)}\ \phi\ (\frac{v_c}{v_0} - 1)(\frac{v_1 + v_2}{v_2})} \tag{2}$$

where $\phi = 1\text{-}v_f$ is the preform porosity; v_1, v_2, v_c, and v_0 are volume fractions of solids in sol, particle suspension, particle compaction, and mixture of suspension and sol, respectively; P is the constant processing pressure; L_f and L_1 are the thicknesses of fiber preform and filter paper, respectively; and x is the thickness of powder compaction layer inside the preform. x is given by

$$x = \frac{d\ (\frac{v_2}{v_1 + v_2})}{\phi\ (\frac{v_c}{v_0} - 1)} \tag{3}$$

and the global permeability of the whole system , κ_e, by

$$\kappa_e = \frac{\kappa_1\ \kappa_2\ \kappa_3(L_f + L_1)}{L_1\kappa_2\kappa_3 + x\kappa_1\kappa_3 + (L_f - x)\kappa_1\kappa_2} \tag{4}$$

where Kozeny-Carman equation [8] has been used for the individual permeabilities. The usual form of the Kozeny-Carman equation for permeability is $\kappa_{CK} = \phi^3/k_0(L_e/L)^2(1\text{-}\phi)^2 S_0^{\ 2}$ where $(L_e/L)^2$ is called "tortuosity", and $k_0(L_e/L)^2$ is called Kozeny constant which equals 5 to fit most experimental data. S_0 is the specific surface area based on the volume of solid which can be related to mean particle diameter D_p or mean fiber diameter D_f. Thus

$$\kappa_2 = \phi\ \kappa_4$$

$$\kappa_3 = \frac{D_f^2}{K}\frac{\phi^3}{(1-\phi)^2}$$

$$\kappa_4 = \frac{D_p^2(1-v_c)^3}{180\, v_c^2}$$

κ_1, κ_2, ,κ_3[9] and κ_4[2] are permeabilities of filter assembly, combination of particle compaction and preform, preform, and particle compaction only, respectively; K = 38.8 is adopted from reference 9 in which 16 layers of 8 harness satin plain weave were used; D_f is fiber diameter; and D_p is mean particle diameter in suspension. κ_1 is 3.43×10^{-14} for the 0.45μm pore size and 6.86×10^{-15} for the 0.1μm pore size nitrate cellulose membrance filter papers used in the experiments[2]. Equations (2), (3), and (4) give

$$Ex = \frac{AD}{B+Cx} \tag{5}$$

Where

$$A = \kappa_1 \kappa_2 \kappa_3$$
$$B = L_1 \kappa_2 \kappa_3 + L_f \kappa_1 \kappa_2$$
$$C = \kappa_1 \kappa_3 - \kappa_1 \kappa_2$$
$$D = \frac{2Pt}{\mu}\frac{v_2}{v_1 + v_2}$$
$$E = \phi\left(\frac{v_c}{v_0} - 1\right)$$

It is to be noted that the model here assumes that the sol has no time to undergo gelation during processing. Solving equation (5) yields

$$x = \frac{-BE+\sqrt{(BE)^2+4ACDE}}{2CE} \tag{6}$$

RESULTS AND DISCUSSION

Figure 3(a) shows the experimental and modeling results for pressure infiltration of sol and suspension with 0.25μm particles through 3-D angle interlock preform under a constant pressure of 123.2 psi. Similarly, figure 3(b) shows the results for infiltrate with 0.5μm particles under a constant pressure of 82.14 psi. Experimental results verify that the processing follows parabolic rate kinetics.

In figure 3(b), the experimental data are lower than those predicted from the model. This may be due to the following reasons. First, sedimentation of the 0.5μm particles was observed and this phenomenon may explain the longer processing times observed in the experiments. Second, application of the pressure may cause the 3-D preform to deform due to a combination of warp and weft yarn deformations. This changes interyarn and intrayarn pores and decreases the rate of infiltration. Third, The real value of κ_3 for our 3-D preform may be lower compared to that in reference 9 due to its complicated fiber architecture. However, the reduction of κ_3 in the theoretical prediction in figure 3(b) has shown slight effect on the processing time which could indicate that the permeability of particle compaction is dominant.

Based on Darcy's law and a series arrangement of the porous media, a global permeability κ_e was provided for the whole system. Permeability of each porous medium was

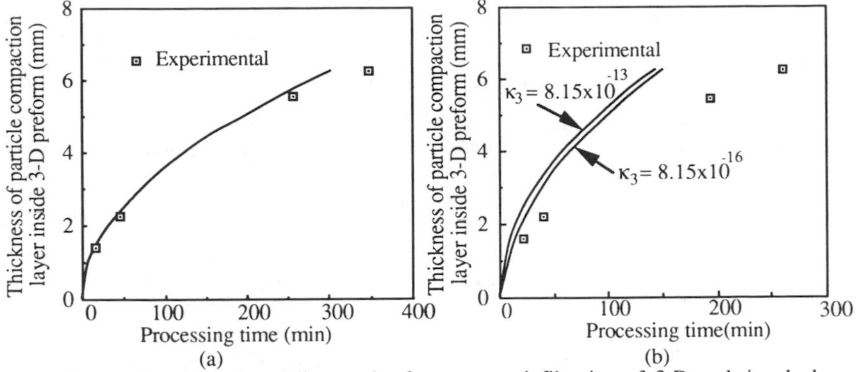

Figure 3 Experimental and modeling results for pressure infiltration of 3-D angle interlock fiber preform with (a) sol and 0.25μm particles suspension under constant pressure at 123.2 psi and (b) sol and 0.5μm particles suspension under constant pressure at 82.14 psi.

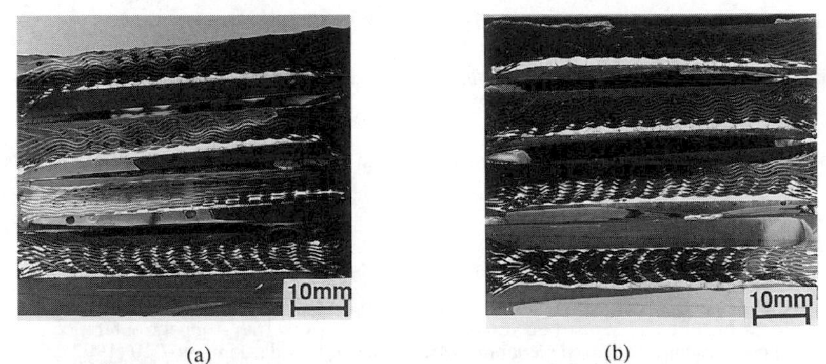

Figure 4 Photographs comparing the thicknesses of the particle compaction layer inside the preform obtained at different processing times in experiment (a) for the case in figure 3(a) and (b) for the case in figure 3(b).

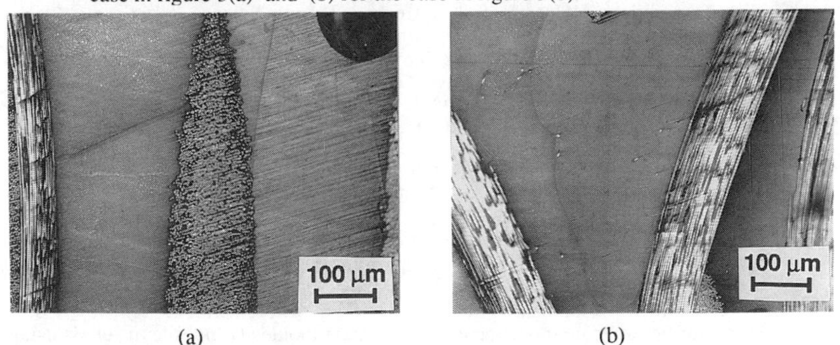

Figure 5 Photographs of (a) material of figure 3(a) after 40 min of infiltration and (b) material of figure 3(b) after 193 min of infiltration. Both cases exhibit good infiltration and a clear and uniform compaction front.

derived from Kozeny-Carman equation. Low pressures were adopted during infiltration in order to prevent flattening of the preform. However, this resulted in long processing times because pressure and time are inversely proportional according to the model. Figure 4 shows the thickness of the particle compaction layer inside the preform as a function of the processing time. Optical micrographs in figure 5 depict that the front of the consolidated layer is clear and uniform and that the infiltration is quite good. Processing at constant pressure provides particles with better packing and longer time to obtain stable packing compared to processing at a constant flow rate.

CONCLUSIONS

The parabolic rate kinetics of pressure infiltration for sol-gel processing of 3-D ceramic composites was verified by experimental and modeling results. There was an excellent correlation between the experimental and theoretical results for the infiltration of composite using sol plus 0.25μm particles. The total processing time for this case was 350 min at a processing pressure of 123.2 psi ($8.5x10^5$ Pa). For the 0.5 μm particles, the experimentally measured thicknesses of the particle compaction layer were somewhat below the theoretical predictions and a total processing time of 262 min was determined at a pressure of 82.14 Psi ($5.66x10^5$ Pa). Optical microscopy also showed better infiltration of the preform with the smaller particle size.

ACKNOWLEDGEMENTS

This work is supported by the Center for Composite Materials at the University of Delaware.

[1] British petroleum Corporation.

[2] Whatman Co. provides flow rate of nitrogen gas through these filter papers under special pressure difference.

REFERENCES

[1] H. K. Liu and A. Parvizi-Majidi, "The effect of solid particles addition in sol-gel processing of ceramic matrix composites," Proceedings of 16th Annual Conference on Composites and Advanced Ceramics, Cocoa Beach, Florida, January 7-10 (1992).

[2] F. F. Lange and K. T. Miller, "Pressure filtration: Consolidation kinetics and mechanics," Am. Ceram. Soc. Bull., 66 [10] (1987), 1498-1504.

[3] F. G. R. Gauthier and S. C. Danforth, "Packing of bimodal mixtures of colloidal silica," Ceramic Transactions, Volume 1, Part B, Ceramic Powder Science, American Ceramic Society, (1988), 709-715.

[4] M. Velazquez and S. C. Danforth, "Pressure filtration of monosized colloidal silica," Surface and Colloid Science in Computer Technology edited by K. L. Mittal, Plenum Publishing Corporation (1987), 86-105.

[5] L. C. Klein, "Sol-gel processing of silicates," Ann. Rev. Mater. Sci. 15 (1985), 227-248.

[6] Werner Stober, Arthur Fink and Ernst Bohn, "Controlled growth of monodisperse silica spheres in the micron size range," J. of Colloid and Interface Science, 26 (1968), 62-69.

[7] Norman J. Wagner and William B. Russel, "Light scattering measurements of a hard-sphere suspension under shear,' Physics fluids, A2 (4) April (1990), 491-502.

[8] F. A. L. Dullien, Porous media:Fluid transport and pore structure, Chapter 4, New York Academic Press (1979).

[9] A. C. Loos, M. H. Weideman, E. R. Long Jr., D. E. Kranbuehl, P. J. Kinsley and S. M. Hart, "Infiltration/cure modeling of resin transfer molded composite materials using advanced fiber architectures," First NASA Advanced Composite Technical Conference, Proceedings of a conference held in Seattle, Washington, Oct. 29-Nov. 1 (1990), 425-442.

PART VI

Hybrid Organic/Inorganic Materials

MACRO-DEFECT-FREE CEMENT: A NOVEL ORGANOCERAMIC COMPOSITE

J. FRANCIS YOUNG AND M. BERG
Center for Cement Composite Materials, University of Illinois, Urbana IL 61801

ABSTRACT

Macro-defect-free cement is a composite material made by processing a hydraulic cement with a water-soluble polymer under high shear roll-milling. This paper examines the principles of processing and the development of microstructure that results in a material with ceramic-like properties. The nature of the chemical interaction between the cement and the polymer is also considered in some detail.

INTRODUCTION

The harnessing of cementitious reactions to create materials with ceramic-like properties by processing at or near ambient temperatures have attracted much interest over the past decade [1-4]. Such materials are known as "chemically bonded ceramics", since the chemical reactions of the inorganic cement produce an assembly of hydrated compounds which create a strongly bonded matrix. Properties can be modified by the incorporation of inert fillers with specific properties. In most cases the cement particles themselves are only partially reacted and act as "micro-fillers".

Macro-defect-free (MDF) cement is an interesting example of a chemically bonded ceramic that is not wholly inorganic but contains a polymer with an inorganic cement. First reported by Birchall and co-workers [5-8] at Imperial Chemical Industries in the early 1980's, MDF cement immediately attracted much attention due to its remarkable mechanical properties (see Table 1). Compressive strengths of 300 MPa were 4-5 times those of conventional cement pastes, while flexural strengths of 150 MPa were 10 times higher. Birchall *et al* [9,10] attributed these improvements to the elimination of large ("macro") defects (> 100 μm) normally found in cast products, due to improved packing. Hence the name MDF cement. The polymer was considered to be primarily a processing aid and not contributing to the properties of the material.

TABLE 1 - Comparison of the Properties of MDF Cement with a Cast Cement Paste and Sintered Ceramics

Property	Sintered Alumina	MDF Cement	Conventional Cement Paste
Compressive Strength (MPa)	2100	300	70
Flexural Strength (MPa)	350	150	20
Modulus of Elasticity (GPa)	40	45	35
Fracture Toughness ($MPa.m^{-1/2}$)	5	3	0.3
Density ($kg.m^{-3}$)	3700	2400	2700
Thermal Expansion ($10^{-6}.°C^{-1}$)	7	10	15

 ©1992 Materials Research Society

FORMULATION AND PROCESSING

Formulation

MDF cement is a mixture of a hydraulic inorganic cement, a water-soluble polymer and water. Birchall *et al.* [6] evaluated many different combinations and found that the best results were given by a combination of a calcium aluminate refractory cement and polyvinyl alcohol/acetate copolymer. Generally calcium silicate (Portland) cements did not provide satisfactory materials, for reasons which will be considered later. The formulation recommend by ICI for the optimum strength is given in Table 2 and has been confirmed in our studies on the roll milling process. [11]

TABLE 2 - Optimum Formulation for MDF Cement

Constitutent	Parts by Weight	Weight %	Volume %
Calcium Aluminate cement#	100	84.3	65.2
Polyvinyl alcohol/acetate*	7	5.9	12.3
Glycerin	0.7	0.6	1.4
Water	11	9.3	21.1

#Secar 71, Lafarge Fondu International; *80% hydrolyzed, Gohsenol KH-17s, Nippon Gohsei, Japan.

Processing

Figure 1 is a schematic representation of the processing sequence. The initial blending of the components is done under low shear in a planetary mixer to provide a damp crumb. The heart of the process is roll-milling under high shear using a conventional twin-roll mill used in polymer compounding. The shear rate is controlled by the relative speed of the rolls and the gap between them:

$$\gamma_{max} = \frac{6U_o}{h_o}\left[0.226 + \frac{(U_1-U_2)}{6U_o}\right] \qquad (1)$$

where U_1, U_2 are the speeds of two rolls, U_o is the average speed and h_o is the nip gap (between the rolls).

Roll-milling can be divided into three regimens, covering a total milling time of about 1 minute. The first is combining under very high rate of shear (1300 s^{-1}) where the damp crumb is transformed into a rubbery dough. During this regimen the PVA is dissolved and the components more fully blended. The second is compounding also under very high rate of shear (1000 cm^{-1}) in which the dough adhers to the fast roll and is drawn repeatedly between the rollers for about 30 s. At this time dense particle packing of the cement particles is achieved. Finally the material is calendered into a sheet at a lower shear rate (200 s^{-1}). The rate of shear during compounding can be varied over a wide range with good results, but a high rate of shear gives a slightly stronger material. Because a large amount of energy is dissipated in the process the rolls must be water-cooled to prevent a rise in temperature which will shorten dough life. Mixing can be done in other high shear mixers that are sufficiently robust to handle the dough [12,13]. Studies using a water-cooled, instrumented Haake-Buehler high shear mixer (100-200s^{-1})have shown [13] that "MDF characteristics" (i.e. the formation of a rubbery dough) can be achieved within two

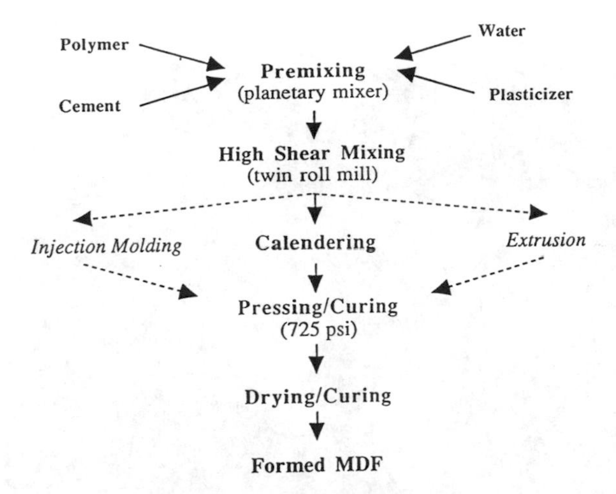

Figure 1. A flow Diagram of the common processing routes to a final MDF product [15].

minutes and dough-like consistency maintained up to about 20 minutes. The rubbery dough can be subsequently extruded[8,12] and processing in a twin screw extruder has been successfully carried out [14].

The pressure step is designed to remove any air from the material, and to heal any lamination defects introduced during the calendering process [6,8]. The heat helps the polymer to initially flow to heal defects and then stiffens the material by promoting further chemical reactions. Processing defects occur during calendering, resulting in surface tears or imperfect lamination. If lamination defects are not removed they will either show up as blisters after drying, or provide the initiation of fracture during testing and lead to low strengths.

The final step is drying in an oven at 80°C to remove residual water. The dry strength is dependent on the amount of evaporable water remaining. Recent work [15,16] has shown that the temperature of pressing and drying can influence the rate at which moisture is adsorbed by the finished product. The hardened sheet can be cut, drilled, tapped, machined, etc. without cracking.

Microstructure

Scanning electron microscopy has shown [7] that the cement particles are close-packed in a polymer matrix. However, roll-milling would be expected to introduce a high degree of anisotropy. Alignment of acicular wollastonite particles in an MDF cement sheet have been observed [17]. Recently it has been shown [18] that the strength of the sheet when tested perpendicular to the planes is about half that of the in-plane strength.

Studies using transmission electron microscopy[19,20] have revealed the presence of an interphase layer (Figure 2) around the cement grains, which is about 0.25 microns thick. High resolution TEM has revealed a nanostructure of very small crystals of hexagonal symmetry (most probably $Ca_2Al_2O_5 \cdot 8H_2O$) embedded in an amorphous matrix. Micronanalysis has revealed the presence of carbon in the interphase region. Special samples have been designed [21,22] to mimic the interphase region, in which a 10nm thick layer of PVA is laid down on a sintered pellet of $CaAl_2O_4$ (the active compound in calcium aluminate cement), by spin casting an aqueous solution of PVA onto the surface. XPS

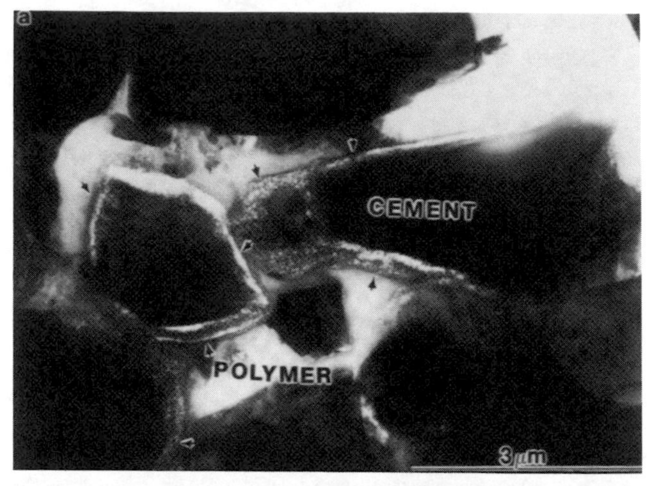

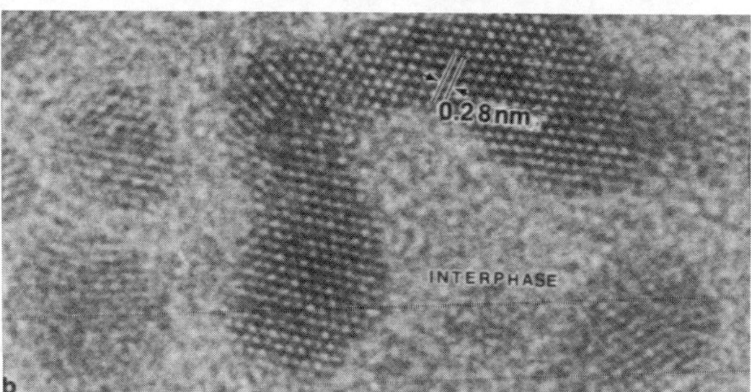

Figure 2. TEM micrographs of MDF cement (ion-thinned) (a) bright field image (arrow show interphase region); (b) high resolution image of interphase region. (After refs. 19,20).

analyses indicate that C-O-Al bonds are present, suggesting that either PVA chains are bonded to the crystals or they are cross-linked by aluminate ions.

Cross-linking Reactions

Cross-linking of PVA chains has long been considered to be an important aspect of the MDF cement composition. Our studies agree with this assessment. Microanalyses [19,20] have shown that aluminum is concentrated preferentially in the PVA matrix [Al:Ca ratio is 4:1 compared to 2.5:1 in the unreacted cement and 1:1 in the crystalline hydrates of the interphase region]. The glass transition temperature (Tg) of the PVA in MDF cement is about 120°C[23] compared to about 50°C for pure PVA films. Since the residual acetate groups on the polymer are hydrolyzed during processing [24], the increase in Tg could possibly be the result of increased crystallinity. However, high resolution TEM shows no evidence of crystalline PVA. Thus, the original suggestion [24] that the PVA chains are cross-linked by metal ions is an attractive hypothesis. The sequence of reactions proposed

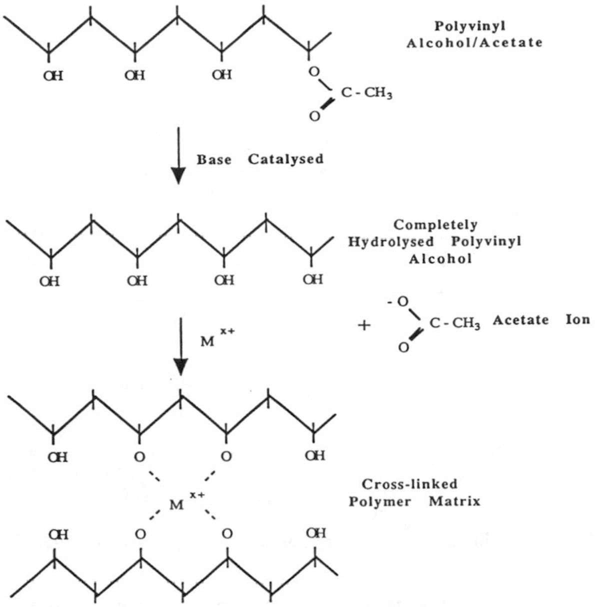

Polyvinyl Alcohol/Acetate

Base Catalysed

Completely Hydrolysed Polyvinyl Alcohol

+ Acetate Ion

M^{x+}

Cross-linked Polymer Matrix

Figure 3. Proposed cross-linking scheme for PVA in MDF cement (Adapted from ref. 24).

is shown schematically in Figure 3. When calcium aluminate cement is used, the principal ions present in solution are Ca^{2+}, $Al(OH)_4^-$ and OH^-, the pH being about 11. Thus M^{x+} in Figure 3 was assumed to be Al^{3+} involved in a covalent rather than an ionic cross-link, analagous to borax which is a well-known cross-linking agent for PVA films [25]. The chemistry is more complex, however, since there is no evidence for interaction between PVA solutions and the $Al(OH)_4^-$ ion [26]. Whereas additions of the borate ion rapidly gel PVA solutions above 3 g/dl in concentration, the equivalent addition of the aluminate ion does not. The current evidence could support an alternative hypothesis that there is an interaction between PVA and precipitated $Al(OH)_3$ [8]. The possibility exists that interactions develop in concentrated gel that form during MDF processing or during heating. It has been shown [11] that even small additions of the borate ion cause premature stiffening during roll-milling, which results in a lower strength material.

When calcium silicate (portland) cement is used, roll-milling is difficult because of rapid reactions that take place on the mill. Flexural strengths in excess of 100 MPa are difficult to achieve, because additional water must be added and lamination defects are common [27,28]. Indeed, MDF cements made with portland cement are generally hydrated subsequently [27,29] to heal "macro-defects" introduced during processing. The reason for this poor performance is due to the reaction of Ca_3SiO_5 [28], which is the principle ingredient of portland cement (see Table 3). When Ca_3SiO_5 is added to water the pH rises rapidly to over 12, representing a calcium hydroxide solution. Only Ca^{2+} and OH^- ions are present, the concentrations of soluble silicate or aluminate ions being negligible. Thus, it was proposed that at this high pH the PVA chains will ionize and form an ionic cross-link. Substituting PVA by polyacrylic acid (PAA) gave an immediate, strongly exotherms reaction and the mass could not be roll-milled. The system is similar to the rapid setting glass-ionomer dental converts developed by Wilson [30], which develop ionic crosslinking of polyelectroytes (see Figure 4).

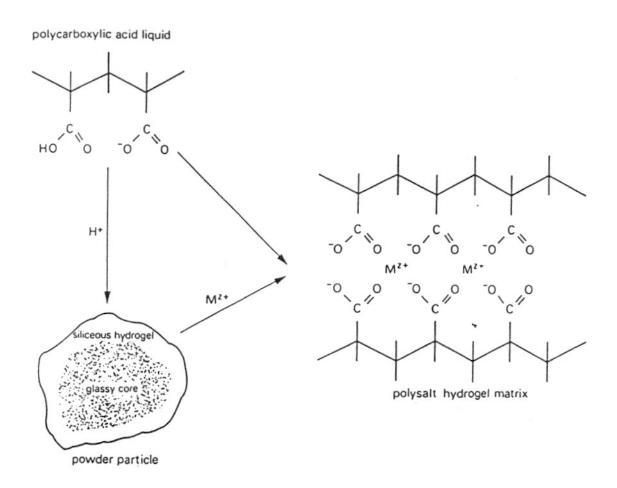

Figure 4. Cross-linking of polymer chains in galssionomer cements [30].

TABLE 3 - Correlation between cement-water interactions and processing characteristics of MDF cement

Cement	Polymer*	Solution Composition(30 min) pH	$[Ca^{2+}]$ (mM)	[sol.Al] (mM)	Processing Characteristics	Flexural Strength
Ca aluminate	PVA	10.69	12	28	Good	250
portland	PVA	12.1	18	negl.	Fair	100
Ca_3SiO_5	PVA	12.6	20	0	Fair	---
Ca_2SiO_4	PVA	11.5	5	0	Did not set	0
portland	PAA	---	---	---	Not possible+	---
C-A-S glass#	PAA	---	5	7	Not possible+	---

*PVA = poly(vinylalcohol); PAA = poly(acrylic acid)
#Calcium aluminosilicate glass
+Mixture highly exothermic shortly after mixing and sets rapidly

This hypothesis is confirmed by substituting Ca_2SiO_4 for Ca_3SiO_5; Ca_2SiO_4 is also a cementing ingredient of portland cement, but it is much less reactive. As a result the pH does not exceed 11.5, although Ca^{2+} and OH^- are still the principal soluble species (see Table 3). The PVA does not ionize appreciably at the lower pH, and cannot be involved in ionic cross-links. The Ca_2SiO_4-PVA combination cannot be properly roll-milled and calendered, because the mass remains soft and sticky, indicating a lack of cross-linking. Therefore, it is concluded that a controlled cross-linking action is need to develop the rubbery characteristics which are suitable for roll-milling. The aluminate ion may be particularly suitable due to the initial cross-linking reaction involving only weak hydrogen bonds, which do not cause too much stiffening. On heating stronger cross-links with covalent character form. The concept is shown schematically in Figure 5.

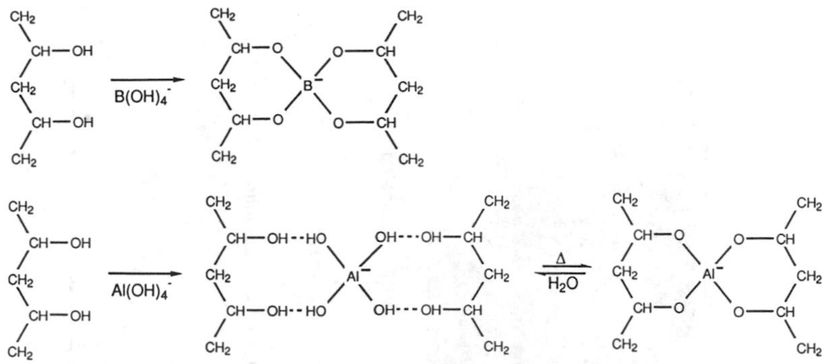

$$\begin{matrix} CH_2 \\ \backslash \\ CH{-}OH \\ / \\ CH_2 \\ \backslash \\ CH{-}OH \\ / \\ CH_2 \end{matrix} \xrightarrow{B(OH)_4^-} \begin{matrix} CH_2 & & & & CH_2 \\ CH{-}O & & & & O{-}CH \\ CH_2 & & B^- & & CH_2 \\ CH{-}O & & & & O{-}CH \\ CH_2 & & & & CH_2 \end{matrix}$$

$$\begin{matrix} CH_2 \\ \backslash \\ CH{-}OH \\ / \\ CH_2 \\ \backslash \\ CH{-}OH \\ / \\ CH_2 \end{matrix} \xrightarrow{Al(OH)_4^-} \begin{matrix} CH_2 & & & & CH_2 \\ CH{-}OH{\cdot\cdot}HO & & & & OH{\cdot\cdot\cdot}OH{-}CH \\ CH_2 & & Al^- & & CH_2 \\ CH{-}OH{\cdot\cdot}HO & & & & OH{\cdot\cdot\cdot}OH{-}CH \\ CH_2 & & & & CH_2 \end{matrix} \xrightarrow[H_2O]{\Delta} \begin{matrix} CH_2 & & & & CH_2 \\ CH{-}O & & & & O{-}CH \\ CH_2 & & Al^- & & CH_2 \\ CH{-}O & & & & O{-}CH \\ CH_2 & & & & CH_2 \end{matrix}$$

Figure 5. Schematic representation of cross-linking of PVA with borate and aluminate ions.

A further set of experiments indicates that the situation is more complex than indicated above. Several studies [7,15,31] have shown that cement can be replaced by significant quantities of an inert filler material without appreciable loss of strength (see Table 4). However, complete replacement of cement is not possible. This situation was explored in more detail[15] by progressively substituting calcium aluminate cement by calcined alumina particles with no significant change in particle size distribution. The results are summarized in Figure 6. Five weight percent of cement was required to calender a sheet having some strength, and 10-20 wt.% needed to get the expected MDF cement characteristics. Addition of only small quantities (5 mmoles/l) of sodium aluminate (adjusted to pH 11.0) provided better processing characteristics, but a good sheet could not be calendered even when the concentration was increased to 1500 mmoles/l). However, by adding aluminate ions to an MDF cement made with a small quantity (6.67 wt.%) of calcium aluminate cement give improved processing and better strength development (see Figure 7). Thus, it appears that the development of the interphase region, as well as the cross-linking of PVA molecules, is necessary for reliable processing and the development of high flexural strengths.

TABLE 4 - MDF Cement with Filler Additions

Filler	Wt.%	Flex. Strength (MPa)	Reference
Quartz	--	290	31
	30	275	31
	50	270	31
	60	260	31
Alumina	0	250	15
	33	230	15
	67	210	15
Iron	--	160	7
	10*	160	7
	30*	150	7

*Vol.%

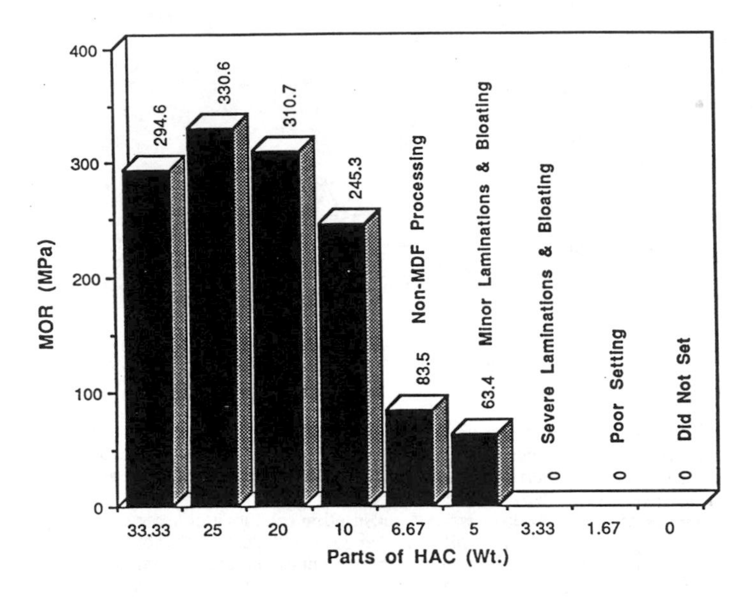

Figure 6. Effect of substitution of calcium aluminte cement (HAC) by calcined alumina on flexural strength (MOR) [15].

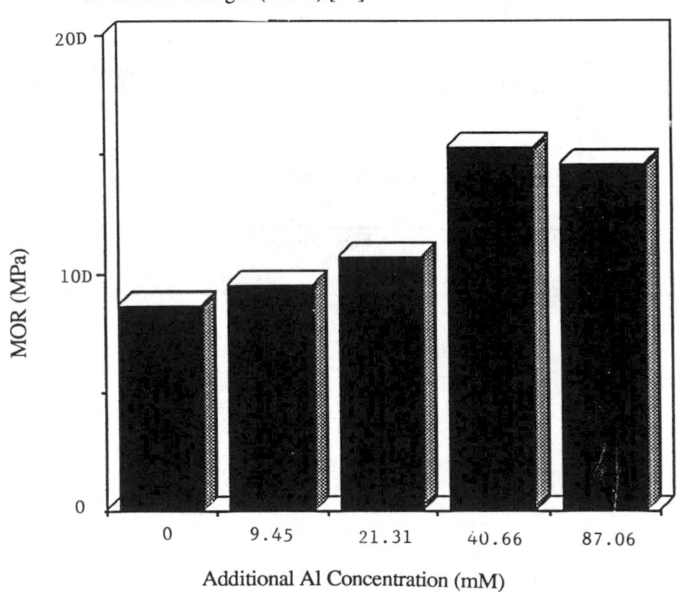

Figure 7. Effect of adding aluminate ions to the flexural strength (MOR) of MDF cement containing 6,.7 wt.% calcium aluminate cement and 93.3 wt.% calcined alumina [15].

Role of Polymer

The exact role of the polymer and interphase region has been explored further. Complete removal of the polymer can be achieved by heating at 500°C [32-34]. Progressive loss of polymer can be achieved by heating at lower temperatures, as shown in Table 5.

TABLE 5 Removal of PVA for MDF Cement

Maximum Temp. °C	Removal of Binder (% of original)	Flexural Strength (MPa)	Reference
25	0	320	33,34
125	13	165	34
200	35	35	34
300	55	35	34
400	70	35	34
500	100	18	33

It can be seen that there is a rapid loss of strength until about one third of the mass of the binder is removed; thereafter the strength remains more or less constant. (The lower strength on full removal of polymer is due to a more rapid rate of heating, which caused additional sample damage). Degradation of the interphase region will also occur on heating since $Ca_2Al_2O_5{\bullet}8H_2O$ and related compounds decompose below 200°C. The high strength loss initially may be due to both loss of interphase and polymer continuity in the binding matrix. It has been shown [21] that the interphase region plays a critical role in fracture. Fracture tends to run through the interphase and polymer ligaments are created to provide a toughening effect. It appears that MDF cement indeed behaves as an effective composite of inorganic and organic components effectively coupled together.

As polymer is removed at higher temperatures capillary porosity is created, as seen in Figure 8, which can be filled by subsequent hydration of residual cement (Figure 9).

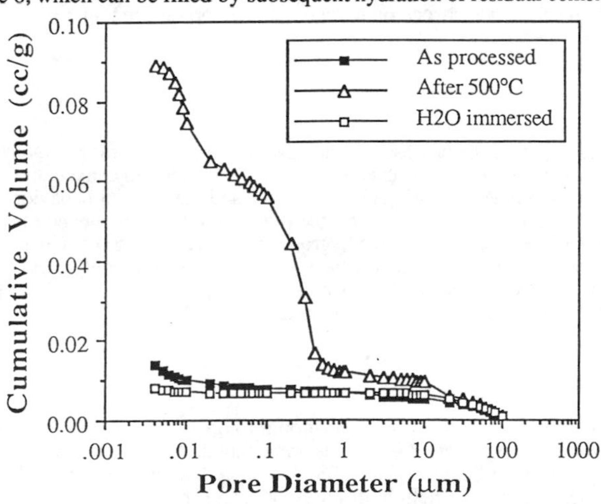

Figure 8. Development of capillary porosity as polymer is removed by burnout [33].

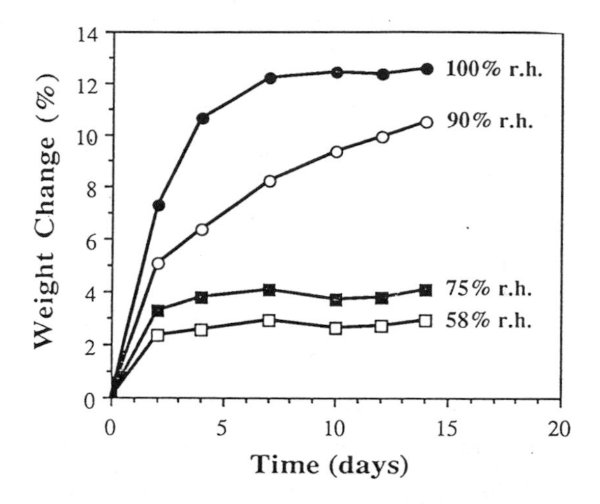

Figure 9. Rehydration of MDF cement after burnout [33].

Hydration is rapid only when >35 wt.% of the binder has been removed, which supports the idea of the polymer providing a percolation path in the matrix. The pore volume occupied by polymer can be completely filled by hydration products (Figure 8) giving a polymer-free MDF cement with a dry flexural strength of about 76 MPa [33]. Alternatively, the pore volume created on burnout can be impregnated with a hydrophobic polymer [35].

Conclusions

This paper has reviewed the processing strategies involved in forming MDF cements and creating their microstructure. The chemical interactions between cement and polymer are responsible for its unique processing characteristics and mechanical behavior. A major concern, not addressed in this paper, concerns the interaction of the hardened material with water. Adsorption of moisture through the hygroscopic polymer matrix can cause additional hydration of the cement particles and significantly degrade polymers. This problem has been addressed in several other publications [11,16,23,27,33,36,37]. Improvements in moisture resistance have been made.

Acknowledgements

The support of the National Science Foundation through the Center for Advanced Cement-Based Materials to prepare this paper is appreciated. Much of the work cited here was supported by the above Center and the U.S. Air Force Office of Scientific Research. We thank Professor J. Lewis for helpful discussions, and Professors J. Lewis and A. J. McHugh for permission to cite unpublished results.

References

1. J. F. Young, ed. Very High Strength Cement-Based Materials MRS Proc. Vol. 42, 317 pp (1985).
2. D. M. Roy, Science 235, 651-658 (1987).
3. M. Doyama, S. Samiya and R. P. H. Chang, eds, Advanced Cements and Chemically Bonded Ceramics, Proc. MRS Intern. Mtg. Adv. Mater. 13, 236 pp (1989).
4. B. E. Scheetz, A. G. Landers, I. Odler and H. Jennings, Specialty Cements with Advanced Properties, MRS Symp. Proc. 179, 298 pp (1991).
5. J. D. Birchall, A. J. Howard and K. Kendall, Nature 289, 388-390 (1981); Mater. Sci. Letters 1, 125-126 (1982).
6. J. D. Birchall et al., U.S. Patents, 4,353,746-8 (1982); 4,410,366 (1983).
7. N. McN.Alford and J. D. Birchall, MRS Symp. Proc. 42, 265-276 (1985).
8. S. D. Tan, A. J. Howard and J. D. Birchall, Phil. Trans. R. Soc. London A322, 479-491 (1987).
9. K. Kendall, A. J. Howard and J. D. Birchall, Phil. Trans. R. Soc. London A310, 139-153 (1983).
10. K. Kendall and J. D. Birchall, MRS Symp. Proc. 42, 143-150 (1985).
11. P. P. Russell, M.S. Thesis, Univ. Illinois, 1991.
12. M. R. Silsbee, D. M. Roy and J. H. Adair, MRS Symp. Proc. 179, 129-144 (1991).
13. A. J. McHugh and L. S. Tan, MRS Symp. Proc. 245, in press.
14. C. L. Sundius, personal communication.
15. R. A. Desai, M.S. Thesis, Univ. Illinois, 1990.
16. P. Lu, J. Kreider, M. Berg and J. F. Young, Abstract No. 38-T-92, Amer. Ceram. Soc. Ann. Mtg., Minneapolis MN (1992).
17. J. M. Bukowski, personal communication.
18. S. Holzgraefe and S. D. Brown, personal communication.
19. O. O. Popoola, W. M. Kriven and J. F. Young, J. Amer. Ceram. Soc. 74, 1928-1933 (1991).
20. O. O. Popoola, W. M. Kriven and J. F. Young, Ultramicroscopy 37, 318-325 (1991).
21. O. O. Popoola, W. M. Kriven and J. F. Young, MRS Symp. Proc. 245, in press.
22. O. O. Popoola and W. M. Kriven, J. Mater. Res., in press.
23. P. P. Russell, J. Shunkwiler, M. Berg and J. F. Young, Ceram. Trans. 16, 501-518 (1991).
24. S. A. Rodger, W. Sinclair, G. W. Groves, S. A. Brooks and D. D. Double, MRS Symp. Proc. 42, 45-51 (1985); J. Mater. Sci. 20, 2853-2860 (1985).
25. C. A. Finch in Polynvinyl Alcohol edited by C. A. Finch (John Wiley & Sons, London, 1973) Chap. 4.
26. P. G. Desai, J. F. Young and R. P. Wool, MRS Symp. Proc. 245, in press.
27. N. McN. Alford, G. W. Groves and D. D. Double, Cem. Concr. Res. 12, 349-358 (1982).
28. R. F. Falkner, M.S. Thesis, Univ. Illinois, 1989.
29. Y. Ohama, M. Endo, Y. Sato and H. Onishi, Proc. 29th Japan Congr. Mater. Res., 137-141 (1986).
30. A. D. Wilson, Cem. Res. Progr., 279-310 (1978).
31. P. P. Russell and M. Berg, ISHM '88 Proc., 141-151 (Intern. Soc. Hybrid Microelectronics, 1988).
32. C. S. Poon, L. E. Wassell and G. W. Groves, Trans. Brit. Ceram. Soc. 86 58-62 (1987); Mater. Sci. Technol., 993-996 (April 1988).
33. J. Shunkwiler, M.S. Thesis, Univ. Illinois, 1989.
34. M. A. Boyer and J. A. Lewis, Abstract No. 37-T-92 Amer. Ceram. Soc. Ann. Mtg., Minneapolis MN, 1992.
35. G. V. Chandrashekhar and M. W. Shafer, J. Mater. Sci. 24, 3353-3355 (1989).
36. C. M. Cannon and G. W. Groves, J. Mater. Sci. 21, 4009-4014 (1986).
37. N. B. Eden and J. E. Bailey, J. Mater. Sci. 19, 2677-2690 (1984); Mater. Sci. Technol., 301-303 (April 1988).

METAL COMPLEXES IN OXIDIC MATRICES

ULRICH SCHUBERT, BORIS BREITSCHEIDEL, HERMANN BUHLER, CHRISTIAN EGGER AND WLODZIMIERZ URBANIAK
Institut für Anorganische Chemie der Universität Würzburg, D-8700 Würzburg, Germany

ABSTRACT

Alkoxides of the type $(RO)_nE\text{-}X\text{-}A$ in which a functional organic group A is connected to the alkoxide moiety via an inert and hydrolytically stable spacer X, are used to bind metal complex moieties. Sol-gel processing of the resulting metal complexes $[(RO)_nE\text{-}X\text{-}A]_yML_m$ under standard conditions results in the formation of metal complex containing gels. If the metal complex moiety ML_m is catalytically active, this method results in the heterogenization of homogeneous catalysts. Another application of the metal complex substituted gels is the preparation of composites containing uniform, nanometer-sized metal particles homogeneously dispersed in a ceramic matrix.

INTRODUCTION

Hydrolyzable silanes having a functional organic substituent bonded to the silicon atom have found wide-spread industrial applications [1]. As shown in formula **1**, there are two reactive centers, A and $Si(OR)_3$, connected by a hydrolytically stable spacer X. Upon addition of water a polysiloxane network is formed, to which the functional organic groups A are bonded via X. The group A can in principle be any organic group, provided that it does not react with water. A $(CH_2)_n$ chain, mostly $(CH_2)_3$, serves as the spacer X in most organofunctional alkoxysilanes.

While Si-C bonds are not hydrolyzed during formation of the inorganic polymer, bonds between carbon and other elements, for instance Ti, Zr or Al, are cleaved. However, other groups X such as carboxylate, ß-diketonate or related groups, can be used to link $E(OR)_n$ and A. Technical applications of compounds of the type **1** with $E \neq Si$ are slowly emerging [2].

$$\underset{\mathbf{1}}{(RO)_nE\text{-}X\text{-}A} \qquad\qquad \underset{\mathbf{2}}{[(RO)_nE\text{-}X\text{-}A]_yML_m}$$

One of the many possible functions of A is that of binding metal ions or metal complex fragments ML_m (**2**). There is an extensive chemistry of $Si(OR)_3$-substituted complexes of the general formula $[(RO)_3Si(CH_2)_nA]_yML_m$ with respect to binding catalytically active metal complex moieties to solid supports [3]. There are four possible ways in which to anchor metal ions or metal complexes to oxidic matrices. As shown schematically in Fig.1 for SiO_2, the different bonds can be formed in a different order.

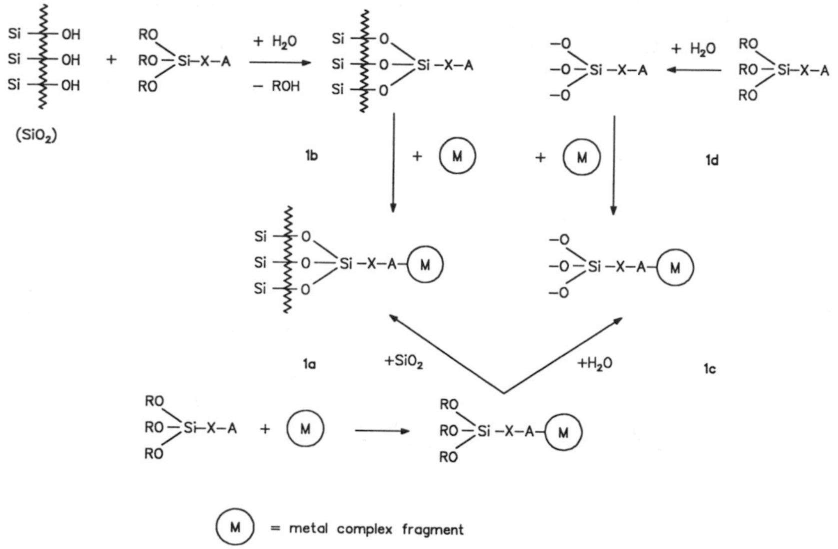

Figure 1. Possible methods (1a-d) in which to anchor metal ions or metal complexes to oxidic matrices.

In method 1a the pre-formed complex $(RO)_3Si\text{-}X\text{-}A\text{-}ML_m$, which is an example of type **2** complexes, is reacted with the surface OH groups of a pre-formed metal oxide support, while in method 1b the support is functionalized by $(RO)_3Si\text{-}X\text{-}A$ (a type **1** compound) and the bond between A and the metal is formed in a later stage. In the alternative approach 1c an oxidic matrix is formed around the metal complex moiety by sol-gel processing [4] using complexes $(RO)_3Si\text{-}X\text{-}A\text{-}ML_m$. It is also possible to process the functionalized alkoxides $(RO)_3Si\text{-}X\text{-}A$ and then to bind the metal complex moieties to the groups A after the gel has been formed (1d). In contrast to the well investigated methods 1a and 1b (Fig.1), very little is known about anchoring metal complexes to inorganic matrices by the sol-gel methods (1c, 1d). The obvious disadvantage of methods 1a and 1b is that the metal loading is directly related to the surface area of the support and the concentration of surface OH groups, which is rather low. Using the sol-gel approach the maximum metal loading depends only on the number of molecules **1** necessary to bind the metal. It cannot however exceed a 1:1 ratio of metal and SiO_2. Another important difference is that in the sol-gel derived materials the metal moieties are not only located at the surface of particles, but are homogeneously distributed through the whole material.

In this article we restrict ourselves to the discussion of method 1c. The drawback of the complementary method 1d is that after sol-gel processing of **1** some of the A groups may be hidden in the matrix and difficult for a metal complex moiety to reach [5]. The metal loading is therefore more difficult to control using this method.

PRECURSORS

We mainly used silicon-containing precursors of the type **1**, particularly $(RO)_3Si(CH_2)_nA$, in which A is NH_2, $NHCH_2CH_2NH_2$, CN, $2\text{-}C_5H_4N$, $CH(COMe)_2$ or PPh_2. Metal complexes of the type **2** can be prepared from these compounds by standard coordination chemistry reactions. Two examples are shown in Eq.1 [6,5]. Complexes **3a** and **3b** are $Si(OEt)_3$-substituted modifications of $Rh(PPh_3)_2(CO)Cl$ and $Rh(acac)(CO)_2$ (acac = acetylacetonate), both being homogeneous catalysts.

$$[RhCl(CO)_2]_2 \xrightarrow{+\ (EtO)_3Si(CH_2)_3PPh_2} (EtO)_3Si(CH_2)_3PPh_2-Rh(Cl)(CO)-PPh_2(CH_2)_3Si(OEt)_3 \quad \textbf{3a}$$

$$[RhCl(CO)_2]_2 \xrightarrow[+\ EtCO_2Na]{+\ (EtO)_3Si(CH_2)_3CH(COMe)_2} (EtO)_3Si(CH_2)_3-C(C(Me)O)_2Rh(CO)_2 \quad \textbf{3b} \qquad (1)$$

For some applications discussed below, isolation of the complexes **2** is either unnecessary or impractical. In these cases they are better prepared in situ from metal salts and $(RO)_3Si(CH_2)_nA$. For instance, when metal salts and $(RO)_3Si\text{-}(CH_2)_3NHCH_2CH_2NH_2$ are stirred in alcoholic solutions, the colour of the solutions and the u.v. spectra are nearly identical with those of the corresponding ethylene diamine complexes ($H_2NCH_2CH_2NH_2$, "en") [7].

Carboxylates or ß-diketonates substituted by functional groups A can be employed in binding metal ions or metal complex moieties to $E(OR)_n$ when $E \neq Si$. For instance, $Cr(CO)_3$ moieties were π-bonded to the phenyl groups of benzoylacetate-substituted alkoxides (where A is a phenyl group) of Ti, Zr or Al [8]. A more general method is the use of functionalized carboxylic acids. It is well known that on reaction of Al, Ti or Zr alkoxides with carboxylic acids some of the alkoxy groups are substituted by carboxylate ligands. The carboxylate-modified alkoxides can be used as precursors in the sol-gel process, during which the alkoxy groups and not the carboxylate ligand are primarily hydrolyzed [9]. The metal alkoxides can also be modified by readily available functionalized acids as shown in Eq.2 [10] (the degree of association of the compounds $(R'O)_3E(OOC\text{-}X\text{-}A)$ is not known in every case).

Phosphino-substituted compounds $(R'O)_nE[OOC(CH_2)_mPPh_2]$ (m = 1,2; E = Ti, Zr: n = 3; E = Al: n = 2) are obtained by reaction of $(R'O)_nE(OOC\text{-}CH{=}CH_2)$ with $HPPh_2$ or of $(R'O)_nE(OOC\text{-}CH_2Cl)$ with $LiPPh_2$, respectively.

The compounds $(R'O)_nE(OOC\text{-}X\text{-}A)$ thus obtained are of the general type **2**. They can be used to coordinate metal ions or metal complex fragments in the same way as the alkoxy silanes $(RO)_3Si(CH_2)_nA$ [10].

$$A\text{-}X\text{-}COOH + E(OR')_4 \longrightarrow R'OH + (A\text{-}X\text{-}COO)E(OR')_3 \qquad (2)$$

(E = Ti, Zr)

A–X–COOH = (structures: methacrylic acid; cinnamic acid (Ph, COOH); sorbic acid-type diene COOH; H_2NCH_2COOH; diaminopropionic acid (COOH, NH_2, NH_2); picolinic acid (N, COOH); 2,2'-bipyridine-3,3'-dicarboxylic acid (COOH, N, N, COOH))

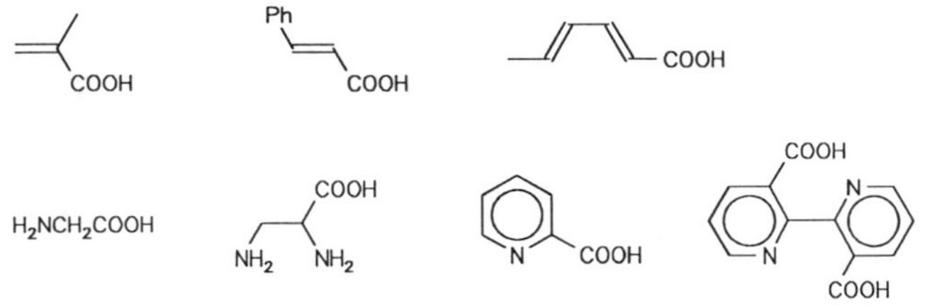

SOL-GEL PROCESSING OF METAL-COMPLEX SUBSTITUTED ALKOXIDES

The compounds **2**, $[(RO)_3Si\text{-}X\text{-}A]_yML_m$ or $[(R'O)_nE(OOC\text{-}X\text{-}A)]_yML_m$ ($E \neq Si$), are processed by the usual sol-gel procedure, as shown in Eq.3 for a metal-complex substituted alkoxysilane. The concentration of the metal ions or metal complex moieties in the gels **4** is controlled by addition of x equivalents of a network forming reagent $E(OR)_m$. The maximum metal concentration in the gels is obtained when x = 0.

$$\underset{\mathbf{2}}{[(RO)_3Si(CH_2)_nA]_yML_m} + x\ Si(OR)_4 \xrightarrow{H_2O/catalyst}$$

$$\longrightarrow \underset{\mathbf{4}}{[O_{3/2}Si(CH_2)_nA]_yML_m \cdot xSiO_2} \qquad (3)$$

The metal-containing entities are homogenously distributed in the resulting gels **4** due to complexation, and the metal complexes are anchored to the oxide backbone via the spacer (Fig.2). In principle, any metal complex moiety ML_m can be processed according to Eq.3, if both the ML_m moiety itself and its bond to the ligating group A are stable against hydrolysis. For example, the i.r. spectrum of $Rh(CO)Cl[PPh_2(CH_2)_2SiO_{3/2}]_2 \cdot xSiO_2$, obtained by acid-catalyzed sol-gel processing of $Rh(CO)Cl[PPh_2(CH_2)_2Si(OEt)_3]_2$ and $Si(OEt)_4$, is very similar to that of $Rh(CO)Cl(PPh_3)_2$. This shows that the complex does not decompose under the conditions of reaction [11]. When a mixture of $Ni(OAc)_2$ (OAc = acetate), three equivalents of $(RO)_3Si(CH_2)_3NHCH_2CH_2NH_2$ and x equivalents of $Si(OEt)_4$ is hydrolyzed under basic conditions, the u.v. spectrum of the resulting colored gel is nearly the same as that of $[Ni(en)_3]^{2+}$. This clearly shows that the concept of in situ formation of metal complexes from metal salts and suitable ligands **1** is valid.

For the remainder of this article, only alkoxysilyl substituted complexes will be discussed, since knowledge of their chemistry is more advanced. Preliminary experiments with E = Al, Ti or Zr indicate, however, that the compounds $(R'O)_nE(OOC\text{-}X\text{-}A\text{-}ML_m)$ can be processed similarly, and give similar results.

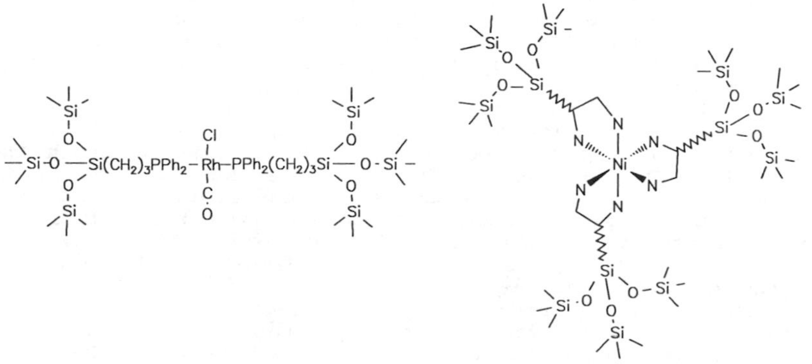

Figure 2. Schematic view of the metal complex moieties $Rh(CO)Cl[PPh_2(CH_2)_2SiO_{3/2}]_2$ or $\{Ni[H_2NCH_2CH_2NH(CH_2)_3SiO_{3/2}]_3\}^{2+}$ anchored to the SiO_2 matrix.

HETEROGENIZED CATALYSTS

The sol-gel process is a promising method for the preparation of heterogenized catalysts. Pioneering work, summarized in Ref.1, was carried out by the Degussa group. They used mainly phosphorus, nitrogen or sulfur containing groups A to coordinate noble metals. Typical heterogenized complexes from their work are $\{[O_{3/2}Si(CH_2)_3]_3N\}_3RhCl_3$, $\{[O_{3/2}Si(CH_2)_3]_2S\}_3RhCl_3$, $\{[O_{3/2}Si(CH_2)_3]_2PPh\}_3IrCl_3$ or $\{[O_{3/2}Si(CH_2)_3]_3NMe\}_2^{2+}[PtCl_4]^{2-}$, which are catalytically active in several technically important reactions.

The most important differences between catalysts heterogenized by methods a and b (Fig.1) and those heterogenized by the sol-gel method (c, d in Fig.1) are the much higher metal loading and the homogeneous distribution of the metal complex moieties throughout the whole material in the latter. The fact that the metal centers are not only located at the outer surface is of no disadvantage with respect to catalytic activity, because the porosity of the support can be tailored to a high degree.

We found that the catalytic activity of sol-gel catalysts is indeed not significantly different if the porosity of the matrix is high enough [12]. Heterogenization of $Rh(CO)Cl[PPh_2\text{-}(CH_2)_2Si(OEt)_3]_2$ (**3a**) by reaction with SiO_2 (method a in Fig.1) gives a compound of approximate composition $Rh(CO)Cl[PPh_2\text{-}$

$(CH_2)_2SiO_{3/2}]_2 \cdot 130\ SiO_2$ (**5a**) (metal only on the surface; metal loading 1.2% Rh; BET surface 485 [$m^3 \cdot g^{-1}$]; mean pore radius 2.8 nm). A heterogenized catalyst of composition $Rh(CO)Cl[PPh_2$-$(CH_2)_2SiO_{3/2}]_2 \cdot 50\ SiO_2$ (**5b**) (metal in the whole material; metal loading 2.7% Rh; BET surface 719 [$m^3 \cdot g^{-1}$]; mean pore radius 1.9 nm) is obtained by sol-gel processing of **3a** and 50 equivalents of $Si(OEt)_4$ (method c in Fig.1). Both heterogenized catalysts have the same activity in the hydrosilylation of 1-hexene with $HSiPh_3$ [12].

One of the problems with heterogenized complexes is leaching of the metal during catalysis. One of the approaches to tackle this problem is to provide multiple anchoring sites, as shown in the Degussa examples. The sol-gel catalyst **5b** turned out to be inherently more stable with respect to leaching than **5a**. In the catalytic hydrosilylation of 1-hexene, catalyst **5a** loses its activity on re-use, while **5b** does not. We attribute this difference to the fact that the metal complex moieties in **5b** are trapped in the SiO_2 matrix (although they are sufficiently accessible for the substrates). Even if the Rh-P bond is cleaved in the course of the reaction, the metal complex moiety cannot easily dissociate into the solution. A similar effect was found for Rh(I) complexes heterogenized via $(EtO)_3Si(CH_2)_2py$ ligands (py = 2-pyridine) by the different methods shown in Fig.1 [13].

The results show that the sol-gel route offers an attractive alternative to more established methods of heterogenizing catalysts. Only silica-based catalysts have been prepared to date. Bearing in mind the co-catalytic effect of certain other metal oxides, an extension of this work to include matrices other than SiO_2, using complexes of the type **2** with $E \neq Si$, may result in even better heterogenized catalysts.

METAL/SiO_2 NANOCOMPOSITES

Composites containing nanometer-sized metal particles of a controllable and uniform size in an insulating ceramic matrix are very interesting materials for use as heterogeneous catalysts and for magnetic and electronic applications. They show quantum size effects [14], particularly the size-induced metal-insulator transition (SIMIT) [15].

There are several approaches in which metal/SiO_2 or metal/Al_2O_3 composites may be prepared by sol-gel processing [16,17]. However, one problem connected with these methods is the control of the metal particle size and the metal distribution. The key to obtain composites with small metal oxide or metal particles is a high, ideally molecular dispersion of the metal compound while the inorganic network is formed. In order to achieve this, our approach is to employ metal complexes of the type **2**. The metal ions cannot aggregate during sol-gel processing because of complexation, and the metal complexes cannot leach because they are linked to the oxidic support.

The metal/SiO_2 composites are synthesized in three steps (Eq. 3 and 4) [7]:

(i) Sol-gel processing of a solution of the metal salt, $(RO)_3Si(CH_2)_3A$ and, optionally, $Si(OR)_4$ (Eq.3). The complexes of type **2** formed in situ are converted to metal complex containing gels **4**, as discussed before.

(ii) Oxidation of the polycondensates **4** at high temperatures, typically 400-550°C, in air to remove all organic components resulting in the composites $MO_m \cdot (x+y)SiO_2$ (**6**) (Eq.4).

(iii) Preparation of the metal/SiO_2 composites **7** (Eq.4) by reduction of the metal oxide particles. The temperature necessary depends on the metal (e.g. 20°C for Pd, 500-900°C for Ni). As discussed before, the metal loading of **6** and **7** is governed by the ratio of the starting compounds.

$$\underset{\mathbf{4}}{[O_{3/2}Si(CH_2)_nA]_yML_m \cdot xSiO_2} \xrightarrow{O_2} \underset{\mathbf{6}}{MO_m \cdot (x+y)SiO_2} \xrightarrow{H_2} \underset{\mathbf{7}}{M \cdot (x+y)SiO_2} \quad (4)$$

The anchoring group A and the spacer X of **1** are only necessary to ensure a high, ideally molecular dispersion of the metal during polycondensation. The chemical composition of these groups and that of the counter-anion or auxiliary ligands of the metal ion must therefore allow their complete removal during the second step. NH_2, $NHCH_2CH_2NH_2$, CN or $CH(COMe)_2$ proved to be suitable A groups, because they form a variety of stable complexes with most transition metals.

The minimum, maximum and mean particle sizes of representative examples of **7** [7] (determined by STEM) are listed in Table I, typical particle size distributions are shown in Fig.3.

Table I. Metal oxide and metal particle sizes

	composition[a)]	minimum and maximum particle diameters [nm]	mean particle diameter [nm]	metal loading (w%)[b)]
7a	$Ag \cdot 4\ SiO_2$	9.0 - 30.9	19.5	28.7
7b	$Co \cdot SiO_2$	11.0 - 24.9	17.4	47.5
7c	$Cu \cdot SiO_2$	1.5 - 7.4	3.9	46.2
7d	$Ni \cdot SiO_2$	37.5 - 62.4	50.3	43.7
7e[c)]	$Ni \cdot 2\ SiO_2$	7.5 - 12.4, 27.5 - 72.4		31.1
7f[c)]	$Ni \cdot 5.5\ SiO_2$	2.5 - 12.4, 32.5 - 57.4		14.0
7g	$Ni \cdot 13\ SiO_2$	12.5 - 47.4	22.9	7.0
7h	$Ni \cdot 33\ SiO_2$	2.5 - 12.4	5.9	2.7
7i	$Ni \cdot 15\ SiO_2$	0.5 - 5.0	2.7	6.1
7j	$Pd \cdot 2\ SiO_2$	1.8 - 4.2	3.0	44.8
7k	$Pd \cdot 32\ SiO_2$	1.3 - 3.7	2.4	5.5
7l	$Pt \cdot 32\ SiO_2$	0.8 - 4.2	2.5	7.9
7m	$Cu \cdot Ru \cdot SiO_2$	0.8 - 3.2	1.6	

a) **7b-h**, **7j-l**: prepared by using $(EtO)_3Si(CH_2)_3NHCH_2CH_2NH_2$; **7a**: prepared by using $(EtO)_3Si(CH_2)_3CN$; **7i**: prepared by using $(EtO)_3Si(CH_2)_3CH(COMe)_2$. b) Determined by elemental analysis. c) Bimodal particle distribution.

The metal particles in **7** are highly and homogeneously dispersed <u>throughout</u> the SiO_2 matrix. They are well separated from each other, and their diameters are very small and uniform, even in the materials with high metal loadings. The particle size distributions are all very narrow. The mean particle sizes depend very strongly on the identity of the metal, as expected. The dependency of the particle sizes on the metal loading was only investigated for Ni and Pd and proved to be different for both metals (Tab.I) [7].

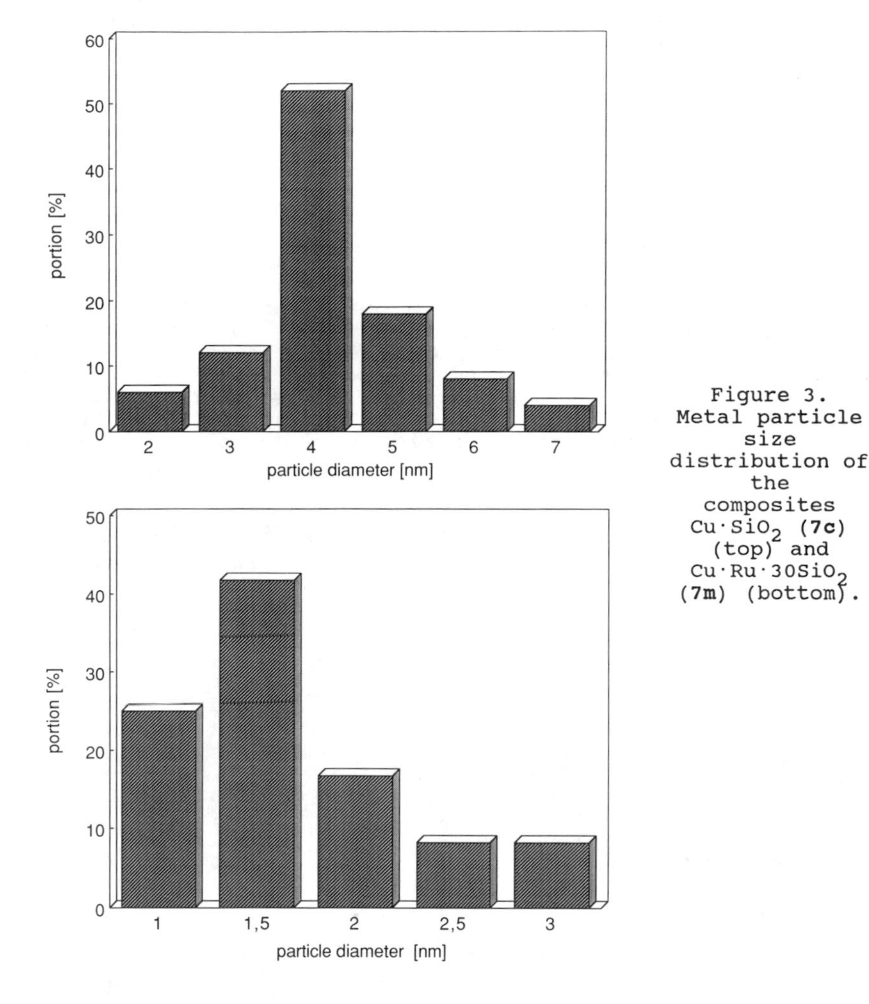

Figure 3. Metal particle size distribution of the composites $Cu\cdot SiO_2$ (**7c**) (top) and $Cu\cdot Ru\cdot 30SiO_2$ (**7m**) (bottom).

Control experiments showed that the use of the compounds **1**, i.e. complexation of the metal ions and fixation of the resulting metal complexes **2** to the SiO_2 matrix during polycondensation, is essential to obtain composites in which the metal particles have the properties described above. Although narrow particle size distributions have been reported for a few metal/SiO_2 nanocomposites prepared from metal salts and TEOS alone [17], these results cannot be generalized. Our method for preparing metal/SiO_2 nanocomposites gives narrow particle size distributions in every case investigated so far.Particle growth occurs mainly during the oxidation step (Eq.4). While isolated metal complexes are present in the polycondensates **4**, corresponding to an atomic dispersion of the metal, metal oxide particles are formed upon oxidation.

The oxidation temperature should be high enough to ensure the complete removal of all organic parts, but not higher than necessary in order to avoid an excessive growth of the metal particles. The optimal temperature for complete oxidation is routinely determined by thermogravimetric analyses. A typical example is shown in Fig.4.

Only water is lost upon heating the gel of composition $[O_{3/2}Si(CH_2)_3NHCH_2CH_2NH_2]_2Pd(acac)_2$ to 200°C. Oxidation of the organic moieties starts above 200°C and is complete at about 390°C. Therefore, the best temperature for the calcination of this gel, resulting in composite **7j** with less than 0.2% C, is about 400°C.

The optimized oxidation temperature are typically in the range of 400-550°C depending on the metal, and on the nature and the amount of the organic moieties used to bind the metal ions. The metals obviously act as oxidation catalysts, because complete oxidation occurs at lower temperatures than for pure, metal-free $O_{3/2}Si(CH_2)_3NHCH_2CH_2NH_2$. Cu and Ag are particularly effective in this respect.

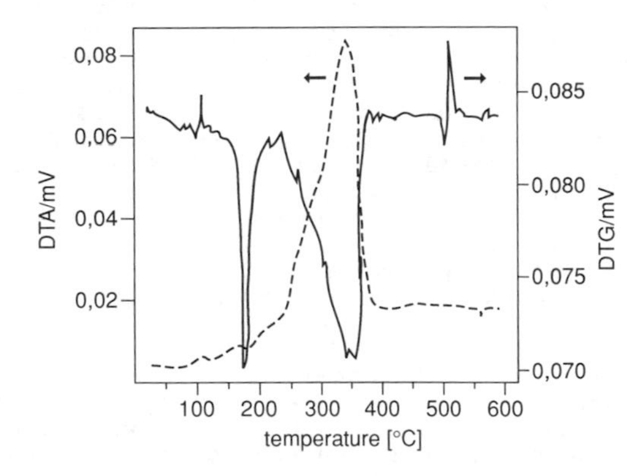

Figure 4. Differential thermal analysis (DTA) and differential thermal gravimetry (DTG) of a gel prepared from $Pd(acac)_2$ and two equivalents of $(EtO)_3Si(CH_2)_3NHCH_2CH_2NH_2$.

In order to obtain the smallest possible metal particles the conditions of the oxidation step have to be carefully optimized, since the final metal particle size is largely determined by this step. Apart from the oxidation temperature, the chemical nature of the groups A used to bind the metal ion is also very important. From $Ni(OAc)_2$, $(EtO)_3Si$-$(CH_2)_3NHCH_2CH_2NH_2$ and $Si(OEt)_4$ (molar ratio 1 : 3 : 10) the composite $Ni \cdot 13\ SiO_2$ (**7g**) with a mean nickel particle size of 22.9 nm is obtained (oxidation: 500°C / 30 min, reduction:

900°C / 2 h). When the composite Ni·15 SiO_2 is prepared by the same method and the same reaction conditions but using $(EtO)_3Si(CH_2)_3CH(COMe)_2$ [18] instead of the ethylene diamine derivative (molar ratio $Ni(OAc)_2$: $(EtO)_3Si(CH_2)_3CH(COMe)_2$: $Si(OEt)_4$ = 1 : 2 : 13. Oxidation: 480°C / 30 min, reduction: 500°C / 2 h), the average nickel particle size is only 2.6 nm! This result indicates that the nature of the organic oxidation products may play an important role in the particle growth mechanism. This aspect needs further investigation.

It is also possible to improve the reaction conditions so that smaller metal particles are formed in the composites **7** by performing the oxidation in a two-step procedure. A temperature of 500°C is necessary for 30 min to remove all organic parts from $[O_{3/2}Si(CH_2)_3NHCH_2CH_2NH_2]_2Ni(OAc)_2$ by air oxidation. In the resulting composite NiO·SiO_2 the NiO particles have an average size of 34.8 nm. Pre-treatment of $[O_{3/2}Si(CH_2)_3$-$NHCH_2CH_2NH_2]_2Ni(OAc)_2$ with 35% H_2O_2 at 100°C results in considerable, but not total removal of the organic components. The carbon content is lowered from 32.7% to 4.2%, and the nitrogen content from 8.5% to 1.7%. A temperature of 350°C for 15 min is then sufficient to remove the remaining organic parts by air oxidation. Due to the lower temperatures during oxidation, the average NiO particle size is reduced to 4.1 nm! H_2O_2 pre-treatment may not be very practical, but the experiment demonstrates the importance of optimizing the oxidation conditions.

Particle sizes do not change very much during reduction with H_2 of the metal oxide containing composites **6**, even if reduction is carried out at rather high temperatures. For instance, when NiO·SiO_2 (average NiO particle size 34.8 nm) is reduced at 500°C, the Ni particles in the resulting composite have an average diameter of 50.3 nm. From NiO·33 SiO_2 (1.4 nm) composite **7h** is obtained (reduction at 900°C), in which the average Ni particle size is 5.9 nm. Preparation of **7d** by reduction of Ni·SiO_2 at different temperatures (500°C and 900°C) shows that the average particle size increases only slightly (57 nm at 900°C, 50 nm at 500°C), but the main effect of the higher temperature is a broadening of the particle size distribution curve (33 - 102 nm at 900°C, 37.5 - 62.4 nm at 500°C).

The composites **7** are obtained as powders. The synthesis of **7d**, which had the most interesting properties in respect to SIMIT activity, in particular a sufficiently high metal loading (16.4 vol%) was successfully upscaled to the preparation of about 500 g composite powder per batch. It is shown elsewhere [19,20] that this composite can be densified to 98% of the theoretical density by hot pressing at 1200°C under a N_2 atmosphere at 30 MPa for 15 min.

The general procedure for the preparation of metal/SiO_2 composites can be extended to the preparation of bimetallic composites, as shown in Eq.5 for the composite Cu·Ru·30 SiO_2 (**7m**). The particle size is extremely small (1.6 nm) (Fig.3). Metals which cannot be alloyed in bulk may form bimetallic particles if the crystallite size is very small [21]. For instance, although ruthenium dissolves to less than 0.1% in copper [22], small bimetallic particles can be prepared. It was concluded from several experiments that copper atoms are preferentially located at the surface of the bimetallic particles [23]. EDX analysis of **7m** showed that the composite indeed contains bimetallic particles with an approximate Cu:Ru ratio of 1:1. It must be stressed that this method does not provide information on the metal distribution within the particles.

$$\begin{array}{c} Cu(OAc)_2 + Ru(acac)_2 + \\ 5\ (EtO)_3Si(CH_2)_3NHCH_2CH_2NH_2 + 25\ Si(OEt)_4 \end{array}$$

1) H_2O, NH_3
2) 400°C / air / 20 min
3) 400°C / H_2 / 1 h

$$Cu \cdot Ru \cdot 30\ SiO_2\ \mathbf{(7m)} \qquad (5)$$

Composites x Ni · y Pd · 15 SiO_2 (x + y = 1) with different Ni : Pd ratios were prepared by the same procedure from $Ni(OAc)_2$, $Pd(acac)_2$, $(EtO)_3Si(CH_2)_3NHCH_2CH_2NH_2$ and $Si(OEt)_4$ [24]. They exhibit bimodal particle size distributions centered around 3.5 nm and 20 nm. For Ni concentrations up to 70% there is a higher percentage of smaller particles. Within the accuracy of EDX, all particles have the same composition. The reason for the bimodal distribution is not yet known.

In summary, the sol-gel route using compounds of type **1** to coordinate metal ions provides metal/ceramic nanocomposites with smaller, more uniform and homogeneously distributed metal particles which have a narrower size distribution than other methods. Both monometallic and bimetallic particles can be prepared. The metal loading can be varied to a high degree. Even with high metal loading, the metal particles remain small, uniform in size and well separated from each other. Preliminary experiments show that this method can also be applied to prepare small and uniform metal particles in ceramic matrices other than SiO_2, when complexes of type **2** with E $\neq$ Si are similarly precessed.

CONCLUSIONS

Sol-gel processing of metal complexes in inorganic matrices offers fascinating new possibilities for the preparation of novel materials. Their advantages and their potential for the heterogenization of homogeneous catalysts and for the preparation of metal/ceramic composites has been stressed in this article. Unexploited possibilities lie in their spectral properties, which may have useful applications in optics, for example for optical coatings.

ACKNOWLEDGEMENTS

We thank the Bavarian Ministry of Economics, the German Ministry of Research and Technology, the Fonds der Chemischen Industrie and the Deutsche Forschungsgemeinschaft for their support of this work.

REFERENCES

1. U. Deschler, P. Kleinschmit and P. Panster, Angew.Chem. 98, 237 (1986); Angew.Chem.Int.Ed.Eng. 25, 236 (1986).
2. C. Sanchez and M. In, J.Non-Cryst.Solids in press.
3. F. R. Hartley, Supported Metal Complexes (D. Reidel Publ. Comp., Dordrecht, 1985); Yu.I. Yermakov, B.N. Kuznetsov and V.A. Zakharov, Catalyses by Supported Complexes (Elsevier Sci. Publ. Comp., Amsterdam, 1981).
4. L.L. Hench and J.K. West, Chem.Rev. 90, 33 (1990); C.J. Brinker, G. Scherer, Sol-Gel-Science, the Physics and Chemistry of Sol-Gel Processing (Academic Press, Boston, 1990).
5. M. Capka, M. Czakoova, W. Urbaniak and U. Schubert, J.Mol. Catal. in press.
6. K.G. Allum, R.D. Hancock, I.V. Howell, S. McKenzie, R.C. Pitkethly and P.J. Robinson, J.Organomet.Chem. 87, 203 (1975).
7. B. Breitscheidel, J. Zieder and U.Schubert, Chem.Mater. 3, 559 (1991).
8. U. Schubert, H. Buhler and B. Hirle, Chem.Ber. in press.
9. C. Sanchez, J. Livage, M. Henry and F.Babonneau, J.Non-Cryst. Solids 100, 65 (1988). C. Sanchez and J. Livage, New J.Chem. 14, 513 (1990).
10. H. Buhler, PhD thesis, Universität Würzburg, 1992. U. Schubert, E. Arpac, W. Glaubitt, A. Helmerich, C. Chau, Chem.Mater. 4, 291 (1992).
11. U. Schubert, K. Rose and H. Schmidt, J.Non-Cryst.Solids 105, 165 (1988).
12. U. Schubert, Ch. Egger, K. Rose and Ch. Alt, J.Mol.Catal. 55, 330 (1989).
13. M. Capka, U. Schubert, B. Heinrich and J. Hjortkjaer, J. Mol.Catal. in press.
14. A. Henglein, Top.Curr.Chem. 143, 113 (1988).
15. P. Marquardt, L. Börngen, G. Nimtz, H. Gleiter, R. Sonnberger and J. Zhu, Phys.Lett. 114A, 39 (1986). P. Marquardt, G. Nimtz and B. Mühlschlegel, Solid State Comm. 65, 539 (1988).
16. G. C. Bond, Surface Sci. 156, 966 (1985).
17. R. A. Roy and R. Roy, Mat.Res.Bull. 19, 169 (1984). R. Roy, S. Komarneni and D. M. Roy, Mat.Res.Soc.Symp.Proc. 32, 347 (1984). G. N. Subbanna and C. N. R. Rao, Mat.Res.Bull. 21, 1465 (1986). F. Orgaz and H. Rawson, J.Non-Cryst.Solids 82, 378 (1986).
18. W. Urbaniak and U. Schubert, Liebigs Ann.Chem. 1991, 1221.
19. U. Schubert, B. Breitscheidel, H. Buhler, J. Petrullat, S. Ray in Proceedings of the 6th European Electromagnetic Structures Conference (Dornier, Friedrichshafen, 1991), p.167.
20. J. Petrullat, S. Ray, U. Schubert, G. Guldner, Ch. Egger and B. Breitscheidel, J.Non-Cryst.Solids in press.
21. D. F. Ollis, J.Catal. 23, 131 (1971).
22. F. Pawlek and K. Reichel, Z.Metallkunde 47, 347 (1956).
23. X. Wu, B. C. Gerstein and T. S. King, J.Catal. 123, 43 (1990).
24. B. Breitscheidel, PhD thesis, Universität Würzburg, 1991.

FORMATION OF HIGHLY DISPERSED METALS ON FUNCTIONALIZED SILICA SUPPORTS

FRANCOIS ROUSSEAU†, Z. DUAN†, M. J. HAMPDEN-SMITH† AND A. DATYE††
†Department of Chemistry, University of New Mexico, Albuquerque NM 87131
††Department of Chemical Engineering, University of New Mexico, Albuquerque NM 87131
Center for Micro-Engineered Ceramics, University of New Mexico, Albuquerque NM 87131

ABSTRACT

The hydrolysis of $H_2N(CH_2)_2NH(CH_2)_3Si(OMe)_3$ with a stoichiometric amount of water results in loss of the -OMe groups after two days and formation of a mixture of species mainly with empirical formula, $[H_2N(CH_2)_2NH(CH_2)_3Si(O)_{1.5}]$. The reaction of representative examples of metal salts, e.g. $Cu(NO_3)_2$, and metal-organic compounds, $[(NBD)RhCl]_2$, with the amino-group of the amine functionalized polysiloxanes $[(H_2N(CH_2)_2NH(CH_2)_3Si(O)_{1.5}]$ has been demonstrated. These species undergo thermal decomposition in air, typically at 200°C, with complete loss of the organic supporting ligands. On heating to 450°C in O_2, sintering occurs to form large, ~1μm, sized silica particles which contain highly dispersed, ~4nm sized metal-containing particles. The surface areas range from 228 to 583 m^2/g. After reduction, the rhodium species is active towards catalytic hydrogenation of pyrene.

INTRODUCTION

There is currently great interest in the preparation of highly dispersed metals on high surface area supports for catalytic applications. The influence of particle size and support interactions on the selectivity and reactivity of catalytic reactions is well known.[1] A variety of methods for the dispersion of metal particles on metal oxide surfaces have been explored, however, few methods are designed to control the dispersion.[2] The use of metal-organic molecules as reagents for site-specific surface reactions is a method in which surface chemical reactions can be used to control the dispersion of the metal-organic complex at the molecular level. With the appropriate complex, decomposition to the final metal under mild conditions may result in retention of the metal as a highly dispersed phase on the metal oxide surface. The reactivity of heterobimetallic metal alkoxide complexes with the surface hydroxyl groups of spherical silica particles was shown to be an effective method for the formation of highly dispersed Rhodium-tin alloy.[3,4] The high dispersion (average particle size 24Å) was retained even on heating to 800°C. As an alternative approach, it has been demonstrated that silicon alkoxide compounds, functionalized with donor substituents, can be reacted with metal complexes and subsequently hydrolyzed and condensed to produce highly dispersed metal particles in a silica matrix according to equation 1.[5]

$$Pd(acac)_2 + 3RHN(CH_2)_3Si(OEt)_3 + xSi(OEt)_4 \xrightarrow{NH_3/H2O} Pd(acac)_2\cdot$$

$$3RHN(CH_2)_3Si(O)_{3/2}\cdot xSiO_2 \xrightarrow{O_2/500^oC} PdO\cdot(3+x)SiO_2 \xrightarrow{H_2/30^oC} Pd\cdot(3+x)SiO_2 \qquad (1)$$

Such highly dispersed conducting materials in a ceramic matrix are interesting because they may show quantum size effects[6] in addition to their catalytic activity. Here we report the results of some preliminary experiments aimed at the formation of highly dispersed metals on silica supports via the reaction of amine derivatized silica gels with low valent metal complexes.

RESULTS AND DISCUSSION

The hydrolysis and condensation of alkyl and aryl silicon trialkoxides has previously been shown to result in the formation of silsequioxanes under optimum conditions.[7] In this work, we chose a silicon monomer which contains an amine functionalized substituent which will subsequently react with a metal-organic complex as illustrated in Scheme 1.

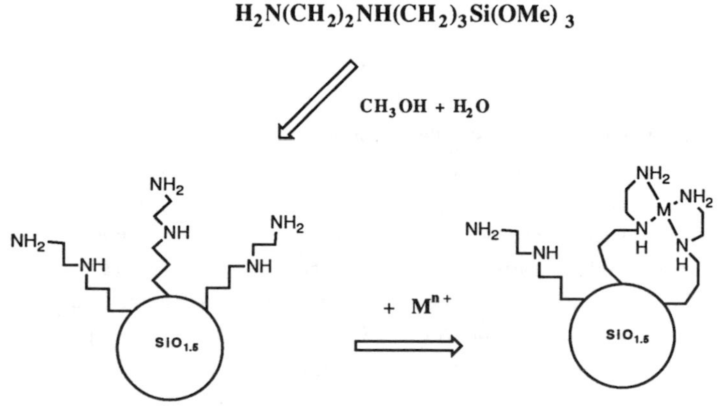

$$H_2N(CH_2)_2NH(CH_2)_3Si(OMe)_3$$

$CH_3OH + H_2O$

$+ M^{n+}$

Scheme 1

The starting material, $(H_2N(CH_2)_2NH(CH_2)_3Si(OMe)_3$ was hydrolyzed with 1.5 equivalents of water in methanol at neutral pH and stirred until the -OMe substituents were completely displaced, as determined by ^1H NMR spectroscopy (~2 days). At this stage, the product possessed approximately seven distinct ^{29}Si NMR chemical shifts. The resonance at -58.5ppm could be assigned to a species of the type $RSi(\text{-}O\text{-})_2(OH)$, while the remaining resonances occurred in a chemical shift region -64.9 to -68.7ppm that is characteristic of the species $RSi(\text{-}O\text{-})_3$. Combustion elemental analysis was consistent with these observations. The sols formed in this way were then reacted with a variety of metal salts. Two illustrative examples, $Cu(NO_3)_2$ and $[(NBD)RhCl]_2$ (where NBD = norbornadiene), will be described here. The metal salts were added to the condensed alkylaminosilica gels in a metal to silicon ratio of ~5% metal by weight. Addition of $Cu(NO_3)_2$ resulted in formation of a deep blue solution consistent with the introduction of amine ligands into the copper(II) coordination environment. The volatile components were then removed *in vacuo* to form a glass. The glass was placed in a pyrex crucible and heated under O_2 to 400°C. For $Cu(NO_3)_2$, with a Cu:Si mol ration of 1:2, TGA data obtained in air showed a sharp weight loss at 200°C which was consistent with the formation of $CuO{\cdot}2SiO_2$, as shown in Figure 1. DTA data showed a large exotherm at this temperature (400μvolts), consistent with the presence of an exothermic reaction.

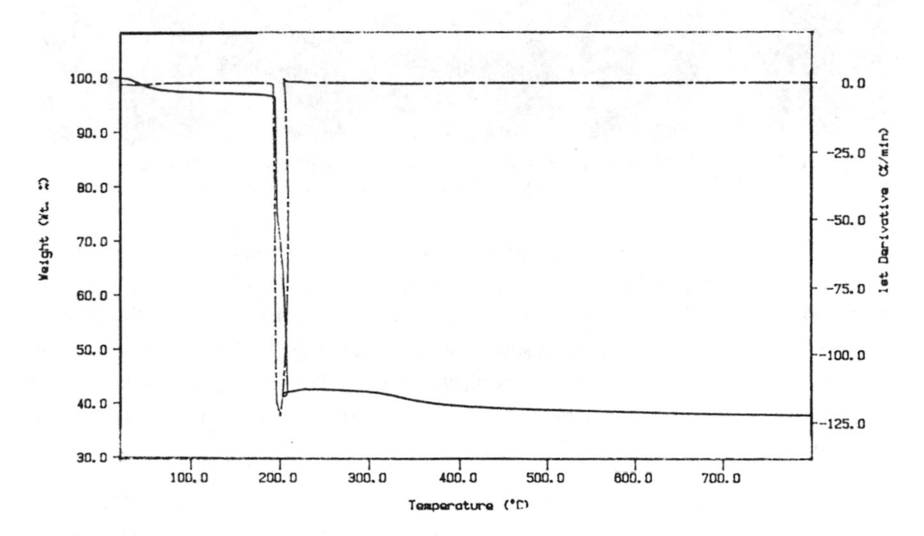

Figure 1: TGA data (continuous line; derivative is the dashed line) for the powder isolated from the reaction between $Cu(NO_3)_2$ and $2(H_2N(CH_2)_2NH(CH_2)_3Si(O)_{1.5}$.

TEM data for methanol solutions of $(H_2N(CH_2)_2NH(CH_2)_3Si(O)_{1.5}$ dried on TEM grids did not show evidence for small particles, which might be expected if silsequioxane and other closo-products were formed, but instead showed that a more fibrous material was present. Close examination revealed a layered structure with an interlayer spacing of ~3.2Å, which was retained after reaction with $[(NBD)RhCl]_2$ as shown in Figure 2. Energy dispersive spectroscopy showed that Rh was present, although no features associated with Rh could be identified at this stage. These observations are consistent with the molecular scale dispersion of the Rh precursor in this material.

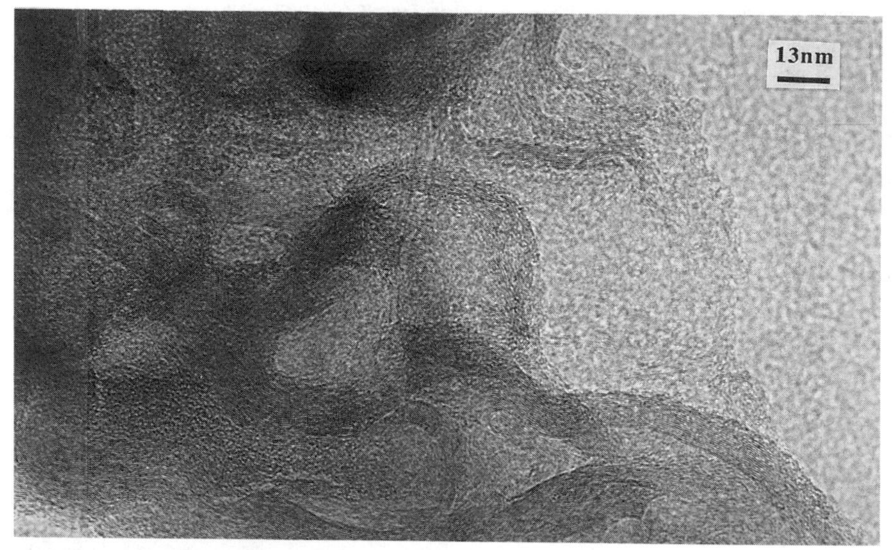

Figure 2: TEM data for the product of the reaction of $[(NBD)RhCl]_2$ with $H_2N(CH_2)_2NH(CH_2)_3Si(O)_{1.5}$ after evaporation of the solvent in air.

Thermolysis of the $[(NBD)RhCl]_2/H_2N(CH_2)_2NH(CH_2)_3Si(O)_{1.5}$ derived material at 400°C in O_2 resulted in a vigorous reaction to produce a black powder. TEM and energy dispersive spectroscopy revealed that large (~1μm) silica particles had been formed which contained a small number of ~100nm sized Rh-containing particles and a homogeneous distribution of much smaller, ~4nm Rh-containing particles throughout the silica matrix, as shown in Figure 3. Whether these Rh-containing particles are rhodium metal or rhodium oxide was not unambiguously established at this stage.

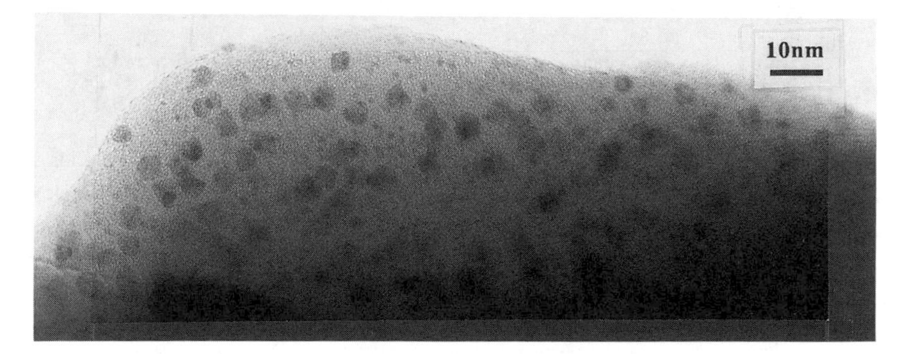

Figure 3: TEM data showing the dispersion of ~ 4nm Rh-containing particles in the silica matrix, derived from the isolated $[(NBD)RhCl]_2/H_2N(CH_2)_2NH(CH_2)_3Si(O)_{1.5}$ after heating in O_2 at 400°C.

BET surface area measurements (N_2 adsorption) revealed that, after thermolysis at 450°C in O_2, the $Cu(NO_3)_2$ derived material, 5% metal by weight, exhibited a surface area of $228m^2/g$ (micropore area 192 m^2/g) and the $[(NBD)RhCl]_2$ derived material, 5% metal by weight, exhibited a surface area of 583 m^2/g (micropore area 445 m^2/g). After reduction of the $[(NBD)RhCl]_2$ derived material in H_2, H_2 adsorption experiments showed a metal surface area of ~100 m^2/g with 22% dispersion of the metal consistent with the 4nm particle size observed by TEM. In pyrene hydrogenation experiments conducted at 100psi H_2, 100°C, this material showed a catalytic activity of 0.42mg conversion/g catalyst/second. This reactivity is similar to the reactivity of Pd dispersed on hydrous titanate supports.

CONCLUSIONS

The reaction of metal salts and metal-organic compounds with amine functionalized polysiloxanes followed by pyrolysis at 450°C in air or oxygen results in formation of highly dispersed metal oxide particles in silica matrices. Reduction under mild conditions results in formation of highly dispersed metal particles. Further experiments are in progress to control the hydrolysis and condensation of $H_2N(CH_2)_2NH(CH_2)_3Si(OMe)_3$ to prepare discrete cage species and determine the reactivity of metal salts with these complexes

ACKNOWLEDGMENTS

M.H.-S. acknowledges funding from Sandia National Laboratories contract # 78-5509, and thanks Johnson-Matthey for the loan of $RhCl_3$ through the Johnson-Matthey Precious Metals Loan Program, We thank the NSF Chemical Instrumentation program for the purchase of a high field NMR spectrometer and the ONR for analytical facilities. Thanks to Dr. Allen Sault for H_2 adsorption experiments, Mr. J. Kowala for Pyrene hydrogenation experiments and Mr. Bill Ackerman for surface area measurements.

REFERENCES

1. M. Che and C.O. Bennett in Advances in Catalysis 36, edited by D.D. Eley, H. Pines and P.B. Weisz (Academic Press, Inc., New York, 1989) p. 55.
2. I.M. Campbell, Catalysis at Surfaces (Chapman Hall, London, 1988).
3. T.A. Wark, E.A. Gulliver, M.J. Hampden-Smith and A.L. Rheingold, Inorg. Chem. 29, 4360 (1990).

4. S.L. Anderson, A.K. Datye, T.A. Wark and M.J. Hampden-Smith, Catalysis Letters 8, 345 (1991).
5. See e.g. B. Breitscheidel, J. Zieder and U. Schubert, Chem. Mater. 3, 559 (1991) and references therein.
6. A. Henglein, Top. Curr. Chem. 156, 113 (1988).
7. F.J. Feher, T.A. Budzichowski, R.L. Blanski, K. Weller and J. Ziller, Organometallics 10, 2562 (1991).

ENERGY TRANSFER BETWEEN Eu^{2+}, Eu^{3+} AND Rh6G IN SILICA, ZIRCONIA AND ALUMINA GELS

W. NIE(1)*, B. DUNN*, C. SANCHEZ** AND P. GRIESMAR**
*UCLA, Department of Materials Science and Engineering, L.A., CA 90024-1595
**Laboratoire de Chimie de la Matiére Condensée, URA 302, Université Pierre et Marie Curie, 4, place Jussieu, 75252 Paris, France

ABSTRACT

The sol-gel process provides new opportunities for the synthesis of photonic materials. Sol-gel chemistry is mainly based on inorganic polymerization reactions performed in solution at lower temperatures than conventional chemical methods. Homogeneous doping by mixing components at a molecular level and synthesis of mixed organic-inorganic materials can then be performed. Luminescent properties of dyes inside silica, aluminosilicates or transition metal oxide based gels have been studied extensively. Much less work has been devoted to the study of energy transfer (ET) inside sol-gel matrices. Sol-gel materials provide the opportunity to investigate dye-dye ET, as well as dye-ion ET. This communication presents an optical study of a variety of sol-gel matrices (SiO_2, Al_2O_3, ZrO_2 based gels) codoped with rhodamine 6G(Rh6G) and europium. Different types of ET are observed among Rh6G, Eu^{3+} and Eu^{2+}. The Eu^{2+}/Eu^{3+} ratio is found to be strongly dependent on the host matrices. The reduction of Eu^{3+} to Eu^{2+} in dried gels may relate to the presence of Rh6G.

INTRODUCTION

The sol-gel method offers an opportunity to incorporate active organic species into inorganic matrices by its low processing temperature [1,2]. Some organic laser dyes are found to exhibit photochemically stable behavior in sol-gel matrices [3,4]. Laser dye-doped sol-gel materials are being actively studied regarding their potential application as tunable solid state laser sources. Lasing effects have been successfully observed in several dye-doped gels [5,6,7].

Codoping of organic dyes and ionic fluorescent activators in an inorganic sol-gel matrix is possible through the sol-gel method. Ion (transition metal and rare-earth) - and defect-activated inorganic materials comprise the two main kinds of solid state lasers. In many cases, low efficiency is the most important obstacle to achieving commercially successful materials due to weak coupling between the pumping source and the absorption band of activators. Energy transfer (ET) between transition metal ions (such as Cr^{3+}) and rare earth laser ions (such as Nd^{3+}) have been studied intensively to improve the lasing efficiency [8,9,10].

Organic dyes generally have wide absorption and fluorescence bands and high energy conversion efficiencies. It is very appealing to use organic dyes in ET applications (either dye$\rightarrow$ dye or dye$\rightarrow$ ion) because of the overlap(a) between the pump source and the absorption band of the sensitizing dyes and (b) between the fluorescence band of the sensitizing dye and the absorption band of the lasing center.

In recent work, ET between the laser dye coumarin 460 and Eu^{3+} or Tb^{3+} in a thorium phosphate xerogel was demonstrated [11a, 11b]. The absorption band of coumarin 460 was observed in the excitation spectrum of Eu^{3+}.

(1) Current address: Laboratoire Science des Matériaux, Kodak-Pathé, 71102 Chalon sur Saone, France.

The present communication addresses some examples of dye-ion ET in other sol-gel matrices. The prospects for obtaining highly efficient new solid state laser materials in this system are excellent.

EXPERIMENTAL

The general approach for preparing SiO_2, ZrO_2 and Al_2O_3-based gels has been described proviously [1,2]. Specific formulations are described below.

i) Preparation of gels

a. SiO_2 based gels: Dimethylaminopyridine (DMAP) is added as a catalyst in the ratio DMAP/Si = 0.003 to a methanolic solution of tetramethoxysilane (TMOS). Europium (III) nitrate and Rh6G are then dissolved in the TMOS solution. Hydrolysis is then performed by adding pure water. The ratio TMOS/MeOH/H_2O is 1/6/7. Transparent gels are obtained within 15 minutes.

b. ZrO_2 based gels: $Zr(OPr^n)_4$ (Fluka) is first diluted in n-propanol (1/9 by mole). The alkoxide is then complexed with acetyl-acetonate (AcacH). The complexation ratio AcacH/Zr is equal to 0.4. Eu(III) nitrate and Rh6G are then dissolved in the complexed $Zr(OPr^n)_4$ solution. The hydrolysis is performed by adding to the acac modified alkoxide, a solution of water diluted in n-propanol (10% by volume). Transparent gels are obtained within 30 minutes.

c. Al_2O_3 based gels: $Al(OBu^s)_2$ (Etac) (Aldrich) is mixed with acetic acid in a ratio of 1/0.3. Eu(III) nitrate and Rh6G are then dissolved in the solution. The hydrolysis is performed by adding a water solution diluted in n-butanol (10% by volume). The H_2O/Al ratio is 5/1. Transparent gels are obtained within 20 minutes.

All gels are codoped with Eu and Rh6G. The concentrations of Rh6G and Eu listed in table I are given by their relative molar ratio to the corresponding metal ion inside the host matrices.

Table I Relative molar concentrations of dopants in the studied xerogels

Name	Si76	Si77	Al7	Al9	Zr92	Zr93	Zr94	Ti66	Ti67
Eu	1%	5%	1%	5%	1%	1%	5%	1%	5%
Rh6G	10^{-5}	10^{-5}	10^{-5}	10^{-5}	10^{-3}	10^{-5}	10^{-5}	10^{-5}	10^{-5}

ii) Optical characterization

Fluorescence and excitation spectra were recorded in front face configuration on a SPEX Fluorolog (model F112A) at room temperature. Absorption spectra were obtained on a Shimadzu double beam spectrophotometer (model No. 260).

RESULTS

The experimental results are classified by the specific sol-gel matrix.

Silica

Two gels labelled Si76 and Si77 with different Eu concentrations are shown below. Fig. 1 indicates the absorption spectra. Sharp lines originate from Eu^{3+}. Apart from the UV band below 325 nm, the other bands are related to the presence of Rh6G. The broad one can be divided into three subbands, two of which may be associated with Rh6G dimers. One of the two dimer bands overlaps with the intense monomer band resulting in the main band peaked at 525 nm. The other one appears as a shoulder at 490 nm. The spectra of Rh6G dimer and monomer species have been well explained in the literature [12]. The presence of dimers may be traced to the hydrophilic nature of the pore surface onto which water molecules are adsorbed leading to a water environment to Rh6G molecules that favors dimer formation.

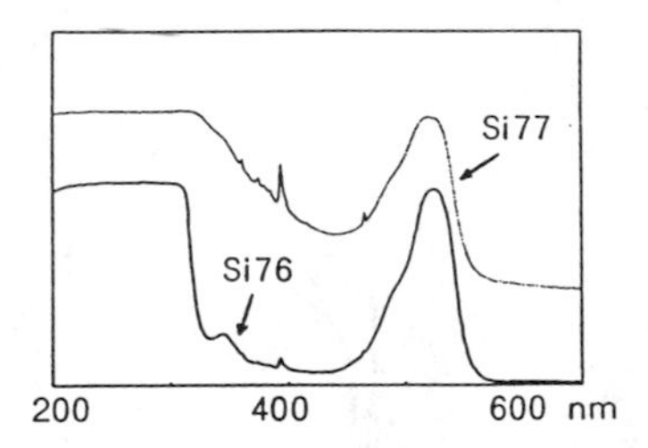

Fig. 1: Absorption spectra of Si76 and Si77

Fig. 2a shows the fluorescence spectra of the two silicate samples when the excitation occurs in a Eu^{3+} absorption line. It is significant to note that fluorescence from both Eu^{3+} and Rh6G is observed and that the Eu^{3+} lines exhibit considerable concentration dependence. Fig. 2b shows fluorescence spectra of Si77 under selective excitations. In this case the Eu^{3+} lines disappear when the sample is excited at the Rh6G absorption band at 345 nm.

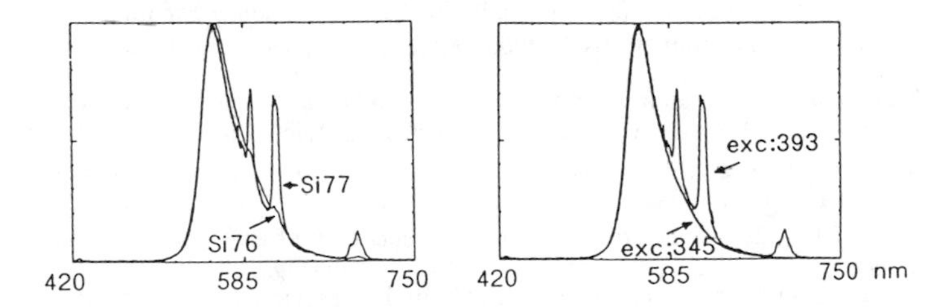

Fig. 2a: Fluorescence spectra of Si76 and Si77 under excitation at 393nm;
b: Fluorescence spectra of Si77 under excitation at 393 and 345 nm.

The optical spectroscopy results also indicate the presence of Eu^{2+} in silica-based gels. The fluorescence band originating from Eu^{2+} is relatively weak in comparison with that of Eu^{3+} causing its disappearance in Fig. 2. The presence of the Eu^{3+} absorption line at 393 nm in the Eu^{2+} luminescence (Fig. 3a) indicates the occurrence of $Eu^{2+} \rightarrow Eu^{3+}$ radiative ET.

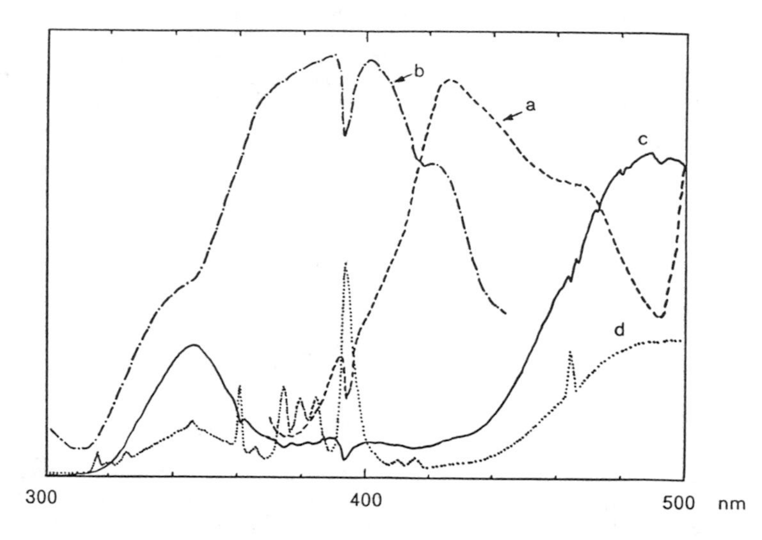

Fig. 3 a: Eu^{2+} emission band of Si76 excited at 350 nm;
b: Excitation spectrum of Si76 by monitoring Eu^{2+} emission at 465 nm;
c: Excitation spectrum of Si76 by monitoring Rh6G emission at 558 nm;
d: Excitation spectrum of Si76 by monitoring Eu^{3+} emission at 694 nm.

Monitoring the excitation spectrum provides additional evidence of ET processes. The absorption at 393 nm observed in the excitation spectrum of Eu^{2+} (Fig. 3b) and Rh6G (Fig. 3c) suggests that ET processes $Eu^{2+} \rightarrow Eu^{3+}$ (resonant non-radiative) and Rh6G$\rightarrow Eu^{3+}$ are occurring (It is also possible that the pump energy is being absorbed at 393 nm). Evidence of Rh6G$\rightarrow Eu^{3+}$ ET is also observed in the excitation spectrum of Eu^{3+} by the presence of two broad Rh6G absorption bands peaked at 345 nm and 490 nm (Fig. 3d).

Similar spectra are also observed for Si77. In this case, the fluorescence and excitation bands of Eu^{2+} are much weaker and less resolved than the corresponding ones in Si76. This may be due to Eu^{2+} concentration quenching.

Alumina and zirconia gels

The optical properties of two alumina gels with different concentrations of Eu were studied. Fig. 4 shows fluorescence spectra of the samples excited at 350 nm at which both Eu^{2+} and Rh6G are excited. There is no absorption line of Eu^{3+} at 350 nm. Two interesting features shown here are the absorption line of Eu^{3+} at 393 nm in the Eu^{2+} fluorescence band (centered at 400 nm) and the Eu^{3+} fluorescence line at 616 nm. Two conclusions can be drawn: i) radiative $Eu^{2+} \rightarrow Eu^{3+}$ ET takes place; ii) Eu^{3+} fluorescence at 616 nm is produced only by ET from other centers, primarily Eu^{2+}. Fig. 4 also indicates Eu^{2+} quenching since its fluorescence band at 400 nm in A19 is much weaker than that in the weakly doped sample A17.

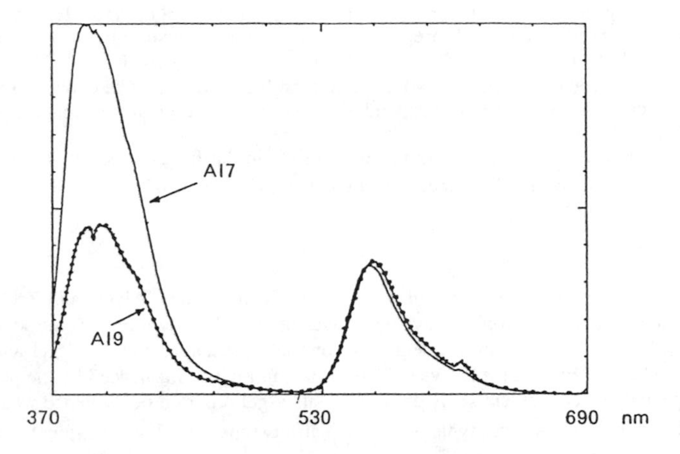

Fig. 4 Fluorescence spectra of A17 and A19 under excitation at 350 nm.

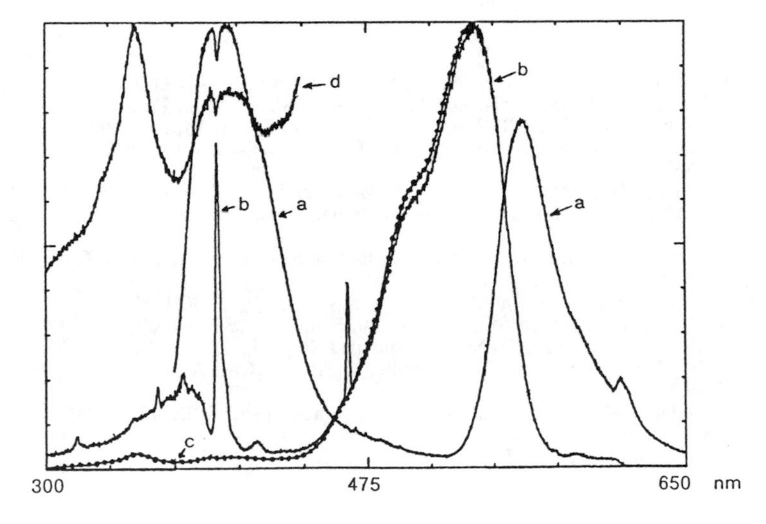

Fig 5 a: Fluorescence spectrum of A19 excited at 350 nm;
b: Excitation spectrum of A19 by monitoring Eu^{3+} emission at 698 nm;
c: Excitation spectrum of A19 by monitoring Rh6G emission at 561 nm;
d: An amplified part of c.

Fig. 5 compares the emission and excitation spectra of A19. ET processes similar to those observed in silica gels are evident. An interesting result obtained in the amplified part (Fig. 5d) of the excitation spectrum for Rh6G emission is that a band identical to that of the Eu^{2+} fluorescence appears. It would seem that this band cannot be explained by $Eu^{2+} \rightarrow$Rh6G ET because it would then correspond to the sum of both the absorption and the fluorescence bands of Eu^{2+}.

For the most part, zirconia gels exhibit similar behavior to that of silica and alumina gels. One interesting difference is that there is no evidence of radiative $Eu^{2+} \rightarrow Eu^{3+}$ ET in the zirconia system.

SUMMARY

The optical properties of the sol-gel matrices SiO_2, Al_2O_3 and ZrO_2 codoped with rhodamine 6G and europium have been investigated. Although Eu^{3+} is introduced in the initial sol, there is a significant concentration of Eu^{2+} present in the dried gel and concentration quenching of Eu^{2+} was observed. This reduction may be influenced by the presence of Rh6G especially since a recent study of Eu^{3+} in silica gel reported no evidence whatsoever of Eu^{2+} formation during sol-gel synthesis [13]. Various types of ET were evident. Rh6G$\rightarrow Eu^{3+}$ ET occurs in all three gel systems while radiative $Eu^{2+} \rightarrow Eu^{3+}$ and Rh6G$\rightarrow Eu^{3+}$ processes are possible ET mechanisms but need to be confirmed by fluorescence decay measurements.

ACKNOWLEDGEMENTS

The partial support of this work by the National Science Foundation (DMR 9003080 and INT 8922345) is greatly appreciated.

REFERENCES

1. C.J. Brinker, G.W. Scherer, Sol-Gel Science, (Acad. Press, 1990)
2. J. Livage, M. Henry, C. Sanchez, Prog. in Solid State Chem., 18, 259(1988)
3. D. Avnir, R. Kaufman, R. Reisfeld, J. Non-Cryst. Solids, 74, 395(1985)
4. B. Dunn, E. Knobbe, J. McKiernan, J.C. Pouxviel, J.I. Zink, Proc. MRS Symp., 121, 331(1988)
5. B. Dunn, J.D. Mackenzie, J.I. Zink, O.M. Stafsudd, Proc. SPIE, 1328, 174(1990)
6. F. Salin, G. LeSaux, P. Georges, A. Brun, C. Bagnall, J. Zarzycki, Opt. Letts., 14, 785(1989)
7. R. Reisfeld, D. Brusilovsky, M. Eyal, E. Miton, Z. Burstein, J. Ivri, Chem. Phys Letts., 160, 43(1989)
8. W. Nie, G. Boulon, A. Monteil, Chem. Phys. Lett., 164, 106(1989)
9. W. Niew, G. Boulon, A Monteil, J. Phys., France, 50, 3309(1989)
10. J.A. Mares, Z. Khas, W. Nie, G. Boulon, J. Phys. I., 1, 881(1991)
11a. M. Genet, V. Brandel, M.P. Lahalle, E. Simoni, C.R. Acad. Sci. Paris, 311, série 11, 1321(1990)
11b. M. Genet, V. Brandel, M.P. Lahalle, E. Simoni, Proc. SPIE, 1328, 194(1990)
12. A. Penkofer, Y. Lu, Chem. Phys., 103, 399(1986)
13. R. Campostrini et. al., J. Mater. Res., 7, 745(1992)

DIFFUSION OF CARBON MONOXIDE IN METALORGANIC THIN FILMS DERIVED FROM TITANIUM ALKOXIDE CARBOXYLATES

CHARLES D. GAGLIARDI*, DILUM DUNUWILA*, C. K. CHANG+, AND KRIS A. BERGLUND*+
Michigan State University, Departments of +Chemistry, *Chemical Engineering, and +Agricultural Engineering, East Lansing, MI 48824.

ABSTRACT

The diffusion of carbon monoxide into thin, transparent, porous film media is studied by doping the film with an iron porphyrin and spectroscopically monitoring the chemical interaction between the reduced iron porphyrin and the diffusing solute. The films are spin-cast from metal alkoxide carboxylates. Both optical absorption and resonance Raman spectroscopies have been used to study the porphyrin-doped film and the diffusion process.

The studies yield information on the effective pore size and homogeneity of the materials, and show that redox chemistry can be performed on incorporated metallo-porphyrins. This investigation also demonstrates the usefulness of porphyrin-doped films as chemical sensors, showing a sensitivity below 70 ppb to aqueous carbon monoxide.

INTRODUCTION

Optically clear, porous, water insoluble films have been produced at ambient temperatures from mixtures of titanium isopropoxide, valeric acid, and water; valeric acid is preferred over other carboxylic acids because it produces films with greater integrity and better adhesion to glass [1]. The production of these film materials can be treated as the sol-gel processing of a chemically modified alkoxide precursor. Although there are significant differences in the behavior of acetic and valeric acids, much of the chemistry is expected to be the same, and there are many relevant examples in the literature in which acetic acid is used to modify a titanium alkoxide sol-gel precursor [2 - 9]. Acetic acid can also be used with zirconium and hafnium alkoxides to produce films [10]. However, films made with acetic acid and processed at ambient temperatures tend to remain soluble in water. The chemistry and processing of the valeric acid-based coating solutions have also been specifically addressed [1, 11 - 12].

Prior to the development of the titanium isopropoxide valerate coatings, a variety of lanthanide cryptate complexes [13] and octaethylporphyrin [14] were successfully incorporated into silica sol-gel materials. Several studies have also been reported in which photoinduced redox chemistry has been performed within silica sol-gel glasses [15 - 17], and silica glasses have been used as matrix supports for optical sensors [18 - 19].

Interest in water insoluble films made with titanium isopropoxide and valeric acid primarily results from their potential use in optically probed chemical sensors. Our metalorganic films, further strengthened by the addition of a small quantity of lauric acid [1], have already been used successfully in a fluorescence-based sensor application with a $Eu^{3+}\subset 2.2.1$ cryptate complex [20]. Due to the combination of adequate solvent resistance, optical clarity, and high porosity, these materials are ideal candidates for sensing applications. When submerged in water, the films produce a very smooth, flat baseline for UV-vis transmission above 350 nm. The material also gives an exceptionally weak Raman scattering signal -- making it an ideal matrix in which to observe the Raman scattering of other molecules.

The current sensing application involves using Fe(II) porphyrins as optical probes incorporated within titanium metalorganic films. Changes in the Soret band occur when the porphyrin binds with CO or various cyano compounds. Due to the reactivity of the reduced porphyrin with dissolved oxygen, the films can also function as low-level aqueous oxygen sensors. Methods of incorporating iron porphyrins, such as those shown in Fig. 1, into the film and methods of *in-situ* porphyrin reduction will be presented, and a CO diffusion study will be shown to yield additional information concerning the structure of the metalorganic film material.

 ©1992 Materials Research Society

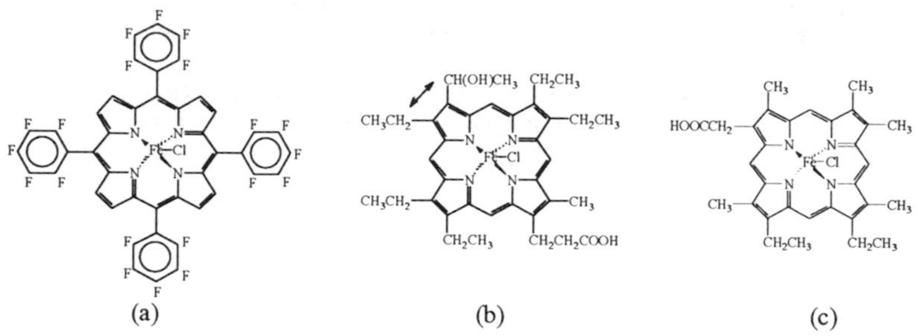

(a) (b) (c)

Figure 1. Structures for three porphyrins (P1, P2, and P3) which have been incorporated into the titanium metalorganic films. (a) P1 is 5,10,15,20-tetrakis(pentafluorophenyl)porphyrin Fe(III) chloride. (b) P2 has both a hydroxyl and a carboxyl group. The structure shown is one of two structural isomers present in a 1:1 ratio of P2; the other isomer is formed by exchanging the two groups indicated by the arrow. (c) P3 has a single carboxyl group.

EXPERIMENTAL

A. Starting Materials

The three porphyrins shown in Fig. 1 were synthesized previously [21-23]. The titanium isopropoxide and valeric acid were obtained from the Aldrich Chemical Company and used without further purification. The argon was 99.99% pure from AGA Gas, Inc., and the CO was Ultra High Purity (99.99% pure) from Matheson Gas Products. The glass substrates for the films where made by cuting standard, pre-cleaned microscope slides (VWR Scientific Inc.) to fit inside 10 mm pathlength quartz cuvettes. The slides were soaked for 24 hours in a 2% solution of Micro (International Products Corporation), rinsed with water, and air dried. All water was de-ionized, with a resistance of 18 Mohm.

B. Coating Solution Preparation and Spin-Casting

The coating solution was prepared in capped glass vials. The valeric acid was added first, followed by the titanium isopropoxide. After thorough mixing, water was added. The molar ratios used of acid/alkoxide (R_a) and water/alkoxide (R_w) were 10 and 1.5, respectively. In order to prevent the aggregation of the porphyrin, 200 proof ethyl alcohol was also added to this mixture in a 8:1 alcohol to alkoxide volumetric ratio [24]. The porphyrin was then added (yielding a 0.002 M solution), and the mixture was aged for three days. The solution was then spin-cast at about 1100 rpm, as previously described [1].

C. Reduction of Fe Porphyrin

The Fe(III) porphyrins were reduced *in-situ*, within the film matrix, by hydrazine, by sodium dithionite, and by photoreduction with UV radiation in the presence of a dilute aqueous solution of isopropanol and acetophenone. The photoreduction of metal porphyrin solutions in the presence of isopropanol and trace amounts of acetophenone or benzophenone has been previously reported [25].

D. Instrumentation

Optical absorption spectra were taken with a Perkin-Elmer Lambda 3A UV-vis spectrophotometer and plotted on a Perkin-Elmer R100A recorder. Raman spectra were collected with a Spex 1877 triple spectrometer equipped with a diode array detector using the 413.1 nm line from a Coherent INNOVA 90 krypton ion laser. Samples were held in septa-sealed quartz cuvettes with 22/14 ground glass joints.

E. CO Diffusion Studies

Deoxygenated water (argon saturated) and CO saturated water were used to prepare the various CO solutions which were used with the diffusion study. The concentration of the saturated CO solution was calculated from temperature and pressure measurements, and from highly accurate data in the form of Henry's Law constants [26]. The reducing solution was then displaced with argon and the cuvette was refilled by syringe with the prepared CO solution.

RESULTS AND DISCUSSION

Porphyrin-Film Interaction

It is often difficult to introduce a sufficient quantity of Fe(III) porphyrin into the coating solution without disrupting the optical clarity or integrity of the film or the chemical reactivity of the porphyrin. For a given coating solution and application, the functional groups substituted on the porphyrin ring must be carefully selected since they effect both the solubility of the porphyrin in the solution and the sensitivity of the absorption bands to porphyrin-solute interaction. Carboxyl, hydroxyl, and fluorinated phenyl substituents can be used to enhance the incorporation of the porphyrin within the film solution. The three porphyrins shown in Fig. 1 represent iron porphyrins which can be introduced into the titanium alkoxide valerate solutions at sufficiently high concentrations, can be reduced *in-situ*, and provide adequate sensitivity to CO. It is certain that porphyrins having one or more hydroxyl or carboxyl groups chemically react with the coating solutions. Porphyrins and other guest molecules having multiple hydroxyl or carboxyl groups have been observed to cause gelation in mixtures which otherwise have never been observed to gel. The extent of reactivity for the fluorinated phenyl groups on P1 is somewhat less certain. The resonance Raman (RR) spectra of the reduced porphyrin in the film and in ethanol solution, shown in Fig. 2 (b, c), are similar but not identical. The RR spectra of tetraphenyl porphyrins are known to show bands associated with the phenyl rings, and this has been attributed to significant conjugation with the porphyrin π-system [27]. The band at 727.1 cm^{-1} is within the expected frequency range for aromatic rings [28] and its disappearance when the porphyrin is in the film probably indicates that phenyl groups are reacting in some way with the film material -- perhaps by the displacement of one or more fluorines on the phenyl ring.

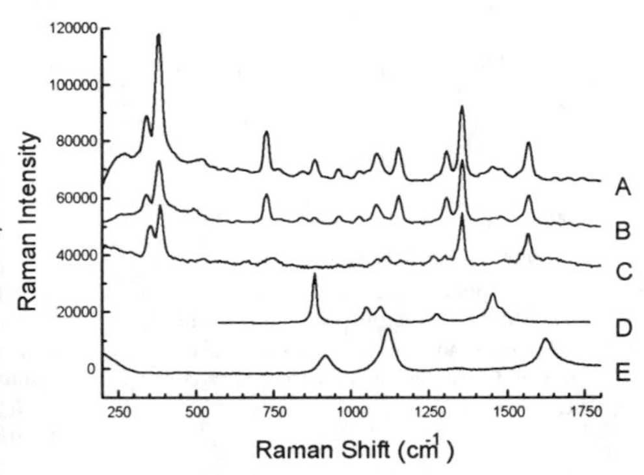

Figure 2. Raman spectra of P1. (A) P1-Fe(III) in EtOH, (B) P1-Fe(II) in EtOH with hydrazine, (C) P1-Fe(II) in film in aqueous hydrazine solution, (D) EtOH, (E) hydrazine.

Reduction to Fe(II) State

The three porphyrins shown in Fig. 1 have all been reduced within the titanium metal-organic film with hydrazine, sodium dithionite, or (photochemically) with UV light in a dilute solution of isopropanol and acetophenone. Hydrazine is the most convenient reducing agent since it is obtained as a ready-to-use aqueous solution. The sodium dithionite is more difficult to use, provided in the form of an air-sensitive solid. These two reducing agents must diffuse into the film before reducing the porphyrin. To photoreduce an Fe(III) porphyrin within the film, acetophenone and isopropanol must also be allowed to permeate the film prior to exposure to UV light. However, by mixing a small amount of acetophenone and benzophenone into the coating solution before the film is even made, the films need only be treated with isopropanol solution during UV exposure. The use of acetophenone, benzophenone, and isopropanol in the photoreduction of metal porphyrins has been previously reported [25]. However, since the film material strongly absorbs in the UV region, the success of the photoreduction within the film was not entirely predictable. The absorption spectra showing the reduction, the formation of the carbonyl ligand, and the re-oxidation are shown in Figures 3 - 4.

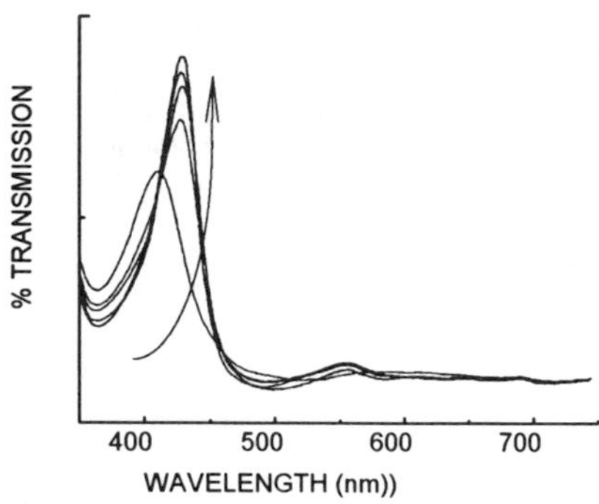

Figure 3. Absorption spectrum of P1 in a titanium metalorganic film showing the photoreduction of P1 by 254 nm light in an aqueous solution of acetophenone and isopropanol. The arrow shows the progression over time, starting with the Fe(III) porphyrin.

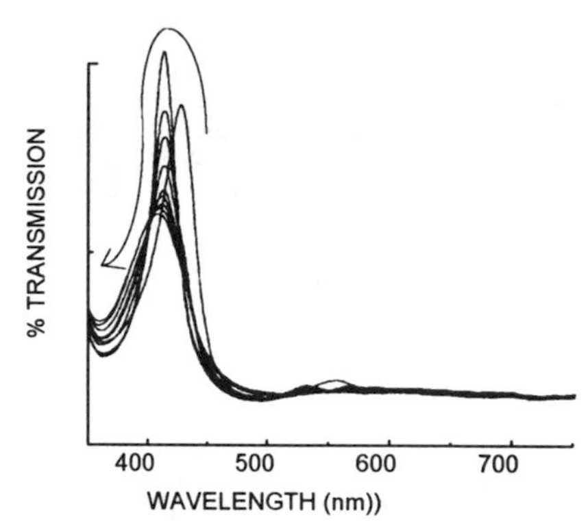

Figure 4. Absorption spectrum of photoreducted P1 in a titanium metalorganic film showing the effects of exposure to CO and re-oxidation by solvated O_2. The arrow shows the progression from the photoreduced Fe(II) porphyrin, to the Fe(II)-CO complex, to the Fe(III) state.

Diffusion Studies

The results of the diffusion study are shown in Figures 5 - 7. The Soret band (~ 415 nm) of P1 in the film is shown for various CO concentrations. The relative intensities and positions of the Soret band for the Fe(III), Fe(II), and Fe(II)-CO states reveal the extent of carbonyl ligand

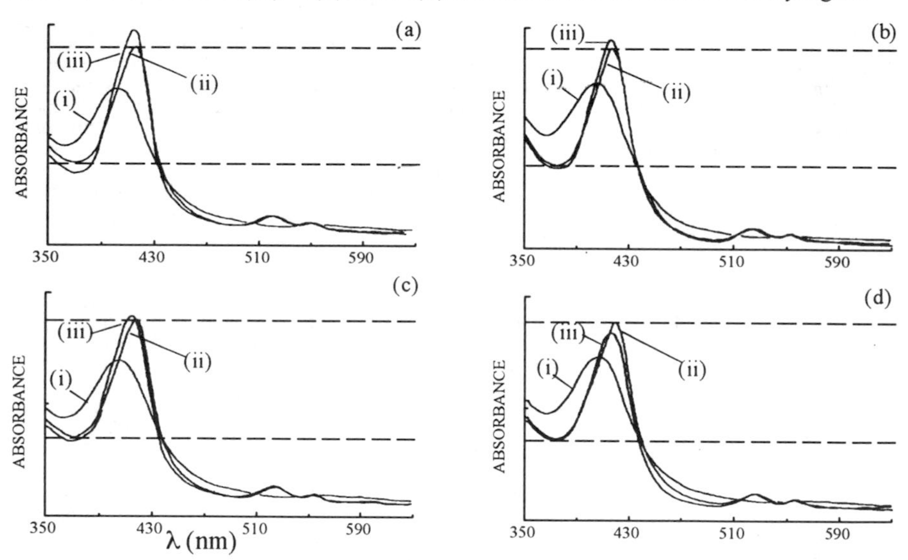

Figure 5. Absorption spectrum showing the response of P1-doped titanium metalorganic films to various concentrations of carbon monoxide in aqueous solution: (a) 1000 ppb, (b) 170 ppb, (c) 70 ppb, (d) 0 ppb. Absorbance spectra were taken (i) before reduction with hydrazine, (ii) after reduction, and (iii) after removal of hydrazine solution and exposure to the carbon monoxide solution.

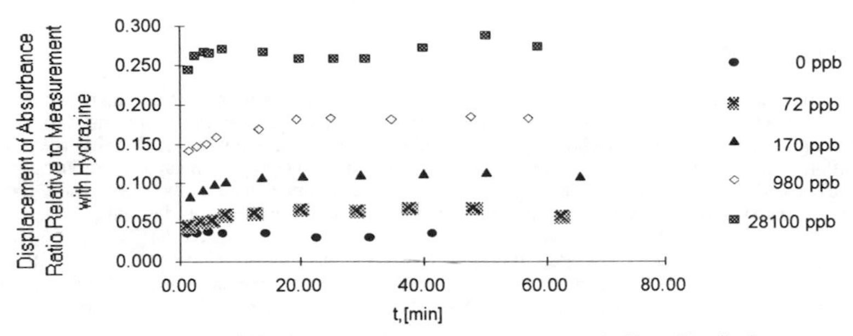

Figure 6. Carbon monoxide diffusion into P1-doped titanium metalorganic films. The displacement ($R(t) = R_{PI}(t) - R_o$) of the absorbance ratio ($R_{PI}(t)$) relative to measurement with hydrazine (R_o) is shown for t between 0 and 60 minutes.

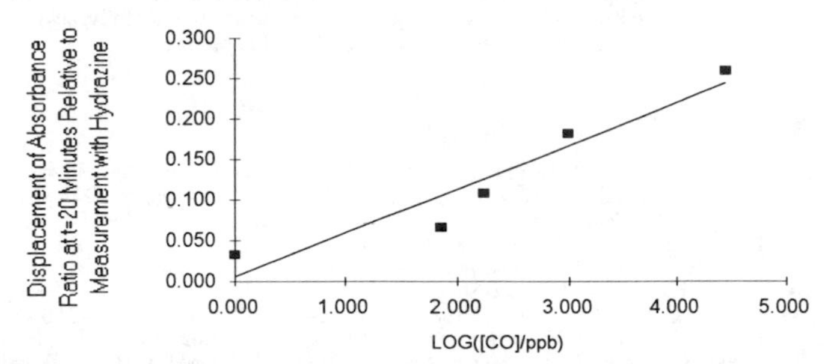

Figure 7. Response of P1-doped titanium metalorganic films to various CO concentrations after 20 minutes of exposure to the CO solution. The displacement ($R(20) = R_{PI}(20) - R_o$) of the absorbance ratio ($R_{PI}(20)$) relative to measurement with hydrazine (R_o) is shown for CO solutions ranging from 0 to 28,100 ppb.

formation after approximately twenty minutes exposure to the CO solutions. The hydrazine not only reduces the Fe(III) P1, it also is weakly retained by the porphyrin as a ligand. Even CO concentrations as low as 70 ppb (Fig. 5 (c)) show an increase in the Soret band magnitude and an easily measurable shift to lower wavelength.

To quantify the changes in the absorption spectrum due to the formation of the carbonyl ligand, it was necessary to normalize the data taken from different films. The baseline corrected absorbance intensity (BCAI) was measured at 414 nm (I^*_{414}) and 419 nm (I^*_{419}) after 30 minutes of film exposure to the hydrazine solution (the asterisk will be used to denote measurements in the presence of hydrazine). Then, after exposure to the CO solution, the BCAI was monitored at 414 nm (I_{414}), the Soret band position for P1 Fe(II)-CO complex. The normalized response to CO was then calculated as $R(t) = R_{PI}(t) - R_o$, where $R_{PI} = I_{414} / I^*_{419}$ and $R_o = I^*_{414} / I^*_{419}$. Figure 6 shows the development of R(t) over a period of ~1 hr for various CO concentrations. From this figure, it is apparent that the initial rate of response to CO changes abruptly prior to the collection of data, and it is also apparent that this initial period accounts for a large percentage of the total response. This would indicate that the porphyrin is not uniformly accessible within the film. The diffusion of benzoate through the film was previously shown to have a diffusivity of no less than 1×10^{-12} [20]. Ignoring the initial response, the diffusivity of CO appears to be on the order of 1×10^{-14}, which is comparable to diffusion in dense solids. A likely explanation for these results is that there is a large pore size distribution within the material; many porphyrin molecules are directly accessible to larger pore channels, while some of the porphyrin remains isolated within less permeable, smaller-pore regions.

In Fig. 7, the CO response after 20 minutes of exposure was compared for various CO concentrations. The response was shown to vary linearly with the log[CO], with a correlation coefficient of 0.91. Although this is only an empirical correlation, it indicates that the films can be calibrated over a very wide concentration range. The correlation also shows that saturation of sensing properties requires relatively high CO concentrations.

CONCLUSIONS

Iron porphyrins can be incorporated into optically clear, porous, metalorganic films cast from titanium isopropoxide valerate solutions. Sufficiently high quantities of the porphyrin can be securely incorporated within the film by choosing suitable substituents on the porphyrin, such as hydroxyl, carboxyl, or fluorinated phenyl groups. However, it must also be noted that the porphyrin carboxyl groups compete with the valeric acid for Ti sites and multiple porphyrin-carboxyl groups tend to reduce rather than enhance compatibility. Some evidence was provided that the fluorinated phenyl groups are also reacting with the coating solution.

Within the film, the iron porphyrin can be reduced to its more reactive Fe(II) state by chemical or photochemical means, and used to detect the presence of a permeating solute such as O_2, CO, or various cyano compounds. The diffusion of CO provided an example of such sensing applications, and gave additional insight into the structure of the metalorganic material. The sensitivity to CO was shown to reach 70 ppb in aqueous solution.

ACKNOWLEDGMENTS

The support for this work by the United States Department of Agriculture (USDA) and the Center for Fundamental Materials Research (CFMR) at Michigan State University is gratefully acknowledged. All Raman spectra were taken at the Michigan State University LASER Laboratory.

REFERENCES

1 C. D. Gagliardi, D. Dunuwila, and K. A. Berglund, in *Better Ceramics Through Chemistry IV*, edited by B. J. J. Zelinsky, C. J. Brinker, D. E. Clark, and D. R. Ulrich (Mater. Res. Soc. Proc. **180**, Pittsburgh, PA, 1990) pp. 801-805.
2 J. Livage, M. Henry, and C. Sanchez, Prog. Solid St. Chem. **18** 259-341 (1988).
3 C. Sanchez and J. Livage, New J. Chem. **14** 513-521 (1990).
4 C. Sanchez, F. Babonneau, S. Doeuff, A. Leaustic, in *Ultrastructure Processing of Advanced Ceramics*, edited by J. D. Mackenzie and D. R. Ulrich (John Wiley and Sons, New York, NY, 1988).
5 S. Doeuff, M. Henry, C. Sanchez, and J. Livage, J. Non-Cryst. Solids **89** 206-216 (1987).
6 C. Sanchez, J. Livage, M. Henry, F. Babonneau, J. Non-Cryst. Solids **100** 65-76 (1988).
7 C. Sanchez, P. Tolando, F. Ribot, in *Better Ceramics Through Chemistry IV*, edited by B. J. J. Zelinsky, C. J. Brinker, D. E. Clark, and D. R. Ulrich (Mater. Res. Soc. Proc. **180**, Pittsburgh, PA, 1990) pp.47-59.
8 R. C. Mehrotra, Inorg. Chim. Acta Rev. **1** 99-112 (1967).
9 R. C. Mehrotra and R. Bohra, *Metal Carboxylates (*Acedemic Press, New York, NY, 1983).
10 C. D. Gagliardi, K. A. Berglund, in *Processing Science of Advanced Ceramics,* edited by I. A. Aksay, G. L. McVay, D. R. Ulrich (Mater. Res. Soc. Proc. **155**, Pittsburgh, PA, 1989) pp.127-135.
11 C. D. Gagliardi, D. Dunuwila, B. Van Vlierberge, and K. A. Berglund, these proceedings.
12 B. Van Vlierberge, J. I. Dulebohn, K. A. Berglund, in *Chemical Processing of Advanced Materials*, ed. by L. L. Hench (Wiley, & Sons, New York, 1992).
13 R. B. Lessard, K. A. Berglund, and D. G. Nocera, in *Processing Science of Advanced Ceramics*, edited by I. A. Aksay, G. L. McVay, D. R. Ulrich (Mater. Res. Soc. Proc. **155**, Pittsburgh, PA, 1989) pp.119-125.
14 R. B. Lessard, M. M. Wallace, W. A. Oertling, C. K. Chang, K. A. Berglund, and D. G. Nocera, in *Processing Science of Advanced Ceramics,* edited by I. A. Aksay, G. L. McVay, D. R. Ulrich (Mater. Res. Soc. Proc. **155**, Pittsburgh, PA, 1989) pp.109-117.
15 A. Slama-Schwok, M. Ottolenghi, and D. Avnir, Nature **355** 240 (1992).
16 A. Slama-Schwok, D. Avnir, and M. Ottolenghi, J. Am. Chem. Soc. **113** 3984 (1991).
17 A. Slama-Schwok, D. Avnir, and M. Ottolenghi, J. Am. Chem. Soc. **93** 7544 (1989).
18 A. R. Zusman, C. Rottman, M. Ottolenghi, and D. Avnir, J. Non-Cryst. Solids **122** 107 (1990).
19 O. Lev, B. I. Kuyavskaya, I. Gigozin, M. Ottolenghi, and D. Avnir, Fresenius J. Anal. Chem. 1 (1992).
20 J. I. Dulebohn, B. Van Vlierberge, K. A. Berglund, R. Lessard, J. Yu, and D. Nocera, in *Better Ceramics Through Chemistry IV*, edited by B. J. J. Zelinsky, C. J. Brinker, D. E. Clark, and D. R. Ulrich (Mater. Res. Soc. Proc. **180**, Pittsburgh, PA, 1990) pp.733-740.
21 C. K. Chang, F. Ebina, J. C. S. Chem. Commun. 778 (1981).
22 C. Sotiriou and C. K. Chang, J. Am. Chem. Soc. **110** 2264 (1988).
23 R. K. DiNello, C. K. Chang in *The Porphyrins* (D. Dolphin, ed.) Vol. I, (Acedemic Press, New York, NY, 1978) 289.
24 D. Dunuwila, B. Van Vlierberge-Torgerson, C. K. Chang, and K. A. Berglund, to be submitted.
25 B. Ward and C. K. Chang, Photochem. Photobiol., **35** 757 (1982).
26 T. R. Rettich, R. Battino, and E. Wilhelm, Ber. Bunsenges. Phys. Chem. **86**, 1128 (1982)
27 T. G. Spiro, in *Iron Porphyrins - Part II, Physical Bioinorganic Chemistry Series,* edited by A. B. Lever and H. B. Gray (Addison-Wesley, New York, NY, 1983).
28 N. B. Colthrop, L. H. Daly, and S. E. Wiberley, *Introduction to Infrared and Raman Spectroscopy,* 3rd edition (Acedemic Press, New York, NY, 1990).

RIGIDOCHROMISM AS A PROBE OF GELATION, AGING, AND DRYING IN SOL-GEL DERIVED ORMOSILS

STEPHEN D. HANNA, BRUCE DUNN*, AND JEFFREY I. ZINK
University of California, Los Angeles, Dept. of Chemistry and Biochemistry, Los Angeles, CA 90024
*University of California, Los Angeles,Dept. of Materials Science and Engineering, Los Angeles CA 90024

The sol-gel derived ORMOSIL (organically modified silicate) composed of copolymerized tetramethylorthosilicate (TMOS), methylmethacrylate (MMA), and 3-(trimethoxysilyl)-propylmethacrylate (TMSPM) is a stable host for organic molecules. Due to the ORMOSILs utility as an optical host matrix, it is important to obtain a more detailed understanding of the changes that occur during gelation, aging and drying. A useful probe molecule for this purpose is $ReCl(CO)_3$-2,2'-bipyridine. The emission band of this molecule shows a large blue shift as the medium is changed from fluid to rigid. In this paper the changes that occur during the gelation, aging, and drying of a sol-gel derived ORMOSIL are studied. The emission maximum shifts gradually from 615 nm in the sol to 575 nm in dried gels. The final wavelength is indicative of an environment that is not fully rigid, and suggests that the molecule is in a flexible organic site. The properties of the ORMOSIL are contrasted with those of silicate and aluminosilicate sol-gel matrices.

INTRODUCTION

The rigidochromism of $ReCl(CO)_3$-2,2'-bipyridine can be used to probe the structural changes which occur during the sol-gel process (gelation, aging and drying) [1,2]. In a rigid medium the luminescence of this molecule undergoes a large blue shift ($\approx$2000 cm^{-1}) and a large increase in quantum yield as compared with fluid solutions [3,4]. Internal changes in rigidity taking place within the evolving polymer network can thus be monitored spectroscopically.

The ORMOSIL precursor CH_2=$CCH_3COO(CH_2)_3Si(OCH_3)_3$, TMSPM, has three hydrolyzable alkoxide groups which will form Si-O-Si bonds upon condensation. TSMPM also has a propylmethacrylate group which may undergo polymerization with other propylmethacrylate groups or with the methylmethacrylate, MMA, which is also added to the ORMOSIL sol. The MMA, $CH_3COOCCH_3C$=CH_2, can undergo polymerization as well. Thus the TMOS, TMSPM and MMA, all of which are present in the ORMOSIL sol, can crosslink and form an organic/inorganic copolymer [5]. The resulting glass is optically transparent and can be used to encapsulate organic and organometallic compounds that are dissolved in the sol as has been done previously with this and other sol-gel systems [6].

In this paper the use of the rigidochromic probe molecule 2,2'-bipyridine-triscarbonylchlororhenium(I) to study the changes in the rigidity of ORMOSIL sols and gels prepared by copolymerization of TMOS, TMSPM, and MMA is reported. The results show that the probe is encapsulated in an area of intermediate rigidity.

EXPERIMENTAL

The rigidochromic molecule $ReCl(CO)_3$-2,2'-bipyridine was synthesized by the literature method [3]. The doped ORMOSIL sol was prepared by stirring a mixture containing 15 mL of TMOS (1 equivalent), 24 mL of TMSPM (1 equivalent) and 6.5 mL of 0.02N hydrochloric acid (3.5 equivalents) for 30 minutes. To this solution was added 0.027 g of $ReCl(CO)_3$-2,2'-bipyridine (5.8 x 10^{-5} mol) dissolved in 11 mL of MMA (1 equivalent). This solution was stirred one hour [5]. The sol (2.5 mL) was then poured into 1 cm x 1 cm x 4.5 cm polystyrene cuvettes which were sealed for 16 days. Gelation occurred approximately 50 hours after the cuvettes were opened and evaporation was allowed to take place.

Fluorescence spectra were recorded at room temperature on a Spex Fluorolog mounted with an RCA C31034 photomultiplier using single photon counting. Samples were excited at 430 nm. The spectra were not corrected for the spectral response of the instrument. Low temperature spectra were taken by cooling the samples in a liquid nitrogen filled quartz dewar. To achieve temperatures below 77 K samples were cooled in an Air Products displex. In these studies samples were excited by using the 351 nm and 457 nm lines of an argon ion laser with an average power of 50 mW.

RESULTS AND DISCUSSION

Spectroscopic Studies of Rigidity

The emission spectrum of $ReCl(CO)_3$-2,2'-bipyridine in freshly prepared ORMOSIL sol was measured at room temperature and 77 K to assess the magnitude of the rigidochromic shift in this system for fluid and rigid environments (see figure 1). At room temperature the emission maximum of the probe occurs at 615 nm (fwhm≈3800 cm^{-1}), while at 77 K the emission is blue shifted by approximately 3000 cm^{-1} to 520 nm (fwhm≈3700 cm^{-1}). On cooling, the band width of the probe in the ORMOSIL sol decreases by only 100 cm^{-1}. The small changes in the fwhm indicate that the rigidochromic effect is not the growth of a new band with a concomitant decrease in the original, but that a shift in the band maximum occurs.

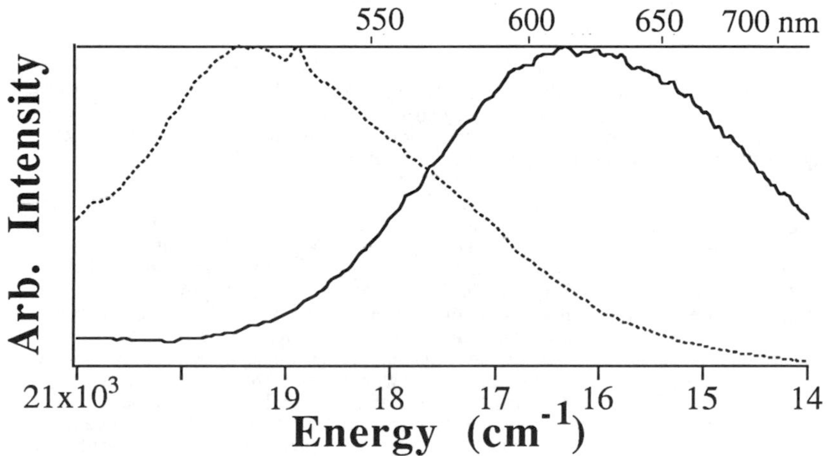

Figure 1. The emission spectra of $ReCl(CO)_3bpy$ in the ORMOSIL sol at room temperature (right) and 77 K (left).

The band maximum of the probe was studied throughout the processing of the ORMOSIL. Following mixing, the emission band of the probe molecule dissolved in the sol is centered at 615 nm (fwhm≈3800 cm^{-1}). No shift is observed during the 16 day aging of the sealed samples. Samples which are allowed to age become increasingly more viscous but do not gel. After the sealed samples are opened to the air, the emission maximum of the probe begins to blue shift as solvent is allowed to evaporate. Gelation occurs approximately 50 hours after the solvent begins to evaporate. A small rigidochromic shift occurs just after the solvent is allowed to evaporate, but prior to gelation of the sample. The changes in the emission maximum and normalized weight as a function of time are shown in figure 2.

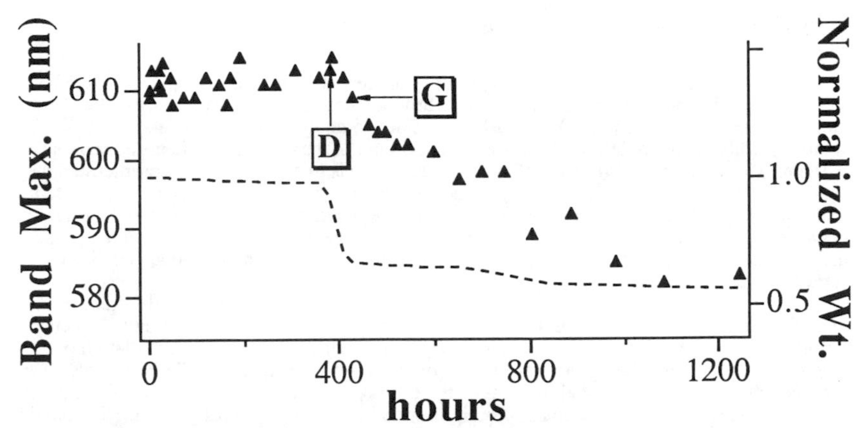

Figure 2. Plot of the wavelength of the emission maxima of the probe in the ORMOSIL system as a function of processing time (triangles). The start of drying is labeled with a "D" and the gelation point is identified by the "G". Normalized weight is shown by the dotted line.

After gelation, the emission maximum continues to blue shift slowly throughout the drying process, reaching its final position at 575 nm (fwhm≈4000 cm^{-1}) after all of the free solvent has evaporated. This value of the emission maximum is approximately midway between that observed in the fluid sol, and the rigid frozen sol. Dried samples were later heated at 60° C for 48 hours to insure the probe was encapsulated in a fully dried gel, and not in excess solvent also trapped in the gel. Within experimental error, no further blue shift beyond 575 nm was noted. Heat-treated samples have their emission band centered at approximately 570 nm. Throughout the study the emission spectra remained similar in shape and band width.

The effect of decreasing temperature on the luminescent properties of the encapsulated probe molecule was studied. As shown by figure 1, a further blue shift in the luminescence of the rigidochromic-doped ORMOSIL can be achieved by cooling the sample. In a series of experiments with the heat-treated ORMOSIL sample, the emission maximum shifted from 570 nm (fwhm≈4000 cm^{-1}) at room temperature to 530 nm (fwhm ≈3500 cm^{-1}) at 15 K. Upon returning the sample to room temperature the band maximum red shifts back to its previous value (see figure 3).

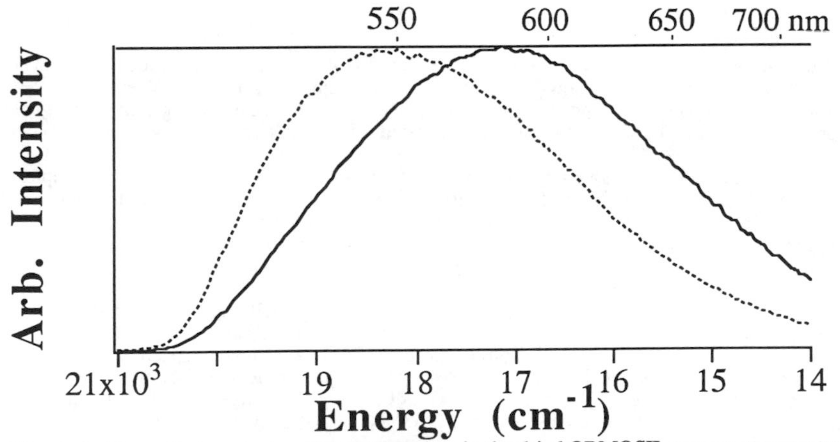

Figure 3. The emission spectra of $ReCl(CO)_3bpy$ in the dried ORMOSIL at room temperature (right) and 70 K (left).

Interpretation of the Spectroscopic Results.

The rigidochromic-doped ORMOSIL does not display full spectroscopic rigidity relative to a frozen sol sample, unlike the silicates and aluminosilicates studied previously. The emission maximum in the solid ORMOSIL occurs at 575 nm (or 570 nm in heat treated samples), while in the frozen sol the emission maximum of the rigidochromic probe occurs at 520 nm. This substantial difference shows that the probe is in an environment of intermediate rigidity. The probe molecule is likely trapped in a flexible "solvent -like" organic site where it feels the effect of a less rigid environment. Alkyl groups surrounding the site of encapsulation could provide the more flexible "solvent-like" medium. The emission maximum of the probe in the heat-treated gel exhibits little change from that of the air dried samples and also indicates a semi-rigid state.

The rigidochromic effect in aluminosilcate and silica sol-gel matrices is different from that of the ORMOSILs. In the TEOS system the probe experiences a non-rigid environment through most of the processing stages. No changes in rigidity are noted during gelation or aging. It is only during the drying stage that increased rigidity is detected. The xerogels become fully rigid upon loss of all solvent. An aluminosilicate composition prepared from the double alkoxide (ASE) also becomes fully rigid upon loss of solvent during drying. Unlike the TEOS gels, however, these compositions reach an intermediate stage of rigidity after aging (see Figure 4) [2].

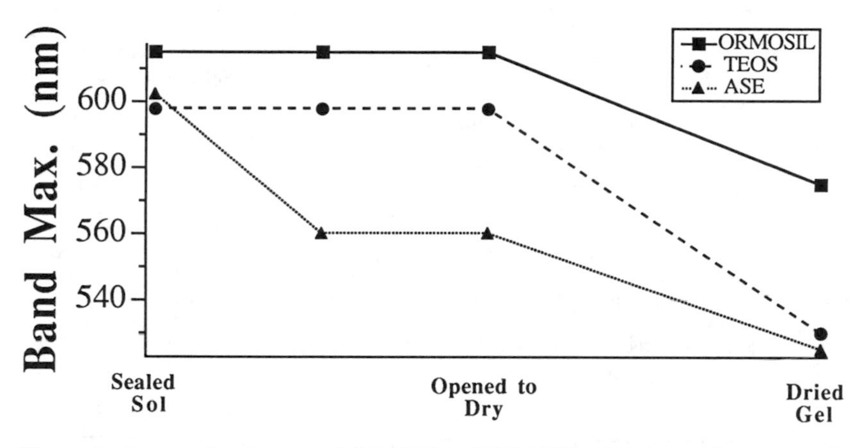

Figure 4. A comparison between ASE, TEOS and ORMOSIL sol gels at various stages of processing. The data points show the observed spectroscopic band maxima. They are connected for illustrative purposes. Squares represent ORMOSIL data, triangles ASE data, and circles TEOS data. The band maximum in the sealed ASE sample shifts after gelation.

Throughout the aging process no rigidochromic shift occurs in the ORMOSIL samples despite the fact that the sol becomes more viscous. Once the evaporation of solvent is allowed to occur, but prior to the formation of a gel, small changes in the rigidity of the sample are detectable in the viscous sol. After gelation, the rigidity of the ORMOSIL increases as the solvent evaporates and the sample loses weight, suggesting that as the sample dries the matrix is constricting. From the luminescence maximum of the fully-dried ORMOSIL at 575 nm, the rigidochromic probe molecule apparently experiences only a partially rigid environment. Flexible alkyl chains in the polymer allow the probe molecule the ability to move in solvent-like surroundings within its site. The motions of these flexible chains, however, can be rigidified at low temperature, thus restricting the mobility of the probe and producing a completely rigid environment. It is interesting to consider that ORMOSIL systems might provide a suitable matrix for the encapsulation of molecules that have special requirements for motion.

In summary, the rigidochromic probe molecule, $ReCl(CO)_3$-2,2'-bipyridine, gives the surprising result that it is trapped in a site that is not fully rigid even though the bulk sample is solid. Heating the sample to forcibly remove trapped solvent does not affect this result. By cooling the sample full rigidity can be achieved.

ACKNOWLEDGEMENTS

This work was made possible by a grant from the National Science Foundation (B.D. and J.I.Z. DMR90-03080) and a graduate research fellowship from the Department of Defense and Air Force Office of Scientific Research (S.D.H.).

REFERENCES

1. C.J. Brinker and G. Scherer, Sol-Gel Science: The Physics and Chemistry of Sol-Gel Processing (Academic Press, Boston, 1990).
2. J.M. McKiernan, J.C. Pouxviel, B. Dunn, and J.I. Zink. J. Phys. Chem. 93, 2129 (1989).
3. M.S. Wrighton, and D.L. Morse, J. Am. Chem. Soc. 96, 998, (1974).
4. P.J. Giordano and M.S. Wrighton, J. Am. Chem. Soc. 101, 2888, (1979).
5. a. C.A. Cappozi and L.D. Pye, SPIE. 970, 135, (1988).
   b. H. Schmidt and B. Seiferling in Better Ceramics Through Chemistry II., (Mater. Res. Soc. Proc. 73, Pittsburgh, PA 1986) pp. 739-751.
6. a. J.C. Pouxviel, B. Dunn, and J.I. Zink, J. Phys. Chem. 93, 2134 (1989).
   b. J.M. McKiernan, S.A. Yamanaka, B. Dunn, and J.I. Zink, J. Phys. Chem. 94, 5652 (1990).
   c. B. Dunn and J.I. Zink, J. Mater. Chem. 1, 903 (1991).
   d. J. M. McKiernan, S.A. Yamanaka, E. Knobbe, J.C. Pouxviel, S. Parveneh, B. Dunn, and J.I. Zink, J. Inorg. Organomet. Polym. 1, 87 (1991).
   e. E.T. Knobbe, B. Dunn, P.D. Fuqua, and F. Nishida, Appl. Opt. 29, 2729 (1990).

SYNTHESIS AND CHARACTERIZATION OF ORGANIC DYE DOPED SILICA GLASSES

ILZOO LEE, JOSEPHINE COVINO AND MICHAEL D. SELTZER
Engineering Sciences Division, Research Department,
Naval Air Warfare Center-Weapons Division,
China Lake, CA 93555-6001

ABSTRACT

The sol-gel process was used to incorporate organic dyes including Rhodamine 6G (Rh6G), 2-(4-pyridyl)-5-(4-phenyl) oxazole (4PyPO) and the n-methyl tosylate salt of 2-(4-pyridyl)-5-(4-methoxy phenyl)oxazole (4PyMPO-MePTS) in silica gel. Thermogravimetric analysis (TGA) and differential scanning calorimeter (DSC) analysis of the dye doped gels showed that the gel structure loses the adsorbed water molecules from room temperature to 150°C and decomposition of the dye molecules followed at the higher temperature. Absorption and emission of the dyes in the sol-gel glass matrix were also studied and compared with the results of the dyes in alcohol solution. The environments of the dye in silica were different than in alcohol solution.

INTRODUCTION

In the past two decades, organic laser dyes have found wide use in different fields of science and technology including spectroscopy, optics, and laser [1]. The lasing medium based on organic dyes may be in the form of solids, liquids or gas. This lasing media causes spectral shifts of both absorption and emission and it affects photochemical stability [2]. It also alters the distribution between processes that the excited states may undergo. For example, a molecule excited to the first excited singlet state may enter the system of triplet states and relax to the lowest level [1]. Although polymers are the earliest and the most common solid state host for organic laser dyes, they have limited photostability and low thermal stability. Polymer hosts also lack the useful chemical, mechanical and optical properties of inorganic hosts. For example, incorporation of Rhodamine 6G in the mixture of MMA (methyl methacrylate) and PMMA (polymethyl methacrylate) showed a laser efficiency of 36%, but lacked photostability [3]. Recently, sol-gel methods, with their low processing temperature, have made it possible to incorporate a dye into a transparent inorganic matrix [2, 4-8]. Embedding organic dyes into silica glass has significant advantages over the other types of matrices. The silica glass is photochemically inert and can enhance the thermal stability of most organic dyes.

Phenyloxazolylpyridinium salts constitute a new class of water soluble low threshold, fluorophoric laser dyes with good lasing characteristics, broad tuning ranges and superior photochemical stability [9]. Since the successful lasing by flash lamp pumping of pyridinium oxazole salts first reported by Lee and Robb [10], Fletcher et. al.[11] reported that the dye modification by the attachment of a methoxy to the phenyl group of 4PyPO to form 4PyMPO showed a great improvement in the lasing performance in a solvent such as ethanol (under argon). This dye showed an extremely long laser lifetime (10,000 MJ/l), laser slope efficiency as high as the best of the coumarin dyes (Coumarin 504) and the lasing threshold was almost as low as the coumarin dyes.

This paper describes the thermal, absorption and emission measurements of organic dyes (Rhodamine 6G, 2-(4-pyridyl)-5-(4-phenyl) oxazole and the n-methyl tosylate salt of 2-(4-pyridyl)-5-(4-methoxy phenyl) oxazole in silica glass from sol-gel processing.

EXPERIMENTAL PROCEDURE

In this study three dyes were used: (1) 4PyMPO-MePTS synthesized by Henry[11], (2) Rh6G and (3) 4PyPO which were both purchased from Aldrich Chemical Co. A 10^{-4}M solution was prepared for each dye in methanol. Silica gels were prepared by hydrolysis of trimethylorthosilicate (TMOS) using a procedure similar to those of Klein [12]. The molar ratio between TMOS and water was 16. The solutions were stirred with a magnetic stirrer for 30 min. and then set aside for 2 weeks to gel. The gel was dried in a convection oven at 60°C.

Thermal analyses of the dyes incorporated in the silica host were performed using a TA 2000 system. Thermogravimetric analysis (TGA, TA 2950) and differential scanning calorimeter (DSC, TA 2910) analysis were performed from room temperature to 600°C at a 5°C/min. and 15°C/min. heating rate, respectively.

Fluorescence spectra of the dried gels were obtained using the harmonic output of a pulsed Nd:YAG laser at 532 nm for Rh6G and at 355 nm for both 4PyPO and 4PyMPO-MePTS. The data was collected using a photomultiplier and stored on a digital oscilloscope (Nicolet 4094). A schematic showing how the measurements were performed is illustrated in Figure 1. Fluorescence life time was measured using the same setup except the boxcar averager was bypassed.

Absorption spectra of the dried gels were measured on a Cary 2390 spectrophotometer with V-block sample holder.

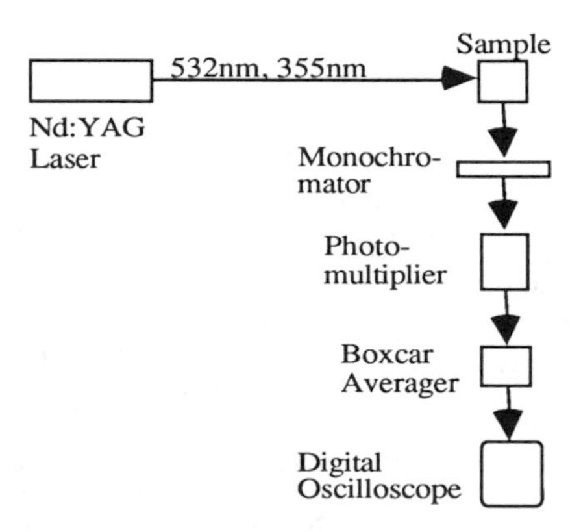

Fig. 1. Schematic of how the fluorescence spectra were obtained.

RESULTS AND DISCUSSION

The thermogravimetric analysis showed that all three of the dye-silica samples lost weight when heated to 600°C (Figure 2). This corresponds to a 20 to 25% weight loss. The approximately 5% weight loss below 150°C was due to the removal of physically adsorbed solvent molecules including surface water. The 15 to 20% weight loss up to ~600°C was due to the decomposition of residual organics and silanols. Above 600°C the weight loss was negligible. Figure 3 shows the DSC curves for the silica xerogels with and without the presence of dyes (the insertion is the DSC curve for 4PyMPO-MePTS itself). It indicates that the xerogels with dyes have a much broader exothermic peak between 250-350°C than silica itself, indicating that organic molecules are being oxidized. It also shows that silica gel has a much narrower endothermic peak at 150°C. In contrast, the silica samples that have the incorporated organic dye

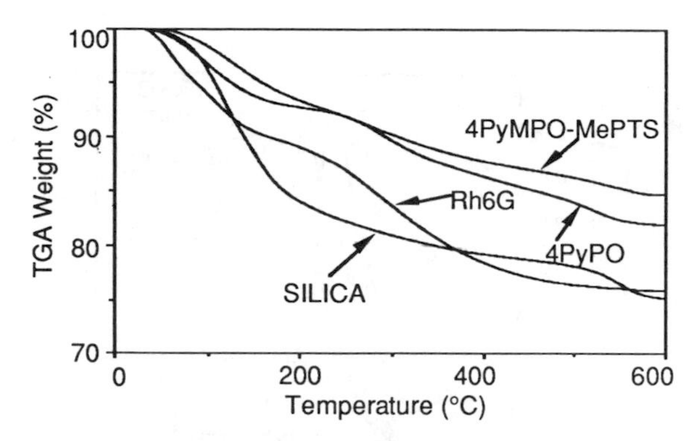

Fig. 2. Thermo-gravimetry of pure SiO_2 gel and SiO_2 gel with incorporated dyes.

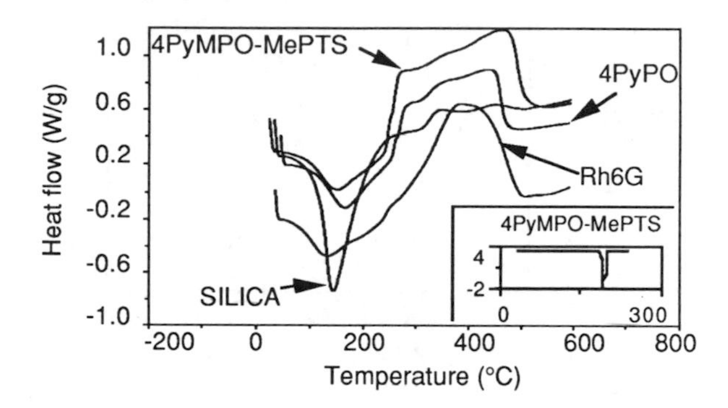

Fig. 3. DSC curves of pure SiO_2 gel and SiO_2 gel with incorporated dyes.

have much broader endothermic peaks. This can be attributed to a combination of dye decomposition and silica dehydration.

Figure 4 shows the absorption and emission spectra of Rh6G, 4PyPO and 4PyMPO-MePTS dyes in silica gel and in solutions. The self absorption of dye, which is indicated by the overlapped portion of absorption and emission spectra, is much smaller in oxazole dyes compared to the Rh6G. Table 1 summarizes the absorption and emission spectra of dyes in silica gel and solution. The experiments performed with Rh6G incorporated in silica gels have

TABLE 1. Emission and excitation characteristics of Rh6G, 4PyPO and 4PyMPO-MePTS in silica gel.

	IN SOL-GEL SILICA MATRIX			IN SOLUTION	
DYE	Fluorescence maximum	Absorption Maximum	Fluorescence lifetime (ns)	Fluorescence maximum	Absorption Maximum
Rh6G	552	522	12.5	559 in Me(OH) [1]	523 in Me(OH) [1]
4PyPO	493	-	12.5	-	325 in Et(OH) [13]
4PyMPO-MePTS	568	408	10	562 in Et(OH)/ H_2O [11]	397 in Et(OH)/ H_2O [11]

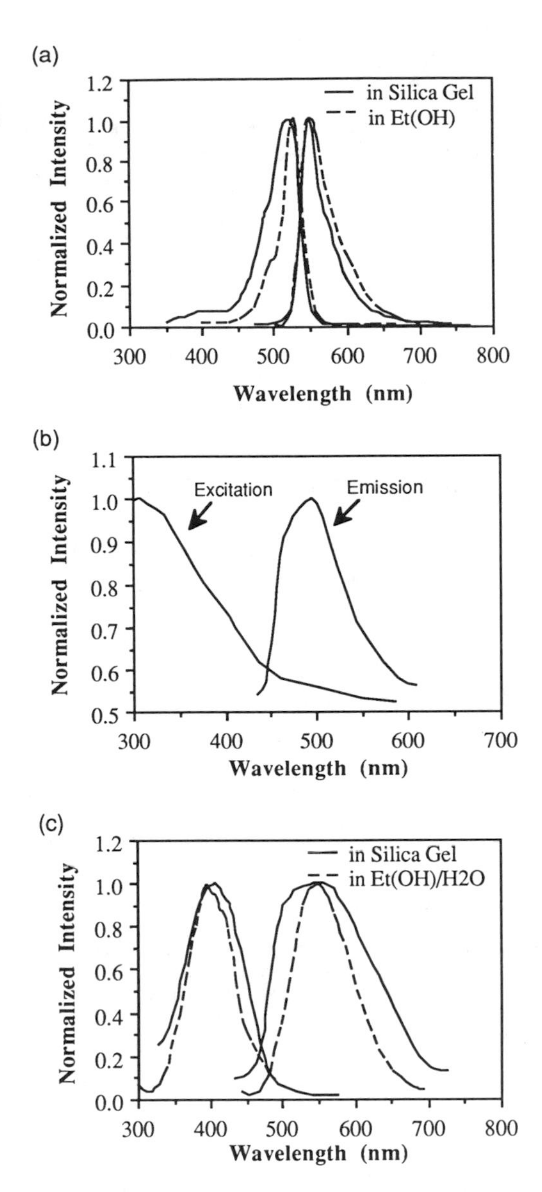

Fig. 4. Excitation and emission spectra of (a) Rh6G in Et(OH) and in silica gel (λ_{ex} = 532 nm) (b)4PyPO in silica gel (λ_{ex} = 355 nm) (c) 4PyMPO-MePTS in Et(OH)/H_2O and in silica gel (λ_{ex} = 355 nm).

demonstrated that the silica matrix does not significantly alter the absorption maximum of the dye molecules when it is compared with the value in methanol. However, a slight blue shift (7 nm) in fluorescence spectra with respect to the methanol solution suggests that the silica gel is more polar. This is probably due to the remaining Me(OH)/H_2O solvent in the gel. Red shifts are observed for both absorption (6 nm) and emission (11 nm) spectra of 4PyMPO-MePTS in silica gel as compared with Et(OH)/H_2O solution. Since similar shifts are often observed in polar organic solvents [4], this may indicate that, in silica gel, the local environment around the dye is slightly less polar than an Et(OH)/H_2O solution. For the 4PyPO dye, Ott et. al. [13, 14] reported that a significant spectral shift of absorption maxima in Et(OH)/HCl with respect to that in Et(OH) is due to the high reactivity of the nonbridging electrons on the pyridyl nitrogen atom for pyridine derivatives. Therefore, significant spectral shifts observed in silica gel compared to the solution spectra are due to the modification of dye chemistry in the initial sol and, subsequently, into oxide gel.

CONCLUSIONS

Thermal analyzers (TGA and DSC) were used to study the thermal reaction of the gels with and without dyes. The TGA data indicated that the samples lost weight continuously until 600°C. The DSC data of the dye doped silica gel showed a broader endothermic peak than the silica gel itself indicating the combination of silica dehydration and dye decomposition. Spectral shifts of the absorption and emission maxima to longer or shorter wavelengths were observed as compared to the dye solution spectra. This is attributed to a modification of the chemistry of the dyes with respect to their local environment during the incorporation of dye molecules into a liquid sol and, subsequently, into an oxide gel.

ACKNOWLEDGMENT

We gratefully acknowledge Dr. R. A. Henry for supplying samples of dye and Dr. C. D. Marrs, Dr. M. E. Hills, Mr. F. H. Hudson, and Mr. C. C. Gilbert for their helpful discussions.

REFERENCES

1. F. P. Schafer, Dye Laser, 2nd ed. (Springer, New York, 1973).
2. D. Avnir, D. Levy and R. Reisfeld, J. Phys. Chem., 88, 5956 (1984).
3. V.V. Rodchenkova, S. A. Tsogoeva, T. M. Muraveva, L.K. Denisov and B. M. Uzhinov, Optical Spectroscopy, 60, 35 (1986).
4. D. Avnir, V. R. Kaufman and R. Reisfeld, J. Non-Cryst. Solids, 74, 395 (1985).
5. J. C. Pouxviel, S. Parvaneh, E. T. Knobbe and B. Dunn, Solid. State Ionics, 32/33, 646 (1989).
6. B. Dunn, E. Knobbe, J. McKiernan, J. C. Zink, Proc. MRS. Symp., 121, 331 (1988).
7. M. Nakamura, H. Nasu and K. Kamiya, J. Non-Cryst. Solids, 135, 1 (1991).
8. R. Gvishi and R. Reisfeld, J. Non-Cryst. Solids, 128, 69 (1991).
9. A. N. Fletcher, R. A. Henry, R. F. Kubin and R. A. Hollins, Optics Comm.. 48 [5], 352 (1984).
10. L. A. Lee and R. A. Robb, IEEE J. Quantm Electronics, QE-16 [7], 777 (1980).
11. A. N. Fletcher, R. A. Henry, M. E. Pietrak, D. E. Bliss, and J. H. Hall, Appl. Phys. B43, 155 (1987).
12. L. C. Klein, Sol-Gel Processing of Silicates, Ann. Rev. Mater. Sci., 15, 227 (1985).
13. D. G. Ott, F. N. Hayes & V. N. Kerr, J. Am. Chem. Soc., 78, 1941 (1956).
14. D. G. Ott, F. N. Hayes, E. Hansbury & V. N. Kerr, J. Am. Chem. Soc., 79, 5448 (1957).

DENSE XEROGEL MATRICES AND FILMS FOR OPTICAL MEMORY

F. CHAPUT*, J.P. BOILOT*, F. DEVREUX*, M. CANVA**, P. GEORGES** AND A. BRUN**
*Groupe de Chimie du Solide, Laboratoire de Physique de la Matière Condensée, Ecole Polytechnique, 91128 Palaiseau (FRANCE)
**Institut d'Optique Théorique et Appliquée, Bat. 503, B.P.147, 91403 Orsay (FRANCE)

ABSTRACT

Optically clear gels doped with organic molecules are prepared in the zirconia-silica system by hydrolysis of metal alkoxides in a wet atmosphere. After drying, dense xerogels are obtained which exhibit a closed porosity. By using the same route, organic-inorganic hybrid xerogels are also produced as films.

The molecule-matrix interactions are evaluated from the absorption recovery times of the $S_0 \rightarrow S_1$ transition for triphenylmethane dyes. Concerning doped xerogels with other organic molecules having polar groups, the application of the strong polarized electric field of an ultrashort optical pulse allows to locally create a birefringence with a memory effect. This type of sample could be used for optical storage and treatment of information.

INTRODUCTION

One of the key problems in the investigation and application of organic molecules in optical devices is the matrix which hosts the molecules. Recently, various organic dyes were embedded in matrices of transparent silica porous glasses prepared at low temperature by the sol-gel technique [1]. Up to now, two main processes have been used to prepare these organically doped materials [2]. In the first route, the dye solution is directly mixed with the sol which is initially formed by adding acidic or basic water to an alcoholic solution of alkoxide. In a second process, porous glasses resulting from partial densification of the gels by thermal treatment are impregnated by the dye solution. In both cases, oxide matrices exhibit high porosity and reactivity leading to poor mechanical properties and chemical instability. These can be partially improved by sol impregnation or by using organic-inorganic hybrid matrices.

When the hydrolysis of metal alkoxides is performed in a wet atmosphere without addition of water in the initial mixture and with sol concentration [3], it is possible to prepare dense gels and films which exhibit a closed porosity and better mechanical properties. We use this technique to prepare organically doped materials in the silica-zirconia system for optical applications. We show that, in some samples, the optical properties may directly result from the interactions between the host matrix and the doping molecules. This type of sample could be used for optical storage and treatment of information [4].

PREPARATION

The hydrolysis of metal alkoxides is carried out by using the moisture in ambient atmosphere (or in a controlled relative humidity). The concentration of the sol is progressively increased by solvent evaporation until the gelation takes place. Four silicon alkoxides have been used: the well-known tetraethoxysilane (TEOS) and three systems having permanent organic groups (methyl, vinyl and amyl triethoxysilane respectively noted MTEOS, VTEOS and ATEOS) in order to prepare organic-inorganic hybrid matrices.

The experimental procedure can be described as follows: raw materials zirconium butoxide $Zr(OBu)_4$ and silicon precursor in the molar ratio 1:4 are mixed in a dry atmosphere using ethanol as a solvent and stirred for several minutes. The volume ratio ethanol: alkoxides is kept at 1:1. The solution is doped

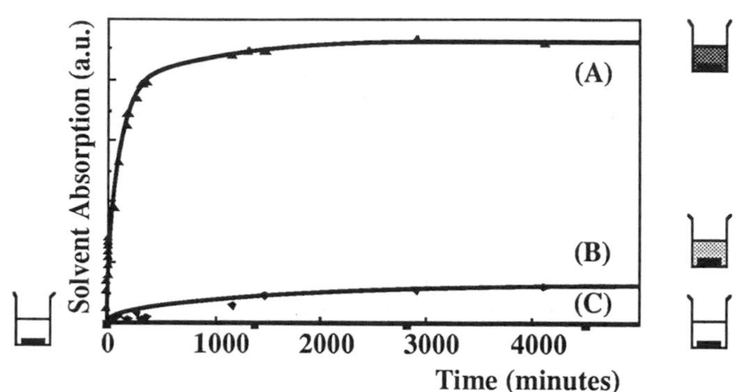

Figure 1: Optical absorption of an ethanol solution in which a dye doped - xerogel is immersed at t=0. Gels are prepared by hydrolysis of alkoxides by using different experimental procedures. The water:alkoxide concentration ratio is 10 (a) or 4 (b). In the case c, the hydrolysis is performed by the moisture in ambient atmosphere and no dye release is observed after 6 months of immersion.

with organic dye (such as Rhodamine 640 at a concentration of 4.10^{-4} mol.l^{-1}) and the hydrolysis is performed in a Petri dish. The mixture becomes viscous and gelatinizes within one week at room temperature. The aging is carried out in the ambient atmosphere for one month. The polycondensation substep leads to a contraction of the gel (90% in volume) which proceeds with the vaporisation of the alcohol until the gel is completely dry and solid. Optically clear and dense xerogels (density of 1.8 g/cm^3 for samples prepared from the TEOS - $Zr(OBu)_4$ mixture) with a glassy aspect are obtained which do not require prior polishing for optical measurements.

Concerning organic-inorganic hybrid gels prepared from a mixture formed of ATEOS and zirconium alkoxide, the hydrophobic character of amyl groups makes easier the drying. The fracture of gels generally observed during the drying step is thus prevented and films of 1-50 μm thickness are then produced.

Xerogel samples are immersed in solvents (water or alcohol) and no detectable leakage of the dye is observed, showing that the dye molecules are completely trapped in a surrounding rigid cage. This behavior highly contrasts with the one of open-porosity gels prepared by classical way (addition of water to the initial mixture) for which the dye release is rapidly observed (figure 1).

STRUCTURAL ASPECTS

Growth of uniform structure

It is now well known that ^{29}Si NMR can be used profitably in the whole condensation process of silicon alkoxides [5]. The relative contribution of the silicon atoms with i siloxanes bridges which are conventionally denoted as Q^i can be quantitatively evaluated. Figure 2 shows typical ^{29}Si NMR spectra recorded at 71.5 MHz as a function of time for TEOS-$Zr(OBu)_4$ and ATEOS-$Zr(OBu)_4$ systems. We note that only a very small fraction of hydrolyzed species with one silanol group is detected at about tg/2 for TEOS and tg/3 for ATEOS (tg is the macroscopic gel time). At this stage a great part of silicon monomers remains in the solution. Moreover Q^1 species are always observed around the gel time showing the presence of small oligomers. This indicates that in these conditions of very slow hydrolysis (neutral medium and no water in the initial mixture) monomer-cluster growth is favoured because the rapid condensation reaction

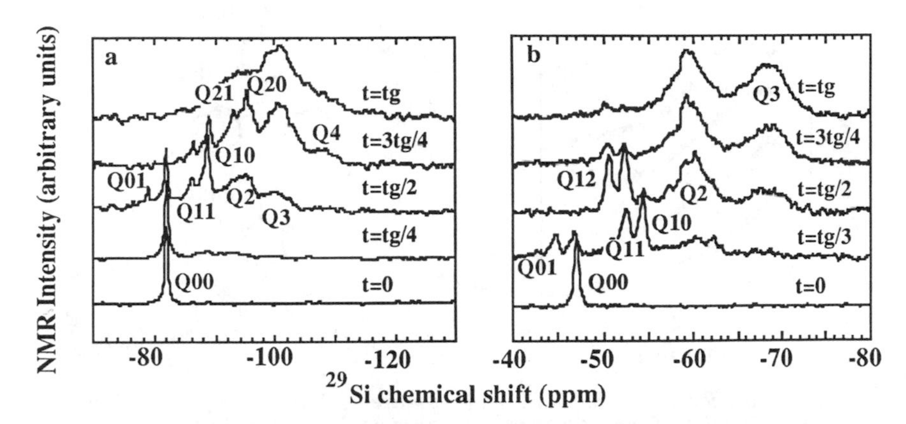

Figure 2: ^{29}Si NMR spectra at different times (in reference to TMS) during the condensation of the TEOS + $Zr(OBu)_4$ (a) and of the ATEOS + $Zr(OBu)_4$ (b) mixtures. Qij denotes a silicon with i siloxanes bondings and j OH side groups. tg is the macroscopic gel time.

consumes hydrolyzed monomers as they are produced. As previously shown [6-7], this growth is consistent either with Eden model which yields compact clusters with smooth surfaces or with the poisoned Eden model which produces uniform structures with fractally rough surfaces. This growth contrasts with the Reaction-Limited-Cluster-Cluster Aggregation (RLCA) observed in the conditions of rapid hydrolysis (acidic medium, water in excess) which leads to mass-fractal aggregates [6].
In fact, a cross-over from the monomer-cluster to the cluster-cluster growth is generally observed in the conditions of slow hydrolysis (basic medium and water in excess) [8]. In our case, the absence of water in the initial mixture and the progressive increase of the sol concentration during the gelation process allow to inhibit a second growth stage and to avoid the formation of open ramified structures.

Local inhomogeneities

A small angle X-ray scattering (SAXS) study has been performed on the TEOS-$Zr(OBu)_4$ system with the 4:1 molar ratio. The variation of the diffuse intensity I(Q) versus the scattering vector Q is taken at different times. During the gelation, there is a gradual decrease in I(Q) which has a form expected for discrete non-interacting particles. After the gel time and during aging and drying processes, curves present a maximum caused by interference effects (Figure 3). This indicates that the particles are not arranged at random but have some short range ordering due to interparticle repulsion. Thus the movement of the maxima to higher values of Q with increasing time reflects a reduction in the equilibrium separation distance between the particles.
It is acknowledged that the hydrolysis rate of TEOS is much slower than the one of zirconium alkoxides. Therefore, one can expect that self polymerizing species based on Zr-O-Zr linkages are initially formed and that they involve local inhomogeneities in the gel. Assuming that these zirconia type clusters contribute in a major part to the scattered intensity, its increasing during the gelation corresponds to their growth. The shift of the maximum towards higher angles during the drying is due to the increase of the concentration of nanometric zirconia particles resulting from the contraction of the gel.

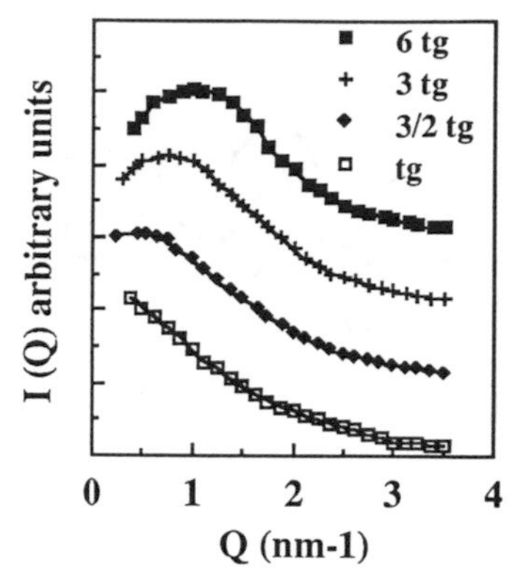

Figure 3: Evolution of the S.A.X.S. curve as a function of the aging time for silica-zirconia gels prepared from the TEOS + $Zr(OBu)_4$ initial mixture.

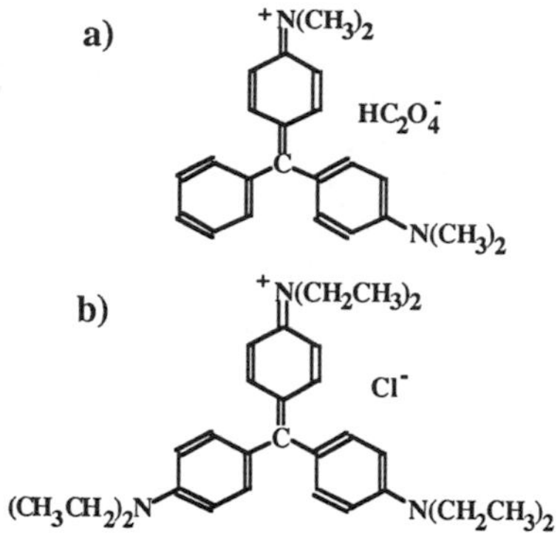

Figure 4: Triphenlymethane dyes a) malachite green, b) ethyl violet

INTERACTIONS BETWEEN THE HOST MATRIX AND THE GUEST MOLECULE

The recovery lifetime of the first excited singlet state S_1 and the ground state S_0 recovery kinetics have been extensively studied for triphenylmethane dyes, such as malachite green (figure 4a) and ethyl violet (figure 4b) in numerous liquids. Optical excitation of the molecules from S_0 to the S_1 state gives rise to a non radiative relaxation process involving rotation of the phenyl rings [9]. To measure the recovery time of the malachite green at 620 nm in oxide gel matrices we have used the previously described pump and probe technique with a 100 fs temporal resolution [10]. For each matrix, the malachite green S_0 state recovery time can be fitted by exponential laws corresponding, as in liquid solvents, to two time constants (Table I). Using the previous work of Ippen et al. [9] which correlates recovery times and solvent viscosity, we can deduce an "apparent viscosity". In fact, the non radiative relaxation process of triphenylmethane dyes involves rotation of the phenyl groups and the "apparent viscosity" can be considered as an evaluation of the interactions between the organic molecule and the oxide matrix. The value is weak (25 cp) for the malachite green in the zirconia-silica matrix showing that the phenyl groups can rotate quite freely within the pores. The relaxation time and consequently the apparent viscosity drastically increases in organic-inorganic matrices suggesting steric hindrance due to the organic groups.

Table I. Recovery times for the malachite green trapped in oxide gel matrices. The apparent viscosity is deduced by using the data of Ippen et al [9].

Precursors	Relaxation times (ps)		Apparent viscosity (cp)
TEOS + $Zr(OBu)_4$	1.4	10	25
MTEOS + $Zr(OBu)_4$	3.5	15	80
VTEOS + $Zr(OBu)_4$	10	23	200

However, a different behavior is noted for ethyl violet. The experiment is destructive whatever the energy sent over the non linear threshold. For this dye, bulky polar groups [N-$(CH_2-CH_3)_2$] are hung on the three phenyl groups. Due to electrostatic interactions with polar groups (such as Si-OH) of the matrices, the amino groups are embedded in the gel and consequently, the energy of rotation of the phenyl groups is transferred to the matrix leading to the fracture of the samples.

OPTICAL MEMORY

The optical set up of a Kerr cell (figure 5) has been used to analyze optical properties of dense xerogels doped with Rhodamine 640. If the sample is optically isotropic no light is transmitted through the analyser. If the medium is birefringent, the transmission is maximum if polarizer and analyser are oriented at 45° from the neutral axes corresponding to the so-called ordinary and extraordinary indices. To induce refractive index change in our materials, the pump beam consists in a sequence of ultrashort light pulses (100 fs, 620 nm and energy of 1 µJ focused on a spot of 50 µm in diameter). The electric field corresponding to such a pulse is about 70MV / cm. By locally applying one hundred pump pulses, the transmission ratio alternately changes from 6% (high level) to 0.5% (low level) depending on the direction of the polarization of the pump beam. The induction of this local birefringence (about 7.10^{-5}) results from the interactions of the optical electric field on the organic molecules via the induced dipole. This corresponds to the alignment of molecules along the direction of the pulse polarization. By changing the polarization direction of the excitation pulses, we may locally control the neutral axis directions
A more surprising fact, observed for the first time, is that zirconia-silica matrices exhibit a memory effect (this effect is weaker in organic-inorganic hybrid matrices). When no pump beam is applied, the output transmission is slightly decreased during the first hours but remains stabilized after at 70% of the maximum value. In fact, the multiple possibilities of electrostatic interactions, such as hydrogen bonds, between the polar groups of the organic molecules and the polar groups of the inorganic host matrix could explain this permanent birefringence. The organic molecules are held by the matrix and the thermal reorientation is prevented when no electrical field is applied. When the torque exerted on the molecules by the optical electric field via the induced dipole is greater than the energy linking the molecules to the pore surface, the molecules are temporarily released and are then aligned along the direction of polarization. This allows to locally control the birefringence and consequently to record and erase information as we like.

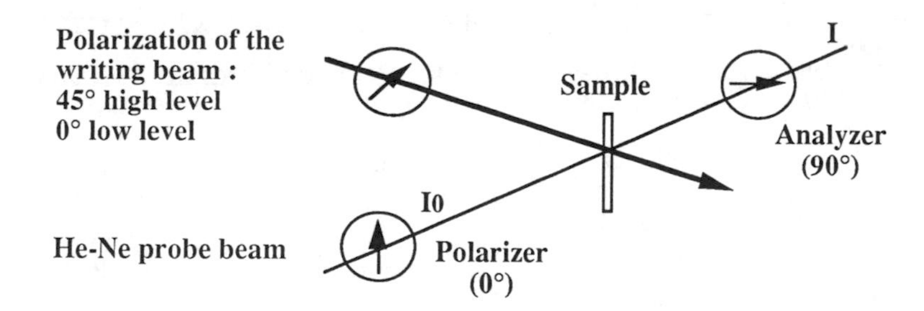

Figure 5: Schematic of the Kerr cell.

CONCLUSION

Optically clear gels have been prepared in the zirconia-silica system by hydrolysis of metal alkoxides in a wet atmosphere. The zirconium alkoxide is rapidly hydrolyzed and self polymerizing species based on Zr-O-Zr linkages initiate the polymerization of the silica network. ^{29}Si NMR data are consistent with a reaction-limited-monomer-cluster growth which produces uniform structures of Eden type. The progressive concentration of the sol during the process prevents the formation of open ramified structures.

These gels can be initially doped with organic molecules. After drying, dense xerogels are obtained which exhibit a closed porosity and some structural inhomogeneities due to the presence of zirconia type nanoclusters. These are real materials directly usable for optical applications. Moreover, by the same process, organic-inorganic hybrid xerogels are also produced as films from a mixture of alkyltrialkoxysilane and zirconium alkoxide.

The measurement of the absorption recovery times of the $S_0 \rightarrow S_1$ transition for triphenylmethane dyes shows that the relaxation process which implies the rotation of phenyl groups can be drastically slowed down: either by steric hindrance due to the methyl or vinyl groups in organic-inorganic hybrid matrices or by polar interactions in inorganic matrices. This indicates the possibility of strong molecule-matrix interactions and suggests that the size of the doping molecule esssentially determines the one of the internal pores of the matrix.

Concerning organic molecules with polar groups, such as Rhodamine 640, the trapped molecules can take numerous orientations for which they are connected to the matrix by hydrogen bonds. These molecules can be freed from the matrix and aligned by the application of a strong optical polarized electric field. Then the new hydrogen bonds prevent any change of the orientation of the molecules allowing to locally control the birefringence. We have named this optical memory matrix OPTOGEL. This is the first potential application in the sol-gel optics field for which the optical effect is based on the interactions between the host matrix and the guest molecule.

REFERENCES

1. D. Avnir, D. Levy and R. Reisfeld, J. Phys. Chem., 88, 5956 (1984).
2. see Sol-Gel Optics, Proceedings of The International Society for Optical Engineering, Eds J.D. Mackenzie and D.R. Ulrich, S.P.I.E. Proc. Ser. 1328 (1990).
3. J.P. Boilot, Ph. Colomban and N. Blanchard, Solid State Ionics, 9/10, 639 (1983).
4. M. Canva, G. Le Saux, P. Georges, A. Brun, F. Chaput and J.P. Boilot, Optics Letters 17 (3), 218 (1992).
5. F. Devreux, J.P. Boilot, F. Chaput and A. Lecomte, Phys. Rev. A, 41 (12), 6901 (1990).
6. D.W. Schaefer and K.D. Keefer, in Better Ceramics Through Chemistry II, edited by C.J.Brinker, D.E.Clark and D.R. Ulrich, (Mater. Res. Soc. Proc. 73, Palo Alto, 1986) p.277.
7. D.W. Schaefer, Science 243, 1023 (1989).
8. D.W. Schaefer, K.D. Keefer, R.A. Shelleman and J.E. Martin, in Advances in Ceramics, Ceramic Powder Science and Technology, The American Ceramic Society, Inc., 21, 561 (1987).
9. E.P. Ippen, C.V. Shank and A. Bergman, Chem Phys. Letters, 38, 611 (1976).
10. M. Canva, G. Le Saux, P. Georges, A. Brun, F. Chaput and J.P. Boilot, Chem. Phys. Letters, 176, (3), 495 (1991).

TAILORING OF TRANSITION METAL ALKOXIDES VIA COMPLEXATION FOR THE SYNTHESIS OF HYBRID ORGANIC-INORGANIC SOLS AND GELS

C. SANCHEZ, M. IN, P. TOLEDANO, P. GRIESMAR
Chimie de la Matière Condensée, Université Pierre et Marie Curie -URA CNRS 1466
4 place Jussieu, 75252 PARIS, FRANCE.

ABSTRACT

The chemical control of hydrolysis-condensation reactions of transition metal alkoxides can be performed through the modification of the transition metal coordination sphere by using strong complexing ligands (SCL). Complexing organic groups can be bonded to the transition metal oxide network in two different ways, as network modifiers or network formers. Different illustrations of the role of complexing ligands on Ti(IV) and Zr(IV) alkoxides are presented. As a network modifier, SCL act as termination agents for condensation reactions allowing a control of particle growth. The complexing ligands being located at the periphery of the oxo core open many opportunities for colloid surface protection. SCL carrying organofunctional groups which exhibit non linear optical (NLO) properties have also been used as probes to study sol-gel transformations. SCL functionalized with organic polymerizable functions act as network formers. Hybrid organic-inorganic copolymers intimately interpenetrated on a nanometer size scale were synthesized from zirconium oxo polymers chemically bonded to polymeric methacrylate chains via a complexing function.

INTRODUCTION

Sol-gel chemistry is based on the polymerization of molecular precursors such as metal alkoxides $M(OR)_n$ [1,2,3]. Hydrolysis and condensation of these alkoxides lead to the formation of oxo-polymers which then transform into an oxide network. Sol-gel chemistry being performed at room temperature offers the possibility to design molecular precursors which retain, after hydrolysis, covalently bonded organic groups. These groups can be bonded to an inorganic network in two different ways, as network modifiers or network formers. Both functions can be easily achieved in the so-called ORMOSILS, ORMOCERS or POLYCERAMS [4,5,6]. The precursors of these compounds are organo-substituted silicic acid esters of general formula $R'_nSi(OR)_{4-n}$, where R' can be any organofunctional group. If R' is a non hydrolyzable group bonded to silicon through a Si-C bond, it will have a network modifying effect ($Si\text{-}CH_3$). If R' can react with itself (R' contains a vinyl group for example) or with additionnal components, it will act as a network former. Chemistry plays the key role for the synthesis of these hybrid materials. Functional coatings for biomedical applications, soft abrasive powders for cosmetics, glues, scratch resistant coatings[4,5] NLO or luminescent coatings [7,8] and ionic conductors gels [9] have recently been produced. These siloxane based materials can be easily synthesized because Si-C bonds are not broken upon hydrolysis. This is no more valid with transition metals for which more ionic M-C bonds are easily cleavable by water. Strongly complexing organic (SCL) ligands have then to be used [10]. This paper reports a discussion on the molecular design of titanium and zirconium alkoxides precursors in the presence of SCL. Such ligands can be functionnalized for any kind of organic reaction. As network modifier, SCL act as termination agents that inhibit further condensation reactions allowing a control of particle growth. The complexing ligands beeing located at the periphery of the oxo core, open many opportunities for colloid surface protection. SCL carrying organofunctionnal groups which exhibit electrochemical or NLO properties are also described. They can be used as probes to follow the sol-gel-xerogel transformations. Moreover SCL carrying an organic polymerizable function can act as network formers leading to the formation of mixed organic-inorganic copolymers.

RESULTS AND DISCUSSION

1 Complexing ligands as termination agents for the control of particle growth.

The chemical tailoring used with siloxane based compounds cannot be directly extended to transition metals because the more ionic M-C bond would be broken upon hydrolysis. Organic modification could however be performed with strong complexing ligands (SCL). The best are β-diketones and allied derivatives, polyhydroxylated ligands such as polyols, and also α or β-hydroxyacids. SCL such as acetylacetone (acacH) react readily with transition metal alkoxides as follows [11,12] :

$$M(OR)_z + acacH \text{ ------> } M(OR)_{z-1}(acac) + ROH \quad (M= Ce, Ti, Zr...)$$

Hydrolysis condensation reaction of acac modified alkoxides are mainly governed by two parameters : the hydrolysis ratio ($h= H_2O/M$) and the complexation ratio x defined as acac/ M (M =Ti, Zr, Ce..)[13,14]. Polymeric sols are obtained with high hydrolysis ratios while for low hydrolysis ratios ($h <1$) acac modified metal alkoxides lead to modified oxo-alkoxide clusters. These compounds are good models for the understanding of the particle growth process. The following examples show that the complexation ratio seems to be the key parameter to avoid precipitation and control particle size.

The hydrolysis of acac modified titanium alkoxides $Ti(OR)_2(acac)_2$ (x= 2) leads to $TiO(acac)_2$. Single crystals have been isolated and X-ray diffraction experiments show dimers where six fold coordinated titanium atoms share corners through oxo-bridges [15] (fig 1a). Upon hydrolysis, all alkoxy groups are quickly removed while SCL are still there. For a smaller complexation ratio larger molecular species can be obtained.

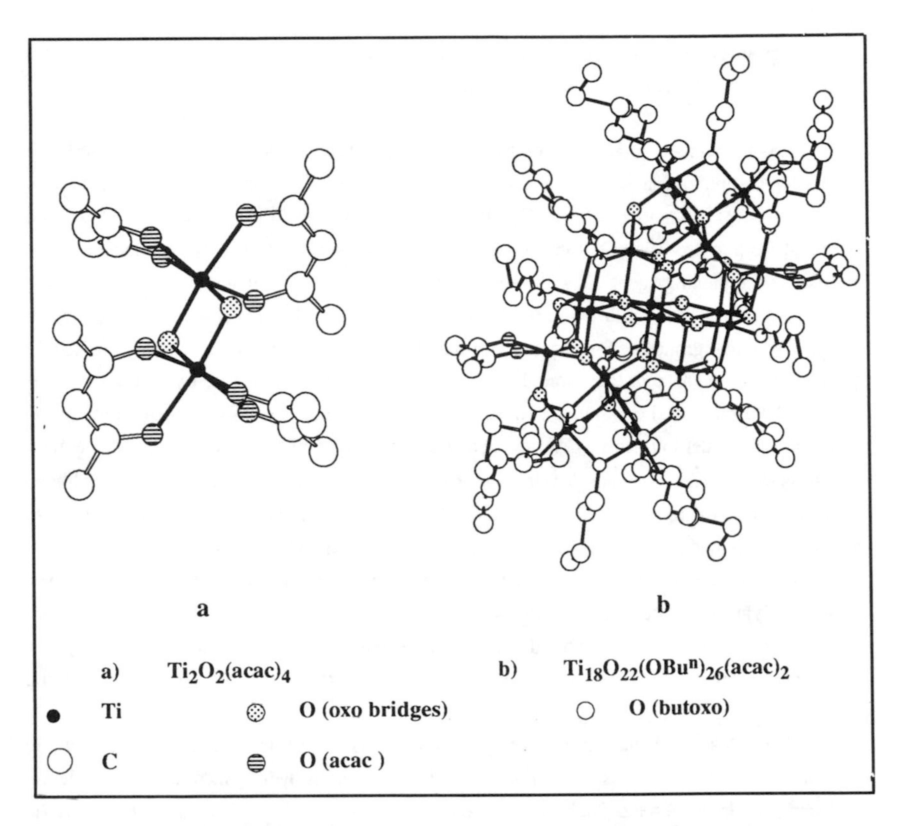

Figure 1 ORTEP view of modified titanium oxo-alkoxo clusters

$Ti_{18}(\mu_5\text{-}O)_2(\mu_4\text{-}O)_2(\mu_3\text{-}O)_{10}(\mu_2\text{-}O)_8(\mu_2\text{-}OBu^n)_{14}(OBu^n)_{12}(acac)_2]$ crystals have been obtained upon hydrolysis of a solution of $Ti(OBu^n)_4$ in the presence of a smaller amount of acacH (x= 0.1). Molecular ORTEP structure is shown in figure 1b [16]. The molecule contains eighteen titanium atoms in distorted octahedral coordination. This molecule is built from a central unit consisting of ten octahedra, linked through edges. Six of these octahedra lead to the formation of a sheet, while the four other ones are located on both sides of this sheet. Two other groups of four octahedra are then linked to this compact decameric central unit, via corner sharing. Only two acac groups are present in the molecule, bonded to the most external free ridges of the compact central unit. The oxide core "$Ti_{18}O_{22}$" is surrounded by the 26 butoxide groups. Among them 14 are bridging groups. Acac ligands act, for low hydrolysis ratios, as a protecting belt towards polymerization reactions.

Similar behavior is also observed for hydrolysis of acac modified zirconium n-propoxide.Figure 2a represents the X-ray structure of a $[Zr_4O(OPr^n)_{10}(Acac)_4]$ cluster obtained upon hydrolysis of the modified precursor $[Zr_2(OPr^n)_6(Acac)_2]$ [17]. Note that many propoxy groups have been removed upon hydrolysis while all acetylacetonato ligands are still there. They prevent further polymerization reactions.

Larger oxo-clusters are obtained when hydrolysis of zirconium n-propoxide is performed with a smaller complexation ratio. The figure 2 b shows the X-ray structure of a decameric zirconium cluste $[Zr_{10}(\mu_4\text{-}O)_2(\mu_3\text{-}O)_4(\mu_3\text{-}OH)_4(\mu_2\text{-}OPr^n)_8(OPr^n)_{10}(\text{allylacetoacetate})_6]$ obtained from hydrolysis of a solution of zirconium propoxide in presence allylacetoacetate (x=0.6) as complexing agent. This cluster is made from ten zirconium atoms. Eight of them are seven fold coordinated while two are in distorted octahedra The (Zr_8) core of the cluster is formed from two groups of four heptacoordinated zirconium atoms located at the corners of a tetrahedron. Each group of four zirconium atoms is connected through a μ_4-oxo bridge as in the $[Zr_4O(OPr^n)_{10}(Acac)_4]$ compound .These two groups are linked via six oxygen atoms (μ_3-O and μ_3-OH) in order to form a a distorted octahedron with one zirconium located at each corner. The last two sixfold coordinated Zr atoms share edges with the Zr_8 core .Six complexing ligand (allylacetoacetate) are bonded to six heptacoordinated Zr atoms, those which are not bonded to sixfold coordinated Zr atoms. Finally, 18 propoxo and 4 hydroxo ligands complete the $Zr_{10}O_6$ core. Note that this cluster is composed by a zirconium oxo core where the six complexing ligands (allylacetoacetate) located at the periphery of the oxo core have a polymerizable organic function. They can be used to extend the cluster network in order to form an hybrid organic-inorganic polymer constituted of an assembly of crosslinked nanoscale crystalline building blocks.

These examples show that strong complexing ligands (acac) are much more difficult to hydrolyse than alkoxy groups. SCL appear to be stable towards hydrolysis because of chelating and steric hindrance effects. Such ligands are located outside the oxide rich part of the core (ie the surface) and act as termination ligands that prevent further condensation. Thus complexing ligands play an important role managing the spatial extension of the transition metal-oxo cores [18]. Similarly short M=O bonds arising from the metal oxygen π-bonding, typical in tungsten, vanadium and molybdenum chemistry, are essentially non basic. They prevent extensive protonation and further polymerization and play a key role in the formation of discrete polyoxoanions[19].

The hydrolysis and condensation behavior of acac modified transition metal alkoxides, was recently studied by NMR, IR, XANES, EXAFS, SAXS, and QELS [13,16,20,21]. When hydrolysis of acac modified transition metal alkoxides $M(OR)_{4-x}(acac)_x$ (M= Ti, Zr, Ce, x =0-2) is performed in the presence of larger amounts of water ($2<h<20$) depending on the complexation ratio and metal concentration sols, gels or precipitates are obtained [13,16]. The size of the colloids was analyzed by light scattering techniques. Within the complexation range x = 0-1, hydrolysis of acetylacetone modified precursors leads to three different domains. For $0.15 \leq x < 1$, sols are obtained. Mean hydrodynamic diameter of the particles measured by light scattering can be modulated from ≈ 400 to ≈ 20 Å when x goes from 0.15 to 1. These sols turn to gels depending on metal concentration. Precipitation is observed for ratio lower than

precipitation can be avoided so that acac ligands have often been reported in sol-gel literature as stabilizing agents for non silicate metal alkoxide precursors [22-25].

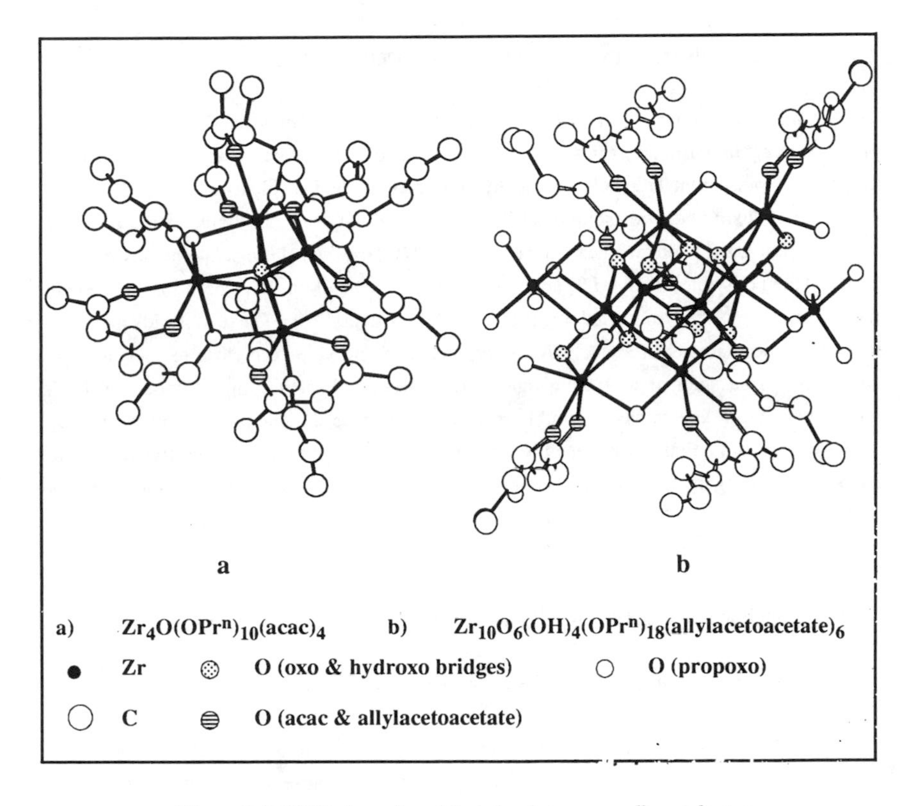

Figure 2 ORTEP view of modified zircoinium oxo-alkoxo clusters

Similar behavior is observed upon hydrolysis of some other metal alkoxides with d or p empty orbitals (Zr(IV), Ce (IV), Ti(IV), Sn(IV), Al(III)) that have the tendency to form oligomeric species in solution and can be complexed by acetylacetone. The key point remains the control of the hydrolysis-condensation of alkoxide based systems by chelated species of lower functionality. The complexation ratio is the key parameter which modulates the mean hydrodynamic diameter of the particles from nano to submicronic range when x decreases from 1 to 0.15 [13,16,20,21].

This synthetic approach to such transition metal oxide (M= Ti, Zr, Ce,..) colloids should be based on the competitive growth / termination mechanism of metal-oxo species in the presence of acac surface capping agents. This way of thinking is general and can be used to control the of acac surface capping agents. This way of thinking is general and can be used to control the size of colloidal particles. Herron et al [26] recently reported the chemical control of cluster size of CdS quantum dots via the use of thiophenols as surface capping agents.

2 Complexing ligands as surface protecting agents for nanosized particles

Complexing ligands can be used to protect the surface of nanosized particles and thus prevent aggregation. But they can also be used to improve chemical stability and photochemical properties. Transparent TiO_2 sols are usually produced via the hydrolysis of $TiCl_4$ in water or $Ti(OPr^i)_4$ in acidic aqueous solutions [27]. Particles about 100Å in diameter are obtained which exhibit good photochemical characteristics [27]. However these sols only absorb U.V. light and are not stable above pH 3. Transparent TiO_2 based colloids have been synthesized via the hydrolysis of the modified precursor $Ti(OPr^i)_3acac$ [28]. These colloids have an hydrodynamic diameter of about 30Å and are made of an oxide core close to TiO_2 anatase surrounded by a shell of protecting ligands (acac) bonded to titanium surface sites [28]. Organic chelating agents have a double role. They increase the chemical stability of TiO_2 modified colloids that remain stable up to pH 10. Moreover, charge transfer from acetylacetonato ligands to titaniums atom gives rise to a strong absorption of visible light that improves the photochemical efficiency of TiO_2 sols. These modified colloids are stronger reducing agents than other TiO_2 colloids [28].

3 Complexing ligands as probes to study the sol-gel transition.

Organic functionalized molecules incorporated inside sol-gel films or fibres present a double interest:

i) They can be used to make sensors [29,32]. Sensors based on optical or electrochemical properties appear to be very promissing because any variation (pH, water content), any chemical or biochemical reaction can be easily detected and monitored [29,32]. Chemically bonded probes present some advantages with respect to simply embedded ones. They are chemically immobilized in the porous texture of the gels and cannot be leached out while at the same time transport of solvent or other any organic molecule into the gel occurs easily. For some optical applications it can be interesting to use chemically bonded dyes that relax slower than embedded ones [33].

ii) They can also be used as probes to monitor textural and structural sol-gel xerogel modifications [30,31,32]. A detailed understanding of the polymer growth process and of the evolution of the gel texture during gelation and aging is essential because the subsequent stages of the glass or ceramic making procedure critically depend on the initial structure and texture of the gel [1,3]. On the course of polycondensation reactions, colloidal solutions (sols) lead to the formation of gels. The sol-gel transition can be followed through the changes of the bulk viscosity of the medium. However rigidification at a macroscopic level does not necessarily imply rigidification at a microscopic scale. Therefore many spectroscopic studies are currently devoted to investigate the physical and chemical transformations that occur upon the sol-gel transformations. Fluorescence polarization [30], ^{2}D NMR [30] or rigidochromic fluorescent probes [35] have been used to characterize silicon oxide based gels. Organic molecules with NLO or electrochemical properties have recently been introduced as probes inside sol-gel systems for the same purpose [31, 34]. They can be simply embedded or chemically bonded to inorganic transition metal oxo backbones through complexing ligands. Both approachs provide complementary informations.

NLO responses such as second harmonics can be generated from organic molecules incorporated in sol-gel matrices [33,34]. However organic molecules inside amorphous sol-gel matrices are in general randomly oriented thus ruling-out the emission of second harmonics. The alignment of nonlinear chromophores incorporated inside sol-gel has been recently performed by using poling techniques by an external electric field. Best results have been obtained with chemically bonded dyes that relax slower than embedded ones[33].

Electric field induced second harmonic (EFISH) is a method for measuring the second harmonic (SH) macroscopic susceptibility Γ. EFISH performed on organic molecules incorporated inside gels appear also to be a convenient method to monitor structural and textural modifications occurring at a microscopic level inside sols and gels. EFISH measurements have been used to determine macroscopic quadratic hyperpolarisabities Γ in sol-gel systems [30]. Time scale of EFISH measurements is about 1 μs. This time is much higher than the reorientational time of organic molecules dissolved in current solvents. The viscosity experienced by optical probes must follow a Stokes-Einstein relationship in which the molecular reorientational correlation time τ_c is given by $\tau_c = 4\pi r^3 \eta / 3kT$. For a given temperature T, an increase of τ_c can be related to an increase of the viscosity η experienced by the NLO probe and/or to an increase of the mean size r of the probe. As soon as τ_c is greater than 1 μs, such molecules should not contribute to the EFISH signal .

A transparent zirconium oxide based gel was prepared as previously described via hydrolysis of an acac modified zirconium propoxide [17,34]. The time evolution of this sol-gel system doped with embedded NLO probes such as methyl-4 nitroaniline (MNA) does not exhibit special variation of the second harmonics when the initial solution turns to a solid amorphous material which does not exhibit macroscopic flow. Thus, in the case of the solidification caused by gelation, EFISH results show that the macroscopic rigidification of the sol is not accompanied by full rigidity at the microscopic level. The solvent phase at the gel point (propanol and water) constitutes the largest volume fraction of the gel (70%). At this stage, the mobility of the embedded MNA molecules is not constrained by the open structure of the gel. Subsequently the reorientational time of MNA is about the same in the gel and in the liquid sol showing that MNA probes are weakly affected by gelation. This results is in agreement with other dynamic studies of the sol-gel transition performed via fluorescence polarisation [30], ^{2}D NMR [30] electrochemical [31] or rigidochromic fluorescent probes [35].

The EFISH signal was also measured as a function of time for a zirconium oxide based gel in which a good chelating agent, aminosalicylic acid (ASA) which present NLO properties was chemically bonded to the zirconium propoxide precursor.
A typical plot of the evolution of the SHG macroscopic susceptibility Γ versus the reduced time t/t_g is shown in figure 3 5 tg is the gelation).

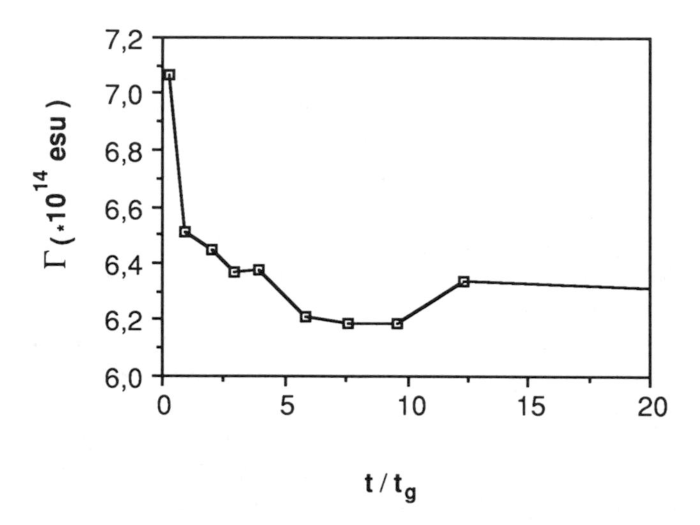

Figure 3 Evolution of the second harmonic susceptibility Γ versus reduced time

The macroscopic susceptibility Γ decreases as soon as polymerisation reaction starts. When the probe (ASA) is chemically bonded to the metallic center (Zr) the onset of the gelation is clearly detected by an immediate decrease of Γ until t/tg=1, contrary to the case of embedded molecules where the decrease of Γ is very smooth. This first transition observed by EFISH must reflect the decrease in mobility of some NLO probes attached to larger polymers. During aging a slight decrease of Γ is observed during a period of time ranging from 1 to 10 t_g. It should reflect the crosslinking of some zirconium oxide based oligomers with the larger ones. However even after the sol-gel transition and upon aging in a closed cell, the relative decrease of Γ is only of about 10-15 % indicating a high degree of mobility even if the NLO probes are chemically anchored to the precursors. It is clear that the NLO molecules that contribute to the EFISH signal are mainly attached to short length polymers or oligomers. Such results are in agreement with a polymerization process that produces a collection of branched polymers with a wide size distribution. The smaller ones, more numerous, control the local densities of chain segment while the few larger ones are responsible for the macroscopic phenomenon of gelation [36].These results, compared to those where the probes are simply embedded, show that a probe anchored on the metallic center experiences more easily and sooner textural and structural changes occurring upon sol-gel transformations. Moreover the comparison of the evolution of diffusion behavior of embedded and chemically bonded probes should give quantitative information on the ratio between large and small polymers (mobil and standingl phase).

4 Complexing ligands for the synthesis of hybrid organic inorganic copolymers with transition metal oxides.

The use of complexing ligands appears to be the key point for the synthesis of organic-inorganic copolymers involving transition metals for which more ionic M-C bonds are easily cleavable by water. Very few papers in this field have been reported [37-39,10].

Prehydrolyzed titanium n-butoxide ($H_2O/Ti=1$) oligomers were complexed with unsaturated organic acid such as cinnamic acid (C_6H_5-CH=CH-COOH). Copolymerization of these clusters was then performed with styrene in presence of benzoyl-peroxide. Transparent brown polymers, insoluble in many organic solvents, and very stable towards the action of water were obtained [37].

Organically modified TiO_2 gels, which give photochromic coatings, were synthesized from an allyl acetylacetone modified $Ti(OBu^n)_4$ alkoxide [10]. A double polymerization process was initiated with partial hydrolysis of the alkoxy groups and radical polymerization of the allyl functions. However polymerization of allyl function is slow and polymerization degree stays low.

Acrylic or metacrylic acid can also be used as a polymerizable chelating ligands. This was recently reported for the sol-gel synthesis of high Tc $YBa_2Cu_3O_{7-x}$ superconducting ceramics [38] and for zirconium oxide based monoliths synthesized by UV copolymerization of zirconium oxide sols and organic monomers [39]. However carboxylic ligands are weak ligands and are largely removed upon hydrolysis [18, 40, 21]. Thus a large amount of the chemical bonds between organic and inorganic networks is lost.

Recently a new approach was chosen with different functionalized chelates which present both a strong chelating part and a highly reactive methacrylate group, such as AcetoAcetoxyEthylMethacrylate (AAEM) [21]. Both organic and inorganic polymerizations were managed simultaneously for the synthesis of hybrid organic-inorganic copolymers. They were performed by hydrolyzing the different AAEM modified precursor solutions at 60°C with an hydrolysis ratio ($h=H_2O/Zr$) of 2, in neutral or basic conditions, with a radical initiator (AIBN=2% w/w). The complexing ratio (AAEM/Zr) was varied between 0.25 and 0.75. Milky sols were obtained.

Local order around zirconium atoms inside hybrid organic-inorganic polymers was investigated with zirconium K-edge X-ray absorption experiments, infra-red, and ^{13}C CP MAS NMR [42]. XANES experiment showed that the zirconium oxo core is made with oxo-alkoxo AAEM modified species where zirconium is in seven fold coordination as in the molecular clusters whose structures are presented in figure 2. The zirconium oxo species are chemically bonded to polymeric methacrylate chains via the β-diketo complexing function as evidenced by infrared and ^{13}C CP MAS NMR. The ratio organic/ inorganic was determined on the hybrid xerogels by TGA and ^{13}C CP MAS NMR[21]. It depends on the complexing ratio. The amount of zirconia can be easily varied from 35 % to 55%.

TEM and SAXS experiments show that bushy scatterers should be a realistic representation of these hybrid organic-inorganic copolymers. The complexation ratio (AAEM/ Zr) seems to be the key parameter which controls the structure and texture of these hybrid materials. A careful adjustment of this parameter leads to zirconium oxo species with more or less open structures and to the tailoring of the ratio between organic and inorganic components. Hybrid organic-inorganic polymers intimately interpenetrated from the nanometer size scale (AAEM/Zr =0.25) to the submicron size scale (AAEM/Zr = 0.75) are obtained.

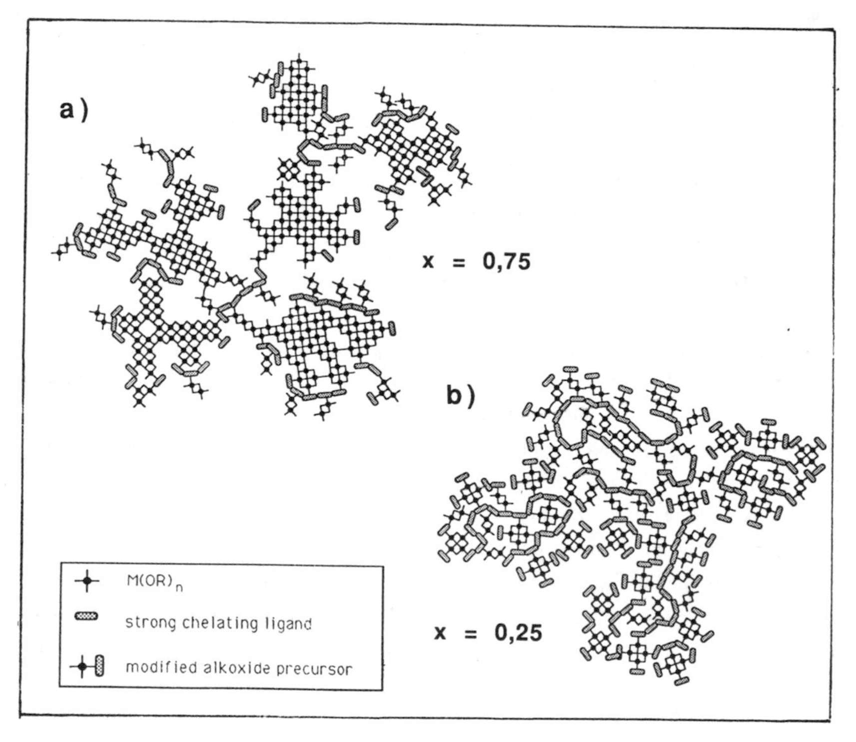

Figure 4 Schematic representation of zirconium-oxo-polymethacrylate copolymers

A simple cartoon of these hybrid zirconium-oxo poly AAEM copolymers based on the local and semi local characterizations [21] is shown in figure 4. For high complexation ratio zirconium-oxo clusters are connected through long polymethacrylate chains while for lower complexation ratio short organic chains crosslink larger zirconium oxo particles.

CONCLUSIONS

Different illustrations of the role of complexing ligands on transition metal alkoxides chemistry were presented. Complexing ligands carrying organofunctional groups which exhibit NLO properties can be used as a probe to study sol-gel transformations. The complexation ratio seems to be the key parameter which controls reactivity and thus the structure and texture. Strong chelating ligands act as termination agents for condensation reactions allowing a control of inorganic particles growth. The complexing ligands beeing located at the periphery of the oxo core they open also many opportunities to protect colloid surface and to control aggregation. SCL functionalized with organic polymerizable functions act as network formers. Hybrid organic-inorganic copolymers intimately interpenetrated at the nanometric scale made with zirconium oxo polymers chemically bonded to polymeric methacrylate chains via a complexing function are synthesized. The ratio organic/ inorganic and the relative spatial expansion of both phases can also be easily varied with the complexation ratio.

Acknowledgements: Rhône Poulenc S.A. is also aknowledged for the financial support of one of the authors.

REFERENCES:

1- C.J. Brinker and G. Scherrer, Sol-Gel Science, the Physics and Chemistry of Sol-gel Processing (Academic press, San-Diego, 1989).
2- J. Livage, M. Henry and C. Sanchez, Progress in Solid State Chemistry 18 , 259 (1988)
3- "Sol-gel technology for thin films, fibers, preforms, electronics and especialty shapes" Edited by. L.C.Klein, (Noyes Pub,.1988).
4- H. Schmidt, H. Scholze and A. Kaiser, J. Non-Cryst. Solids 63 ,1 (1984).
5- H. Schmidt and B. Seiferling, Mat. Res. Soc. Symp. Proc. 73 739 (1986).
6- B. Wang, G.L. Wilkes, C.D. Smith and J.E. McGrath, Polymer Communication Vol.32 13 , 400 (1991).
7- E. Toussaere, J. Zyss, P. Griesmar and C. Sanchez, NLO 3 (1991)
8- S. Dire, F. Babonneau, C. Sanchez and J. Livage, Journal of Material Chemistry, 2,(2) 239 (1992).
9- Y. Charbouillot, D. Ravaine, M. Armand and C. Poinsignon, J. Non-Cryst. Solids, 103, 325 (1988).
10- C. Sanchez, J. Livage, M. Henry and F. Babonneau, J. Non-Cryst. Solids 100, 650 (1988).
11- D.C. Bradley, R.C. Mehrotra and D.P. Gaur, Metal Alkoxides (Academic Press, London, 1978).
12- R.C. Mehrotra, R. Bohra and D.P. Gaur, Metal β-diketonates and Allied Derivatives (Academic Press, London,1978).

13- F. Ribot, P. Toledano and C. Sanchez, Chem. Mater. 3 , 759 (1991).

14- J. Livage and C. Sanchez , J. Non-Cryst. Solids (1992) (in press)

15- G.D. Smith, C.N. Caughlan and J.A. Campbell, Inorg. Chem. 11(12) 2989 (1972).

16- P. Toledano, M. In and C. Sanchez, C. R. Acad. Sci. Paris Serie II 313, 1247 (1991)

17- P.Toledano, M. In and C. Sanchez, C. R. Acad. Sci. Paris Serie II 311 1161 (1990).

18- C.Sanchez, F. Ribot, S.Doeuff, in Inorganic and Organometallique Polymers with Special Properties, Edited by R.M. Laine (Nato,ASI series, Kluwer Academic Pub, vol 206,1992) p 267.

19- M.T. Pope, in Heteropoly and Isopoly Oxometalates, Inorg. chem. concepts 8, (Springer Verlag, Berlin 1983).

20- P. Papet, N. LeBars, J.F. Baumard, A. Lecomte and A. Dauger, J. Mater. Sci. 24, 3850 (1989).

21- C.Sanchez and M. In, J. Non-Cryst. Solids (1992), in press.

22- J.C. Debsikar, J .Non-Cryst. Solids 87, 343 (1986).

23- D. Kundu and D. Ganguli, J. Mater. Sci. Lett. 5 , 293 (1986).

24- W.C. Lacourse and S. Kim, in : Science of Ceramic Processing, eds L.L. Hench and D.R. Ulrich (Wiley, New-York, 1986) p.304.

25- H. Unuma, T. Tokoda, T., Y. Susuki, T. Furusaki, K. Kodaira and T. Hatsushida,J. Mater. Sci. Lett., 5, 1248 (1986).

26- N. Herron, Y. Wang and H. Eckert, J. Am. Chem. Soc.,112,1322 (1990)

27- D. Duonghong, E. Borgarello and M. Grätzel, J. Am. Chem. Soc., 103, 4685 (1981).

28- A. Leaustic, F. Babonneau and J. Livage, Chem. Mater. 1, 240 and 248 (1989).

29- D. Avnir, S. Braun, and M. Ottolenghi, in Supramolecular Architecture in 2 and 3 dimensions, Edited by T Bein , ACS Symp. Ser.(1992)

30- R. Winter, D. W. Hua, X.Song, W. Mantulin, and J. Jonas, J. Phys. Chem, 94,N° 6, 2706 (1990)

31-P. Audebert, P.Griesmar and C.Sanchez, J Mater. Chem.,1,4, 699 (1991).

32- B. Dunn and J.I. Zink, J. Japan Ceram, Soc, (1992), in press.

33- J.M.Boulton, J.Thompson, H.H.Fox, I.Gorodisher, G.Teowee, P.D.Calvert, and D.R.Uhlmann, Mat. Res. Soc. Symp. 180, 987.(1990)

34- P. Griesmar, C. Sanchez, G. Pucetti, I. Ledoux and J. Zyss, Molecular Engineering,1, 3, 205 (1991).

35- J.M. McKierman, J.C. Pouxviel, B. Dunn, and J.I. Zink J. Phys. Chem.,93, 2129 (1989).

36- B. Cabane, M.Dubois and R. Duplessix, J. Physique, 48, 2131 (1987).

37-.L. Suvorov and S.S. Spasskii, Proc. Acad. Sci. USSR 127 , 615 (1959).

38- I. Valente, C. Sanchez, M. Henry and J. Livage, Industrie Céramique 836, 193, (1989).

39- R. Naβ and H. Schmidt, in : Sol-Gel Optics, Edited by J.D. Mackenzie and D.R. Ulrich (Proc. SPIE 1328, Washington, 1990) p.258.

40- S. Doeuff, M. Henry, C. Sanchez and J. Livage, J. Non-Cryst. Solids 89, 206 (1989).

STRUCTURE-RELATED MECHANICAL PROPERTIES OF ORMOSILS BY SOL-GEL PROCESS

Yi Hu and John D. Mackenzie
Department of Materials Science and Engineering,
University of California-Los Angeles, CA 90024

ABSTRACT

The mechanical properties corresponding to the different structures of organically modified silicates (Ormosils) prepared by the sol-gel process were studied. Tetraethoxysilane (TEOS) and polydimethyl siloxane (PDMS) were used as the inorganic and polymeric components, respectively. Large variations in the microstructure of the Ormosils were found by changing the reaction temperature, and the content of PDMS, acid catalyst (HCl),and water. The structure of the Ormosils and the mechanical properties were correlated by the edge connectivity. Modulus of elasticity and fracture strength were used to study their relations Ormosils structures.

INTRODUCTION

The synthesis of new hybrid materials incorporating polymeric/oligomeric species into inorganic ceramics and glasses by the sol-gel process is currently an area of intense investigation [1-5]. The sol-gel technique provides great possibilities for developing the new or better ceramics and glasses at low temperatures without decomposition the organic components. A few examples are contact lenses with good oxygen-permeability [6], hard coatings and diffusion barriers [7], and the incorporation of organic dyes in gels for optical applications [8]. Some of these are materials known as Ormosils (organically modified silicates) and "Ceramers".

The studies on the Ormosils were concentrated on synthesizing new hybrid systems and their structure recently. The introduction of an organic group into an inorganic matrix may act as a network modifier or as network former. The bridge between organic group and inorganic network has been demonstrated by different methods [9-11]. The structure-property behaviors have also been reported for these sol-gel derived hybrid materials. One important structure-related property of the hybrid materials is the rubber-like behavior and has been found in special Ormosils [11,12]. With increasing demands for flexible and elastic glasses or ceramics, mechanical properties of these hybrid microcomposites become very important. In this paper, we relate the effects of certain factors on the microstructure and the mechanical properties correlated to the structure of the investigated Ormosils.

EXPERIMENTAL PROCEDURE

Silanol-terminated polydimethyl siloxane (PDMS) with molecular weight of about 1,700 from Petrach System was used as the prepolymer. Tetraethoxysilane (TEOS) from Fluka Co. was used as the inorganic component of the hybrid gels. The TEOS and PDMS were mixed into an isopropanol/THF solvent, and a solution containing HCl, isopropanol and water was subsequently added to the same mixture. The mixture was then immediately refluxed in a water bath at certain temperature and cast into a prepared mold and sealed before it gelled at same temperature. The gelation time was measured by titling the glass tube. After a long gelation time of two weeks, the wet gels were dried at room temperature for one week, followed by further drying at 70°C for 48 hours in air. The procedure for preparing samples have been described in more detail in a previous report [12]. In this study, the weight ratio of PDMS/TEOS is 2/3. The reaction conditions and the properties of the gels are shown in Table 1.

The microstructure of the completely dried gels was examined by electron scanning microscopy (SEM). Compressive loading experiments were carried out with an Instron machine using an initial strain rate of 0.5 mm/min. The cylinder-shaped samples for testing are 5 cm long and 2.5 cm in diameter. The bulk densities were determined by an autopycnometer and also by the Archimedes method.

Table 1 The reaction condition for and the properties of several Ormosils

Sample Number	Reaction Temperature(°C)	HCl/TEOS molar ratio	H_2O/TEOS molar ratio	Average Bulk Density(g/cm^3)	Open Porosity (%)
1	25	0.3	3	1.09	6.5
2	40	0.3	3	0.91	17.3
3	50	0.3	3	0.83	26.8
4	60	0.3	3	0.65	40.9
5	70	0.3	3	0.56	49.5
6	60	0.1	3	1.10	1.9
7	60	0.2	3	0.88	19.2
8	60	0.3	3	0.65	40.9
9	60	0.4	3	0.50	54.5
10	70	0.3	2	0.72	33.9
11	70	0.3	3	0.56	49.5
12	70	0.3	4	0.50	54.1
13	70	0.3	5	0.48	56.4
14	70	0.3	6	0.49	55.8

RESULTS AND DISCUSSION

It is well known that the final properties of gels are governed by their microstructures. There are many interdependent parameters which control the microstructure of a gel. In the reactions of Ormosils solutions, the effect of these parameters are far more complex. The overall reactions of Ormosils may be represented by following simplified reactions.

Hydrolysis:

$$Si(OR)_4 + 4H_2O \rightarrow Si(OH)_4 + 4ROH \quad (\text{R is } C_2H_5, CH_3, \text{etc.}) \qquad (1)$$

Polycondensation:

$$2Si(OH)_4 + HO\text{-}(Si(CH_3)_2\text{-}O)_x\text{-}Si(CH_3)_2\text{-}OH \rightarrow \text{-}Si(O)_2\text{-}O\text{-}(Si(CH_3)_2\text{-}O)_x\text{-}Si(CH_3)_2\text{-}O\text{-}Si(O)_2\text{-} + 2H_2O \quad (2)$$

Self-condensation:

$$Si(OH)_4 \rightarrow SiO_2 + 2H_2O \quad (3)$$

The extent of reactions depends on the amount of water added and the conditions under which the reactions are carried out. The relative rates of these reactions are crucial in determining the final bulk structure, and, thus the properties of the resultant materials. However, it is difficult to completely separate these reactions in the entire gelation process. Gelation time can be defined as the point when the sol does not flow upon tilting the tube under gravity, and when the viscosity of the solution had reach about 10^4 poise to represent the overall reaction time. Figure 1 shows the variation of gelation time varied by temperature, H_2O/TEOS ratio, and HCl/TEOS ratio. The gelation time as shown decreases with increasing temperature, H_2O , and HCl.

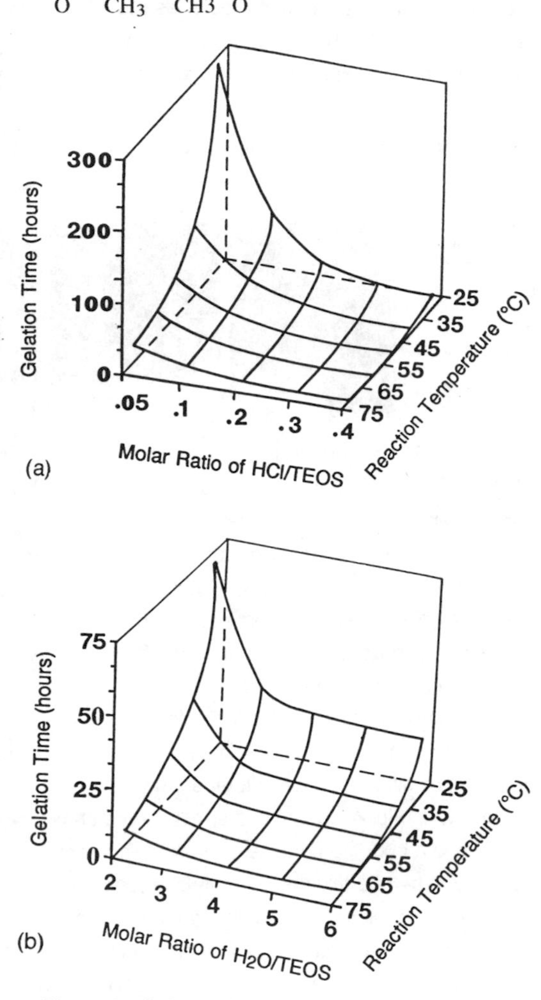

Figure 1. Gelation time versus varied reaction temperature, HCl concentration, and H_2O content.

The microstructures of Ormosils gels made by various parameters are shown in Fig. 2. When the reaction temperature, water content, or the concentration of HCl is increased, the structure of Ormosils becomes more open and the porosity, as shown in Table 1, also increases. The cellular structure which characterizes the high porous

Ormosils has been observed in some other system. The structure of the cells was found to change with different reaction conditions. The average edge connectivity, Zc, which can be used to characterize the structure of Ormosils by SEM photography, increased as the gelation time

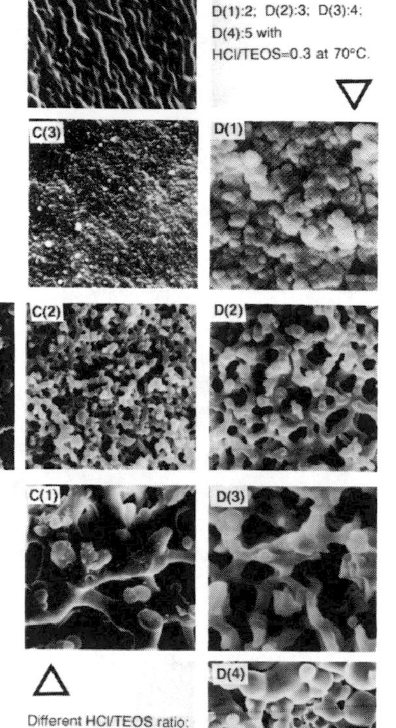

Figure 2. Microstructures of Ormosils with different reaction conditions.

Different reaction temperature:
A:30°C; B:50°C;
C:60°C; D: 70°C

HCl/TEOS=0.3
H_2O/TEOS=3.

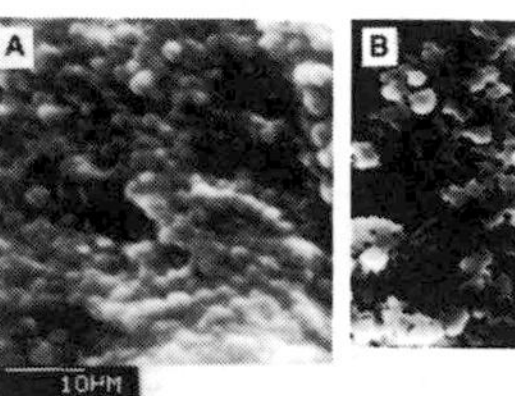

increased. The schematic illustration of the ideal cellular structure based on edge connectivity corresponds to gelation time is shown in Fig. 3. The structure of Ormosils transforms from cellular to the solid structure when the gelation rate was decreased. The porosity decreased significantly as the gelation time decreased by lowering HCl, H_2O, or temperature as shown in Table 1.

The stress-strain curves of samples prepared by different reaction temperatures are shown in Fig. 4. All stress-strain curves show two distinct regions: a linear elastic region and a buckling non-linear region. The sample with higher porosity shows higher flexibility. The modulus of elasticity corresponding to different edge connectivity of the Ormosils is shown in Fig. 5. When the edge connectivity increases, the elastic modulus of the Ormosils increases also. The elastic modulus of the solid material, Es, was estimated to be about 114 MPa for PDMS/TEOS weight ratio of 2/3.

The strength of Ormosils is difficult to correlated with structure. However, we can use the strength per unit solid volume to show this relationship. The strength per unit solid volume, (F/V), as a function of porosity, P, can be expressed as follows:

$$F/V=F/AL=F/[(A^*L^*)P]= S /L^*P \quad (4)$$

where F is the compression strength, A is the cross sectional area of the

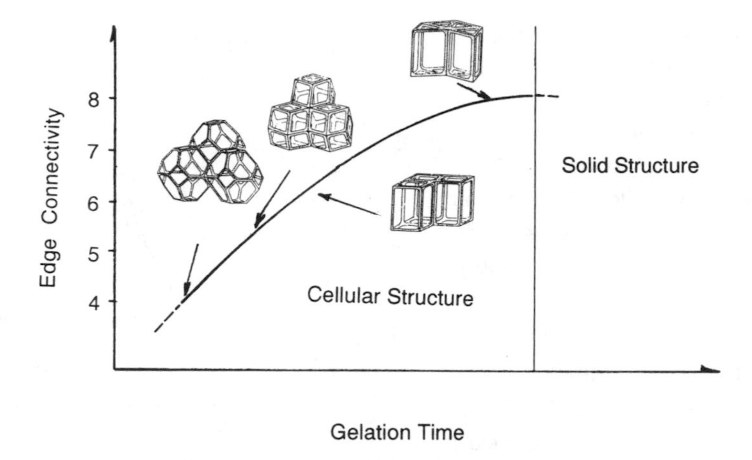

Figure 3. Ideal cellular structure of Ormosils varies with gelation time.

solid Ormosils, A* is the cross section area of the cellular Ormosils, L is the length of solid Ormosils, L* is the length of the cellular Ormosils, and S is the fracture strength of cellular Ormosils. Size effect can be neglected, since the sizes of the samples tested were constant. The strength per unit volume vs. edge connectivity is shown in Fig. 6. It was found that the strength/unit volume of Ormosils increased with increasing edge connectivity, and the strength/unit volume of solid material , Fs/V, was estimated to be about 9.5 MPa/cm.

The mechanical properties as shown apparently depend on the edge connectivity of the Ormosils structure. When the structure becomes more open and the edge connectivity decreases the high free volume allows a chain movement of PDMS and rearrangement of structure so that the "rubbery" behavior becomes more apparent.

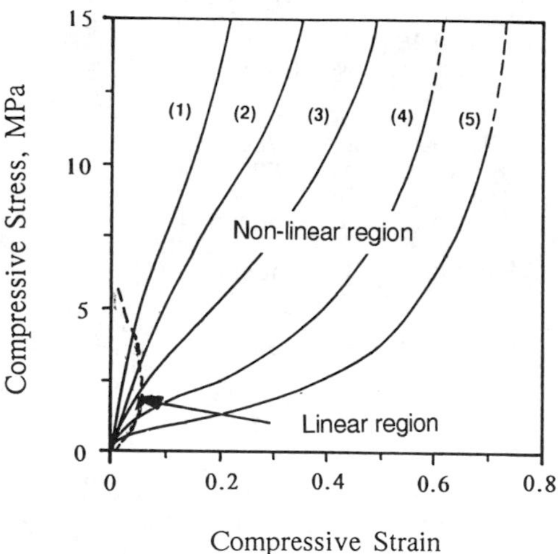

Figure 4. Compressive stress-strain curves of different resultant Ormosils.

CONCLUSIONS

The structure of Ormosils varies with gelation time and can be controlled by adjusting the reaction temperature, water content, and

HCl concentration. Using edge connectivity to characterize the structure of Ormosils shows a good way to correlate gelation time with mechanical properties. Edge connectivity was found increase as the gelation time increased. The elastic modulus and fracture strength/unit volume both increase as edge connectivity increases.

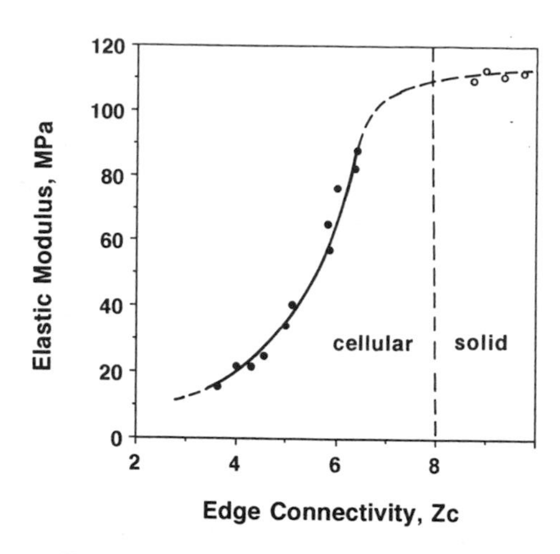

Figure 5. Elastic modulus versus the edge connectivity of Ormosils.

ACKNOWLEDGEMENT:
We are grateful to the Airforce Office of Scientific Research, Directorate of Chemistry and Materials Science for the Support of this work. A research grant donated by the Scientific Laboratory of the Ford Motor Company is also gratefully acknowledged.

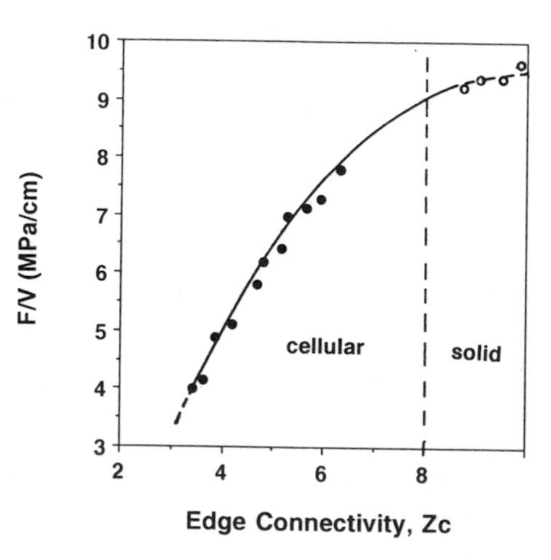

Figure 6. Compressive fracture strength/unit volum versus the edge connectivity of Ormosils.

References

1. H. Schmidt, Mat. res. Soc. Symp. Proc. **32** (1984) 327.
2. G. L. Wilkes, B. Orler, and H. Huang, Polym. Prepr. (Am. Chem. Soc., Div. Polym. Chem.) **26** (1985) 300.
3. H. Schmidt, J. Non-Cryst. Solids **73** (1985) 681.
4. D. Ravaine, A. Seminel, Y. Charbouillot, and M. Vincens, J. Non-Cryst. Solids, **82** (1986) 210.
5 B. Wang, A. B. Brennan, H. Huang and G. L. Wilkes, J. Macromol. Sci.-Chem., **27** (1990) 1447.
6. G. Philipp and H. Schmidt, J. Non-Cryst. Solids, **63** (1984) 283.
7. H. Schmidt and H. Wolter, J. Non-Cryst. Solids, **121** (1990) 428.
8. E. J. A. Pope and J. D. Mackenzie, Mater. Res. Soc. Bulletin (1987) 29.
9. H. Huang, B. Orler and G. Wilkes, Polym. Bull., **14** (1985) 557.
10. R. H. Glaser and G. L. Wilkes, Poly.r Prep., **25** (1987) 236.
11. Y. J. Chung, S. Ting, and J. D. Mackenzie, in: *Better Ceramics through Chemistry*, Mater. Res. Soc. Symp., 180 (1990)
12. Y. Hu and J. D. Mackenzie, to be published in J. Mater. Sci. (1992)

MONITORING OF HETEROPOLYSILOXANE THIN FILM PROPERTIES USING RHEOLOGICAL AND ACOUSTIC CHARACTERIZATION METHODS

C. GUIZARD*, P. LACAN**, J.M. SAUREL***, F. HAJOUB*** and L. COT*
*Laboratoire de Physicochimie des Matériaux CNRS-URA 1312
ENSC - 8, rue de l'Ecole Normale. 34053 Monpellier Cedex 1 (France).
**Kodak-Pathe, Centre de Recherches et de Technologie,
Zone Industrielle, 71102 Chalon sur Saône Cedex (France).
***Laboratoire de Microacoustique, Université de Montpellier II,
Place Eugène Bataillon, 34060 Montpellier Cedex (France).

ABSTRACT

This paper emphasizes on the mechanical characterization of heteropolysiloxane (HPS) derived materials using rheology and acoustic microscopy methods. Dynamic rheological measurements have been performed at the sol and the gel stages with two different precursors. An organic gel-like behaviour has been brought into evidence related to the fact that several microns thick coated layers can be obtained with these materials. On the contrary acoustic analysis revealed a glass-like behaviour for the cured layers. These results confirm the interest of organic/inorganic polymers to design new materials with tailored properties.

INTRODUCTION

Heteropolysiloxanes (HPS) have been emphasized as belonging to a new class of material, ORMOSIL, from which very versatile coating systems can be prepared [1,2]. Most of HPS precursors are based on modified silicon alkoxides offering the possibility to use the sol-gel route to synthesize tailored materials. They can be processed at relatively low temperature yielding dense or microporous layers with thicknesses being classed in a one to several microns range. Related to their mixed organic/inorganic nature an intermediate viscoelastic behaviour between purely organic and purely inorganic polymers is expected. Moreover HPS thin layers exhibit improved physical properties compared to pure organic polymers. The aim of this paper is to quantify the basic properties of HPS in terms of viscoelastic behaviour at the sol and gel states and mechanical properties on coated thick films.

In gelation processes, molecules or particles connect into a three-dimensional network. The molecular architecture determines the rheological properties during the evolution of the system. As far as sol-gel processing of simple alkoxides is concerned, the sol-gel transition consists in clusters growing through condensation reactions until these clusters collide, producing a gel. A dynamic rheological investigation of these systems will differ from an organic polymer gel-like system in which typical characteristics are based on molecular size, chain stiffness and cross-linking level [3]. HPS offer the possibility to prepare a three-dimensional network in which either the inorganic part or the organic part can be qualified as the network backbone. On that account, two kinds of gels have been prepared in this work, one with organic links formed at the precursor stage, the other with organic links formed during the gelation step. This approach gave us the opportunity to investigate the relationship which can exist between the structural organization and the rheological properties of these HPS derived materials.

In this paper, rheological measurements concern properties of HPS derived sols and gels. At the solid state and specially with coated materials, difficulties arise to go further in the characterization of mechanical properties. At this stage of material processing, acoustic microscopy can be substituted to rheological measurements. The application of the technique called "Acoustic Material Signature" (AMS) allowed calculation of Young modulus and Shear modulus.

EXPERIMENTAL

HPS crosslinked materials were prepared by using sol-gel routes with two different organically modified silicon alkoxide precursors. Precursor n°1, $CH_3(OC_2H_5)_2Si(CH_2)_3NCHCO(C_6H_4)NHCO(CH_2)_3Si(OC_2H_5)_2CH_3$, was prepared by reacting at 25°C stoichiometric quantities of aminopropylmethyldimethoxysilane with terephtalate chloride in anhydrous chloroform (4). Hydrochloric acid resulting from the reaction was eliminated using pyridine as a trapping reagent. Once the precursor had been purified a one step gelling process was performed in ethanol with a precursor concentration of 30 w.%. The water/alkoxide molar ratio, h, was equal to 3 and hydrochloric acid (1 w%) was used as reaction catalyst. Precursor n° 2, $(CH_3O)_3Si(CH_2)_3OCO(CH_3)=CH_2$, was purchased from Fluka and differs from the previous one with the presence of a methacrylic group fixed on the silicon atoms. It was reacted in a two step process in order to obtain an organic/inorganic interpenetrated network. This precursor was dissolved (12 w.%) in THF and methacrylic groups fixed on silicon were reacted to form primary organic polymer chains. Then the alkoxo groups were hydrolysed and condensed using similar conditions to precursor n°1 in order to form the secondary -Si-O-Si- inorganic network. The evolving crosslinking processes for the two precursors were investigated by gel permeation chromatography, liquid and solid NMR spectroscopies in order to check that proper reactions occured in each step of gel preparations.

Rheology measurements on sols and gels were performed using a Haake dynamic rheometer equipped with both coaxial cylinders and plate sensors. Acoustic measurements were performed on thick films spincoated on glass plates and cured at 80°C for one hour. HPS film thicknesses were respectively 4.6 μm and 15 μm for precursor n°1 and 2 μm and 4.9 μm for precursor n°2. A reflexion acoustic microscope equipped with a single lens was used for both transmitting and receiving the acoustic signal. Acoustic measurements were performed with a coupling liquid (water) placed between the lens and the analysed material. Acoustic frequencies of 58, 140 and 600 MHz were used to characterize the different samples.

RESULTS AND DISCUSSION

Dynamic rheological characterization

Rheological behaviour of alkoxide derived sols and gels have been described in the literature [5,6,7,8] but very little is known about mixed organic/inorganic gels. In dynamic measurements, the storage (G') and loss (G") moduli as well as the phase shift δ between stress and strain are defined by the theory of linear viscoelasticity and contained in the real part and the imaginary part of the complex modulus G* :

$$G^* = G' + iG'' \quad (1)$$
$$G' = [G^*] \cos \delta \quad (2)$$
$$G'' = [G^*] \sin \delta \quad (3)$$

with tang δ = G"/G'(loss tangent), defined as the ratio of energy lost to energy stored.

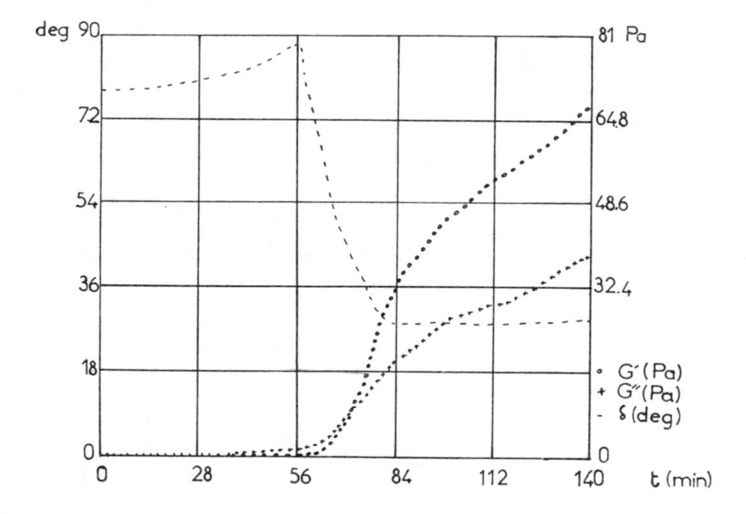

Figure 1. Plots of storage (G') and loss (G") moduli versus aging time for a sol-gel system prepared from HPS precursor n°1 with used frequency $\omega = 6.3\ s^{-1}$ and amplitude $\gamma = 3°$.

Figure 1 shows the evolution of G' and G" moduli as well as the δ angle evolution during the isothermal sol-to-gel transition of precursor n°1. The crossover of the storage and loss moduli during the evolving crosslinking process can be defined as the gel point. At the beginning of the process G" was higher than G' showing a liquid-like behaviour while above the gel transition a solid-like behaviour was evidenced. One can see that in this case, expectations of a maximum in tang δ or G" at the gel point are not in agreement with the observed behaviour. The relative evolution of G' and G" versus frequency at different stages of the sol-to-gel transition has also been investigated. Two main features of these organic/inorganic systems can be pointed out from the curves on Figure 2. Before reaching the gel point the shape of G' and G" curves corresponds to the behaviour of tetraethoxysilane derived gels described in references [6] and [7]. This is typical of a cluster based system in which the energy, W, dissipated in a viscous flow is affected by the excitation frequency. The maximum rate in dissipated energy is observed at the natural frequency for cluster motion. One can see that this maximum for G" is shifted to lower frequencies when clusters grow up in the evolving sol. Above the gel point the value for storage modulus is always higher than for the loss modulus for all the frequencies investigated. This state corresponds to the transition from a material without permanent elasticity to a material with permanent elasticity. This result confirms an organic gel-like behaviour previously described with polydimethylsiloxane systems [9]. Quite similar results were observed with precursor n°2 except that the loss modulus exhibits a higher value at the beginning of the process due to the presence of the methacrylic chains.

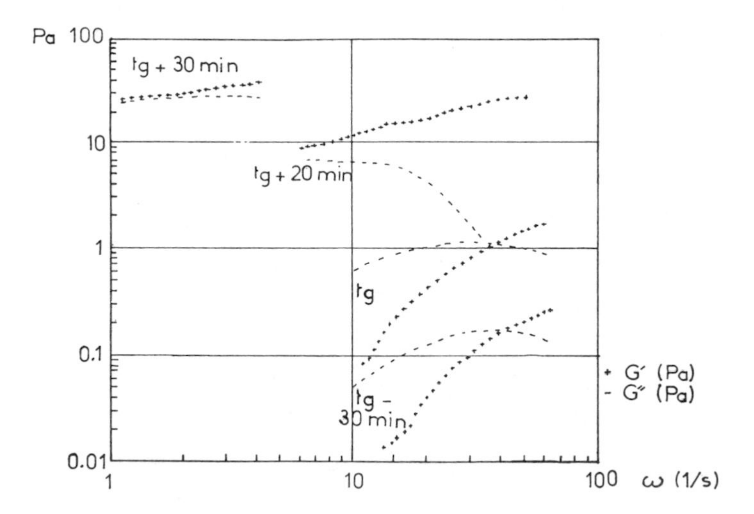

Figure 2. Evolution of G' and G" versus frequency at successive stages of the sol-to-gel transition (system n°1), $\gamma = 3°$.

Acoustic characterization

Quantitative informations on mechanical properties can be obtained when measuring the output signal intensity V (volts) generated at the tranducer versus the defocus z (μm) of the lens. The V(z) response is also dependent of the reflexion coefficient of the material and the opening angle of the lens. The simplest application of the so-called Acoustic Material Signature (AMS) technique is the direct calculation of the surface velocity by measuring the periodicity of the V(z) curves. A numerical analysis of these curves using fast Fourier transform (FFT) can improve the determination accuracy of the Rayleigh velocity V_R.

$$V_R = V_{liq} / [1 - (1 - V_{liq} / 2 F \Delta z)^2]^{1/2} \quad (4)$$

where V_{liq} is the velocity in the coupling liquid, F the frequency of the acoustic wave and Δz the oscillation periodicity [10,11].

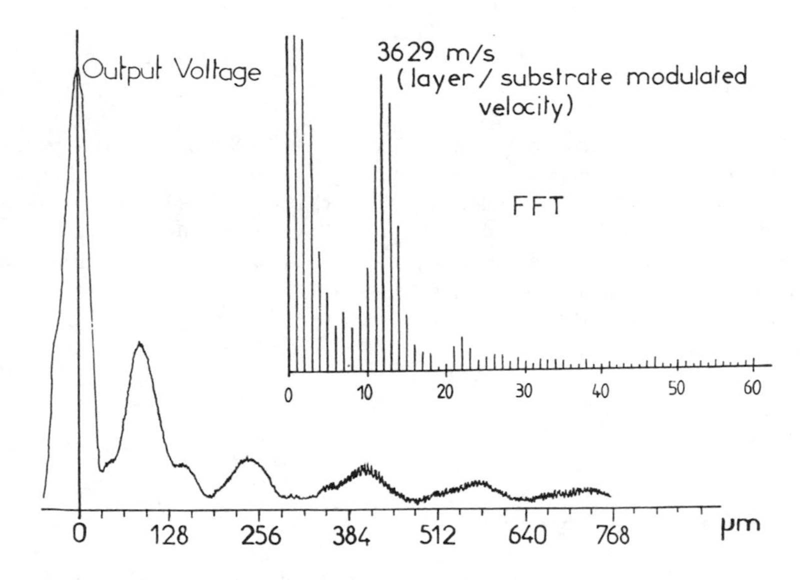

Figure 3. V(z) curve and its FFT for an HPS layer (system n°2), thickness 4. 9 μm, frequency 140 MHz .

An example of a V(z) experimental curve and its FFT is given in Figure 3. From V_R we can determine a longitudinal velocity V_L and a shear velocity V_T used in the calculation of the Young modulus E and the shear modulus G. A comparison of the mechanical properties of the two sorts of material prepared in this work with a pure silica glass and an organic polymer film (polyethylene) is provided in Table I. One can see that these HPS materials, which behave as organic polymers at the gel state, exhibit mechanical characteristics close to glass materials once cured at relatively low temperatures.

Table I. Comparison of Young modulus E and shear modulus G of HPS thin films to glass and organic materials measured by reflective acoustic measurements.

Sample	Glass	HPS type1	HPS type2	Organic polymer
V_L(m/s)	5660	6000	5680	2800
V_T(m/s)	3420	2740	2590	1600
E (GPa)	58	43	37	8
G (GPa)	22	15	14	3

CONCLUSION

Rheological and acoustic investigations on HPS materials confirmed the unusual mechanical properties expected for organic/inorganic polymers. The unique advantage of this new class of material is that it can be processed as organic polymers with a glass-like behaviour at the final stage. This is a very profitable approach in coated materials because of the possibility to obtain tailored coatings with several microns thicknesses, on various substrates and at low temperature. Finally we want to point out acoustic microscopy as a non-destructive powerful method to characterize mechanical properties of coated materials in the micron range.

References

1. H. Sholze, J. Non-Cryst. Solids, 73, 669 (1985).
2. H.K. Schmidt, ACS Symp. Ser. 360 (27), 333 (1988).
3. H. Henning Winter, MRS Bulletin, XVI (8), 44 (1991).
4. C. Guizard, N. Ajaka, M.P. Besland, A. Larbot and L. Cot, in Polyimides and other High-Temperature Polymers, edited by M.J.M. Abadie and B. Sillion, (Elsevier Science Pub. B.V., Amsterdam, 1991) , p. 537.
5. B. Gauthier-Manuel, E. Guyon, S. Roux, S. Gits and F. Le Faucheux, Journal de Physique, 48 (5), 869 (1987).
6. M.D. Sacks and R.S. Sheu in Science of Ceramic Chemical Processing, edited by L.L. Hench D.R. Ulrich (Wiley, New York, 1986), p. 100.
7. M.D. Sacks and R.S. Scheu, J. Non Cryst. Solids, 92, 383 (1987).
8. C. Guizard, J.C. Achddou, A. Larbot and L. Cot, to be published in J. Non-Cryst. Solids, Special issue on the Six International Workshop on Glasses and Ceramics from Gels, Seville 1991.
9. F. Chambon and H. Henning Winter, Polymer Bull. 13, 499 (1985).
10. A. Atalar, J. Appl. Phys., 50, 8237 (1979).
11. J. Attal, L. Robert, G. Despaux, R. Caplain and J.M. Saurel, to be published in "19th Int. Symp. Proceed. on Acoustical Imaging", Bochum (Germany), April 3-5, 1991.

THE EFFECT OF ULTRASONIC RADIATION ON GELATION AND PROPERTIES OF ORMOSILS

Kazuki Morita, Yi Hu and John D. Mackenzie
University of California Los Angeles, Department of Materials Science & Engineering, Los Angeles, CA 90024

ABSTRACT

Ultrasound has been applied to organically modified silicates (ORMOSILs) solutions during gelation. A series of silica and polydimethylsiloxane (PDMS) composites have been prepared by the sol-gel method. Tetraethoxysilane (TEOS), PDMS, water and a HCl catalyst were mixed and exposed to 20 kHz ultrasonic radiation under different conditions. Ultrasonic radiation made it possible to obtain wet gels in 5 to 30 min. without solvent (e.g. alcohol, and THF). And exothermic reaction was observed due to hydrolysis and polycondensation of TEOS. The reaction process was examined by liquid state ^{29}Si NMR. Densities of obtained sono-ORMOSILs were observed to be higher than those produced by ordinary processes. Some sono-ORMOSIL samples demonstrated extremely low porosity and low specific surface areas.

INTRODUCTION

Sol-gel processing has been used to synthesize organic and inorganic compounds such as ORMOSILs, and organic dye doped glasses. [1, 2, 3] However, obtained gels are porous and therefore have low densities.

Ultrasonic radiation is known to enhance some chemical reactions, such as polymerization and decomposition. [4, 5, 6] The effects of ultrasound on the gelation process were examined with regard to TEOS and TMOS based solutions. [7,8,9,10] It was found that denser dry gels could be obtained without solvent addition.

In this study, the effects of ultrasonic radiation on gelation and subsequent properties of ORMOSILs were investigated. The physical properties and microstructure of the obtained samples were characterized to demonstrate the advantages of using ultrasonic radiation.

EXPERIMENTALS

TEOS, PDMS, water and HCl were used as starting materials. TEOS was mixed with 0, 10, 20, 40 and 70 wt% PDMS. The molar ratio of H_2O/TEOS and HCl/TEOS was held constant for these samples at 2 and 0.05, respectively. Other samples were made using 10 wt% PDMS with H2O/TEOS varying from 2 to 8 and the HCl/TEOS varying from 0.025 to 0.2.

All solution were prepared with a total volume of 60 ml and were placed into a 100 ml beaker before being subjected to 20 kHz ultrasonic radiation. All the samples demonstrated phase separation before the ultrasonic radiation. Ultrasound wave was supplied by a Fisher Scientific Sonic Dismembrator 550 with an 8 mm diameter probe that was placed into the solution. The supplied power was kept constant at 70 W throughout the experiment. The ultrasonic radiation was supplied long enough to accomplish homogeneous mixing but shorter than time needed for gelation. Temperature measurements were taken using a thermometer until they leveled off at a constant value.

The ultrasound solution was poured into petri dishes and covered with a paraffin film. Samples were kept at room temperature (22 ± 3 °C) until 10 days after gelation. Small pin holes were then placed into the paraffin film. Some samples were further dried at 60 °C for 7 days.

Reaction mechanisms were studied by ^{29}Si NMR (Bruker AM360/Wb) at ambient temperatures with the reference of Tetramethylsilane. A series of solution that had been quenched in liquid nitrogen at different stages of ultrasound were analyzed by NMR spectroscopy shortly after melting. Bulk densities of the gels were measured by the archimedes method and porosity was determined by nitrogen sorption.

RESULTS AND DISCUSSION

Temperature Behavior

Temperatures of TEOS, PDMS, water and HClaq. were observed to increase constantly when the ultrasound was irradiated to them separately. The temperature increase is considered to be due to the collapse of vapor bubbles in the liquid during ultrasonic irradiation. When the components were mixed and subjected to ultrasonic irradiation, the solution temperature increased drastically until leveling off at 78 °C. (Fig. 1) This temperature corresponds to the boiling point of ethanol which is produced during the hydrolysis of the TEOS. This increase in temperature was much larger than that calculated from observed temperature increases of individual components during ultrasonic irradiation and each heat capacity, indicating that this excess heat is generated due to the exothermic reactions of hydrolysis and polycondensation of TEOS.

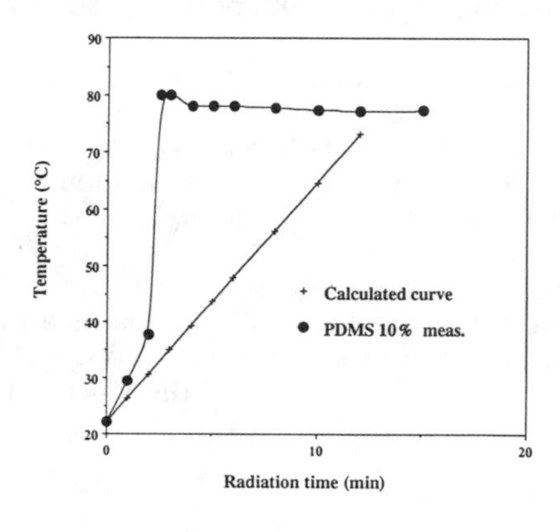

Fig. 1 Change in temperature of the solution during sono-gel reactions.

(The calculated curve is based on the observed temperature increases of individual components during the ultrasonic irradiation and their heat capacities.)

Gelation Time

Gelation time is dependent on the PDMS, water, and HCl concentration along with ultrasound radiation time. Increasing the amount of acid and water gives higher reaction rates. However, the amount of PDMS did not have a simple relationship to gelation time. It was found that the gelation time is negatively proportional to the length of the ultrasound period. (Fig. 2) This result is not consistent with that for pure TEOS, where the logarithm of gelation time is negatively proportional to the irradiation time. [10]

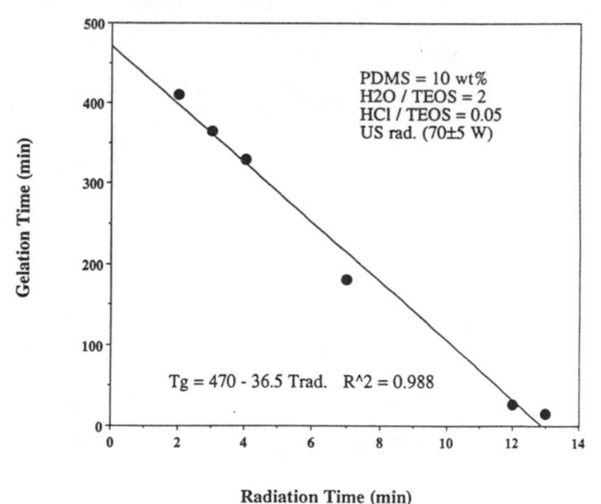

Fig. 2 Ultrasonic radiation time v.s. gelation time.

Reaction Mechanism

The ^{29}Si NMR data for samples with 10 wt% PDMS, a H_2O to TEOS ratio of 2, and a HCl to TEOS ratio of 0.05 are given in Fig. 3 and can be characterized as follows:

- For the first 30 seconds, all TEOS monomers were hydrolyzed and condensation started leading to the growth of the Q^2 and Q^3 peaks.
- For the next 2.5 minutes further condensation occurs leading to the growth of the Q^3 and Q^4 peaks.
- After Q^3 and Q^4 peaks level off the PDMS peak, D, begins to decrease and new peaks, which include that of a four member ring $Si_4O_4(CH_3)_8$ (19.3 ppm), appear and begin to grow.

These results mean that the hydrolysis and polymerization of TEOS begins immediately followed by breaking of PDMS chains and then formation of other structures including a four member ring $Si_4O_4(CH_3)_8$. Unidentified peaks could include that of the silicon in PDMS bonded to hydrolyzed TEOS.

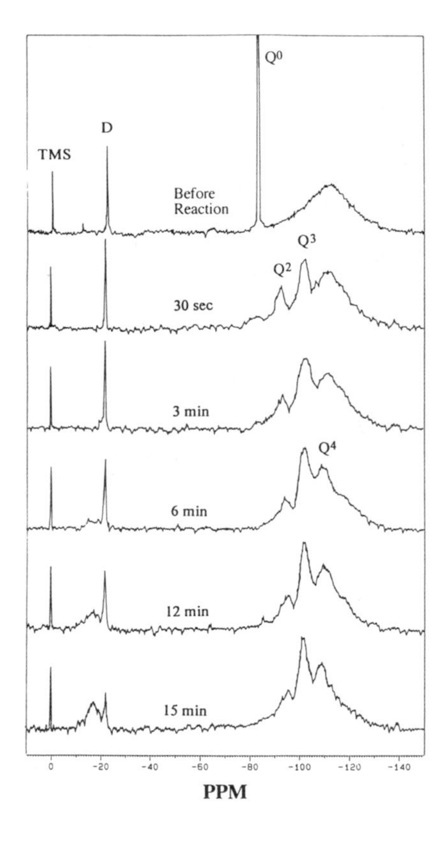

Fig. 3 Silicon - 29 spectra of 10 wt% PDMS - 90 wt% TEOS solutions during sol-gel reactions under the radiation of 70 W, 20 kHz ultrasound. (H_2O/ TEOS = 2, HCl/TEOS = 0.05)

Physical Properties

Since the sono-gel process does not need a solvent, the gels obtained do not contain as much alcohol as those made by traditional processes. As shown in Fig. 4, the bulk densities of sono-gel samples were 10 % higher than those made using a solvent. Fig. 5 shows the effect of water on the bulk density. High concentrations of water decrease the bulk density slightly because water enhances the hydrolysis reaction of TEOS leading to increased polymerization before reaction with PDMS. This can be seen visually as increasing the water concentration leads to decreasing transparency. As a result, the structure of the ORMOSIL becomes less rigid. However, there was no detectable difference in specific surface area between low and high concentration water samples. All the samples had low surface areas of less than 1 m^2/g and low porosities of less than 0.1 %. For the same compositions using a solvent and without ultrasound the surface areas were found to be between 200 and 800 m^2/g, and the porosities between 20 and 50 %. Differences in the amount of catalyst did not seem to have much effect on the density.

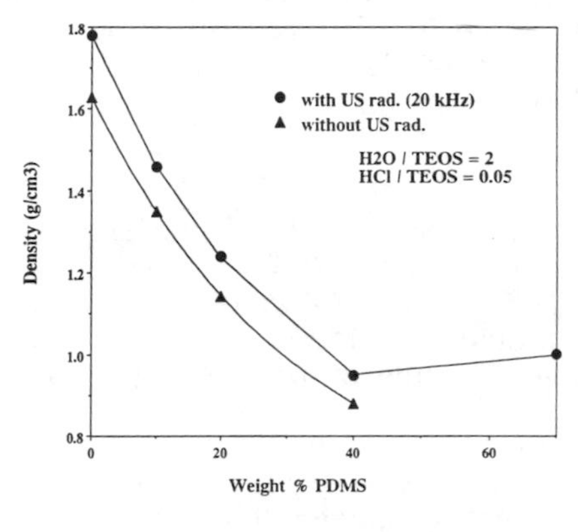

Fig. 4 Densities of ORMOSILs prepared with ultrasound and without ultrasound.

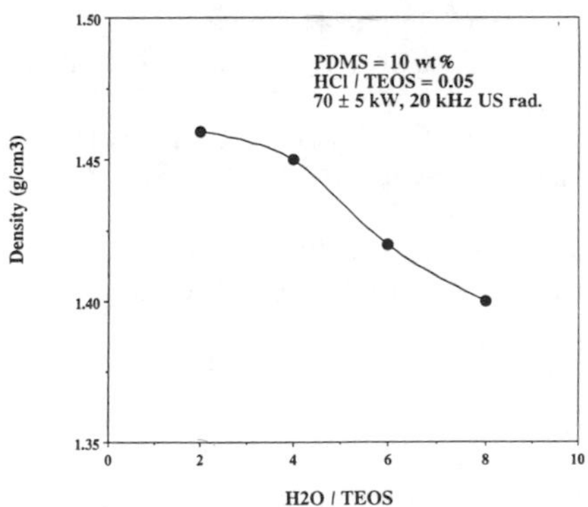

Fig. 5 Effect of H_2O/TEOS on the densities of ORMOSILs.

CONCLUSIONS

1. Ultrasonic radiation (20 kHz, 70 W) enables the synthesis of ORMOSILs without a solvent.
2. Ultrasonic radiation enhanced hydrolysis and polycondensation of TEOS. These exothermic reactions caused the thermal runaway of solutions.
3. ORMOSILs obtained with ultrasound have higher densities than those without ultrasound. These dense samples, which were prepared from 10 wt% PDMS solutions, had surface area lower than 1 m^2/g.
4. ORMOSIL composition and ultrasonic radiation time significantly affected the gelation rate and was found to be negatively proportional to the radiation time.

ACKNOWLEDGMENT

We are grateful to the Airforce Office of Science Research, Directorate of Chemistry and Materials Sciences for support of this work. A research grant donated by the Scientific Laboratory of the Ford Motor Company is also gratefully acknowledged.

REFERENCES

1. B. Lintner, N. Arfsten, H. Dislich, H. Schmidt, G. Philipp, and B. Seiferling, J. Non-Cyst. Solids, 100, pp.378-382 (1988)
2. Y. Hu, and J. D. Mackenzie, J. Mater. Sci. (1992) to be published
3. A. Makishima, and T. Tani, J. Amer. Ceram. Soc., 69, c72 (1986)
4. M. E. Fitzgerald, V. Griffing, and J. Sullivan, J. Chem. Phys., 25, No.5, pp.926-932 (1956)
5. P. Kruus, Ultrasonics, Sept. pp.201-204 (1983)
6. K. S. Suslick, ed. Ultrasound, Its Chemical, Physical, and Biological Effects, VCH Publishers, Weinheim RFA (1988)
7. M. Tarasevich, Ceram. Bull., 63, p.500 (1984)
8. L. Esquivias, and J. Zarzycki, Proc. 1st. Int. Workshop on non-Cryst. Solids, pp.409-414 (1986)
9. L. Esquivias, and J. Zarzycki, 3rd Int. Conf. on Ultrastructure Processing of Ceramics, Glasses and Composites, pp.255-270 (1987)
10. N. de la Rosa Fox, L. Esquivias, and J. Zarzycki, 2nd Int. Conf. The Effects of Modes of Formation on the Structure of Glasses, pp.363-374 (1987)

PREPARATION OF ARYL-BRIDGED POLYSILSESQUIOXANE AEROGELS

Douglas A. Loy,[◊] Kenneth J. Shea,[*] and Edward M. Russick.[◊]

◊ Org. 1812, Sandia National Laboratories, Albuquerque, NM 87185

*Department of Chemistry, University of California Irvine, Irvine, CA 92717

ABSTRACT

We report the preparation of a new class of organic/inorganic hybrid aerogels from aryl-bridged polysilsesquioxanes [1]. 1,4-Bis(triethoxysilyl)benzene and 4,4'-bis(triethoxysilyl)-biphenyl were sol-gel processed to form phenyl- and biphenyl-bridged polysilsesquioxane gels (Figure 1, structures **1** and **2**, respectively). The gels were then dried using supercritical carbon dioxide extraction. It was discovered that aryl-bridged polysilsesquioxane aerogels are indeed formed, but with a pronounced influence on surface area from the reaction conditions used in preparing the initial gels. Specifically, high surface aerogels (up to 1750 m^2/g) are obtained from gels prepared with either acid or base catalysts. With reduced concentrations of base catalyst, however, supercritical processing afforded phenyl-bridged *xerogel-like materials*. The materials were characterized by nitrogen sorption surface analysis, and by transmission electron microscopy.

INTRODUCTION

Aerogels are gels which have been dried using a supercritical extraction technique thereby avoiding shrinkage and fracturing associated with simple evaporative drying[2]. Since Kistler's discovery of aerogel formation [3] there have been numerous reports of aerogels derived from metal oxides [4], crosslinked organic polymers [5], and even biological polymers[3]. We have used supercritical carbon dioxide extraction to prepare aerogels of aryl-bridged polysilsesquioxanes, a class of hybrid organic-inorganic materials [1]. These aryl-bridged polysilsesquioxanes differ from silica and other metal oxide materials in that a rigid organic (aryl) spacer is inserted at regular intervals into the silica network. The gels are prepared by the hydrolysis and condensation of bis(triethoxysilyl)-phenyl **1** and -biphenyl **2** under either acid or basic conditions in tetrahydrofuran prior to supercritical extraction. Gels were prepared with varying amounts of catalyst with constant amounts of H_2O in order to determine if aerogel structure varied with catalyst concentration.

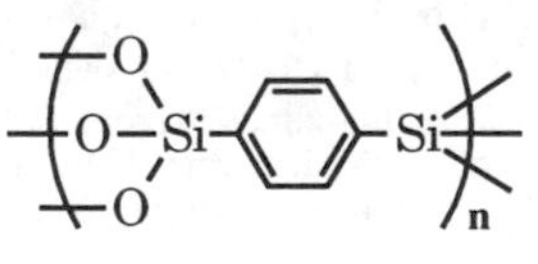

Figure 1. Phenyl- and Biphenyl-Bridged Polysilsesquioxanes.

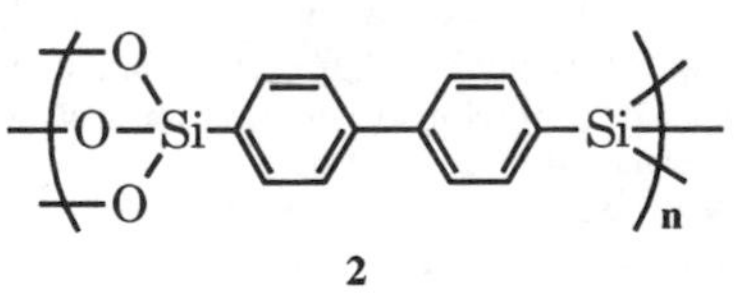

Mat. Res. Soc. Symp. Proc. Vol. 271. ©1992 Materials Research Society

EXPERIMENTAL PROCEDURES

1,4-Bis(triethoxysilyl)benzene **1** and 4,4'-bis(triethoxysilyl)biphenyl **2** were prepared from the aryldibromides using Barbier-Grignard chemistry and are described elsewhere [1]. Gels were prepared by both acid (HCl) and base (NH_4OH) catalyzed reactions at monomer concentrations between 0.1-0.4 M in tetrahydrofuran (Table 1). The time required for gelation to occur was dependent upon both concentration and type of catalyst used. The gels were allowed to stand in sealed containers under excess tetrahydrofuran for periods up to one year without any visible changes. The gels were dried by first exchanging the THF with methanol at room temperature for at least three days. The samples were then supercritically extracted at 1400 psig carbon dioxide at 35 °C for three days with intermittent purging of solvent ladden carbon dioxide and simultaneous recharging with fresh carbon dioxide. Upon depressurizing the vessel, the gels are completely dry.

RESULTS

Hydrolysis and condensation of 1,4-bis(triethoxysilyl)benzene and 4,4'-bis(triethoxysilyl)biphenyl (0.2-0.4 M in tetrahydrofuran) result in the formation of transparent gels with a blue tint upon reflection, and yellow with transmission. The gels are firm, but can be sliced with a razor. With air drying, the materials lose between 90-95% of their volume from shrinkage to afford transparent xerogels. Extraction of a 0.2 M biphenyl-bridged polysilsesquioxane gel with supercritical CO_2 afforded an opaque white gel (Figure 2). A series of phenyl-bridged polysilsesquioxane gels in tetrahydrofuran were prepared using different amounts of either HCl or NH_4OH with the monomer: water ratio held constant over most of the series. The gels (three identical gels for each formulation) were originally prepared to examine the effect of catalyst concentration on the time required for gelation to occur. Formulations with 0.054 mol% HCl and 0.54 mol% NH_4OH failed to gel, even after 1500 hours. The gels prepared with 0.54 mol% HCl and 5.4 mol% NH_4OH each required more than 100 hours to gel, yet appeared identical to the other gels in the series. The only visible differences noticed were that the base catalyzed gels, as a rule, contracted slightly from the walls of the glass more quickly than acid catalyzed gels.

After supercritical extraction, we discovered that the amount of shrinkage experienced by the drying gels was related to the concentration of catalyst used in their preparation. The yield of the polymerizations, based on moles monomer, were all between 100-112% indicating that the lower surface areas were not simply the product of lower % conversion with less catalyst. Product yields are typically in this range because 100% condensation is never achieved and there are residual silanol and ethoxysilyl groups adding to the mass of the gels prepared [1c-f].

Surface area analysis using nitrogen sorption porosimetry [6] revealed an interesting trend (Table 2). With NH_4OH as catalyst, the surface area was directly proportional to the concentration of catalyst used. The phenyl-bridged polysilsesquioxane prepared with 540 mol% NH_4OH was found to have a surface area of 1670 m^2/g, over 500 m^2/g higher than previously reported for aryl-bridged polysilsesquioxanes (xerogels) [1] and higher than any silica aerogel surface area we could find in the literature [7]. The phenyl-bridged gel prepared with only 0.54 mol% NH_4OH not only took longer to reach gel point, but had a surface area of only 269 m^2/g.

Table 1 Preparation of Phenyl-Bridged Polysilsesquioxane Gels: Gelation Time.

Catalyst (mol%)	H_2O (equiv.)	**Gelation Time** (hrs)[a]
HCl 10.0	4.0	5.0
" 5.40	3.0	31.0
" 0.54	3.0	232
" 0.054	3.0	none[b]
NH_4OH 0.54	3.0	none[b]
" 5.40	3.0	720
" 54.0	3.0	151
" 540	15.0	4.3

a. Average for three identical 0.4 M solutions (5.0 mL) in THF. b. No gelation after 1500 hours.

Figure 2. Photograph of Biphenyl-Bridged Polysilsesquioxane Aerogel Prepared by Supercritical CO_2 Extraction.

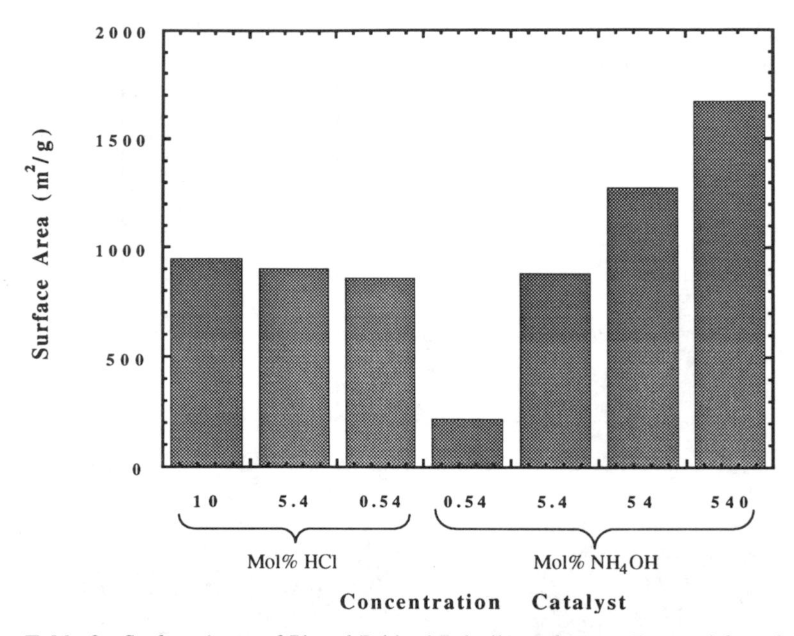

Table 2. Surface Areas of Phenyl-Bridged Polysilsesquioxanes Prepared from 0.4 M 1,4-Bis(triethoxysilyl)benzene in THF with HCl or NH_4OH catalysts. Each bar represents an average value obtained for two identically processed gels.

Gels prepared with different amounts of HCl as catalyst exhibited a similar but less dramatic decrease in surface area with concentration of catalyst from 962 m^2/g for the gel made with 10 mol% HCl to 846 m^2/g for the gel made with 0.54 mol% catalyst. The calculated mean pore diameter for the gels followed a similar trend. Average pore sizes ranged from 160 Å for the 540 mol% NH_4OH gel down to 35 Å for the 0.54 mol% gel and from 53 Å for the 0.54 mol% HCl gel to 76 Å for the 10 mol% HCl gel. Total pore volume for the gels also displayed a minimum with 0.54 mol% NH_4OH catalyst. The one measurement which remained relatively constant was the micropore volume. The differences in the volume and surface areas seem to be primarily dependent on the presence and amount of mesopores. Isotherms of the dried gels prepared with 0.54 mol% base were more type I or microporous. Similar isotherms were observed with the air-dried xerogels prepared from the same materials [1]. The effect of catalyst concentration on surface area was also reported for silica-siloxane copolymers [8]. Surface areas of these materials were observed to increase with increasing catalyst concentrations and the increases were attributed to increasing mesoporosity. The biphenyl-bridged polysilsesquioxane aerogels (Figure 2), analyzed by nitrogen sorption, were shown to have surface areas as high as 1765 m^2/g.

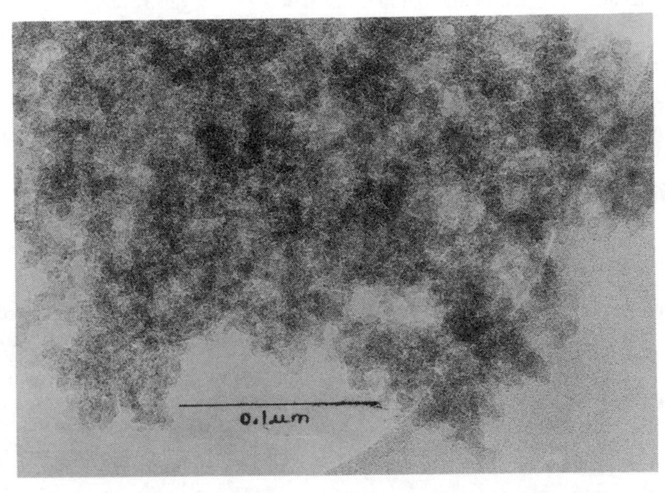

Figure 3a Transmission Electron Micrograph (600,000 X) of Phenyl-Bridged Polysilsesquioxane Prepared with 10 Mol% HCl.

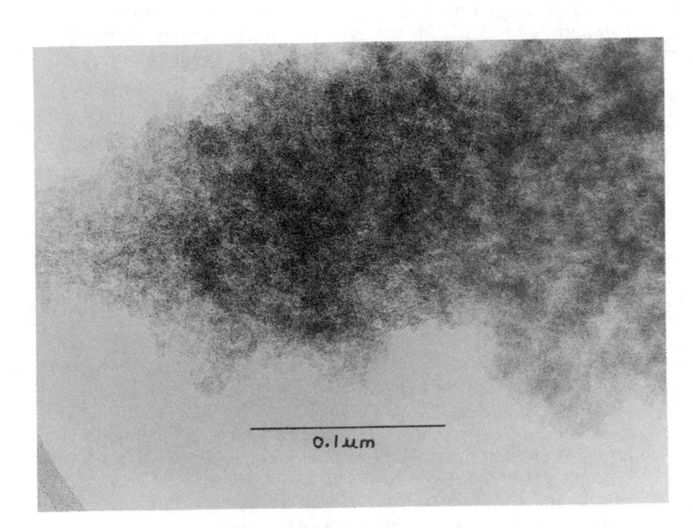

Figure.3b. Transmission Electron Micrograph (600,000 X) of Phenyl-Bridged Polysilsesquioxane Prepared with 540 Mol% NH_4OH.

Transmission electron micrographs taken of the dried gels reveal that both the acid (10 mol% HCl, Figure 3a) and base (540 mol% NH_4OH, Figure 3b) formulations give rise to similar sized structures. There are no radical differences visible in the micrographs corresponding to the differences in pore structure as revealed earlier by gas porosimetry. The size scale on the micrographs suggests that the grain-like structures are on the order of 100 nm in diameter or less [1]. Surprisingly, it would appear that the structures or grains in the aerogel prepared with 10 mol% HCl are slightly larger (60 nm) than those in the base catalyzed material (30 nm).

CONCLUSIONS

Supercritical CO_2 extraction of aryl-bridged polysilsesquioxanes has been shown to permit the preparation of aryl-bridged aerogels. Formation of aerogels does not, however, seem to be solely dependent on the mode of drying. Perhaps more important was the concentration of catalyst used, particularly under alkaline conditions. Creation or retention of mesoporous structure would appear to be the critical goal for preparing the high surface area aerogels.

Closer examination of the dried phenyl-bridged polysilsesquioxanes structure using small angle x-ray and neutron scattering, coupled with atomic force microscopy may help to illuminate the differences in the dried gel structure in these unusual materials. In addition, experiments are planned to examine the degree of hydrolysis and condensation in the gels to determine if and how important the contribution of the sol-gel chemistry may be for these systems.

ACKNOWLEDGEMENTS

We would like to thank Robert Dosch for the use of his gas sorption porosimeter, and Linda McLaughlin for her invaluable assistance in running the surface area analyses. We must also gratefully acknowledge the expertise of Thomas Headley, who provided us with transmission electron micrographs. This work was supported by the United States Department of Energy under contract # DE-AC04-76DP00789.

REFERENCES

[1] a) K.J. Shea; D.A. Loy; O.W. Webster, Chem. Materials, 1, 572, (1989). b) Shea, K.J.; Webster, O.; Loy, D.A. in Better Ceramics Through Chemistry IV, edited by B.J.J. Zelinski, C.J. Brinker, D.E. Clark, D.R. Ulrich (Mater. Res. Soc. Proc., 180, Pittsburgh, PA 1990, pp 975-980. c) K.J. Shea; D.A. Loy; O.W.Webster J. Am. Chem. Soc. In Press, 4/92. d) K.J. Shea; D.A. Loy; James H. Small; O.W.Webster. in preparation. f) D.A. Loy, Ph.D. Dissertation, University of California at Irvine, 1991.

[2] C.J. Brinker and G.W. Scherer, Sol-Gel Science: The Physics and Chemistry of Sol-Gel Processing, Academic Press, Boston, 1990, pp. 453-515.

[3] S.S. Kistler, Nature 127, 741 (1931).

[4] For reviews of aerogels see: a) J. Fricke, J. Non-Cryst. Solids 100, 169 (1988). b) H.D. Gesser and P.C. Goswami, Chem. Reviews, 89, 765 (1989).

[5] R.W. Pekala , J. Mater. Sci., 24(9), 3221 (1989).

[6] Nitrogen sorption experiments were carried using a Quantachrome Autosorb Automated Gas Adsorption System; Surface areas were obtained using the five point BET method ($0.02 < p/p_o < 0.30$).

[7] C.A.M. Mulder and J.G. Van Lierop reported a silica aerogel with a surface area of 1590 m^2/g in, Aerogels (Proceedings of the First International Symposium. Wurzburg, FRG, Sept. 1985), ed. by J. Fricke, Springer-Verlag, Berlin, 1986, p. 68.

[8] A. Kaiser, H. Schmidt, and H. Bottner, J. Membrane Sci. 22, 257 (1985).

CHARACTERIZATION OF MICROPOROUS SILICA GELS PREPARED FROM MODIFIED SILICON ALKOXIDES

W.G. Fahrenholtz and D.M. Smith
UNM/NSF Center for Micro-Engineered Ceramics
University of New Mexico, Albuquerque, NM 87131

ABSTRACT

Organically modified silica gels were prepared from mixtures of TEOS and methyl-substituted silicon alkoxides. The pore structure of dried gels was investigated in order to determine the effect of the organic additions on gel structure. For base catalyzed gels, surface area, pore volume, and skeletal density all showed dramatic decreases at high methyl contents. The surface area dropped by nearly three orders of magnitude, from over 800 m^2/g for normal silica gels to less than 1 m^2/g for gels containing over 60 mole percent modified alkoxide. The pore structure of acid catalyzed gels showed less dependence on composition. The nature of the surface of both acid and base catalyzed gels varied with modified alkoxide content. Higher organic contents produced materials that were more hydrophobic. In addition, the mechanical strength of acid catalyzed gels dropped in direct proportion to methyl content.

INTRODUCTION

Organically modified silica gels (ORMOSILS) can be prepared from any number of modified silicon alkoxides [1-4]. Gels can be prepared by several methods including a standard two step hydrolysis/condensation procedure using a mixture of tetraethoxysilane (TEOS) and an organic-substituted silicon alkoxide [5,6]. With this method it is possible to control the organic content in the gels by varying the ratio of TEOS to the modified silicon alkoxide or by changing the number or size of organic groups on the modified alkoxide. A number of important properties may be tailored by the controlled addition of organic groups. For example, it may be possible to alter mechanical properties through controlled addition of modified alkoxides since the methyl groups are unreactive and do not participate in network formation. In addition, the nature of the surface, i.e. hydrophobic vs. hydrophilic, can also be affected by the number of organic groups present. Finally, the pore structure itself can be altered by the addition of these species. Since the methyl sites do not participate in network formation, addition of methyl-substituted alkoxides forms terminal bonds within the gel structure. Terminal sites may be considered surface even though the area may be as small as the group size. This additional surface may not be accessible to the gases commonly used to probe pore structure, but the resultant drop in network connectivity can be detected as a drop in the skeletal density of the gel.

This paper will discuss the formation and characterization of gels from mixtures of TEOS and two modified silicon alkoxides, methyltriethoxysilane (MTEOS) and dimethyldiethoxysilane (DMDEOS). Gels were prepared containing one or both of the modified alkoxides. Properties of gels were studied in order to determine the effect of methyl content for both acidic and basic gelation conditions.

 ©1992 Materials Research Society

EXPERIMENTAL

Organically modified silica gels were prepared from mixtures of TEOS and modified silicon alkoxides in a method similar to that described for pure TEOS gels by Brinker et. al. [7]. Gels ranging in composition from 0 to 75 mole percent MTEOS (M = 0 to 75) and 0 to 60 mole percent DMDEOS (D = 0 to 60) were prepared by mixing various ratios of stock solutions containing 50 or 75 mole percent modified alkoxide (balance TEOS) with a standard TEOS (M = D = 0) stock solution. The composition is based on a total silicon content in solution, thus M + D + %TEOS = 100. Stock solutions of identical total silicon concentration were prepared by diluting stoichiometric amounts of the alkoxides to a standard volume with ethanol. Water and HCl were added to bring the molar ratio of silicon to water to HCl to 1 to 1 to 0.0007. The mixture was refluxed at approximately 353 K for 4 hours. Stock solutions were aged for at least 24 hours prior to use. Gels were formed under either acidic or basic conditions. Acid catalyzed gels were formed by adding 1 M HCl. The final ratio of silicon to water to HCl was nominally 1 to 5 to 0.05. After addition of the acid/water solution, the gelation vials were capped and placed in a 323 K oven. Gelation occurred in approximately 48 hours and gels were aged at 323 K for 24 hours after gelation. Base catalyzed gels were formed by the addition of 0.5 M NH_4OH. The final ratio of silicon to water to base was 1 to 4 to 0.02. Gels were prepared in sealed vials at room temperature (~ 295 K). Gelation time varied from approximately 1 hour for M = D = 0 gels to almost 72 hours for M = 75 and D = 50 gels. Base catalyzed gels were aged for 24 hours at room temperature after gelation. Both acid and base catalyzed gels were dried in the same manner. Gels were uncapped and exposed to ambient atmosphere for 5 days and then oven dried at 373 K for 24 hours. After drying, gels were lightly ground to produce a free flowing powder.

The surface area of all gels was determined by nitrogen adsorption at 77 K. At least five adsorption points were taken at relative pressures between 0.05 and 0.3. Data was reduced using the BET method [8]. Micropore area was determined from a plot of statistical layer thickness as a function of volume adsorbed, the t-plot method [8]. The adsorption cross-section of nitrogen was taken as 0.162 nm^2. Total pore volume was determined from a single adsorption point at a relative pressure of 0.995. Skeletal density of all gels was determined by helium pycnometry. Single point water adsorption data was collected at a relative pressure of approximately 0.23. This corresponds roughly to a single statistical monolayer. The cross section of water was taken as 0.108 nm^2. Weight adsorbed was normalized by the nitrogen surface area to obtain the fractional monolayer coverage. Mechanical properties of the wet gels were determined in a three point bend experiment similar to the method reported by Scherer [9]. Data is reported for specimens 7 cm long and 0.78 cm in diameter. The span of the bend fixture was 5.1 cm and the cross head speed was 750 μm/minute. Mechanical properties were measured in a bath containing pore fluid (ethanol, water, and NH_4OH) at a temperature of 305 K.

RESULTS AND DISCUSSION

The pore structure of gels formed from solutions of TEOS and MTEOS was a function of both modified alkoxide content and catalyst. For acid catalyzed gels, as shown in Figure 1, the total surface area and pore volume increased from around 400 m^2/g for the M = 0 gel to over 800 m^2/g for the M = 40 gel. Above this organic content, the surface area and pore volume were nearly constant. The micropore area, that is the surface area of pores less than 1 nm in radius, was constant from M =

0 to M = 40, then it dropped. All of the acid catalyzed gels contained a significant fraction of micropores. The observed increase in the area of the acid catalyzed gels with increasing methyl content may be an indication that the organic additions prevent collapse of the gel structure upon drying. Acid catalyzed gels formed from DMDEOS showed a similar behavior. For base catalyzed gels [5], the results were significantly different, as illustrated in Figure 2. Surface area was relatively high, over 800 m^2/g, for gels ranging in composition from M = 0 to M = 50, but fell dramatically to less than 1 m^2/g as MTEOS content approached 70 mole percent. This loss in surface area has been attributed to differences in gel structure, not simply collapse of the pores during drying [5]. The structure of several wet gels was probed by small angle x-ray scattering. Results showed a significantly smaller feature size for low surface area gels such as M = 70 compared to gels with lower methyl contents and higher surface areas, i.e. M ≤ 50. In addition base catalyzed gels were much less mircoporous than acid catalyzed gels of similar composition. For gels prepared using DMDEOS, a similar drop in surface area was noted, but at approximately half of the modified alkoxide content, D = 40 as compared to M = 70. This difference is reasonable since each DMDEOS molecule provides two methyl groups while each MTEOS provides only one. Surface area data for both acid and base catalyzed DMDEOS gels is shown in Figure 3. Interestingly, the hydraulic radius, that is the ratio of pore volume to surface area, remained nearly constant for both acid and base catalyzed MTEOS gels regardless of composition. The hydraulic radius was approximately 1.25 nm for base catalyzed MTEOS gels and was around 1.05 nm for acid catalyzed MTEOS gels. The hydraulic radius of acid and base catalyzed DMDEOS gels were similar to those of the MTEOS gels.

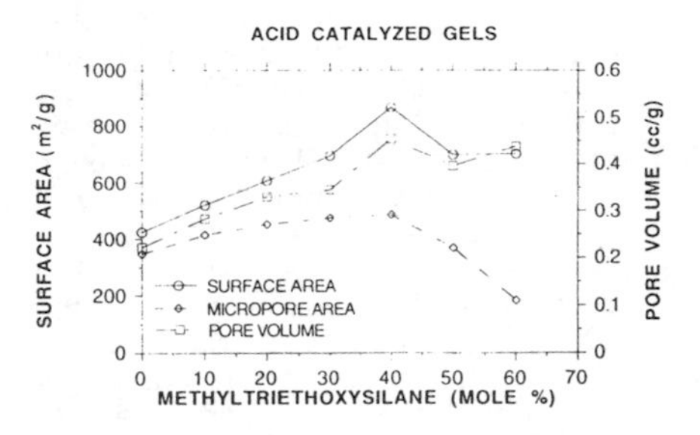

Fig. 1. Pore structure of acid catalyzed gels as a function of MTEOS content.

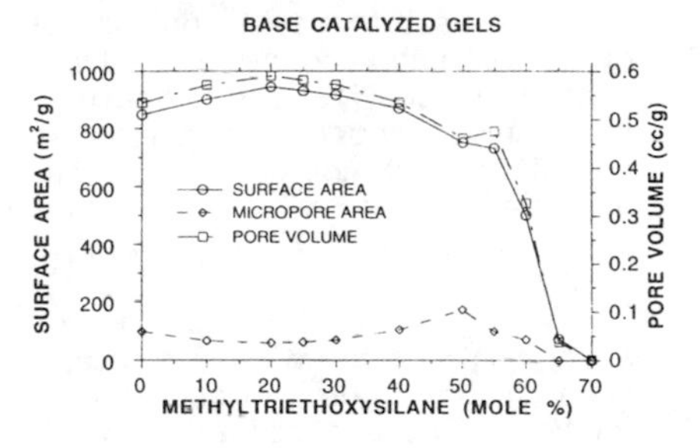

Fig. 2. Pore structure of base catalyzed gels as a function of MTEOS content [5].

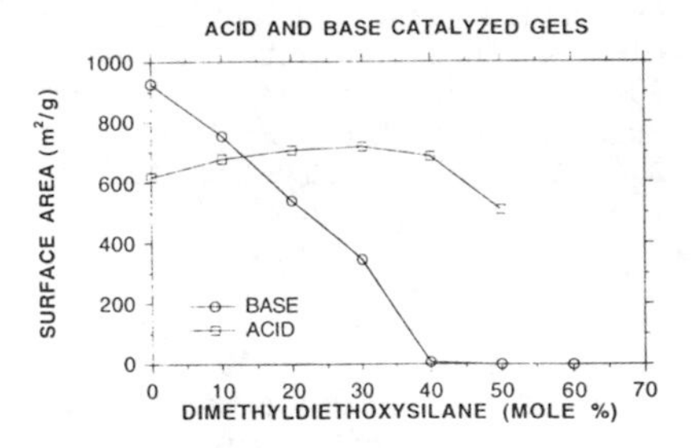

Fig. 3. Surface area of acid and base catalyzed gels as a function of DMDEOS content.

The skeletal density of both acid and base catalyzed gels was a function of composition. The density of all gels decreased with increasing organic content, diagramed in Figure 4. The density of MTEOS and DMDEOS base catalyzed gels decreased more rapidly as a function of composition than did the density of acid catalyzed gels. As with the surface area measurements, the decrease in density is magnified in the DMDEOS gels because twice as many methyl groups are incorporated per alkoxide ligand as compared to MTEOS. The steady decrease in density with the addition of modified alkoxide can be attributed to two causes. First, the density will decrease solely because of the organic content, but the skeletal density will also decrease as the average number of network-forming bonds per silicon atom decreases [10]. A decrease in the average number of network bonds per silicon with increased modified alkoxide content was confirmed for base catalyzed MTEOS gels using ^{29}Si MAS NMR [5].

Figure 5 shows water adsorption data for MTEOS acid and base catalyzed gels. In general, the fractional coverage decreased as M increased. This data indicates that the nature of the surface is changing as methyl content increases. For the base catalyzed gels above M = 60, the monolayer coverage appears to rise, but this is an artifact of the calculation which normalizes adsorbed weight by surface area. The lower values of surface coverage for the base catalyzed gels as compared to acid catalyzed gels is a result of a higher fraction of surface OR groups for silicons not terminating with a methyl group.

Base catalyzed gels were also prepared from mixtures of MTEOS, DMDEOS, and TEOS. Two series of gels were prepared in order to study the effect of incorporating both alkoxides into the gel network. For one series, MTEOS content was held

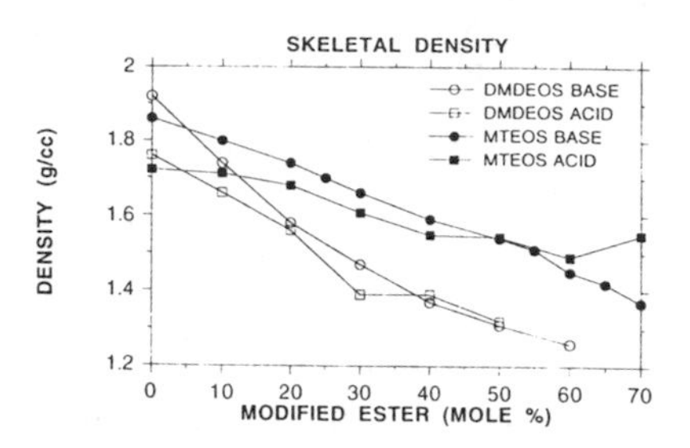

Fig. 4. Skeletal density of acid and base catalyzed gels as a function of DMDEOS or MTEOS content.

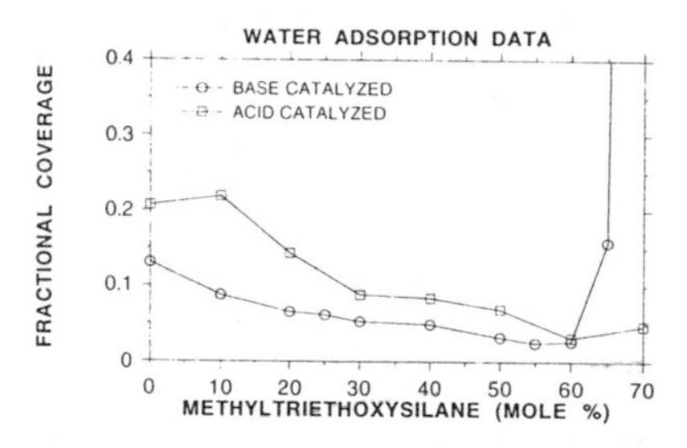

Fig. 5. Water adsorption as a function of MTEOS content for acid and base catalyzed gels.

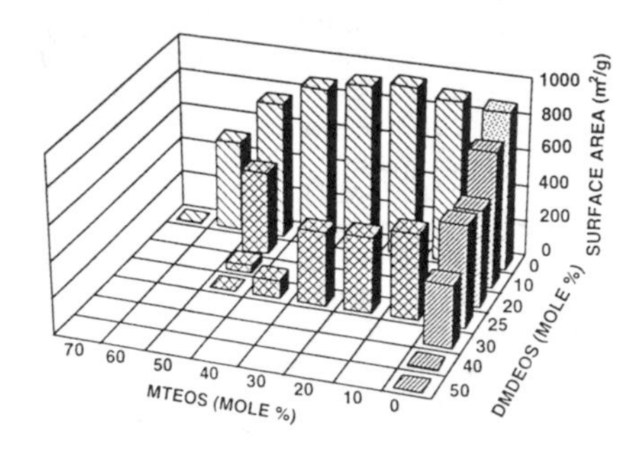

Fig. 6. Surface area of base catalyzed gels containing mixtures of MTEOS and DMDEOS.

constant at M = 50 while DMDEOS content was varied from D = 0 to D = 25. For the other series, DMDEOS content was held constant at D = 25 and MTEOS content was varied from M = 0 to M = 50. As shown in Figure 6, the surface area for each series was a maximum for the gel containing none of the second alkoxide. As the amount of the second alkoxide increased, the surface area dropped. For the M = 50, D = 25 gel, surface area was less than 1 m^2/g. From this data, it seems that the pore structure of the gels is controlled by both the total number of methyl groups incorporated into the structure and the total amount of the individual modified alkoxide added. If the pore structure were simply a function of total organic content, the M = 50 and D = 25 gels would have the similar surface area.

Shear modulus and modulus of rupture were measured for acid catalyzed MTEOS wet gels. Both properties, as illustrated in Figure 7, decreased as MTEOS content increased. Loss of mechanical strength in the wet gels was expected since the addition of modified alkoxide, essentially addition of unreactive sites, should lead to a less fully developed gel network. According to Brinker and Scherer [11], acid catalyzed gels consist of weakly branched polymers cross linked at various points. If the MTEOS containing gels have similar structure, the methyl groups may act to prevent cross linking thus lowering the gel strength by significantly decreasing the integrity of the network.

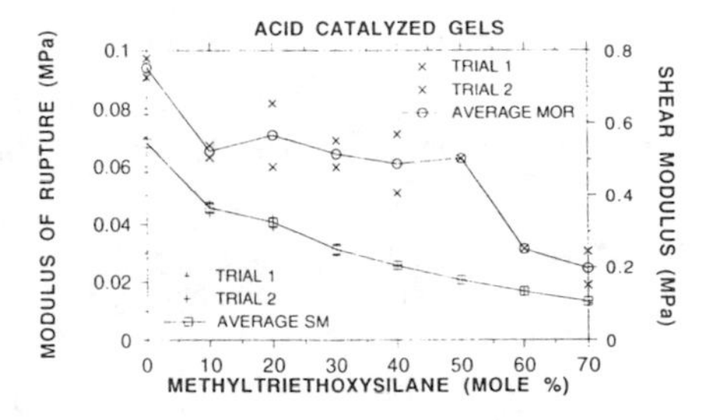

Fig. 7. Mechanical properties of acid catalyzed MTEOS gels as a function of composition.

CONCLUSIONS

Both acid and base catalyzed gels were prepared from mixtures of TEOS and organically modified silicon alkoxides. Gels could be prepared using one or both of the modified species with TEOS. For the base catalyzed gels, surface area decreased from around 800 m^2/g to less than 1 m^2/g as the modified alkoxide content increased. Acid catalyzed gels showed no drastic loss in surface area with higher organic content. In addition to pore structure, both the surface properties and the mechanical strength of the gels varied with organic content. Gels became more hydrophobic with higher organic contents while the mechanical strength of wet gels decreased significantly as modified alkoxide content increased. Overall, the results indicate that the addition of modified silicon alkoxides changes the structure of the gels.

ACKNOWLEDGEMENTS

This work was supported by the UNM/NSF Center for Micro-Engineered Ceramics (CMEC). CMEC is supported by NSF (CDR 8803512), Los Alamos and Sandia National Laboratories, and the ceramics industry. The authors would like to thank the following people, all of CMEC, for assistance; R. Desphande for performing mechanical property measurements, J. Anderson for sample preparation and mechanical testing, and W. Ackerman for gas adsorption measurements.

REFERENCES

1. S.Kohjiya, K. Ochiai, and S. Yamasita, J. Non-Cryst. Solids 119 132-135 (1990).

2. K. Kamiya, T. Yoko, T. Sano, and K. Tanaka, J. Non-Cryst. Solids 119 14-20 (1990).

3. K. Kamiya, T. Yoko, K. Tanaka, and M. Takeuchi, J. Non-Cryst. Solids 121 182-187 (1990).

4. H.K. Schmidt, in Better Ceramics Through Chemistry IV, edited by B.J.J. Zelinski, C.J. Brinker, D.E. Clark, and D.R. Ulrich, (Mater. Res. Soc. Proc. 180, Pittsburgh, PA 1990) pp. 961-973.

5. W.G. Fahrenholtz, D.M. Smith, and D.W. Hua, "Formation of Microporous Silica Gels From a Modified Silicon Alkoxide," J. Non-Cryst. Solids, in press.

6. R.H. Glaser, G.L. Wilkes, and C.E. Bronnimann, J. Non-Cryst. Solids 113 73-87 (1989).

7. C.J. Brinker, K.D. Keefer, D.W. Schaefer, and C.S. Ashley, J. Non-Cryst. Solids 48 47-64 (1982)

8. S.J. Gregg and K.S.W. Sing, Adsorption, Surface Area, and Porosity (Academic Press, London, 1967).

9. G.W. Scherer, S.A. Parandek, and R.W. Swiatek, J. Non-Cryst. Solids 107 14-22 (1988).

10. C.J. Brinker and G.W. Scherer, Sol-Gel Science (Academic Press, Boston, 1990) p. 112.

11. C.J. Brinker and G.W. Scherer, Sol-Gel Science (Academic Press, Boston, 1990) p. 536.

HYPERVALENT SILICONATE MATERIALS. SYNTHESIS AND CHARACTERIZATION OF NOVEL LADDER AND NETWORK IONOMERS

KENNETH J. SHEA*, DOUGLAS A. LOY**, AND JAMES H. SMALL*
*Department of Chemistry, University of California, Irvine, California 92717
**Sandia National Laboratories, Albuquerque, New Mexico 87185

ABSTRACT

The first representatives of novel materials that contain five- and six-coordinate anionic silicon are reported. Linear (or cyclic) ionomers incorporating pentacovalent siliconate were synthesized by condensation of 1,2,4,5-tetrahydroxybenzene (THB) with alkyl and aryl trialkoxysilylanes. Network ionomers that incorporate pentacovalent siliconates were formed from THB and alkyl and aryl bistrialkoxysilanes. In addition, network ionomers with hexacoordinate silicon result from condensation of TEOS and THB. Spectroscopic evidence is presented to verify the identity of the novel structural features of these materials.

INTRODUCTION

Hypervalent siliconates have been under intense scrutiny and have been implicated as intermediates in many nucleophilic substitution reactions at silicon [1-8]. Stable five- and six-coordinate siliconates are isolated when strongly electron withdrawing (fluoride) or chelating (catechol) groups are attached to silicon. Structural data for these compounds have been reviewed [9]. Stishovite, a polymorph of quartz, and thaumasite, a hydrated calcium silicate sulfate carbonate material, are two *rare* examples of naturally occurring hexacoordinate siliconate minerals formed under extremely high pressure [10,11].

Synthetic materials which incorporate hypervalent silicon are limited to siloxane-phthalocyanines, a remarkable family of linear polymers with unique optical and electrical properties (Figure 1). In this paper, we report the synthesis and characterization of three new families of hypervalent siliconates and present spectroscopic evidence in support of the structural assignments.

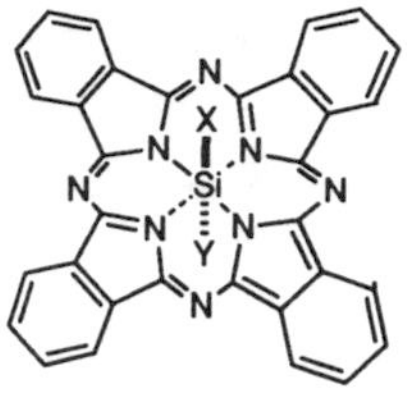

EXPERIMENTAL

General Procedures

Catechol and phenyltriethoxysilane were purchased from Aldrich Chemical Company and were used as received. Tetraethoxysilane was purchased from Aldrich and distilled from CaH_2 before using. 1,4-Bis(triethoxysilyl)benzene was prepared from TEOS and 1,4-dibromobenzene by a Barbier-Grignard reaction [12-14]. 1,6-Bis(trimethoxysilyl)hexane was purchased from Hüls Inc. and distilled from CaH. The reactions were run in flame dried glassware under nitrogen atmosphere. Model siliconates and oligomeric materials were characterized by ^{1}H, ^{13}C, and ^{29}Si NMR on a GN-500 Nuclear magnetic resonance spectrometer. Materials were also characterized by solid-state ^{13}C and ^{29}Si CP/MAS NMR on a Chemagnetics 200 MHz instrument. All samples were characterized by elemental analysis. A representative procedure for the synthesis of a ladder ionomer, PTES-THB-ET, is given below.

A solution of equimolar amounts of tetrahydroxybenzene and phenyltriethoxysilane in ethanol solvent was heated to just below reflux. Triethylamine was then added dropwise by syringe over 5 min. The solution turned a reddish color almost instantly. Within 1/2 to 1 h, a pinkish solid began to adhere to the walls of the reaction vessel. After refluxing for 11 h, the solution was cooled slowly to room temperature. The solids were filtered and washed (CH_3CN, 100 mL) under an inert atmosphere. The solid was dried at 70°C (15 mTorr) for 12 h (73%).

RESULTS AND DISCUSSION

The synthesis of five- and six-coordinate silicon derivatives can be achieved by condensation of suitable silicon precursors with fluoride or chelating bidendate oxygen ligands. Two examples from the earlier literature by Rosenheim [15] and by Frye [16] are illustrated in equations 1 and 2 and involve

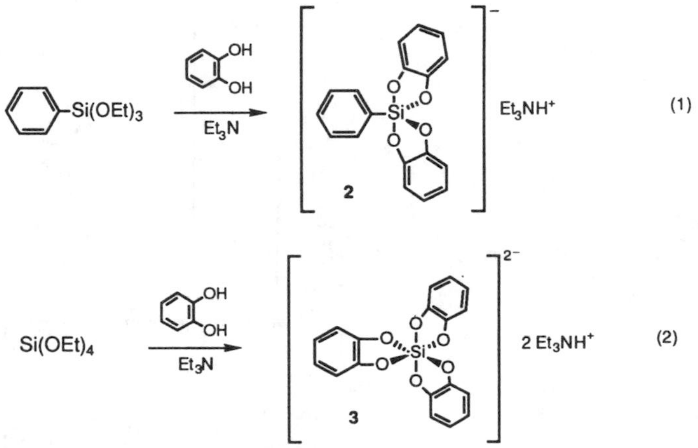

Figure 1

condensation reactions of catechol with either triethoxysilylbenzene or TEOS.

The preceding synthetic technology was employed for the preparation of three new families of materials. The first representatives are *ladder ionomers*, linear (or perhaps cyclic) structures that contain an anionic pentacovalent siliconate in the polymer backbone. The ionomer (PTES-THB-ET) is prepared by condensation of equimolar amounts of 1,2,4,5-tetrahydroxybenzene with triethoxysilylbenzene. The solvent systems include CH_3CN/MeOH, THF, EtOH/THF mixtures, and diglyme (Figure 2).

The materials, which are air-sensitive due to the oxidative lability of the pendent catechol groups, become discolored when exposed to air. They are soluble in DMSO and acetone. The ^{29}Si NMR chemical shift is very sensitive to the coordination state of silicon [17]. As a result, both solution and solid-state ^{29}Si NMR spectroscopy were employed to verify the coordination state of silicon and the integrity of repeat unit.

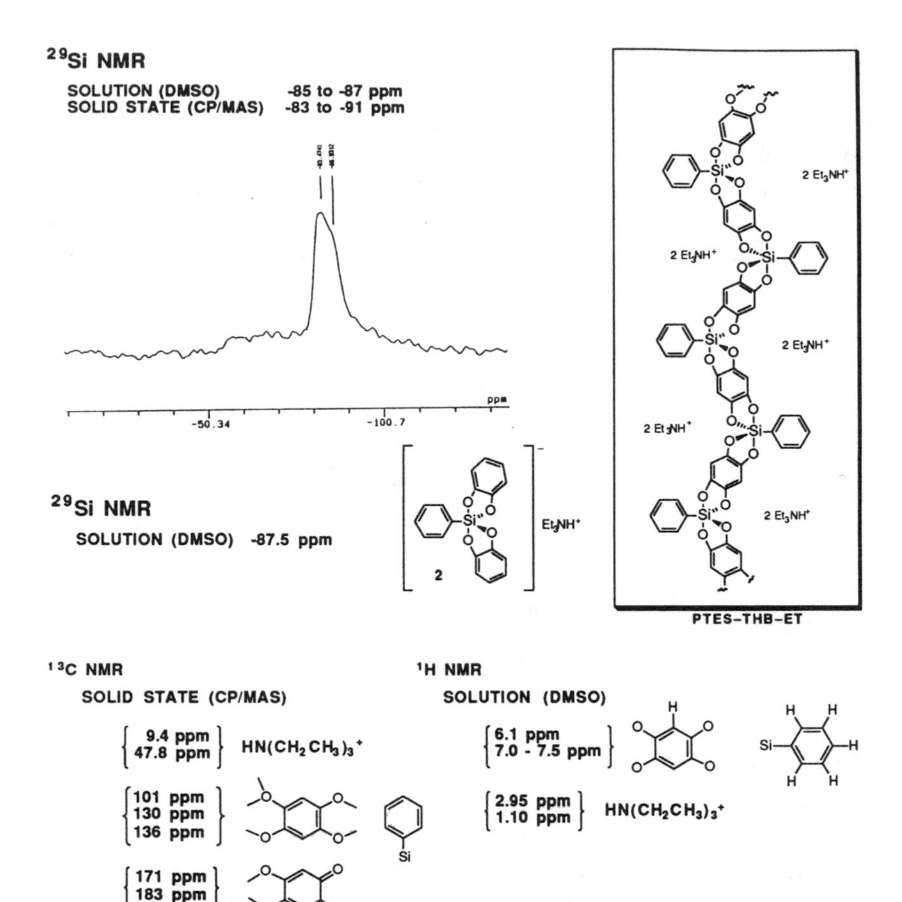

Figure 2

As a benchmark, a reference pentacoordinate silicon compound, the monomeric bis-catecholate salt **2**, exhibits a ^{29}Si chemical shift in DMSO of -87.5 ppm. DMSO solutions of the ladder material exhibit peaks from -85 to -87 ppm, consistent with a pentacovalent silicon. A solid-state ^{29}Si NMR spectra (cross-polarization, magic angle spinning) is shown in Figure 2. There is an absorption envelope from -83 to -91 ppm, all within the pentacovalent chemical shift region. The similarity of the solution and solid-state ^{29}Si NMR suggests that in coordinating solvents as well as in the solid-state, silicon is in the pentacovalent state. Despite the mildness of the reaction conditions, there was concern regarding the integrity of the carbon-silicon linkage. The spectroscopic evidence

(^{29}Si - absence of chemical shifts corresponding to four- or six-coordinate Si), and ^{13}C NMR (solution and solid), and ^{1}H NMR (solution), were consistent with the absence of cleavage of the silicon-carbon bond.

^{13}C NMR resonances are observed for the triethyl ammonium ion as well as THB and phenyl groups. Minor absorption at 171 and 183 ppm are also observed which are tentatively assigned to the oxidized catechol end groups. The proton NMR gives the correct ratio of triethyl ammonium groups to the two aromatic signals. Satisfactory analytical data for most elements was achieved by combustion analysis.

In contrast to the linear ionomers, *siliconate networks* are produced upon condensation of tetrahydroxybenzene with bistrialkoxysilanes such as 1,6-bistriethoxylsilylhexane. Solid ionomeric materials are produced from these condensation reactions that feature a pentacovalent siliconate as part of the main chain. Fully condensed materials will be three-dimensional networks. Since each pentacovalent silicon is anionic, the solid is a polyion with triethyl ammonium counter ions. Key NMR absorptions are summarized for the hexyl-bridged bis-siliconate materials in Figure 3. A reference compound, a

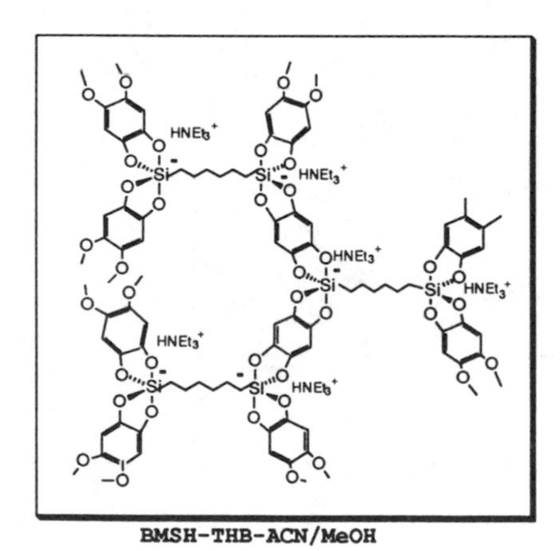

BMSH-THB-ACN/MeOH

BMSH-CAT-THF

NMR ANALYSIS

	^{13}C Catechol (ppm)	^{13}C Hexyl-bridge (ppm)	^{13}C $HN(CH_2CH_3)_3$ (ppm)	^{13}C $HN(CH_2CH_3)_3$ (ppm)	^{29}Si (ppm)
BMSH-CAT-THF (solution NMR) DMSO-d_6	150.72, 117.01, 109.48	33.47, 24.55, 18.73	45.82	8.75	-71.90
BMSH-THB-ACN/MeOH (solid-state NMR)	141.95, 102.01	33.49, 25.09	48.63	9.95	-73.12

Figure 3

monomeric *bis*-silicon catecholate, is given in the top entry. Key data include: ^{29}Si absorption in the pentacovalent region (-71 ppm), and a solid-state ^{13}C NMR that is consistent with the assigned structure.

Finally, *hexacoordinate siliconate networks* are produced by condensation of TEOS with tetrahydroxylbenzene (Figure 4). The networks contain a dianionic silicon as the key building block, each silicon is associated with two triethyl ammonium ions. For comparision, the spectroscopic properties of a model *monomeric* siliconate dianion is given on the top row of Figure 4. The characteristic chemical shifts of a hexacoordinate silicon dianion is approximately -140 ppm. Both solution ^{29}Si NMR of oligomeric material and solid-state ^{29}Si NMR of fully condensed siliconates exhibit characteristic chemical shifts in the hexacoordinate region (*c.a.* -135 ppm). The ^{13}C adsorptions of soluble oligomeric material and of the solid, fully condensed networks are reconciled in a straightforward manner from carbon absorptions associated with either triethyl ammonium counter ions or the two chemically distinct carbon residues of the THB component.

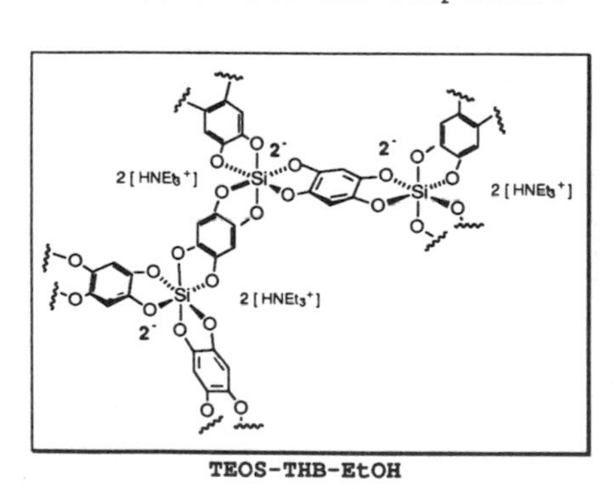

TEOS-THB-EtOH

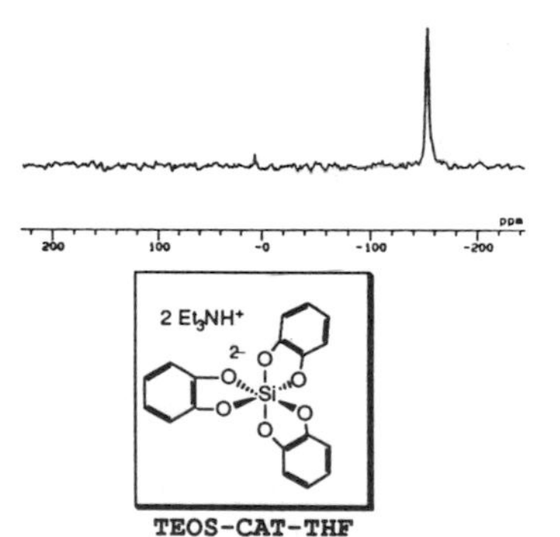

TEOS-CAT-THF

NMR ANALYSIS

	^{13}C Catechol (ppm)	^{13}C $HN(CH_2CH_3)_3$ (ppm)	^{13}C $HN(CH_2CH_3)_3$ (ppm)	^{29}Si (ppm)
TEOS-CAT-THF (solution NMR) DMSO-d_6	152.2, 115.1, 108.8	45.81	8.94	-139.99
TEOS-THB-EtOH (solution NMR) DMSO-d_6	139.48, 102.38	48.03	8.93	-135.00
TEOS-THB-EtOH (solid-state NMR)	142.37, 104.67	45.88	9.79	-135.53

Figure 4

The preceding examples provide representatives of three new families of hypervalent siliconates. The building blocks from which they are assembled offer considerable flexibility with regard to the organic component. These materials are being examined as novel ion conducting materials, and for their nonlinear optical properties. The solution and mechanical properties are also of interest, particularly in view of the dynamic stereochemistry at silicon.

ACKNOWLEDGMENT

We are grateful to the Materials Research Division of the National Science Foundation and to the Air Force Office of Scientific Research for financial support of this research.

REFERENCES

1. Holmes, R. R. *Chem. Rev.* **1990**, *90*, 17.
2. Corriu, R. J. P.; Guerin, C. *Adv. Organomet. Chem.* **1982**, *20*, 265.
3. (a) Corriu, R. J. P.; Guerin, C.; Moreau, J. J. E. *The Chemistry of Organic Silicon Compounds*; Patai, S.; Rappoport, Z., Eds., Wiley, New York, **1989**, Part 1, pp. 305-370. (b) Corriu, R. J. P.; Young, *ibid.*, Part 2, pp. 1241-1288.
4. Fleming, I. in *Comprehensive Organic Chemistry*, Barton, D. H. R.; Ollis, D. W., Eds., Pergamon Press, Oxford, **1979**, *Vol. 3*, p. 555.
5. (a) Corriu, R. J. P.; Guerin, C.; Moreau, J. J. E. *Topics in Stereochemistry* **1984**, *15*, 43. (b) Johnson, S. E.; Day, R. O.; Holmes, R. R. *Inorg. Chem.* **1989**, *28*, 3182. (c) Johnson, S. E.; Payne, J. S.; Day, R. O.; Holmes, J. M.; Holmes, R. R. *Inorg. Chem.* **1989**, *28*, 3190. (d) Damraver, R.; O'Connell, B.; Denahey, S. E.; Simon, R. *Organometallics* **1989**, *8*, 1167.
6. Shklover, V. E.; Struchkov, Yu T.; Voronkov, M. G. *Russian Chemical Reviews* **1989**, *58*, 211.
7. Johnson, S. E.; Deiters, J. A.; Day, R. O.; Holmes, R. R. *J. Am. Chem. Soc.* **1989**, *111*, 320.
8. Sakurai, H. *Selectivities in Lewis Acid Promoted Reactions*; Schinzer, D., Ed., Kluwer, **1989**, p. 203.

9. Tandura, S. N.; Voronkov, M. G.; Alekseev, N. V. *Top Stereochem.* **1986**, *131*, 99.

10. (a) Thomas, J. M.; Gonzales-Calbert, J. M.; Fyfe, C. A.; Gobbi, G. C.; Nicol, M. *Geophysical Research Letters* **1983**, *10*, 91. (b) Edge, R. A.; Taylor, H. F. W. *Nature* **1969**, *224*, 363.

11. Edge, R. A.; Talor, H. F. W. *Nature* **1969**, *224*, 363.

12. Shea, K. J.; Loy, D.; Webster, O. *Chemistry of Materials* **1989**, *1*, 572.

13. Shea, K. J.; Webster, O. W.; Loy, D. A. *Better Ceramics Through Chemistry IV*, MRS Symposium Proceedings, **1990**, *180*, 975.

14. Shea, K. J.; Loy, D. A.; Webster, O. *J. Am. Chem. Soc.* **1992**, *114*, 0000.

15. Rosenheim, A.; Baibmann, B.; Schendel, G. *Z. Anorg. All. Chem.* **1931**, *196*, 160.

16. Frye, C. L. *J. Am. Chem. Soc.* **1964**, *86*, 3170.

17. Cella, J. A.; Cargiol, J. D.; Williams, E. A. *J. Organomet. Chem.* **1980**, *186*, 13.

MULTIFUNCTIONAL (METH)ACRYLATE ALKOXYSILANES A NEW TYPE OF REACTIVE COMPOUNDS

H. WOLTER, W. GLAUBITT and K. ROSE
Fraunhofer-Institut für Silicatforschung, Neunerplatz 2, D-8700 Würzburg, Federal Republic of Germany

ABSTRACT

A new class of (meth)acrylate substituted alkoxysilanes has been developed. These are synthesized by addition of the H-S- or H-Si-unit of thiosilanes (e.g. $HS-(CH_2)_3-SiCH_3(OCH_3)_2$) or hydrosilanes to one (meth)acrylate C=C bond of commercially available di-, tri-, tetra- or penta-(meth)acrylate compounds. The alkoxysilyl groups allow the formation of an inorganic Si-O-Si-network by a hydrolysis and polycondensation reaction (sol-gel process), and the (meth)acrylate groups are available for thermally or photochemically induced organic polymerisation. The elegant procedure for the synthesis of a wide variety of multifunctional alkoxysilanes and the preparation of optical lenses will be discussed.

INTRODUCTION

A large number of silanes are known with modifications resulting from different inorganic and organic groups and methods of preparation [1]. Such compounds are used for the preparation of inorganic-organic copolymers, the so-called ORMOCERs [2]. These copolymers have found successful applications in a wide variety of fields, one example being coating materials with various properties (scratch resistance [3], corrosion resistance [4], special dielectric properties [5]).

The alkoxysilyl groups can be used for the construction of an inorganic network (-Si-O-Si- units) by the sol-gel process. An additional organic network can be formed for example by thermally or photochemically induced organic polymerisation of C=C-bonds, of the organic substituent of the silane. The commonly used alkoxysilanes with polymerisable C=C-bonds are methacryloxypropyl-trimethoxysilane 1 and vinyltrimethoxysilane 2.

$$CH_2{=}C(CH_3){-}C({=}O){-}O{-}(CH_2)_3Si(OCH_3)_3 \quad \textbf{1} \qquad\qquad CH_2{=}CH{-}Si(OCH_3)_3 \quad \textbf{2}$$

There is a constant need to modify the existing silanes in order to open up a new field of applications and to optimize the properties for specific purposes. The essential disadvantage of the precursors mentioned above is the fact that there is only one polymerisable group. This means that only a linear organic structure can be formed. A fast curing step, which is sometimes necessary for coating systems, is hardly possible. Furthermore appropriate silanes with additional special functionalities or structures resulting in adjustable mechanical [6] and optical properties [6], for example, are generally not available. A new family of reactive compounds, the multifunctional (meth)acrylatealkoxysilanes (Fig. 1), has been developed to overcome this disadvantages.

EXPERIMENTAL

The preparation of the new silanes by a mercaptoaddition or a hydrosilylation process results in a new type of inorganic/organic copolymers after polycondensation and photochemically or thermally induced curing. The method used is described below.

0.1 mole (18.03 g) mercaptopropylmethyl-trimethoxysilane was added to 0.1 mole (29.63 g) trimethylol-propanetriacrylate (TMPTA) dissolved in 100 ml ethylacetate under a nitrogen atmosphere. Under basic conditions the thiol addition was finished after 5 minutes, as monitored by an iodine test. 1.8 g of 6 n HCl was added for the porpose of hydrolysis and condensation, whereupon the mixture was stirred at room temperature for 20 hours. After working up a colorless, sometimes slightly yellowish transparent resin was obtained. The viscosity varied over a wide range (9500 - 20000 mPas) and depended strongly on the exact conditions during the preparation.

It is possible, but not necessary to isolate the silane intermediate. The preferred method is to prepare these silanes first in a one-pot-process, immediately followed by the sol-gel process to build up the inorganic backbone.

The polymerisation of the remaining acrylate groups can be carried out using UV radiation. The solvent free resin was therefore mixed with a photoinitiator, put into a curing mould and irradiated from the front and the back for approximately 1 minute (100 W/cm, medium pressure Hg-lamp) resulted in a transparent colorless raw lense. Similar results were obtained with dibenzoylperoxid after thermal curing at 70 to 80 °C for one hour.

0.1 mole (29.5 g) TMPTA was reacted with 0.1 mole (16.5 g) triethoxysilane in 100 ml of solvent (e.g., ethanol, benzene, cyclohexane, diethylether or methyl-tert.-butyl.-ether). 0.3 mmol of a rhenium catalyst was added to this solution, which was stirred in the dark at 40 °C until (in the IR-spectrum) the Si-H vibrations could no longer be observed (48 to 72 hours). Following completion of the reaction the catalyst was filtered off, and the solvent was removed under vacuum. A yield of 43.5 g (94 %) of the yellowish, light-sensitive oil was obtained.

The isolated silane can be used as shown above for the preparation of inorganic/organic condensates by the sol-gel process.

The course of thiol addition can be examined by a iodometric titration, IR and NMR-spectroscopy. The second step, the inorganic condensation, can be examined by the determination of water consumption during hydrolysis using Karl-Fischer titration [8]. The curing process of the acrylate groups, can be examined by following the intensity and relation of the C=C and C=O bands in the IR-spectra. The viscosities of the resins were determined using a rotating viscosimeter.

GENERAL ASPECTS

The novel multifunctional (meth)acrylate alkoxysilanes open a wide field of variations as shown in Fig. 1.

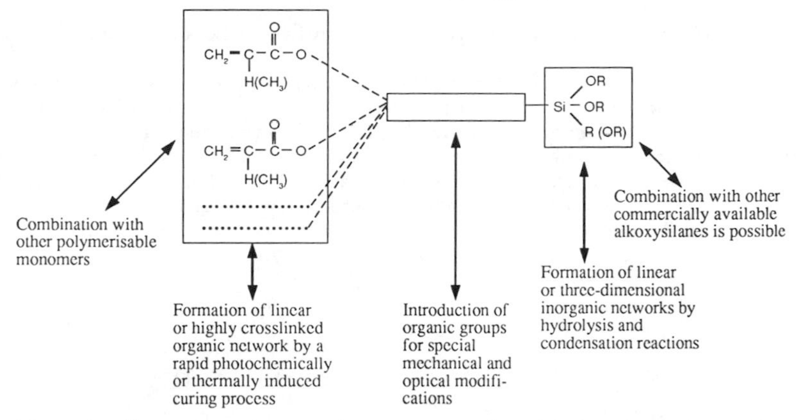

$CH_2=C-C-O$
$H(CH_3)$
$CH_2=C-C-O$
$H(CH_3)$
OR
Si - OR
R (OR)
Combination with other polymerisable monomers
Formation of linear or highly crosslinked organic network by a rapid photochemically or thermally induced curing process
Introduction of organic groups for special mechanical and optical modifications
Formation of linear or three-dimensional inorganic networks by hydrolysis and condensation reactions
Combination with other commercially available alkoxysilanes is possible

Fig. 1 The general formula of a novel multifunctional (meth)-acrylate alkoxysilane and possible variations concerning the physical properties

The inorganic network is formed by hydrolysis and condensation of the pure compounds or in combination with other commercially available alkoxysilanes. The remaining (meth)acrylate groups can be used to build up an additional organic network. Incorporation of other reactive monomers is also possible, for example those used for the preparation of silanes mentioned above.

The following structural modifications can be achieved:

- variation of the number of (meth)acrylate groups <—> variation of the organic crosslinking density
- variation of the number of alkoxy groups in the silyl unit <—> variation of the inorganic crosslinking density
- variation of chain length between two (meth)acrylate groups <—> variation of the organic network density
- variation of the chain connecting the alkoxysilyl group and the (meth)-acrylate unit <—> variation of the overall density
- variation of the kind of connecting unit (e.g. aliphatic or aromatic units) <—> variation for example the optical properties (refractive index, Abbe-number)

The manyfold kinds of variations demonstrates the number of ways in which the desired properties of the resulting material may be adapted to specified fields of application.

Synthesis of the multifunctional (meth)acrylate alkoxysilanes

Two possible methods for the preparation of the new silane type will be described. The thiol group of an appropriate silane, in this case 3-mercaptopropylmethyldimethoxysilane added by a Micheal reaction to one C=C-bound of commercially available multifunctional (meth)acrylates, in this case TMPTA, forms a thioether unit (Fig. 2). This reaction takes place very rapidly, using basic conditions and can be monitored by an iodine test or by IR- and ^{1}H-NMR spectroscopy (Tab. I).

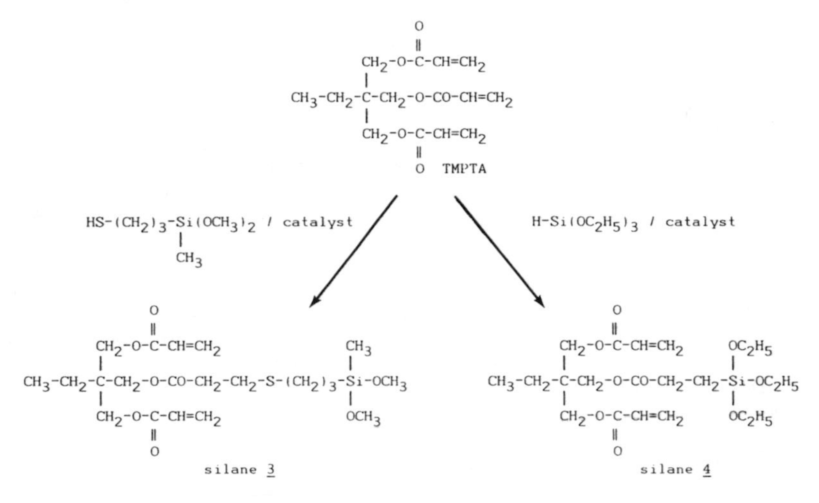

Fig. 2 Reaction scheme for the preparation of a novel silane type

Table I IR- and ^{1}H-NMR data characterizing the thiol addition

ν(S-H) = 2560 cm^{-1} (thiosilane),	disappears during the addition
ν(C=C) = 1620/1635 cm^{-1} (TMPTA),	decreases during the addition
ν(C=O) = 1730 cm^{-1} (TMPTA),	C=O band becomes broader caused by the production of a saturated ester group
δ(S-H) = 1.37 ppm (thiosilane),	disappears during the addition
δ(-$\underline{CH}$=CH_2) = 6.12 ppm and	
δ(-CH=$\underline{CH_2}$) = 5.86/6.41 ppm (TMPTA),	decreases during the addition and δ(-CH_2-CH_2-S-) signal arises at 2,55 - 2.85 ppm

The hydrosilylation is also a possible method (Fig. 2) of synthesis for the highly reactive silane precursor.

Some of the precursors used for the synthesis of the new type of silane are listed below:

(mercaptomethyl)methyldiethoxysilane, mercaptopropyltrimethoxysilane, (mercaptomethyl)dimethylethoxysilane, methyldiethoxysilane [9] combined with 1,6-hexanedioldiacrylate, trimetylolpropanetrimethacrylate, ethoxylated bisphenol-A-diacrylate, pentaerythritoltetraacrylate, pentaerythritoltriacrylate, dipentaerythritolpentaacrylate [10].

Synthesis of the inorganic condensates

In order to prepare the inorganic poly-condensate, silane 3 shown in Fig. 2 is hydrolysed by addition of water (0.8 mol/mol $-OCH_3$), e.g. under conditions of acid catalysis (Fig. 3). The water consumption and therefore the progress of this reaction step can be monitored by Karl-Fischer titration.

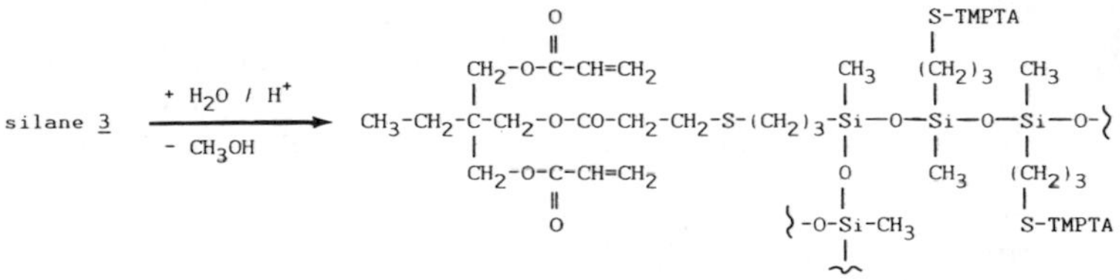

```
                                        O                                   S-TMPTA
                                        ||                                  |
                                CH2-O-C-CH=CH2                  CH3      (CH2)3   CH3
             + H2O / H+                 |                       |          |       |
silane 3  --------------->  CH3-CH2-C-CH2-O-CO-CH2-CH2-S-(CH2)3-Si---O---Si---O---Si-O-}
             - CH3OH                    |                       |          |       |
                                CH2-O-C-CH=CH2                  O         CH3   (CH2)3
                                        ||                      |                  |
                                        O                  }-O-Si-CH3           S-TMPTA
                                                                |
                                                                ~
```

Fig. 3 Reaction scheme for the preparation of the inorganic polycondensate

Another method for monitoring the reaction is to use IR- and ^{1}H-NMR spectroscopy. The alkoxy group of the silane results in a $\nu(OCH_3)$-band at 2840 cm^{-1} and a $\delta(OCH_3)$ signal at 3.52 ppm. Both decrease during course of the reaction (Fig. 3). When the reaction is complete a solvent free resin, used for the organic polymerisation step, is obtained after work up.

Curing of the inorganic/organic copolymer

Curing the (meth)acrylate containing siloxane (a solvent free resin) based on silane 3 irradiation with UV-light in the presence of a initiator is achieved in up to 0.1 sec. for particular applications, for example thin layers. Another application is the manufacturing of bulk materials e. g. for optical purposes with high refractive index and/ or high Abbe number [7] (Fig. 4). The resin was filled in a glass mold shaped as a lens and cured by UV-light irradiation on both sides for approx. 1 minute.

Refractive indices from 1.52 to 1.60 were obtained by variation of the connecting unit between the acrylates and alkoxysilyl group in the copolymers of the new type of silane described here.

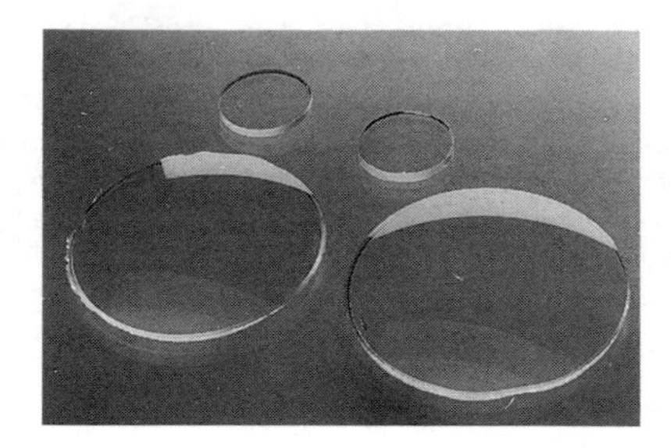

Fig. 4 Examples of optical lenses manufactured on base of the new silane type

ACKNOWLEDGEMENT

We gratefully thank Ms. H. Bäuerlein for her technical assistance.

REFERENCES

[1] E. P. Plueddemann, Silane Coupling Agents, Plenum Press, New York (1991)

[2] H. Schmid, B. Seiferling, G. Philipp, K. Deichmann in: Proc. Int. Congres on Ultrastructure Processing of Ceramics, Glasses and Composites, San Diego, J.D. Mackenzie, D.R. Ulrich (Eds.), Mat. Res. Soc. Symp. Proc. (1987) 651

[3] S. Amberg-Schwab, E. Arpac, W. Glaubitt, K. Rose, G. Schottner and U. Schubert in: High Performance Films and Coatings, Ed. P. Vincenzini, Elservier Science Publ. (1991)

[4] K. Greiwe, Farbe und Lacke 97 (1991) 968

[5] H. Wolter and H. Schmidt, DVS-Berichte 129 (1990) 80
H. Schmidt and H. Wolter, J. Non-Cryst. Solids 121 (1990) 428
M. Popall, J. Kappel, M. Pilz and J. Schulz, VDI-Berichte 933 (1991) 139

[6] H. Wolter, W. Glaubitt and K. Rose, Mat. Res. Soc. Symp. Proc. (1992), in print

[7] H. Wolter, W. Glaubitt; Berlin, 4. Polymertage (1991) Berliner Verband für Polymerforschung

[8] E. Scholz, Karl Fischer Titration, Springer Verlag, New York (1984) 92

[9] Petrach Systems Silanes & Silicons, USA

[10] Sartomer Company, Oakland, USA

CHARACTERISATION OF HYDROLYSED ALKOXYSILANES AND ZIRCONIUMALKOXIDES FOR THE DEVELOPMENT OF UV-CURABLE SCRATCH RESISTANT COATINGS

K. GREIWE, W. GLAUBITT, S. AMBERG-SCHWAB and K. PIANA
Fraunhofer-Institut für Silicatforschung, Neunerplatz 2, D-8700 Würzburg, Federal Republic of Germany

ABSTRACT

The development of a new scratch resistant UV-curable coating system is described. The material can be applied to optical data discs by spin coating. The coating shows superior adhesion even without priming and good weathering performance especially on polycarbonate.

The material is prepared from $Zr(OPr)_4$ and 3-Methacryloxipropyltrialkoxysilanes via sol-gel processing. Complexation of $Zr(OPr)_4$ with methacrylic acid yields complexes which can be copolymerized with the prehydrolyzed silane. The hydrolysis and self-condensation of the trialkoxysilanes were studied under different conditions. The consumption of water was studied by Karl-Fischer titration and the formation of oligomeric and polymeric species by GPC-measurements.

INTRODUCTION

Scratch resistant hard coatings on transparent plastics for optical applications are well known [1]. Most of these coating materials are manufactured from polysiloxanes and colloidal silica. After application of thin lacquer films to polymeric substrates, these coating materials have to be thermally cured to display their scratch resistance. After the successful development of plastic data discs for optical information storage, it was necessary to develop radiation curable hard coating materials for the surface protection of these data discs [2]. The material of choice for the production of these data discs is the very scratch sensitive polycarbonate (PC), which offers several advantages compared to other high quality plastics or glass [3]. Most UV-curable coating materials for PC are based on purely organic methacrylate/acrylate-systems [4]. Colloidal silica was added to some multifunctional acrylate containing systems to increase the scratch resistance of the coating [5]. Despite all the developments in this field, so far none of the materials fulfills all demands required for hard coatings on optical data discs.

In order to obtain good adhesion of radiation curable coatings on polymeric substrates, more requirements have to be met than those attained by use of thermally curable coatings [6]. It seems feasible that the adhesion of radiation curable coatings on PC can be improved by use of reactive monomeric diluents, which can form a strong interpenetrating bonding layer after diffusion into the substrate [6, 7]. The conditions for the UV-curing must be optimized to avoid the formation of a weak boundary layer (WBL) [8]. An overlap between the UV-emission of the Hg-lamp and the absorption spectra of the polymeric substrate may lead to a reduced degree of polymerisation or incomplete curing of the coating material in the boundary phase and, thus to the formation of a WBL.

The controlled hydrolysis and cocondensation of metal

alkoxides and organically modified alkoxysilanes offers an extremely attractive route for the development and synthesis of inorganic-organic polymers. One striking feature of these sol-gel derived coating materials (ORMOCERs = ORGanically MOdified CERamics) is their scratch resistance [9]. It is difficult to imagine systems for a fast curing step other than those involving free radicals which are compatible with sol-gel conditions. We therefore investigated first the hydrolysis and condensation of 3-methacryloxipropyltrialkoxysilanes under different conditions. Reaction of $Zr(OPr)_4$ with methacrylic acid yields methacrylate (MAS) complexes [10]; these can be copolymerized with the prehydrolyzed silane to obtain a new UV-curable sol for the development of a scratch resistant coating system for PC. Transparent bulk materials based on this system have been described very recently [11].

EXPERIMENTAL

Different samples of 3-methacryloxipropyltrimethoxysilan (MEMO and MEMO-E), both commercially available (HÜLS AG), were studied. The two products are prepared by different synthetic routes [12] and contain different additives [13]. $Zr(OPr)_2(MAS)_2$ was prepared by the reaction of $Zr(OPr)_4$ (Hüls AG, commercially available) with two equivalents of distilled methacrylic acid (MAS-H). The synthesis of the UV-curable coating material is described elsewhere [14].

GPC-measurements were performed in THF at 38°C (20 µl sample, flow rate: 1ml/min; UV-detection at 220 nm). A combination of columns (PSS-Gel-SVD 5 µm (8*300 mm)) with different pore sizes (100 A, 1000 A and 100 000 A) were used. Commercially available polystyrene standards in the molecular weight range between 162 and 1 550 000 were used for calibration [15].

UV-curing was performed with commercially available equipment (Fa. Beltron; type 22/III).

RESULTS AND DISCUSSION

The hydrolysis of 3-methacryloxipropyltrimethoxysilane can be carried out with 1,5 mol water/1 mol silane to minimize the

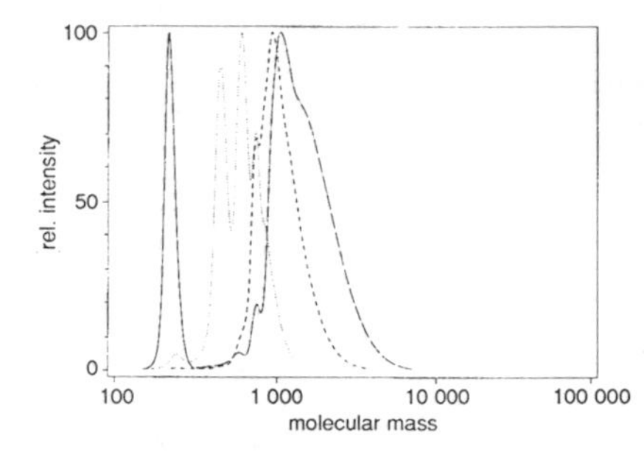

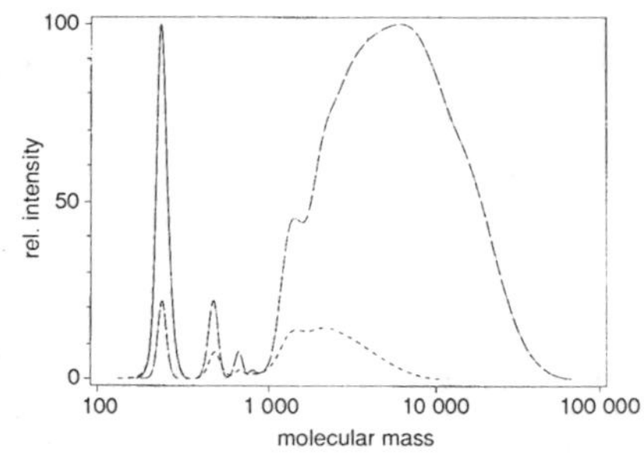

Fig. 1 Molecular weight distribution in MEMO- (left) and MEMO-E- (right) solutions determined by GPC-measurements after ——1 hour, ·········1 day, - - - - -3 weeks and —·— 13 weeks of hydrolysis with 1,5 mol water/1 mol silane

residual water content in the resulting sol [11]. GPC-measurements were performed on the reaction mixtures in order to elucidate the nature of the silane species in the resulting sol. Figure 1 displays the time dependance of the molecular weight distribution in the MEMO- and MEMO-E-solutions after the addition of 1,5 mol of deionized water (pH-value of around 5,5). The results are summarized in figure 2.

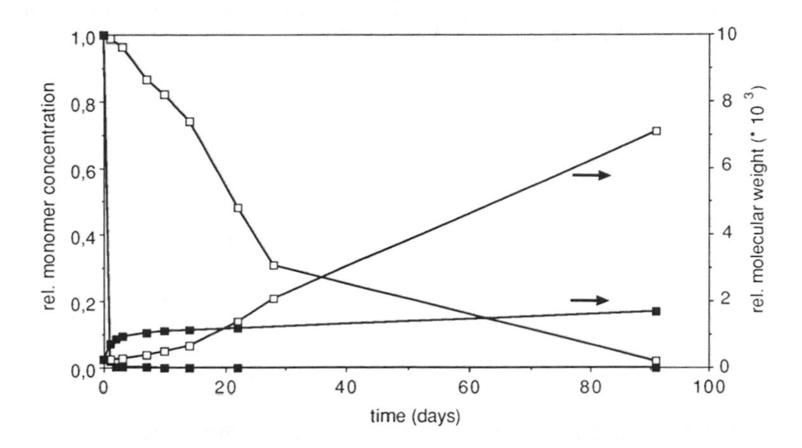

Fig. 2 Time dependance of the monomer concentration (left side) and the molecular weight (Mw) (right side) in the MEMO-(■) and MEMO-E-(□)hydrolysates

The two samples display totally different hydrolysis and condensation characteristics. In the MEMO-hydrolysate the unhydrolyzed monomers react very rapidly with water, yielding oligomeric (di-, tri- and tetrameric) species which condense to form polymeric compounds with a low molecular weight. In the MEMO-E-hydrolysate the first hydrolysis step is very slow, and the formation of a polymeric compound with a higher molecular weight is favoured. This is reflected macroscopically by the determination of the clear point of the silanes: MEMO clears up within a day, while MEMO-E has a clear point of almost 3 weeks under the same conditions.

It was not obvious up to this time whether the impurities in the samples influence the hydrolysis/condensation reactions or, whether the different pH-values of the starting compounds are the crucial factors. MEMO has a lower pH-value (ca. 4) than MEMO-E (ca. 6) [13]. It is well established that the pH in alkoxysilane solutions governs the reaction pathways for hydrolysis and condensation. It has been observed in dilute aqueous solutions of 3-methacryloxipropyltrimethoxysilane, that at lower pH-values polysiloxanes with a lower molecular weight are formed, while at higher pH-values polysiloxanes with higher molecular weight are found [16]. This is in agreement with our results from the GPC-measurements. Spectroscopic evidence for the existence of trimeric, tetrameric and cyclic polysiloxanes in these dilute solutions has been presented [16, 17]. The reaction of $Si(OCH_3)_4$ with an understoichiometric amount of water is characterized under acidic conditions by a fast hydrolysis of the monomer, and under basic conditions by a selective consumption of water for the hydrolysis of higher order intermediate products [18]. We studied the influence of catalysts on the hydrolysis and condensation of MEMO and MEMO-E by GPC to find a more convenient method than pH-adjustment to reduce the differences between the two commercially available 3-methacryloxipropyltrimethoxysilanes in the sol-gel processing. With respect to the development of UV-curable coatings, our work has been mainly aimed at compounds which could be copolymerized by UV-radiation with the methacrylic groups in the alkoxysilanes.

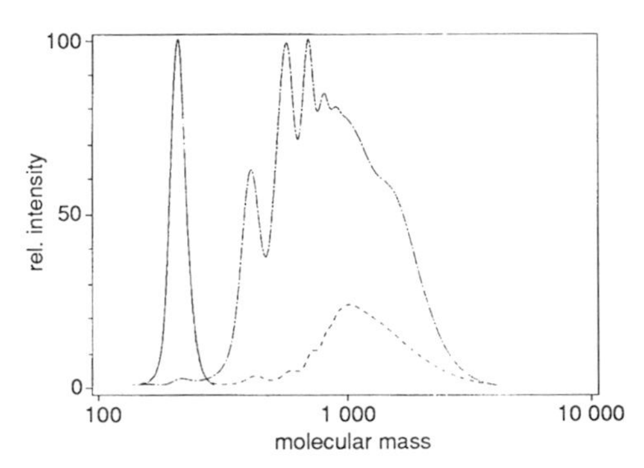

Fig. 3 Molecular weight distribution in different MEMO-E solutions (MEMO-E ——, with 0,1 mole methacrylic acid - - -, with 0,02 mole $Zr(OPr)_2(MAS)_2$)—·—, as determined by GPC measurements after 1 day of hydrolysis with 1,5 mol water/1 mol silane

Figure 3 shows the molecular weight distribution of MEMO-E solutions with different catalysts after 1 day of hydrolysis. The addition of $Zr(OPr)_2(MAS)_2$ very rapidly induces the formation of oligomeric species, while the influence of the methacrylic acid is much lower. The influence of $Zr(OPr)_2(MAS)_2$ in the reaction with water seems to be so dominant, that the MEMO-E-hydrolysate now features the same characteristics in the GPC-measurements as the MEMO-hydrolysate without catalyst (s. Fig. 1). Whether this catalytic effect of $Zr(OPr)_2(MAS)_2$ is limited to the trimethoxy-derivatives of the 3-methacryloxipropylsilane or whether other alkoxy groups can also be influenced in their hydrolysis and condensation reactions is under investigation. To date we have not determined whether the zirconium compound will catalyze transesterification reactions between the methoxy and propoxy groups. Very recently it has been reported that $Mg(OMe)_2$ rapidly catalyzes ligand exchange in the TMOS/-TEOS-system [19].

The storage stable hydrolysate of 3-methacryloxipropyltrimethoxysilane can be used as one building block for a UV-curable hard coating. Multifunctional acrylates can be added to this solution as desired. The other component for preparing scratch-resistant coatings can be taken from the aforementioned reaction product of $Zr(OPr)_4$ and methacrylic acid. It has been discovered that the 2:1 product ($Zr(OPr)_4$:2 methacrylic acid) $Zr(OPr)_2(MAS)_2$ can not only be added to the MEMO-hydrolysates as a catalyst for sol-gel processing as described earlier (vide supra) but also in amount from 10 to 30% to obtain a lacquer, which can be applied by common application methods to polycarbonate. Depending on the initiator used the coating can be cured thermally, by UV-radiation, or by both methods in succession.

lacquer composition:	10 MEMO/2 $Zr(OPr)_2(MAS)_2$
photoinitiator:	2% Irgacure 184 (Ciba Geigy)
adhesion:	cross cut test* : Gt 0
	tape test* : B 5
scratch resistance:	haze** (100 cycles): 2 - 3 %
humidity resistance:	no change of adhesion or scratch resist.

Table 1 Characteristics of hard coated UV-cured PC samples
(* DIN 53151 and ASTM D 3359, respectively)
(** taber abraser test (ASTM D 3359))

lacquer composition:	10 MEMO/2 $Zr(OPr)_2(MAS)_2$/2 ditrimethylolpropantetraacrylate
photoinitiator:	2% Irgacure 184 (Ciba Geigy)
adhesion:	cross cut test* : Gt 0 (PC)
	tape test* : B 5 (PC)
scratch resistance:	haze** (100 cycles): 2 - 3 %
humidity resistance:	no change of adhesion or scratch resist.
solvent resistance:	stable towards ethanol and acetone

Table 2 Characteristics of hard coated UV-cured compact discs

The results displayed in Tables 1 and 2 were obtained on PC-sheets or CDs without priming or pretreating the surface. The incorporation of multifunctional acrylates with the mixed hydrolysate of MEMO and $Zr(OPr)_2(MAS)_2$ significantly improves the adhesion of the coating on the CDs. While the lacquer described in Table 1 features irreproducible adhesion properties on the PC-side of the CDs, an improved adhesion is observed on this side after the addition of the tetraacrylate.

In order to meet the optical quality required for data discs, we observed that the wetting and levelling properties of the lacquer could be improved if instead of the methacryloxipropyltrimethoxysilane other alkoxy derivatives of this silane are processed. It has been calculated that much better results can be obtained by the spin coating technology when two different solvents with highly different volatilities are used [20]. We are investigating how less volatile solvents change the sol-gel processing of the coating material. By use of another solvent system it was possible for us to develop a scratch resistant coating material which can be applied by using a spray coating technology. Only acrylate/methacrylate systems filled with colloidal silica [5] can compete with the scratch resistance of our sol-gel derived coating material, while unfilled systems do not match [21]. But it is difficult to apply these scratch resistant acrylate/methacrylate systems to planar data discs by a spin coating technology due to their high viscosity [22].

ACKNOWLEDGEMENTS

We like to thank the Philips and DuPont Optical Deutschland GmbH (PDO) and the Bayerisches Staatsministerium für Wirtschaft und Verkehr for financial support.

REFERENCES

[1] J. Hennig, Kunststoffe 71, 103 (1981).

[2] J. Hennig, Kunststoffe 75, 425 (1985).

[3] W. Siebourg, Kunststoffe 76, 917 (1986).

[4] K. P. Murray, W. E. Hoffmann and D. C. Thompson, Offenlegungsschrift DE 39 32 460 A1.

[5] L. J. Cottington, and A. Revis, European Patent Application 0 437 327.

[6] A. Priola, G. Gozzelino and F. Ferrero, Adv. Org. Coat. Sci. Technol. Ser. 11, 156 (1989).

[7] L. N. Lewis and D. Katsamberis; J. Appl. Polym. Sci. 42, 1551 (1991).

[8] S. Jönsson, C. G. Gölander, A. Biverstedt, S. Göthe and P. Stenius, J. Appl. Polm. Sci. 38, 2037 (1989).

[9] H. K. Schmidt in Inorganic and Organometallic Polymers, edited by M. Zeldin, K. J. Wynne, and H. R. Allcock (ACS Symposium Series 360, 1988) pp. 333 - 344.

[10] U. Schubert, E. Arpac, W. Glaubitt, A. Helmerich and C. Chau, Chem. Mater. 4, 291 (1992)

[11] R. Nass, H. Schmidt, and E. Arpac, in Sol-Gel Optics (Proc. SPIE-Int. Soc. Opt. Eng. 1328, 1990) pp. 258.

[12] K. Winnacker and L. Küchler, Chemische Technologie 6 (Carl Hanser Verlag, München, 1982) p. 836

[13] H. Peeters, Hüls AG (private communication).

[14] K. Greiwe, W. Glaubitt, E. Arpac and S. Amberg-Schwab, German Patent Application No. 4 122 743 (22. July 1991).

[15] Handbook of Plastics Testing Technology, edited by V. Shah (John Wiley & Sons, Inc., 1984) p. 178

[16] S. Savard, L.-P. Blanchard, J. Leonard and R.E. Prud'homme Polm. Comp. 5, 242 (1984).

[17] N. Nishiyama and K. Horie, J. Appl. Polm. Sci. 34, 1619 (1987).

[18] I. Matsuyama, S. Satoh, M. Katsumoto and K. Susa, J. Non-Cryst. Sol. 135, 22 (1991).

[19] K. E. Yeager and J. M. Burlitch, Chem. Mater. 3, 387 (1991).

[20] D. E. Bornside, C. W. Macosko and L.E. Scriven, J. Appl. Phys. 66, 5185 (1989).

[21] S. R. Kerr, European Patent Application 0 323 560.

[22] K. P. Murray, W. E. Hoffamnn and D. C. Thompson, German Patent Application No. 3 932 460 (30. September 1988).

MULTIFUNCTIONAL ACRYLATE ALKOXYSILANES FOR POLYMERIC MATERIALS,

K. ROSE, H. WOLTER, W. GLAUBITT, Fraunhofer-Institut für Silicatforschung, Würzburg, Germany.

ABSTRACT

Multifunctional acrylate alkoxysilanes synthesized from commercial acrylate compounds and mercapto-substituted alkoxysilanes were used as precursors in the development of materials for various purposes. The alkoxysilyl groups are available for the construction of an inorganic backbone by the sol-gel process, and the acrylate groups for building an organic polymeric matrix by thermal or photochemical curing.

Tailoring the material properties is achieved by use of specific molecular structures and functionalities within the monomeric compounds. Linear and branched structural units influence the mechanical properties (modulus of elasticity, abrasion resistance, flexural strength). Rapidly UV-cured materials have been prepared for use as primary or secondary coatings for silica optical fibres with improved adhesion under humid conditions . Further organic modification results in polymers which can be used for manufacturing lenses of high optical quality. Preliminary attempts to draw fibres have also been successful.

INTRODUCTION

In recent years sol-gel technology has been well established in the manufacture of ceramic materials [1], glasses [2] or organically modified ceramics (ORMOCERs) [3]. ORMOCERs are inorganic organic copolymers synthesized from monomeric hydrolysable compounds, for example silicon alkoxides.

In the first reaction step an inorganic Si-O-Si backbone is built up via hydrolysis and condensation. Using substituted alkoxy compounds containing non-hydrolysable organic groups the inorganic network can be modified in many ways. Whereas non-reactive organic groups act as network modifiers, olefinic moieties can build an additional organic polymeric chain by UV- or thermal induced vinyl or a methyl methacrylate polymerization reaction.

The resulting material combines properties of ceramics or glasses with those of organic polymers. These properties depend on the inorganic or organic components of the precursor compound or of a mixture derived from different compounds.

Due to the inorganic network these materials are mostly used as scratch and abrasion resistant hard coatings for plastics [4, 5]. Other applications in which they show promise are in corrosion protection of metals [6] and for isolation layers in microelectronics [7].

In order to increase variety of applications of sol-gel derived ORMOCERs we have developed a new class of fast curable organo-alkoxysilanes, based on oligofunctional acrylate compounds.

Based on these alkoxy compounds which contain more than one reactive acrylate moiety per molecule, a higher degree of organic crosslinking is achieved.

In this paper some mechanical and optical properties of the new materials are described, and two examples of preliminary developments are featured.

MATERIAL PROPERTIES AND APPLICATION

Formation of the monomeric acrylate alkoxysilane is achieved by base catalysed addition of a mercapto-substituted alkylalkoxysilane to a commercially available compound with more than one acrylate moiety [8].

Bulk and coating materials with variable flexibility, variable thermal expansion, high strength and low shrinkage on curing are required for a number of different purposes. These properties are influenced to a great degree by the molecular structural units of the monomeric precursor compound. Furthermore there are several possible areas of chemical modification by which the mechanical and physical properties of the resulting material can be greatly varied [8].

Incorporation of linear moieties, for example derived from hexandiol, $-O-(CH_2)_6-O-$ or ethoxylated bisphenol-A-, $-CH_2CH_2-O-C_6H_4-C(CH_3)_2-C_6H_4-O-CH_2CH_2$ act as flexible spacer between the inorganic and the organic network. This as well as the degree of organic crosslinking based on the amount of reactive acrylate groups influences flexibility and rigidity. Further combination with a variable inorganic network - glasslike or siliconelike - allows the synthesis of materials with defineable mechanical properties like elasticity, abrasion resistance and flexural strength. New possibilities in material development are offered by mixing various acrylate alkoxysilanes or combination of alkoxysilanes containing different functional groups.

Following hydrolysis and polycondensation of the alkoxysilyl groups the very reactive olefinic groups attached to the resulting siloxane backbone can undergo a thermally or UV-induced polymerisation reaction.

The parameters and reaction conditions for hydrolysis and polycondensation are described elsewhere [9].

Application for fibre coating

Coating materials for silica optical fibres must fulfill many different requirements. The fibres are drawn from silica rods and must be protected from mechanical damage even during manufacturing.

A double coating system generally is used to achieve optimal protection of the fibre, both during handling and usage:

1. a primary coating of low modulus of elasticity to absorb mechanical stress or different thermal expansion of coating and fibre, which would result in microbending and signal loss

2. a secondary coating of high modulus of elasticity to preserve the strength and performance of the glass fibre, and for protection against abrasion.

Based on acrylate alkoxysilanes we have synthesized a fast curable coating material via the sol-gel process which allows UV curing in $< 0,1$ s, and can be used for fibre coating on a technical scale.

In this case the required moduli of elasticity for the primary and secondary coatings have been adjusted by incorporation of different structural units (figure 1):

- a linear, flexible bisphenol-A-moiety with one reactive group for UV-curing resulting in only linear polymer chains (component x) and

- a branched, aliphatic moiety with two reactive groups, resulting in crosslinked polymer chains (component y).

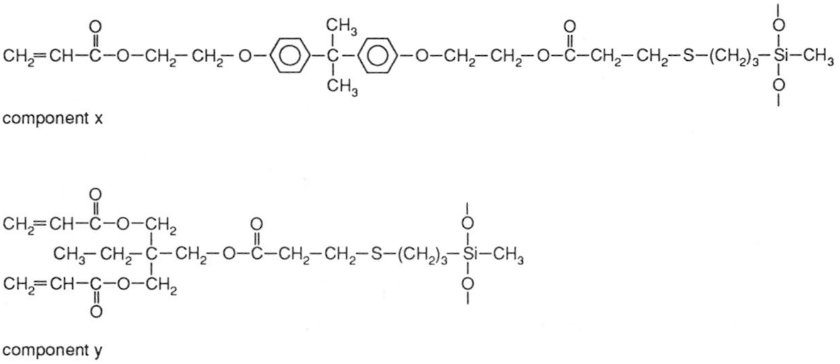

$$CH_2{=}CH{-}\overset{O}{\overset{\|}{C}}{-}O{-}CH_2{-}CH_2{-}O{-}C_6H_4{-}\underset{CH_3}{\overset{CH_3}{C}}{-}C_6H_4{-}O{-}CH_2{-}CH_2{-}O{-}\overset{O}{\overset{\|}{C}}{-}CH_2{-}CH_2{-}S{-}(CH_2)_3{-}\underset{O}{\overset{O}{Si}}{-}CH_3$$

component x

$$\begin{array}{l} CH_2{=}CH{-}\overset{O}{\overset{\|}{C}}{-}O{-}CH_2 \\ \qquad CH_3{-}CH_2{-}C{-}CH_2{-}O{-}\overset{O}{\overset{\|}{C}}{-}CH_2{-}CH_2{-}S{-}(CH_2)_3{-}\underset{O}{\overset{O}{Si}}{-}CH_3 \\ CH_2{=}CH{-}\underset{O}{\underset{\|}{C}}{-}O{-}CH_2 \end{array}$$

component y

Figure 1 Polysiloxane backbone with different acrylate side groups

Adjustment of modulus of elasticity in the range from 20 MPa to 1500 MPa for layers of 30 µm thickness after 0,5 s UV-curing is achieved using resins consisting of either one of the two components x and y (figure 2) or a mixture of the two components. For use as primary coating the modulus of elasticity should be < 10 MPa for reasons mentioned above. In this case long chained aliphatic moieties, such as those derived from hexandiol or ethylene glycol, should be used as a spacer between the organic and inorganic network.

After curing times of 0,5 s and 0,1 s respectively, abrasion resistance is high enough to protect the freshly drawn and coated fibres during winding and further handling. 30 µm thick films of various coatings based on different mixtures of components x and y have been tested after 0,5 s UV-curing on glass plates with a taber abraser (figure 2).

Whereas the mechanical data are comparable to the commercially available optical fibre coatings, the acrylate siloxane coating shows much better adhesion to silica in the presence of water. After 2 hours at 100 % relative humidity the common materials detach from the SiO_2-substrate. The ORMOCER materials, however, show no loss in adhesion even after 24 h. The formation of a water film on the silica surface is suppressed because of the good adhesion of the ORMOCER coating and thus corrosion of the fibre, which can cause signal loss, is prevented. In order to connect fibres the layer can be removed within a few seconds by treatment with CH_2Cl_2.

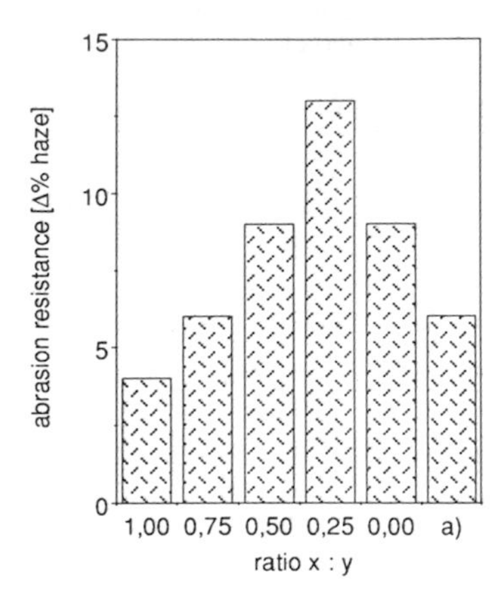

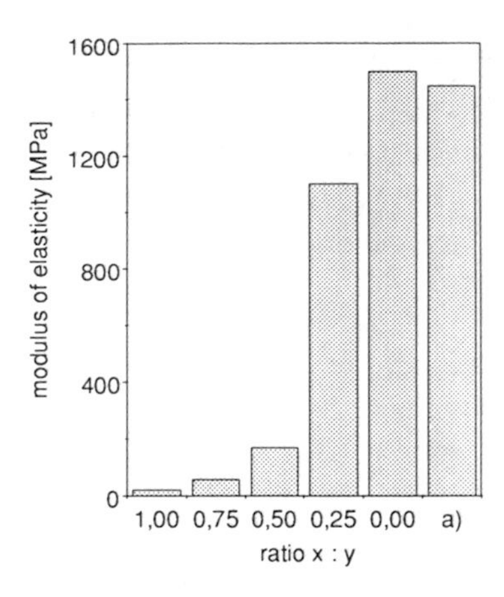

Figure 2 Mechanical properties of ORMOCER coating in comparison to a commercial secondary coating a) (films of 30 μm thickness)

Application to optical lenses

The optical industry is seeking low weight abrasion resistant materials which have the mechanical properties of organic polymers, and high optical quality, namely a high refractive index and a high Abbe-Number.

We synthesized UV- or thermally cured optical lenses of Abbe-Number > 35 based on acrylate alkoxysilanes and the siloxane resins produced from them. The refractive index depends on various organic moieties in the precursor compound, which connect the inorganic with the organic network (table 1).

Table 1. Refractive index depending on different organic moieties

organic moiety	refractive index
linear aliphatic	1,52
aromatic	1,56
halogenated aromatic	1,60

The modulus of elasticity can be varied by use of structural units which influence the degree of organic and inorganic crosslinking or the chain length between the two networks. The modulus of the resulting material varies in a range of 20 - 3600 MPa, thus the production of flexible contact lenses seems to be possible. The density of the bulk materials, in the range 1,16 - 1,28 g/cm^{-3}, is comparable to that of organic polymers like Polycarbonate or Polymethylmethacrylate.

Commonly available polymer optical lenses made from CR 39 are scratch sensitive and must be protected by a hard coating. As described above, ORMOCER materials have sufficient abrasion resistance due to their silicate network. For this reason an additional coating may be avoided. Abrasion resistance of ORMOCER lenses in comparison to common optical materials is shown in figure 3.

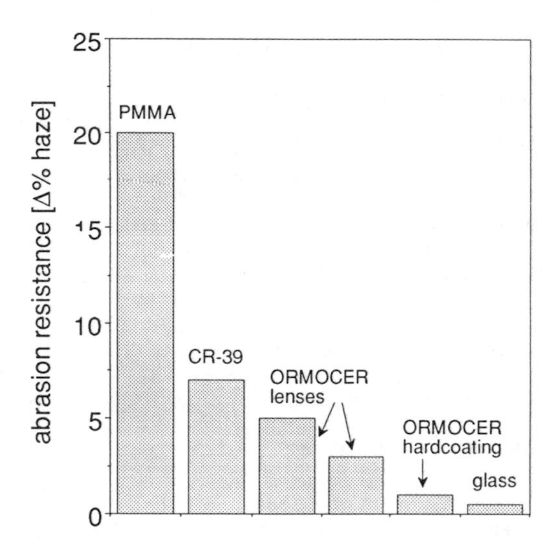

Figure 3 Abrasion resistance (100 cycles Taber Abraser) of different materials

CONCLUSION

We have developed new inorganic-organic polymers via sol-gel processing, which are based on a new class of acrylate alkoxysilanes as precursor compounds. The degree of organic or inorganic crosslinking can be varied to a great extent by varying the number of free acrylate moieties or alkoxy groups.

The flexibility, strength, abrasion resistance and thermal expansion of the resulting material is influenced by the degree of crosslinking, and the structural units which connect the inorganic and organic networks on a molecular scale.

After hydrolysis and condensation the materials may be used in solution or as resin. Organic crosslinking as curing reaction is achieved by thermal treatment or in a fraction of a second by UV-curing.

Up to now the materials have been tested in the coating of silica fibres and for the manufacture of optical lenses with high refractive indices. Preliminary investigations in drawing fibres have also been successful. In the latter case the optical properties in combination with the flexibility and abrasion resistance of the materials open the way for their use in the manufacture of polymer optical fibres. The materials have also been tested successfully as an adhesive to connect silica optical fibres with themselves or with electronic devices.

The wide variety of possible applications of the new materials has been further demonstrated in the preparation of flexible foils which also show better abrasion resistance than PMMA, PET or PVC, or in coating systems which demonstrate a self-healing effect.

REFERENCES:

[1] D. W. Johnson, Jr., Am.Ceram.Soc. Bull 64 (1985) 1597

[2] S. Sakka, Am.Ceram.Soc. Bull. 64 (1985) 1463

[3] H. Schmidt, ACS Symp. Ser. No 360, Inorganic and Organometallic Polymers M. Zeldin, K.J. Wynne, H.R. Allcock, Eds., 333 (1988)

[4] H. Schmidt, B. Seiferling, G. Philipp, K. Deichmann, Development of Organic-Inorganic Hard Coatings by the Sol-Gel Process, in: Ultrastructure Processing of Advanced Ceramics, J.D. Mackenzie, D.R. Ulrich, eds., (J. Wiley & Sons, New York, 1988), 651

[5] S. Amberg-Schwab, E. Arpac, W. Glaubitt, K. Rose, G. Schottner, U. Schubert, Protective Coatings for Organic Polymers by Sol-Gel Techniques, in: High Performance Films and Coatings, P. Vincenzini, ed., Elsevier Sci. Publ., 1991, 203

[6] K. Greiwe, Farbe + Lack 97 (1991) 368

[7] H. Schmidt, H. Wolter, J. Non-Cryst. Solids 121 (1990) 428
M. Popall, J. Kappel, M. Pilz, J. Schulz in: Proc. Eurogel 91, in print

[8] H. Wolter, W. Glaubitt, K. Rose, Multifunctional (Meth)acrylate Alkoxysilanes A new Type of Reactive Compounds, Mat. Res. Soc. Symp. Proc. (1992), in print

[9] S. Sakka, Y. Tanaka, T. Kokubo, J. Non-Cryst. Solids 82 (1986) 24
B. E. Yoldas, J. Non-Cryst. Solids 83 (1986) 375

PART VII

Novel Routes to Non-Oxide Ceramics

PREPARATION OF NON-OXIDE CERAMICS BY PYROLYSIS OF ORGANOMETALLIC PRECURSORS

LEONARD V INTERRANTE[§], WAYDE R. SCHMIDT[§], PAUL S. MARCHETTI[¥], and GARY E. MACIEL[¥], [§]Department of Chemistry, Rensselaer Polytechnic Institute, Troy, NY 12180-3590; [¥]Department of Chemistry, Colorado State University, Fort Collins, CO 80523.

INTRODUCTION

Chemical processing routes to advanced ceramic materials are gaining importance as a convenient approach to control the stoichiometry, purity, microstructure and final form of the ceramic products [1]. The pyrolytic conversion of organometallic molecules and polymers is one such chemical processing route that has been widely applied in ceramic fiber technology [1,2], in coating processes [1,2], and in the sintering of bulk ceramic objects [3]. Despite these advances in practical applications, there is a continuing need in this area for a better fundamental understanding of the chemistry involved during the precursor-to-ceramic conversion process and for the development of new precursors which yield the desired ceramic(s) in high yield and purity.

The recent discovery of superplasticity for nanocrystalline SiC/Si_3N_4 obtained by pyrolysis of a methylsilazane [4], along with the potential advantages afforded by nanocrystalline materials in terms of mechanical, optical, and electronic properties [5], has focused attention on the preparation of these materials by pyrolysis of molecular precursors. Finally, the prospect of developing ceramic powders and/or films with controlled nanoporosity for use as catalysts, catalyst supports and as gas separation media offers additional incentive to develop further the range of processes and new systems available through polymer precursor pyrolysis [6].

This paper focuses on our own efforts over the past *ca.* 10 years to explore polymer/molecular precursor pyrolysis as a route to both known and potentially new non-oxide ceramic compositions with unique properties.

Organometallic Precursor Routes to AlN

Our efforts in this area began with an attempt to obtain AlN powder in high yield and purity from organoaluminum-ammonia derivatives. These experiments led to a process for AlN powder production that employs Et_3Al and NH_3 as the Al and N sources, respectively (Scheme I), and which gives a high surface area, pure AlN powder in nearly quantitative yield at a low net

$$R_3Al + NH_3 \longrightarrow R_3Al{:}NH_3$$

$$R_3Al{:}NH_3 \longrightarrow R_2AlNH_2 + RH$$

Scheme I

$$R_2AlNH_2 \longrightarrow RAlNH + RH$$

$$RAlNH \xrightarrow{NH_3} AlN + RH$$

materials cost [7]. A key step in this procedure is the use of an NH_3 atmosphere in the final stage of conversion to promote loss of residual alkyl groups as volatile hydrocarbons, thereby yielding stoichiometric AlN with both C and O contents less than 0.1 wt. % and negligible amounts of other impurities. This is to be contrasted with alternative processing methods that start with Al or Al_2O_3 where control of product stoichiometry and/or C and O content is problematic.

The use of Me_3Al as the Al source in this same sequence of reactions involving NH_3 yields $[Me_2AlNH_2]_3$ as a crystalline intermediate. This compound is relatively volatile and serves as an excellent single-source precursor for AlN CVD yielding high quality, polycrystalline AlN films at relatively low deposition temperatures (*ca.* 500 °C) compared to alternative mixed Al and N precursor systems, such as AlX_3 (X = Cl, Br) or Me_3Al and NH_3 [8]. In this case, the loss of C from the growing AlN surface is relatively efficient, leading to low levels of C incorporation in the product films even without the use of additional NH_3 as a co-reactant.

These applications of the $R_3Al + NH_3$ system as a source of AlN powder and thin film have been aided and enhanced by our continuing studies of the chemistry represented in the sequence of reactions shown in Scheme I. This work has included kinetic studies of the $Me_3Al{:}NH_3$ decomposition to $[Me_2AlNH_2]_3$, which provided the first indirect evidence for a trimer (T) <=> dimer (D) <=> monomer (M) equilibrium for the Me_2AlNH_2 product species [9]. The existence of at least the T <=> D equilibrium in hydrocarbon solution was evidenced in further NMR studies of the R = Me, Et, i-Bu and t-Bu derivatives which yielded the respective ΔH and ΔS values [10]. Our isolation of the trimeric form of the $[(t\text{-}Bu)_2AlNH_2]_3$ compound led to a crystal structure study that revealed a novel planar arrangement for the $(AlN)_3$ six-membered ring [11]. In this study, the factors that determined the relative stabilities of the dimer and trimer forms for the $[R_2AlNH_2]_n$ compounds, as well as the conformation of the $[R_2AlNH_2]_3$ ring system, were noted and discussed [11]. A subsequent theoretical study indicated a probable planar structure for the monomeric Me_2AlNH_2 species and led to a prediction of its bond lengths and angles [12]. Time of flight studies of the $[Me_2AlNH_2]_3$ system have recently confirmed the existence of dimeric and trimeric forms in the vapor phase [13]; however, the corresponding monomeric form currently remains elusive.

An alternative solution-based route to AlN films and coatings which employs a mixture of R_3Al and ethylenediamine (en) has also been investigated yielding, in the case of the 2:1 R_3Al:en system, a soluble polymeric intermediate $[(R_2Al)_2(en\text{-}2H)]_n$ which was used to make AlN films on Si [14]. Again, a detailed study of the chemistry involved in the conversion of the xR_3Al:en (x = 2,1) adducts to polymeric products was carried out which has revealed a series of novel aluminum/(en-2H) cluster intermediates on thermolysis of the 1:1 adducts [15].

Precursors to SiC and Si_3N_4

Our efforts to synthesize new SiC precursors were prompted by the problem of excess C in SiC derived from most of the organosilicon polymers that are currently employed as precursors. These precursors typically contain a 2/1 or greater ratio of C/Si initially, and the excess C is not eliminated cleanly as hydrocarbons on pyrolysis. Residual carbon can lead to decreased oxidative stability for the resulting SiC ceramics, as well as excessive creep and other undesireable effects on the microstructure and/or properties. Moreover, in the presence of oxygen derived from curing, handling, and/or partial oxidation of the precursors or from oxide components in a composite, loss of CO_x species becomes a significant problem for these materials beyond about 1200 °C [16].

For these reasons, we have directed our attention to the preparation of polycarbosilanes that have a nominal "SiH_2CH_2" composition and have explored two approaches to such polymers that yield structurally quite different products having either highly branched or linear -Si-CH_2-Si- backbones. Both polymers were found to have high ceramic yields after thermal processing and gave near-stoichiometric SiC on pyrolysis in N_2.

The first approach developed employs Mg as a coupling agent and Cl_3SiCH_2Cl as the starting material. The resultant "chloropolymer" (nominally, "$[SiCl_2CH_2]$") was converted to the hydridopolycarbosilane (HPCS) "$[SiH_2CH_2]$" by reduction with $LiAlH_4$. A side reaction with the diethylether solvent leads to the incorporation of a small amount of ethyl groups, giving the actual formula, $[SiH_{1.85}Et_{0.15}CH_2]$ [17].

A structural analysis of this polymer by solution NMR methods has revealed a complex, highly branched structure, resulting from the ternary functionality of the Cl_3Si end of the monomer unit. Despite this branching, the polymer is not appreciably crosslinked and remains liquid and highly soluble in hydrocarbon solvents. This "$[SiH_2CH_2]$" polymer undergoes crosslinking through loss of H_2 above *ca.* 200 °C. Additional crosslinking by thermal treatment produces a gel at 200 °C (12 hrs) and an insoluble rubber at 400 °C (2 hrs), which have significantly increased ceramic yields (53% and 91%, respectively) on pyrolysis in N_2 to 1000 °C (Figure 1), as compared to the initially isolated, untreated polymer (30-40% ceramic yield) [18].

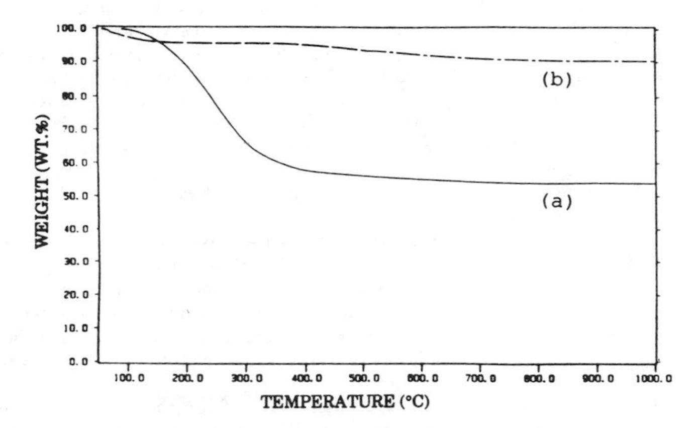

Figure 1. TGA curves for highly branched hydridopolycarbosilane (HPCS) after thermal crosslinking a) 200 °C (12 hrs), b) 400 °C (2 hr).

We have used this polymer to obtain amorphous "SiC" coatings on glassy carbon substrates and C fiber that inhibit oxidation of these substrates at 600 °C in air. Adherent coatings have also been applied to various metal surfaces, including C-steels. These coatings can be applied as polymer solutions through dip or spin coating techniques, followed by pyrolysis to 1000 °C in N_2. We have also used this hydridopolycarbosilane to prepare particulate and fiber-reinforced SiC/SiC composites by liquid phase impregnation. The advantages over alternative precursors for SiC include the effective elimination of excess C and O, thereby providing a SiC product that is more thermally and oxidatively stable than those obtained using alternative precursors [19]. This polymer also does not require a catalyst or an oxidation step for thermosetting and exibits a high ceramic yield of essentially pure SiC.

The second approach that we have developed to obtain a "$[SiH_2CH_2]$" polymer involves the ring-opening polymerization (ROP) of tetrachlorodisilacyclobutane, $[SiCl_2CH_2]_2$ [20]. The resultant "chloropolymer" is again reduced to the hydridopolymer with $LiAlH_4$. In this case, however, the product polycarbosilane is a high molecular weight linear polymer with $[SiH_2CH_2]_n$ as the actual repeat unit. This polymer is also a liquid a room temperature, is soluble in hydrocarbons, and undergoes pyrolysis under N_2 to give stoichiometric SiC in yields approaching the theoretical value (up to 91%) by 1000 °C. This polymer is currently available only in small quantities by a relatively low yield process and is not expected to be competitive in the short term with the branched "$[SiH_2CH_2]$" polymer as a source of SiC for practical applications.

Both of these synthetic methods are fundamentally new approaches for preparing useful polymeric precursors to SiC. These polycarbosilanes undergo a novel, thermally-induced crosslinking process that involves elimination of H_2 from SiH_n groups on the polymer backbone. They show considerable promise as high yield precursors to stoichiometric SiC and, in the case of the branched polymer, as a viable source of SiC coatings and matrices for composites.

In addition to these ongoing investigations of the pyrolysis of our own "SiH_2CH_2" polymers, we have carried out parallel studies of a commercial polysilane of the approximate composition,

$$[[Si(Me_3)]_{0.32}[Si(CH{=}CH_2)Me]_{0.35}[Si(H)Me]_{0.18}[SiMe_2]_{0.07}[CH_2SiMe_3]_{0.08}]_n$$

obtained from Union Carbide (Y-12044) [19,21]. This liquid polymer provides a predominantly polysilane $(-RR'Si-)_n$ chain structure with reactive vinyl and H groups for possible thermosetting through radical coupling of the vinyl groups and hydrosilylation. It is known to yield a C-rich (*ca.* 17 wt.% excess C) SiC on pyrolysis in N_2 or other inert atmospheres. We have studied the crosslinking process and the evolution of the SiC network structure in this polymer by a combination of methods, but of particular value has been the use of solid state NMR methods, where the Si, C and even H environments in the precursor were followed through the different stages of the thermal conversion process [21]. The chief conclusion in the case of the pyrolysis in N_2 was that cleavage of the end groups from the polymer backbone to give $R_3Si^\bullet$ radicals at relatively low temperature is a key step in both the thermosetting and the initial part of the polymer conversion process, during which the -Si-Si- backbone converts to $-Si-CH_2-Si-$ via a radical-initated $-CH_2$ insertion process [21].

Following other studies in which NH_3 has been used to convert polycarbosilanes and polysilazanes to Si_3N_4, we have found that the vinylic polysilane and our "SiH_2CH_2" polymer can be used to obtain high surface area powders of largely amorphous Si_3N_4 at 1000 °C [22]. The Si_3N_4 eventually crystallizes with reduction in surface area on further heating above 1200 °C; however, below this temperature, in the absence of certain impurities such as B or BN [23], this high surface area Si_3N_4 is relatively stable and conceivably could be used as a catalyst support [6].

We have recently completed a detailed study of the chemistry occurring during the conversion of this vinylic polysilane to Si_3N_4 in NH_3 that has included esr and dynamic nuclear polarization, as well as solid state NMR measurements on the intermediate and final solid products [24]. A key difference here with respect to the analagous conversion in N_2, in addition to the eventual loss of all C and its replacement by N, is the relatively low spin concentrations for the samples pyrolyzed in NH_3 (Figure 2). Whereas the "inert" atmosphere pyrolysis generates a significant concentration of paramagnetic centers and unpaired spins in the solid products, in NH_3 the unpaired spins are effectively quenched, presumably by radical transfer processes involving NH_3 (Scheme II).

$$NH_3 + {}^\bullet Si\lessgtr \text{ (or } R^\bullet) \longrightarrow NH_2^\bullet + HSi\lessgtr \text{ (or RH)}$$

Scheme II

$$NH_2^\bullet + {}^\bullet Si\lessgtr \longrightarrow \gtrless Si\text{-}NH_2$$

The resultant $NH_2^\bullet$ radicals generated can couple with Si-based radical sites on the polymer backbone. This process, along with nucleophilic attack of NH_3 directly on the backbone Si atoms of the polymer, are believed to be responsible for the efficient removal of C, as hydrocarbons, and its replacement by NH_x species. Condensation processes that eliminate NH_3 and silylamine species then lead to the extended -Si-N- network structure of Si_3N_4 [24].

A study of the condensation polymerization and pyrolysis of silicon tetradiethylamide, $[Si(NEt_2)_4]$, was also carried out with the aid of solid state ^{29}Si, ^{13}C, and ^1H (CRAMPS) NMR spectroscopy [25]. Among the key observations here was that the initial SiN_4 bonding of the precursor was maintained throughout the conversion of the tetraamide to the polymeric silazane and then to the amorphous preceramic product at 800 °C, albeit with the formation of a considerable amount of free C by-product. On heating further to 1500-1600 °C, this amorphous "silicon nitride" plus C mixture is converted under an Ar atmosphere largely to SiC and $N_2(g)$, whereas under one atmosphere of N_2, crystalline α-Si_3N_4, along with free C, are the main products.

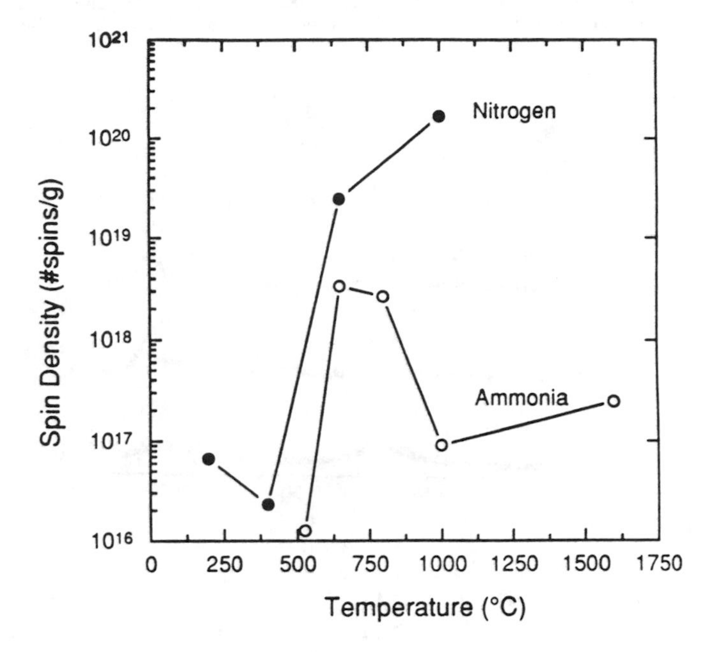

Figure 2. Spin densities as a function of temperature for solids isolated by pyrolysis of the vinylic polysilane in either nitrogen or ammonia.

Preparation of AlN/SiC Solid Solutions and Ceramic Nanocomposites Using Organometallic Precursors

We have found that both homogeneous solid solutions and nanoscale composite mixtures of SiC with AlN can be obtained on pyrolysis of mixtures of the polycarbosilanes or polysilane (as the SiC source) and dialkylaluminum amides (as the AlN precursor) and that the ceramic yield of the polycarbosilane is greatly enhanced in the presence of the AlN precursor [26].

In this work the use of miscible liquid precursors to both AlN and SiC allowed the attainment of molecular-level homogeneity in the initial precursor mixture. Subtle differences in the extent to which this molecular-level homogeneity was maintained after pyrolysis of the precursor mixture were evidenced by X-ray diffraction studies and solid state ^{27}Al and ^{29}Si NMR spectroscopy. Figure 3 shows a series of X-ray diffraction patterns of a 50/50 mole % mixture of the completely miscible liquid HPCS and $[Me_2AlNH_2]_3$ precursors which were pyrolyzed gradually to 1000 °C (GP) (Fig. 3a) or rapidly pyrolyzed at 500 °C (RP) prior to further heating (Fig. 3b). These samples were further heated in N_2 to temperatures as high as 1800 °C to crystallize the solid solution. In both cases the X-ray diffraction patterns obtained at 1800 °C are consistent with expectations for the *2H* solid solution [26,27]; however, the sharper lines exhibited by the rapidly pyrolyzed sample suggest that the crystallinity is somewhat higher in this case as compared to the gradually pyrolyzed sample. These XRD patterns contrast markedly with that obtained on pyrolysis, and subsequent annealing at 1800 °C, of the HPCS polymer alone. In this case the diffraction pattern is consistent with expectations for the cubic 3*C* form of SiC.

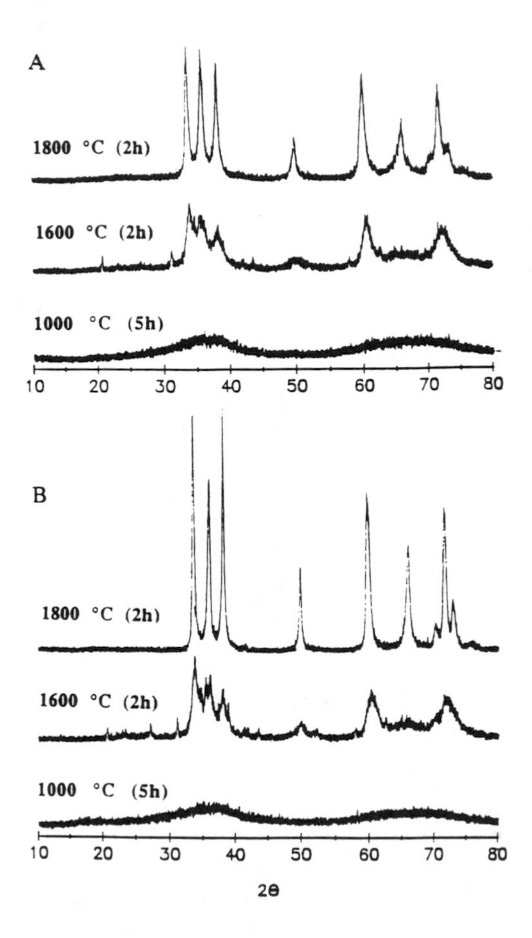

Figure 3. X-ray diffraction patterns of a) gradually pyrolyzed (GP) and b) rapidly pyrolyzed (RP) SiC/AlN precursor mixture.

The solid state NMR spectra of the solid solution samples (Figures 4-6) show the effects of the miscibility gap in the SiC/AlN solid solution below 1950 °C [27]. Thus, both 1000 °C samples show spectra consistent with expectations for "averaged" $Al(Si)N_xC_y$ environments whereas, after crystallization of these samples at 1600-1800 °C, the the Al and Si NMR peaks shift to chemical shift values more nearly representative of the "pure" components, AlN and SiC. On the other hand, there are clearly differences in these spectra that support the presumption of solid solution formation and suggest further that the processing method employed plays a role in determining the atomic-level homogeneity of the crystallized SiC/AlN solid solution phases obtained. In particular, after heating to 1000 °C the gradually pyrolyzed (GP) sample exhibits a two-peak ^{27}Al NMR spectrum, with one relatively sharp peak close to the expected position for pure AlN (Figure 4). This suggests that partial phase separation occurs during the slow heating of the precursor mixture, as would be expected on the basis of the lower thermal stability of $[Et_2AlNH_2]_3$ relative to HPCS [26]. Moreover, in the ^{29}Si NMR spectrum obtained for the rapidly pyrolyzed (RP) sample after heating to 1800 °C, the main Si peak is broader and shifted to a more shielded position relative to that observed for the corresponding GP sample (Figures 5 and 6). These data suggest that on an atomic level, the crystallized solid solution phases are slightly different, with the RP sample exhibiting a greater degree of homogeneity and a more "solid-solution like" ^{29}Si NMR spectrum.

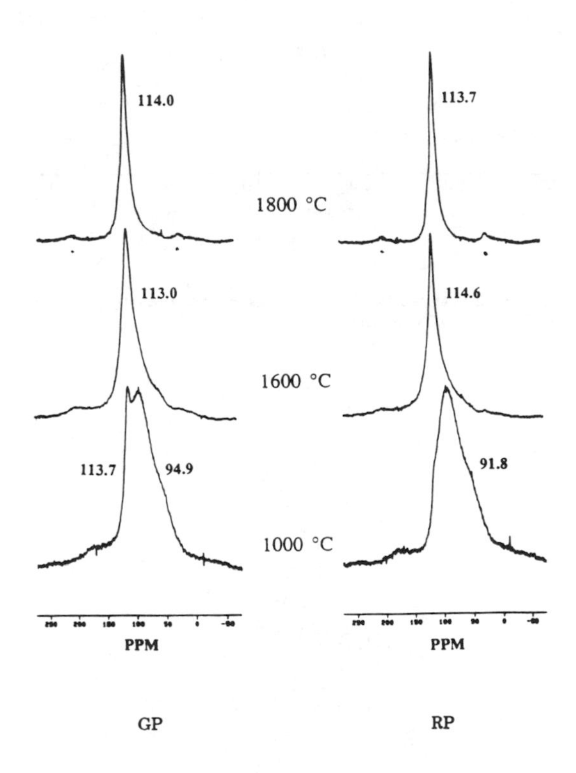

Figure 4. Solid state 27Al MAS NMR spectra of the gradually pyrolyzed (GP) or rapidly pyrolyzed (RP) HPCS/[Et2AlNH2]3 precursor mixture.

Despite these subtle processing-dependant differences in composition and homogeneity and the presence of a miscibility gap in the phase diagram, the successful generation of 2*H* solid solutions of SiC and AlN from HPCS and $[Me_2AlNH_2]_3$ mixtures represents a significant advance in ceramic processing by polymer precursor pyrolysis. This opens up the possibility of using liquid phase, thermally curable precursors such as HPCS not only as a source of SiC coatings and matrices for ceramic composites, but also for the preparation of other mixed-component ceramic

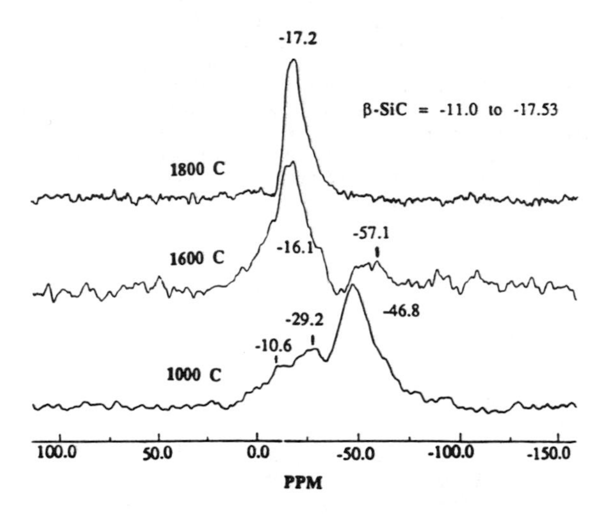

Figure 5. Solid state 29Si MAS NMR spectra of a gradually pyrolyzed HPCS/[Et2AlNH2]3 precursor mixture.

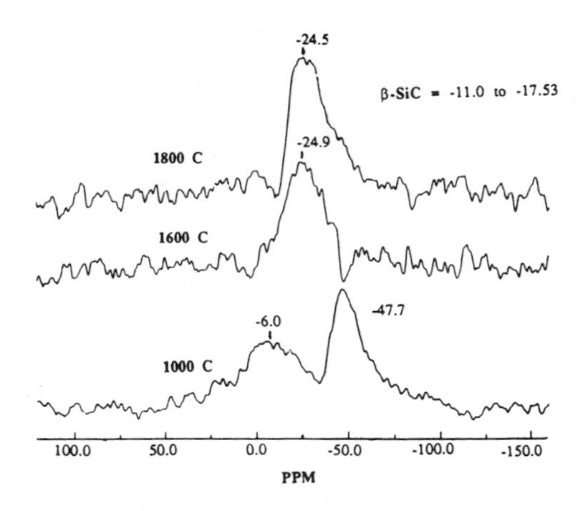

Figure 6. Solid state 29Si MAS NMR spectra of a rapidly pyrolyzed HPCS/[Et2AlNH2]3 precursor mixture.

systems that may be more advantageous than SiC alone. In particular, in the case of SiC/AlN, the alloying of SiC with AlN has been reported to improve its microstructural stability, inhibiting the exaggerated grain growth that is known to contribute to the loss of strength typically observed for commercial "SiC" fibers at high temperatures (1200-1800 °C) [27]. The SiC/AlN solid solution has also been reported to have superior strength and oxidation resistance compared to either SiC or AlN alone [28].

We have also explored the use of both precursor mixtures and specially designed single-component precursors to prepare nanocrystalline composites of AlN/SiC, AlN/Si_3N_4, BN/Si_3N_4, AlN/BN, and BN/TiN [29]. We have demonstrated that such precursors can be used to obtain extremely fine-grained (typically less than 100 nm), homogeneously mixed composite powders of these crystalline ceramic phases. Moreover, these fine-grained mixtures resist coarsening during exposure to temperature in excess of 1500 °C. Such mixed-phase ceramics may be useful as matrices for SiC-reinforced composites, in applications requiring superplasticity, or as tough monolithic materials in their own right.

ACKNOWLEDGEMENTS

Funding for this research has been provided in part by the National Science Foundation, the Air Force Office of Scientific Research, the Office of Naval Research, and a DARPA/ONR URI program. Solid state NMR spectra were obtained at the Colorado State University NMR Center, which was funded by the NSF.

REFERENCES

1. a) K. J. Wynne and R. W. Rice, *Annu. Rev. Mater. Sci.*, **14** (1984) 297; b) D. R. Ulrich, "Chemical Processing of Ceramics", *Chemical and Engineering News*, Jan. 1 (1990) 28; c) D. Segal, Chemical Synthesis of Advanced Ceramic Materials, Cambridge University Press: New York, (1989).

2. H. E. Fischer, D. J. Larkin, and L. V. Interrante, *MRS Bull.*, 26[4] (1991) 59-65; b) T. F. Cooke, *J. Am. Ceram. Soc.*, **74**[12] (1991) 2959.

3. a) S. T. Schwab and C. R. Blanchard-Ardid, *Mat. Res. Soc. Symp. Proc.*, **121** (1988) 581; b) W. H. Atwell, G. T. Burns, and G. A. Zank, "Silicon Carbide Preceramic Polymers as Binders for Ceramic Powders", in Inorganic and Organometallic Oligomers and Polymers, J. F. Harrod and R. M. Laine, eds., Kluwer Academic Publishers, (1991) 147-159.

4. F. Wakai, Y. Kodama, S. Sakaguchi, N. Murayama, K. Izaki, and K. Niihara, *Nature*, **344**[29 March] (1990) 421.

5. a) Y. Maehara and T. G. Langdon, *J. Mat. Sci.*, 25 (1990) 2275; b) I.-W. Chen and L. A. Xue, *J. Am. Ceram. Soc.*, 73[9] (1990) 2585; b) R. W. Siegel, *MRS Bull.*, **25**[10] (1990) 60.

6. a) R. W. Chorley and P. W. Lednor, *Adv. Mater.*, **3**[10] (1991) 474; b) S. Yamanaka, *Ceram. Bull.*, **70**[6] (1991) 1056.

7. L. V. Interrante, L. E. Carpenter II, C. Whitmarsh, and W. Lee, *Mat. Res. Soc. Symp. Proc.*, **73** (1986) 359.

8. L. V. Interrante, W. Lee, M. McConnell, N. Lewis and E. Hall, *J. Electrochem. Soc.*, **136**[2] (1989) 472-478.

9. F. C. Sauls, L. V. Interrante and Z. Jiang, *Inorg. Chem.*, **29** (1990) 2989-96.

10. F. C. Sauls, C. L. Czekaj, and L. V. Interrante, *Inorg. Chem.*, **29**[23] (1990) 4688-4692.

11. a) L. V. Interrante, G. A. Sigel, C. Hejna and M. Garbauskas, *Inorg. Chem.*, **28** (1989) 252-257; b) L. V. Interrante, G. A. Sigel, C. Hejna and M. Garbauskas, *Phosphorus, Sulfur and Silicon*, **41** (1989) 325-334.

12. M. M. Lynam, L. V. Interrante, C. H. Patterson, and R. P. Messmer, *Inorg. Chem.*, **30** (1991) 1918-1922.

13. a) C. C. Amato, J. B. Hudson and L. V. Interrante, *Mat. Res. Soc. Symp. Proc.*, **168** (1990) 119-124; b) C. C. Amato, J. B. Hudson, and L. V. Interrante, *Mat. Res. Soc. Symp. Proc.*, **204** (1991) 135-140.

14. Z. Jiang and L. V. Interrante, *Chem. Mater.*, **2** (1990) 439-446.

15. Z. Jiang, L. V. Interrante, D. Kwon, F. S. Tham, and R. Kullnig, *Inorg. Chem.*, **30** (1991) 995-1000.

16. a) D. J. Pysher, K. C. Goretta, R. S. Hodder, Jr., and R. E. Tressler, *J. Am. Ceram. Soc.*, **72**[2] (1988) 244; b) Z.-F. Zhang, F. Babonneau, R. M. Laine, Y. Mu, J. F. Harrod, and J. A. Rahn, *J. Am. Ceram. Soc.*, **74**[3] (1991) 1670.

17. C. Whitmarsh and L. V. Interrante, *Organometallics*, **10** (1991) 1336-1344.

18. C.-Y. Yang and L. V. Interrante, "Thermally and Chemically Induced Crosslinking of a Hydridopolycarbosilane for Optimization of Silicon Carbide Yield", submitted to *Polymer Preprints*.

19. L. V. Interrante, C. K. Whitmarsh, T. K. Trout, and W. R. Schmidt, "Synthesis and Pyrolysis Chemistry of Polymeric Precursors to SiC and Si_3N_4", Kluwer Academic Publishers, Dordrecht, (1991); Proceedings, NATO Workshop on Organometallic Polymers with Special Properties, Cap D'Agde, France, Sept. 10-15, 1990.

20. a) H.-J. Wu and L. V. Interrante, "Preparation of A Linear Polycarbosilane via Ring-Opening Polymerization", Polymer Preprints, 4th Chem. Congress, NY, Aug. 1991; b) H.-J. Wu and L. V. Interrante, *Macromolecules*, **25** (1992) 1840.

21. W. R. Schmidt, L. V. Interrante, R. H. Doremus, T. K. Trout, P. S. Marchetti, and G. E. Maciel, *Chem. Mater.*, **3** [2] (1991) 257-267.

22. W. R. Schmidt, V. Sukumar, W. J. Hurley, Jr., R. Garcia, R. H. Doremus, and L. V. Interrante, *J. Am. Ceram. Soc.*, **73**[8] (1990) 2412.

23. V. Sukumar, W. R. Schmidt, R. Garcia, R. H. Doremus and L. V. Interrante, *Mater. Letters*, **9** (1990) 117-120.

24. W. R. Schmidt, P. S. Marchetti, L. V. Interrante, W. J. Hurley, Jr., R. H. Lewis, R. H. Doremus, and G. E. Maciel, "Ammonia-Induced Pyrolytic Conversion of a Vinylic Polysilane to Silicon Nitride", submitted to *Chem. Mater.*

25. D. M. Narsavage, L. V. Interrante, P. S. Marchetti, and G. E. Maciel, *Chem. Mater.*, **3**[4] (1991) 721.

26. a) C. L. Czekaj, M. L. Hackney, W. J. Hurley, Jr, L. V. Interrante and G. A. Sigel, *J. Am. Ceram. Soc.*, **73**[2] (1990) 352-357; b) L. V. Interrante, W. R. Schmidt, S. N. Shaikh, R. Garcia, P. S. Marchetti, and G. Maciel, "Silicon Carbide Aluminum Nitride Solid Solutions by Pyrolysis of Organometallics", Proc. 5th Intl. Conf on Ultrastructure Processing of Ceramics, Glasses, Composites, Ordered Polymers and Advanced Optical Materials, Series E: Applied Sciences, Vol. **206**, J. Wiley and Sons, publs., (1991) 243-254.

27. a) A. Zangvil and R. Ruh, *J. Am. Ceram. Soc.*, **71**[10] (1988) 884; b) R. Ruh and A. Zangvil, *J. Am. Ceram. Soc.*, **65**[5] (1982) 260; c) W. Rafaniello, M. R. Plichta, and A. V. Virkar, *J. Am. Ceram. Soc.*, **66**[4] (1983) 272; d) W. Rafaniello, K. Cho, and A. V. Virkar, *J. Mater. Sci.*, **16**[12] (1981) 3479.

28. G. Ervin, Jr., US Patent 3,492,153, Jan. 27, 1970.

29. a) D. Kwon, W. R. Schmidt, L. V. Interrante, P. Marchetti, and G. Maciel, "Preparation and Microstructure of Organometallic Polymer Derived AlN-BN Composites", Inorganic and Organometallic Oligomers and Polymers, J.F. Harrod and R.M. Laine, eds., Kluwer Academic Publishers, (1991) 191-197; b) W. R. Schmidt, W. J. Hurley, Jr., V. Sukumar, R. H. Doremus, and L. V. Interrante, *Mat. Res. Soc. Symp. Proc.*, **171** (1990) 79-84; c) L. V. Interrante, W. J. Hurley, Jr., W. R. Schmidt, D. Kwon, R. H. Doremus, P. S. Marchetti, and G. Maciel, "Preparation of Nanocrystalline Composites by Pyrolysis of Organometallic Precursors", Proceedings, Symposium on Composites, American Ceramic Society Meeting, Orlando, FL, Nov. 1990, Advanced Composite Materials-Ceramic Transactions, Vol. 19, M.D. Sacks, ed. American Ceramic Society: Westerville, Ohio, (1991) 3-17; d) W. R. Schmidt, W. J. Hurley, Jr., R. H. Doremus, L. V. Interrante, and P. S. Marchetti, "Novel Polymer Precursors to Si-C-Al-O-N Ceramic Composites", Proceedings, Symposium on Composites, American Ceramic Society Meeting, Orlando, FL, Nov. 1990, Advanced Composite Materials-Ceramic Transactions, Vol. 19, M.D. Sacks, ed. American Ceramic Society, Westerville: Ohio, (1991) 19-25.

CHEMICALLY DESIGNED, UV CURABLE POLYCARBOSILANE POLYMERS

KEVIN J. THORNE*, STEPHEN E. JOHNSON**, HAIXING ZHENG*, JOHN D. MACKENZIE* AND M.F. HAWTHORNE**
*UCLA Materials Science and Engineering Department, Los Angeles, CA, 90024.
**UCLA Department of Chemistry and Biochemistry, Los Angeles, CA, 90024.

ABSTRACT

To prepare new polycarbosilane polymer precursors with high solubility and the capability of UV cross-linking, commercial polycarbosilane was modified by a chemical route. These modifications involved $AlCl_3$ catalyzed chlorination reactions of polycarbosilane's Si-H bonds. The resultant Si-Cl bonds were substituted by a reaction with sodium acetylyde to form Si-C≡CH ligands. These ligands are suitable for controlled, free radical initiated cross-linking of the polycarbosilane polymers. The increase in molecular weight should allow for increased Tg's and the retention of polymer pre-forms. In this report, the chlorination of the polycarbosilane polymer and the substitution reactions of polycarbosilane were examined with IR, ^{29}Si and ^{13}C NMR spectroscopy. In addition, the retention of polymer pre-forms were analyzed after UV exposure and inert atmosphere pyrolysis.

INTRODUCTION

The inability of polycarbosilane polymeric precursors to maintain structural rigidity during pyrolysis severely limits the macroscopic SiC ceramics which can be produced. Direct pyrolysis of the commercial Nicalon polycarbosilane polymer into silicon carbide results in extremely porous and weak monolithic ceramic products. The low Tg of the polymer precursor (MW_{ave}=1200-1450) accounts for its liquification at approximately 300°C [1,2,3]. The gaseous evolution of CH_4 and H_2 products during its polymerization and conversion into a macromolecular solid at temperatures between 300-500°C explain the extreme porosity in the ceramic product [4].

Regardless of this limitation, polycarbosilane (PCS) has been successfully used as a polymer precursor for the production of Nicalon silicon carbide fibers. Their production requires controlled surface oxidation of the polymer fibers [5]. This oxidation results in a silica layer that provides structural rigidity during pyrolysis and allows the retention of the fiber form. Our recent investigations in the polymer route production of advanced ceramic carbides and nitrides have suggested a solution that will not require surface oxidation for the retention of the monolithic polymer pre-forms.

With commercial polycarbosilane, it is only required that the polymers be structurally rigid to temperatures of 500°C. This is because the polymerization and cross-linking of the polycarbosilane molecules into a macromolecular solid occurs between 300-500°C [6]. If the molecular weight of the polycarbosilane polymer can be increased, it should be possible to provide structural rigidity during pyrolysis to these temperatures since the Tg of the polymer would be increased.

 ©1992 Materials Research Society

Through chemical modification of the polycarbosilane polymers, organic ligands capable of free radical initiated polymerization have been introduced into polycarbosilane's polymer structure. The introduction of silicon bonded ethynyl ligands (Si-C≡CH) was made possible by the chlorination of polycarbosilane. The Si-Cl ligands of the chlorinated polymer were substituted by a reaction with sodium acetylyde, NaC≡CH. In this report, these reactions, the products and the UV cross-linking of the ethynyl substituted polycarbosilane polymer were examined using ^{13}C, ^{29}Si NMR and IR spectroscopy.

EXPERIMENTAL PROCEDURES

All reactions and processing were conducted inside a nitrogen atmosphere glove box [$P(O_2)<3$ ppm] or by means of standard Schlenk techniques. Polycarbosilane was purchased from DOW Corning. Sodium acetylyde, trimethylchlorsilane, vinyl magnesium bromide and aluminum chloride were purchased from the Aldrich Chemical Co. and used without further purification.

^{29}Si and ^{13}C NMR spectra were recorded using a Bruker AM-360 FT-NMR spectrometer. Transmission IR spectra were obtained using a Beckman FT-1100 photospectrometer over a range of 4000-500 cm^{-1}. The polymeric samples were exposed to ultraviolet radiation inside a nitrogen atmosphere glove box. Pyrolysis was conducted in a flowing argon atmosphere (99.999% Ar), graphite refractory furnace (Thermal Technology - Series 1000).

Preparation of Poly(alkynyl)carbosilane

Polycarbosilane was dissolved in an excess of trimethylchlorosilane [$ClSi(CH_3)_3$]. Upon dissolution, 5 weight percent of aluminum chloride [$AlCl_3$] was added to the solution. The $ClSi(CH_3)_3$ mixture spontaneously reacted and continuously boiled. The solution was refluxed for two hours and then the excess solvent was distilled. The $AlCl_3$ catalyst was then removed by vacuum sublimation at 50°C. The chlorinated PCS polymer was then dissolved in oxygen-free tetrahydrofuran. To this solution, dry sodium acetylyde [NaC≡CH] was added in excess. An exothermic reaction occurred and white insoluble particles precipitated from the solution. The modified polycarbosilane polymer was obtained by filtration of the insoluble reactants and salt products followed by distillation of the THF solvent.

RESULTS AND DISCUSSIONS

The reaction scheme utilized for the chlorination of polycarbosilane is outlined in Figure 1. This reaction mechanism had originally been proposed by Kumada for the chlorination of bis(trimethylsilyl)methane [7]. The reaction mechanisms previously proposed were different then actually observed according to the IR and NMR spectroscopic results.

In Figure 2, the IR spectra during the chlorination reaction are presented. The most obvious change during the reaction is the significant reduction in the relative intensity of the Si-H stretching vibration at 2150 cm^{-1}. After a two hour reflux, the Si-H vibration is no longer present. It is also important to notice that the relative intensity of the $Si-CH_3$ stretching vibration at 1250 cm^{-1} didn't change. This information suggests that the reaction primarily occurred through the consumption of the Si-H bonds and the formation of trimethylsilane. The previous studies had suggested that the reaction primarily occurs through the consumption of the $Si\text{-}CH_3$ bonds [8].

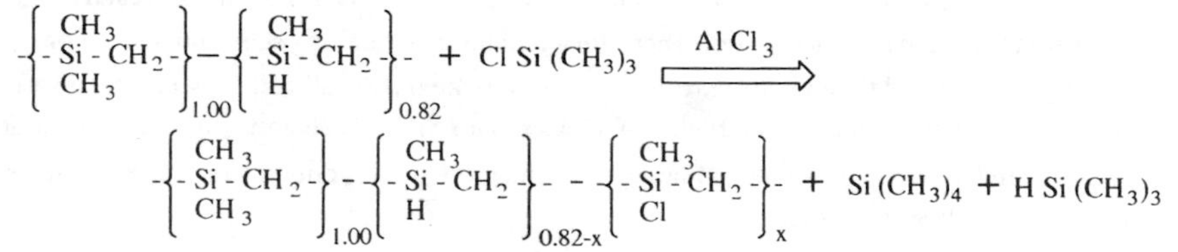

$$\left\{ -\underset{CH_3}{\overset{CH_3}{Si}}-CH_2- \right\}_{1.00} - \left\{ -\underset{H}{\overset{CH_3}{Si}}-CH_2- \right\}_{0.82} + Cl\,Si\,(CH_3)_3 \xrightarrow{Al\,Cl_3}$$

$$\left\{ -\underset{CH_3}{\overset{CH_3}{Si}}-CH_2- \right\}_{1.00} - \left\{ -\underset{H}{\overset{CH_3}{Si}}-CH_2- \right\}_{0.82-x} - \left\{ -\underset{Cl}{\overset{CH_3}{Si}}-CH_2- \right\}_{x} + Si\,(CH_3)_4 + H\,Si\,(CH_3)_3$$

Figure 1: Reaction mechanism for the chlorination of Nicalon polycarbosilane.

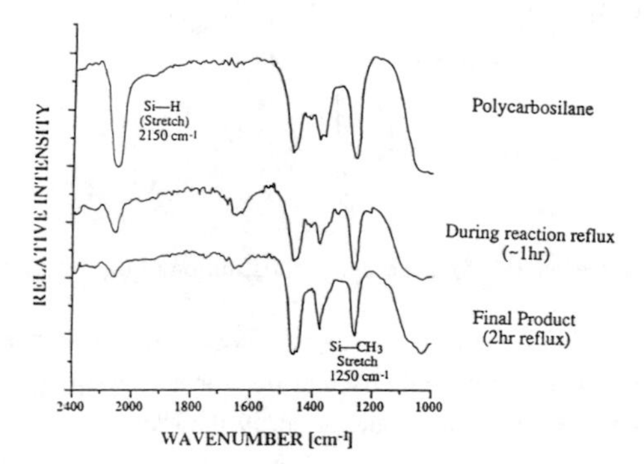

Figure 2: Infrared spectra during the chlorination reaction.

The ^{29}Si NMR spectra during the reaction are presented in Figure 3. The ^{29}Si NMR data shows a dramatic decrease in the intensity of the peak at -18 ppm which corresponds to a silicon environment of 1 hydrogen and 3 carbon atoms. Simultaneously, a new peak at approximately 26 ppm develops. This peak is due to silicon environments with either 3 carbon and 1 chlorine or 2 carbon and 2 chlorine atoms. The peak is centered close to 26 ppm which suggests the primary formation of a single chlorine, 3 carbon silicon environment. Integration of the NMR peaks allow a rough estimate of the chlorine concentration in the modified polymer. Concentrations for x ranging between 0.2 and 0.62 were obtained depending upon the time of reaction reflux.

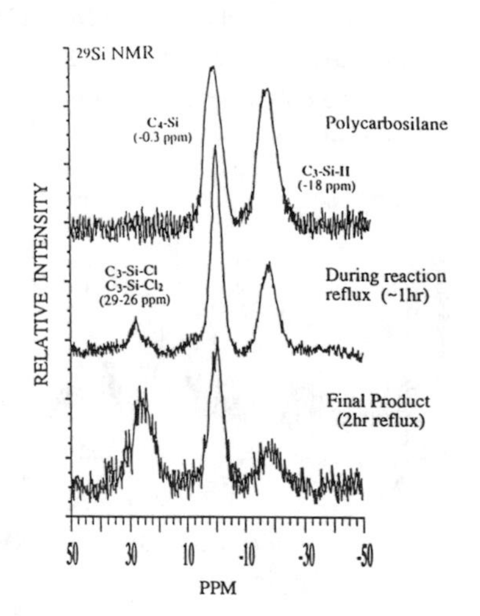

Figure 3: ^{29}Si NMR spectra during the reflux chlorination of polycarbosilane.

The chlorinated polymer product is highly reactive. Exposure to air quickly oxidizes the polymer as does contact with water or alcohols. The polymer is less soluble in non-polar organic solvents, but it is very soluble in THF or chlorinated solvents. Because of its high reactivity, this polymer product can have many uses. There are a number of interesting substitution reactions that can be easily induced through the reaction of the chlorinated polycarbosilane with organic salts. We have successfully introduced both Si-CH=CH_2 and SiC≡CH ligands through substitution reactions similar to that presented in Figure 4. At the present time, only the results of the alkynyl substitutions will be discussed.

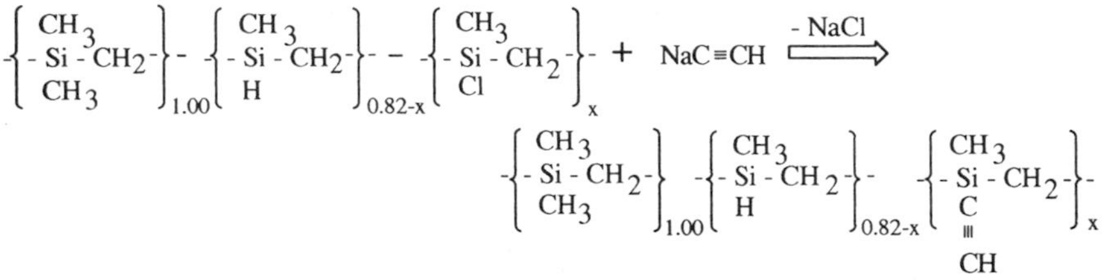

$$\left\{ -\overset{CH_3}{\underset{CH_3}{Si}}-CH_2- \right\}_{1.00} - \left\{ -\overset{CH_3}{\underset{H}{Si}}-CH_2- \right\}_{0.82-x} - \left\{ -\overset{CH_3}{\underset{Cl}{Si}}-CH_2- \right\}_{x} + NaC{\equiv}CH \xrightarrow{-NaCl}$$

$$\left\{ -\overset{CH_3}{\underset{CH_3}{Si}}-CH_2- \right\}_{1.00} - \left\{ -\overset{CH_3}{\underset{H}{Si}}-CH_2- \right\}_{0.82-x} - \left\{ -\overset{CH_3}{\underset{C\equiv CH}{Si}}-CH_2- \right\}_{x}$$

Figure 4: Reaction mechanism for the ethynyl substitution of the chlorinated polycarbosilane.

Following the reaction presented in Figure 4, the modified polymer was cleaned to remove the excess sodium acetylyde and the sodium chloride product. The proposed substitution reaction for the formation a Si-C≡CH ligand is supported by the ^{29}Si and ^{13}C NMR data presented in Figure 5. The formation of Si-C≡CH is not directly evident as its characteristic ^{29}Si NMR peak is hidden in the broad polycarbosilane [Si-C_4] peak at -0.3 ppm. However, the consumption of the Si-Cl bonds indicated in the ^{29}Si NMR (26 ppm) occurred simultaneously with the formation of the characteristic ethynyl peaks at 92 and 96 ppm in the ^{13}C NMR. As the silicon bonded chlorine sites are the only reactive site, it is probable that a direct substitution reaction resulted in the formation of Si-C≡CH ligands.

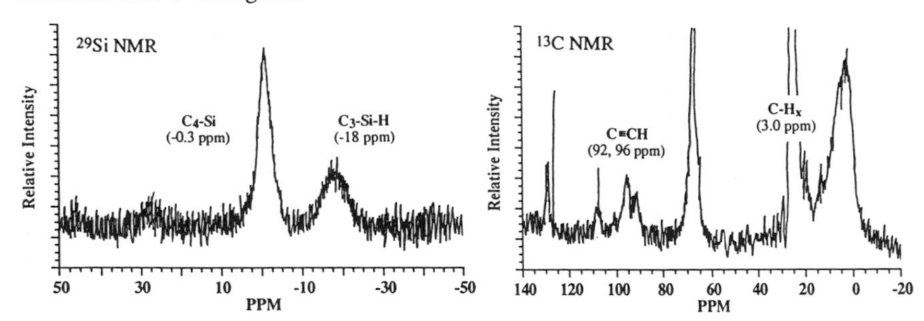

Figure 5: ^{29}Si and ^{13}C NMR spectra of poly(alkynyl)carbosilane.

<u>UV Cross linking of Poly(alkynyl)carbosilane</u>

As the polymer would quickly react with oxygen , all UV exposures were conducted inside a nitrogen atmosphere glove box. The poly(alkynyl)carbosilane was dissolved in THF. To this solution, 2wt% Daracur [C_6H_5-COC$(CH_3)_2$OH] was added as photo initiator. The THF was allowed to evaporate until a tacky gum formed that was then molded into 1 cm diameter round balls. The samples were then sealed in a silica jar for UV exposure. The UV lamp emits radiation at variable intensities between 200 and 600 nm and heats the sample to approximately 140°C. After a 1 hour exposure, the yellow polymer precursor changed to an orange color.

In Figure 6, the IR spectra of the polymer before and after 1 hour exposure to the UV radiation suggests that free radical polymerization occurs. This is primarily supported by the formation of a significant concentration of C=C vibrational peaks between 1600 and 1500 cm^{-1}. These absorption peaks are very broad because the polymerized vinyl bonds can assume a number of orientations. A fraction of these peaks, especially the strong absorption at 1700 cm^{-1}, are due to the presence of the photo initiator.

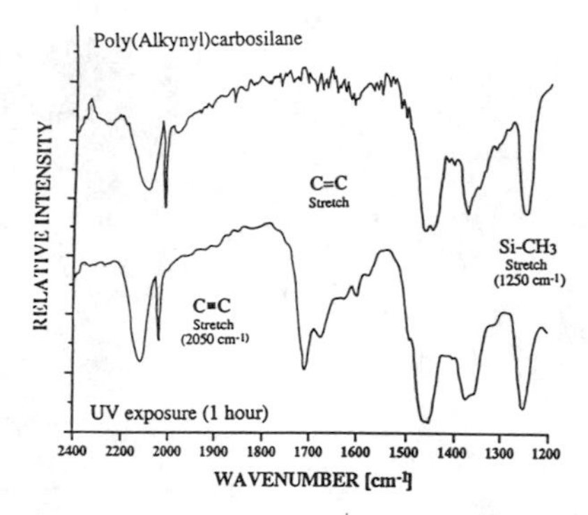

Figure 6: IR spectra of poly(alkynyl)carbosilane before and after UV exposure.

During the formation of the C=C bonds, a reduction in the characteristic alkynyl stretch at 2050 cm-1 was observed. Since no reduction in the Si-CH_3 stretch (1250 cm^{-1}) occurred, it can be assumed that polymerization only occurs through the ethynyl ligands. It is of interest to notice that only a small percentage of the alkynyl bonds have been consumed during the UV exposure.

The primary goal of this polymer modification was to prepare a polymer that could be cross-linked to provide structural rigidity during pyrolysis. In Figure 7, a comparison between the pyrolysis of polycarbosilane and poly(alkynyl)carbosilane are presented. With the incorporation of alkynyl ligands into the polymer structure, it is possible with UV cross-linking of the polymer to maintain polymer shapes during pyrolysis into SiC as we were able to maintain the dimensions of the molded polymer ball. The chemical modification of polycarbosilane is a significant improvement for maintaining structural rigidity as evidenced by the drastic shape change that occurred during a similar pyrolysis of the commercial polycarbosilane polymer.

CONCLUSIONS

Chemical modification of the commercial polycarbosilane polymer is possible as alkyl groups can be introduced through controlled chlorination of the polymer and subsequent substitution with alkyl metallic salts. Through an $AlCl_3$ catalyzed reaction of polycarbosilane and trimethylchlorosilane [$(CH_3)_3SiCl$], Si-Cl groups were introduced into the polycarbosilane polymer network. These chlorine bonds were substituted by a reactions with sodium acetylyde for the formation of Si-C≡CH ligands. This substitution does not significantly change the molecular weight of the polymer precursor allowing for high polymer solubility. The advantage is that these organic ligands can be polymerized after shaping of the polymer. UV initiated polymerization of the ethynyl ligands can cross-link the polycarbosilane polymers through the formation of organic bridges. These organic bridges can increase the molecular weight and the T_g of the polymer to provide increased structural rigidity during pyrolysis. Through controlled cross-linking of this type, it was possible to prepare a polymer that maintained its shape and dimensions during inert atmosphere pyrolysis into SiC without the use of surface oxidation.

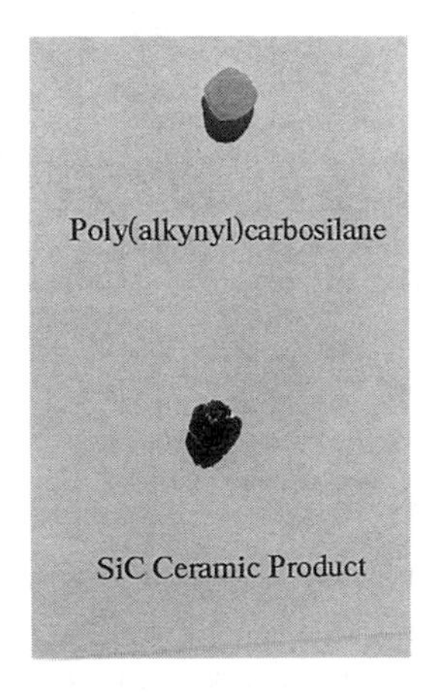

Figure 7: Pyrolysis of polycarbosilane and UV cross linked, poly(alkynyl)carbosilane into SiC.

Acknowledgments The authors gratefully acknowledge the National Science Foundation for financial support of this research through contract grant DMR-90-14487

References

[1] S. Yajima, J. Hayashi, M. Omori. Chemistry Letters. (1975), 931.
[2] S. Yajima, K. Okamura, T. Matsuzawa, Y. Hasegawa,. Nature. **279** (1979) 706.
[3] K. Okamura, M. Sato, Matusuzawa and Y. Hasegawa. Third International Conference on Ultrastructure Processing of Ceramics, Glasses and Composites. San Diego, CA. Feb 23-27, 1987.
[4] S. Yajima, T. Iwai, T. Yamamura, K. Okamura, Y. Hasegawa. J. Mater. Sci. 16 (1981) 1349.
[5] S. Yajima, K. Okamura, J. Hayashi and M. Omori. J. Am. Ceram. Soc., **59**, 7-8, (1976) 324.
[6] F. Babonneau, G. Soraru and J.D. Mackenzie. J. Mater. Sci., **25** (1990) 3664.
[7] M. Ishikawa, M. Kumada and H Sakurai. J. Organometal. Chem., 23, 63 (1970).
[8] E. Bacque, J.P. Pillot, M. Birot and J. Dunogues in Transformation or Organometallics into Common Exotic Materials: Design and Activation, edited by R.M. Laine (Martinus Nijhoof Publishers, 1988) pp. 116-132.

PROCESSING OF SiC-BASED FIBRES FROM POLYCARBOSILANE

VASILIOS.KALYVAS*, J.C.KO**, G.C.EAST*, J.E.McINTYRE*, B.RAND** AND F.L.RILEY**
*Department of Textile Industries, **School of Materials, The University of Leeds, Leeds LS2 9JT, United Kingdom.

ABSTRACT

A commercial polycarbosilane, thermolysis product of polysilastyrene, was spun into fibres. The self-curing character of the polymer permitted both oxidative and thermal curing of the precursor fibres . Pyrolysis of the latter produced ceramic fibres in high yields, which were fully characterised. Optimum tensile strength was attained after heat treatment at 1100°C. Oxidative reactions and crystallisation caused strength degradation above this temperature.

INTRODUCTION

Continuous ceramic fibres play a key role for the advancement of ceramic- and metal-matrix composites for the high technology industries. Further improvement of the properties of the existing ceramic fibres, especially the retention of their mechanical properties at temperatures higher than the present limit of 1100°C in oxidising atmospheres, in combination with the understanding of the structure-property-performance relationship, are the current subjects of vigorous research. The current status of these issues has been reviewed recently[1-3].

Non-oxide ceramic fibres have been produced mainly via the polymer precursor route, although other techniques have also been employed[1].

Polycarbosilastyrene (PCSS) has been shows to produce SiC fibres with good mechanical properties ($\sigma \leq 4.1$GPa and $E \leq 190$GPa), via the route outlined in Scheme 1[4-5].

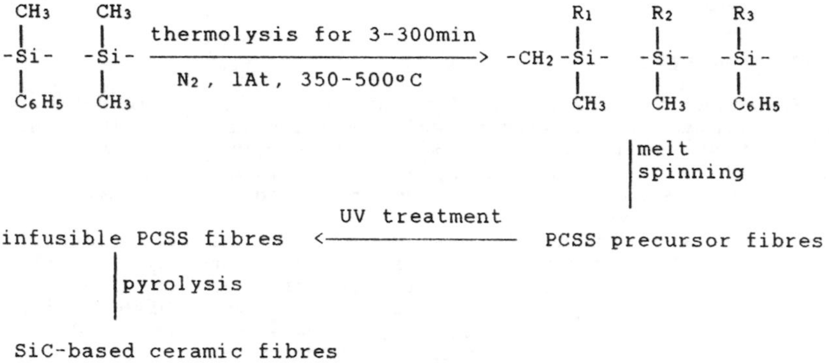

```
CH3    CH3                                      R1      R2      R3
 |      |     thermolysis for 3-300min          |       |       |
-Si-   -Si-  -------------------------->  -CH2-Si-    -Si-    -Si-
 |      |        N2, 1At, 350-500°C             |       |       |
C6H5   CH3                                     CH3     CH3     C6H5

                                                        |melt
                                                        |spinning
                                                        |
                              UV treatment
infusible PCSS fibres    <----------------    PCSS precursor fibres
        |
        |pyrolysis
        |
 SiC-based ceramic fibres
```

R_1, R_2, R_3 = H, CH_3, C_6H_5

Scheme 1

It is claimed that the fibres can retain up to 90-100% of the original tensile strength after heat treatment at 1200°C in air for 1hr.

A similar, high molecular weight (MW) PCSS is marketed under the commercial name PSS-400. However, no information has been disclosed in the open literature about the properties of this polymer, especially about its suitability as a ceramic fibre precursor. Thus the aim of the present contribution is to study the evolution of the ceramic structure and its performance at high temperatures in relation to microstructure developed under these conditions.

EXPERIMENTAL

Starting materials

PSS-400 was supplied by Nissho Iwai Corp., with the following specifications: melting point (m.p.)=150-180°C, molecular weight (MW)=8000-12000, bulk specific gravity =0.3-0.4Kg.m^{-3}, soluble in toluene, THF etc., and sensitive to UV/X-ray irradiation.

Spinning and pyrolysis of the precursor fibres

Preliminary work showed that this polymer could be thermoset if the heating rate was sufficiently low. Consequently, attempts at melt-spinning of the polymer at temperatures in the range of 240-260°C were unsuccessful due to the cross-linking reactions. The melt spun fibres showed poor surface finish and inconsistent fibre diameter and the fibre line was frequently disrupted. Melt-spun fibres were not converted into the ceramic form. However, the solubility of the polymer allowed the use of dry spinning from concentrated solutions in THF or toluene, through a single hole spinneret. The precursor fibres were pyrolysed in an argon atmosphere at temperatures up to 1700°C, at a heating rate of 4°C/min. No bond-fusion of the fibres was observed.

Characterisation

FTIR spectra were recorded in air on a 1725X Perkin-Elmer FTIR spectrometer. UV spectra were recorded on a Perkin-Elmer 402 UV/VIS spectrometer. TGA and DSC were carried out on a Du Pont 951 Thermogravimetric Analyser and 910 DSC cell, equipped with a 2000 Thermal Analyser. Number-Average MW was measured by Vapour Phase Osmometry (VPO). Element analysis for carbon and hydrogen was provided by the Microanalytical Laboratory of the Chemistry Department of the University of Leeds, using the Carlo Erba Elemental Analyser. Silicon was measured by atomic absorption spectroscopy after fusing the sample with sodium carbonate at 1000°C in air. Oxygen was calculated by difference. Density was measured by the floating method, using toluene / 1,1,2,2 tetrabromoethane mixture. SEM was performed in a Camscan 3-30BM microscope. X-ray diffraction patterns were recorded in a Philips PW 1840 diffractometer, equipped with a Ni-filtered CuKa radiation. Tensile properties of single fibres were measured in an Instron Tensile Tester, Model 1122.

RESULTS

PSS-400 is a light brown, brittle solid, soluble in THF, toluene and similar solvents. It was fully characterised by FTIR, ^1H NMR, UV, dilute solution viscosity, GPC, TGA DSC and element analysis.

The FTIR absorption bands were assigned as follows (in cm^{-1}).: 3067, 3048, 1590, 1487, 1428, 1100, 697 ($-C_6H_5$); 2955, 2893, 1408, 1255 (C-H in CH_3 and $-CH_2-$); 771, 731, 697 ($Si-CH_3$); 1354, 1019 ($Si-CH_2-Si$); 2104 (Si-H); 472 ($Si-Si-C_6H_5$); 3445 (H-bonded -OH); 3600 (free -OH); Residual Si-Si bonds were present, as revealed by UV spectroscopy. Number average MW, MW=2300, was lower than that reported in the specifications. Similar values were obtained by GPC measurements. The polymer, heated at 45°C/min under nitrogen, melted at 220-230°C. At lower heating rates it thermoset rather than melted. The TGA thermogram (nitrogen, 100°C/min) showed a very gradual weight loss. The rate increased between 380 and 470°C with a maximum at 430°C. The final residue was 89%. The main feature of the DSC thermogram was a broad and strong exothermic peak with a maximum at 340°C and a shoulder at 420°C. Element analysis of the as-received polymer gave the following results: C=44.8% H=6.1% Si=45.85% O=3.15%.

The weight percentages of the residue of the as-received polymer and of a sample pre-oxidised at 160°C for 10hr PSS-400 as a function of the temperature of heat treatment are recorded in Fig.1. Heating was in argon at 4°C/min up to the specified temperature which was held isothermally for 1hr. The precursor fibres were pyrolysed under the same conditions. The effect of the maximum temperature of the heat treatment on the tensile strength is recorded in Fig.2 and on the x-ray diffraction patterns in Fig.3. SEM of the ceramic fibres obtained by pyrolysis of the precursor fibres up to 1100 and 1700°C are shown in Fig.4 and 5.

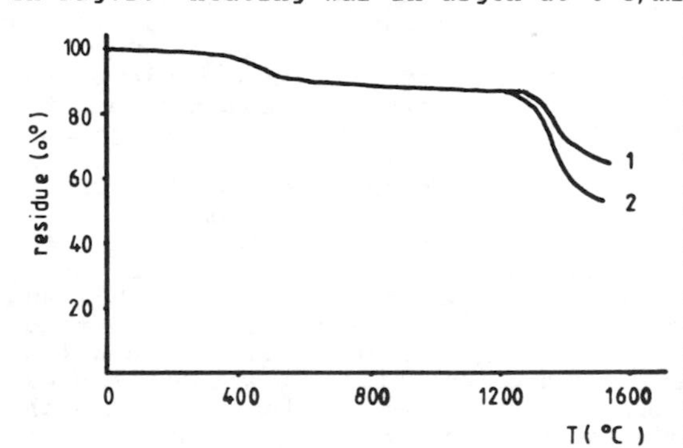

Fig.1 Residue of PSS-400 as function of the temperature of the heat tratment; (1) as-received (2) pre-oxidised

DISCUSSION

The chemical changes taking place in the pyrolysis step of the PCSS were followed by FTIR. The residual hydroxyl groups can be condensed at relatively low temperatures (Eqs. 1-2)

$$\text{Si-OH} + \text{HO-Si} \longrightarrow \text{Si-O-Si} + H_2O \quad (1)$$
$$\text{Si-OH} + \text{H-Si} \longrightarrow \text{Si-O-Si} + H_2O \quad (2)$$

and this accounts for the self-curing character of the polymer. Above approx. 250°C, scission of the residual Si-Si bonds takes place, generating radicals which can yield Si-C-Si moieties, according to a mechanism proposed by Yajima et. al. [6-8]. At even higher temperatures, methane and benzene are produced in accordance with Eqs. 3-5.

$$\text{Si-}CH_3(C_6H_5) \longrightarrow \text{Si}\cdot + \cdot CH_3(C_6H_5) \quad (3)$$
$$\text{Si-}CH_3 + \cdot CH_3(C_6H_5) \longrightarrow \text{Si-}CH_2\cdot + CH_4(C_6H_6) \quad (4)$$
$$\text{Si}\cdot + \cdot CH_2\text{-Si} \longrightarrow \text{Si-}CH_2\text{-Si} \quad (5)$$

An initial weight loss is attributed to the evaporation of the water produced via the above reactions (Eqs. 1-2). Volatilisation of low MW compounds could also explain further weight loss at intermediate temperatures. At these and higher temperatures, methane, benzene and hydrogen are evolved. Above 800°C no organic moieties could be detected by FTIR. Above about 1200°C degradation of the fibres takes place, involving oxidation of carbon and of silicon carbide by the oxygen contained in the structure. The greater the oxygen content of the samples the higher is the weight loss to be expected and this is observed. Optimum strength ceramic PSS-400 fibres, obtained by heating the precursor fibres up to 1100°C under the conditions specified above, do not show any appreciable weight loss when heated in an argon atmosphere up to 1200°C or isothermally at 1100°C for up to 10hr. When they were heated in air under the same conditions, a weight loss was observed. This implies that the weight loss due to the oxidation of carbon is higher than the weight gain due to oxidation of silicon carbide to yield silica. In both cases, the weight loss was higher in the early stages of the heat treatment, which can be explained by the formation of silica at the surface of the fibre, retarding the rate of oxidation (passive oxidation).

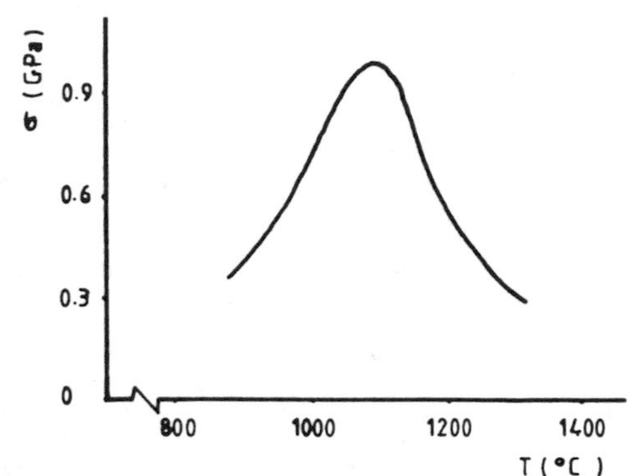

Fig.2 Tensile strength of PSS-400 fibres as function of the temperature of the heat treatment

The strength of the fibres, as a function of temperature, initially increases up to 1100°C and decreases above this temperature. The ceramic fibres obtained after heat treatment at temperatures ≥1400°C are too weak to be tested. Densification of the fibres via elimination of hydrogen can account for the increase of the strength and the onset of the carbothermal reduction and of the oxidation of carbon for its decrease.

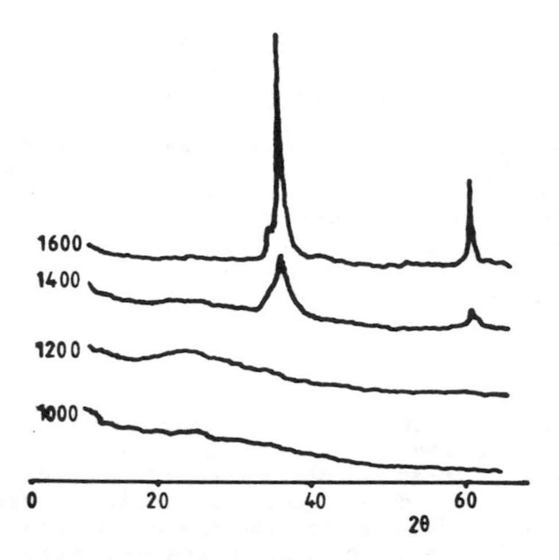

Fig.3 X-ray diffraction patterns of PSS-400 fibres as function of the temperature of the heat treatment

The strength of the fibres is governed mainly by the flaw population. It is also known [9] that the latter are strongly affected by the diameter of the fibres. The PSS-400 ceramic fibres of a diameter of approximately 25μ and of tensile strength of the order of 1GPa are rather better than those derived from PCS [9-10].

Wide angle X-ray diffraction patterns of PSS-400 ceramic fibres as function of the temperature of the heat treatment (Fig. 3) show that no crystallisation of β-SiC takes place below 1300°C. The apparent crystallite size, calculated by the Scherrer line-broadening method, increases continuously, as the temperature of the heat treatment increases. Below 1250°C the crystalline size is <2nm and the structure of the fibres is considered as amorphous. Peaks due to silica are not observed as a separate phase and, instead, as chemically bonded to silicon and carbon in a more or less continuous phase of silicon carbide/silicon oxycarbide.

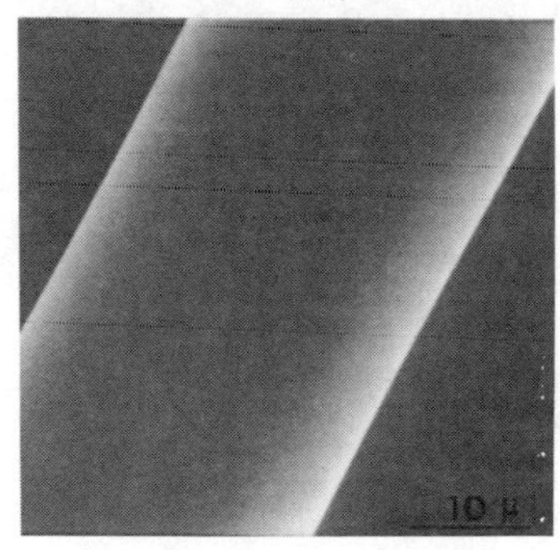

Fig.4 SEM of PSS-400 ceramic fibre at 1000°C

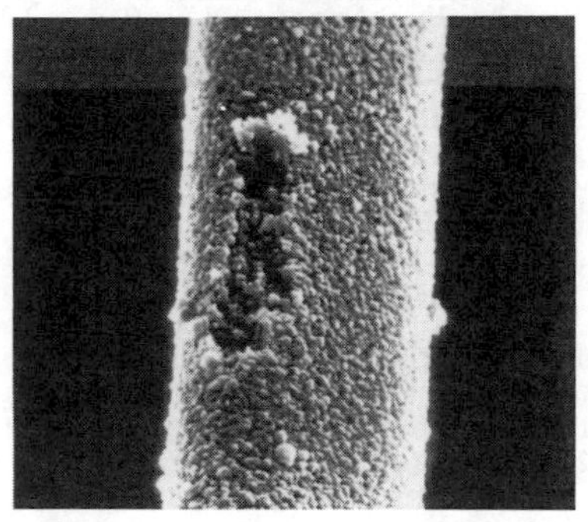

Fig.5 SEM of PSS-400 ceramic fibre at 1700°C

The density of the PCSS fibres increases as the temperature of the heat treatment increases. The rate of the increase reaches a maximum at approx. 1350°C which corresponds fairly well with the maximum rate of the weight loss of the PCSS and the maximum rate of gas evolution observed in PCS and other systems [11-12]. The resulting composition has higher content of the more dense silicon carbide and lower content of the less dense silica and carbon. The crystallisation of silicon carbide also contributes to the increase of the density.

The density of ceramic fibres obtained by heat treatment up to 1000°C is 2.30Kgm^{-3}. Taking into account the elemental analysis of the PCSS ceramic fibres, i.e. C=45.0%, Si=48.2% and O=6.8%, then the theoretical density is 2.62kg•m^{-3}. Consequently, the calculated porosity is 0.12. This value is below the estimated porosities of four out of five ceramic fibres examined by Langley et al. [3].

CONCLUSIONS

Polycarbosilastyrene, the thermolysis product of polysilastyrene, was characterised. It was spun into continuous precursor fibres by dry-spinning. These fibres were stabilised by controlled heating in an inert atmosphere, via self-condensation of residual hydroxyl groups and/or hydroxyl and silyl hydride groups. The ceramic fibres resulting from the pyrolysis of the stabilised precursor fibres were characterised. The mechanical properties as a function of temperature exhibit a maximum at 1100°C. Above this temperature the reduction of the mechanical properties is due to crystallisation and degradation reactions involving oxygen.

REFERENCES

1. J.K.Weddell, J. Text. Inst., 81(4), 333 (1990).
2. J.Lipowitz, Am.Ceram.Soc., Ceram.Bull., 70,1888 (1991).
3. N.R.Langley,G.E.LeGrow and J.Lipowitz,in Fibre Reinforced Ceramic Composites, edited by K.S.Mazdiyasni (Noues Publications, NJ, USA, 1990), pp.63-92; L.C.Sawyer, ibid., pp. 141-181.
4. EP 212 485, Teijin Ltd (4 March 1987).
5. EP 363 509, Teijin Ltd (18 April 1990).
6. Y.Hasegawa and O.Okamura, J.Mat.Sci., 18, 3633 (1983).
7. T.Taki, M.Inui, K.Okamura and M.Sato, J.Mat.Sci. Lett., 8, 918 (1989).
8. S.Yajima, Y.Hasegawa, J.Hayashi and M.Iimura, J.Mat.Sci., 13, 2569 (1978).
9. S.Yajima, Phil.Trans.R.Soc.Lond., A294, 419 (1980).
10. L.C.Sawyer, M.Jamieson, D.Brikowski, M.I.Haider and R.T. Chen, J.Am.Ceram.Soc., 70, 798 (1987).
11. Y.Hasegawa, J.Mat.Sci., 24, 1177 (1989).
12. (a) F.C.Montgomery and H.H.Streckert, Pol.Preprints, 28, 258 (1987). (b) L.F.Chen, unpublished results.

POLYMER-DERIVED SILICON CARBIDE FIBERS WITH IMPROVED THERMOMECHANICAL STABILITY

W. TOREKI, C.D. BATICH, M.D. SACKS, M. SALEEM, AND G.J. CHOI
Department of Materials Science and Engineering, University of Florida, Gainesville, FL 32611

ABSTRACT

Continuous silicon carbide fibers ("UF fibers") with low oxygen content (~2 wt%) were prepared by dry spinning of high molecular weight polycarbosilane solutions and subsequent pyrolysis of the polymer fibers. Room temperature mechanical properties were similar to those of commercially-available Nicalon™ fibers, as average tensile strengths as high as 3 GPa were obtained for some batches with fiber diameters in the range ~10-15 μm. Furthermore, UF fibers showed significantly better thermomechanical stability compared to Nicalon™, as indicated by lower weight losses, lower specific surface areas, and improved strength retention after heat treatment at temperatures up to 1700°C. UF fibers were also characterized by elemental analysis, X-ray diffraction, and scanning Auger microprobe. Strategies were suggested for achieving further improvements in thermomechanical stability.

INTRODUCTION

Ceramic-matrix composites reinforced with continuous ceramic fibers are of considerable interest for applications requiring improved mechanical properties at high temperatures. Numerous studies have been reported concerning fabrication and properties of composites prepared with Nicalon™ SiC fibers (Nippon Carbon Co., Tokyo, Japan).[1-5] Although excellent mechanical properties (e.g., high strength and high fracture toughness) have been achieved in these composites, applications are limited by degradation of the Nicalon™ fibers at elevated temperatures. Fiber degradation depends on the heat treatment atmosphere, but significant decreases in strength have been observed after exposure to temperatures at or below 1000°C.[5-10] Previous studies have shown that the poor thermomechanical stability of Nicalon™ is associated with the presence of oxygen (~8-15 wt%) and excess carbon in the fibers. Volatile reaction products (primarily CO and SiO) form during heat treatment at elevated temperatures, resulting in large weight losses, formation of porosity, and the growth of SiC grains and other strength-degrading flaws in the fiber microstructure.[5,8-13]

The development of SiC fibers with improved thermomechanical stability requires better control over the chemical composition of the fibers, especially the oxygen content.[14-16] Nicalon™ fibers are prepared by the method of Yajima et al.[17,18] which involves melt spinning of low-molecular-weight (<2,000) polycarbosilane (PC). As-spun fibers are then heated at low temperature (~150-200°C) in air in order to cure (cross-link) the polymer. This treatment incorporates a large amount of oxygen in the fiber, but it is necessary to prevent the fibers from melting during the subsequent pyrolysis step (in which the polymer is decomposed at higher temperatures to a SiC-rich ceramic). In contrast to the Nicalon™ process, we recently reported[19] a method in which PC fibers were pyrolyzed without an oxidative cross-linking step, thereby allowing SiC fibers (designated "UF fibers") with low oxygen content (<2 wt%) to be produced. The processing steps for this method are outlined in Fig. 1. Fibers were spun using higher molecular-weight (~5,000-10,000) PC's which remained highly soluble in appropriate solvents, but did not melt during subsequent heat treatment. Fibers were formed at room temperature by extruding concentrated polymer solutions through stainless steel spinnerets under nitrogen pressure. (Vinylic polysilazane was added to the PC solutions as a fiber spinning aid.) As-spun fibers were directly pyrolyzed in nitrogen at 750-1000°C. Pyrolyzed fibers (see Fig. 2) were produced with low oxygen content, round cross-sections, and controlled diameters in the range ~10-50 μm. Fine-diameter fibers (~10-15 μm) had room temperature properties similar to Nicalon™, with average tensile strengths as high as 3 GPa in some batches. In addition, UF fibers showed superior thermomechanical stability

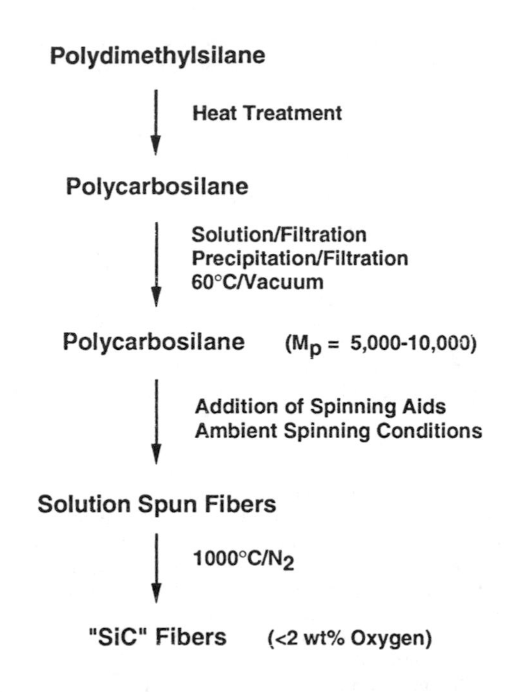

FIG. 1. Processing steps for preparation of SiC fibers with low oxygen content.

FIG. 2. Scanning electron micrograph of pyrolyzed UF fibers.

compared to Nicalon™. Significant reduction in weight loss and improved strength retention were observed in UF fibers subjected to heat treatments at temperatures up to 1700°C.[19] In this paper, additional information is reported concerning the structure and properties of pyrolyzed and heat-treated UF fibers.

EXPERIMENTAL

Fibers were characterized by a variety of techniques. Weight loss behavior of pyrolyzed fibers was followed by thermogravimetric analysis, TGA (model STA-409, Netzsch Co., Exton, PA). Specific surface areas of heat-treated fibers were determined by the BET method using N_2 or Kr gas adsorption measurements (model ASAP 2000, Micrometrics, Atlanta, GA). Phase analysis of heat-treated fibers was carried out by X-ray diffraction, XRD (model APD 3720, Phillips Electronic Instruments Co., Mt Vernon, NY) using Ni-filtered CuKα radiation. Fibers were also examined by scanning and transmission electron microscopies, SEM and TEM (models JSM-6400 and 200CX, respectively, JEOL, Tokyo, Japan).

The overall elemental compositions of heat-treated fibers were determined by outside analytical laboratories (IRT Corp., San Diego, CA; Radio-Analytical Services, University of Kentucky, Lexington, KY; Galbraith Laboratories Inc., Knoxville, TN) using neutron activation analysis (for oxygen, nitrogen, and silicon), atomic absorption (for silicon), and LECO combustion analysis (for carbon, hydrogen, and nitrogen). Compositional analyses of individual pyrolyzed fibers were also carried out using scanning Auger microprobe, SAM (model PHI 660, Perkin Elmer Corp., Edin Prairie, MN). Fibers were mounted horizontally and a 10 kV electron beam (30 nA) was focused on a small area ($<0.01\ \mu m^2$) on the center portion of the fiber surface. A detailed survey spectrum (10 scans) was collected over the entire kinetic energy range (50-2050 eV). The analyzed surface was then etched by sputtering for 6 sec with an Ar^+ ion beam; this was followed by re-analysis of the same fiber area, but only over the spectral regions for C, Si, and O. This process was repeated in rapid sequence in order to obtain a depth profile for the fiber elemental composition. A detailed survey spectrum of the fiber interior was then collected.

Fiber tensile strengths were measured using ASTM procedure D-3379,[20] as described in detail previously.[19] Elastic modulus values were obtained from the stress-strain data collected during the strength measurements.

RESULTS AND DISCUSSION

We have previously reported that UF fibers show improved retention of strength (i.e., compared Nicalon™ fibers) after heat treatment at elevated temperatures.[19] Figure 3 shows average tensile strength values for UF and Nicalon™ fibers that were heat treated in an argon atmosphere at temperatures in the range 1000-1700°C.† The data labeled "UF-80" were collected from one lot of fibers and are considered representative of UF fiber batches prepared with low oxygen content (~2 wt%). In comparison with Nicalon™, UF-80 fibers show considerably higher strengths at heat treatment temperatures above 1300°C. For example, after 1.0 h at 1400°C, UF-80 fibers have an average tensile strength of ~1.2 GPa compared to only ~0.3 GPa for Nicalon™.* In addition, the UF-80 fibers retained

† Heat treatment times were 1.0 h except that the 1300°C Nicalon™ sample was heated for 2.0 h and the 1500°C UF-80 and Nicalon™ samples were heated for 0.5 h.

* The results in Fig. 3 for Nicalon™ fibers heated to high temperatures (e.g., ≥1400°C) are probably not representative of the true average strengths because the values reported were determined on fibers that had sufficient handling strength to be tested. However, many fibers failed during preparation for testing, suggesting that true average strengths for the Nicalon™ fibers are probably less than indicated in Fig. 3.

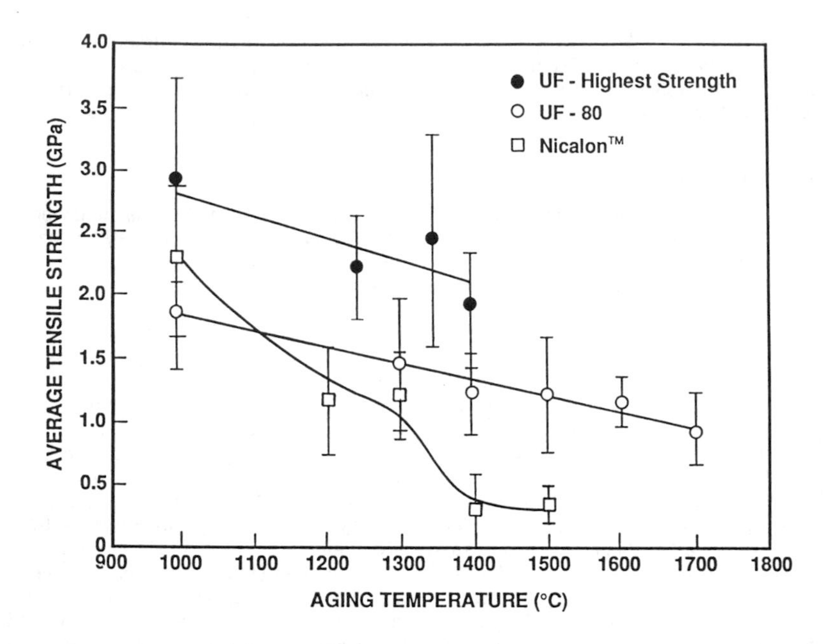

FIG. 3. Plots of average tensile strength vs. heat treatment temperature for UF fibers and Nicalon™ fibers. Values for UF fibers are given for batch no. 80 and for the batches with the highest average strengths.

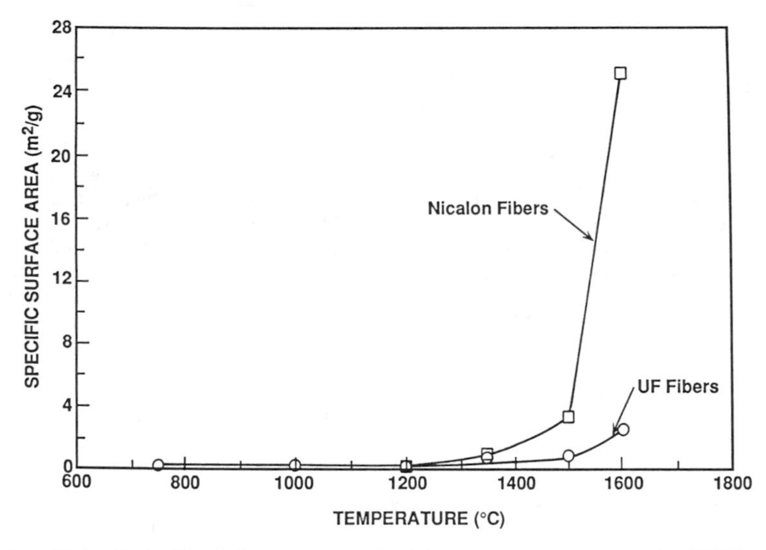

FIG. 4. Plot of specific surface area vs. heat treatment temperature for UF fibers and Nicalon™ fibers.

substantial strength (~0.9 GPa) after heat treatment at 1700°C (1.0 h), while the Nicalon™ fibers were too weak to be tested. Figure 3 also shows data for samples designated "UF - highest strength." This data represents the highest average strength values obtained from ~15 fiber lots that were processed after the UF-80 lot. Maximum average strengths approached ~3 GPa and ~2 GPa for 1000°C and 1400°C heat-treated fibers, respectively. It may be possible to obtain these high strength levels on a consistent basis if processing conditions are more rigorously controlled. In particular, it is suspected that the observed variations in strength arise, at least in part, from batch-to-batch variations in the characteristics of the polymer (i.e., amount and type of impurities, molecular weight distribution, etc.) and the spinning dope (i.e., polymer concentration, rheological properties, amount and type of impurities, etc.).

The superior strength retention of UF fibers (i.e., compared to Nicalon™) can be attributed to the relatively small weight loss during the heat treatment. TGA experiments showed that UF fibers (previously pyrolyzed to 1000°C) lost only ~3 wt% when heated in argon at 10°C/min to 1550°C and subsequently held for 1.0 h at temperature.[19] In contrast, Nicalon™ fibers lost ~25 wt% when subjected to the same heat treatment.[19] Large weight losses - which are directly related to the formation of pores, large grains, and other microstructural flaws in the fibers - lead to dramatic decreases in tensile strength for Nicalon™ fibers after heat treatment at elevated temperature.[5-13,19] In this study, the development of porosity in heat-treated Nicalon™ and UF fibers was assessed by specific surface area measurements. Figure 4 shows that the surface area for Nicalon™ fibers increased sharply above ~1350°C. In contrast, the surface area did not increase significantly in UF fibers until the temperature exceeded 1500°C.

It was noted earlier that the large weight loss during heat treatment of Nicalon™ fibers is associated with its high oxygen content, as silica (SiO_2) reacts with excess carbon to form volatile products (primarily CO and SiO).[5,8-13] In contrast, the relatively low weight loss observed for UF fibers can be attributed to the low oxygen content. Table 1 shows that UF fibers have been produced with oxygen contents less than 2 wt%, while Nicalon™ fibers have oxygen contents in the range of 10-15 wt%. Other elemental analyses show that UF and Nicalon™ fibers have similar Si, N, and H contents. These results indicate that UF fibers have higher carbon content compared to Nicalon™. It should be noted that the carbon content (~42 wt%) listed in Table 1 for UF fibers was determined by difference, i.e., 100% minus the combined concentrations for Si, O, N, and H. Combustion analysis on UF fibers consistently gave lower carbon contents (~37 wt%) than the value listed in Table 1; however, these results were considered unreliable since oxidative combustion of the UF fibers was probably incomplete under the experimental conditions used in the measurements. In contrast, measurement of the carbon content of Nicalon™ by combustions analysis is considered reliable because the fibers degrade more rapidly at elevated temperatures. The development of porosity and high specific surface area in heat-treated Nicalon™ fibers (such as shown in Fig. 4) is expected to accelerate the oxidative combustion reaction rate. The measured carbon content is in good agreement with the value reported by the manufacturer. In addition, the concentrations of Si, C, O, N, and H total to ~100% for the Nicalon™ fibers.

Despite the excess carbon content, oxidation resistance of UF fibers is relatively good. This is indicated by tensile strength measurements that were carried out at elevated temperature in an air atmosphere and at room temperature after high temperature oxidizing heat treatments. For example, after heat treatments of 1.0 h in air at 1400 and 1500°C, room temperature average tensile strengths for three different lots of UF fibers were in the range of ~1.3-1.7 GPa and ~1.2-1.4 GPa, respectively. In addition, an average strength of ~1.1 GPa was obtained on a UF fiber lot that was tested at 1400°C in air.[19] (The fiber batch used in this experiment was initially pyrolyzed at 1250°C in nitrogen, i.e., prior to testing at high temperature.) A TGA experiment also showed that UF fibers had relatively good oxidation resistance. UF fibers (initially pyrolyzed at 1000°C) showed only ~1.5% weight gain when heated in air at 10°C/min to 1550°C and subsequently held for 1 hour at temperature. Neutron activation analysis showed that the oxygen content in this particular sample increased from ~2.6 wt% (as-pyrolyzed) to ~5.5 wt% (after oxidation treatment).

TABLE 1. ELEMENTAL ANALYSIS

Sample	O (wt%)	Si (wt%)	C (wt%)	N (wt%)	H (wt%)
UF Fiber	1.5-2.0†	55†,‡	42▼	1-2▲,†	<0.5▲
Nicalon	15†	55‡	29▲	<0.5▲	<0.5▲
Nicalon*	10	58	31		

* Reported by manufacturer
† Determined by neutron activation analysis
‡ Determined by atomic absorption
▲ Determined by LECO combustion method
▼ Determined by difference

It is presumed that the oxygen weight gain is greater than the overall weight gain due to an oxidation reaction that forms silicon dioxide (non-volatile) and carbon monoxide (volatile), i.e., a reaction such as:

$$SiC_{x(s)} \quad + \quad (1+\tfrac{x}{2})O_{2(g)} \quad \Rightarrow \quad SiO_{2(s)} \quad + \quad xCO_{(g)}\uparrow$$

It is well known that PC-derived fibers develop an amorphous, carbon-rich structure after low temperature pyrolysis.[5,7,10,13] Thus, it is not surprising that XRD analysis of heat-treated UF fibers shows only weak crystallinity. Figure 5 gives typical diffraction patterns for the fibers after pyrolysis at 750°C and after subsequent heat treatments in the range 1000-1600°C. There is increased crystallization of β-SiC as the heat treatment temperature increases, but the broad diffraction peaks indicate that the average crystallite sizes remain small. (This was confirmed by TEM observations.) The XRD patterns in Fig. 5 are very similar to results reported for Nicalon-type fibers,[13] except that diffraction peaks associated with silica are absent in UF fibers due to their low oxygen content.

Nicalon™ and UF fibers have low elastic modulus values because of the substantial amounts of amorphous carbon and weakly crystalline SiC. Figure 6 shows plots of elastic modulus vs. heat treatment temperature for both types of fibers. (This plot includes data for the UF-80 fibers and for the UF fiber batches having the highest <u>average</u> elastic modulus values.) Nicalon™ fibers show a significant decrease in modulus after heat treatment above 1300°C. This presumably reflects the formation of porosity in the fibers as weight loss occurs at high temperatures. In contrast, UF fibers undergo a much smaller weight loss during heat treatment, so only a small amount of porosity develops. In fact, the UF fibers actually show small increases in elastic modulus during heat treatment due to the increasing degree of crystallinity (β-SiC), as indicated in Fig. 5.

Although UF fibers show better thermomechanical properties compared to Nicalon™, development of SiC fibers with improved thermal stability and substantially higher elastic modulus is still desired. A fully crystalline fiber is needed in order to meet the latter goal. This would require the development of fibers that have an Si:C stoichiometry closer to 1:1 after pyrolysis (i.e., as opposed to the carbon-rich stoichiometry of current UF fibers). Improvements in the high temperature stability of UF fibers may also require better control

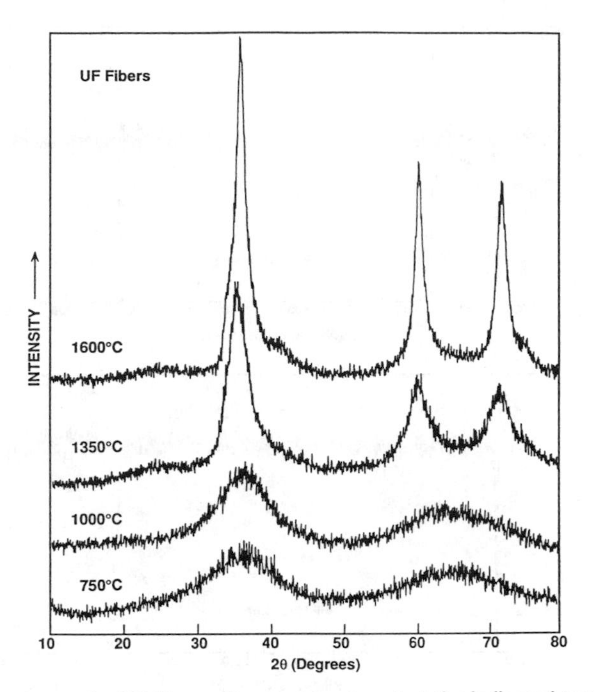

FIG. 5. XRD patterns for UF fibers after heat treatment at the indicated temperatures.

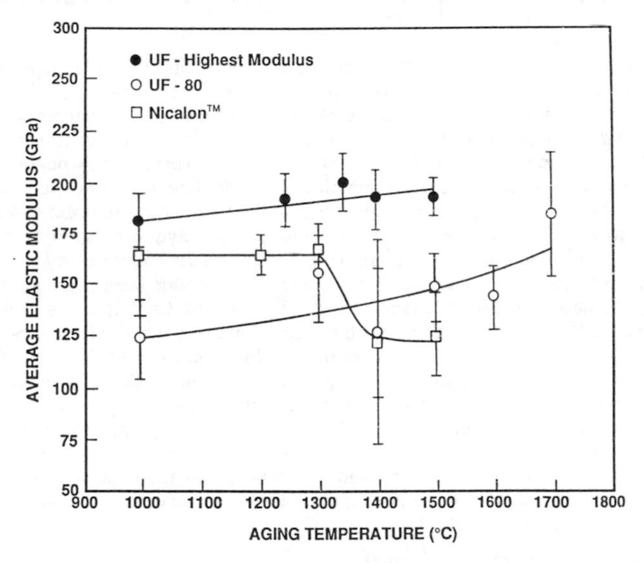

FIG. 6. Plots of average elastic modulus vs. heat treatment temperature for UF fibers and Nicalon™ fibers. Values for UF fibers are given for batch no. 80 and for the batches with the highest average moduli.

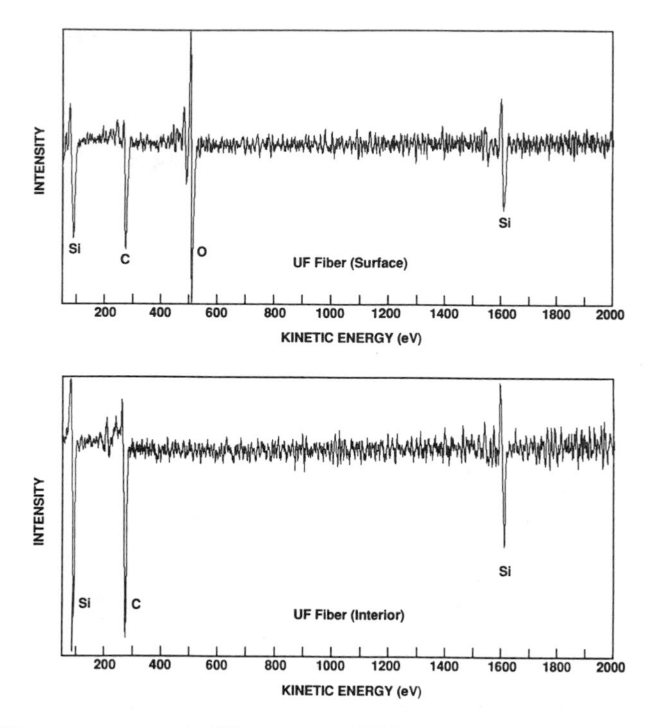

FIG. 7. SAM survey scans at the (A) surface and (B) interior of a 1000°C-pyrolyzed UF fiber.

over the fiber chemical composition. The decreases in tensile strength after high temperature heat treatments may be associated, at least in part, with high concentrations of oxygen at the fiber surfaces. Specific surface measurements (Fig. 4) and SEM observations show that microstructural degradation (i.e., porosity formation and grain growth) occurs in some fibers heated at elevated temperatures ($\geq$1500°C). This degradation presumably occurs by reaction between free carbon and siliceous (Si-O containing) material near the fiber surface. Oxygen enrichment at fiber surfaces was confirmed by scanning Auger microprobe (SAM) analysis on 1000°C-pyrolyzed UF fibers with relatively low overall oxygen content (~1.5 wt% as determined by neutron activation analysis). Figures 7A and 7B show survey scans obtained from the fiber surface (i.e., before sputtering) and fiber interior (i.e., after sputtering to a depth of approximately 180 nm), respectively. Fig. 7A shows that there is a very large oxygen content at the fiber surface, while no oxygen signal is observed in the spectrum from the fiber interior (Fig. 7B).* Atomic concentration depth profiles of Si, C, and O for this sample showed that the oxygen content is significantly higher in the outer several hundred angstroms of the fiber. It should be possible to reduce the oxygen content in UF fibers by controlling the spinning and pyrolysis atmospheres so that less exposure to oxygen and water vapor occurs during processing. (Note that UF fibers are currently spun into an ambient air atmosphere. In addition, pyrolysis is carried out in an ungettered furnace using standard-grade tank nitrogen. Thus, it is not surprising that fiber surfaces were enriched with oxygen.)

---

* Based on atomic sensitivities, magnitude of C and Si signals, and signal-to-noise ratio, it is estimated that the detection limit for oxygen for the survey scan shown in Fig. 7B is ~1 at%.

CONCLUSIONS

Continuous SiC fibers ("UF fibers") were prepared by dry spinning of high-molecular-weight (5,000-10,000) polycarbosilane (PC) solutions and subsequent pyrolysis of the polymer fibers. These fibers showed excellent thermomechanical stability in comparison to commercially-available Nicalon™ fibers, as indicated by lower weight losses, lower specific surface areas, and improved strength retention after heat treatment at elevated temperatures. This was attributed to the relatively low oxygen content for UF fibers (i.e., ~2 wt%).

As observed in other PC-derived ceramics, heat-treated UF fibers have a carbon-rich, weakly-crystalline structure. This results in relatively low elastic modulus (i.e., compared to fully-crystalline, stoichiometric SiC), even after heat treatment at temperatures up to 1700°C. Scanning Auger microprobe analysis showed that the residual oxygen in UF fibers is concentrated near the fiber surfaces, presumably due to exposure to atmospheric oxygen and water vapor during the spinning and pyrolysis operations. Reaction between the siliceous and carbon-rich materials apparently results in some surface degradation of the fibers after high temperature (≥1500°C) heat treatment. Thus, ongoing investigations are directed toward (i) more rigorous control over the atmosphere during processing in order to minimize oxygen contamination in the fibers and (ii) adjustment of the overall fiber chemical composition in order to approach an Si:C stoichiometry closer to 1:1.

ACKNOWLEDGEMENTS

This work was supported by the DARPA under Contract Nos. MDA 972-88-J-1006 and N00014-91-J-4075. The authors gratefully acknowledge Mr. Eric Lambers of the University of Florida's Major Analytical Instrumentation Center for assistance with the SAM analysis of UF fibers.

REFERENCES

1. K. Prewo and J. Brennan, J. Mater. Sci. 15 (2), 463 (1980).
2. J. Brennan and K. Prewo, J. Mater. Sci. 17 (8), 2371 (1982).
3. K.M. Prewo, J. J. Brennan, and G.K. Layden, Am. Ceram. Soc. Bull. 65 (2), 305 (1986).
4. P.J. Lamicq, G.A. Bernhart, M.M. Dauchier, and J.G. Mace, Am. Ceram. Soc. Bull. 65 (2), 336 (1986).
5. K.S. Mazdiyasni (ed.), Fiber Reinforced Ceramic Composites: Materials, Processing, and Technology, (Noyes Publications, Park Ridge, NJ, 1990).
6. D.J. Pysher, K.C. Goretta, R.S. Hodder, Jr., and R.E. Tressler, J. Am. Ceram. Soc. 72 (2), 284 (1989).
7. G. Simon and A.R. Bunsell, J. Mater. Sci. 19 (11), 3649 (1984).
8. T.J. Clark, R.M. Arons, J.B. Stamatoff, J. Rabe, Ceram. Eng. Sci. Proc. 6 (7-8), 576 (1985).
9. M.H. Jaskowiak and J.A. DiCarlo, J. Am. Ceram. Soc. 72 (2), 192 (1989).
10. B.A. Bender, J.S. Wallace, and D.J. Schrodt, J. Mater. Sci. 26 (4), 970 (1991).
11. T. Mah, N.L. Hecht, D.E. McCullum, J.R. Hoenigman, H.M. Kim, A.P. Katz, and H. Lipsitt, J. Mater. Sci. 19 (4), 1191 (1984).
12. S.M. Johnson, R.D. Brittain, R.H. Lamoreaux, and D.J. Rowcliffe, Comm. Am. Ceram. Soc., 71 (3), C-132 (1988).
13. Y. Hasegawa, J. Mater. Sci. 24 (4), 1177 (1989).
14. M. Takeda, Y. Imai, H. Ichikawa, T. Ishikawa, Ceram. Eng. Sci. Proc. 12 (7-8), 1007 (1991).
15. L.A. Silverman, W.D. Hewett, Jr., T.P. Blatchford, and A.J. Beeler, J. Appl. Polymer Sci.: Appl. Polymer Symp. 47, 99 (1991).
16. F. Frechette, B. Dover, V. Venkateswaran, and J. Kim, Ceram. Eng. Sci. Proc. 12 (7-8), 992 (1991).
17. S. Yajima, J. Hayashi, M. Omori, and K. Okamura, Nature 261, 683 (1976).
18. S. Yajima, K. Okamura J. Hayashi, and M. Omori, J. Am. Ceram. Soc. 59 (7-8), 324 (1976).
19. Wm. Toreki, G.J. Choi, C.D. Batich, M.D. Sacks, and M. Saleem, Ceram. Eng. Sci. Proc. 13 (7-8), 198 (1992).
20. ASTM Designation D3379, American Society for Testing and Materials, Philadelphia, PA.

SYNTHESES, STRUCTURES AND PROPERTIES OF POLYCARBOSILANES FORMED DIRECTLY BY POLYMERIZATION OF ALKENYLSILANES

JOHN MASNOVI*, XIN Y. BU*, KASSAHUN BEYENE*, PAULA HEIMANN*, TERRENCE KACIK*, A. HARRY ANDRIST* AND FRANCES I. HURWITZ**

* Cleveland State University, Cleveland, OH 44115
** NASA Lewis Research Center, Cleveland, OH 44135

ABSTRACT

Vinylsilane polymerizes to form predominantly a carbosilane polymer using dimethyltitanocene catalyst. This is in contrast to alkylsilanes, which afford polysilanes under the same conditions. The mechanism of polymerization of alkenylsilanes has been shown to be fundamentally different from that for the polymerization of alkylsilanes. The silyl substituent apparently activates a double bond to participate in a number of polymerization processes in this system, particularly hydrosilation. Isotopic labeling indicates the involvement of silametallocyclic intermediates, accompanied by extensive nuclear rearrangement. Polymers and copolymers derived from alkenylsilanes have relatively high char yields even for conditions which afford low molecular weight distributions. Formation of crystalline ß-SiC is optimum for a copolymer of an alkylsilane and an alkenylsilane having a silane/carbosilane backbone ratio of 85/15 and a C/Si ratio of 1.3/1 .

INTRODUCTION

Preceramic polymers represent a useful strategy for formation of fibers, coatings and composite matrices. Much effort has been devoted to development of novel polymeric precursors which can be pyrolyzed to form silicon carbide in high yield and purity [1,2]. The major commercial routes to silicon carbide are based on polymerization of dichlorosilanes, such as dichlorodimethylsilane (Eq. 1, R = CH_3) [3]. Pyrolysis of the polysilane formed by coupling of the dichlorosilane is believed to involve insertion of carbon into the polymer backbone to form a polycarbosilane [4].

$$n\ Cl\text{-}\underset{R}{\overset{CH_3}{Si}}\text{-}Cl \xrightarrow{\ 2n-2\ \ Na\ } Cl\text{-}[\text{-}\underset{R}{\overset{CH_3}{Si}}\text{-}]_n\text{-}Cl \longrightarrow \text{-}[\text{-}\underset{R}{\overset{H}{Si}}\text{-}CH_2\text{-}]_n\text{-} \qquad (1)$$

polyalkylsilane polycarbosilane

Subsequent loss of small fragments (such as R-H and H_2 following Eq. 1) results in formation of silicon carbide (SiC).

Generally, linear polysilanes exhibit low ceramic yield due to cleavage of silicon-silicon bonds, leading to breakup of the polymer backbone and formation of small volatile fragments. Consequently, most such precursors require crosslinking in order to avoid this material loss [5]. The presence of hydrogen attached to silicon (Eq. 1, R = H) provides a thermosetting mechanism which minimizes mass loss due to depolymerization processes [6].

Another strategy receiving recent attention is the formation of polycarbosilanes directly. Polycarbosilanes may be formed from alkenylsilanes by hydrosilation (Eq. 2) and ring-opening polymerization (Eq. 3) [7,8]. Although low carbon/silicon ratios in the polymer can be realized using derivatives of dichlorosilane, such monomer precursors may be difficult to obtain, the yields of soluble polymer usually are low, and a subsequent reduction step is required to produce the halogen-free polymer.

$$\begin{aligned} (n+1)\ CH_2{=}CH{-}SiHCl_2 &\xrightarrow{Pt} CH_2{=}CH{-}[-SiCl_2{-}CH_2{-}CH_2{-}]_n{-}SiHCl_2 \\ &\xrightarrow{LiAlH_4} CH_2{=}CH{-}[-SiH_2{-}CH_2{-}CH_2{-}]_n{-}SiH_3 \end{aligned} \qquad (2)$$

$$n\ \begin{matrix} Cl_2Si - CH_2 \\ | \quad\quad | \\ CH_2{-}SiCl_2 \end{matrix} \xrightarrow{Pt} -[-SiCl_2{-}CH_2{-}]_{2n}- \xrightarrow{LiAlH_4} -[-SiH_2{-}CH_2{-}]_{2n}- \qquad (3)$$

Harrod has reported a method for polymerization of alkylsilanes to form polysilanes under mild conditions (Eq. 4). This technique involves dehydrogenative coupling of primary alkylsilanes ($R{-}SiH_3$) using dimethyltitanocene (cp_2TiMe_2) and related catalysts [9]. The polymers are predominantly linear, although of rather short persistence lengths (n = 10-20) [10]. Nevertheless, the polymer derived from methylsilane (R = CH_3, with n > 30) can be converted directly to silicon carbide in high yield [11]. We wish to report here some of the properties of polymers, and their pyrolysis products, derived from the polymerization of alkenylsilanes such as vinylsilane ($CH_2{=}CH{-}SiH_3$) using dimethyltitanocene catalyst.

$$n\ R{-}SiH_3 \xrightarrow{cp_2TiMe_2} R{-}SiH_2{-}[-\overset{R}{\underset{H}{Si}}-]_{n-2}{-}SiH_2{-}R \ + \ (n-1)\ H_2 \qquad (4)$$

RESULTS AND DISCUSSION

Polymer characteristics and ceramic conversion

Vinylsilane polymerized at room temperature for 30 days as described previously [12] affords a soluble, low molecular weight polydisperse polymer (M_n = 540, M_w = 1800) in 82% yield following removal of the solvent. Analysis by ^{1}H, ^{13}C and ^{29}Si NMR indicated that the polymer contained 74% polycarbosilane and 26% polysilane. The polymer was not (<10%) cross-linked. Despite the relatively low molecular weight distribution, vinylsilane exhibited a 60% char yield. The char was slightly carbon-rich, and analyzed as 58% Si and 39% C (3% titanium originating from the catalyst would be expected to remain in the polymer). Pyrolysis to 1400°C in flowing argon indicated formation of both amorphous silicon carbide and ß-SiC crystallites together with a phase which was X-ray amorphous.

Shorter reaction times could be achieved practically without decrease in polymer yields by either increasing the concentration of catalyst or employing higher temperatures (50-80°C). Interestingly, introduction of up to 10 mole % of the potential cross-linking agent 1,2-disilylethane ($H_3Si-CH_2-CH_2-SiH_3$), BSE, during polymerization did not significantly increase polymer molecular weight of the vinylsilane polymer as determined by GPC (Table I). This is in contrast to increase in the molecular weight observed on addition of BSE to methylsilane polymerization.

An optimum composition of such monomers for conversion into silicon carbide was found to derive from the copolymer 70% methylsilane and 30% vinylsilane. This mixture corresponds to a C/Si ratio of 1.3/1 . An 82% yield of copolymer, comprised of 80% polysilane and 20% polycarbosilane

Table I. Polymer and Copolymer Properties

Monomer Composition	Yield (%)	Polymer Composition (ratio silane: carbosilane)	M_n	M_w	Char Yield (%)
100% Vinylsilane	82	26:74	540	1800	60
98% Vinylsilane:2% BSE	80	26:74	480	1200	64
95% Vinylsilane:5% BSE	90	30:70	560	1900	67
30% Vinylsilane: 70% Methylsilane	86	85:15	580	1300	73
100% Methylsilane	90	100:0	680	1960	65
95% Methylsilane:5% BSE	100	100:0	920	7400	71

linkages, was derived from this mixture of monomers. Pyrolysis of the copolymer to 1400 °C afforded 73% char with essentially stoichiometric analysis for SiC. The char was primarily amorphous, with some small (95 Å) crystallites of ß-SiC.

<u>Mechanism of polymerization of vinylsilane</u>

Alkenylsilanes polymerize by a different process than is seen for alkylsilanes [12]. Vinylsilane, for example, affords a polymer which is predominantly a polycarbosilane. A carbosilane type polymer would form by hydrosilation of the vinyl group of one monomer by the SiH_3 group of another. This process may lead to two types of linkages, -Si-C-Si-C- and -Si-C-C-Si-C-C- (Eq. 5). Both types are observed in roughly equal amounts.

```
R3Si  H                    R3Si-H
 |    |      <-----<          +
 CH2-CH-SiH3               CH2=CH-SiH3                                    (5)
                              +          >--->  CH2-CH-SiH3
                           H-SiR3               |    |
                                                H    SiR3
```

The novel results obtained with vinylsilane have a mechanistic basis. Photolytic decomposition of dimethyltitanocene in the presence of silanes ($R\text{-}SiH_3$) affords titanium complexes in reduced oxidation states, such as titanocene (cp_2Ti) and titanocene hydrides, as the apparent active catalytic species [10]. Isotopic labeling studies involving silane-SiD_3 ($R\text{-}SiD_3$) demonstrate that reaction of vinylsilane differs fundamentally from reactions of alkylsilanes. In particular, extensive nuclear rearrangements accompany polymerization of vinylsilane, unlike the alkylsilanes. This evidence is inconsistent with a mechanism for the polymerizations which is free radical in nature, although odd-electron species involving the catalyst are formed. Instead, the polymerization of alkenylsilanes likely proceeds by way of silatitanocyclic intermediates (Eq. 6), from which the rearrangements can proceed. Similar silylmetallic species are known to undergo alkene insertion and other reactions [13], and the metallocycles depicted in Eq. 6 also would be expected to be susceptible to such processes.

Other evidence demonstrates intimate participation by the catalyst in

```
CH2=CH-SiH3           H                        CH2                CH-CH3
                     /                        /  \               /|
    +      -->  cp2Ti     CH=CH2   <-->  cp2Ti    CH2  +   cp2Ti  |         (6)
                     \   /                    \  /               \|
  cp2Ti               SiH2                     SiH2               SiH2
```

reactions of alkenylsilanes. Several processes, including hydrosilation, hydrogenation, dehydrogenative coupling, desilylation, trans-silylation and metathesis, were observed to occur with exceptional facility. Thus, the silyl substituent is highly activating for reaction of double bonds with this catalyst. Reactions such as desilylation and trans-silylation involve translocation of the heavy atoms, and are responsible for formation of small amounts of various side-products, including silane and disilylethylene, which incorporate into the polymer more slowly. The polymer of vinylsilane appropriately should be considered a copolymer of vinylsilane, ethylsilane, and these other materials. Nevertheless, incorporation of <10% of such materials into the polymer was found to have little effect on the overall structure and molecular weight of vinylsilane polymer (again contrary to the alkylsilane polymers), as demonstrated by deliberate addition of ethylsilane and disilylethane.

This result probably is due to the high reactivity of vinylsilane (compared to alkylsilanes), and hydrosilation (as opposed to crosslinking) of vinylsilane in the presence of saturated silanes such as SiH_4 and $H_3SiCH_2CH_2SiH_3$ (BSE). Therefore, vinylsilane rapidly forms homopolymer, even in the presence of alkylsilanes, which are incorporated into the polymer more slowly. Copolymer, when formed during the early stages of polymerization, probably also is formed largely by hydrosilation; thus the disilylalkane BSE tends to afford linear polymer (with two $-SiH_3$ termini, which may be available for subsequent crosslinking) rather than altering the structure of the vinylsilane polymer. Apparently, the high char yields obtained, despite the relatively low molecular weights of the polymers, should be due at least in part to the structure of the polymer, which consists of predominantly carbosilane linkages with free H-Si-H groups [3].

CONCLUSIONS

Whereas polyalkylsilanes (except for methylsilane) produce low ceramic yields, alkenylsilanes yield polycarbosilanes which are suitable as high char yield precursors. In particular, vinylsilane polymers are suitable as precursors to carbon-rich material which may remain amorphous to higher temperature. If desired, however, stoichiometric SiC may be produced by copolymerization of vinylsilane with methylsilane.

ACKNOWLEDGEMENT

The authors would like to thank D. A. Scheiman and A. M. Modock for technical assistance and the NASA Lewis Research Center for support of this work under cooperative agreement NCC 3-63.

REFERENCES

[1] R. West, D. D. David, P.I. Djurovich, H. Yu, and R. Sinclair, Am. Ceram. Soc. Bull. 62, 899 (1987).

[2] D. Seyferth and H. Lang Organometallics 10, 551 (1991).]

[3] S. Yajima, J. Hayashi, M. Omori, and K. Okamura, Nature 261, 683 (1976); S. Yajima, J. Hayashi, and M. Omori, Chem. Lett. 1975, 931.

[4] T. Ishikawa, M. Shibuya, and T. Yamamura, J. Mater. Sci. 25, 2809 (1990).

[5] D. Seyferth in Inorganic and Organometallic Polymers, edited by M. Zeldin, K. J. Wynne, and H. R. Allcock (ACS Symposium Series, 1988), Vol. 360, p. 89.

[6] H. J. Wu and L. V. Interrante, Chem. Mater. 1, 564 (1989).

[7] B. Boury, L. Carpenter, and R. J. P. Corriu, Angew. Chem., Int. Ed. Engl. 29, 785 (1990); B. Boury, R. J. P. Corriu, D. Leclercq, P. H. Mutin, J.-M. Planeix, and André Vioux, Organometallics 10, 1457 (1991).

[8] H.-J. Wu and L. V. Interrante, Macromolecules 25, 1840 (1992).

[9] E. Samuel and J. F. Harrod, J. Am. Chem. Soc. 106, 1859 (1984).

[10] J. F. Harrod in Inorganic and Organometallic Polymers, edited by M. Zeldin, K. J. Wynne, and H. R. Allcock (ACS Symposium Series, 1988), Vol. 360, p. 89, and references therein.

[11] Z. Zhang, F. Babonneau, R. M. Laine, Y. Mu, J. F. Harrod, and J. A. Rahn, J. Am. Ceram. Soc. 74, 670 (1991).

[12] J. Masnovi, X. Y. Bu, P. Conroy, A. H. Andrist, F. I. Hurwitz, and D. Miller, Mat. Res. Soc. Symp. Proc. 180, 779 (1990).

[13] J. Arnold, M. P. Engeler, F. H. Eisner, R. H. Heyn, and T. D. Tilley, Organometallics 8, 2284 (1989); R. H. Heyn and T. D. Tilley, J. Am. Chem. Soc. 114, 1917 (1992).

NMR CHARACTERIZATION OF SILICON CARBIDES AND CARBONITRIDES. A METHOD FOR QUANTIFYING THE SILICON SITES AND THE FREE CARBON PHASE.

GERARDIN C., HENRY M. and TAULELLE F.
Laboratoire de chimie de la Matière Condensée, University P. et M. Curie, T54-55
4 Place Jussieu, 75252 Paris Cedex 05, FRANCE

ABSTRACT

With ceramics prepared from organic precursors, polymers are often formed by thermolysis, and the ceramics are obtained by pyrolysis of the polymer. This paper will cover the NMR methods used to characterize the polymerization as well as the pyrolysis of such polymers. The reticulation rate of such materials is measured on the silicon atoms and on the carbon atoms, leading to a silicon reticulation state and a carbon reticulation state. For the ceramics, the quantification of the SiC_xH_{4-x} or SiC_xN_{4-x} sites is performed using a priori chemical shift calculation. A measure of the H/C of the free carbon phase at the different temperatures of pyrolysis is presented. This measure uses a absolute determination of hydrogen content in ceramics, more accurate than a classical chemical analysis of hydrogen. The evolution of the reticulation, of the free carbon amount and of the H/C ratio in the free carbon phase will be followed with the temperature.

INTRODUCTION

Since the discovery of the Nicalon fiber new routes to silicon carbides have been constantly developed[1,2,3]. The goals are to reach a high molecular weight with a high ceramization yield combined in a high purity material. A starting material called PPCS with a linear formula $-(-SiH_2-C_2H_4-)_n-$ has been used to form a SiC ceramics[4]. The polymerization process has been a catalytic hydrosilylation[5]. This route leads to a ceramic precursor well defined, less complex than the Yajima's precursor PCS. Another family of compounds named PVSZ of general formula $-(-SiHVinyl-NH-)_n-$ lead to SiCN ceramics.
The goal of this contribution is not to describe the details of the materials condensation polymerization and ceramization but to show how NMR can be used to measure the parameters quantifying at each stage the evolution of the material.

CHEMICAL SHIFT IN AMORPHOUS PHASES

It has been shown that the screening constant of silicon can be calculated as follows[6]:

$$\sigma = \sigma_{dia} + \sigma_{para} = \sigma_{dia} - \frac{e^2\hbar^2<r^{-3}>_p P_u}{3m_e c^2 \Delta E} \qquad (1)$$

Using the SDPCM (structure dependant partial charge model) [6] leads to a simplified expression of the chemical shift.

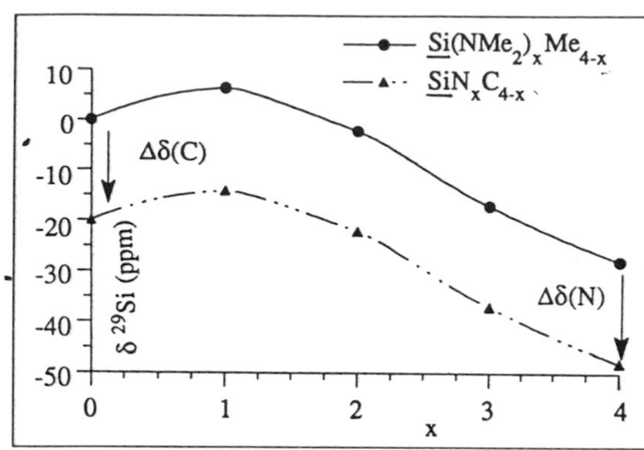

Figure 1: Chemical shift transposition from molecular to amorphous state in the SiN_xC_{4-x} system.

$$\delta = A - \frac{B(1+fq)^3 P_u}{C + Dq} \quad (2)$$

With A = 701 ppm, B = -1605 ppm, f = 0.873, C = 1.76eV and D = 9.25eV.

In the case of silicon (with the usual conventions) chemical shift noted as δ and screening constant noted σ have the same numerical values. In order to predict the chemical shift of a molecular or a structure compound, a knowledge of the positions of atoms in the structure or the distances and bond angles in a molecular entities is sufficient. However, one cannot estimate such a chemical shift in an amorphous material by such an approach. Furthermore, the chemical shift measured by NMR will always be the center of gravity of a distribution of chemical shifts related to sites in slightly different conformations.

Chemical shift estimation in amorphous material is therefore performed as follows, presented below for the SiN_xC_{4-x} system. First, the first sphere of coordination in which one is interested is taken on a family of molecular compounds. Here the family $Si(NMe2)_x(Me)_{4-x}$ has been considered.The chemical shift of these species can be calculated and measured. Both set of values are in excellent agreement. Then the series of solid samples can be measured, this is the case of SiC and Si_3N_4. In the case of amorphous SiC an approximate shift of -20 ppm is observed relative to liquid TMS. In the case of a first sphere of SiN_4 type in Si_3N_4, there is also a shift of -20 ppm compared to $Si(NMe)_4$. The chemical shift estimation for the three solid "unknown" mixed first spheres proceeds by translation of the complete series of molecular compounds chemical shifts by -20 ppm. Actually the estimation procedure is governed by the actual expression of the chemical shift.

$$\delta_n = 701 - \frac{1065\,(\,1 + 0.873\,q_n)^3\,P_u}{\Delta E_n}$$

For the mixed spheres SiN_xC_{4-x} the different terms can be treated as follows:

SiC4	SiC3N	SiC2N2	SiCN3	SiN4
-20 ppm	-14 ppm	-22 ppm	-37 ppm	-48 ppm
SiC4	SiC3O	SiC2O2	SiCO3	SiO4
-20 ppm	-10 ppm	-44 ppm	-73 ppm	-109 ppm
SiN4	SiN3O	SiN2O2	SiNO3	SiO4
-47 ppm	-58 ppm	-74 ppm	-90 ppm	-109 ppm
Si(CH2)4	SiHC3	SiH2C2	SiH3C	SiH4
20 ppm	6 ppm	-20 ppm	-51 ppm	-90 ppm

Table I: selected characteristic positions of mixed first spheres occuring in silicocarbides, silicocarbonitrides and silicocarboxides amorphous materials.

$\Delta E_n = (n\Delta E_4 + (4\text{-}n)\Delta E_0)/4$, the charges are estimated by $q_n = (nq_4 + (4\text{-}n)q_0)/4$ and the population unbalance is given by $P_u = 1 - q_n^2/16$ [6].
This relation of the three parameters governing the chemical shift allows for a generalization of the chemical shift estimation starting with a knowledge of the molecular species and an "educated" transposition of those shifts to the solid state. The accuracy of the positions thus obtained is in the order of 2 to 5 ppm. This is lose enough to localize the center of gravity of a ditribution of sites in amorphous materials. This "assignment" provides a good way to attribute a peak to its correct first sphere. The final positioning is done on the experimental spectrum itself.

LOCAL SITE ENVIRONNEMENT OF SILICON

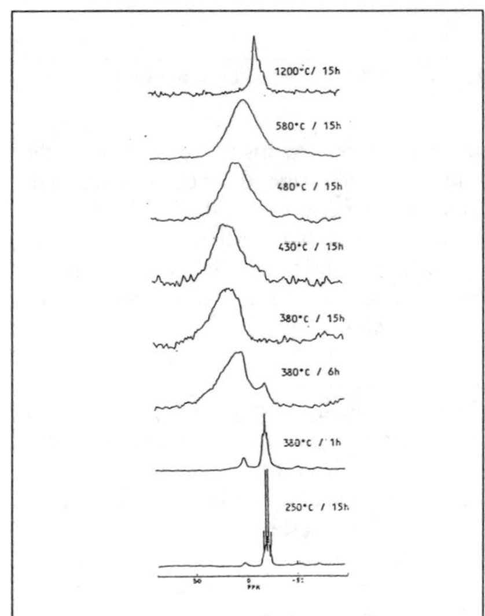

Figure 2: ^{29}Si MAS NMR of PPCS.

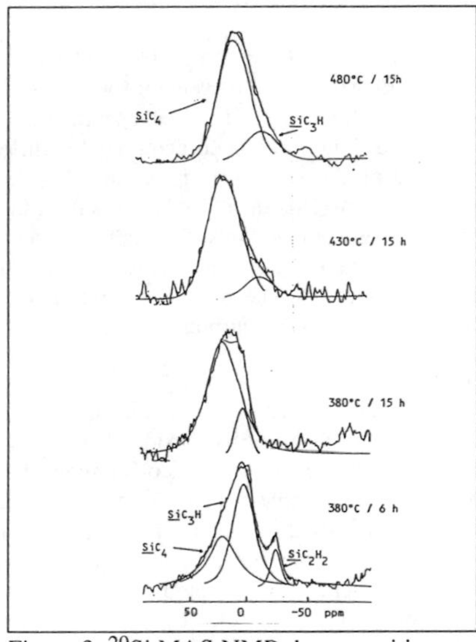

Figure 3: ^{29}Si MAS NMR decomposition.

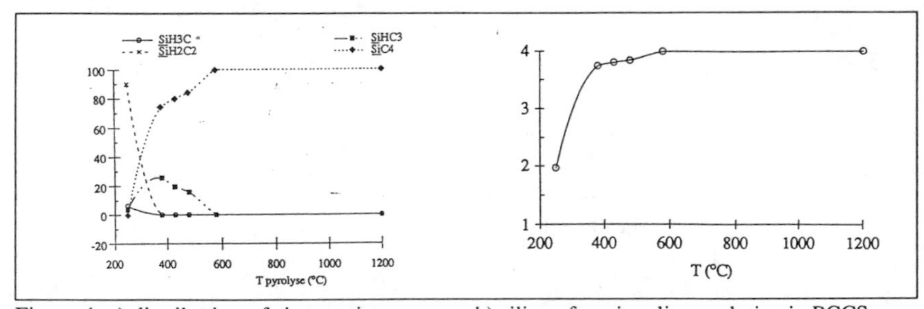

Figure 4: a) distribution of sites environnement; b) silicon functionality evolution in PCCS.

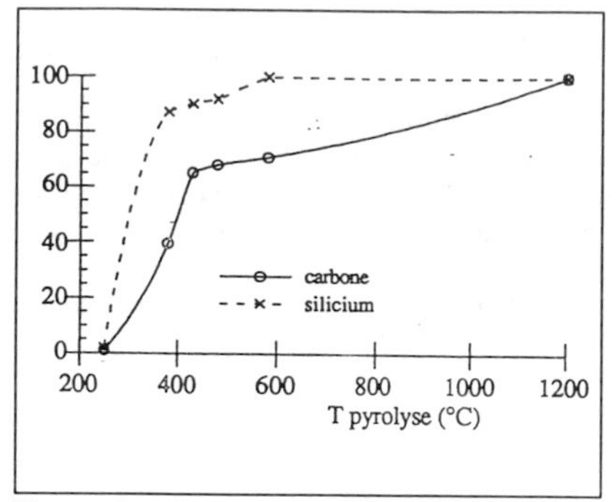

Figure 5: reticulation rate of PPCS;

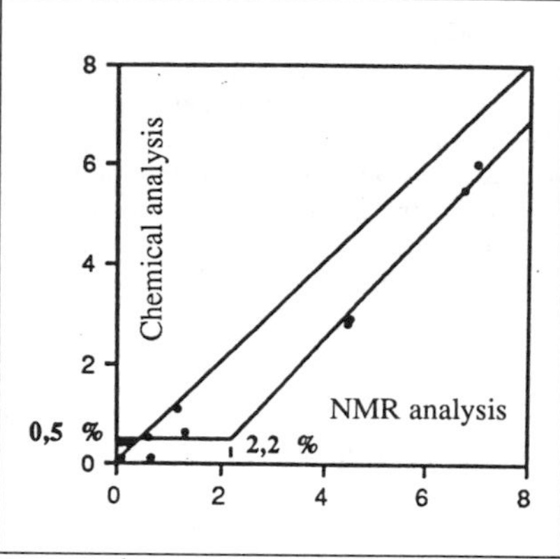

Figure 6: ^{1}H NMR and chemical analysis.

In the case of the PCCS (polycarbocarbosilane) of general formula -SiH2-C2H4-Si- the ^{29}Si NMR spectra in MAS conditions have been acquired for different temperatures treatments. In Figure 2 the ensemble of the spectra is presented. Figure 3 presents a selection of spectra, with their decomposition. This decomposition is done with positions fixed at their calculated values by the above procedure. Only the intensities and widths are fitted. The proportion of sites are therefore plotted as a function of temperature in Figure 4a, and the functionality "n" of silicon deduced figure 4b. The reticulation rate is calculated as n/4. By an analogous procedure the reticulation rate of the carbon has been measured and is given in figure 5.

HYDROGEN CONTENT MEASURE

While the ceramic is being formed, the C-H and the Si-H bonds are progressively replaced by C-Si and Si-C bonds. The amount of hydrogen in the sample is related to the reticulation of carbon and silicon. Usually hydrogen measurent is done through a chemical analysis of the sample. At low amounts of hydrogen, the chemical analysis is not reliable. A NMR measure of the areas of the ^{1}H peaks has been done on a series of PVSZ compounds, and the NMR results are plotted versus the chemical analysis ones in Figure 6. Under 2% in weight the chemical analysis gives erratic results. Above 2% they contain a systematic error, underestimating the results by 2.2%. For the PVSZ series of compounds the amount of hydrogen has been plotted on figure 7 as a function of the temperature. In the chemical analysis that will be used later on, the hydrogen elemental analysis has been systematically replaced by the NMR ^{1}H analysis.

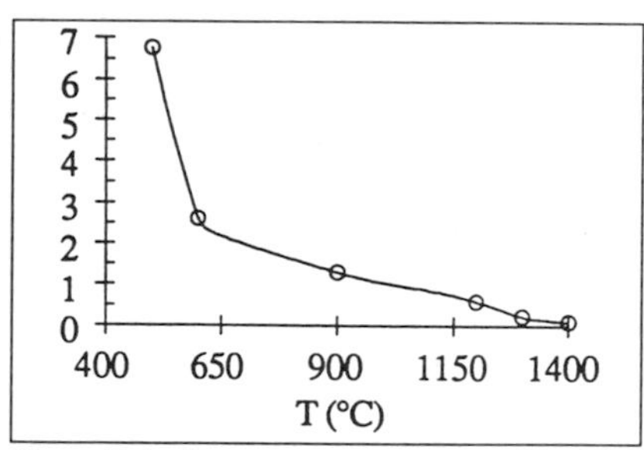

Figure 7: hydrogen content measured by NMR.

FREE CARBON PHASE MEASURE

As the results obtained in ^{29}Si MAS NMR allow measurement of the environment of the silicon, if the material is single phased, the analysis of the distribution sites in ^{29}Si is equivalent to performing a chemical analysis. A plot of this NMR elemental analysis versus chemical analysis appears in figure 8a for the PVSZ compounds. It is obvious that a large discrepancy between

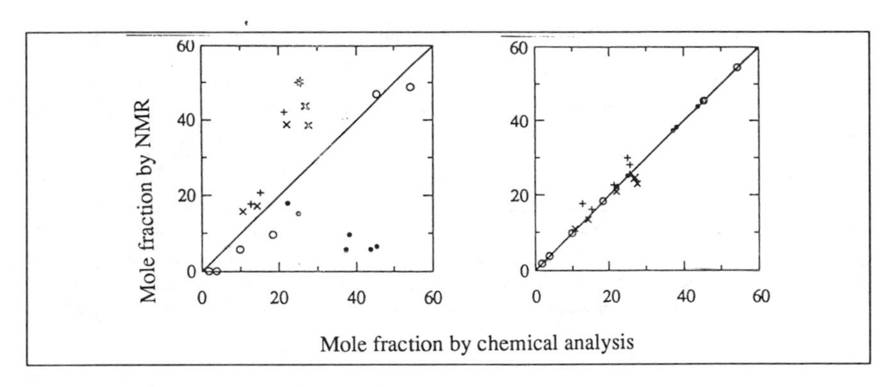

Figure 8: a) on left correlation NMR chemical shift without free carbon phase; b) on right with free carbon.

the two sets of values exists. This discrepancy is attributed to the fact that not all the atoms in the phase are connected to the silicon network. The material is not single phased, shown by electronic microscopy, it contains a "free carbon" phase. The relations between the chemical analysis and ^{29}Si NMR analysis are:

(3) $1 = x_C + x_H + x_{Si} + x_N$ — ^{29}Si NMR

(4) $1 = X_C^{Si} + X_C^{free} + X_H^{Si} + X_H^{free} + X_{Si} + X_N$ — Chemical analyis

Renormalization of equation (3) leads to:

(5) $1 + x_C^{free} + x_H^{free} = x_C + x_C^{free} + x_H + x_H^{free} + x_{Si} + x_N$

and combining (4) and (5) gives:

$$X_C = (x_C + x_C^{free}) / (1 + x_H^{free} + x_C^{free})$$
$$X_H = (x_H + x_H^{free}) / (1 + x_H^{free} + x_C^{free})$$
$$X_{Si} = x_{Si} / (1 + x_H^{free} + x_C^{free})$$
$$X_N = x_N / (1 + x_H^{free} + x_C^{free})$$

with X_i are the molar fraction measured by chemical analysis and x_i the molar fraction measured by NMR. x_H^{free} and x_C^{free} are the fractions of hydrogen and carbon not measured by ^{29}Si NMR. The comparison of the two sets of results allows therefore for a complete determination of x_H^{free} and x_C^{free}. Both sets are now in agreement as plotted in Figure 8b. The amount of free carbon has been plotted in Figure 9 and the ratio H/C of the free carbon

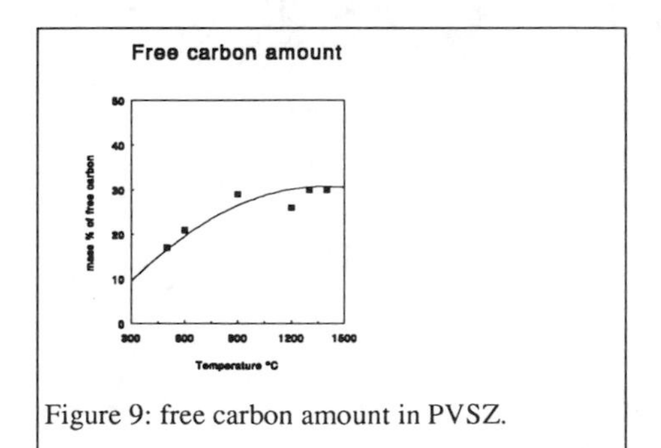

Figure 9: free carbon amount in PVSZ.

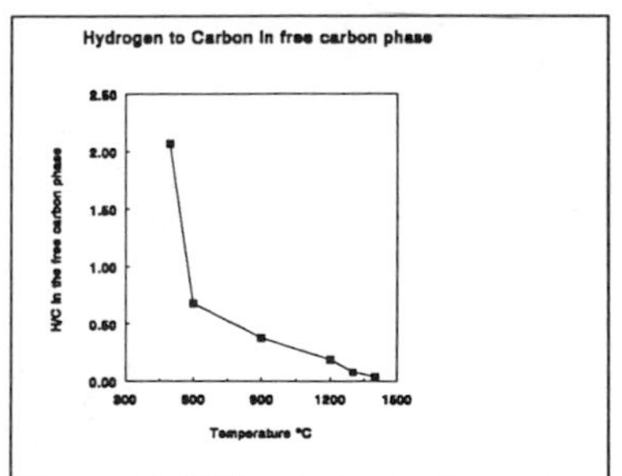

Figure 10: H/C ratio in the free carbon phase of PVSZ.

has been plotted in Figure 10. The H/C ratio evolution confirms previous work[7] on the formation of the free carbon phase, involving coronene formation. The coronene H/C ratio is 0.5. As the material has a starting H/C ratio of 2, the evolution through coronene and condensation of these basic structural units by sides to form the free carbon phase, is followed quantitatively with the present measurements.

CONCLUSION

Though the details of the mecanisms of the material transformation have not been treated in this paper, due to space limitations, it has been shown that most of the quantitative parameters describing the evolution of a series of polymeric, amorphous or crystalline materials can be identified through NMR combined with chemical analysis.The concepts used to describe the evolution through the amorphous state, up to the crystalline state, follow from polymer science.

REFERENCES

[1] Bacque E., Pillot J.P., Birot M. and Dunoguès J., Macromolecules, 21, 30 (1988)

[2] Zhang Z. F., Babonneau F., Laine R.M., Mu Y.,Harrod J.F. and Ramm J.A. J. Am. Ceram. Soc. 74, 670 (1991)

[3] West R., David L.D., Djurivich P.I., Yu H., Sinclair R. Am. Ceram. Soc. Bull. 62, 899 (1983)

[4] Boury B., Carpenter L., Corriu R., and Mutin H. New J. Chem. 14, 535 (1990)

[5] Speier J.L. Adv. Organomet. Chem. 17, 407 (1979)

[6] Gerardin C. Ph. D. Dissertation, University P. and M Curie, Paris 1991
and This MRS session 1992 Poster N11.1

[7] Oberlin A. in Chemistry and Physics of Carbon, (Ed. P.A. Thrower, Dekker New York), 22,, 1 (1989)

HIGH TEMPERATURE STABILITY OF OXYCARBIDE GLASSES

HANXI ZHANG AND CARLO G. PANTANO
Department of Materials Science and Engineering, The Pennsylvania State University, University Park, Pa 16802

ABSTRACT

The stability of silicon oxycarbide glasses has been studied at temperatures up to 1500°C. The silicon oxycarbide glasses were synthesized using a sol/gel process. The pyrolysis treatment in argon influenced the structure and composition of the synthesized glasses, and in turn, their high temperature stability in oxidizing atmosphere. The oxycarbide glasses pyrolyzed at ≥ 1000°C had lower hydrogen concentration and a more polymerized network structure, and thereby were more resistant to oxidation and crystallization at higher temperatures.

I. INTRODUCTION

The thermochemical and thermomechanical stability of glasses has always been a critical issue in their high temperature applications. Ordinarily, oxide glasses crystallize and soften at elevated temperatures. There has been great interest in enhancing the stability of the glasses by incorporating carbon into glass structures[1-2]. The sol/gel process has made it practical to synthesize these glasses[3-5]. Carbon offers the possibility of 4 coordinate bonds to replace the oxygen anion which is only 2 coordinate, and this is expected to strengthen the molecular structure of the glasses. Chi[2], Zhang and Pantano[4], and Runland[5] have independently reported that there was limited crystallization of SiO_2 from the oxycarbide glasses. But these oxycarbide glasses were processed and evaluated in very different ways. Thus, the goal of this study was to systematically examine the relationships between processing and high temperature stability. The gels were synthesized using an established procedure and they were pyrolyzed to the glassy state in argon over the temperature range 800°C to 1400°C. Solid state Magic Angle Spinning ^{29}Si Nuclear Magnetic Resonance (MAS ^{29}Si NMR), ^{13}C NMR and chemical analysis were used to characterize these glasses. The decomposition and oxidation resistance was examined by thermogravimetric analysis (TGA). The role of glass structure and composition in the thermochemical stability will be discussed.

II. EXPERIMENTAL PROCEDURES

The oxycarbide glasses were synthesized by a sol/gel process[4]. Methyltrimethoxysilane was the starting material. 1 mole of $MeSi(OMe)_3$ was mixed with 4 moles of H_2O in a beaker, and the pH value of the solution was adjusted to 1 or 2 by adding 1M HCl solution. After 2 hours of stirring, the pH value of the solution was raised to 6.5 by addition of 2M NH_4OH. The solution gelled within about 24 hours. The gels were further dried in petri-dishes for 48 hours during which time the gels shrank by syneresis. The solid gels were dried at 65°C to 105°C for about a week. The dry gels were pyrolyzed to the glassy state in flowing argon at 800°C, 1000°C, 1200°C and 1400°C for 60 minutes. The oxycarbide glasses obtained were pulverized into 100 mesh powder for solid state ^{29}Si NMR+, TGA++ and chemical analysis+++. In the TGA analysis, the oxycarbide glasses were heated at 10°C/min up to 1500°C and held for 60 minutes. The powders studied in the TGA were subsequently examined by x-ray diffraction.

III. RESULTS AND DISCUSSION

The carbon and hydrogen contents of the oxycarbide glasses are shown in Figure 1. There is a big difference in composition between the glass heat treated at 800°C relative to those heat treated at ≥1000°C. The carbon content increases from 12.5% at 800°C to 14% at 1000°C and remains almost constant at higher temperatures. The hydrogen content

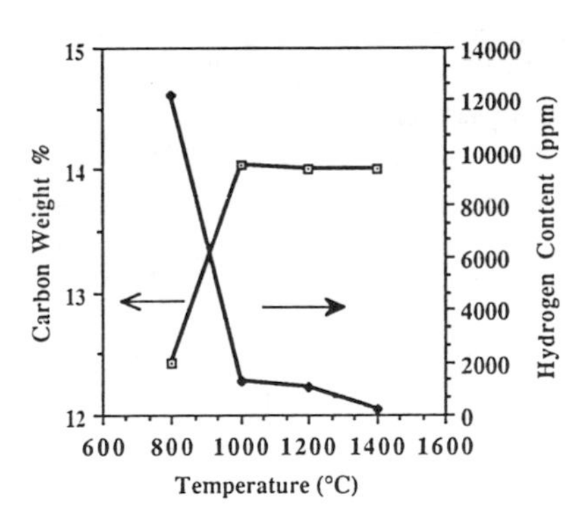

Figure 1. The variation of the carbon and hydrogen contents of the oxycarbide glasses with pyrolysis temperature.

+ Chemagnetic CMX-300A operating at 59.08MHZ

++Netzsch Simultanous Analyzer 429

+++Leco Chemical Analysis

decreases most drastically between 800°C and 1000°C and continues to decrease up to 1400°C.

Figure 2 is the Cross Polarization ^{13}C NMR of the glass pyrolyzed at 800°C. CP NMR depends upon the fast relaxation of H (to shorten the spectra acquisition time) and hence is only sensitive to those carbons that are bonded with or in the close proximity of, hydrogen. The glasses heat treated at higher than 1000°C exhibited weak spectra and this verified their low concentration of hydrogen. The strong signals in Figure 2 indicate that there is a large amount of elemental carbon and $\equiv SiCH_3$ in the structure of the oxycarbide glass pyrolyzed at 800°C.

The ^{29}Si NMR spectra of the oxycarbide glasses pyrolyzed at progressively higher temperatures are shown in Figure 3. The ^{29}Si NMR spectrum for the gel is also shown for comparison. The assignment of the peaks was made using standard compounds[6]. There are two species in the gel, $[O_3SiCH_3]$ and $[HOO_2SiCH_3]$[4]. The hydroxyl group left in the gel is the result of incomplete polymerization of the gel structure. The spectra for all the glasses show the presence of $[SiO_4]$, $[O_3SiC]$ and $[O_2SiC_2]$ species, but the amount of the oxycarbide species is quite different. With the increase of pyrolysis temperature, the relative concentration of $[O_3SiC]$ decreases. Above 1200°C, $[SiC_4]$ (~ -15 ppm) appears in the ^{29}Si NMR spectra. At 1400°C, most of the $[O_3SiC]$ species in the glass decompose and $[SiC_4]$ species increase in concentration.

It should be noted that the $[O_3SiC]$ peak shifts from -65ppm in the spectrum of the glass pyrolyzed at 800°C to ~ -75ppm in these higher temperature glasses. The broad $[O_3SiC]$ peak at -65 ppm in the spectrum of the 800°C glass probably represents a distribution of $[O_3SiCH_3]$ and $[OHO_2SiCH_3]$; these species are clearly resolved in the spectrum of the gel at -68 and -58 ppm, respectively. At ≥ 1000°C, the downfield shift of this broad peak denotes a relative loss of the terminal OH species (-58 ppm). This is consistent with the ^{29}Si NMR spectra of silicate minerals where polymerization of the silicate structure leads to a down field shift of the peaks[7]. Here the $[O_3SiC]$ peak shift from -65 ppm to -75 ppm corresponds to the removal of network terminating OH and CH_3 groups through polymerization of the network structure.

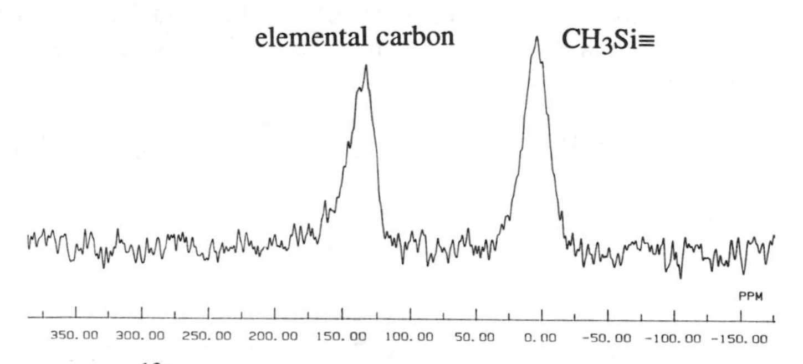

Figure 2. CP ^{13}C MAS NMR of the oxycarbide glass pyrolyzed at 800°C.

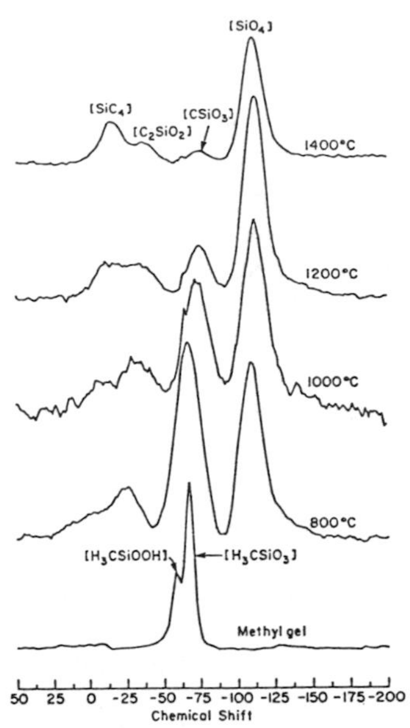

Figure 3. ^{29}Si MAS NMR spectra of the gel and the oxycarbide glasses pyrolyzed at various temperatures

Reactions which result in the polymerization of the glass structure can be shown as:

$$H_3C\underset{|}{\overset{|}{Si}}OH + HO\underset{|}{\overset{|}{Si}}CH_3 \rightarrow H_3C\underset{|}{\overset{|}{Si}}\text{-O-}\underset{|}{\overset{|}{Si}}CH_3 + H_2O \qquad 1)$$

$$\equiv SiCH_3 + H_3CSi\equiv \rightarrow \equiv Si\text{-}CH_2\text{-}Si\equiv + 2H_2 + C \text{ (free carbon)} \qquad 2)$$

The evolution of H_2O and H_2 accounts for the chemical analysis in Figure 1 where a relative increase in total carbon, and decrease in hydrogen, is revealed.

TGA of the glasses in oxygen up to 1500°C is shown in Figure 4. The weight change values are listed in Table I. The TGA results in Ar are included for reference.

Table I. Weight Change Behavior of the Oxycarbide Glasses up to 1500°C

Atmosphere	O_2	Ar
Pyrolysis Temperature(°C)		
800	-8%	-13%
1000	+2%	-11%
1200	+2%	-7%
1400	+2%	-6%

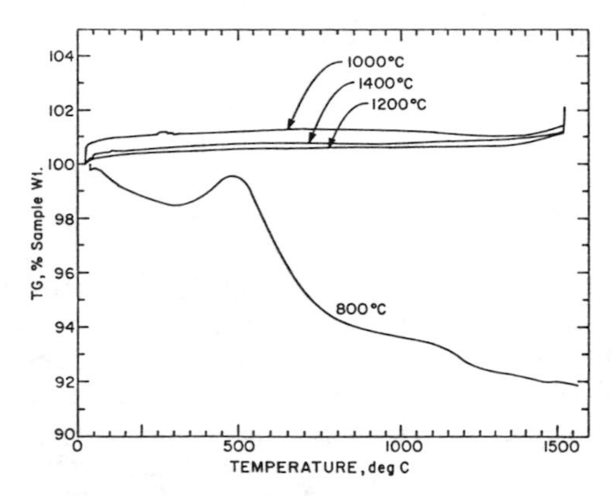

Figure 4. TGA of the oxycarbide glasses in Oxygen up to 1500°C

The behavior of the oxycarbide glass pyrolyzed at 800°C is very different from the other glasses. The glass loses weight up to 400°C and then gains weight between 400°C and 600°C. Above 600°C, the sample loses weight again. The residual material left after the TGA runs is almost white, and shows a strong peak of crystalline SiO_2 in the x-ray diffraction pattern. Conversely, the glasses heat treated in argon loses weight continuously in this temperature range.

Weight loss in O_2 has two possible sources:

1. Oxidation of the methyl group in the silicate network:

$$2 \equiv Si\text{-}CH_3 + 3O_2 \rightarrow 2 \equiv Si\text{-}O_{1/2} + 2CO\uparrow + 3H_2O\uparrow \qquad 3)$$

Weight loss is the result of the mass difference of CH_3 and $O_{1/2}$.

2. Oxidation of free (aromatic) carbon:

$$2C + O_2 \rightarrow 2CO\uparrow \qquad 4)$$

Since free carbon undoubtedly exists in the oxycarbide glasses, its oxidation contributes a weight loss.

Weight gain can be explained by replacement of network carbon by oxygen. One possible reaction is :

$$2 \equiv Si\text{-}CH_2\text{-}Si\equiv + 2O_2 \rightarrow 2\equiv Si\text{-}O\text{-}Si\equiv + 2CO\uparrow \qquad 5)$$

The sample gains weight from the mass difference of CH_2 and O.

Because of the terminal CH_3 and OH groups in the network of the glass pyrolyzed at 800°C, it has much less resistance to oxidation and crystallization. The terminal sites enhance the mobility of the network. The CH_3 is more readily oxidized than the network carbon that forms at high pyrolysis temperature. The OH groups, which can condense to water at high temperatures, lower the viscosity and provide internal oxidants.

IV. CONCLUSION

It has been shown that there is a relationship between pyrolysis, structure/composition, and stability of oxycarbide glasses synthesized through the sol/gel process. Although the oxycarbide glass pyrolyzed at 800°C has abundant Si-C bonds in the glass structure, the network is terminated by $Si-CH_3$ and SiOH groups. These terminating groups increase the mobility and reactivity of the glass at high temperatures. Pyrolysis at ≥ 1000°C in argon is found to create a polymerized oxycarbide glass structure that is more stable upon further oxidation and decomposition at even higher temperatures.

ACKNOWLEDGEMENT:
The authors gratefully thank the Air Force Office of Scientific Research under grant No. AFOSR-89-0446 for their support of this research.

REFERENCES

1. T.H. Elmer and H.E. Meissner "Increase of Annealing point of 96% SiO_2 Glass on Incorporation of Carbon" J. *Amer.* Ceram. *Soc.*, 59(5-6), 206-09 (1976).
2. F.K. Chi, "Carbon-Containing Monolithic Glasses via Sol/gel Process" Ceram. Eng. Sci Proceedings, 4, 704-17 (1983).
3. Dennis N. Coon, "Effect of Silicon Carbide Additions on the Crystallization Behavior of a Magnesia-Lithia-Alumina-Silica Glass", J. *Amer. Ceram. Soc.*, 72(7), 1270-73 (1989).
4. Hanxi Zhang and C.G. Pantano "Synthesis and Characterization of Silicon Oxycarbide Glasses", J. *Amer. Ceram. Soc.*, 73(4), 958-963 (1990).
5. G. M. Runland, S.Prochazka and R. Doremus, "Silicon Oxycarbide Glasses: Part I. Preparation and Chemistry" *J. Materials Research.*, 6, (12) 2716-2722 (1991).
6. H. Marsman, NMR, Basic Principles and Progress, P. Diehl, E.fluck and R. Kosfeld, eds., New York, Springer-verlag, 65-235, (1981).
7. E. Lippman, M.Magi, A. Samoson, G. Engelhardt and A. R. Grimmer, "Structural Studies of Silicates by Solid-State High-Resolution ^{29}Si NMR", *J. Amer. Chem. Soc.* 102 (15), 4889-4893 (1980).

SILICON OXYCARBIDE GLASSES FROM SOL-GEL PRECURSORS

F. BABONNEAU**, G.D. SORARU*, G. D'ANDREA*, S. DIRE* and L. BOIS**
*Dipartimento di Ingegneria dei Materiali, Università di Trento, 38050, Mesiano-Trento, Italy.
**Chimie de la Matière Condensée, Université P. et M. Curie, 4 place Jussieu, 75252 Paris, France.

ABSTRACT

Silicon oxycarbide glasses have been prepared from sol-gel precursors containing not only Si-CH_3, but also Si-H bonds. Three systems have been chosen containing various Si units but the same C/Si ratio. Their pyrolysis process has been mainly followed by ^{29}Si MAS-NMR and the composition of the final glass extracted. This study shows that a suitable choice of Si-CH_3 and Si-H functionnalized silicon alkoxides can lead to a strong decrease in the free carbon content and to an almost pure silicon oxycarbide phase.

INTRODUCTION

Numerous works have been dedicated to the preparation of silicon oxynitride glasses either by conventionnal melting process or sol-gel approach. Nitrogen is usually introduced via a heat treatment in flowing ammonia or nitrogen. The anionic substitution of divalent oxygen atoms by trivalent nitrogen atoms improves mechanical properties and thermal stability of the glasses [1]. Following the same idea, it was attempted to substitute oxygen atoms by tetravalent carbon atoms in the glass network. The use of conventionnal processing was not very successfull. Silicon carbide powder was used as a carbon source [2], but lead only to a very low C content. Even if no structural characterization was performed to prove the existence of an oxycarbide phase, the samples presented improved properties.

More recently, the sol-gel aproach was considered as a suitable way to get silicon oxycarbide glasses [3-5]. The hydrolysis of modified silicon alkoxides, $R'_nSi(OR)_{4-n}$ leads to the formation of silica-based network containing Si-C bonds which can be pyrolyzed under an inert atmosphere to form "black glasses". The use of several precursors can even allow the variation of the Si-C bond content. The most common modified alkoxides contain terminal alkyl groups. Pantano et al.[4] studied the influence of the R' chain length starting from trifunctionnal $R'Si(OR)_3$ precursors and found that the maximum amount of C retained after the pyrolysis under inert atmosphere is one C per Si atoms. ^{29}Si MAS-NMR studies on systems derived from methyltriethoxysilane [4] or mixtures of tetraethoxysilane and dimethyldiethoxysilane [6] have shown that Si-C bonds are retained after pyrolysis under inert atmosphere at ≈900-1000°C. However, the comparison of the ^{29}Si MAS-NMR and chemical analysis results, show the formation of a large percentage of free carbon in the glass : the free carbon content can be estimated to ≈60-70% of the total carbon content, assuming that C atoms in the glass are bonded to four Si atoms. Ideally, the transformation of the silica-based network into a silicon oxycarbide network could occur via crosslinking steps at the C atoms as represented in Figure 1. Unfortunately, for silica network containing only terminal Si-CH_3 groups, a large number of Si-C bonds seems to be cleaved during the pyrolysis with formation of carboneous residues.

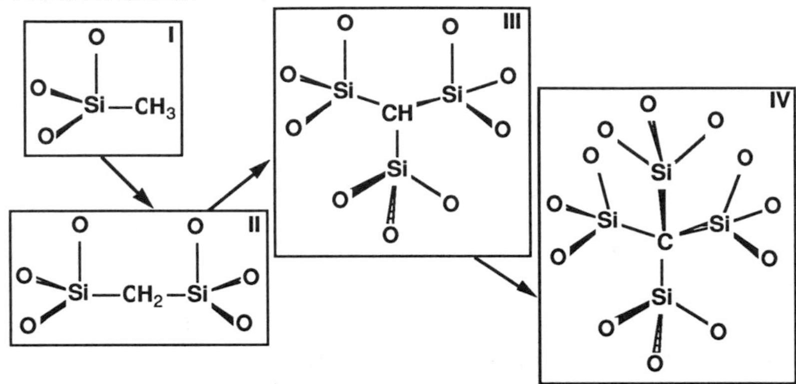

Figure 1: Ideal transformation of a silica gel modified with methyl groups into a silicon oxycarbide network

The specific objective of this work was to minimize the formation of free carbon during pyrolysis in order to get a pure silicon oxycarbide network. Precursors containing Si-H groups were used to favor at low temperature the following reaction:

$$\equiv Si\text{-}CH_3 + H\text{-}Si\equiv \text{ ----> } \equiv Si\text{-}CH_2\text{-}Si\equiv + H_2 \quad (1)$$

that will ensure a better incorporation of bridging C atoms in the silica-based network. The pyrolyzed samples were characterized by ^{29}Si MAS-NMR and the percentages of free carbon estimated. Influences of the structure of the starting gels on the composition of the final oxycarbide glasses will be discussed.

EXPERIMENTAL

Three systems were chosen with a constant C/Si ratio of 0.5 as summarized in Table 1 :

Gels	Precursors	Composition
T/T^H	T : methyltriethoxysilane $CH_3Si(OEt)_3$ T^H : triethoxysilane $HSi(OEt)_3$	$T/T^H/H_2O/EtOH$ 1/1/6/4
D^H/Q	D^H : diethoxymethylsilane $(CH_3)HSi(OEt)_2$ Q : tetraethoxysilane $Si(OEt)_4$	$D^H/Q/H_2O/EtOH$ 1/1/6/4
D^H/T^H	D^H : diethoxymethylsilane $(CH_3)HSi(OEt)_2$ T^H : triethoxysilane $HSi(OEt)_3$	$D^H/T^H/H_2O/EtOH$ 1/1/5/4

Table 1: Compositions of the gels, precursors for silicon oxycarbides

The precursors were mixed with ethanol and then co-hydrolyzed with a stoichiometric amount of water (pH≈6). After stirring, the solutions were poured into plastic open tubes.

Monolithic transparent gels are obtained within few days. The samples were pyrolyzed under flowing argon in a tubular furnace. TGA experiments were run with a Netszch apparatus under argon. ^{29}Si MAS-NMR spectra were recorded on a MSL 400 Bruker spectrometer with pulsewidth of 2.5 μs and relaxation delays of 60 s. Peaks will be labelled with the X, M_n, D_n, T_n and Q_n notation. X, M, D, T and Q respectively refer to $SiC_{4-x}O_x$ units with x=0, 1, 2, 3 or 4. n is the number of bridging O atoms surrounding Si. D^H and T^H will represent $SiCHO_2$ and $HSiO_3$ units.

RESULTS

The ^{29}Si MAS-NMR spectra of the three gels have been simulated and the results are summarized in Table 2:

Gel	Q	T^H	T	D^H	Average Functionnality	Degree of Condensation
	Chemical Shift in ppm (Percentage)					
T/T^H	-112.5 (12)	-86.0 (36)	-67.0 (30) -58.5 (22)		3.12	93%
D^H/Q	-109.5 (39) -100.5 (17)			-34.0 (44)	3.12	95%
D^H/T^H	-110.2 (4)	-85.3 (46)	-65.0 (2)	-34.8 (48)	2.56	100%

Table 2 : ^{29}Si MAS-NMR results of the various gels.

They reveal that Si-H bonds are retained in the gel to a large extent. Only a small fraction of Q units in the T/T^H gel, and Q and T units in the D^H/T^H gel are formed due to the hydrolysis of these groups. Simulation shows that the D^H/T^H system contains only fully condensed Si sites while the other gels present sites with terminal groups, such as T_2 (δ=-58.5 ppm) in T/T^H and Q_3 (δ=-100.5 ppm) in D^H/Q. A relation between the condensation degree of gels prepared from methyl-modified silicon alkoxides and the average functionality of the precursors has already been pointed out [7]. For an average functionality of ≈2.5, the condensation degree was found to be over 95% while for a functionality of 3, it decreases around 90%. Similar results are observed in the present case (Table 2).

TGA experiments show similar behaviors for the T^H/T and D^H/Q systems with a total weight loss around 7%, divided into three steps: from 200 to 400°C, then from 400 to 600°C and finally from 600 to 800°C. On the contrary, the D^H/T^H system exhibits only one main weight loss (16%) from 400 to 500°C.

For each system, the ^{29}Si MAS-NMR spectra of samples pyrolyzed at 400°C, 600°C and 1000°C for 1 hour have been recorded and simulated. The percentages of the various Si sites present in the samples are represented in Figure 2. For all the systems, the samples pyrolyzed at 400°C display a lower content of Si-H bonds and a higher content of Si-O bonds compared to the starting gels. This can be due to redistribution reactions between the various Si-X bonds (X=H, C or O) that are known to take place at these temperatures in similar systems [8]. These reactions usually occur with a loss of silanes and this could explain the relative increase of the Si-O bonds in the samples. Another mechanism can be the reaction of

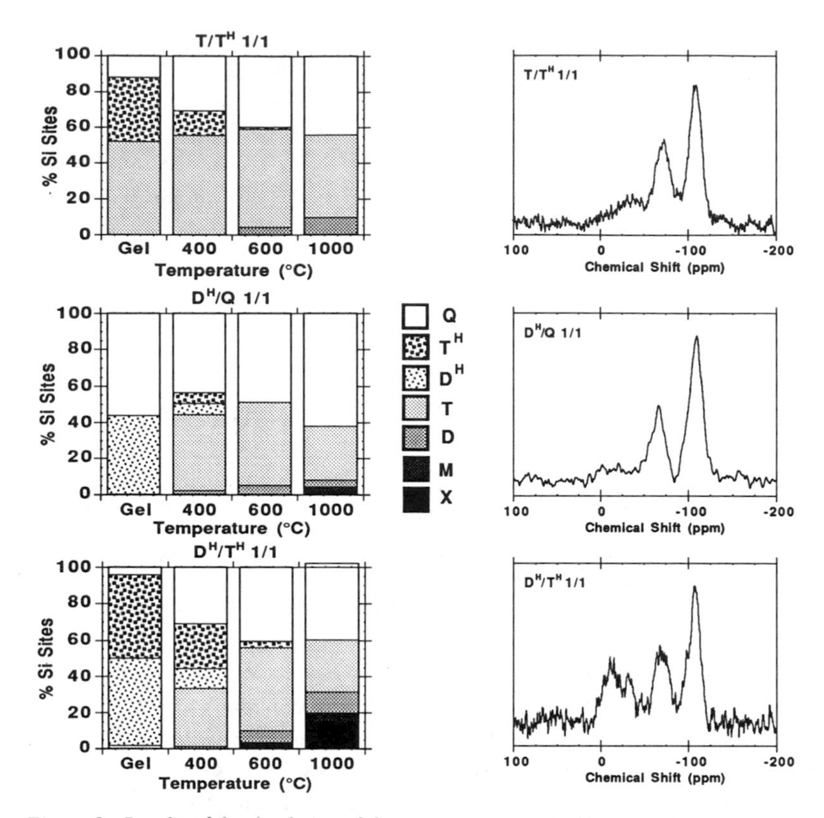

Figure 2 : Results of the simulation of the ^{29}Si MAS-NMR spectra of pyrolyzed samples.

Figure 3 : ^{29}Si MAS-NMR spectra of the oxicarbide glasses pyrolyzed at 1000 °C for 1 hour under argon.

Si-H bonds either with terminal Si-OH groups present in the gel or with water produced by condensation reactions between these terminal groups. The latter reactions are certainly responsible for the first weight loss only observed in the T/T^H and D^H/Q systems that are not fully condensed according to the previous NMR study. The second weight loss in the temperature range, 400-600°C, common to all systems could be related to redistribution reactions. Indeed, an experiment performed on the D^H/Q gel with a TGA equipment coupled with a mass spectrometer reveals a loss of H_2 and H_2O around 200°C and a loss of CH_3SiH_3 around 450°C. Despite the loss of Si-H bonds in these first stages of pyrolysis, a certain amount of Si sites still have such a bond at 400°C : 12% for D^H/Q, 14% for T/T^H and even 35% for D^H/T^H. The percentage of Si sites having a Si-C bond does not change compared to the starting gels and varies from 45 to 55% for the different systems. All the reactions occuring during this first stage lead to a decrease of the terminal groups, Si-OH, Si-H and eventually $Si-CH_3$ and thus to a more cross-linked network.

From 400°C to 600°C, the main feature is the almost complete consumption of the Si sites containing Si-H bonds (D^H and T^H) and the formation of Si sites containing Si-C bonds such as T, D or even M. This is particularly clear for the D^H/T^H system that contains at 400°C the highest amount of Si-H bonds and suggests that the reaction (1) mentioned in the introduction of this paper could be very efficient in this case. Indeed, TGA-MS experiments on the D^H/T^H gel have shown that the weigth loss at 400-500°C is associated with a noticeable evolution of H_2. This second stage corresponds to the starting point of condensation reactions occuring between the remaining terminal groups, Si-H and Si-CH_3.

From 600°C to 1000°C, the systems are completly converted into inorganic materials. The number of Q units does not vary too much in the different systems. On the contrary, mixed SiC_xO_{4-x} units containing a higher amount of Si-C bonds (D, M and X) are created suggesting that a replacement of the Si-O bonds by Si-C bonds within these units occured. The importance of these reactions depends on the systems. The spectra of the various oxycarbide glasses are presented in Figure 3. For the T/T^H system, only T (46%) and D (10%) units are present at 1000°C while for the D^H/T^H system T (29%), D (11%), M (5%) and even X (13%) units are clearly visible. The C/Si ratio in the oxycarbide phase revealed by ^{29}Si MAS-NMR has been extracted from the precentages of the various SiC_xO_{4-x} units assuming that the C atoms were bonded to 4 Si atoms. A comparison with the C/Si ratio found by chemical analysis allows to calculate the amount of free carbon phase (Table 3). Results for the T/T^H and D^H/Q systems are not different from previously studied systems that did not contain any Si-H bonds : the amount of free carbon is ≈70 mole% of the total carbon. On the contrary, the D^H/T^H system really shows a decrease of the free carbon content (≈36%). Because of this encouraging result, D^H/T^H gels with a lower amount of D^H units, and thus of Si-C bonds, were prepared : D^H/T^H 0.5/1 and D^H/T^H 0.1/1 and pyrolyzed at 1000°C. The C/Si ratio extracted from the ^{29}Si MAS NMR spectra and from chemical analysis are similar suggesting that the amount of free carbon phase is very small, if any.

Samples	C/Si ratio from NMR	C/Si ratio from Chemical Analysis	Free Carbon/Total Carbon (mole%)
T/T^H 1/1	0.16	0.54	70
D^H/Q 1/1	0.12	0.52	76
D^H/T^H 1/1	0.31	0.47	36
D^H/T^H 0.5/1	0.37	0.32	0
D^H/T^H 0.1/1	0.11	0.11	0

Table 3 : Comparison of the C/Si ratios obtained from ^{29}Si MAS-NMR and chemical analysis for the various oxycarbide phases.

DISCUSSION

Three sol-gel systems containing Si-CH_3 and Si-H bonds have been investigated as potential precursors for pure silicon oxycarbide glasses. This study shows clearly the influence of the composition of the starting gels on the composition of the final glasses. For the T/T^H and

D^H/Q systems, the amount of free carbon is quite high (≈70%) and the presence of Si-H bonds does not seem to increase the amount of C atoms bonded to Si atoms. A strong improvement comes from the D^H/T^H system. This difference in behavior seems related to the amount of Si-H bonds retained in the polymeric material heated at 400°C that can lead via reaction (1), to the formation at 600°C of a larger amount of bridging C atoms within the silica based network. Indeed, in the D^H/T^H system, the amount of Si sites with a Si-H or a Si-CH_3 bond are respectively ≈35% and ≈45%, while for the other two systems, these amounts are respectively around 12-14% and 50-55%.

At higher temperature, during the conversion of the polymeric material into an oxycarbide glass, this study points out an interesting feature : the formation of Si-C bonds occurs with almost no change in the Q unit content and thus seems related to the cleavage of Si-O bonds belonging to mixed SiC_xO_{4-x} ($x \neq 0$). This shows that the Si-O bonds in a Q unit are certainly more stable than in a mixed unit. As a consequence, a large amount of carbon present in the oxycarbide phase will favor the formation of units with a high Si-C bond content. This will certainly prevent an homogeneous distribution of the Si-C bonds within the network.

This study shows that with a suitable choice of precursors, an almost pure silicon oxycarbide phase can be obtained. The ^{29}Si MAS-NMR study performed on the pyrolyzed samples allowed a better understanding of the transformation of the gels into the oxycarbide phases and pointed out the important role of Si-H bonds during this process. A study is under progress to get a better description of the glass phase and to study its thermal stability.

ACKNOWLEDGEMENTS

Dr H. Mutin (University of Montpellier, France) is greatly acknowledged for the TGA-MS experiment of the D^H/Q gel. Dr D. Bahloul (University of Limoges, France) is greatly acknowledged for the TGA-MS experiment of the D^H/T^H gel. NATO and CNR are also acknowledged for financial support.

REFERENCES

1. R.E. Loehman, J. Non Cryst. Solids 56, 123 (1983).
2. J. Homeny, G.G. Nelson, S.H. Risbud, J. Am. Ceram. Soc. 71, 386 (1988).
3. F. Babonneau, K. Thorne, J.D. Mackenzie, Chem. Mater. 1, 554 (1989).
4. H. Zhang, C. Pantano, J. Am. Ceram. Soc. 73, 958 (1990).
5. G.M. Renlund, S. Pochazka, R.H. Doremus, J. Mater. Res. 6, 2716 (1991).
6. F. Babonneau, L. Bois, J. Livage, J. Non Cryst. Solids (in press).
7. R.H. Glaser, G.L. Wilkes, C.E. Bronnimann, J. Non Cryst. Solids 113, 73 (1985).
8. V. Belot, R. Corriu, D. Leclercq, P.H. Mutin, A. Vioux, Chem. Mater. 3, 127 (1991).

THE ROLE OF Si-H FUNCTIONALITY IN OXYCARBIDE GLASS SYNTHESIS

ANANT K. SINGH AND CARLO G. PANTANO
Department of Materials Science and Engineering, The Pennsylvania State University, University Park, Pennsylvania 16802

ABSTRACT

Silicon oxycarbide gels and glasses were synthesized using various ratios of methyldimethoxysilane and TEOS. These gels and glasses were compared with those made from methyltrimethoxysilane and TEOS. The effect of the Si-H functionality in the methyldimethoxysilane was of primary interest. Hydrolysis and condensation processes were monitored using ^{1}H and ^{29}Si-NMR spectroscopy. The structures and the oxycarbide fractions of the glasses, obtained after heating the gels to 900°C in flowing argon, were investigated with ^{29}Si-MAS NMR. The total carbon and silicon contents of the glasses were determined using chemical analysis. The glasses covered the range of carbon concentration from ~1 to 15%, while the silicon concentrations were constant at about 40%. The concentration of the oxycarbide species was enhanced in the glasses synthesized with the Si-H functionality in the precursor.

INTRODUCTION

The synthesis of silica glasses with partial substitution of oxygen atoms by carbon or nitrogen atoms has attracted a lot of attention as a way to improve the mechanical and thermal properties. Silicon oxycarbide glasses, i.e., those in which carbon atoms enter the silicate network, have been the focus of quite a few studies recently [1-7]. Conventional glass melting is not a feasible synthesis route for these glasses because, among other reasons, the decomposition reactions occurring at the high temperatures involved in melting make carbon-retention irreproducible. Therefore, these glasses have been synthesized by the sol-gel process. Alkyl-substituted alkoxysilanes are used as starting materials so that Si-C bonds can be introduced into the precursors themselves. The idea is to maintain these bonds throughout the processing, especially through the pyrolysis. It has been shown that the Si-C bonds persist in the glasses made after heat treatment at 900°C [1, 2, 3]. One of the key issues is the ability to maximize the retention of the Si-C bonds introduced in the precursor, which would therefore minimize the free carbon phase in the glasses.

In this study, methyldimethoxysilane (henceforth referred to as MDMS) was mixed in various ratios with TEOS and formed into gels and glasses. Gels and glasses were also formed with mixtures of methyltrimethoxysilane (henceforth referred to as MTMS) and TEOS. MDMS differs from MTMS only in that it has an Si-H functionality substituted for one of the methoxy groups. The Si-H bonds are expected to enhance the reactivity of the gels - especially during the pyrolysis where condensation of Si-H and Si-CH_3 could enhance cross-linking of the network and the retention of carbon. ^{29}Si MAS NMR spectra were used to compare the proportions of the Si-C bonds in the glasses made from the two precursors.

EXPERIMENTAL

TEOS and either MDMS or MTMS, in the molar ratios shown in Figure 2, were mixed in ethanol for 6 hours. The ratio of the total silane to ethanol was 1:2. Water acidified with 1M-HCl was then added to the mixtures. The amount of water was adjusted so as to have one mole for every alkoxy group of the silanes. The pH of the mixture was about 2 in all instances. The solutions were left to gel in covered petri-dishes. The gels were dried at 80°C for 7 days and then pyrolyzed in flowing argon with a 2°C/min ramp to 900°C and held there for 30 minutes. Carbon was determined by LECO CHN 600, while lithium metaborate fusion in Pt crucibles followed by Plasma Emission Spectrometry was used to determine the silicon contents. Liquid-state ^{1}H NMR spectra were recorded for MDMS hydrolyzed with deuterated water. Solid-state ^{29}Si MAS NMR was used to characterize the silicon bonding environment in the gels and the glasses.

RESULTS AND DISCUSSION

Since the Si-H functionality in the MDMS was the essential difference between the two precursors, the first question addressed was whether the Si-H bond survived the hydrolysis process. Figure 1(a) shows the liquid-state ^{1}H NMR spectra of MDMS, before and after hydrolysis with deuterated water.

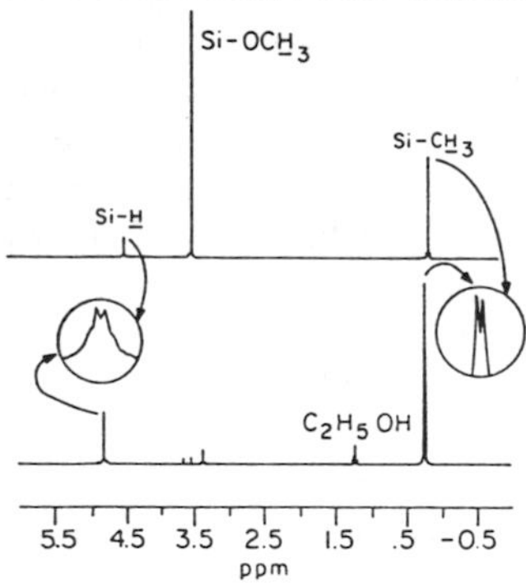

Fig. 1(a)
^1H NMR spectra of unhydrolyzed MDMS (upper) and hydrolyzed MDMS (lower).

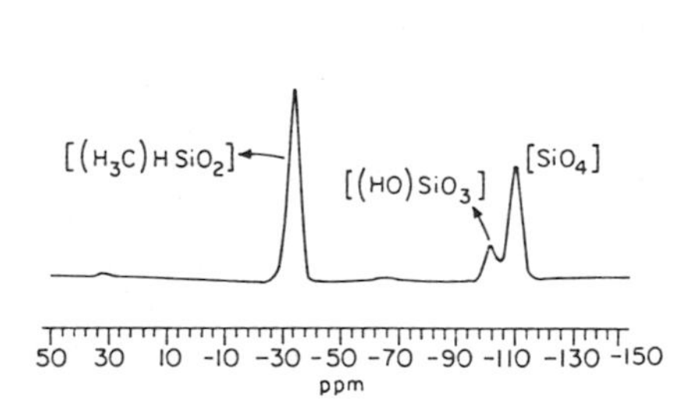

Fig. 1(b)
^{29}Si CP MAS NMR spectrum of dried gel made from 50% MDMS concentration.

The extent of hydrolysis can be estimated from the drastic reduction in the Si-$OC\underline{H}_3$ peak. In contrast, the Si-H̲ and the Si-$C\underline{H}_3$ peaks are seen to persist in the spectrum of hydrolyzed MDMS. Due to coupling between the 3 protons in the Si-$C\underline{H}_3$ and the one in the Si-H̲, there is fine structure (shown in insets) associated with these two peaks. The three protons of the methyl group split the Si-H̲ peak into a quartet, whereas the one proton in the Si-H̲ splits the Si-$C\underline{H}_3$ peak into a doublet. This verifies the peak assignments. The two peaks are slightly shifted to higher chemical shift values because the hydrolysis of the methoxy groups has resulted in a slight deshielding of the protons. Thus, Si-H bonds are seen to survive the hydrolysis under these conditions.

Figure 1(b) shows the ^{29}Si CP MAS NMR spectrum of the dried gel made with 50% MDMS. The three peaks at –110 ppm, –101 ppm and –35 ppm have been assigned to [SiO_4], [(HO)SiO_3] and [(H_3C)HSiO_2] silicon-configurations respectively [7]. The [(H_3C)HSiO_2] peak confirmed the presence of Si-H bonds in the MDMS-gels. Thus, the unique chemistry of the MDMS precursor was maintained in the gel state. The two sets of gels - MDMS and MTMS, were subsequently pyrolyzed in argon at 900°C to create glasses.

Figure 2 is a plot of the total carbon and silicon concentrations in the glasses, as determined by chemical analyses. As the ratios of MDMS:TEOS and MTMS:TEOS were increased in the starting solution, there was a steady rise in the carbon contents of both sets of glasses, but the carbon contents were approximately equal for equal starting concentrations of MDMS and MTMS. Thus, the total carbon concentration was unaffected by the choice of the Si-C containing silane. The silicon contents of the glasses remained constant over the entire range.

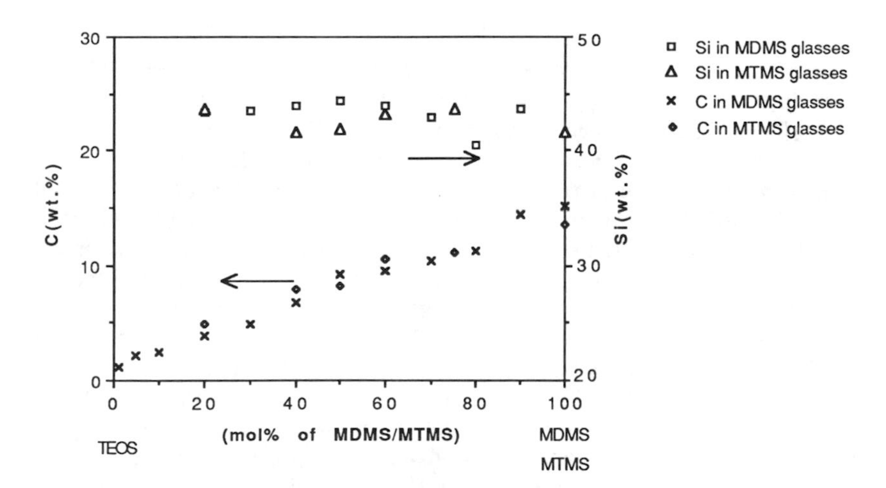

Fig. 2. *Plot of the total weight percentages of carbon and silicon in MDMS- and MTMS-glasses.*

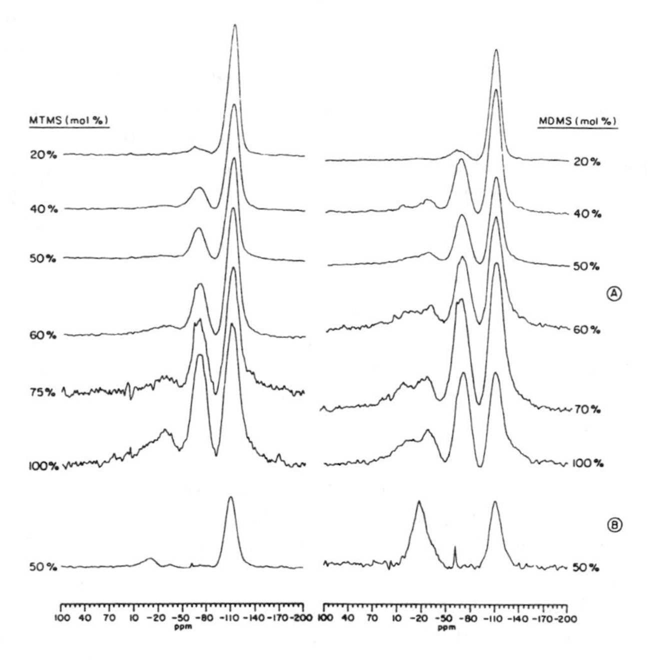

Fig. 3. ***(A)*** *^{29}Si MAS NMR spectra of glasses made at 900°C*
(B) *^{29}Si MAS NMR spectra of glasses made at 1400°C for 50% MDMS/MTMS concentrations.*

Figure 3 shows the ^{29}Si MAS NMR spectra of the two sets of glasses. All the spectra show peaks at −110 ppm and around −67 ppm, which have been assigned to [SiO_4] and [$CSiO_3$] species respectively [7]. The denominations of the structural species here indicate only the first-neighbor bonding environment of the silicon atoms. A third peak, assigned to [C_2SiO_2] species, appears at around −27 ppm for MDMS concentrations of 30% (not shown in the figure) or higher, and for MTMS concentrations higher than 50%. A broad fourth peak,convoluted with the third, can be seen for MDMS-glasses made with MDMS concentrations higher than 50% and for the MTMS-glasses made with 100% MTMS concentration. It can be assigned to [C_4Si] species [8] but the presence of [C_3SiO] species, with a peak between 6.1 and 9.8 ppm [7], can not be ruled out.

The spectra show that silicon species with 1, 2, 3 and 4 carbon bonds are present in the glasses. A careful examination of the spectra reveals that the formation of these 1-, 2-, 3- and 4-coordinated Si-C species is enhanced in the case of MDMS-glasses. This is depicted in Figure 4 which is a plot of the ratios of the areas of the oxycarbide to the oxide peaks for MDMS- and MTMS-glasses. Thus, these ratios are a relative measure of the oxycarbide contents of the glasses. The MDMS- glasses have consistently higher ratios and hence higher oxycarbide fractions than the corresponding MTMS-glasses. In fact, the fourth peak, corresponding to the formation of $[C_3SiO]$ and $[C_4Si]$ species, can be observed for MDMS-glasses made with as low as 40% MDMS concentration, while it appears only for those MTMS-glasses made with 100% MTMS.

One reaction which could be responsible for this increase in the formation of higher-order $[C_xSi]$ species in MDMS-glasses is the following:

$$H-\overset{|}{\underset{|}{Si}}-CH_3 + H-\overset{|}{\underset{|}{Si}}-CH_3 \rightarrow H-\overset{|}{\underset{|}{Si}}-CH_2-\overset{|}{\underset{|}{Si}}-CH_3 + H_2$$

The remaining Si-H and Si-CH_3 bonds in the product could lead to further cross-linking through similar reactions. Such a reaction has been reported to be responsible for the increase in the molecular weight of polycarbosilane after its initial formation [9]. In the MTMS system, this polymerization can occur only through a direct reaction of the methyl groups which, of course, could be occurring in the MDMS system as well.

The verification of this reaction mechanism will require further study, but it seems clear that the reactive Si-H bonds enhance cross-linking and therefore, lead to higher fractions of 1-, 2-, 3- and 4- coordinated Si-C species. The impact of this more polymerized structure is evident in Fig. 3(B) which shows the ^{29}Si MAS NMR spectra of the MDMS- and MTMS-glasses made with 50% starting concentrations, and pyrolyzed at 1400°C. In both cases, the $[CSiO_3]$ species has been consumed and higher-order $[C_xSi]$ species have been formed, but the proportion of these higher-order $[C_xSi]$ species is much higher in the MDMS-glass. The peak centered at −17 ppm for the MDMS-glass is quite broad; it envelops the crystalline SiC peaks (ß-SiC at −18 ppm, and α-SiC at −14, −20 & −25 ppm), and the $[C_2Si_2O_2]$ peak [10]. The crystallization of the SiC in these

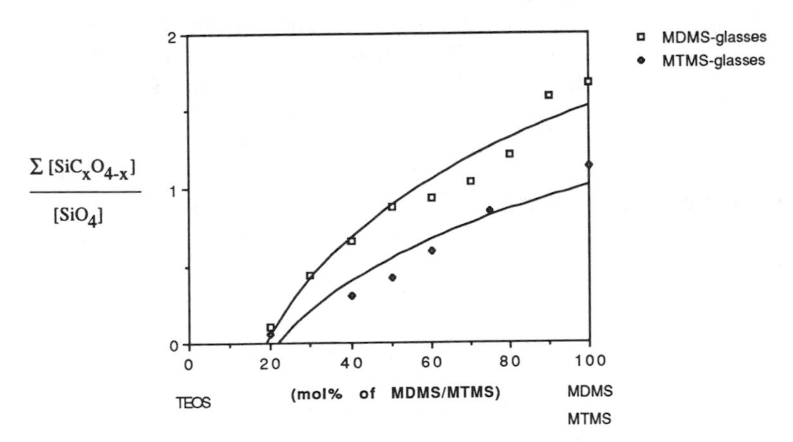

Fig. 4. Comparison of the oxycarbide fractions in MDMS- and MTMS- glasses made at 900°C, where x = 1, 2, 3 & 4.

higher temperature glasses has been confirmed by x-ray analysis. In view of this much higher oxycarbide fraction in the MDMS-glass and the fact that the weight percentages of C and Si are approximately equal for the 900°C- and the 1400°C-glasses, it is suggested that MDMS-glasses contain less free carbon phase than the corresponding MTMS-glasses.

CONCLUSIONS

Methyldimethoxysilane (MDMS) and methyltrimethoxysilane (MTMS) were mixed in various ratios with TEOS to form oxycarbide glasses at 900°C in an argon atmosphere. The Si-H functionality in the MDMS precursor influenced the molecular structure of the glass. The glasses retained approximately equal amounts of total carbon, but MDMS-glasses contained higher oxycarbide fractions and less free carbon phase, than the corresponding MTMS-glasses.

ACKNOWLEDGEMENTS

We gratefully acknowledge the financial support of ONR under contract N00014-90-J-1392. We also thank Alan Benesi for his help with NMR. Carlo G. Pantano also acknowledges Giando Soraru for his contribution to the initiation of this study.

REFERENCES

1. G. M. Renlund , "Silicon Oxycarbide Glasses", PhD thesis, Rensselaer Polytechnic Institute, Dec. 1989.
2. H. Zhang and C. G. Pantano, J. Amer. Ceram. Soc., 73 (4), 958 (1990).
3. F. Babonneau, K. Thorne and J. D. Mackenzie, Chem. Mater., 1, 554 (1989).
4. J. Homeny, G. G. Nelson and S. H. Risbud, J. Amer. Ceram. Soc., 71 (5), 386 (1988).
5. F. K. Chi, Ceram. Eng. Sci. Proc., 4, 704 (1983).
6. K. C. Chen, K. J. Thorne, A. Chemseddine, F. Babonneau and J. D. Mackenzie in Better Ceramics Through Chemistry III, edited by C. J. Brinker, D. E. Clark and D. R. Ulrich (Mater. Res. Soc. Proc. 121, Pittsburgh, PA 1988) pp. 571–574.
7. V. Belot, R. J. P. Corriu, D. Leclercq, P. H. Mutin and A. Vioux, to be published
8. K. Thorne, E. Liimatta and J. D. Mackenzie, J. Mater. Res., 6 (10), 2199 (1991).
9. S. Yajima, Y. Hasegawa, J Hayashi and M. Iimura, J. Mat. Sci. 13, 2569 (1983).
10. J. Lipowitz, H. A. Freeman, R. T. Chen and E. R. Prack, Advanced Ceramic Materials, 2 (2), 121 (1987).

CHARACTERIZATION OF HIGH SURFACE AREA SILICON OXYNITRIDES

PETER W. LEDNOR, RENE DE RUITER and KEES A. EMEIS

Koninklijke/Shell-Laboratorium, Amsterdam (Shell Research B.V.), Badhuisweg 3, 1031 CM Amsterdam, The Netherlands

ABSTRACT

High surface area silicon oxynitrides have been prepared by nitridation of silica with ammonia. Characterization by Fourier-transform infrared spectroscopy has allowed quantitative determination of hydroxyl, amido and imido groups. Data obtained by X-ray photoelectron spectroscopy show that the nitrogen is well distributed in the surface of the materials.

INTRODUCTION

In heterogenous catalysis, liquid or gaseous feedstocks are converted over a solid catalyst into more desirable products. Such processes form an essential part of the oil and petrochemical industries. The solid catalyst usually consists of an inorganic phase, with or without metal particles on the surface. Examples include platinum particles on gamma alumina (a reforming catalyst used in oil processing), chromium particles on silica (an ethylene polymerization catalyst) and zeolites or amorphous silica-aluminas (used as solid acids).

Oxides have been widely investigated in catalysis, and silica, alumina, and aluminosilicates find application commercially on a large scale. On the other hand, non-oxide materials such as nitrides, carbides and borides have been relatively little investigated. The main reason for this has been the lack of routes to the high surface area forms usually required in catalysis. However, this situation has changed significantly in recent years, due to the interest in high surface area non-oxides as precursors to fully dense ceramics; we have reviewed synthetic routes to high surface area non-oxides [1].

With the aim of extending the range of inorganic materials used in catalysis we have investigated the preparation and properties of high surface area silicon oxynitride. Of various preparative routes explored, the reaction of Aerosil silica with flowing gaseous ammonia at 1100°C proved convenient, giving an amorphous material of about 150 $m^2 . g^{-1}$, with a surface and bulk composition corresponding to Si_2N_2O [2]. The surface nitrogen content showed good stability under both oxidizing and reducing conditions [3], and the material can be used as a solid, basic catalyst [4]. The present paper describes further characterization of high surface area silicon oxynitride.

EXPERIMENTAL

Nitridations were carried out by passing ammonia (30 $l.h^{-1}$) over Aerosil silica (Degussa, 300 $m^2 . g^{-1}$, typically 2 g) at temperatures of 500°C (4 h), 800°C (4 h) or 1100°C (24 h).

IR spectra were recorded on a Bio-rad Qualimatic spectrophotometer. Samples of the above silicon oxynitride powders, which had been stored in screw-top bottles, were prepared as self-supporting discs, by pressing at 1-2 tons cm^{-2}. The sample discs were placed in a closed IR cell which allowed exposure of the sample to heat, vacuum or various gas atmospheres. After evacuation (10^{-5} torr, 500°C, 2 h) to remove physisorbed water, spectra were recorded at ambient temperature.

Hydrolysis of surface amido groups and measurement of the resulting ammonia concentration in solution was carried out as follows. A silicon oxynitride sample (50-100 mg) was suspended in 100 ml 10^{-4} M HCl. After stirring for at least 1 h, a 20 ml portion of this solution was taken, and to this was added 200 μl of a 10 M NaOH solution. An ammonia gas electrode (Methrom) was suspended in this solution, and the ammonia concentration measured. Solution samples with or without the solid material (removed by centrifuging) gave the same result. The gas electrode was calibrated regularly using standard solutions of ammonium chloride.

X-ray photoelectron spectroscopy (XPS) was carried out with a Kratos XSAM-800 spectrometer using Al K_α incident radiation of 1484 eV.

MOLECULAR GROUPS: QUALITATIVE ANALYSIS BY FOURIER-TRANSFORM INFRARED (FT-IR) SPECTROSCOPY

Identification of the groups present was facilitated by carrying out the reaction of silica with ammonia at different temperatures. Fig. 1 shows the IR spectra of Aerosil silica nitrided at 500°C, 800°C and 1100°C; the surface nitrogen content, determined by XPS, in these three samples was 0.8, 1.8 and 27 %wt respectively.

The principal bands shown are assigned to ≡Si-OH* (3743-5 cm^{-1}), ≡Si-NH_2* (3522-24, 3447-48 and 1550-1 cm^{-1}) and ≡Si)$_2$NH* (3367-89 cm^{-1}). These assignments are supported by the following values: (i) 3743 cm^{-1} for the hydroxyl group in Aerosil silica (this work); (ii) 3500-10, 3430-40, and 1550 cm^{-1} for ≡$SiNH_2$ groups formed in the zeolite ZSM-5 on reaction with ammonia [5], and 3526, 3446 and 1555 cm^{-1} for ≡$SiNH_2$ groups on silica aerogel [6]; (iii) 3372 cm^{-1} for the N-H vibration in hexamethyldisilazane, $[(CH_3)_3Si]_2NH$, (this work), and 3340-3380 cm^{-1} for the NH band in silicon oxynitride thin films, with the peak maximum in this range depending on the nitrogen content [7]. It is clear from Fig. 1 that, in the absence of the spectra for the samples nitrided at 500°C or 800°C, the presence of ≡$SiNH_2$ in the sample nitrided at 1100°C can easily be missed. Indeed, an IR analysis of silicon nitride [8] did not find the two high wavenumber bands for ≡$SiNH_2$, and attributed a band at 1550 cm^{-1} to the NH bending vibration of ≡Si)$_2$NH; this latter assignment must be incorrect in view of the above data and the assignment of the ≡Si)$_2$NH bending vibration to a band at 1150 [9], 1175 [10] or 1180 [11] cm^{-1}.

MOLECULAR GROUPS: QUANTITATIVE ANALYSIS BY FT-IR SPECTROSCOPY

The concentrations of the groups identified in the previous section can be determined from the IR spectra if extinction coefficients are known. These coefficients were determined as follows.

---

* This notation indicates molecular units in the silicon oxynitride solid.

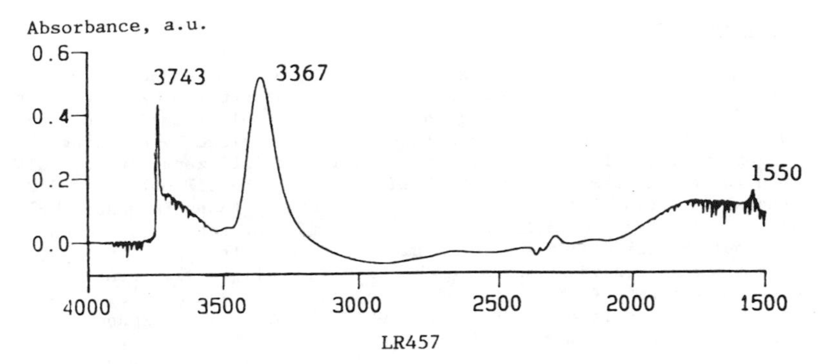

Aerosil 300 silica NH_3 treated at 1100°C for 24 h

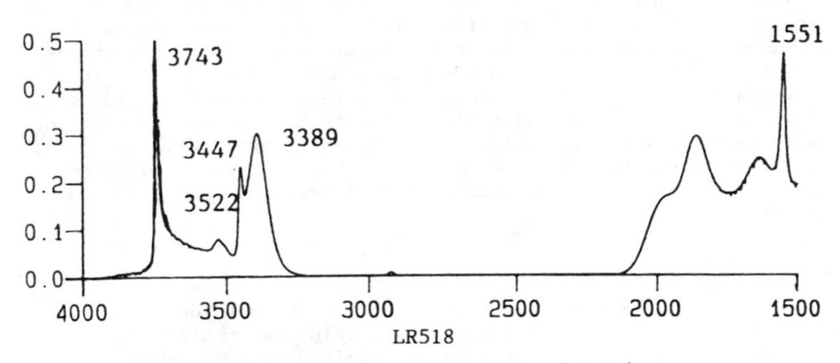

Aerosil 300 silica NH_3 treated at 800°C for 4 h

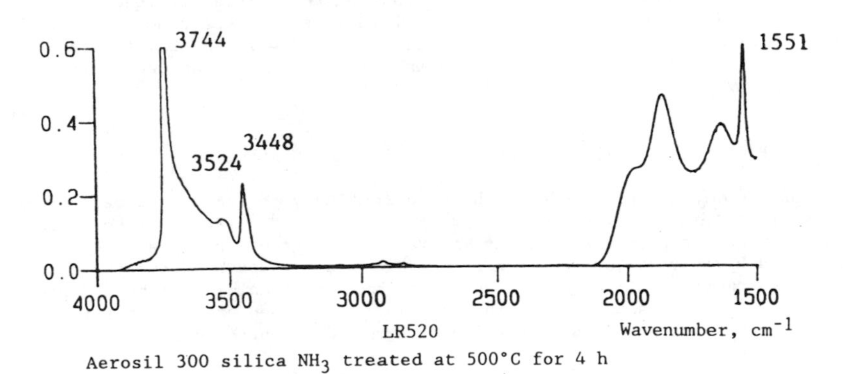

Aerosil 300 silica NH_3 treated at 500°C for 4 h

Fig. 1. The FT-IR spectra of Aerosil silica nitrided with ammonia under various conditions

Hydroxyl groups, Si-OH

During the pretreatment (see experimental), hydrogen-bonded silanol groups are removed, and all samples showed a sharp absorption band at about 3743 cm^{-1} (see Fig. 1) corresponding to isolated silanol groups. An extinction coefficient for these isolated silanols was determined by allowing them to react, in the infrared cell, with hexamethyldisilazane vapour: this was found to lead to complete trimethylsilylation of the silanols at 250°C. Using the known extinction coefficient for the methyl groups in hexamethyldisilazane, the integrated molar extinction coefficient for the isolated silanols could be calculated as 3.8 cm μmol^{-1}. Using this value, the concentration of isolated silanols in the untreated silica was found to be 0.8 OH groups per nm^2. This is in good agreement with a value of ca. 1 OH per nm^2 given for a silica dehydrated at 400°C [6]. The values found for the nitrided silicas are shown in Table I.

Amido groups, $Si-NH_2$

In situ reaction of a silica wafer with gaseous ammonia (100 mbar) at 500°C led to partial (ca. 50%) substitution of silanol groups by $Si-NH_2$ groups. This reaction is selective, in the sense that Si-O-Si groups are not attacked by ammonia, which is evident from difference spectra. The silanol groups could be completely restored by hydrolysing the amido groups with water vapor (ca. 40 mbar) at 500°C. With these data, and knowing the extinction coefficient for the silanol groups, an integrated molar extinction coefficient for the amido 1550 cm^{-1} band can be calculated: 2.0 cm μmol^{-1}. This value allows calculation of the amido concentrations shown in Table I.

Imido groups, $Si)_2NH$

In work on silicon oxynitride thin films, in which imido groups were the only proton-containing species present, a value for the imido group extinction coefficient (2.9 cm μmol) was determined by independent analysis of the proton concentration, and this value was used to calculate the concentrations shown in Table I. In the case of the sample nitrided at 500°C, a shoulder at 3426 cm^{-1} is discernible in the IR spectrum; it has been assumed that this represents a small amount of $\equiv Si)_2NH$, shifted to a higher wavenumber by the low N/O ratio in this sample (as has been found in silicon oxynitride thin films [7]).

MOLECULAR GROUPS: QUANTITATIVE ANALYSIS USING A GAS ELECTRODE

An ammonia gas electrode can be used routinely to measure concentrations of ammonia in solution in the range 10^{-2} - 10^{-6} M. It occurred to us that it might be possible to cleave surface amido groups using acid-catalysed hydrolysis. This will replace the amido groups by hydroxy groups, releasing ammonia into solution, which, in excess of acid, will be trapped as ammonium ion. Measurement of this species could then provide confirmation of the amido group concentration determined by infrared spectroscopy.

For a sample nitrided at 500°C, a gas electrode determination (see experimental section above) was carried out on a wafer which had been used for infrared analysis. The gas electrode measurement showed an amido concentration of 0.25 $mmol.g^{-1}$, in good agreement with a value of 0.32 $mmol.g^{-1}$ determined by IR.

In the case of samples which had been nitrided at 1100°C, a study using both a gas electrode and IR analysis showed that amido groups were generated by contact with dilute acid, presumably by hydrolysis of imido and nitrido groups.

Table I

Hydroxyl, amido and imido group concentrations in nitrided silicas[a]

Nitridation temp. °C	≡Si-OH		≡Si-NH_2		≡Si-NH-Si≡	
	$mmol.g^{-1}$	$OH.nm^{-2}$	$mmol.g^{-1}$	$NH_2.nm^{-2}$	$mmol.g^{-1}$	$NH.nm^{-2}$
500	0.46	0.9	0.23	0.5	0.02	0.1
800	0.24	0.5	0.29	0.6	0.9	2.0
1100	0.23	0.8	0.15	0.5	4.1	13.0[b]

[a] Spectra of samples shown in Fig. 1. Surface areas after nitridation measured as 175 $m^2.g^{-1}$ (1100°C), 286 $m^2.g^{-1}$ (800°C); assumed to be 300 $m^2.g^{-1}$ (as in starting material) after nitridation at 500°C.
[b] This value implies that at least some of the imido groups are sub-surface; see text.

MOLECULAR GROUPS: DISCUSSION

Table I summarizes the data on the concentrations of the molecular groups determined by IR. The values are believed to be accurate to +/- 15%. The accuracy is limited by sample preparation (homogeneity of the IR wafer) and measurement of peak area. The following conclusions can be drawn:

- the hydroxyl concentration is roughly constant in all three samples,
- the amido concentration is also fairly constant, i.e. it does not increase with the nitridation temperature,
- the imido concentration increases very significantly with nitridation temperature, and
- the imido concentration in the sample nitrided at 1100°C is equivalent to 13 groups per square nanometer; this is too high to be realistic, and the value thus means that at least some of the imido groups are sub-surface.

It should also be noted that the composition Si_2N_2O implies a nitrogen concentration of 20 gram-atom N per gram of material. This is much higher than the concentration of the nitrogen-containing groups reported in Table I, and shows that most of the nitrogen is present as nitride. Analysis of the IR spectrum in the region of the Si-O and Si-N bending vibrations is in progress [12].

SURFACE NITROGEN DISTRIBUTION

For silicon oxynitride films prepared by the reaction of ammonia with porous silicon dioxide films (ca. 100 nm thick), it has been reported [13] that the silicon 2p binding energy, determined by XPS, decreases linearly with increasing nitrogen content; this has been rationalized on the basis of the increasing charge on silicon which occurs on substituting oxygen by the less electronegative nitrogen [13]. We have found the same correlation for a variety of nitride powders prepared by nitriding Aerosil silica with ammonia. Early XPS work on silicon oxynitride thin films, prepared by chemical vapor deposition, led to the conclusion that Si-O-Si and Si-N-Si units were mixed on a molecular level, and the SiN_xO_y film was not a mixture of silicon dioxide and silicon nitride [14]. This conclusion is also valid for our samples, since the Si 2p peak is symmetrical, and of constant width, over a wide range of nitrogen contents: the peak width at half height was 3.13 eV for a sample with 8.8 %wt surface N, and 3.19 eV for a sample with 28 %wt surface N.

CONCLUSIONS

Characterization by FT-IR spectroscopy of high surface area silicon oxynitrides, prepared by nitriding silica with ammonia, has allowed determination of hydroxyl, amido and imido group concentrations. The amido group concentration was confirmed by an independent measurement using a gas electrode. Only the concentration of the imido groups changes significantly when the nitridation temperature is raised, increasing up to a value which implies that at least some of these groups are sub-surface in a material prepared at 1100°C. X-ray photoelectron spectroscopy has shown that the nitrogen is well distributed in the surface of the materials.

REFERENCES

1. R.W. Chorley and P.W. Lednor, Adv. Mater. 3, 475 (1991).
2. P.W. Lednor and R. de Ruiter, in Better Ceramics Through Chemistry III, edited by C.J. Brinker, D.E. Clark, and D.R. Ulrich (Mater. Res. Soc. Proc. 121, Pittsburgh, PA 1988) pp. 497-502; J. Chem. Soc., Chem. Commun., 320 (1989).
3. P.W. Lednor and R. de Ruiter in Inorganic and Metal-Containing Polymeric Materials, edited by J.E. Sheats, C.E. Carraher, C.U. Pittman, M. Zeldin and B. Currell (Plenum, New York, 1990), p. 187.
4. P.W. Lednor and R. de Ruiter, J. Chem. Soc., Chem. Commun., 1625 (1991).
5. P. Fink and J. Datka, J. Chem. Soc. Faraday Trans. I, 85, 3079 (1989).
6. J.B. Peri, J. Phys. Chem. 70, 2937 (1966).
7. C.M.M. Denisse, J.F.M. Janssen, F.H.P.M. Habraken and W.F. van der Weg, Appl. Phys. Lett. 52, 1308 (1988).
8. G. Busca, V. Lorenzelli, G. Porcile, M.I. Baraton, P. Quintard, and R. Marchand, Mater. Chem. Phys. 14, 123 (1986).
9. K. Murakami, T. Takeuchi, K. Ishikawa and T. Yamamota, Appl. Surf. Sci. 33/34, 742 (1988).
10. D.V. Tsu, G. Lucovsky and M.J. Mantini, Phys. Rev. B, 33, 7069 (1986).
11. S. Narikawa, Y. Kojima and S. Ehara, Jpn. J. Appl. Phys. 24, L861 (1985).
12. M.I. Baraton, P.W. Lednor and P. Quintard, in preparation.
13. R.K. Brow and C.G. Pantano, J. Am. Ceram. Soc. 69, 314 (1986).
14. S.I. Raider, R. Flitsch, J.A. Aboaf and W.A. Pliskin, J. Electrochem. Soc. 123, 560 (1976).

ISOCYANATE-MODIFIED POLYSILAZANES: CONVERSION TO CERAMIC MATERIALS

JOANNE M. SCHWARK and MARK J. SULLIVAN
Hercules Incorporated, Research Center, Wilmington, DE 19894-0001

ABSTRACT

A tailorable viscosity, liquid polysilazane has been developed which thermosets to a solid ceramic precursor for both silicon nitride (Si_3N_4) and silicon carbide (SiC). Conversion of the thermoset, crosslinked polysilazane to a ceramic material has been examined. Preferential formation of silicon carbide or silicon nitride is controlled by the pyrolysis conditions employed.

INTRODUCTION

Many polysilazane precursors to silicon nitride or mixed silicon nitride/silicon carbide ceramic materials have been developed [1]. Both polymer structure [2] and pyrolysis conditions [3] are important in determining the ultimate ceramic product generated. While most polysilazanes provide Si_3N_4 or a mixed Si_3N_4/SiC ceramic upon pyrolysis, a thermosetting, isocyanate-modified polysilazane has been developed which affords either Si_3N_4 or SiC [4].

The precursor polymer may be prepared with a broad range of viscosities by the reaction of cyclic silazanes with mono- and multi-functional isocyanates [5]. The liquid isocyanate-modified polysilazanes which contain vinyl substituents are readily thermoset by heating with a free radical source such as dicumyl peroxide. The thermoset solid is ideal for binder, coating or infiltration applications because it has no detectable glass transition temperature (T_g) and does not deform upon subsequent heating. Both Si_3N_4 and SiC have been produced from the thermoset polysilazane using appropriate pyrolysis conditions.

EXPERIMENTAL PROCEDURES

Preparation of Isocyanate-Modified Polysilazane: A nitrogen-sparged, two liter, three-necked, round-bottomed flask equipped with an overhead stirrer, reflux condenser and a septum was charged with 1451.85 g of a poly(methylvinyl)silazane, $[(CH_3SiHNH)_{0.8}(CH_3SiCH{=}CH_2NH)_{0.2}]_x$, prepared by the coammonolysis of methyldichlorosilane and methylvinyldichlorosilane (4:1 mol ratio). Phenylisocyanate (Aldrich, 6.62 ml, 7.26 g) was added by syringe. The reaction mixture was heated to 70°C for one hour and then cooled to room temperature. The clear, liquid polymer had a viscosity of 48 cp (Brookfield cone-and-plate viscometer).

Thermosetting of Isocyanate-Modified Polysilazane: A sample of the isocyanate-modified polysilazane was transferred to a glass jar containing 0.1 wt% of dicumyl peroxide. The jar was capped with a septum, sparged with nitrogen, and placed in a preheated oil bath (160°C). When the polymer/peroxide mixture reached ≈130°C, an exotherm occurred and the liquid cured in ≈30 sec. A solid exhibiting no T_g was obtained in 100% yield.

Pyrolysis of Thermoset Polysilazane: The solid polysilazane was ground with a MC-17B Miracle Mill. All pyrolyses were conducted in a CM Model 1700-HTF tube furnace equipped with a mullite tube and alumina dee insert with Coors alumina boats. Gas flows were set at 1.0 L/min. Pyrolyses under Ar were ramped at 10°C/min from 25°C to the maximum T with an 8 hour hold at the maximum T. Pyrolyses under NH_3 were conducted from 25-600°C at 10°C/min, and from 600-1000°C at 5°C/min. No heat soak was used after reaching the maximum T. Subsequent pyrolysis from 1000-1600°C under Ar was performed at 10°C/min.

Analysis of Pyrolysis Products: X-ray powder diffraction (XRD) patterns were recorded on a Philips APD3600/02 diffractometer with a Cu-Kα source. Solid state ^{29}Si NMR (SSNMR) spectra were obtained on a Bruker MSL-200 operating at 39.7 MHz (4 kHz spin rate). Spectra were obtained using a single pulse, gated ^1H decoupling sequence with a 50 μsec spin-locking pulse to eliminate baseline artifacts. The spectra contained 16-40 transients that were collected using a 75° flip pulse and a 3600 sec pulse delay.

RESULTS AND DISCUSSION

Pyrolysis to Silicon Nitride

Pyrolysis of both carbon-free [1] and carbon-substituted polysilazanes [3] in a reactive ammonia atmosphere has been successfully used to generate Si_3N_4 with little or no excess carbon. When the thermoset, isocyanate-modified polysilazane was pyrolyzed in ammonia to 1000°C, followed by pyrolysis under Ar to 1600°C, a mixture of α- and ß-Si_3N_4 was produced. The transformation to silicon nitride is shown in the ^{29}Si SSNMR spectra of the ceramic products at 600-1600°C (Figure 1).

The unfired polysilazane had a peak at -3.3 ppm assigned to a (-CH_3Si(-$CHCH_2$-)NH-) unit containing an ethylene group. This was generated during the thermoset cure by the free radical polymerization of vinyl substituents. The peak at -21.7 ppm corresponds to (-CH_3SiHNH-) units. After pyrolysis to 600°C under NH_3, the -3.3 ppm peak was gone and new peaks appeared at -23.4 and -42.2 ppm indicating substantial carbon loss. By 800°C, only one broad peak centered at -46.2 ppm remained, indicative of a product with mainly Si-N bonding. The peak narrowed upon

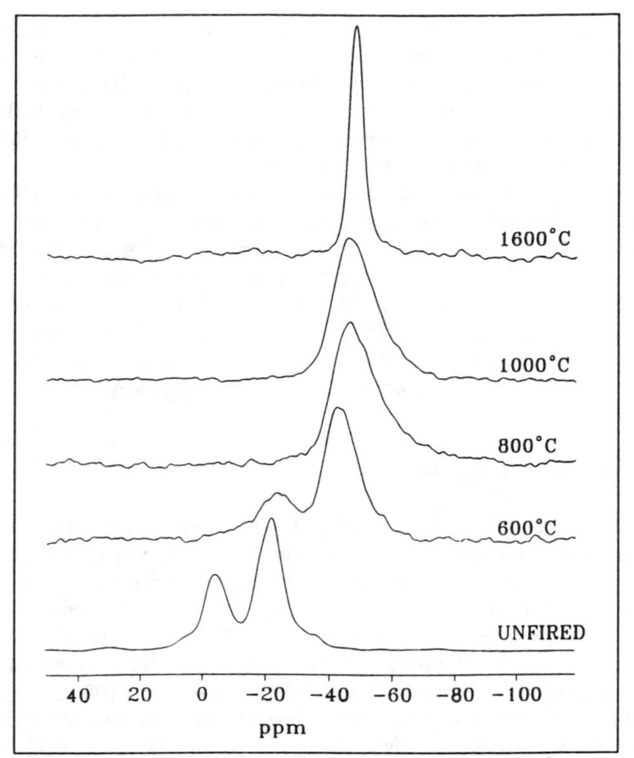

Figure 1 29Si SSNMR Spectra of Polysilazane Pyrolyzed Under NH_3 to 1000°C, and Ar to 1600°C

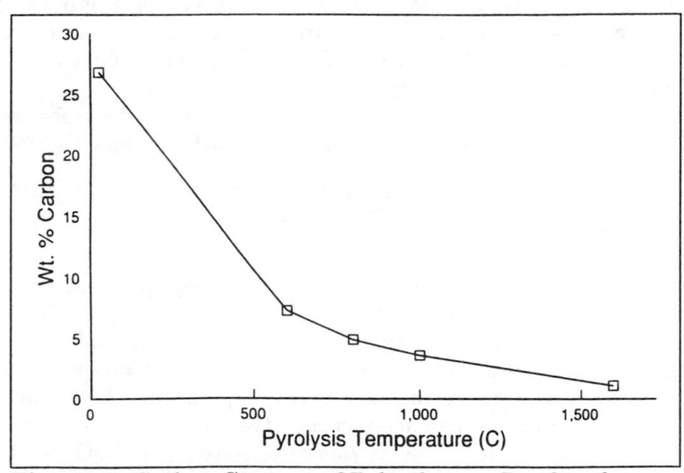

Figure 2 Carbon Content of Polysilazane Pyrolyzed Under NH_3 to 1000°C, and Ar to 1600°C

heating to 1000°C and, after further pyrolysis to 1600°C under Ar, the ceramic product had a sharp peak (-47.7 ppm) in the expected shift range [6] for α- and ß-Si_3N_4--no SiC was observed. The XRD pattern confirmed the formation of a mixture of crystalline α- and ß-Si_3N_4 at 1600°C. Only amorphous material was observed in the XRD patterns for material pyrolyzed at temperatures below 1600°C.

Elemental analysis of the pyrolysis product generated at each temperature showed that as the pyrolysis temperature increased, the carbon content of the resultant ceramic char decreased (Figure 2). The unfired polysilazane had 26.8% C. After pyrolysis to 600°C under ammonia, the carbon content of the char had decreased to 7.3%. Less than 4% carbon remained after heating to 1000°C in ammonia. Carbon removal continued during higher temperature pyrolysis under Ar. After pyrolysis to 1600°C, the silicon nitride contained only 1.1% residual carbon.

Pyrolysis to Silicon Carbide

While the production of Si_3N_4 from the precursor was anticipated, more surprising was the transformation of the thermoset polysilazane to *silicon carbide*. Few polysilazanes have been described as effective SiC precursors [7]. When our thermoset polysilazane was pyrolyzed under Ar to 1600°C, ß-SiC with ≈8% Si was obtained. The evolution of crystalline SiC from the polysilazane is shown in the ^{29}Si SSNMR spectra of Figure 3. When the polysilazane was fired to 1000, 1200, and 1400°C under Ar, the ceramic materials produced had very broad NMR peaks covering the shift range for both SiC [8] and Si_3N_4 [6]. After pyrolysis at 1600°C for 8 hours, the SSNMR spectrum for the ceramic product showed a major peak at -18.0 ppm (ß-SiC [8]) and a minor peak at -81.3 ppm (Si° [6c])--no Si_3N_4 was observed. The XRD patterns for the pyrolysis products (Figure 4) confirm the formation of elemental Si (2θ=28.4 and 47.3°) in the ceramic product at 1600°C. Elemental analysis of the 1600°C pyrolysis product showed only 1.0% N content.

Conversion of the polysilazane to SiC occurs because Si_3N_4 is thermodynamically unstable in the presence of carbon above 1440°C (Eq. 1) [9].

$$Si_3N_4 + 3\,C \xrightarrow{>1440°C} 3\,SiC + 2\,N_2 \qquad (1)$$

A crosslinked Si-C network is present in the thermoset polysilazane, formed during the free radical polymerization of the Si-vinyl groups. This network provides substantial nonvolatile carbon upon decomposition of the precursor in a nonammonia atmosphere. At 1400°C, the pyrolysis product still shows both Si-C and Si-N bonding in the SSNMR spectrum. Upon heating to 1600°C, the Si-N bonds are replaced with Si-C species as N_2 is expelled from the material. Because the pyrolysis was conducted in an Al_2O_3 boat, Si was produced during the high temperature reaction. Pyrolysis in a carbon crucible, which provides a carbonaceous pyrolysis atmosphere, affords SiC.

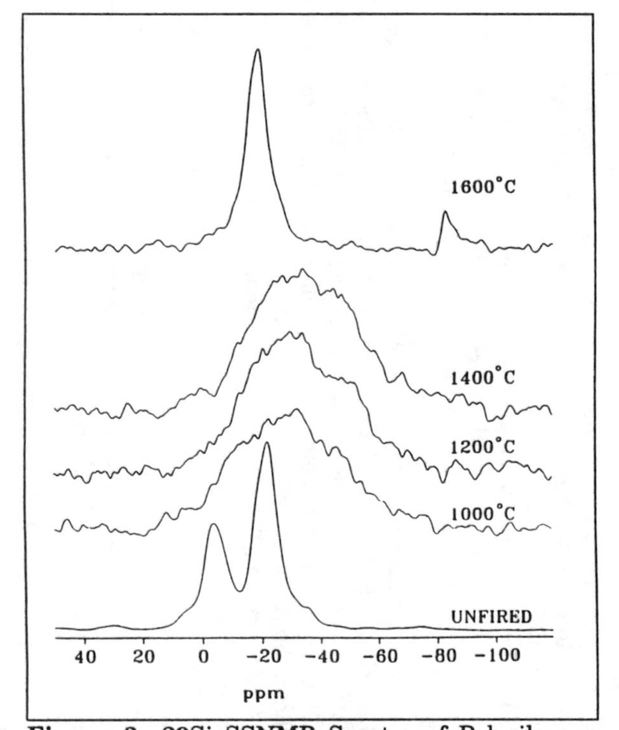

Figure 3 29Si SSNMR Spectra of Polysilazane Pyrolyzed Under Ar to 1600°C

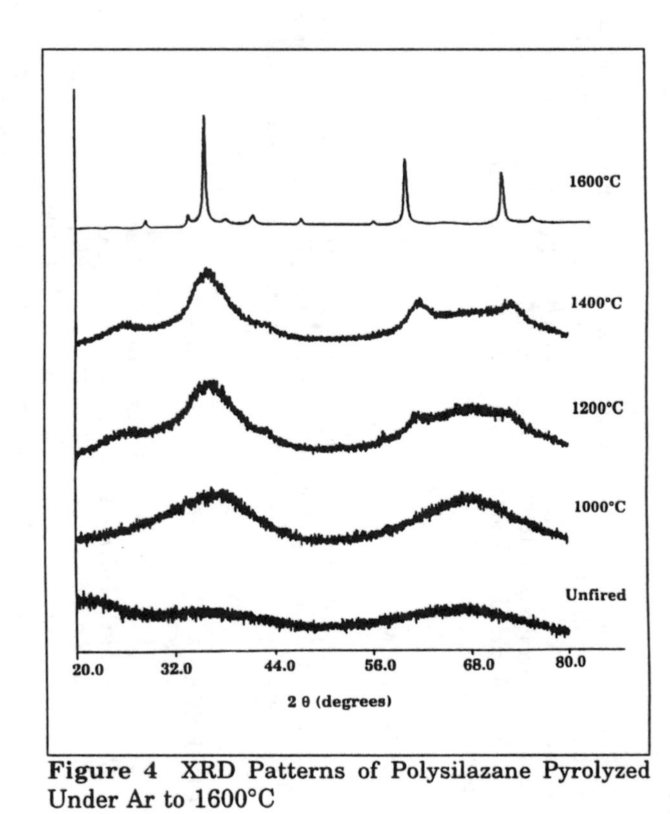

Figure 4 XRD Patterns of Polysilazane Pyrolyzed Under Ar to 1600°C

SUMMARY

The conversion of a thermoset polysilazane to either Si_3N_4 or SiC has been demonstrated. Pyrolysis under ammonia to 1000°C, followed by pyrolysis under Ar to 1600°C, produced α- and ß-Si_3N_4. Silicon carbide was generated by pyrolysis to 1600°C under Ar. The crosslinked nature of the thermoset polymer is responsible for the ability to form SiC at high temperatures.

ACKNOWLEDGMENTS

We acknowledge D.A. Moroney for preparing the polysilazanes, J.A. Mahoney and M.J. Centuolo for the pyrolysis work, R.H. Kridler for the X-ray diffraction data, and R.L.K. Matsumoto for invaluable discussions.

REFERENCES

1. (a) R.M. Laine, Y.D. Blum, D. Tse, and R. Glaser, in *Inorganic and Organometallic Polymers*, M. Zeldin, K.J. Wynne, and H.R. Allcock, Eds., ACS Symposium Series **360**, 124-42 (1988); (b) K.J. Wynne and R.W. Rice, Ann. Rev. Mater. Sci., **14**, 297-334 (1984).
2. (a) Y.D. Blum, G.A. McDermott, R.B. Wilson, and A.S. Hirschon, Polymer Preprints, **32**(3), 548-49 (1991); (b) Y.D. Blum, K.B. Schwartz, and R.M. Laine, J. Mater. Sci., **24**, 1707-18 (1989).
3. (a) G.T. Burns and G. Chandra, J. Am. Ceram. Soc., **72**(2), 333-37 (1989); (b) F.K. van Dijen and J. Pluijmakers, J. Eur. Ceram. Soc., **5**, 385-90 (1989).
4. J.M. Schwark, Polymer Preprints, **32**(3), 567-68 (1991).
5. J.M. Schwark, U.S. Patent No. 4 929 704 (May 29, 1990); U.S. Patent No. 5 001 090 (March 19, 1991); U.S. Patent No. 5 021 533 (June 4, 1991).
6. (a) K.R. Carduner, C.S. Blackwell, W.B. Hammond, F. Reidinger, and G.R. Hatfield, J. Am. Chem. Soc., **112**, 4676-79 (1990); (b) G.R. Hatfield and K.R. Carduner, J. Mater. Sci., **24**, 4209-19 (1989); (c) K.R. Carduner, R.O. Carter III, M.E. Milberg, and G.M. Crosbie, Anal. Chem., **59**, 2794-97 (1987).
7. A.A. Morrone, W. Toreki, and C.D. Batich, Materials Letters, **11**(1,2), 19-25 (1991).
8. (a) D.C. Apperley, R.K. Harris, G.L. Marshall, and D.P. Thompson, J. Am. Ceram. Soc., **74**(4) 777-82 (1991); (b) K.R. Carduner, S.S. Shinozaki, M.J. Rokosz, C.R. Peters, and T.J. Whalen, J. Am. Ceram. Soc.,**73**(8), 2281-86 (1990); (c) J.S. Hartman, M.R. Richardson, B.L. Sherriff, and B.G. Winsborrow, J. Am. Chem. Soc., **109**, 6059-67 (1987).
9. K.G. Nickel, M.J. Hoffmann, P. Greil, and G. Petzow, Adv. Ceram. Mater., **3**(6), 557 (1988).

REACTION SINTERING MECHANISM OF SUBMICROMETER SILICON POWDER IN NITROGEN

YOSHIYUKI YASUTOMI and MASAHISA SOBUE
Hitachi Research Laboratory, Hitachi Ltd., Ibaraki 319-12 JAPAN
JIRO KONDO
Advanced Materials & Technology Research Lab., Nippon Steel Corp., Kawasaki 211 JAPAN

ABSTRACT

Knowing the nitridation mechanism of Si is important to obtain near-net-shape high strength reaction bonded ceramics. In this report, comparison of nitridation mechanism of small spherical Si powder produced by plasma arc method and large faceted Si powder, is discussed on the basis of microstructural analyses by SEM and TEM. From the analyses, nitrogen diffuses through the oxide film of Si powder and forms Si_3N_4 under the oxide films during the initial stage of sintering. At higher temperature, the oxide film transform into fine Si_3N_4 grains when heated to the 1350°C. The Si_3N_4 has morphology of hollow shell and a size, which corresponds to that of the original Si particles.

1. INTRODUCTION

Reaction bonding of pure Si powder into Si_3N_4 ceramics is well known as a favored route to obtain near-net-shape ceramics [1]. The reaction between Si and nitrogen has been extensively studied and reviewed [2] - [5]. Figure 1 shows a schematic of the reaction bonding model of Si_3N_4 for typical a ceramic composite. The green body is a mixture of metal Si powder and ceramic powder. During the heating process, Si metal reacts with nitrogen and transforms into Si_3N_4. The dimensional change of this composite is measurably smaller than monolithic reaction bonded Si_3N_4 ceramics. With this process, we can control the strength, electroconductivity, and resistivity by properly selecting the ceramic particles and developed ceramic composites with wide ranging properties with excellent near-net-shape control [6] - [12]. But these new Si_3N_4 bonded ceramic composites have a typical porosity of 12vol% and maximum pore size of 30μm which is detrimental to strength.

Prof. Haggerty's group successfully made small pore distribution reaction bonded Si_3N_4 ceramics also by using fine Si powder (average size: 2μm) [13] - [17] produced by lazer method. The surface of the Si powder was covered with hydrogen (Si-H) as opposed to (Si-O) in our powder. Their work showed the reaction bonded Si_3N_4 can be formed rapidly and Si_3N_4 forms on Si powders by the nucleation and growth of discrete crystallites on the Si particle surfaces [17].

Figure 2 shows the pore distributions of Si_3N_4 bonded SiC ceramic composites in our study. Using small spherical Si powder (plasma arc method; Nippon Steel), the pore size distribution peaks ranged from 0.15μm to 5μm. In contrast, faceted large Si gives a pore size distribution which ranged from 0.15μm to 30μm. As suported previously by Prof. Haggerty, small spherical Si powder has the effect of making pore size smaller.

Aside from effect on poresize, we also investigated the nitridation mechanism of faceted large Si powder and small spherical Si powder's mechanism in nitrogen between 1100°C to 1350°C for Si_3N_4 bonded SiC ceramic composite. The comparison is discussed on the basis of the microstructural analyses from SEM and TEM.

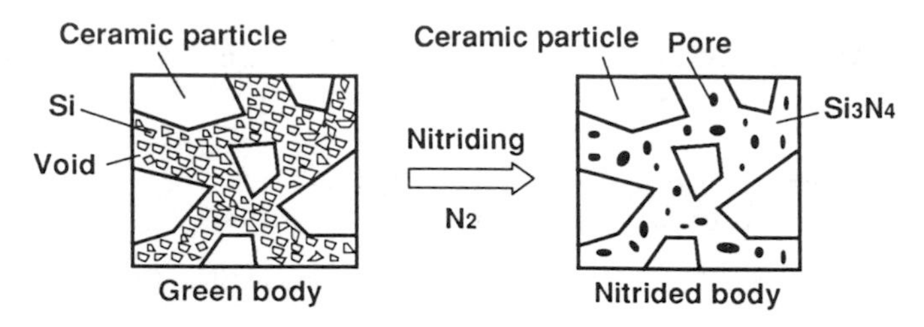

Fig.1 Bonding model of Si_3N_4 bonded ceramics

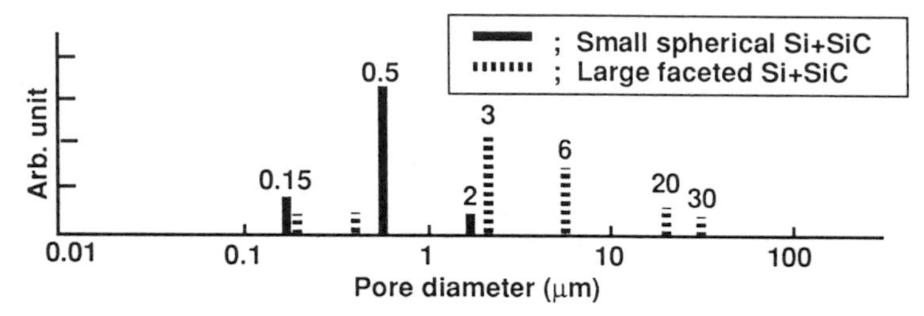

Fig.2 Comparison of pore distributions of the nitrided bodies

2. EXPERIMENTAL PROCEDURE

Figure 3 shows the micrographs of starting powders. Large faceted Si powder and small spherical Si powder had an average size of about 3μm and 0.5μm respectively, and SiC powder, 16μm. The weight percents of Si and SiC powders used were 80 and 20. Surface of the Si powders were covered with oxide (Si-O) as opposed to the hydrogen (Si-H) covered surface on the laser powder used by Prof. Haggerty. Powders are mixed with distilled water, citric acid and sodium metaphosphate for 24 hours in a pot mill together with Si_3N_4 balls. The green body was made in a die by slip casting into a 40 x 5x 20mm bars. The samples are dried, dewaxed and heated stepwise from 1100°C for 20h, 1200°C for 20h, 1250°C for 10h, 1300°C for 10h and to the highest temperature of 1350°C for 10h. All heating was done in nitrogen atmosphere.

Fig.3 Micrographs of starting powders

3. RESULT

3.1 Nitridation mechanism of large faceted Si / SiC

Figure 4 shows the microstructure of the fracture surfaces of a partially nitrided sample of faceted Si and SiC at 1100°C for 5 hours. Whisker-like reaction products grew on the surface of the SiC particle. From TEM and X-ray analysis, these reaction products were amorphous Si-O or amorphous Si-N-O. These resulted possibly from the reactions of the oxide film on the particle surfaces with nitrogen in the atmosphere and vaporized silicon.

Figure 5 shows the nitrided microstructure of faceted Si and SiC at the highest temperature of 1350°C in the vicinity of the Si_3N_4 and SiC particle interface. The whisker-like reaction products changed into very fine Si_3N_4 crystals, which connected SiC particles. The Si_3N_4 crystals are very small, about 200 nanometer. It appeared that Si_3N_4 is generated on the surfaces of the SiC particles by the process of chemical vapor deposition because the particle size of the Si_3N_4 is much smaller than that of starting faceted silicon powder. Further in figure 6, ß-Si_3N_4 whiskers appeared to have grown via a vapor-liquid-solid reaction route. This ß reaction which requires exothermic melting of the Si particles, is beleived to be root cause of the formation of the large pores observed in the samples. If the ß reaction is suppressed, the formation of large pores should be suppressed. Detailed analyses of nitridation mechanism of this composite as described in the following section [9].

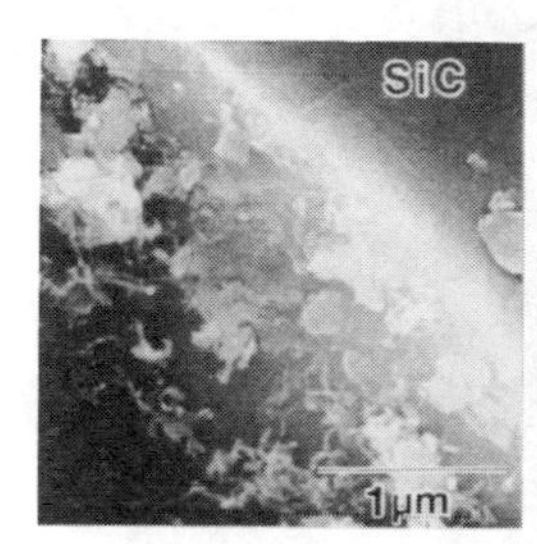

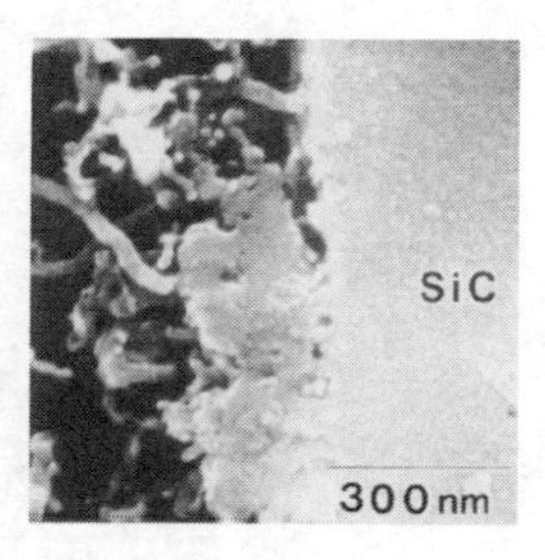

Fig.4 SEM of the fracture surfaces of a partially nitrided sample of faceted Si and SiC at 1100°C for 5hours

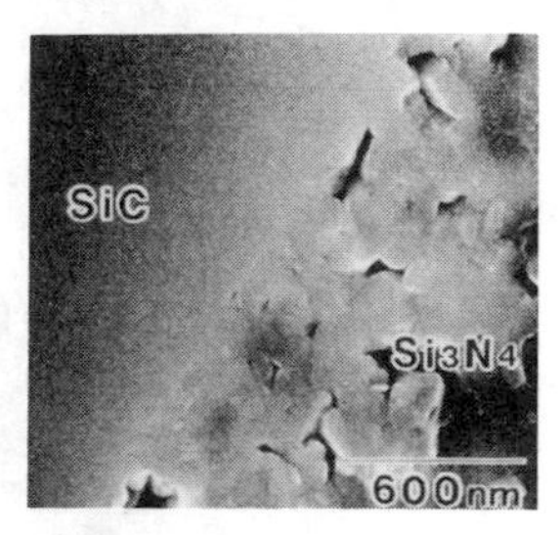

Fig.5 SEM of the ntrided sample of faceted Si and SiC at highest 1350°C

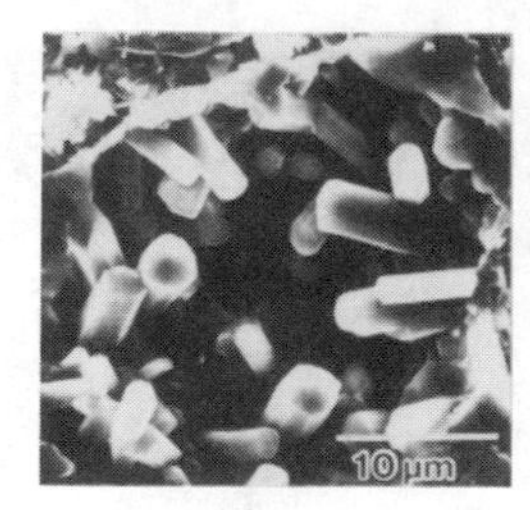

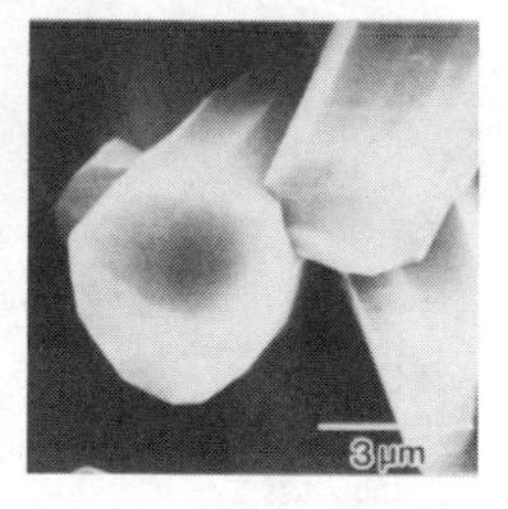

Fig. 6 ß-Si_3N_4 whisker in the nitrided body of faceted Si and SiC

3.2 Nitridation mechanism of small spherical Si / SiC

With smaller particle size, the powder is more reactive and therefore will transform into nitride at temperature significantly lower than the melting point of Si. Figure 7 shows the microstructure of the partially nitrided sample of fine spherical Si and SiC at 1100°C for 20 hours showed that whisker-like reaction products were not observed on the Si particles and SiC particles. Instead, the nitride was formed via a gas phase reaction via diffusion of nitrogen in the Si particle across the SiO_2 layer.

Figure 8 shows the nitrided body of small spherical Si and SiC at the highest temperature of 1350°C. The Si_3N_4 grains resembled a hollow spherical shell. In a few areas, Si_2N_2O grains appeared. The morphology of the Si_3N_4 corresponds to that of the starting Si particles which further suggests no melting had occurred. From this fact, we believe that within the Si, compressive stress was produced and temperature was increased by covering Si_3N_4, consequently Si diffused outside and reacted with nitrogen. Figure 9 shows nitridation mechanism of small spherical Si powder. At the initial stage nitrogen diffused the inside and produced Si_3N_4. Next step, the Si_3N_4 layer cracked by thermal stress. Si diffused outside and reacted with nitrogen. Finally, a Si_3N_4 hollow shell, which corresponds to that of the original Si particles.

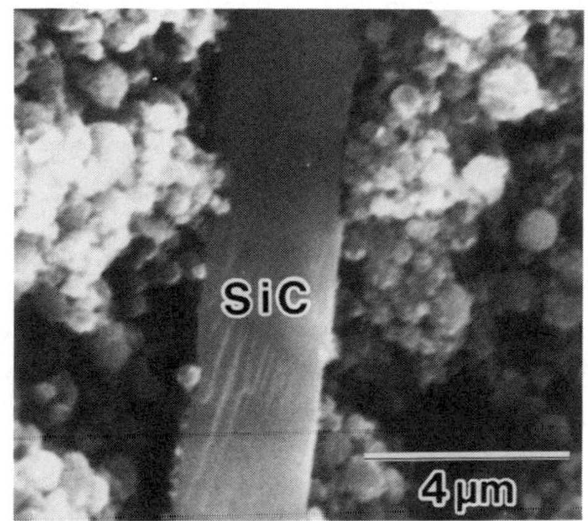

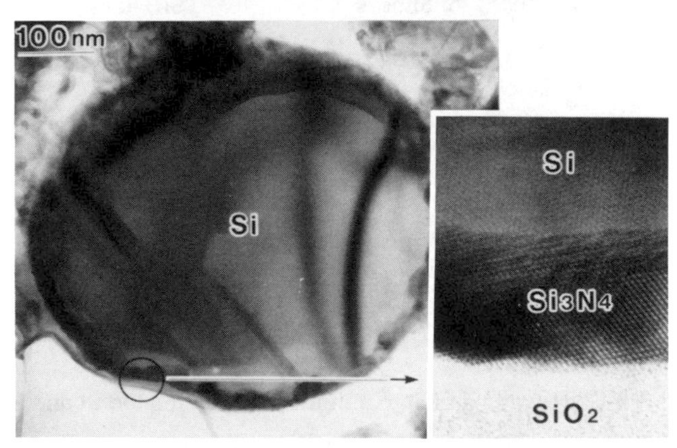

Fig.7 SEM of the fracture surface of a partially nitrided sample of spherical Si and SiC at 1100°C for 20 hours

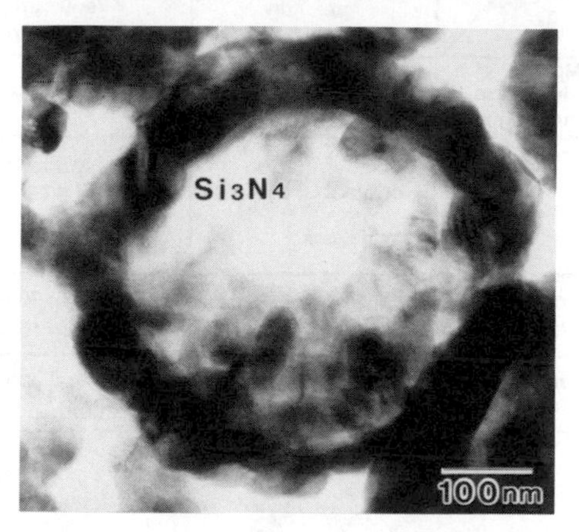

Fig.8 SEM of the ntrided sample of spherical Si and SiC at highest 1350°C

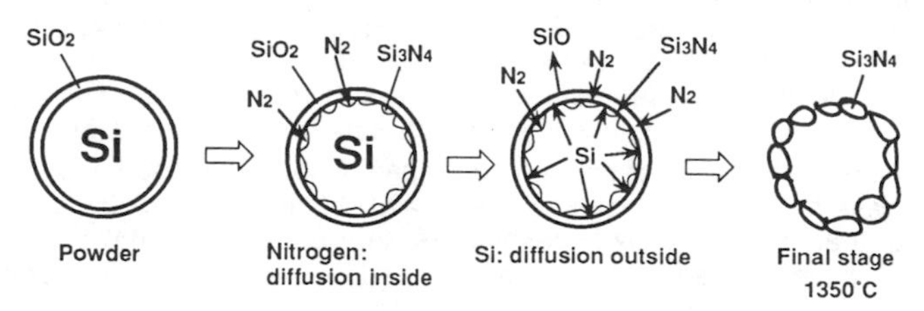

Fig.9 Nitridation mechanisim of small spherical Si powder

Table 1 summarizes the results. Small spherical Si powder produced bodies with smaller pore size. Average pore size of nitrided body is 0.5μm and maximum is 5μm which is better than bodies from faceted Si powder. Also, far faceted Si powder, the main nitridation mechanisms are chemical vapor deposition and liquid phase reaction. For that reason, Si_3N_4 grains were smaller than original Si particles, whiskers were produced in the composite, and crystal structure of composite have large quantity of ß-Si_3N_4. On the other hand, in case of small spherical Si powder, the main nitridation mechanism is nitrogen diffusion in the Si particles. The transformed Si_3N_4 has the morphology of a hollow shell with a size which corresponded to the original Si powder. The crystal structure of composite have almost 90% α-Si_3N_4.

Table 1 Comparison of nitridation mechanism of Si powders

Si powders / Item	Faceted Ave. 3μm (0.1μm - 10μm)	Spherical Ave. 0.5μm (0.1μm - 2μm)
Main nitridation mechanism	CVD and VLS	Nitrogen diffusion
Si_3N_4 grains	1. Smaller than original Si particles 2. Whiskers	1. Correspondence of oliginal Si particles
Crystal structure	α-Si_3N_4 : 60% β-Si_3N_4 : 40%	α-Si_3N_4 : 90% β-Si_3N_4 : 10%
Pore size of nitrided body	Average; 5μm Maximum; 30μm	Average; 0.5μm Maximum; 2μm

CONCLUSIONS

Comparison of the nitridation mechanism of small spherical Si powder and large faceted Si powder, multi-day firing schedule between 1100°C to 1350°C in nitrogen led to the following conclusions,

(1) Nitridation mechanism depends on the Si particle size.

(2) In case of small spherical Si powder, main nitridation mechanism is nitrogen diffusion in Si particles. At the initial stage nitrogen diffused inside and produced Si_3N_4. Next step, Si diffused outward and reacted with nitrogen. Finally, the Si_3N_4 has the morphology of a hollow shell, which corresponds to that of the original Si particles. Crystal structure of this composite had almost 90% α-Si_3N_4.

(3) In case of large faceted Si powder, main nitridation mechanisms are CVD and VLS. For that reason, Si_3N_4 grains were smaller than original Si particles and whiskers were produced in the composite, and crystal structure of composite had large quantity of ß-Si_3N_4.

(4) Small spherical Si powder had the effect making of pore size smaller.

REFERENCES

[1[R.Pompe, L.Hermansson and R.Carlsson, Sprechsaal, 115, 1098-101 (1982)
[2] H.M.Jenings, J. Mater. Sci., 18, 951-67 (1983)
[3] A.J.Moulson, Ibid., 14, 1017-51 (1979)
[4] A.Atkinson, P.J.Leatt, A.J.Moulson and E.W.Roberts, Ibid., 9, 981-84 (1974)
[5] J.W.Evans and S.K.Chatterji, J. Phys. Chem., 62, 1064-67 (1958)
[6] Y.Yasutomi, H.Kita, K.Nakamura and M.Sobue, J. Ceram. Soc. Japan, 96, 783-8 (1988)
[7] Y.Yasutomi, K.Nakamura, M.Sobue and Y.Kubo, Ibid., 97, 148-54 (1989)
[8] Y.Yasutomi, M.Sobue and Y.Kubo, Ibid., 97, 721-27 (1989)
[9] Y.Yasutomi, M.Sobue, S.Shinozaki and J.Hangas, Ibid., 99, 429-38 (1991)
[10] Y.Yasutomi, A.Chiba and M.Sobue, Ibid., 99, 407-10 (1991)
[11] Y.Yasutomi, A.Chiba and M.Sobue, J. Am. Ceram. Soc., 74, 950-57 (1991)
[12] Y.Yasutomi, M.Sobue, S.Shinozaki and J.Hangas, J. Ceram. Soc. Japan, 99, 692-98 (1991)
[13] S.C.Danforth and J.S.Haggerty, J. Am. Ceram. Soc., 64, C-58 (1983)
[14] Y-M.Chiang, J.S.Haggerty, R.P.Mesner and C.Demetry, Am. Ceram. Soc. Bull., 68, 420-28 (1989)
[15] J.S.Haggerty, A.Lightfoot, J.E.Ritter, P.A.Gennari and S.V.Nair, J. Am. Ceram. Soc., 72, 1675-79 (1989)
[16] J.E.Ritter, S.V.Nair, P.A.Gennari, W.A.Dunlay, J.S.Haggerty and G.J.Garvey, Advanced Ceram. Mater., 3, 415-17 (1988)
[17] B.W.Sheldon and J.S.Haggerty, to be published in the proceedings of the 13th Annual Conference on composites and Advanced Ceramics, Am. Ceram. Soc., Cocoa Beach, FL, Jan., 1989

SYNTHESIS OF ADVANCED CERAMICS IN THE SYSTEMS SI-B-N AND SI-B-N-C EMPLOYING NOVEL PRECURSOR COMPOUNDS

H.-P. BALDUS*, O.WAGNER**, AND M. JANSEN**
*Bayer AG, 5090 Leverkusen, FRG
**Institut für Anorganische Chemie, Universität Bonn, 5300 Bonn, FRG

ABSTRACT

The reaction of Cl_3Si-NH-$Si(CH_3)_3$ with BCl_3 yields the novel compounds Cl_3Si-NH-BCl_2 and $(Cl_3SiNH)_2$-BCl. The ammonolysis of these monomeric precursors in liquid ammonia gives rise to highly crosslinked borosilicon imides. Pyrolysis of the polymeric imides in a stream of ammonia at 1250°C yields amorphous borosilicon nitrides with formal compositions of $Si_3B_3N_7$ and $Si_6B_3N_{11}$ respectively. MAS-NMR, DTA/TG, SEM and TEM investigations reveal that the resulting materials are single phase.
The reaction of the novel precursors with primary amines yields meltable and soluble polymers suitable for producing fibers or coatings. Subsequent pyrolysis in NH_3 leads also to amorphous borosilicon nitride. The physical and chemical properties of the new materials will be discussed.

INTRODUCTION

Composite ceramics consisting of two or more binary nitrides and carbides respectively have witnessed increased interest in recent years, because this type of ceramics shows improved properties compared to the pure binary nitrides [1-3]. Synthesis of these composites is usually performed by co-milling the binary compounds, which, however leads to an inhomogeneous elemental distribution and impurities in the material.
These observations led us to investigate the synthesis of molecular precursors having exactly the stoichiometry and molecular structure desired in the final ceramic.
Resulting materials should therefore exhibit a homogenous elemental distribution on a molecular scale, which leads to improved mechanical properties of the shaped body. Recently, following this approach, Niihara et al. developed a Si_3N_4/SiC nanocomposite with superiour bending strength and toughness [4].

In the present paper we report the synthesis of Cl_3Si-NH-BCl_2 and $(Cl_3SiNH)_2$-BCl. These novel precursor molecules contain the Si-N-B-bond structure required for the synthesis of a multinary ceramic. Furthermore, the conversion of Cl_3Si-NH-BCl_2 into a new borosilazane polymer and pure borosilicon nitride respectively by ammonolysis and pyrolysis will be presented.

EXPERIMENTAL

Preparation of (trichlorosilyl)-amino-dichloroborane (TADB) and bis-((trichlorosilyl)-amino)-chloroborane (TACB)

The reaction of 1,1,1 trichloro-3,3,3 trimethyldisilazane (TTDS), a compound easily prepared by stirring a mixture of $SiCl_4$ and hexamethyldisilazane at room temperature for 8 h [5], with boron trichloride (1 : 1) at - 78 °C led to a novel molecule with the formula Cl_3Si-NH-BCl_2 ((trichlorosilyl)-amino-dichloroborane) (yield 89%):

$$Cl_3Si\text{-}NH\text{-}Si(CH_3)_3 + BCl_3 \text{---->} Cl_3Si\text{-}NH\text{-}BCl_2 + ClSi(CH_3)_3 \quad (1)$$

The preparation of TACB was performed in the same manner as the preparation of TADB with the exception that the ratio of TTDS : BCl_3 was changed from 1:1 to 2:1 (yield 51%):

$$2\ Cl_3Si\text{-}NH\text{-}Si(CH_3)_3 + BCl_3 \text{---->} (Cl_3Si\text{-}NH)_2\text{-}BCl + 2\ ClSi(CH_3)_3 \quad (2)$$

Both compounds were characterized by GC-MS and multi-nuclei NMR-measurements.

Analytical data of TADB: MS (EI) m/z = 229 (M^+), ^{1}H-NMR (TMS, $CDCl_3$) $\delta_{N\text{-}H}$ = 5.15 ppm, ^{11}B-NMR (BF_3*Et_2O) δ_B = 37 ppm, ^{29}Si-NMR (TMS) δ_{Si} = -23 ppm.

Analytical data of TACB: MS (EI) m/z = 342 (M^+), ^{1}H-NMR (TMS, $CDCl_3$) $\delta_{N\text{-}H}$ = 4.15 ppm, ^{11}B-NMR (BF_3*Et_2O) δ_B = 30 ppm, ^{29}Si-NMR (TMS) δ_{Si} = -27 ppm.

Preparation of the borosilicon nitride (from TADB)

A polymeric borosilicon imide amide was prepared by ammonolysis of TADB, dissolved in toluene, at -78°C. The ammonium chloride produced by the ammonolysis reaction was washed out with liquid ammonia for four days. A white powder was obtained with an oxygen content of 0.6 wt.%. The chlorine content of the powder amounted to 0.4 wt.%. For pyrolysis, samples were heated under flowing anhydrous NH_3 from room temperature to 1000°C with a heating rate of 15°C per minute, then held at that temperature for an additional 10 - 15 h and finally cooled slowly to room temperature.
The pyrolyzed material was pressed at 750 MPa. The pellets were subsequently annealed at 800, 1250, 1400, and 1650 °C respectively for 6 h in a boron nitride crucible.
Sintering of the amorphous borosilicon nitride was conducted at 1800°C with 10% oxidic sintering additives using a pressure sintering furnace (KCE Sondermaschinen GmbH, FRG).

Preparation of polymers by reaction of TADB with methylamine

TADB, dissolved in hexane was added dropwise to liquid methylamine under vigorous stirring at - 20°C. The precipitated solid consisting mainly of methylamine hydrochloride was filtered off, the solvent was evaporated and finally the volatiles were removed by vacuum destillation. The reaction yielded a yellow glass-like residue with a softening point of about 130°C. Pyrolysis of the polymeric materials was performed under Argon, N_2, and NH_3.

Analytical data: ^{1}H-NMR data: SiBNH 2.1 ppm, CH_3NHSi 0.6 ppm, CH_3NHB 1.6 ppm, NCH_3 2.3 -2.9 ppm; ^{29}Si-NMR: 33 ppm; ^{11}B-NMR: 26 ppm; ^{13}C-NMR: 27-28 ppm; IR data: N-H stretch 3450 cm^{-1}, 3416 cm^{-1}, C-H-stretch 2886 cm^{-1}, 2809 cm^{-1}, Si-N stretch 933 cm^{-1}, B-N-stretch 1357 cm^{-1}, Si-N deformation 820 cm^{-1}, 775 cm^{-1}.

All operations were carried out in an inert gas atmosphere either in a dry box or with standard

Schlenk techniques.

RESULTS AND DISCUSSION

A) Synthesis and characterization of the ceramic material obtained by ammonolysis of TADB with NH_3 and susequent pyrolysis

Reaction of TADB with NH_3 resulted in a white powder with the following chemical composition: Si: 28.3 %, B: 10.6 %, N: 54.2 %, O: 1.1 %, H: 5.05 %, Cl: 0.1 %.

The corresponding reaction equation is given below:

$$Cl_3Si\text{-}NH\text{-}BCl_2 + 10\ NH_3 \dashrightarrow (NH_2)_3Si\text{-}NH\text{-}B(NH_2)_2 + 5\ NH_4Cl \quad (4)$$
$$\dashrightarrow SiB(NH)_3(NH_2) + NH_3$$

^{11}B-NMR spectra obtained from the polymeric borosilicon imide and from the pyrolyzed borosilicon nitride show quadrupole interactions also observed in the borosilazane derived materials, indicating the absence of a tetrahedral environment. The chemical shift of the ^{11}B-nucleus in borosilicon nitride is similar to the ^{11}B-resonance of borosilicon carbonitride. The ^{29}Si resonance of the polymeric borosilicon imide is observed at -33 ppm, indicating silicon atoms tetrahedrally co-ordinated to four nitrogen atoms[10].

The pyrolyzed SiBN-samples exhibit a chemical shift at - 40 ppm and -43 ppm respectively. This is close to the corresponding Si-resonance of silicon nitride (- 48 ppm)[10]. The IR-spectra of the untreated polymer and of 800°C-, 1000°C-, and 1250°C-samples show three distinct regions. An N-H-stretch at 3396 cm^{-1}, which disappears at 1250°C, indicates the loss of residual hydrogen at this temperature. The B-N and Si-N stretches are seen as expected at 1339 cm^{-1} and 934 cm^{-1} respectively. Due to the reaction of Si with NH_3 in the temperature range 800°C - 1250°C an Si-H stretch emerges temporarily.

The pyrolysis of the borosilicon imide was studied with the aid of DTA/TG between 50°C and 1600°C in a nitrogen atmosphere. A continuous loss of mass was detected between 100°C and 1250°C, between 1250°C and 1600°C the mass remained nearly constant. The total weight loss within this range of temperature amounted to ca. 28.5 %. Neither exothermic nor endothermic effects could be observed. The composition of the powder annealed at 1250°C is given below:
Si: 38.7 % (39.06 %), B: 15.1 % (15.3 %), N: 44.2 % (45.56 %), O: 1.5 %, Cl: 0.01 %. This corresponds to the expected formula $Si_3B_3N_7$ (the calculated values are given in parentheses).

DTA-TG investigations between 1000°C and 1800°C in helium atmosphere reveal that the sample decomposition begins at 1550°C and is completed at 1780°C resulting in Si, BN and N_2. The narrow DTG-peak width indicates a pure compound. A comparison of the TG-data of borosilicon nitride with the corresponding data of a Si_3N_4/BN composite of the same nominal composition and Si_3N_4 respectively shows the enhanced thermal stability of the SiBN-powder (Fig. 1).

SEM and TEM studies on the powder, annealed at 1000°C exhibit a grain size of primary particles between 20 and 50 nm. The agglomerates have an overall grain size between 100 and 500 nm. EDX measurements exhibit a homogenous elemental distribution of silicon, boron and nitrogen in all particles which gives further evidence to a pure compound. The powders, pyrolyzed at 1400°C, 1550°C, and 1750°C respectively have formed loose agglomerates with a size of ca. 200 µm. The original particle size and shape as well as the distribution of the elements did not change. The specific surface area of the powder calcined at 1650°C amounts to 60 m^2/g. All the SiBN-samples pyrolyzed in nitrogen up to 1750°C are amorphous to X-rays and to electron beams.

Preliminary sintering experiments with $Si_3B_3N_7$ in comparison with a 65/35 composite of Si_3N_4 and BN reveal a strongly improved sintering activity of the new material (Fig. 2). X-ray powder diffraction of the sintered body exhibited only the presence of crystalline ß-silicon nitride. No evidence of boron nitride could be detected.

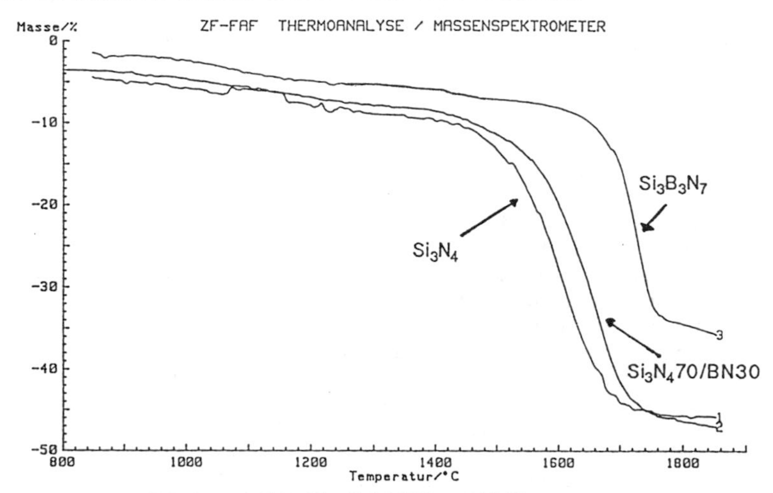

Figure 1: DTA/TG data of $Si_3B_3N_7$, Si_3N_4/BN and Si_3N_4

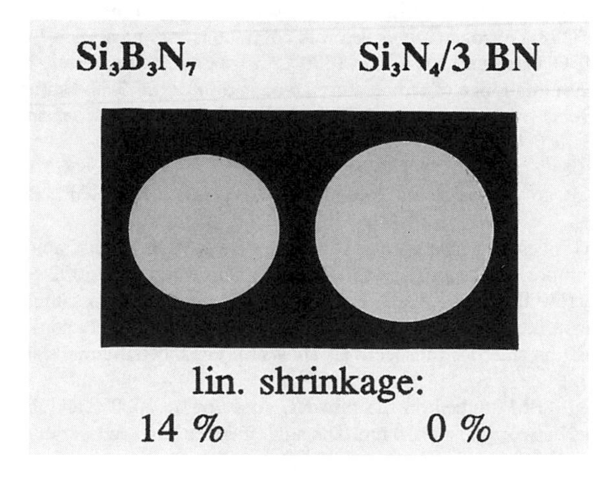

Figure 2: Sintering behaviour of $Si_3B_3N_7$ in comparison with a Si_3N_4/3 BN composite (sintering temperature 1800°C; sintering additives: 5 wt.-% Y_2O_3 + 5 wt-% Al_2O_3)

B) Borosilazane polymers

Reaction of TADB with methylamine and subsequent heating in vacuum at 160°C gave a soluble and meltable polymeric borosilazane in about 80% yield. Converting the elemental analytical data (C: 27.5 %, H: 7.18 %, O: 0.7 %, N: 38.6 %, Cl: 0.06 %, B: 6.87 %, Si: 17.9 %) of the polymer in an empirical formula, $Si_2B_2(NH)_2(NHCH_3)_3(NCH_3)_4$, suggests that the following reaction takes place:

$$10\ n\ CH_3NH_2 + n\ Cl_3Si\text{-}NH\text{-}BCl_2 \dashrightarrow n\ (CH_3NH)_3Si\text{-}NH\text{-}B(NHCH_3)_2 + 5\ n\ CH_3NH_3^+Cl^- \dashrightarrow [(SiBNH)_2(NHCH_3)_2(NCH_3)_4]_{n/2} + 2\ n\ CH_3NH_2 \quad (3)$$

NMR data (^1H, ^{29}Si, ^{11}B, ^{13}C), IR data as well as the elemental analysis data suggest a polymer structure of $Si_3(NCH_3)_3$ six-membered rings connected via HN-B- and $N(CH_3)$-B-units respectively (see Scheme 1).

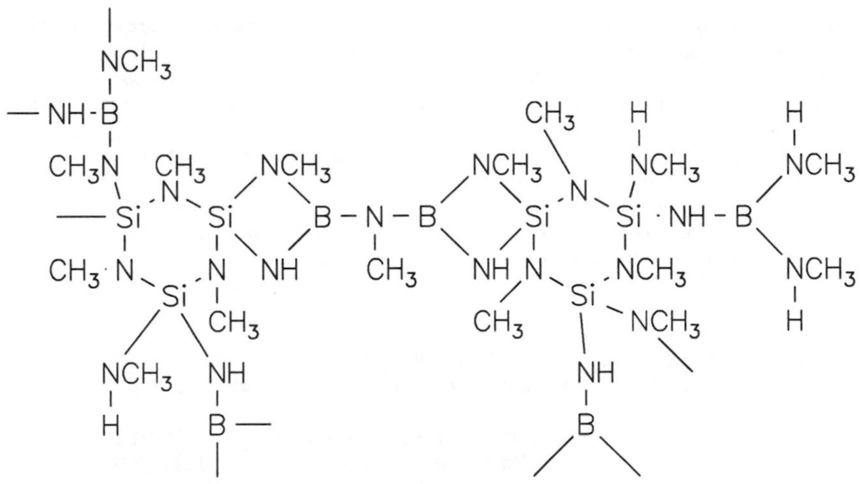

Scheme 1: Proposed structure of the borosilazane

We can exclude the existence of Si_2N_2-four-membered rings because ^{29}Si-signals at ca. - 50 ppm and Si-N-deformation modes at ca. 860 cm^{-1} indicating such a structural unit are not detected [6-7]. Furthermore, due to the absence of ^{11}B-signals in the range 29-36 ppm there is no evidence for the existence of borazine like ring structures [8-9]. In summary, the analytical data reveal that there is no difference in the Si : B ratio of the polymeric material compared with the ratio in the precursor molecules, and indicate that the Si-N-B-bonds of the precursor molecule are still present in the polymer.

The yellow brownish glass-like polymer has an average molecular weight of 2500, melts at 130°C and is soluble in methylene chloride and toluene. Fibers can be drawn from this material either by a dry spinning process from solution or simply by melt spinning at 140°C.

Pyrolysis of the polymer in NH_3 at 1000°C leads to a white powder with the same composition as the pyrolyzed ammonolysis product of TADB. Pyrolysis in Ar or N_2 at 1650°C leads to a black amorphous borosilicon carbonitride in 70% yield. During the initial stages of pyrolysis further elimination of methylamine takes place, resulting in further crosslinking. According to TG/MS measurements the elimination of CH_4 and H_2 begins at 450°C. IR-spectra indicate that the organic functionalities are no longer present at 800°C. IR also reveals the disappearence of N-H bonds in samples pyrolyzed at 1250°C.

^{29}Si- (- 45 ppm), ^{13}C- (122 ppm) as well as ^{11}B-solid state MAS-NMR (19 - 25 ppm) and TEM - investigations of the material pyrolyzed at 1650°C give evidence for the formation of an Si-N-B-C quaternary borosilicon carbonitride with silicon connected to four nitrogen atoms and boron connected to three nitrogen atoms. No graphitic carbon is detected by TEM measurements even though ^{13}C-NMR indicates sp^2 hybridized carbon.

CONCLUSIONS

We have demonstrated a simple and low cost synthesis of the novel precursor molecules Cl_3Si-NH-BCl_2 and $(Cl_3Si\text{-}NH)_2BCl$. Cl_3Si-NH-BCl_3 is converted to a meltable borosilazane polymer by reaction with methylamine and to a pure single phase borosilicon nitride by reaction with ammonia and subsequent pyrolysis. In both cases the Si:B ratio in the ceramic material does not differ from the ratio in the precursor molecule.

Therefore, these novel precursors represent examples in which the bond structure of the precursor determines the bond structure of the ceramic.

Further work concerning the conversion of $(Cl_3Si\text{-}NH)_2$-BCl to ceramic material, spinning of fibers from borosilazane melts as well as the ceramic processing and characterisation of borosilicon nitride material is in progress.

REFERENCES

1. K.S. Mazdiyasni, R.Ruh, J. Am. Chem. Soc. 64, 415 (1981).
2. J.D. Lee, H.H. Moeller, D.R. Petrak, P.L. Berneburg, Am. Ceram. Soc. Bull., 63 (3), 422 (1984).
3. A. Bellosi, S. Guicciardi, A. Tampieri, J. Euro. Ceram. Soc. 9, 83 (1992).
4. A. Sawaguchi, K. Toda, K. Niihara, J. Am. Ceram. Soc., 74 (5), 1142 (1991).
5. J.P. Moser, H. Nöth, W. Tinhof, Z. Naturforsch.B 29, 166 (1974).
6. D.M. Narsavage, L.V. Interrante, P.S. Marchetti, G.E. Maciel, Chem. Mater. 3, 721 (1991).
7. H. Burger, W. Geschwandtner, G.R. Liewald, J. Organomet. Chem. 259, 145 (1983).
8. D. Seyferth, H. Plenio, J. Am. Ceram. Soc. 73 (7), 2131 (1990).
9. H. Nöth, H. Vahrenkamp, Chem. Ber. 99, 1049 (1966).
10. G.R. Hatfield, K.R. Carduner, J. Mater. Sci. 24, 4209 (1989)

DESIGN AND SYNTHESIS OF CHEMICAL PRECURSORS TO BORON CARBIDE

XIAOGUANG YANG, STEPHEN JOHNSON AND M. FREDERICK HAWTHORNE*
*Department of Chemistry and Biochemistry, University of California at Los Angeles, 405 Hilgard Ave., Los Angeles, CA 90024.
HAIXING ZHENG, KEVIN THORNE AND J.D. MACKENZIE**
**Department of Materials Science and Engineering, University of California at Los Angeles, 405 Hilgard Ave., Los Angeles, CA 90024.

ABSTRACT

The design of molecular and preceramic polymer precursors to boron carbide based ceramics is described. The goal was to design nonvolatile, tractable precursors to boron carbide with high ceramic yields. Molecular precursors containing carborane cages and acetylenic groups were synthesized and converted to nonvolatile, soluble preceramic polymers. Pyrolysis of these polymers gave boron carbide based ceramics in 60-70% ceramic yield. A B_4C/SiC film has been fabricated from one of the polymers and the film is uniform and crack-free.

INTRODUCTION

It is well known that there are difficulties in fabricating metal and non-metal carbides into useful forms such as films and fibers by conventional techniques. However, recent work has shown that the conversion of preceramic polymers to ceramics provides access to ceramic films, fibers, and complex shapes [1]. Even though boron carbide is an attractive material for structural ceramics, fibers and coatings, there are few reports on the pyrolysis of preceramic polymers to boron carbide [2]. It was reported that $B_4C/SiC/SiO_2$ and B_4C/SiC composites can be obtained from carboranesiloxane polymers in 60% ceramic yield [3]. Sneddon has recently shown that 2-$(H_2C{=}CH)$-B_5H_8 undergoes thermal polymerization to air-sensitive vinylpentaborane oligomers which can be converted to bulk pure boron carbide in 77% ceramic yield [4]. We recently reported that the DEXSIL 202 carboranesiloxane polymer can be converted to B_4C/SiC ceramics, but the poor rheological behavior of the polymer makes it very difficult to fabricate fibers or thin films [5]. Here we present our recent effort in designing molecular and polymeric carborane-containing precursors and converting and fabricating them into B_4C-based ceramics and films.

RESULTS AND DISCUSSION

To design and construct preceramic polymers which would give high ceramic yields, it is desirable to have the elemental composition of the preceramic polymer as close as possible to that of the target ceramic and have the preceramic polymers which are highly branched, crosslinked, or contain rings, cages and unsaturated functional groups such as vinyl and acetylenic groups [6]. The synthesis and reactions of well-defined low molecular weight compounds that are terminated with ethynyl or acetylenic groups have received a great deal attention as a means of thermally chain extending, rigidizing, and cross-linking polymers over the past 20 years [7]. We accordingly decided to synthesize disubstituted acetylenic derivatives of icosahedral carboranes as molecular precursors for B_4C ceramics. The synthesis of 1,2-bis(trimethylsilylethynyl)-1,2-$C_2B_{10}H_{10}$ was unsuccessful and we turned our attention to 1,7-$C_2B_{10}H_{10}$ derivatives.

1,7-bis(trimethylsilylethynyl)-1,7-$C_2B_{10}H_{10}$ (**1**) was synthesized according to the literature procedure [8] and is shown in Scheme I. The molecule was characterized by ^1H,

Scheme I

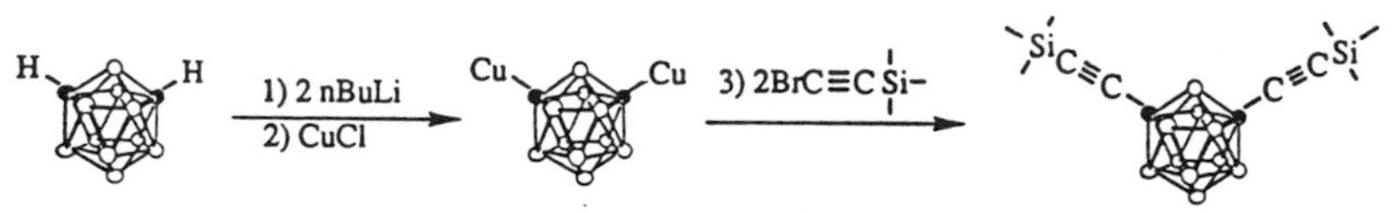

H H 1) 2 nBuLi 2) CuCl Cu Cu 3) 2BrC≡C Si–

^{11}B, ^{13}C NMR spectroscopy and X-ray crystallography. The structure of (**1**) is shown in Figure 1 [9]. (**1**) contains two unsaturated trimethylsilylethynyl functional groups and a carborane icosahedron cage, which promise to give polymers with tractability, latent reactivity and high ceramic yield.

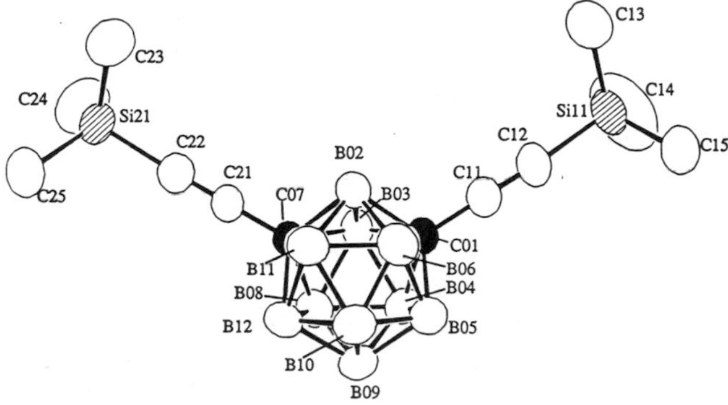

Figure 1. The molecular structure of 1,7-bis(trimethylsilylethynyl)-1,7-$C_2B_{10}H_{10}$ (**1**).

The pyrolysis of (**1**) was investigated. Unfortunately, (**1**) is volatile at high temperature and the TGA result indicated complete weight loss by 240 ^oC (Figure 2(a)).

High temperature and high pressure are commonly used to initiate cross-linking in polymers to increase their average molecular weight and to improve their ceramic yields. This process was therefore used to convert (**1**) to nonvolatile polymers. (**1**) was dissolved in hexane and the autoclave was heated at 300 ^oC for three hours under a pressure of 1400 psi. The dark brown resin obtained was viscous and soluble in most common organic solvents.

The dark brown polymers were characterized by ^1H, ^{11}B, ^{13}C NMR and FT-IR

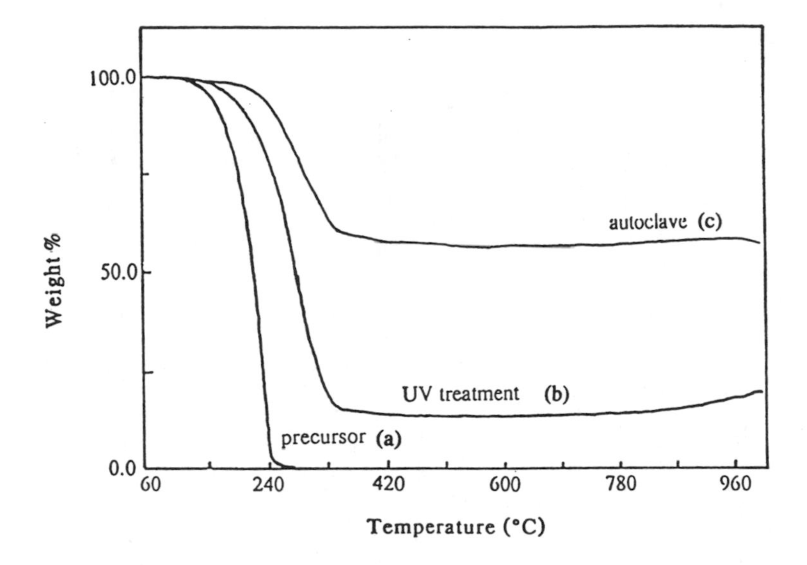

Figure 2. Thermogravimetric analyses of (a) molecular precursor (**1**), (b) polymers from UV-irradiation of (**1**), (c) polymers from autoclave treatments of (**1**).

spectroscopy . The infrared spectrum (Figure 3(b)) of the polymer still contains an absorption at 2198 cm^{-1} ($\upsilon_{C\equiv C}$) but at half the intensity of that of (**1**). Strong absorptions at 2923 and 2903 cm^{-1} indicated that new types of C-H groups were formed. No peaks due to C=C double bonds were observed in the infrared spectrum of the polymer. The ^{11}B NMR spectrum of the polymer indicates that the carborane cages are still intact after the polymerization (Figure 4(b)). The carbon-carbon triple bonds still intact, but not equivalent, since the ^{13}C NMR spectrum of the polymer in acetone-d_6 exhibits three types of acetylenic groups with a ratio of 1:2:4 around 98.6 and 86.1 ppm which are observed for (**1**) and assigned to trimethylsilyl-ethynyl carbons. No peaks higher than 100 ppm were observed in the ^{13}C NMR spectrum of the polymer, which further confirms that no C=C double bonds are present in the polymer. The polymer formed from UV irradiation of (**1**) has similar spectroscopic features. Based on the spectroscopic data, we believe that one possible structure of the polymers is the structure with carborane acetylenic moieties linked through carbosilane groups as shown in Scheme II.

Scheme II

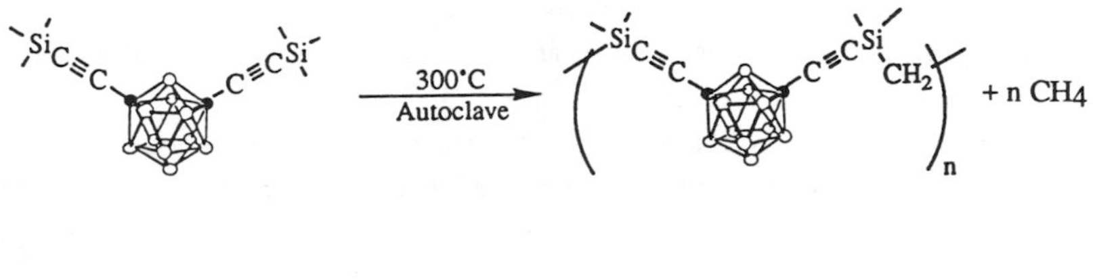

Si–C≡C–(carborane)–C≡C–Si —300°C, Autoclave→ (Si–C≡C–(carborane)–C≡C–Si–CH_2)$_n$ + n CH_4

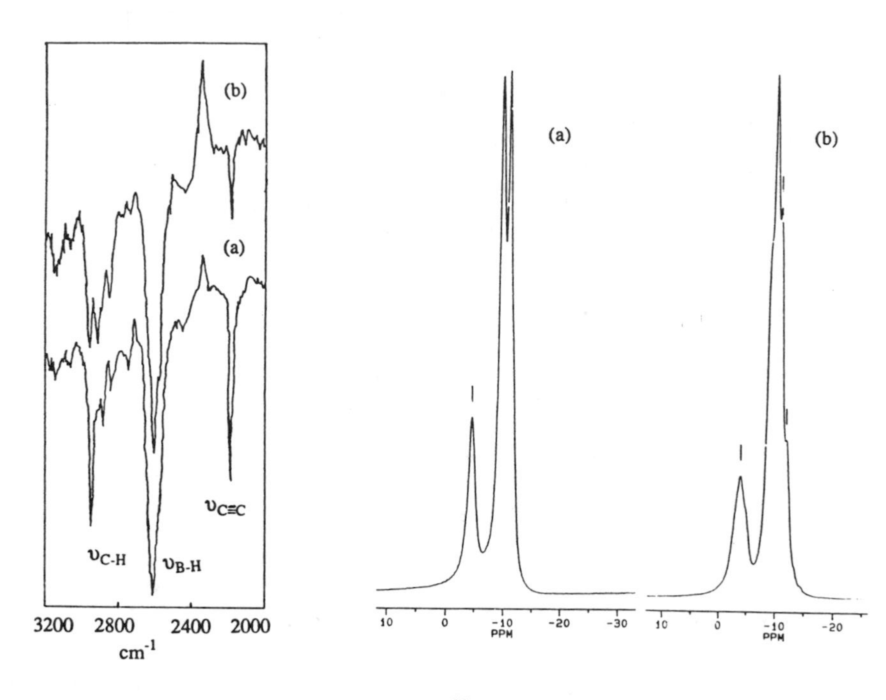

Figure 3. Infrared spectra of (**1**) (a) and its polymer (b).

Figure 4. ^{11}B NMR spectra of (**1**) (a) and its polymer (b)

The presence of carbon-carbon triple bonds in the polymer apparently provides the latent reactivity of the polymers which can then be cured and crosslinked during the pyrolysis to give B_4C/SiC with a high ceramic yield. In addition, carborane cages are in the polymer chain, which enhances the chemical and thermal stability of the polymer and prevents the initial chain degradation observed in the polymers with carborane cages as pendant groups [10].

The lack of reactivity of acetylenic groups of (**1**) in the initial thermolysis treatment can be due to the rigidity of (**1**) and bulky substitutents around the carbon-carbon triple bonds. We observed ethynyl groups in the ^{13}C NMR spectrum of the polymer formed from (**1**). We believe that at higher temperatures (pyrolysis conditions), Cross-linking occurs through the addition reaction between the acetylenic groups.

The TGA results indicate that ceramic yields as high as 60 wt% have been achieved (Figure 2(c)). The ceramic yield of the polymer from (**1**) exposed to UV-light is 20% (Figure 2(b)). It is assumed that the non-transparency of the polymer formed on the surface layer to UV-radiation is the primary problem. Lack of cross-linking after UV-irradiation may be due to the bulky substitutents of carbon triple bonds and the rigidity of (**1**).

Since the polymers from (**1**) after the autoclave treatment are soluble in organic solvents and give high ceramic yields, films from these polymers have been fabricated. The polymers

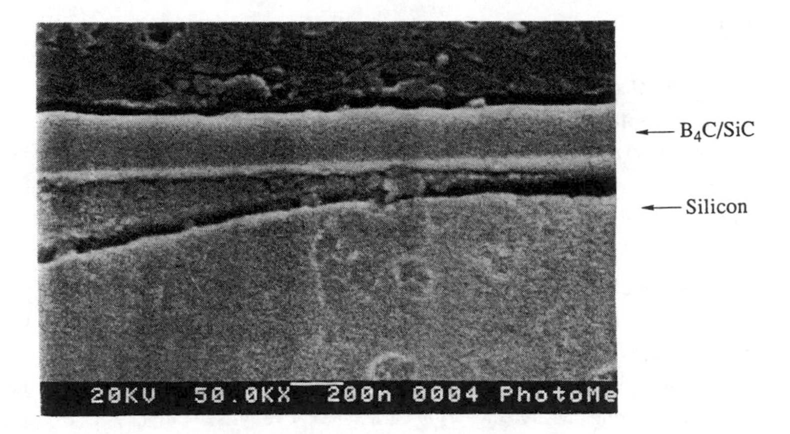

Figure 5. Cross-section SEM of the film fabricated from the polymer derived from (**1**).

formed from (**1**) were dissolved in *n*-hexane at a concentration of 10 wt%. The resulting clear solution was used for dip coating on silicon substrates. After each dip, the polymer film was thermally cross-linked at 500 ^oC in N_2 for 5 minutes to prevent the redissolution of the polymer film. After three dips, the film was fired at 1150 ^oC in Ar for one hour. The X-ray diffraction of this thin film indicated that the film is a B_4C/SiC crystalline film. Further examination of the film by scanning electronic microscopy (SEM) (Figure 5) showed that the film is dense, without cracks, and the thickness of the film is about 0.2 μm.

CONCLUSION

The polymers derived from 1,7-bis(trimethylsilylethynyl)-1,7-$C_2B_{10}H_{10}$ (**1**) have latent reactivity and enable the formation of branched or ring structures giving good ceramic yields. Such latent chemical reactivity provides a particular advantage for the preparation of ceramic films and fibers. Good quality films have been fabricated and preliminary work has shown that B_4C/SiC fibers can also be fabricated. The structural characterization and properties of thin films and fibers are currently under active investigation.

EXPERIMENTAL SECTION

Standard Schlenk line technique was employed for all manipulation of air- and moisture-sensitive compounds. All solvents were dried and distilled before use. Photolysis and thermolysis were performed under Ar or N_2 atmospheres. CuCl was purchased from CERAC Inc. and Bis(trimethylsilyl)acetylene was purchased from the Aldrich Chemical Co. Trimethylsilylethynylbromide and 1,7-bis(trimethylsilylethynyl)-*m*-carborane were prepared according to the literatures [11,8].

IR spectra of KBr pellets were recorded with a Beckman FT-1100 instrument. ^{1}H, ^{13}C and ^{11}B NMR spectra were recorded with Bruker AF 200, AM 360 and AM 500 spectrometers. Chemical shifts for ^{1}H and ^{13}C NMR spectra were referenced to the residual protons and carbons in deutrated solvents. Chemical shift values for ^{11}B spectra were referenced relative to external $BF_3 \cdot OEt$ (0.00 ppm). The thermogravimetric analysis (TGA) results were obtained with a Perkin Elmer Thermostation using model TGS-2. X-ray powder diffraction patterns were obtained with a Phillips Diffractometer using Cu $K_{\alpha-1}$ radiation.

Preparation of poly(1,7-bis(trimethylsilylethynyl)-*m*-carborane).

1,7-bis(trimethylsilylethynyl)-*m*-carborane (**1**) was dissolved in 10 mL hexane and the autoclave was heated at 300 °C for three hours under a pressure of 1400 psi. The brown viscous material was transferred to a 50 mL round-bottom flask and hexane was removed under vacuum to afford poly(1,7-bis(trimethylsilylethynyl)-*m*-carborane). Pyrolysis of poly(1,7-bis(trimethylsilylethynyl)-*m*-carborane) was conducted in a flowing argon atmosphere (99.999% Ar) graphite refractory furnace (Thermal Technology- Series 1000).

Acknowledgment. This research was supported by Grant DMR-90-14487 from the National Science Foundation.

REFERENCES

[1] K.J. Wynne and R.W. Rice, Ann. Rev. Mater. Sci 14, 297 (1984).

[2] W. Toreki, Polymer News 16, 6 (1991)

[3] (a) R.W. Rice, K.J. Wynne, W.B. Fox, U.S. Patent No.4,097,294 (1978). (b) B.E.Jr. Walker, R.W. Rice, P.E. Becher, B.A. Bender, W.S. Coblenz, Am. Ceram. Soc. Bull. 62, 916 (1983).

[4] M.G.L. Mirabelli and L.G. Sneddon, J. Am. Chem. Soc. 110, 3305 (1988).

[5] H. Zheng, K. Thorne, J.D. Mackenzie, X. Yang, M.F. Hawthorne, presented at the 1991 MRS Fall Meeting, Boston, MA.

[6] K.J. Wynne, in Transformation of Organometallics into Common and Exotic Materials: Design and Activation, edited by R.M. Laine, Matrinus Nijhoff Publisher, 1988, p 89-96.

[7] Reactive Oligomers, edited by F.W. Harris and H.J. Spinelli (ACS Symposium Series 282, St. Louis, MO, 1984) pp17-29.

[8] L.I. Zakharkin, A.I. Kovderov, V.A. Ol'Shevskaya, Izv. Akad. Nauk SSSR, Ser. Khim. 35, 1388 (1986).

[9] X. Yang, C.B. Knobler, M.F. Hawthorne, unpublished results.

[10] R. Grimes, Carboranes, Academic Press, New York 1970, p181.

[11] J.A. Miller and G. Zweifel, Synthesis 129 (1983).

ALKYNYL SUBSTITUTED CARBORANES AS PRECURSORS TO BORON CARBIDE THIN FILMS, FIBERS AND COMPOSITES

STEPHEN E. JOHNSON*, XIAOGUANG YANG*, M. F. HAWTHORNE*, KEVIN J. THORNE**, HAIXING ZHENG** AND JOHN D. MACKENZIE*
* UCLA Department of Chemistry and Biochemistry, Los Angeles, CA, 90024.
**UCLA Materials Science and Engineering Department, Los Angeles, CA, 90024.

ABSTRACT

The use of alkynyl substituted derivatives of *o*-carborane as precursors to boron containing ceramics is described. These compounds undergo a thermally or photochemically induced polymerization to afford cross linked polyakynyl-*o*-carborane derivatives. The increase in molecular weight should allow for increased Tg's and the retention of modelled polymer pre-forms. In this report, these modification reactions are described. In addition, the retention of molded polymer preforms were analyzed after UV exposure and inert atmosphere pyrolysis.

INTRODUCTION

There has been considerable interest in non-oxide ceramics such as B_4C, BN and Si_3N_4 derived from metal-organic precursors [1-4]. Boron carbide is an important engineering material, having a wide variety of commercial applications [5,6]. Moreover, boron carbides have interesting mechanical and electrical properties [7]. Several polymeric precursors to B_4C [8-10] and B_4C/SiC composites have have been reported only within recent years [1,11,12]. For example, pyrolysis of a polymer derived from 2-(CH_2=CH)-B_5H_8 affords boron carbide in a high ceramic yield [8]. Boron carbide is representative of ceramic whose preparation from a polymeric precursor might offer advantages over conventional synthetic methods [13]. One aspect of the work in these laboratories has focused on the design and synthesis of functionalized carboranes to serve as molecular precursors to non-oxide ceramics of boron [14-16]. Herein we report the synthesis, structural characterization, and materials chemistry of alkynyl substituted carborane derivatives.

RESULTS AND DISCUSSION

We have recently described a novel protecting group for *o*-carborane which has greatly aided in the design and construction of rational molecular routes to inorganic and organometallic materials as opposed to simply screening a variety of potential compounds [17]. These are versatile synthetic reagents in that monofunctional and linked carborane cages are realized. The protecting group is easily removed and additional carboranes can be introduced or further synthetic transformations can be performed. We have recently extended this synthetic methodology to *m*-carborane and *p*-carborane derivatives [18,19]. The synthesis of some alkynyl *o*-carborane derivatives are outlined in Scheme I. The compound 1-ethynyl-*o*-carborane (**1**) was prepared by the addition of diacetylene to a solution of $B_{10}H_{12}(CH_3CN)_2$ in a 70% yield [20]. The acetylene function can be selectively protected and addition of trimethylsilylbromide gives the protected dialykynyl-*o*-carborane. Deprotection using nBu_4NF in THF affords 1,2-diethynyl-*o*-carborane (**2**). Compound (**2**) is also obtained directly using similar methods. All compounds have been spectroscopically characterized and we are currently trying to optimize the yield of (**2**) [21].

Scheme I

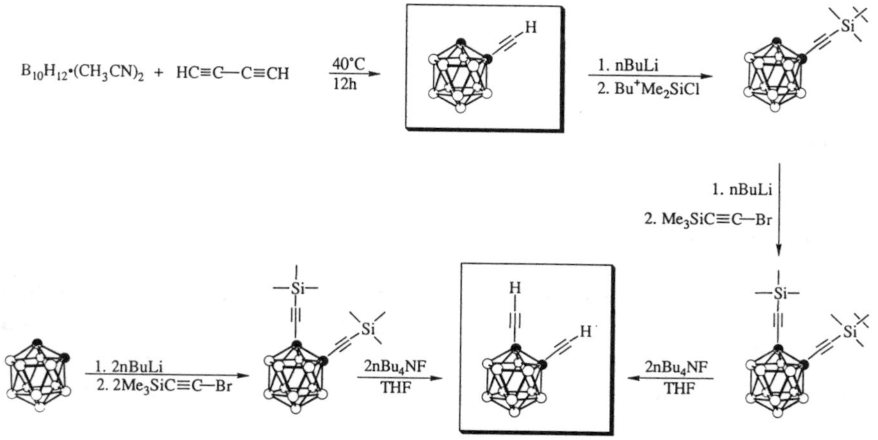

$B_{10}H_{12}\cdot(CH_3CN)_2$ + HC≡C—C≡CH —(40°C, 12h)→ [product] —(1. nBuLi; 2. Bu^tMe_2SiCl)→ [product] —(1. nBuLi; 2. $Me_3SiC≡C—Br$)→ [product] —($2nBu_4NF$, THF)→ [product]

[carborane] —(1. 2nBuLi; 2. $2Me_3SiC≡C—Br$)→ [product] —($2nBu_4NF$, THF)→ [product]

These compounds are attractive precursors for two reasons: they contain the nearly ideal ratio of boron to carbon found in B_4C. Secondly, the alkynyl substituents can undergo a thermal or photoinduced polymerization reaction that gives rise to cross-linked polyalkynyl-*o*-carborane derivatives. In the case of the UV photolysis of compound (**1**), polymer formation was followed spectroscopically. The infrared spectrum of the monomer (**1**) exhibits strong absorptions which are assigned to alkynyl C-H stretching and B-H fundamental stretching modes [Figure 1a]. The fundamental carbon-carbon triple bond vibration is a moderately intense absorption near 2000 cm^{-1}. After exposure to UV irradiation, the alkynyl C-H band has decreased in intensity and new bands have formed between 1670-1640 cm^{-1}, characteristic of C=C fundamental stretching modes, while the B-H fundamental remains essentially unchanged. The data are consistent with polymerization through the carbon skeleton rather than through carborane cages. Comparison of ^{13}C NMR data for (**1**) and the carboranyl polymer is in line with the formulation of a poly(yne) structure with a pendant carborane ligands [Figure 1b]. The observation of a number of signals centered near 128 ppm are characteristic of sp^2 type carbons in the polymer backbone and indicates a number of different stereochemical configurations. In addition, there were no significant changes in the ^{11}B NMR spectrum of the polymer as compared to the monomer.

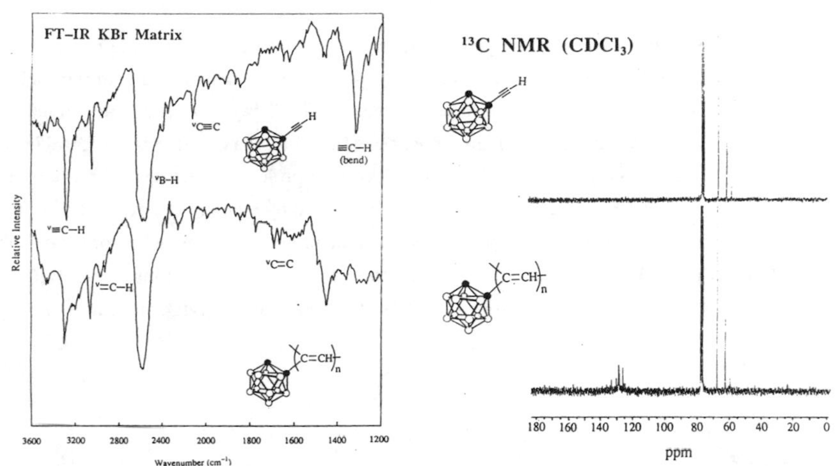

Figure 1: Spectroscopic data for (**1**) and resulting polymer obtained by UV irradiation, FT-IR in KBr matrix (a) and ^{13}C NMR spectra in $CDCl_3$(b).

Thermogravimetric data for the monomer (**1**), the polymer after UV photolysis and the polymer generated from the autoclave reaction were obtained [Figure 2]. The monomer exhibits a steep slope, indicating a nearly complete weight loss, consistent with the volatile nature of (**1**). The polymer generated by exposure to UV radiation exhibits a sharp slope to 240° C, indicating loss of monomer and incomplete polymerization. There is a more gradual weight loss up to 420° C attributed to residual cross-linking after which essentially no weight loss occurs. Longer exposure times should increase the degree of polymerization. The autoclaved monomer is most completely polymerized, exhibits significantly reduced weight loss and was subsequently used for pyrolysis.

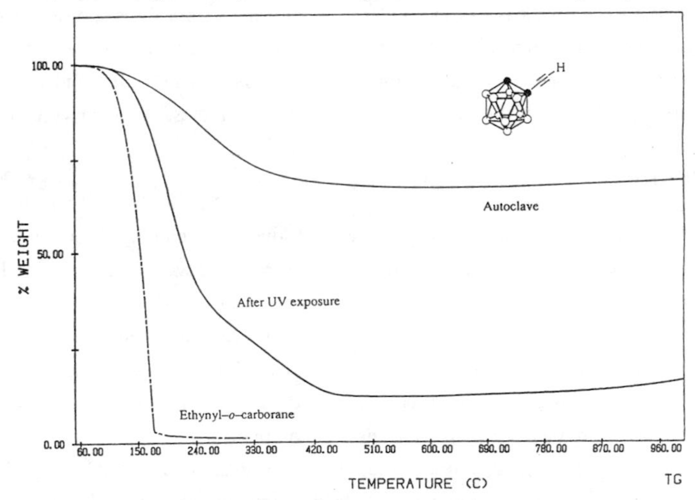

Figure 3: TGA data for (**1**) and polymers resulting from UV irradiation and autoclave reaction.

Bulk pyrolyses were performed on poly(1-ethynyl-*o*-carborane) in a graphite oven under a flow of argon gas. Boron carbide prepared in this manner was obtained in a ceramic yield of 68%, consistent with a mechanism proposed in Scheme II. The X-ray powder diffraction patterns exhibited no sharp lines, consistent with an amorphous structure. Crystalline boron carbide was obtained by pyrolysis at temperatures higher than 1300° C or prolonged heating of the amorphous material. Prelimnary work indicates films could be fabricated by evaporation of solutions of the polymeric alkynylcarborane derivatves and fibers could be drawn from the melted polymer. The films had a shiny, black appearance and appeared to retain some structural integrity. B_4C/SiC composites, as well could be prepared from pyrolysis of the polymeric silylakynyl carboranes.

SCHEME II

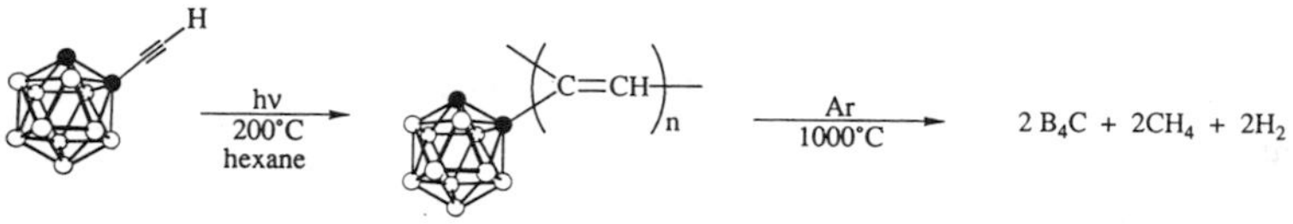

H

hv / 200°C / hexane → C=CH, n → Ar / 1000°C → $2\,B_4C + 2CH_4 + 2H_2$

CONCLUSIONS

These initial studies indicate that the akynyl functionalized carboranes described can serve as low temperature precursors to boron carbide. The synthetic methodology employed demonstrates that polymeric carborane derivatives are acccessible. The fact cross-linking occurs at relatively low temperatures minimizes losses of molecular fragments during pyrolysis and suggests it might have utility in applications where thermosetting behavior is desired. The method appears to be favorable for the fabrication of boron-containing ceramics such as thin films and composites of boron carbide [22,23]. Efforts towards processed ceramics from these precursors are in progress.

EXPERIMENTAL PROCEDURES

Standard glovebox, Schlenck and vacuum line techniques were employed for manipulation of air- and moisture-sensitive compounds. Photolysis and pyrolysis were performed in nitrogen or argon atmospheres. Reaction solvents were reagent grade and were distilled from appropriate drying agents under argon before use. THF was distilled from sodium benzophenone ketyl. The reagents tetra-*n*-butylammonium fluoride, *t*-butyldimethylchlorosilane and *n*-butyl lithium were purchased from the Aldrich Chemical Co. and used without further purification. Trimethylsilylethynylbromide was prepared using a literature method.

^{1}H and ^{13}C NMR spectra were recorded using a Bruker AF-200 (200.133 MHz, 50.324 MHz respectively) FT-NMR spectrometer. ^{11}B NMR spectra were recorded using a Bruker AM-

500 (160.46 MHz) FT-NMR spectrometer. Transmission IR spectra were obtained in Nujol or KBr matrices using a Beckman FT-1100 spectrometer over a range of 4000-500 cm^{-1}. The thermogravimetric analysis (TGA) results were obtained with a Perkin Elmer Thermostation using model TGS-2. X-ray diffraction patterns were obtained with a Phillips Diffractometer using Cu $K_{\alpha-1}$ radiation at 40 kV and 30 mA. Elemental analysis was conducted by Galbraith Laboratories, Knoxville, Tennessee.

The alkynyl carborane samples and polymers were exposed to ultraviolet radiation (200 - 700 nm) inside a nitrogen atmosphere glove box. Pyrolysis was conducted in a flowing argon atmosphere (99.999% Ar), graphite refractory furnace (Thermal Technology - Series 1000). All samples were heated at a rate of approximately 10°C/min.

Preparation of poly(1-ethynyl-*o*-carborane)

A soluble polymer was prepared by heating a neat sample of 1-ethynyl-*o*-carborane at 250-300° C for several hours under vacuum. As the reaction proceeded, a brown, viscous material was observed. At this point the flask was freeze-thawed to remove any non-condensable gas(es), leaving the viscous polymer, identified as poly(1-ethynyl-*o*-carborane). Autoclaved monomer afforded the same polymer in better yield. Treatment of a hexane solution of the monomer (**1**) under UV irradiation produces a polymer of similar composition. All polymers were characterized by a combination of ^{1}H, ^{13}C, ^{11}B NMR and IR spectral data.

Acknowledgements

The authors gratefully acknowledge the National Science Foundation for financial support of this research through contract grant DMR-90-14487. S. E. J. thanks NSF for a post-doctoral fellowship (CHE-90-01819).

REFERENCES

[1] K. Wynne and R. W. Rice, Ann. Rev. Mater. Sci. 14, 297 (1984).

[2] R. W. West, in Ultrastrructure Processing of Ceramics, Glasses, and Composites, edited by L. Hench and D. L. Ulrich (J. Wiley and Sons, New York, 1984).

[3] S. Yajima, Amer. Ceram. Soc. Bull. 62, 892 (1983).

[4] R. R. Willis, R. A. Markle, S. P. Mukherjee, Amer. Ceram. Soc. Bull. 62, 904 (1983).

[5] D. R. Richerson, Modern Ceramics Engineering, (Marcel Dekker, New York, 1983).

[6] N. N. Greenwood, The Chemistry of Boron, Vol. 8, (Pergamon, Oxford, 1973).

[7] C. A. Wood, Phys. Rev. B 29, 4582 (1984).

[8] M. G. Mirabelli and L. G. Sneddon, J. Amer. Chem. Soc. 110, 3305 (1988);

Sol. St. Ionics, 32-33, Pt. 2, 655 (1989).

[9] W. S. Rees, D. Seyferth, Ceram. Eng. Soc. Proc. 9 (7-8), 1009 (1983), ibid., 9 (7-8), 1021 (1983); Mater. Res. Symp. Proc. 121, 449 (1988).

[10] R. S. Johnson, U. S. Patent 4,810,436 (1989).

[11] R. W. Rice, K. J. Wynne, W. B. Fox, U. S. Patent 4,097,294 (1978).

[12] B. E. Walker, R. W. Rice, P. F. Becher, B. A. Bender, W. S. Coblenz, Am. Ceram. Soc. Bull., 62, 916 (1983).

[13] Gmelin Handbuch der Anorganische Chemie, Boron; Suppl., Vol. 2 (1981).

[14] T. D. Getman, P. M. Garrett, C. B. Knobler, M. F. Hawthorne, K. Thorne, J. D. Mackenzie, Organometallics, in press (1992).

[15] H. Zheng, K. Thorne, J. D. Mackenzie, X. Yang, M. F. Hawthorne, Mater. Res. Symp. Proc., in press (1991).

[16] S. E. Johnson, F. A. Gomez, M. F. Hawthorne, K. J. Thorne, J. D. Mackenzie, Eur. J. Sol. St. and Inorg. Chem., in press (1992).

[17] F. A. Gomez, S. E. Johnson, M. F. Hawthorne, J. Am. Chem. Soc. 113, 5915 (1991).

[18] F. A. Gomez and M. F. Hawthorne, J. Org. Chem. 57, 1384 (1992).

[19] S. E. Johnson and M. F. Hawthorne, unpublished results.

[20] T. E. Paxson, K. P. Callahan, M. F. Hawthorne, Inorg. Chem. 12, 708 (1973).

[21] See ref. 19.

[22] J. Jamet, J. R. Spann, R. W. Rice, D. Lewis, W. S. Coblenz, Am. Ceram. Eng. Sci. Proc. 5, 677 (1984).

[23] C. Wood, in Boron Rich Solids, edited by D. Edmin, T. Aselage, C. L. Beckel, I. A. Howard, C. Wood (Amer. Inst. of Physics, New York, 1986), pp. 362-372.

BORON CARBIDE NITRIDE DERIVED FROM AMINE-BORANES

JOACHIM BILL AND RALF RIEDEL
Max-Planck-Institut für Metallforschung (PML), Heisenbergstr. 5, 7000 Stuttgart 80, Germany, FRG

ABSTRACT

Amine-boranes such as pyridine- or piperazine-borane can be converted into infusible polymers by thermal crosslinking at temperatures up to 420°C. Further rise of the temperature up to 1050°C in argon results in transformation of the polymers into black residues. Microstructural (TEM/EELS, ESCA) and chemical investigations indicate the presence of single phase boron carbide nitrides which exhibit a graphite-like, turbostratic structure with a homogeneous distribution of the elements B, N, and C. Subsequent annealing at 2200°C in argon gives rise to crystallization of the pyrolytic material generating the thermodynamically stable phases BN, C, and B_4C.

The semiconducting properties of the X-ray amorphous boron carbide nitride synthesized at 1050°C depend on the B/N/C-ratio which can be influenced a) by the type of amine-borane-precursor and b) by the applied atmosphere (Ar or NH_3) during pyrolysis.

The amine-boranes can be converted into boron carbide nitride- and BN-monoliths at 1050°C under argon or reactive gas (NH_3), respectively. The monoliths are transformed into composites with 91% rel. density containing BN, C, and B_4C when heated up to 2200°C.

INTRODUCTION

Due to the similar crystal structures of α-BN and graphite and due to the isoelectronic properties of BN- and C_2-units the formation of solid solutions in the ternary system B-N-C can be expected. Boron carbide nitrides, i. e. carbon materials in which carbon atoms are partially substituted by boron and nitrogen can be prepared by the reaction of amorphous boron with carbon black in nitrogen at 1900 °C [1]. Alternatively, boron carbide nitrides have been synthesized in form of thin solid films deposited from a gaseous mixture containing BCl_3, C_2H_2, and NH_3 at 400-700 °C [2]. The boron and nitrogen containing graphite-like compounds exhibit interesting features such as semiconducting properties with a much lower band gap than α-BN [2]. However, the preparation of monolithic materials with sheet-like structures is difficult and requires the addition of binders or hot-pressing techniques. This observation is due to the strong covalent bonds within the layers and the weak van der Waals-forces between them. Very recently we found that pyrolysis of amine-boranes at low temperatures (1050 °C) results in the formation of boron carbide nitrides (BC_2N, BC_4N) in high yields [3, 4]. Boron carbide nitride-powder as well as monolithic materials can be prepared. In this paper the thermal decomposition of pyridine-borane ($C_5H_5N{\cdot}BH_3$) and piperazine-borane ($HNC_4H_8NH{\cdot}BH_3$) is discussed with respect to the preparation of metastable B-C-N-solid solutions. The crystallization behaviour and the electrical properties of the synthesized graphite-like materials are discussed.

EXPERIMENTAL PART

For the decomposition of pyridine- and piperazine-borane to boron carbide nitride the amine-boranes were heated with 5 °C/min in argon or ammonia up to 1050 °C with an isothermal hold for 80 min at this temperature. Isothermal heat treatments of the boron carbide nirides at $T > 1100$ °C were performed in a graphite furnace. For the preparation of the pyridine-borane derived BC_4N-monoliths pyridine-borane was heated under argon at 5 °C/min up to 110 °C with isothermal hold for 4 h. The resulting unmeltable foamed polymer was milled with ZrO_2-balls, sieved through a 80 μm screen in Ar, and cold-isostatically pressed into a cylindric green body. The polymeric green sample was heated under argon at 0.5 °C/min up to 1050 °C with isothermal hold for 2 h at this temperature. Heating piperazine-borane at 420 °C yields a yellow infusible polymer which can be converted to BC_2N monoliths according to the processing described in the case of pyridine-borane.

RESULTS AND DISCUSSION

Synthesis and characterisation of the boron carbide nitrides

The synthesis of boron carbide nitride by thermal decompositon of the molecular $BH_3 \cdot C_5H_5N$-complex proceeds via the formation of a highly crosslinked infusible and insoluble polymeric intermediate. The polymer is formed by hydroboration of the pyridine-ring occuring at a reaction temperature of 110 °C without loss of mass [7]. Further heating of the polymer up to 1050 °C in argon atmosphere results in the formation of a black residue. The overall yield has determined to be 57 wt.-%. The thermal decompositon is accompanied by the simultaneous loss of H_2 and CH_4 as detected by mass spectroscopy (MS) in the thermal gravimetric analysis (TGA/MS). The analytical data of the ceramic residue are given in Table I and correspond to the composition $B_{1.1-1.3}\ C_{3.8-4.1}\ N$. Equation 1 gives a schematic description of the overall decompositon reaction:

$$BH_3 \cdot C_5H_5N \xrightarrow{Ar,1050\ ^\circ C} BC_4N + 2H_2 + CH_4 \tag{1}$$

If the pyrolysis of the pyridine-borane-derived polymer is carried out in a reactive NH_3-atmosphere most of the carbon is removed and α-BN with overall yield of 26 wt.-% is formed (Eq. 2, Table I):

$$BH_3 \cdot C_5H_5N \xrightarrow{NH_3,1050^\circ C} BN + H_2 + RH\ (CH_4) \tag{2}$$

The thermal decomposition of piperazine-borane in argon at 1050 °C yields boron carbide nitride with the stoichiometry BC_2N (Table I). The overall yield is 64 wt.-%.

TABLE I
B-, C-, N-content and O-contamination of the pyridine- and piperazine-borane derived boron carbide nitrides prepared at 1050 and 2200 °C.

Amine-Borane	T[°C]/ Atmosphere	B	C [wt.-%]	N	O
$BH_3 \cdot C_5H_5N$	1050 °C/Ar	16.6-17.5	62.0-62.3	17.8-19.3	0.8
$BH_3 \cdot C_5H_5N$	1050 °C/NH_3	n. d.	0.74	53.0	2.6
$BH_3 \cdot C_4H_{10}N_2$	1050 °C/Ar	22.4	47.0	28.0	1.7
$BH_3 \cdot C_5H_5N$	2200 °C/Ar	17.7	65.8	15.9	$\leq$0.5
$BH_3 \cdot C_4H_5N$	2200 °C/Ar	28.5	46.4	24.3	$\leq$ 0.5

The X-ray diffraction patterns of the amine-borane derived boron carbide nitrides and boron nitride reveal only (hk0)- and (00l)- diffraction lines indicating the presence of a graphite-like turbostatic structure with roughly parallel layers which are otherwise completely disordered with respect to each other. Wave-length dispersive X-ray analysis (WDX) in the scanning electron microscope (SEM) of BC_2N and BC_4N exhibits a homogeneous distribution of B, N and C.

In the case of BC_4N a homogenous distribution of the elements with a resolution of +/- 20 Å is found by electron spectroscopic imaging (ESI) in the transmission electron microscope. Electron energy loss spectroscopy of the BC_4N-sample clearly reveals π^*-features corresponding to the sp^2-hybridization of the elements which is consistent with the layered structure of the boron carbide nitride [3]. The sp^2-character of the elements is also supported by ^{13}C-solid state-NMR spectroscopy. Electron spectroscopy for chemical analysis (ESCA) reveals the presence of B-N-, B-C-, C-C-, and C-N-bonds [4]. Therefore, BC_4N has to be considered a single phase metastable boron carbide nitride. The conversion of the pyridine-borane derived polymer into boron carbide nitride is schematically shown in equation 3.

1000 °C, $-CH_4$, $-H_2$ (3)

Thermal decomposition of boron carbide nitride into the thermodynamical stable phases

Pyridine-borane derived BC_4N subsequently heat-treated at 1800 °C for 2h in Ar still contains (hk0)- and (00l)-diffraction lines exclusively in the XRD pattern. However, the inter-layer spacing in BC_4N is reduced from d = 3.609 Å at 1400 °C to d = 3.457 Å in the sample annealed at 1800 °C for 2h in Ar as could be determined by the diffraction angle of the (002)-reflex. ESI in TEM-mode reveals the presence of a still homogenous distribution of the elements B, N, and C. Additionally, small amounts of elemental boron are detected, which is consistent with the results of Saugnac et al. [7]. At temperatures exceeding 1800 °C, BC_4N decomposes according to the reaction given in equation (4):

$$BC_4N \xrightarrow{Ar,\ 2200^\circ C} 0.068\ B_4C + 0.729\ BN + 3.932\ C + 0.136\ N_2 \qquad (4)$$

Similarly the piperazine-borane derived composition BC_2N starts to decompose at T > 1800 °C giving a mixture of BN, C, and B_4C as can be determined by X-ray powder diffraction. The analytical data of BC_4N and BC_2N heat-treated at 2200 °C in Ar for 2h are summarized in Table I.

Ceramic monoliths derived from amine-boranes

The pyridine- and piperazine-borane derived BC_4N-monoliths could be prepared as described in the experimental part. The overall process is shown in a flow diagram given in Fig 1.

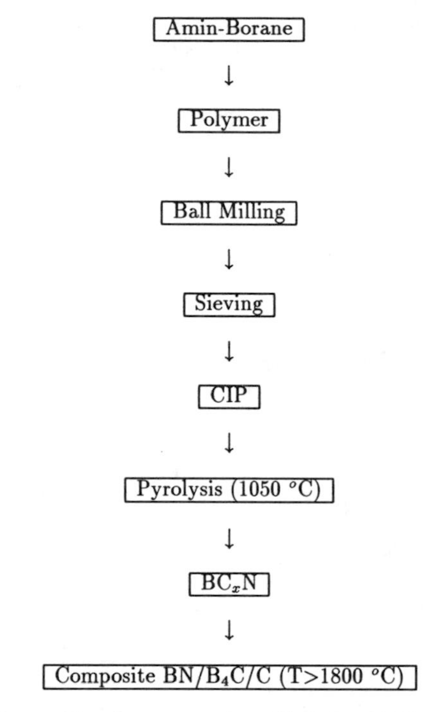

Fig. 1 Flow diagram for the preparation of boron carbide nitride monoliths.

In the case of pyridine-borane derived boron carbide nitride (BC_4N) monoliths obtained at 1050 °C an open porosity of 20-25 % is determined by means of mercury-pressure-porosimetry. The pore volume comprises pore radii smaller than 400 nm. The average pore radius is 300 nm. Pyrolysis of the pyridine-borane derived polymeric green compact under ammonia instead of argon results in pale yellow monoliths containing boron nitride. The open porosity of the boron nitride- and the piperazine-borane derived BC_2N-monoliths is determined to be ca. 30 %. The BC_4N-parts crystallize into ceramic composites containing BN, C, and B_4C by heating the samples up to 2200 °C which is combined with a linear shrinkage of 15% (Fig. 2). The densification of the monoliths during the heat treatment up to 1800 °C is associated with an increase of the bulk density from 1.59 g/cm^3 at 1050 °C to 1.82 g/cm^3 at 1800 °C. At temperatures exceeding 1800 °C the sample starts to sinter and the open porosity is reduced from 20% at 1800 °C to 9% at 2200 °C.

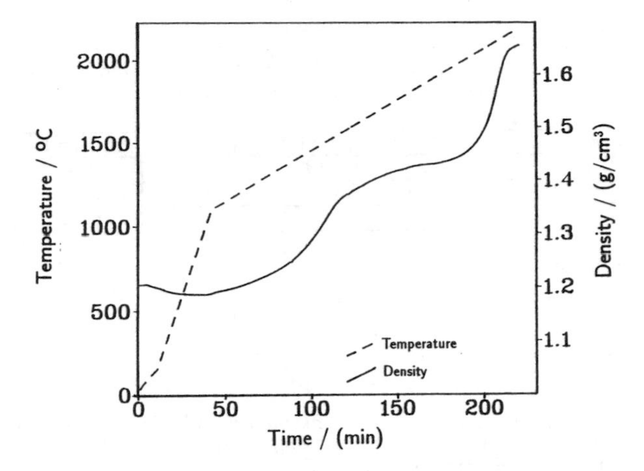

Fig. 2 Dilatometry of BC_4N.

Physical properties of boron carbide nitride

Electrical conductivity two-probe measurements reveal that the as-synthesized boron carbide nitride monoliths (BC_2N and BC_4N) exhibit semiconducting properties. In Fig. 3 the logarithm of the electrical conductivity of BC_4N is plotted versus the reciprocal temperature. The slope of the measured line in the range from 20 to 500°C implies a thermal activation energy for the transport of the charge carriers of 0.07 eV. In contrast to the BC_xN materials the pure BN monoliths show insulating properties. The determined electrical and thermal properties of BC_4N, BC_2N, and BN are summarized in Table II.

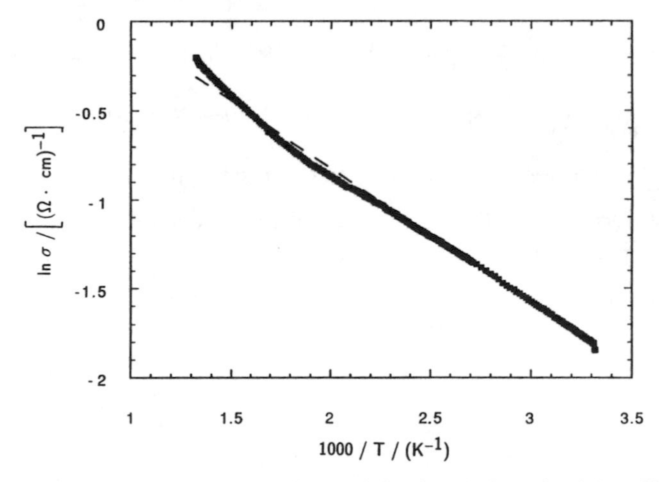

Fig. 3 Temperature dependence of the electrical conductivity of BC_4N.

TABLE II
Open porosity, electrical conductivity at room temperature (σ_{RT}), activation energy for the transport of the charge carriers (E_a), and thermal coductivity at room temperature (λ) of the prepared monoliths.

Sample	Porosity/%	σ_{RT} / $(\Omega \cdot cm)^{-1}$	E_a [eV]	λ [W/mK]
BC_4N	20-25	$1.6 \cdot 10^{-1}$	0.07	0.4
BC_2N	30	$5.5 \cdot 10^{-4}$	0.11	n. d.
BN	30	10^{-10}	n. d.	n. d.

n. d.: not determinded

CONCLUSIONS

Pyridine- and piperazine-borane are suitable precursors for the synthesis of single phase boron carbide nitrides (BC_xN) at low temperatures (1050 °C). The BC_xN decompose above 1800 °C forming crystalline B_4C, BN, and C.

Ceramic powders and monoliths comprised of semiconducting boron carbide nitride or insulating BN as well as ultrafine B_4C/BN/C-composites can be prepared.

ACKNOWLEDGEMENTS

The authors wish to thank V. Szabo, Max-Planck-Institut für Metallforschung Stuttgart, for the ESI measurement in the TEM-investigations. This work was financially supported by the Keramikverbund Karlsruhe Stuttgart (KKS). J. Bill also gratefully acknowledges the financial support of the KSB-Stiftung, 7000 Stuttgart 80, FRG.

REFERENCES

1. E. V. Prilutskii, G. N. Makarenko, T. I. Serebryakova, Vysokotemperatur. Karbidy, 84-89(1975).
2. R.B. Kaner, J. Kouvetakis, C.E. Warble, M.L. Sattler, N. Bartlett, Mater. Res. Bull. 22, 399(1987).
3. R. Riedel, J. Bill, and G. Passing, Adv. Mater. 3, 551(1991).
4. J. Bill, R. Riedel und G. Passing, Z. anorg. allg. Chem., in press.
5. J. Bill, M. Frieß, and R. Riedel, Eur. J. Solid State Inorg. Chem., (in press).
6. F. Saugnac, F. Teyssandier, and A. Marchand, J. Am. Ceram. Soc. 75, 161(1992).

POLYMER PRECURSORS FOR ALUMINUM NITRIDE CERAMICS

JAMES A. JENSEN
Hercules Incorporated, Research Center, Wilmington, DE 19894-0001.

ABSTRACT

The aluminum imine complexes $[RCH{=}NAlR'_2]_2$ undergo thermal condensation polymerization to yield alazane [poly(alkylaminoalane)] polymers. The aluminum imine complexes are prepared in quantitative yield by reduction of nitriles, such as acetonitrile and benzonitrile, with alkylaluminum hydride reagents. These alazane polymers are thermoplastic, soluble in aprotic solvents, and display low moisture and oxygen sensitivity. The alazanes convert to aluminum nitride ceramics in non-oxidizing atmospheres above 500°C.

INTRODUCTION

Aluminum nitride is a refractory, low dielectric ceramic which possesses high intrinsic thermal conductivity. AlN also has good mechanical properties making it suitable for both electronic and structural applications. These valuable properties have resulted in research activity on a variety of AlN processing techniques [1]. There is currently a great deal of interest in utilizing polymer precursors which can be pyrolyzed to yield ceramic materials, including aluminum nitride [2-4]. The use of polymer technology to prepare ceramic materials offers certain processing advantages over conventional ceramic processing routes. Polymer precursors can be fabricated into complex shapes using a variety of standard plastic forming techniques such as molding, fiber spinning, film casting, coating, and infiltration.

The present study describes thermoplastic alazane polymers prepared by thermal polymerization of alkylaluminum imine complexes. Alkylaluminum imine complexes can be prepared by aluminum hydride reduction of nitriles [5].

EXPERIMENTAL

Synthesis of aluminum imine complexes $[RCH{=}NAl(i\text{-}C_4H_9)_2]_2$ with a variety of R groups has been described [5,6]. In a typical experiment, a 250 ml Schlenk round bottom flask was fitted with a pressure equalized dropping addition funnel, purged, and cooled to 0°C. Acetonitrile (20 ml, 479 mmol) was added to the flask. Diisobutylaluminum hydride (100 ml, 1.0 M in toluene, 100 mmol) was charged to the funnel and added dropwise over 30 minutes. The colorless solution was warmed to room temperature and stirred four hours. The solvent was removed under vacuum leaving 18.0 g (98%) $[CH_3CH=NAl(i\text{-}C_4H_9)_2]_2$. [$^1H$ NMR (C_6D_6, 300 MHz): δ 8.08 (q, 1H, $\underline{H}C=N$), 2.11 (m, 2H, >C$\underline{H}$), 1.74 (d, 3H, $C\underline{H}_3C=N$), 1.25 (dd, 12H, $C\underline{H}_3$-CH), 0.37 (dd, 4H, $C\underline{H}_2Al$) ppm.]

Thermolysis of the neat imine $[CH_3CH=NAl(i\text{-}C_4H_9)_2]_2$ (75 g, 409 mmol) under nitrogen from room temperature to 300°C for 8 hours results in evolution of isobutene and isobutane and formation of the alazane polymer $\{[(C_2H_5)NAl(C_4H_9)]_x[(C_2H_5)NAl(R')]_y[(C_2H_5)NAl]_z\}_n$. Upon cooling to room temperature, a clear red-orange solid forms. The softening point of the solid polymer is 100°C. [^1H NMR (C_6D_6, 300 MHz): δ 3.2 (m, 2H, $C\underline{H}_2N$), 3.0 (t, 3H, $C\underline{H}_3CH_2N$) 2.1 (m, 1H, -CH<), 1.9 (m, 6H, $C\underline{H}_3CH$), 0.5 (m, 2H, $C\underline{H}_2Al$) ppm.] Elemental analysis provides an empirical formula of $C_{4.6}H_{11}NAl$. The value of n determined from size exclusion chromatography and vapor phase osmometry is 6 to 20.

Thermal gravimetric analyses were conducted on a DuPont 951 instrument using platinum boats in a nitrogen atmosphere with a temperature ramp of 20°C per minute. Molecular decomposition FT-IR was conducted in a closed stainless steel cell fitted with KCl windows. The sample was heated on a tungsten filament in a static nitrogen atmosphere and the volatile products observed as a function of temperature.

The alazane polymers were pyrolyzed to AlN in an Astro Model 1000 furnace using graphite heating elements and crucibles. Pyrolysis gases were purified through a Nanochem® Gas Purification System. Gas flows were set at 1 l/min. Pyrolyses were conducted at 10°C/min from room temperature to the maximum temperature with a 2 hour hold at the maximum temperature. Pyrolysis in an ammonia-containing atmosphere was conducted in ammonia from 25-1000°C and in nitrogen above 1000°C. X-ray powder diffraction patterns (XRD) were obtained on a Philips APD3600/02 diffractometer with a Cu-Kα source.

RESULTS AND DISCUSSION

Dialkylaluminum imines were prepared at room temperature by reduction of a nitrile with a dialkylaluminum hydride reagent. The reaction is general for a variety of nitriles. Low molecular weight nitriles, such as acetonitrile and benzonitrile, form liquid products upon reduction with diisobutylaluminum hydride.

Thermolysis of the neat imines from room temperature to 300°C in an inert atmosphere yields an alazane polymer with concomitant release of both isobutane and isobutene. Thermolysis at 300°C removes 1.3 equivalents of the aluminum bound isobutyl groups in the form of isobutane and isobutene. The isobutane/isobutene ratio varies with temperature. Below 150°C isobutane is the major by-product, but above 150°C isobutene is the main off-gas.

The resulting polymer has a softening point near 100°C and is readily soluble in aprotic solvents. The polymer is represented by the formula

$$\{[(C_2H_5)NAl(C_4H_9)]_x[(C_2H_5)NAl(R')]_y[(C_2H_5)NAl]_z\}_n \qquad \mathbf{I}$$

Under the conditions described above, x + y = 0.65, z = 0.35, the value of n ranges between 6 and 20 and R' is an octyl group bridging two aluminum centers formed by C_4-C_4 coupling [6]. The values of x and y are related to the quantity of isobutyl groups evolved as either isobutane or isobutene and vary with polymerization time and temperature. There are two types of gross aluminum environments in the polymer: those retaining one Al-C bond as in units x and y, and those bonded only to nitrogen as in unit z. Alazane polymer **I** shows very low moisture and oxygen sensitivity; the polymer be handled for short periods of time in air without appreciable degradation of properties.

Polymer **I** can be processed using typical thermoplastic polymer fabrication techniques. In addition, coatings and films may be cast from solutions of the polymer in hydrocarbon solvents.

The "char yield" of alazane polymer **I** in nitrogen, as determined by thermal gravimetric analysis, is 65% at 1000°C, see Figure 1. Only slight (≈ 6%) weight loss was observed below about 250°C; primary loss of volatiles is complete by 500°C. Figure 2 displays the infra-red absorbance trace of the evolved volatile products as a function of temperature as obtained by molecular decomposition FT-IR (MD/FT-IR) analysis. The specific molar absorbance of each compound is not accounted for in the figure so the relative absorbance is not directly representative of the absolute concentration of gas in the cell. The ethylene observed above 300°C results from decomposition of the N-bound ethyl group and does not result from secondary reactions involving cleavage of C_4 hydrocarbons. This was

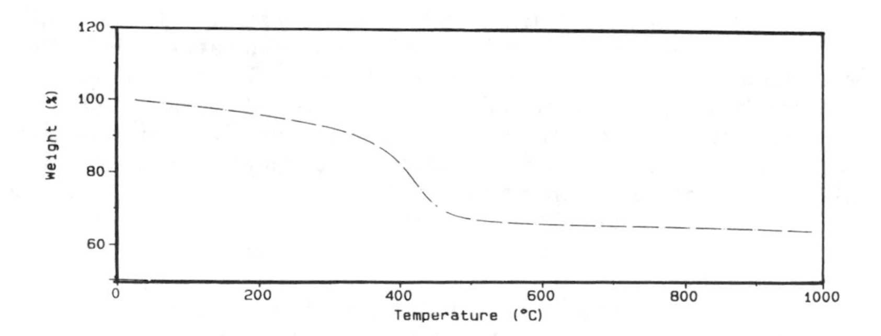

Figure 1. Thermal gravimetric analysis of polymer **I** in nitrogen.

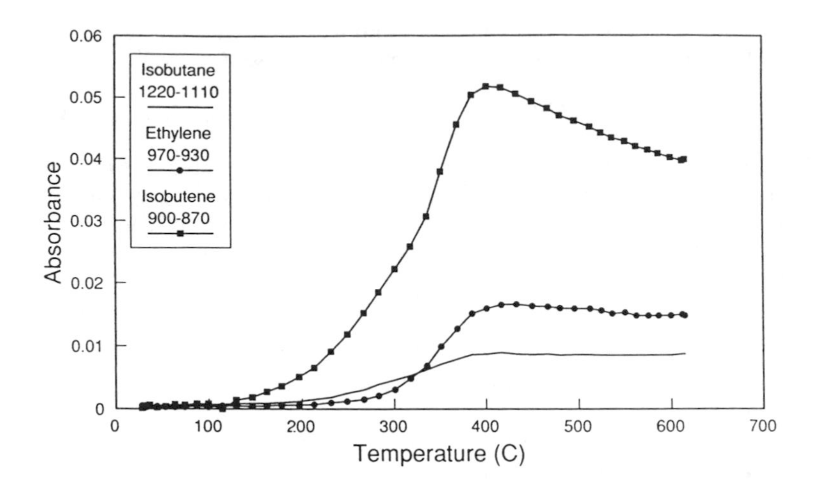

Figure 2. Molecular decomposition FTIR of polymer I.

confirmed by substitution and labeling of the nitrogen substituent: polymers containing N-bound benzyl groups showed no ethylene, while polymers containing CD_3CH_2-N groups evolved deuterium-labeled ethylene. Notably, ethylamine was not observed on pyrolysis indicating N-C and Al-C bond cleavage is favored over Al-N cleavage under the pyrolysis conditions.

Figure 3 shows the X-ray powder diffraction patterns of alazane polymer I pyrolyzed in ammonia or ammonia/nitrogen atmospheres as a function of temperature. The evolution of AlN crystallization is evident. At 500°C a primarily amorphous pattern is observed. By 1000°C all five AlN peaks in the $20 < 2\theta < 60$ degree region are identifiable. By 1200°C baseline separation of the (100), (002), and (101) peak envelope is evident. A fully crystalline AlN product results at 1600°C. Samples pyrolyzed in ammonia-containing atmospheres are grey below 800°C and white above 800°C. Bulk elemental analysis indicate pure AlN is obtained (< 0.5 wt% carbon and < 0.5 wt% hydrogen impurities) from samples fired at or above 1000°C.

For comparison, alazane polymer **I** was also pyrolyzed in nitrogen at various temperatures. The X-ray patterns are shown in Figure 4. It is evident that crystallization requires a higher temperature in nitrogen pyrolysis vis à vis ammonia. At 800°C the pattern is amorphous. Individual peaks are evident in the 1300°C pattern. Samples fired in nitrogen are black.

Application of the Scherrer equation [7] to the XRD patterns provides an estimation of the average crystallite size as a function of pyrolysis temperature. Calculations using the (110) reflection yield average crystallite sizes ranging from 18 Å at 500°C to about 70 Å at 1000°C in ammonia. In nitrogen the values range from 32 Å at 1000°C to 92 Å at

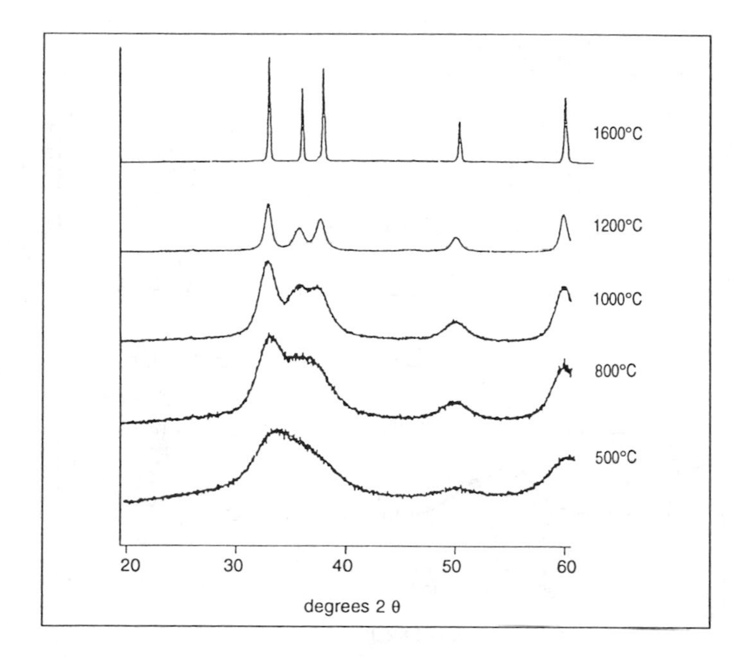

Figure 3. X-ray patterns of polymer **I** pyrolyzed in ammonia.

1300°C. The average crystallite sizes determined by the Scherrer equation compare favorably with microscopy observations which indicate pyrolysis in nitrogen at 1000°C results in crystallites under 100 Å in size, while nitrogen pyrolysis to 1800°C yields micron sized crystals.

CONCLUSIONS

Alazane polymers of the form $\{[(C_2H_5)NAl(C_4H_9)]_x[(C_2H_5)NAl(R')]_y$-$[(C_2H_5)NAl]_z\}_n$ are prepared by thermolysis of aluminum imine complexes, $[RCH{=}NAlR'_2]_2$. The polymers have low moisture and oxygen sensitivity, are thermoplastic and are soluble in aprotic solvents allowing facile processing into useful shapes such as coatings and films. Pure AlN is formed by pyrolysis of the alazane polymer in ammonia atmospheres. Bulk elemental analysis indicates <0.5 wt% C and <0.5 wt% H are present.

ACKNOWLEDGEMENTS

I wish to acknowledge A. Lukacs for invaluable guidance, L. J. Hanna and J. A. Mahoney for technical support, D. E. Pivonka for obtaining the pyrolysis FTIR data, and R. H. Kridler for collecting the XRD data.

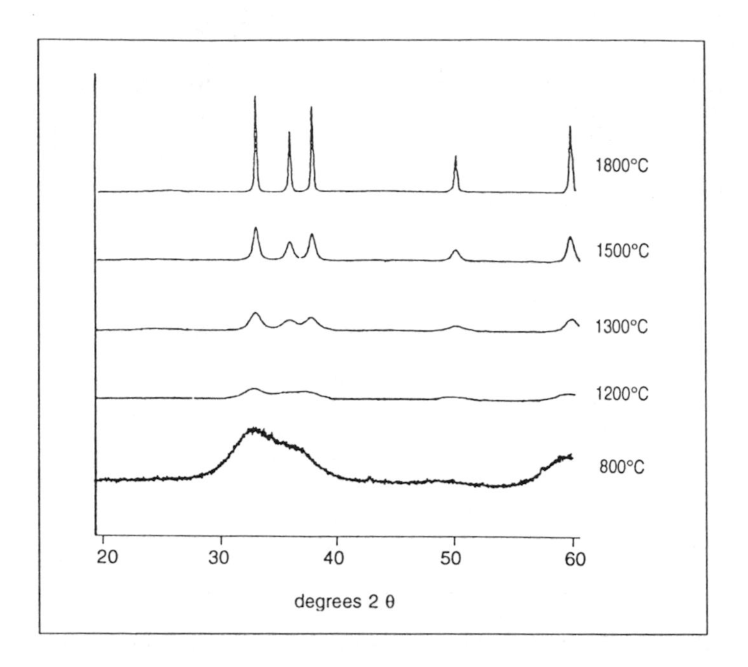

Figure 4. X-ray patterns of polymer **I** pyrolyzed in nitrogen.

REFERENCES

1. G. W. Prohaska and G. R. Miller in Mater. Res. Soc. Proc., 167, Pittsburgh, PA, (1990), pp. 215-227.
2. G. Pouskouleli, Cer. Int. 15, 213, (1989) and references therein.
3. R. T. Baker, J. D. Bolt, G. S. Reddy, D. C. Roe, R. H. Staley, F. N. Tebbe, and A. J. Vega in Better Ceramics Through Chemistry III, edited by C. J. Brinker, D. E. Clark and D. R. Ulrich (Mater. Res. Soc. Proc. 121, Pittsburgh, PA 1988) pp. 471-476.
4. M. Seibold and C. Russel in Better Ceramics Through Chemistry III, edited by C. J. Brinker, D. E. Clark and D. R. Ulrich (Mater. Res. Soc. Proc. 121, Pittsburgh, PA 1988) pp. 477-482.
5. L. I. Zakharkin, and I. M. Khorlina, Proc. Acad. Sci. USSR, 112, 879, (1957).
6. J. A. Jensen (submitted for publication).
7. P. Scherrer, Ges. Wiss. Gottingen, 1918, 98.

OPTIMIZATION OF ALUMINUM NITRIDE THERMAL CONDUCTIVITY VIA CONTROLLED POWDER PROCESSING

THERESA A. GUITON, JAMES E. VOLMERING, KRIS K. KILLINGER
Dow Chemical, Advanced Ceramics Laboratory, 1776 Building, Midland, MI 48674

ABSTRACT

The emergence of aluminum nitride ceramics in electronic and structural applications has been largely due to the increased availability of high quality powders. Much of the process development has been concentrated in capturing markets within the electronics industry, mainly high thermal conductivity substrates. The thermal conductivity of AlN substrates is strongly dependent on oxygen chemistry and sintering parameters. A key source of oxygen is AlN powder impurities. In this study, powders with differing oxygen contents were formulated with various Y_2O_3 levels. The results indicate that the distribution of sintered oxygen is strongly correlated with the relative concentration of the powder's lattice and surface oxygen. Under the sintering parameters of the analysis, optimal thermal conductivities were achieved when $Y_4Al_2O_9$ was present as the grain boundary phase. As predicted, the powders with the lowest lattice oxygen concentrations exhibited the highest thermal conductivities. Hence, depending upon the surface oxygen content, different yttria levels were required for optimal thermal conductivities, i.e., higher surface oxygen contents required higher sintering additive levels than did low surface oxygen powders.

INTRODUCTION

Aluminum nitride (AlN) has generated increasing interest as a microelectronic substrate material [1-3]. With a thermal conductivity approaching that of BeO and a thermal expansion coefficient well matched to silicon, AlN represents an attractive alternative in high power or multi-chip module applications [3-4]. The emergence of aluminum nitride ceramics in electronic and structural applications has been largely due to the increased availability of high quality powders. There remains, however, some concern as to the consistency of commercial AlN substrate properties, e.g., thermal conductivity, appearance, and metallization, as compared to traditional ceramic materials.

At room temperature, single crystal AlN has a theoretical thermal conductivity of 319 W/m-K [5-6]. Polycrystalline ceramics have lower thermal conductivities due to random orientation of grains, impurities dissolved in the crystalline lattice, and lower thermal conductivity grain boundary phases [7]. It is well known that the thermal conductivity of AlN substrates is strongly dependent on oxygen concentration, namely lattice oxygen, and sintering parameters [7-13]. A key source of oxygen impurities is the AlN starting material [13]. The distribution of the oxygen-containing impurities is a function of the powder's synthesis method. For powders prepared by carbothermal reduction of aluminum oxide, oxygen remains from incomplete conversion of Al_2O_3 and/ or impurities such as silica. This residual oxygen is identified as lattice oxygen. As a result of the incorporation of oxygen, charge compensating aluminum vacancies are created in the AlN lattice. It is the aluminum vacancies which cause phonon scattering and hence lower the thermal conductivity [5-6, 8].

The second source of powder oxygen is an artifact from processing "crude" AlN into finished powder [13]. For example, the carbothermal process involves an additional oxidation step to remove excess carbon. This step generates an oxygen rich surface layer on the powder. Enhanced densification of AlN is achieved via reaction of this oxide surface with select sintering additives, typically alkaline earth or rare earth oxides [14-16]. These sintering additives not only promote densification via formation of intermediate liquid phases, but they also enhance the thermal conductivity of AlN.

Previous studies have demonstrated that the AlN powder with the lowest oxygen generates the highest thermal conductivity substrate [12-13, 17]. In each study, the sintering additive level was held constant. Separate studies have demonstrated that for any given powder there appears to be an optimal sintering additive level [7-8, 17] where maximum thermal conductivity is achieved. However, the previous studies did not examine the interdependence of powder oxygen chemistry and optimization of sintering additives. This study will examine this interdependence and its impact on the thermal conductivity of AlN.

EXPERIMENTAL PROCEDURE

The aluminum nitride powders utilized in this investigation were prepared via the carbothermal reduction process (Dow Chemical). In total, four powders were evaluated and are hereafter designated as powders "A" (XUS 35544), "B" (XUS 35544),"C" (XUS 35548) and "D" (XUS 35548). Oxygen and carbon levels were determined via combustion analysis (LECO TC-136 and LECO IR-212 respectively). The distribution of the surface and lattice oxygen was estimated using combustion analysis of powders from which the surface oxygen had been removed via a reductive thermal treatment. The metallic impurities were determined via x-ray fluorescence. Surface areas were determined via single point BET measurements. A summary of the chemical impurities and surface areas is given in Table I.

Table I. Analyses of Aluminum Nitride Powders

Powder Identification	"A"	"B"	"C"	"D"
Oxygen (wt%)	1.08	0.95	0.76	0.67
Carbon (wt%)	0.03	0.05	0.04	0.04
Iron (ppm)	12	23	40	9
Calcium (ppm)	201	134	46	54
Silicon (ppm)	42	77	77	51
Surface Area (m^2/g)	3.21	3.37	2.66	2.49

Each powder was formulated using a standard composition of AlN with 1 to 5 wt% Y_2O_3 (Molycorp, 99.99%) in an organic solvent compounded with 6 wt% organic binder. The specimens were uniaxially pressed at a pressure of 104 MPa. The binder was removed thermally in a nitrogen atmosphere at 550°C. The greenware were sintered using a boron nitride crucible in a graphite resistance furnace. Two sintering cycles were employed and are given in Table II.

Table II. Sintering Parameters

Sintering Cycle	Cycle 1	Cycle 2
Heating Rate	5°C/min	2.5°C/min
Soak Temperature	1850°C	1850°C
Soak Time	2 hours	3 hours
Cooling Rate	50°C/min	1°C/min
Cooling Temperature	1000°C	1500°C

Densities were measured via the Archimedes method. Thermal conductivities were determined using a CO_2 laser flash device [23]. Crystalline phases within the sintered parts were identified using powder x-ray diffraction (Cu-Kα radiation).

RESULTS AND DISCUSSION

A series of formulations using powders "A", "B", and "C" with 1 to 5 wt% Y_2O_3 was sintered following cycle 1. The measured thermal conductivities are shown in Figure 1. The powder with the low oxygen level ("C") exhibits a maximum thermal conductivity at 1.5-2.0

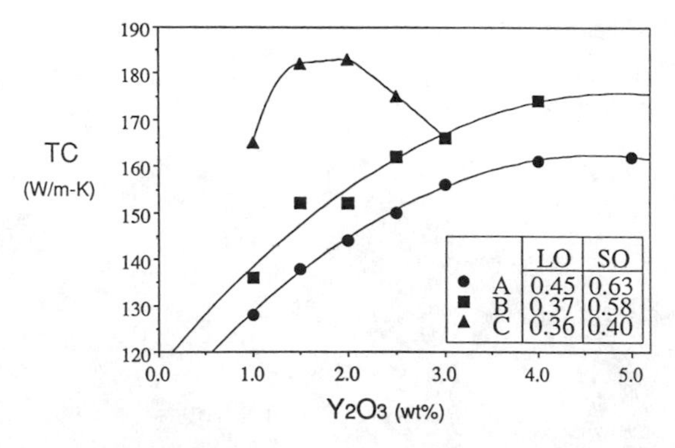

Figure 1. Thermal conductivity as a function of Y_2O_3 content for powders "A", "B", and "C".

wt% Y_2O_3 whereas the higher oxygen levels associated with "A" and "B" result in a shift in the maximum thermal conductivity to 4-5 wt% Y_2O_3. It was also noted that the density of parts sintered with "C" decreased as the Y_2O_3 level exceeded 2.5 wt%.

The crystal structure and phase chemistry of the grain boundary phases were studied by powder x-ray diffraction. The grain boundary phases were determined to be single-phase or two-phase mixtures of Y_2O_3 [1:0], $Y_4Al_2O_9$ [2:1], $YAlO_3$ [1:1], and/ or $Y_3Al_5O_{12}$ [3:5]. The phase compositions were estimated as follows. Reference peaks were selected for each of the phases: (222) for Y_2O_3, $(\bar{1}22)$ for $Y_4Al_2O_9$, (121) for $YAlO_3$, and (400) for $Y_3Al_5O_{12}$. After correcting for scattering factors, ratios of the reference peaks were obtained. The resulting relative phase chemistry of the residual yttrium aluminates is shown in Figure 2. As the Y_2O_3 content was increased, the stoichiometry of the second phases shifted toward more yttrium-rich aluminates. In addition, at constant Y_2O_3 levels, the highest oxygen "A" powder exhibited a more alumina-rich chemistry relative to "B", while the lowest oxygen "C" powder exhibited a more yttria-rich chemistry relative to "B". When the thermal conductivity is examined as a function of relative phase chemistry, Figure 3, it is evident that peak thermal conductivity is observed when $Y_4Al_2O_9$ is the primary second phase.

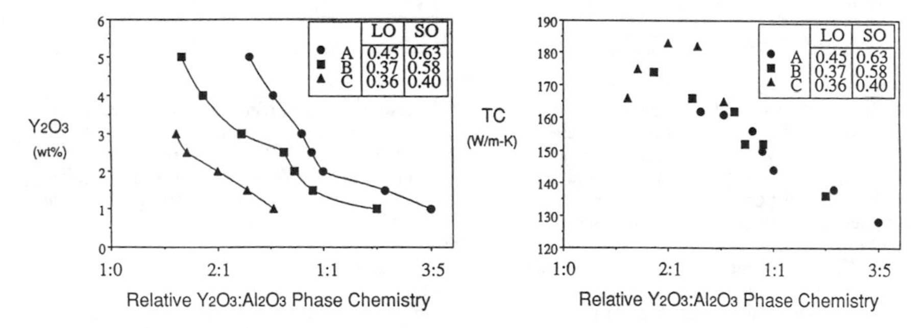

Figure 2. Relative yttrium aluminate phase chemistry measured as a function of Y_2O_3 levels.

Figure 3. Thermal conductivity versus relative yttrium aluminate phase chemistry.

These data can be interpreted in light of the distribution of oxygen in the original powder and in the sintered specimens. A high resolution transmission electron micrograph of an AlN particle, displayed in Figure 4, shows that a native "AlON" layer is found on the surface of the powders. Through a thermal treatment, this layer can be removed. Analysis of the oxygen levels before and after this treatment provide an upper estimate of the concentration of oxide in this surface layer. The remaining oxygen is, then, assumed to be dissolved in the AlN crystal lattice. The derived surface and lattice oxygen contents for the powders are given in Table III.

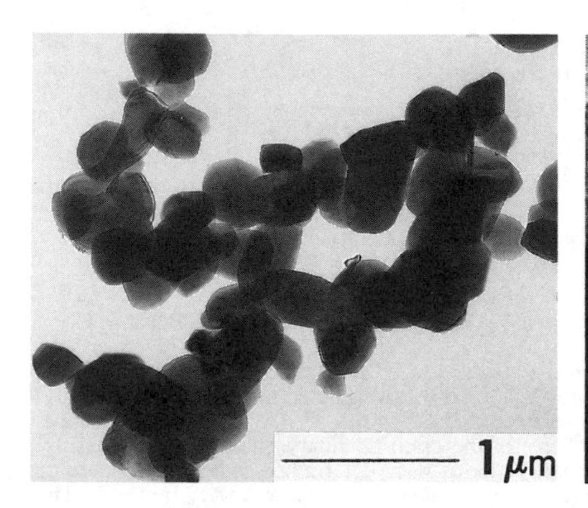

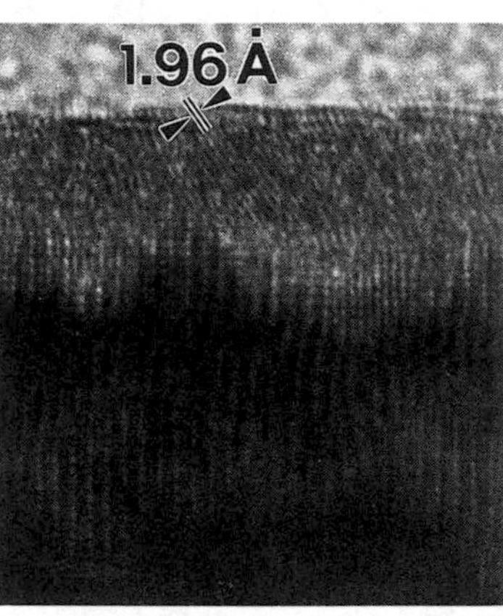

Figure 4. Transmission electron micrograph (TEM) of powder "A" (30,000x) (left). High resolution TEM (4,400,000x) recorded along the [0001] zone axis, showing the edge of an AlN particle. The thickness of the AlON surface coating varies from 20-40 Å (right).

Table III. Estimated Surface and Lattice Oxygen Levels

Identification	"A"	"B"	"C"	"D"
Total Oxygen (wt%)	1.08	0.95	0.76	0.67
Lattice Oxygen (wt%)	0.45	0.37	0.36	0.30
Surface Oxygen (wt%)	0.63	0.58	0.40	0.37

The sintering results are consistent with this difference in the surface oxygen levels. The formation of an yttrium aluminate liquid phase which promotes consolidation involves a reaction between Y_2O_3 and the surface oxide. In addition, high thermal conductivity is associated with a low oxygen activity, i.e., yttria-rich phase chemistry. Thus, higher Y_2O_3 additions are required with elevated powder oxygen content to maintain a low lattice oxygen concentration in the sintered part. Excess Y_2O_3 however, will impede densification. As the Y_2O_3 level increased beyond the $Y_4Al_2O_9$ phase region, the melting point of the composition increases. Thus, at 1850°C, the concentration of liquid necessary to facilitate densification is reduced.

In addition to powder chemistry, sintering conditions have a dramatic effect on thermal conductivity. Numerous studies have verified that higher sintering temperatures and longer sintering times result in higher thermal conductivities [7-12, 17]. These higher thermal conductivities are generally attributed to lattice oxygen purification, migration of second phases to isolated triple points, and reduction in second phase content. In this study, it has also been demonstrated that cooling rate has a significant impact on the thermal conductivity of polycrystalline AlN.

To demonstrate the impact of sintering conditions, a second sintering cycle was conducted on a low and high oxygen powder: "A" and "D". Figure 5 compares the thermal conductivities for cycles 1 and 2 as a function of Y_2O_3 content. Again, the lowest lattice oxygen powder generated a higher maximum thermal conductivity than did the higher lattice oxygen powder. Furthermore, the low surface oxygen powder required lower sintering additive levels for optimal conductivity. X-ray diffraction analysis of samples from sintering cycles 1 and 2 indicated nearly equivalent relative phase composition. However, the concentration of the yttrium aluminates was reduced as was evident by yttrium x-ray fluorescence analysis (Table IV).

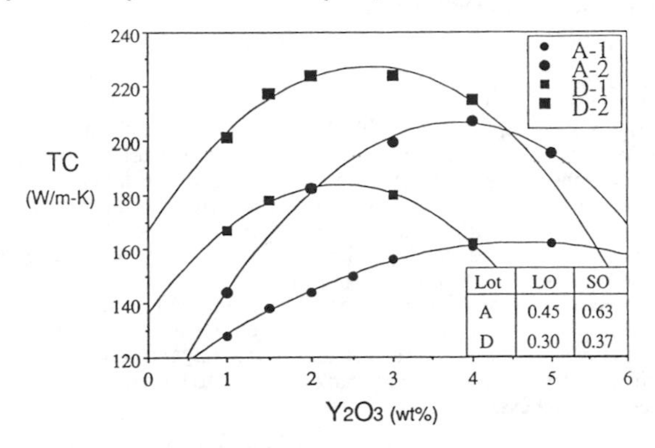

Figure 5. Thermal conductivity as a function of Y_2O_3 content for powders "A" and "D" sintered using cycles 1 and 2. Powder "A", cycle 1 (A-1); powder "A", cycle 2 (A-2); powder "D" cycle 1 (D-1), and powder "D", cycle-2 (D-2).

Table IV. Y_2O_3 Content of Parts Sintered Using Powder "D"

Target Y_2O_3 (wt%)	1.0	1.5	2.0	3.0	4.0
Cycle 1- Y_2O_3 (wt%)	0.93	1.38	1.83	2.96	3.96
Cycle 2- Y_2O_3 (wt%)	0.94	1.19	1.59	2.67	3.76

At constant Y_2O_3 levels, microstructural analysis of the specimens indicated equivalent AlN grain sizes for cycles 1 and 2. In contrast, the yttrium aluminate grain boundary microstructure had shifted from fine, evenly distributed grains in cycle 1 to coarse, isolated triple point grains in cycle 2, see Figure 6. Thus, the elevated thermal conductivities measured for cycle 2 specimens

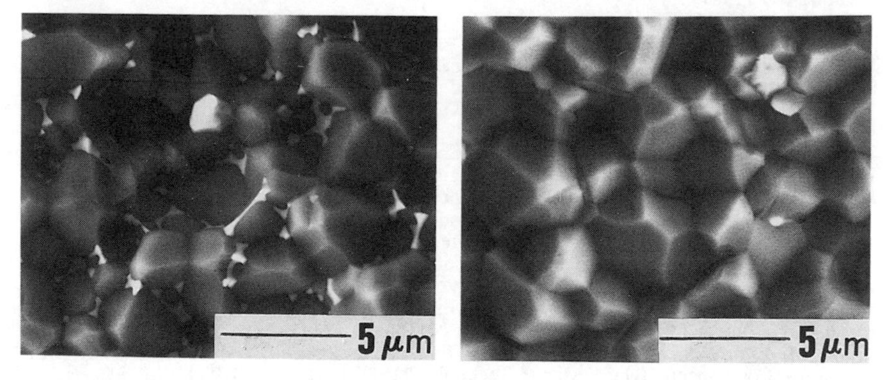

Figure 6. Scanning electron micrographs of sintered AlN microstructures for cycle 1 (left) and cycle 2 (right) (magnification = 5000x).

are attributed to removal of low thermal conductivity intergranular yttrium aluminate phases. Although direct lattice oxygen measurements have not been obtained for these specimens, it is hypothesized that the lattice oxygen of the sintered specimens is reduced due to extended annealing times at elevated temperatures.

CONCLUSIONS

In this study, a series of powders with differing surface and lattice oxygen contents were obtained and formulated using a range of yttrium oxide levels. The resulting components demonstrated the importance of optimizing sintering additive levels as a function of powder chemistry. It was found that optimal thermal conductivities were achieved when $Y_4Al_2O_9$ was present as the primary second phase. As predicted, the lowest lattice oxygen powders produced the highest absolute values. Hence, depending upon the surface oxygen content of the powders, different yttrium oxide levels are required for optimal thermal conductivities, i.e., higher surface oxygen contents require higher sintering additive levels than do low surface oxygen powders. Thus, it was determined that the distribution of oxygen in sintered products is strongly correlated with the relative concentration of the surface and lattice oxygen content in the raw powder. As a result, AlN substrate thermal conductivity can be related to the quality of the powder source used in the manufacturing process.

ACKNOWLEDGEMENTS

The authors would like to thank Don Beaman, David Susnitzky, Mike Wiznerowicz, Harold Klassen, Frank Wong and Terry Hiller for the electron microscopy and x-ray fluorescence analysis. We also extend thanks to Arne Knudsen for his review of this investigation.

REFERENCES

1. N. Kuramoto, H. Taniguchi and I. Aso, IEEE Trans. Components Hybrid, Manuf. Technol. CHMT-9 (4), 386-390 (1986).
2. W. Werdecker and F. Aldinger, IEEE Trans. Components Hybrid, Manuf. Technol. CHMT-7 (4), 402-406 (1984).
3. E. Y. Luh, J. H. Enloe, J. W. Lau, and L. E. Dolhert, IEPS 1991, 366-372.
4. H. Farzanehfard, A. Eishabini-Riad and V. Vorperian, ISHM Proceedings 1991, 472-477.
5. G. A. Slack, J. Phys. Chem. Solids 34, 321-335 (1973).
6. G. A. Slack, R. A. Tanzilli, R. O. Pohl, J. W. Vandersande, J. Phys. Chem. Solids 48 (7), 641-647 (1987).
7. H. Buhr, G. Muller, H. Wiggers, F. Aldinger, P. Foley, and A. Rossen, J. Am. Ceram. Soc. 74 (4), 718-723 (1991).
8. A. V. Virkar, T. B. Jackson, R. A. Cutler, J. Am. Cer. Soc. 72 (11), 2031-2042 (1989).
9. K. Watari, M. Kawamot, and K. Ishizaki, J. Materials Sci. 26, 4727-4732 (1991).
10. T. Yagi, K. Shinozaki, N. Mizutani, M. Kato, and Y. Sawada, J. Ceram. Soc. Jpn. Inter. Ed. 97, 1374-1381 (1989).
11. N. Mizutani and K. Shinozaki, Bull. Cer. Soc. Japan 26 (8), 738-743 (1991).
12. N. S. Raghavan, Materials Science and Engineering A148, 307-317 (1991).
13. M. S. Paquette, J. L. Board, C. N. Haney, A. K. Knudsen, D. W. Susnitzky, P. R. Rudolf, D. R. Beaman, R. A. Newman, and S. W. Froelicher, in Ceram. Trans., Ceramic Powder Science III Vol. 12, ed. G. L. Messing, S. Hirano, and H. Hausner (American Ceramic Soc., Westerville, OH, 1990) pp. 885-893.
14. K. Komeya, H. Inoue, and A. Tsuge, Yogyo-Kyokai-Shi 89 (6), 330-336 (1981).
15. K. Komeya, A. Tsuge and H. Inoue, U.S. Patent No. 4 435 513 (1984).
16. I. C. Huseby and C. F. Bobik, U.S. Patent No. 4 478 785 (1984).
17. F. Miyashiro, N. Iwase, A. Tsuge, F. Ueno, M. Nakahashi, and T. Takahashi, IEEE Trans. Components Hybrid, Manuf. Technol. CHMT-13 (2), 313-319 (1990).

ELECTROCHEMICAL SYNTHESIS OF ALUMINUM NITRIDE IN LIQUID AMMONIA ELECTROLYTE SOLUTIONS

TRAVIS WADE, JONGMAN PARK, GENE GARZA, CLAUDIA B. ROSS,
DOUGLAS M. SMITH, AND RICHARD M. CROOKS
UNM/NSF Center for Micro-Engineered Ceramics, University of New Mexico, Albuquerque, NM 87131

ABSTRACT

An electrochemical method for the preparation of high purity metal nitride ceramic precursors is described. Constant current electrolysis of an electrolyte solution containing NH_3 and NH_4Br at an Al electrode yields a solid mixture consisting of $Al(NH_3)_6Br_3$ and $[Al(NH_2)(NH)]_n$ after evaporation of excess NH_3. Calcination of this mixture above 800 C in flowing NH_3 results in sublimation of $Al(NH_3)_6Br_3$ and conversion of the ceramic polymer precursor, $[Al(NH_2)(NH)]_n$, to pure, high surface area AlN. Here we discuss some electrochemical aspects of the polymer precursor synthesis, precursor processing parameters, and materials characterization of the AlN powder before and after sintering.

BACKGROUND

Metal nitride ceramics represent an important area of research at the present time, because they often have superior electronic, thermal, or mechanical characteristics compared to the corresponding oxides. For example, compared to Al_2O_3, AlN has higher mechanical strength, much higher thermal conductivity, a better thermal expansion match to silicon, lower electrical conductivity, and materials properties that are suitable for optoelectronics applications in some cases.[1-3] Despite the obvious advantages of metal nitride ceramics, they have not as yet found wide-spread commercial applications, principally because they are usually more challenging and costly to synthesize and process than oxides. As a result, it is desirable to develop new synthetic procedures that yield low cost, high purity metal nitride ceramics for use in electronic packaging, composite fibers, and as high temperature wear coatings.

Commercial AlN powder is presently made by direct nitridation of AlN, eq 1, or by carbothermal reduction of alumina, eq 2.[1-2] Thin films are made by polymer pyrolysis or by chemical vapor deposition (CVD).[1-5] Thin film syntheses have advanced significantly

$$N_2 + 2Al \dashrightarrow 2AlN \qquad (1200\text{-}1400\ C) \tag{1}$$

$$Al_2O_3 + 3C + N_2 \dashrightarrow 2AlN + 3CO \qquad (1200\text{-}1750\ C) \tag{2}$$

in recent years, but synthetic routes to powders that have both desirable materials properties and low levels of contamination are still elusive.

We recently reported an electrochemical route for the preparation of metal nitride ceramic precursors in NH_3/NH_4Br electrolyte solutions.[6] The process is similar to previously reported homogeneous[7] and heterogeneous[8-10] metal nitride ceramic precursor syntheses, but the electrochemical approach described here is somewhat more easily implemented, requires only inexpensive starting materials, yields nitride powders containing lower oxygen impurity levels, and does not result in the formation of any unrecoverable or noxious by-products. Scheme I illustrates the processes that we believe are responsible for AlN precursor formation. The anode and cathode reactions shown in Scheme I are known to occur in liquid NH_3, as are the other reactions shown in Table I.[7,11,12] The product that results from

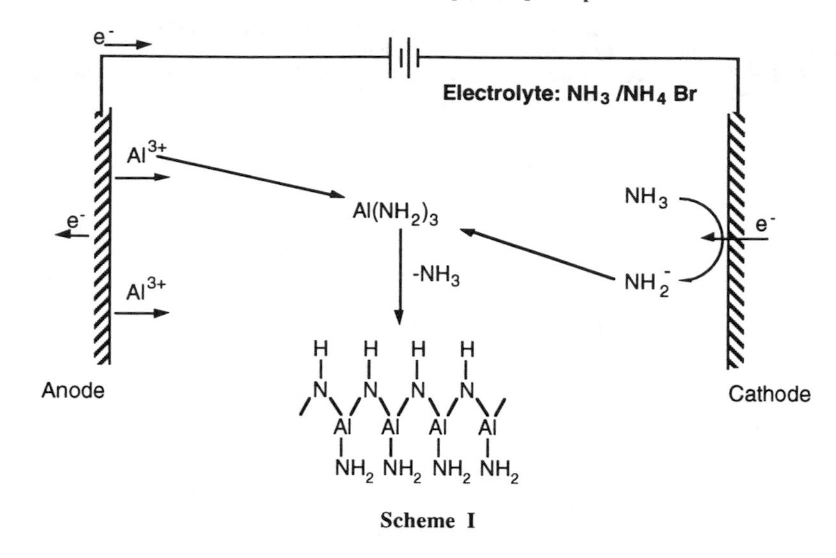

Scheme I

Table I Electrode Reactions in the Cell: $Al/NH_3, NH_4^+, Br^-/Al$

Cathode Reactions:	$e^-(Al) \rightarrow e_s^-$	(3)
	$NH_4^+ + e^- \rightarrow NH_3 + 1/2H_2$	(4)
	$NH_3 + e^- \rightarrow NH_2^- + 1/2H_2$	(5)
Anode Reactions:	$NH_3 \rightarrow 1/2N_2 + 3H^+ + 3e^-$	(6)
	$Br^- \rightarrow 1/2Br_2 + e^-$	(7)
	$Al \rightarrow Al^{3+} + 3e^-$	(8)

reaction between Al^{3+} and NH_2^- in NH_3 solvent, $Al(NH_2)_3$, is also known, eq 9.[7,11] $Al(NH_2)_3$ can undergo NH_3 condensation at room temperature to produce oligomers of the general form $[Al(NH_2)NH]_n$, and further NH_3 condensation at elevated temperature results in AlN, Scheme II.[7,13]

$$AlBr_3 + 3KNH_2 \rightarrow Al(NH_2)_3 + 3\ KBr \qquad (9)$$

In this paper, we provide an overview of the critical synthesis and processing parameters that lead to ceramics with desirable chemical and morphological properties. The results show that careful attention to electrochemical and processing parameters results in inorganic polymer precursors that, when calcined, decompose to form high purity metal nitride

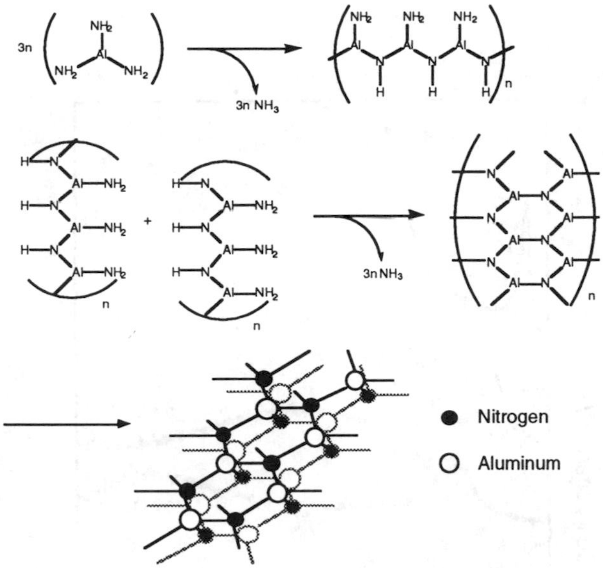

Scheme II Formation of AlN from an electrosynthesized polymer precursor.

powders. Finally, we provide information about the chemical structure of the ceramic precursors and the materials properties of the ceramic powders before and after sintering.

RESULTS

We are interested in the effect of electrochemical and processing variables on electrosynthesized ceramic polymer precursors and AlN purity, yield, and morphology. Information about these relationships will permit us to identify the electrosynthetic reaction mechanism responsible for the ceramic polymer precursor, maximize the cell current efficiency, and control the chemical and morphological characteristics of the ceramic powders. The first part of this paper addresses issues related to the electrochemical synthesis of the precursor, and the second part is concerned with powder processing. Details of the experimental procedures used to obtain the results reported here have been discussed previously [6a].

Electrochemical Synthesis of Metal Nitride Ceramic Precursor Polymers

The magnitude of the cell potential as a function of electrolysis time for different concentrations of NH_4Br electrolyte is shown in Figure 1. These data were obtained for constant current electrolyses, 260 A/m^2, at -80 C in a 350 mL cell containing two concentric Al electrodes. This cell is significantly larger than the one we have previously used,[6] and it allows us to synthesize about 5-10 times more powder per electrolysis cycle. For 0.05 M

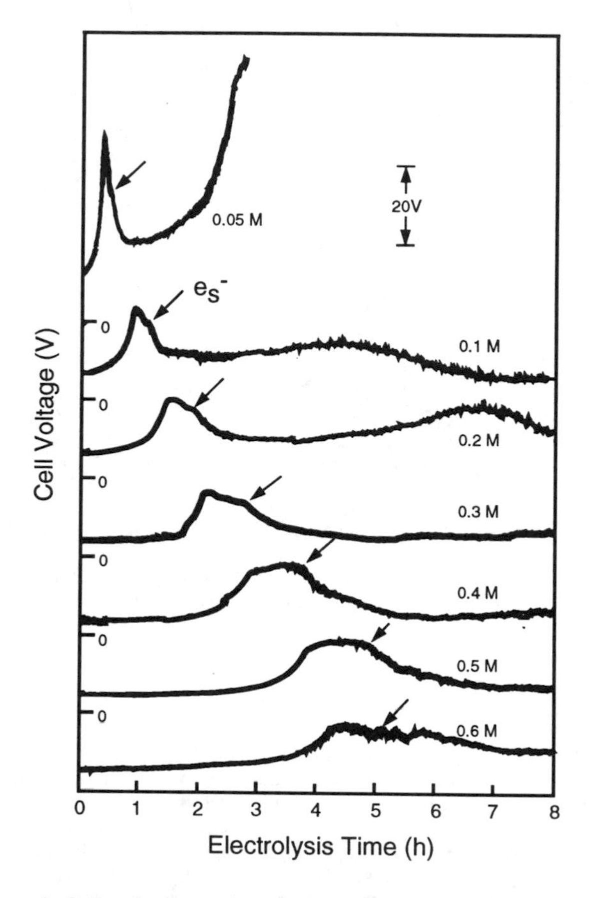

Figure 1 Cell voltage vs. electrolysis time for an AlN polymer precursor synthesized using different concentrations of NH_4Br electrolyte. The arrows indicate the onset of solvated electrons, e_s^-.

NH_4Br, the potential versus time profile indicates rapid attainment of a maximum voltage after about 30 min of electrolysis, which is followed by a steady increase to the maximum output voltage of the current source after 3 h. Concentrations of 0.1 - 0.6 M NH_4Br also yield voltage versus time curves that have a pronounced voltage maximum, but in these cases the maxima are lower and there is no rapid voltage increase associated with the electrolysis. Each of the principal voltage maxima are accompanied by a shoulder which typically appears about 30 min after the time corresponding to the peak potential. Concurrent with the onset of the shoulders, we noted that the electrolyte solution changed from colorless to deep blue. The blue color is characteristic of solvated electrons, e_s^-, which can be generated and are stable in liquid NH_3 electrolyte solutions, eq 3.[14]

We interpret the voltage versus time plots as follows. The slow initial increase in voltage results primarily from the electrolytic conversion of NH_4Br to neutral species, eq 4 and 7, and a concomitant decrease in solution conductivity. When the current demand is exceeded by the flux of electrolyte to the electrodes, additional heterogeneous reactions occur. Thermodynamically, the lowest energy cathode reactions in this electrochemical cell are the generation of e_s^-, eq 3, and formation of NH_2^-, eq 5.[11,14] The likely anode reaction corresponds to solvent oxidation, eq 6.[11,14] When these reactions commence, the characteristic blue color of e_s^- is observed, and the cell potential drops, because the newly formed, highly mobile e_s^- ions increase the solution conductivity. This argument is strongly supported by the approximately linear relationship between the maxima in the voltage versus time curves and the electrolyte concentrations, Figure 2.

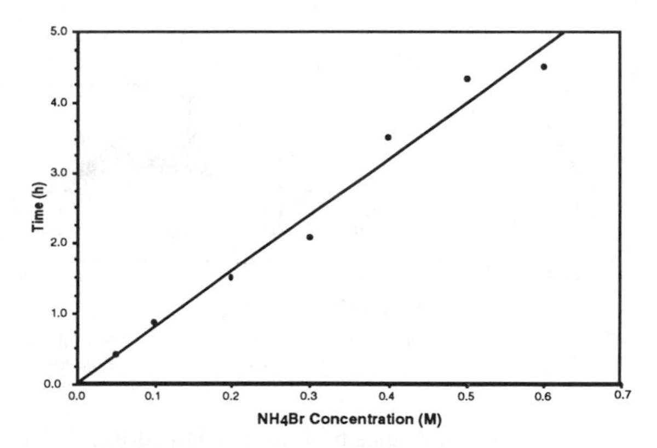

Figure 2 Time of the cell voltage maxima vs. [NH_4Br] for the data shown in Figure 1.

Two other features of Figure 1 are worth noting. First, the voltage versus time curves for 0.1 and 0.2 M NH_4Br electrolyte concentrations have a second local maximum. We speculate that these maxima may be related to electrode passivation, rather than to a significant change in the solution conductivity, because significant amounts of deposits (probably the insoluble polymer precursor, $[Al(NH_2)(NH)]_n$) are clearly visible on the electrode at the end of each electrolysis. These deposits may increase the cell voltage by reducing the active electrode area. Second, the lowest electrolyte concentration, 0.05 M NH_4Br, is not sufficient to sustain the electrolysis, and the cell voltage increases dramatically after about 2 h. Although we do not fully understand the electrochemistry that leads to this result at the present time, the high cell voltage probably arises from high solution resistance.

The interpretation of Figure 1 has two important ramifications. First, the electrolysis conditions which result in formation of the AlN polymer precursor also consume electrolyte. This effect lowers the overall current efficiency of the electrolysis; however, the steady state solution resistance does not change appreciably because parasitic electrode reactions generate mobile charge carriers such as e_s^-. The absence of NH_4Br after electrolysis is confirmed by FTIR and XRD results discussed later. Second, e_s^- formed at the cathode are consumed at the anode, thereby reducing current efficiency. We are presently altering the cell configuration to minimize these seemingly unnecessary side-reactions. However, it is important to note that e_s^- may be an integral part of precursor formation, and we are presently investigating this possibility.

Processing of Metal Nitride Ceramic Precursor Polymers

The mass of AlN resulting from calcination of the precursor is related to the time of electrolysis, Figure 3. These data were obtained using a 75 mL electrolysis cell and a 0.05 M

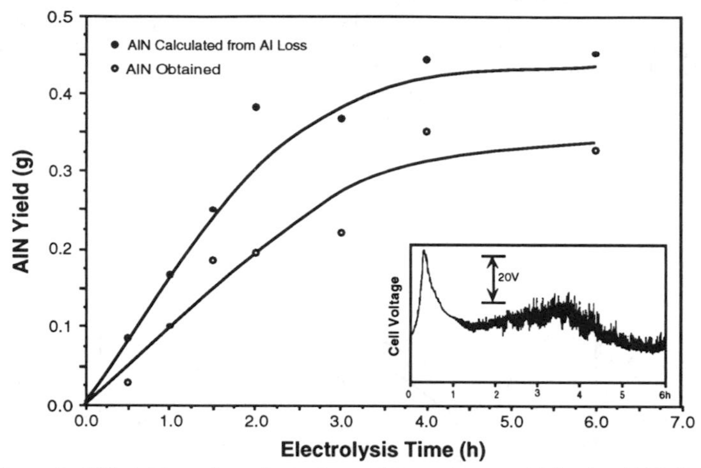

Figure 3 AlN yield vs. electrolysis time. AlN powder was calcined at 1100 C in flowing NH_3. The inset shows the voltage vs. time profile for this cell.

NH_4Br electrolyte solution so that limiting behavior could be attained on a reasonable time scale. Under these conditions, the mass of AlN synthesized is a linear function of the electrolysis time between 0 and ~ 2.5 h. However, prolonged electrolysis does not result in additional yield. This result suggests that some critical reactant is not present in the cell after prolonged electrolysis. Since it is not likely that the concentration of NH_2^- is diminished (it is formed at about the same potential as e_s^-), we conclude that Al oxidation is suppressed since the anodic current demand is satisfied by e_s^- oxidation. That is, e_s^- is more easily oxidized than Al.

When the mass of electrogenerated AlN is calculated from the Al-electrode mass loss, rather than directly from the mass of powder remaining after calcination, the shape of the yield versus time curve is similar to that discussed above, but the calculated yield of AlN is about 25 % higher. The difference between the two curves indicates that not all of the electrolyzed Al takes part in the formation of the polymer precursor. FTIR, elemental analysis, and TGA/DTA data suggest that the parasitic by-product is ammoniated $AlBr_3$, $Al(NH_3)_6Br_3$.

Figure 4 shows the transmission FTIR spectra obtained for the precursor after calcining at different temperatures in flowing NH_3. The bottom spectrum, obtained prior to calcination, shows two unresolved bands centered at 1513 cm^{-1}, which are characteristic of the NH_2 and NH bending modes of $[Al(NH_2)NH]_n$, and peaks at 1620 and 1307 cm^{-1}, which are characteristic of the NH_3 degenerate and symmetric deformations, respectively, in $Al(NH_3)_6Br_3$.[7,15] After heating to 200 C, however, the $[Al(NH_2)NH]_n$ band intensities diminish, consistent with NH_3 condensation from the polymer precursor according to Scheme II, but the $Al(NH_3)_6Br_3$ band intensities remain essentially constant, consistent with its boiling temperature of ~260 C. However, at 400 C, $Al(NH_3)_6Br_3$ differentially sublimes

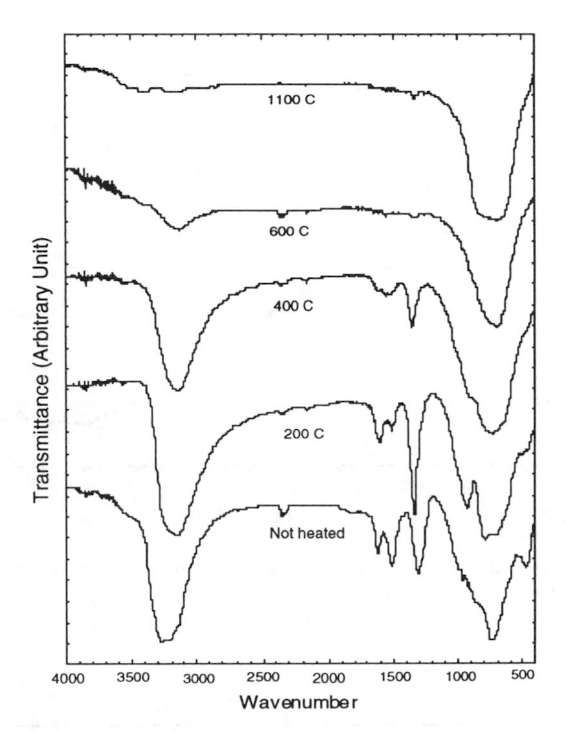

Figure 4 Transmission FTIR spectra for an electrosynthesized precursor calcined at different temperatures in flowing NH_3.

from the AlN precursor powder and the peaks at 1620 and 1307 cm^{-1} disproportionately decrease. Essentially all N-H absorptions are absent at 1100 C, including the large N-H stretching feature originally present at about 3250 cm^{-1}. These data, along with the broad AlN stretching peak (~800 cm^{-1}) present at 1100 C, are consistent with formation of H-free AlN powder. Importantly, most of the chemistry responsible for the precursor-to-ceramic transformation is over by 600 C. Moreover, comparison of the spectra shown in Figure 4 with a spectrum of pure NH_4Br indicates that all of the electrolyte originally present in the cell is consumed by electrochemical reactions under these reaction conditions.

Thermal analysis of the ceramic precursor is also consistent with electrochemical formation of species that react in solution to form $Al(NH_3)_6Br_3$ and $[Al(NH_2)NH]_n$ followed by $Al(NH_3)_6Br_3$ sublimation and $[Al(NH_2)NH]_n$ decomposition, Figure 5. TGA/MS data indicate that the mass decrease commencing at about 80 C is due to NH_3 condensation. The inflection point in the TGA data at 250 C corresponds to sublimation of an Al-containing species, probably $AlBr_3$ or $Al(NH_3)_6Br_3$.

X-ray diffraction (XRD) spectra, Figure 6, are consistent with the FTIR and thermal analyses. These data show the onset of a crystalline AlN phase at temperatures as low as 600 C, but no evidence for NH_4Br or $AlBr_3$. The XRD line widths are also consistent with our previous finding that the calcined precursor results in nanophase AlN particles.[6] The elemental and morphological analyses of AlN powder calcined at 1100 C in flowing NH_3 are shown in Table II.

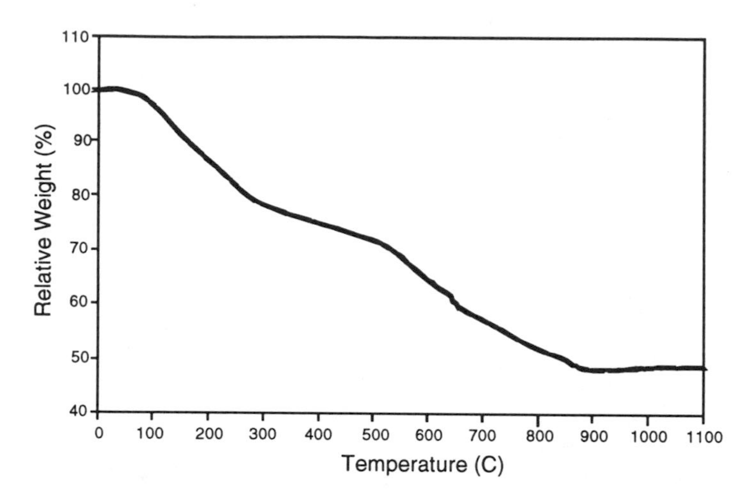

Figure 5 TGA data for an electrogenerated AlN polymer precursor obtained in flowing N_2.

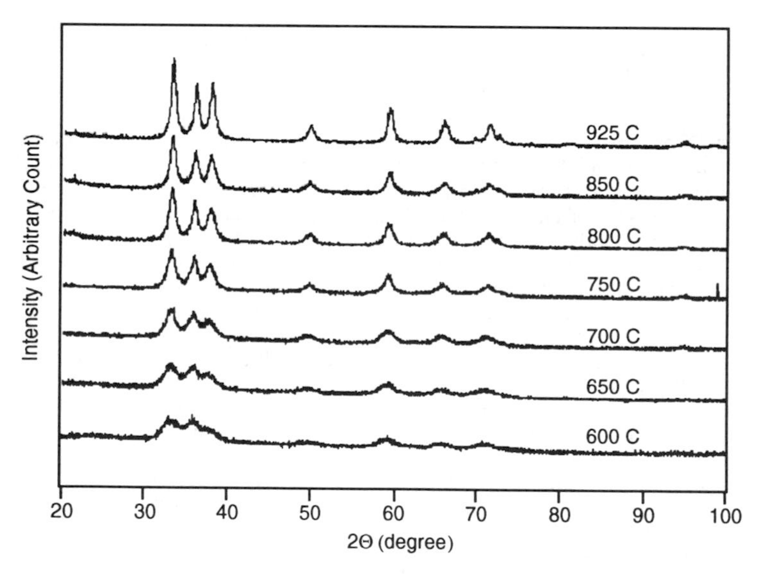

Figure 6 XRD data for AlN polymer precursors calcined at different temperatures in flowing NH_3.

Table II Summary of the elemental composition and morphological characteristics of AlN prepared in NH_4Br/NH_3 electrolyte solution and calcined at 1100 C. Theoretical values, where appropriate, are listed in parentheses.

	AlN
Elemental Analysis (wt %)	
Al	66.12 (65.9)
N	33.92 (34.1)
O	1.4*
C	<0.5
Br	<0.5
H	<0.5
Total	101.4 (100)
Mole ratio, Al/N	1.01 (1.00)
BET Surface Area (m^2/g)	87
Helium Density (g/cm^3)	2.97 (3.26)
Mean Crystallite Size (nm)	12

*Depending on conditions, the oxygen impurity level ranges from 0.70 - 1.4 wt.%.

We have also examined the effect of the calcining ambient on the purity of AlN powder. Figure 7 shows XRD spectra for AlN powders calcined at 1000 C under various flowing gases.

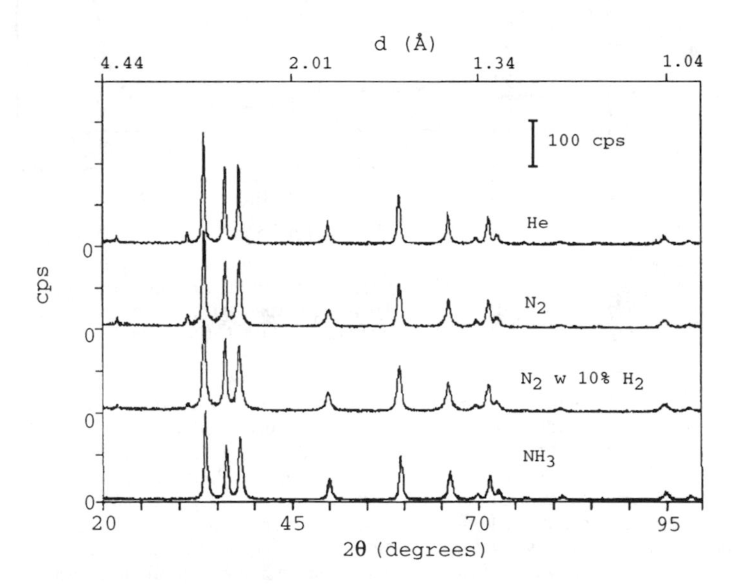

Figure 7 XRD data for AlN polymer precursors calcined in different atmospheres at 1000 C.

The bottom spectrum was obtained from a precursor sample calcined in flowing NH_3. The diffraction maxima in this spectrum are identical to those reported in the literature for pure AlN.[16] Spectra obtained from samples calcined in He, N_2, and 10% H_2/90% N_2 are very similar to the spectrum of pure AlN, except for small peaks at $2\theta=22^o$ and 31° that are due to NH_4Br impurities which may be entrained within AlN crystallites. Elemental analyses of the powders used to obtain the spectra shown in Figure 7 provide the following weight percent oxygen impurities: NH_3 (0.87), 90%N_2/10%H_2 (1.01%), N_2 (1.25), He (0.69).

We have examined the stability of electrochemically prepared and calcined AlN powders in wet and dry O_2 and N_2. These experiments were carried out by exposing a freshly calcined powder to a flowing gas stream of either dry N_2 or wet O_2, and then measuring the mass change of the powder as a function of time and temperature. The results indicate that the powders are stable under a wide variety of conditions. Most other commercially available, high surface area powders degrade upon exposure to air as a result of surface oxidation.[3,17] Figure 8 shows the result obtained when, first N_2, and then wet O_2, are passed over AlN powder at 25 C. Importantly, no mass changes result except for a small increase during the period that the sample was exposed to wet O_2. When the powder is heated in dry O_2, the water adsorbed during the previous cycle is desorbed at about 100 C. The AlN powder is stable to about 800 C before it is quantitatively converted to Al_2O_3 at about 1065 C.

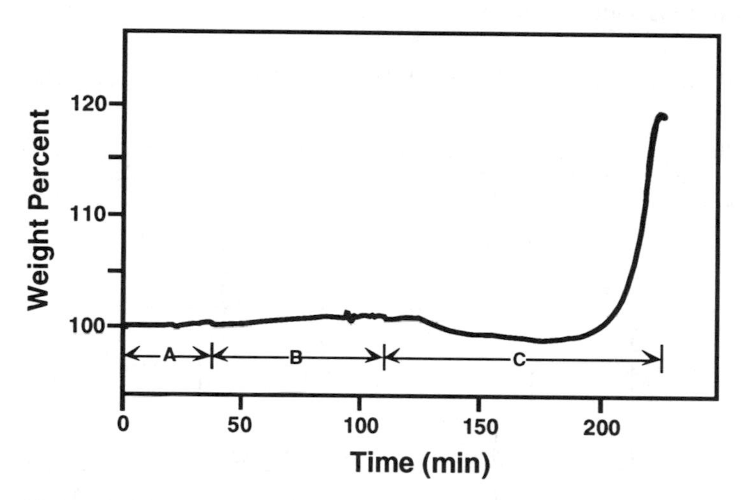

Figure 8 TGA of AlN powder calcined at 1000 C and then exposed to flowing N_2 at 25 C (part A), $O_2 + H_2O$ at 25 C (part B), and finally heated in flowing O_2 at 10 C/min to a final temperature of 1100 C (part C).

Calcined AlN powders were hot-pressed at 1800 C and 1500 psi for 2 h in gettered Ar. After pressing, the material was 97% densified and translucent. More recent data indicate that hot pressing at 1600 C and 5000 psi for 2 h in gettered Ar results in fully dense material. The narrow XRD lines of the material sintered at 1800 C, Figure 9, indicate that the crystallites increase in size, but that no impurities are present in the sintered powder within the accuracy of XRD. SEM micrographs, Figure 10, confirm the crystallite size. Elemental analysis shows that the O_2 impurity level after calcining, without sintering aids, amounts to 0.9%. Thermal conductivity measurements of the sintered material are presently underway.

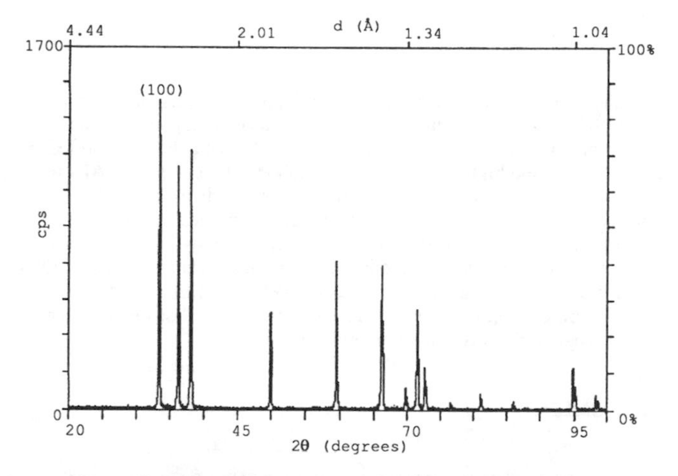

Figure 9 XRD data for AlN powder hot pressed at 1800 C and 1500 psi in gettered Ar for 2 h without sintering aids.

Figure 10 SEM micrographs of AlN powders prepared by calcining the polymer precursor in flowing NH_3 at 1100 C: (top) before sintering; (bottom) after hot pressing at 1800 C and 1500 psi in gettered Ar.

SUMMARY

We have shown that a simple electrochemical apparatus consisting of two Al electrodes and an NH_4Br/NH_3 electrolyte solution can be used to generate an inorganic AlN polymer ceramic precursor. Calcination of this material in flowing NH_3 at 600 C results in an amorphous AlN phase, and higher temperatures result in fully crystalline AlN with tetrahedral symmetry. Elemental analysis of the crystalline material indicates that it is largely free of C and O impurities. The high surface area powder fully densifies when hot pressed at 1600 C and 5000 psi to yield 5 -10 μm crystallites. Finally, we have shown previously that this technique is appropriate for synthesizing other metal nitride ceramics such as TiN.[6a]

Experiments are presently underway to fully reveal the mechanistic chemistry that leads to AlN, improve cell current efficiency, and construct a small-scale flow reactor. Results obtained from these experiments will be reported shortly.

ACKNOWLEDGMENTS

This work has been partially supported by the UNM/NSF Center for Micro-Engineered Ceramics, a collaborative effort supported by NSF (CDR-8800352), Los Alamos and Sandia National Laboratories, the New Mexico Research and Development Institute, and the ceramics industry. Acknowledgment is made to the donors of The Petroleum Research Fund, administered by the ACS, for partial support of this research. We thank Dr. Kevin Howard and Ms. B. Lynn Stiehl of the Dow Chemical Company for providing some of the elemental analysis data.

REFERENCES

1. L.M. Sheppard, Ceram. Bull. 69, 1801 (1990).
2. N. Kuramoto and H. Taniguchi, J. Mater. Sci. Lett. 3, 471 (1984).
3. D.D. Marchant and T.E. Nemecek, Adv. Ceram. 26, 19 (1989).
4. L.V. Interrante, W. Lee, M. McConnell, N. Lewis and E. Hall, J. Electrochem. Soc. 136, 472 (1989).
5. Z. Jiang and L.V. Interrante, Chem. Mater. 2, 439 (1990).
6. (a) C.B. Ross, T. Wade, R.M. Crooks and D.M. Smith, Chem. Mater. 3, 768 (1991). (b) R.M. Crooks, D.M. Smith and C.B. Ross, U.S. Patent Application.
7. L. Maya, Adv. Ceram. Mater. 1, 150 (1986).
8. C. Rüssel and M.M. Seibold, J. Am. Ceram. Soc. 72, 1503 (1989).
9. M. Seibold and C. Rüssel, Mat. Res. Soc. Symp. Proc. 121, 477 (1988).
10. C. Rüssel, Chem. Mater. 2, 241 (1990).
11. D. Nicholls, Inorganic Chemistry in Liquid Ammonia (Elsevier, Amsterdam, 1979).
12. L. Maya, Aluminum Electrochemistry in Liquid Ammonia, Report ORNL/TM-9762 (Oak Ridge National Laboratory, Oak Ridge, TN, 1985).
13. Results from Al MAS NMR indicate that the Al coordination symmetry is octahedral prior to calcining, but fully tetrahedral after calcining in NH_3 for 2 h at 1000 C. This suggests that the Al in $[Al(NH_2)(NH)]_n$ is further ligated to three NH_3 molecules. See reference 11, p. 158.
14. T. Teherani, K. Itaya and A.J. Bard, Nouv. J. Chim. 2, 481 (1978).
15. K. Nakamoto, Infrared and Raman Spectra of Inorganic and Coordination Compounds, 3rd ed. (Wiley, New York, 1978) pp. 197-199.
16. Powder Diffraction File Alphabetical Index, Inorganic Phases, edited by W.F. McClune (JCPDS Internation Center for Diffraction Data, Swarthmore, 1988).
17. M.S. Paquette, J.L. Board and C.N. Haney, Ceram. Trans. 12, 855 (1990).

ELECTROCHEMICALLY PREPARED POLYMERIC PRECURSORS FOR THE FORMATION OF NON-OXIDE CERAMICS AND COATINGS

RALPH ZAHNEISEN and CHRISTIAN RÜSSEL
Institut für Werkstoffwissenschaften III (Glas und Keramik), Universität Erlangen-Nürnberg, Martensstr. 5, 8520 Erlangen, Germany

ABSTRACT

The electrochemical preparation of polymeric precursors and the subsequent formation of non-oxide powders, ceramics and coatings is described. The metals were anodically dissolved in an electrolyte consisting of a primary organic amine, acetonitrile as a solvent, and a tetraalkylammonium salt. This procedure led to the formation of polymeric precursor solutions. Removal of the excess organic compounds resulted in the formation of polymeric amorphous solids. Pyrolysis was carried out at temperatures in the range of 750 to 1100°C. In an atmosphere of ammonia, metal nitrides were formed, while calcination under nitrogen or argon led to carbonitrides or to carbides, depending on temperature and the metal used. Up to now, this route has been applied to Al, Ti, Zr, Cr, Ta, Mg, Ca and Y, and it is supposed, that this route is applicable to the formation of many metal carbides and nearly all metal nitrides relevant for materials science.

INTRODUCTION

In the last few years, many polymeric routes to non-oxides have been developed. Although most of them describe the preparation of silicon nitride [1] and carbide [2], other non-oxides, such as BN, B_4C, TiN, TiC, WC and TiB_2, have also been investigated. All polymeric routes for the formation of non-oxides mentioned in literature are not generally applicable and mostly restricted to the production of only one metal nitride or carbide [3, 4]. For the formation of composites, composed by two or more non-oxides, special routes have to be used (e. g. [5]). Simple mixing of different precursor solutions, in the most cases, is not applicable. The electrochemical preparation of polymeric precursors by an anodic dissolution process was at first developed for the formation of aluminum nitride powders and coatings. It is a route using commercially available and comparably cheap starting materials. In this paper the preparation of various non-oxides is described.

ELECTROCHEMICAL PREPARATION OF THE PRECURSOR

The metals were anodically dissolved in a purely organic electrolyte. It consisted of an organic amine, acetonitrile to increase the polarity, and a tetrabutylammonium salt as a supporting electrolyte to achieve sufficient electric conductivity. A double walled glass vessel contained the electrodes and the electrolyte. Both cathodes and anodes were formed by metal sheets (thickness: 1-2 mm). The distance between two electrodes was about 1 mm. A condenser, fixed at the top of the vessel, recovered organic compounds which were vaporized or carried along with the gas stream. The electrolysis was carried out applying a current density of 60-70 mA/cm^2 and a DC-voltage in the range of 6-8 V. The polarity of the DC-voltage was reversed every 15 min in order to achieve a uniform dissolution of all electrodes. For about 4-6 h, the current remained nearly constant and then decreased. Then the electrolysis was stopped, and the solution was drawn off to another vessel. The apparatus used was the same as already described for the anodic dissolution of metallic aluminum and titanium [6-12].

At the anode the metal is oxidized. At the cathode NH_2R is reduced and the corresponding anion and gaseous hydrogen (gas bubbles) are formed. The assumed chemical reactions at the electrodes are as follows:

$$\begin{array}{lrcl} \text{anodic reaction:} & Me & \longrightarrow & Me^{x+} + x\,e^- \\ \text{cathodic reaction:} & x\,NH_2R + x\,e^- & \longrightarrow & x\,NHR^- + x/2\,H_2 \\ \hline & x\,NH_2R + Me & \longrightarrow & Me(NHR)_x + x/2\,H_2 \\ & & & \text{(I)} \end{array} \qquad (1)$$

In order to study the chemistry, the obtained polymeric solutions were dried by applying vacuum to the vessel and subsequent heating. The evaporation of excess amine and solvent led to a further increase in viscosity and subsequently to the formation of a solid of gellike consistency. On further heating under vacuum, a solid foam full of gas bubbles was obtained. For the formation of powders with mean grain sizes in the submicron range, the fluid electrolyte was dropped in paraffin, previously heated up to 450°C. For the preparation of coatings on various substrates, a dip-coating procedure was carried out. Figure (1) shows the flow diagram of the different steps and procedures used [11, 13-16].

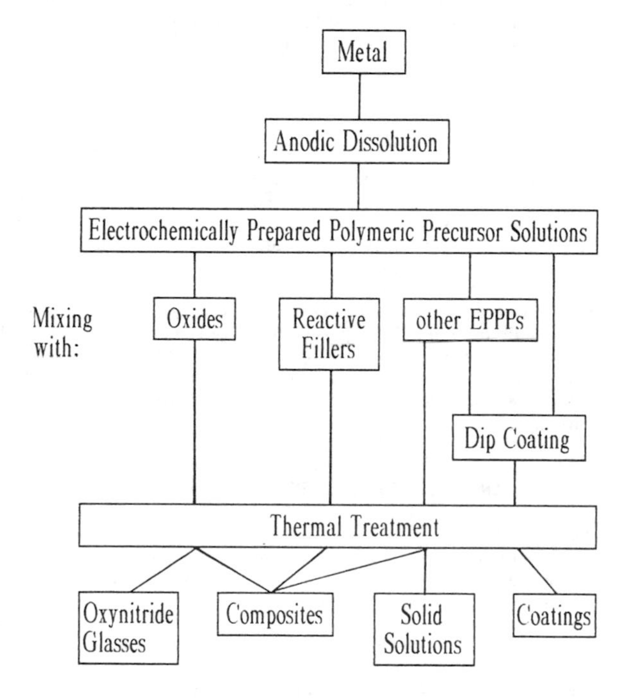

Figure (1) Flow diagram of different procedures applied.

The compound $Me(NHR)_x$ which is supposed to be formed, according to equation (1), as an intermediate, polycondenses. FTIR-spectra of the dried precursors (see figure (2)) show strong absorptions in the N-H regions (3000-3400 and 1500-1600 cm^{-1}) and the C-H region (2800-2900 cm^{-1}). The broad peak

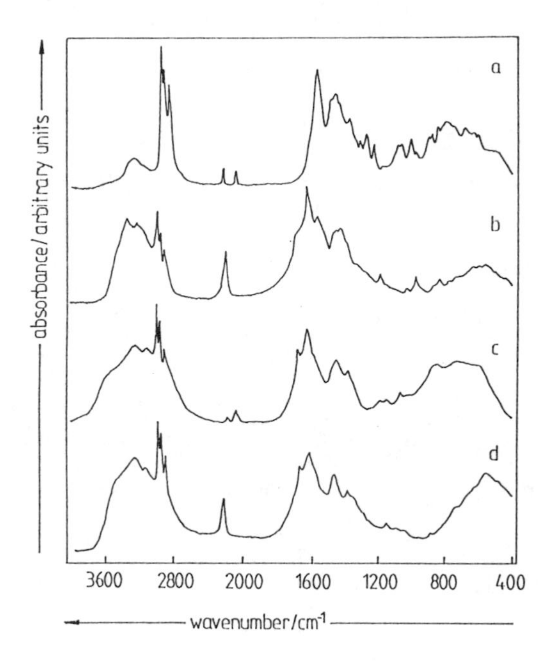

Figure (2) FTIR-spectra a: fluid Al-containing electrolyte; b-d: dried precursors: b: Ca; c: Al; d: Cr.

in the range of 400-800 cm^{-1} indicates the presence of Me-N-Me bonds. The occurrence of ν(N-H) vibrations indicates that a complete polycondensation according to equation (2) did not take place:

$$Me(NHR)_x \longrightarrow \underset{(II)}{Me(NR)_{x/2}} + x/2\ NH_2R \qquad (2)$$

The empirical formulae of dried precursors are summarized in Table I column 2 and prove that the chemical composition cannot sufficiently be described by formula (II) and that the occurring chemical reactions are much more complicated. XRD-patterns of dried precursors showed that all products formed were completely amorphous.

CALCINATION

Pyrolysis of the dried precursors was carried out in nitrogen, argon or anhydrous ammonia atmospheres over the temperature range of 600-1200°C. Table I, column 3 summarizes the weight loss obtained from TGA-profiles. The main weight loss was observed in the temperature range between 150 and 350°C. Further slight weight loss occurred up to 550°C, then the weight remained nearly constant. It should be noted that the TGA-profiles were fairly similar and were only slightly influenced by the type of metal. The ceramic yield, however, strongly depended on the type of metal used and were in the range of 20 to 50 wt %.

Tab. I Empirical formulae, of various polymeric precursors and products formed during calcination

metal dissolved	empirical formula	ceramic yield /%	calcination gas	product
Al	$AlN_{1.43}C_{5.5}H_{12.6}$	42	NH_3	AlN
Ti	$TiN_{1.14}C_{3.94}H_{8.24}$	20	NH_3	TiN
			N_2	Ti(C,N)
Zr	$ZrN_{0.67}C_{1.59}H_{4.44}$	41	NH_3	ZrN
Cr	$CrN_{2.16}C_{6.83}H_{12.44}$	40	NH_3	CrN
			N_2	Cr_3C_2
			Ar	Cr_3C_2
Ta	$TaN_{5.87}C_{19.3}H_{34.4}$	31	NH_3	TaC
Ca	$CaN_{1.18}C_{4.51}H_{8.33}$	34	NH_3	$CaCN_2$
Mg	$MgN_{1.07}C_{3.88}H_{7.71}$	21	NH_3	$MgCN_2$
Y	$YN_{3.84}C_{11.38}H_{26.42}$	39	NH_3	YN, $Y_{0.45}C_{0.55}$

Figure (3) shows XRD-patterns of some calcined products. Graph a presents an aluminum containing precursor calcined at 750°C in ammonia. All lines observed are attributable to aluminum nitride and are notably broadened, indicating small crystallite sizes in the range of 30 nm. Graph b shows a titanium containing precursor, calcined at 1100°C in ammonia. All lines are related to titanium nitride which, however, cannot be distinguished from TiC using XRD. Carbon and nitrogen analysis of calcined titanium containing precursors indicate that always TiN/TiC solid solutions are obtained. The quantity of carbon strongly depends on the calcination atmosphere applied (argon: 31 wt %, ammonia: 5.1 wt %). By contrast, chromium containing precursors are pyrolysed in ammonia to CrN (see graph c), while calcining in argon leeds to the formation of Cr_3C_2 (see graph d). Other products formed are ZrN, TaC (graph e), YN, $MgCN_2$ and $CaCN_2$ (graph f). Due to small crystallite sizes, the obtained powders possess very high sinterability. As already shown for aluminum nitride, pressureless sintering can be carried out without the addition of sintering additives already at 1750°C resulting in densities > 98 % of theory [8]. The oxygen contents of the obtained powders were in the range of 1.5 to 2.0 wt % for aluminum nitride, calcium cyanamide and the transition metal compounds investigated.

All prepared precursor solutions can be mixed with one another. Two examples highlight the possibilities of this procedure: mixed Ti- and Al-containing precursors were dried and subsequently calcined in ammonia. XRD-patterns of products calcined at temperatures up to 800°C show solely lines attributable to cubic TiN, at temperatures of at least 1000°C, lines attributable to both TiN and AlN can be seen. This behaviour should be interpreted in terms of the formation of a metastable solid solution of AlN in TiN and its thermal decomposition at higher temperatures. Similar solid solutions are wellknown from physical vapour deposition and possess superior mechanical properties. Although the obtained powders cannot be densified below the decomposition temperature, the formation of dense cubic coatings is possible.

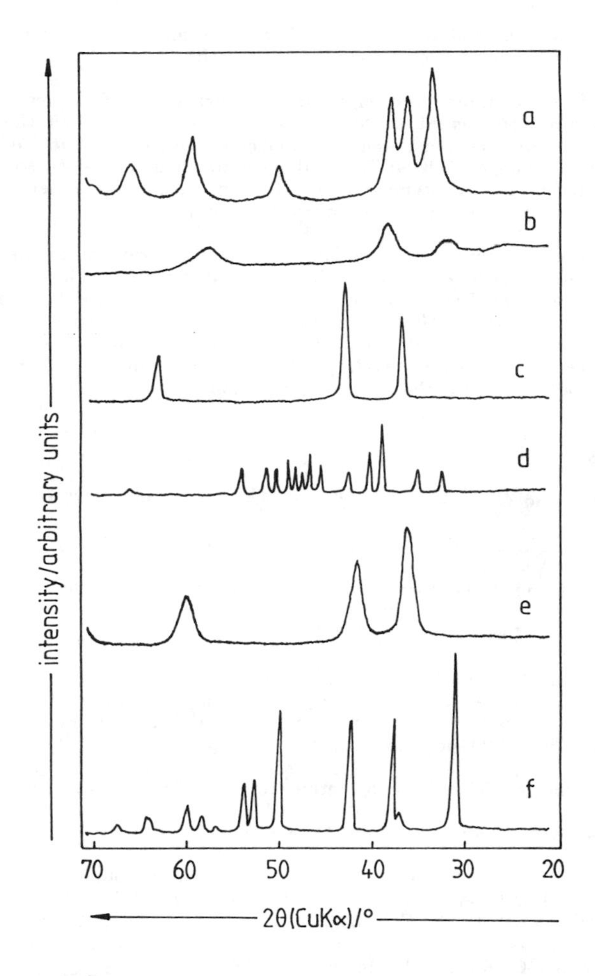

Figure (3) XRD-patterns of calcined products: a: Al-containing (NH_3 at 750°C); b: Ti-containing (NH_3 at 1100°C); c: Cr-containing (NH_3 at 800°C); d: Cr-containing (Ar at 800°C); e: Ta-containing (NH_3 at 1000°C); f: Ca-containing (NH_3 at 1000°C).

In a second example, Al-, Ca- and Y-containing precursors were mixed together. Drying and subsequent calcining resulted in the formation of AlN doped with around 7 wt % calcium and 1.5 wt % yttrium. Subsequent sintering was carried out at a temperature of 1850°C using a soaking time of 1 h. The oxygen content of the obtained fully dense AlN ceramic was 0.3 wt % and the thermal conductivity measured was 155 W/m·K. Although pyrolytically derived aluminum nitride powder can also pressureless be densified without addition of Ca- and Y-compounds, the resulting ceramics possess only poor thermal conductivities in the range of 40 W/m·K. Therefore, Y- and Ca-compounds do not primarily act as sintering additives but as oxygen getters. The homogeneous mixing of these compounds, however, enables a more effective removal of oxygen from the aluminum nitride grains due to short diffusion paths.

Another fairly different procedure is the formation of oxynitride glasses. Sol-gel derived amorphous $MgO \cdot x SiO_2$-powders were mixed with the aluminum containing precursor solutions, and subsequently calcined and melted. Nitrogen contents in the range of 5-8 wt % could be realized using short soaking times and comparably low melting temperatures. The obtained glasses were transparent with a slightly grey colour at higher nitrogen contents.

The greatest advantage of the electrochemical preparation of polymeric precursors is the versatility of this method. Various metals can be electrolytically dissolved and transferred into polymeric precursors and this method seems to be nearly generally applicable. All precursor solutions can be mixed with one another and therefore the formation of composites or solid solutions is enabled. In the future, however, and for a large scale production, alloys should be electrolytically dissolved instead of mixing the precursor solutions.

REFERENCES

1. D. Seyferth and G. H. Wiseman, J Am Ceram. Soc. 64, C-132 (1984).
2. S. Yajima, J. Hajashi and M. Omori, Chem Lett. 931 (1975).
3. K. J. Wynne and R. W. Rice, Ann. Rev. Mater. Sci. 14, 297 (1984).
4. G. Pouskopeli, Cerram. Int. 15, 213 (1989).
5. C. L. Czekaj, M. L. J. Hackney, W. J. Hurley, L. V. Interrante, G. A. Sigel, P. J. Schields and G. A. Slack, in Better Ceramics Through Chemistry III, edited by C. J. Brinker, D. F. Clark, and D. R. Ulrich (Mat. Res. Soc. Proc. 121 Pittsburgh, PA 1988) pp. 465.
6. C. Rüssel and M. Seibold, U. S. Patent No. 07/ 316, 914 (4 February 1989).
7. M. Seibold and C. Rüssel, J. Am. Ceram. Soc. 72, 1503 (1989).
8. M. Seibold and C. Rüssel in Better Ceramics Through Chemistry III, edited by C. J. Brinker, D. F. Clark, and D. R. Ulrich (Mat. Res. Soc. Proc. 121 Pittsburgh, PA 1988) pp. 477-482.
9. P. Distler and C. Rüssel, J. Mater. Sci. 27, 133 (1992).
10. M. Seibold, U. Vierneusel and C. Rüssel, in Ceramic Powder Processing Science, edited by H. Hausner, G. C. Messing and S. Hirano (Deutsche Keramische Gesellschaft, Köln 1989) pp. 173-179.
11. I. Teusel and C. Rüssel, J. Mater. Sci. 25, 3531 (1990).
12. R. Jaschek and C. Rüssel, Surf. Coat. Technol. 45, 99 (1991).
13. R. Jaschek and C. Rüssel, Thin Solid Films 208, 7 (1992).
14. C. Rüssel, Chem. Mater. 2, 241 (1990).
15. R. Jaschek and C. Rüssel, J. Non-Cryst. Solids 135, 236 (1991).
16. C. Rüssel, J. Mater. Sci. Lett. in press.

ORGANOMETALLIC PRECURSORS TO VANADIUM AND TITANIUM CARBONITRIDE

FRANCOIS LAURENT, CHRISTOPHE DAURES, LYDIE VALADE, ROBERT CHOUKROUN, JEAN-PIERRE LEGROS, AND PATRICK CASSOUX
CNRS Laboratoire de Chimie de Coordination, 205 Route de Narbonne, 31077 Toulouse Cedex, France

ABSTRACT

We report on the syntheses, structural and thermoanalytical studies of two titanium complexes $CpTiCl_2N(SiMe_3)_2$ and $FTi[N(SiMe_3)_2]_3$ and one vanadium complex $V(NEt_2)_4$. $CpTiCl_2N(SiMe_3)_2$ has been used in a hot wall CVD reactor as a precursor to titanium nitride. Thin films of typically 1 µm in thickness have been deposited at 600°C on glass, silicon and tool steel. XPS analyses of the deposits show titanium carbonitride to be formed. The new $FTi[N(SiMe_3)_2]_3$ complex can also be used as a precursor to titanium nitride. $V(NEt_2)_4$ led at 500°C to the deposition of vanadium carbonitride characterized by XPS analysis. Films tended to contain excess free carbon.

INTRODUCTION

Titanium (and possibly vanadium) nitride coatings have a wide area of applications related to their good mechanical, thermal, chemical and conductive properties [1]. The typical precursors of titanium nitride used in CVD processes are titanium tetrachloride $TiCl_4$ [2,3] and tetrakis-dialkylammino-titanium complexes $Ti(NR_2)_4$ [4,5,6]. In order to get good quality titanium nitride, authors often mention the necessary use of ammonia in the process [5,6]. Otherwise, the thin films contain impurities such as free carbon or have a poor adhesion due to corrosion by hydrogen chloride HCl. On the contrary, vanadium nitride coatings have not been often studied, but have been recently obtained by Hoffman [5] using $V(NMe_2)_4$ as CVD precursor .

We have explored a series of titanium and vanadium complexes bearing ligands that could be, on a thermal process, easily and cleanly separated from the M-N core (M = Ti or V). This search led to the three following compounds: $CpTiCl_2N(SiMe_3)_2$, $FTi[N(SiMe_3)_2]_3$ and $V(NEt_2)_4$.

EXPERIMENTAL SECTION

While this work was in progress, Bai *et al.* [7] described the synthesis and crystal structure of $CpTiCl_2N(SiMe_3)_2$. We prepared this complex by a similar procedure that described by these authors but we purified it by recrystallization in toluene [8] instead of sublimation [7].

$FTi[N(SiMe_3)_2]_3$ was prepared by the reaction of $LiN(SiMe_3)_2$ in hexane with TiF_4 in toluene at room temperature. Single crystals can be collected either by crystallization from the reaction mixture at -30°C or by sublimation at 120°C under 0.01torr.

$V(NEt_2)_4$ has been prepared as in [9].

Crystal structure data of $CpTiCl_2N(SiMe_3)_2$ and $FTi[N(SiMe_3)_2]_3$ have been recorded on a four circle CAD-4 Enraf-Nonius automatic diffractometer using the Mo Kα radiation (λ = 0.71069Å).

Thermoanalytical studies have been run on a Setaram, Model TG-DTA 92, thermoanalyzer allowing simultaneous recording of TGA and DTA data. This thermoanalyzer is connected to a Leybold-Heraeus, Model QX 2000, quadrupolar mass spectrometer for the analysis of the gases evolving during the thermal analysis. Studies have been made with various heating rates and by using hydrogen or nitrogen as carrier gases.

Chemical vapor deposition experiments have been run with controlled quality nitrogen gas, at ambient or reduced pressure, in a hot wall type reactor consisting of a 30 mm diameter quartz tube inserted in a tubular furnace.

The substrates used for the experiments were glass, silicon plates (orientation 100), and tool steel. These substrates were cleaned before use following reported procedures [6].

Surface analysis of the deposits were made by X-ray Photoelectron Spectroscopy using a VG Escalab, Model MK2, spectrometer working with the Mg Kα radiation (1253.3 eV). The binding energies of the core levels of the various elements were recorded before and after etching the surface contamination layer by bombardment with Ar^+ ions (4kV, 100μA, 12min).

RESULTS AND DISCUSSION

Structural studies

Single crystals of $CpTiCl_2N(SiMe_3)_2$ have been studied by X-ray crystallography. Details on the determination will be reported elsewhere [10]. $CpTiCl_2N(SiMe_3)_2$ crystallizes in the triclinic system. The unit cell parameters at 185 K are a = 8.831(4), b = 14.270(8), c = 6.774(8)Å, α = 91.98(8), β = 99.33(9), γ = 85.13(5)°, V = 839.2(8)$Å^3$. Final conventional agreement factors were R(F) = 0.042 and R_w(F) = 0.53 [8]. The room temperature unit cell is also triclinic with consistent parameters. The crystal structure of $CpTiCl_2N(SiMe_3)_2$ reported by Bai *et al.* [7] is different. Their compound crystallizes in the monoclinic system, space group $P2_1/n$. This difference is probably related to the method used for crystallizing the compound. Nevertheless, no marked difference is observed in the two molecular structures. As shown in Figure 1, the titanium atom lies in a classical tetrahedral environment. The four positions of the tetrahedron are occupied by two chlorine atoms, the nitrogen atom of the $N(SiMe_3)_2$ group and the center of the cyclopentadienyl group in the η^5 bonding mode. The geometry around the nitrogen atom is planar. This and the relatively short titanium to nitrogen bond length (Ti-N = 1.883(6)Å) which is among the shortest reported Ti-N σ-bond length (1.85-1.98Å [11]) suggest the existence of a nitrogen-to-metal π-donation [12].

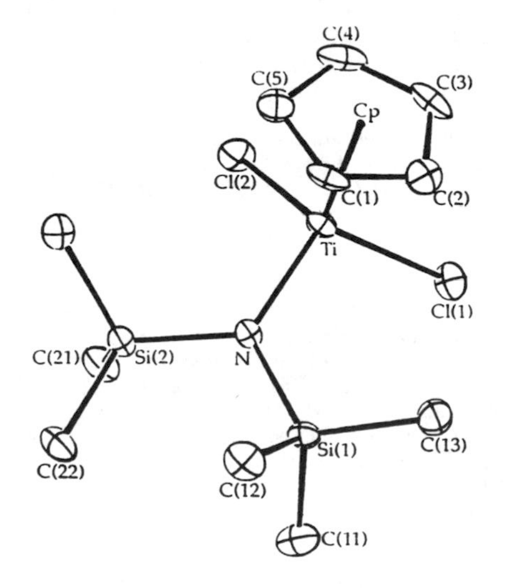

Figure 1. Molecular structure of $CpTiCl_2N(SiMe_3)_2$

$FTi[N(SiMe_3)_2]_3$ has been characterized by elemental analysis, ^1H and ^{19}F NMR, and X-ray crystallography. X-ray structural data recorded at room temperature show that this complex crystallizes in the orthorombic system, space group $Pna2_1$. The unit cell parameters are a = 19.935(4), b = 10.758(2), c = 15.118(2)Å, V = 3242(1)$Å^3$. Final conventional agreement factors were R(F) = 0.0317 and R_w(F) = 0.0347. As shown in Figure 2 the titanium atom lies in a classical tetrahedral environment and bears the fluorine atom and the three nitrogen atoms of the $N(SiMe_3)_2$ groups. The tetrahedron is slightly distorted due to sterical interaction between the $(SiMe_3)$ groups. Each $TiNSi_2$ fragment of the molecule is planar. As in the previous $CpTiCl_2N(SiMe_3)_2$ complex, this could result from a retro π-donation of the nitrogen atoms

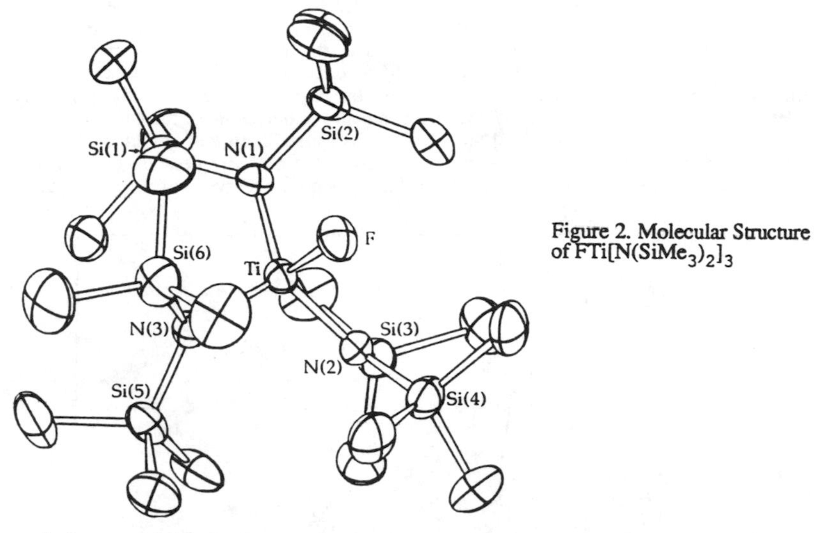

Figure 2. Molecular Structure of $FTi[N(SiMe_3)_2]_3$

towards the empty orbitals of the titanium atom. However, the Ti-N bond is significantly longer than that encountered in $CpTiCl_2N(SiMe_3)_2$ (average value = 1.93Å *versus* 1.883(6)Å, respectively).

Thermoanalytical studies

Figure 3 shows the thermal analysis of $CpTiCl_2N(SiMe_3)_2$ run at normal pressure under a flow of nitrogen gas. We observe that: (i) this complex melts at 130°C (DTA peak). The 2% weight loss observed at this temperature is due to the vaporization of the compound as no low-weight chemicals can be detected by mass spectrometry; (ii) the decomposition then occurs at 250°C and leads to the formation of CpH, Me_3SiCl and $HN(SiMe_3)_2$, all species being detected by mass spectrometry. The overall weight loss corrected from vaporization is 73.5%

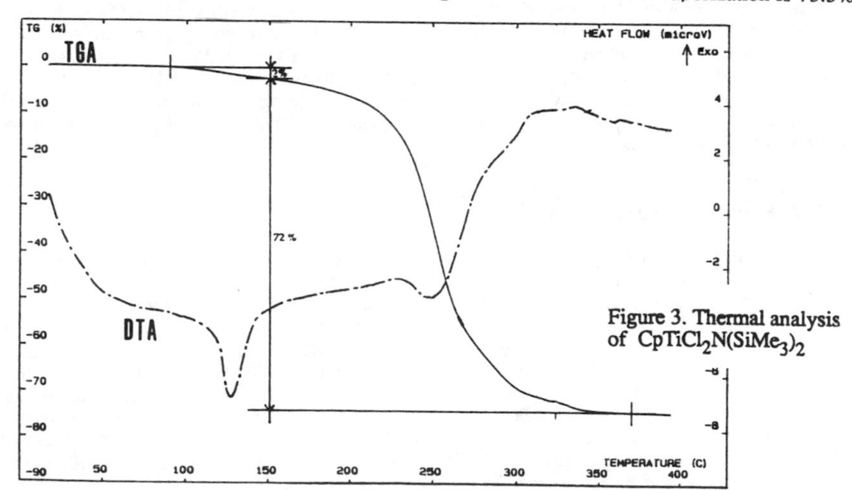

Figure 3. Thermal analysis of $CpTiCl_2N(SiMe_3)_2$

under a nitrogen flow and has been found to be 81.5% under hydrogen. The weight loss expected for the formation of titanium nitride is 82% and compares well with that obtained under H_2. However, elemental analyses of the residues indicate the formation of titanium carbonitride under both N_2 and H_2 conditions [13]. As could be expected, the amount of carbon in the residue is lower under H_2 than under N_2. The decomposition of $CpTiCl_2N(SiMe_3)_2$ is completed at a temperature as low as 350°C, independently of the nature of the gas.

The thermal analysis of $FTi[N(SiMe_3)_2]_3$ has also been run under nitrogen and hydrogen. Figure 4 shows the thermal behavior of this complex under hydrogen. $FTi[N(SiMe_3)_2]_3$ melts at 120°C as evidenced by the DTA peak observed at this temperature. Its

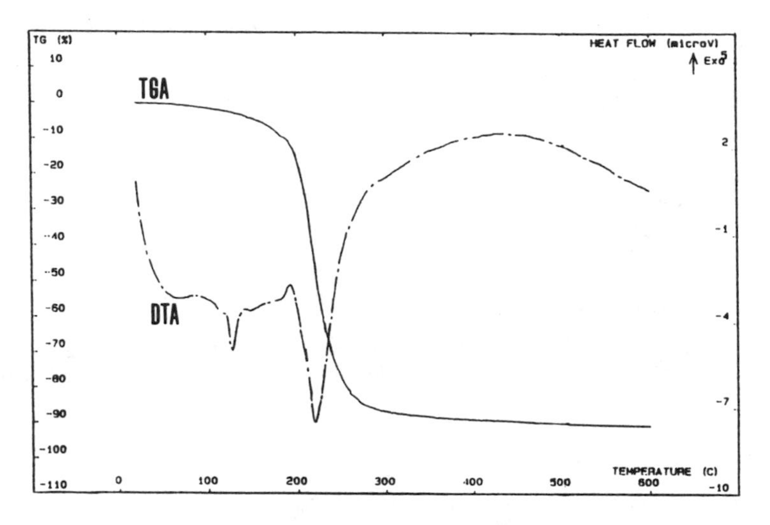

Figure 4. Thermal analysis of $FTi[N(SiMe_3)_2]_3$

decomposition occurs at 220°C and leads to the evolution of Me_3SiF and $(Me_3Si)_2NH$ as evidenced by simultaneous mass spectrometry analysis. The decomposition of $FTi[N(SiMe_3)_2]_3$ is completed at 280°C and corresponds to a 88.6% weight loss in good agreement with the formation of titanium nitride (calculated weigth loss for TiN = 88.7%). $FTi[N(SiMe_3)_2]_3$ appears to be an excellent candidate for the formation of titanium nitride by CVD.

Because of difficulties of handling a highly oxygen and moisture sensitive liquid compound in our TGA equipment, the thermoanalytical analysis of $V(NEt_2)_4$ could not be realized before the CVD experiments.

CVD experiments and characterization of the coatings

Thin films have been obtained by chemical vapor deposition using $CpTiCl_2N(SiMe_3)_2$ as precursor. The precursor was heated at 100-120°C and the substrates at 600°C. The pressure of the equipment was kept at 10torrs. In all experiments, shiny, bronze-black metal-like deposits were obtained on glass, silicon and steel. Typically 1 µm thick coatings could be formed within two hours of deposition. Deposits are reasonably adherent (Scotch tape test) on glass or silicon, but not on steel. XPS analyses have been carried out on deposits obtained on silicon substrates, and show (Figure 5) the presence of titanium, nitrogen and carbon but also of oxygen and traces of chlorine. After Ar etching the surface contamination layer, the amount of oxygen decreases. Figure 5a shows the binding energies of the $Ti_{2p3/2}$ and $Ti_{2p1/2}$ levels. The two peaks at 455.5eV and 461.0eV may be attributed to Ti-N and/or Ti-C bonds [14]. In addition, the peaks arising at 457.9eV and 463.2eV may be attributed to the $Ti_{2p3/2}$ and $Ti_{2p1/2}$ levels of TiO_2 respectively. The existence of titanium oxide is supported by the observation of a peak at

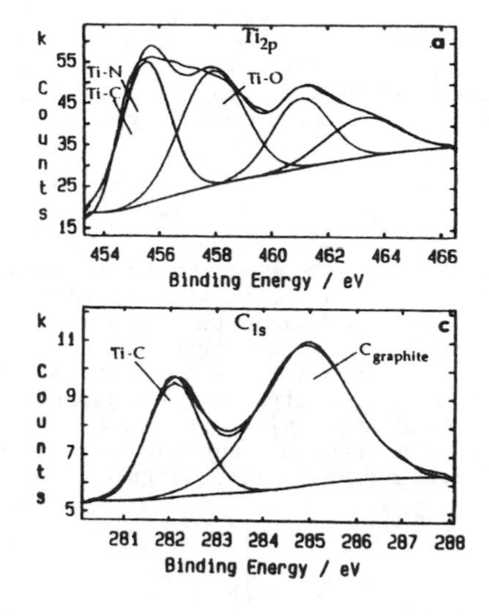

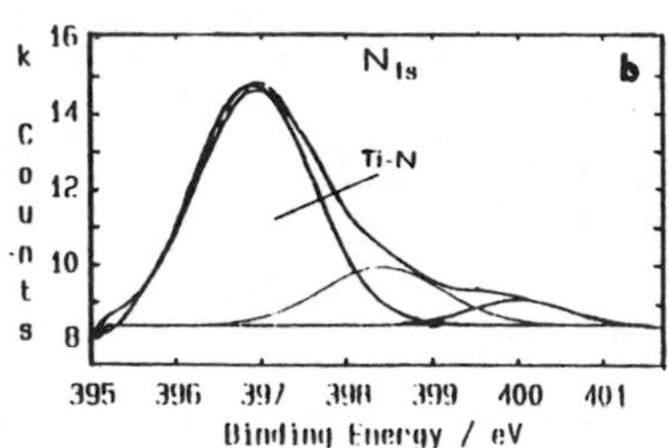

Figure 5. (a) Ti 2p, (b) N 1s, (c) C 1s regions of the XP Spectrum

530.6eV corresponding to the O_{1s} level [14]. The N_{1s} level binding energy at 396.8eV confirms the presence of titanium nitride [14] (Figure 5b) and that of the C_{1s} level (Figure 5c) indicates two kinds of carbon, carbide at 282.1eV and free carbon at 284.9eV [14]. These results show that $CpTiCl_2N(SiMe_3)_2$ can form titanium carbonitride by CVD. Deposition conditions remain to be optimized in order to reduce the free carbon and the chlorine contents of the films. The presence of chlorine is probably responsable for the poor adherence on steel.

Brown, adherent and metal-like coatings have been obtained by CVD from $V(NEt_2)_4$. The precursor was heated at 80°C, the substrates at 500°C and the pressure was kept between 1-10torrs. XPS analysis (Figure 6) shows the presence of vanadium, nitrogen, carbon and oxygen. The oxygen content is considerably decreased after Ar etching the surface contamination

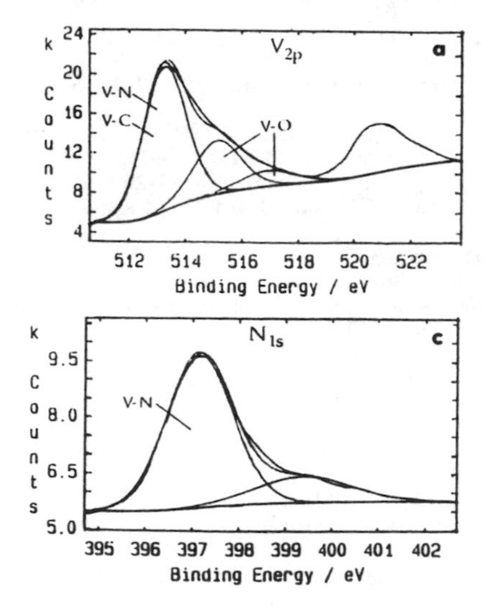

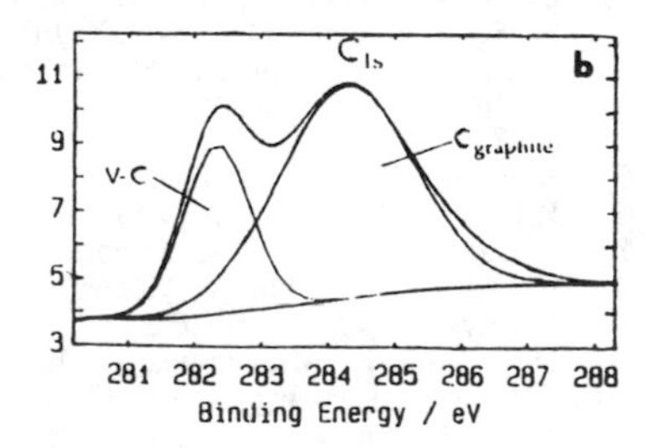

Figure 6. (a) V 2p, (b) C 1s, (c) N 1s regions of the XP Spectrum

layer. The energy value of the $V_{2p3/2}$ level (513.3eV, Figure 6a) is very close to that reported for vanadium carbide (512.6eV) and vanadium nitride (514.2eV) [14]. The formation of vanadium carbonitride is confirmed by the binding energies of the C_{1s} and N_{1s} levels on Figure 6b and 6c respectively, which correspond to vanadium carbide (282.3eV) and vanadium nitride (397.2eV) [14]. Additional peaks also reveal the presence of free carbon and traces of nitrogen containing species in the film.

CONCLUSION

Titanium carbonitride and vanadium carbonitride can be deposited by CVD on various substrates from $CpTiCl_2N(SiMe_3)_2$ and $V(NEt_2)_4$ respectively. According to thermoanalytical studies, $FTi[N(SiMe_3)_2]_3$ also appears to be a possible candidate for the CVD formation of titanium nitride. All these precursors can form the desired ceramic at temperature lower than 700°C. This allows them to be used in processes devoted to the protection of steel against abrasion without any risk of degradation of its intrinsic mechanical properties. Modulation of the M/C/N ratio remains to be realized whenever pure metal nitride coatings are desired. This may be done either by running the CVD experiments under ammonia which proved to be effective in many cases, or through modifications of the ligands in these complexes, resulting in a lower carbon content in the precursor.

AKNOWLEGMENTS

This work has been carried out within a collaborative CNRS program (SDI 5578). We thank the DGA-DRET (groupe 8) of the French National Defense Ministry as well as the Conseil Régional Midi-Pyrénées for the support of this work. We also thank R. Feurer and R. Morancho for assistance and tutoring in XPS measurements and data analyses.

REFERENCES

1. L.E. Toth, in Transition Metal Carbides & Nitrides, Refractory Materials, Edited by J.L. Margrave, (Academic Press, New-York, 1971), Vol 7; R. Juza, in Advances in Inorganic Chemistry and Radiochemistry, Edited by H.J Emeleus and A.G. Sharpe, (Academic Press, New-York and London, 1966), Vol IX, p. 81.
2. W. Schintlmeister, O. Pacher, K. Pfaffinger, J. Electrochem. Soc.123, 924 (1976).
3. S.R Kurtz, R.G. Gordon, Thin Solid Films 140, 277-290 (1986).
4. K. Sugiyama, S. Pac, Y. Takahashi, S. Motojima, J. Electrochem. Soc. 122, 1545-1549 (1975).
5. R.M. Fix, R.G. Gordon, D.M. Hoffman, Chem. Mat. 2, 235-241 (1990).
6. R.M. Fix, R.G. Gordon, D.M. Hoffman, Chem. Mat. 3, 1138-1148 (1991).
7. Y. Bai, H.W. Roesky, M. Noltemeyer, Z. Anorg. Allgem. Chem. 595, 21-26 (1991.
8. F. Laurent, J.S. Zhao, L. Valade, R. Choukroun, P. Cassoux, J. Anal. and Appl. Pyrolysis 1992, submitted.
9. D.C. Bradley, I.M. Thomas, J. Chem. Soc. 1960, 3857.
10. F. Laurent, J. P. Legros, Acta Crystallogr. 1992, submitted.
11. M. Bochmann, L.M. Wilson, M. B. Hursthouse, M. Motevalli, Organometallics 7, 1148-1154 (1988); S. Gambarotta, C. Floriani, A. Chiesivilla, C. Guastini, J. Amer. Chem. Soc. 105, 7295-7301 (1983); 104, 1918-1924 (1982); D. L. Thorn, W.A. Nugent, R.L. Harlow, J. Amer. Ceram. Soc. 103, 357-363 (1981); H. Burger, K. Wiegel, J. Thewalt, D. Schomberg, J. Organomet. Chem. 187, 301-309 (1975); J. Fayos, D.Z. Mootz, Z. Anorg, Allgem. Chem. 380, 196-201 (1971).
12. M.F. Lappert, A.R. Sanger, R.C. Srivastava, P.P. Power, Metal and Metalloid Amides (J. Wiley & Sons Publishers, New-York, 1980).
13. Elemental analyses of the TGA residues of $CpTiCl_2N(SiMe_3)_2$: under N_2, C 25.3, H 0.9, N 9.0; under H_2, C 20.2, H 0.5, N 11.0; these values correspond approximatively to the following ratios, Ti/C/N = 2/3/1 (N_2), 2/2/1 (H_2).
14. C.D. Wagner et al. Handbook of X-ray Photoelectron Spectroscopy (Perkin-Elmer Corporation, Eden Prairie).

NEW ROUTES TO GROUP IVA METAL-NITRIDES

CHAITANYA K. NARULA, Department of Chemistry, Research Laboratory, Ford Motor Company, P.O.Box 2053, MD 3083, Dearborn, MI 48121

ABSTRACT

The reactions of $TiCl_4$ with $[(CH_3)_3Si]_2NH$ have been examined under several reaction conditions. One of the reaction products, $(CH_3)_3Si(H)NTiCl_3$, can be crystallized in 60% yield on reacting $TiCl_4$ with $[(CH_3)_3Si]_2NH$ in a 1:1 molar ratio in dichloromethane at -78°C. $[(CH_3)_3Si(H)NTi(Cl_2)(NH)]_2TiCl_2$ is the primary product on increasing the amount of $[(CH_3)_3Si]_2NH$ to two equivalents. $(CH_3)_3Si(H)NTiCl_3$ and $[(CH_3)_3Si(H)NTi(Cl_2)(NH)]_2TiCl_2$ form titanium nitride on pyrolysis at 600°C in an ammonia atmosphere.

In absence of solvents, a variety of powders are formed on reacting $TiCl_4$ with $[(CH_3)_3Si]_2NH$ in 1:1 to 1:2 ratios. These powders are mixtures of several species, insoluble in common organic solvents, and form titanium nitride on pyrolysis.

TiN coatings on MgO are prepared by adding MgO particles to a solution of $(CH_3)_3Si(H)NTiCl_3$ in dichloromethane, warming and pyrolysis. Electron microscopic studies indicate that MgO particles are uniformly coated with titanium nitride.

INTRODUCTION

In recent years, preceramic precursor processing has been used to prepare several advanced materials. This route offers numerous advantages over conventional methods. For example, low temperature syntheses, control on the purity and ease of preparation of materials in the desired forms of fibers, coatings and films [1] are achieved easily by preceramic precursor processing. Materials such as silicon nitride [2], silicon carbide [2], and boron nitride [3] have been synthesized by this method.

Preparation of Group IVb metal-nitrides from preceramic precursors offers advantages over classical high-temperature methods, [4] but require high-temperature processing. Maya et.al. [5] reacted titanium or zirconium dialkylamides with ammonia to obtain insoluble polymers. The polymers furnish partially crystalline TiN or ZrN on pyrolysis in an ammonia atmosphere at 800°C. Seyferth et.al. [6] prepared polymers by reacting bifunctional amines with $Ti[N(CH_3)_2]_4$. These polymers require pyrolysis temperatures of 800-1200°C to form amorphous titanium nitride. Calcining at 1500°C is needed for crystallization. Recently, Rhine et.al. have prepared polymers by reacting furfuryl alcohol with titanium alkoxides. The polymers can be pyrolyzed in ammonia to prepare titanium nitride [7].

We became interested in methods to prepare Group IVb metal-nitride coatings on amorphous metal-oxides for applications in 'Sodium Heat Engine' [SHE]. One SHE developed at Ford contained sodium ion conducting sodium-ß-alumina membrane coated with TiN [8]. TiN on ß-alumina was reactively sputtered with a magnetron sputtering system [9]. It was considered necessary to find alternate low temperature, economical methods for the preparation of TiN coatings.

Recent results from our laboratory describe the photolytic decomposition of $Ti[N(CH_3)_2]_4$ on exposure to infra-red radiation from a pulsed ND-YAG laser. The decomposition products can be pyrolyzed to TiN and this method can also be used to prepare titanium nitride coatings on Al_2O_3, TiO_2 and MgO [10]. In this report, we will describe studies on the reactions of $TiCl_4$ with $[(CH_3)_3Si]_2NH$, characterization of reaction products, and their pyrolysis to form titanium-nitride. Preliminary work has also been carried out on the preparation of coatings on metal-oxide particles with a view to investigate the microscopic structure of the coated materials. The preparation of coatings on ß-alumina membranes and its evaluation is under progress.

EXPERIMENTAL

Preparation of $[(CH_3)_3Si(H)NTiCl_3$: A 250 ml three-neck flask fitted with a condenser was charged with $TiCl_4$ (6.9 g, 36.3 mmol) and CH_2Cl_2 (100 ml) and cooled to -78°C. To this solution $[(CH_3)_3Si]_2NH$ (5.87 g, 36.3 mmol) was slowly added with a syringe with stirring. After complete addition, the reaction mixture was slowly warmed to room temperature and the precipitated orange compound (1.0 g) was separated by filtration. The filtrate was concentrated and cooled to obtain red-orange crystals. Elemental Analysis for $C_3Cl_3H_{10}NSiTi$ Calc. C 14.8, H 4.16, N 5.8, Ti 19.7; Found C 15.1, H 4.2, N 6.7, Ti 19.1. Yield 5.16g, 58%.

Preparation of $[(CH_3)_3Si(H)NTi(Cl_2)(NH)]_2TiCl_2$: $TiCl_4$(6.90g, 36.4mmol) and $[(CH_3)_3Si]_2NH$ (11.75g, 72.8mmol) were reacted in CH_2Cl_2 at -78°C. The reaction mixture was allowed to warm slowly to room temperature. Initially, a clear solution was obtained from which a yellow precipitate separated out. The product was isolated by filtration and repeated washing with CH_2Cl_2. Elemental analysis for $C_6H_{22}N_4Si_2Ti_3$ Calc. C 12.8, H 3.94, N 9.95, Ti 25.52; Found C 13.04, H 4.57, N 9.87, Ti 25.97 yield 5.73, 84%.

$TiCl_4$ was also reacted with $[(CH_3)_3Si]_2NH$ in the absence of solvents under the following reaction conditions.

Method a. $[(CH_3)_3Si]_2NH$ (5.87g, 36.37 mmol) was added to $TiCl_4$ (6.9g, 36.3mmol) at room temperature dropwise. An orange colored powder forms immediately. After complete addition, volatile material was removed in vacuum to obtain a free flowing orange powder. Yield: residue, 6.31g; volatiles: $(CH_3)_3SiCl$, 5.22 g, 48 mmol.

Method b. $TiCl_4$(6.9g, 36.3mmol) and $[(CH_3)_3Si]_2NH$ (11.75g, 72.8mmol) were reacted at room temperature. The reaction mixture was heated to 70°C for 2 hours, cooled to room temperature and volatiles were removed to obtain a free flowing yellow powder. Yield: residue, 7.60 g; volatiles: $(CH_3)_3SiCl$, 9.92g, 66.6mmol; $[(CH_3)_3Si]_2NH$, 1.0g, 6.2mmol.

Deposition on Magnesium Oxide Particles: To a solution of $TiCl_4$ (0.48g) in CH_2Cl_2 was added hexamethyldisilazane(0.73g) at -78°C. After warming to -20°C, magnesium oxide particles (0.25g) were added to the reaction mixture. The reaction mixture was slowly warmed to room temperature and the volatiles were removed in vacuum.

RESULTS AND DISCUSSION:

The reaction chemistry of $TiCl_4$ with $[(CH_3)_3Si]_2NH$ was first examined by Andrianov et. al. [11] during their investigation of new manufacturing methods for $(CH_3)_3SiCl$. These authors prepared $(CH_3)_3SiCl$ in 88% yield and described titanium containing residue to be "$NHTiCl_2$". Wannagat et.al. [12] also studied the reaction of $TiCl_4$ with $[(CH_3)_3Si]_2NH$ and isolated $[(CH_3)_3Si]_2NH.2TiCl_4$ and $(CH_3)_3Si(H)NTiCl_3$ by varying the reaction conditions. The characterization of the compounds was limited to elemental analyses.

Our studies show that $NHTiCl_2$ is not formed in the above reaction under the conditions described by Andrianov et.al. Instead, an orange, free flowing powder, henceforth called I, separates out with the evolution of $(CH_3)_3SiCl$ (identified by ^1H and ^{13}C NMR) in 66% yield when equimolar amounts of $[(CH_3)_3Si]_2NH$ with $TiCl_4$ are reacted in the absence of solvents. The infrared spectrum of the powder shows a broad strong band at 3148 cm^{-1} for νNH, a sharp medium intensity band at 1251 cm^{-1} for δs CH_3 of $(CH_3)_3Si$ and a sharp band at 805 cm^{-1} for νas Si-N-Ti. These data suggest the presence of trimethylsilylamino groups in the residue and mass balance calculations suggest it to have a ratio of 3:3:1 for Ti:N:Si. The material is largely insoluble in inert organic solvents, but a small fraction dissolves in dichloromethane and can be separated. The spectral properties of the soluble fraction suggest it to be $(CH_3)_3Si(H)NTiCl_3$.

Two equivalents of $[(CH_3)_3Si]_2NH$ react with $TiCl_4$ to furnish a free flowing yellow powder, henceforth called II. The volatile byproducts are a mixture of $(CH)_3)_3SiCl$ and $[(CH_3)_3Si]_2NH$ in 10:1 ratio as determined from ^1H NMR. The yield of evolved $(CH_3)_3SiCl$ is calculated to be 74% based on reacted $[(CH_3)_3Si]_2NH$ and mass balance calculations suggests the ratio of Ti:N:Si to be 1.15:1.92:1. The infrared spectrum of the yellow powder residue is similar to that of the orange powder described in the preceding paragraph except that the band at 1251 cm^{-1} due to δs CH_3 of $(CH_3)_3Si$ is stronger suggesting higher ratio of Si to Ti.

Reproducible elemental analysis of these compounds could not be obtained, and these powders are probably mixtures of compounds. It should be pointed out that Wannagat et. al. also suggested that the powders that precipitate out from the solutions on reactions of $TiCl_4$ with $[(CH_3)_3Si]NH$ in varying ratios of 1:0.6 to 1:2 are mixtures of compounds.

Pure compounds can be prepared if the reactions are carried out under controlled conditions. A red solution is formed on the addition of $[(CH_3)_3Si]_2NH$ to a dichloromethane solution of $TiCl_4$ at -78°C. On warming to 25°C, an orange powder separates outs. This powder dissolves readily in MeCN and on removal of MeCN the powder is recovered suggesting that MeCN does not react with the powder. Elemental analysis is not reliable and suggest it to be a mixture of compounds. The filtrate forms orange-red crystals on cooling at -20°C. Elemental analysis of the crystals agrees with the formula $Me_3Si(H)NTiCl_3$. ^1H and ^{13}C NMR spectra of the crystals show single peaks at 0.42 and 1.98 ppm respectively originating from protons and carbons of trimethylsilyl groups.

$$TiCl_4 \quad + \quad [(CH_3)_3Si]_2NH \ \text{-->} \ [(CH_3)_3Si(H)NTiCl_3] \quad + \quad Me_3SiCl$$

A yellow powder separates from the clear solution obtained on reacting $TiCl_4$ with two equivalents of $[(CH_3)_3Si]_2NH$ in dichloromethane at -78°C on warming to room temperature. Elemental analysis of the materials is consistent with the formula $C_6H_{22}Cl_6N_4Si_2Ti_3$ with a probable structure shown below.

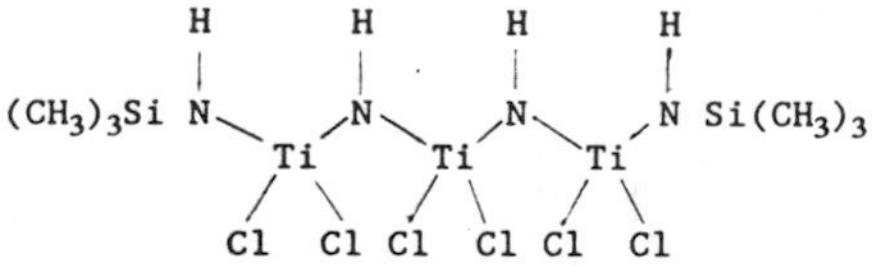

```
              H         H         H         H
              |         |         |         |
(CH3)3Si  N\       /N\       /N\       /N  Si(CH3)3
            Ti        Ti        Ti
           /  \      /  \      /  \
         Cl    Cl  Cl    Cl  Cl    Cl
```

The compound is insoluble in inert organic solvents but dissolves readily in CH_3CN. A red powder separates out from the clear CH_3CN solution on storing it at room temperature. ^1H NMR of the freshly prepared CD_3CN solution of the material shows singlets at 0.42, 0.27, 0.02 ppm. The 0.27 peak gradually decreases and completely disappears within 60 minutes. This clearly indicates an irreversible reaction with acetonitrile. In a similar reaction, Roesky et.al. found that $(i\text{-}Pr)_2P(S)N[Si(CH_3)_3]_2$ reacts with $TiCl_4$ in the presence of CH_3CN with the complete elimination of $(CH_3)_3SiCl$ to forms oligomers incorporating CH_3CN [13].

Unlike $TiCl_4$, $[Me_3Si]_2NH$ does not react with $[(CH_3)_2N]_{4-x}TiCl_x$, x = 1, 2 or 3, [14] in the presence or absence of solvents or on heating to 100°C. It is possible that $[(CH_3)_3Si]_2NH$ is too weak a base to form adducts with dimethylamino substituted titanium halides which seems to be the first step in these reactions.

Pyrolysis

Prior to pyrolysis, thermogravimetric analyses [TGA] of $(CH_3)_3Si(H)NTiCl_3$, $[(CH_3)_3Si$-$(H)NTi(Cl_2)(NH)]_2TiCl_2$, and powders I and II was carried out to select optimum temperature conditions. $(CH_3)_3Si(H)NTiCl_3$ shows a weight loss of 44% in the 120-200°C, a gradual weight loss of 11% in the 220-550°C and of 7% 550-600°C region. $[(CH_3)_3Si(H)NTi(Cl_2)$-

(NH)]$TiCl_2$ shows a weight loss of 37% in 180-240°C and 24 % in 220-420°C regions respectively. Powder I shows a weight loss of 30% in 50-100°C region and an additional weight loss of 37% in 100-400°C region. Powder II, on the other hand shows a weight loss of 12% in the 50-100°C region, 37% in the 100-180°C region and 14% in the 180-400°C region. Since the weight loss for all samples is complete below 600°C, we fired the samples at this temperature.

Pyrolyses of $(CH_3)_3Si(H)NTiCl_3$, $[(CH_3)_3Si(H)NTi(Cl_2)(NH)]TiCl_2$ and powders I and II were carried out by the following procedure. A weighed amount of the sample was placed in a quartz tube connected to a system to maintain nitrogen atmosphere. The temperature of the furnace was slowly raised to 600°C and maintained at that temperature for 5 hours. The X-ray powder diffraction [XRD] patterns of black powders match with that reported (JCPD# 38,1420) for titanium nitride. No additional diffraction peaks are present in the XRD patterns, which could be assigned to silicon containing species.

Pyrolyses of the precursors in an ammonia atmosphere at 600°C furnishes a dark-olive-brown titanium nitride powders. The diffraction peaks in XRD patterns of the powders are somewhat broad suggesting small particle size. After heating the samples at 800°C, the resulting powder shows sharp peaks in its XRD patterns [Fig. 1]. Transmission electron micrographs [TEM] show the particle size to be 40-200 nm range. Energy dispersive spectra [EDS] of several particles show Kα peaks for titanium and nitrogen. No peak for silicon is observed suggesting that trimethylsilyl groups are eliminated at relatively low temperatures in the presence of ammonia without leaving silicon residues. Titanium nitride yields are close to quantitative based on titanium.

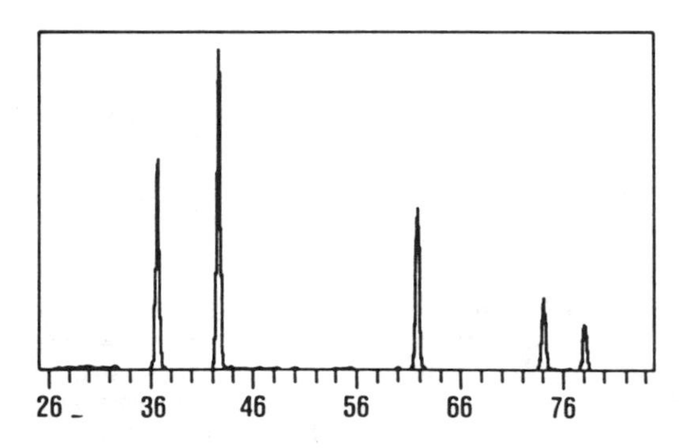

Fig. 1: XRD Pattern of the Pyrolysis Product of $(CH_3)_3$ Si(H)NTiCL$_3$

Applications

Solubility of $[CH_3)_3Si(H)NTiCl_3$ and its low temperature conversion to TiN makes it an attractive precursor for the preparation of TiN coatings or the composite materials with TiN as one of the components. Magnesium oxide was selected as a representative substrate because its convenient preparation as non-porous uniform particles, crystallized on hkl=100 axis, makes it suitable for HREM studies. TiN/MgO prepared by a variety of methods has also been the subject of extensive microscopic studies, providing an opportunity to compare preceramic coated MgO particles with those obtained by other methods [15].

In efforts to simplify the process, we (i) prepared the reaction mixture in dichloromethane by mixing $[CH_3)_3Si]_2NH$ with $TiCl_4$ at -78°C, (ii) added MgO oxide particles to the solution (iii) warmed the reaction mixture to room temperature and collected the particles (iv) and carried out pyrolysis under an ammonia atmosphere at 600°C for 4 hours. The X-ray powder diffraction of the material matched with that of MgO (JCPD # 4,829) and TiN (JCPD # 38,1420) and from peak intensities the ratio of TiN to MgO was found to be 1:4. TEM shows two varieties of particles: single particles and agglomerates. MgO particles either fuse on exposure to high temperature, or TiN growth over several particles gives them the appearance of agglomerates. EDS of several particles showed the expected Ti, Mg, O and N $K\alpha$ peaks.

Further simplification of the process was attempted by (i) grinding magnesium oxide particles (0.25g) with $[(CH_3)_3Si(H)NTi(NH)(Cl_2)]_2(NH)TiCl_2$ (1.0 g) and (ii) pyrolysis at 600°C for one hour in ammonia atmosphere. The sample required grinding and firing at 1000°C for 4 hours to form a homogeneous material. The XRD pattern of the material matches with those of MgO (JCPD # 4,829) and TiN (JCPD # 38,1420) as far as peak positions are concerned. EDS on several particles shows magnesium and titanium $K\alpha$ peaks although a few particles exhibit either magnesium or titanium $K\alpha$ peak. TEM shows that a large number of particles have joined to form agglomerates.

CONCLUSIONS

Reactions of titanium tetrachloride with hexamethyldisilazane form free flowing powders, $(CH_3)_3Si(H)NTiCl_3$, or $[(CH_3)_3Si(H)NTi(Cl_2)(NH)]_2TiCl_2$ depending on the reaction conditions. All these products form titanium nitride on pyrolysis in an ammonia atmosphere at 600°C. Since $(CH_3)_3Si(H)NTiCl_3$ dissolves in halocarbons, it is an attractive precursor for the preparation of coatings or composite materials. Preliminary studies on the application of this material in the preparation of coatings on MgO particles support its suitability as a precursor for titanium nitride coatings or composites.

ACKNOWLEDGEMENT

The use of WM-360 Bruker NMR at Chemistry Department and Phillips CR/2 ASTEM and JEOL 4000 HREM at Center for Microbeam Analysis of University of Michigan, Ann Arbor is also gratefully acknowledged.

REFERENCES

[1] Better Ceramics Through Chemistry, edited by C.J. Brinker, D.E. Clark, D.R. Ulrich, (Mater. Res. Soc. Proc. 32, Pittsburgh, PA 1984). Better Ceramics Through Chemistry II, edited by C.J. Brinker, D.E. Clark, D.R. Ulrich, (Mater. Res. Soc. Proc. 73 Pittsburgh, PA 1986). Better Ceramics Through Chemistry III, edited by C.J. Brinker, D.E. Clark, D.R. Ulrich, (Mater. Res. Soc. Proc. 121 Pittsburgh, PA 1988). Better Ceramics Through Chemistry IV, edited by B.J.J. Zelinski, C.J. Brinker, D.E. Clark, D.R. Ulrich, (Mater. Res. Soc. Proc. 180 Pittsburgh, PA 1990).

[2] K.J. Wynne, R.W. Rice, Ann. Rev. Mater. Sci., 14 (1984) 297.

[3] R.T. Paine, C.K. Narula, Chem. Rev., 90 (1990) 73.

[4] Kirk-Othmer Encyclopedia of Chemical Technology, Third Edition, (Wiley, New York, 23).

[5] G.M. Brown, L. Maya, J. Am. Ceram. Soc., 71 (1988) 78.

[6] G. Mignani, D. Seyferth, Gov. Rep. Announce. Index (U.S.), 88 (1988) 827,109. CA 109 (1988) 214,916.

[7] Z. Jiang, W.E. Rhine, Chem. Mater. 3 (1991) 1132.

[8] J.V. Lasecki, R.F. Novak, J.R. McBride, J.T. Brockway, T.K. Hunt, Proceedings of the 22nd Intersociety Energy Conversion Engineering Conference (1987) 1407.

[9] J.R. McBride, D.J. Schmatz, T.K. Hunt, R.F. Novak, Proceedings of the Second Symposium on Electrode Materials and Processes for Energy Conversion and Storage, Electrochemical Society, 87 (1987) 594.

[10] C.K. Narula, M.M. Maricq, Am.Chem. Soc., Div. Poly. Chem., 32 (1991) 499. M.M. Maricq, C.K. Narula, Chem. Phys. Lett. 187 (1991) 220.

[11] K.A.Andrianov, V.V. Astakhin, D.A. Kuchkin, I.V. Sukhanova, Žh. Obsh. Khim., 31 (1961) 3410.

[12] H. Bürger, U. Wannagat, Mh. Chem., 94 (1963) 761.

[13] H.W. Roesky, T. Raubold, M. Witt, R. Bohra, M Noltemeyer, Chem. Ber. 124 (1991) 1521.

[14] E. Benzing, W. Kornicker, Chem. Ber. 94 (1961) 2263.

[15] L. Hultman, D. Hesse, W.-A. Chiou, J. Mater. Res., 6 (1991) 1744.

TITANIUM CARBIDE FROM CARBOXYLIC ACID MODIFIED ALKOXIDES

Tom Gallo, Carl Greco, Claude Peterson, Frank Cambria and Johst Burk,
Akzo Chemicals Inc., 1 Livingstone Ave, Dobbs Ferry, NY 10522, USA.

ABSTRACT

Transition metal carbide precursors have been made in the past by the reaction of alkoxides with polymeric materials to form gels and resins. A new route to transition metal carbide precursors has been developed using alkoxides polymerized with dicarboxylic acids. (Dicarboxylic acid precursors have the advantage of precipitating as powders that can be removed from solvents by filtration and that are not very air sensitive.) Precursors were pyrolyzed under inert or reducing conditions to form metal carbides.

The choice of ligand(s) determined the carbon content after pyrolysis. Unsaturated ligands tended to increase carbon content. Materials from oils to fine powders were produced by varying the stereochemistry of the ligands. The morphology of the pyrolyzed product mimicked that of the precipitated powder. Pyrolysis was typically carried out under Ar/H_2 at 1200-1600°C. X-ray diffraction (XRD) was used to follow the incorporation of carbon into the lattice.

INTRODUCTION

There are four criteria for choosing a route to materials preparation: price, purity, morphology and phase assemblage. Direct carbothermal reduction of the oxide will always be the most economical route to carbide formation. The metal to carbon ratio is easiest to control when working with pure metals and carbon. Other routes are used to increase purity and control morphology and phase assemblage. Routes based on metal alkyls are generally very costly due to the use of expensive organometallic reagents. Oxygen containing precursors must go through some form of carbothermal reduction which may lead to residual oxygen.

One choice of a starting material for non-oxides is metal alkyls. H. Lang and D. Seyferth [1] used precursors such as $TiCp_2(CCC_6H5)_2$ and $TiCp_{2}$-μ-CH_2 to form TiC. TiC from these precursor materials always have excess carbon. J.R. Fox et. al. [2] used a phenyl modified siloxane polymer to form silicon carbide. T. Gallo et. al. [3] found that a similar titanium structure, $Ti(C_6H_5)_2(OC_3H_7)_2$, only forms titanium oxides because of the weak Ti-C bond. A simple way of producing titanium carbide is the pyrolysis of titanocene dichloride or titanocene dimethyl in Ar [3].

Alkoxides polymerized with polyalcohols have also been used for the production of carbides. This was first reported in a Japanese patent application by Ube Industries [4]. $TiCl_4$ was reacted with catechol (1,2 dihydroxy benzene) to form a polymeric precursor and HCl. Pyrolysis of this precursor gave TiC. Although not reported this precursor must have yielded an excess of carbon. In a series of European and US patents J.D. Birchall et. al. [5] formed polymeric alkoxides as precursors for non-oxide ceramics. Their precursors were alkoxides or halides reacted with polyalcohols. Birchall typically reacted one mole of tetraethyl orthosilicate, TEOS, with one mole of furfuryl alcohol and then reacted this "new" alkoxide with one mole of glycerol. Glycerol, being a tri-functional alcohol, forms a nonvolatile resin or gel. Another way of forming polymerized alkoxides was developed by S-J Ting et. al. [6]. They reacted alkoxides with diacetates to form polymeric alkoxides and esters. From XRD data, they claimed that the TiC formed was substoichiometric in carbon. Most likely this means the material, which contained 3% oxygen, showed XRD line shifts due to the presence of TiO and not a substoichiometric TiC.

Metal alkoxides will react with hydroxylated polymers/ resins to form pendant metal alkoxy groups. These modified resins are carbide precursors. Furfuryl alcohol polymerizes under acidic conditions to form a resin. Z. Jiang and W. Rhine[7] and Krumbe et. al. [8] used the polymerization of furfuryl alcohol as a resin system for carbide precursor formation. Krumbe also used the polyalcohol formed by the polymerization of phenol with formaldehyde. M.A. Janey [9] used phenolic resins or methylcellulose reacted with alkoxides to form carbide precursors. Although his examples only showed the use of polyalcohols, Janey claimed the use of polymers having active hydrogens on a hydroxyl, amino, amido, carboxyl, or thio groups.

EXPERIMENTAL WORK

Reactions were carried out in 1L triple neck flasks equipped with a mechanical stirrer, Dean-Stark trap with a condenser, addition funnel, and heating mantle or oil bath. Tetraisopropyl titanate (TIPT) was reacted with dicarboxylic acids with the removal of isopropanol/solvent using the Dean-Stark trap. Typically TIPT was added slowly to a solution of acids in a refluxing solvent. Heating was continued until no -OH peak was detected in the distillate by infra-red spectroscopy. Powders were separated from solvents by distillation on a rotoevaporator or filtration. The powders were vacuum dried at 110°C prior to pyrolysis.

Precursor samples from 2 to 15 grams were placed in an alumina tray and placed on a "D" tube in the middle of a horizontal tube furnace. End plates equipped with gas tight purge fittings were placed on the ends of the furnace. Purge gas flow was controlled using a rotameter. The exit end of the furnace was equipped with an oil bubbler so that the furnace was always under a pressure of 1 cm oil. Pyrolysis was typically carried out using a heating rate of 5 to 20°C/min, a soak time of 0.5 to 4 hour, a cooling rate of 10°C/min, under an atmosphere of Ar with 0-10% H_2 at 0.1 to 4.5 l/min. Because of the pyrophoric nature of some of these materials, samples were cooled to less than 50°C prior to removal from the furnace. LECO Corporation preformed carbon analysis. Particle size distribution was measured using a Granulometre. X-ray Diffraction was carried out using a Cu X-ray source of 1.54056 Å.

RESULTS AND DISCUSSION

The largest challenge in the formation of carbides from organometallic precursors is forming a product with the correct amount of carbon. A good example of varying the precursor chemistry to control the final carbon content is the use of two acids. One acid which yields high carbon, *e.g.*, isophthalic acid, and a second acid which yields a low carbon content, *e.g.*, fumaric acid. Table I shows the effect of ligand(s) on the carbon content of TiC, as well as, the weight gain on oxidation of the TiC formed. The theoretical carbon content of TiC is 20.05%, the theoretical weight gain on oxidation of TiC is 33.37%. A weight gain of less than 33.37% on oxidation of TiC shows the presence of free (uncombined) carbon and residual oxygen. Varying the phthalic to fumaric acid ratio shows how the carbon content of TiC may be controlled.

Fumaric acid is the trans- dicarboxylic acid and maleic acid is the cis- acid. Fumaric acid cannot chelate to the metal, whereas maleic acid may chelate to the metal. Succinic acid potentially should chelate the metal half the time. It is interesting to see that the carbon content decreases from fumaric to maleic to succinic acid. The use of acids that can chelate titanium, such as maleic or cis-muconic acids, yields oils that are interesting as liquid carbide precursors. Succinic acid forms a gel depending upon the choice of solvents. The use of nonchelating

Table I: The effect of ligands on carbon content and morphology

Precursor formula	Carbon Leco	Wt. gain ox.	Morphology
$(OiPr)_2Ti(O_2CC_6H_4CO_2)_1$	32.8	10.1	fine powder
$(OiPr)_2Ti(O_2CC_6H_4CO_2)_{.5}(O_2CCH{=}CHCO_2)_{.5}$	28.3	12.3	fine powder
$(OiPr)_2Ti(O_2CC_6H_4CO_2)_{.4}(O_2CCH{=}CHCO_2)_{.6}$	21.3	21.4	fine powder
$(OiPr)_2Ti(O_2CC_6H_4CO_2)_{.3}(O_2CCH{=}CHCO_2)_{.7}$	20.6	27.5	fine powder
$(OiPr)_2Ti(O_2CC_6H_4CO_2)_{.2}(O_2CCH{=}CHCO_2)_{.8}$	19.1	16.1	fine powder
$(OiPr)_2Ti(O_2CCH{=}CHCO_2)_1$	12.0	29.2	fine powder
$(OiPr)_2Ti(O_2CCH{=}CHCO_2)$ [fumarate]	12.0	29.2	fine powder
$(OiPr)_2Ti(O_2CCH_2CH_2CO_2)$ [succinate]	8.0	25.8	yellow gel
$(OiPr)_2Ti(O_2CCH{=}CHCO_2)$ [maleate]	6.7	24.6	oil
$(OiPr)_2Ti(O_2CCH{=}CHCH{=}CHCO_2)$ [muconate]	21.4	27.2	oil/gel
$(O_2CC_6H_4CO_2)Ti(OiPr)_2$	32.8	10.1	fine powder
$(O_2CC_6H_4CO_2)Ti(O\text{-}2ethylhexyl)_2$	28.3	18.3	fine powder
$(O_2CCH_2CH_2CO_2)Ti(OiPr)_2$	8.0	25.8	yellow gel
$(O_2CCH_2CH_2CO_2)Ti(O\text{-}2ethylhexyl)_2$	6.8	18.8	powder

acids such as fumaric are better for the formation of fine powders. The effects of the stereochemistry of the acid on precursor morphology, *i.e.*, the formation of less condensed species leading to oils and gels, can also be explained by the low second dissociation constant of the "chelating" acids.

Contrary to our expectations, the use of 2-ethylhexyl titanate as opposed to isopropyl titanate actually lowers the carbon content. Propene is such a good leaving group that we expected 2-ethylhexyl to increase the carbon content.

Heating rate, firing temperature, soak time, and sample size all affect the carbon content, in a predictable fashion. Decreased heating rates, increased firing temperature, increased hold times, and decreased sample size decrease the carbon content of the final material. For titanium carbide, each of these parameters can affect the final total carbon content by 1-3%.

According to Schwarzkopf and Kieffer [10], there is a solid solution between TiC and TiO. The work of Krainer and Konopieky [11] shows that for TiC carbon contents ranging from 16 to 18%, the lattice parameter shifts by 0.00036 nm for a 1% change in carbon. The data had a correlation coefficient, R^2, of 99%. For this study, shifts in the 60.5° two theta peaks were used to estimate the combined carbon content. Different precursors were fired to 1600°C and held for 4 hours. At this point it was assumed that all carbon, up to 20.05%, was combined. Carbon contents were measured by LECO and an empirical formula, R^2 92%, was generated for all Akzo data. Data from Krainer and Konopieky and this study are shown in Figure 1.

Figure 1: Change in lattice constant with carbon. (Krainer and this study data).

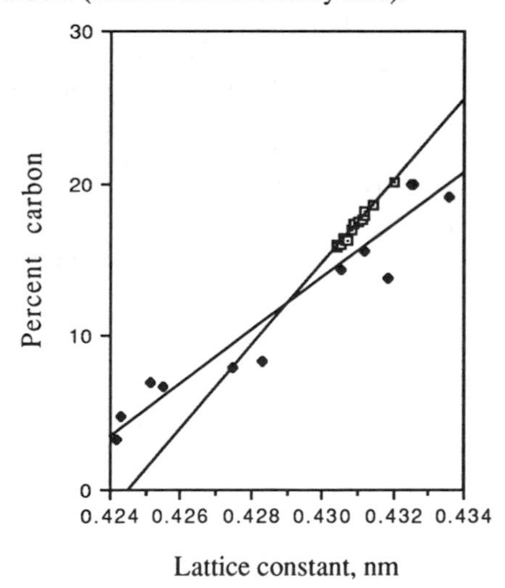

Figure 2: Combined carbon vs time (measured by XRD).

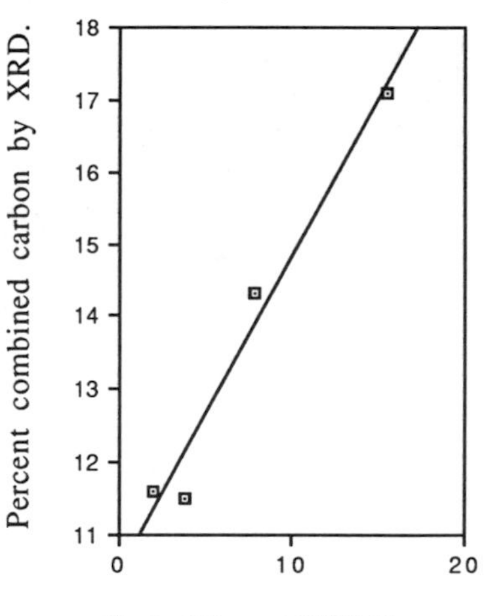

In general, the precursors are amorphous up to 1000°C. By 1200°C they crystallize into a Ti-O-C solid solution, containing only a few percent carbon. After an hour at 1400°C the materials are essentially TiC, assuming there is enough carbon present. Figure 2 shows that the amount of combined carbon increases linearly with the square root of time, min^{-2}, at 1350°C for the 50/50 isophthalic acid/fumaric acid precursor.

Table II shows the effect of gas composition and flow rate on total and XRD carbon content of the 50/50 isophthalic acid/fumaric acid precursor. Gas composition and flow rate have little affect on total carbon content for these high carbon samples. The data also suggest that combined carbon is affected by flow rate only and not gas composition. Increased flow rates probably remove oxygen containing volatile species, thereby driving carburization toward completion.

Solvent exchange and surfactants can be used to control the agglomerate size of precursor powders. Figures 3 and 4 shows the particle size distribution of 30/70 isophthalic acid/fumaric acid precursors after pyrolysis. The precursor (Figure 3) was precipitated by solvent exchange and a surfactant, Duomeen® TDO, was used. Controlled agglomeration during precipitation was used to produce a material (Figure 4) that could be filtered, had good flow properties, and low dust formation for easy calcination. This material was run through an air-mill at inlet pressures less than 0.3 MPa (40 psig) to form 1-5 micron TiC.

CONCLUSIONS

Transition metal carbides have been produced by the pyrolysis of alkoxides that have been polymerized using dicarboxylic acids. The morphology of the precursor was changed from oils to powders by varying the stereochemistry and/or second dissociation constant of the acids. Carbon content was controlled by the ratio of aliphatic to aromatic acids. Increasing the flow rate of the firing atmosphere had a stronger influence on the amount of combined carbon than the hydrogen content of the firing atmosphere. XRD was used to follow the incorporation of carbon into the lattice.

Table II: Effects of Gas Composition and Flow Rate on Carbon Content

At 1 l/min:	XRD Carbon	Leco Carbon
Ar	17.1	28.1
Ar/4%H_2	16.5	28.3
Ar/10%H_2	17.2	29.3

Ar/4%H_2 at:	XRD Carbon	Leco Carbon
0.1 l/min	13.2	30.0
1.0	16.5	28.3
4.5	18.2	28.8
Stark	18.2	19.1

Figure 3: PSD of a precipitated TiC after pyrolysis.

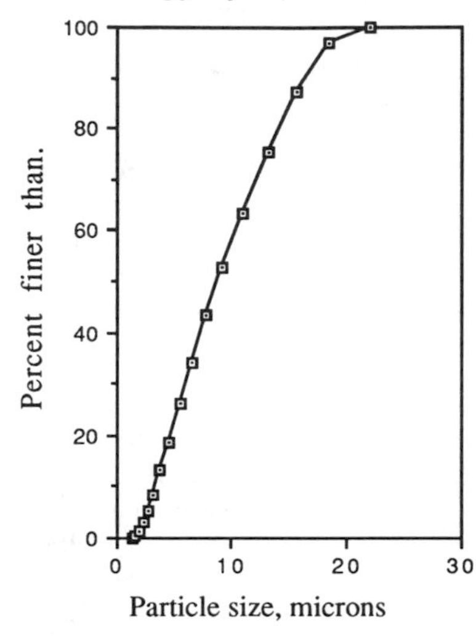

Figure 4: Effect of air-milling pressure, MPa, on PSD.

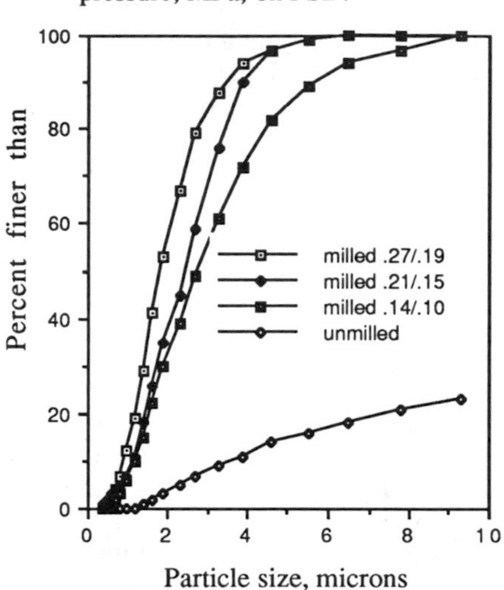

REFERENCES

1. H. Lang and D. Seyferth, Appld. Organomet. Chem. 4 (6), 599-604 (1990).
2. J.R. Fox, D.A. White, S.M. Oleff, R.D. Boyer and P.A. Budinger, in Better Ceramics Through Chemistry II, edited by C.J. Brinker, D.E. Clark, D.R. Ulrich (Mater. Res. Soc. Symp. Proc. 73, Palo Alto, CA 1986) 395-400.
3. T. Gallo, C. Greco, B. Simms and F. Cambria, in Covalent Ceramics, edited by G.S. Fischman, R.M. Spriggs and T.L. Aselage (Mater. Res. Soc. Ext. Ab. 23, Boston, MA1990) pp. 29-32.
4. Ube Industries KK J. Patent Pub. 56 155 013 (1 December 1981).
5. J.D. Birchall et al, E. Patent App. 239 301 (13 March 1987); E. Patent App. 266 104 (16 October 1987); E. Patent App. 266 105 (16 October 1987)/U.S. Patent 4 861 735 (29 August 1989).
6. S-J Ting, C-J Chu, E. Limatta, J.D. Mackenzie and T. Getman, M.F. Hauthorne, in Better Ceramics Through Chemistry IV, Edited by B.J.J. Zeninski, C.J. Brinker, D.E. Clark, D.R. Ulrich (Mater. Res. Soc. Symp. Proc. 180, 1990) pp. 457-460.
7. Z. Jiang and W.E. Rhine, Chem. Mater. 3 1132-1137 (1991).
8. W. Krumbe, B. Lauback, G. Franz, U.S. Patent 4 948 762 (14 August 1990).
9. M.A. Janey, U.S. Patent 4 622 215 (11 November 1986).
10. Paul Schwarzkopf and Richard Kieffer, Refractory Hard Metals (The Macmillan Company, New York, 1953), p. 87.
11. H. Krainer and K. Konopieky, Berg- und. Hüttenmännische Monatshefte, Band 92, Heft 10/11, 166-178 (1947).

PREPARATION OF TITANIUM CARBONITRIDE COATINGS BY THE SOL-GEL PROCESS

Haixing Zheng, Kris Oka and John D. Mackenzie
Department of Materials Science and Engineering, University of California at Los Angeles, 405 Hilgard Ave., Los Angeles, CA 90024.

ABSTRACT

$Ti(NPr^i_2)_4$ has been synthesized and used as a precursor for aminolysis and polymerization by reaction with isopropylamine. The resultant dark red solution was utilized for dip coating of silicon substrates. Firing of these coatings in N_2 or Ar produced dense and crack-free titanium carbonitride coatings.

I. INTRODUCTION

Titanium nitride and titanium carbide are two hard refractory materials with microhardness values of 2000 and 2900 kg/mm^2, respectively. Titanium nitride and titanium carbide have the same fcc (B1) crystal structure. They are miscible and form a complete solid state solution of Ti(C,N) [1]. TiO is also soluble in TiN, TiC and Ti(C,N) over a wide range of compositions. Since TiN and TiC have high melting temperatures (~3000°C) and are extremely hard, they have been used for coating industrial materials in order to reduce friction and to increase wear and corrosion resistance. TiN, Ti (C, N) and (Ti, Al)N coated tools fabricated by PVD are already available on the market. In addition, the golden color of TiN has made it desirable for a decorative coating.

At present, most of the TiN and Ti(C, N) coatings are made by vapor deposited techniques [2]. However, these techniques have many inherent limitations that are nearly impossible to overcome such as the difficulty in coating large and complex shaped materials, a large capital investment and high operation costs. The sol-gel process has been proven to be successful in making oxide coatings on various materials [3]. Recently, preparation of non-oxide coatings via the pyrolysis of soluble polymers has been investigated [4 -7]. By utilization of polycarbosilane solution, silicon carbide coatings have been derived on various materials such as silicon [4], and steel [5]. Jaschek and Russel [7] coated silica glass with electrolysis derived TiN- and (Al, Ti)N-containing polymers to yield TiN and (Al, Ti)N coatings. By pyrolysis in ammonia and N_2 at 800°C or higher, (Al, Ti)N coatings were obtained. There are also a few reports of preparation of TiN coatings from pyrolysis of alkanolamine modified titanium alkoxide or titanium dialkylamides [8-10]. The synthesis of the oxygen-free chemical precursors $Ti(NR_2)_4$ and the fabrication of titanium carbonitride coatings from these precursors are reported in this work.

II. Synthesis of the Chemical Precursors

Titanium tetrakisdialkylamides, $Ti(NR_2)_4$, have been synthesized by following Bradley's reaction [11].

$$TiCl_4 + 4LiNR_2 \dashrightarrow Ti(NR_2)_4 + LiCl \qquad (1)$$

99% pure $TiCl_4$ and $LiNR_2$ (10 wt.% suspension in hexane) from Aldrich were used as source materials. The solution was stirred for 6 hours in a dry box. The resultant mixture was centrifuged to remove LiCl precipitates. The color of the clear solution was dark red.

It was known that $Ti(NR_2)_4$ is a monomer, is volatile and is highly susceptible to water and alcohol to yield titanium hydroxide and titanium alkoxide, respectively.

$$Ti(NR_2)_4 + 4H_2O \dashrightarrow Ti(OH)_4 + NR_2H$$

$$Ti(NR_2)_4 + 4R'OH \dashrightarrow Ti(OR')_4 + NR_2H$$

It is clear from these two reactions that oxygen can be easily incorporated. Therefore, precautions were taken to minimize the introduction of oxygen during handling and processing.

III. Aminolysis and Polymerization

The tendency for large coordination number in transition metal ions such as titanium always causes the metal-organics to form polymers. The formed polymer is normally insoluble and nonvolatile. The absence of coordination-polymerization of $Ti(NR_2)_4$ is attributed to steric hindrance. However, it is expected that aminolysis of $Ti(NR_2)_4$ with less steric effects of primary amines will lead to the polymerization reactions:

Aminolysis:

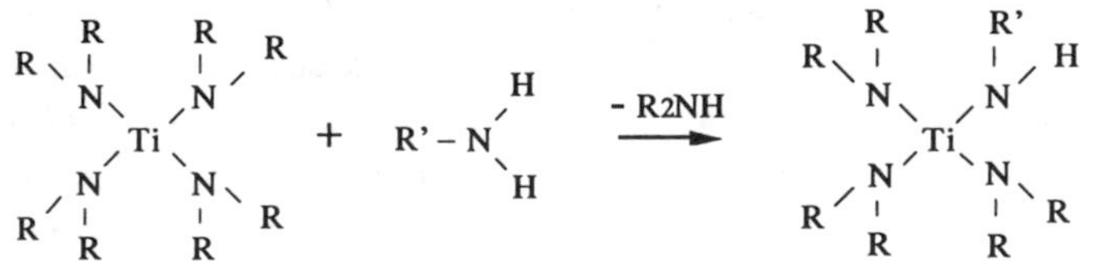

```
   R   R                                      R    R'
R\ |   |  R                H                R\ |    |  H
   N\  N/             /       - R2NH          N\  N/
    Ti       +   R'- N       ------->          Ti                (2)
   N/  N\             \                       N/  N\
R/ |   |  R            H                   R/ |    |  R
   R   R                                      R    R
```

Polymerization:

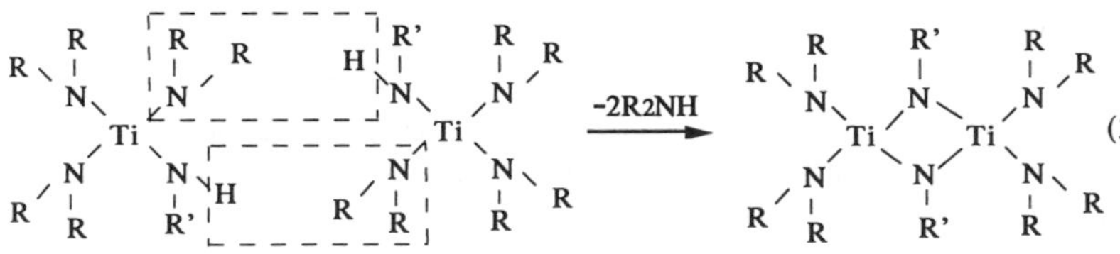

```
   R    R       R'   R                        R    R'    R
R\ |    |  R  H |    |  R                  R\ |    |     | R
   N\   N/      N\   N/        -2R2NH         N\   N\   N/
    Ti              Ti         ------->        Ti      Ti         (3)
   N/   N\ H    N/   N\                       N/   N/   N\
R/ |    |     R/|    |  R                  R/ |    |     | R
   R    R'      R    R                        R    R'    R
```

These reactions are similar to the reactions of hydrolysis and polymerization of metal-alkoxides; therefore, can be controlled through partial aminolysis, or modification of $Ti(NR_2)_4$ in order to make soluble polymers. Bradley, et al. [12] have reported that partial aminolysis of $Ti(NEt_2)_4$ with n-butylamine yields soluble polymers. In this work, $Ti(NPr^i_2)_4$ was used to react with excess Pr^iNH_2. Since isopropylamino groups are branched, it is expected that the aminolysis reaction of $Ti(NPr^i{}_2)_4$ will not be completed even with excess Pr^iNH_2 and the resultant polymer should be soluble. The experimental results proved the prediction: reaction of $Ti(NPr^i_2)_4$ with excess Pr^iNH_2 produced no precipitation.

$$(n+1)Ti(NPr^i_2)_4 + 2nPr^iNH_2 \rightarrow Ti_{n+1}(NPr^i)_{2n}(NPr^i_2)_4 + 4n\ Pr^i_2NH \quad (4)$$

This result agreed with Bradley's observation that treatment of $Ti(NPr^i)_2$ with an excess of n-butylamine only caused slow replacement of isopropylamine by n-butylamino groups and a soluble n-butylamino-titanium polymer was formed.

IV. Preparation of Ti(C, N) Coatings

The clear solution with titanium concentration of 0.05 M resulted from reaction (4) was used to coat silicon substrates. The silicon substrate was dipped into the solution and then was fired at 500°C in N_2 for each dip to avoid redisolvation of polymer coatings in the subsequent dipping. Finally, the coatings were fired at elevated temperatures in an inert atmosphere. Fig.1 shows the x-ray diffraction pattern of Ti (C, N) coatings fired at 500°C, 800°C, 1000°C in N_2 for one

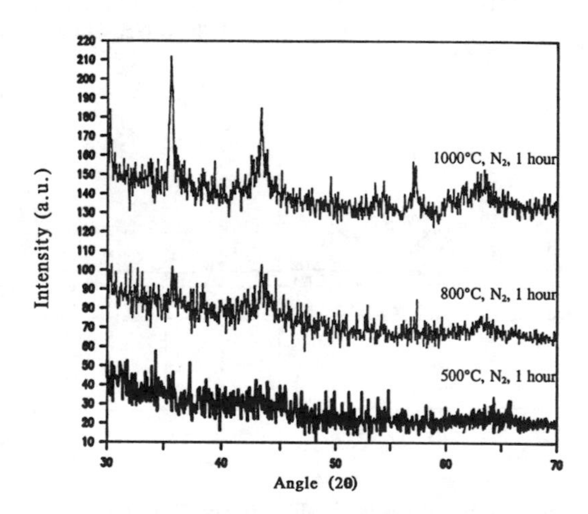

Fig.1 X-ray diffraction pattern of Ti(C, N) coatings fired at different temperature

Table 1 The (200) X-ray diffraction Peaks of Ti(C, N, O)

Material	Angle (2θ)	Lattice Parameter (Å)
TiC	41.41°	4.328
Coating fired at 1200°C	41.95°	4.304
TiN	42.60°	4.242
Coating fired at 1000°C	42.90°	4.213
Coating fired at 800°C	42.90°	4.213
TiO	43.29°	4.177

hour each. At 500°C, the coating was amorphous; at 800°C and 1000°C the coatings were crystalline and broad x-ray diffraction peaks were observed. The diffraction peaks could be indexed to the diffraction of Ti(C, N, O) with an exception of one unidentified peak at 57°. Table 1 listed values of the (200) x-ray diffraction peaks of Ti(C, N, O). The diffraction peaks of the coatings fired at 800°C and 1000°C located between the peaks of TiN and TiO. This implies that the coating may contain a certain amount of oxygen. This prediction was confirmed by Auger electron spectroscopy (Fig.2) which showed the coexistance of Ti (32 at.%), C (28 at.%), N (20 at.%) and O (20 at.%). The ratio between

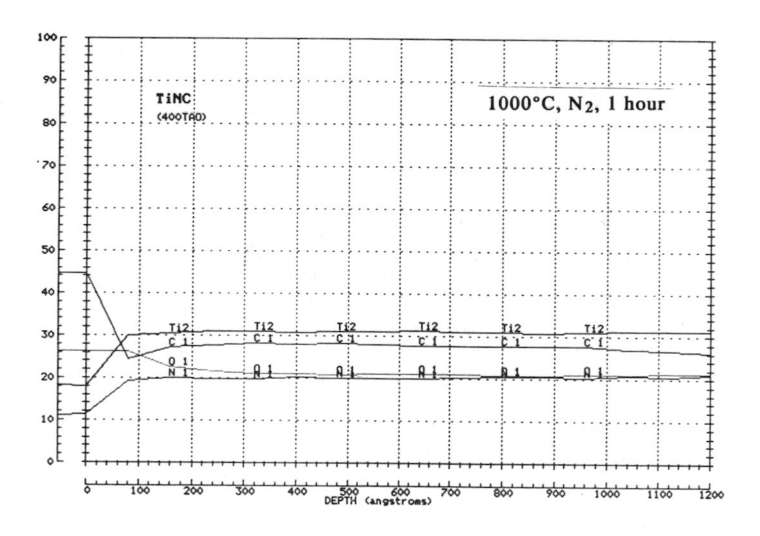

Fig.2 Auger electron spectroscopy sputter depth profile of Ti(C, N) coating on silicon

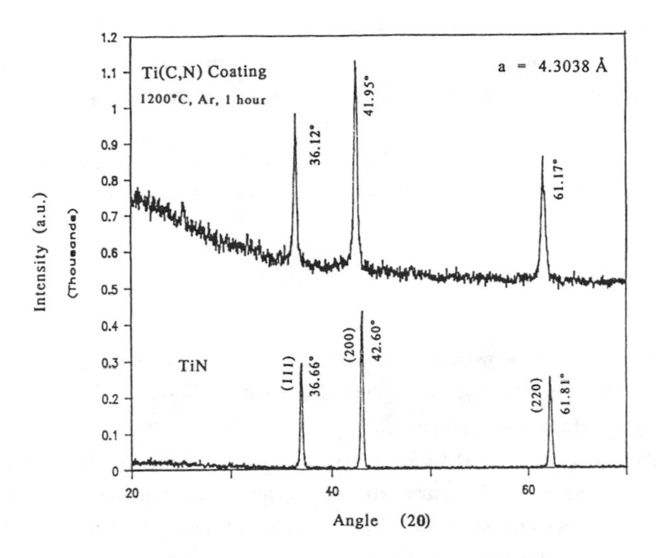

Fig.3 X-ray diffraction pattern of Ti(C, N) coating and TiN

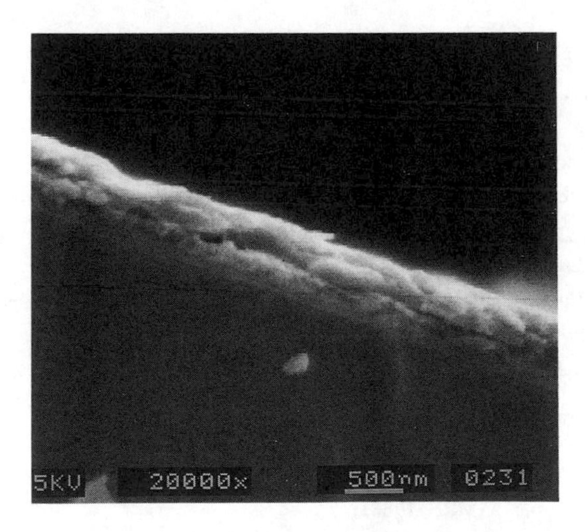

Fig. 4 Cross-section of $TiC0_{0.8}N0_{0.2}$ coating on silicon substrate

(C + N + O) (68 at.%) and Ti (32 at.%) is more than 2, indicating that there were free C or other unknown phases in the coating. The coating fired at 1200°C in Ar for one hour produced sharp x-ray diffraction peaks of Ti(C, N) without any extra peaks (Fig 3). The lattice parameter was calculated to be 4.3038Å. Application of Vegard's law [1] yielded that the coating had the composition of $TiC_{0.8}N_{0.2}$ providing that there was no oxygen present.

Fig. 4 showed the microstructure of $TiC_{0.8}N_{0.2}$ coating. The coating was about 0.6 µm with a smooth surface. No pores or cracks were observed in the coatings.

V. Summary

In this work, it was demonstrated that $Ti(NR_2)_4$ can be used as the precursor for processing Ti(C, N). The aminolysis and polymerization of $Ti(NR_2)_4$ are controllable and partial aminolysis yields a soluble polymer.

The resultant coatings from $Ti(NR_2)_4$ precursor are amorphous after being fired at 500°C. The coatings are easily incorporated with oxygen, and the amount of oxygen is reduced at higher firing temperatures. The Ti(C, N) coatings derived from sol-gel process are dense and smooth without any cracks.

Acknowledgements. This work was financially supported by NSF under DMR-9014487.

References

1. P. Duwez and F. Odell, J. Electrochem. Soc., **97**, 299(1950)
2. B.D. Sartwell and A.Mattews, Surface & Coatings Technology, **49**, Nos.1-3 (1991)
3. Sol-Gel Technology for Thin Films, Fibers, Preforms, Electronics, and Specialty Shapes, edited by Lisa Klein (Noyes, Park Ridge, N.J., 1988)
4. C.J.Chu, Haixing Zheng, S. Park, S.J. Ting and J.D.Mackenzie, to be published
5. C.J.Chu, Ph.D Disertation, University of California-Los Angeles, 1992
6. Haixing Zheng, K. Thorne and J.D.Mackenzie, et al., to be published in Synthesis and Processing of Ceramics: Scientific Issues, MRS Proceedings, Vol. 249
7. R.Jaschek and C.Russel, J. Non-Cryst. Solids, **135**, 236(1991)
8. K. Kuroda, Y. Tanaka, Y.Sugahara and C. Kato, Better Ceramics Through Chemistry III, Vol.121, pp.575 (MRS, 1988)
9. K.Sugiyama, et al. J.Electrochem. Soc., **122**, 1545(1975)
10. L.M.Dyagileva, et.al., J. Gen. Chem. USSR (Engl.Trans.) **54**, 538(1984)
11. D.C.Bradley and I.M. Thomas, J.Chem.Soc., 3857(1960)
12. D.C.Bradley and E.G.Torrible, Canadian J. Chem., **41**, 134(1963)

XPS STUDY OF THE NITRIDATION PROCESS OF A POLYTITANOCARBOSILANE INTO Si-Ti-N-O CERAMICS

G. GRANOZZI*, A. GLISENTI*, R. BERTONCELLO*, G.D. SORARU**

*Dipartimento di Chimica Inorganica, Università di Padova, Via Loredan 4, 35131 Padova, Italy.

**Dipartimento di Ingegneria dei Materiali, Università di Trento, 38050, Mesiano-Trento, Italy.

ABSTRACT

The nitridation process of a polytitanocarbosilane, leading to the formation of a Si-Ti-N-O ceramics, has been investigated mainly by means of X-ray photoelectron spectroscopy (XPS). Ti2p spectra collected on samples fired at various stages of the transformation process clearly shown that Ti-N bonds of TiO_xN_y mixed units start to form at 500 °C. By increasing the firing temperature, the Ti2p peak shifts toward values typical of titanium nitride ceramics, so indicating the progressive nitrogen enrichment of the mixed units up to the formation of a TiN phase.

INTRODUCTION

Non oxide ceramics can now be prepared with a powder-free processing technique, *via* controlled pyrolysis of metal-organic polymer precursors. This new synthetic route to advanced ceramics has been mainly applied to the development of highly refractory ceramic fibers [1,2]. Following this preceramic approach, the structure and properties of the resulting material are strongly influenced by the polymer architecture and by the conversion mechanisms during the pyrolysis process under controlled atmosphere [3]. Due to the amorphous nature of the polymer-derived ceramics, spectroscopic techniques have been currently applied to follow the structural evolution from the starting metal-organic compound to the resulting ceramic material. In this context, X-Ray Photoelectron Spectroscopy (XPS) has been already successfully applied to study silicon carbide ceramics obtained from polycarbosilane (PCS) and SiC/TiC ceramics from polytitanocarbosilane (PTC) [2,4,5]. In the latter case, XPS analysis provided relevant information about the transformation of the starting Ti-O bonds of the polymer precursor into the final Ti-C bonds. The same PTC polymer has been recently used as precursor for Si-Ti-N-O ceramics when fired in NH_3 atmosphere [6].

This paper will report XPS spectra collected on PTC samples at various stages of the nitridation process up to the formation of a Si-Ti-N-O ceramic phase. Experimental results will be discussed together with complementary information, obtained from XRD analysis, concerning the ordering process that leads, at high temperature (T >1500°C), to the formation of a microcrystalline silicon nitride/titanium nitride ceramic.

EXPERIMENTAL

The preparation of the Ti-modified polycarbosilane (PTC), was performed reacting PCS (Dow Corning) and titanium n-butoxide (Strem Chemicals) following a procedure described in details in ref. [5]. Weighted amounts of the two reagents were selected to give a nominal Ti/Si = 0.25. Ammonolysis of preceramic polymers were carried out under ammonia flow (100 cc/min) in a silica tube at various temperatures up to 1000°C using the following heating schedule: 5°C/min up to 500°C, 2°C/min from 500°C to 700°C, 5°C/min up to 1000°C with a maintenance step of 1 hour at the maximum temperature. The lower heating rate step has been selected according to published results [7] showing that nitridation of polycarbosilane begins at 500 °C and it is complete at 700 °C. In order to follow the evolution at higher temperatures, the powders obtained after the ammonolysis process at 1000°C were further heated at 10°C/min (1 hour at the maximum temperature) under N_2 up to 1600°C using a graphite furnace. XPS spectra were collected using a VG ESCALAB MKII spectrometer operating at 10^{-7} Pa with Mg-Kα radiation (hν = 1253.6 eV). To minimize surface contamination and with the aim to provide fresh bulk surfaces, the examined materials were ground into fine powders and pressed to form pellets just before XPS measurements. The spectrometer was calibrated by assuming the binding energy (BE) of the Au4$f_{7/2}$ line at 83.9 eV with respect to the Fermi level. As a second reference, the C1s peak of hydrocarbon contamination has been observed at 285.0 eV [8]. Survey scans (50 eV pass energy; 1 eV step; 0.5 s. step^{-1} dwell time) were obtained in the BE range between 0 and 1000 eV. Detailed scans were recorded for the C1s, O1s, Ti2p, Si2p and N1s regions (20 eV pass energy; 0.1 eV step; 1 s. step^{-1} dwell time). The standard deviation in the BE values of the XPS lines is 0.15 eV. Electron flood gun facilities and the BE of C1s peak as internal standard were used to evaluate charging effects and the reported BE values were corrected accordingly. XRD analysis were performed on a Philips instrument using Cu-Kα radiation and a Ni filter.

RESULTS AND DISCUSSION

Nitridation of PCS resulted always in white powders while ammonolysis of PTC leads to a dark bluish product even at the lowest temperature used in this study (500°C). Si2p and Ti2p spectra recorded on the PTC precursor and at various temperatures during the nitridation process are reported in Figure 1. Concerning the PTC precursor, XPS results shows a single Si2p peak centered at 101.7 eV (FWHM=2.1 eV) and a spin doublet $2p_{1/2}$-$2p_{3/2}$ for the Ti2p signal at 465.2 ad 459.5 eV. These values reveal that silicon atoms are mainly engaged in forming Si-CH_3 and Si-CH_2 bonds while titanium atoms are present into an oxygen environment [5]. At 500°C, the Si2p signal shifts to 102.3 eV and becomes broader (FWHM=2.7 eV). By increasing the temperature up to 1000°C the XPS Si2p spectra show the same peak position and only a slight decrease of the linewidth (FWHM=2.2 eV at 1000°C). The binding energy of Si2p at 500°C is typical of silicon atoms in silicon nitride materials [9]. Indeed, XPS analysis of a reference Si_3N_4 powders (Cerac) gave a BE of Si2p of 102.0 eV (FWHM=2.5 eV). The small shift observed in the PTC sample at 500°C compared to the one of Si_3N_4 (+0.3 eV) could be due to the presence of some silicon oxynitride sites such as $SiN_{4-x}O_x$. Indeed, ^{29}Si solid state NMR investigations of PTC powders nitridated at 1000°C have clearly shown the presence of SiN_4 and $SiN_{4-x}O_x$ sites [10]. XPS analysis of Si2p spectra (Figure 1) indicates that the evolution of the local environment of the Si atoms during the ammonolysis process of PTC polymer is close to that one observed for pure PCS [11] and for Al or Zr modified PCS [10].

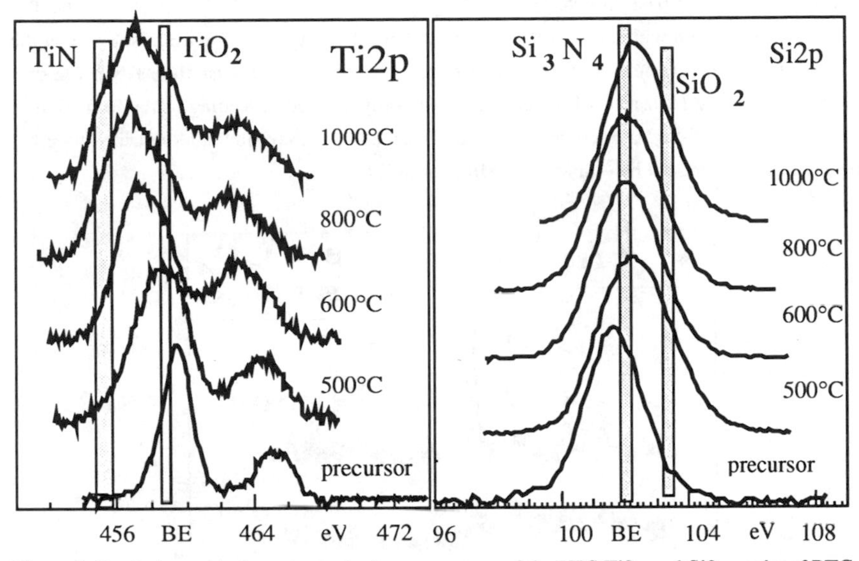

Figure 1. Evolution with the ammonolysis temperature of the XPS Ti2p and Si2p peaks of PTC.

The nitridation process starts at temperatures as low as 500°C and it is complete around 600°C when a silicon nitride-based network is already formed; a further increase of the nitridation temperature up to 1000°C does not seems to greatly modify the local environment of the silicon atoms.

The analysis of the Ti2p spectra (Figure 1) gives information on the corresponding evolution of the local environment of Ti atoms during the nitridation process. At 500°C the spin doublet $2p_{1/2}$-$2p_{3/2}$ is still observed but the positions of the two maxima are shifted toward lower values of BE (464.3 and 458.7 eV, respectively). At 600°C a further shift can be observed up to 463.1 and 457.5 eV respectively. From 600°C to 1000°C the main evolution of the XPS spectra leads to a considerable increase of the linewidth while the position of the maxima shows minor changes being, at 1000°C, 463.0 and 457.3 eV, respectively. These results suggest that, starting from 500°C, Ti atoms are not any more engaged in forming only Ti-O bonds but some Ti-N bonds start to appear. Actually, the measured BE of $Ti2p_{3/2}$ peak in TiN reference (Cerac) ceramics is 455.2 eV. Accordingly, the observed progressive shift of the spectra toward lower BE values can be assigned to formation of Ti-N bonds. Up to 1000°C the position of the $Ti2p_{3/2}$ peak is intermediate between BE values typical of titanium nitride and titanium oxide. This evidence suggests that the nitridation of the titanium oxide-based phase present in the PTC precursor proceed *via* the formation of mixed titanium sites such as TiO_xN_y, where Ti atoms are engaged in both Ti-O and Ti-N bonds. According to these results, the ceramic phase obtained from the ammonolysis process at 1000°C can be described as an amorphous silicon oxynitride network containing mixed TiO_xN_y sites.

The evolution with the firing temperature of the XRD spectra (Figure 2) shows that the disordered phase is stable up to 1500°C where very broad peaks of a crystalline TiN phase start to appear. The crystallization of the system occurs only at 1600°C leading to the formation of TiN, Si_2N_2O and β-Si_3N_4 microcrystals. The presence of a glassy silicon oxynitride phase has been already pointed out by ^{29}Si MAS NMR analysis [10].

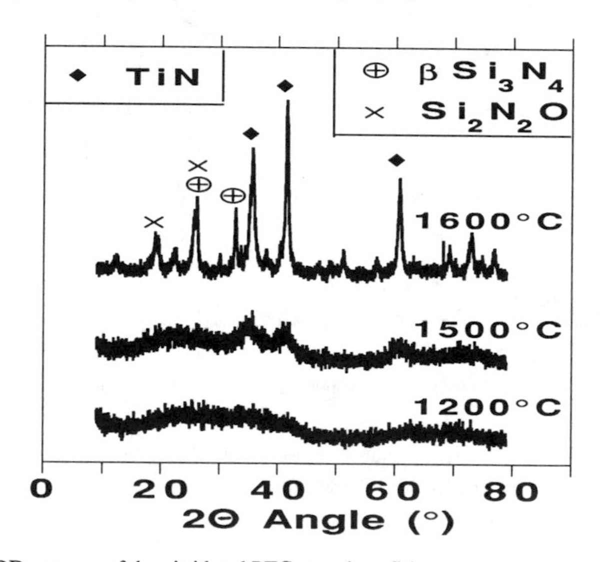

Figure 2. XRD patterns of the nitridated PTC at various firing temperatures.

Si2p and Ti2p XPS spectra collected on PTC samples fired at various temperatures from 1000°C to 1600°C are reported in Figure 3. From 1000°C to 1200°C the Si2p peak shifts toward higher BE values (102.7 eV); by increasing the firing temperature up to 1600°C Si2p peak does not cause any further evolution. According to these results the amount of mixed $SiN_{4-x}O_x$ sites of the amorphous Si-Ti-N-O ceramic seems to increase up to 1200°C and to be stable up to 1600°C. The corresponding evolution of the Ti2p spectra is more progressive: the component around 455 eV is continuously growing from 1200°C to 1600°C indicating the nitrogen enrichment of the mixed TiO_xN_y units finally approaching the composition of TiN.

The N1s peaks are found within the 397.2-397.9 eV interval, which is the typical value found in Si_3N_4. Bearing in mind that silicon sites are prevailing, these data confirms the above discussed pattern.

CONCLUSION

The ammonolysis in NH_3 flow of a polytitanocarbosilane leading to the formation of a Si-Ti-N-O ceramic phase has been investigated by means of XPS analysis. Both Si2p and Ti2p peaks indicate that the nitridation process starts at 500°C. At this temperature, the BE value of Si2p suggests the formation of a silicon nitride network in which some Si-O bonds are present. The oxygen content of the silicon nitride-based network seems to increase in the temperature range from 1000°C to 1200°C and to be stable at higher firing temperatures in agreement with XRD analysis showing the eventual crystallization of a Si_2N_2O phase at 1600°C.

The evolution of Ti2p peaks during the nitridation process shows that, at 500°C, Ti-O bonds start to be cleaved to form Ti-N bonds of mixed TiO_xN_y units. TiN phase seems to be produced, from these units, by a progressive enrichment in nitrogen during the firing process. Therefore, the observed mechanism for the formation of TiN phase by ammonolysis of PTC seems different compared to that one leading to TiC microcrystals by the pyrolysis in inert atmosphere of the same PTC precursor.

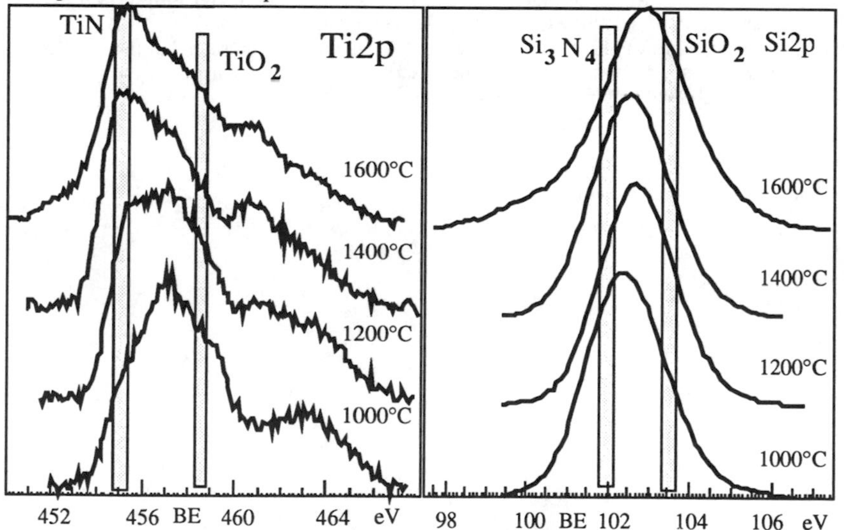

Figure 3. Evolution with the firing temperature of the XPS Ti2p and Si2p peaks of the nitridated PTC.

Actually, in the latter case Ti2p band envelope suggests a minor contribution from TiO_xC_y mixed units and, even at the first firing temperature when Ti-C bonds start to appear (840°C), the presence of a pure TiC phase mixed with the precursor titanium oxide phase. In the present case the transformation from Ti-O to Ti-N based units occurs *via* intermediate TiO_xN_y species, where x and y change smoothly with the firing temperature.

Acknowledgements

NATO is acknowledged for financial support. This work has been partially funded by "Progetto Finalizzato Materiali Speciali per Tecnologie Avanzate" of CNR, Rome.

REFERENCES

1. S. Yajima, M. Omori, J. Hayashi, K. Okamura, T. Matsuzawa and C-F. Liaw, Chem. Lett. 551 (1976).
2. J. Lipowitz, Bull. Am. Ceram. Soc. 70, 1888 (1991).
3. Z. Zhang, F. Babonneau, R.M. Laine, Y. Mu, J.F. Harrod and J.A. Rahn, J. Am. Ceram. Soc. 74, 670 (1991).
4. E. Bouillon, D. Mocaer, J. F. Villeneuve, R. Pailler, R. Naslin, M. Monthioux, A. Oberlin, C. Guimon and G. Pfister, J. Mat. Sci. 26, 1517 (1991).
5. G. D. Soraru, A. Glisenti, G. Granozzi, F. Babonneau and J. D. Mackenzie, J. Mat. Res. 5, 1958 (1990).
6. Y. Hasegawa, C. Feng, Y. Song and Z. Tan, J. Mat. Sci. 26, 3657 (1991).
7. T. Taki, M. Inui, K. Okamura and M. Sato, J. Mater. Sci. Lett. 8, 1119 (1989).
8. M. P. Seah, G. C. Smith in Practical Surface Analysis - Vol 1 2nd Edition, Ed. by D. Briggs and M. P. Seah, Wiley Chichester (UK) 1990 App. 2.
9. M. P. Seah, G. C. Smith in Practical Surface Analysis - Vol 1 2nd Edition, Ed. by D. Briggs and M. P. Seah, Wiley Chichester (UK) 1990 App. 5.
10. G. D. Sorarù, A. Ravagni, R. Dal Maschio and F. Babonneau, presented at the 2nd ECerS Meeting, Ausburg, Germany, 1991 (unpublished).
11. K. Oamura, M. Sato and Y. Hasegawa, Ceram. Int.13, 55 (1987).

PART VIII

Vapor Phase Routes to Oxide Ceramics

MULTICOMPONENT CERAMIC POWDER GENERATION BY SPRAY PYROLYSIS

SHIRLEY W. LYONS*, J. ORTEGA*, L. M. WANG**, T. T. KODAS*
*Chemical Engineering Department, Center For Micro-Engineered Ceramics, University of New Mexico, Albuquerque, 87131
**Geology Department, University of New Mexico, Albuquerque, 87131

ABSTRACT

We have examined methods for controlling the morphology and microstructure of ceramic particles produced by spray pyrolysis. A variety of materials were examined including $SrTiO_3$ and $BaTiO_3$ and the oxides of Al, Mg, Zn, Pd, V, Mo, and Bi. The morphology of the particles was influenced by using colloidal precursors in combination with molecular precursors for particle generation. Slow drying rates obtained by using high relative humidities and controlled axial temperature gradients did not influence particle morphology for the systems and conditions studied. The microstructure of Al_2O_3, Bi_2O_3, V_2O_5, and PdO particles was controlled by varying the temperature to provide nanocrystalline or single-crystal particles. Evaporation and condensation of volatile species such as MoO_3 and V_2O_5 dramatically modified particle microstructure and morphology.

INTRODUCTION

Aerosol Decomposition or Spray Pyrolysis is a powerful route for ceramic powder production. This process has been called aerosol decomposition [1], spray pyrolysis [2], evaporative decomposition [3], spray roasting [4], spray calcination [5], Aman process [6], aerosol pyrolysis [7], atomizing burner technique [8], and other names. A solution of precursors is atomized and passed through a hot-wall reactor in which the solvent evaporates and the precursors react to form a powder. This process has several attractive features. Inexpensive metal salts and colloidal materials can be used as precursors. The particles formed can be solid, submicron, high-purity, unagglomerated, multicomponent, and have a controlled microstructure. The primary advantage over solid-state chemical reaction, sol-gel, and precipitation techniques is that submicron, unagglomerated particles can be produced without milling which can result in introduction of impurities. For these reasons, the spray pyrolysis approach has been scaled up by several European and Japanese Companies; one is currently producing 60 tons/year of micron-sized mullite powder.

The main problem with using spray pyrolysis to generate ceramic powders is the lack of control over particle morphology. The particles produced are sometimes hollow or porous. This leads to difficulties in packing green bodies which in turn causes difficulties in sintering. Hollow or porous particles require higher sintering temperatures or longer sintering times than solid particles to achieve near-theoretical density [9]. This leads to lack of control over micro-structure resulting in degraded mechanical, electrical, and magnetic properties. The final particle morphology is a result of many physicochemical processes. Some of these include solvent evaporation from liquid and solid/liquid particles, precipitation, polymerization or gelation of precursors in the droplets, reactions to form gaseous and solid products, phase changes, intraparticle transport, and evaporation/condensation of volatile precursors.

Solvent evaporation is one of the most critical steps in spray pyrolysis since it directly influences product particle morphology. Some insight into the role of solvent evaporation can be obtained by considering the times required for droplet evaporation and solute diffusion in the droplet. The evaporation time for a 1 μm water droplet at 20°C and 50% relative humidity is roughly 2×10^{-3} sec [10]. Thus, droplet evaporation can occur before the droplets have entered the reactor if the gas is not near 100% relative humidity. The characteristic time for diffusion of a low molecular weight species through a droplet is $t \sim R^2/D$. Using a droplet radius, R, of 10^{-4} cm and a diffusion coefficient, D, of 10^{-5} cm^2/s, a characteristic time of 10^{-3} sec is obtained. Therefore, the processes of solute diffusion and droplet evaporation occur over similar time scales. As a result, evaporation of solvent

 ©1992 Materials Research Society

proceeds until a sufficiently high solute supersaturation develops in the droplet, at which point the solute begins to precipitate out of the solution. Because solute diffusion is not rapid relative to droplet evaporation, precipitation occurs at the surface of the droplet where the solute concentration is a maximum. A more complex situation occurs for cases where polymerization or gelation occur, and will not be discussed here.

The equation that describes the diffusion of a soluble species in a solution droplet of initial radius r_o is:

$$\frac{\partial C}{\partial t} = \frac{D_L}{r_o^2}\frac{\partial}{\partial r}\left(r_o^2\frac{\partial C}{\partial r_o}\right) \quad (1)$$

where C(r) is the concentration of the soluble species, D_L is the liquid phase diffusion coefficient and the boundary moves as the result of solvent evaporation (i.e. $\partial r_o/\partial t<0$). If the assumption that $r_o(\partial r_o/\partial t) = K$ (constant) is made, an analytical solution to the equation can be obtained [11]. From Hinds [10], this constant is:

$$K = \frac{D_v(P_\infty - P_S)M}{\rho RT} \quad (2)$$

where D_V is the vapor phase diffusion coefficient of the solvent, r and M are the density and molecular weight of the solvent, P_∞ is the partial pressure of the solvent far away from the droplet and R is the universal gas constant. P_S is the vapor pressure of the solvent above the curved surface of the droplet which is a function of droplet size and temperature, T, because of the Kelvin effect. P_S is also affected by the presence of a solute. Using Gardner's solution to Eq.1 and Eq. 2, the solution which describes the ratio of the surface concentration, C_S, to the concentration at the droplet's center, C_o, is:

$$\frac{C_S}{C_o} = \exp\left[\frac{D_V M(P_S - P_\infty)}{2RT\rho D_L}\right] \quad (3)$$

The solution shows the exponential dependence of the surface concentration on the difference in vapor pressure and partial pressure of the solvent. By increasing the relative humidity to near 100% ($P_\infty/P_S \approx 1$), the solute concentration at the surface would be approximately the same as at the center of the drop. In addition, increasing the relative humidity to near 100% would extend the lifetime of a 1μm water droplet from 0.001 to 0.1 sec. This would allow enough time for diffusion of the solute inside the droplet to lead to a nearly uniform solute concentration profile thus potentially avoiding solute precipitation at the droplet's surface and crust formation.

Few studies have examined the dependence of particle morphology on the humidity and temperature of the drying environment of aerosol droplets. In one study, Leong [12] demonstrated limited control over the morphology of salt particles (dried 10 - 100 μm sized solution droplets) by controlling the humidity and temperature of the drying environment. Precursor characteristics, such as solubility, largely influenced particle morphology and it was concluded that the final morphology must be determined empirically. In this study, however, precursors useful for ceramic powder production were not examined.

An alternative approach for control of particle morphology that also has not been investigated systematically is the use of seed particles in the solution to avoid precipitation at the droplet's surface. The most favorable site for solute nucleation within an aerosol droplet is the surface because of the high solute concentration. Precipitation onto seed particles could provide uniform precipitation throughout the solution droplet. By avoiding precipitation at the droplet's surface, crust formation could be eliminated resulting in a solid salt particle which could subsequently react to form a solid oxide particle. This approach would be most useful when combined with control of the droplet drying rate which could cause the solute concentration profile within the droplet to become more uniform.

Spray pyrolysis can also be carried out using colloidal precursors in suspension in a liquid. This is an extension of using seed particles in which the precursor for the product is in the seed particles rather than in solution. In this case, evaporation can proceed in a qualitatively different manner from evaporation with dissolved molecular precursors [13,14]. If two or more colloidal precursors are used with no molecular precursors, there are no solutes to precipitate and it is possible to form solid particles during the drying phase. An example is the generation of mullite from boehmite and silica [15]. With one or more colloidal species and one or more molecular species, the latter can precipitate onto the colloidal particles in the droplet as for the seed particles. An example is formation of mullite from $Al(NO_3)_3$ and SiO_2 [15]. Other attempts at formation of mullite powder from molecular precursors have resulted in formation of hollow particles [16]. Nucleation of crystallites at the surface of the droplet can be avoided since the activation barrier for heterogeneous nucleation onto the colloidal particles is lower than that for nucleation at the droplet surface. Colloidal approaches for control of particle morphology have not been examined in a systematic manner.

The considerations outlined above suggest several approaches for generation of solid particles by spray pyrolysis. These include control of droplet evaporation rates, addition of seed particles, and use of colloidal precursors. Associated problems that have not been examined in detail are (1) producing particles with controlled microstructures with crystallite sizes ranging from the nanometer to micrometer scale and (2) determining how evaporation of volatile components at elevated temperatures influences particle morphology. We have examined these problems for a variety of precursors and product materials. $SrTiO_3$, $BaTiO_3$, ZnO, and MgO formation from several precursors was examined because hollow particles are formed under most conditions, providing model systems for attempts to control morphology. Oxides of Pd, V, Mo, and Bi were studied because of the low melting points and high vapor pressures of the oxides. These systems have been used to study the (1) influence of temperature and relative humidity on morphology, (2) control of particle morphology by addition of seed particles as nucleation sites or use of colloidal precursors, (3) control of particle microstructure via control of temperature and residence time for generation of nanophase materials, and (4) the influence of evaporation and condensation of volatile species on particle morphology.

EXPERIMENTAL

The experimental apparatus consisted of an ultrasonic aerosol generator, drying tube with humidity (0-100%) and axial temperature gradient (25 - 100°C in 60 sec) control, reactor tube in a 3-zone furnace, and a filter for particle collection (Figure 1). A similar experimental system was used for studies of vapor wall losses and control of particle microstructure, but without the controlled drying system and with a TSI Inc. Constant Output Atomizer Model 3076. A variety of precursors were examined including: nitrate solutions of Pd, Bi, Mg, Zn, and Al, acetate solutions of Mg and Zn, ammonium paramolybdenate hydrate, and $ZnSO_4$. $BaTiO_3$ and $SrTiO_3$ were made from barium acetate/titanium lactate and strontium nitrate/titanium citrate respectively. The morphology and microstructure of the particles were examined by Transmission Electron Microscopy (TEM) using a TEM-2000FX microscope operated at 200keV. The morphologies of the

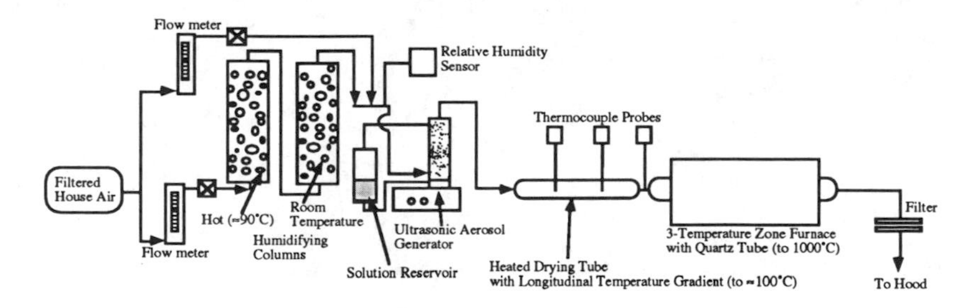

Figure 1. Experimental apparatus including Ultrasonic Aerosol Generator and drying tube with temperature gradient (25°C-110°C in 60 sec).

particles were investigated using a Hitachi Model S-800 Field Emission Scanning Electron Microscope (SEM) which was operated between 10 - 25keV. The phase composition of the powders was determined by a Scintag/USA CuKα Diffractometer (XRD). A Quantachrome Autosorb-1 was used to measure specific surface areas using Brunaur, Emmett, and Teller of nitrogen adsorption isotherms at 77K.

INFLUENCE OF DRYING RATE, SEED PARTICLES AND COLLOIDAL PRECURSORS ON PARTICLE MORPHOLOGY

Attempts were made to modify particle morphology by using a temperature gradient in the drying tube (Figure 1) to obtain controlled drying rates. A wide variety of particle morphologies were observed for drying with completely dry air depending on the nature of the precursor. Nitrate derived MgO and ZnO particles appeared to be hollow as indicated by holes on the particle surfaces (Figures 2 and 3). TEM confirmed that these particles were hollow. Nitrate derived Al_2O_3 particles were solid as indicated by TEM. Acetate derived MgO and ZnO particles were aggregates and had the appearance of "crumpled paper." Barium titanate and strontium titanate powders were hollow and/or porous as shown by TEM. Similar results have been reported before for Al_2O_3 [17-19], MgO [20,21], and ZnO [22-24]. The morphology of particles obtained by drying droplets containing colloidal and molecular precursors has been examined for a variety of systems [25-29], but only for very large droplets (500 μm) and mainly with species that are not precursors for ceramic particle generation.

The precursors described above which produced hollow particles were used as model systems for experiments in modifying particle morphology. Magnesium and zinc nitrate solution droplets were dried in a humid environment with a temperature gradient, and with addition of colloidal seed particles (ultrafine MgO and ZnO). The $BaTiO_3$ and $SrTiO_3$ precursor droplets were dried in an initially nearly saturated (100% relative humidity) environment with a temperature gradient. In addition, colloidal TiO_2 particles were added to the lactate and citrate precursor solutions to give varying fractions of titanium from colloidal TiO_2.

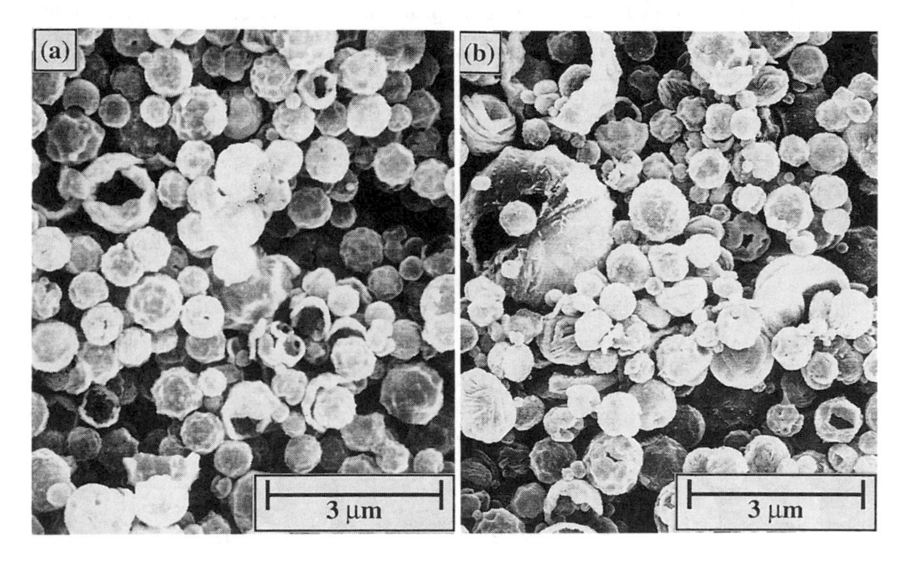

Figure 2. ZnO particles produced in (a) dry air at 25°C and (b) 100% R.H. air with temperature gradient from 25 to 100°C in 1 min. residence time and subsequently reacted in gas phase at 800°C reactor temperature.

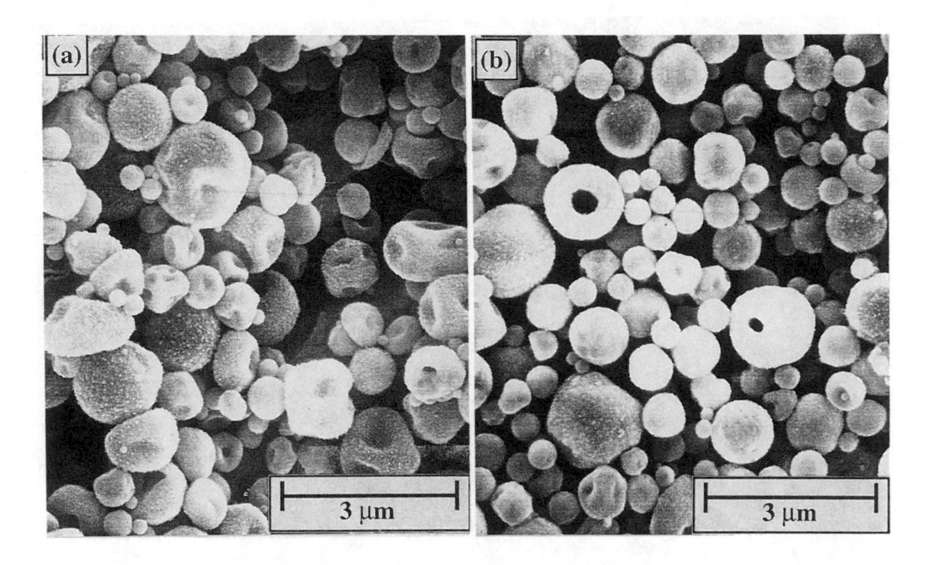

Figure 3. MgO particles produced in (a) dry air at 25°C and (b) 100% R.H. air with temperature gradient from 25 to 100°C in 1 min. residence time and subsequently reacted in gas phase at 800°C reactor temperature.

No change in particle size or morphology was observed in nitrate derived ZnO or MgO by generating the droplets in a 100% R.H. environment and introducing them into a flow system with a temperature gradient (25 to 100°C in 1 min. residence time). The addition of small amounts of seed particles (0.1 - 10% by weight) and the addition of seed particles combined with an initially saturated drying environment and temperature gradient also had no effect. Small amounts of TiO_2 added to the $BaTiO_3$ and $SrTiO_3$ precursors did not affect the particle morphology. Adding large percentages of colloidal TiO_2 to the $BaTiO_3$ and $SrTiO_3$ systems strongly influenced particle morphologies and specific surface areas. Specific surface areas for $SrTiO_3$ made with 0%, 50% and 100% of the titanium from colloidal TiO_2 were 30.7 m^2/g, 21.4 m^2/g, and 8.4 m^2/g respectively. Particle morphologies changed from hollow and porous spheres to aggregates as the amount of TiO_2 added was increased from 0 to 100% (Figure 4). X-ray diffraction revealed extraneous phases such as Sr_2TiO_4, and TiO_2 in $SrTiO_3$ and Ba_2TiO_4 and TiO_2 in $BaTiO_3$ (Figure 5). Although the morphology of the particles was changed by adding TiO_2, the stoichiometry was altered because of losses of TiO_2 particles in the aerosol generator and incomplete reaction. Thus, the successful application of this approach will require better stoichiometry control and higher temperatures for complete reaction to form the products.

Although many attempts were made to form solid particles by controlling the drying rate and by adding seed particles, particle morphology could not be easily modified for the systems studied. The reason for this may be that rates of precipitation, gelation and polymerization of solutions depend too strongly on solute concentration which depends on the drying rate. Similarly, the seed particles used in this study were not effective either. Again, this may be due to preferential precipitation at the droplet's surface as well as the strong dependence of the solute precipitation, gelation or polymerization rates on solution concentration. Another difficulty is that true seed particles that are composed of the same material as the precipitate cannot be used because they would dissolve. Thus, the oxide seed particles provide surface area, but act more like inert surfaces. In addition, only a few out of the many possible precursors were studied. It is possible that the slow drying of other

Figure 4. SEM micrographs of $BaTiO_3$ produced at 1000°C using (a) 40% TiO_2 and (b) 100% TiO_2.

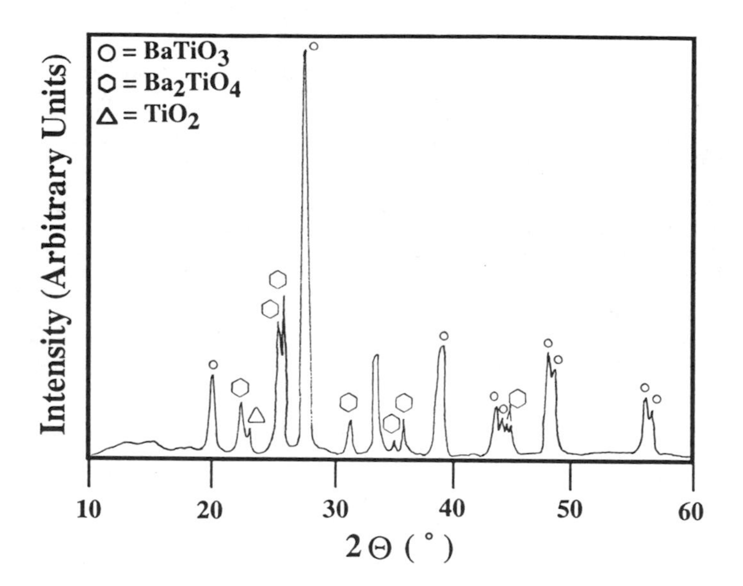

Figure 5. X-ray diffraction pattern of powder produced at 1000°C using barium acetate and colloidal TiO_2 as precursors.

precursor solutions, the addition of seed particles, or the use of other drying conditions could have an affect on particle morphology. However, it appears as though the choice of precursor is of much greater importance in controlling particle morphology.

CONTROL OF PARTICLE MICROSTRUCTURE: GENERATION OF NANOPHASE MATERIALS

Few methods are available for generation of nanophase materials, especially multicomponent materials. The most successful approach is an aerosol method involving metal evaporation followed by gas-phase nucleation and subsequent oxidation [30]. However, this method is limited to volatile materials. Also, multicomponent materials generation has not been demonstrated. We are examining an alternative method for nanocrystalline materials generation where particles are produced by aerosol decomposition at a sufficiently low temperature where grain growth is minimized in each particle, but full reaction still occurs. This approach produces submicron-sized particles which consist of nanometer sized crystallites.

In order to investigate this approach to nanophase materials generation, simple metal oxides (Mo, Bi, V, Pd, and Al) were investigated. The Al_2O_3 [17-19] and V_2O_5 [31] systems have been studied previously. Nanophase materials were produced in the cases of Al, V, and Pd. Bi and Mo produced either single-crystal particles or incompletely reacted material for all temperatures examined. Previous results with the $YBa_2Cu_3O_{7-x}$ system have shown that multicomponent nanophase materials can also be produced [32].

The β-Bi_2O_3 and MoO_3 systems did not provide nanophase materials. β-Bi_2O_3 was formed at furnace temperatures of 400-900°C. Powder formed at 400°C showed broad peaks by XRD, but was not fully reacted as indicated by TGA. β-Bi_2O_3 particles examined by TEM showed that solid single-crystal particles were formed at 500-900°C. Thus, generation of nanocrystalline β-Bi_2O_3 may require temperatures between 400 and 500°C. MoO_3 was produced at furnace temperatures of 400-600°C. MoO_3 particles formed at 400°C were spherical single-crystal particles. Nanocrystalline particle generation may require a temperature below 400°C.

The Al_2O_3, V_2O_5, and PdO systems allowed generation of nanophase materials. Particles were generated using aluminum nitrate at temperatures up to 1000°C. However, complete reaction to form alumina was not obtained under any of the conditions studied as indicated by TGA and XRD. Figure 6 shows TEM micrographs of a Al_2O_3 particle produced at 1000°C. The insert is an electron diffraction pattern taken from the particle indicating the particle contains nanocrystallites. Higher temperatures are required for complete crystalline particle formation. V_2O_5 was formed at temperatures between 300-800°C. Extensive peak broadening was observed by XRD for the powder produced at 300°C. At higher temperatures (700, 800°C) XRD results could not be obtained because of low yields due to vapor wall losses of the volatile vanadium oxide. TEM micrographs showed (Figure 7a) spherical V_2O_5 particles formed at 300°C where the insert is an electron diffraction pattern taken from a particle showing nanocrystallites within the particle. Single-crystal V_2O_5 rods were formed at 600°C (Figure 7b). PdO powders were formed at temperatures between 400-800°C in air. Powders produced at all temperatures were phase-pure PdO as indicated by XRD. XRD spectra of powders produced at (a) 400, (b) 500, and (c) 800°C (Figure 8) showed a decrease in peak broadening as the reactor temperature was increased, consistent with the formation of nanocrystalline material (Figure 9) at lower temperatures and grain growth at higher temperatures. TEM of the powder produced at 500°C (Figure 9) showed that each spherical particle was composed of many small crystallites with dimensions on the order of 10nm. The surface area of the powder showed a decrease from $56.4m^2/g$ to $3.2m^2/g$ as the furnace temperature was increased from 400 to 800°C. This suggested that the grain size in the particles increased as the temperature was increased. These results for several different materials indicate that the microstructure of particles produced by spray pyrolysis can be controlled by varying reactor temperature at a fixed residence time.

The evaporation of metal oxides during spray pyrolysis occurs in numerous systems where volatile species such as oxides of Pb [33], Mo, V, and Bi are present. Evaporation/condensation of volatile species can lead to dramatic changes in particle morphology. Figure 7b shows rods of V_2O_5 formed by this process. Similar results were

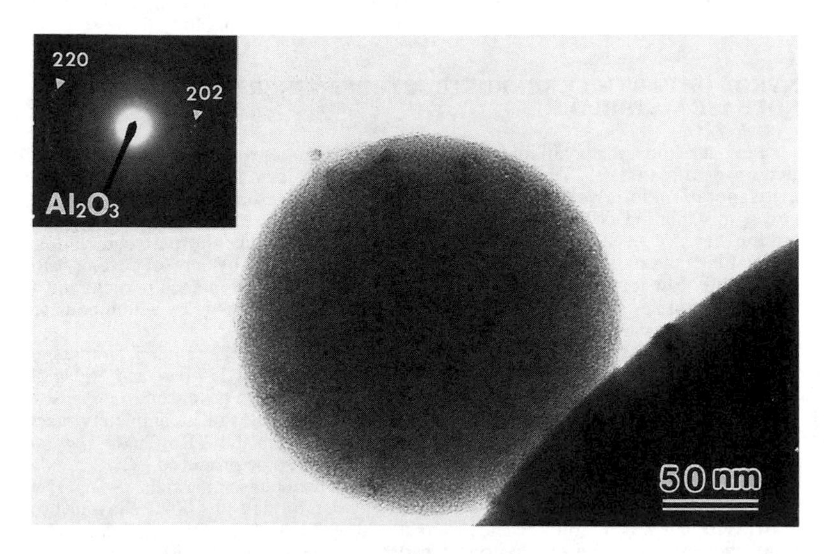

Figure 6. TEM micrographs of an Al_2O_3 particle produced at 1000°C which contains nanocrystallites. The insert is an electron diffraction pattern taken from the particle.

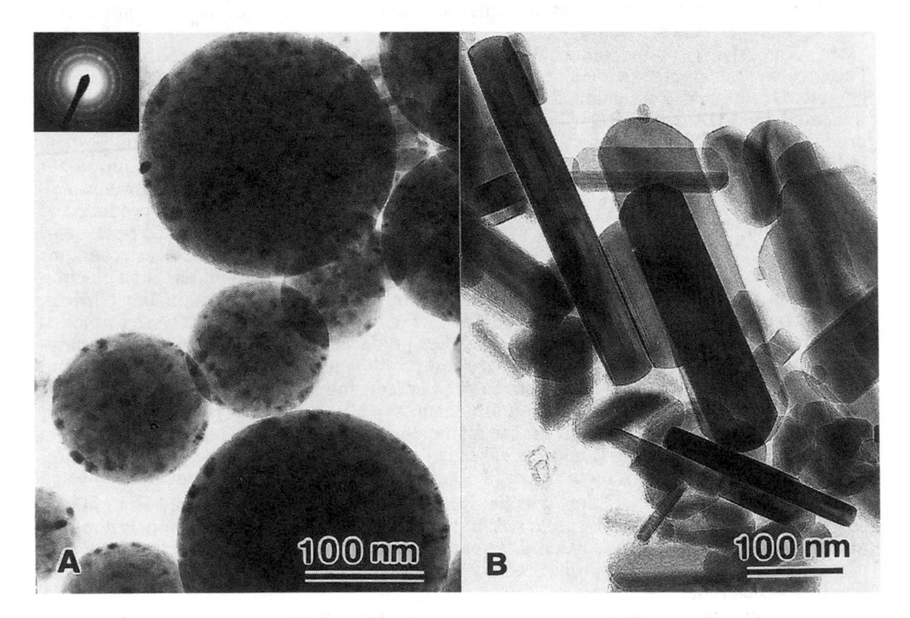

Figure 7. TEM micrographs of (a) Nanocrystalline, spherical V_2O_5 particles which contain nanocrystallites made at 300°C. The insert is an electron diffraction pattern taken from a single particle and (b) single-crystal V_2O_5 particles formed at 600°C.

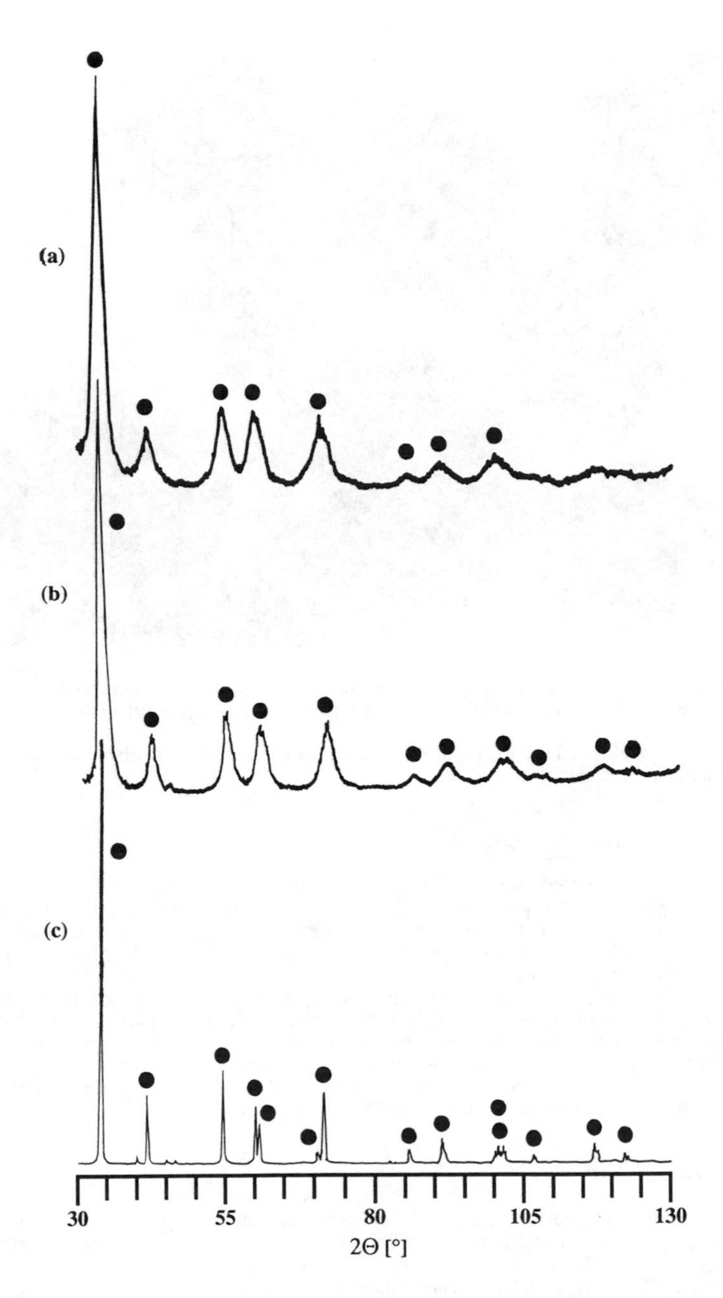

Figure 8. XRD spectra for PdO powders formed at (a) 400, (b) 500, and (c) 800°C.

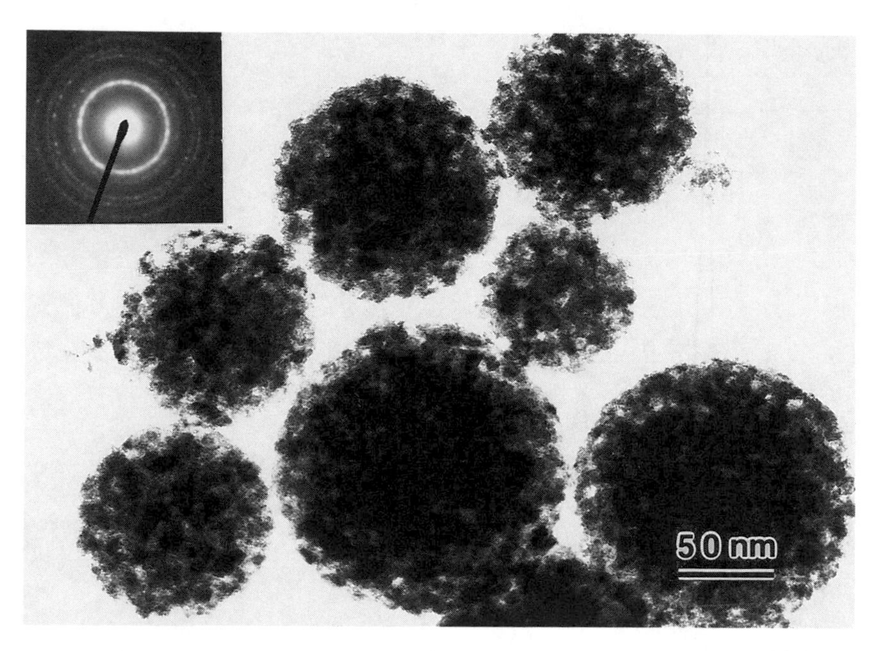

Figure 9. TEM micrographs of nanocrystalline PdO particles produced at 500°C. The insert is an electron diffraction pattern from a single spherical particle.

obtained for MoO_3. This evaporation/condensation process can also play a role in densification of particles.

SUMMARY AND CONCLUSIONS

We have examined methods for controlling the microstructure and morphology of particles produced by Spray Pyrolysis. Controlled drying rates and the addition of small amounts of seed particles did not influence particle morphology. The use of colloidal precursors provided control over particle morphology, but altered phase purity and stoichiometry of the powders.

The intraparticle reaction approach of spray pyrolysis is a useful method for the generation of nanocrystalline materials. The microstructure of the particles can in many cases be controlled to provide particles with crystallite sizes ranging from the nanometer to the micron scale to single-crystal particles. Further characterization of the powders, generation of other simple and mixed metal oxides, and studies of the processing of these powders are required to evaluate the potential of this approach.

ACKNOWLEDGMENTS

We acknowledge the support of the Center for Micro-Engineered Ceramics, the DuPont Co., and Texas Instruments. TEM was carried out in the Electron Microbeam Analysis Facility of the Department of Geology of the University of New Mexico, which is supported by NSF, NASA, DOE-BES and state of New Mexico.

REFERENCES

1. T. T. Kodas, Angew. Chemie: Int. Ed. in English 28, 794 (1989).
2. S. Zhang and G. Messing, J. Amer. Ceram. Soc. 73, 61 (1990).
3. T. Gardner, D. Sproson, and G. Messing, Mat. Res. Soc. Symp. Proc. 32, 227 (1984).
4. M.I. Ruthner, Science of Sintering 11, 203 (1979); M.I. Ruthner, Ceramic Powders, ed. by P. Vincenzini, (Elsevier, Amsterdam, 1983) p.515; M.I. Ruthner, Science of Sintering 6, 82 (1974).
5. D. Vollath, J. Mater. Sci. 25, 2227 (1990).
6. J. Epstein, Hydrometallurgy 2, 1 (1976).
7. A. Pebler and R. Charles, Mat. Res. Bull. 24, 1069 (1989).
8. J. Wenkins and W. Leavitt, Magnetism and Magnetic Materials Conference, Boston, MA, 1956, Amer. Inst. Elect. Engs.(1957).
9. L. Edelson and A. Glaser, J. Amer. Ceram. Soc. 71, 225 (1988).
10. W. Hinds, Aerosol Technology, (Wiley Interscience, N.Y., 1982).
11. G. Gardner, Int. J. Heat Mass Transfer 8, 667 (1965).
12. K. Leong, J. Aerosol Science 12, 417 (1981); K. Leong, J. Aerosol Science 18, 511 (1987); K. Leong, J. Aerosol Science 18, 525 (1987).
13. W. Ranz and W. Marshall, Chem. Eng. Prog. 48, 141 (1952).
14. J. McDuffie and W. Marshall, Chem. Eng. Prog. 49, 417 (1953).
15. K. Moore, D. Smith, T. T. Kodas, J. Amer. Cer. Soc. 75, 213 (1992).
16. S. Kanzaki, H. Tabata, T. Kumazawa, and S. Ohta J. Am. Ceram. Soc. 68, C6 (1985).
17. D. Roy, R. Neurgonkar, T. O'Holleran, and R. Roy, Amer. Ceram. Soc. Bull. 56, 1023 (1977).
18. W. Baukal, H. Beck, W. Kuhn, and R. Sieglen, Power Sources 6, 655 (1977).
19. R.R. Ciminelli, and G.L. Messing, Ceramica 30, 131 (1984).
20. A. Kato, H. Ishimatsu, and Y. Suyama, J. Japan Soc. Powder and Powder Metallurgy 26, 131 (1979).
21. T. Gardner and G. Messing, Amer. Cer. Soc. Bull. 63, 1498 (1984).
22. T. Liu, O. Sakurai, N. Mizutani, and M. Kato, J. Mat. Sci. 21, 3698 (1986).
23. K. Seitz, E. Ivers-Tiffee, H. Thomann, and A. Weiss, in High Tech. Ceramics, edited by P. Vincenzini (Elsevier, Amsterdam, 1987) p. 1753.
24. Z. Szabo, B. Jover, and J. Juhasz, Mater. Sci. Monographs 28B, React. Solids Pt. B, 659 (1985).
25. D. Charlesworth and W. Marshall, AIChE J. 6, 9 (1969).
26. S. Nesic and J. Vodnik, Chem. Eng. Sci. 46, 527 (1991).
27. K. Masters, Spray Drying, (John Wiley and Sons, NY, 1972).
28. Y. Sano and R. Keey, Chem. Eng. Sci. 37, 881 (1982).
29. H. Cheong, G. Jeffreys, and C. Mumford, AIChE J. 32, 1334 (1986).
30. R. Siegel, Material Research Bulletin 25, 60 (1990).
31. R. Sullivan, T. Srinivasan, and R. Newnham, J. Amer. Cer. Soc. 73, 3715 (1990).
32. A. H. Carim, P. Doherty, T. T. Kodas, and K. Ott, Materials Letters 8, 335 (1989)
33. S. Lyons, T. Ward, T. Kodas, R. Prasad, S. Pratsinis, submitted to J. Mater. Res. (1992).

STRUCTURAL AND THERMAL INVESTIGATIONS ON BARIUM, COPPER AND YTTRIUM MOCVD PRECURSORS

ALAIN GLEIZES*, DOMINIQUE MEDUS AND SANDRINE SANS-LENAIN
Institut National Polytechnique de Toulouse, E. N. S. Chimie, Laboratoire des Matériaux, URA-CNRS 445, 118 route de Narbonne, 31077 Toulouse cedex, France.

ABSTRACT

Structural and thermal results concerning four barium tetramethylheptanedionates, $Ba(thd)_2(CH_3OH)_3.CH_3OH$, amorphous $Ba(thd)_2$, $[Ba(thd)_2]_4$, and $Ba(thd)_2(H_2O)_2(CH_3OH)_2$, a methoxo bridged dimer of copper tetramethylheptanedionate, $[Cu(\mu\text{-}OCH_3)(thd)]_2$, and two yttrium tetramethylheptanedionates are presented and discussed.

INTRODUCTION

Y, Ba and Cu tetramethylheptanedionates (thd) have been the most widely used molecular precursors for deposition of thin films of $YBa_2Cu_3O_{7-x}$ by the MOCVD technique. $Cu(thd)_2$ and $Y(thd)_3$ are monomers sufficiently volatile around 140°C under a few torrs. The Ba derivative is less volatile. This is a consequence of its oligonuclearity. Barium requires a high coordination number. The coordination number of Ba in $Ba(thd)_2$ nominally is only 4. Therefore, "$Ba(thd)_2$" adopts an oligonuclear structure with oxygen sharing Ba atoms to secure the metal coordinative saturation. Turnipseed et al. [1] have shown that the most frequently used form is a pentanuclear species of formula $Ba_5(thd)_9(OH)(H_2O)_3$. At the use temperature of ~230°C, partial decomposition occurs yielding non-volatile products. The ensuing lowering of the barium molar fraction in the gas phase with respect to Cu and Y generates inhomogeneous films. A physical remedy would be to act on the flows to keep the gas phase composition constant. This implies that one is able to measure any change in composition throughout the deposition process. Médus [2] has shown that it is possible to monitor the transport rates of precursors by using a mass spectrometer fitted to the reactor.

Chemical approaches to stable "$Ba(thd)_2$" transport also exist. They have been reviewed in previous papers [3, 4]. Addition of free ligand Hthd or of various Lewis ligands to the carrier gas was found to reduce decomposition. As oligonuclearity lowers volatility, one may also try to stabilize mononuclear species by saturating the cation coordination with neutral ligands in addition to the two β-diketonato ligands. Polyethers were used successfully, but only with Ba fluorinated acetylacetonates [5]. This is not satisfactory since fluorinated precursors yield BaF_2 in the films.

We currently deposit thin films of $YBa_2Cu_3O_{7-x}$ [2] using a Ba carrier quite different from the pentanuclear species characterized by Turnipseed et al. [1]. It is a white, amorphous solid (cpd **2**), deriving from transparent single crystals (cpd **1**) grown from a methanol solution of the product of the reaction between Ba hydroxide and Hthd in water [3]. The formula of **1** is $Ba(thd)_2(CH_3OH)_3.CH_3OH$. It is a quasi mononuclear compound with methanol bound to Ba. When separated from the mother solution, crystals of **1** spontaneously lose most of the methanol yielding the amorphous compound **2**, which is oligonuclear. We have also established that crystals grown by subliming **2** are tetrameric and can be formulated as $[Ba(thd)_2]_4$ (cpd **3**) [6]. This set of results has shown that methanol is not a suitable auxiliary ligand to make thermally stable monomers of Ba tetramethyl-heptanedionate. We decided to try a stronger coordinating ligand such as water. We obtained a new compound, $Ba(thd)_2(H_2O)_2(CH_3OH)_2$ (cpd **4**), which is definitely monomeric and higly volatile. We report the structural and thermal works on Ba compounds **1**, **2**, **3** and **4**.

Paradoxically, oligonuclear species may be interesting CVD precursors because in them the metal centers are already close to each other. We present a new copper dimer, $[Cu(thd)(\mu\text{-}OCH_3)]_2$. The paper will end with a brief mention to the yttrium compounds $Y(thd)_3(H_2O)$ and $Y(thd)_3$.

BARIUM TETRAMETHYLHEPTANEDIONATES

$[Ba(thd)_2(CH_3OH)_3].CH_3OH$ (1) - The reaction between Ba hydroxide and Hthd in water yielded a white precipitate which was separated by filtration and dissolved in methanol. Transparent crystals formed on standing at -18°C. They are stable in solution only. The structure was determined from a crystal sealed wet in a capillary tube. The crystal sytem is triclinic, space group P1, with a = 10.693 (3) Å, b = 10.720 (3) Å, c = 15.627 (3) Å, α = 109.26 (1)°, β = 91.9 (1)°, γ = 94.86 (1)°.

The unit-cell contains two units of formula $Ba(thd)_2(MeOH)_3$ (Fig. 1a) and two non coordinated methanol molecules. An important feature is the wide range of Ba-O distances. The chelating oxygens O11, O21, O12 and O22 are at distances ranging from 2.58 to 2.74 Å. The methanol oxygens $O1_m$, $O2_m$ and $O3_m$ are slightly farther at around 2.80 Å. The next neighbor, at 3.32 Å, is the methanol oxygen $O1_m$ of the unit related by inversion through center at (1/2, 0, 0). Both $O1_m$ atoms form a centrosymmetrical double bridge with two sets of Ba-$O1_m$ distances differing by 0.5 Å. A weak dimer association $[(\mu\text{-}MeOH)Ba(MeOH)_2(thd)_2]_2$ results. It is strengthened by H-bonds from methanol oxygens $O1_m$ and $O2_m$ to ketone oxygens O11 and O21 respectively (Fig. 1b). The intradimer Ba-Ba distance is 4.937 (1) Å.

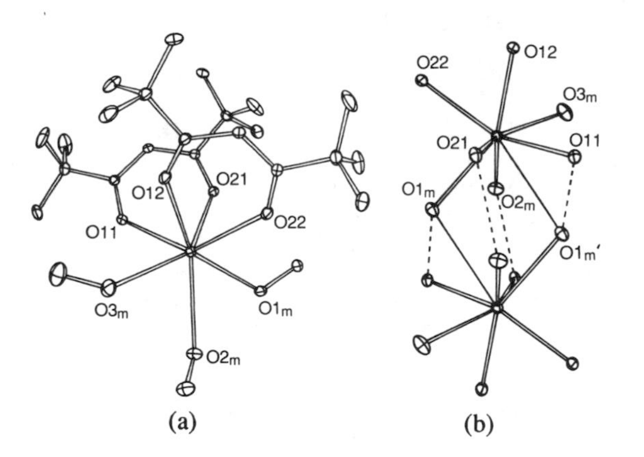

Fig.1 - a) ORTEP view of **1**; b) details of the dimer association.
Ba-O11=2.739(5) -O12=2.578(5) -O21=2.675(5) -O22=2.686(5) -$O1_m$ = 2.822(5) -$O1_m'$ = 3.321(5) -$O2_m$=2.789(6) -$O3_m$=2.775(8) Å.

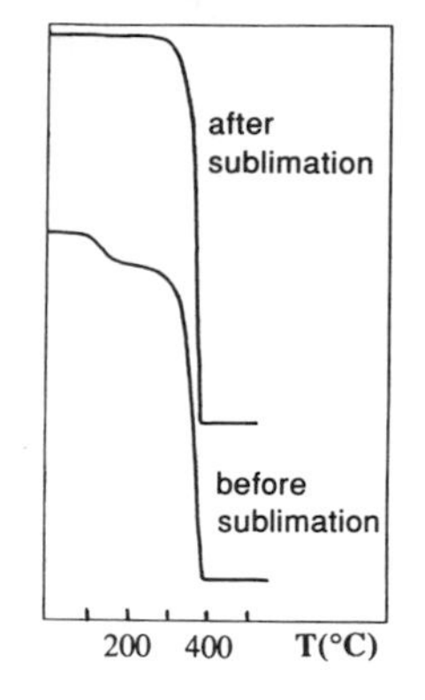

Fig.2 - TGA plots for **2** (He atm.; 4.5°/min)

The dimer association, and the presence of the non coordinated methanol oxygen $O4_m$ at 3.70 Å, give barium a (7+1+1) coordination number. The atom $O4_m$ is also H-bounded to atoms $O3_m$ and O22 not belonging to the same dimer entity.

Amorphous $Ba(thd)_2$ (2) - Compound **2** is formed when crystals of **1** are separated from the mother solution. Wether this is done in air or under argon, they turn white and opaque in a few minutes. The product is amorphous on X-ray powder diffraction. TGA shows a two-step loss of weight (Fig. 2). The first step may be due to the release of one methanol molecule per barium, and the second one results from sublimation. As IR bands representative of methanol showed up before and desappeared after the first loss of weight, compound **2** has been given the formula $Ba(thd)_2(CH_3OH)$. However the presence of methanol is questionable. Samples are often found free of it when fresh and always after staying some weeks in a glove box.

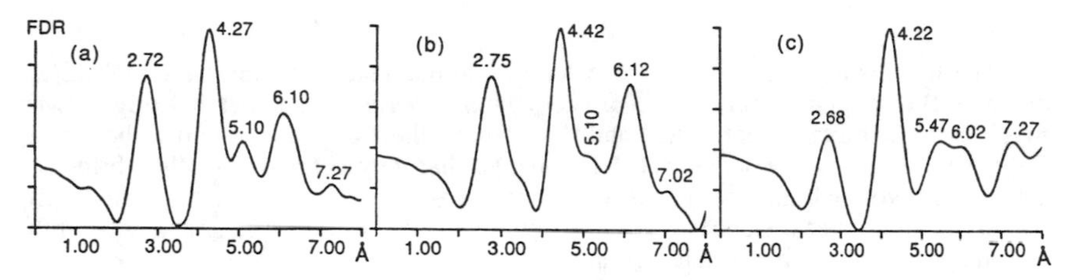

Fig.3- Experimental RDF for **2**: a) freshly prepared, b) 6-month aged; c) calculated RDF for **3**.

Large angle X-ray scattering measurements were performed for a freshly prepared sample and a six-month aged one. The experimental radial distribution function (RDF) curves shown in Fig. 3a and 3b, reveal that the structure has been altered on aging. However both RDF's show peaks at ~1.4 Å, ~2.7 Å and ~4.3 Å: the first one corresponds to C-C and C-O bonds, and the second one mainly to Ba-O bonds. The intense peak at ~4.3 Å is signature of an oligonuclear structure. The RDF calculated from the tetramer structure of $[Ba(thd)_2]_4$ (**3**) (see below) (Fig. 3c) does not fit satisfactorily the experimental ones. Several dinuclear models with a Ba-Ba distance of 4.3 Å do not fit either. Moreover the structure of $[Ba(thd)_2(NH_3)_2]_2$ recently described by Rees [4] suggests that for a dimer a Ba-Ba distance of about 3.8 Å is more likely than 4.3 Å. Trinuclear models issued from molecular mechanics calculation are to be tested.

$[Ba(thd)_2]_4$ (3) - Transparent crystals were grown by sublimating **2** at 180°C, in a Pyrex glass tube, under a dynamic vacuum of ~ 10^{-1} torr. Crystals formed within a few hours on the cold part of the tube. The crystal used for the structure determination was sealed in a capillary tube. The crystal system is orthorhombic, space group Pbca, with a = 26.226 (5) Å, b = 21.887 (3) Å, c = 18.203 (9) Å.

The structure consists of centrosymmetric cyclic tetramers of composition $Ba_4(thd)_8$ (Fig. 4a). Due to high atomic thermal motions, the bond lengths and angles were not accurately determined. Two symmetrically independant sets of bariums, Ba(1) and Ba(2), form rhombuses made of nearly equilateral triangles with three Ba-Ba distances equal to 4.153 (3) Å, 4.191 (3) Å and 4.199 (4) Å, and one equal to 7.211(5) Å (Fig. 4b).

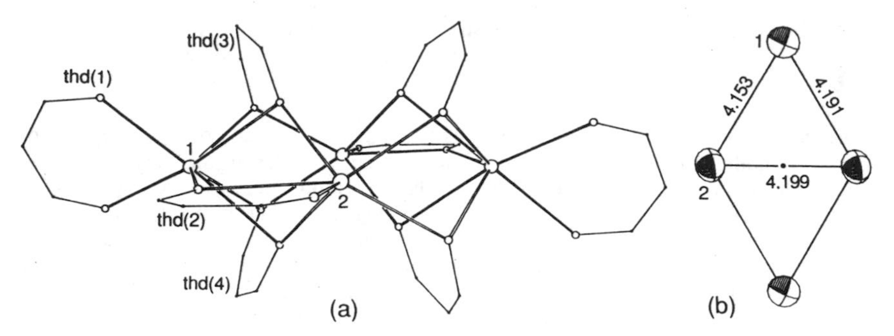

Fig.4 - a) ORTEP view of **3**; b) the rhombus formed by the Ba atoms.

The ligands exhibit three modes of coordination. Thd(1) chelates Ba(1). Thd(2) chelates Ba(2), and has also one oxygen bound to Ba(1), so that it bridges two neighboring cations. Either thd(3) and thd(4) have their oxygens bound to both one Ba(1) cation and two separate Ba(2) cations, so that they act as two-arched bridges. Ba(1) has seven neighbors at distances ranging from 2.53 (2) Å to 2.95 (2) Å. Ba(2) has six neighbors at distances ranging from 2.50 (2) Å to 2.98 (2) Å and a seventh one at 3.43 (2) Å from the same tetramer.

$Ba(thd)_2(H_2O)_2(MeOH)_2$ (4) - Since it was not possible to stabilize a monomeric species using methanol as an auxiliary ligand, we thought that use of a stronger, small-sized Lewis base such as water should improve stability. We used a mixture of water and methanol to avoid hydrolysis and subsequent formation of the pentanuclear species $Ba_5(thd)_9(OH)(H_2O)_3$ [1]. Transparent crystals were obtained by dissolving **2** in a 12% mixture of water in methanol followed by standing at -18°C. The structure was determined from a crystal sealed wet in a capillary tube. The crystal system is monoclinic, space group C2/c, with a = 27.801 (4) Å, b = 10.569 (1) Å, c = 11.132 (9) Å, β = 107.11 (4)°.

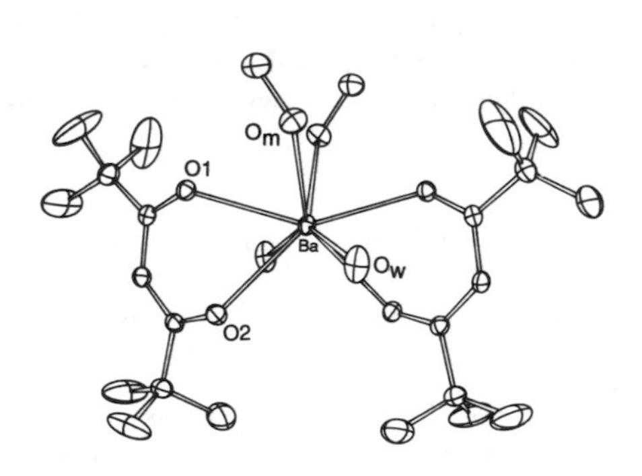

Fig.5 - ORTEP view of **4**. Ba-O1 = 2.725(2) Å, Ba-O2 = 2.761(2) -O_m = 2.777(3) -O_w = 2.720(4) Å

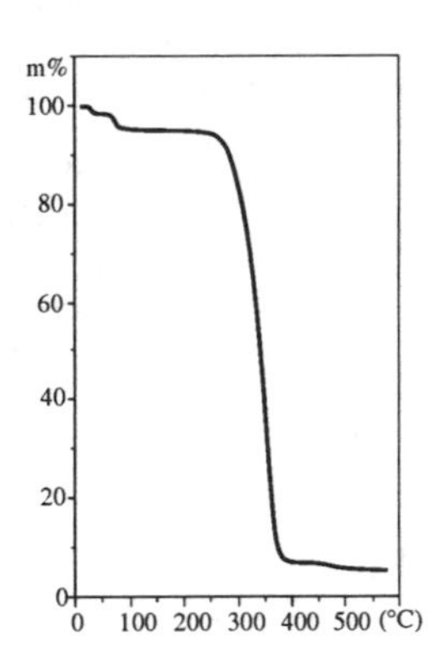

Fig.6 - TGA plot for 4. (He atmosphere, 4.5°/min)

The structure is made of well separated molecules of $Ba(thd)_2(H_2O)_2(CH_3OH)_2$ possessing a twofold symmetry axis (Fig. 5). Eight oxygens around Ba form a square antiprism: the square basis atoms O1, O2, O_m and O_w do not deviate by more than 0.09 Å from their mean least-squares plane; the square bases are related through the twofold axis; their normals make an angle of 1.8°. The narrow range of Ba-O distances (2.72 to 2.78 Å) contrasts with the wide one observed for **1**. This cofirms that an important ligand size effect prevents methanol from stabilizing a monomeric species. This effect is reduced when the methanol molecules are partly replaced by smaller molecules such as water. This and the higher polarization of water make $Ba(thd)_2(H_2O)_2(CH_3OH)_2$ more stable than $Ba(thd)_2(CH_3OH)_3.(CH_3OH)$.

The TGA performed in an He stream under a light overpressure (about 3.10^3 Pa) (Fig. 6) shows that the product sublimes between 260 and 380°C which is only about 10°C lower than observed for **2**. The two weight losses below 100°C are difficult to interpret because of the use of partial vacuum to settle the helium atmosphere. More promising than the result of TGA in view of potential use in OMCVD is the fact that **4** sublimes efficientely at 100°C under ~10^{-1} torr, which is nearly a hundred degrees lower than for any of the polymetallic species. Measurements of vapor pressure are currently in progress. First results have confirmed the high volatility.

COPPER AND YTTRIUM TETRAMETHYLHEPTANEDIONATES

$[Cu(thd)(\mu\text{-}CH_3OH)]_2$ - This compound was prepared according to the method of Bertrand and Caine [7] for the acetylacetonate derivative. The crystal system is monoclinic, space group $P2_1/c$, with a = 9.656 (2) Å, b = 30.168 (5) Å, c = 10.095 (7) Å, β = 103.26 (4)°.

The structure is made of dimers $(thd)Cu(\mu\text{-}OCH_3)_2Cu(thd)$ with double methoxo bridges (Fig. 7a). The dimers are paired off into centrosymmetric tetranuclear units by weak apical Cu-O bonds (Fig.7b). Each Cu atom has local square pyramidal coordination geometry, with four basal Cu-O bond lengths ranging from 1.883 (6) Å to 1.982 (6) Å , and a larger apical bond of 2.386 (6) Å for Cu(1) and 2.922 (6) Å for Cu(2). The dimers exhibit a boat configuration due to Cu elevation above basal square planes and bending along axes O11···O21, O12···O22 and $O1_m$···$O2_m$, (Fig. 7b).

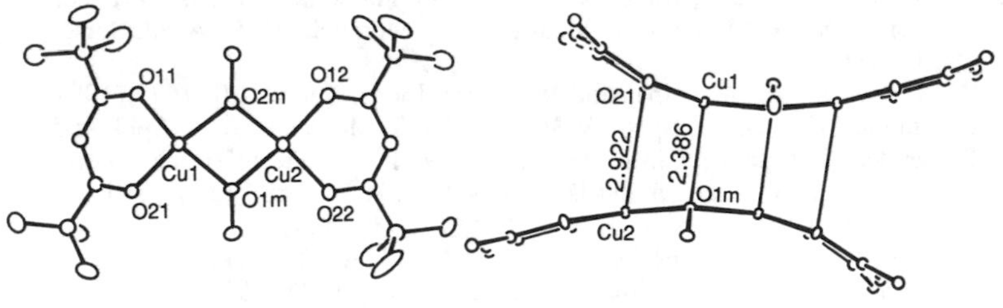

Fig.7 - ORTEP view of a) the dimer $[Cu(thd)(\mu\text{-}CH_3OH)]_2$; b) the tetramer association.
Cu1-O11 = 1.914(6) Cu1-O21 = 1.918(6) Cu1-$O1_m$ = 1.982(5) Cu1-$O2_m$ = 1.920(8) Å
Cu2-O12 = 1.914(6) Cu2-O22 = 1.900(6) Cu2-$O1_m$ = 1.934(6) Cu2-$O2_m$ = 1.883(6) Å.

Intermolecular van der Waals interactions involve methyl groups from both methanol and *t*-butyl fragments. A similar structure has been described by Andrew and Blake for a derivative of acetylacetone and phenylmethanol [8].

Thermogravimetric analyses carried out under helium at 1 atmosphere showed contrasted results which did not allowed to conclude definitively on the use of $Cu_2(thd)_2(\mu\text{-}OMe)_2$ as a suitable copper carrier for OMCVD. In one experiment, sublimation occured between 130°C and 185°C that is about 60°C lower than for $Cu(thd)_2$ in the same conditions [2], and this was the only weight loss. In other experiments, sublimation took place in about the same range of temperature as for $Cu(thd)_2$, after a small amount of weight loss still not interpreted.

$Y(thd)_3(H_2O)$ and $Y(thd)_3$ - Yttrium tetramethylheptanedionate prepared according to the method of Eisentraut and Sievers [9], and recristallized in hexane has proved to be $Y(thd)_3(H_2O)$ isostructural with $Dy(thd)_3(H_2O)$ previously described by Erasmus and Boeyens [10]. The crystal system is monoclinic, space group P1, with a = 11.566 (4) Å, b = 14.571 (5) Å, c = 14.811 (3) Å, α = 119.01 (2)°, β = 99.98 (2)°, γ = 106.78 (3)°. Anhydrous $Y(thd)_3$ forms on sublimation. It is monomeric and has the structure of $Er(thd)_3$ previously determined by de Villiers and Boeyens [11]. The crystal system is orthorhombic, space group $Pmn2_1$, with a = 17.874 (3) Å, b = 10.625 (2) Å, c = 9.966 (2) Å.

ACKNOWLEDGEMENTS

This work has been supported by the French CNRS (PIRMAT). We thank Professor A. Mosset (UPR-CNRS 8011, Univ. Toulouse 3) for help in studying amorphous compounds, Dr. M. Heughebaert (URA-CNRS 445, INP Toulouse) for recording TGA, Drs. J.-P. Petitet, P. Tobaly and G. Lanchec (UPR-CNRS 1311, Univ. Paris 11) for measuring vapor pressures.

REFERENCES

1. S. B. Turnipseed, R.M. Barkley and R. E. Sievers, Inorg. Chem. 30, 1164 (1990).
2. D. Médus, Thèse de l'Institut National Polytechnique de Toulouse, France, 1992.
3. A. Gleizes, S. Sans-Lenain, D. Médus and R. Morancho, C.R. Acad. Sci. 312(II), 983 (1991).
4. W. S. Rees Jr., M. W. Carris and W. Hesse, Inorg. Chem. 30, 4479 (1991).
5. K. Timmer, K. I. M. A. Spee, A. Mackor, H. A. Meinema, A. L. Spek and P. van der Sluis, Inorg. Chim. Acta 190, 109 (1991), and references therein.
6. A. Gleizes, S. Sans-Lenain and D. Médus, C.R. Acad. Sci. 313(II), 761 (1991).
7. J. A. Bertrand and D. Caine, J. A. C. S. 86, 2298 (1964).
8. J. E. Andrew and A. B. Blake, J. C. S., Dalton Trans. 1973,1102.
9. K. E. Eisentraut and R. E. Sievers, Inorg. Synth. XI, 95 (1965).
10. C. S. Erasmus and J. C. A. Boeyens, J. Cryst. Mol. Struct. 1, 83 (1971).
11. J. P. R. de Villiers and J. C. A. Boeyens, Acta Cryst. B28, 2335 (1972).

ELECTRONIC PROPERTIES OF OMVPE GROWN FILMS OF $YBa_2Cu_3O_{7-\delta}$ ON 1" $LaAlO_3$ SUBSTRATES

William S. Rees, Jr.,[*,‡,§] Yusuf S. Hascicek,[§,¶] and Louis R. Testardi[§,¶]
Department of Chemistry,[‡] Department of Physics,[¶] and Materials Research and Technology Center,[§] Florida State University, Tallahassee, FL., 32306-3006.

ABSTRACT

Films of $YBa_2Cu_3O_{7-\delta}$ have been grown on 1" $LaAlO_3$ by OMVPE utilizing $M(tmhd)_n$ (M = Ba, Cu: n = 2; M = Y: n = 3; tmhd = 2,2,6,6-tetramethylheptane-3,5-dionato) as the source materials in a cold wall, vertical rotating disk reactor. The resultant films were characterized by SEM, XRD, T_c, J_c, and surface profilometry measurements. Relative to laser ablated thin films, the surface morphology was determined to be virtually featureless. *In-situ* depositions at substrate temperatures of <700°C, employing nitrous oxide as the oxidizing reagent, produced annular irregularities in the electronic properties of these films. The highest quality was observed near the film's center, with a marked decay evident toward the exterior 7 mm perimeter of the coated wafer.

BACKGROUND

The initial reports of CVD of $YBa_2Cu_3O_{7-\delta}$ emanated almost simultaneously from three independent laboratories.[1-3] Since these early investigations, significant improvements have been reported steadily by researchers at Northwestern,[4] EMCORE,[5] Tohoku,[6] Georgia Tech,[7] H-P,[8] and TRW.[9] The area was reviewed in its early days,[10] and is the subject of an upcoming book chapter.[11] We have chosen to focus, in this, our initial report on this topic, on the issue of uniformity of electronic properties across rather somewhat larger substrate areas than those previously discussed.

EXPERIMENTAL

The source chemicals employed were $Y(tmhd)_3 \cdot H_2O$, $Ba_5(tmhd)_9OH \cdot 3\ H_2O$, and $Cu(tmhd)_2$ (tmhd = 2,2,6,6-tetramethylheptane-3,5-dionato). The yttrium[12] and copper[13] complexes were prepared in-house, and twice purified by vapor transport. The barium complex was purchased from Strem Chemical Co., and used as received without additional purification.[14] The substrate utilized was highly twinned 1" circular $LaAlO_3$ from ATT Bell Laboratories. It was washed with electronic grade MeOH, deionized H_2O, and a second time with electronic grade MeOH, blown dry with ultrapure He, and used for deposition experiments. A custom designed vertical flow CVD system constructed

 ©1992 Materials Research Society

by EMCORE was used throughout the growth runs. Rotation of the substrate at 100 rpm in a chamber pressure of 2.2 torr was effected by a sealed stem drive. Thin films were characterized by XRD on a Siemens cassette loaded generator. SEM results were obtained from a JEOL instrument equipped with an EDAX accessory. Surface profilometry was measured with a Tencor instrument. The experimental set-up for determining photo-induced voltages in ambient temperature thin films has been described in earlier contributions.[15] T_c and J_c measurements were obtained from a home-built H-P transport property workstation.

The temperature of the yttrium source was 108°C, of the barium source was 203°C, and of the copper source was 105°C. The flow rates of ultrapure $N_{2(g)}$ through the sources were 50 sccm for yttrium and copper, and 200 sccm for barium. The films were deposited under a growth ambient of 200 sccm of N_2O at a substrate temperature of 600°C. The lowering of Tg, relative to earlier work, was the result of this oxidizing ambient. All reported properties derive from these as-deposited films, after cooling to ambient temperature under 600 torr O_2. No post deposition processing was performed on any of these samples.

RESULTS AND DISCUSSION

The films exhibit c-axis preferential epitaxial growth (**Figure 1**). The computer simulation (error limits $<\pm 0.05\%$) of the spectra indicate 100% texturing. Careful examination of the 001 peak reveals a line width that is comparable to that for single crystalline samples. The films examined were approximately 0.75 μ thick. Profilometry data indicate no gross surface features. On a separation of *ca.* 0.8 μ one can observe some pinholes in the films; however, there is an unusually high degree of smoothness in these films, when compared to those we previously have produced by laser ablation. There are no discernable surface features on the SEM size scale. There does not appear to be any obvious surface difference across the coated wafer, as found by SEM. The structures of the films are indicative of less than fully dense construction (**Figure 2**). We attribute this observation to the rather slow growth rates present within the system. A film of ~800Å requires ~10 hours to deposit. Obviously, these growth rates are well outside of the conventional realm of CVD.[16] This reproducible difficulty is not understood fully at the present moment; however, some data indicate it to be a hydrodynamic problem within the reactor design (**Figure 3**). Experiments in progress are aimed at further elicitation of the source of the low growth rates. In any event, the results reported herein are for films deposited at this level.

As with films produced by laser ablation, the current films exhibit a voltage response to photo-stimulation.[15] This phenomenon has not yet been explained. However, in a direct comparison of CVD grown films with laser ablated films, it has been observed that the voltage response is several times (~30) larger and is reproducible for a significantly higher number (>500) of pulses without an observable diminution in peak intensity. While it is premature to speculate on the full physical explanation behind such photo-induced voltage observations, one current line of reasoning, comparing the CVD films with the laser ablated films, invokes the above-mentioned slow growth rate as the source of the increased sensor performance. A concomitant fallout of the CVD low accretion speed is the higher order of resultant crystallinity present in the ultimate thin film. In the absence of a feasible method for performing the photo-induced voltage experiments on *single crystals* of $YBa_2Cu_3O_{7-\delta}$, the present data lend some confirmation to this suggestion.

Figure 1: XRD of film of $YBa_2Cu_3O_{7-\delta}$ on 1" $LaAlO_3$ grown by CVD.

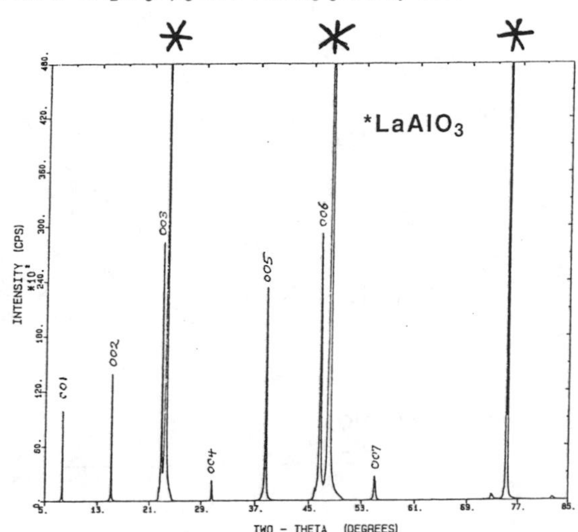

Figure 2: SEM of film of $YBa_2Cu_3O_{7-\delta}$ on 1" $LaAlO_3$ grown by CVD. The scale is the same for both views. A: Exterior. B: Interior.

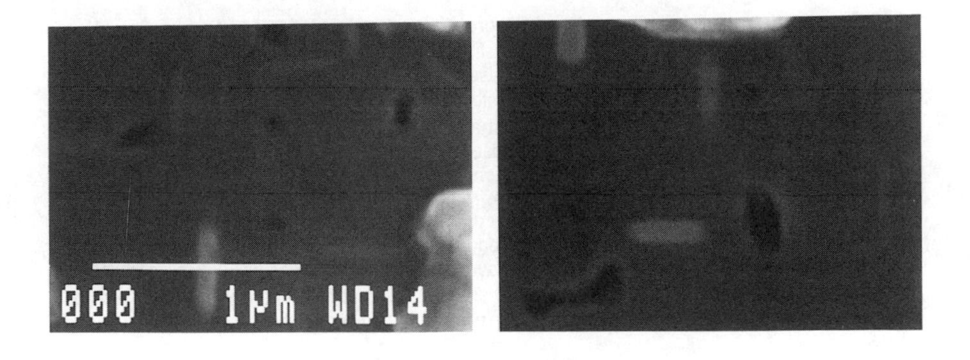

Figure 3: Scale representation of the deposition chamber utilized in CVD growth of $YBa_2Cu_3O_{7-\delta}$ on 1" $LaAlO_3$.

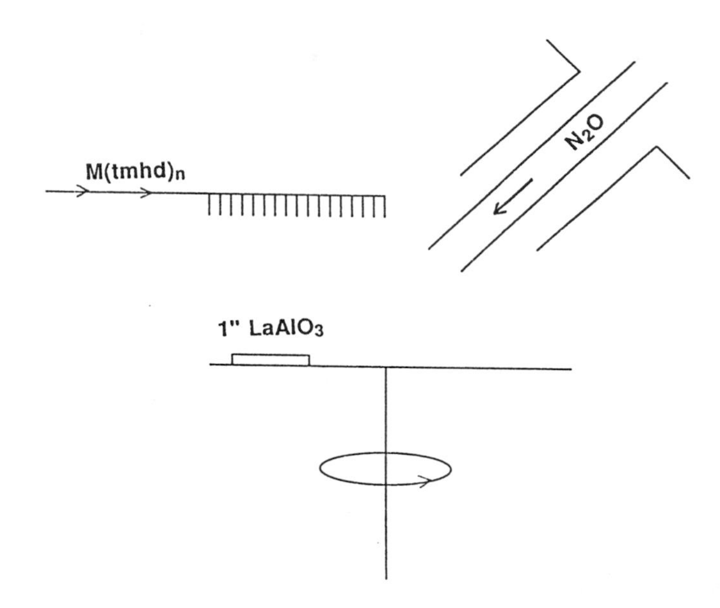

Prior to efforts to expand the scope of the present work to include 2" $LaAlO_3$ substrates, we examined the electronic properties of the current films as a function of location on the wafer. The full details of these experiments will be reported elsewhere; however, a summary of T_c data (**Figure 4**) and J_c data (**Figure 5**) reveal annular distributions in the properties of interest. Such a radial tailing may not be a problem with larger substrates, where some external edge loss is to be anticipated in any event.[16] As with the low observed growth rates, it is our working model that these irregularities result from hydrodynamic problems associated with the design of the growth chamber.

One of our main objectives has been to lower the temperature of the growth process and bring it into a regime compatible with conventional semiconductor processing (400 - 500°C). The solid state diffusion of oxygen out of the $YBa_2Cu_3O_{7-\delta}$ overlayer into the oxophillic underlying semiconductor is rapid at elevated temperatures. Likewise, the lowered temperature of the barium source is perhaps the single most important chemical objective.[17] We will report on our results in these areas at a later time.

Figure 4: T_c of $YBa_2Cu_3O_{7-\delta}$ thin film grown by CVD on 1" $LaAlO_3$ as a function of location on the surface.

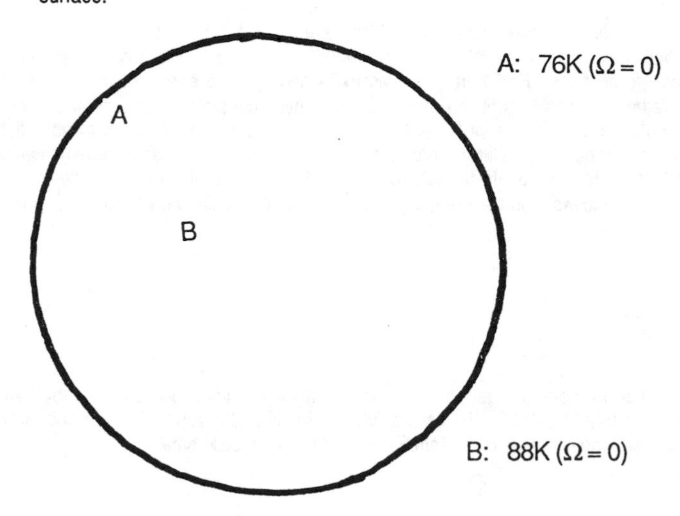

Figure 5: J_c of $YBa_2Cu_3O_{7-\delta}$ thin film grown by CVD on 1" $LaAlO_3$ as a function of location on the surface.

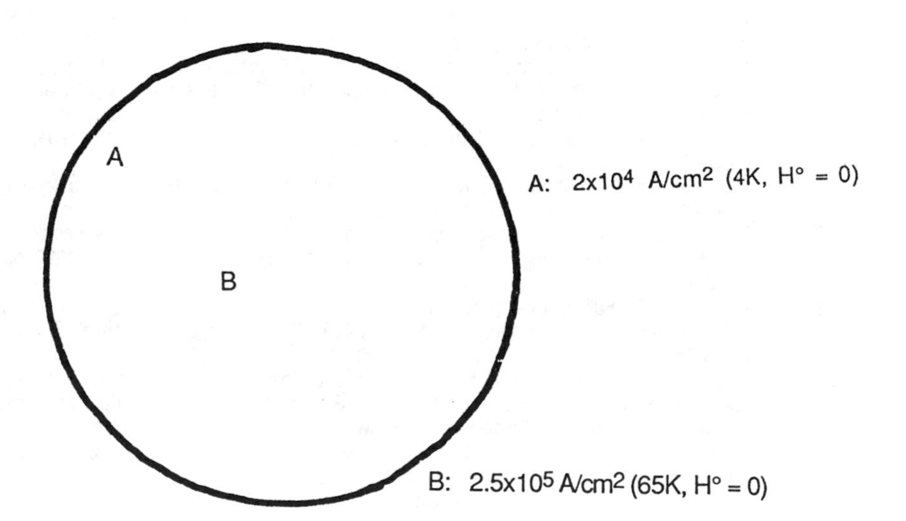

SUMMARY

Excellent c-axis oriented $YBa_2Cu_3O_{7-\delta}$ thin films have been deposited on 1" $LaAlO_3$ by CVD. The surface smoothness of these CVD films is improved significantly over those previously reported for laser ablation growth films. Radial irregularities in T_c and J_c have been observed and these, together with a rather low growth rate, are attributed to a hydrodynamic reactor geometry design difficulty. The photo-induced voltage of these highly crystalline, oriented thin films exhibits a high response at ambient temperature. Current problems include the lack of suitable, readily available, lattice-matched 2" substrate materials for $YBa_2Cu_3O_{7-\delta}$, the absence of a realistic flow model for the growth chamber employed, and the on-going search for highly stable, increased vapor pressure barium source compounds.

ACKNOWLEDGMENTS

We gratefully acknowledge the generous financial support provided to this project by DARPA contract number MDA 972-88-J-1006. Professors Edwin Hilinski and Stephen Foster performed the photo-induced voltage experiments and Mr. Tom Fellers obtained the SEM data.

REFERENCES

1. A. D. Berry, K. D. Gaskill, R. T. Holm, E. J. Cukauskas, R. Kaplan, and R. L. Henry, *Appl. Phys. Lett.*, **1988**, *52*, 1743.
2. J. Zhao, K. H. Dahman, H. O. Marcy, L. M. Tonge, T. J. Marks, B. W. Wessels, and C. R. Kannewurf, *Appl. Phys. Lett.*, **1988**, *53*, 1750.
3. H. Yamane, H. Kurosawa, and T. Hirai, *Chem. Lett.*, **1988**, 1515.
4. S. J. Duray, D. B. Buchholz, S. N. Song, D. S. Richeson, J. B. Ketterson, T. J. Marks, and R. P. H. Chang, *Appl. Phys. Lett.*, **1991**, *59*, 1503.
5. J. A. Ladd, B. T. Collins, J. R. Matey, J. Zhao, and P. Norris, *Appl. Phys. Lett.*, **1991**, *59*, 1368.
6. N. Kobayashi, Y. Minagawa, K. Watanabe, S. Awaji, H. Yamane, H. Kurosawa, and T. Hirai, *Physica C*, **1991**, *185-189 (Pt. 4)*, 2353.
7. K. Zhang, B. S. Kwak, E. P. Boyd, A. C. Wright, and A. Erbil, in *Science and Technology of Thin Film Superconductors*, R. D. McConnell and S. A. Wolf, Eds.; Plenum: NY; 1989, pp. 271-279.
8. R. Hiskes, MRS Fall 1991 Meeting, Abstract H 13.7.
9. J. H. Takemoto, C. M. Jackson, H. M. Manaseuit, D. C. St. John, J. F. Burch, K. P. Daly, and R. W. Simon, *Appl. Phys. Lett.*, **1991**, *58*, 1109.
10. L. M. Tonge, D. S. Richeson, T. J. Marks, J. Zhao, J. Zhang, B. W. Wessels, H. O. Marcy, and C. R. Kannewurf, in *Electron Transfer in Biology and the Solid State: Inorganic Compounds With Unusual Properties, Part III*, M. K. Johnson, R. B. King, D. M. Kurtz, Jr., C. Kutal, M. L. Norton, and R. A. Scott, Eds.; ACS Advances in Chemistry Series 226; American Chemical Society: Washington, D.C., **1990**; pp. 351-368.
11. W. S. Rees, Jr. and A. R. Barron, manuscript in preparation.

12. a. W. S. Rees, Jr. and M. W. Carris, *Inorg. Syn.*, submitted for publication.
    b. W. S. Rees, Jr., H. A. Luten, M. W. Carris, E. J. Doskocil, and V. L. Goedken, accompanying manuscript, this volume.
13. a. W. S. Rees, Jr. and M. W. Carris, *Inorg. Syn.*, submitted for publication.
    b. W. S. Rees, Jr. and C. R. Caballero, in *Chemical Vapor Deposition of Refractory Metals and Ceramics*, T. M. Besmann, B. M. Gallois, and J. Warren, Eds., *Materials Research Society Proceedings*; Materials Research Society: Pittsburgh, Pennsylvania, accepted for publication.
    c. W. S. Rees, Jr. and C. R. Caballero, *Advanced Materials for Optics and Electronics*, **1992**, *1(1)*, accepted for publication.
    d. W. S. Rees, Jr. and C. R. Caballero, manuscript in preparation, to be submitted.
14. a. Although sold as "anhydrous $Ba(tmhd)_2$," in reality this compound is $Ba_5(tmhd)_9OH{\bullet}3H_2O$.
    b. S. B. Turnipseed, R. M. Barkley, and R. E. Sievers, *Inorg. Chem.*, **1991**, *30*, 1164-1170.
    c. W. S. Rees, Jr., M. W. Carris, and W. Hesse, *Inorg. Chem.*, **1991**, 4479-4481.
15. a. C. L. Chang, A. Kleinhammes, W. G. Moulton, and L. R. Testardi, *Phys. Rev. B*, **1990**, *41*, 11564.
    b. K. L. Tate, R. D. Johnson, C. L. Chang, E. F. Hilinski, and S. C. Foster, *J. Appl. Phys.*, **1990**, *67*, 4375.
    c. K. L. Tate, E. F. Hilinski, and S. C. Foster, *Appl. Phys. Lett.*, **1990**, *57*, 2407.
16. G. B. Stringfellow, *Organometallic Vapor Phase Epitaxy: Theory and Practice*; Academic Press: New York, 1989.
17. a. See reference 14c, and references cited therein.
    b. W. S. Rees, Jr. and D. A. Moreno, *J. Chem. Soc., Chem. Commun.*, **1991**, *1759*.
    c. W. S. Rees, Jr., C. R. Caballero, and W. Hesse, *Angewandte Chemie*, **1992**, accepted for publication.
    d. W. S. Rees, Jr. and K. A. Dippel, in *Ultrastructure Processing of Ceramics, Glasses, Composites, Ordered Polymers, and Advanced Optical Materials V*, L. L. Hench, J. K. West, and D. R. Ulrich, Eds.; Wiley: New York, **1992**, in press.
    e. W. S. Rees, Jr., K. A. Dippel, M. W. Carris, C. R. Caballero, D. A. Moreno, and W. Hesse, accompanying manuscript, this volume.

REACTION OF BARIUM BIS(ß-DIKETONATE) COMPLEXES WITH THE SURFACE OF MAGNESIUM OXIDE

R. A. GARDINER and P. S. KIRLIN
Advanced Technology Materials, Inc., 7 Commerce Drive, Danbury, CT
G. A. M. HUSSEIN and B. C. GATES
Center for Catalytic Science and Technology, Department of Chemical Engineering, University of Delaware, Newark, DE

ABSTRACT

Second generation barium MOCVD precursors have been investigated for the growth of high quality superconducting and ferroelectric thin films. Novel barium ß-diketonate polyether adducts have been synthesized and characterized by X-ray crystallography, ^1H and ^{13}C nmr spectroscopies, elemental analysis, melting and sublimation points. Metalorganic chemical vapor deposition experiments have been performed using this new class of barium ß-diketonate polyether adducts. The as-deposited barium films were characterized by scanning electron microscopy and energy dispersive x-ray spectroscopy. These results were correlated with reaction intermediates observed by FT-IR in temperature programmed oxidation of barium bis(2,2,6,6-tetramethyl-3,5-heptanedionate) on magnesium oxide. Conditions under which the barium source reagents proceed to BaO, $BaCO_3$ or BaF_2 were elucidated.

INTRODUCTION

The preparation of oxide thin films has been a central theme in the material science community for the past decade. A wide array of technologically important applications are driving the thin film development ranging from superconducting microwave components to ferroelectric capacitors in ULSI DRAMs. Several deposition techniques are being developed and MOCVD is exceptional in that it offers superior conformality and scalability and can be carried out at higher effective oxygen partial pressures relative to PVD methods. The implementation of MOCVD requires source reagents which are pure, can be reproducibly transported into the reactor and decompose to the desired products. Results from the MOCVD of YBaCuO, BiSrCaCuO, and $Ba_{1-x}Sr_xTiO_3$ indicate that group II carbonates/fluorides are deposited from metal ß-diketonates source reagents at growth temperatures below 650°C. Our investigations have concentrated on delineating the major factors affecting the volatility, stability and deposition chemistry of the barium complexes which determine their potential as MOCVD source reagents. In addition, the nucleation step has a profound impact on film quality and we report preliminary data on the effect of the MgO surface on the decomposition pathway(s) of barium bis(2,2,6,6-tetramethyl-3,5-heptanedionate) [Ba(thd)$_2$].

 ©1992 Materials Research Society

EXPERIMENTAL

The 2,5,8,11,14-pentaoxaheptadecane (tetraglyme) adducts of barium complexes of 1,1,1,5,5,5-hexafluoro-2,4-pentanedione (H(hfacac)), 1,1,1-trifluoro-2,4-pentanedione (H[tfacac]) were synthesized by literature procedures. Commercially available $Ba(thd)_2$ has been identified as an oligomer, $Ba_5(thd)_9(OH)\cdot 3H_2O$ [1]. $Ba(thd)_2$ used in our studies was synthesized via a non aqueous route[2], to avoid the inclusion of a hydroxy group and water molecules. All of the barium complexes were purified by reduced pressure sublimation.

Reduced pressure MOCVD experiments of the barium complexes were carried out in a standard horizontal hot wall reactor described previously[3]. The precursor (0.5 g) was loaded into the reagent bubbler in an inert atmosphere and connected to the system without exposing the compound to air. The system was evacuated to 0.05 Torr and bubbler temperature was raised to 175°C for the barium ß-diketonate polyether adducts and 225°C for $Ba(thd)_2$. Total system pressure was varied from 4 and 10 Torr and the barium reagents were transported in an Ar carrier through heated lines to the quartz reactor tube. A typical growth run lasted 2 hrs and the films deposited on MgO powder were characterized either by scanning electron microscopy (SEM) and energy dispersive x-ray analysis (EDX) or by IR spectroscopy. Powder samples for the IR spectroscopy were pressed into thin porous wafers held between KBr windows in a controlled-atmosphere cell. Spectra of the solids were obtained by absorption subtraction of the spectrum of the MgO support or by transmission ratioing with the background of the empty cell. Temperature programmed decomposition studies were carried out by heating a sample made by vapor phase condensation of $Ba(thd)_2$ on MgO at room temperature; these experiments were done in a cell that enabled the measurement of both solid state and gas phase spectra.

RESULTS AND DISCUSSION

Films resulting from the decomposition of the barium ß-diketonate/polyether adducts on quartz slides at 923 K were barium fluoride contaminated with a trace of carbon as determined by EDX. The carbon presumably arose from the fragmentation and incorporation of the organic ligands $Ba(hfacac)_2$ reacts to give barium fluoride under similar conditions and the as-deposited films grown using the tetraglyme adducts do not differ in composition i.e. tetraglyme did not affect the formation of barium fluoride films [Figure 1]. The formation of barium fluoride films from the decomposition of fluorine containing ß-diketonates at temperatures between 773 and 973 K is well documented [4], [5], [6]. In, contrast, decomposition of $Ba(thd)_2$ on quartz leads to the formation of films containing barium, carbon and oxygen. $BaCO_3$ films have previously been obtained from the decomposition of $Ba(thd)_2$ [7], [8], [9].

The MgO is almost completely dehydroxylated with only weak peaks at 3677, 3667, and 3590 cm^{-1} [Figure 2(a)]. The IR spectrum of the surface species resulting from the deposition of $Ba(thd)_2$ on MgO at room temperature include bands indicating thd; the bands indicative of surface OH groups disappeared as a result of the deposition [Figure 2(b)]. These observations suggest a reaction involving $Ba(thd)_2$ and surface functional groups.

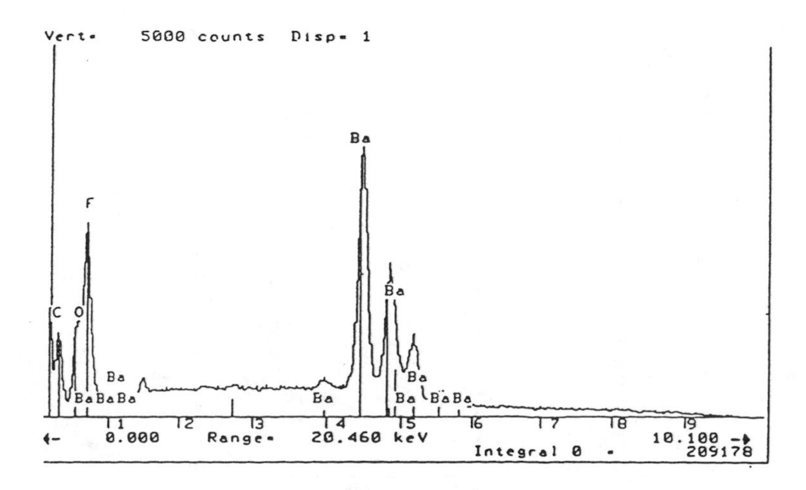

Figure 1 Energy Dispersive X-ray analysis of film deposited using $Ba(CF_3COCHCOCF_3)_2 \cdot (CH_3O(CH_2CH_2O)_4CH_3)$.

The IR spectra of the surface species formed from $Ba(thd)_2$ on MgO were independent of the deposition temperature up to 373 K, but as this temperature increased to 673 K, the structure of resultant surface species changed, as indicated by the appearance of peaks at 1580 and 1410 cm^{-1}, likely indicative of acetates [Figure 2(c)] [10]. When the samples containing these species were heated to 873 K, carbonates were observed on the MgO surface and CO_2 in the gas phase, inferred to be formed from the decomposition of surface carboxylates [8], [11]. When the deposition temperatures were in the range of 873-1073 K, surface carbonates were observed [Figure 2(d)], in the absence of surface carboxylates; the carbonates were characterized by peaks at 1560 and 1420 cm^{-1} and are attributed to monodentate surface species.

Infrared spectra of the gas phase when adsorbed $Ba(thd)_2$ was heated to 373 K include weak bands at 3970, 2970, 2910, 2870, 1710, 1620, 1480, 1386, 1240, 1140, 1070, 950, and 870 cm^{-1}, indicative of thd. These were also observed when the sample was evacuated to 10^{-2} Torr prior to heating and may indicate the vapor pressure of the thd.

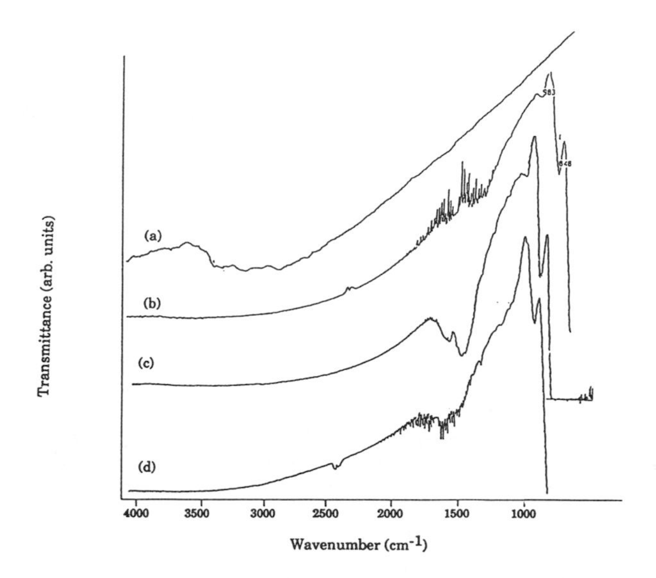

Figure 2 (a) MgO support calcined at 800 K for 4 h; (b). Infrared spectra of MgO supported species formed by adsorption of Ba(thd)$_2$ at room temperature (c) Infrared spectrum of (b) after treatment in N_2 at 673 K; (d) sample of spectrum (b) after treatment of N_2 at 873 K

When the temperature was increased, the amount of thd in the vapor phase increased, and water was also detected (ν_{OH} 3600, δ OH 1620 cm $^{-1}$). As the temperature reached 673 K, a non conjugated ketone (**1**) was detected in the gas phase (1750 cm^{-1}) along with CO (2130 cm^{-1}) and CO_2 (2345 cm^{-1}). Unambiguous identification of **1** was not made. When the temperature reached 673 K, the gas-phase products included CO, CO_2, CH_4 (1310 cm $^{-1}$), and **1**, (Figure 3). At 723 K and 773 K, they included CO, CO_2, CH_4, **1**, and isobutylene. One possible mechanism of isobutylene formation is ß-hydride elimination from 2,2-dimethyl-3-butanone.

The gas phase analyses are consistent with the IR spectra of the surface species and imply that the Ba(thd)$_2$ was chemisorbed on the MgO. We speculate that when the temperature was increased, the absorbed Ba complex decomposed, forming a ketone, which readsorbed on the MgO to give CH_4 and surface acetate [Eqn. 1 and 2] [10], [12].

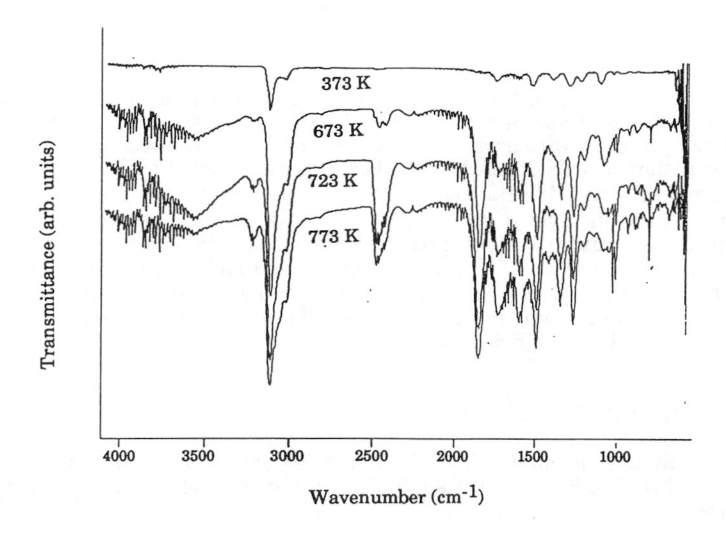

Figure 3. Infrared spectrum of gas phase species formed from 373 to 773 K.

The data suggest that the surface acetate decomposed to give surface carbonates, with the liberation of CH_4. The carbonate bands, 1560 and 1420 cm $^{-1}$, are assigned to magnesium carbonate[13]; there is no evidence of barium carbonate, 1550-1300 cm^{-1} (ν_3 antisymmetric stretching vibration mode) and 1890-800 cm^{-1} (ν_2 out of plane bending vibration mode) [13], [14]. Bulk $BaCO_3$ is stable at temperatures up to 1423 K and the evolution of CO and CO_2 at temperatures above 673 K is consistent with this peak assignment. The difference between the IR spectra obtained on quartz and MgO indicates that the surface has a dramatic effect on the decomposition chemistry of the barium source and that the destruction of carbonate intermediates, which presumably is a critical step in the formation of the perovskite phase, is promoted by the MgO surface.

$$R{-}CO{-}CH_3\ (g) + OH^-(s) \longrightarrow CH_4\ (g) + RCOO^-\ (s) \tag{1}$$

$$(CH_3)_3C{-}CO{-}CH_3\ (g) + OH^-(s) \longrightarrow (CH_3)_2C{=}CH_2 + CH_3COO^-\ (s) \tag{2}$$

CONCLUSION

Films resulting from the decomposition of partially fluorinated barium ß-diketonate/polyether adducts on quartz slides at 923 K were barium fluoride which is consistent with earlier work done with the parent complexes. The decomposition of $Ba(thd)_2$ on MgO and quartz was studied by IR spectroscopy. The results indicate that $BaCO_3$ is formed on quartz whereas $Ba(thd)_2$ decomposes on MgO via surface acetate and carbonates intermediates. These results demonstrate that the substrate surface has a dramatic effect on the decomposition of the MOCVD source reagents and suggests that the MOCVD of Ba-containing perovskites should be possible on MgO at temperatures as low as 550 °C subject to the constraint that the other temperature dependent parameters such as the sticking coefficient and the ad-atom mobility have appropriate values.

ACKNOWLEDGMENTS

The authors which to thank NASA (NAS8-38485) and NSF (ISI 9002399) for their support. Dr. Hussein was supported by a Peace Fellowship provided by the Egyptian government.

REFERENCES

1. S. B.. Turnispeed, R. M. Barkley, R. E. Seivers, Inorg. Chem. 30, 1164 (1991).
2. R. A. Gardiner, P. S. Kirlin, unpublished results.
3. R. A. Gardiner, P. S. Kirlin, D. W. Brown, A. L. Rheingold, Chem. Mater., 3, 1053 (1991).
4. P. S.Kirlin, R. Binder, R. A. Gardiner, D. W. Brown, Proc. SPIE: Processing of Films for High Tc Superconducting Electronics, 115, 1187 (1989).
5. A. P. Purdy, A. D. Berry, R. T. Holm, M. Fatemi, K. K. Gaskill, Inorg. Chem. 28, 2799 (1989).
6. J. Zhao, H. O. Marcy, L. M. Tonge, T. J. Marks, B. W. Wessels, App. Phys. Lett. 53, 1750 (1988).
7. H. Busch, A. Fink, A, Muller, J. Appl. Phys. 70(4), 2449 (1991).
8. H. Yamane, T. Hirai, J. Appl. Phys. 69(11), 7948 (1991).
9. K. H. Young et al, J. Mater. Res. 6 (11), 2259 (1991).
10. R. O. Kagel, R. G. Greenler, J. Phys. Chem. 49, 1638 (1968).
11. M. J. D. Low, H. Jacobs, N. Takezawa, Water Air Soil Pollut. 2, 61 (1973).
12. M. I. Zaki, N. Sheppard. J. Catal. 80, 114 (1983).
13. J. A. Gadsen, Infrared Spectra of Minerals and Related Inorganic Compounds, (Butterworths, London, 1975) page 15.
14. R. B. Fahim, M. I. Zaki, G. A. M. Hussein, Powder Technol. 30, 161 (1982).

OMCVD ELABORATION AND CHARACTERIZATION OF SILICON CARBIDE AND CARBON-SILICON CARBIDE FILMS FROM ORGANOSILICON COMPOUNDS

R. MORANCHO*, A. REYNES*, M'b. AMJOUD* AND R. CARLES **
*Ecole Nationale Supérieure de Chimie de Toulouse, Laboratoire des Matériaux (URA 445 CNRS), 118 Route de Narbonne 31077 Toulouse Cédex France
**Université Paul Sabatier, Laboratoire de Physique des Solides (URA 074 CNRS), 118 Route de Narbonne 31062 Toulouse Cedex France

ABSTRACT

Two organosilicon molecules tetraethysilane (TESi) and tetravinylsilane (TVSi) were used to prepare thin films of silicon carbide by chemical vapor deposition (C.V.D.). In each of the molecule, the ratio C/Si = 8, the only difference between TESi and TVSi is the structure of the radicals ethyl ($.CH_2-CH_3$) and vinyl ($.CH=CH_2$). This feature induces different thermal behavior and leads to the formation of different materials depending on the nature of the carrier gas He or H_2. The decomposition gases are correlated with the material deposited which is investigated by I.R. and Raman spectroscopy. The structure of the starting molecule influences the mechanisms of decomposition and consequently the structure of the material obtained.

INTRODUCTION

The work presented in this paper is devoted to the thermal behavior of two organosilicon compounds, tetraethylsilane [$Si(CH_2-CH_3)_4$],(TESi) and tetravinylsilane [$Si(CH=CH_2)_4$], (TVSi) used to elaborate thin films of silicon carbide at low temperatures. These two precursors, which present the same C/Si atomic ratio (C/Si = 8) were selected to compare the effects of the :

-nature of the radicals ethyl ($.CH_2-CH_3$) and vinyl($.CH=CH_2$)
-carrier gas, helium or hydrogen

on the nature (composition and structure) of the materials deposited and on the composition of the pyrolysis gas phase.

EXPERIMENTAL

The pyrolysis of the two precursors was studied in a classical horizontal cold-wall reactor which has been presented elsewhere [1]. The apparatus was interfaced with a gas phase chromatograph (G.P.C.) in order to follow the variations of the gas phase (precursor non-decomposed and decomposition gases) with the temperature and to identify from their retention time the nature of the decomposition gases. The experimental conditions used were the same for the two precursors :

-temperature of decomposition	:	500-1000 °C
-molar fraction	:	2.10^{-3}
-carrier gas flow rate	:	$10\ l.h^{-1}$
-total pressure	:	760 torr

The quantity of precursor input to the reactor was controlled by the vapor pressure through variation of the temperature of the bubbler containing the liquid organometallic precursor through which the carrier gas was blown. (4.5 °C for TESi and -13 °C for TVSi) The substrates used were silica and alumina.

GASEOUS PHASE DECOMPOSITION STUDIES

Figure 1, reports the variations of the percentages of decomposition of the two precursors against the temperature and the carrier gas (He or H_2). An identical percentage of pyrolysis (50%) was reached for different pyrolysis temperatures (T_{50}). Under He, for TESi, T_{50} = 845 °C while for TVSi T_{50} = 765 °C.

The use of H_2 as carrier gas lowered the temperature of decomposition to T_{50} = 785 °C for TESi and 740°C for TVSi.

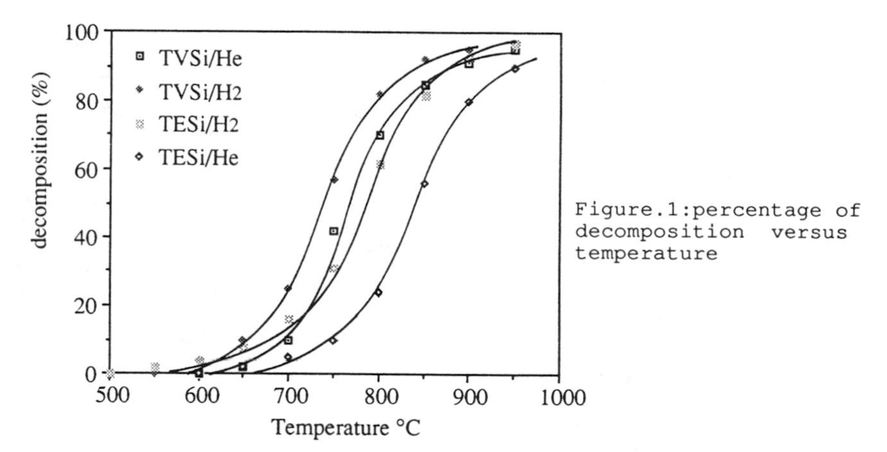

Figure.1: percentage of decomposition versus temperature

The decomposition gases were the same for the two precursors, their formation is summarized as follows :

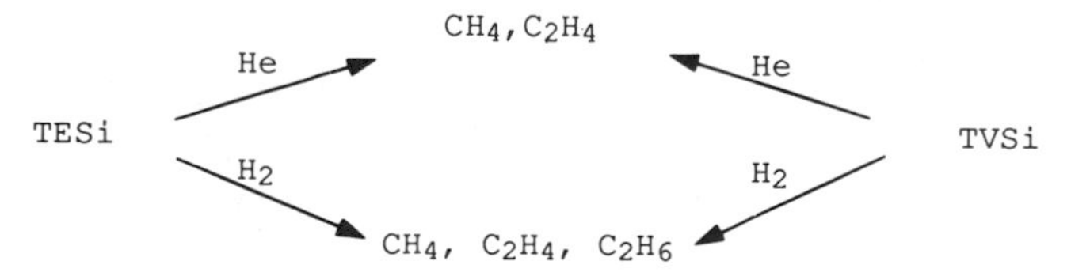

and their variations are displayed figure 2. Though the decomposition gases are the same, their concentrations are very different (depending on the starting molecule, the carrier gas and the temperature) and consequently the thermal behavior of the precursors appears different. Particularly, the pyrolysis of TESi with He gives a greater quantity of gases than the pyrolysis of TVSi. Moreover, the use of H_2 in place of He also leads to an increase of the percentages of the gases and to the formation of ethane.

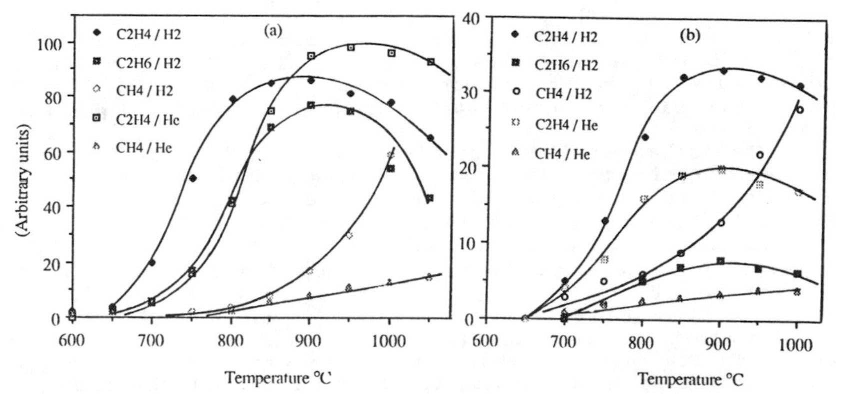

Figure.2 : Variation of the concentration of the decomposition gases (a) for TESi, (b) for TVSi against temperature for the two precursors

SOLID PHASE INVESTIGATION

The solid phase was investigated by different techniques. We report here the results obtained by electron microprobe analysis, (EMPA), infra-red and Raman spectroscopy. The surfaces of the coatings were examined by scanning electron microscopy.

The morphology of the films changed with the carrier gas and the nature of the substrate. The use of hydrogen induced smoother surfaces than helium whichever precursor was decomposed. The coatings obtained on silica presented some cracks as a consequence of the thermal stresses coming from the differences in the thermal expansion coefficients (α = 5.4 10^{-7} K^{-1} for SiO_2, α = 5.5 10^{-6} K^{-1} for SiC). The coatings made on alumina were uniform, the α coefficient of this substrate being very near of that of SiC (α = 5.6 $10^{-6}K^{-1}$ for Al_2O_3).

Only the atomic percentage of silicon was determined by EMPA using monocrystalline silicon as standard. The values are gathered in table I for the different precursors, carrier gases and the temperatures.

Table I : silicon atomic percentage of the coatings deposited from TESi and TVSi precursors. The accuracy was ± 2 %

Precursor	TESi		TVSi	
Carrier gas / Temperature	He	H_2	He	H_2
760 °C	39	46	29	34
800 °C	43	54	32	37
840 °C	44	53	30	42
880 °C	40	54	28	45

The percentage of carbon was estimated by the difference from 100 % considering the formula Si_xC_{1-x}. The nature of the precursor and of the carrier gas influence the percentage of silicon, the system TESi/H_2 led to a silicon-rich deposit while the system TVSi/He gave a carbon-rich coating. The decomposition temperature has a weak effect except for the system TVSi/H_2 where the silicon percentage increases with the deposition temperature.

The infra red (I.R.) studies carried out on the coatings coming from the decomposition of TESi/H_2, TESi/He, TVSi/H_2 and TVSi/He showed a very broad and intense band located at about 800 cm^{-1} characteristic of a Si-C stretching mode [2]. In the case of TESi/H_2, we also noted the presence of a Si-H bond (2100 cm^{-1})[3,4] while with He this feature did not appear. The systems TVSi/He and TVSi/H_2 displayed Si-H bands and also C-H bands at around 3100 cm^{-1} [5]. The Raman spectra are complementary and particularly the C-C bonds which are not very active in I.R. are well resolved by Raman spectroscopy and give useful information concerning the surroundings of the elements. The Raman spectroscopy of the four coatings made under He and H_2 from TESi and TVSi, at 800 °C are gathered in figure 3.

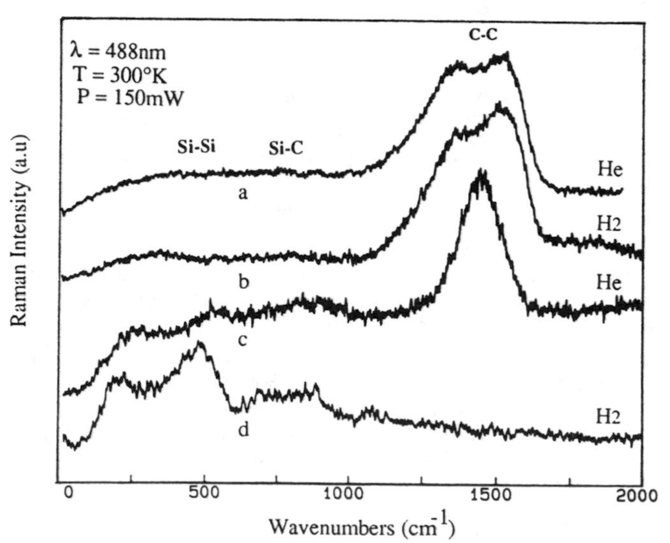

Figure. 3 : Raman spectra of coatings obtained at 800 °C from TVSi/He(a), TVSi/H_2(b), TESi/He(c) and TESi/H_2(d)

The two spectra resulting from TVSi are characterized by two broad intense peaks at 1380 and 1570 cm^{-1} typical of graphite and disordered carbon [6,7]. From the system TESi/He, only one peak at 1420 cm^{-1} argues in favour of C-C bond incorporation in the Si_xC_{1-x} phase. When He was replaced by H_2 no C-C bond was detected. We can only see the presence of Si-Si (~500 cm^{-1}) and Si-C (~760 cm^{-1}) [7] bands which suggests a totally tetrahedrally coordinated Si_xC_{1-x} phase. All the carbon atoms are linked to silicon atoms.

DISCUSSION CONCLUSION

From the overall results, some comments can be made concerning the materials deposited and the mechanisms of pyrolysis.

The materials deposited can be represented by the general formula Si_xC_{1-x}. However the chemical bonds between these two elements are different. From I.R., we see that silicon carbide is present whichever the precursor and the carrier gas used. With this in mind, as the silicon carbide does not present any deviation from the stoichiometry, its quantity is defined by the element which has the lowest percentage. The coatings where an excess of carbon is present $x < 0.50$ can be represented by the formula $(SiC)_x\ C_{1-2x}$ and the coatings having an excess of silicon ($x > 0.50$) by the formula $(SiC)_{1-x}\ Si_{2x-1}$. According to this remark, the solid phase obtained at for example 800 °C can be formulated as follows (see table I) for the different systems :

- TESi/He $Si_{0.43}\ C_{0.57}$ $\longrightarrow$ $(SiC)_{0.43}\ C_{0.14}$
- TESi/H_2 $Si_{0.54}\ C_{0.46}$ $\longrightarrow$ $(SiC)_{0.46}Si_{0.08}$
- TVSi/He $Si_{0.32}\ C_{0.68}$ $\longrightarrow$ $(SiC)_{0.32}\ C_{0.36}$
- TVSi/H_2 $Si_{0.37}\ C_{0.63}$ $\longrightarrow$ $(SiC)_{0.37}\ C_{0.26}$

The excess of carbon with respect to the silicon carbide stoichiometry was identified by Raman spectroscopy. and in the case of TESi/H_2 Si-Si bonds were shown to be present.

The discussion of the pyrolysis mechanisms can be based on the nature of the gas and solid phases. The two molecules gave the same decomposition gases (CH_4,C_2H_4 with He and CH_4, C_2H_4 and C_2H_6 with H_2).

Under He carrier gas.

TESi gave more C_2H_4 than TVSi and led to solids having less carbon. The formation of ethylene in the case of TESi can be explained by ß hydride elimination.

$$>Si(C_2H_5)(C_2H_5) \longrightarrow >Si(H)(C_2H_5) + C_2H_4$$

On the other hand, with the vinyl radical ($.C_2H_3$), an extra hydrogen atom is necessary to form ethylene. We must imagine the dehydrogenation of a neighbouring radical leading to the formation of reactive, weakly hydrogenated organosilicon species responsible for the introduction of carbon into the solid phase.

Under H_2 carrier gas.

With either precursor, H_2 favours the departure of carbon by hydrogenation of the radicals : ethyl radical gave ethane and vinyl produced more ethylene and ethane. The role of H_2 is not as yet well defined, but the presence of Si-H bonds seems to indicate the involvement of the surface of the solid. In this case, the formation of C_2H_6 may come from a reductive elimination between an ethyl radical and a hydrogen atom bonded to a silicon atom of the surface.

In order to pursue these studies, experiments with deuterium as carrier gas and/or deuterated precursors are planned.

References

1 P. Mazerolles, A. Reynes, S. Séfiani and R. Morancho, J.Anal. and Appl. Pyrolysis, 22, 95-105 (1991).

2. Y. Katayama, K. Usami and T. Shimada, Phil. Mag. B 43, 283 (1981).

3. A. Hiraki, Timura, K. Mogi and M. Tashiro, J. Phys. C4, 277 (1981).

4. M.H. Brodsky, M. Cardona and J.J. Cuomo, Phys. Rev. B 16, 3556 (1977).

5. K. Mui, D.K. Basa and F.W. Smith, Phys. Rev. B 35, 8089 (1987).

6. A. Morimoto, T. Kataoka, M. Kumeda and T. Shimizu, Phil. Mag. B 50, 517 (1984).

7. G. Constant, J.P. Gérault, R. Morancho and J.J. Ehrhardt in Studies in Inorganic Chemistry, edited by R. Metselaar, H.J.M. Heijiligers and J. Schoonman (Solid State Chemistry Proc. of the Second European Conference, 3, 1983) pp 267-270.

COMPUTATION, VISUALIZATION, AND CHEMISTRY OF ELECTRIC FIELD-ENHANCED PRODUCTION OF CERAMIC PRECURSOR POWDERS

MICHAEL T. HARRIS, WARREN G. SISSON, AND OSMAN A. BASARAN
Chemical Technology Division, Oak Ridge National Laboratory, Oak Ridge, TN 37831

ABSTRACT

The stability of a liquid emanating from a nozzle is profoundly affected by an electric field. This electric-field induced instability is used here to form ceramic precursor powders having well-controlled particle-size distribution and morphology. Moreover, a hybrid boundary element/finite element method is used to determine the shapes and stability of a drop hanging from a nozzle. The efficiency of two different electrode configurations is considered: in one configuration, the nozzle is attached to the top plate of a parallel-plate capacitor and in the other, the nozzle is surrounded by a concentric cylindrical electrode. The computational results show that such pendant drops lose stability at turning points with respect to field strength. The experimental and computational results reported here are of importance not only in the development of electrodispersion apparatus, but in fields as diverse as capillarity and separations.

INTRODUCTION

The scientific study of the effects of electric fields on ***shapes and stability*** of drops that are pendant from a nozzle and ***dispersion*** of drops from a nozzle can be traced to the research of Zeleny [1] and Wilson and Taylor [2]. However, it was not until Vonnegut and Neubauer [3] rediscovered the phenomenon of electro-atomization of drops that the subject became of wide-spread interest due to applications in fields as diverse as [4] propulsion, spraying, painting, and printing, ***and*** separations. More recently, this technique has been used for atomization of droplets into air [5] and into a second liquid phase [6] to form micron-size ceramic powders.

Harris *et al.* [6] reported the formation of ceramic precursor particles in a device called the Electrical Dispersion Reactor (EDR) where the dispersed phase is a highly conducting aqueous solution and the continuous phase is a non-conducting organic solvent (e.g. 2-ethyl-1-hexanol): see Figure 1. Thus, a microreactor is created in the form of a micron-size droplet within a liquid continuous phase. Chief among the attractive features of the EDR is that some of the reactants may be placed in the continuous phase and others in the drops. Harris *et al.* [6] demonstrated the flexibility of the EDR by synthesizing micron-size porous shells and dense spherical particles of a single component as well as mixed hydrous oxide particles in organic solvents.

One of the requirements for the EDR to be commercially successful is that the electrical energy needed to disperse drops from a nozzle should be low. Harris *et al.* [6] studied the chemistry of particle making, but did not investigate optimal electrode configurations for destabilizing drops. Providing this insight is a goal of this paper.

Although there have been numerous studies of electrified pendant drops recently, only two of these works have been detailed enough to draw any conclusions on the effect of geometrical parameters on drop stability [7, 8]. In [7], shapes and stability of drops pendant from a face of a parallel-plate capacitor are determined by the Galerkin/finite element method (GFEM). In [8], shapes and stability of drops pendant from a nozzle

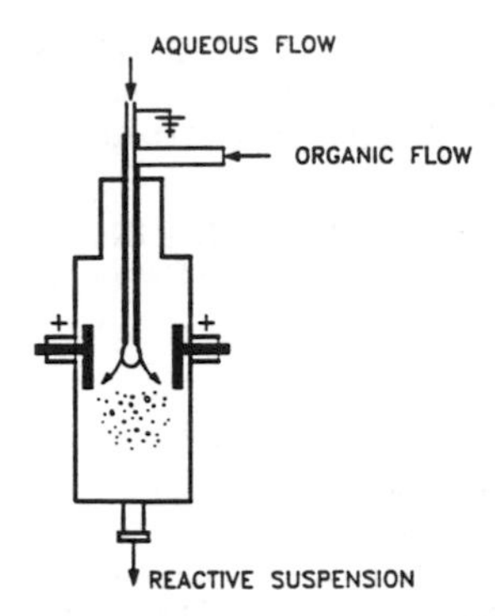

Figure 1. Electric dispersion reactor (EDR).

or a pipe that is attached to the top plate of a parallel-plate capacitor — Figure 2 — are analyzed by a hybrid method that solves the *augmented Young-Laplace equation* that governs the drop shape by the GFEM and the integral statement of the *Laplace equation* that governs electric field distribution outside the conducting drop by the boundary element method (BEM). Here we analyze the shapes and stability of drops hanging from a nozzle that is surrounded by a so-called cup electrode — Figure 3 — by the same hybrid method (see, also, [9]).

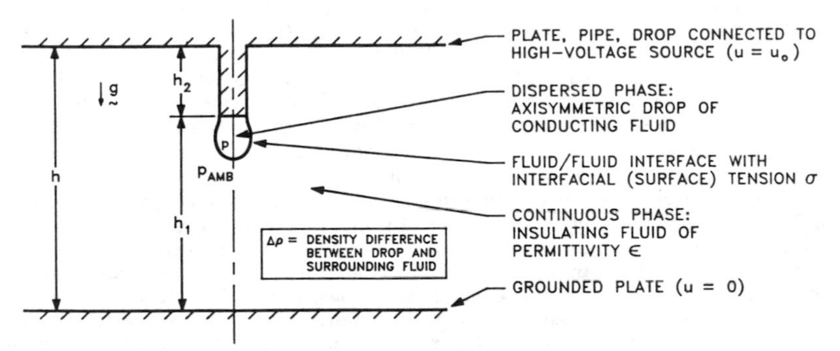

Figure 2. Parallel-plate capacitor. Here $\underset{\sim}{g}$ is the acceleration due to gravity. As the distance measured from the axis of symmetry $l \to \infty$, the electric field $\underline{\mathbf{E}}$ between the two plates approaches a uniform vertical field.

The mathematical statements of the drop shape and electrostatics problems are presented next, followed by predictions of the potential required to destabilize pendant drops. Then, experimental means for synthesizing hydrous aluminum oxide particles are described and the flexibility of the new electrode configuration in creating particles having different morphologies is brought out.

MATHEMATICAL FORMULATION

If length is measured in units of R, a characteristic length (see below), and electrostatic potential U outside the drop is measured in units of u_o, Table I shows that

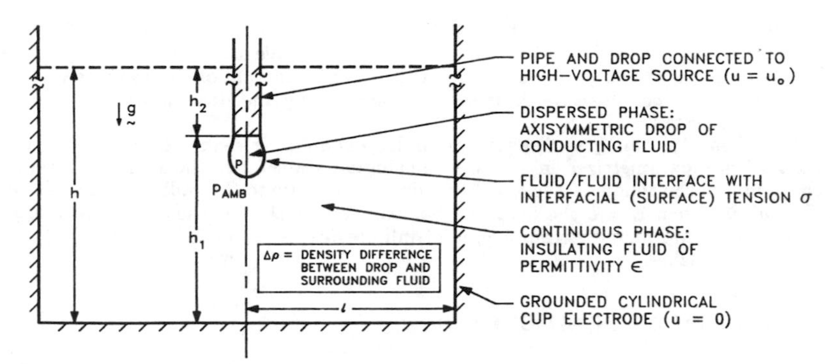

Figure 3. Cup electrode. For distances exceeding h_2 above the tip of the pipe, the electric field between the pipe and the cup approaches a uniform radial field.

equilibrium shapes and stability of electrified pendant drops are governed by six independent dimensionless groups: N_e, G, any two among the three geometrical parameters H, H_1, and H_2, L and D. The meaning of the drop size parameter D is clarified below. It is sometimes convenient to employ two dimensionless groups that are related to N_e; these two parameters are defined in Table I.

Table I. Governing dimensionless groups.

DIMENSIONLESS GROUP	
N_e = ELECTRICAL BOND NUMBER = $\frac{\text{ELECTROSTATIC}}{\text{SURFACE ENERGY}}$	$\frac{\epsilon u_o^2}{2\sigma R}$ *
G = GRAVITATIONAL BOND NUMBER = $\frac{\text{GRAVITATIONAL}}{\text{SURFACE ENERGY}}$	$\frac{gR^2\Delta\rho}{\sigma}$
H = PLATE SEPARATION OR CUP HEIGHT	h/R
H_1 = DISTANCE BETWEEN PIPE AND BOTTOM ELECTRODE	h_1/R
H_2 = LENGTH OF PIPE	h_2/R
L = ASYMPTOTIC BOUNDARY LOCATION OR CUP RADIUS	l/R
D = DROP-SIZE PARAMETER	d/R

*$\mathcal{P} = \sqrt{N_e} = \sqrt{\frac{\epsilon}{2\sigma R}}\, u_o$ = DIMENSIONLESS "EFFECTIVE" POTENTIAL

$\mathcal{E} = \mathcal{P}/H = \sqrt{\frac{\epsilon R}{2\sigma}}\, \frac{u_o}{h}$ = DIMENSIONLESS PARALLEL-PLATE FIELD

The drop shape and the electrostatic field $\underline{\mathbf{E}} \equiv -\nabla U$ are governed by the (dimensionless) augmented Young-Laplace and Laplace equations, respectively

$$-2\mathcal{H} = K + Gz + N_e E_n^2 \qquad \text{on} \qquad S_{\text{DROP}}, \tag{1}$$

$$c(\underline{\mathbf{r}})U(\underline{\mathbf{r}}) + \int_{\partial V'} U(\underline{\mathbf{r}}')\frac{\partial \mathcal{G}}{\partial n}(\underline{\mathbf{r}},\underline{\mathbf{r}}')dS = \int_{\partial V'} \frac{\partial U}{\partial n}(\underline{\mathbf{r}}')\mathcal{G}(\underline{\mathbf{r}},\underline{\mathbf{r}}')dS. \tag{2}$$

Here S_{DROP} is the drop surface, $2\mathcal{H}$ is twice the local mean curvature of the interface, and z is the distance measured in the direction of gravity. Reference pressure K is the pressure difference between the drop and the ambient fluid in the horizontal plane

$z = 0$. $E_n \equiv -\underline{n} \cdot \nabla U = -\frac{\partial U}{\partial n}$ is the normal component of the electric field at the drop surface and $\underline{n}$ is the unit normal to S_{DROP}. $\partial V'$ is the union of the drop and solid electrode surfaces and $\underline{r}, \underline{r}' \in \partial V'$. $\mathcal{G}$ is the Green's function of the Laplace operator.

The reference pressure K is set by constraining the drop volume to be a fixed amount V_o, *viz.* $V = V_o$.

When $G \to 0$, $N_e \to 0$, equilibrium drop shapes are segments of spheres and are conveniently parametrized in terms of the single parameter d, the signed distance from the center of the sphere to the tip of the pipe, or its ratio to the radius of the sphere, *viz.* $D \equiv d/R$. When $D = 0$ the drop is a hemisphere; as $D \to 1$, the situation approaches a spherical drop hanging from a pipe of infinitesimally small radius; and as $D \to -1$, the drop vanishes.

THEORETICAL PREDICTIONS

Figure 4 shows the variation of drop aspect ratio, a measure of the drop deformation that equals the ratio of the length of the pendant drop to the radius of the pipe, with increasing effective potential. Along each curve, the drop shapes are stable up to the last point shown, which is a stability limit or a turning point [10]. Plainly, for the situation depicted in Figure 4, the potential required to destabilize a drop is lower by 25% to 30% with the cup electrode than with the parallel-plate (pp) capacitor.

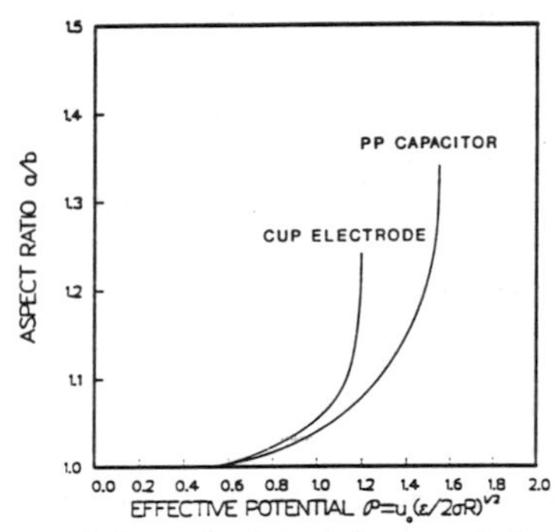

Figure 4. Effective potential required to deform and destabilize a pendant drop. $G = 0$, $D = 0$, $H_2 = 5$, $H_1 = 10$, and, for the cup electrode, $L = 5$.

Figure 5 shows the effect of cup radius on the variation with cup height of the critical value of the effective potential to destabilize a pendant drop. When the cup height exceeds a critical value, Figure 5 demonstrates that the potential required to destabilize a drop decreases as cup radius decreases. Figure 6 shows the effect of cup radius on drop shapes at the stability limit. Evidently, the drops in Figure 6(a) and the one corresponding to $H = 12$ in Figure 6(b) would jet from their tips. However, Figure 6(b) shows and experiments confirm that when L is small and $L \ll H_1$, a pendant drop would jet from its periphery.

EXPERIMENTAL OBSERVATIONS

Figure 7 shows particles produced by two different modes of operation. In the first mode, the precipitation agent was placed in the disperse phase (Figure 7(a)), whereas

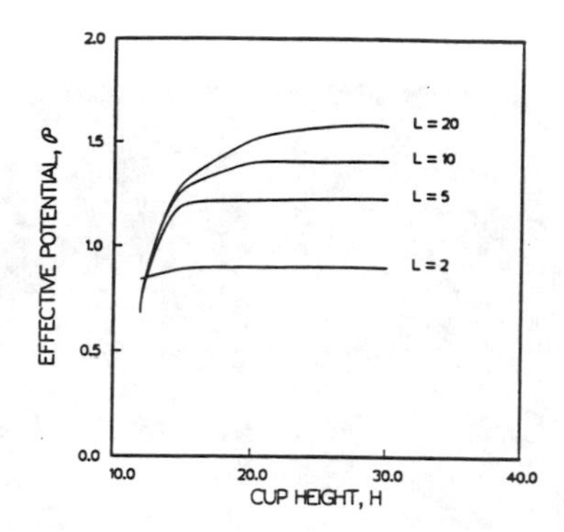

Figure 5. Effect of cup radius, L, on effective potential required to destabilize pendant drops. $G = 0$, $D = 0$, and $H_2 = 10$.

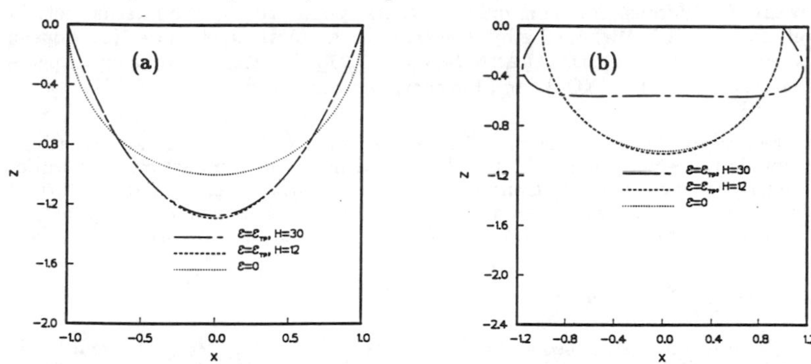

Figure 6. Effect of cup height, H, on drop shapes at their stability limits: (a) case of large cup radius, $L = 20$, and (b) case of small cup radius, $L = 2$. $\mathcal{E} = 0$ corresponds to the situation when the field is off and $\mathcal{E} = \mathcal{E}_{TP}$ corresponds to the situation at the turning point or stability limit. $G = 0$, $D = 0$, and $H_2 = 10$.

in the second mode, it was placed in the continuous phase. After the particles were formed, they were removed by centrifugation, washed several times with ethanol, and were then dried for 24 hours. The particles were imaged before and after the washing and drying steps by electron microscopy.

The particles shown in Figure 7(a) were dense and their size ranged between a few tenths of a micron and several microns. The particles shown in Figure 7(b) had holes in them. When a drop forms from a nozzle, it undergoes oscillations that are damped at a rate governed by the physical properties of the two phases and the local electric field. These oscillating drops display multi-lobed shapes that are characteristic of nonlinear oscillations [11]. When the particles gel rapidly, a consequence of mode coupling is the dimple-like, multi-lobed morphology shown in Figure 7(b).

CONCLUSIONS

Computational techniques have been developed that can predict the electric field strength needed to disperse or destabilize a liquid at a nozzle. Experiments have shown

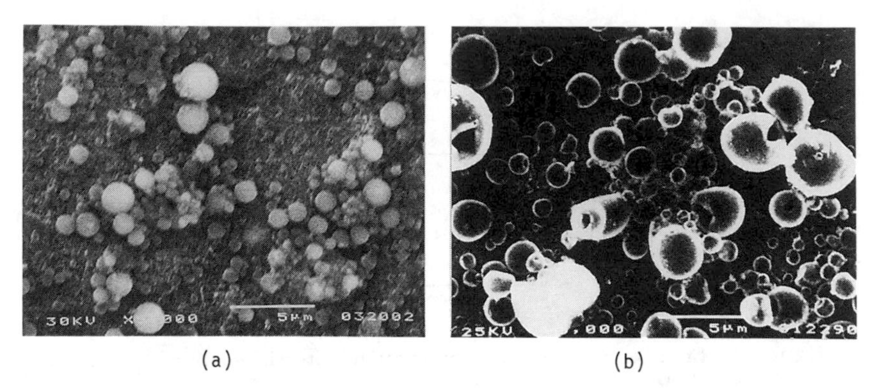

Figure 7. Hydrous aluminum oxide particles synthesized by the internal gelation method in the EDR where 2-ethyl-1-hexanol is the continuous phase. (a) Aqueous dispersed phase contains 0.5M $Al(NO_3)_3$ plus 1.25M HMTA. (b) Aqueous dispersed phase contains 1M $Al(NO_3)_3$ and the continous phase contains 0.3M NH_3.

that the characteristics of the particles can be changed by strategically locating the reactants in the proper phase. Future theoretical work will emphasize incorporation of the underlying chemistry into transient models to predict the entire particle formation process. Future experimental work will entail detailed studies of powder synthesis in the EDR and comparative evaluation of these powders as ceramic precursors.

ACKNOWLEDGMENTS

This research was sponsored by the Advanced Technology Programs and the Division of Chemical Sciences, Office of Basic Energy Sciences, U.S. Department of Energy under contract DE-AC05-84OR21400 with Martin Marietta Energy Systems, Inc.

REFERENCES

[1] Zeleny, J., Proc. Camb. Phil. Soc. 18, 71 (1915).
[2] Wilson, C. T. R. and Taylor, G. I., Proc. Camb. Phil. Soc. 22, 98 (1925).
[3] Vonnegut, B. and Neubauer, R. L., J. Colloid Sci. 7, 616 (1952).
[4] Bailey, A. G., Electrostatic Spraying of Liquids, (John Wiley and Sons, New York, 1988).
[5] Slamovich, E. B. and Lange, F. F., Mater. Res. Soc. Proc. 121, 257 (1988).
[6] Harris, M. T., Scott, T. C., Basaran, O. A., and Byers, C. H., Mater. Res. Soc. Proc. 121, 257 (1988).
[7] Basaran, O. A. and Scriven, L. E., J. Colloid Interface Sci. 140, 10 (1990).
[8] Harris, M. T. and Basaran, O. A., Bull. Amer. Phys. Soc. 36 (10), 2690 (1991).
[9] Pelekasis, N. A., Tsamopoulos, J. A., and Manolis, G. D., Phys. Fluids A 8, 1328 (1990).
[10] Iooss, G. and Joseph, D. D. Elementary Stability and Bifurcation Theory, 2nd edition (Springer-Verlag, New York, 1990).
[11] Basaran, O. A., J. Fluid Mech., in press (1992).

EFFECT OF DOPANTS IN VAPOR PHASE SYNTHESIS OF TITANIA POWDERS

M. KAMAL AKHTAR*, S. E. PRATSINIS*, AND S. V. R. MASTRANGELO**
*Department of Chemical Engineering, Center for Aerosol Processes, University of Cincinnati, Cincinnati, OH 45221.
** DuPont Chemicals, Edge Moor, DE 19809.

ABSTRACT

Gas phase synthesis of titania from titanium tetrachloride ($TiCl_4$) oxidation in the presence of dopants ($SiCl_4$ and $POCl_3$) was systematically investigated in an aerosol reactor as a function of temperature (1300-1700 K) and dopant concentration (0-15 mole % of $TiCl_4$). The particle morphology was dramatically altered in the presence of dopants from polyhedral to spherical. Energy dispersive analysis indicated that the powders were homogeneous and that the dopants were not segregated at the surface or at the grain boundaries. Lattice parameter measurements from X-ray diffraction indicated that the dopant oxide was present in solid solution in titania. While titania synthesized in the absence of dopants was ≈80% anatase, the introduction of Si^{4+} and P^{5+} resulted in greater than 98 % anatase. The effects of foreign ions on titania phase composition, aggregate size and gas phase coalescence are explained by the creation of oxygen vacancies and reduction/enhancement of the titania sintering rates.

INTRODUCTION

Gas phase routes provide the advantage that they do not involve the handling of large volumes of liquids, multiple steps and surfactants used in conventional wet chemical processes [1]. Dopants or additives are routinely used in the gas phase manufacture of titania to control particle size and phase composition. Though there are several patents dealing with the formation of titania particles, the fundamentals of the effect of dopants on titania particle characteristics are still not understood [2].

The vapor synthesis of titania powders doped with different metal halides was first studied by Suyama and Kato [3]. The particles formed in the presence of all additives were significantly smaller and had a narrower size distribution than those produced in the absence of additives. The phase composition of the product powders was not reported. Mezey [4] reported that if 0.01 to 10 percent of a volatile silicon compound is present in flame oxidation of $TiCl_4$, it leads to >90 percent anatase; if 0.01 to 10 percent of an aluminum compound is present, the rutile phase is predominantly formed.

While some of the properties of doped titania in the solid state have been studied, contradictions and differing results may arise due to the different sample preparation conditions. In this work, a systematic experimental investigation is conducted into the gas phase synthesis of titania powder in the presence of Si and P halide vapors in a externally heated flow reactor.

EXPERIMENTAL

Titania particles were formed by $TiCl_4$ oxidation in an aerosol flow reactor in the presence of dopants: $SiCl_4$, and $POCl_3$. All experiments were carried out at constant reactor residence time (1.6 s) but at various dopant : $TiCl_4$ ratios at reactor temperatures of 1300, 1500 and 1700 K.

Titanium tetrachloride (Aldrich, 99.9%) was reacted with oxygen (Linde, 99.9%) in argon at a high temperature in the presence of dopants in an aerosol flow reactor [5]. Particle-free, dry argon gas (Wright Brothers, 99.98%) was bubbled through a fritted glass outlet into a gas washing bottle containing $TiCl_4$ at room temperature (296 K). The Ar-$TiCl_4$ vapor stream was diluted with additional argon and mixed with about 10 times the stoichiometric amount of oxygen. Vapors of liquid $SiCl_4$ (Aldrich, 99.9%) and $POCl_3$ (Alfa, 99.9%) were entrained into the carrier gas similarly to $TiCl_4$ vapor. The argon-dopant vapor stream was mixed with the Ar-$TiCl_4$-O_2 mixture and introduced into the reactor. The vapor pressure of $SiCl_4$ at room temperature is much higher than that of $TiCl_4$ and in order to introduce only small amounts (up to 15 mole % of the $TiCl_4$ content in the gas phase) of $SiCl_4$, the temperature of liquid $SiCl_4$ was maintained at 273 K. The vapor pressure of $POCl_3$ is low enough so that controlled amounts can be introduced into the reactor without any difficulty. The concentration of $TiCl_4$, $SiCl_4$ and $POCl_3$ in the gas stream was determined by recording the weights of the halide containing bottles before and after each experiment carried out at constant argon flow rate through the bottle. The detailed apparatus description, experimental procedure, reactor flow rates, and measurements of precursor $TiCl_4$ vapor concentrations and temperature profiles in the reactor can be found in Akhtar et al. [5].

The morphology of the powders was obtained by transmission electron microscopy (Phillips CM 20). Elemental analysis of aggregates was performed using a PGT System4 Plus EDS system. The elemental composition of the primary particles comprising the aggregate was determined by an energy dispersive x-ray spectroscopy unit (9800 Plus, EDAX International) attached to the TEM.

X-ray diffraction was used to determine the phase composition and lattice parameters of the titania powders (D500, Siemens; using CuKa radiation). The weight fraction of the anatase and rutile phases in the samples were calculated from the relative intensities of the strongest peaks corresponding to anatase and rutile as described by Spurr and Myers [6].

The powder specific surface area was measured by nitrogen adsorption (Quantasorb, Quantachrome Inc.) at 77 K using the BET equation. Assuming uniformly sized spherical primary particles within an aggregate, an equivalent average primary particle diameter was calculated ($d_p = 6/\rho A$, where ρ is the titania density and A is the specific surface area).

RESULTS AND DISCUSSION

(1) Powder Morphology and Chemical Composition

Transmission electron micrographs showed that pure titania powders produced in an aerosol reactor were polyhedral particles that were presumably dense since no surface connected porosity was observed (Figure 1). In addition, SEM micrographs showed that changing the reactor temperature or the reactant concentration did not significantly affect the morphology of pure titania particles [5].

Introduction of $SiCl_4$ as a dopant, resulted in mainly spherical primary particles within the aggregates and the degree of aggregation increased with increasing $SiCl_4$ concentration in the gas phase (Figure 2). Suyama and Kato [3] have also reported that silica doped titania particles are more round than those produced in the absence of dopants. Pure silica synthesized under similar conditions is extensively aggregated and the primary particles are considerably smaller with an irregular shape. Elemental analysis (EDS) on titania particles produced in the presence of $SiCl_4$ vapor showed that silicon was also present in the particles.

Titania powder produced in the presence of $POCl_3$ consisted of aggregates and much smaller particles which were interspersed among the aggregates. Increasing the gas phase concentration of $POCl_3$ led to an increase in the number of smaller particles. The corresponding TEM pictures show that the primary particles comprising the aggregate were more round than in the case of Si doped titania; in fact they were nearly perfect spheres (Figure 3). EDS analysis of these particles showed that the molar P/Ti ratio in the particles were slightly higher than the relative amounts of Ti and P introduced into the reactor as $TiCl_4$ and

$POCl_3$. No particles were found by EDS analysis on a TEM to be pure TiO_2 or pure P_2O_5. These results suggest that the oxides are nearly simultaneously incorporated from the gas phase into the particle and that there is strong interaction between the gas phase reactions of the two halides.

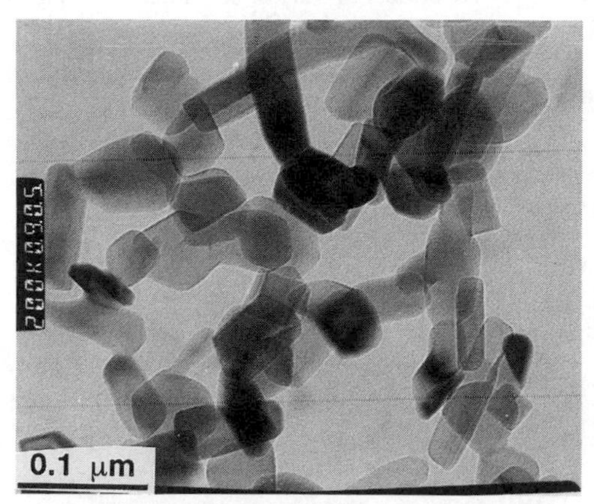

Figure 1: TEM micrograph of pure titania powders synthesized at 1500 K.

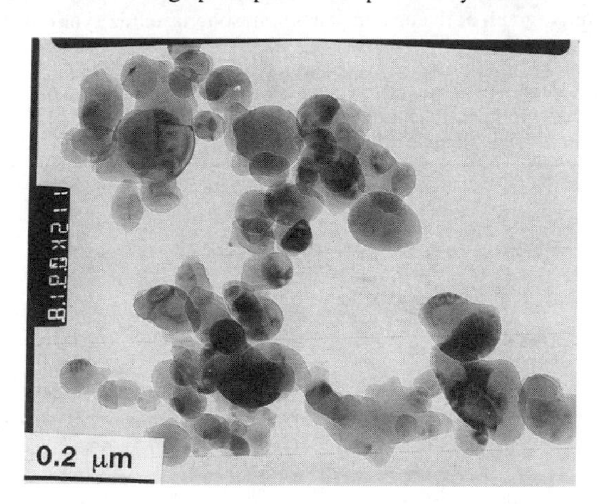

Figure 2: TEM micrograph of titania doped with silica synthesized at 1500 K with a gas phase $SiCl_4$ loading of 10%.

(2) Phase Composition

Pure TiO_2 produced by vapor phase oxidation of $TiCl_4$ is primarily anatase though the rutile phase content begins to increase as the reactor temperature is raised above 1200 K [5]. Figure 4 shows the effect of increasing $SiCl_4$ content in the gas phase on the titania phase composition for particles produced at 1500 K. With increasing $SiCl_4$ concentration the rutile content was consistently reduced as evidenced by the progressive loss in intensity of the characteristic rutile peak at $2\theta = 27.5°$. At a $SiCl_4$ concentration of 5 mole % of gas phase $TiCl_4$ concentration the rutile completely disappeared and pure anatase powder was obtained. It was also observed that with an increase in gas phase $SiCl_4$ concentration, the (004), (200) and (105) anatase peaks ($2\theta = 37.8°$, 48.1° and 53.9°) were broadened and lost definition indicating the formation of polycrystalline, highly aggregated titania powder. This is in agreement with the SEM

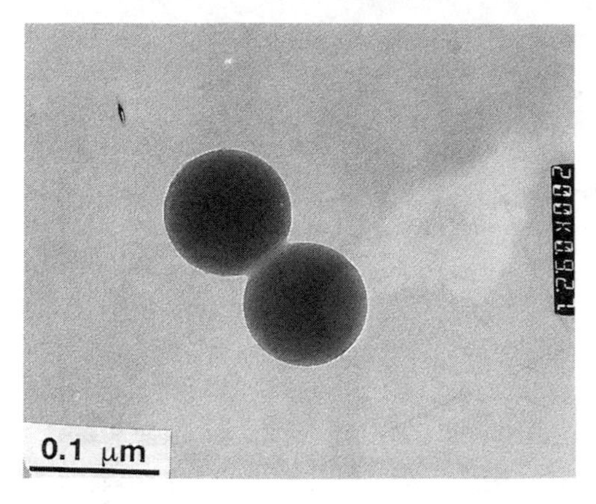

Figure 3: TEM micrograph of titania doped with phosphorus oxide synthesized at 1500 K.

observations that the degree of aggregation increases with the introduction of $SiCl_4$ as a dopant which indicates that amorphous silica probably slows down sintering and hinders the formation of single crystalline titania particles.

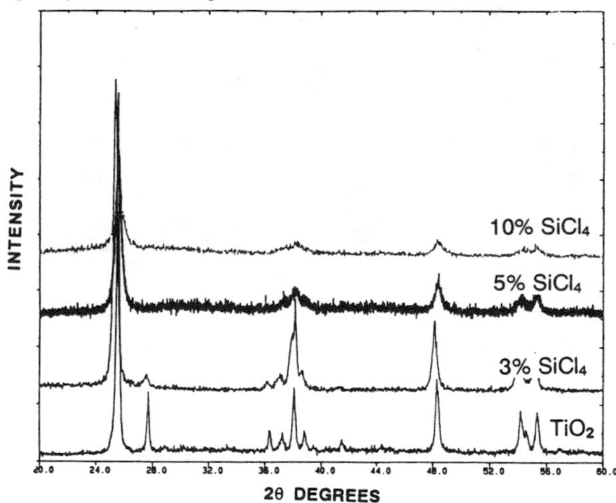

Figure 4: X-ray diffraction patterns for titania doped with silica for different amounts of dopants. All the powders were synthesized at 1500 K.

The rutile content decreased with increasing gas phase $POCl_3$ concentration though larger concentrations of the dopant were required to produce the same effect than with $SiCl_4$. The determination of the cell parameters from XRD data for anatase titania produced in the presence of $POCl_3$ showed an increase in a_0 from 3.780 Å for pure titania to 3.791 Å for titania produced in the presence of $POCl_3$ at 15 mole % of $TiCl_4$ concentration in the gas phase The c_0 axis increased in length from 9.498 Å to 9.504 Å. This increase in the unit cell parameters indicates that phosphorus is incorporated into the titania lattice.

The co-ordination of Ti is 6-fold in both anatase and rutile but the octahedral sites in anatase are highly distorted. The defect structure of TiO_2 is very complex; the various point defects for which evidence exists in anatase are oxygen vacancies, interstitial or substitutional Ti^{3+} ions, and interstitial Ti^{4+} ions [7]. Shannon and Pask [8] considered defect formation by foreign ions in the titania lattice and suggested that ions which enter the system interstitially inhibit the transformation from anatase to rutile. On the other hand, ions which can substitute directly for Ti^{4+} may either enhance or retard the transformation depending on whether the number of oxygen vacancies is increased or decreased. Ti^{4+} has an ionic crystal radius of 0.61 A while the ionic radii of Si^{4+}, and P^{5+} are 0.40 A, and 0.35 A, respectively; all small enough to enter the titania lattice interstitially thus inhibiting the transformation to rutile. Silicon is the most effective in inhibiting the transformation as the addition of these ions to interstitial sites does not involve any associated charge balance. The extra valance on phosphorus reduces oxygen vacancies retarding, thus, the transformation to rutile. Furthermore, the difference in the ionic radii between Si^{4+}, and P^{5+} explains the difference in their effectiveness in inhibiting the anatase to rutile transformation .

(3) Surface Area Measurements

Surface area measurements provide an estimate of the average primary particle size within an aggregate. As long as sintering is fast and complete coalescence takes place there is no difference between a primary particle and an aggregate; the primary particles constituting the aggregate lose their identity and their size coincides with that of the particle itself. For incomplete coalescence (slow rates of sintering), however, irregular aggregates are obtained and the individual primary particles comprising the aggregate retain their identity.

Figure 5 shows the variation in specific surface area for titania/silica powders synthesized at different temperatures and at various gas phase $SiCl_4$:$TiCl_4$ ratios. Titania particles produced in the absence of dopants have the lowest specific surface area at all temperatures. As the gas phase $SiCl_4$ concentration is increased, the specific surface area of silica doped titania powder increases resulting in a decrease in the BET equivalent primary particle diameter. Pure silica aggregate particles have the highest specific surface area containing primary particles that are about 5 times smaller than pure titania primary particles. The specific surface area data corroborate the TEM observations.

Anderson [9] concluded that oxygen diffusion was rate limiting in titania sintering and that decrease in oxygen diffusion led to slower sintering of titania. The reduction of oxygen vacancies will reduce oxygen diffusion which in turn will retard sintering. This reduced sintering will affect only intra-aggregate densification in that the primary particles will become smaller.

CONCLUSIONS

$SiCl_4$ was the most effective in producing predominantly anatase titania and these results were explained in terms of the ionic radii and valence of the species. Significant changes in particle morphology and degree of aggregation was obtained by the addition of dopant $SiCl_4$. These results were interpreted in terms of partial coalescence and reduced rate of sintering. The addition of $POCl_3$ resulted in the formation of smaller titania aggregates with nearly spherical primary particles. Chemical additives (dopants) in the gas phase synthesis of titania not only alter the phase composition and crystallinity of the product powder but may also drastically alter the morphology and specific surface area of these powders. The latter result is of significant importance not only in pigment manufacture but also in the synthesis of catalytic substrates and in separations applications involving titania and other ceramic materials.

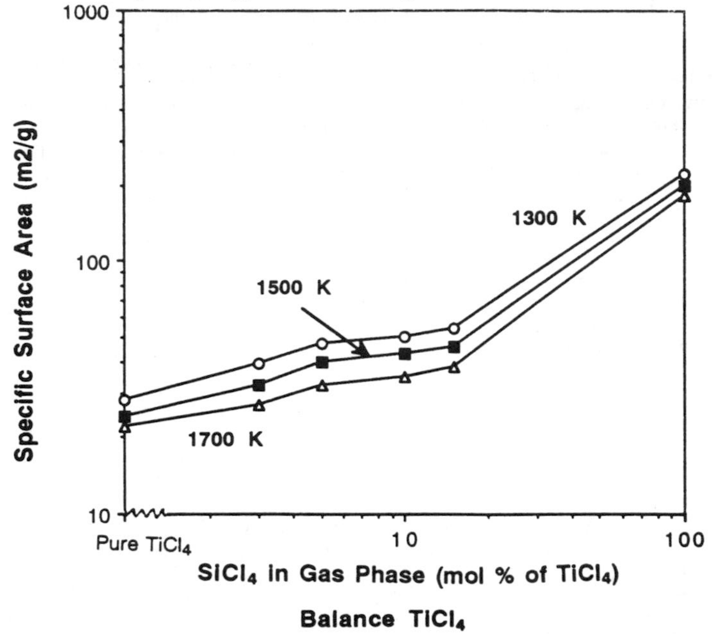

Figure 5: Specific area of silica doped titania powders as a function of temperature and gas phase $SiCl_4$ concentration.

Acknowledgements

This research was supported by the National Science Foundation, Grant CTS-8908197 and E. I. DuPont de Nemours and Co. Inc.

REFERENCES

[1] S. E. Pratsinis and S. V. R. Mastrangelo, Chem. Eng. Prog., 85, 62 (1989).
[2] H. K. Bowen, Matl. Sci. Engg, 44, 1 (1980).
[3] Y. Suyama and A. Kato, J. Amer. Ceram. Soc., 68, C154 (1985).
[4] E. J. Mezey, in Vapor Deposition , eds. C. F. Powell, J. H. Oxley and J. M. Blocher, Jr., (John Wiley & Sons, New York, 1966) p 423.
[5] M. K. Akhtar, Y. Xiong and S. E. Pratsinis, AIChE J., 37, 1561 (1991).
[6] R. A. Spurr and H. Myers, Analytical Chem., 29, 760 (1957).
[7] J-L. Hébrard, P. Nortier, M. Pijolat and M. Soustelle, J. Amer. Ceram. Soc., 73, 79 (1990).
[8] R. D. Shannon and J. A. Pask, J. Amer. Ceram. Soc., 48, 39 (1965).
[9] H. A. Anderson, J. Am. Ceram. Soc., 50, 235 (1967).

DEPOSITION OF CERAMIC FILMS BY A NOVEL PULSED-GAS MOCVD TECHNIQUE

KENNETH A. AITCHISON, JAMES D. BARRIE, AND JOSEPH CIOFALO
The Aerospace Corporation, M2/248, P.O. Box 92957, Los Angeles, CA, 90009.

ABSTRACT

Metal-Organic Chemical Vapor Deposition (MOCVD) is a versatile technique for the deposition of thin films of metals, semiconductors and ceramics. Commonly used hot wall flow-reactor designs suffer from a number of limitations. Chemical processes occurring in these reactors typically include a combination of homogeneous (gas-phase) and heterogeneous (gas-surface) reactions. These complex conditions are difficult to model and are poorly understood. In addition, flow reactors use large quantities of expensive precursor materials and are not well suited to the formation of abrupt interfaces. We report here a novel MOCVD technique which addresses these problems and enables a more thorough mechanistic understanding of the heterogeneous decomposition pathways of metal-organic compounds. This technique, the low-pressure pulsed gas method, has been demonstrated to provide high deposition rates with excellent control over film thickness. The deposition conditions effectively eliminate homogeneous processes allowing surface-mediated reactions to dominate. This decoupling of gas-phase chemistry from film deposition allows a better understanding of reaction mechanisms and provides better control over film growth. Both single metal oxides and binary oxide systems have been investigated on a variety of substrate materials. Effects of precursor chemistry, substrate surface, temperature and pressure on film composition and morphology will be discussed.

INTRODUCTION

The fabrication of thin films by Metal Organic Chemical Vapor Deposition (MOCVD) is an extremely versatile technique, finding application in many diverse technologies and devices. In addition to the well-known advantages of CVD (such as conformal coverage of irregular or non-planar substrate topographies and high growth rate), the use of metal-organic compounds as source materials has been shown to confer even greater benefits. The large and still-growing body of literature on synthesis and characterization of metal-organic compounds has, from the earliest days of MOCVD research, inspired the possibility of altering film deposition conditions by "molecular engineering", that is, the deliberate choice or design of source molecules with advantageous properties for the peculiar needs of a particular film/substrate system. Based on the large variety of structures and physicochemical properties found in this family of compounds, design criteria have included vapor pressure, decomposition temperature and pathways, chemical reactivity, purifiability and photosensitivity. The earliest work in this field focussed on developments in III-V semiconductor compounds, which at the time were assuming increasing technological importance coupled with a growing realization of the difficulty of growing films of adequate quality with the combination of halide-hydride chemistries then in use [1]. Recent advances in knowledge of electronic, electro-optic and optical properties of oxides and other ceramic materials has prompted increased interest in the possibility of fabricating those materials in thin film form. The pioneering and rigorous work of D.C. Bradley and co-workers [2] in the 1950's and 60's sparked the realization in the ceramics community that metal alkoxides would serve as excellent source materials, considering their exceptionally high vapor pressures and ready decomposition to the parent oxides at relatively modest temperatures or under hydrolytic conditions. This realization led to a rejuvenation of research in that area of metal-organic chemistry, and recent years have seen an explosion in the number of new compounds and detailed studies of their physicochemical properties. At the same time, there has been an increasing body of literature addressing the use of those compounds as MOCVD source materials.

Of equal importance to source material properties and deposition chemistry is the design of MOCVD reactors. Most CVD work has been carried out in flow type reactors at or near atmospheric pressure. This is in the viscous flow regime, and a very complex set of hydrodynamic conditions apply which include thermal, concentration and reaction-rate gradients within the reactor as well as at the substrate/gas boundary layer. Depletion effects can occur, leading to non-uniform growth along the flow axis. In order to compensate for this, large excesses of expensive source materials are used, most of which is wasted. At these pressures, gas phase nucleation is often a problem, especially when multiple reacting species are used, so high dilution by inert gases is also practiced. The relationship of the hydrodynamics of such systems with film deposition chemistry has been studied experimentally and theoretically for decades, but our understanding of these phenomena is still poor. Conventional methods of delivering source material vapors to MOCVD reactors also suffer from problems. Typically, a flow of inert gas is passed over or through the source material in a bubbler to entrain vapor. Pickup rates are variable, due to a lack of temperature uniformity within the bubbler resulting from the cooling effect of large amounts of flowing gas, and mist or particle entrainment lead to defects in growing films as well as large concentration excursions within the reactor. Since it is difficult to switch gas flows abruptly in such a system, film thickness control and interface sharpness are difficult to achieve.

As a result of the disparate nature of these two disciplines, two parallel sets of effort are discernible in the MOCVD literature, one based on the development and analysis of reactor designs, and the other focussing on the development and understanding of source material chemistries. Neither approach is sufficient by itself. At the outset of this work, it appeared that a more integrated approach was necessary. In particular, we sought a technique that would enable us to decouple hydrodynamic considerations from the surface catalyzed reactions leading to film growth. Our goal was to develop an MOCVD technology which combined the best reactor design with appropriate source chemistry to grow multicomponent oxide films for optical, electro-optic or electronic applications, while maintaining close control over film composition and thickness.

Recently, a reactor design was published which addresses many of the problems identified earlier [3,4]. That technique was developed for the growth of III-V semiconductor materials for quantum well device applications which, in many ways, is similar to the problem of growing multicomponent oxides for optical applications. Briefly, it consists of pulsing small quantities of source material into a chamber containing a heated substrate. Reactions occur at low pressures (approximately 0.1 Pa) for periods of approximately 1 sec at which time the chamber is evacuated and the cycle repeated. There are considerable advantages to such a system. Since only the vapor of the source vessel is being sampled, the equilibrium vapor pressure of the compound can be used to predict the exact amount of material introduced into the chamber. The deposition chamber is always at a pressure in the molecular flow regime, and viscous boundary layer effects are absent. Having filled a known volume with a known amount of gas at these pressures, kinetic molecular gas theory can be used to model properties of the reaction such as substrate-molecule collision frequencies, theoretical deposition rates and deposition kinetics. In addition, theoretical concepts of surface science and heterogeneous catalysis (sticking coefficients, surface diffusion, catalytic poisoning) can be used with much greater confidence to explain observations, as these experimental conditions are much closer to the "ideal" conditions under which those theories were derived. Many practical advantages also accrue from this method. Firstly, the minuscule consumption rate of source materials will tend to make this process much more economical to run than processes which consume large amounts of chemicals. Secondly, the small amount of effluent can be efficiently treated without requiring huge disposal systems. Finally, this technique offers the promise of virtually digital control over film thickness, an important aspect of many thin film technology requirements.

The first system we chose to study was the deposition of TiO_2 from titanium isopropoxide. This system has been well studied [5,6] and is amenable to characterization by many techniques. The various crystalline phases of TiO_2 and their ready interconversion at moderate temperatures provide a useful tool for studying the state of the final product. TiO_2 is commonly used in multilayer optical devices, and is also a component of a number of technologically important titanates. We have also prepared composite films of TiO_2 /SiO_2 using a combination of titanium isopropoxide and silicon tetraethoxide as source materials.

EXPERIMENTAL

The reactor design is illustrated schematically in figure 1 and can be described as follows. The source material is held at constant temperature while connected to a computer-controlled valve system leading to the reaction chamber. This valve system contains a precisely known small volume. At periodic intervals, the source vessel is opened to this volume, and a portion of the vapor in the vessel diffuses into the volume, which will then be at the vapor pressure of the pure compound, in the 10^2 - 10^4 Pa region. The vessel is then closed off, and allowed to re-attain its equilibrium vapor pressure. After closing the vessel, the small volume is quickly opened to the much larger volume of the reactor chamber which contains the heated substrate. This chamber is maintained at approximately 10^{-3} -10^{-2} Pa. When the source vapor enters the chamber, the pressure increases moderately, and the system is held thus for a period of approximately 1 second, during which time the source material or materials diffuse into the chamber and interact with the substrate. At the end of this period, the chamber exit valve is opened and the chamber pumped out. This cycle is repeated numerous times, a very thin film being grown during each cycle, until the desired film thickness is attained.

The reactor is constructed of a stainless steel high vacuum cross with 6" flanges. The cross is wrapped with several turns of heater tape, to maintain the walls at approximately 60°C. Access to the interior is provided by a combination viewport/door assembly, which also aids in viewing during experiments. The substrate heater is constructed of quartz with a boron nitride core which is heated by a current carrying molybdenum wire (based on an earlier published design [7]. With the substrate heater installed, the unoccupied volume of the chamber is 3500 cm^{-3}. The chamber is isolated from the vacuum system by means of a pneumatically actuated 50 mm bellows valve with opening and closing times in the 100 msec range. Pressure in the reactor is measured by a 10^3 Pa high accuracy capacitance manometer (Baratron, MKS Instruments) held at 100°C to prevent condensation of source materials. The response time of this device is 25 msec.

The pulsing valve assembly is constructed of Nupro air actuated bellows valves connected by Cajon VCR gasket fittings. The volume of the sampling elements between valves Vn1 and Vn2 is 4.60 cm^{-3}. The valve assembly is controlled by a Macintosh Plus computer. The valve train is in an insulated enclosure heated to a temperature higher than the highest source material temperature. When multiple source materials are used, they are passed from their respective valve trains through a mixing tube before entering the reaction chamber. The source materials are held in stainless steel sample cylinders equipped with Nupro manual bellows valves. Pumping is accomplished by means of a Leybold 360 L/sec turbomolecular pump backed by a mechanical roughing pump. Interposed between the turbo pump and the roughing pump is a section of Inconel pipe held in a tube furnace at ≈900°C, the purpose of which is to decompose unreacted source materials before they are exhausted. The base pressure achievable with this system is approximately $4x10^{-4}$ Pa.

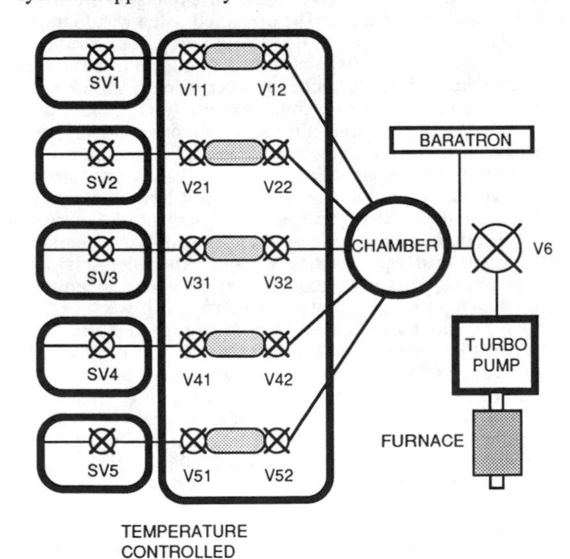

Figure 1: Schematic diagram of the pulsed-gas MOCVD reactor. Valves SVn are the manual valves on source vessels. The shaded regions between valves Vn1 and Vn2 represent the sampling elements.

In order to determine the reproducibility of this process, we calibrated the pressure fluctuations from cycle to cycle using ethylene glycol (Fisher, AR grade) as a model compound, since its vapor pressure is in the same range as many desirable source materials. For these experiments, degassed ethylene glycol was held at a temperature of 53°C (vapor pressure=$1.3x10^3$ Pa). Using the high accuracy baratron, the pressure maximum and rise time were recorded for 1000 consecutive cycles using a 4.6 cm^{-3} sampling element. The maximum reaction chamber pressure was shown to be 0.18 ± 0.01 Pa, and was reached in 150 msec in all cases. Pumpdown to base pressure (6.5×10^{-3} Pa) required 742 ± 90 msec.

For the deposition experiments, titanium isopropoxide (Alfa Chemicals) was distilled under vacuum into a previously dried and evacuated source cylinder. After collection, the cylinder was disconnected from the distillation apparatus and the contents degassed by repeated freeze-pump-thaw cycles before being installed in the system. Silicon tetraethoxide (Alfa Chemicals, 97%) was treated in an identical fashion.

Most films were deposited on 25 mm single crystal silicon wafers (100 orientation, Virginia Semiconductor Corporation), with selected films for special analyses being deposited on sapphire (00·1 and 11·0 orientations, Union Carbide) or fused silica. All substrates were cleaned by immersion in freshly prepared sulphuric acid/hydrogen peroxide solutions, rinsed with deionized water and dried by nitrogen blowoff. Silicon wafers were given an additional immersion in a 100:1 HF:water mixture followed by rinsing with deionized water in order to remove native oxide. All growth experiments were performed within 8 hours of cleaning the substrates. Deposition experiments were performed in the temperature range of 400°C to 730°C.

Stylus profilometry was performed using a Dektak IIA profilometer (Sloan Technology). UV-Visible spectra were obtained using a Perkin-Elmer Lambda-9 spectrophotometer. Ion Microprobe Mass spectral measurements were made in both positive and negative ion modes. Raman spectra were performed using a Coherent model 305 Argon ion laser operating at 488.0 nm and an Instruments SA U-1000 spectrometer with an Olympus microscope attachment.

RESULTS AND DISCUSSION

TiO_2 growth experiments were performed with titanium isopropoxide held at 60°C, and the sampling element at 110°C. Based on the data of Bradley [8] the vapor pressure in the source vessel should be $0.61x10^2$ Pa at this temperature, and the pressure on expansion from the sampling element to the reaction chamber should be 0.078 Pa. The measured value was found to be 0.069 Pa (with the chamber walls and substrate heater at 55°C). During growth experiments with a hot substrate holder, the pressure in the chamber was observed to rise to a maximum pressure plateau of approximately 0.31 Pa within 300 msec. Similar experiments with inert gas showed the pressure rise due to simple heating of the gas to be much slower (on the order of 5 seconds) and to account for only a few percent of this magnitude of increase. Under these conditions, the actual amount of source material delivered to the chamber is 2.5×10^{-5} grams per pulse. Pressure measurements using a high accuracy capacitance manometer have demonstrated the reproducibility of the process as well as the attainable degree of control in the amount of material delivered to the chamber during each pulse. If one accepts the attainment of a pressure maximum as indication of completion of reaction, we have shown that deposition is essentially complete by 300 msec. The 4.5-fold pressure increase during growth experiments is also indicative of rapid and complete decomposition. This is in good agreement with a theoretical analysis of the time domain of GaAs growth in a pulse reactor of similar design [4].

In general, films of pure TiO_2 grown by this technique were of very high quality. The absence of significant impurities in the films has been demonstrated by IMMA. Positive ion spectra of films on silicon showed the presence of sodium in sub ppb quantities, but no other impurities. Negative ion spectra showed the presence of peaks corresponding to C^- and small hydrocarbon fragments, but in concentrations below 0.1%. The levels of these peaks were fairly constant regardless of the deposition temperature, and depth profiling did not reveal any dependence of these peaks on the film thickness. Films on silicon that were thicker than 50 nm displayed interference colors which were very reproducible in multiple experiments run under the same conditions.

Thickness measurements by profilometry indicate that growth rates ranging from 0.27 - 0.75 nm per cycle are possible. Figure 2 shows the growth rate of films as a function of substrate temperature. The observation of a maximum in the apparent deposition rate at 500°C has been reported previously [9] in a hot-walled flow system at atmospheric pressure. Films studied in that work were much thicker, and the investigators reported a change in microstructure as thickness increased, with a tendency to form rutile gradually taking over. X-ray diffraction results on our TiO_2 films deposited on 100 silicon between 400°C and 600°C showed the films to be polycrystalline, with anatase being the only phase observed in this temperature range. Films deposited on silicon at 650°C showed some evidence for the presence of rutile, however, and a slight degree of preferred orientation was found. Raman spectra also showed anatase to be the dominant phase at all deposition temperatures. Rutile was not detected by this technique even in films deposited at 650°C. Since we have not measured the density of these films, we cannot positively assert whether this phenomenon is due to the growth of denser films at temperatures higher than 500°C, or to inefficient reaction conditions (i.e., deposition on the hotter portions of the substrate heater). However, SEM micrographs of our films indicate that films grown at higher temperatures possess finer grain structure than those deposited at lower temperatures. This indicates a higher density of nucleation sites at higher temperatures, and that grain growth is not significant at these temperatures.

TiO_2/SiO_2 film growth experiments were performed with titanium isopropoxide under conditions identical to those used for pure TiO_2 film growth, and TEOS held at 45°C. This provided an approximately 20-fold excess of TEOS over titanium isopropoxide. Films were grown in the 400 - 650°C temperature range. Under identical conditions, but in the absence of titanium isopropoxide, no film growth from pure TEOS was observed. Energy dispersive x-ray fluorescence (EDX) analyses were performed on TiO_2/SiO_2 composite films to determine the ratio of Si to Ti, and the results are summarized in figure 3. These films were deposited on sapphire to minimize interference from substrate contributions. There is a rise in the ratio of Si to Ti (atomic percent) as the temperature increases from 400°C to 700°C.

Low-temperature, chemically catalyzed decomposition of TEOS by metal alkyls has previously been reported [10,11]. To our knowledge, this is the first report of the apparent catalytic effect of titanium isopropoxide upon the decomposition of tetraethoxysilane to deposit films of variable TiO_2/SiO_2 ratios. Earlier work in hot-walled flow systems at higher pressures have shown that gas-phase intermediates play an important role in the decomposition of pure TEOS to form SiO_2 [12]. These intermediates are no doubt formed as a result of molecule-wall and intermolecular gas-phase collisions in portions of the hot-walled reactors prior to their reaching the substrate. These processes are absent under our experimental conditions and we conclude that the reactions leading to formation of composite TiO_2/SiO_2 films are dominated by reactive surface intermediates which may consist of partially decomposed organic or organometallic species or of particularly reactive surface sites generated during film growth. Detailed studies to elucidate the nature of these intermediates are underway.

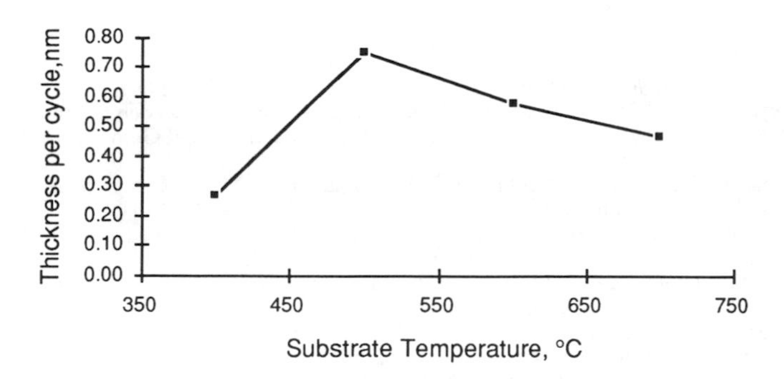

Figure 2: Measured thickness of TiO_2 films deposited per cycle at various temperatures. Each experiment consisted of 500 cycles with a residence time of 1 sec per cycle.

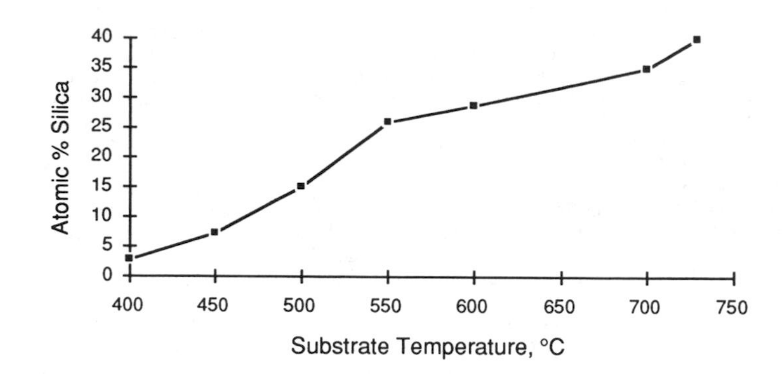

Figure 3: Atomic percentage of SiO_2 in TiO_2/SiO_2 films prepared at various substrate temperatures.

SUMMARY

We have successfully demonstrated a novel technique for the preparation of single and multicomponent oxide thin films by ultra-low pressure pulsed gas MOCVD. This technique has been shown to provide digital control over film thickness without sacrificing high growth rates. We have reported, for the first time, the apparent catalytic effect of titanium isopropoxide upon the decomposition of tetraethoxysilane to deposit films of variable TiO_2/SiO_2 ratios. This technique will prove valuable for the deposition of other mixed oxides where control of stoichiometry and thickness are critical to film properties and device performance.

ACKNOWLEDGMENTS

The authors wish to thank P. D. Chaffee and T. Park for assistance and helpful discussions. This work was supported by the Aerospace Sponsored Research Program.

REFERENCES

1. Manasevit, H. and W.I. Simpson, J. Electrochem. Soc. 116, 1725 (1969).
2. Bradley, D.C., R.C. Mehrotra, and D.P. Gaur, Metal Alkoxides, (Academic Press, 1978).
3. Van Suchtelen, J., Hogenkamp, J.E.M., Van Sark, W.G.J.H.M., and Giling, L.J., J. Crystal Growth, 93, 201 (1988).
4. Van Sark, W.G.J.H.M., Van Suchtelen, J., Hogenkamp, J.E.M., De Croon, M.H.J.M, Velthuis, R.A. and Giling, L.J., J. Crystal Growth, 102, 1 (1990).
5. E.T. Fitzgibbons, K. J. Sladek, and W.H. Hartwig, J.Electrochem. Soc., 119, 92 (1972).
6. Siefering, K.L. and Griffin, G.L., J. Electrochem. Soc., 137, 814 (1990).
7. Boldish, S.I., Ciofalo, J.S., and Wendt, J.P., J. Electron. Mater., 14, 587 (1985).
8. Bradley, D.C., J. Chem. Soc. A, 5020 (1952).
9. Takahashi, Y., Suzuki, H., and Nasu, M., J. Chem. Soc. Faraday Trans. 1, 81, 3117 (1985).
10. Eversteijn, F.C., Phillips Res. Repts., 21, 379 (1966).
11. Li, P.C., and Tsang, P.J., J. Electrochem. Soc., 129, 165 (1982).
12. Kalidindi, S.R., and Desu, S.B., J. Electrochem. Soc., 137, 624 (1990).

Author Index

Subject Index